INDUSTRIAL AND ENGINEERING APPLICATIONS OF ARTIFICIAL INTELLIGENCE AND EXPERT SYSTEMS

INDUSTRIAL AND ENGINEERING APPLICATIONS OF ARTIFICIAL INTELLIGENCE AND EXPERT SYSTEMS

IEA/AIE
96

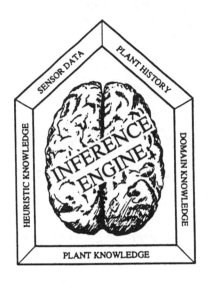

Proceedings of the Ninth International Conference
Fukuoka, Japan, June 4-7, 1996

Edited by

Takushi Tanaka
Fukuoka Institute of Technology
Fukuoka, Japan

Setsuo Ohsuga
Waseda University
Tokyo, Japan

Moonis Ali
Southwest Texas State University
San Marcos, Texas, USA

GORDON AND BREACH PUBLISHERS
Australia • Canada • China • France • Germany • India • Japan • Luxembourg • Malaysia •
The Netherlands • Russia • Singapore • Switzerland • Thailand • United Kingdom

Emmaplein 5
1075 AW Amsterdam
The Netherlands

British Library Cataloguing in Publication Data

A catalogue record for this book is available from the British Library.

CONTENTS

Preface   xvii

Conference Organization   xix

List of Sponsors   xxi

INVITED PAPERS

Micromachines as Unconventional Artifacts   3
Naomasa Nakajima

Multi-Strata Modeling to Automate Problem Solving Including Human Activity   9
Setsuo Ohsuga

The Current State of AI Planning Research   25
Drew McDermott

DCSG: A Logic Grammar for Knowledge Representation of Hierarchically Organized Objects   35
Takushi Tanaka

Pattern Separation by Chaotic Networks Employing Chaotic Itinerancy   43
Takeshi Yamakawa, Tsutomu Miki and Masayoshi Shimono

AUTOMATED REASONING

Design and Implementation of Multiple-Context Truth Maintenance System
with Binary Decision Diagram   47
Hiroshi G. Okuno, Osamu Shimokuni and Hidehiko Tanaka

PARCAR: Parallel Cost-Based Abductive Reasoning System   57
Shohei Kato, Chiemi Kamakura, Hirohisa Seki and Hidenori Itoh

CAD/CAM

Automatic Placement Using Static and Dynamic Groupings   65
Makoto Hirahara, Natsuki Oka and Kunio Yoshida

A Domain-Specific Conceptual Model in Building Design   71
Masaaki Hashimoto, Toyohiko Hirota, Isao Nagasawa, Keiichi Katamine and Yoshiaki Tegoshi

The Elements of Programming Style in Design Calculations   77
Masanobu Umeda, Isao Nagasawa and Tatsuji Higuchi

CASE-BASED REASONING

Knowledge-Based Expert System Combining Case-Based Reasoning and Rule-Based
Reasoning to Select Repair and Retrofitting Methods for Fatigue Damage on Steel Bridge   89
Shigenori Tanaka, Ichizou Mikami, Hidenori Maeda and Atsushi Kobayashi

Multiple Domain Feature Mapping Utilizing Case-based Reasoning and Blackboard Technology   95
S.S. Lim, B.H. Lee, E.N. Lim and K.A. Ngoi

Plant Fault Diagnosis by Integrating Fault Cases and Rules   103
Yoshio Nakatani, Makoto Tsukiyama and Tomo-omi Wake

DATABASE

Characteristics of Query Responses on Fuzzy Object-Oriented Databases   111
Shyue-Liang Wang

A Knowledge Media Approach Using Associative Representation for Constructing Information Bases   117
Harumi Maeda, Kazuto Koujitani and Toyoaki Nishida

A Query Procedure for Allowing Exceptions in Advanced Logical Database   127
Kouzou Ohara, Noboru Babaguchi and Tadahiro Kitahasi

DECISION SUPPORT

Decision Support System for Water-Supply Risk Management   135
Xu Zongxue, Akira Kawamura, Kenji Jinno and Kazumasa Ito

Suggestions on Developing "Creative Decision Stimulation Systems (CDSS)"   141
Ronggui Ding and Wenxiu Han

DIAGNOSIS

A Method of Diagnosis Using Possibility Theory   149
Koichi Yamada and Mitsuhiro Honda

Model-Based Program Debugging and Repair   155
Markus Stumptner and Franz Wotawa

An Object-Oriented Method for Control Loop Diagnosis and Monitoring   161
P.A. Collier and M.G. Casey

A Self-Learning Fuzzy System for Automotive Fault Diagnosis   167
Yi Lu

DISTRIBUTED AI

An Automated Reasoning System for Multi-Agent Cooperation   175
Lifeng He, Hirohisa Seki and Hidenori Itoh

GUARDNET: A Distributed and Concurrent Programming Environment for Multi-Agent Systems 181
Motoyuki Takaai, Hideaki Takeda and Toyoaki Nishida

FUZZY LOGIC AND CONTROL

Assigning Weights to Rules of an Expert System Based on Fuzzy Logic 189
Mehdi R. Zargham and Leishi Hu

Detection of Chaos in Process Control Systems Using a Julia Set Methodology 195
David W. Russell and James J. Alpigini

Fuzzy Johnson's Algorithm for Two-Machine Flow Shop 201
Tzung-Pei Hong and Tzung-Nan Chuang

Prediction of Properties of Mixture by Integration of Pattern Recognition and Logic Procedure by Using Fuzzy Clustering 207
Hiroyuki Tanaka and Yuji Naka

Self-Learning Fuzzy Logic Control of Non-Linear Processes 213
S.H. Ghwanmeh, K.O. Jones and D. Williams

Using Fuzzy Logic in Feedrate Control for Free-Form Surface Machining 223
Ralph W.L. Ip and Felix T.S. Chan

GENETIC ALGORITHM

GA-Based Reconstruction of Plane Images from Projections 235
Zensho Nakao, Yen-Wei Chen and Fath El Alem Fadlallah Ali

Genetic Algorithm Processor for the Frequency Assignment Problem 241
Mehrdad Salami and Greg Cain

Selection of Representatives Using Genetic Algorithms 247
Y. Hamamoto, Y. Tsuneta, C. Kaneyama and S. Tomita

Using the Baldwin Effect to Accelerate a Genetic Algorithm 253
John R. Podlena and Tim Hendtlass

INTELLIGENT TUTORING

Generating Courses in an Intelligent Tutoring System 261
R. Nkambou, M.C. Frasson and C. Frasson

Object-Oriented Model of Knowledge for the Intelligent Tutoring Shell 267
Ljubomir Jerinic and Vladan Devedzic

KNOWLEDGE ACQUISITION

A Data Mining Technique for Time Series Data 275
Yoshinori Satoh and Akira Maeda

Development of a Knowledge Based System Using a Combined Knowledge Acquisition
and Knowledge Modeling Approach 283
Glenn D. Aguilar, Hiroyuki Yamato and Takeo Koyama

A Human-Intelligence-Imitating Question-Asking Strategy for Propositional Knowledge-Based Systems 289
Jinchang Wang

Incremental Rule Production: Towards a Uniform Approach for Knowledge Organization 295
Mondher Maddouri and Ali Jaoua

Ontology-Based Information Gathering and Categorization from the Internet 305
Michiaki Iwazume, Hideaki Takeda and Toyoaki Nishida

KNOWLEDGE-BASED SYSTEMS

A Knowledge-Based Mesh Generation System for Forging Simulation 317
*Osamu Takata, Koukichi Nakanishi, Nariaki Horinouchi, Hiroshi Yano, Tadao Akashi
and Toyohide Watanabe*

A Knowledge Based Tool for Checking Large Knowledge Bases 327
Rym Nouira and Jean-Marc Fouet

OKCFTR: Translators for Knowledge Reuse 333
Kunhuang Huarng and Ming-Cheng Chen

Integration of Forward and Backward Inferences Using Extended Rete Networks 339
Yong H. Lee and Suk I. Yoo

VAREX: An Environment for Validating and Refining Rule Bases 345
Heribert Schimpe and Martin Staudt

KNOWLEDGE REPRESENTATION

A Formalism for Defeasible Inheritance 357
Nadim Obeid

Knowledge Compilation for Interactive Design of Sequence Control Programs 363
Yasuo Namioka and Toshikazu Tanaka

Shape Reasoning in CAD Systems 369
Jose C. Damski and John S. Gero

LOGIC PROGRAMMING

Applications of Logic Programs with Functor Set to Automated Problem Solving Under Uncertainty 377
Hiroshi Sakai and Akimichi Okuma

Channel Routing with Constraint Logic Programming and Delay 383
Neng-Fa Zhou

Design of Database Interface to ILP for Building Knowledge Base 389
Keiko Shimazu and Koichi Furukawa

An Incremental Solver for Constraint Hierarchies over Real Intervals 395
Duong Tuan Anh and Kanchana Kanchanasut

MACHINE LEARNING

Coin Classification Using a Novel Technique for Learning Characteristic Decision Trees
by Controlling the Degree of Generalization 403
Paul Davidsson

Evaluation of Attribute Selection Measures in Decision Tree Induction 413
TuBao Ho and TrongDung Nguyen

Feature Selection and Classification - A Probabilistic Wrapper Approach 419
Huan Liu and Rudy Setiono

Knowledge Discovery from Numerical Data 425
Chie Morita and Hiroshi Tsukimoto

A Knowledge Revision Learner Using Artificially Generated Examples 431
Megumi Ishii, Hussein Almuallim and Shigeo Kaneda

Learning Probabilistic Networks from Data 439
Bozena Stewart

A Practical Object-Oriented Concept Learning System in Clinical Medicine 449
Gou Masuda, Norihiro Sakamoto and Kazuo Ushijima

Sampling Effectiveness in Discovering Functional Relationships in Databases 455
Atsuhiro Takasu, Tatsuya Akutsu and Moonis Ali

Signal Interpretation in Two-Phase Fluid Dynamics Through Machine Learning
and Evolutionary Computing 461
Bogdan Filipic, Iztok Zun and Matjaz Perpar

MANUFACTURING

Hybrid AI Solutions and Their Application in Manufacturing 469
L. Monostori, Cs. Egresits and B. Kadar

Performance Optimization of Flexible Manufacturing Systems Using Artificial Neural Networks 479
S. Cavalieri

Real-time Intelligent Monitoring System in Manufacturing System 487
Sun Yu, Yu Zhigang and Zhang Shiqi

Simulation of Scheduling Rules in a Flexible Manufacturing System Using Fuzzy Logic 491
A. Kazerooni, F.T.S. Chan, K. Abhary and R.W.L. Ip

MONITORING

An Expert System for the Monitoring and Simulation of the AXAF-1 503
Kai H. Chang, Mark Rogers and Richard McElyea

Knowledge-Enhanced CO-monitoring in Coal-Mines 511
Wolfram Burgard, Armin B. Cremers, Dieter Fox, Martin Heidelbach, Angelica M. Kappel
and Stefan Luttringhaus-Kappel

Medi-View - An Intelligent ICU Monitoring System 521
Kevin Kennedy, Michael Mckinney and Yi Lu

NEURAL NETWORK

Comparison of Two Learning Networks for Time Series Prediction 531
Daniel Nikovski and Mehdi Zargham

An Evolutionary Algorithm for the Generation of Unsupervised Self Organizing Neural Nets 537
Tim Hendtlass

Exploiting Don't-care Information in Neural Network Learning 543
Chung-Yao Wen

Extended Tree based Regression Neural Networks for Multifeature Split 547
Sook Lim and Sung Chun Kim

Mixing Stochastic Exploration and Propagation for Pruning Neural Networks 553
Eric Fimbell

A Note on the Generalization Error in Neural Networks 559
Yoshihiko Hamamoto, Toshinori Hase, Yoshihiro Mitani and Shingo Tomita

A Proposal of Emotional Processing Model 563
Kaori Yoshida, Masahiro Nagamatsu and Torao Yanaru

Recursive Prediction Error Algorithm for the NUFZY System to Identify Nonlinear Systems 569
B.T. Tien and G. van Straten

Reflective Learning of Neural Networks 575
Yoshiaki Tsukamoto and Akira Namatame

NEURAL NETWORK APPLICATIONS

Collision Avoidance Using Neural Networks Learned by Genetic Algorithms 585
Nicolas Durand and Jean-Marc Alliot

CONTENTS

Identifying Chemical Species in a Plasma Using a Neural Network 593
Phil D. Picton, Adrian A. Hopgood, Nicholas St. J. Braithwaite and Heather J. Phillips

A Method for Selecting Learning Data in the Prediction of Time Series with Explanatory Variables
Using Neural Networks 599
Hisashi Shimodaira

A New Second-Order Adaptation Rule and Its Application to Electrical Model Synthesis 605
Jan Wilk, Eva Wilk and Bodo Morgenstern

Real Option Valuation with Neural Networks 611
Alfred Taudes, Martin Natter and Michael Trcka

NATURAL LANGUAGE

Automatic Segmentation and Tagging of Hanzi Text Using a Hybrid Algorithm 621
An Qin and Wing Shing Wong

Computer-aided News Article Summarization 627
Hisao Mase, Hiroshi Tsuji and Hiroshi Kinukawa

A Connectionist/Symbolic Dependency Parser for Free Word-Order Languages 633
Jong-Hyeok Lee, Taeseung Lee and Geunbae Lee

Intelligent Support for Construction and Exploration of Advanced Technological Information Space 639
Toshiyuki Matsuo and Toyoaki Nishida

PLANNING AND SCHEDULING

Adaptation of a Production Scheduling Framework to Distributed Work Environment 647
Taketoshi Yoshida and Masahiro Hori

Application of Genetic Algorithm to Alloy Casting Process Scheduling 653
Masuhiro Ishitobi

Design and Development of an Integrated Intelligent Process Planning System
Taking CAD Data Files as Input 659
Shu-Chu Liu

Encapsulation of Actions and Plans in Conditional Planning Systems 665
Marco Baioletti, Stefano Marcugini and Alfredo Milani

On-line Optimization Techniques for Dynamic Scheduling of Real-Time Tasks 673
Yacine Atif and Babak Hamidzadeh

PRACTICAL APPLICATIONS

Application-Specific Configuration of Telecommunication Systems 681
Andreas Bohm and Stefan Uellner

Character Design Based on Concept Space Formation   687
Takenao Ohkawa, Kaname Kakihara and Norihisa Komoda

Customizing A* Heuristics for Network Routing   693
M. Hitz and T.A. Mueck

Feature Selection in Automatic Transmission Shift Quality Classification   699
Robert Williams and Yi Lu

KOA: General Affairs Expert System with Easy Customization   705
*Shigeo Kaneda, Katsuyuki Nakano, Daizi Nanba, Hisazumi Tsuchida, Megumi Ishii
and Fumio Hattori*

Using Constraint Technology for Predictive Control of Urban Traffic Based on Qualitative
and Temporal Reasoning   711
F. Toledo, S. Moreno, E. Bonet and G. Martin

ROBOTICS

Fuzzy Behavior Organization and Fusion for Mobile Robot Reactive Navigation   719
Jiancheng Qiu and Michael Walters

Learning and Classification of Contact States in Robotic Assembly Tasks   725
Enrique Cervera and Angel P. del Pobil

A Neural Approach for Navigation Inspired by the Human Visual System   731
J. Fernandez de Canete and I. Garcia-Moral

Ultrasonic Perception: A Tri-aural Sensor Array for Mobile Robots Using a Competition
Neural Network Approach   737
J. Chen, H. Peremans and J.M. Van Campenhout

VISION

A Genetic Algorithm for Image Segmentation   745
A. Calle, W.D. Potter and S.M. Bhandarkar

A Hierarchy of Detail for Representing Non-Convex Curved Objects   751
Begona Martinez and Angel P. del Pobil

A Qualitative Traffic Sensor Based on Three-Dimensional Qualitative Modeling
of Visual Textures of Traffic Behavior   761
E. Bonet, S. Moreno, F. Toledo and G. Martin

Towards an Automatic Determination of Grasping Points Through a Machine Vision Approach 767
P.J. Sanz, J.M. Inesta and A.P. del Pobil

ABSTRACTS FOR POSTER SESSION

Case-Based Reasoning

A Case-Based Reasoning Approach in Injection Modeling Process Design 775
K. Shelesh-Nezhad and E. Siores

SMARTUSA: A Case-Based Reasoning System for Customer Services 776
Pradeep Raman, Kai H. Chang, W. Homer Carlisle and James H. Cross

Decision Making

Modeling and Implementing Aspects of Holistic Judgment in Industrial Decision Making 777
Philip A. Collier and Stewart A. Leech

Diagnosis

Computer Intrusion Detection and Incomplete Information 778
Mansour Esmaili, Reihaneh Safavi-Naini and Josef Pieprzyk

TURBOLID: Time Use in a Rule-Based, On-Line Industrial Diagnoser 779
C. Alonso Gonzalez, B. Pulido Junquera and G.G. Acoste Lazo

Distributed AI

An Approach to a Multi-agent Based Scheduling System Using a Coalition Formation 780
Takayuki Itoh and Toramatsu Shintani

Integrating Agent and Object-Oriented Programming Paradigms for Multi-Agent
Systems Development 781
Agostino Poggi, Giovanni Adorni and Paola Turci

Necessary Knowledge Sharing for Cooperative Learning Among Agents 782
Akira Namatame and Yoshiaki Tsukamoto

Genetic Algorithm

Optimal Selection of Cutting Length of Bars by Genetic Algorithms 783
Toshihiko Ono and Gen Watanabe

Intelligent Tutoring

A Descriptive Model of Student Program Generation Based on a Protocol Analysis 784
Tsuruko Egi and Kazuoki Osada

An Intelligent Tutoring System for the Teaching of the Industry Equipment
and Its Student Modeling 785
Lidia L. Hardy and Jose M. Yanez Prieto

Interactive System

Skill Training at a Distance 786
Shuichi Fukuda, Yoshifusa Matsuura and Premruedee Wongchuphan

Knowledge Based System

K-Tree: An Efficient Structure for Verification and Inference in Rule-Based Systems   787
T. Rajkumar and H. Mohanty

Software Fault Reduction Methodology for Reliable Knowledge Bases   788
Yasushi Shinohara

Knowledge Representation

Macro Scale Object Complexity Measurement and Its Relation to a Semantic Model   789
John W. Gudenas and C.R. Carlson

Logic Programming

B-Prolog: A High Performance Prolog Compiler   790
Neng-Fa Zhou, Isao Nagasawa, Masanobu Umeda, Keiichi Katamine and Toyohiko Hirota

Machine Learning

A Cascade of Neural Networks for Complex Classification   791
David Philpot and Tim Hendtlass

PAC-Learning of Weights in Multiobjective Function by Pairwise Comparison   792
Ken Satoh

Natural Language

A Machine Translating Approach from Japanese to Tamil   793
Sivasundaram Suharnan and Hiroshi Karasawa

Neural Network

Annealed Hopfield Network Approach for Boundary-Based Object Recognition   794
Jung H. Kim, Eui H. Park and Celestine A. Ntuen

Better Neurons for Biological Simulations   795
Howard Copland and Tim Hendtlass

Canonical Form of Recurrent Neural Network Architecture   796
N. Selvanathan and Mashkuri Hj. Yaacob

Equalisation of Digital Communication Channel Using Hartley-Neural Technique   797
Jitendriya K. Satapathy, Ganapati Panda, and Laxmi N. Bhuyan

A Nonlinear Optimization Neural Network for Optimum Multiuser Detectors   798
Guiqing He and Puying Tang

Parallel Computer Implementation for Feature Extraction via Answer-in-Weights Neural Network   799
Iren Valova and Yukio Kosugi

Serial and Parallel Neural Networks for Image Compression   800
Ryuji Hamabe and Ho Chun Kuo

Planning and Scheduling

Explore the Operational Problems of FMSs through Analytical Hierarchy Process
and Simulation Approach   801
Felix T.S. Chan and Ralph W.L. Ip

Practical Applications

Spectacle Designing and Advice Computer Graphics System Using Artificial Intelligence   802
Ryuto Fujie, Hiroyuki Fujie, Kunie Takeuchi, Oskar Bartenstein and Kosaku Shirota

Study of a Consultation System for Railway Safety Countermeasure   803
Hisaji Fukuda

Robotics

A Step Toward Human-Robot Cooperative System   804
Masaru Ishii and Hironori Saita

Index of Authors   805

PREFACE

The editors present the Proceedings of the Ninth International Conference on Industrial and Engineering Applications of Artificial Intelligence and Expert Systems (IEA/AIE-96) with a great deal of pride and pleasure. This international forum for AI researchers and practitioners attempts to maintain a balance between theory and practice. Participants share their experience, expertise, and skills in developing and enhancing intelligent systems technology.

Since the establishment of this conference in 1988, intelligent systems have been successfully applied in most countries. Over the last few years, the number of countries submitting papers has gradually increased, with 30 countries being represented in the 170 submissions we received this year. Each submission was sent to three referees, and the best 100 were selected for presentation as regular papers.

This volume contains these regular papers along with the invited papers and the abstracts of the poster sessions presented at this conference. The selected papers in this volume represent a broad spectrum of new ideas in the field of applied artificial intelligence and expert systems, and serve to disseminate information regarding intelligent methodologies and their implementation as applied in solving various problems in industry and engineering.

We would like to thank the Organizing Committee for their efforts, and note that the committee members have changed over the four years since the decision was made to hold this conference in Asia for the first time in its history. We are especially grateful to Professor Mamoru Nakajima, former president of the Fukuoka Institute of Technology, who invited this conference to Fukuoka, Japan. We would like to thank the Program Committee and the external referees listed on the following pages. Without them, we would not have been able to evaluate the papers or even hold the conference itself. We would like to thank the Japan Executive Committee for local arrangements. Finally, we would like to thank the staff members who worked so diligently, both in the United States and Japan.

Takushi Tanaka, *Program Co-Chair*

Setsuo Ohsuga, *Program Chair*

Moonis Ali, *General Chair*

CONFERENCE ORGANIZATION

Organizing Committee

Moonis Ali, Southwest Texas State University, General Chair
Setsuo Ohsuga, Waseda University
Masanori Akazaki, Fukuoka Institute of Technology
Takushi Tanaka, Fukuoka Institute of Technology

Advisor

Keiji Ito, Fukuoka Institute of Technology
Tsuneo Tamachi, Fukuoka Institute of Technology
Susumu Tanaka, Kyushu Electric Power Co.

Program Committee

Setsuo Ohsuga, Waseda University, Chair
Takushi Tanaka, Fukuoka Institute of Technology, Co-Chair

Makoto Amamiya Kyushu Univ., Japan	**Toshio Fukuda** Nagoya Univ., Japan	**Drew McDermott** Yale Univ., USA	**Franz. J. Radermacher** Univ. Ulm, Germany
Frank Anger NSF, USA	**Andreas Günter** Univ. Hamburg, Germany	**John Mitchiner** Sandia Natl. Lab., USA	**Rita Rodriguez** NSF, USA
Bechir el. Ayeb Univ. Sherbrooke, Canada	**Mehdi T. Harandi** Univ. Illinois, USA	**Riichiro Mizoguchi** Osaka Univ., Japan	**Erik Sandewall** Linkoping Univ.,Sweden
Fevzi Belli Univ. Paderborn, Germany	**Tim Hendtlass** Swinburne UT, Australia	**Hiroshi Motoda** Osaka Univ., Japan	**Nigel Shadbolt** Univ. Nottingham,UK
Gautam Biswas Vanderbilt Univ., USA	**Koichi Hori** Univ. Tokyo, Japan	**Laszlo Monostori** HAS, Hungary	**Mildred L.G. Shaw** Univ. Calgary, Canada
Ivan Bratko Univ. Ljubljana, Slovenia	**Masumi Ishikawa** KIT Iizuka, Japan	**Masao Mukaidono** Meiji Univ., Japan	**Katsuhiko Shirai** Waseda Univ., Japan
Bruce G. Buchanan Univ. Pittsburgh, USA	**Shun Ishizaki** Keio Univ., Japan	**Yi Lu Murphey** Univ. Michigan-Dbn, USA	**Yoshiaki Shirai** Osaka Univ., Japan
Kai H. Chang Auburn Univ., USA	**Masatoshi Ito** Tokyo Inst. Tech., Japan	**Isao Nagasawa** KIT Iizuka, Japan	**Reid Simmons** CMU, USA
Hon Wai Chun City Univ., Hong Kong	**Shuichi Iwata** Univ. Tokyo, Japan	**Toyoaki Nishida** NAIST, Japan	**Motoi Suwa** ETL, Japan
Paul Chung Loughborough Univ., UK	**Lakhmi C. Jain** UNISA, Australia	**Gordon S. Novak** Univ. Texas, USA	**Michio Suzuki** CRIEPI, Japan
Tony Cohn Univ. Leeds, UK	**Kazuhiko Kawamura** Vanderbilt Univ., USA	**Norio Okino** Kyoto Univ., Japan	**Swee Hor Teh** UVSC, USA
Alain Costes LAAS-CNRS, France	**Shigenobu Kobayashi** Tokyo Inst. Tech., Japan	**Dick Peacocke** BNR, Canada	**Takao Terano** Tsukuba U, Japan
Roberto Desimone Defence Res. A.,UK	**Michael Magee** Univ. Wyoming, USA	**Francois Pin** Oak Ridge N. Lab., USA	**Spyros Tzafestas** NTU Athens, Greece
Ken Ford Univ. West FL, USA	**Wolfgang Marquardt** RWTH Aachen, Germany	**Angel P. d. Pobil** Jaume-I Univ., Spain	**Ian H.Witten** U. Waikato, New Zealand
Graham Forsyth DSTO, Australia	**Gen Matsumoto** ETL, Japan	**Don Potter** Univ. Georgia, USA	**Takeshi Yamakawa** KIT Iizuka, Japan
Shuichi Fukuda Tokyo Metro. IT, Japan	**Manton Matthews** U. South Carolina, USA	**Henri Prade** Univ. Paul Sabatier, France	

External Referees

Ruth Aylett
Univ. Salford, UK
Kozo Bannai
CRIEPI, Japan
Lonnie Chrisman
CMU, USA
S.J. Cunningham
U. Waikato, New Zealand
Bogdan Filipic
Josef Stefan Inst., Slovenia
Matjaz Gams
Josef Stefan Inst., Slovenia
R. Goodwin
CMU, USA
Masaaki Hashimoto
Kyushu Inst.Tec., Japan
Toyohiko Hirota
Kyusyu Inst.Tec., Japan
G. Holmes
U. Waikato, New Zealand
Kho Hosoda
Osaka Univ., Japan

Thomas Kämpke
FAW Ulm, Germany
Christoph Klauck
Bremen Univ., Germany
Sven Koening
CMU, USA
Sabine Kokskaemper
Univ. Hamburg, Germany
Jutta Kreyss
Univ. Bremen, Germany
Matjaž Kukar
Univ. Ljubljana Slovenia
Yosihinori Kuno
Osaka Univ., Japan
Akeo Kuwahata
CRIEPI, Japan
Umeda Masanobu
Kyusyu Inst. Tech., Japan
Jun Miura
Osaka Univ., Japan
H. Jurgen Muller
Univ. Bremen, Germany

Chikahito Nakajima
CRIEPI, Japan
Takao Ohya
CRIEPI, Japan
Takashi Onoda
CRIEPI, Japan
Uros Pompē
Univ. Ljubljana, Slovenia
Wolf-Fritz Riekert
FAW Ulm, Germany
Thomas Rose
FAW Ulm, Germany
Klaus Rotter
FAW Ulm, Germany
Ashraf S. Saad
Austin Peay State U.,USA
Shigeo Sagai
CRIEPI, Japan
Serdkamp
U. Hamburg, Germany
Yasushi Shinohara
CRIEPI, Japan

Marko Robnik Šikonja
Univ. Ljubljana, Slovenia
James Soutter
Loughborough Univ., UK
Dorian Šuc
Univ. Ljubljana, Slovenia
Kenichi Suzaki
Fukuoka Inst. Tech., Japan
Fujio Tsutsumi
CRIEPI, Japan
Edger Velez
Northern Telecom, Canada
S. Wang
Univ. Sherbrooke, Canada
Kenichi Yoshida
Hitachi Ltd., Japan
Neng-Fa Zhou
Kyushu Inst. Tech., Japan
D. Ziou
Univ. Sherbrooke, Canada

Japan Executive Committee

Masanori Akazaki, Fukuoka Institute of Technology, Chair
Takushi Tanaka, Fukuoka Institute of Technology, Co-Chair
Ryohei Tamura, Fukuoka SRP Corp., Exhibition Chair
Toshihiko Ono, Fukuoka Institute of Technology, Poster Session
Ryuji Hamabe, Fukuoka Institute of Technology, Treasurer
Masaru Ishii, Fukuoka Institute of Technology, Registration
David J. Littleboy, Qtech Translation, Interpretation
Masanobu Umeda, Kyushu Inst Tech., Computer Network
Yoshio Nishikawa, AINEC Co., Exhibition

Makoto Amamiya
Kyushu Univ.
Sinji Araya
Fukuoka Inst. Tech.
Setsuo Arikawa
Kyushu Univ.
Toru Hidaka
Kyushu Univ.
Kazuo Ichimura
Mitsubishi Heavy Ind.

Masumi Ishikawa
Kyushu Inst. Tech.
Kazunori Kakemoto
Kyushu Electric Pow. Co.
Hiroki Kondo
Saga Univ.
Tadashi Kuroiwa
Kyushu Matsushita E. Co.
Kazuhisa Matsuo
Fukuoka Inst. Tech.

Naoyuki Motomura
Yaskawa Electric. Co.
Isao Nagasawa
Kyushu Inst. Tech.
Tadashi Nagata
Kyushu Univ.
Setsuo Sagara
Fukuoka Inst. Tech.
Takatoshi Tsuchie
Nippon Steel Corp.

Teruyoshi Ushijima
Fukuoka SRP Corp.
Takeshi Yamakawa
Kyushu Inst. Tech.
Masao Yokota
Fukuoka Inst. Tech.
Tokio Yoshida
Saibu Gas Co.

Staff

Cheryl Morriss,
SW Texas State Univ.

Mito Inoue
Fukuoka Inst. Tech.

Sachiko Kojima
Fukuoka Inst Tech.

Akemi Onuki
Fukuoka Inst Tech.

SPONSORS

International Society of Applied Intelligence
Fukuoka Institute of Technology

Cooperating Organizations

American Association for Artificial Intelligence
Association for Computing Machinery /SIGART
Canadian Society for Comp. Studies of Intelligence
European Coordinating Committee for Artificial Intelligence
IEEE Computer Society
Information Processing Society of Japan
Institute of Electronics, Info. and Comm. Engineers
Institute of Measurement & Control
Institution of Electrical Engineers
Institution of Electrical Engineers of Japan
International Neural Network Society
Japan Society for Fuzzy Theory & Systems
Japan Society for Precision Engineering
Japan Society of Mechanical Engineers
Japanese Cognitive Science Society
Japanese Neural Network Society
Japanese Society for Artificial Intelligence
Society of Instrument & Control Engineers

Support Organizations

Fukuoka City
Fukuoka Conv. & Visitors Bureau
Fukuoka Prefectural Government
Fujitsu Ltd.
Fukuoka SRP Corp.
Japan Airlines
Kyushu Electric Power Co.
Kyushu Matsushita Electric Co.

Mitsubishi Heavy Industries
Nippon Steel Corp.
Nippon Telegraph & Telephone Corp.
Research Found. for Electro Tech. of Chubu
Saibu Gas Co.
Support Center for Adv. Telecom. Tech. Res.
Yaskawa Electric Corp.

Exhibitors

Daewoo Electronics Co.
Fujitsu Ltd.
Hitachi Ltd.
IBM Japan Ltd.
IF Computer Japan Ltd.
Matsushita Electric Industrial Co.

Mitsubishi Electric Corp.
NEC Corp.
Nippon Telegraph & Telephone Corp.
Nippon Steel Corp.
SONY Corp.

Invited Papers

MICROMACHINES AS UNCONVENTIONAL ARTIFACTS

Naomasa Nakajima

Research into Artifacts, Center for Engineering, The University of Tokyo
7-3-1 Hongo, Bunkyo-ku, Tokyo, 113 JAPAN
Email: nakajima@mech.t.u-tokyo.ac.jp

ABSTRACT

Micromachines are extremely novel artifacts, with a variety of special characteristics which can be applied to people and nature. Micromachines do not affect the object or environments as much as the conventional machines do. Many Japanese researchers see micromachines as the ultimate in mechatronics, and the typical targets of them are autonomous machines which can be put on a fingertip, composed of parts the smallest sized of which is a few dozen micrometers. The development of micromachines requires a number of generic technology including those of design, materials, processing, assembly, energy supply, electric circuits, minute functional elements, information processing, and control. It is useful to categorize micromachine technologies from the two perspective of performance and diversity. Micromachine technology is essential not only for the development of micromachines themselves, but also for improving the convenience of machines in general. In using micromachines, their distinguished features, and their user- and environment-friendliness should be fully applied.

1. INTRODUCTION

From the beginning, mankind seems instinctively to have desired large machines and small machines. That is, "large" and "small" in comparison with human-scale. Machines larger than human are powerful allies in the battle against the fury of nature; smaller machines are loyal partners that do whatever they are told.

If we compare the facility and technology of manufacturing larger machines versus smaller machines, common sense tells us that the smaller machines are easier to make. Nevertheless, throughout the history of technology, larger machines have always stood out. The size of the restored models of the watermill invented by Vitruvius in the Roman Era, the windmill of the Middle Ages, and the steam engine invented by Watt is overwhelming. On the other hand, smaller machines in the history of technology are mostly tools. If smaller machines are easier to make, a variety of such machines should exist, but until modern times, no significant small machines existed except for guns and clocks.

This fact may imply that smaller machines were actually more difficult to make. Of course, this does not mean simply that it was difficult to make a small machine; it means that it was difficult to invent a small machine that would be significant to human beings.

Some people might say that mankind may not have wanted smaller machines. This theory, however, does not explain the recent popularity of palm-size mechatronics products.

The absence of small machines in history may be due to the extreme difficulty in manufacturing small precision parts.

2. WHY MICROMACHINES NOW?

The dream of the ultimate small machine, or micromachine, was first depicted in detail 30 years ago in the 1966 movie "Fantastic Voyage." At the time, the study of micromachining of semiconductors had already begun. Therefore, manufacturing of minute mechanisms through micromachining of semiconductors would have been possible, even at that time. There was, however, a wait of over 20 years before the introduction, 8 or 9 years ago, of electrostatic motors and gears made by semiconductor micromachining.

Why didn't the study of micromachining and the dream of micromachines meet earlier? A possible reason for this is as follows. In addition to micromachining, the development of micromachines requires a number of technologies including materials, instrumentation, control, energy, information processing, and design. Before micromachine research and development can be started, all of these technologies must reach a certain level. In other words, the overall technological level, as a whole, must reach a certain critical point, but it hadn't reached that point 20 years ago.

Approximately 20 years after "Fantastic Voyage," the technology level for micromachines finally reached a critical point. Micromotors and microgears made by semiconductor micromachining were introduced at about that time, triggering the research and development of micromachines.

The background of the micromachine boom which started 8 or 9 years ago can be explained by the above.

3. MICROMACHINES AS GENTLE MACHINES

How do micromachines of the future differ from conventional machines? How will they change the relationship between nature and humans?

The most unique feature of a micromachine is, of course, its small size. Utilizing its tiny dimensions, a micromachine can perform tasks in a revolutionary way that would be impossible for conventional machines. That is, micromachines do not affect the object or the environment as much as conventional machines do. Micromachines perform their tasks gently. This is a fundamental difference between micromachines and conventional machines.

The medical field holds the highest expectations for benefits from this feature of micromachines. Diagnosis and treatment will change drastically from conventional methods, and "Fantastic Voyage" may no longer be a fantasy. If a micromachine can gently enter a human body to treat illnesses, humans will be freed from painful surgery and uncomfortable gastro-camera testing. Furthermore, if micromachines can halt the trend of ever-increasing size in medical equipment, it could slow the excess growth and complexity of medical technology, contributing to the solving of serious problems with high medical costs for citizens.

Micromachines are gentle also in terms of machine maintenance, since they can be inspected and repaired without difficulty in reaching and overhauling the engine or plant. The more complex the machine, the more susceptible it is to malfunction due to overhaul and assembly. In addition, there have been more instances of human errors during overhaul and assembly. It is good for the machine if overhaul is not necessary. It is even better if maintenance can be performed without stopping the machine. Repeated stop-and-go operation will accelerate damage of the machine due to excess stress caused by thermal expansion.

Such gentleness of a micromachine is an advantage, as well as a weakness, in that a micromachine is too fragile to resist the object or the environment. This is the drawback of the micro-scale objects.

For example, a fish can swim freely against the current, but a small plankton cannot. This is a result of physical laws and nothing can change it. Still, the plankton can live and grow in the natural environment by conforming to the environment.

Unlike conventional machines which fight and control nature, micromachines will probably adapt to and utilize nature. If a micromachine cannot proceed against the current, a way will be found to proceed with the flow, naturally avoiding collisions with obstacles.

4. MICRO-ELECTRONICS AND MECHATRONICS

The concept of micromachines and related technologies is still not adequately unified, as these are still at the development stage. The micromachines and related technologies are currently referred to by a variety of different terms. In the United States, the accepted term is "Micro Electro Mechanical Systems" (MEMS); in Europe, the term "Microsystems Technology" (MST) is common, while the term "micro-engineering" is sometimes used in Britain. Meanwhile in Australia "micro machine". The most common term if it is translated into English is "Micromachine" in Japan. However "micro-robot" and "micro-mechanism" are also available case by case.

The appearance of these various terms should be taken as reflecting not merely diversity of expression, but diversity of the items referred to. Depending on whether the item referred to is an object or a technology, the terminology may be summed up as follows:

Object: micro-robot, micro-mechanism
Technology: micro-engineering, MST
Object & technology: MEMS, micromachine, micro machine

With regard to technology, if we summarize the terms

according to (1) where the technology for micro machine systems branched off from, and (2) whether the object dealt with by the technology in question is an element or a machine system, the terms can be organized as follows. That is, MEMS, MST and "micro-engineering"stem from micro-electronics, and have developed with targeting elements. "Micromachines" stem from mechatronics, and have developed dealing mainly with machine systems ("micromachines" are regarded in Japan as the ultimate mechatronics). In this sense, MEMS, MST and "micro-engineering" on the one hand and "micromachines" (including "micro machines") on the other hand form two separate groups, but as the former has started to move in the direction of machine systems, while the latter has already incorporated microelectronics, the differences between the two groups are gradually disappearing.

Looking at the areas included in the two groups, given that the machine systems which are the main concern of "micromachines" include elements, and given also that "micromachines" include microelectronics, it would be natural to assume that micromachines include MEMS and MST. The same relationship holds true for the actual objects treated under each category.

5. THE DEFINITION AND DEVELOPMENT AIM

It is difficult at present to give a unified definition of micromachines, but if these are taken to be limited to machine systems as output of micromachine technologies, the scope for variation of the definition narrows slightly.

The micromachine technology project being promoted under the ISTF (Industrial Science and Technology Frontier) Program Agency of Industrial Science and Technology (AIST) of MITI, and the Micromachine Center, define micromachines as follows:

Micromachines are small machines composed of sophisticated functional elements less than a few millimeters in size, constructed to perform complex tasks on a small scale.

The above definition of micromachines is in fact inseparable from the development aims for micromachines. At present, debate on the definition of micromachines is exactly the same as debating development aims, that is, the diversity of definitions of micromachines reflects the diversity of development aims.

6. THE EVOLUTION OF MACHINES AND MICROMACHINES

Many Japanese researchers see micromachines as the ultimate in mechatronics, developed out of machine systems. I should like to discuss this point further.

Ever since the Industrial Revolution, machine systems have grown larger and larger in the course of their evolution. Only very recently has evolution in the opposite direction begun, with the appearance of mechatronics. Devices such as video cameras, tape recorders, portable telephones, portable copiers which at one time were too large to put ones arms around, now fit on the palm of ones hand.

Miniaturization through mechatronics has resulted mainly from the development of electronic controls and control software for machine systems, but the changes to the structural parts of machine systems have been minor compared to those in the control systems. Even today, the smallest parts of machine systems measure about one millimeter; this applies equally to watches and to construction machinery. This is probably because existing production systems are geared to produce parts only down to this size. As a result, there are very few practical machine systems smaller than an insect; the only example that comes to mind is watches.

The next target in miniaturization of machine systems is miniaturization of the structural parts left untouched by present mechatronics. These are the micromachines which are seen as the ultimate in mechatronics.

Seen in this light, the aim of micromachines can be expressed as follows:
——"Micromachines are autonomous machines which can be put on a fingertip, composed of parts the smallest sized of which is a few dozen micrometers"

That is, since micromachines which can be put on a fingertip have to perform operations in spaces inaccessible to humans, they are required to be autonomous and capable of assessing situations independently, as are intelligent robots. To achieve this kind of functionality, a large number of parts must be assembled in a confined space. This factor determines the size of the smallest parts, and given the resolution of micro-machining systems, a target size of several dozen micrometers should be achievable.

7. PERFORMANCE AND DIVERSITY OF TECHNOLOGY

However the development targets for micromachines are set, it will probably be generally agreed that silicon-based microelectronics will be important in their achievement. Even if we kick up an approach different from MEMS to form structural elements and the structural parts of machine systems by processing techniques other than micro-machining of silicon, the control of micromachines will depend on the mounting of sensors and integrated circuits based on micro-electronics.

There will probably be general agreement also that whatever the development aims for micromachines, other technologies besides micro-electronics and silicon micro-machining are also important, such as materials, measurement, control systems, energy supply, communications, data processing, design and manufacture technologies, etc.

That is, micromachines will not come about through the completion of one particular key technology; various technologies such as those mentioned above will have to be developed to a certain level. In this sense, micromachines can be said to require comprehensive technology, as do mechatronics devices.

It is useful to categorize micromachine technologies from the two perspective of performance and diversity. Firstly, performance has been given top priority in research and development, for instance the precision, speed, torque, power, efficiency, durability, reliability of functional elements and machine systems.

However, even if these factors are brought up to the target levels, this will still not be enough to produce practical, commercial micromachines. This is because the incorporation of actuators and energy sources in complete micromachines must fulfill other conditions, such as special product conditions stemming from the environment the machine is to be used in; these may be prohibitions on electrical power, limitations on the materials used. Fulfillment of these conditions will require adequate diversification of related technologies as well as of performance.

8. THE POSITION OF MICROMACHINES

I should like here to compare micromachines as machine systems with conventional machine systems, molecular machines, and nano-machines, from the standpoint of design and composition.

Firstly, comparing the differences in the design of micromachines and conventional machines, several problems emerge which are not relevant to conventional machines, such as the need for scale analysis and analysis of phenomena unique to the microscopic world, and the impossibility of drawing a clear distinction between research and development and design processes. On the other hand, the design requirements of conventional machines mostly apply to micromachines also, and the design of micromachines can therefore be said to include the design of conventional machines.

Next, comparing micromachines and molecular machines, whereas micromachines achieve the required function through integration of element parts formed into the required shape by processing of lumps of material, molecular machines aim to achieve the required function through direct control of atoms and molecules. In micromachines, however, we can take it that there are no restrictions on the processing techniques used to form the element parts. If therefore molecular machines are incorporated into micromachines as component parts, the result would also class as a micromachine.

The definition of nano-machines is vague, but the term seems in general to refer either to molecular machines or to micromachines whose component parts are measured in nano-meters. In the latter case, there is no meaningful distinction to be made between nano-machines and micromachines, except where quantum effects are a problem, and nanomachines can apparently therefore be included in the category of micromachines. The "micro" of micromachine does not necessarily refer to micrometers, and there is therefore no lower limit on the size of the component parts.

9. CHANGING VIEW OF MACHINES

It is very important that a machine is easy for people to use. Since the beginning, engineers have worked toward the goal of developing such a machine. Even a classic machine, which may look simple, was the result of an effort to make an easy-to-use device.

For example, it is well known that Watt's steam engine had a speed regulating mechanism called a governor, but the windmills of the Middle Ages are also said to have included automatic brakes that prevented excessive rotation speed. The brakes were needed to avoid high-speed rotating, which would cause the turning stone mill to float, thus reducing its efficiency.

When ease of use is required of a machine, the number of mechanisms generally increases, and such mechanisms must be arranged efficiently within the space inside the machine. The placement of the control mechanism mentioned above was probably not a problem because of the device's large size, but the problem is not so easily solved in a small machine.

One of the reasons for the difficulty in developing smaller machines may have been the problem just mentioned.

When there is no more room for additional mechanisms in the machine, it is necessary, to reduce the size of the parts. In general, the size of the smallest part possible to process with current technology is approximately 1mm. This size limit is even larger if productivity and economy must be considered.

To make a machine easier to use, the degree of integration of the parts must be increased. Thus, the current size of parts must be drastically reduced; micromachine technology makes such size reduction possible. Micromachine technology can probably reduce the part size to 1/100 to 1/1,000 of current sizes. This will make future versions of current machines far more convenient and useful.

Micromachine technology is essential not only for the development of micromachines themselves, but also for improving the convenience of machines in general.

When micromachine technology brings gentle machines into existence and makes machines in general far more convenient than they are today, our view of machines will also change. The new world opened up by micromachines may also contribute to guiding the development of modern machine civilization in desirable directions.

10. CONCLUSIONS

Micromachines are unconventional artifacts with respects to their gentle features to people and nature. The current diversity of the definition of them are originating from development objectives and technological starting points. Micromachine technologies, categorized from the two perspective of performance and diversity, are expected as generic technologies for the twenty-first century to support industry and medicine as well as daily life. Micromachine technologies are essential also for improving the conventional machines in general.

Micromachines are artifacts in tiny size, but they will exert a strong influence on our lifestyles and society.

MULTI-STRATA MODELING TO AUTOMATE PROBLEM SOLVING INCLUDING HUMAN ACTIVITY

Setsuo Ohsuga

Department of Information and Computer Science,

Waseda University,

3-4-1 Ohkubo Shinjuku-ku Tokyo 169, Japan

Email: ohsuga@ohsuga.info.waseda.ac.jp

ABSTRACT

The objective of this paper is to discuss a formal method of designing automatic systems and also an intelligent system to realize this idea. This method is to build a model of objects including human activity and then transform the model into an automatic problem solving system. In other words, this is to generate a special purpose problem solving system for a given problem. This problem specific system is extracted from a general purpose problem solving system using the object model as a template. Thus in this paper a way of developing general-purpose problem solving systems is discussed first, then concept of multi-strata model is introduced as a new modeling scheme to represent problem specific models including human activity. Finally a way of generating a problem-specific problem solving system is discussed. This idea leads us to automatic programming.

1. INTRODUCTION

The objective of this paper is to discuss a scheme for modeling a problem solving process including persons and for automating the process in intelligent systems. A concept of multi-strata model is introduced. It is argued that a model has two aspects corresponding to the different stage of problem solving; one is a representation of human idea and requirement to be met by problem solving process and the other is a representation of a problem specific object which meets the user's requirement. Problem solving then consists of an incipient model building to represent the first aspect and goal finding by modifying the incipient model to represent the second aspect. The first process is called in this paper 'externalization' denoting human activity to externalize one's idea. Usually an idea created in a person's brain is quite nebulous and the person cannot externalize it clearly. Some methods of aiding to externalize one's idea are being studied and are expected effective for fast problem definition [HOR94]. The second process is called 'exploration'. This is an activity to find the goal in a possible solution space. Since the goal is not foreseen in advance for most problems, exploring in the space is necessary. Externalization and exploration use the different technologies.

From the human-computer relation's point of view, externalization needs an intrinsic human-computer interaction because its objective is to transmit information from human users to computers. Contrarily, if a problem is clearly defined through externalization and computers are provided with enough knowledge to solve problems in advance as well, then, in principle, exploration can be automated within computers. Practically however to meet this condition is often difficult. Even with methods for aiding externalization it is difficult for persons to externalize their idea completely and some part of information which could not be externalized remains in one's brain. This information must be excavated later. Computer may find during explorations process that information on a problem is incomplete and ask the person to supply the information. Until this time person may not note that the information has not been externalized completely yet. Other than this case, computers may need person's help because knowledge for solving problems is not enough. These human-interactions in exploration are however not intrinsic. These are necessary because technology level is not high enough. With the progress of computer-aided externalization and with the use of very large knowledge base, the frequency of this type of interactions may reduce and the chance of automatic exploration increases.

In this paper computer aided externalization is first discussed and then attention is paid on a way of achieving the autonomy of the exploration. In particular, the paper intends to automate complex problem solving process involving persons. It includes to model an activity including person and represent it explicitly so that formal method of automation system design can be attained as a method of processing this model.

2. MODEL-BASED METHOD OF KNOWLEDGE UTILIZATION

2.1 Basic Method of Problem Solving

In the early expert systems knowledge was used directly to generate answer for user's question. It was assumed there that knowledge can be made universal. In reality knowledge is specific to each individual and its interpretation can be different for every person. This difference of interpretations by individuals causes users confusions and as the result makes people distrustful of the system. Moreover it is difficult to build up complex objects in the systems and to acquire knowledge. These were shortcomings in the way of using knowledge in computers.

As an alternative of this style a problem solving scheme as shown in Figure 1 is proposed in which user's problem is represented first in a special form in the system such that a solution can be derived by transforming it formally. The result of transformation is evaluated in the system based on user's

criterion given to the system or directly by the user. The transformation is repeated until the goal is reached. Knowledge is used for this transformation in this process. Thus it takes part in the solution process merely indirectly. Let this representation of problem be called conceptual object model or simply object model by the reason discussed below. Problem solving is a process of changing the model representation successively to arrive at the final form to generate solution. Model must be able to be evaluated and changed not only by part but as a whole. This approach is adopted in Black Board model partly which showed a wider applicability than any other expert systems but is still problem dependent. Knowledge sources and their management must be specified by human experts to each specific problem.

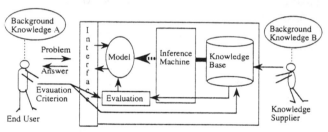

Fig. 1 Model-based method of problem solving

If a model is evaluated completely from the user's point of view every time it is changed, any knowledge can be used without worrying much about the differences in interpretations by individuals. This scheme is robust enough for the variation of inputs and knowledge. This is a necessary condition of making large knowledge base to be shared by users who can't be specified in advance. It should also be noted that this scheme enables users to construct a big model incrementally from small components.

Figure 2 shows diagrammatically the basic configuration of this scheme. The different blocks represent the different operations. In reality these are performed by the same deductive operations. The different knowledge sources are assigned to the deductive operation to define the different operations by a strategic and control operation in the higher level. This is to classify a knowledge base into the different chunks such as for analyzing and modifying model, and assign different chunks to the deductive operation in a certain order in order to define the different operations. This is also a knowledge-based operation in the meta-level. It is possible then that a model is built such that it reflects at every instance all decisions ever made until the current model representation is reached during problem solving so that the next decision can be made on the basis only of the current model without referring to the history of decisions. This is an important advantage expected to this style of model based problem solving. In this paper this model-based problem solving scheme is considered as the basis. It is combined with the other issue such as the structuring knowledge bases.

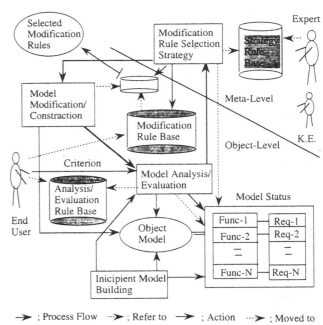

; Process Flow ⋯▶ ; Refer to ➔ ; Action ⋯▶ ; Moved to

Figure 3 A realistic representation of system

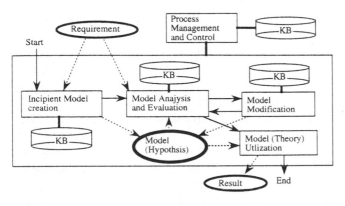

Figure 2 Basic method of problem solving

A more realistic representation of model-based problem solving system is shown as Figure 3. Problem solving in this system proceeds as follows. A model state is created by analyzing the current model to reveal the distance to the goal therefrom. One or some model modification rules are selected from a large modification rule base by a selection subsystem. The selected rule(s) is used to modify the model. This selection is carried out on the basis of strategic knowledge. As the selection implies the evaluation of knowledge, this knowledge-selection knowledge is meta-knowledge in the meta-level.

Different strategy may be used to select knowledge. A typical way is to classify the knowledge by its characteristics and give information to each class to represent its characteristics in advance. For example, some knowledge modify the model to strengthen one of its properties while the other weaken.

These are classified into different classes and each class is given a comment to denote the characteristic. This is meta-knowledge. The selection subsystem selects one of them referring to the model state. An example of application of this system to chemical compound design is shown in [SUS93]. If the knowledge for selecting knowledge is so made as to select only a best modification knowledge, this process progresses autonomously. If more than one rule is selected for modification, more than one models are created as new candidates. One is selected among them either by means of the different strategy in the system or by human user. In order to manage more than one different meta-level operations, meta-meta-level operation becomes necessary. Thus an architecture with more than one meta levels becomes necessary. The necessary number of levels is dependent on the problem and user's requirement. System must be able to define an arbitrary number of description levels to meet user's requirement. It is necessary to study the concepts of modeling from this point of view and also to define multiple meta-level operations.

2.2 Model Manipulation

Various model manipulations are necessary for attaining to solution for the given problems. These are classified into Externalization and Exploration. The major objective of this paper is to discuss the way of automating exploration.

2.2.1 Group of Models

With a proper method of representing a conceptual model, a single problem solving scheme as was shown in Figure 2 can cover a very wide class of problems. In this scheme, a model is seen from different points of views depending on problems. A view is defined by an operation applicable to a specific part of model and is referred to as an aspect hereafter. This is in fact an operation specific model representation. For example, a model must be analyzed in many aspects and a specific operation is necessary for each of them which requires an operation specific model representation. Thus a model must be able to represent different aspects. Since the aspects are problem dependent these cannot be prepared in advance but must be generated in each case dynamically from a kernel model. This condition can be met by giving systems domain specific or problem specific knowledge which generates an operation specific model from a kernel model. This implies that not only a kernel model but in reality number of its variations are involved in the representation of a problem solving. Some researchers note that multiple models are necessary [TOP94] for knowledge based operations. Thus the model in Figure 2 is in reality a group of models (Figure 4).

2.2.2 Two Classes of Operations

Operations needed for problem solving are classified into two classes; externalization and exploration. Externalization is operations for representing person's idea/problem explicitly. An external model is obtained through externalization. The process of making the external model is called the Externalization Phase.

Various externalization methods may be used to define the different aspects of an external model. These aspects must be combined to form an object model in the canonical form (kernel model) by the use of knowledge prepared for each externalization method.

Problems (external model) brought in by

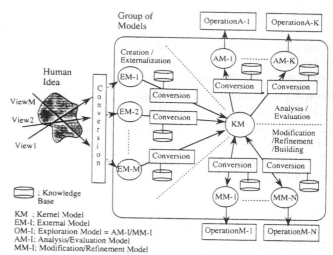

; Knowledge Base
KM ; Kernel Model
EM-I; External Model
OM-I; Exploration Model = AM-I/MM-I
AM-I; Analysis/Evaluation Model
MM-I; Modification/Refinement Model

Figure 4 A group of models

this process must be solved by exploration. This is called the Exploration Phase. Various methods must be used for exploration. Every method must be represented explicitly as a set of rules (knowledge) involving a model as an object. Some rules may include procedural programs. Each method is defined to deal with a specific aspect of model and requires its own operational model (aspect) representation as mentioned above. For example it may be a graph with predefined functions specified to nodes. An arc may denote an interrelation between nodes or a path of message passing.

Thus a kernel model must be converted to an operational model which is specific to each method used there. This conversion is necessary not only for exploration but for producing a product from the model and is achieved by domain/problem specific knowledge. This knowledge is defined and given to the system together with the method. To generate an operational model which can be processed in the system is called Operationalization.

In this manner, a group of models must be generated to solve a problem. If the system is provided with a modeling scheme by means of which (1) every model can be represented by means of a language and (2) every operation involved in the Externalization Phase and the Exploration Phase can be represented by the same language including this modeling scheme, then the whole process can be processed by this system. In order to design such a language as to meet this condition, a modeling scheme must be defined first.

2.3 Externalization

Every problem must be represented by person. This is an activity to represent human idea explicitly in the form of model. It can be a multi-strata object model. It is not an easy task due to ambiguities included therein, especially to unclearness of idea in the human side [BIJ94]. Persons come to vague ideas and cannot represent them clearly. Computer can help them to externalize the ideas in various ways. Externalization therefore includes to help persons to clarify their idea. Therefore it is closely related with creativity. Intelligent systems can take part in this process and stimulate persons to clarify their ideas as is shown in Figure 5 below.

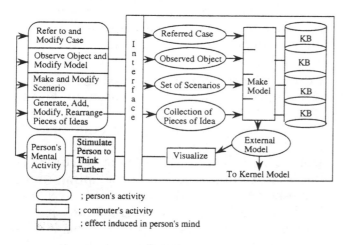

Figure 5　Model building methods

There are some possible ways of making external models depending on the case. Some of them are referred here. These are (1) using past case, (2) referring to an existing object, (3) prototyping, (4) making a scenario of achieving the given goal with not yet existing object and (5) forming a model from a collection of fragments of idea. The brief comments are given as follows. (1) is well known as case-based method. (2) is useful when an object is observable. For example, an enterprise model is made by observing an existing enterprise organization. Requirement specification for a business information system is generated from this model

[BUB93]. (3) is well known and is effective if it is possible. (4) is used for starting an innovative design, especially in engineering. For example, in designing a new type of rocket vehicle, its expected missions are first made and represented in the form of a scenario. This scenario defines an outline of a model representation. (5) aids a person to build up his idea. For example, computer visualizes person's idea by arranging the keywords representing fragments of the idea in a multi-dimensional space on the display. Users manipulate these keywords so that these are arranged according to their closeness with the idea. As the result the person can note or find something unnoticed before. Hori developed such a system in [HOR94]. Requirement or even the specification for the thing which he/she wishes to do, or in other words, a problem representation may be made clear through these processes. These functions form the important parts of human-oriented interfaces. The detail is abbreviated.

2.4 Model Exploration

Problem solving is in general to explore aiming a goal. The search space is vast and in general unforeseen. Exploration begins by creating an incipient model as a starting point in this space. The point is examined whether it satisfies the requirements. Thus exploration is composed of such operations as (1) creating an incipient model and guiding the direction in which to proceed and to select a next point (model) as a candidate, (2) making or changing the model and (3) analyzing and evaluating the model. The second and third are the operations to manipulate an object model directly while the first is to control the second and the third operations. Thus the first is a meta-level operation for the second and the third operations in the object level.

Among the object level operations, an analysis-and-evaluation is defined for a specific model. It can be used when a specific aspect of a model is fixed tentatively. On the other hand, there are different ways of building and changing a model depending on the case. These are (1) lateral expansion, (2) bottom-up model construction, and (3) top-down model refinement. The lateral expansion is to use the model of an existing object as the reference. It is the same as the using past case ((1)) in the externalization. The bottom-up model construction is to make a structure making the structural relations between components. This can be used for rather the small scale problems and when a set of components are fixed. A top-down problem solving is closely related to problem decomposition and is discussed in [OHS94] in detail. This is the almost only method of solving very large scale problems. The further details are abbreviated because these are not the main objective of discussion in this paper. Usually in exploration, only a small portion of the exploration space, i.e. the neighborhood of the current model, can be really explored. Guiding the process properly is very important.

2.5 Autonomous Process and Interactive Process

There are two type of human-computer interactions in problem solving; intrinsic interaction and interaction because of lack of problem solving capability of systems. Human-computer interaction is intrinsic to externalization not only because it is to transfer information from human to computers but because ambiguity involved in human thinkings is the intrinsic nature.

On the other hand, even if problems could be specified clearly and even with systems with well designed architecture, human decision making becomes necessary in solving them because of the lack of problem solving capability of the systems due to the lack of knowledge in the system. Unknown problems always arise to which enough knowledge cannot be provided in advance. In particular, making and modifying models in non-deterministic problem solving require interactions. As was mentioned above, explorations are apt to be confined to the neighbor of the current model. In order to achieve innovation, it is necessary to change the context and make an abrupt change. Usually such an abrupt change cannot be anticipated beforehand but need human decision. Thus interaction is indispensable though it is not intrinsic. In this case human beings behaves as one of knowledge sources to the system. If the system is added with knowledge to

make the same decision as is made by person or accumulates the human responses obtained through the human-computer interactions, then the system can make decisions itself afterward. The system can glow and the problem solving capability gets large approaching the more autonomous system.

An ideal system may be such that externalization and exploration are clearly separated and human intervention because of lack of problem solving capability is the least. Since an intrinsically interactive activity cannot be included in automation, the objective of system design is to automate exploration. As the matter of course, the system must be so designed that persons can intervene the system's operation any time.

2.6 Meta-Operation for Controlling Problem Solving Process

It is necessary to represent a problem solving process explicitly in order to perform certain operations in a specified order. A control is also necessary to guide exploration in order to reach the goal as fast as possible. As an operation is an application of knowledge to a model through the inference mechanism, problem solving process and its control can be represented as the certain methods of selecting knowledge. In order to give the system a capability of adapting to the changing environment and also the freedom of adding, deleting or changing knowledge, these meta-operations should also be knowledge-based. This knowledge is to describe about other knowledge. For example it is to decide the preference of some knowledge to the other at the object level. This is a meta-knowledge.

Meta-level representation is necessary in various other cases. In particular, it is necessary for representing multi-strata models. Thus the concept of meta-level representation is indispensable for developing intelligent systems. This is basically the matter of language design to allow user to define any kind of meta-level operation.

3. APPROACH TO AUTOMATIC EXPLORATION

An approach taken in this paper for automatic exploration is to extract only the necessary functions to solve the specific problem from a general structure of intelligent facilities when the problem is given or, in other words, to generate a problem specific system from a very general purpose problem solving system. There are two big problems to be solved. The first problem is to provide computer systems with capability to generate a large solution space to cover different type of problems which an agent can explore to find solutions. Unless the space is large enough, the limit of the space is reached before reaching the goal. It implies to develop very general problem solving systems and, considering that a real complex problems relate with the different problem domains, it must be able to cover wide problem areas. However in general the larger the exploration space becomes, the longer the time is spent for reaching the goal. There must be some method to reach the goal in as small time as possible yet allowing to use huge knowledge bases. To generate a problem specific system from a very general system is an approach to resolve this conflicting requirements. As a cue for extracting a necessary functions from the general purpose system, the concept of multi-strata modeling is introduced.

4. EXPLORATORY SPACE GENERATION

The first difficulty to resolve is to provide computer systems with a large solution space. In reality what is represented in the system is not the space of instance solutions but the space of knowledge to generate possible solutions. It is necessary to find a way of making such a large knowledge space as well as a way of rapid access to the necessary knowledge. In the early expert systems, it was believed that a knowledge base formed as a collection of rules and some heuristic search algorithm can be effective for producing answer for user's question. This is too simple compared to the way of persons use knowledge. In many cases an intelligent activity like design includes various other intelligent methods such as model building, mathematical analysis, decomposing a problem into smaller problems, and so on. There

are dependency relations among intelligent activities such that a complex activity is defined using simpler ones. A structure composed of intelligent functions can be built on the basis of this relation. This issue has been discussed by the author in [OHS95] and is mentioned only briefly in this paper with slight modifications. A basic method of problem solving has been shown in Figure 1. This method is available however only when the scale of problems is rather small. A very large problem needs to be decomposed into a set of small and less complex problems before this method is used [OHS94]. Figure 6 shows a way of problem decomposition. A large set of requirements to define a large problem are classified into a set of smaller ones. Each small set of requirements defines a small subproblem. This classification is continued top-down until the every new subproblem is small enough to be processed conveniently by the basic problem solving scheme of Figure 1. Decomposition is to find a best way of classifying the set so that every subproblem can be solved in the easiest way. Since the subproblems cannot always be independent completely from each other, the objective of decomposition is to decompose an original problem into a structure of small and less complex subproblems with the least interactions among them. This is to find a best structure of subproblems and is a kind of design problem. This problem is therefore represented in terms of basic method of problem solving of Figure 1 and is ranked upper to the basic problem solving method to indicate that the former can be defined by using the latter.

In this way a hierarchy of different intelligent functions are made. The upper layer functions are defined on the basis of the lower layer functions. This structure must or can be defined by analyzing various activities such as design, diagnosing, decision making, planning, automatic programming, etc. For example, design involves different functions such as (i) clarify the design requirement and externalize designer's idea, (ii) to decompose large design to a set of smaller ones so that each small design can be managed by each individual or terminal computer, (iii) to solve small design problems, (iv) to generate

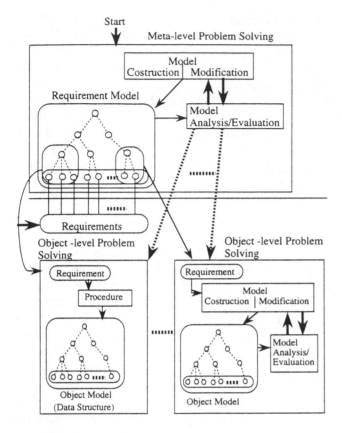

Figure 6 An illustration of decomposition

telligent functions through education, starting from understanding the easy subjects then proceeding to the more complex issues represented on the basis of the easy functions. This is a way students can glow and acquire such high level capabilities as to solve complex problems. The term 'layer' is used here to represent the dependency relation of the functions. It is different from the term 'level' which is used to represent description level such as object-level, meta-level etc.

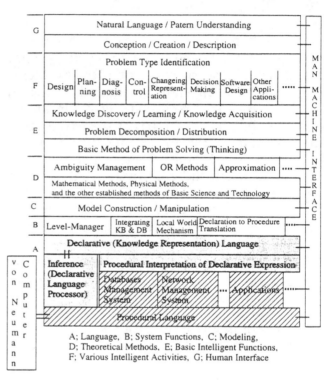

A; Language, B; System Functions, C; Modeling,
D; Theoretical Methods, E; Basic Intelligent Functions,
F; Various Intelligent Activities, G; Human Interface

Figure 7 Structure of intelligent functions

information for producing the object from the resulting object model. Different activity requires the different ways of using intelligent functions. To find such structures for the various other activities as for the case of design is necessary. This top-down approach is not easy to do nor can be made complete. But it must be made once as far as possible in order to find necessary set of functions with which various human activities can be covered.

Figure 7 is a structure which the author proposes. This figure shows only a part of the large functions map because of the lack of space. In this structure intelligent functions are arranged in such a way that a function in a layer can be decomposed to and represented by the functions arranged in the lower layers than at this layer. The function structure made up in this way is called the global function structure. This structure shows the ways of various intelligent functions being realized with no concern with their applications to the specific problem solving. Making this structure is similar to making curriculum in a school where students get the structure of in-

In problem solving, various theoretical methods which have been established in mathematics, operations research and the other domains must be fully utilized. Each method requires its own model representation and the different model representation is used in the different method. There must be a canonical form of model and the specific models required by these methods are generated as the transformation from the canonical form as has been discussed before. The functions for manipulating model must be at the lower layer of the theoretical methods.

In the next higher layers of these functions, there are different function layers including basic problem solving, decomposition and so

on. These are used directly for exploring the solutions to the given problems.

On the other hand, functions at the top layers in Figure 7 is to aid human mental activity for externalization. As has been discussed in 2.5, the operations in these layers require intrinsically the human machine interaction.

Between the layers related with externalizing and exploring respectively is a layer for identifying problem types. Through the externalization process at the highest layers of this structure, problems are defined and represented explicitly. The problems are accepted by the function in this layer for their problem types being identified. There can be the different types of problems such as design type, diagnostic type, planning type, and so on. Each one requires the inherent problem solving method which is represented by a specific structure of intelligent functions in the global function structure. Let this problem oriented structure of functions be called a problem-specific function structure. In contrast to the global function structure being made in such a way that every function in a layer can be realized as a composition of some functions in the lower layers, this problem specific structure represents the specific way of using functions to solve the required type of problems. The 'problem-type-identification' function located below the human interface layer binds the given question to a corresponding structure of functions. This function is represented by a set of rules for identifying problem-type. Externalization and exploration are separated here. This does not mean that these two operations can be separated at execution but to-and-fro operations are often necessary. Systems must be able to facilitate this process.

Declarative language is necessary for representing these functions and the ways of organizing the functions to represent various activities. At the bottom of the structure are the functions which can be represented in the procedural form and therefore can be processed by the current computers. Declarative language is put above the procedural language layer and made executable because its language processor (inference) is realized by procedural language. Hence the functions in the upper layer are expanded finally to a set of procedures at the bottom layer via the declarative language expressions. Some parts of this ideas have already been implemented by the author's group and applied to various problems. The system named KAUS (Knowledge Acquisition and Utilization System) has been developed for the purpose [YAM93].

5. METHOD TO REDUCE PROCESS TIME

The more general a system becomes, the longer the time is needed for reaching the goal. There must be some methods to resolve this conflict. First, it is necessary to limit the search space before going into exploration. The scope of exploration must be specified to each problem. Model is used as a cue for limiting the exploration space because a model is a representation of an instance problem. Second, it is required to guide the exploration in this limited area so that faster processing is achieved. Strategic knowledge may become necessary which gives general principle to guide exploration. It is not problem specific but domain specific. There must be some scheme to acquire strategic knowledge as well as object knowledge. The scope of exploration is narrowed by these methods and finally problem specific goal is found by trial-and-error process in this scope.

The author has discussed in [OHS95] the second problem, i.e. an architecture of intelligent systems to control the exploration by means of control knowledge in the meta-level in contrast to the object-level in which exploration is achieved. Thus the first problem, i.e. a way of generating a problem specific system is discussed hereafter.

6. MULTI-STRATA STRUCTURE OF MODEL

Conceptual modeling must be defined in the context of a relation between subject and object. The subject has an intention to obtain some product from an object. The subject can be either a person or a computer and the object is anything in which the subject has an interest. There can be the different combinations of subjects and objects and a different conceptual model is created for each combination (Figure 8).

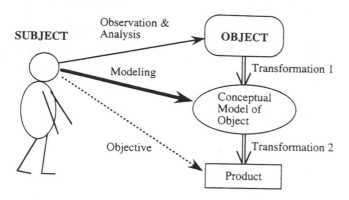

Figure 8   Conceptual Modeling

Everything in the world can be an object of human thinking. It can be a real object or an abstract concept. It can even be other problem solving processes. An object has many aspects or characteristics, but only some of them related with a person's interest are included in a conceptual model. Hence, different models may be created depending on an personal interests even if the object is the same. In the same way, a subject can be other than a person. We have an interest in the case where the subject is a computer.

There must be different methods of building a model and generating a solution. When the subject is a person, these methods are created in the human brain. People are not always conscious of the method of building this model by themselves, if their objective is to obtain a final product. However, when the subject is a computer, it is necessary to formalize this method of building a conceptual model and give it to the computer in advance. Human way of building a model must be studied and represented explicitly. This study is itself another problem consisting of a subject and an object in which the object is a human mental activity to build a model and generate a product and the subject is a person who studies and models this mental activity. This problem involves another problem solving and is represented as a double process; the inside process for obtaining a product directly from the given object and the outside process for obtaining a way of obtaining the product in the inside process.

Studying human activity is not easy but once after the model building activity has been clarified, its implementation in computers becomes possible. Thus this double process represents a process of computerizing problem solving (Figure 9).

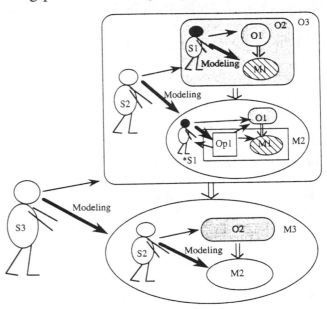

Figure 9   Multi-strata object

Sometimes a model with the more number of strata is necessary. The author discussed it in [OHS96] referring to the case of intelligent tutoring system design. Another example of many strata modeling is seen in modeling big enterprises. Thus the modeling scheme can be different depending on the situation and characteristics of the constituents in this context. If a taxonomy of schemes of modeling to cover all different situations can be defined, then it enables us to design intelligent information processing systems which can accept diverse problems with the different characteristics. It resolves some difficulties involved in information processing. For example, it will help us to formalize software design. To discuss conceptual modeling, therefore, is to discuss the basis of information processing. It means that a general purpose information processing system which is completely different from the ordinary computer systems is realizable.

It is possible to assume always the existence of a still outer stratum for every existing problem solving. Accordingly, in principle, an infinite strata scheme may be assumed resulting in an infinite taxonomy of problem solving. However, practically this is nonsense. In practice the number of strata is limited and almost all problems can be represented by models of within a certain finite strata because the effect of an inner process to the outer processes reduces rapidly and it is of no use to take too many strata into account. Contrarily sometimes the number of strata which a person can take into consideration is less than practically necessary because of the limit in the human capability of dealing with very complex problems. Intelligent system being discussed here can help him/her by including the necessary number of strata in a scope. The depth of strata which is taken into account is included in the rules to process this multi-strata model as discussed in Chapter 9. In this way diverse human mental activities can be formalized and represented in the form of conceptual model in this paper. It is necessary to find a modeling scheme which can cover this diversity.

7. MODELING SCHEME

7.1. Basic Modeling Scheme

It is desirable that a single modeling scheme can generate different conceptual models. Since different models are necessary for representing not only different problems but even a single problem (refer to 2.2.1), the generative power of the modeling scheme must be large enough. It is well known that a model must include structural information and functional aspect of the object as its basic components. Here various concepts such as attribute, property, behavior, function, relation with the other entity(ies), i.e. characteristics of an object which need to be represented in the form of a predicate, are referred inclusively by the term functional aspect. Its basic form of representation is a predicate composed of the basic atom predicates of the form Functionality (Object-Structure, Value). On the other hand, object structure is represented as a data structure. Thus the modeling scheme is an object structure in which functional aspects are arranged. This scheme will be referred to as a standard modeling scheme. Figure 10 shows an example.

In order for the modeling scheme to have a large descriptive power, expressive power of both data structure and functional aspect must be large enough. In most cases a functional aspect can be well represented by a first order predicate. But in representing multi-strata models and strategic rules, it requires the higher level expressions. Data structure must also be rich enough in order to represent every necessary structural concept. Moreover, in order to give systems creative power, it is desirable to be self-generative. For the practical purpose the ordinary database scheme may be used to save a part of model. Then database model must be represented by the same modeling scheme. The characteristics of this modeling scheme is discussed in [OHS96].

When some part of this model representation, either object structure or functional aspect, is unknown and someone wishes to know the unknown, a problem is created. Some quantity can be substituted into the unknown to form a solution. Thus the same modeling scheme covers the internal representation of problems and their solutions.

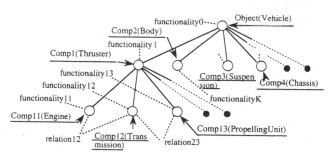

Figure 10 Basic Modeling Scheme

7.2 Problem Type

Different type of problems can be represented in this modeling scheme depending on the unknown part in the form of Functionality (Object-Structure, Value). First, if Value is the unknown, it represents a deterministic (analytic) problem. For example, a value of an attribute of an object is uniquely determined when its structure is fixed. Second, if Object-Structure is the unknown while the other items are known, it represents a nondeterministic problems. A typical example is the design problem in which requirement is given in the form of Functionality (x, Value) and Object-Structure is to be obtained as a structure to be substituted into x. Third, it is possible to assume that Functionality is the unknown. It is represented x(Object-Structure, Value). Formally it is read as "to what Functionality the given Object-Structure has the given Value ?". Discovery and learning problems can be included as the variation of this type of problem.

Different exploring methods are necessary for the different type of problems because of the different direction to follow the relation between structure and functionality. In general, structure-to-functionality relation is one-to-one but the functionality-to-structure relation is one-to-many. In the first case above in which a value is unknown, a solution is uniquely determined because of the one-to-one relation of structure-to-functionality and if some theoretical or the other method to obtain the solution has already been found, it is obtained straight. Such a method is stored as knowledge at the layer of the established methods in the function structure of Figure 3. If any theoretical method is not yet discovered for some given problem, then simulation method or physical experimental method become necessary. Any attempt to find such theoretical method forms another problem solving of either the second or the third type depending on the problem formulation.

In the second case above, problem is non-deterministic because of one-to-many relation of the functionality-to-structure relation and a heuristic method becomes necessary for problem solving. Figure 1 is the standard problem solving scheme of this type. It includes the first type problem as a part for analyzing a model. If model modification is not necessary in this scheme as a special case, then it reduces to the first case.

In the third case, it is required to create a general relation from a set of instance relations. The scheme of Figure 1 is also used effectively in this case because the general relation cannot be found in the deterministic way. A hypothetical relation is created and tested if it implies all the given specific relations. If it does not, it is modified and this process is repeated [ZHO94]. In this sense the basic method of problem solving as shown in Figure 1 is very

general. The knowledge bases which are used in this scheme is different by the cases.

7.3 Multi-Strata Modeling Scheme

A stratified model is made if there is an outside process to an existing problem solving. It means to model human activity when the subject of the inside process is human being. A special scheme including a subject node is used for multi-strata object modeling. Figure 11 illustrates an example. In a multi-strata object as shown in Figure 9, a base level problem arises from some object (O1). In the ordinary problem solving, a person (S1) makes its model. In the next level this problem solving is an object (O2) and another subject (S2) has a problem of making its model. Requirements given to O1 define a problem in this level. The requirement given to S1 defines the task he/she should do. For example, it may be a command given to S1 such as "find a model to meet the given object-level requirements", "check the legality of the object-level requirements" and so on. These specify the activities of S1 which are formalized in the next stratum and are considered as the requirement to define the model of O2. Similarly some command is given to S2 to specify the task the S2 should do. For example, "automate the activity of lower stratum subject" is given. If there is still the outer process, then this command becomes the requirement to define the model of the higher stratum object. This stratified requirements define a multi-strata object and are used to make a multi-strata model. Upper strata requirement includes very often the lower strata requirement. In order to represent it, a higher (meta) level representation scheme is necessary.

8. MULTI-LAYER LOGIC (MLL)

In order to represent the concepts discussed so far a special knowledge representation is necessary. It must be able to represent multi-strata model and its operation as well as multi-level representation architecture. A language to meet this condition is discussed in [OHS83, OHS95]. This is presented here in brief because it is necessary to understand the following discussion. This language is an

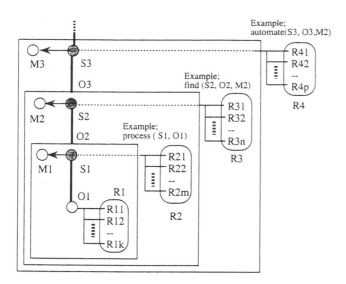

Figure 11 Multi-strata object model

extension of first order logic. The main extensions are (1) introduction of a method of manipulating data structure, (2) high order predicate under certain restrictions, and (3) their combination. The data structure is based on the axiomatic (Z-F) set theory and the language has a capability to define a new data structure as a compound of the primitive relations by a set of composing operations. The primitive relations are; element-of (is-a), component-of, product-of, power-of, pair-of. The basic form to represent knowledge is,

$$(Q_x x/X)(Q_y y/Y)--(Q_z z/Z)[R(x, y, --, z) :-$$
$$P_1(x, y,--,z) * P_2(x, y, --,z) *$$
$$-- * Pn(x, y, --,z)]$$

in which Q_x etc. denotes either a universal quantifier (A) or an existential quantifier (E), x/X etc. means x in a set X, $*$ denotes a logical connector, i.e., either conjunction or disjunction. This expression is read as "For all /some x in X, all /some y in Y,--, all /some z in Z, if the relation P1 and/or P2 and/or --- Pn hold, then R". In the following however, the prefix is often abbreviated for the sake of simplicity of expressions.

An evaluation of predicate is a mapping from a set of variables (x, y,--,z) to either True or False (T, F). This mapping can be performed by a procedure. A predicate to which a procedure for evaluation is provided is named a Procedural Type Predicate (PTP). The others are NTP (Normal Type Predicate). When a PTP is to be evaluated the corres-

ponding procedure is evoked and executed.

Any variable (say x) can be a structural variable to represent a structure. In this case X is a set of structures. For example, if X is a product set of the sets D1, D2, --, Dm, then x represents a tuple (d1, d2, --, dm) of which di $\in$ Di. Moreover any variable can be a closed sentence which means a predicate without free variable. For example a predicate of the form P(x,--, y, S(u, --,w)) can be used. Any of the variables in S, i.e. u, -- ,w, must not be the same with the other variables in the predicate P, i.e. x, --,y.

As a combination of these two expansions, any variable can be a structural variable of which the domain is a structure of predicates. This language is named MLL and illustrated in Figure 12. Using MLL as a knowledge representation language a knowledge based system named KAUS (Knowledge Acquisition and Utilization System) has been developed by the author's group [YAM93] and used for many problems [GUA88, SUZ15, YAM90]. The system is designed to accept any PTP, i.e. a pair of predicate and its evaluation procedure. The ideas on intelligent systems discussed so far are represented by this language. The function structure as shown in Figure 7 can be represented in reality as a structure of the separate knowledge sources arranged in the different levels as shown in Figure 13.

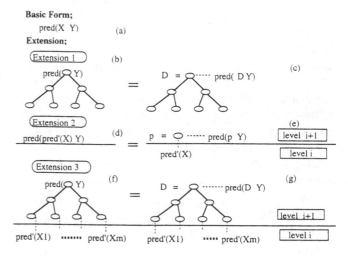

Figure 12 Syntax of MLL

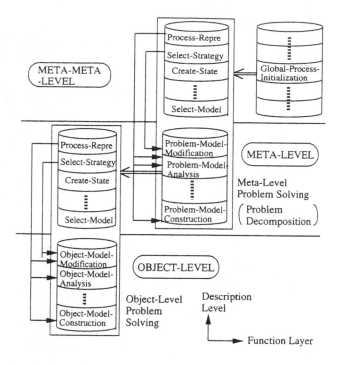

Figure 13 Structure of knowledge sources to represent function structure

9. PROBLEM SOLVING INCLUDING MULTI-STRATA MODEL

The way of problem solving is different depending on the requirements given to the subject even if the object is the same in a multi-strata model. For example, the higher stratum requirement may be to restructure the enterprise or to automate the activity of human subject in the lower stratum in the same enterprise model. The problems are completely different by these cases. Among a number of possible cases, an attention is paid on designing automatic systems. It is formalized as a process of making a multi-strata model and then generating a system to replace the subjects in the lower strata. This leads us to an idea of automatic design of automatic system. Automatic programming is a typical example because a program is an automatic system to replace an activity which has been performed by persons.

9.1 Multi-Strata Model as a Cue of Automatic System Generation

A problem solving including a multi-strata object begins by making a multi-strata model to represent the object as shown in Figure 11, then the stratified requirements are processed from the highest stratum. The highest stratum requirement is sent to 'problem type identification' layer in the function structure of Figure 7. The requirement is interpreted there and a specific operation corresponding to the requirement is activated. Let the requirement be "automate the activity of lower stratum subject". Then it is resolved into a three-step operation. The following rule may be prepared in the knowledge source to represent the activity for this requirement.

```
automateActivity (h-node, ($req), system) :-
    generate (h-node, ($req), system),
                            -- * first step
    tailor(system,($$req)),
                            -- * second step
    replaceNode (*h-node, system).
                            -- * third step
```

In this rule, 'h-node' represents the node to which this highest level requirement is given, *h-node is a lower stratum subject node of h-node, 'system' represents a system to be generated for this specific problem, '($req)' represents the requirement to the next lower stratum object of h-node, and '($$req)' represents those at the two levels lower stratum object.

The first step is to generate a system to satisfy the requirement given to the lower stratum object. For example, let us assume that the lower stratum requirement is "find a model to meet the given requirements" as given before. This is a design type problem and requires a problem solving by exploration. Then a system to solve the problem must be generated. Since every operation to achieve a required activity is represented by a specific substructure of functions, the one for achieving the required operation is extracted from a global function structure and a system provided with this special function structure is created. Since every function is represented in the form of a structure of separate knowledge sources (Figure 13), it is to extract a specific structure of the knowledge sources from the global one. This system is capable of solving the given problems of the lower-stratum object O1, if it is executed.

The second step is to tailor the produced system to the given problem by deleting irrelevant knowledge. If large irrelevant knowledge remains in the system, then the system operation is inefficient, while if necessary knowledge is lost, the system cannot solve given problems. A rough selection of knowledge has been made at the 'problem-type-identification'. But still a lot of irrelevant knowledge can remain in the object-level. For example, knowledge sources belonging to the different problem domains are related to the basic problem solving scheme. It is necessary to filter only useful knowledge for solving given problems. For this purpose the generated system is executed tentatively to meat the requirement at the two-levels lower stratum (($$requirement)). For example, a set of instance problems are given to the system and the system's operation to solve them are traced to form a new knowledge base including only knowledge used there. Selection of instance problems is important because the produced system depends on them. Tracing is a function at the higher level than the functions to solve problems. Finally the generated system replaces the subject at the lower stratum.

9.2 Automatic Programming

Computer program is an automatic system to generate a solution to a given input. Thus automatic programming is an automatic design of automatic system. At an automatic programming, a specific multi-strata model as shown in Figure 11 is made. In this case O1 is an object which is processed by a subject S1. For example it is a transaction in an office information processing. Computer program is made to replace the S1. Object-level requirements are made by users to define tasks performed there. S1 is given the requirement R2 such as "to process this task". A subject S2 represents a programmer who is required to automate the activity of S1 by making the object model of O2 meeting the requirements. A subject S3 is an entity which intends to formalize and computerize the activity of S2. The procedure similar to that discussed above but with an additional

step to generate a procedural code is executed to generate a system. It replaces S2.

In tailoring the system the requirements to O1 are taken into account. A set of the requirements R1 is classified and the original large problem is decomposed into a set of small problems holding the relations to each others. This decides the global program structure.

If some necessary meta-knowledge is lacking, then the system cannot design the program structure autonomously. Then a human designer must intervene the system in order to decompose a large problem into a set of smaller problems and, if necessary, to distribute the problems to the workstations or PCs in a network environment. Then the generated system is executed. Instance problems are solved in order to filter the relevant set of knowledge sources. As the result an object model is created as a combination of an object structure and predicates linked to the nodes in the structure.

This structure is expanded further until the termination condition is reached as follows [LI93]. If some of the predicates in this structure is NTP (non-procedural type predicate), this model is not yet complete but expanded by deduction. An NTP at the highest level in the object model representation is selected and a deduction step is performed, by looking for a rule with a conclusion part unifiable with the NTP and by replacing the NTP by the premise of the rule. In this way the object model is created and expanded while NTP remains in the structure and finally a tree having only PTPs (procedural type predicate) is made. This is the termination condition. The PTPs are threaded traversing the tree in a certain order. This is an incipient program and then it is refined and finally coded.

It is assumed here that enough object-level knowledge exists as well as PTPs as program components. If some of them are lacking, then the system cannot generate a program autonomously. Human programmer's intervention is necessary.

10. CONCLUSION

A method for modeling a problem solving process including persons and for automating the process in intelligent systems has been discussed. In this paper problem solving is separated into two stages; externalization and exploration. The objective of this paper was to automate the latter. The approach taken in this paper is first to develop a general purpose problem solving system which can accept different type of problems, then to generate a problem-specific problem solving system automatically from the general purpose system. As a cue for extracting only knowledge which is necessary for solving given problems from a very general knowledge sources a concept of multi-strata model is introduced. This is a model of problem solving process including persons. A way of making multi-strata model and also processing it to generate a required system has been discussed. Through this discussion a way of automating the design of automatic system was opened. One of its most important applications is automatic programming.

REFERENCES

[BIJ94] Bijl, A.; Showing Your Mind, Proc. Fourth European-Japanese Seminar on Information Modelling and Knowledge Bases., 1994

[BUB93] Bubenko, J.A., Jr, and Wangler, B. Objectives Driven Capture of Business Rules and of Information System Requirements, IEEE Systems Man and Cybernetics '93 Conference. 1993

[GUA88] J. Guan, J. and S. Ohsuga, An Intelligent Man-Machine Systems Based on KAUS for Designing Feedback Control Systems, Artificial Intelligence in Engineering Design, Elesevier Science Pub. Co. 1988

[HOR94] K. Hori, A system for aiding creative concept formation, IEEE Transactions on Systems, Man and Cybernetics, Vol.24, No.6, 1994

[LI93] C.Y.Li, and S.Ohsuga, A MetaKnowledge Structure for Program Development Support, Pro. 5th Intn'l Conf. on Software Engineering and Knowledge Engineering (SEKE), 1993

[OHS85] S. Ohsuga and H.Yamauchi, Multi-Layer Logic - A Predicate Logic Including Data Structure As Knowledge Representation Language, New Generation Computing, 1985

[OHS94] S.Ohsuga, How Can Knowledge Based Systems Can Solve Large Scale Problems - Model Based Decomposition and Problem Solving, Knowledge Based Systems, Vol. 5, No.3, 1994

[OHS95] S.Ohsuga. A Way of Designing Knowledge Based Systems, Knowledge BasedSystems, Vol.6, No.1, 1995

[SUZ93] E.Suzuki, T.Akutsu, and S. Ohsuga, Knowledge Based System for Computer-Aided Drug Design, Knowledge Based Systems, Vol.6, No.2, 1993

[TOP94] Toppano, E. Chittaro, L.& Tasso, C.. Dimensions of Abstraction and Approximation in the Multimodeling Approach, Proc.Fourth European-Japanese Seminer on Information Modelling and Knowledge Bases, 1994

[YAM90] H.Yamauchi, and S. Ohsuga, Loose Coupling of KAUS with Existing RDBMSs, Data and Knowledge Engineering, Vol.5, No.3, 1990

[YAM93] H. Yamauchi, KAUS6 User's Manual, RCAST, Univ. of Tokyo, 1993

[ZHO94] N.Zhong, and S.Ohsuga, Discovering Concept Clusters by Decomposing Data-bases, Data and Knowledge Engineering, Vol.12, 1994

THE CURRENT STATE OF AI PLANNING RESEARCH

Drew McDermott
Yale Computer Science Department
51 Prospect Street, P.O. Box 808285
New Haven, CT 06520-8285
e-mail: mcdermott@cs.yale.edu
phone: 203-432-1281 fax: 203-432-0593

Abstract

Planning is designing the behavior of an agent. The classical theory of planning assumes that the planner has perfect information, and needs to bring about a definite result, by finding a series of actions that brings it about. These classical assumptions have been questioned, but are quite reasonable for many engineering applications. The problem is that traditional algorithms, while elegant, are not effective on realistic-sized problems. Recently some more promising algorithms have been developed, based on avoiding costly searches by doing a more exhaustive analysis at each search state. However, these algorithms may still not be ready for the kinds of problems that arise in manufacturing and other domains where real geometry and kinematics are important.

Introduction

Planning is the process of designing the behavior of an agent of some kind. The agent can be a robot, a factory, an organization, or even some kind of software entity. The question immediately arises: Why *design* behavior? Why not simply behave?

The answer should be obvious to engineers: Whenever the environment permits "looking before leaping," it makes sense to do it. If one course of action can be shown in advance to be better than another, it should be selected. Situations of this kind arise in manufacturing, in mobile-robot applications, in transportaiton schedling, and in process planning, plus many other areas.

There has been a tremendous amount of work in planning in the last ten years. The result has been a big increase in our understanding of the theoretical underpinnings of the subject. However, what many people want to know is when this theory will translate into practice, and here we find frustration.

Some of the work described here was supported by ARPA and administered by ONR under Contract Number N00014-93-1-1235

Let me start by laying out a framework for talking about planning. We can make several different distinctions:

- General-purpose vs. special-purpose: As in other areas of AI, the textbooks talk about very general algorithms, but an examination of practice would show that these are almost never used. Instead, there is an array of more or less specialized algorithms.

- Off-line vs. on-line: Is planning a separate process that takes a certain amount of time, to be followed by an execution phase? Or does it occur at the same time as execution?

- Atemporal or temporal: In atemporal planning the planner does not need to model the dynamics of the world, but only select actions. In temporal planning, it must think of its actions as inserting control pulses into an otherwise autonomous system.

- Perfect information or imperfect information: Must the planner worry about gathering further information or coping with its absence?

What the textbooks talk most about is the case of general-purpose, off-line, atemporal, perfect-information planning, which we summarize with the phrase "classical planning." In this paradigm we make the following assumptions:

1. Nothing happens unless the agent makes it happen.

2. Only one action may be taken at a time.

3. The agent knows the state of the world, and knows the effects of its actions.

It is often said that these assumptions are unrealistic, and that therefore classical planning is of no interest. Actually, there are lots of domains where the assumptions are quite reasonable, especially in engineering applications. For example, in manufacturing, it is reasonable to assume that all operations on a product are under the control of the factory managers. Of course, there will inevitably be events that fall outside the model. But if these events are rare, and do not depend on planning decisions, it is acceptable to handle them by a separate error-recovery process.

The main problem with classical planning is that it is intractable. What I would like to do is to explain why that is, and talk about some hopes for fixing the problem.

I will not say too much about the nonclassical case. It's not that I don't think it's important, or that I don't have something to say. But in my opinion most engineering situations really do fit the classical case, and talking about other paradigms mainly looks like changing the subject to avoid unpleasantness.

Classical Planning and its Classical Headaches

We begin by talking about notation. It is fairly standard in the AI literature to present classical planning problems by giving a set of *action schemata* that define what actions are possible, an *initial situation*, and a *goal formula*. The problem then is to find a sequence of actions that, if executed in the initial situation, will get to a situation in which the goal formula is true.

Here is an example of an action schema:

Action:　　　　fix(?ob, ?orientation)
Preconditions:　free(?ob)
Effects:　　　*Add:*　fixed(?ob,?orientation)
　　　　　　　　Del:　free(?ob)

This schema defines an action fix(*object, orientation*) that might be used in a manufacturing process to fix an object to work on. It has as precondition that the object is not already fixed (it is "free"), and as effect that the object is now oriented at the given orientation. (The question marks flag free variables, which behave like Prolog variables.)

This example is very simple, but it is not much simpler than the typical example given in AI talks. As we will see, even with this simple kind of action, planning gives rise to serious search problems. But even ignoring that issue, the representation may be too simple. I will come back to this problem later.

Action schemas bear a superficial resemblance to predicate-calculus descriptions of actions, using the situation calculus [MH69]. However, we are not going to use them deductively. Instead, we treat them, not as axioms for use in deducing properties of situations, but as specification of how to *edit situation de-*

scriptions. We assume that we are given a complete description of an initial situation. To produce the description of the situation following an action, we delete the formulas flagged with "*Del:*," and add the formulas flagged with "*Add:*" (with variables suitably substituted). All other formulas remain the same. This idea, of course, goes back to the Strips planner [FN71].

We are given an initial-situation description and a goal formula. We can find an action sequence from one to the other by working forward or backward. That is, we can try feasible action sequences in the initial situation to see where they lead, and we can try action sequences that would cause the goal to be true and see if they can be executed in the initial situation. But working backward and forward are not done symmetrically. In the forward direction, we start with a complete situation description, and after computing the result of every action, we still have a complete description. In the backward direction, we have a goal statement that merely constrains one aspect of the final situation. So we have to solve the following problem:

Given a proposition P, and an action specification for action A, compute the weakest precondition R such that if R is true before the execution of A, then A will be feasible, and P will be true afterward. We call this precondition $[A]^R(P)$.

This is called the *regression problem*, and $R = [A]^R(P)$ is called the *regression of P through A*.

There is good news and bad news about the regression problem. The good news is that for the kind of action-specification language we sketched above, the regression problem is easy to solve. The bad news is that for more realistic action specifications, it may not be so easy. I will defer discussion of the bad news until later, and focus on computing

Computing the regression of a literal (an atomic formula or its negation) through an action specification is quite easy. If the formula is negated, we look for entries of the form *Del:* ... in the effects list of the action spec; otherwise, we look for an entry of the form *Add:* .... The preconditions, with appropriate substitutions, then are the regression we're looking for.

The picture is a bit more complicated if we allow *context-dependent effects* in action specifications, that is, if effects are allowed to depend on the truth of propositions other than those that determine the feasibility of an action. Suppose we have a spray-painting application. Then the action spray_paint(x, c) must paint everything in x, if x is an open box:

Action:	spray_paint(?object,?color)
Preconditions:	...
Effects:	*Add:* color(?object,?color)
	box(?object)
	$\wedge$ open(?object)
	$\wedge$ in(?y,?object)
	$\Rightarrow$*Add:* color(?y,?color)

This complexity has two ramifications. First, if you want to change the color of an object, you can do that by putting into another object that is going to be spray-painted. Second, if you want to *avoid* changing the color of an object, you may want to take it out of a container that is going to be spray-painted. Following [Ped89], we say that in(y,x) is a *causation precondition* of color(y,c) before spray_paint(x,c); and ¬in(y,x) is a *preservation precondition* of ¬color(y,c) before that same action. We use the term *secondary preconditions* to denote both kinds.

For all but the most trivial domains, finding a plan to solve a problem requires an extensive search. For our purposes, a plan is just a sequence of actions. (There is no need to have conditionals, because there is nothing to test if the planner has perfect knowledge at planning time.) But just knowing we're looking for a sequence of actions does not tell us what the proper search space should be. As I said above, intuitively we might want to work forward or backward, from the initial situation to the goal or vice versa. But we must careful not to assume without further thought that the search space we want is the set of action sequences.

In fact, it is traditional to work with search spaces that combine two elements:

1. A *plan prefix* that represents a sequence of actions the planner has tentatively committed to, starting from the initial situation.

2. A *goal network* that represents a set of actions the planner has tentatively committed to in order to achieve the given goals, or preconditions of actions the planner has committed to.

See Figure 1.

Some planners include only one of these elements, and some include both. The two dominant traditions might be characterized thus:

- The GPS-Strips-Prodigy tradition[NS61, FN71, FV94]: A search state has both elements, as in Figure reffig:plansearch. There are two kinds of

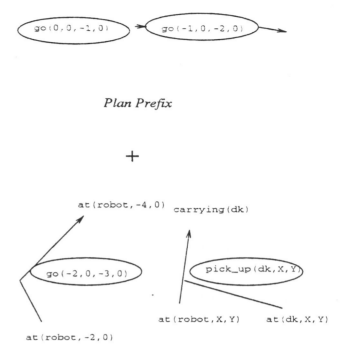

Plan Prefix

+

Goal Network

Figure 1: Bipartite Planning Search State

operator in the search space: committing to an action in the plan prefix, and committing to an action (and new precondition goals) in the goal network.

- The Nonlin-SNLP-Ucpop tradition[Tat77, MR91, PW92]: A search state consists only of a goal network. There are two kinds of operator: Committing to an action to achieve a goal, and imposing ordering relationships among steps to exploit or avoid interactions.

It is agued in [Kam95] that the difference between these two traditions is superficial, but I believe there is a profound distinction between the two, based on the fact that in the first style of planning, there is a *current situation*, resulting from the plan prefix already committed to, and in the second style there are only *partial descriptions* of situations before and after plan steps.

The second style is illustrated in Figure 2, where the planner has decided that the step go(-2,0,-3,0) must follow the step pick_up(dk,2,1), to make sure that it can't spoil the precondition at(robot,-2,0) of pick_up(dk,2,1). In the figure, the planner is considering a three-step plan (not counting the dummy START and END steps), but it has not picked a total ordering for the steps. In some cases [BW94], this

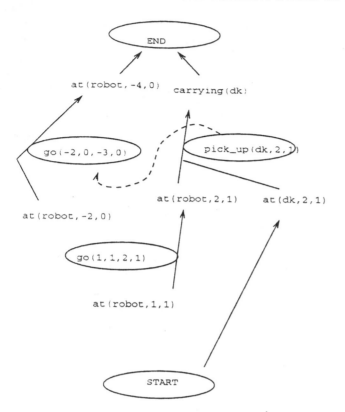

Figure 2: Partial-Order Planning Search State

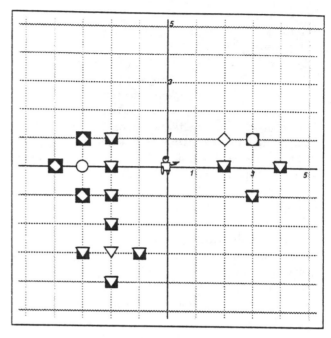

Figure 3: The Manhattan World

ability to postpone ordering decisions can make a big difference to efficiency.

Each of these traditions is alive and well. Each camp is enthusiastic about its own approach, and critical of the other camp's. Each type of planner has been implemented in several variations, and the implementations have been refined to the point where they are exquisitely elegant. The only problem is that they are not terribly efficient. There is little hope that they will ever be able to solve realistic-sized problems.

Let's look at an example domain, in fact the one that the last two figures were drawn from. I call this the "Manhattan world," because it consists of a grid of intersections that a robot moves among, using the action $go(X_1,Y_1,X_2,Y_2)$. If we allow context-dependent effects, we can simplify this action to $go(X_2,Y_2)$. As the robot moves, it can carry at most one object. It can pick_up and put_down objects. It can move from one intersection to another provided the destination is not locked. If it's locked, it can be opened with a key, which is another object to be carried. This domain is simple, but is bordering on being realistic. One could imagine a factory-delivery robot whose world model would not be too different from this.

A problem in this domain is as shown in Figure 3,

which is borrowed from [McD96]. The robot needs to get dk, the diamond-shaped key at location $\langle 2,1 \rangle$ to location $\langle 3,0 \rangle$. Unfortunately, the destination is surrounded by locked intersections. So the robot will have to use a key to open one of them. But which one, and which key?

If that problem seems too hard, then let's simplify it even further. Get rid of the keys, and leave only the grid, with no locks or other obstacles. Problems just involve goals of the form at(robot,x,y). Now, it seems as though, if any domain was straightforward, this one is. After all, if we were to write a special-purpose algorithm to solve problems in this domain, one would imagine that the task would not be too hard, and the resulting program would work pretty well. So what happens if we turn a standard classical planner loose on this kind of problem? The answer is that it would run into a terrible combinatorial explosion, and die. The problem is in the goal network. There are four ways to get to a point in the grid, so for each goal of the form at(robot,x,y) there are four possible subgoals. But each of these has four subgoals, and so forth, as shown in Figure 4. Furthermore, it is not possible to delete a goal if it occurs as a subgoal of itself (as it would be in a deductive context), because there are cases where one has to do exactly that. For example, to achieve the goal "Be here with your paycheck cashed," you might

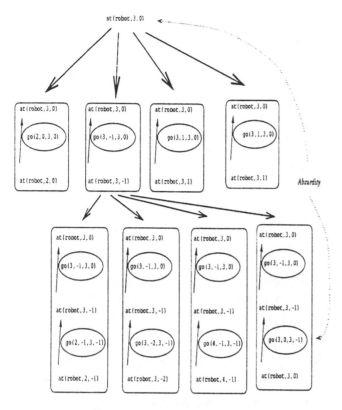

Figure 4: A Goal-Network Explosion

have to achieve the goal to "Be in the car with your paycheck cashed," and that might eventually lead to the subgoal "Be in your car" as a part of getting to the bank. So the goal "Be in your car" can occur as a legitimate subgoal of itself.

Hence the number of goal networks for the goal of getting to an intersection n blocks away is about 4^n.

Note that goal networks are the element that both main traditions have in common. So arguments about "totally ordered" vs. "partially ordered" (or "linear" vs. "nonlinear") planning are really beside the point.

Some New Ideas

So the question arises whether we can find a different way of organizing the search space. What I would like to do from here on out is talk about a couple of ideas that depart fairly radically from the "tradition" in this area. One of these is due to me [McD96], the other to Blum and Furst [BF95]. Each seems to show some promise in overcoming the combinatorial explosion that arises in classical planning. (Of course, the explosions cannot be eliminated completely, because the planning problem is NP-complete [ENS95].) Although they were developed independently, they bear some interesting similarities. Both seem to take off in directions that are counterintuitive to AI researchers,

which may say something about our intuitions.

My own work starts from the observation that the standard planning algorithms are curiously deficient in a key area: the heuristic estimator. One can read many planning papers without ever seeing a mention of search control. Even then, the focus is often on strategies for ordering alternative operators (e.g., [KKY95]). Almost no attention is given to the question: *How do you measure whether one partial plan is more promising than another?* That's why our planners get so lost on problems like those in the Manhattan world. There are so many possibilities, and they all look alike. When a person looks at one of these problems, his eye is immediately drawn to issues of distance and direction. One's inclination is to go east if the goal is to the east. Of course, people are "cheating," and using built-in spatial-navigation skills. But suppose there were a way of estimating the distance without using anything but the action specifications. It turns out that there is. If an atomic goal can be achieved in one action, we can detect that by looking for an action instance whose preconditions are true in the current situation. Call that an *estimated effort* of 1. Now suppose that the goal requires two actions, A_1 followed by A_2, in order to be achieved. In other words, A_2 achieves the goal, and would be feasible if one more precondition were true, and A_1 achieves it. Unfortunately, in general there is an indeterminacy in applying this idea. For a given goal g and action A, $[A]^R(g)$ contains free variables, even if g doesn't. If we allow these variables to proliferate, then finding bindings for them all becomes as difficult as the original planning problem. So we adopt the following heuristic: *Eliminate free variables by binding them in such a way as to make as many preconditions of A true in the current situation as possible.*

For example, suppose we have the following action specification:

Action:	`turn(?ob,?lathe)`
Preconditions:	`in(?ob,?lathe)` $\land$ `on(?lathe)`
Effects:	`cut(?ob)`

and the goal `cut(B)`. Regressing through the action specification gives the precondition `in(ob,?lathe)` $\land$ `on(?lathe)`. If the object is already known to be in a certain lathe, L1, then we can bind *lathe* = L1, and treat `on(L1)` as a *difference* to be reduced (using GPS terminology [NS61]).

It's at this point that the GPS-Strips-Prodigy tradition continues by considering all possible ways to bind the variables, and by exploring them using goal networks. But we have a more humble aim: we want merely to estimate how many actions it will take to achieve the goal. Hence we will focus only on the

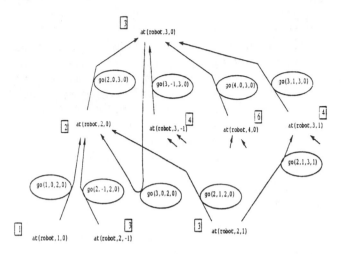

Figure 5: Greedy Regression-Match Graph (Fragment)

maximal matches, variable bindings that bind *all* the variables, and make as many conjuncts true as possible. For each maximal match, we then take the differences, if any, and repeat the process. It eventually stops at one of two places: either we reach a precondition that has no differences, or we reach a precondition that we have already encountered. We can then construct a graph of all the goal literals that are reached in this process. We call it the *greedy regression-match graph*. The size of the subgraph below the original goal in this graph is then our estimate of the effort required to achieve it.

Figure 5 shows a fragment of the greedy regression-match graph for the same example as in Figure 4. At first glance it looks similar, until you realize that the nodes are not goal networks, but simple literals. Hence it is easy to detect repetitions, which have already begun to appear at the third level. The numbers in boxes in the figure are the estimated effort numbers for each literal. In this case the estimated effort for the top goal is 3, which is correct. In other cases, the estimate can be too high or too low.

I don't have space in this paper to go into details about this structure. Please see [McD96] for those details. The key ideas in the algorithm, called "Unpop," are these:

- We dispense with the goal network, and treat search spaces as just plan prefixes.

- We use the greedy regression-match graph to provide a heuristic estimate for the number of steps that will have to be added to the current prefix.

- We don't worry about exploring a large number of possible literals, because the number is quite small compared to the number of possible situations or possible goal networks.

For some domains, this idea works remarkably well. For example, in the problem described above, of getting dk to $\langle 3, 0 \rangle$, the planner finds a plan around 48 steps (43 is optimal), after exploring only about 60 plan prefixes. Of course, it spends more time than a traditional planner on each partial plan. But it marches inexorably toward the goal without generating millions of silly partial plans.

There are cases, however, where the Unpop algorithm does not work. It deals better with addition than with deletion of literals, and in particular is fooled when a deletion becomes inevitable, but it requires an exploration of an exponential number of plan prefixes to prove that it is inevitable.

The algorithm of Blum and Furst [BF95] is even more divergent from the planning mainstream, and it is not yet clear just which planning problems it works well on. One feature that puts it out of the mainstream is that it does not simply search a space of partial plans. Instead, its outer loop looks like this:

```
For l=1 by 1 to maxlength
    Generate a ''progression graph''
                of length l;
until a plan can be extracted
        from the progression graph
```

The progression graph[1] is similar to the greedy regression-match graph, in that it is basically a structure of literals and action terms. However, the key difference is that it represents all propositions that might conceivably be made true starting from the current situation rather than all propositions whose truth might be relevant to achieving the goal. In the first layer of the graph, we put all the literals that are true in the initial situation. In the next layer we put all actions that are feasible in the initial situation. In the third layer we put all propositions affected by those actions. See Figure 6, where *Add:* effects are indicated by solid arrows and *Del:* effects by dashed arrows.

Suppose that the original goal is a conjunction with three conjuncts, g_1, g_2, and g_3, and that all three of them are now found in the third layer, that is, that there are actions A_1, A_2, and A_3, not necessarily distinct, that achieve the three goals. Then, if we're lucky, we're done. *Any* ordering of the three actions will succeed. To be lucky, it must be the case that no A_i deletes a precondition or effect of some A_j, $i \neq j$.

---

[1] Blum and Furst call this a "planning graph," which has all the wrong connotations.

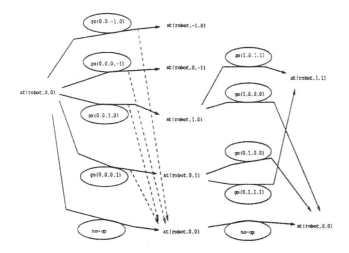

Figure 6: Progression Graph (Fragment)

To formalize this idea, they introduce the notion of *mutual exclusion* among actions and propositions. Two propositions are mutually exclusive if they are effects of mutually exclusive actions (that is, if every action that causes one is exclusive of every action that causes the other). Two actions at the same level are mutually exclusive if one deletes a precondition or effect of the other; or if they depend on mutually exclusive preconditions. There are no mutually exclusive propositions in layer 1 of the graph, but they arise in layers 3 and beyond. The construction of the progression graph can be extended as long as required. In the example, all the go actions are mutually exclusive, because they all delete their precondition at(robot,0,0).

Note that the progression graph approximates forward search in a way that is oddly similar to the way the regression-match graph approximates backward search. It doesn't generate situations, but structures of literals representing batches of situations. The literals in a layer can't all be made true, but various non-exclusive subsets are good candidates for situations that *can* be made true. It also has in common with the regression graph the fact that at first glance it looks too huge to be interesting, until you realize that it is actually quite manageable compared to the search space for traditional algorithms.

After extending the progression graph to a given length, the algorithm tries to extract a plan from it, by carrying out a backward-chaining process. Starting at the last layer, it attempts to make all the goals true by choosing non-mutually-exclusive actions the previous level, such that all their preconditions can be achieved at the layer before that, and so forth.

In the Manhattan domain, the graph would be built

out to a length of 43 before the goal at(dk,3,0) would appear. Prior to that point the algorithm would quickly reject any attempt to extract a plan from the progression graph. At that point it would see that there was only one way to get to ⟨3, 0⟩ that is compatible with carrying(robot,dk), and that is to have at(robot,3,1) and carrying(robot,dk), and execute action go(3,1,3,0). It would then recurse to think about getting to ⟨3, 1⟩ while carrying dk, and see that the only way is to come from ⟨2, 1⟩, and so forth.

Real-World Applications

In the previous section, I examined some of the good news about classical planning, that after several years of disappointment, we seem to have some promising new algorithms. Now we can ask the question, Are these techniques ready for practical use? That is, can the theory of classical planning now be brought to bear on problems of interest to engineers?

As I said earlier, the classical planning assumptions are not in themselves very confining. In a manufacturing application, it is plausible that we can break processes down into steps that are atemporal and fully under our control. For example, in a milling application, we might have as goal transforming a block of material into a certain shape, as shown in Figure 7. Getting the object will require making various kinds of cuts and drilling holes through various faces, and these operations will require fixing the object in jigs and mounts; and these fixing operations will require that there be parallel faces to grip the work with, and so on. It all seems to cohere nicely into the framework of actions, goals, and subgoals.

The difficulties arise when we attempt to define the legal actions using the Strips-style formalism that is so standard in AI planning research. This formalism works well when propositions have a clear-cut status, true or false. But in working with real objects, the facts are often stated using complex and messy data structures.

For example, we would like to say that whenever an object is milled along a certain plane, a face is created in that plane. More formally:

If part of an object is a volume V_o, and a volume V_c is cut away, then

- any face of $V_c \cap V_o$ that lies entirely in V_o becomes a face of the object;
- any face of V_o that intersects V_c ceases to be a face of the object;
- if A is a face of V_c, and $A - V_c$ is nonempty, then $A - V_c$ becomes a face of the object.

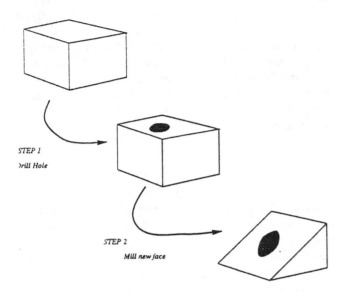

STEP 1
Drill Hole

STEP 2
Mill new face

Figure 7: Process Planning is Classical

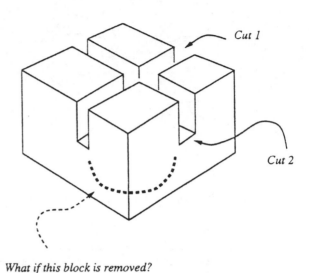

Cut 1

Cut 2

What if this block is removed?

Figure 8: Milling Can Make a Face Larger

This definition may seem formal, but it doesn't capture all the cases. Figure 8 shows a case where a face could get *bigger* when a volume is milled away. Figure 9 shows a case where it is not clear whether a face is present or not. (Could it be used to fix the object for further milling operations?)

From the point of view of geometric modeling, these problems are not that difficult. It is well known how to compute volumes and surfaces of polyhedral objects, so it may seem as if we should just make use of those techniques. After every modification of a shape, we can use whatever specialized algorithms are necessary to infer what the geometric and physical properties of the facets of the shape are.

Unfortunately, to apply the standard planning algorithms, we need not just to be able to predict the effects of an operation, but to perform regression on it, that is, as I said in an earlier section, to specify the weakest precondition of an operation that will cause it to have a desired effect. So what's the weakest precondition of the action mill(*object, volume*) that will cause it to have a face in a certain direction? At the very least, it will have to spell out exactly the set of volumes that intersect the given cut volume in just the right way.

The Graphplan algorithm of Blum and Furst, described above, relies less on regression than most planners do. Their algorithm would construct a progression graph representing all the possible milling

Is this a face?

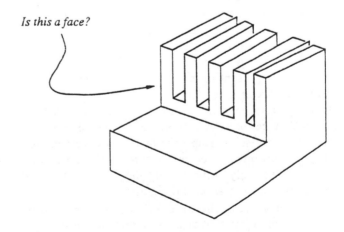

Figure 9: Faces Can be Complicated

operations, then try to extract a plan from it. The extraction phase does use regression, but of a "propositional" variety; no variables are bound during this phase. Unfortunately, Graphplan does not allow actions' effects to be context-dependent. Hence the action mill(*object, volume*) would have to be replaced with a variant in which every condition that can influence an effect is encoded as an argument to the action: mill(*object, volume, existing-volumes, existing faces, ...*).

Assuming that can be done, the next problem that will arise is that Graphplan requires enumerating all possible operations on the object. This sounds crazy, but our intuitions on this matter have been wrong before. It's possible that the large space of possible manufacturing operations is quite manageable for a modern computer. It would be foolish to prejudge the issue.

Conclusion

There are several points I have tried to make in this paper:

1. Classical planning, while it has been rejected as a research paradigm by many workers in the field, is still quite relevant to real-world applications.

2. Traditional algorithms are elegant, but don't perform very well.

3. Newer algorithms show some promise. They are based on the idea of enumerating all possible effects and goals at once, and exploring the resulting data structures as clues to how to solve the overall problem. These algorithms have been overlooked because they are willing to generate what may seem like unmanageably large data structures; but the resulting structures are still much smaller than the search spaces traditional algorithms inhabit.

4. There is still a gap between theory and practice. Closing it is a challenging but worthwhile project.

Acknowledgements: Many of my ideas on planning have been worked out in collaboration with Jim Hendler [MH95]. Some of the figures in this paper were used in an unpublished talk at the AI Planning Systems Conference, 1994.

References

[BF95]   Avrim L. Blum and Merrick L. Furst. Fast planning through planning graph analysis. In *Proc. Ijcai*, 1995.

[BW94]   Anthony Barrett and Daniel S. Weld. Partial-order planning: evaluating possible efficiency gains. *Artificial Intelligence* , 67(1):71–112, 1994.

[ENS95]   Kutluhan Erol, Dana Nau, and V.S. Subrahmanian. Complexity, decidability and undecidability results for domain-independent planning. In Drew McDermott and James Hendler, editors, *Artificial Intelligence* **76**, *Special Issue on Planning and Scheduling*, pages 75–88. NIL, 1995.

[FN71]   Richard Fikes and Nils J. Nilsson. Strips: A new approach to the application of theorem proving to problem solving. *Artificial Intelligence 2*, pages 189–208, 1971.

[FV94]   Eugene Fink and Manuela Veloso. Prodigy planning algorithm. Technical Report 94-123, CMU School of Computer Science, 1994.

[Kam95]   Subbarao Kambhampati. Universal classical planner: an algorithm for unifying state-space and plan-space planning. In *Proc. AAAI-95*, 1995.

[KKY95]   Subbarao Kambhampati, Craig A. Knoblock, and Qiang Yang. Planning as refinement search: a unified framework for evaluating design tradeoffs in partial-order planning. In Drew McDermott and James Hendler, editors, *Artificial Intelligence* **76**, *Special Issue on Planning and Scheduling*, pages 167–238. NIL, 1995.

[MH69]   John McCarthy and Patrick Hayes. Some philosophical problems from the standpoint of artificial intelligence. In Bernard Meltzer and Donald Michie, editors, *Machine Intelligence 4*, pages 463–502. Edinburgh University Press, 1969.

[McD96]   Drew McDermott. A Heuristic Estimator for Means-ends Analysis in Planning. In *Proc. International Conference on AI Planning Systems 3*, 1996.

[MH95]   Drew McDermott and James Hendler. Planning: What it is, what it could be. *Artificial Intelligence (1-2)*, 76:1–16, 1995.

[MR91]   David McAllester and David Rosenblitt. Systematic nonlinear planning. In *Proc. AAAI 9*, pages 634–639, 1991.

[NS61]   Allen Newell and Herbert Simon. GPS:
           a program that simulates human thought.
           In *Lernende Automaten*, pages 279–293. R.
           Oldenbourg KG. Reprinted in Feigenbaum
           and Feldman 1963, 1961.

[Ped89]   Edwin Peter Dawson Pednault. ADL: Ex-
           ploring the middle ground between Strips
           and the situation calculus. In *Proc. Conf.
           on Knowledge Representation and Reason-
           ing 1*, pages 324–332, 1989.

[PW92]   J. Scott Penberthy and Daniel S. Weld.
           Ucpop: A sound, complete, partial order
           planner for ADL. In *Proceedings, Conf.
           on Knowedge Representation and Reason-
           ing 3*, 1992.

[Tat77]   Austin Tate. Generating project networks.
           In *Proc. IJCAI 5*, pages 888–893, 1977.

DCSG: A LOGIC GRAMMAR FOR KNOWLEDGE REPRESENTATION OF HIERARCHICALLY ORGANIZED OBJECTS

Takushi Tanaka

Fukuoka Institute of Technology
3-30-1 Wajiro-Higashi, Higashi-ku, Fukuoka 811-02, Japan
E-mail: tanaka@fit.ac.jp

ABSTRACT

Most artifacts designed by engineers have a hierarchical structure of functional blocks. In order to formalize knowledge of these artifacts and to construct knowledge based systems, we need a method of knowledge representation for hierarchically organized objects. Using electronic circuits as examples, we introduce a logic grammar called **DCSG** (Definite Clause Set Grammar), which not only represents these hierarchical structures but also analyzes these hierarchically organized objects. Circuit structures are defined by **DCSG** rules. Each circuit is viewed as a sentence and its elements as words. A given circuit is decomposed into parse trees by the **DCSG** top-down parsing mechanism. These parse trees represent hierarchical structures of functional blocks.

1. INTRODUCTION

Most artifacts designed by engineers, computers, cars, home appliances, desks, chairs, even software, have a hierarchical structure of functional blocks. Every artifact consists of functional blocks. Each functional block also consists of sub-functional blocks. Each sub-functional block is also decomposed into sub-sub-functional blocks until the individual parts are reached. In other words, each functional block, even a single screw, has a special goal for its containing functional block. Each artifact as a final product itself can also be viewed as a functional block designed for a special goal for users. In order to formalize knowledge of these artifacts, we need a method of knowledge representation for hierarchically organized objects.

Using electronic circuits as examples, we introduce a logic grammar called **DCSG** (Definite Clause Set Grammar), which not only represents these hierarchical structures but also analyzes these hierarchically organized objects. The hierarchical structures of electronic circuits are defined by **DCSG** rules. Each circuit is viewed as a sentence and its elements as words. Given circuits are decomposed into parse trees by the **DCSG** top-down parsing mechanism. These parse trees represent hierarchical structures of functional blocks.

2. DEFINITE CLAUSE GRAMMARS FOR FREE WORD-ORDER LANGUAGES [Ta91]

Definite Clause Grammars (**DCG**s) [PW80] are a method for expressing context-free grammars in logic programming. A set of grammar rules itself forms a logic program which implements top-down parsing. Inspired by the method of **DCG**, we have developed Definite Clause Set Grammars (**DCSG**s) [Ta91] for representation and analysis of hierarchically organized objects. In contrast with the **DCG** formalism, **DCSG** is based on the concept of using difference sets (complementary sets), as opposed to the difference lists used by **DCG**, to define grammar rules. The **DCSG** can be viewed as **DCG** for free word-order language.

2.1 Free-Word-Order Languages

A free-word-order language L(G') can be defined by modifying the definition of a formal grammar. We define a context-free free-word-order grammar **G'** to be a quadruple $<V_N, V_T, P, S>$ where: V_N is a finite set of non-terminal symbols, V_T is a finite set of terminal symbols, P is a finite set of grammar rules of the form:

$$A \rightarrow B_1, B_2, ..., B_n. \qquad n \geq 1$$
$$A \in V_N, \quad B_i \in V_N \cup V_T \quad i = 1, ..., n$$

The above grammar rule means rewriting a symbol A not with the string of symbols $B_1, B_2, ..., B_n$, but with the set of symbols $\{B_1, B_2, ..., B_n\}$. A sentence in the language **L** (G') is a set of terminal symbols which is derived from S by successive application of grammar rules. Here the sentence is a multi set which admits multiple occurrences of elements taken from V_T. Each non-terminal symbol used to derive a sentence can be viewed as a name given to a subset of the multi set. This modified formal grammar is the base of **DCSG** formalism.

2.2 Definite Clause Set Grammars

A free-word-order sentence is, therefore, a multi set of

elements taken from V_T. Each non-terminal symbol used to derive the sentence can be viewed as a name given to a subset of the multi set. Therefore, grammar rules represent relationships between these subsets and their elements. Using the predicates *subset* and *member*, we can translate grammar rules directly into a Prolog program. As with **DCG** parsing, this results in a Prolog program that will parse the **DCSG** defined language in a top-down manner. Each grammar rule is translated into a definite clause by the **DCSG** translation procedure.

Both terminal and non-terminal symbols in grammar rules are written as strings of characters beginning with a lower case letter. Each terminal symbol in a grammar rule is surrounded by "*[*" and "*]*", so that the translation procedure can distinguish terminal from non-terminal symbols. The general form of the translation from a grammar rule to a definite clause is shown by formulas (1) and (1)'.

$$A \to B_1, B_2 ..., B_n. \tag{1}$$

$$
\begin{aligned}
subset(A,S_0,S_n) :- \; & subset(B_1,S_0,S_1), \\
& subset(B_2,S_1,S_2), \\
& ... \\
& subset(B_n,S_{n-1},S_n).
\end{aligned}
\tag{1}'
$$

Here, all symbols in the grammar rule (1) are assumed to be non-terminal symbols. That is, A and $B_1, ..., B_n$ are names given to subsets of elements composing a free-word-order sentence. The definite clause (1)' explicitly represents relationships between these subsets. The arguments $S_0, S_1, ..., S_n$ in (1)' are substituted by elements composing a free-word-order sentence. The predicate subset is used to refer to a subset of an object set which is given as the second argument, while the first argument is the name of its subset. The third argument is a complementary set which is the remainder of the second argument less the first; e.g. $subset(A,S_0,S_n)$ states that A is a subset of S_0 and that S_n is the remainder. That is, the predicate subset defines A as the difference between two sets S_0 and S_n. The clause (1)' states if B_1 is a subset of S_1, B_2 is a subset of S_1, ..., and B_n is a subset of S_{n-1}, then A is a subset of S_0.

If B_i ($i \le n$) is found in the right hand side of grammar rules, where B_i is assumed to be a terminal symbol, then $member(B_i,S_{i-1},S_i)$ is used instead of $subset(B_i,S_{i-1},S_i)$ in the translation, because the terminal symbol B_i is an element composing a free-word-order sentence. The predicate member is defined by the following definite clauses:

$$member(M,[M|X],X).$$
$$member(M,[M|X],[A|Y]) :- member(M,X,Y). \tag{2}$$

2.3 The Parsing Process

Parsing consists of the iterative application of grammar rules controlled by the Prolog backtracking mechanism. Grammar rules, such as (1)' above, contains only non-terminal symbols, and grammar rules that generate terminal symbols differ slightly. When the clause (1)' is used in parsing, an object set (multi set) is substituted into S_0, and the non-terminal symbol A is identified as a subset of S_0. During parsing, the elements in S_0 are successively replaced to form the sequence $S_1, ..., S_n$. Procedurally, the clause (1)' can be read as: In order to show A to be a subset of S_0, show B_1 to be a subset of S_0, show B_2 to be a subset of S_1 which is the remainder of B_1 from S_0 ..., show B_n to be a subset of S_{n-1}, and set the remainder to be S_n.

Non-terminal symbols finally generate terminal symbols such as:

$$B \to [C].$$

This type of rule is translated into the following definite clause:

$$subset(B,S_0,S_1) :- member(C,S_0,S_1).$$

This clause can be read as: In order to show that B is a subset of S_0, show C to be a member of S_0, and set the remainder to S_1. Here, the variable S_0 will be bound to a current context, which is the unidentified part of the object set. The goal $member(C,S_0,S_1)$ searches the context S_0 for the terminal symbol C using the rule (2). If C is found, the goal returns the remainder in the variable S_1. Thus, the non-terminal symbol B is identified as a subset of the current context. When the starting symbol S is finally identified as a subset of the object set, the top-down parsing succeeds.

2.4 Extensions to DCSG

Several extensions to the basic **DCSG** formalism are necessary for actual use. One extension introduces context-dependent features into grammar rules by extending **DCSG**-syntax. The context-dependent conditions "*test C*" and "*not C*" in the grammar rules

$$A \to B_1, ..., B_i, test \; C, B_{i+1}, ..., B_n.$$
$$A \to B_1, ..., B_i, not \; C, B_{i+1}, ..., B_n.$$

are translated into

$$subset(C,S_i, \_) \quad and$$
$$not \; subset(C,S_i, \_),$$

respectively. The "_"s are anonymous variables which may be ignored. In the process of parsing the non-terminal symbol "*A*", "*test C*" and "*not C*" respectively demand the existence and the absence of subset *C* in the current parsing context. These conditions will apply to S_i after identifying B_1, ..., B_i. When the condition succeeds, the process of identifying B_{i+1}, ..., B_n continues on S_i. However, the translation differs for terminal symbols. Conditions of the form "*test [C]*" and "*not [C]*" will be translated into

$$member(C, S_i, \_)$$ and
$$not \ member(C, S_i, \_),$$

which respectively demand the existence and the absence of element *C* in the current context S_i. These conditions are used in grammar rules for parsing context-dependent circuits.

Another extension which enable coupling syntactic and semantic information is discussed in another [Ta93]. The predicate *member* has three arguments. The first is an element of a set. The second is the whole set. The third is the complementary set of the first. That is, both terminal and non-terminal symbols are represented as differences of the last two arguments of these predicates.

DCSG syntax allows terminal and non-terminal symbols with variables. When the grammar rules are used in parsing sentences, grammar rule variables are instantiated with specific terms in the object sentences.

3. REPRESENTATION OF HIERARCHICAL STRUCTURES

Electronic circuits are designed as goal oriented compositions of basic circuits with specific functions. We formalize these design rules for circuit structures as **DCSG** rules. That is, we assume circuits as sentences generated by circuit grammar.

3.1 Circuits Represented as Sentences

In order to represent the circuit "*ya88*" in FIGURE 1 which is a fuzzy membership function circuit* by T. Yamakawa [Ya88], we introduce six terminal symbols with variables:

V_T = { *resistor(Name,Node1,Node2)*,

  diode(Name,Anode,Cathode),
  npnTr(Name,Base,Emitter,Collector),
  pnpTr(Name,Base,Emitter,Collector),
  terminal(Name,Node), ground(Node) }

The first argument of each terminal symbol is a name of specific element, and the others are connected nodes of the element. Connections of circuit elements are defined by

----
* The original circuit is modified by T. Tanaka.

binding terminal symbol variables to shared nodes.

The circuit topology of "*ya88*" is represented by the following assertion:

ya88([resistor(r1,9,0),resistor(r2,8,7),resistor(r3,11,12),
  resistor(r4,1,10),resistor(r5,1,15),resistor(r6,14,0),
  resistor(r7,5,2),npnTr(q1,6,2,6),npnTr(q2,6,2,13),
  npnTr(q3,15,5,1),pnpTr(q4,9,1,9),pnpTr(q5,9,1,8),
  pnpTr(q6,3,8,2),pnpTr(q7,4,7,6),pnpTr(q8,4,11,2),
  pnpTr(q9,9,1,11) pnpTr(q10,3,12,6),
  pnpTr(q11,9,10,13),pnpTr(q12,14,15,2),
  diode(d1,13,14),terminal(t1,3),terminal(t2,4),
  terminal(t3,1),terminal(t4,5),terminal(t5,2),
  ground(0)]).   (3)

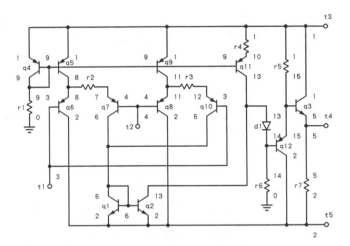

FIGURE 1 Fuzzy membership-function circuit

The terminal symbol *resistor(r2,8,7)* denotes a resistor named *r2* connecting node *8* and node *7*. The node order is arbitrary because a resistor does not have polarity. *npnTr(q3,15,5,1)* denotes an NPN-transistor named *q3* with the base connected to node *15*, the emitter to node *5*, and the collector to node *1* respectively. *diode(d1,13,14)* denotes a diode named *d1* with the anode connected to node *13* and the cathode connected to node *14* respectively. *terminal(t1,3)* denotes an external-terminal named *t1* connected to node *3*. The list of these terminal symbols surrounded by "*[*" and "*]*" is a free-word-order sentence denoting the circuit. As the list is used to represent a multiset, the order of the elements is not important. The assertion "*ya88([...])*" states that the list "*[...]*" is the circuit *ya88*.

3.2 Grammar Rules

Such functional blocks as "*diode-connected transistor*", "*voltage regulator*", "*current source*", "*current mirror*", and "*emitter-follower*" are found in the circuit "*ya88*". The circuit "*ya88*" itself can also be viewed as the functional block "*membership-function circuit*" for fuzzy logic. Each functional block is defined as a non-terminal symbol, with the circuit elements as terminal symbols.

The simplest functional block, the *"diode-connected transistor"*, is defined by (4) and (5) as a non-terminal symbol. The rule (4) means that an NPN-transistor Q with the base and the collector connected to the same node A works functionally as a diode (FIGURE 2). Here, "ndtr(Q)" is a name given to the diode-connected transistor. The rule (5) is defined for PNP-transistor similar to the rule (4).

$$dtr(ndtr(Q),A,C) \rightarrow [npnTr(Q,A,C,A)]. \qquad (4)$$
$$dtr(pdtr(Q),A,C) \rightarrow [pnpTr(Q,C,A,C)]. \qquad (5)$$

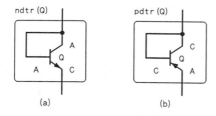

FIGURE 2   Diode-connected transistor

Translating the grammar rules (4) and (5) into definite clauses gives (4)' and (5)', as follows:

$$subset(dtr(ndtr(Q),A,C),S0,S1) :- \\ member(npnTr(Q,A,C,A),S0,S1). \qquad (4)'$$
$$subset(dtr(pdtr(Q),A,C),S0,S1) :- \\ member(pnpTr(Q,C,A,C),S0,S1). \qquad (5)'$$

When the clause (4)' is used in parsing, an object circuit is substituted into $S0$, and the functional block $dtr(ndtr(Q),A, C)$ is identified in the circuit, and the remainder of the circuit is bound to the variable $S1$. Procedurally, the clause (4)' can be read as: In order to identify the non-terminal symbol $dtr(ndtr(Q),A,C)$ in $S0$, find the terminal symbol $npnTr(Q,A,C,A)$ in the circuit $S0$, then set the remainder into $S1$. More formally, it can also be read as: $npnTr(Q,A, C,A)$ being a member of the multiset $S0$ implies $dtr(ndtr (Q),A,C,A)$ to be a subset of the multiset.

The following grammar rule (6) enables us to refer to resistors independently of their node order. e.g. A resistor represented by a terminal symbol either $resistor(r2,8,7)$ or $resistor(r2,7,8)$ can be identified by a non-terminal symbol $res(r2,8,7)$.

$$res(X,A,B) \rightarrow [resistor(X,A,B)]; \\ [resistor(X,B,A)] . \qquad (6)$$

Here, the symbol ";" is used for abbreviation of two grammar rules with the same left-hand side. This grammar rule is translated into the following clause:

$$subset(res(X,A,B),S0,S1) :- \\ member(resistor(X,A,B),S0,S1); \\ member(resistor(X,B,A),S0,S1). \qquad (6)'$$

The symbol ";" in the clause is the logical connective "or". When an object circuit is substituted into S0, the first subgoal is to identify $resistor(X,A,B)$ as an element of the circuit. If it succeeds, this subgoal returns the remainder of the circuit in S1. If it fails, the second subgoal tries $resistor(X,B,A)$ instead of $resistor(X,A,B)$. This kind of rule is defined for all non-polar elements such as capacitors and inductors.

The following grammar rule defines a non-terminal symbol "*nVbe-voltage regulator*" which has negative Vbe (-0.6 ~ -0.7 volts) as its output (FIGURE 3).

$$nVbeReg(nVreg(D,R),Vm,Out,Com) \rightarrow \\ dtr(D,Com,Out), \\ res(R,Vm,Out). \qquad (7)$$

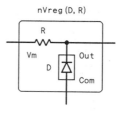

FIGURE 3   nVbe-voltage regulator

The grammar rule is translated into a definite clause as:

$$subset(nVbeReg(nVreg(D,R),Vm,Com,Out),S0,S2) :- \\ subset(dtr(D,Com,Out),S0,S1), \\ subset(res(R,Vm,Out),S\_1,S2). \qquad (7)'$$

Procedurally, the clause (7)' can be read as: in order to find a *nVbe-voltage regulator* in the circuit $S0$, fist find a diode-connected transistor $dtr(D,Com,Out)$ in $S0$, then find a resistor $res(R,Vm,Out)$ in $S1$ which is the remainder of $dtr(D,Com,Out)$ from $S0$. The remainder of the circuit is held in $S2$ from which the *nVbe-voltage regulator* has been removed.

A simple current source is defined by the following grammar rule (8). The current source consists of a *nVbe-voltage regulator* and an *PNP-transistor* (FIGURE 4).

$$currentSource(cSorc(VR,Q),Out,Com) \rightarrow \\ nVbeReg(VR,\_,B,Com), \\ [pnpTr(Q,B,Com,Out)]. \qquad (8)$$

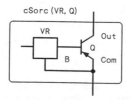

FIGURE 4   Simple current source

However, this definition is incomplete. Since the second

argument of *nVbeReg* is an anonymous variable, the definition does not specify the connections of the input node of the *nVbe-voltage regulator*, though the input node must be connected to a power supply.

One method to make this rule complete is to include a power supply as a component of the current source. As a result, a single power supply in an actual circuit is shared with many parts of the circuit as a voltage source. This method is developed for context-dependent circuits in another paper [Ta93]. More sophisticated methods would use electrical conditions as semantic information [Ta93], but in this section we focus on grammar rules specifying topological conditions.

If an electronic circuit consists only of hierarchically organized functional blocks, its structure is defined by context-free rules. However, the circuit ya88 has a context dependent structure which can not be analyzed by context-free rules defined so far. If a circuit goal generates two current sources, each current source generates voltage source as subgoal using rule (8), then two voltage sources are generated. When one of the voltages is derived from the other, an engineer may combine two voltage sources into one voltage source for simplicity. That is, he has ability to use context dependent circuit generation rules.

The *nVbe-voltage regulator* (*q4, r1*) is shared with three current sources. We define the following rule for parsing this type of context dependent structure [Ta93].

$$currentSource(cSorc(VR,Q),Out,Com) \rightarrow$$
$$test\ nVbeReg(VR,\_,B,Com),$$
$$[pnpTr(Q,B,Com,Out)]. \qquad (9)$$

Another simple current source with an adjustable resistor is defined as follows.

$$currentSource(cSorc2(VR,Q,R),Out,Com) \rightarrow$$
$$nVbeReg(VR,\_,B,Com),$$
$$[pnpTr(Q,B,E,Out)],$$
$$res(R,E,Com),$$
$$not\ anyElm(\_,E). \qquad (10)$$

The last condition *not anyElm(_,E)* demands no element other than PNP-transistor *Q* and resistor *R* must be connected to the node *E*. The non-terminal symbol *anyElm (Name,Node)* which represents any element connected to Node is defined as follows.

$$anyElm(Name,Node) \rightarrow$$
$$[terminal(Name,Node)];$$
$$res(Name,Node,\_);$$
$$[npnTr(Name,Node,\_,\_)];$$
$$[npnTr(Name,\_,Node,\_)];$$
$$[npnTr(Name,\_,\_,Node)];$$
$$[pnpTr(Name,Node,\_,\_)];$$
$$[pnpTr(Name,\_,Node,\_)];$$
$$[pnpTr(Name,\_,\_,Node)];$$
$$[diode(Name,Node,\_)];$$
$$[diode(Name,\_,Node)]. \qquad (11)$$

An emitter-follower using a current source is defined as follows.

$$pnpEF(pnpEF2(Q,CS),In,Out,Vm,Com) \rightarrow$$
$$[pnpTr(Q,In,Out,\ Vm)],$$
$$currentSource(CS,Out,Com). \qquad (12)$$

Figure 5 shows a *voltage-current converter* used in the circuit *ya88*. If the node voltage of *Vi1* is grater than the one of *Vi2*, it generates the collector current of transistor *Q* proportional to the difference of these node voltages.

$$viConverter(viConv(EF,Q,R),Vi1,Vi2,Io,Vp,Vm) \rightarrow$$
$$pnpEF(EF,Vi1,A,Vm,Vp),$$
$$res(R,A,E),$$
$$[pnpTr(Q,Vi2,E,Io)]. \qquad (13)$$

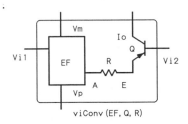

FIGURE 5   Voltage-current converter

We further define non-terminal symbols for functional blocks found in *ya88* as follows:

$$npnEF(npnEF(Q,R),In,Out,Vp,Com) \rightarrow$$
$$[npnTr(Q,In,Out,Vp)],$$
$$res(R,Out,Com). \qquad (14)$$
$$pnpEF(pnpEF(Q,R),In,Out,Vm,Com) \rightarrow$$
$$[pnpTr(Q,In,Out,Vm)],$$
$$res(R,Out,Com). \qquad (15)$$
$$cascadeEF(cEF(PEF,NEF),In,Out,Vp,Vm) \rightarrow$$
$$pnpEF(PEF,In,E,Vm,Vp),$$
$$npnEF(NEF,E,Out,Vp,Vm). \qquad (16)$$
$$currentMirror(cMirr(D,Q),In,Com,Sink) \rightarrow$$
$$dtr(D,In,Com),$$
$$[npnTr(Q,In,Com,Sink)]. \qquad (17)$$

The whole circuit of *ya88* is defined by (19) as shown in Figure 6. The composite voltage-current converter *CVI* generates a current proportional to the absolute value of voltages across two inputs (*Vi1, Vi2*). The difference current of *CVI* from *J* flows into *D* and *R* and forms a shape of fuzzy membership-function.

$$cVIconverter(cVIconv(VI1,VI2,CM),$$
$$Vi1,Vi2,Sink,Vp,Vm) \rightarrow$$
$$viConverter(VI1,Vi1,Vi2,S,Vp,Vm),$$
$$viConverter(VI2,Vi2,Vi1,S,Vp,Vm),$$
$$currentMirror(CM,S,Vm,Sink). \qquad (18)$$

$$fzMembership(fzMember(CVI,J,D,R,CEF),$$
$$Vi1,Vi2,Out,Vp,G,Vm) \rightarrow$$
$$cVIconverter(CVI,Vi1,Vi2,A,Vp,Vm),$$
$$currentSource(J,A,Vp),$$
$$[diode(D,A,B)],$$
$$res(R,B,G),$$
$$cascadeEF(CEF,B,Out,Vp,Vm). \qquad (19)$$

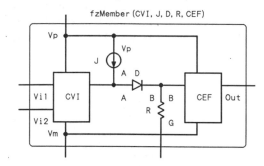

FIGURE 6   Structure of membership-function circuit

4. TOP-DOWN PARSING

In top-down parsing of free-word-order languages, grammar rules are applied to a starting symbol until it generates terminal symbols, and the terminal symbols are found in the object sentence. Since all non-terminal symbols are functional blocks, any non-terminal symbol may be viewed as a starting symbol. When all of the terminal symbols are found, and no unknown terminal symbols remain in the sentence, then the top-down parsing succeeds. If a given circuit is a grammatical one, the following goal clause successfully parses the circuit. Although we have not introduced a single starting symbol, the variable S in the following goal clause works as the starting symbol which generates all of the non-terminal symbols.

$$?\text{- } subset(S, ObjectCircuit, [\ ]).$$

Since the third argument of subset is a null circuit, the goal clause asks whether the whole *ObjectCircuit* is identified as a non-terminal symbol. Since the variable S matches all of the non-terminal symbols defined by grammar rules, the system tries all of the grammar rules one after another until the remainder part of subset becomes the null circuit. In this case, the top-down parsing mechanism does not work efficiently, because the first argument S does not contain any information about how to parse the object circuit. However, if the object circuit is a grammatical one, the goal clause eventually parses the circuit.

Even if the given circuit is not grammatical, namely, even if the structure of the whole circuit is not defined as a non-terminal symbol, the following goal clause identifies all known structures in the given circuit.

$$?\text{- } subset(S, ObjectCircuit, Rest).$$

The goal clause tries to identify all kinds of the non-terminal symbols in the object circuit. If a non-terminal symbol is identified in the object circuit, the remainder of the circuit is bound to the variable *Rest*. The goal clause separates the object circuit into a known structure and an unknown part.

The top-down mechanism works efficiently when a specific non-terminal symbol is given as the starting symbol. The following goal clause (20) executes the top-down parsing for circuit *ya88*.

$$?\text{- } ya88(Circuit),$$
$$subset(fzMembership(X,\_,\_,\_,\_,\_,\_),Circuit,\_). \qquad (20)$$

The first subgoal *ya88(Circuit)* binds circuit *ya88* to the variable Circuit. The second subgoal *subset(...)* identifies a functional block *fzMembership(...)* (fuzzy membership function circuit) in the circuit *ya88*. Here, the non-terminal symbol *viConverter(...)* works as the starting symbol. According to these grammar rules, the initial goal (20) is repeatedly decomposed into sub-goals until each *subset* goal reaches *member* goals which identify terminal symbols. Each time a *member* goal succeeds, an element is identified as a part of a functional block, and the unknown part of the object circuit decreases.

When the goal clause (20) succeeds, we will acquire the following values for the variable X in (13):

$$X = fzMember(cVIconv(viConv(pnpEF2(q6,$$
$$cSorc(*,q5)),$$
$$q7,$$
$$r2),$$
$$viConv(pnpEF2(q8,$$
$$cSorc(*,q9)),$$
$$q10,$$
$$r3),$$
$$cMirr(ndtr(q1),q2)),$$
$$cSorc2(*,q11,r4),$$
$$d1,$$
$$r6,$$
$$cEF(pnpEF(q12,r5),npnEF(q3,r7)))$$
$$* = nVreg(pdtr(q4),r1)$$

The value of the variable X in the starting symbol keeps track of successful goals. It forms a hierarchical structure of functional blocks and can be viewed as a parse tree for the circuit corresponding to a syntactic structure of a sentence (FIGURE 7).

5. GENERATING CIRCUITS

As data flows to predicates are bilateral in logic programming, generating circuits can be done by changing the input and the output of the predicate "*subset*". Namely, the first argument of "*subset*" is its input to specify circuits and the second argument is its output from which the generated circuit appear.

Using the grammar rules defined for C-MOS circuits, following goal clause generates a master-slave JK-flipflop [TJ95]. The starting symbol "*jkff(Name,j,k,q,nq)*" specifies the JK-flipflop with its inputs connected to nodes " *j*" and "*k*", its clock connected to node " *c*", and its output connected to node " *q*" and "*nq*" respectively.

?- subset(jkff(Name,j,k,c,q,nq),CT,[ ]).

Name = jkff(dff(gl(inv(p(_53),n(_54)),
inv(p(_71),n(_72)),
tg(p(_89),n(_90)),
tg(p(_108),n(_109))),
gl(inv(p(_139),n(_140)),
inv(p(_157),n(_158)),
tg(p(_175),n(_176)),
tg(p(_194),n(_214))),
inv(p(_213),n(_214))),
and(nand(p(_239),p(_240),n(_241),n(_242)),
inv(p(_273),n(_274))),
nor(p(_291),p(_292),n(_293),n(_294)),
nor(p(_326),p(_327),n(_328),n(_329)))

CT = [pmos(_53,_50,vdd,_38),nmos(_54,_50,gnd,_38),
pmos(_71,_38,vdd,_37),nmos(_72,_38,gnd,_37),
pmos(_89,_36,_24,_50),nmos(_90,_36,_24,_50),
pmos(_108,c,_37,_50),nmos(_109,c,_37,_50),
pmos(_139,_136,vdd,nq),nmos(_140,_136,gnd,nq),
pmos(_157,nq,vdd,q),nmos(_158,nq,gnd,q),
pmos(_175,_36,_37,_136),
. . .
nmos(_329,_25,gnd,_334)]

The variable *"Name"* is substituted by the parse tree of generated circuit (FIGURE 8). The tree shows that the starting symbol *JK-flipflop* generates a *D-flipflop*, a logical *AND*, and two *NORs* as non-terminal symbols. Then, *D-*

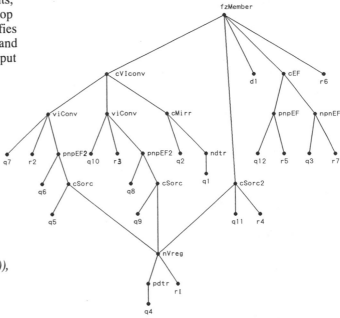

FIGURE 7   Parse tree for the circuit *ya88*

flipflop generates two *Gated feedback-loops* and an *Inverter*. The *Gated feedback-loop* generates two *Inverters* and two *Transmission gates*. Each *Inverter* generates *P-MOS* and *N-MOS* transistors as terminal symbols. Other non-terminal symbols also finally generate *P-MOS* and *N-MOS* transistors as terminal symbols.

The variable *CT* is substituted by the circuit generated. Here, the symbol such as " *_53*", " *_54*", "*_71*", " *_72*", ... are names given to *FET*s and nodes in the generated circuit.

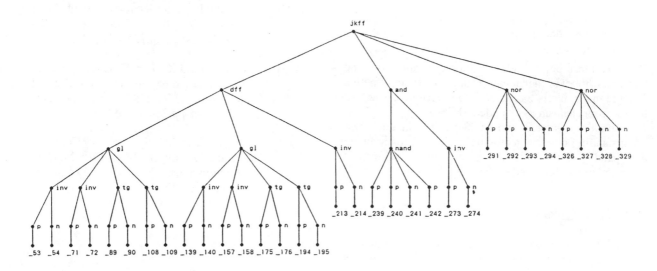

FIGURE 8   Parse tree for generated JK-flipflop

6. SEMANTIC INFORMATION

We introduce semantic terms into grammar rules, so that information other than circuit topology is available in parsing. The semantic terms are surrounded by " $\{$ " and " $\}$ " in any position of the right hand side of grammar rules. For example, the following rule has a semantic condition:

$$A \rightarrow B_1, ..., B_n, \{[C]\}. \tag{21}$$

When this rule is used in parsing, not only sub-circuits B_1, ..., B_n must be identified, but also the semantic condition C must be filled. A **DCSG** rule using this extension is translated into the following clause.

$$ss(A, S_0, S_n, E_0, E_n) :- ss(B_1, S_0, S_1, E_0, E_1),$$
$$\cdots$$
$$ss(B_n, S_{n-1}, S_n, E_{n-1}, E_n),$$
$$member(C, E_{n\_}). \tag{21'}$$

The first three arguments of the predicate ss(...) are the same arguments as *subset(...)*. That is, an object circuit is bound to S_0, then, the sub-circuit A is identified in that circuit. The remainder of the circuit is substituted into S_n.

The last two arguments of the predicate *ss(...)* are used for semantic information. The fourth argument is used as an input and the last argument as an output of the predicate. Usually, electrical environments or constraints on the circuit are given in the fourth argument. The information is used by semantic terms in the right hand side of the grammar rules. Similar to ordinary terminal symbols in grammar rules, the semantic term *{[C]}* requires that the current semantic information E_n contains C as a member.

But, unlike ordinary terminal symbols, it does not remove C from the current semantic information in transferring to the next goal.

When we define another free-word-order language to represent semantic information, semantic terms of the form *{C}* become available. It is translated into $ss(C, E_{n\_}, \_\_)$.

Here, C must be defined by grammar rules as a non-terminal symbol of the language for semantic information. The grammar rule is translated into a definite clause using the same mechanism as extended **DCSG**.

Coupling syntactic and semantic information toward understanding circuits, removing undesired interpretation, and parsing context dependent circuits using semantic information are discussed in the paper [Ta93].

7. CONCLUSIONS

In this paper, we introduced a logic grammar called **DCSG** using electronic circuits as a sample domain. We viewed electronic circuits as sentences in a formal language, and developed grammar rules for hierarchically organized circuit structures. The set of grammar rules itself can function as a logic program that implements top-down parsing. The parsing performance depends on the defined grammar rules. As more grammar rules are defined, more circuits can be parsed. The system cannot analyze circuits which consist of arbitrary connected circuit elements. As is true for natural language, it is not possible to understand structures not included in the grammar. That is, all circuits which can be parsed are sentences defined by the given circuit grammar. If an object circuit has unknown structures, our system cannot parse the whole circuit, but it can separate the object circuit into known parts and unknown sections.

DCSG can be used to represent and analyze a wide range of hierarchical organized objects.

To use **DCSG** to represent such objects, first view the hierarchically organization of the objects as a generalized sentence. Then, identify the functional blocks and individual parts as non-terminals and terminals respectively. Next, define symbolic expressions for these non-terminals and terminals. Finally, using these symbols, define the hierarchical structures as grammar rules.

REFERENCES

[PW80]   F.C.N. Pereira and D.H.D. Warren, Definite Clause Grammars for Language Analysis, *Artificial Intell.*, vol. 13, pp. 231-278, 1980.

[Ta85]   T. Tanaka, Parsing Circuit Topology in A Deductive System, Proc. *IJCAI-85*, pp. 407-410, 1985.

[Ta91]   T. Tanaka, Definite Clause Set Grammars: A Formalism for Problem Solving, *Journal of Logic Programming*, vol.10, pp. 1-17, 1991.

[Ta93]   T. Tanaka, Parsing Electronic Circuits in a Logic Grammar, *IEEE-Trans. Knowledge and Data Eng.*, Vol.5, No.2, pp.225-239, 1993.

[TJ95]   T. Tanaka and L.C. Jain, Circuit Representation in A Logic Grammar, in Electronic Technology Directions to the Year 2000, IEEE Computer Society Press, pp.28-34, 1995.

[Ya88]   T. Yamakawa, High-Speed Fuzzy Controller Hardware System: The Mega FIPS Machine, *Information Science*, Vol.45, pp.113-128, 1988.

PATTERN SEPARATION BY CHAOTIC NETWORKS EMPLOYING CHAOTIC ITINERANCY

Takeshi Yamakawa, Tsutomu Miki and Masayoshi Shimono
Department of Control Engineering and Science
Kyushu Institute of Technology
Iizuka, Fukuoka 820, Japan
Phone : +81-948-29-7712
Fax : +81-948-29-7742
E-mail : yamakawa@ces.kyutech.ac.jp

ABSTRACT

An ordinary neural network can store information in the form of distributed connective weights and thresholds, and facilitates readout of the information in the form of training output data by applying input data similar to but different from the training input data. Although storing behavior in the training mode in the network is dynamical, the readout behavior is static. Accordingly, there cannot be searching of alternative candidates to be read out. This is quite different from information retrieval in a biological brain. The brain retrieves events one by one, by chaining a train of thought or by analogy of different fields.

A group of biological neurons connected to each other often exhibits chaotic itinerancy. This is regarded as the primary behavior of chaining retrieval of stored information in the recurrent neural network.

Interesting behavior of a chaotic network is presented here, which is constructed with chaotic units connected to each other including self-feedback. The chaotic unit is constructed with the delay block and the nonlinear block which a linear combination of a sigmoidal function and an inverted-N type nonlinear function. This unit can be implemented with a chaos chip developed by the authors [1]. The combinatory coefficient (one of the nonlinear parameters) can be determined to make the chaotic unit to come up to fixed point attractor, periodic attractor and strange attractor (chaos) . The connective weights in the chaotic network are assigned by Hebbian learning rule. When a distorted pattern is applied to the chaotic network, the stored pattern most closed to the distorted pattern is retrieved. When an overlapped pattern is applied, separated patterns are retrieved sequentially, the scheme of which is presented in Fig.1. When a mixed pattern or noisy pattern is applied, all the stored patterns are retrieved sequentially. This interesting behavior cannot be obtained in the ordinary mutually connected networks, e.g. Hopfield network, agitated by noisy signal.

This chaining retrieval of information is inherent to the chaotic network proposed here and is possibly a model of associative readout of stored information in a human brain.

REFERENCES

[1]Takeshi Yamakawa, Tsutomu Miki and Eiji Uchino, "A Chaotic Chip for Analysing Nonlinear Dynamical Network Systems," Proc. 2nd Int'l Conf. on Fuzzy Logic and Neural Networks, Iizuka, Japan, July 17-22, pp.563-566,1992.

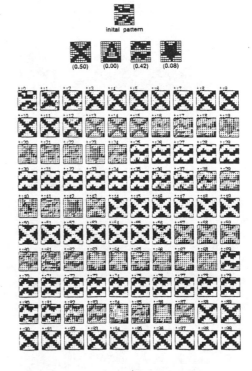

Fig.1 Overlapped two patterns similar to two of four stored paptterns in the chaos network are separated by sequential retrieval from the network.

43

Automated Reasoning

DESIGN AND IMPLEMENTATION OF MULTIPLE-CONTEXT TRUTH MAINTENANCE SYSTEM WITH BINARY DECISION DIAGRAM

Hiroshi G. Okuno, **Osamu Shimokuni,** and **Hidehiko Tanaka**

NTT Basic Research Laboratories

Nippon Telegraph and Telephone Corporation

3-1 Morinosato-Wakamiya, Atsugi

Kanagawa 243-01, JAPAN

Graduate School of Engineering

The University of Tokyo

7-3-1 Hongo, Bunkyo-ku

Tokyo 113, JAPAN

okuno@nuesun.brl.ntt.jp, osamus@ipl.t.u-tokyo.ac.jp, and tanaka@mtl.t.u-tokyo.ac.jp

ABSTRACT

Implicit enumeration of prime implicates in Truth Maintenance System (TMS) is investigated. CMS (Clause Management System), an extension of Assumption-based TMS (ATMS), that accepts any type of justification has a burden to compute all prime implicates, since its complexity is NP-complete. To improve the performance of multiple-context TMS such compact representation of boolean functions. In this paper, we propose a BDD-based Multiple-context TMS (BMTMS) and present the design and implementation of interface between TMS and BDD. The interface provides high level specifications of logical formulas, and has mechanisms to schedule BDD commands to avoid combinatorial explosions in constructing BDDs. In BMTMS, most TMS operations are carried out without enumerating all prime implicates.

1 INTRODUCTION

The capability of thinking with explicit multiple alternatives is required by sophisticated problem solving systems such as qualitative simulation, multi-fault diagnosis or non-monotonic reasoning. Since a consistent database (data collection) is referred to as *context*, the above requirement can be paraphrased as *multiple context reasoning*. Multiple-context reasoning is, in general, superior to single-context reasoning in switching contexts or comparing the results between context [deK86a].

A problem solving system consists of an inference engine and a truth maintenance system (TMS). The inference engine introduces hypotheses and makes inferences, while TMS records hypotheses as *assumptions* and inferences as *justifications*. TMS maintains the inference process and enables the inference engine to avoid futile or redundant computations [FdK93].

TMSs are classified according to two properties. One property is whether they provide a single or a multiple context and the other is whether they accept Horn clauses or general clauses as justifications. This classification is summarized in Table 1.

Justification-based TMS (JTMS) is a single-context TMS that accepts only a Horn clauses as a justification. JTMS is very popular so far, because JTMS runs very efficiently with backtracking mechanism. The satisfiability of a set of Horn clauses, or the assignment of variables that satisfies the set of clauses, can be solved in a linear time of the number of variables [Dow84]. However, JTMS cannot perform efficiently reasoning with alternatives.

Logic-based TMS (LTMS) is a single-context TMS that accepts general clauses including disjunction or negation. An efficient algorithm for LTMS is a *Boolean constraint Propagation (BCP)*, a kind of intelligent backtracking mechanism. First, the conjunction of justifications is converted to a conjunctive normal form (CNF, or Product of Sum, POS) and decomposed to a set of clausal forms. BCP takes a set of clausal forms and labels each node consistently by backtracking. BCP with a set of clausal forms is complete for Horn clauses, but not for general clauses. This incompleteness is caused by being information dropped in conversion of a clause to a set of clausal forms. To

Table 1 Classification of TMS

context	single context	multiple context
JUSTIFI-CATION		
HORN CLAUSES	Justification-based TMS (JTMS)	Assumption-based TMS (ATMS)
GENERAL CLAUSES	Logic-based TMS (LTMS)	Clause Management System (CMS)

recover completeness, all prime implicates should be added to the set of clausal forms, but the whole performance of such a BCP deteriorates because the total number of prime implicates is generally very large. For example, the modeling of two containers connected by a valve with seven observation points produces 2,814 prime implicates [FdK93]. In addition, the computational complexity of enumerating prime implicates is NP-complete.

A multiple-context TMS is, in general, superior to a single-context TMS in its capabilities and efficiency in seeking all solutions [deK86a]. In particular, *Assumption-based TMS (ATMS)*, a multiple-context TMS that accepts only Horn clauses as justifications can be implemented quite efficiently by compiling justifications into a network [deK86a; Oku90]. A datum of inference engine is represented by a node and its belief status is maintained by a label. The labels are computed incrementally by using the network. However, ATMS has a limited power of expression. When the problem solving system wants to handle a general clause, it must be encoded to a set of Horn clauses with special constructs. And some additional routines are needed to validate the completeness of TMS operations [deK86b]. This type of encoding sometimes gives rise to the degradation of the total performance of the system.

Clause management system (CMS) is a multiple-context TMS that accepts general clauses including disjunction and negation [RdK87]. In CMS, the label of a node is computed via minimal support [RdK87]. Let Σ be the set of all the justifications and PI be a prime implicates of Σ. The minimal support for the clause C is computed as follows:

$$MinSup(C, \Sigma) = \{S | S \in \Delta(C, \Sigma), \text{ S is minimal}\} \quad (1)$$
$$\Delta(C, \Sigma) = \{PI - C | PI \cup C \neq \{\}\}$$

Note that the computation of the minimal support requires the enumeration of all prime implicates. This is the main factor of intrinsic poor performance of CMS.

We have been studying an implicit enumeration of prime implicates. Recently, a Binary Decision Diagram (BDD) is proposed, which is a compact representation and provides efficient manipulations of boolean functions [Bry92]. Recent techniques with BDDs can generate more than 10^{10} prime implicates [LCM92]. Since BDDs can represent all solutions simultaneously, it is reasonable to use BDDs to implement a multiple-context TMS. In this paper, we propose a *BDD-based Multiple-context TMS (BMTMS)* to exploit two ways:

1) Implementing efficient methods for enumerating all prime implicates, and

2) Implementing TMS operations without enumerating prime implicates explicitly.

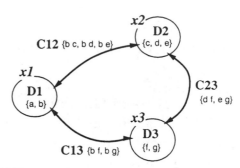

FIGURE 1   **A simple constraint satisfaction problem**

Madre *et al.* implemented ATMS by a variant of BDD called TDG (Typed Decision Graphs), but only the first issue was addressed [MC91].

The rest of the paper is organized as follows: In Section 2, the issues of CMS are identified. In Section 3, BDDs are explained and the issues in applying BDDs to multiple-context TMS is investigated. In Section 4, the BMTMS is proposed and its details are described. In Section 5, some capabilities of the BMTMS are demonstrated. The evaluation results of the BMTMS, and conclusions are given in Section 6 and Section 7, respectively.

2   IDENTIFYING THE ISSUES OF CMS

In this section, the drawbacks of the ATMS and the problems of CMS are demonstrated by using a a simple problem [deK89] (hereafter, *the simple problem*) shown in Figure 1. It has three variables, x_1, x_2, and x_3. The domain of each variable is $D_1 = \{a, b\}$, $D_2 = \{c, d, e\}$, and $D_3 = \{f, g\}$, respectively. For a pair of variables, x_i and x_j, a constraint on them, $C_{i,j}$, is given as the set of permissible combinations of values. They are $C_{12} = \{b\,c, b\,d, b\,e\}$, $C_{13} = \{b\,f, b\,g\}$, and $C_{23} = \{d\,f, e\,g\}$.

2.1   Encoding by ATMS

In ATMS, an inference-engine's datum is represented by a *TMS-node*, and an assumption is a special kind of TMS-node. A justification has a set of *antecedents* and a single *consequence*, all of which are TMS-nodes. In other words, a justification is a propositional Horn clause. An *environment* is represented by a set of assumptions. A contradictory environment is called *nogood*. Each TMS-node has a *label*, or a set of *environments*, under which the node is proved valid.

The simple problem cannot encoded by ATMS itself, because it contains disjunctions and thus is not a Horn clause. The encoding techniques proposed by de Kleer [deK89] is used for the encoding. The inference engine gives the following data to ATMS:

- A propositional symbol, $x_{i:v}$. A symbol $x_{i:v}$ means that a variable x_i has a value v. Thus, $x_{1:a}$, $x_{1:b}$, $x_{2:c}$, $x_{2:d}$, $x_{2:e}$, $x_{3:f}$, $x_{3:g}$.

- A set of justifications, each of which encodes the domain of a variable.

  $x_{1:a} \vee x_{1:b}$, $x_{2:c} \vee x_{2:d} \vee x_{2:e}$, $x_{3:f} \vee x_{3:g}$.

  Since these clauses are not Horn ones, a disjunction is encoded by using the choose predicate as follows:

  $$\text{choose}\{x_{1:a}, x_{1:b}\}, \quad \text{choose}\{x_{2:c}, x_{2:d}, x_{2:e}\},$$
  $$\text{choose}\{x_{3:f}, x_{3:g}\}.$$

- A set of justifications, each of which states that each variable have only one value.

  $\neg x_{1:a} \vee \neg x_{1:b}$, $\neg x_{2:c} \vee \neg x_{2:d}$, $\neg x_{2:c} \vee \neg x_{2:e}$,

  $\neg x_{2:d} \vee \neg x_{2:e}$, $\neg x_{3:f} \vee \neg x_{3:g}$.

  These are encoded by using the nogood predicate as follows:

  $$\text{nogood}\{x_{1:a}, x_{1:b}\}, \quad \text{nogood}\{x_{2:c}, x_{2:d}\},$$
  $$\text{nogood}\{x_{2:c}, x_{2:e}\}, \quad \text{nogood}\{x_{2:d}, x_{2:e}\},$$
  $$\text{nogood}\{x_{3:e}, x_{3:f}\}.$$

- A set of justifications, each of which encodes an inhibited pair specified by a constraint.

  $\neg x_{1:a} \vee \neg x_{2:c}$, $\neg x_{1:a} \vee \neg x_{2:d}$, $\neg x_{1:a} \vee \neg x_{3:e}$,

  $\neg x_{2:d} \vee \neg x_{3:g}$, $\cdots$.

  These are encoded as follows:

  $$\text{nogood}\{x_{1:a}, x_{2:c}\}, \quad \text{nogood}\{x_{1:a}, x_{2:d}\},$$
  $$\text{nogood}\{x_{1:c}, x_{3:e}\}, \quad \text{nogood}\{x_{2:d}, x_{3:g}\}, \quad \cdots.$$

To find solutions that satisfy these justifications, ATMS uses a label update algorithm. Since ATMS is complete for Horn clauses but not for arbitrary logical formula, ATMS fails to calculate the correct labels for the above encodings. To attain completeness, meta rules concerning choose and nogood are needed. These meta rules perform hyperresolutions for these predicates. Some examples of hyperresolutions are listed below:

$$\begin{array}{l} \text{choose}\{x_{3:f}, x_{3:g}\} \\ \text{nogood}\{x_{2:c}, x_{3:f}\} \\ \text{nogood}\{x_{2:c}, x_{3:g}\} \\ \hline \\ \text{nogood}\{x_{2:c}\} \end{array}$$

$$\begin{array}{l} \text{choose}\{x_{2:c}, x_{2:d}, x_{2:e}\} \\ \text{nogood}\{x_{2:c}\} \\ \hline \\ \text{choose}\{x_{2:d}, x_{2:e}\} \end{array}$$

The final solutions are

$$(x_{1:b} \wedge x_{2:d} \wedge x_{3:f}) \vee (x_{1:b} \wedge x_{2:e} \wedge x_{3:g}).$$

This encoding works well for the simple problem, but does not work for all CSPs. de Kleer pointed out that the meta rules correspond to local consistency algorithms [deK89]. Applying a meta rule that corresponds to arc-consistency algorithm is very expensive, and thus the completeness may be fulfilled not at all time.

2.2 Encoding by CMS

The simple problem is directly encoded by logical expressions. In CMS encoding, the same propositional symbols as ATMS are used. The domain of each variable is encoded as $X1$, $X2$, and $X3$. The select-one type constraint that each variable has only one value is encoded by $C1$, $C2$, and $C3$. The constraints between two variables are directly encoded.

$$\begin{aligned} X1 &= x_{1:a} \vee x_{1:b} \\ X2 &= x_{2:c} \vee x_{2:d} \vee x_{2:e} \\ X3 &= x_{3:f} \vee x_{3:g} \\ C1 &= (x_{1:a} \wedge \neg x_{1:b}) \vee (\neg x_{1:a} \wedge x_{1:b}) \\ C2 &= (x_{2:c} \wedge \neg x_{2:d} \wedge \neg x_{2:e}) \vee (\neg x_{2:c} \wedge x_{2:d} \wedge \neg x_{2:e}) \\ &\quad \vee (\neg x_{2:c} \wedge \neg x_{2:d} \wedge x_{2:e}) \\ C3 &= (x_{3:f} \wedge \neg x_{3:g}) \vee (\neg x_{3:f} \wedge x_{3:g}) \\ C12 &= (x_{1:b} \wedge x_{2:c}) \vee (x_{1:b} \wedge x_{2:d}) \vee (x_{1:b} \wedge x_{2:e}) \\ C13 &= (x_{1:b} \wedge x_{3:f}) \vee (x_{1:b} \wedge x_{3:g}) \\ C23 &= (x_{2:d} \wedge x_{3:f}) \vee (x_{2:e} \wedge x_{3:g}) \\ F &= X1 \wedge X2 \wedge X3 \wedge C1 \wedge C2 \wedge C3 \\ &\quad \wedge C12 \wedge C13 \wedge C23 \qquad\qquad (2) \end{aligned}$$

The simple problem is encoded as F. The prime implicates of F are

$\neg x_{2:d} \vee \neg x_{3:g}$, $x_{2:d} \vee x_{3:g}$, $\neg x_{3:f} \vee x_{2:d}$,

$x_{3:f} \vee \neg x_{2:d}$, $\neg x_{2:e} \vee \neg x_{2:d}$, $x_{3:f} \vee x_{2:d}$,

$\neg x_{2:e} \vee \neg x_{2:d}$, $\ldots$

Let $Goal$ be the goal literal. The simple problem is thus encoded as $F \supset Goal$. The label of the $Goal$ literal is computed by Equation (1) and the following results are obtained:

$$(x_{1:b} \wedge x_{2:e} \wedge x_{3:g} \wedge \neg x_{1:a} \wedge \neg x_{2:c} \wedge \neg x_{3:d} \wedge \neg x_{3:f}) \supset Goal$$

$$(x_{1:b} \wedge x_{2:d} \wedge x_{3:f} \wedge \neg x_{1:a} \wedge \neg x_{2:e} \wedge \neg x_{2:c} \wedge \neg x_{3:g}) \supset Goal$$

Thus, the solution is obtained as the label of the literal $Goal$, that is,

$$\{x_{1:b} \wedge x_{2:e} \wedge x_{3:g} \wedge \neg x_{1:a} \wedge \neg x_{2:c} \wedge \neg x_{3:d} \wedge \neg x_{3:f},$$
$$x_{1:b} \wedge x_{2:d} \wedge x_{3:f} \wedge \neg x_{1:a} \wedge \neg x_{2:e} \wedge \neg x_{2:c} \wedge \neg x_{3:g}\}.$$

Since the labels in ATMS satisfies four properties — soundness, completeness, consistency, and minimality —, they can be computed incrementally and efficiently

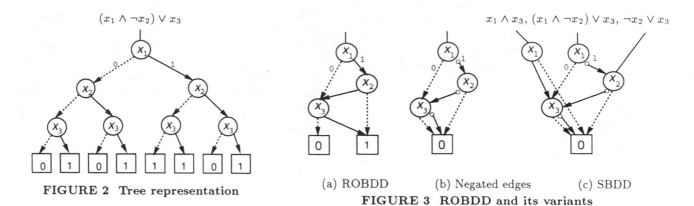

FIGURE 2 Tree representation

(a) ROBDD   (b) Negated edges   (c) SBDD

FIGURE 3 ROBDD and its variants

[deK86a]. On the other hand, the labels are computed in batch in CMS by enumerating all prime implicates, because no algorithms are proposed so far to get all prime implicates incrementally. The label of a node plays a key role for efficient processing in ATMS, while it is one source of inefficiency in CMS. To cope with CMS's inefficiency due to enumeration of prime implicates, we exploit BDDs in the following two ways in the next section.

1) TMS operations without using labels, and

2) efficient enumeration of prime implicates.

3 Binary Decision Diagram (BDD)

3.1 Representing logical functions

Akers proposed a Binary Decision Diagram (BDD) [Ake78] as a compact representation of boolean (logical) functions, and Bryant invented efficient algorithms for manipulating them [Bry86]. Boolean functions can be represented by a tree. For example, the logical function, $((x_1 \wedge \neg x_2) \vee x_3)$, is represented by a tree, where a node represents a variable and two kinds of leaves represent 0 and 1 (Figure 2). Two edges of each node are called *0-edge* and *1-edge*, which represent the value when the variable associated with the node takes 0 and 1, respectively.

By fixing the order of variables, say, x_1, x_2, x_3 (where x_1 is the uppermost variable), sharing leaves and duplicate nodes, and removing redundant nodes, this tree can be transformed into an ROBDD (Reduced Ordered BDD) (Figure 3-(a)). An ROBDD gives a canonical form of a boolean function, and thus equivalent functions are represented by the same ROBDD. Various techniques have been invented to reduce the size of BDDs. One is a negated edge, which is indicated as a small circle attached to an edge (Figure 3-(b)). Another example is a Shared BDD (SBDD), which shares BDDs among many functions. Three functions

share BDDs (Figure 3-(c)). Hereafter, BDD refers to SBDD with negated edges.

A BDD is constructed incrementally by the *apply* function instead of reducing a tree representation. Suppose *var-order* returns the order of a variable. Then, the apply function is defined recursively as follows:

$apply(\text{bdd1,bdd2,operation}) =$

- if *var-order*(bdd1's root) = *var-order*(bdd2's root), create a node by performing the operation to get a 0-edge and 1-edge.

- if *var-order*(bdd1's root) > *var-order*(bdd2's root), create a node with *apply*(bdd1's 0-edge, bdd2, operation) for the 0-edge, and *apply*(bdd1's 1-edge, bdd2, operation) for the 1-edge.

- otherwise,
  apply(bdd2, bdd1, operation).

This definition can be explained as performing logical operations on the Shannon expansion. Suppose that the Shannon expansions of functions f and g are as follows:

$$f = (\neg x \wedge f_{|_{x=0}}) \vee (x \wedge f_{|_{x=1}}), \text{ and}$$
$$g = (\neg x \wedge g_{|_{x=0}}) \vee (x \wedge g_{|_{x=1}}).$$

The logical AND of f and g, $f \wedge g$, is computed as

$$(\neg x \wedge (f_{|_{x=0}} \wedge g_{|_{x=0}})) \vee (x \wedge (f_{|_{x=1}} \wedge g_{|_{x=1}})).$$

The computational cost of the logical AND is in the order of the product of the numbers of nodes of both BDDs. Most functions can be executed in the same order. In addition, the *apply* function can be implemented efficiently by using a node hash table and a result cache (hash table). Since the result cache stores recent results of *apply* functions, redundant computations may be avoided if the cache hits. Since BDD is a compact representation of boolean functions and many usual boolean operations are performed efficiently with BDDs, BDD is commonly used in VLSI CAD systems.

Table 2 Comparison of size of BDDs

Constraint ordering	(2)	(3) [*]
Variable ordering	Best	Best [*]
Maximum size of intermedi-ate BDDs	101	31
Size of final BDD	24	30

[*] Both constraint and variable ordering are determined by the CCVO heuristics.

The simple problem of the previous section can be solved directly by creating the BDD for F by using Equation (2). However, this BDD is an implicit representation of solutions and all the paths from the root of the BDD to the leaf of 1 should be enumerated to get the explicit representations of solutions.

3.2 Issues in using BDD

Applying BDDs to TMS has four main issues:

1) **Variable ordering**

2) **Constraint ordering**

3) **Encoding by logical functions**

4) **Compatibility with existing systems**

The number of nodes of a BDD, or *the size* of BDD, is determined by variable order. The minimum (best) size of 2-level AND-OR circuit of n variables is in the linear order of n, while the worst size is in the order of 2^n. Therefore the optimal variable order under which the size of BDD is minimum is very important. However, the computational complexity of finding the optimal variable order is NP-complete. This is well-known problem and many heuristics to find a near-optimal variable ordering is proposed.

However, constraint order is essential to avoid combinatorial explosions [Oku94]. In creating the BDD for F, nine constraints, $X1$, ..., $C23$, are used. Consider two constraint orderings below:

$$X1 \wedge X2 \wedge X3 \wedge C1 \wedge C2 \wedge C3 \wedge C12 \wedge C13 \wedge C23 \quad (2)$$

$$X1 \wedge C1 \wedge X2 \wedge C2 \wedge C12 \wedge X3 \wedge C3 \wedge C13 \wedge C23 \quad (3)$$

The size of BDDs created by (2) and (3) is shown in Table 2. The size of intermediate BDDs is affected by constraint order. The importance of constraint ordering is discovered by applying BDDs to solve combinatorial problems. Okuno proposed the CCVO (Correlation-based Constraint and Variable Ordering) heuristics to find a near-optimal constraint ordering. The CCVO first calculates the correlation of variables between constraints and then determines constraint ordering. Variable ordering is determined the first time a variable is used by constraint ordering.

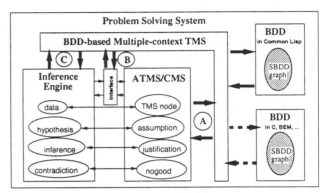

FIGURE 4 Structure of BMTMS

Another problem of constraint ordering occurs in applying BDDs to TMS, because constraints or justifications are given incrementally. Therefore, a new method of determining constraint ordering is required.

The third issue is concerning the expressive power of logical functions. In this paper, arithmetic boolean functions are used [Min93a]. A select-one type constraint can be easily encoded by them. For example, a constraint that variable x_2 has one and only one value of its domain, that is $X2 \wedge C2$, can be encoded by

$$x_{2:c} + x_{2:d} + x_{2:e} == 0 \quad (4)$$

The final issue is that the interface between BDD and TMS should be compatible with existing systems, since many applications are developed and it costs very expensive to re-implement them from scratch.

4 Design of BMTMS

In this section, the *BDD-based Multiple-context TMS (BMTMS)*, is proposed to cope with the above issues. The whole system is depicted in Figure 4. BMTMS provides three interfaces to various kinds of existing systems:

1) interface to ATMS/CMS (Ⓐin Figure 4),

2) interface to extensions of ATMS such as consumer architecture [deK86b] (Ⓑin Figure 4), and

3) interface to inference engine (Ⓒin Figure 4).

These interfaces are only conceptual and thus treated uniformly within the BMTMS, since it can manipulate any logical formulas.

4.1 Interface functions

The criteria of designing the primitive functions is whether they can be implemented efficiently with BDDs.

1) **BDD primitive functions:**

   bddand, bddor, bddnot, bddrstr0, bddrstr1, ···

Each function corresponds to a primitive operation of BDDs, and take BDDs as its arguments. For example, (bddrstr0 BDD_f x) computes a BDD restricted to the case that x has value of 0, that is, $f_{|x=0}$.

2) **Variable ordering**:

variable-order specifies the variable order.

3) **Logical operations**:

land, lor, lxor, imply, lnot, $\cdots$
These functions accept any number of arguments.

4) **Arithmetic logical operations**:

1* (multiplication), 1+ (addition), 1/ (quotient), 1% (remainder), 1- (subtraction), 1<< (left-shift) 1== (equal to), ...

These operations are implemented by simulating arithmetic logic unit (ALU) operations [Min93a]. Equation (4) is encoded by

$$(\texttt{1==} (\texttt{1+} x_{2:c} \; x_{2:d} \; x_{2:e}) \; \texttt{0}).$$

5) **BDD graph operations**:

forall, exists, 1-path, $\cdots$

forall is universal quantifier and exists is existential quantifier. Consider a function, f, is expressed in Shannon expansion,

$$f = (\neg x \wedge f_{|x=0}) \vee (x \wedge f_{|x=1}).$$

Universal quantifier $\forall x f$ is computed by

$$\forall x f = f_{|x=0} \wedge f_{|x=1}.$$

Or (bddand (bddrstr0 f x) (bddrstr1 f x)). Existential quantifier $\exists x f$ is computed by

$$\exists x f = f_{|x=0} \vee f_{|x=1}.$$

Or (bddor (bddrstr0 f x) (bddrstr1 f x)). 1-path enumerates all the paths from the root of a BDD to the leaf of 1.

6) **Output functions**:

sop returns a sum of products, which is a set of irredundant prime implicants. And pos returns a product of sums, which is a set of irredundant prime implicates.

7) **User defined data types**:

In addition to system-defined binary and integer data, the user can define any data type by defentity.

```
(defentity datatype
    (slot_1 slot_2 ⋯ )
    (:exclusive-p) )
```

A slot can be accessed by a function *datatype-slotname*. If :exclusive-p is specified, elements of a data defined by defentity hold exclusively. An operation for a data type is defined as follows:

```
(defop (operation datatype)
    argument-list . body )
```

Its name is *datatype-operation*.

8) **Select function**:

(choice-of p$_1$ $\cdots$ p$_n$) selects one element exclusively from the set. This function can be encoded by (lor p$_1$ $\cdots$ p$_n$) and (1== (1+ p$_1$ $\cdots$ p$_n$) 1).

9) **Addition of Constraints**

(add-constraint l C) adds a constraint C to the logical formula l. This is conceptually the same as (setq C (land C l)).

4.2 TMS Data Representation

Four kinds of nodes in TMS are encoded as follows:

1) **Premise**: the variable itself.

2) **Contradiction**: This node is used to represent a nogood relation in ATMS. It is not needed in the BMTMS, since a nogood relation can be represented by a logical formula.

3) **Assumption**: An assumption variable is introduced to express environments in the BMTMS, but is not discriminated from other nodes in BDDs.

4) **Normal node**: a variable itself.

Constraints between nodes are represented straightforward as follows:

- **Justification**, $n_1, \cdots, n_k \rightarrow c$, (c is an arbitrary clause), is encoded by
  (imply (land n$_1$ $\cdots$ n$_k$) c).

- **nogood**$\{n_1, \cdots, n_k\}$ is encoded by
  (lnot (land n$_1$ $\cdots$ n$_k$)).

- **class**$\{n_1, ..., n_k\}$, that selects one element from a set of nodes is encoded as follows:
  (choice-of n$_1$ $\cdots$ n$_k$).

4.3 TMS operations

Let Π be a logical AND of all justifications and d be a node. An environment E is expressed by an AND of assumption variables, $a_1 \wedge ... \wedge a_k$. Implementations of some main TMS operations are listed below:

1) **Check whether all justification are not consistent.**

If the BDD for Π reduces to 0, all justifications are not consistent.

2) **Check whether node d is consistent with environment E**: node-consistent-with(d, E)

If the BDD for $\Pi \wedge d \wedge a_1 \wedge \cdots \wedge a_k$ reduces to 0, node d is not consistent with environment E.

3) **Compute the label of node** d: `tms-node-label`(d).

Let $a_1, a_2, \dots$ be assumption variables, and $d_1, d_2, \dots$ be normal variables. The label of d is computed as follows:

(1) Construct the BDD for $\Gamma = (\forall d_1, d_2, \dots (\Pi \supset d))$. (2) Enumerate all prime implicates of Γ that contain d. This enumeration is described in Section 4.4. (3) Compute $\{X | X \supset d$ of a prime implicate$\}$, and this set is the label of d.

If the justifications are restricted to Horn clauses (as in ATMS), the step (2) can be replaced with $\text{pos}(\Gamma)$, which runs much faster.

4) **Check the status of node** d: `in-node?`, `out-node?`, `true-node?`, `false-node?`

Let $a_1, a_2, \dots$ be assumption variables, and $d_1, d_2, \dots$ be normal variables. Construct the BDD for $(\Pi \supset d)$. If the BDD reduces 1 or 0, the node status of d is `:TRUE` and `:FALSE`, respectively. Otherwise, construct the BDD for $\forall d_1, d_2, \dots (\Pi \supset d)$. If the BDD is not 0, the node status of d is `:IN`. Otherwise it is `:OUT`. (In addition, if the node status of $\neg d$ is `:OUT`, the node status of d is `:UNKNOWN`.)

5) **Force an assumption true or false.**

These functions are implemented by `retract-assumption`$(a_i$ or $\neg a_i)$, which are encoded as follows: (`add-constraint` a_i or $\neg a_i$ Π).

4.4 Prime implicates

In the research of boolean and switching functions, a *prime implicant* which is in the form of product of literals is usually used. (A literal is either a symbol or a negated symbol). Efficient algorithms for obtaining all prime implicants are proposed [CM92].

On the other hands, in truth maintenance, constraint satisfaction and logic programming, a *prime implicate* which is in the form of sum of literals is usually used. Since a conjunctive normal form is dual to a disjunctive normal form, a set of prime implicates can be computed by performing the duality operation on a set of prime implicants.

In other words, given a logical function $f(x)$, let $\bar{f}(\bar{x})$ be a function gotten by exchanging $\vee$ with $\wedge$ and vice versa, and by replacing every literal with its negation, simultaneously. Then all the prime implicates for $\bar{f}(\bar{x})$ are computed. Finally, exchanging $\wedge$ with $\vee$ and replacing every literal with its negation simultaneously computes every prime implicate.

The resulting set of prime implicates is enough for computing the label of a node, although it is not complete in the sense that any tautology such that $(y \vee \neg y)$ is a prime implicate,

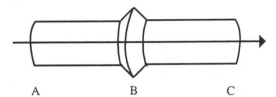

FIGURE 5 Two-pipe system ([FdK93, p.466])

In the BMTMS, all the prime implicates are computed directly from BDDs. The key point of the algorithm is that a BDD can be interpreted as another expansion on f.

$$f = (x_k \vee f_{|x_k=0}) \wedge (\neg x_k \vee f_{|x_k=1}),$$

The prime implicates of the function f can be obtained by combining the prime implicates of two subfunctions $f_{|x_k=0}$ and $f_{|x_k=1}$. The algorithm recursively computes prime implicates. This algorithm is a dual version of Coudert's [CM92]. Intermediate sets are maintained by Zero-Suppressed BDDs [CMF93], which are well suited for the set representation [Min93b].

4.5 Label enumeration

In the BMTMS, it has been shown that all TMS operations concerning labels except `tms-node-label` can be implemented without enumerating prime implicates. Although such enumeration is needed at least by abduction reasoning or dependency-directed search [RdK87], we guess that it is not so often in other applications. If this is true, many applications can be implemented by BDDs very efficiently.

4.6 Constraint and variable ordering

The constraint ordering algorithm used in the BMTMS is simpler than CCVO. The BMTMS does not construct a BDD for class constraints immediately when they are given to the system. Instead, adding those to the system is delayed until some variable of the class is used by some constraint. Variable ordering is determined the first time when it is used by such a constraint similar to that of CCVO.

5 APPLICATIONS OF BMTMS

In this section, some capabilities of the BMTMS are demonstrated in qualitative simulation.

Consider the system that has two pipes, A and C, which are connected by a joint, B (Fig.5). The pressure and flow in the system can be modeled by the following qualitative equations [deK91]:

$$[dP_A] - [dP_B] = [dQ_{AB}],$$
$$[dP_B] - [dP_C] = [dQ_{BC}],$$
$$[dQ_{AB}] = [dQ_{BC}],$$

where $[dx]$ denotes the sign $(+, 0, -)$ of $\frac{dx}{dt}$.

First, the sign data type and its operations are defined as follows:

```
(defentity sign
   (positive zero negative)
   (:exclusive-p t) )

(defop (- sign) (s)
   (make-sign (sign-minus s)
      (sign-zero s) (sign-plus s) ))

(defop (+ sign) (s1 s2)
  (let ((xp (sign-plus s1))
        (x0 (sign-zero s1))
        (xm (sign-minus s1))
        (yp (sign-plus s2))
        (y0 (sign-zero s2))
        (ym (sign-minus s2)) )
   (make-sign
    (bddnot
     (lor (bddand x0 ym) (bddand x0 y0)
        (bddand xm y0) (bddand xm ym)))
    (bddnot
     (lor (bddand xp yp) (bddand xp y0)
        (bddand x0 yp) (bddand x0 ym)
        (bddand xm y0) (bddand xm ym)))
    (bddnot
     (lor (bddand x0 yp) (bddand x0 y0)
        (bddand xp yp) (bddand xp y0)))))))

(defop (sign =0) (s) (sign-zero s))
```

Since the `sign` is defined with `:exclusive-p`, the operation `=0` is specified quite simply instead of the following complicated definition:

```
(land (lnot (sign-plus s))
      (sign-zero s)
      (lnot (sign-minus s)) )
```

The logical formula that express $(x + y = 0)$ is defined as follows: (we use logical formula for economy of space.)

$$(y_m \wedge \neg y_p \wedge \neg y_0 \wedge \neg x_m \wedge x_p \wedge \neg x_0) \vee$$
$$(\neg y_m \wedge y_p \wedge \neg y_0 \wedge x_m \wedge \neg x_p \wedge \neg x_0) \vee$$
$$(\neg y_m \wedge \neg y_p \wedge y_0 \wedge \neg x_m \wedge \neg x_p \wedge x_0)$$

where indexes, $p, 0, m$, indicate three slots of the sign data type, respectively.

Let `C` be a Lisp variable that holds the constraint set. The above qualitative equations are encoded as follows:

```
(setq Pa (make-sign) Pb (make-sign)
      Pc (make-sign) Qab (make-sign)
      Qbc (make-sign) )
```

```
(add-constraint
   (sign-= (sign-- Pa Pb) Qab) C)
(add-constraint
   (sign-= (sign-- Pb Pc) Qbc) C)
(add-constraint
   (sign-= Qab Qbc) C)
```

Now, consider the situation that the pressure at A is increasing and one at C is not changing. This fact is encoded as follow:

```
(setq Constraint
   (land Constraint
      (sign-plus Pa)
      (sign-zero Pc) ))
```

```
(add-constraint Constraint C).
```

Finally, the result shows that the BDDs for
   (imply C (sign-plus Qab)), and
   (imply C (sign-plus Qbc))
reduce to 1. This means that both $[dQ_{AB}]$ and $[dQ_{BC}]$ are $+$, that is, the pressures at the interfaces between A and B, and between B and C are increasing. Note that this computation does not need enumerating prime implicates.

On the other hand, in QPE, every qualitative equation is expanded to a set of clausal form and all the prohibited combination are enumerated [deK91]. Then, all the prime implicates are computed. The final step is to apply the BCP to the set of all the prime implicates, which computes the labels of $[dQ_{AB}]$ and $[dQ_{BC}]$.

6 EVALUATION OF BMTMS

The current BMTMS system is implemented in Lisp and has two BDD packages (see Figure 4). One BDD package is implemented in C, which is called Boolean Expression Manipulator, BEM-II [Min93a]. This implementation of the BMTMS is referred to *BMTMS with BDD in C*. Another BDD package is written in Lisp and this implementation of the BMTMS is referred to *BMTMS with BDD in Lisp*. The timing data is measured on the SPARCStation10 with 128 MBytes of main memory.

6.1 Evaluating interface Ⓐ

N-Queens problem Two implementations of the BMTMS are compared with the ATMS which is implemented in Lisp. Two kinds of N-Queen problem programs by ATMS are used; one is by label update, and the other is by interpretation construction. Both programs computes column by column. The timing results of N-Queen problem programs are shown in Figure 6. The BMTMS with BDD in Lisp is much

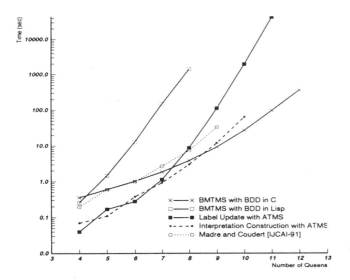

FIGURE 6   N-Queen problem results

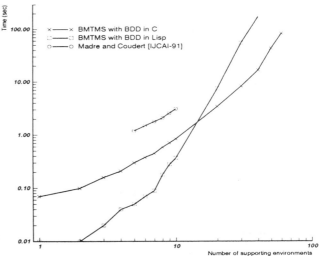

FIGURE 7   Minimal Cover results

slower than the ATMS. The reasons are twofold. (1) The algorithm of N-Queens is different. If the most fair algorithm by ATMS, which creates only one goal, installs all the justifications that justify the goal, and then computes labels, is used, it runs much slower than the BMTMS with BDD in Lisp and can get solutions only up to 6-Queen problem [Oku90]. (2) The BDD is not considered suitable for Lisp due to Lisp's memory management. Usually efficient implementations of BDD use only main memory, while such memory management is difficult in Lisp systems [Bry92; MIY91]. The BMTMS with BDD in C shows a good performance for large n.

Minimal cover   This benchmark was used by Madre and Coudert [MC91]. The timing results of minimal cover are shown in Figure 7. The figure shows that the overheads of the BMTMS with BDD in Lisp exceed those of the BMTMS with BDD in C. This is caused by the overheads of Lisp runtime system.

6.2   Evaluating interface Ⓑ: ATMS trace file

We use an ATMS trace file generated by consumer architecture [deK86b], for which de Kleer's ATMS could not compute the labels. Both BMTMS implementations succeed in constructing BDDs by the constraint ordering algorithm. However, they fail in constructing BDDs due to combinatorial explosions, when the constraint and variable ordering mechanism is not used. The ordering algorithm is very simple, but proved effective.

6.3   Evaluating interface Ⓒ: Qualitative simulation

As described in the previous section, qualitative simulation of the two pipe system can be effectively computed by the BMTMS without enumerating all the prime implicates. However, this is a preliminary result and full assessment should be done by implementing the full set of QPE by the BMTMS.

7   CONCLUSIONS

In this paper, a new multiple-context truth maintenance system called BMTMS is proposed and its design and implementation is presented. The key idea of the BMTMS is to use BDDs to avoid enumerating prime implicates in manipulating general clauses. The two issues in applying BDDs to multiple-context TMSs are pointed out; variable ordering, and constraint ordering. To cope with these issues, the BMTMS schedules the constraint ordering by considering the dependency of variables and justifications (or constraints). The BMTMS also provides three level interface to existing problem solving and TMS systems so that BDD can be used in "plug-and-play" manner. Since the logical relations are stored in BDDs and most TMS operations can be performed without enumerating all prime implicates, the BMTMS runs efficiently. The BMTMS enumerates all prime implicates directly from BDDs. In addition, the capability of defining a data type and its operations makes it easy to implement applications with the BMTMS.

Future work includes implementation of the full QPE system with the BMTMS to demonstrate it power and open a new load to intelligent systems. Applying the BMTMS to various sophisticated expert systems is also an interesting area.

Acknowledgments

The authors thank Mr. Yuji Kukimoto of UCB, Dr. Shin'ichi Minato of NTT LSI Laboratories, and Mr. Hideki Isozaki of NTT Basic Research Laboratories for their discussions. The first author also thanks Dr. Ken'ichiro Ishii and Dr. Norihiro Hagita of NTT Basic Research Laboratories for their supports.

References

[Ake78] S.B., Akers. Binary Decision Diagrams. *IEEE Transactions on Computer*, Vol. C-27, No. 6, pages 509–516, 1978.

[Bry86] R.E. Bryant. Graph-based algorithm for Boolean function manipulation. *IEEE Transactions on Computer*, Vol. C-35, No. 5, pages 677–691, 1986.

[Bry92] R.E. Bryant. Symbolic Boolean Manipulations with Ordered Binary Decision Diagrams. *Computing Surveys*, Vol. 24, No. 3, pages 293–318, ACM, 1992.

[CM92] O. Coudert, and J.C. Madre. Implicit and Incremental Computation of Primes and Essential Primes of Boolean Functions. In *Proceedings of the 29th Design Automation Conference (DAC)*, pages 36–39, ACM/IEEE, 1992.

[CMF93] O. Coudert, J.C. Madre, and H. Fraisse. A New Viewpoint on Two-Level Logic Minimization. In *Proceedings of the 30th Design Automation Conference (DAC)*, pages 625–630, ACM/IEEE, 1993.

[deK86a] J. de Kleer. An Assumption-based TMS. *Artificial Intelligence*, **28**:127-162, 1986.

[deK86b] J. de Kleer. Extending the ATMS. *Artificial Intelligence*, **28**:163-196, 1986.

[deK89] J. de Kleer. A Comparison of ATMS and CSP Techniques. In *Proceedings of the Eleventh Internatioal Joint Conference on Artificial Intelligence (IJCAI)*, pages 290–296, 1989.

[deK91] J. de Kleer. Compiling Devices: Locality in a TMS. In Faltings, B. and Strauss, P. (eds): *Recent Advances in Qualitative Physics*, MIT Press, 1991.

[Dow84] W.F. Dowling and J.H. Gallier. Linear time algorithms for testing the satisfiability of propositional horn formulas. *Journal of Logic Programming,* vol.3, pages 267–284, 1984.

[FdK93] K. Forbus, and J. de Kleer. *Building Problem Solvers*, MIT Press, 1993.

[LCM92] B. Lin, O. Coudert, and J.C. Madre. Symbolic prime generation for muliple-valued functions. In *Proceedings of the 29th Design Automation Conference (DAC)*, paages 40–44, ACM/IEEE, 1992.

[MC91] J.C. Madre, and O. Coudert. A logically complete reasoning maintenance system. In *Proc. of the Twelfth Internatioal Joint Conference on Artificial Intelligence (IJCAI)*, pages 294–299, 1991.

[MIY91] S. Minato, N. Ishiura, and Y. Yajima. Shared Binary Decision Diagram with Attributed Edges for Efficient Boolean Function Manipulation. In *Proceedings of the 27th Design Automation Conference (DAC)*, pages 52–57, IEEE/ACM, 1990.

[Min93a] S. Minato. BEM-II: An arithmetic Boolean expression manipulator using BDDs. *IEICE Transactions of Fundamentals*, Vol. E76-A, No. 10, pages 1721–1729, 1993.

[Min93b] S. Minato. Zero-Suppressed BDDs for Set Manipulation in Combinatorial Problems. In *Proceedings of the 30th Design Automation Conference (DAC)*, pages 272–277, ACM/IEEE, 1993.

[Oku90] H.G. Okuno. AMI: A New Implementation of ATMS and its Parallel Processing (*in Japanese*). *Journal of Japanese Society for Artificial Intelligence*, Vol.5, No.3, pages 333–342, 1990.

[Oku94] H.G. Okuno. Reducing Combinatorial Explosions in Solving Search-Type Combinatorial Problems with Binary Decision Diagrams (*in Japanese*). *Transactions of Information Processing Society of Japan* Vol.35, No.5, pages 739–753, 1994.

[OST96] H.G. Okuno, O. Shimokuni, and H. Tanaka. Binary Decision Diagram based Multipli-Context Truth Maintenance System BMTMS (*in Japanese*). *Journal of Japanese Society for Artificial Intelligence*, Vol.10, No.2, 1996.

[RdK87] R. Reiter, and J. de Kleer. Foundations of Assumption-Based Truth Maintenance System. In *Proceedings of National Conference on Artificial Intelligence (AAAI)*, pages 183–188, 1987.

PARCAR: A PARALLEL COST-BASED ABDUCTIVE REASONING SYSTEM

Shohei Kato   **Chiemi Kamakura**   **Hirohisa Seki**   **Hidenori Itoh**

Department of AI and Computer Science
Nagoya Institute of Technology,
Gokiso, Showa-ku, Nagoya 466, Japan.
E-mail:{ shohei@juno.ics, chiemi@juno,ics, seki@ics, itoh@ics}.nitech.ac.jp

ABSTRACT

We propose an efficient parallel first-order cost-based abductive reasoning system, which can find a minimal-cost explanation of a given observation. In this paper, we introduce a search control technique of parallel best-first search into abductive reasoning mechanism. We also implement a PARallel Cost-based Abductive Reasoning system, PARCAR, on an MIMD distributed memory parallel computer, Fujitsu AP1000, and show some performance results.

1 INTRODUCTION

In recent years, the demands of AI for industrial and engineering applications become more and more strong. The research for expert systems and knowledge-base systems is indispensable for practical AI applications. The reasoning mechanisms in the systems, complying with the demand for more intelligent inferences, come to shift from deductive to non-deductive, such as default, abductive, and inductive reasoning and reasoning under uncertainty. We are taking our stand on that the mechanisms should be also parallelized so as to deal with large scale knowledge-bases within realistic time. We, therefore, adopt abductive reasoning as a reasoning mechanism of the systems and aim to propose an efficient abductive reasoning system by means of its parallelization.

Abductive reasoning, a form of non-deductive inference which can reason suitably under the incomplete knowledge (e.g., [Poo88]), has attracted much attention in AI and also has many interesting application areas such as diagnosis, scheduling and design. Thus abductive reasoning may hold the key to improve the power of the industrial and engineering applications.

It is, however, known to be computationally very expensive for large problems [SL90], thus will require sophisticated heuristic search strategies. Then, several query processing techniques studied in logic programming and deductive databases are applied to abductive reasoning (e.g.,[Sti94], [OI93], [KSI93]).

In general, abductive reasoning might find more than one solution. We, however, do not always require all the solutions. We often need the most preferable solution instead. For example, it would be often natural in some planing problems that our required solution is not all orders of processes to accomplish a given task, but the minimal-cost (or fastest) one among them. Some work has been reported to solve such problem, by giving costs to hypotheses as the criterion judging which hypothesis to be selected preferably (e.g.,[Poo93],[CS94]). Then, we [KSI94] have proposed an abductive reasoning system which can find efficiently the most preferable solution of a given query, by introducing a search control technique of A* into abductive reasoning mechanism.

In addition to the search control technique in [KSI94], we propose a method for distributing the search space to multiple processors and exploring each distributed search space with synchronizations and communications. We introduce a search control technique of parallel best-first heuristic search into abductive reasoning mechanism, thereby obtaining much more efficiently the most preferable explanation of a given observation. We then implemented parallel cost-based abductive reasoning system, PARCAR on an MIMD distributed memory parallel computer, Fujitsu AP1000 [IHI+91]. The performance results of PARCAR on AP1000 are also shown.

The organization of the paper is as follows. In Section 2, we give a brief description of our framework of abductive reasoning. Section 3 describes PARCAR algorithm. In section 4 we discuss related work. Section 5 shows some performance results of PARCAR on AP1000.

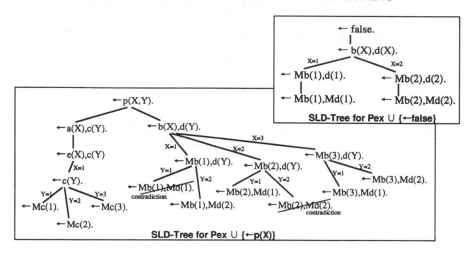

Figure 2   An Example of Abductive Reasoning for P_{ex}

2 COST-BASED ABDUCTION

In this paper, we consider our abductive reasoning to find the most preferable explanation of an observation, by giving each hypothesis a non-negative real number as its cost for the criterion of selection of preferable hypotheses. For a set of hypotheses D, we suppose that a cost of D is the sum of costs of hypotheses in $D^{\dagger}$. We suppose further that an explanation of O is *the most preferable* (or *optimal*) if the sum of costs of the hypotheses constructing the explanation is minimum in all the explanations of O.

Definition 2.1 Suppose that a set of first-order Horn clauses F, called *facts*, and a set of atoms (unit clauses) H, called the set of *hypotheses*, are given. Suppose further that an existentially quantified conjunction O of atoms, called *an observation* or simply a *goal*, is given. Then, an *optimal explanation* of O from $F \cup H$ is a set D of instances of elements of H such that

$F \cup E \vdash O$ (O can be proved from $F \cup E$)   (AR1)

$F \cup E \nvdash false$ ($F \cup E$ is consistent)   (AR2)

$cost(E) \leq cost(D)$ for all D :

$F \cup D \vdash O, F \cup D \nvdash false$

(*cost*(D) is minimum among all sets of hypotheses which satisfy AR1 and AR2),   (AR3)

where $cost(H)$ is the sum of costs of hypotheses in H. Abductive reasoning is now defined to be a task of finding an optimal explanation E of O from $F \cup H$. In this framework, F is assumed to be consistent and treated as always true.   □

Definition 2.2 A headless clause in F is called a *consistency condition*. A consistency condition is denoted by "$false \leftarrow A_1, \cdots, A_n$", where A_i ($1 \leq i \leq n$; $n \geq 1$)

is an atom and *false* designates falsity.   □

Example 2.1 Figure 1 shows a simple example P_{ex}. Suppose that an observation "p(X,Y)" is given. Knowledge-base P_{ex} and goal "$\leftarrow p(X,Y)$" correspond to $F \cup H$ and O respectively.   □

Facts	Hypotheses : Costs	
p(X,Y) ← a(X),c(Y).	c(1). : 5	c(2). : 6
p(X,Y) ← b(X),d(Y).	c(3). : 7	
a(X) ← e(X).	b(1). : 2	b(2). : 3
e(1).	b(3). : 2	
false ← b(X),d(X).	d(1). : 2	d(2). : 4

Figure 1   An Example P_{ex}

Any first-order proof procedure works as an abductive reasoning system if it distinguishes hypotheses from facts, satisfying the conditions given in Definition 2.1.

Example 2.2 Figure 2 shows SLD-trees [Llo84] for $P_{ex} \cup \{\leftarrow p(X,Y)\}$ and $P_{ex} \cup \{\leftarrow false\}$.

In the trees, an atom L in the form of ML means that L is assumed to be true. It is called *assumed atom* [KSI93].

It follows from SLD-tree for $P_{ex} \cup \{\leftarrow false\}$ that it causes inconsistency to assume that any of $\{b(1), d(1)\}$ or $\{b(2), d(2)\}$ is true (we call each set of hypotheses *set of incompatible hypotheses*). We also know that $p(3,1)$, for example, has the most preferable explanation $\{b(3), d(1)\}$, since the cost of the set of hypotheses is a minimum among all sets of hypotheses obtained from all successful leaves in the SLD-tree for $P_{ex} \cup \{\leftarrow p(X,Y)\}$, while maintaining the consistency.   □

---

† We suppose that cost of D is 0 if $D = \phi$.

In this paper, if some consistency conditions exist in a given knowledge-base, we suppose that all sets of incompatible hypotheses are derived in advance, by constructing SLD-derivation for $\leftarrow false$.

The above SLD-tree for $P_{ex} \cup \{\leftarrow p(X, Y)\}$, however, shows that the unnecessary search to finding an optimal explanation is very large. We therefore apply a parallel heuristic search technique to our abductive reasoning. Preliminary to our reasoning system, we define a following function to evaluate search spaces explored by the system.

Definition 2.3 Let P be given facts and hypotheses (knowledge-base), and O be an observation. To each goal g, "$\leftarrow ML_1, \cdots, ML_i, A_{i+1}, \cdots, A_n$", in SLD-tree for $P \cup \{\leftarrow O\}$, the evaluation function $f(g)$ is defined as follows:

$f(g) = cost(\{L_1, \cdots, L_i\}) + \hat{h}(g)$,

where $\hat{h}(g)$ is an estimated cost satisfying the following conditions,

$0 \leq \hat{h}(g) \leq h(g)$,

$h(g) = cost(\bigcup_{j=i+1}^{n}$ an optimal explanation of $A_j)$.

$cost(\{L_1, \cdots, L_i\})$ means the cost of the derivation from $\leftarrow O$ to g. $h(g)$ also means an actual cost of an optimal derivation from g to any successful leaf. We assume that A_j has a makeshift explanation whose cost is ∞ if there is no explanation of A_j. □

The evaluation function defined as above enables heuristic search to be applied to finding an optimal explanation of given observation. In [KSI94], we have actually implemented an efficient abductive reasoning system, by introducing a search control technique of A* into abductive reasoning mechanism. This paper proposes a parallelization of the abductive reasoning system. In addition to the search control technique in [KSI94], we propose a method for distributing the search space to multiple processors and exploring each distributed search space with synchronization and communication.

3 PARCAR ALGORITHM

The section describes A PARallel Abductive Reasoning system, PARCAR. It incorporates a parallel heuristic search algorithm into the reasoning mechanism. Abductive reasoning imposes much computation on expanding nodes in search space. The computational cost largely depends upon the number of successor nodes generated by node expansion. Then, PARCAR performs the load balancing based on the number of successor nodes. The Complexity of data structure of goals labeled nodes enlarges communication overhead for node distribution. PARCAR is de-

signed so that it can keep the communication overhead down, by giving adequate independence from communications to each processor.

PARCAR proceeds iteratively. Figure 3 shows the algorithm. In the figure, n is the number of processors, and $OPEN_i^j, CLOSED_i^j, Gmin^j, SG^j, RS^j$, and ANS^j are sets of goals dealt to processor j ($1 \leq j \leq n$) at i-th iteration. $OPEN_i^j$ and $CLOSED_i^j$ contain goals which are not expanded yet and were already expanded by processor j at i-th iteration, respectively. SG^j contains goals labeled leaves in sub tree constructed by j. ANS_j contains goals labeled successful leaves obtained by j. These sets are implemented as ordered sets where each goal g is in order of a value of its evaluation function $f(g)$.

The algorithm consists of iterations of series of three procedures, "goal expansion", "sub-goal distribution", and "sub-goal reception". Let $T_{i,l}^j (l \geq 1)$ be a sub-SLD-tree constructed by processor j at i-th iteration, and $|T_{i,l}^j|$ be number of goals in $T_{i,l}^j$. In each processor j, these procedures at i-th iteration behave intuitively as follows:

goal expansion procedure goes on in either of following 2 ways according to its phase:
  in *progress* phase,
    by best-first expanding goals gs in order of $f(g)$s, the procedure constructs $T_{i,l}^j$s whose roots are labeled goals in $OPEN_i^j$ until $\sum_l |T_{i,l}^j| \geq K$, where K is a parameter given by user. The procedure, however, halts immediately when it finds a successful leaf (i.e., an answer), or has no goal to expand.
  in *confirmation* phase,
    by best-first expanding goals gs in order of $f(g)$s, the procedure constructs $T_{i,l}^j$s whose roots are labeled goals in $OPEN_i^j$ until $f(g)s \geq t$ for all leaves gs in $\bigcup_l T_{i,l}^j$ or it finds a new answer whose cost is less than t, where $t = min\{f(g_{ans}) \mid g_{ans}$ is labeled a successful leaf obtained in *progress* phase $\}$. Then, PARCAR terminates in success.

Figure 4 shows a model of a sub-SLD-tree constructed by processor j at i-th iteration.

subgoal distribution procedure checks whether an answer has been found by *goal expansion procedure*. if so, it broadcasts the request for shifting to *confirmation* phase with the cost of the answer. Then, it distributes the goals gls labeled leaves in $\bigcup_l T_{i,l}^j$ in order of $f(gl)$s. The goal g, such that $f(g)$ is m-th least among the leaves, is put to processor $((j + m) \bmod n)$.

subgoal reception procedure shifts to *confirmation*

PARCAR: A PARalell Abductive Reasoning system
  Input: a program: P, an observation: O
  Output: a solution: (answer, an optimal explanation) (with success) or false (with failure)

1 **begin**
2     $OPEN_0^1 := \{\leftarrow O\}$; $i := 0$; *phase* := **progress**; $t := \infty$;
3     $OPEN_0^{j(1<j\leq n)} := \phi$; $CLOSED_0^{j(1\leq j\leq n)} := \phi$; $ANS^{j(1\leq j\leq n)} := \phi$;
4   **repeat**
5      $OPEN' := OPEN_i^j$; $SG^j := \phi$; *escape* := false; *shifted* := false;
6   **repeat**   % Goal Expansion Procedure
7        $Gmin^j := \{^\forall g \in OPEN' \cup SG^j \mid$ for all $g' \in OPEN' \cup SG^j : f(g)\leq f(g')\}$;
8   Generate all resolvents (called subgoals) of $^\forall g_{min} \in Gmin^j$ and $^\exists C \in P$
          (the set of the subgoals is denoted by RS^j). ;
9   Check the consistency of goals in RS^j †, and remove the inconsistent goals from RS^j †. ;
10       $ANS^j := \{^\forall g \in RS \mid g$ is an successful leaf, and $f(g)$ is less than any other successful leaves} ;
11       $SG^j := SG^j \cup RS^j$;
12       $CLOSED_{i+1}^j := CLOSED_{i+1}^j \cup \{g\}$;
13       $OPEN' := OPEN' \setminus Gmin^j$; $SG^j := SG^j \setminus Gmin^j$;
14   **if** $(ANS^j \neq \phi)$ **then** *escape* := true ;
15   **if** (*phase* = **progress** and $(|SG|\geq K$ or $OPEN' \cup SG^j = \phi)$) **then** *escape* := true ;
16   **if** (*phase* = **confirmation** and $(^\forall g \in OPEN' \cup SG^j \mid f(g) > t)$) **then** *escape* := true ;
17   **until** (*escape*)
18      $OPEN_{i+1}^j := OPEN'$;
19   *distribution_goals*(ANS^j, SG^j) ;   % Subgoal Distribution Procedure
20   *reception_goals*$(OPEN_{i+1}^j, phase, shifted)$;   % Subgoal Reception Procedure
21      $i := i+1$;
22   **until** ((*phase* = **confirmation** and *not shifted*) or $(OPEN_{i+1}^{j(1\leq j\leq n)} = \phi)$)
23    $ANS := \{^\forall g \in ANS \cup ANS^{j(1\leq j\leq n)} \mid$ for all $g' \in ANS \cup ANS^{j(1\leq j\leq n)} : f(g)\leq f(g')\}$
24   **if** $(^\exists \leftarrow ML_1 \cdots ML_m \in ANS)$ **then return** (answer,$\{L_1 \cdots L_m\}$)
25   **else return** false
26 **end**.

   † A goal $\leftarrow ML_1, \cdots, ML_i, A_{i+1}, \cdots, A_n$ is inconsistent if any set of incompatible
     hypotheses is a subset of $\{L_1, \cdots, L_i\}$.

Figure 3   The PARCAR Algorithm

phase and assigns *shifted* "true" and gives t the cost as a threshold for termination in **confirmation** phase, if the procedure receives a request and a cost from any processors. Then, it merges goals distributed from all processors into $OPEN_{i+1}^j$.

The algorithm is admissible for all finite SLD-trees, if $\hat{h}(g) \geq h(g)$ for all goals g in the SLD-trees.

4   RELATED WORK

Troya and Benjumea [BT93] have proposed an OR-parallel Prolog model. The depth-first constructing an SLD-tree in Prolog complicates OR-parallelization. On the other hand, Our cost-based abduction constructs an SLD-tree in a breadth-first manner because of necessity for obtaining a minimum-cost explanation. The breath-first manner promotes more parallelism and simplifies the division of search space into processors. each processor carries out best-first search in a part of the space.

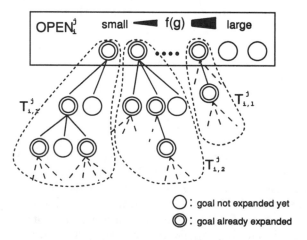

: goal not expanded yet

: goal already expanded

Figure 4   A Model of a Sub-SLD-tree Constructed by Processor j at i-th iteration

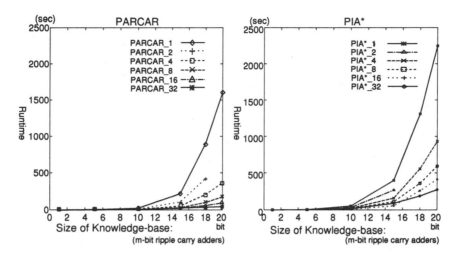

Figure 5   Experimental Results

Huang and Davis [HD89] have proposed parallel iterative A* search (PIA*). In PIA*, each processor j at i-th iteration usually expands one node g in a priority queue WL_i^j (it corresponds to $OPEN_i^j$ in PARCAR) such that $f(g) \leq t$, where $t = min\{f(g)$ for all $g \in \bigcup_{j=1}^n WL_i^j\}$. So, PIA* is a simple parallelization of A*. PIA*, however, has a few defects with respect to applied to cost-based abduction. PIA* performs the load balancing in goal expansion procedure based on the number of goals expanded by processor j at a single iteration. We think that it depends upon the assumption that the number of subgoals generated by expanding a goal is nearly constant. The number of them, however, is not constant in general. For example, in the SLD-tree in figure 1, PIA* makes processor PE1 expand $\leftarrow a(X), c(Y)$ and processor PE2 expand $\leftarrow b(X), d(Y)$ (see the branches from depth 2 to 3 in figure 1). PE1 generates 1 subgoal while PE2 generates 3 subgoals. Cost-based abduction imposes several tasks on the generation of subgoal: unification, propagation of binding for variables, and so on. The computational costs of these tasks cannot be negligible. Therefore, the difference of the number of subgoals generated by each processor may become an obstacle to the load balancing in PIA*. PARCAR performs the load balancing in goal expansion procedure based on the number of subgoals generated by processor j at a single iteration.

The total number of goals expanded for finding an optimal solution is nearly same between PIA* and PARCAR. Let G_i^j be a quantity of task in processor j at i-th iteration, and suppose that G_i^j is approximated as the number of goals expanded by j at i-th iteration. In PIA*, G_i^j is nearly equal to 1 in case there are not so numerous goals gs in an SLD-tree such that $f(g)$s are the same. This means that PIA* imposes a large amount of iterations on j. As a result, each processor needs to communicate with other processors for many times. The communication cost may become a bottleneck with the number of processors increasing. In PARCAR, G_i^j can be adjusted by changing the parameter K to keep the communication cost down.

5   EXPERIMENTAL RESULTS

We have made some experiments on an MIMD distributed memory parallel computer Fujitsu AP1000. We have considered a diagnostic problem to diagnose an m-bit ripple carry adder circuit. We have represented each hypothesis as a state in which a gate x is, (i.e., $ok(x)$, $stuck0(x)$, or $stuck1(x)$), and have given each hypothesis $|log_e \mathcal{P}(state(x))|$ as its cost, where $\mathcal{P}(state(x))$ is a probability of x being in $state$[†]. Figure 5 shows the experimental results on the problem, by changing the size m of the circuit[††]. Our experimental system is written in C. In the experiments, a parameter for PARCAR, $K = 50$ and for all goals gs, $\hat{h}(g)$s= 0. In this particular example, the results indicate that PARCAR is about 2.5 ~ 5.8 times faster than PIA* as the problem becomes larger. Figure 6 shows the speedups. The speedup of PARCAR comes nearer to linear when the problem size becomes larger, while the speedup of PIA* falls below linear.

---

† It follows from $0 \leq \mathcal{P}_i \leq 1$ and $\prod_i \mathcal{P}_i = exp(\sum_i |log_e \mathcal{P}_i|)$ that $\prod_i \mathcal{P}_i$ is maximum if $\sum_i |log_e \mathcal{P}_i|$ is minimum.

†† The scale of knowledge-base which represents the problem is in proportion to m: facts consist of $17n + 24$ first-order Horn clauses and set of Hypotheses contains $15n + 3$ unit clauses.

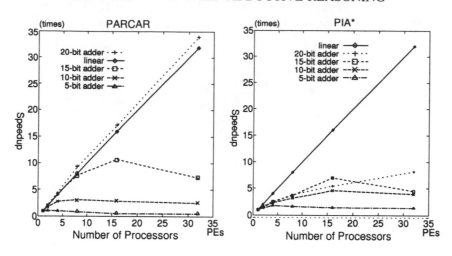

Figure 6   Speedups

6   CONCLUDING REMARKS

This paper proposed a parallelization method of cost-based Horn abduction. We introduced a search control technique of parallel best-first heuristic search into abductive reasoning mechanism, thereby obtaining much more efficiently the optimal explanation of the given observation. We then implemented a parallel cost-based abductive reasoning system PARCAR on an MIMD distributed memory parallel computer Fujitsu AP1000. The well performance results were obtained by some experiments on AP1000. In the experiments, we supposed that $\hat{h}(g) = 0$ for all goals g (see definition 2.3), because the experiments were to evaluate the parallelism of the algorithms. We have already proposed the pre-analysis to discover effective admissible heuristic function $\hat{h}(g)$ in [KSI94]. It can also make PARCAR more efficient, by pruning the search space.

In this paper, we mainly described the method for dividing the search space to processors and constructing the search-tree by each processor. We have also performed loop-check function, which is required in case knowledge-base is recursively defined, by pipelining. In the case, we have promoted the parallelism of pipelining, by equalizing the number of leaves in a search tree constructed by each processor.

REFERENCES

[BT93]   V. Benjumea and J. M. Troya. An OR Parallel Prolog Model for Distributed Memory Systems. In *Proc. of the Intl. Symp. PLILP'93*, *LNCS-714*, pages 291–301. Springer-Verlag, 1993.

[CS94]   E. Charniak and S. E. Shimony. Cost-based abduction and MAP explanation. *Artificial Intelligence*, 66:345–374, 1994.

[HD89]   S. Huang and L. S. Davis. Parallel Iterative A^* Search: An Admissible Distributed Heuristic Search Algorithm. In *Proc. of the IJCAI-89*, pages 23–29, 1989.

[IHI+91]   H. Ishihara, T. Horie, S. Inano, T. Shimizu, and S. Kato. An Architecture of Highly Parallel Computer AP1000. In *Proc. of the IEEE Pacific Rim Conf. on Communications, Computers and Signal Processing*, pages 13–16, 1991.

[KSI93]   S. Kato, H. Seki, and H. Itoh. An Efficient Abductive Reasoning System Based on Program Analysis. In *Static Analysis, Proc. of the 3rd Intl. Workshop, LNCS-724*, pages 230–241, Padova, 1993. Springer-Verlag.

[KSI94]   S. Kato, H. Seki, and H. Itoh. Cost-based Horn Abduction and its Optimal Search. In *Proc. of the 3rd Intl. Conf. on Automation, Robotics and Computer Vision*, pages 831–835, Singapore, November 1994.

[Llo84]   J. W. Lloyd. *Foundations of Logic Programming*. Springer, 1984. Second, extended edition, 1987.

[OI93]   Y. Ohta and K. Inoue. Incorporating Top-Down Information into Bottom-Up Hypothetical Reasoning. *New Generation Computing*, 11:401–421, 1993.

[Poo88]   D. Poole. A Logical Framework for Default Reasoning. *Artificial Intelligence*, 36:27–47, 1988.

[Poo93]   D. Poole. Probabilistic Horn abduction and Bayesian networks. *Artificial Intelligence*, 64:81–129, 1993.

[SL90]   B. Selman and H. J. Levesque. Abductive and Default Reasoning: A Computational Core. In *Proc. of the AAAI-90*, pages 343–348, 1990.

[Sti94]   M. E. Stickel. Upside-Down Meta-Interpretation of the Model Elimination Theorem-Proving Procedure for Deduction and Abduction. *Journal of Automated Reasoning*, 13:189–210, 1994.

CAD/CAM

AUTOMATIC PLACEMENT USING STATIC AND DYNAMIC GROUPINGS

Makoto Hirahara, Natsuki Oka, and Kunio Yoshida
Matsushita Research Institute Tokyo, Inc.
3-10-1, Higashimita, Tama-ku, Kawasaki 214,Japan
Email: mhira@mrit.mei.co.jp

Abstract

An automatic placement system for printed circuit boards is presented. The system detects repetitions of similar structures in a target circuit to be designed, and recognizes the similar structures as sub-circuits. After the sub-circuit recognition, using simulated annealing, placements are repeatedly modified so as to reflect the similarities of the recognized sub-circuits. To accelerate modifications, the system generates two kinds of component groups: static and dynamic ones. Components in each of the sub-circuits form a static group. Although the constitutions of the static groups are fixed during the modifications, those of the dynamic groups are adaptively rearranged depending on the modification histories. Relocating not only each component but also each of the static and dynamic groups as a unit, and reflecting the sub-circuit similarities in placements, our system quickly produces high-quality placements comparable to expert's ones.

1 Introduction

Although various CAD systems have been developed to reduce the design period, automatic placement systems satisfying practical requirements scarcely exist. One of approaches to the placement problems is to make use of expert systems which have knowledge-bases consisting of many know-hows of skillful experts [Yo+ 90]. Though these systems work well if they have enough know-hows, knowledge acquisition to build the knowledge-bases is time-consuming and difficult. Know-hows for pure analog circuits are known to be more complicated than those for digital-analog mixed circuits, and acquisition of the former is quite difficult.

There also exists approaches where systems do not require such knowledge-bases, one of which solves the placement problems as optimization ones. In this approach, systems need cost functions to evaluate placements instead of knowledge-bases. These systems repeatedly modify placements so as to minimize the values of the cost functions, where simulated annealing (SA) is frequently used [SS 85]. Though SA is a powerful technique to search global minima of optimization problems [GG 84, KGV 83], time to be consumed increases exponentially as the problem size becomes larger.

Although our system modifies placements using SA, it automatically generates two kinds of component groups, called static and dynamic ones, in order to reduce the problem size, to accelerate placement modifications, and to improve the placement quality. Before placement modifications, the system detects repetitions of similar structures in a target circuit to be designed, and recognizes the similar structures as sub-circuits. For example, RGB sub-circuits in a circuit often have similar structures to each other. Components in each of the recognized sub-circuits form a static group. Although the constitutions of the static groups are fixed during placement modifications, those of the dynamic groups are adaptively rearranged depending on the modification histories. Relocating not only each component but also each of the static and dynamic groups as a unit, placements are quickly modified so as to reflect the similarities of the recognized sub-circuits. Resultant placements are comparable to expert's ones.

Our system is introduced in Section 2. Sub-circuit recognition and static group generation are described in Section 2.1. In Section 2.2, an initial placement is generated using the static groups. Placement modifications by simulated annealing are described in Section 2.3. In Section 2.4, dynamic groups are generated by learning histories of placement modifications. To confirm the behavior of the system, we perform computer simulations using a circuit of a television in Section 3.

2 Automatic placement system

Figure 1 demonstrates an example of a placement designed by an expert, where the six dotted rectangles enveloping some components show sub-circuits (recognized by Static grouping module described below). The three sub-circuits in the left (right) side have similar structures to each other. Experts, thus,

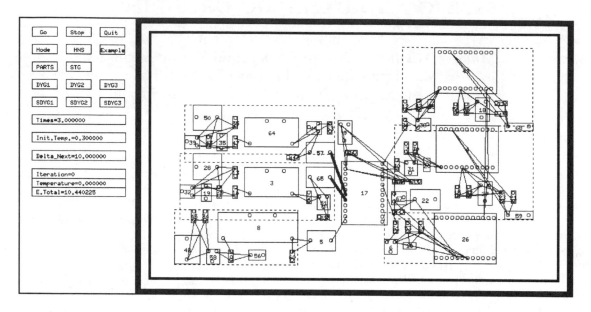

Figure 1: Placement designed by an expert. Experts, thus, tend to reflect the similarities of sub-circuits in placements. The dotted rectangles show the sub-circuits (the static groups) recognized by our system. The three sub-circuits in the left (or right) side have similar structures to each other.

tend to reflect the sub-circuit similarities in placements. In order to produce such a placement quickly, our system consists of the four modules: Static grouping, Initial placement, Modification and Dynamic grouping modules. Figure 2 shows the constitution of our system.

First, Static grouping module receives a specification of a circuit to be designed. The module detects repetitions of similar structures in the target circuit, and recognizes the similar structures as sub-circuits. Components in each of the sub-circuits form a static group. Secondly, Initial placement module produces an initial placement using the static groups. If an initial placement is given by a user, the module emits it instead of the generated one. Thirdly, using simulated annealing [GG 84, KGV 83, SS 85], Modification module repeatedly modifies placements so as to reflect the sub-circuit similarities. During placement modifications, Dynamic grouping module learns the modification histories, and generates dynamic groups adaptively according to the situations. Relocating not only each component but also each of the static and dynamic groups as a unit, our system quickly produces high-quality placements comparable to expert's ones.

2.1 Static grouping module

Before placement modifications, Static grouping module recognizes sub-circuits in a target circuit, and

generates static groups automatically.

Using the specification of a target circuit to be designed, Static grouping module detects repetitions of similar structures in the circuit, and recognizes the similar structures as sub-circuits. The algorithm of sub-circuit recognition is as follows:

1. Similarities between components are calculated by comparing them in terms of shape, the number of pins, and so forth. If the kinds of two components (IC, register, etc.) are different, the similarity between them is set to zero.

2. Similarities between nets are calculated by comparing them in terms of the number of components, component similarities calculated above, and so forth. If two nets hold some components in common, the similarity between them is set to zero [1].

3. Nets, whose similarities to the other nets are above a threshold, are detected.

---

[1] The two thick lines (nets) shown in Figure 1 is a typical example. If the similarity between the two nets is above the threshold introduced in the third process, they are connected in the forth process, so that the above two sub-circuits in the left side and the component 17 are grouped together. If the similarity can be set to zero (below the threshold), Static grouping module divides into the two sub-circuits as shown in Figure 1, which division meets our expectation.

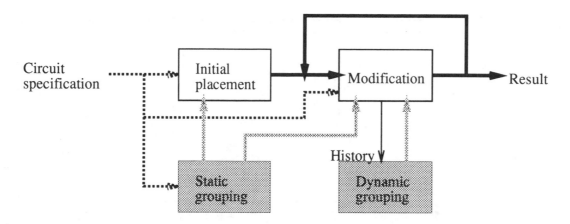

Figure 2: Automatic placement system consisting of the four modules. Relocating not only each component but also each of static and dynamic groups as a unit, and reflecting sub-circuit similarities in placement, the system quickly designs placements comparable to expert's ones.

4. Net groups are constructed by grouping the detected nets which hold some components in common.

5. Similarities between net groups are calculated by comparing them in terms of the number of nets belonging to a net group, net similarities calculated above, and so forth.

6. If similarities between net groups are above a threshold, these net groups are recognized as sub-circuits having similar structures to each other.

The effectiveness of this algorithm is shown in Figure 1. The dotted rectangles which envelope some components show the sub-circuits (the static groups) recognized by Static grouping module. In this example, Static grouping module recognizes six sub-circuits. The three sub-circuits in the left (or right) side have similar structures to each other. Components in each of the sub-circuits form a static group whose constitutions are fixed during placement modifications. Relocating each static group as a unit, placement modifications by SA are accelerated.

2.2 Initial placement module

If an initial placement is not given by users, Initial placement module produces it using the static groups generated above. Figure 3 shows an example of an initial placement produced from the same data as in Figure 1.

In Figure 3, the static groups, whose structures are similar to each other, are arranged in a line. The

inner layouts within the static groups having similar structures are also similar to each other. Since our system starts to modify from such an initial placement, our system have the advantage over conventional systems starting from random placements. The algorithm to generate an initial placement is as follow:

1. A static group which is not placed is selected.

2. A point on the printed circuit board is randomly determined. The components of the static group are randomly placed around the point.

3. The inner layout within the static group is copied to those in the others having similar structures, following which the static groups are arranged in a line.

4. If there are static groups which are not placed, go back to the first process.

5. Components which do not belong to any static groups are randomly placed.

2.3 Modification module

When an initial placement is entered, Modification module starts to modify placements using simulated annealing. Simulated annealing is a powerful technique to search the state of the minimum cost [GG 84, KGV 83].

In our system, the cost function of a placement is defined mainly by the following four terms:

1. Total wiring length of all nets.

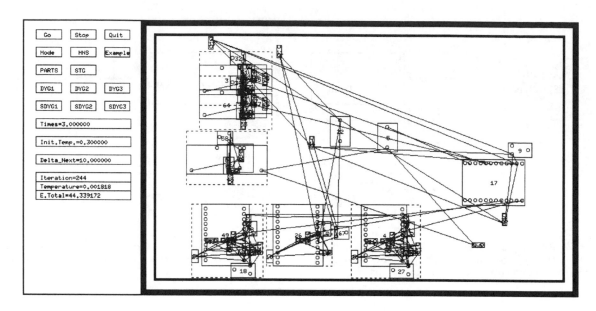

Figure 3: Example of an initial placement generated by our system. The static groups, whose structures are similar to each other, are arranged in a line. The inner layouts within the static groups having similar structures are also similar to each other. Our system has the advantage over conventional systems starting from random placements.

2. Overlaps between components.

3. Disagreement of inner layout between static groups having similar structures.

4. Distance among the centers of static groups having similar structures.

Though the first and second terms are generally used in conventional systems [SS 85], the third and forth terms are the characteristics of our system since they are related to static groups. The reflections of the sub-circuit similarities in placements are realized by the minimization of the third and forth terms.

Relocating not only each component but also each of static groups and dynamic groups (described below) as a unit, Modification module modifies placements repeatedly. The relocation of each unit is performed by one of modification procedures. Some of the procedures are as follows:

1. Position modification of the unit.

2. Angle modification of the unit.

3. Swapping the unit for one of the other units.

When the unit is a static group, the following procedures are also available:

1. Copy the inner layout of the unit to those of the other units having similar structures.

2. Arrangement of the units and the other units having similar structures in a line.

These two procedure facilitate the reflections of the sub-circuit similarities in placements, which facilitation is one of the reasons for the acceleration of placement modifications.

2.4   Dynamic grouping module

In addition to static groups, our system automatically generates component groups, called dynamic groups, to accelerate modifications further. The constitutions of dynamic groups are adaptively rearranged depending on the situations. The dynamic groupings are realized by learning the histories of modifications performed by Modification module.

The situations, where relative positions of components belonging to same nets are stable, are often observed during placement modifications. Dynamic grouping module generates component groups based on such the stability of each net. One of simple criteria to evaluate the stability of a net is the change in the wiring length of the net. If relative positions of components belonging to a net are stable, the change in the wiring length becomes small, so that the components in the net are grouped together. Relocating each dynamic group as a unit, placement modifications by SA are further accelerated.

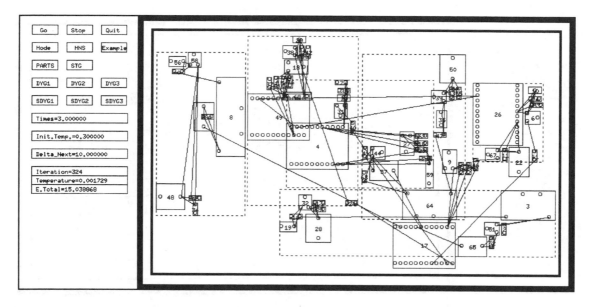

Figure 4: Placement designed by a conventional system. Since the conventional system takes no account of the similarities of sub-circuits, they are not reflected in the resultant placement. It is difficult to identify the sub-circuits.

3 Computer simulation

Computer simulations were performed in order to compare our system with a conventional one. The data to be used in our simulations is one of blocks in a circuit of a television, and is shown in Figure 1 where the placement was designed by an expert. Experts tend to reflect the similarities of sub-circuits in placements.

If Static and Dynamic grouping modules are excluded from our system, and if the system starts to modify from a random initial placement, our system is reduced to a conventional one which relocates each component one by one so as to minimize the costs of total wiring length and overlaps (the first and second terms of the cost function) such as in [SS 85]. Hence, we regard our system without Static and Dynamic grouping modules as a conventional one. The placement designed by the conventional system is shown in Figure 4. Since the conventional system takes no account of the similarities of the sub-circuits, they are not reflected in placements at all. Furthermore, it is difficult for us to identify sub-circuits in the resultant placement.

The placement designed by our system is shown in Figure 5. The result was obtained for about two or three minutes by SUN work-station (CPU: micro SPARC II, 110MHz). Our system recognizes sub-circuits having similar structures before placement modifications by simulated annealing. Reflecting the similarities of the recognized sub-circuits in place-ments as experts do, and relocating not only each component but also each of the static and dynamic groups as a unit, our system quickly produces high-quality placements comparable to the expert's one shown in Figure 1. It is easy to identify sub-circuits in our resultant placement.

4 Conclusions

An automatic placement system for printed circuit boards is presented. The system detects the repetitions of similar structures in a target circuit to be designed, and recognizes the similar structures as sub-circuits. After the sub-circuit recognition, using simulated annealing, placements are repeatedly modified so as to reflect the similarities of the recognized sub-circuits in placements, which process improve the placement quality effectively. To accelerate placement modification by simulated annealing, the system performs the two kinds of component groupings: static and dynamic groupings. Components in each of the recognized sub-circuits form a static group. Although the constitutions of the static groups are fixed during the placement modifications, those of the dynamic groups are adaptively rearranged depending on situations. The dynamic groupings are realized by learning the histories of the placement modifications. Reflecting the similarities of the recognized sub-circuits, and relocating not only each component but also each of the static and dynamic groups as a

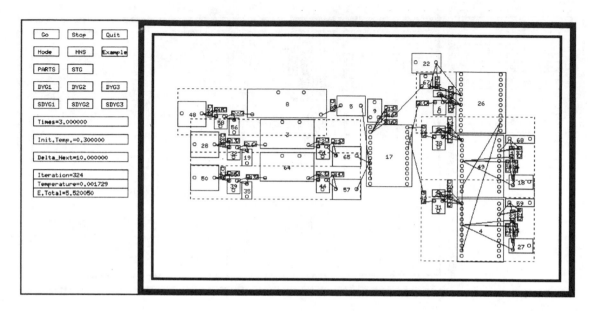

Figure 5: Placement designed by our system. Before placement modifications, the system recognizes sub-circuits having similar structures to each other. Reflecting the sub-circuit similarities in placements, the resultant placement is comparable to the expert's one shown in Figure 1.

unit, our system quickly produces high-quality placements. Resultant placements are comparable to expert's ones.

In the future, we should further analyze and improve our system as follows:

- In order to evaluate our system, we have used only the circuit shown in Figure 1. We should also use the other circuits, and analyze the benefit and limitations of static and dynamic groupings.

- To generate dynamic groups, wiring length is used to evaluate the stability of each net. This implies that it is impossible to divide components belonging to a net into more than one group. In our simulations, we have observed the situations where such the grouping has been useful to accelerate placement modifications. In order to realize such the grouping, we will further improve our criterion to generate dynamic groups.

- Our system has some parameters such as the weights of the terms of the cost function. In this present, we have determined the values of such the parameters carefully in order to get the result shown in Figure 5. In the future, we would like to realize automatic setting of the values of these parameters by learning.

Acknowledgements

The authors would like to thank S. Fukuhara and H. Yoshimura, Production Engineering Laboratory, Matsushita Electric Industrial Co., Ltd., for their useful comments and helpful discussion. We would also like to express our appreciation to H. Morimoto, AVC Products Development Laboratory, Matsushita Electric Industrial Co., Ltd., for providing us with the data to be used in our simulations.

References

[GG 84] Geman,S., Geman,D.: Stochastic relaxation, Gibbs distributions, and the Bayesian restoration of images. IEEE Trans. Pattern Analysis and Machine Intelligence, PAMI-6, 6, pp.721-741, 1984.

[KGV 83] Kirkpatrick,S., Gelatt, C.D., Vecchi, M.P.: Optimization by simulated annealing. Science, 220, 4598, pp.671-680, 1983.

[SS 85] Sechen,C., Sangiovanni-Vincentelli,A.: The TimberWolf placement and routing package. IEEE J. Solid-State Circuits, SC-20, 2, pp.510-522,1985.

[Yo+ 90] Yoshimura,H. et al.: Knowledge-based placement and routing system for printed circuit board. Proc. PRCAI'90, pp.116-121, 1990.

A DOMAIN-SPECIFIC CONCEPTUAL MODEL IN BUILDING DESIGN

Masaaki HASHIMOTO[†]   Toyohiko HIROTA[†]   Isao NAGASAWA[†]
Keiichi KATAMINE[†]   Yoshiaki TEGOSHI[‡]
[†]Kyushu Institute of Technology   [‡]Hiroshima Institute of Technology
Kyushu Institute of Technology, 680-4 Kawazu, Iizuka, 820 JAPAN
Tel:+81-948-29-7619   Fax:+81-948-29-7601   e-mail:hasimoto@ai.kyutech.ac.jp

ABSTRACT

This paper proposes a domain-specific model of building structures and operations for editing the structures. The authors have developed an experimental compiler for generating building CAD systems from the specification description language based on the domain-specific model. The experiment has shown the followings: The building structure model has minimality, constructibility and comprehensibility of specification descriptions. Therefore, building designers would describe CAD systems specifications by themselves. On the other hand, the operation model requires a lot of specification descriptions because the model is procedural.

Key words – CAD, building structure design, domain-specific conceptual model, specification description language, CAD generator.

1 INTRODUCTION

Software requirements should be exactly analyzed in the first phase of software life cycle. For aiding the requirements analysis, the object-oriented methodologies such as OMT (Object Modeling Technique) [RBP91] and CASE(Computer Aided Software Engineering) tools [Bel94] have been developed and applied. These methodologies and tools were designed not for a specific domain but for various domains. That is, their conceptual models for describing software requirements specifications are considerably general.

In order to obtain exact requirements specifications, domain experts such as building designers should describe their requirements specifications by themselves. However, they hardly use general methodologies, tools nor computer languages since they are not software engineers. Therefore the domain-specific conceptual models, methodologies, tools and computer languages, which domain experts can use, should be developed[Pri87].

The authors have been studying domain analysis and modeling techniques in various domains such as building design, business data processing and so on. Concerning the building design, we already proposed an idea of IBDS(Integrated Building Design Systems) [NTM89] and a conceptual model specific to building structural design.

IBDS would integrate individual CAD systems: architectural design, structural design, equipment installation design and estimation by means of the domain-specific conceptual model. The domain-specific conceptual model has an object-oriented framework specialized in existence dependency constraints between building structural parts. We already designed the conceptual model-based specification description language, and evaluated it [HHN95].

The authors have recently extended the conceptual model for expressing operations to edit building structures, and have experimentally implemented a compiler for generating building CAD systems from the specification description language based on the model. This paper describes the extended conceptual model and discusses the model on the experimental implementation. Section 2 refers to related works. Section 3 describes the extended conceptual model. Section 4 explains the specification description language. Section 5 describes the experimental implementation. Section 6 discusses them.

2 RELATED WORK

The requirements analysis methodologies applied to several CASE tools are based on data flow and functional structure [YC78]. On the other hand, object-oriented methodologies [RBP91] are based on objects and their associations in a domain. However, such general methodologies hardly acquire actual domain knowledge necessary for developing application systems. Therefore, domain analysis for extracting domain knowledge, and domain modeling [IWA91] for describing the extracted knowledge are recently studied in various domains.

The extracted domain knowledge should be expressed with a model. There are two kinds of models: formal and conceptual models. If the extracted domain knowledge can be expressed with a formal model [HH93], a computer program can easily been generated from the model. However, the formal model is hardly applied in the requirements analysis because of their restrictions. Therefore, conceptual models are usually applied.

The data flow model is sometimes applied as a conceptual model to the domain of business data processing. In the domain, flows of orders, products and money are easily modeled as data flows. However, data flows are not easily detected in various domains such as CAD(Computer Aided Design) systems.

The ER(Entity-Relationship) model and object-oriented model are other conceptual models. The models mainly concern entities (or objects) and their associations in a domain. Relations among attribute values are expressed with data flows or constraints.

3 CONCEPTUAL MODEL

3.1 MAINTENABILITY AND INTEGRITY

Conceptual models are essential for building CAD systems from the following view points:

- Maintenability of the systems.
- Integrity of the systems.

Maintenability
Building designers think of figures of buildings which are represented on X, Y and Z plane drawings. Then they think of attributes of the building parts, such as their volume, material, weight, and relationship between parts.

Therefore, the building CAD systems which edit only the geometric drawings can not help the designers sufficiently. To support the designers sufficiently, the attributes of the building parts should be introduced into the systems. However, this will make the systems complicated, and make maintenance of the systems more difficult.

The conceptual models which directly express the attributes of the building parts would be essential for the clear specifications to satisfy the designer's requirements.

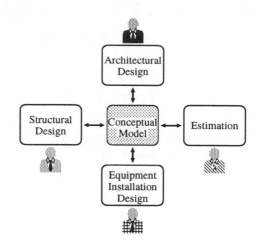

FIGURE 1: Integrated CAD system.

Integrity
Most existing CAD systems are developed for architectural design, structural design, equipment installation design and estimation respectively. Therefore, the designers who use a certain CAD system must exchange design information with other CAD systems by hand.

The individual CAD systems must be integrated for the designers to access the same design information as shown in Fig. 1. Conceptual models, which directly express the attributes of the building parts, would provide the information with frameworks.

3.2 STRUCTURE

The conceptual model of building structures is a framework for representing structural design information as attributes of the building parts. The information is necessary for each divided design work. The following three concepts are important in the model:

- Part as object, and its attributes.
- Existence dependency constraint between parts.
- Attribute dependency constraint among attributes.

Existence dependency constraint determines dependency between parts in a building structure. For example, a girder is suspended by two columns. The girder can not exist in the structure if one or both of the columns do not exist. Therefore, the girder must be removed when one or both of the columns are removed. In the same manner, the both columns must be exist in the structure when the girder is inserted.

Attribute dependency constraint determines dependency among attributes, and describes a computation expression for obtaining a value of one of the attributes

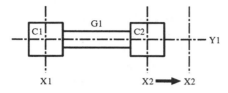

(1) Modify position of base line

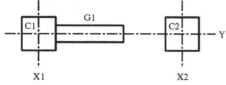

(2) Modify position of column

(3) Modify figure of girder

FIGURE 2: Attribute dependency constraint.

from other attributes. If attribute A depends on attribute B and C, value of attribute A must be calculated again when attribute B or C is modified.

For example, girder G1 depends on two columns C1 and C2 in Fig. 2. The both end positions of the girder are determined by the positions of both columns. When the position of column C2 is modified, the right position of the girder must be calculated again. Attributes, such as volume, weight and cost, of the girder also should be calculated again.

3.3 OPERATION

This subsection describes operations for designers to edit the building structure expressed with the above-mentioned conceptual model. Each operation is a sequence of procedure for solving a problem of structural design in a typical manner, or for editing the structure efficiently. The operations are composed of the following primitive functions:

- Inserting a part into the structure according to the existence dependency constraint.
- Removing a part from the structure.
- Modifying a connection relationship between parts.
- Modifying an attribute value of part.

A complicated operation can be composed of these primitive functions. For example, pulling out a column shown in Fig. 3 is composed as follows:

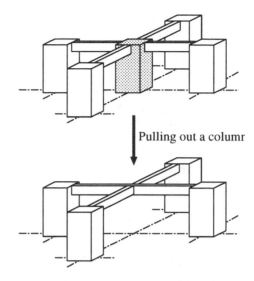

Pulling out a column

FIGURE 3: Pulling out a column.

1. Remove a column in the room.
2. Remove one of four girders depending on the column.
3. Extend the girder in opposite side of the removed girder.
4. Connect other two girders to the extended girder.
5. Connect beams and slabs, which depend on the removed girder, to the extended girder.

4 SPECIFICATION DESCRIPTION LANGUAGE

The authors have designed a specification description language based on the conceptual model mentioned in the previous section. The language is composed of following three languages:

- Structure description language.
- Structure editing operation description language.
- Geometry description language.

The last one describes a geometric figure of each part type. The figure is defined with functions such as rectangle and line. The language also describes a region of each part type for detecting a mouse position which indicates the part. The region is defined with a functions such as rectangle or polygon.

The first two languages are explained in the following:

```
PART column
 PLUG
  x_con INTO x_line::col_con;
  y_con INTO y_line::col_con;
  f_con  INTO z_line::col_con;
  c_con INTO z_line::col_con;
  p_col  INTO column::s_col;
 END
 SOCKET
  s_col TAKE column::p_col;
  x_w_gird TAKE x_girder::x_e_con;
  x_e_gird  TAKE x_girder::x_w_con;
  y_n_gird  TAKE y_girder::y_s_con;
  y_s_gird  TAKE y_girder::y_n_con;
 END
 ATTRIBUTE
  x_diam INT DEFAULT default_x_diam;
  y_diam INT DEFAULT default_y_diam;
  x_pos  INT
   WHERE xpos := x_con->x_pos+x_disp;
```

FIGURE 4: Example of structure descriptions

4.1 STRUCTURE

The structure description language describes part types and their associations in the similar way of object class and association definitions in OMT. The language has the following description elements:

- Name of part.
- Existence dependency constraint which is an association connecting two parts to each other.
- Name of attribute, its value type and measure.
- Attribute dependency constraint which determines computation of attribute values.

Fig. 4 shows an example of the structure descriptions.

PLUG and SOCKET are used for describing an existence dependency constraint. A part instance, except for foundation, depends on other part instances in a building. A plug of the former part instance is connected with a socket of the latter part instance. All plugs of a part instance must be connected with sockets of other part instances for existing in a building. However, a socket is not necessarily connected with a plug.

ATTRIBUTEs are classified into the following two groups:

- Basic attribute: A designer determines its value.

- Derived attribute: Its value is obtained from basic attributes and other derived attributes whose values are already decided.

An attribute having DEFAULT value is a basic attribute. For example, diameter and material are basic attributes of a column. On contrary, its position is a derived attribute since the column is located depend on the base lines.

An attribute dependency constraint is described with a substitution expression in succession of the attribute definition.

4.2 OPERATION

The structure editing operation description language describes procedures for editing structures composed of parts. Since the editing procedures are used through graphical user interface (GUI), the operation description language also describes GUI. The operation description language has procedure descriptions of two types: menu and operation.

- Menu describes a procedure which is called through GUI.
- Operation describes a precondition and procedure. The procedure run if the precondition is satisfied. The operation is called from menus and operations.

Fig. 5 shows an example of the menu and operation descriptions.

A MESSAGE function outputs a message through GUI when the menu is selected. A GETPART function substitutes an instance of the part into a variable. The part instance is specified through GUI by a designer. After the function evaluations, the variables can be referred to in the procedure.

A precondition statement after PREMISE describes a condition which must be satisfied on calling the procedure. The precondition also includes definitions of variables. The values of the variables are substituted only once, and are fixed. If the condition is not satisfied or if one of the variable substitutions fails, the procedure does not run.

A procedure statement after THEN describes a procedure for editing structure using the following functions and variables.

A NEW statement creates a new instance, then PLUGIN, plug and socket connection, must be described for inserting the part instance.

The language has also the following statements:

- Removing a part.
- Plug and socket disconnection.
- Modifying an attribute value.
- Calling operation procedure.
- Selection of procedure.

```
MENU 'Basic creation of Slab'
   MESSAGE( 'Select 1st part (E-W Girder or Beam)' );
   GETPART( gird1: e_w_gird );



   CALL( 'BasicCreationOfSlab', [gird1, gird2, gird3, gird4] );
END MENU
OPERATION 'BasicCreationOfSlab'
   PREMISE
      n_gird:  x_girder := INPUT( 1 );
      s_gird:  x_girder := INPUT( 2 );
      e_gird:  y_girder := INPUT( 3 );
      w_gird:  y_girder := INPUT( 4 );
      n_gird->n_e_y_pos > s_gird->n_e_y_pos;
      e_gird->n_e_x_pos > w_gird->n_e_x_pos;
   THEN
      NEW slb: slab;
      PLUGIN( slb->n_con, n_gird->SET( s_con ) );
      PLUGIN( slb->s_con, s_gird->SET( n_con ) );
      PLUGIN( slb->e_con, e_gird->SET( w_con ) );
      PLUGIN( slb->w_con, w_gird->SET( e_con ) );
END OPERATION
```

FIGURE 5: Example of menu and operation descriptions

5   EXPERIMENTAL IMPLEMENTATION

The authors has experimentally implemented a compiler for generating CAD systems from specifications described with the language mentioned in the previous section. For maintenability, the compiler is coded with SICStus Prolog, SICStus objects library and graphics manager. The system is composed of the following four subsystems:

- Building structure management subsystem.
- Building structure editor.
- Geometry management subsystem.
- GUI subsystem.

Building structure management subsystem
This subsystem manages a building and its parts with OMT object models. Two object classes are used for each type of parts. One contains fixed information of the type, such as method and constant data, and is called control object. The other contains variable information of part instance, and called data object. Control and data prototype objects are also generated for an association between plug and socket.

Building structure editor
This subsystem has two kinds of predicates. One is predicate 'callBack'. The other is predicate 'operation'. The former predicate is called when a menu

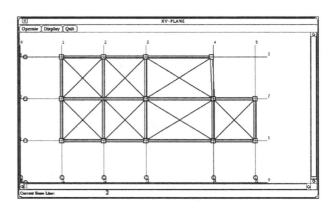

FIGURE 6: GUI

is selected. Then, the predicate calls predicate 'construct' or 'getAttribute' of a part instance in the building structure management subsystem.

The predicate 'operation' calls a precondition predicate and procedure predicates. If the precondition predicate fails, the procedure predicates are not called.

Geometry management subsystem
This subsystem generates structural drawings to be displayed using the predicates of the building structure management subsystem, and send the drawings to GUI subsystem. The subsystem also detects a part instance which is specified with a mouse by a designer.

GUI subsystem
This subsystem is composed of 'windowManager' object for managing the whole GUI subsystem, and of 'plane' objects for controlling the display of X, Y and Z plane drawings individually. When a designer selects a structure editing operation in a menu on a plane window, the 'windowManager' object calls predicates 'getFigure' and 'getPartObj'. Fig. 6 shows an example of GUI.

6   DISCUSSION

6.1   STRUCTURE

The conceptual model of building structures has been obtained from the object model and function model of OMT by restricting them. One of the main restrictions is as follows: No association can be expressed except for ones which have the same meaning as the existence dependency constraints between building parts. Besides, associations are always binary. Therefore, general associations which are expressed in OMT can not be described. This restriction has achieved the minimality which does not require a lot of descriptions for describing the same specifications.

Moreover, the object-oriented model elements such as part, attribute and constraint correspond to concepts which designers have in their mind. Therefore, the model is favorable to the constructibility and comprehensibility of specifications.

As mentioned in the previous section, the compiler automatically generates Prolog programs of CAD systems from specifications described with the conceptual model-based language. This proves that the model is also favorable to formality.

The minimality, constructibility and comprehensibility would encourage building designers to describe specifications by themselves. The formality, for generating programs automatically, makes it possible that the designers test the specifications in short turn around time. Therefore, the domain-specific conceptual model would be useful for requirements analysis and specification.

6.2　OPERATION

The authors have tried a procedural model for expressing the structural editing operations. The language based on the model is useful for describing the typical operations such as pulling out a column in order to make a wide lobby.

However, the language requires a lot of descriptions, because it is procedural. A nonprocedural operation model should be studied. An operation would be nonprocedurally expressed by defining the precondition and postcondition of the operation. Then, architectural space concepts such as lobby might be necessary.

7　CONCLUSION

A domain-specific conceptual model for expressing building structures and structural editing operations has been proposed, and discussed on the experimental implementation.

The conceptual model of building structures is favorable to the minimality, constructibility, comprehensibility because it is specialized from a object-oriented model by the existence dependency constraints between structural parts. The conceptual model of operations can express typical operations for editing the structures. These features would encourage building designers to describe requirements specifications of CAD systems by themselves. Therefore, domain-specific conceptual models might be one of the excellent means of obtaining exact requirements specifications.

Moreover, the model has the formality for generating programs automatically. This feature together with the above-mentioned ones means that building designers could maintain CAD systems by themselves.

In the future, experiments for describing the specification description languages should be tried. Nonprocedural operation description language should be studied. The structural part management subsystem should be improved in efficiency. Architectural space concepts such as lobby should be studied. All individual subdomains of building design should be studied and integrated.

REFERENCES

[RBP91]　Rumbaugh, J., Blaha, M., Premerlani, W., Eddy, F. and Lorenson, W.: *Object-Oriented Modeling and Design*, Prentice Hall (1991).

[Bel94]　Bell, R.: Choosing Tools for Analysis and Design, *IEEE Software*, 11(3), pp. 121–125 (1994).

[Pri87]　Prieto-Diaz, R.: Domain Analysis for Reusability, *proc. IEEE Computer Software and Application Conference*, pp. 23–29 (1987).

[NTM89]　Nagasawa, I., Tegoshi, Y. and Makino, M.: IBDS: An Integrated Building Design System, *Trans. IPS Japan*, 30(8), pp. 1058–1067 (1989). (In Japanese)

[HHN95]　Hirota, T., Hashimoto, M. and Nagasawa, I.: A Dicussion on Conceptual Model Description Language Specific for an Application Domain, *Trans. IPS Japan* 36(5), pp. 1151–1162 (1995). (In Japanese)

[YC78]　Yourdon, E., and Constantine, L. L.: *Structured Design, Fundamentals of a Discipline of Computer Program and Systems Design*, Prentice Hall (1979).

[IWA91]　Iscoe, N,. Williams, G. B. and Arango, G.: Domain Modeling for Software Engineering, *Proc. 13th International Conference on Software Engineering*, pp. 340–343 (1991).

[HH93]　Hashimoto, M, and Hirota. T.: A Case Study on Conceptual and Formal Modeling of Software, *Proc. Joint Conference on Software Engineering '93*, pp. 333–340 (1993).

THE ELEMENTS OF PROGRAMMING STYLE IN DESIGN CALCULATIONS

Masanobu Umeda*   Isao Nagasawa*   Tatsuji Higuchi**

Faculty of Computer Science and System Engineering, Kyushu Institute of Technology*,
Iizuka, Fukuoka, Japan, 820.
Email: umerin@mse.kyutech.ac.jp
DI Business Development Department, OLYMPUS Optical Co., Ltd.**,
Hachioji, Tokyo, Japan, 192.

ABSTRACT

DSP is a programming language specific for design calculation problems, and is intended for designers themselves to describe and maintain their design knowledge. This paper describes the fundamental programming concepts of DSP and its programming style. It clarifies how to formalize a given problem in design calculations and which concept can be applied to what kinds of problems. It also shows the actual stage of the programming using examples of design calculation problems, and discusses the usefulness of this approach to design calculation problems based on the experiences from the applications of DSP.

1. INTRODUCTION

Intellectual activities of human being have become complicated due to advancements of sciences and technologies and changes of social environments. In the example of product developments, design work has become difficult and complicated because of varieties of user requirements and social requirements from the exhaustion of resources and the destruction of environments. Since the improvement of intellectual productivity of experts such as a product designer is essential to solve these kinds of problems, knowledge-based systems such as intelligent CAD systems [Nag89, NS95] are expected heavily.

It is essential to formally analyze, systematize, and describe knowledge of an application domain in the development of a knowledge-based system. The description of the knowledge is conceptually possible in any conventional programming language. However, a knowledge base described in a conventional language such as FORTRAN is difficult to develop, maintain and manage since most of the domain experts do not have enough techniques to use such a language without help from computer programmers. This problem causes the systems to be stiffened and makes the transfer of the technologies to succeeding generations difficult [Nag88]. Therefore, design support systems should have the capability of enabling the designers themselves to develop and maintain the knowledgebase. It is perceived that a description language that is specific to an application domain and is designed so as to be constructive by the domain experts is superior in terms of the minimality, constructibility, comprehensibility, extensibility, and formality of the language [HHN95].

DSP [TNMM89, NMTM90, UN90] is a programming language specific for design calculation problems, and is intended for designers themselves to describe and maintain their design knowledge. It is a functional programming language based on attribute grammars [DJ90] with capability of describing generate and test algorithm. The concepts of data flow and generate and test in DSP are suitable for the formal description of design knowledge where trial and error are involved. DSP and its programming environment have been applied for over 5 years to a wide range of industrial application domains, such as cameras design, copying machines design, architectural design, and plant elements design.

This paper describes the fundamental programming concepts of DSP and its programming style. It also shows the actual stage of the programming using examples of design calculation problems, and discusses the usefulness of this approach to design calculation problems based on the experiences from the applications of DSP.

2. FUNDAMENTAL PROGRAMMING CONCEPTS IN DSP

2.1 DATA FLOW

The most fundamental work in a design is to decide the values of design variables from other design variables and to verify if the specified constraints hold on the design variables. This work can be formally described as the dependency on design variables, that is the data flow.

Let's consider a problem for calculating the profit from the purchase cost and the retail price as an example of a program based on the data flow. In Figure 1, items (a) ∼ (f) are the declarations of integer (int) type variables indicating the cost, price, count, payment, sale and profit. Items (g) and (h) are two expressions for computing the payment and sale from the products of the cost by the count and of the price by the count, respectively. Item (i) is a expression for computing the profit by subtracting the payment from the sale. Here, if we change the calculation order between (g) and (h), both programs are equivalent in the meaning and they can be expressed as equivalent in the data flow.

As above mentioned, the representation in data flow is more abstract than the procedural representation when the calculation order has no meaning, therefore, even those representations which are different in procedure can often be represented as equivalent with the data flow.

```
Cost : int;                    --(a)
Price : int;                   --(b)
Count : int;                   --(c)
Payment : int;                 --(d)
Sale : int;                    --(e)
Profit: int;                   --(f)

Payment := Cost * Count;       --(g)
Sale := Price * Count;         --(h)
Profit := Sale - Payment;      --(i)
```

FIGURE 1   An example program based on the data flow

2.2 GENERATE AND TEST

There are some difficult problems to be solved analytically in design. In this case, designers repeatedly perform such work as assuming the values of design variables to verify if they satisfy the specified constraints through trial and error. The design work where trial and error are involved can be generalized as the generate and test algorithm.

Generate and test algorithm is weak to solve a complex problem of a large search space since it is too general. However, the actual search space of most design calculation problems is likely to be bounded so as to be solved manually if design knowledge is well systematized. Thus, generate and test algorithm is strong enough for such problems.

For example, let's consider a problem of acquiring the set of X, Y values that satisfy the expression $X > Y^2$, where the domains of variables X, Y are both $\{1, 2, 3\}$. This problem can be represented as the expressions shown in Figure 2.

$$
\begin{aligned}
X &\in \{1, 2, 3\} & (1) \\
Y &\in \{1, 2, 3\} & (2) \\
X &> Y^2 & (3)
\end{aligned}
$$

FIGURE 2   An example of search problem

Variables X, Y are both discrete and finite, so the number of results are also finite. Therefore, this problem can be solved with the program given in Figure 3. The items (a) ∼ (c) of the program have the following meanings:

(a) Assume X is either of 1, 2 or 3.

(b) Assume Y is either of 1, 2 or 3.

(c) Evaluate the expression X>Y². If this expression is true, then the pair X, Y is a result.

Where, select(L) is a multi-valued function that returns one element in the list L nondeterministicly, and is also referred to as a generator.

select([1,2,3]) is a multi-valued function, so that the values of variables X, Y are nondeterministicly determined. Therefore, the entire program can be considered as a multi-valued function. As the result of the execution, {2,1} and {3,1} can be acquired as solutions.

```
X : int;
Y : int;

X := select([1, 2, 3]);        --(a)
Y := select([1, 2, 3]);        --(b)
test(X > Y^2);                 --(c)
```

FIGURE 3   An example of generate and test program

2.3   ABSTRACTION USING MODULES

When using a number of similar mechanical elements, it is not effective to repeatedly describe the similar calculation procedures in the design program. If any processing unit which is considered as the same functional unit is repeatedly used in one program, this program can be simply expressed by arranging these units as one program unit to perform abstraction allowing it to be reference with the name. Such an abstracted program unit is called a module and the abstraction of a program unit as a module is referred to as modularization.

Let's discuss a problem of generalizing the domains of the variables X, Y to $\{1, ..., N\}$ for a given positive integer N in Figure 3. Figure 4 shows the definition of the module combination with the vector $\{N\}$ of integer type variable N as the input and the vector $\{X, Y\}$ of integer type variables X, Y as the output.

(a) Assume X is either of 1, ..., N.

(b) Assume Y is either of 1, ..., N.

(c) Evaluate the expression X>Y^2. If the expression is true, then the pair X, Y is a result.

Where, for(B,E,S) is a multi-valued function that nondeterministicly returns one of the elements of the following set defined with integers (or real numbers) B, E, S:

$$\{i \mid \quad i = B + n * S, n : \text{integer}; \text{where} \\ 0 \leq n, B \leq i \leq E, \text{or } n \leq 0, E \leq i \leq B\}$$

Therefore, the module combination can be considered as a multi-valued function which returns a pair of the values of variables X, Y with an integer N as the input. If N is equal to 3, the values of the module combination is either $\{2,1\}$ or $\{3,1\}$.

```
combination({N : int},
            {X : int, Y : int})
  method
    X := for(1, N, 1);          --(a)
    Y := for(1, N, 1);          --(b)
    test(X > Y^2);              --(c)
end;
```

FIGURE 4   An example of the module combination

2.4   PROBLEM REDUCTION

Problem reduction is one of the effective techniques for solving complex problems. If a design problem can

be expressed as a combination of several partial problems, it can be solved as a set of less complex partial problems. For example, when the design object consists of some mechanical elements, problem reduction can be applied if the design problem is able to be divided into each mechanical element problems.

As an example of problem reduction, let's consider a pressure vessel design problem [JIS81]. For simplification, let's assume the pressure vessel consists of a cylindrical body and two heads as shown in Figure 5. In this case, by applying problem reduction, the pressure vessel design can be considered as three partial design problems: a body design and two head designs. Figure 6 shows an example program of the pressure vessel design module vessel. The module vessel has the vector of head types Type1, Type2; material Mat; design pressure P; design volume V as the input and that of diameter D; plate thickness and volume of each head T1, V1, T2, V2; plate thickness and length of body Tb, Lb as the output. The modules head and body that are submodules of vessel are omitted here.

(a) Diameter D is assumed by the unit of 50mm within the range of 200mm ∼ 3000mm.

(b) One of the heads is designed from head type Type1, material Mat, design pressure P and diameter D to acquire the plate thickness T1 and volume V1. Where, T1 is assumed in the small-to-large plate thickness order in the module head.

(c) The other head is designed from head type Type2, material Mat, design pressure P and diameter D to acquire the plate thickness T2 and volume V2. Where, T2 is assumed in the small-to-large plate thickness order in the module head.

(d) The difference between the design volume V and the volumes of two heads is acquired as the design volume of body Vb.

(e) The body is designed from the material Mat, design pressure P, diameter D and design volume of body Vb to acquire the plate thickness Tb and length Lb.

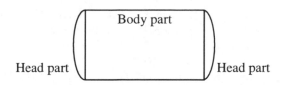

FIGURE 5   A simple example of pressure vessel

```
vessel({Type1 : atom, Type2 : atom ,
        Mat : atom , P : real , V : real},
       {D : int,
        T1 : real, V1 : real,
        T2 : real, V2 : real,
        Tb : real, Lb : real})
  method
    Vb : real;

    D := for(200, 3000, 50);          --(a)
    call(head,{Type1, Mat, P, D},
             {T1, V1});               --(b)
    call(head,{Type2, Mat, P, D},
             {T2, V2});               --(c)
    Vb := V - V1 - V2;                --(d)
    call(body,{Mat, P, D, Vb},
             {Tb, Lb});               --(e)
end;
```

FIGURE 6   An example of problem reduction

2.5   RECURSION

As a special case of problem reduction, there is recursive technique for solving partial problems using the module itself directly or indirectly. Recursion is effective for a design work in which the same design procedures can be applied to its partial problems.

Let's consider a problem for computing the factorial of positive integer N in Figure 7 as an example of recursion. Figure 8 shows the definition of the module fact with the vector $\{N\}$ of integer type variable N as the input and the vector $\{M\}$ of integer type variable M as the output.

(a) The expression N=0 is evaluated. If this expression is true, then the value of M is the result.

(b) The value of M is set to 1. Therefore, if the expression in (a) is true, $\{1\}$ will be the value of the module fact.

(c) Integer type variable M1 is declared.

(d) The expression N>0 is evaluated. If this expression is true, then the value of M is the result.

(e) The module fact is called with $\{N-1\}$ as the input and $\{M1\}$ as the output.

(f) The value of M is set to that of N*M1. If the expression is true, then $\{M\}$ will be the value of the module fact.

The module fact consists of two methods: (a) $\sim$ (b) and (c) $\sim$ (f). The expressions (a) and (d) give the domains for each method. In this example, the domains of two methods have no overlaps, so that these two methods can be exclusively applied and method 1 is the termination condition of the recursion.

$$fact(N) = \begin{cases} 1 & \text{when } N = 0 \\ N * fact(N - 1) & \text{when } N > 0 \end{cases}$$

FIGURE 7   Factorial problem

```
fact({N : int},{M : int})
  method                         --Method 1
    when(N = 0);                 --(a)
    M := 1;                      --(b)
  method                         --Method 2
    M1 : int;                    --(c)

    when(N > 0);                 --(d)
    call(fact,{N-1}, {M1});      --(e)
    M := N*M1;                   --(f)
end;
```

FIGURE 8   An example of the module fact

2.6   ITERATION

Loop constructs in conventional procedural languages, such as DO in FORTRAN, are often used for representing iterations in convergence computations. Conceptually, most of programs which use loops can be described using recursion. However, the use of recursion is often a difficult task [Rob86] for designers. It is helpful if iteration can be represented without using an explicit loop construct.

Let's consider a problem computing a square root of a given number using Newton method. The module sqrt in Figure 9 computes the square root S of the input X by calling the module sqrt_step which does the following:

(a) The next approximate square root Xn of the number X is calculated using the previous approximate value Xi.

(b) A relative error En is calculated for Xn and Xi.

(c) If the relative error En is still greater than E, the call of the module sqrt_step from the module sqrt will be continued.

repeat iterates a module call until continue in the module is evaluated to false. During the iteration, the output of the module in one step is used in turn as the input in the following step. Thus, the structure of the input of the module sqrt_step is same as that of the output.

```
sqrt({X : real}, {S : real})
  method
    Xo : real;
    En : real;

    repeat(sqrt_step,{X, 1.0, 1.0E-10},
                     {Xo, S, En});
end;

sqrt_step({X : real, Xi : real, E : real},
          {X : real, Xn : real, E : real})
  method
    En : real;

    Xn := (Xi+X/Xi)/2.0;        --(a)
    En := abs((Xn-Xi)/Xi);      --(b)
    continue(En > E);           --(c)
end;
```

FIGURE 9   An example of convergence computation

2.7   NONDETERMINISTIC PROCESSING

In the process of design work, there can be several different choices of calculation procedures and the information enough to deterministicly choose one of them may not be available. In case that it is impossible to deterministicly perform a processing, the nondeterministic calculation procedure will be necessary. The nondeterministic calculation procedure is expressed with multiple methods so defined that their domains overlap within one module.

Figure 10 shows the definition of the module for that nondeterministicly returns one of the elements of the following set defined with integers B, E, S:

$$\{i \mid \quad i = B + n * S, n : \text{integer}; \text{where}$$
$$0 \leq n, B \leq i \leq E, \text{or } n \leq 0, E \leq i \leq B\}$$

Method 1 If the expression $(B \leq E \land S > 0) \lor (B \geq E \land S < 0)$ is true, the value of the module shall be set to $\{B\}$.

Method 2 If the expression $(B+S \leq E \land S > 0) \lor (B+S \geq E \land S < 0)$ is true, the value $\{N\}$ in case of executing the module for with the vector $\{B+S,E,S\}$ as the input shall be the value of the module.

Since method 1 is always applicable when method 2 is, the domains of method 1 and 2 overlap. Therefore, the module for acts nondeterministicly on the inputs B, E and S. For example, when B, E and S are 1, 3 and 1 respectively, $\{1\}$, $\{2\}$ and $\{3\}$ will be the result values.

The description order of methods has a meaning on the order of values returned from a module. For example, if the order between method 1 and 2 is reversed, then the values are returned in the order: $\{3\}$, $\{2\}$, $\{1\}$ for the same input.

```
for({B : int, E : int, S : int},{N : int})
  method                     --Method 1
    when((B =< E and S > 0) or
         (B >= E and S < 0));
    N := B;
  method                     --Method 2
    when((B+S =< E and S > 0) or
         (B+S >= E and S < 0));
    call(for,{B+S, E, S}, {N});
end;
```

FIGURE 10   An example of nondeterministic module

2.8   LIST PROCESSING

When several results are obtained by design work, the analysis and evaluation of the results will be necessary for determining which is the most suitable. It is also the case that a set of the results itself may be important rather than individual results as how many solutions are acquired. If a set of the results is expressed as a list of results, the selection of a result and the statistical processing of the results can be realized as a list processing.

Let's discuss a problem for computing the number π using Monte Carlo method as an example of handling solution sets. The module monte as shown in Figure 11 returns the vector consisting of a pair of real numbers X, Y satisfying the expression $X^2+Y^2<1.0$ with a vector of integer N as the input. Where, random(N, {uniform, uniform}) is a multi-valued function generating up to N vectors $\{X,Y\}$ ($0 \leq X < 1$, $0 \leq Y < 1$) that are pseudo uniform random numbers.

The module pi for computing the number π can be defined as shown in Figure 11 using the module monte.

(a) The module monte is called with the vector of integer N as the input, and the value of variable A is the list of all results of the call.

(b) The value of variable L is the length of list A.

(c) The value of variable Pi, the number π, is the ratio of the length of list L and the number of random numbers N.

3.   PROGRAMMING PROCESS

The process of analyzing and systematizing design problems to describe complete programs in DSP can be summarized as follows:

```
monte({N : int},{X : real, Y : real})
  method
    {X, Y} := random(N,{uniform,uniform});
    test(X^2 + Y^2 < 1.0);
end;

pi({N : int},{Pi : real})
  method
    A : [{real, real}];
    L : int;

    find(monte,{N}, A);        --(a)
    L := length(A);            --(b)
    Pi := 4.0*L/N;             --(c)
end;
```

FIGURE 11   Computing π using Monte Carlo method

Problem Definition   This step is to define the problem by analyzing and systematizing the given design problems. For instance, the design conditions and requirements are summarized as constraints and the inputs and outputs of the program are decided.

Formulation for Generate and Test Algorithm
This step is to rearrange the problem so as to apply generate and test algorithm. For instance, variables with continuous values are changed into discrete variables if they are treated as assumptions. Constraints over such variables are modified accordingly.

Data Flow Representation   This step is to represent design procedure as the data flow. Those which can be considered as same functional groups are separated as modules. In case of complex data flow, it is useful to use the data flow diagram [TNMM89, NMTM90].

Programming   This step is to describe programs based on the data flow.

Debugging   This step is to run the program and verify its correctness. If there is any logical error, the modification is done after returning to the data flow representation.

4.   DESIGN EXAMPLES

This section explains the actual steps of the programming in DSP using the practical design problems. We only show a few examples in the experiments as space is limited.

4.1   FLOOR PARTITION DESIGN

The floor partition design problem in the architectural structure design is to decide the structure of beams spanned on the floor and the structure of slabs. The thickness of slabs are determined by restrictions, such as the purpose, finish weight, loading weight and bending. Beams support the loading weight applied on slabs. When the design cannot be performed with one slab, the floor is divided into multiple slabs by beams. The thickness of a slab and the size of beams depend on the floor partition method, which also affects the height, weight and price of an entire building.

In floor partition, several dividing points and directions are possible as shown in Figure 12, so the partition method cannot be deterministicly selected. Therefore, the nondeterministic processing is necessary for the partition method. The partition procedure can be similarly applied to those slabs which have already separated, so that recursion is also applicable. Figure 13 shows an example of the module floorplan. It returns the vector of the slab weight Wt [1] and the number of partitions N as well as longitudinal and lateral sizes and the thickness of slab as the dividing result with the vector of the floor size Lx, Ly, purpose Purpose and finish weight Wf as the input. Where, the module slab is for calculating the slab weight and thickness with the size, purpose and finish weight as the inputs.

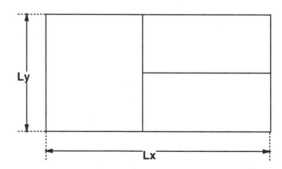

FIGURE 12   Floor partition problem

4.2   CONTACT TOLERANCE ANALYSIS OF SPUR GEAR

Design standards such as mechanical engineering handbooks usually cannot be applied to those plastic gears made with molding process like spur gears used for cameras. Therefore, in the actual design, the

---

[1] Beam weight is ignored for simplicity's sake.

```
floorplan({Lx : real, Ly : real,
          Purpose : atom, Wf : real},
         {Wt : real, N : int,
          Plan : [{real, real, int}]})
  method                  --Design slab part
    call(slab,{Lx, Ly, Purpose, Wf}, {Wt, T});
    N := 1;               --The number of partitions
    Plan := [{Lx, Ly, T}]; --Result of partitioning

  method                  --Swap long side and short side
    when(Lx > Ly);
      call(floorplan,{Ly, Lx, Purpose, Wf},
                     {Wt, N, Plan});

  method                  --Short side partitioning
    Lx1 : real;           --Length of partition 1
    Lx2 : real;           --Length of partition 2
    Wt1 : real;           --Weight of partition 1
    Wt2 : real;           --Weight of partition 2
    N1 : int;             --Number of partitions
    N2 : int;             --Number of partitions
    Plan1 : [{real, real, int}]; --Partition result 1
    Plan2 : [{real, real, int}]; --Partition result 2

    when(Ly >= Lx and
        (Lx > 4.2 or Ly > 7.5 or Lx*Ly > 36));
      Lx1 := for(floor(Lx)/2, Lx - 2, 0.5);
      Lx2 := Lx - Lx1;
      call(floorplan,{Lx1, Ly, Purpose, Wf},
                     {Wt1, N1, Plan1});
      call(floorplan,{Lx2, Ly, Purpose, Wf},
                     {Wt2, N2, Plan2});
      Wt := Wt1 + Wt2;
      N := N1 + N2;
      Plan := append(Plan1, Plan2);

  method                  --Long side partitioning
    Ly1 : real;
    Ly2 : real;
    Wt1 : real;
    Wt2 : real;
    N1 : int;
    N2 : int;
    Plan1 : [{real, real, int}];
    Plan2 : [{real, real, int}];

    when(Ly >= Lx and
        (Lx > 4.2 or Ly > 7.5 or Lx*Ly > 36));
      Ly1 := for(floor(Ly)/2, Ly - 2, 0.5);
      Ly2 := Ly - Ly1;
      call(floorplan,{Lx, Ly1, Purpose, Wf},
                     {Wt1, N1, Plan1});
      call(floorplan,{Lx, Ly2, Purpose, Wf},
                     {Wt2, N2, Plan2});
      Wt := Wt1 + Wt2;
      N := N1 + N2;
      Plan := append(Plan1, Plan2);
end;
```

FIGURE 13   Floor partition program

tolerances are assigned by worst case analysis or statistical analysis. However, because the backlash or contact ratio is solved with general contact equation, it is necessary for exact sensitivity analysis to verify the tolerance distribution by assigning tolerances in random numbers. For example, with the spur gear as shown in Figure 14, the backlash or contact ratio can be calculated from tolerances of the outer diameters of gears D1 and D2; bearing diameters S1 and S2; spindle hole diameters of gears H1 and H2; distance between centers A; coaxial (eccentricity) degree formed with bearing and diameters of D1 and D2.

Figure 15 shows an example of the module **gear** for calculating the backlash Sn and contact ratio E with the inputs of the number of trials N, module M, the number of teeth Z1 and Z2, the driver outer diameter and its minimum and maximum tolerances D1, D1l and D1u, the follower outer diameter and its minimum and maximum tolerances D2, D2l and D2u, the distance between centers in the direction of X and its minimum and maximum tolerances Xa, Xal and Xau, the distance between centers in the direction of Y and its minimum and maximum tolerances Ya, Yal and Yau, the coaxial degrees of driver and follower E1in and E2in, minimum and maximum tolerances of bearing outer diameter Dhmin and Dhmax, the tolerances of bearing inner diameter Dsmin and Dsmax and the factor Dp that depends on the forming method. In case of plastic gears, the tolerances of gear outer diameters D1 and D2, bearing diameters S1 and S2 and gear spindle hole diameters H1 and H2 will show a normal distribution for a short period, however, it is natural to treat it as an uniform distribution taking account of the wear. On the other hand, the distance between centers A and the coaxial (eccentricity) degree formed with bearing and outer diameters D1 and D2 will show a normal distribution.

In this example, the modules **uniform** and **normal** calculate the values including tolerances with the random numbers of uniform and normal distribution, values, minimum and maximum tolerances as the inputs. The function inv(θ) is an involute function: $tan(\theta) - \theta$ [Ohn89]. Figure 16 shows the backlash Sn in the horizontal axis and its occurrence frequency in the vertical axis for the results of 10000 trials with one spur gear sample.

5.   DISCUSSIONS

5.1   EXPERIMENTAL RESULTS

A number of experts in the design of various industries such as cameras, copying machines, architectures

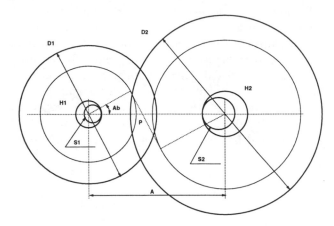

FIGURE 14　An example of spur gear

and plant elements have actually described their design problems in DSP. These include mechanical elements designs such as springs and gears; floor partition and planning of temporary platform stage in the architectural design; and pressure vessel design.

These applications clarify that the concepts based on the data flow and generate and test algorithm in DSP are more appropriate for describing design procedures used by designers than the conventional procedural languages, therefore it is easier to learn the language and its programming and to understand the written programs than that of the conventional languages. For example, in case of a designer who has an experience with programming in FORTRAN, it was possible to describe a spring design program only with one week training, and an high appraisal of maintenance and productivity of DSP was given.

DSP is also applied to elevator configuration design [Yos92] that is a well-known example in AI community. Configuration design is basically a combinatorial problem, and the elevator configuration has about 10^{17} possibilities in case of simple combination. VT [MSM88] reduces this search space by implementing failure recovery knowledge that imitate behaviors of expert designers in case of failures. This conventional knowledge-based approach may obtain a solution with a short time, but is neither guaranteed to reach a solution nor the best solution because of unreliableness and individual variations of the knowledge. On the other hand, the system developed with DSP reduces the search space by using the ontologies for the search control, such as stepwise refinement, bounding by linearity, and bounding by minor parameters. The system can enumerate all significant solutions including the optimal solution at some engineering or economical

```
gear({N : int, M : real, Z1 : int, Z2 : int,
      D1 : real, D1l : real, D1u : real,
      D2 : real, D2l : real, D2u : real,
      Xa : real, Xal : real, Xau : real,
      Ya : real, Yal : real, Yau : real,
      E1in : real, E2in : real,
      Dhmin : real, Dhmax : real,
      Dsmin : real, Dsmax : real,
      Dp : real}, {Sn : real, E : real})
  method                -- Declarations omitted
    {R0, R1, R2, R3, R4, R5, R6, R7, R8, R9} :=
      random(N, {uniform, uniform, normal, normal,
                 normal, normal, uniform, uniform,
                 uniform, uniform});
    call(uniform,{R0, D1, D1l, D1u},{Dk1});
    call(uniform,{R1, D2, D2l, D2u},{Dk2});
    call(normal,{R2, Xa, Xal, Xau},{X});
    call(normal,{R3, Ya, Yal, Yau},{Y});
    call(normal,{R4, 0.0, -E1in*0.5, E1in*0.5},{E1in2});
    E1 := abs(E1in2);           --Eccentricity
    call(normal,{R5, 0.0, -E2in*0.5, E2in*0.5},{E2in2});
    E2 := abs(E2in2);           --That of follower
    call(uniform,{R6, 0.0, Dhmin, Dhmax},{Dh1});
    call(uniform,{R7, 0.0, Dsmin, Dsmax},{Ds1});
    call(uniform,{R8, 0.0, Dhmin, Dhmax},{Dh2});
    call(uniform,{R9, 0.0, Dsmin, Dsmax},{Ds2});
    Pi := 2*asin(1.0);          --Constant π
    An := Pi*20.0/180.0;        --Tool pressure angle
    Te := Pi*M*cos(An);         --Normal line pitch
    Gata1 := 0.5*(Dh1 - Ds1);   --Driver bearing space
    Gata2 := 0.5*(Dh2 - Ds2);   --Follower bearing space
    A := sqrt(X^2+Y^2)-(E1+E2)-(Gata1+Gata2);
    X1 := (Dk1/M-Z1-2)/2;       --Modification coefficient
    X2 := (Dk2/M-Z2-2)/2;       --That of follower
    G1 := M*Z1*cos(An);         --Base circle diameter
    G2 := M*Z2*cos(An);         --That of follower
    Ab := acos((G1+G2)/2/A);    --Contact pressure angle
    B1 := G1/cos(Ab);           --Pitch circle diameter
    B2 := G2/cos(Ab);           --That of follower
    Sn := M*(Z1+Z2)*cos(An)     --Normal backlash
          *(tan(Ab)-Ab-tan(An)+An
            -2*tan(An)*(X1+X2)/(Z1+Z2));
    L1 := sqrt((Dk1/2)^2-(G1/2)^2) --Contact length
          -sqrt((B1/2)^2-(G1/2)^2);
    L2 := sqrt((Dk2/2)^2-(G2/2)^2) --Contact length
          -sqrt((B2/2)^2-(G2/2)^2);
    E  := (L1+L2)/Te;           --Contact ratio
    S1 := Pi*M/2+2*X1*M*tan(An);
    S2 := Pi*M/2+2*X2*M*tan(An);
    THf1 := S1/M/Z1+inv(An);
    THf2 := S2/M/Z2+inv(An);
    THs1 := sqrt(((Dk1-Dp)/G1)^2-1.0)
            +Dp/G1-acos(G1/(Dk1-Dp));
    THs2 := sqrt(((Dk2-Dp)/G2)^2-1.0)
            +Dp/G2-acos(G2/(Dk2-Dp));
    test(Sn >= 0);              --Verify backlash
    test(E >= 1);              --Verify contact ratio
    test(L1 >= 0);             --Verify contact length
    test(L2 >= 0);             --Verify contact length
    test(X1-(1-Z1*sin(An)^2/2) >= 0); --Verify undercut
    test(X2-(1-Z2*sin(An)^2/2) >= 0); --Verify undercut
    test(THf1 - THs1 >= 0);    --Verify sharpness
    test(THf2 - THs2 >= 0);    --Verify sharpness
end;
```

FIGURE 15　Tolerance analysis program of spur gear

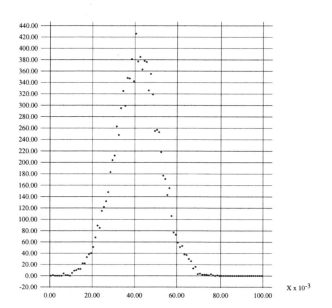

FIGURE 16   Backlash distribution of spur gear

point of view in a practical time. The system consists of 23 modules in basic design phase and 17 modules in detail design phase. 850 solutions are found in the basic design phase in about 17 seconds on SparcStation 2. Finally, 844 solutions are found in about 30 hours. The knowledge described in DSP is also easier to understand than that of the conventional approach.

Practical experiments using DSP is also performed for 38 students as an education program in the faculty of computer science. The students had a training of information processing in the faculty. They were taught the fundamental programming concepts of DSP using a pillar layout problems of parking garage (90 minutes) and given a practical training (90 minutes), then they were required to submit the reports. As the result of this training course, the students who had received the education of information processing were able to complete programming after getting the lecture and practical training for about 3 hours in total. They were asked for the comparison of the description in DSP with that in C language in the reports. Many of them listed an advantage of DSP, i.e., DSP enables the straightforward description of data flow, while C language needs the conversion of data flow into control flow. Furthermore, they also pointed out that the assumption of the values which can be described declar-

atively in DSP should be converted into multiple loops with C language.

5.2   RELATIONS WITH ENGINEERING INFORMATION

The engineering information, such as laws and regulations about the application domains and products catalogs used in a product, is needed in design [IUN93]. DSP has a feature handling the engineering information as functions or multi-valued functions in the program. For this information, refer to [UN91].

5.3   PROGRAMMING ENVIRONMENT

Programming and debugging steps in programming process are strongly influenced by its programming environment. DSP has an interactive programming environment shown in Figure 17. The environment provides facilities of editing, compilation, execution, and debugging of the programs, and management of program libraries.

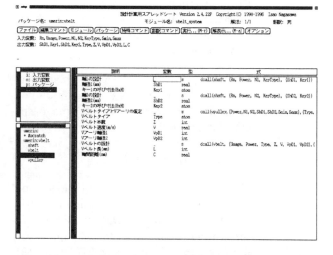

FIGURE 17   A snapshot of DSP programming environment

6.   CONCLUSIONS

DSP is a programming language specific for design calculation problems, and is intended for designers themselves to describe and maintain their design knowledge. It is suitable to the descriptions of design knowledge where trial and error are involved. However, it may not be clear how to formalize the given problem and which concept can be applied to what kinds of problems.

This paper shows the fundamental programming concepts of DSP such as the data flow and generate and test, and shows the five steps of a typical programming process in DSP. Then, examples of design calculation problems are represented in DSP based on the concepts. The experiments by the experts in various industries and the students show that the language is easy to learn and has advantages in productivity and maintenance of the programs. Effectiveness of DSP in configuration design is also shown using the elevator design example.

DSP is going to be applied to new application domains, such as the heat exchanger design system [YNU+95] and the parametric drawing system. In the heat exchanger design system, DSP is used as a subsystem for solving sub-problems or providing possible solutions to the superior system. The details about the system will be reported in the future.

REFERENCES

[DJ90]   Pierre Deransart and Martin Jourdan, editors. *Attribute Grammars and their Applications*, number 461 in Lecture Notes in Computer Science,. Springer-Verlag, September 1990.

[HHN95]   Toyohiko Hirota, Masaaki Hashimoto, and Isao Nagasawa. A discussion on conceptual model description language specific for an application domain. *Transactions of Information Processing Society of Japan*, 36(5):1151–1162, 1995.

[IUN93]   Masatoshi Ito, Masanobu Umeda, and Isao Nagasawa. Standardization activities for standard parts library in japan. *International Journal of the Japan Society for Precision Engineering*, 27(3):185–191, September 1993.

[JIS81]   JIS Pressure Vessels Editing Committee, editor. *JIS Pressure Vessels – Interpretations and Examples of Calculation –*. Japanese Standards Association, 1981.

[MSM88]   Sandra Marcus, Jeffrey Stout, and John McDermott. Vt: An expert elevator designer that uses knowledge-based backtracking. *AI Magazine*, pages 95–111, 1988.

[Nag88]   Isao Nagasawa. Feature of design and intelligent CAD. *Journal of The Japan Society for Precision Engineering*, 54(8):1429–1434, 1988.

[Nag89]   Isao Nagasawa. *Paradigms in Design Support Methodologies and Intelligent CAD*, volume 1 of *Intelligent CAD –Idea and Paradigm–*, chapter 5, pages 113–141. Asakura Shoten, June 1989.

[NMTM90]   Isao Nagasawa, Junji Maeda, Yoshiaki Tegoshi, and Minoru Makino. A programming technique for some combination problems in a design support system using the method of generate-and-test. *Journal of Structural and Construction Engineering*, (417), 1990.

[NS95]   Isao Nagasawa and Hiromasa Suzuki. Roles in the development of intelligent cad systems. *Journal of Advanced Automation Technology*, 7(1):13–16, 1995.

[Ohn89]   Kiyoshi Ohnishi. *Mechanical Design and Drafting Handbook based on JIS*. Rikougakusha, 6 edition, 1989.

[Rob86]   Eric S. Roberts. *Thinking Recursively*. John Wiley & Sons, 1986.

[TNMM89]   Yoshiaki Tegoshi, Isao Nagasawa, Junji Maeda, and Minoru Makino. An information processing technique for a searching problem of an architectural design. *Journal of Architecture, Planning and Environmental Engineering*, (405), 1989.

[UN90]   Masanobu Umeda and Isao Nagasawa. A design support system for pressure vessels. *IEICE Technical Report, AI90-65*, 1990.

[UN91]   Masanobu Umeda and Isao Nagasawa. The design of public engineering knowledge-base library. In *9th Design Symposium*, pages 49–58, 1991.

[YNU+95]   Hideyuki Yamaguchi, Isao Nagasawa, Masanobu Umeda, Naoko Sakurai, and Tsutomu Katoh. Standardization of basic design process of heat exchanger in power plant using design process modeling. In *13th Design Symposium*, pages 215–222, July 1995.

[Yos92]   Gregg R. Yost. Configuring elevator systems, December 1992.

Case-Based Reasoning

KNOWLEDGE-BASED EXPERT SYSTEM
COMBINING CASE-BASED REASONING AND RULE-BASED REASONING
TO SELECT REPAIRING AND RETROFITTING METHODS
FOR FATIGUE DAMEGE OF STEEL BRIDGES

Shigenori Tanaka[1], Ichizou Mikami[2], Hidenori Maeda[3] and Atsushi Kobayashi[2]

[1] Faculty of Informatics, Kansai University, Osaka, Japan.
[2] Department of Civil Engineering, Kansai University, Osaka, Japan.
[3] Toyo Information Systems Co., Ltd., Osaka, Japan.
E-mail: tanaka@kutc.kansai-u.ac.jp

ABSTRACT

We have studied the knowledge-based expert system (KBES) to select repairing and retrofitting methods for fatigue damage of steel bridges. The case-based reasoning (CBR) system can solve the problem of knowledge acquisition and perform efficient inference based on similar cases, whereas the rule-based reasoning (RBR) system has the difficulty in acquiring knowledge bases.

In the present paper, we constructed a KBES, in which CBR and RBR were combined, to select repairing and retrofitting methods for steel bridges' fatigue damage. The inference results obtained from this system are discussed to prove its usefulness, and subjects for the future are identified.

1 INTRODUCTION

Recently, maintenance and administration of existing civil engineering structures have become an important subject, but their maintenance and administration services are complex and extensive. This situation holds true of steel bridges, too. On the other hand, the selection of repairing and retrofitting methods for fatigue damage of steel bridges is at present dependent upon expertise and experiential knowledge of experts. In steel bridge engineering fields, KBES can be an effective tool. Accordingly, the authors have been developing KBES's [MTK94 and TMM94] to select repairing and retrofitting methods for fatigue damage of steel bridges. The first system [MTK94] was developed by using causal relation knowledge, which is a kind of RBR, and an inference method based on modal logic. The last one [MTK94] was constructed by using CBR with retrieval functions such as abstraction matching and part matching. We described in the paper [MTK94] that RBR could acquire general rules from experiential knowledge of experts and perform efficient inference based on the rules. In the paper [TMM94], it was presented that CBR could produce exceptional inference results, using knowledge acquired as actual cases, as well as experts' knowledge.

In case of RBR, however, there remain problems such as knowledge acquisition and refinement of acquired knowledge, while in case of CBR, problems such as reliability about solution space and rightness of analogous solutions [Koba92]. Accordingly, recently proposed were the methods using both the CBR and RBR [RS89 and GR91].

Presented in this paper are a KBES using both CBR and RBR developed recently by the authors and the verification of its usefulness through actual bridge damage cases.

2 PLAN OF SYSTEMATIZATION

2.1 Solution of problems in domain

The domain dealt with in this study is of maintenance and administration to select repairing and retrofitting methods for fatigue cracks in superstructures of steel bridges. As shown in Fig. 1, there exist the causal relations from causes to phenomena and from phenomena to countermeasures, as for damage by fatigue cracks. In usual maintenance and administration services, expert engineers are solving their problems in the following three ways: (i) observing the damaged areas and damage phenomenon and remembering his experienced cases in the past, an engineer performs repair and retrofit work,

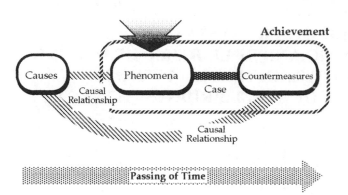

FIGURE 1   Relationship of Times in Problem-Solving

(ii) grasping the damage situation and condition from a damage phenomenon, an engineer searches and finds out the causes based on his expert domain knowledge and executes countermeasures such as removing the causes directly, and (iii) an engineer performs repair and retrofit work, considering both the damage phenomenon and its causes, which is the combination of the first two approaches. Systematization methods of KBES conceivable from these three approaches are as follows.

In case of the approach (i), CBR fits in here because cases from phenomena to countermeasures are easily acquired from the past actual results. In case of the approach (ii), RBR fits in with the first half of the process (from phenomenon to cause) and CBR with the latter half (from cause to countermeasure). Cases from causes to phenomena can not be acquired as past actual results. Even if such cases could be acquired as actual results, they are inferred ones. Thus, CBR does not fit in here, but RBR does, in which causes can be inferred from phenomena in reverse direction by using causal relation knowledge from causes to phenomena. On the other hand, cases from causes to countermeasures in the latter half are easy to obtain as actual results. And hence CBR is preferable to RBR in the latter half of the process. But, since there also should be causal relation from causes to countermeasures, RBR may be used in addition to CBR in the latter half. In case of the approach (iii), the first two approaches are combined and hence it is effective to use both CBR and RBR.

In this study, the third approach was adopted and therefore a KBES using both CBR and RBR was developed.

2.2   Domain knowledge

According to the attributes shown in Fig. 2, 205 cases of fatigue damage of steel bridges were acquired from the literature [Miki82 and Fisher84].

Considered as the "attributes of a phenomenon" were the bridge's outline, the damaged location, and the

damage condition. The "bridge's outline" and the "damaged location" contained eight attributes each, and the "damage condition" four. Then, as the "attributes of causes," external and internal causes and their joint action were considered.

As the "attributes of a countermeasure," repair and retrofit were considered. As the "attributes of repair," pretreatment of repair, repair methods, posttreatment of repair, and their purposes were considered. As the "attributes of retrofit," methods and purposes of retrofit were considered.

2.3   Specifications of KBES

The specifications of this system are divided into "Process 1: retrieval, verification, and modification [Fig. 3 (a)]" and "Process 2: evaluation, restoration, and storage [Fig. 3 (b)]." The description of each process will follow.

2.3.1 Process 1: retrieval, verification, and modification

The attributes shown in Fig. 2, which represent a subject, or phenomenon, are inputted into this system. Once the subject is inputted, four inference results (① - ④) shown in Fig. 3 (a) are obtained in the retrieval process. ① and ② are inference results of CBR, ③ of CBR and RBR, and ④ of RBR. The retrieval in CBR uses the abstraction matching and part matching presented in the paper [TMM94]. But the calculation method of degree of similarity was extended as follows. For example, as shown in Fig. 4, assume the attributes of a subject, A (its value: a), B (its values: b and c), and C (its value: unknown), and the attributes of case data, A (its values: a and d), B (its value: b), and C (its value: e). Use the equation (1) to calculate the degree of agreement (0 - 1) of each attribute between the subject and the case data.

$$
\begin{aligned}
(degree\ of\ agreement)_i = {} & \\
& \frac{number\ of\ agreed\ attri.val.}{number\ of\ subject's\ attri.val.} \\
& - C \times \Bigg[\frac{(number\ of\ case\ data's\ attri.val.)}{number\ of\ agreed\ attri.val.} \\
& - \frac{(number\ of\ agreed\ attri.val.)}{number\ of\ agreed\ attri.val.} \Bigg]
\end{aligned}
\tag{1}
$$

where the coefficient C is an optional variable (0 - 1).

Then, calculate the degree of similarity of the case data to the subject with the equation (2). W stands for the weight of each attribute (optional value of 0 - 1).

$$
\begin{aligned}
(degree\ of\ similarity) = {} & \\
& \frac{(number\ of\ attri.known\ attri.val.)}{(number\ of\ subject's\ attri.)^2} \\
& \times \sum_{}^{\substack{number\ of\ attri.\\ known\ attri.val.}} W_i \times (degree\ of\ agreement)_i
\end{aligned}
\tag{2}
$$

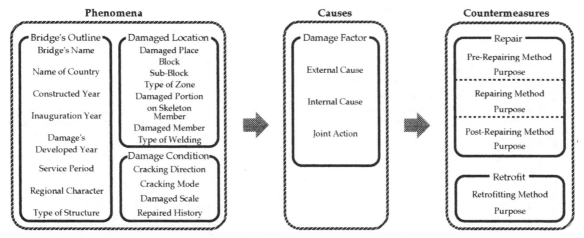

FIGURE 2 Attributes of System

As shown in Fig. 4, the number of the attributes of the subject is 3, and the number of the attributes of which the values are known (the attributes A and B) is 2. If $C = 0.5$, the degree of agreement of the attribute A is 0.5 [= 1/1 - 0.5 (2 - 1)/1] and that of the attribute B 0.5 [= 1/2 - 0.5 (1 - 1)/2]. The agreement degree of the attribute C is not calculated since its attribute value is unknown. If the weight of the attribute A (W_A) is 1.0 and that of the attribute B (W_B) is 0.75, the degree of similarity of the case data to the subject is 0.19 [= $2/3^2 \times (1.0 \times 0.5 + 0.75 \times 0.5)$].

The inference result ① [Fig. 3 (a)] is exceptional countermeasures obtained from the case database by analogical inference retrieval for the damage phenomenon (the subject). This solution is lacking in reliability because it does not take account of the damage causes which are importance in deciding upon countermeasures for repair and retrofit. Then, exceptional damage factors are automatically extracted from the cases of the inference result ①, and exceptional countermeasures (inference result ②) are obtained from the guided analogical inference of the damage factors. Because even with this result we are not sure if all the solution space is covered, further exceptional countermeasures (inference results ③) have to be obtained by using both RBR and CBR. Through the reasoning process of RBR, damage factors can be inferred from the damage phenomenon (the subject), and then exceptional countermeasures can be inferred from the damage factors through CBR. With these three exceptional analogous solutions, it would be safe to say that all the solution space have been covered.

On the other hand, RBR deduces the damage factors, and from the damage factors the inference result ④ is inferred as exceptional countermeasures. Therefore, no special results are included in the inference result ④.

Fig. 5 shows the reasoning procedure in the RBR unit.

Firstly, the subject inputted is classified into attributes of damaged location and damage condition. Next, damage factors are inferred through two routes: "damaged location → damage factor" and "damage condition → damage factor." The logical product of the inference results from these two routes is calculated [Fig. 5 (a)]. Furthermore, these damage factors are classified into three categories, i.e., external factors, internal factors, and joint action. Countermeasures for repair and retrofit are inferred through three routes: "external factor → repair and retrofit," "internal factor → repair and retrofit," and "joint action → repair and retrofit." The logical product of the inference results from these three routes is calculated [Fig. 5 (b)].

To single out the most appropriate one from the above four inference results, the following method was adopted. The result of ① [Fig. 3 (a)] is given the highest priority, and after it the result of the logical product of ② and ③ comes. ④ is used as a supplement to the results of ① and the logical product. Thus, the logical sum of the inference result ①, the logical product of ② and ③, and the inference result ④ are the final result of the retrieval process.

In the verification process, the countermeasures' purposes are determined from the final result of the retrieval process. The purposes are used as domain knowledge to identify the inference result. If there is any inconsistency in the final result, it is forwarded to the modification process. If not, the modification process is skipped.

In the modification process, a prospective subject is automatically formed by referring to the subject of the inference result and the information obtained in the verification process. Then, the new subject is forwarded to the retrieval process.

2.3.2 Process 2: evaluation, restoration, and storage

See Fig. 3 (b). In the evaluation process, the inference

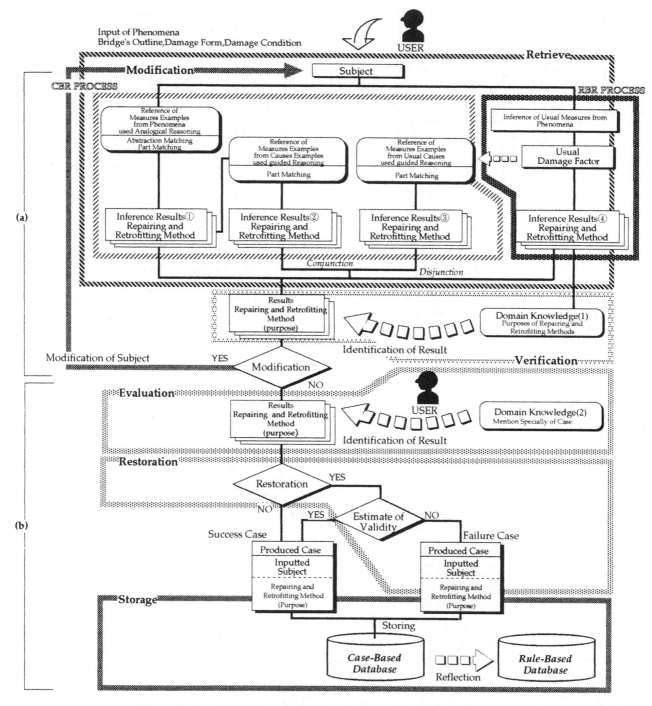

FIGURE 3 Process of Retrieve / Verification / Modification and Process of Evaluation / Restoration / Storage

result is identified by using the domain knowledge about executionality of work, economical efficiency, and appearance. Here, if restoration of the result is necessary, it is forwarded to the restoration process. If not, it is forwarded as a success case to the storage process.

In the restoration process, the inference result is restored based on the information obtained in the verification and evaluation processes, and a prospective result is formed. If it is proper, it is forwarded as a success case to the storage process. If not, as a failure case to the storage process.

In the storage process, success and failure cases are stored as new cases into the case database. Failure cases are administered separately form success cases. If a case, whether success or failure, need not be stored in the case database, it is not stored. As rule knowledge can easily be acquired from success cases newly stored in the case database, it is reflected into the rule database.

3 EXECUTION OF SYSTEM

This system is verified by using a damage case of an actual bridge. Let us take Polk County (Des Moines) Bridge as an example, with which damage was found in 1979. 17 years after its construction, cracks occurred in the connection areas between cross beams and main girders, running from fillet welds between web plates of main girders and vertical stiffeners mainly to the web plates. In this system, the input items for the bridge's outline were made as follows: the service period, 16 - 20 years; and the type of structure, composite plate girder bridge. Input for the damaged locations: blocks, girders; subblocks, connection areas between cross beams and main girders; damaged spots, welds and main-girder web plates; members, main girders and vertical stiffeners; and the type of weld, fillet welding. Input for the damage condition: the damage direction, slanting; the cracking mode, (c); and the damage scale, middle.

Firstly, in the retrieval process, inferred from the similar cases of the inference result ① were damage factors, i.e., lateral distribution and out-of-plane deformation as external factors, secondary stress concentration as an internal factor, and T joint action "③" as working force. Then, based on this inference result, inference result ② was deduced.

On the other hand, deduced from the inputted subject through the RBR unit were damage factors, i.e., lateral distribution as an external factor, secondary stress

concentration as an internal factor, and T joint force "③" as joint action. Based on this information, inference result ③ was obtained. Furthermore, from the logical product of the inference results ② and ③ and the inference result ①, the five similar cases and RBR's inference result ④ were obtained.

In fact, repair work by "stop holes (prevention of cracks from advancing)" and retrofit work by "increase of gaps of web plates (reduction of out-of-plane deformation)" were conducted to Polk County Bridge.

In this system, "stop holes" are presented as the repair method for all the cases. Besides, "increase of gaps of web plates" is also recommended by RBR. Thus, CBR and RBR complement each other in their inference results, when both are used in combination.

4 EVALUATION OF SYSTEM

This system was applied to further five bridges, and the Inference results of the six bridges including Polk County Bridge are summarized in Table 1. The meanings of the symbols used in the table are as follows: ○, actual measure; ◎, actual and more valid measure; △, actual measure uncompleted; ×, unsatisfactory.

In Case 1, CBR inferred valid measures but missed the retrofit method of "increase of gaps of web plates" actually used, which RBR inferred. Therefore, the result of the total

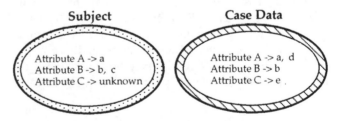

Weight of Attribute -> $A : W_A = 1.0$, $B : W_B = 0.75$, $C : W_C = 1.0$

FIGURE 4 Calculating Example for Degree of Agreement

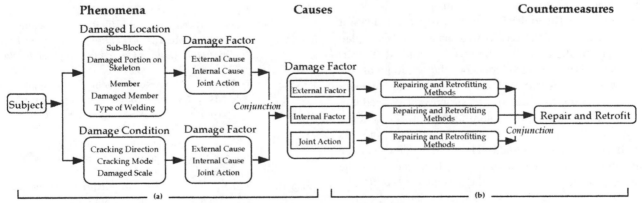

FIGURE 5 Process of RBR

Table 1   Examination of Inference Results

Case	Input Case of Damage		Examination of Inference Results		
	Bridge's Name	Crack Detected Point	CBR	RBR	Total
1	Polk County Bridge	Welds between main girder web and vertical stiffener at main girder to floor beam joint	△	△	○
2	Mukaijima Bridge	Welds between main girder flange and gusset plate at hanger connecting	◎	○	◎
3	Gulf Outlet Bridge	Welds of pin plate at tie girder connecting	○	○	○
4	I-480 Cuyahoga River Bridge	Welds between main girder web and vertical stiffener at web gap	○	×	○
5	Yellow Mill Pond Bridge	Welds between main girder bottom flange and cover plate at cover plate connecting	○	△	○
6	Poplar Street Bridge	Welds between main girder web and vertical stiffener at main girder to floor beam joint	△	△	△

evaluation is ○. In Case 2, actual measures and more valid ones were obtained with CBR and RBR, grading the system as ◎. In Case 3, both the CBR and RBR showed valid inference results, the system scoring ○. In Cases 4 and 5, while CBR presented valid measures, RBR failed to do so, the system scoring ○ again. In Case 6, both the CBR and RBR presented valid repair methods but failed to bring about any valid retrofit method, the system scoring only △. The retrofit method applied to the actual bridge was to connect main girders to floor beams.

Thus, it was made clear that the system using both CBR and RBR brings about more reliable results than systems using either CBR or RBR only. A KBES should be constructed by combining CBR and RBR, because CBR's merits can cover RBR's demerits, and vice versa. In other words, RBR's results may be substantiated by CBR and CBR's results may be complemented by RBR.

5   CONCLUSIONS

A KBES for selecting repairing and retrofitting methods for steel bridges' fatigue damage was developed by using both CBR and RBR. In this paper, a policy for the development of the KBES was made clear. The plan of systematization to use both CBR and RBR was described based on such concept, the solution of the domain problems and the expansion of domain knowledge being considered. The system was divided into the retrieval / verification / modification process and the evaluation / restoration / storage process and its specifications were designed. The system, of which the main unit is the retrieval unit, was built based on the design, and the usefulness of the system was verified by using a damage case of an actual steel bridge.

The performance of this system can be enhanced to the practical level by expanding its case database and expanding and refining its rule database. In the future, this system will be expanded along its detailed design to materialize the processes of verification, modification, evaluation, restoration, and storage. Incidentally, this system was developed on SUN S-4/2, C++ being used.

REFERENCES

[MTK94]   Mikami, I., Tanaka, S., and Kurachi, A. Expert system with learning ability for retrofitting steel bridges, *Journal of Computing in Civil Engineering*, ASCE, Vol.8, No.1, pp.88-102, 1994.

[TMM94]   Tanaka, S., Mikami, I., and Maeda, H. Case-based reasoning system to select retrofitting methods for fatigue damages in steel highway bridge, *Proceeding of Application for civil engineering using fuzzy sets*, Fuzzy Society of Japan, pp. 37-42, 1994. (in Japanese)

[Koba92]   Kobayashi, S. Present and future of case-based reasoning, *Journal of Japanese Society for Artificial Intelligence*, Vol.7, No.4, pp.559-565, 1992. (in Japanese)

[RS89]   Rissland, E.L. and Skalak, D. B. Combining case-based reasoning and rule-based reasoning; A heuristic approach, *Proceedings of the 12th International Joint Conference on Artificial Intelligence*, AAAI, pp.524-530, 1989.

[GR91]   Golding, A. R. and Rosenbloom, P. S. Improving rule-based systems through case-based reasoning, *Proceedings of the 9th National Conference of Artificial Intelligence*, AAAI, pp.22-27, 1991.

[Miki82]   Miki C. Damage and maintenance of road bridges in USA, *Road*, Japan Road Association, pp.26-30, 1982. (in Japanese)

[Fisher84]   Fisher, J. W. *Fatigue and fracture in steel bridges - Case Study -*, John Wiley and Sons, Inc., 1984.

Multiple Domain Feature Mapping Utilizing Case-based Reasoning and Blackboard Technology

S.S. Lim, B.H. Lee

Gintic Institute of Manufacturing Technology, Nanyang Technological University,
71 Nanyang Drive, Singapore 638075
Email: gsslim@ntuvax.ntu.ac.sg

E.N. Lim, K.A. Ngoi

School of Mechanical and Production Engineering, Nanyang Technological University,
Nanyang Avenue, Singapore 639798

ABSTRACT

This paper describes a methodology for multiple domain feature mapping. Graph techniques are used to represent concisely the geometric semantics of features needed to support reasoning in feature mapping. Based on the proposed methodology, a knowledge-based feature mapping shell has been implemented using blackboard technology to facilitate the integration of different reasoning techniques as required by the methodology. A case-based reasoning (CBR) system is employed to represent past experiences related to feature mapping. Such experiences are crucial for determining emergent features together with the relevant parameters. The use of CBR simplifies the identification of possible emergent features in a specific mapping situation.

1. Introduction

The engineering intent of the geometry of a design can be concisely represented as features [Sha92]. A feature-based design representation can encompass form, tolerances, material, functionality or assembly characteristics. Two main approaches have been proposed for the derivation of feature-based representations, namely feature recognition and feature-based design. The first approach attempts to extract domain-dependent feature information from a geometric model using algorithms [WCJ93], rule-based expert systems [HA84] or graph-based heuristics [JC88]. The process can be interactive or fully automatic. The second approach, feature-based design, employs features directly in a design. This approach is effective for capturing knowledge associated with a design during the design process itself rather than relying on complex extraction or recognition algorithms to re-create a feature-based model later.

Since interpretation of features tends to be domain specific, a product design can often be described with different sets of features depending on the target domain. While it is possible to construct a design with features that are directly related to the functionality of a product, these features can be meaningless to downstream applications such as process planning or design for assembly (DFA) analysis. The need to transform a feature-based representation from one application domain to another, therefore, arises. This process, known as feature mapping, feature transformation, feature conversion or feature transmutation [Pra93], is the focus of discussion in this paper.

Further information on various feature recognition methodologies can be found in [Woo88]. Please refer to [LLL95] for detailed information on existing feature mapping methodologies.

2. The Feature Mapping Methodology

In this research, an initial feature model is constructed using neutral features or design features. Neutral features, called Feature Oriented Generic Shapes (FOGS), represent generic shapes with information about form but without any non-geometric semantics. A feature modeling system with FOGS has been developed using the Wisdom Systems Concept Modeller [Con92]. It is fully integrated with a geometric kernel called CV DORS, as provided by Concept Modeller. Design features can be customized from FOGS by including more specific knowledge such as the functionality pertaining to an application domain.

The proposed feature mapping approach consists of three stages of reasoning, namely:

i) filtration and selection;
ii) construction of an enhanced representation; and
iii) matching.

In the first stage, an initial feature-based design model is systematically searched to determine collections of faces that form certain patterns of interest to a particular application domain. The search is constrained by utilizing the information contained in the FOGS or design features of an initial design. The relationships among these features can provide some clues about faces that are meaningfully related.

In contrast, the feature recognition technique suffers a major drawback since feature information is not easily available in an initial model. A large amount of knowledge is often needed to establish meaningful relationships between faces for a target domain in an unconstrained search space. Failure to identify related faces will render the whole recognition process useless in later stages. It is also impossible to extract non-available information from an original pure geometric model. Such information includes tolerances, geometric constraints and material properties. The proposed filtration and selection stage improves the opportunity of finding a solution.

Next, an enhanced representation of the selected faces is constructed to complement the underlying boundary representation (Brep) and constructive solid geometry (CSG) techniques, which cannot be easily employed for the determination of emergent features and symbolic reasoning in feature mapping. With the enhanced representation, information relating to the collection of faces and their specific relationships as identified in the first stage, can be captured explicitly. Such explicit information is critical to the subsequent matching process. The enhanced representation is described further in the next section.

In the final stage, the enhanced representation is matched against a library of pre-defined cases within the system to identify a similar feature. The case library effectively functions as a repository of mapping knowledge that allows the relevant parameters of emergent features to be easily extracted. It contains not only the descriptions of individual features in a target domain, but also the descriptions of collections of faces that can be decomposed into more than one feature. This

extension is particularly important for mapping complex features.

3.   Graph Representation Of Feature Semantics

Two types of feature semantics have been identified in this research to support feature mapping. The first type, geometric semantics, is defined by a face connectivity graph (FCG). A FCG is an augmented face-edge semantics network that allows the geometric knowledge of a feature to be uniquely specified. Information such as the types of geometric edges and faces constituting a particular feature, and their inter-relationships, can be effectively represented by a graph. With richer semantics, it complements existing solid modeling techniques, such as Brep and CSG. Figure 1 illustrates a FCG describing the necessary conditions for identifying a rectangular slot geometry in a concise manner. The FCG technique has been implemented in a prototype feature mapping shell, described in Section 4.0. Please refer to [LLL95] for further information on its derivation.

Non-geometric knowledge of features constitutes the second type of semantics represented within the system. The knowledge associated with non-geometrical attributes is useful for representing semantics that differentiate between geometrically similar but functionally different features in a particular domain. Differentiating factors can include feature dimensions, locations, material properties or machining processes needed. Non-geometric semantics is also required for features that cannot be represented conveniently with geometric information alone. An undercut feature in an injection molding is an example of such features. It is formed when placement of other features hinders the ejection of a component from a mold. It emerges as a result of interaction between the direction of mold closure and the regions protruding from other features in that direction. The shape of an undercut is therefore determined by factors that make it impossible to associate a form to its definition.

The explicit representation of both geometric and non-geometric knowledge facilitates feature validation, and avoids confusion in its interpretation. In this respect, the explicit representation of feature semantics is important to feature mapping. Figure 2 summarizes the types of semantics that define a feature. The two primary types constitute a deep model of a feature, providing a basis for feature reasoning and mapping in a feature modeling system. The enriched semantics will further provide a reference model for determining whether feature interaction will render a feature invalid.

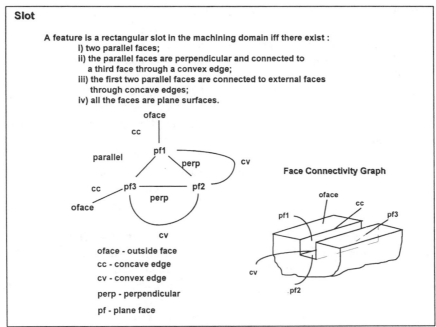

Figure 1 Definition of a slot and its face connectivity graph.

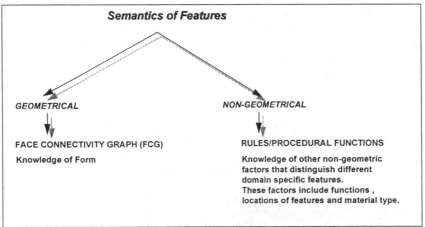

Figure 2 Semantics of features.

4. Implementation Of A Knowledge-Based Feature Mapping Shell

A feature mapping shell has been implemented to demonstrate the viability of the proposed feature mapping methodology. The blackboard architecture [Nii88] has been employed to support the needs of multiple knowledge representations and reasoning techniques in deriving solutions during feature mapping. These techniques include case-based reasoning [Pu93], graphical methods, geometric reasoning and algorithmic analytical methods. As shown in Figure 3, the blackboard has been structured into four independent layers acting as a global repository for information required by the methodology. Three functionally independent knowledge sources (KS) are attached to the corresponding layers of the blackboard. These are the:

i)   Filtration and Selection Knowledge Source (labeled 'Identifier' in the figure);

ii) Face Connectivity Graph Constructor Knowledge Source ('FCG Constructor');

iii) Matcher Knowledge Source ('Matcher').

Appropriate control knowledge can be incorporated into the control shell of the blackboard, allowing it to activate a suitable KS in response to information posted onto the blackboard system. The type of information and the layer onto which it is posted are considered dynamically during the activation of a KS.

The top layer of the blackboard contains a primary feature model that is essentially a tree structure with leave nodes representing neutral features or design features. The functionality of each KS is described in the following subsections.

4.1 Filtration and Selection Knowledge Source

This KS systematically examines an evaluated boundary model in the first layer, grouping together related faces which may form features in a target application domain. By using the heuristics shown in Table 1 to guide filtration and selection tasks, neutral features or design features that correspond directly with process features can be mapped easily. This type of mapping is known as identity mapping. Faces of original features are compared with faces in an evaluated boundary model. Those residing in the same space are grouped together. In the case of feature interactions, faces are grouped together based on the conditions of remnant faces in an evaluated boundary model. Interactions may establish the size and identity dependencies among features, further decompose a feature, or invalidate the identity of original design features.

More complex conjugate mapping requires further analysis to determine faces that should be grouped together. A procedure has been developed to handle simple cases involving conjugate mapping. The related faces that have been identified are posted onto the second layer of the blackboard for the next stage of reasoning.

4.2 FCG Constructor Knowledge Source

This KS constructs an enhanced description of the relationships among faces which have been grouped together during filtration and selection. Face Connectivity Graphs (FCGs) are employed to explicitly capture information and relationships such as convex edges, concave edges, parallelism and perpendicularity between faces.

The FCG Constructor KS systematically examines all grouped faces and edges between them to extract relevant information about their types and relationships as described above. A FCG is implemented as structures with pointers. Figure 4 gives an illustration of the various component structures used to represent a feature graph. A structure named Feature_Graph is constructed for each FCG at the top level. This structure contains components that point to a list of all *nodes*, a list of all graph *edges* and a list of features or case that the graph represents. *A_Node* is created to represent a face while *An_Edge* maintains its relationship with another node. All the *edges* and *nodes* are kept at the top level feature graph structure to facilitate their ease of access.

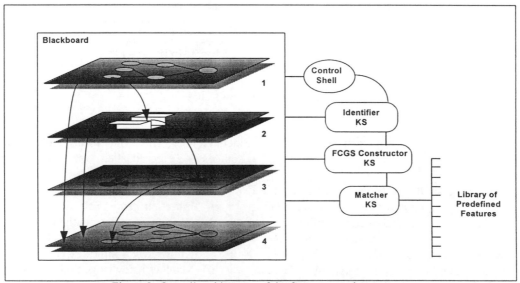

Figure 3 Overall architecture of the feature mapping system.

Table 1   Procedure for filtration and selection without conjugate mapping

Procedure Filter&SelectWithoutConjugate
(1) select a feature from the feature model as parent feature.
(2) determine whether any other feature interacts with the selected feature.
(3) if no feature interaction exists, then identify and group all the faces in the evaluated boundary model that belong to the original FOGS of the features.
(4) if there are interactions,
for each feature that interacts with the parent feature, perform the following checks :
(4.1) when an inner-loop is created on a face of the parent feature, or when additional edges are created on the outer-loop of a face, then set up dependency information between this feature and the parent feature
(4.2) when this feature sub-divides the faces of the parent feature into two or more faces, then group all the faces belonging to this feature and the parent feature together. Repeat steps 2 and 4 recursively with this feature as the parent feature.

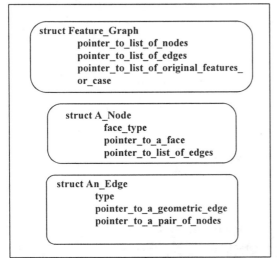

Figure 4 Components of structures developed for representing feature graphs

4.3 Matcher Knowledge Source

The Matcher KS determines the corresponding cases from a library of pre-defined cases that are potential matches to a given FCG. As described in Section 2, the case library contains pre-defined FCGs of individual features and their associated knowledge in a target application domain. In addition, graphs that describe the semantics of faces created from interaction and the presence of features that require conjugate mapping are also maintained in the system.

Each case library is always associated with two types of knowledge. These are subsets of non-geometric semantics. The first type, matching knowledge, determines a match between a given FCG and a particular case in the library. The second, known as decomposition knowledge, is responsible for determining emergent features from a collection of related faces, and for extracting the relevant parameters. No restriction is

imposed on their implementation; rules, methods and procedural techniques can be used.

A case-based reasoning (CBR) system is employed to index pre-defined cases. Since it uses information from an initial feature model to select potential cases for matching, the search space is significantly reduced in the initial phase. An exhaustive matching of potential cases is then performed with the matching knowledge. When a match is found, its decomposition knowledge is activated to extract the relevant parameters. A schematic diagram of the Matcher KS is shown in Figure 5.

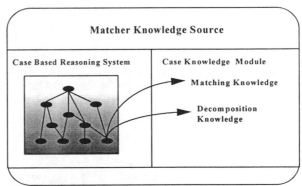

Figure 5 Schematic diagram of the Matcher KS

The CBR system has been developed based on Schank's dynamic memory concept [Sch82]. A hierarchy of Memory Organization Packages (MOPs) in the memory represents knowledge related to classes of events. Each MOP in the hierarchy can have specializations that are more specific versions of itself. A MOP instance is then used to represent a specific case in the memory. The CBR system uses the hierarchy links to get from one MOP to another. Abstraction links between MOPs provide a network connecting very specific instances at a lower level to very abstract general knowledge at the top. Each MOP can also have 'roles' and 'fillers' representing its distinct characteristics. The MOP-based abstraction hierarchy facilitates the retrieval of previous cases and the addition of new cases into the memory. For further details, please refer to [RS89].

The mapping shell has been further customized with knowledge to allow the mapping of generic shapes used in a design into the corresponding machining features for process planning. The CBR system contains cases that capture past mapping experiences related to individual features as well as those dealing with feature interaction and conjugate mapping. A subset of the abstraction hierarchy for indexing cases involving feature interactions is shown in Figure 6. The figure shows the 'roles' and

'fillers' of MOPs, where 'fillers' are related to other parts of the abstraction hierarchy.

Nearest neighbour matching is a technique that allows the retrieval of cases by comparing a collection of weighted features in the input case to cases in the library. The technique can be implemented within the Matcher Knowledge Source to find cases whose features match the input case the closest. Such a technique has been implemented within ReMind, a commercial cased-based reasoning development shell. Please refer to [Rem92] for details on the implementation of the nearest neighbour matching technique.

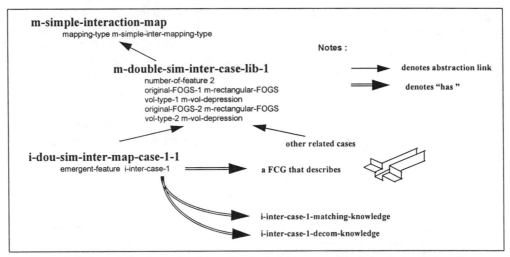

Figure 6   Subset of the abstraction hierarchy for indexing cases involving feature interactions.

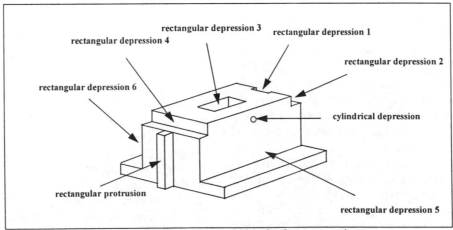

Figure 7   An example part design for feature mapping.

Table 2 Results of feature mapping for the example part shown in Figure 7.

	Neutral features used in example part	Emergent machining features after mapping
a	rectangular depressions 1 & 2	2 through steps & 1 through slot
b	rectangular depression 3	1 rectangular pocket
c	rectangular depression 4	1 through step
d	rectangular depression 5	1 through step
e	rectangular depression 6	1 through step
f	rectangular protrusion	2 through steps
g	rectangular depression	2 through holes

4.4 An Example

Figure 7 shows an example part with 8 neutral features designed using FOGS in the prototype system to test the customized mapping shell. The emergent machining features have been identified as listed in Table 2. It can be seen that conjugate mapping is necessary for the sole rectangular protrusion in the part, while interaction occurs between rectangular depressions 1 and 2.

5.0 CONCLUSIONS

A feature mapping methodology has been developed to handle feature interactions and conjugate mapping more effectively than other existing techniques. The methodology has been implemented within a prototype feature mapping shell that can be easily customized for different application domains by including appropriate cases in a case library. Further research is still needed to extend its conjugate mapping capability.

The identification of different stages of reasoning has contributed to an improved understanding of the issues and mechanisms involved in feature mapping. In order to represent the diverse types of knowledge required by the methodology, different artificial intelligence technologies, including blackboard technology and case-based reasoning, have been adopted within the prototype feature mapping shell.

REFERENCES

[Con92] The Concept Modeller, *Concept Modeller, Release 2.0, Reference Manual*, Wisdom Systems, Pepper Pike, Ohio, USA, 1992.

[HA84] Henderson, M.R. and Anderson, D.C., Computer Recognition and Extraction of Form Features: A CAD/CAM Link, *Computers in Industry*, Vol. 4, No. 5, Elsevier Science Publishers B.V., pp. 329-339, 1984.

[JC88] Joshi, S. and Chang, T.C., Graph-based Heuristics for Recognition of Machined Features from a 3-D Solid Model, *Computer Aided Design*, Vol. 20, No. 2, Butterworth & Co. (Publishers) Ltd., UK, pp. 58-64, 1988.

[LLL95] Lim, S.S., Lee I.B.H., Lim L.E.N. and Ngoi, B.K.A., Multiple Domain Feature Mapping: A Methodology Based on Deep Models of Features, *International Journal of Intelligent Manufacturing*, Vol. 6, Chapman and Hall, UK, pp. 245-262, 1995.

[Nii88] Nii, H. Penny, Introduction, *Blackboard Architecture and Applications,* Jagannathan, V., Dohiawala, R., Baum, L.S. (eds.), Academic Press, Inc., Boston, USA, 1988.

[Pra93] Pratt, M.J., Automated Feature Recognition and its Role in Product Modeling, *Geometric Modelling*, Farin, G., Hagen, H., Noltemeir, H. (eds.), Computing Supplementum 8, Springer Verlag, Austria, pp. 241-250, 1993.

[Pu93] Pu, P., Introduction: Issues in case-based design systems, AI EDAM, Vol. 7, No. 2, Academic Press Limited, UK, 1993.

[RS89] Riesbeck, C.K. and Schank, R.C., *Inside Case-Based Reasoning*, Lawrence Erlbaum Associates, New York, USA, 1989

[Rem92] ReMind, Developer's Reference Manual, Cognitive Systems, Inc., USA, 1992.

[Sch82] Schank, R.C., *Dynamic memory: A theory of reminding and learning in computers and people*, Cambridge University Press, UK, 1982.

[Sha92] Shah, J.J., Features in Design and Manufacturing, *Intelligent Design and Manufacturing*, edited by Andrew Kusiak, John Wiley & Sons, Inc., pp. 39-71, 1992

[WCJ93]   Wang, M.T., Chamberlain, M.A., Joneja, A., Chang, T.C., Manufacturing Feature Extraction and Machined Volume Decomposition in a Computer-Integrated Feature-Based Design and Manufacturing Planning Environment, *Computers In Industry*, Vol. 23, Elsevier Science Publishers B.V., pp. 75-86, 1993.

[Woo88]   Woodward, J.R., Some Speculation on Feature Recognition, *Computer-Aided Design*, Vol. 20, No. 4, Butterworth & Co. (Publishers) Ltd., UK, pp. 189-196, 1988.

PLANT FAULT DIAGNOSIS BY INTEGRATING
FAULT CASES AND RULES

Yoshio Nakatani*   Makoto Tsukiyama*   Tomo-omi Wake**
Mitsubishi Electric Corp.
*Industrial Electronics & Systems Lab.   **Power & Industrial Systems Center
1-1, Tsukaguchi-Honmachi 8, Amagasaki,   1-1-2 Wadasaki cho, Hyogo ku, Kobe,
Hyogo 661, Japan   Hyogo 652, Japan
E-Mail: nakatani@soc.sdl.melco.co.jp

ABSTRACT

This paper proposes a method of combining diagnosis rules and fault cases to support computer-aided fault diagnosis of plants. This method helps users to maintain and use the rules and the cases by representing them in a unified format. The rules and the cases are used by analogical retrieval. The rules and the cases are represented in two modes: the natural language/index mode. The latter is made from the former by abstraction. The natural language mode is used for the detailed explanation, while the index mode for easier understanding of the contents. The effectiveness of this method is demonstrated through the prototype system for the control and maintenance computer of the water treatment plant. The customer support engineers, accepting the claims from the plant operators, abstract the keywords and input them into the system. The system diagnoses by searching for the relevant rules and cases.

1 INTRODUCTION

This paper proposes a method of combining diagnosis rules and fault cases to support fault diagnosis of plants. When the plant operators observe unusual symptoms in the plant, they generally take the following procedures: (1) observation of the symptoms, (2) reasoning of the causes, (3) decision and execution of the countermeasures, (4) evaluation of the results, (5) writing a trouble-shooting report. The steps from (1) to (4) are repeated until the satisfactory evaluation can be obtained in the step (4). The steps (2) and (3) are based on the trouble-shooting manuals and the trouble-shooting cases in the past. The cases are stored in the form of paper as the trouble-shooting reports. The operators sometimes ask advice of customer support engineers of the plant makers.

One major problem on this kind of diagnosis is that it is difficult to retrieve the relevant pages from the paper form reports and manuals in case of emergency. In order to solve this problem, many methods have been proposed. We can classify them into four kinds of method: model-based, rule-based, case-based, and combination method.

(1) Model-based method: It diagnoses the faults based on causal models of the plant actions [e.g., DW87]. Only the predictable faults in the design phase can be explained. The external environment of the plant is difficult to be completely modeled.

(2) Rule-based method: It is based on general rules about the plant [e.g., CM84]. A rule consists of a condition-action pair. When the same symptoms as the observed symptoms are found in a condition of a rule, its action is thought to be the cause. If the same symptoms as the observed symptoms cannot be found in any rules, diagnosis is impossible. Generality of the rules makes it difficult to consider characteristics of each plant.

(3) Case-based method: It diagnoses the faults based on the trouble-shooting cases in the past [e.g., KF94]. A case consists of the symptoms, the causes, the countermeasures, and so on. When the similar symptoms to the observed symptoms are found in a case, the case is adopted. It is impossible to obtain the cases from the newly designed plants. Diagnosis is difficult when there is no very similar cases.

(4) Combination method: It combines two or more methods above [e.g., GR91, Man+93]. The mechanism is needed to use the different types of representation and usage of the model, the rules, and the cases.

We adopt the combination method of the rule-based method and the case-based method. The following sections explaines our method and demonstrates its effectiveness.

2 CHARACTERISTICS

The customer service engineers use this system to process the claims from the plant operators through the telephones. The characteristics of our method are as follows:

(1) It uses the rules and the cases complementarily. The rules are collected from the trouble-shooting manuals, and the cases from the trouble-shooting reports. The cases are collected from many plants which have similar composition in order to collect sufficient number and variety of cases. We call both the rules and the cases the "diagnosis knowledge."

(2) A rule and case have the same representation including the symptom, cause, and countermeasure columns in order to diagnose the fault by itself. These two kinds of knowledge are used by analogical retrieval. This makes it easy to maintain the rule base and the case base, and the rules can be used flexibly.

(3) We adopt two modes of representation of the diagnosis knowledge: the natural language mode and the index mode. There are two reasons. First, the simple representation is required because the diagnosis knowledge must be easily understood for quick solution of the claims, while the detailed representation is required because the users often want to understand the cases which are written by the other users. The natural language mode is for detailed understanding and the index mode is for quick understanding. Second, the continual addition of the new cases to the case base causes delay of case retrieval and difficulty of case maintenance. Our system must keep all cases because it is required to maintain the trouble-shooting reports. By forgetting the verbose index mode cases, we can keep retrieval speed.

(4) When the user inputs symptoms as the keywords, the system searches for the rules and the cases, in parallel, whose symptoms are most similar to the keywords. The index mode knowledge is retrieved in default, and the user can refer to the natural language mode knowledge to select the best one of the retrieved knowledge.

(5) Different from the general framework of case-based reasoning, our method does not adapt the retrieved cases to match for a new faults, because it is not easy to acquire knowledge of how to adapt them. Our alternative is to provide the relevant information of the cases, which includes the equipment data, their plans and photos, and the figures and the videos of the fault cases, along with the diagnosis knowledge.

(6) The system can abstract new rules from the cases (Figure 1). After repairing a fault, the user writes a trouble-shooting report as a natural language mode case. Next, the user create the index mode case by extracting the indices from this new case. This is done interactively by using the natural language processing tool. The indices are organized into the index templates including such attributes as the troubled equipment, the troubled part, and the content of the trouble. The system abstracts the troubled equipment name by one level and creates the index mode rule. The user manually decodes this into

the natuarl language mode rule. Some rules are implemented initially from the trouble-shooting manual as the natural language mode rules, and the index mode rules are created from them by the same method of creating the index mode cases.

(7) The system can diagnose the complex faults caused by multiple causes.

We propose a method to solve the problems described above. Our idea is to combine the rules and the cases to complement each other. Any combination methods neither mix the rules and the cases nor retrieve them by analogical retrieval.

3 USAGE OF THE SYSTEM

When the user receives the claims from the plant operator, he/she extracts the keywords based on their own knowledge, and inputs them into the system through keyword templates of the same format as the symptom of the index mode knowledge (Figure 2). The user can select the multiple faults mode, assuming that the faults are caused by the multiple causes. The system searches for the index mode knowledge which have the similar symptoms to the keyword templates.

When there are many rules and cases of the same similarity, the system gives priority to the cases than the rules, to the cases of the same plant than the cases of the other plants, to the cases of the more frequently occurring fault of the same plant, and to the more recent cases of the same frequency. As for the rules that are abstracted from the cases, it is verbose to list up the original cases. So the system does not present the original cases. For this purpose, the abstracted rules store the list of the original cases.

When the user selects one of the best matching knowledge, the system presents its causes and countermeasures. After the fault is fixed, the user writes and stores a natural language mode case.

The system uses the natural language processing tool to extract the words for the index mode case from the symptoms, the causes, the countermeasures of this new case. The user selects the best words and fills the templates of the symptom, the cause, and the countermeasure with them. The user manually fills out the blanks. When there is already the same faults occured at the similar equipments, the system abstracts the troubled equipment names by one level based on the dictionary and creates a index mode rule.

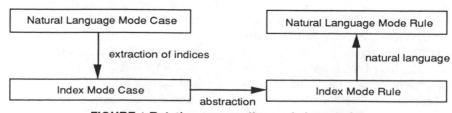

FIGURE 1 Relation among diagnosis knowledge

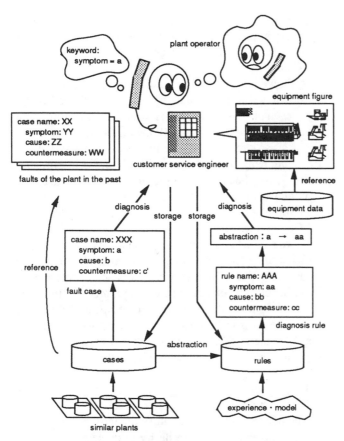

FIGURE 2 Usage of this system

The user manually decodes the index mode rule into the natuarl language mode rule.

4 ARCHITECTURE

This section shows the components of our framework.

4.1 Subsystems

Figure 3 shows the software architecture of our system. It consists of six kinds of databases and seven kinds of knowledge units.

Databases:

(1) Natural language mode knowledge — It stores the rules and the cases of the natural language mode.

(2) Index mode knowledge — It stores the rules and the cases of the index mode.

(3) Hyperscript — It organizes the words used to represent the contents of the index mode knowledge. It is also used as a dictionary for natural language processing.

(4) Parts-interference database — It stores the existence of influence from one equipment to another, such as causal relation. These data are induced mainly from the connections between equipments, partly from the experiences of the engineers.

(5) Statistical database — It stores statistical analysis of the faults occured at each plant. It involves fault frequency of each equipment and time series changes of the faults. It is used to decide the priority among the retrieved diagnosis knowledge.

(6) Equipment database --- It stores the specifications, the plans, and the other data of the plant components.

Knowledge unit:

(1) Knowledge retrieval unit —- It provides the keyword template to search for the relevant diagnosis knowledge. From this template, the user can refer to the diagnosis knowledge and the various plant information.

(2) Natural language mode case editor — It provides the templates to create, delete, and modify the natural language mode case.

(3) Index mode case editor — It provides the templates to create, delete, and modify the index mode case.

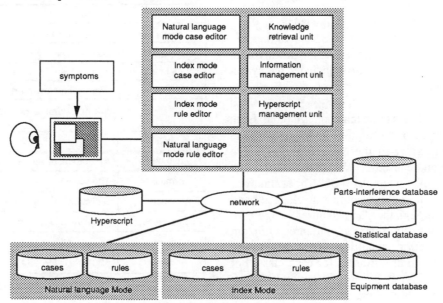

FIGURE 3 Software architecture of this system

(4) Index mode rule editor — It provides the template to support creation of the index mode rule from the index mode case, and deletion and modification of the index mode rule.

(5) Natural language mode rule editor — It provides the template to support creation of the natural language mode rule from the index mode rule, and deletion and modification the natural language mode rule.

(6) Hyperscript management unit — It manages the hyperscript by providing the template for adding, deleting, modifying the words. From this template, the user can refer to the diagnosis knowledge and the various plant information.

(7) Information management unit — It receives all requests from the various templates to show the diagnosis knowledge and the various plant information, and presents them in their proper form (e.g., text, video).

4.2 Natural Language Mode Knowledge

The natural language mode case consists of nine attributes including the case ID, the plant name, the order name, the fault occurrence date, the engineer name, the symptom column, the cause column, the countermeasure column, and the memo. The symptom, the cause, the countermeasure, and the memo can be written in the natural language. An example of the symptom column is "The sealing water of the No.2 washing sludge pump was clogged and caused the sludge to leak through the gland."

The natural language mode rule consists of six attributes including the rule name, the symptom column, the cause column, the countermeasure column, the memo, and the case list (Figure 4). The case list is the list of original cases of the rule. The fault equipment name of the natural language mode rule is more abstract than the case.

4.3 Representation of Index Mode Knowledge

The index mode case and rule have the same format, except that the case involves the case ID and the rule involves the rule name and the case list. Both includes the symptoms, the causes, and the countermeasures.

(1) Symptoms

A fault generally includes more than one symptom. Each Symptom is written in one symptom template. The example "The sealing water of the No.2 washing sludge pump was clogged and caused the sludge to leak through the gland" is written in two templates of "The sealing water of the No.2 washing sludge pump was clogged" and "The sludge to leak through the gland is caused" (Figure 5). The user first classifies each symptom into three symptom classes: (abnormal) action, condition, and location. The template provides different attributes for each class. For example, when the class is "abnormal action," the attributes consist of the class, the equipment, the part, the kind of action, the cause number, and the case ID (or the

rule name). The kind of action includes the emergency stop, the impossibility of operation, and so on. The cause number is for the multiple causes.

(2) Causes

A fault can be caused by multiple causes. Each cause is written in a cause template. The user classifies each cause into seven classes: inferior part, loss, wrong setting, wrong operation, foreign matters, unknown, and the others. When the class is "inferior part," the attributes consist of the class, the equipment, the part, the kind of inferiority, the cause number, and the case ID (or the rule name). The user gives any integer to each cause template and the corresponding symptom templates and countermeasure templates.

(3) Countermeasures

More than one countermeasure may be executed against each cause. Each countermeasure is written in one countermeasure template. The user classifies each countermeasure into nine classes: diversion, exchange, modification, addition, buy, removal, cleaning, reporting, and the others. When the class is "modification," the attributes consist of the class, the equipment, the part, the kind of modification, the cause number, and the case ID (or the rule name).

4.4 Hyperscript
(1) Concept definition

The hyperscript serves to unify the terminology used by the users, decide on the similarity of concepts, and extract the indices.

Attribute	Example
case ID	FB92####
plant name	xxxxxx
order name	BV45XXX
date	920806
engineer	ooooo
symptoms	The sealing water of the No.2 washing s ...
causes	Sufficient amount of water was not supplied ...
countermeasures	The sleeve was replaced with a new one ...
memo	The Teflon-immersed gland packing is ...

(a) Example of Natural language Mode Case

Attribute	Example
rule name	#152: Leak of sludge through the gland
symptoms	The sealing water of the washing sludge pump ..
causes	Sufficient amount of sealing water was not ...
countermeasures	The sleeve is to be replaced with a new one ...
memo	......
case list	FB92###1, FB92##13, FB94###7

(b) Example of Natural language Mode Rule

FIGURE 4 Example of natural language mode knowledge

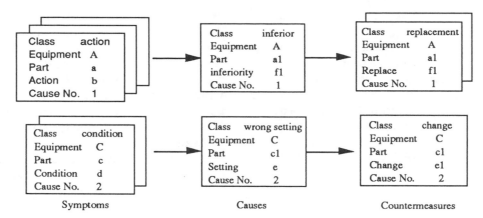

FIGURE 5 Templates of Index mode knowledge

Relation among concepts is represented as a network [Tul72]. We analyze the relation further and adopt two relation identifiers: the "relation type" and the "viewpoint." With these, we can define the relation type T and the viewpoint V of a concept C1 and a concept C2 in the form: [C1, T, V, C2], which means that a concept C2 has the relation type T with a concept C1 from the viewpoint of V.

Four relation types are considered based on the analysis of the relations [e.g., JSAI 90]: "an instance of," "a kind of," "a part of," and "has a value of."

The viewpoint is important in defining similarity of concepts. Although a "daily report" has the same relation type of "a kind of" with a "form" and a "creation of documents," the former sees the "daily report" from the viewpoint of its period, and the latter sees it from the viewpoint of object. The "an instance of" type do not have the viewpoint.

Figure 6 shows a part of the hyperscript. Each arrow is from the concept 1 to the concept 2.

(2) Indexing

When the system extracts the indices from the natural language mode case, it extracts words from the symptom, cause, and countermeasure columns of the case by natural language processing, and selects the concepts stored as C2s in the hyperscript.

(3) Abstraction

The system abstracts the concepts by using the relation type "a kind of" and the viewpoint "functional." An abstract concept A of the concept B is decided by searching the hyperscript for the concept definition [A, a kind of, functional, B].

5 DIAGNOSIS

5.1 Diagnosis in index mode

When the user specifies the keywords through the keyword template, the system searches for the diagnosis knowledge by similarity matching. The keyword template is the same as the symptom template without the cause number and the case ID. Similarity is calculated by how many keyword templates match the symptom templates.

Matching between the templates are judged by similarity matching between the values of the corresponding attribute.

When two values are different, Similarity between the values is judged. Two values are similar when they share the same C1, the same relation type, and the same viewpoint in the hyperscript. For example, "monthly report" can match "daily report." Similarity is weighted by the priority of the relation type. The priority is as follows:

an instance of > a kind of > a part of > has a value of.

The user can use the wild card as the keyword, which can match any values.

5.2 Diagnosis in natural language mode

In diagnosis in the natural language mode, the system checks if the keywords are involved in the symptom column of each knowledge. The wild cards are automatically added before and after each keyword. The cases that involve the keywords most are retrieved.

6 RULES FROM CASES

The index mode rules are created from the index mode cases by abstracting the equipment names in the symptom, cause, and countermeasure columns. When the case in-

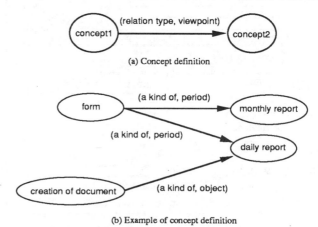

(a) Concept definition

(b) Example of concept definition

FIGURE 6 Concept definition

cludes more than one cause, the system creates a rule for each cause number.

After creation of the rule, the system checks if there is already the same rule or the rules having the same symptoms but different causes. This is done by searching for the rules by using a new rule as the keywords. If such rules are found, the user modifies the new rule because the retrieved rules have been already checked. There is, however, the case where the user must modify the retrieved rules. So the system provides the original cases of the retrieved rules for reference.

7 EVALUATION

7.1 Prototype system

To evaluate the effectiveness of this method, we implement the prototype system for the diagnosis of the control and maintenance computer of the water treatment plant on the workstation. The architecture of the system is the same as Figure 3. The knowledge unit is written in the C language, and the databases are managed by the conventional relational database management system. The natural language processing tool is also conventional [Nic94]. We collect the rules from the manuals. We collect more than two hundred latest cases from more than ten plants. Other previous cases are stored in another case base, which can be loaded if necessary.

Figure 7 shows an example of the display. The left top window is the keyword template, the right window is the retrieved natural language mode case, and the left bottom window is the picture of the relevant equipment referred from the left top window.

7.2 Example

We show an example of diagnosis by using the symptom "A report was not printed out by print request." The keywords are "Print of the report was requested" and "The report was not printed out." As a result, the first candidate

was a case that the date data management file was broken. This case had the same symptoms as the test fault, and occurred in the same plant. This case diagnosed the correct cause. The second candidate was a case that had the same symptoms but occurred in a different plant. The third candidate was a rule that the report was not printed out because a certain data management file was broken. According to this result, the system correctly diagnosed the fault.

7.3 Evaluation

So far we get the evaluation as follows:

(1) The rules are often referred to because about two hundred cases are not enough. The rules make the users feel easy.
(2) The users are anxious about the trivial differences among the plants. So we must provide more information about the differences
(3) Keyword extraction from the natural language mode case mentally costs much.
(4) Extraction of the keywords from the claims is not easy for the users. We must consider some help.

8 CONCLUSION

This version of the system is only the first step to the end goal. Now we are evaluating the system and there are some open problems. They include automatization of index extraction, more effective similarity matching, and so on.

REFERENCES

[CM84] B. Chandrasekaran and S. Mittal : Deep v.s. Compiled Knowledge Approaches to Diagnositic Problem-Solving. In Coombs, M.J. (Ed.), *Developments in Expert Systems*, Academic Press, 1984.

[DW87] J. de Kleer and B.C. Williams : Diagnosis with Multiple Faults, *Artificial Intelligence*, Vol.32, 1324-1330, 1987.

[GR91] A.R.Golding and P.S. Rosenbloom : Improving Rule-Based Systems through Case-Based Reasoning, *Proc. of AAAI-91*, 22-27, 1991.

[JSAI90] Japanese Society for Artificial Intelligence(Ed.): *Artificial Intelligence Handbook*, OHM, 1990 (in Japanese).

[KF94] S.T. Karamouzis and S. Feyock : Cases, Models, and Rules for Fault management: An Evaluation, *AAAI Workshop on Case-Based Reasoning*, 62-66, 1994.

[Man+93] M. Manago et al. : Induction and Reasoning from Cases, *EWCBR-93*, 313-318, 1993.

[Nic94] Nichigai Associates Incorporated: *NICE/UNIX Manual*, 1994.

[Tul72] E. Tulving : Episodic and Semantic Memory. In Tulving, E. and Donaldson, W. (Eds.), *Organization of Memory*, Academic Press, 1972.

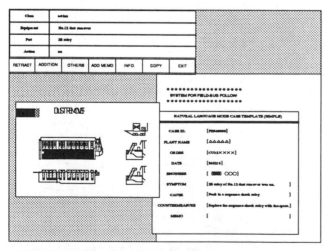

FIGURE 7 Example of display of the prototype system

Database

CHARACTERISTICS OF QUERY RESPONSES ON FUZZY OBJECT-ORIENTED DATABASES

Shyue-Liang Wang

Department of Information Management, Kaohsiung Polytechnic Institute
Kaohsiung, Taiwan, ROC
Email: slwang@nas05.kpi.edu.tw

ABSTRACT

This work describes an approach to model the characteristics of query responses on a fuzzy object-oriented database. The fuzzy object-oriented database, originated by Buckles et al., is based on similarity relation and subsumes the traditional object-oriented data model. The characteristics of query responses can be measured by membership values for each object in the query responses as well as overall entropy for the entire query responses. A new membership function is presented here in order to induce the membership value for each object. In addition, an approach for determining the entropy of query response is introduced.

Keywords: fuzzy object-oriented database, fuzzy query, similarity relation, fuzzy entropy.

1. INTRODUCTION

Object-oriented database technology emerged in recent years as an effective tool for complex applications such as CASE, CAD/CAM, and office automation [1,2,18]. In contrast to traditional database models, it provides powerful mechanisms to support both semantic abstractions and behavioral abstractions. It also combines the expressive power of both programming languages and database management systems. Nevertheless, these database systems basically deal with only precise and well-defined data. There are still many real world applications that require the management of imprecise, incomplete, and uncertain information.

Extending current database models to represent uncertain information has been investigated under many directions [3,15-17,22]. Two main approaches can be classified, (1) extending the data model utilizing fuzzy logic theory while minimizing the changes to query language, or (2) extending the query language to a fuzzy query format while maintaining the same data model. In extending the data models, an approach based on similarity relation [21] for both relational data model and object-oriented data model has been proposed [4-9,11].

The similarity relation based fuzzy object-oriented database has been extended to representing the fuzzy object attribute values [9] and modeling the fuzzy class hierarchy [11]. In this work, we describe an approach to model the characteristics of the query responses on this similarity-based fuzzy object-oriented data model. The characteristics will be examined from two aspects. The first aspect is to obtain a membership value for each object in the query response. These values can be used to evaluate which objects best match the specified query. The second aspect is to obtain an entropy value for the entire query response. The entropy value can be used to establish the confidence in the results of queries.

We will first review the similarity-based object-oriented data model originated by Buckles et al. in section 2. The characteristics of query responses will be modeled in section 3. A simple example illustrating these characteristics will be given in section 4 and followed by a brief discussion in section 5.

2. THE FUZZY OBJECT-ORIENTED DATA MODEL BASED ON SIMILARITY RELATION

A similarity-relation-based fuzzy object-oriented data model is an extension to an object-oriented data model that permits (1) class/object attribute values to be fuzzy predicates and numbers, and (2) class/object and class/subclass hierarchies to be fuzzy hierarchies.

The domain of an attribute, *dom*, is the set of all values the attribute may take, irrespective of the class it falls into. The range of an attribute, *rng*, is defined as the set of allowed values that a member of a class, i.e., an object, may take for the attribute and in general dom $\supseteq$ rng. For instance, assume that *salary* is an attribute and the domain is between 0 and 1,000,000. If there is a class Employee, the range of *salary* for this class may be 0 to 200,000, which is most appropriate for the employees.

Unlike the regular attribute domains, a simple extension is made on these domains. For each attribute domain D, a similarity relation s is defined over the domain elements:

$$s : D \times D \to [0, 1].$$

A similarity relation is a generalization of an equivalence relation in that if a, b, c $\in$ D, then s is

reflexive: $s(a, a) = 1$
symmetric: $s(a, b) = s(b, a)$
transitive: $s(a, c) \geq \max\{\min(s(a, b), s(b, c))\}$
                for all b $\in$ D

The similarity relation introduces different degrees of similarity to the elements in each domain and this is one mechanism for the representation of "fuzziness" in this fuzzy object-oriented data model.

Another mechanism for modeling uncertainty in the fuzzy object-oriented database is to accommodate uncertainty in class hierarchies. For instance, objects which are members of a class may have a degree of membership or a class may not be contained in a superclass but may be only somewhat of a subclass.

To model object-class relationships, a membership function is proposed [11]. Let C be a class, and

$$Attr(C) = \{a_1, a_2, ..., a_n\},$$
$$SClass(C) = \{C_1, C_2, ..., C_n\}$$

where C_1, C_2, ..., C_n are the direct superclasses of the class C. The membership of an object O_j in a class C with attributes a_1, a_2, ..., a_n, is defined as

$$\mu_c(O_j) = g[f(RLV(a_i, C), INC(rng(a_i)/O_j(a_i)))]$$

where $RLV(a_i, C)$ indicates the relevance of the attribute a_i to the class C, and $INC(rng(a_i)/O_j(a_i))$ denotes the degree of inclusion of the attribute values of O_j in the range of a_i pertaining to the class C. The degree of inclusion measures the similarity between a value (or a set of values) in the denominator with respect to the

value (or a set of values) in the numerator. It can be calculated by considering three cases as shown in [11]. The value of $RLV(a_i, C)$ may be supplied by the user or computed as in [10]. The function f represents the aggregation over the n attributes in the class and g reflects the type of link existing between an object and a class.

A query $Q(a'_1, a'_2, ..., a'_m)$, as defined by Buckles and Petry[6], is an expression of one or more factors combined by disjunctive or conjunctive Boolean operators: V_i op V_h op ... op V_k. Each factor V_j must be 1) a domain element a'_j, $a'_j \in D_j$, where D_j is an attribute domain of a class C, 2) a domain element modified by one or more linguistic modifier, e.g.

NOT, VERY, MORE-OR-LESS.

The linguistic modifiers such as VERY or MORE-OR-LESS are usually defined by primitive operations[20], e.g., VERY is treated as CON(F), and MORE-OR-LESS as DIL(F). The CON means concentration and DIL means dilation, where F is a fuzzy set and μ is a membership function over F:

$$\mu : F \to [0, 1]$$

and

$$CON(F) = \{ \mu^2(a) \mid a \in F \}$$
$$DIL(F) = \{ \mu^{1/2}(a) \mid a \in F \}$$

In fact, we have relaxed the condition of domain element a'_j in the query factor from singleton to a domain subset in another work [19], i.e., $a'_j \subseteq D_j$. As a simple illustration, consider a class called Market (see Fig. 1), which has three attributes, a_1 = Size, a_2 = Stability, and a_3 = Forecast. The attribute domains with their similarity relations for corresponding domain elements are shown in Fig. 2. Assume further that three objects O_1 = North, O_2 = Central, O_3 = South have been instantiated with attribute ranges shown in Fig. 3. Therefore, a trivial query would look like Q(Size = {Large}, Stability = {Very Stable}, AND Forecast = MORE-OR-LESS {Excellent, Good}), which is used to find objects with three attributes, Size, Stability, and Forecast with specified values.

Class Market
a1: Size
a2: Stability
a3: Forecast

Figure 1   Class Market

SIZE	Large	Medium	Small
Large	1	0.6	0.3
Medium	0.6	1	0.7
Small	0.3	0.7	1

SIZE={Large, Medium, Small}

STABILITY	Very Stable	Stable	Not Stable
Very Stable	1	0.8	0.2
Stable	0.8	1	0.6
Not Stable	0.2	0.6	1

STABILITY={Very Stable, Stable, Not Stable}

FORECAST	Excellent	Good	Moderate	Poor
Excellent	1	0.8	0.5	0
Good	0.8	1	0.6	0.2
Moderate	0.5	0.6	1	0.5
Poor	0	0.2	0.5	1

FORECAST={Excellent, Good, Moderate, Poor}

FIGURE 2 Similarity relations for attribute domains

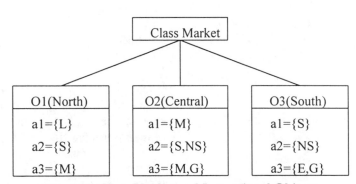

Figure 3 Class Market and Instantiated Objects

3. CHARACTERISTICS OF QUERY RESPONSES

The responses to a query may be a single object or a set of objects with possible attribute values not completely matching the query factors. It is therefore important to measure the degree of preciseness of the query responses. Two characteristics are considered here. First, if a set of responses is obtained, a measure should be provided to discriminate among all responses to indicate which response best matches the specified query. A membership function inducing membership value for each query response will be described. Second, if the responses match completely against the specified query or are partially matched, a measure should be provided to establish the confidence on the entire results obtained.

Entropy measures based on membership value of each object will be presented here.

Let $Q(a'_1, a'_2, ..., a'_n)$ be a query and objects O_j be the query responses with attributes $a_1, a_2, ..., a_n$. We define a membership function of O_j with respect to query $Q(.)$ as

$$\mu_Q(O_j) = g[f(RLV(a'_i, Q), INC(rng(a'_i)/O_j(a_i)))]$$

where $RLV(a'_i, Q)$ indicates the relevance of the query attribute a'_i to the query $Q(.)$, and $INC(rng(a'_i)/O_j(a_i))$ denotes the degree of inclusion of the attribute values a_i of O_j in the range of a'_i in the query $Q(.)$. The function f represents the aggregation over the n attributes in the object and g reflects the type of link existing between an object and the class it instantiates from.

The membership function defined above induces membership values for each object in the query response. It characterizes how well the object matches the query. In practice, we can consider the object with the highest membership value as the best matching to a query.

In order to measure how well the entire query response matches the query, an entropy measure for fuzzy sets proposed by Loo[14] is adopted. It has been shown that the Shannon-function entropy and Minkowski-distance entropy are special cases of this class of entropies[12].

DEFINITION Let $O = \{O_1, O_2, ..., O_n\}$ and define

$$E(O|Q) = F(\sum_{j=1}^{n} c_j f_j (\mu_Q(O_j))$$

Then $E(O|Q)$ is an entropy on $F(O)$, where $c_j \in R^+=[0, +\infty)$, f_j is a function from $[0, 1]$ to $[0, 1]$, $F(O)$ is the class of all fuzzy sets of O, such that
(1) $f_j(0) = f_j(1) = 0$;
(2) $f_j(u) = f_j(1 - u)$; $\forall u \in [0, 1]$;
(3) f_j is strictly increasing on $[0, 1/2]$;
and F is a positive increasing function from R^+ to R^+ with $F(0) = 0$.

For example, given any $w \in [1, \infty]$, let

$$f_j(\mu_Q(O_j)) = \mu_Q^w(O_j) \quad \text{for } \mu_Q(O_j) \in [0, 1/2]$$
$$(1 - \mu_Q(O_j))^w \quad \text{for } \mu_Q(O_j) \in (1/2, 1]$$

for all $O_j \in O$, $c_j = 1$, $\forall j$ and

$$F(b) = b^{1/w}$$

then Minkowski fuzzy entropy measures are obtained.

4. APPLICATION

To demonstrate how membership values for each object and fuzzy entropy for the query responses are calculated, the class Market and its attribute domains with similarity relations shown in Fig. 1 and Fig. 2 are used. Suppose we wish to select a market which is Large in Size, is Very Stable, and its forecast is MORE-OR-LESS Excellent or Good. A corresponding query would look like

Q(Size = {Large}, Stability = {Very Stable}, AND Forecast = MORE-OR-LESS {Excellent, Good})

Let $p_i = RLV(a'_i, Q)$, $q_i = INC(rng(a'_i)/O_j(a_i))$, and assume that [11]

$$\mu_Q(O_j) \quad = g[f(RLV(a'_i, Q), INC(rng(a'_i)/O_j(a_i)))]$$
$$= g[f(p_i , q_i)]$$
$$= g[p_i / RLV_{Max} * q_i]$$
$$= Max [p_i / RLV_{Max} * q_i]$$
$$1 \le i \le n$$

where p_i's are given by users to indicate the relevance of a'_i to the query $Q(.)$, and the inclusion q_i is induced by a similarity measure between a'_i and a_i. A similarity measure of two fuzzy sets is a measure that describes the similarity between fuzzy sets and is defined as following:

DEFINITION Let a'_i and a_i be subsets of fuzzy set D_i, i.e., $a'_i \subseteq D_i$ and $a_i \subseteq D_i$. The similarity measure is defined to be

$$SM(a'_i, a_i) = max \; s_i(x, y) \quad \text{for all } x \in a'_i \text{ and}$$
$$y \in a_i$$

where s_i is the similarity relation on attribute domain D_i. It can be shown that the similarity measure we define here satisfies the axiom definition of similarity measure of fuzzy sets given by Liu [13].

Assuming that MORE-OR-LESS is interpreted as DIL(F) and $p_1 = RLV(a'_1, Q) = 1$, $p_2 = RLV(a'_2, Q) = 0.5$, $p_3 = RLV(a'_3, Q) = 0.75$, and objects "North", "Central", and "South" in Fig. 3 are query responses. The object corresponding to "South", e.g., has membership value $\mu_Q = 0.75$ derived from

$$q_1 = INC(rng(a'_1)/O_3(a_1))$$
$$= SM_1(\{Large\}, \{Small\})$$
$$= max (s_1(Large, Small))$$
$$= 0.3$$
$$q_2 = INC(rng(a'_2)/O_3(a_2))$$
$$= SM_2(\{Very Stable\}, \{Not Stable\})$$
$$= max (s_2(Very Stable, Not Stable))$$

$$= 0.2$$

$$q_3 = INC(rng(a'_3)/O_3(a_3))$$
$$= SM_3(\{Excellent,Good\}, \{Excellent,Good\})$$
$$= max(s_3(Excellent,Excellent),s_3(Excellent,$$
$$Good),s_3(Good,Excellent),s_3(Good,Good))$$
$$= max (1, 0.8, 0.8, 1)$$
$$= 1$$
$$MORE\text{-}OR\text{-}LESS (q_3) = (1)^{0.5} = 1$$

$$\mu_Q(O_3) = Max [1 * 0.3, 0.5 * 0.2, 0.75 * 1] = 0.75$$

Objects "North" and "Central" can be calculated similarly with membership values 1 and 0.75 respectively. The highest membership value belongs to the object "North" which matches the first query factor completely and has the highest relevance to the query specified by the user.

To determine the fuzzy entropy of the query result, using Minkowski fuzzy entropy with w = 1, we obtain

$$E(O|Q) = [1 - 1] + [1 - 0.75] + [1 - 0.75] = 0.5$$

Note that Minkowski fuzzy entropy achieves its maximum value when each object has membership value of 0.5 and minimum value when $\mu_Q(O_j) = 0$ or 1, for every j, $1 \leq j \leq n$, i.e., $0 \leq E(O|Q) \leq 1.5$ in this example.

5. CONCLUSION

Extending object-oriented databases to accommodate uncertain information has been widely studied recently. Querying fuzzy databases may result in fuzzy responses. Characteristics of query responses should be well-investigated before their utilization. In this work, two characteristics of query responses have been examined. Membership values induced for each object in the query responses is proposed. These values suggest a mechanism to discriminate, among all responses, which response best matches a query. In addition, a fuzzy entropy measure based on these membership values is introduced to provide a mechanism for determining the overall confidence on the query responses.

A simple example illustrating these characteristics has been described. The results seem to confirm with our expectations in that objects best matching the query would lead to maximum membership values and minimum entropy. Further study should be carried out on applying these mechanisms to constructive applications as well as elaboration on query mechanism.

REFERENCES

1. Bertino, E., and Martino, L. Object-oriented database management systems: concepts and issues, IEEE Computer, April 1991, 33-47.
2. Bertino, E., et al., Object-oriented query languages: the notion and the issues, IEEE Trans. on Knowledge and Data Engineering, Vol. 4, No. 3, June 1992, 223-237.
3. Bosc, P. and Pivert, O. Fuzzy querying in conventional databases. In Fuzzy Logic for the Management of Uncertainty, Zadeh, L. and Kacprzyk, J. Eds, John Wiley, New York, 1992, 645-671.
4. Buckles, B.P. and Petry, F.E. A fuzzy representation of data for relational databases, Fuzzy Sets and Systems, 7, 1982, 213-226.
5. Buckles, B.P. and Petry, F.E. Fuzzy databases and their applications, in Fuzzy Information and Decision Process, Gupta, M. and Sanchez, E., Eds, North-Holland, New York, 1982, 361-371.
6. Buckles, B.P. and Petry, F.E. Information-theoretic characterization of fuzzy relational databases, IEEE Trans. SystemsMan Cyernet, 13, 1983, 74-77.
7. Buckles, B.P. and Petry, F.E. Extending the fuzzy database with fuzzy numbers, Inform. Sci. 34, 1984, 145-155.
8. Buckles, B.P. and Petry, F.E. Query languages for fuzzy databases, in Management Decision Support Systems using Fuzzy Sets and Possibility Theory, Kacprzyk, J. and Yager, R.R., Eds,Verlag TUV Rheinland, Cologne, 1985, 241-252.
9. Buckles, B.P., George, R., and Petry, F.E. Towards a fuzzy object-oriented model, Proc. of the NAFIPS-91 Workshop on Uncertainty Modeling in the 90's, 1991, 73-77.
10. Dutta, S., Approximate reasoning by analogy to answer null queries, Int. J. of Appr. Reasoning, 5, 1991, 373-398.
11. George, R., Buckles, B.P., and Petry, F.E., Modeling class hierarchies in the fuzzy object-oriented data model, Fuzzy Sets and Systems, 60, 1993, 259-272.
12. Klir, G.J. Where do we stand on measures of uncertainty, ambiguity, fuzziness and the like, Fuzzy Sets and Systems, 24, 1987, 141-160.
13. Liu, X. Entropy, distance measure and similarity measure of fuzzy sets and their relations. Fuzzy Sets and Systems, 52, 1992, 305-318.
14. Loo, S.G. Measures of fuzziness. Cybernetica, 20, 1977, 201-210.
15. Prade, H. Lipski's approach to incomplete information databases restated and generalized in

the setting of Zadeh's possibility theory, Inform. Systems, 9, 1984, 27-42.

16. Prade, H. and Testmale, C. Generalizing database relational algebra for the treatment of incomplete or uncertain information and vague queries, Inform. Sci. 34, 1984, 115-143.

17. Ruspini, E.H. Possibility theory approaches for advanced information systems, IEEE Computer., 9, 1982, 83-91.

18. Vossen, G. Bibliography on Object-Oriented Database Management, SIGMOD RECORD, Vol. 20, No. 1, March 1991, 24-46.

19. Wang, S.L., Measuring fuzzy query results in uncertainty management, Proceedings of the Six International Conference on Information Management, 1995, Taipei, Taiwan, 213-218.

20. Zadeh, L.A. PRUF - a meaning representation language for natural language. In Fuzzy Reasoning and its Application, E.H. Mamdani and B.R. Gaines, Eds., Academic Press, New York, 1981, 1-66.

21. Zadeh, L.A. Similarity relations and fuzzy orderings. Inform. Sci., vol 3, no. 1, Mar. 1971, 177-200.

22. Zemankova-Leech, M. and Kandel, A. Fuzzy Relational Databases - A Key to Expert Systems, Verlag TUV Rheinland, Cologne, 1985.

A KNOWLEDGE MEDIA APPROACH USING ASSOCIATIVE REPRESENTATION FOR CONSTRUCTING INFORMATION BASES

Harumi Maeda, Kazuto Koujitani and Toyoaki Nishida
Graduate School of Information Science,
Nara Institute of Science and Technology
8916-5 Takayama, Ikoma, Nara, 630-01 Japan
Email: harumi-m@is.aist-nara.ac.jp

Abstract

In this paper, we present a new approach based on *knowledge medium* using *associative representation* as a framework of information representation to gather raw information from vast information sources and to integrate it into information bases cost-effectively.

We then present a knowledge media information base system called CM-2 which provides users with a means of accumulating, sharing, exploring and refining conceptually diverse information gathered from vast information sources. We describe the system's four major facilities; (a) an *information capture facility*, (b) an *information integration facility*, (d) an *information retrieval facility* and (d) an *information refinement facility*. We discuss the strength and weakness of our approach by analyzing results of experiments.

keywords: associative representation, knowledge media, knowledge media system, CM-2, information base

1 Introduction

There exist various kinds of information sources around us. For instance, personal memoranda, research notes, hypertexts, image files and so on. Most of such information is conceptually diverse in the sense that its semantics is not rigorously defined.

In addition, widespread access to the Internet and WWW has led to a new phase in information acquisition. There already exist large scale information resources and they are increasing rapidly. We need to integrate a wide variety of information into personal information space from our point of view. However, it seems almost impossible to design a well-defined conceptual structure for organizing diverse information obtained from heterogeneous information sources.

In this paper, we present a new approach based on *knowledge medium* [Stefik86] using *associative representation* as a framework of information representation. The basic recognition behind this research is a trade-off between the benefit from conceptually well-structured information space and the cost for organizing information space. The more well-structured information representation becomes, the more useful it is for computational manipulation, however, the more expensive the cost of information acquisition becomes. Associative representation is a plain and weakly structured knowledge medium which is visible and manipulatable to humans and computers. We use associative representation to gather raw information from vast information sources and to integrate it into information bases cost-effectively.

We then present a knowledge media information base system called CM-2 which provides users with a means of accumulating, sharing, exploring and refining conceptually diverse information gathered from vast information sources. We describe the system's four major facilities;

- an *information capture facility* which helps users gather information from multiple information sources

- an *information integration facility* which allows users to integrate heterogeneous information into personal information space from the user's point of view

- an *information retrieval facility* which gives users access to multimedia information stored in the information base through associative indexing mechanisms

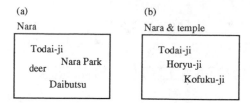

Figure 1: Example associations in CM-2

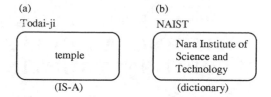

Figure 2: Example associations (special types)

- an *information refinement facility* which helps users reorganize the information base to be more comprehensive

We discuss the strength and weakness of our approach by analyzing results of experiments.

In what follows, we first describe the role of a plain indexing method using associative representation and overview the CM-2 information base system. We then present the system's four major facilities. Finally, we show experimental results and make discussion.

2 Associative Representation in CM-2 Information Base System

2.1 An Indexing Method Using Associative Representation

In this paper, we focus on *associative representation*, which allows the user to explore a way of articulating conceptually diverse information by aggregating conceptually relevant information. The basic entities of associative representation are (a) a *unit* which represents either a concept or an external datum, and (b) an *association* which connects a collection of key concepts (hereafter *keys*) with a collection of units (hereafter *values*) which is normally reminded by the given keys. Figure 1 shows a couple of associations. Figure 1(a) says that given a concept "Nara", one may be reminded of "Todai-ji", "Nara Park", "deer", and "Daibutsu". Figure 1(b) is an example of association with more than one key. It says that "Todai-ji", "Horyu-ji", and "Kofuku-ji" are reminded when "Nara" and "temple" are given as keys.

Users can define special types of associations to be used in information integration and refinement facility. Figure 2(a) is a "IS-A" relation which connects a unit with other units which are reminded as a class of the given unit. Figure 2(b) is a "dictionary" relation which can be used for translation.

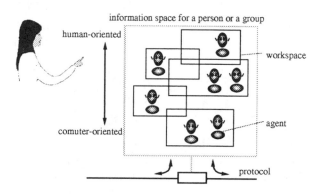

Figure 3: The Architecture of the CM-2 Information Base System

2.2 Overview of the CM-2 Information Base System

CM-2 [1] is a knowledge media information base system which provides users with a means of accumulating, sharing, exploring, and refining conceptually diverse information gathered from vast information sources.

CM-2 consists of a collection of information bases. Each CM-2 information base is possessed by an individual person or a group (Figure 3) and it consists of a collection of *workspaces* and *agents*. Each workspace provides a particular view of multimedia information stored in the information base.

Each agent manipulates information tasks and interacts with the user. The user or the agents in CM-2 can interact with other, or incorporate information from other kinds of information sources connected to the Internet.

Figure 4 shows an example screen of CM-2.

In what follows, we describe four major facilities of CM-2.

---

[1] "CM" stands for "Contextual Media" which stands for our long term theoretical research goal.

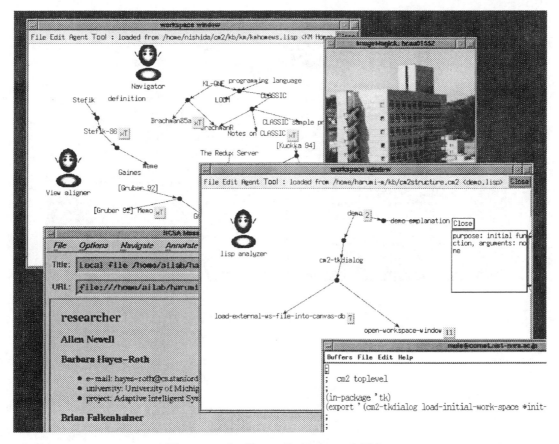

Figure 4: An Example Screen of CM-2

3   Information Capture Facility

The structure of information may well differ according to authors, and people might give different names to same concept and same names to different concepts.

Compare Figure 5(a) and (b). These are WWW pages of two famous AI researchers [2] [3] displayed on WWW browsers. They are different in structure and use of words. We need to obtain useful information from these kinds of diverse information sources.

Information capture facility helps users gather information from multiple information sources and generate CM-2 associations.

It is easy to generate associative representations from various information sources using a simple keyword extraction and text analysis algorithm.

We have implemented capture programs for digitized information, such as UNIX file system, program files written in Lisp, Nikkei newspaper full-text database and HTML documents on WWW.

We have also implemented those for capturing undigitized information, such as ideas. Users can input units using keyboard and connect them using mouse through workspaces.

3.1   Information Capture for WWW pages

We focus on capture facility for HTML documents on WWW. The general procedure of the facility is composed of the following steps.

1. generation of raw CM-2 units and associations

    (a) collection of HTML documents by analyzing URL

    (b) extraction of noun phrases and generation of units using Rule Based Tagger by Eric Brill [Brill68]

    (c) generation of associations by analyzing the structure of HTML documents

2. generation of "IS-A" relations and modification of units using domain knowledge

---

[2] http://www.cs.rochester.edu/u/james/
[3] http://www-ksl.stanford.edu/people/bhr/index.html

(a) James Allen's Home Page   (b) Barbara Hayes-Roth's Home Page

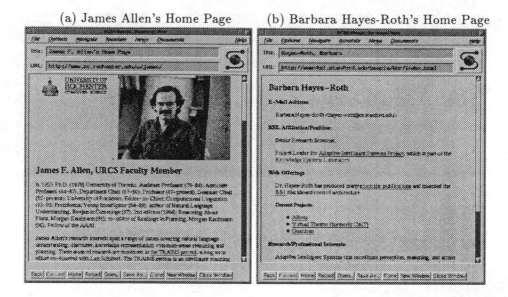

Figure 5: Example Home Pages

(a) extraction and generation of "IS-A" relations from units

(b) removal of unnecessary units and modification of associations

An example of domain knowledge used in step2 (a) is as follows:

> it is inferred that the class of a unit "James F. Allen" is a "people", because the unit's label contains "James" which is one of common English names.

Figure 6 shows the overview of the algorithm when a URL of James Allen's Home Page is given.

4   Information Integration Facility

Information integration facility allows users to integrate heterogeneous information into personal information space from the user's point of view. When a user input items for reorganizing information, it generates new associations in accordance with users request and itemizes them.

The following describes the general procedure of the facility.

1. unification of units and associations using several heuristics

2. generation of new associations according to user input

   (a) extraction of keys by path finding

   (b) extraction of values by path finding

   (c) generation of new associations

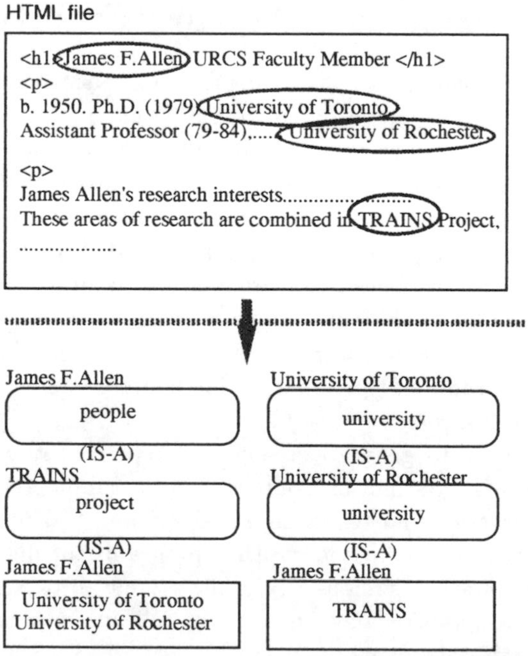

Figure 6: Information Capture Facility

Some examples of heuristics used in step1 are stated below.

- unification of units whose labels are the same

- unification of units referring to user dictionaries

- generation of associations between units when a unit's label is included in another unit's label

- unification of associations whose keys are the same

Let us think a case that when a user wants to know AI researchers and their contact information and projects concerning your research interest but there is no such database available. The user may search WWW pages about AI, read appropriate encyclopedia and organize the information using his/her knowledge. Information integration facility helps user's such process.

Figure 7 shows how the facility answer the following question against the sets of CM-2 associations which are mixtures of associations generated by information capture facility and those obtained by other information sources.

"Display a list of researchers and related projects concerning 'reasoning' ?"

Figure 8 and figure 9 illustrate example results of the facility.

5 Information Retrieval Facility

Information retrieval facility gives users access to multimedia information stored in the information base through associative indexing mechanisms. The system has three information retrieval facilities: (a) keyword search, (b) neighbor search and (c) intelligent associative retrieval. The rest part of this section describes neighbor search and intelligent associative retrieval.

5.1 Neighbor Search

Neighbor search enables users to search and display units which are linked to the selected unit by associations. For example, when an association shown in Figure 1 is given and the user selects "Nara", linked units such as "temple", "Todai-ji", "Horyu-ji" and "Kofuku-ji" will be displayed. Users can execute neighbor search

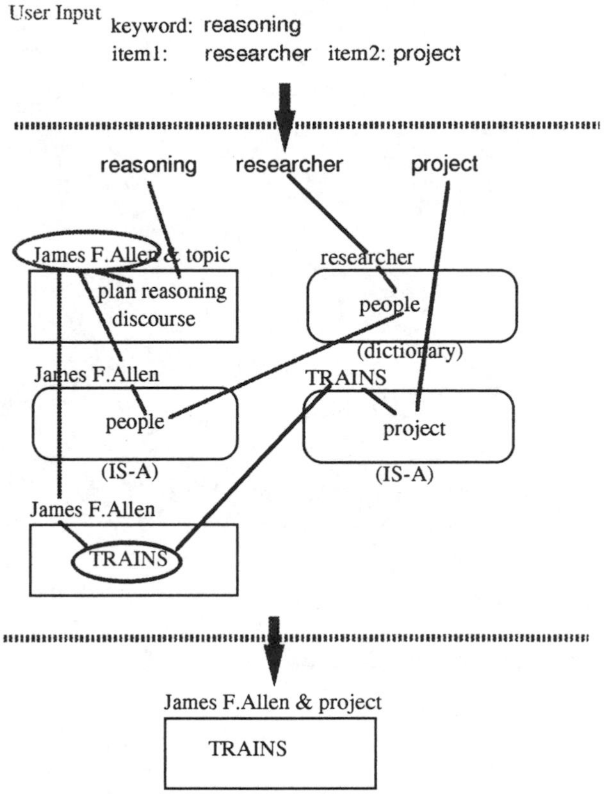

Figure 7: Information Integration Facility (step2)

by pressing buttons displayed nearby units on workspaces [4].

Neighbor Search causes a problem when there are too many values associated with the selected unit; it is very difficult to identify the displayed units. Figure 10 shows an example of workspace in such a case. To remedy this problem, we need more intelligent and dynamic search facility to obtain the desired information and it will be described in the next section.

5.2 Intelligent Associative Retrieval

Path finding is a powerful means of retrieving information, in particular when what is contained in an information base is structurally different from the presupposition of a given query.

Figure 11 illustrates how the algorithm works to answer a question:

"are there any places in Nara that are famous for rhododendron?"

---

[4]These buttons are displayed when units have some values undisplayed on workspaces. A number displayed within buttons describes the number of values of the unit.

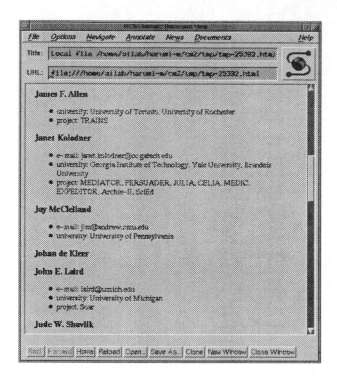

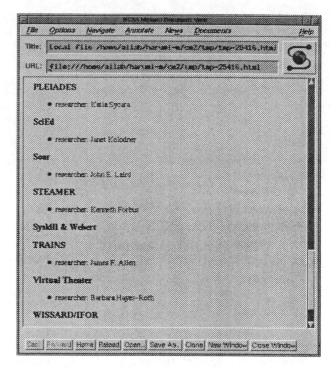

Figure 8: An Example Screen of Information Integration Facility 1 (A List of AI Researchers with their Related Information)

Figure 9: An Example Screen of Information Integration Facility 2 (A List of AI Projects with their Related Information)

Intelligent associative retrieval is based on the idea of "spreading activation" on semantic networks [Quillian68] and composed of the steps below.

1. extraction of units by analyzing the query

2. extraction of answers by path finding

3. display of the extracted units, answers, paths which links these concepts, and concepts which are useful for understanding the paths

Figure 12 shows the result of the intelligent associative retrieval.

6 Information Refinement Facility

Information refinement facility helps users reorganize the information base to be more comprehensive.

Compare two sets of associations in Figure 1 (a) and (b). The association in Figure 1(b) is more comprehensive and useful than that in Figure 1(a) because various kinds of entities are mixed up in the association in Figure 1(a).

We present a couple of heuristic techniques which will detect inappropriate associations from CM-2 information base and suggest a possible way of remedying them.

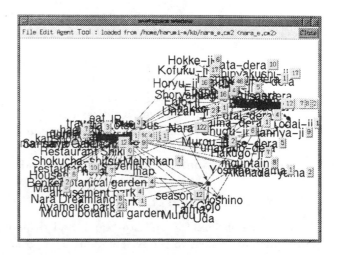

Figure 10: An Example Screen of Neighbor Search (Problematic Case)

Are there any places in Nara which are famous for rhododendron ?

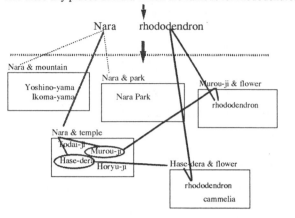

Figure 11: Intelligent Associative Retrieval

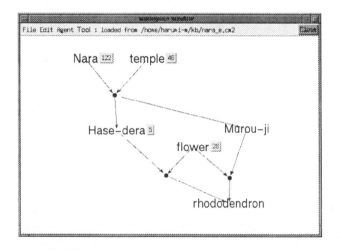

Figure 12: An Example Screen of Intelligent Associative Retrieval

6.1 Orthogonal Decomposition

Orthogonal decomposition attempts to decompose a given information base into coherent groups of associations, by analyzing how the user intersects associations. It is a technique of refining CM-2 information base using diagnosis rules shown in Figure 13.

An example of orthogonal decomposition is illustrated in Figure 14.

6.2 Analogical Refinement

Analogical refinement is a less efficient but more powerful technique for further elaborating information base based on the measurement of similarity.

Given a couple of non-orthogonal keys x and y, we

if
$\langle$ for concepts x and y:
$$V^*[\{x\}] \cap V^*[\{y\}] \neq V^*[\{x,y\}] \rangle$$
then
$penalty \leftarrow \dfrac{|(V^*[\{x\}] \cap V^*[\{y\}]) - V^*[\{x,y\}]|}{|V^*[\{x,y\}]|}$;

$suggestion \leftarrow$ "resolve the difference between $V^*[\{x\}] \cap V^*[\{y\}]$ and $V^*[\{x,y\}]$, by adding z to $V[\{x,y\}]$ if $z \notin V^*[\{x,y\}]$ and $z \in (V^*[\{x\}] \cap V^*[\{y\}])$ "

if
$\langle$ for two sets of concepts $\alpha, \beta, \ \alpha \subset \beta$:
$$\exists z[\, z \in V[\alpha] \wedge z \in V[\beta]\,] \rangle$$
then
$penalty \leftarrow \infty$;

$suggestion \leftarrow$ "remove z from $V[\alpha]$."

Figure 13: Diagnosis rules for orthogonal decomposition

define the similarity $\mathrm{Sim}[x,y]$ as shown in Figure 15. Based on that definition, we define the key similarity $\mathrm{Sim}^*[\alpha, \beta]$ between keys α and β as the sum of maximal pairwise similarities of units in α and β. Namely,

$$\mathrm{Sim}^*[\alpha, \beta]$$
$$= \max\left[\sum_{x \in \alpha} \max_{y \in \beta}[\mathrm{Sim}[x,y]], \sum_{y \in \beta} \max_{x \in \alpha}[\mathrm{Sim}[x,y]]\right]$$

For concepts x, y, and a threshold $\theta > 0$, we denote $x \sim y$ if $\mathrm{Sim}[x,y] \geq \theta$. Similarly, for keys α, β, and a threshold θ, $\alpha \sim \beta$ if $\mathrm{Sim}^*[\alpha, \beta] \geq \theta$.

The analogical refinement heuristic suggests to refine a CM-2 information base according to the following diagnosis rule:

if
$x \in V^*[\alpha]$,
$y \in V^*[\beta \cup \{a\}]$, and
$x \notin V^*[\alpha \cup \{a\}]$
then
$penalty \leftarrow \mathrm{Sim}[x,y] + \mathrm{Sim}^*[\alpha, \beta]$
$suggestion \leftarrow$ add x to $V[\alpha \cup \{a\}]$.

There are several interesting suggestions. For example, from "cherry blossom" $\in V[\{$ "Ikoma park" $\}]$ and, "cherry blossom" $\in V[\{$ "flowers" $\}]$, we obtained

"cherry blossom" $\in V[\{$ "Ikoma park", "flowers" $\}]$,

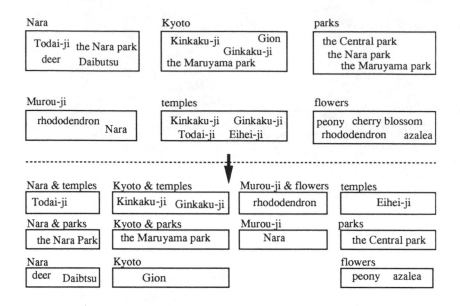

Figure 14: Orthogonal decomposition of CM-2 information base

Given a couple of non-orthogonal keys x and y, we define the similarity $\text{Sim}[x,y]$ between x and y from three perspectives and let it:

$$\text{Sim}[x,y] = \frac{\text{Sim}^{(a)}[x,y] + \text{Sim}^{(b)}[x,y] + \text{Sim}^{(c)}[x,y]}{3} \in [0,1].$$

$\text{Sim}^{(a)}[x,y]$ measures the similarity between x and y by comparing concepts in $V^*[\{x\}]$ and those in $V^*[\{y\}]$. The definition is as follows:

$$\text{Sim}^{(a)}[x,y] = \frac{1}{|V^*[\{x\}] \cup V^*[\{y\}]|}$$
$$\cdot (\ |\{z \mid z \in V^*[\{x\}] \wedge z \in V^*[\{y\}]\}|$$
$$+ |\{z \mid z \in V^*[\{x\}] - V^*[\{y\}] \wedge \exists u[\, u \in V^*[\{y\}] \wedge (K^*[z] \cap K^*[u] \neq \{\})]\}|$$
$$+ |\{z \mid z \in V^*[\{y\}] - V^*[\{x\}] \wedge \exists u[\, u \in V^*[\{x\}] \wedge (K^*[z] \cap K^*[u] \neq \{\})]\}|\ ).$$

$\text{Sim}^{(b)}[x,y]$ measures the rate of common keys of associations containing x and y as values. Namely,

$$\text{Sim}^{(b)}[x,y] = \frac{|\{z \mid z \in K^*[x] \wedge z \in K^*[y]\}|}{|K^*[x] \cup K^*[y]|}.$$

$\text{Sim}^{(c)}[x,y]$ measures the rate of keys orthogonal both to x and to y. Thus,

$$\text{Sim}^{(c)}[x,y] = \frac{|\{z \mid \langle z \text{ is orthogonal to } x\rangle \wedge \langle z \text{ is orthogonal to } y\rangle\}|}{|\{z \mid \langle z \text{ is orthogonal to } x\rangle\} \cup \{z \mid \langle z \text{ is orthogonal to } y\rangle\}|}.$$

Figure 15: Defining similarity between concepts

from which we in turn obtained

  "iris" $\in V[\{$"Ayameike park", "flowers"$\}]$

based on

  "iris" $\in$ $V[\{$"Ayameike park"$\}]$,

"Ikoma park" $\sim$ "Ayameike park", and

"cherry blossom" $\sim$ "iris",

as shown in Figure 16.

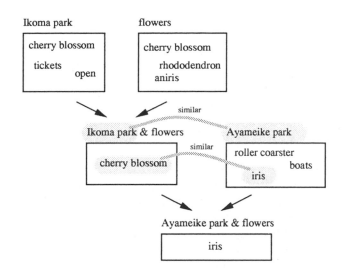

Figure 16: An interesting suggestion obtained by orthogonal decomposition and analogical refinement

7 Experiments

We have implemented CM-2 on top of Common Lisp and tcl/tk. We are evaluating CM-2 against accumulating various kinds of information such as research memoranda, technical surveys, regional guide, personal diary, and so on. Besides testing against these small examples and the examples described so far, we have made a couple of experiments with a nontrivial scale.

Experiment 1: Information Capture Facility We have gathered 22 WWW pages concerning AI researchers. CM-2 has extracted units about 7 classes such as researchers, topics and universities and generated associative representations. We have evaluated that 222 units out of 272 are appropriate.

Experiment 2: Information Integration Facility After having modified CM-2 associations generated in Experiment 1, we have tested information integration facility. A sample question is: "Display AI projects with their related researchers and universities." 60 units out of 72 are extracted properly.

Experiment 3: Information Retrieval Facility We have manually constructed a CM-2 information base for regional guide of Nara, Japan. It contains about 1,850 units and 870 associations. We have tested intelligent associative retrieval against the above information base. We have obtained appropriate 24 answers out of 30 queries.

Experiment 4: Information Refinement Facility We have tested orthogonal decomposition and analogical refinement against the information base constructed in Experiment 3. As a result of orthogonal decomposition, CM-2 has produced 212 revisions, about 80 of which have been found useful. Others are uninteresting. On the other hand, the analogical refinement heuristic has generated 65 suggestions, 20 of which are found useful.

8 Related Work and Discussion

The work reported in this paper is part of the **Knowledgeable Community** [Nishida93] project which aims to develop a computational framework of collecting, accumulating, systematizing, sharing, and creating knowledge by human-computer interaction. Crucial issues in the Knowledgeable Community are (a) knowledge media, (b) ontology and (c) agent-assisted mediation technology. We focus on knowledge media and have built an information base system using associative representation.

Our work is related to recent work on information gathering from heterogeneous sources on Internet ([Levy94],[Armstrong95],[Balabanovi'c95],[Li95]). Instead of focusing on the strategies and heuristics for information gathering, we concentrate on how to classify information obtained from multiple information sources and integrate it into personal information base.

The basic recognition behind this research is a tradeoff between the benefit from conceptually well-structured information space and the cost for organizing information space. The more well-structured information representation becomes, the more useful it is for computational manipulation, however, the more expensive the cost of information acquisition and integration becomes.

Our approach is to provide a framework of information representation with a low structural facilities and to facilitate raw information from vast information sources to be incorporated without much labor and gradually refined and elaborated as more insights are obtained.

How successful is our approach? Experiment 1,2 and 3 have ended up in very promising results. Members of our group have been able to use associative representation to accumulate and access varieties of information taken from vast information sources and access relevant information.

However, Experiment 4 shows that there is much

space to improve heuristics about information refinement facility, since the rate of useful suggestions from the heuristics seems to be low. To improve the quality of heuristics, we are currently looking at introduction of other kinds of heuristics and domain knowledge.

9 Conclusions

In this paper, we have proposed a new approach based on *knowledge medium* using *associative representation* as a framework of information representation to facilitate raw information from vast information sources to be incorporated without much labor.

We have presented CM-2 information base system which provides users with a means of accumulating, sharing, exploring and refining conceptually diverse information gathered from vast information sources. We have described the system's four major facilities: (a) an *information capture facility* which helps users gather information from multiple information sources, (b) an *information integration facility* which allows users integrate heterogeneous information into personal information space from the user's point of view, (c) an *information retrieval facility* which gives users access to multimedia information stored in the information base through associative indexing mechanisms, and (d)an *information refinement facility* which helps users reorganize the information space to be more comprehensive. We have discussed the strength and weakness of the method on the analysis of experimental results.

We have implemented a kernel of CM-2 on top of Common Lisp and tcl/tk. The system currently operates on the UNIX platform.

References

[Armstrong95] Robert Armstrong and Dayne Freitag and Thorsten Joachims and Tom Mitchell. A learning apprentice for the World Wide Web. In *Working Notes of the AAAI Spring Symposium on Information Gathering from Heterogeneous, Distributed Environments*, pages 6–12, 1995.

[Balabanovi'c95] Marko Balabanovi'c and Yoav Shoham. Learning information retrieval agents: Experiments with automated web browsing. In *Working Notes of the AAAI Spring Symposium on Information Gathering from Heterogeneous, Distributed Environments*, pages 13–18, 1995.

[Brill68] Eric Brill, Some Advance in Transformation-Based Part of Speech Tagging. In *Proceedings of the Twelefth National Conference on Artificial Intelligence (AAAI-94)*,1994.

[Levy94] Alon Y. Levy and Yehoshua Sagiv and Divesh Srivasava. Towards efficient information gathering agents. In *Working Notes of the AAAI Spring Symposium on Software Agents*, pages 64–70, 1994.

[Li95] Wen-Syan Li. Knowledge gathering and matching in heterogeneous databases. In *Working Notes of the AAAI Spring Symposium on Information Gathering from Heterogeneous, Distributed Environments*, pages 116–121, 1995.

[Nishida93] Toyoaki Nishida and Hideaki Takeda. Towards the knowledgeable community. In *Proceedings of International Conference on Building and Sharing of Very Large-Scale Knowledge bases 93*,pages 157–166. Japan Information Processing Development Center, 1993.

[Quillian68] M.R.Quillian, Semantic memory. In Marvin Minsky edition, *Semantic Information Processing*, MIT Press, 1968.

[Stefik86] Mark Stefik, The next knowledge medium, *AI Magazine*, 7(1):34–46,1986.

A QUERY PROCEDURE FOR ALLOWING EXCEPTIONS IN ADVANCED LOGICAL DATABASE

Kouzou Ohara, Noboru Babaguchi and Tadahiro Kitahashi

The Institute of Scientific and Industrial Research, Osaka University,

8-1, Mihogaoka, Ibarakishi, Osaka 567, Japan

Email : ohara@am.sanken.osaka-u.ac.jp

ABSTRACT

In this paper, we discuss a query procedure allowing exceptions in Advanced Logical DataBase, ALDB. It is capable of handling the incomplete knowledge containing exceptions as well as the complete knowledge. The rules in ALDB represented by logical formulas are translated into relational algebraic formulas by making use of the previously proposed *exc-representation*, and are evaluated by the method in Stratified Database, which stratify all predicates; however in ALDB it is not always possible. Thus we propose the *multiple answer set* as an answer in that irregular case and add a rule to acquire it to the original rule set. Since the additional rules allow us to stratify the predicates in the set of rules, we can get an answer by the ordinary method in any case.

1. INTRODUCTION

In recent years, Deductive DataBase(DDB)[Min87] has provoked a great deal of controversy as one of new database models. DDB is a logical database on the mathematical basis of the formal logic. The existing database such as Relational DataBase(RDB) can only handle given facts, whereas DDB can deal with not only facts but some rules about them. In DDB the rule and the facts are represented by first-order logic or related logics, and DDB can offer the conclusions derived from them as answers.

Toward constructing a large-scale DDB for data in real world, there is a problem that exceptions may be included in a data set, which will cause inconsistencies. Since arbitrary facts can be derived from an inconsistent theory in first-order logic, DDB will not work rationally under inconsistencies. Thus many efforts have been made to maintain database consistency[Nis93].

From the above background, we proposed an Advanced Logical DataBase, ALDB, which is capable of handling the incomplete knowledge containing some exceptions as well as the complete knowledge. The incomplete knowledge is not always true while the other is always true, and is interpreted by nonmonotonic logic. Hence the manipulation of a query in ALDB corresponds to nonmonotonic reasoning.

For ALDB we developed its query procedure using SLD-refutation [IB93] and its consistency maintenance mechanism[IBK95]. However the query procedure does not always work efficiently for a query which requires all answers satisfying each condition, because it has to produce a great deal of similar refutation trees with different constants. To improve the efficiency, we proposed a query procedure using relational algebra[1][Ull88,89] in ALDB [OBK95a]. Relational algebra regards some related data as a set called the *relation*, and the procedure includes no backtracking.

In the procedure using relational algebra, to decide the order to evaluate the relational algebraic formulas, all predicates are stratified as similar to Stratified Database(SDB)[Ull88,89], which is one of extensions of DDB. If the predicates are unstratifiable, the procedure can make no answer since it can not decide the order uniquely. This corresponds to a problem called *multiple extensions*[2] in nonmonotonic reasoning.

In this paper, we propose the *multiple answer set* as an answer in the case of multiple extensions, which consists of two kinds of answer sets: the *definite answer set* and the *indefinite answer set*. The elements of the former set are certainly correct as an answer, but those of the latter may be correct. To acquire these answer sets, we introduce an rule which defines the indefinite answer set. The rule set added the rule become stratifiable; it is evaluated by the ordinary method.

---

[1]The relational algebraic operators are the basic operators in RDB.

[2]An extension is a logical consequence in nonmonotonic logic.

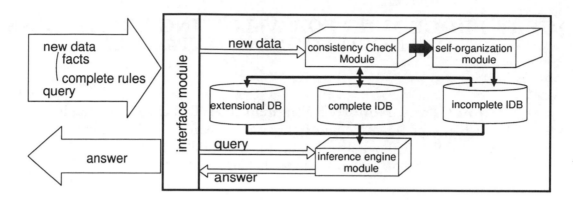

FIGURE 1   Basic architecture of ALDB

2. OUTLINE OF ALDB

The basic architecture of ALDB is shown in Figure1. ALDB has both extensional DB and intensional DB(IDB) same as DDB, which are sets of given facts and given rules, respectively. Furthermore in ALDB, IDB consists of both complete IDB and incomplete IDB, which are sets of the complete knowledge and the incomplete knowledge, respectively. When adding new data to ALDB, the consistency check module investigates whether inconsistencies occur due to it. If an inconsistency occurs, the self-organization module converts the complete knowledge causing it into incomplete knowledge containing exceptions. Thus ALDB can maintain its own consistency [IBK95]. The inference engine module produces an answer set excluded exceptions.

The knowledge representation in ALDB is based on Horn clauses without function symbols. A fact and a piece of knowledge are represented by a ground atom and which of the rules defined below, respectively.

Definition 1 ALDB allows the following three types of rules.

- *Complete-rule* (for the complete knowledge)

$$r \leftarrow s_1, \cdots, s_m \qquad (1)$$

This rule means that "the predicate r is true if the predicates $s_1, \cdots, s_m$ are true ".

- *Incomplete-rule* (for the incomplete knowledge)

$$r \Leftarrow s_1, \cdots, s_m, \qquad (2)$$

This rule means that "the predicate r is normally true if the predicates $s_1, \cdots, s_m$ are true ".

- *Restriction-rule* (definition of an inconsistency)

$$\perp \leftarrow s_1, \cdots, s_m \qquad (3)$$

This rule means that "it is inconsistent to satisfy the predicates $s_1, \cdots, s_m$ at the same time ".

The symbols, $\leftarrow$, $\Leftarrow$ and $\perp$ denote the logical implication, the special logical implication allowing exceptions and an invalid proposition representing an inconsistency, respectively.   □

We define the logical semantics of the incomplete-rule that "**an incomplete-rule entails its head if and only if its body is true, and the negation of its head is not proved**". From this definition, we can recognize it by means of general nonmonotonic logic such as default logic[Rei80], autoepistemic logic[Moo85]; ALDB is a logical database based on nonmonotonic logic.

3. RELATIONAL ALGEBRA IN ALDB

In this section, we discuss the translation of an incomplete-rule into a relational algebraic formula. Before moving to the main topic, a brief description will be necessary to the translation of the facts and the complete-rules[Ull88,89]. Here a *relation* is a set of instances which satisfy a certain predicate: an instance may be either a constant or a set of constants. The following definition is based on one in DDB.

Definition 2 Translations of the facts and the complete-rule into the relational algebraic formulas are defined as follows.

- Facts with same predicate name are a relation whose relation name is the predicate name.
- A complete-rule,

$$r \leftarrow s_1, \cdots, s_m \qquad (4)$$

is translated into the following formula.

$$R = \pi_{\boldsymbol{\alpha}}(\sigma_{\theta_1}(S_1) \bowtie \cdots \bowtie \sigma_{\theta_m}(S_m)) \qquad (5)$$

where $R, S_1, \cdots, S_m$ denote relations corresponding to predicates $r, s_1, \cdots, s_m$, respectively; π, σ and $\bowtie$ denote operators of relational algebra projection, selection and natural join, respectively; α denotes a set of variables in argument of r; and $\theta_i (1 \leq i \leq m)$ is a condition for each selection. $\square$

Although generally a relational algebraic formula can be translated into another equivalent formulas according to some laws such as the commutativity, we use equation(5) as a normal form in this paper.

In the following, predicate names in a rule are represented by lowercase characters, while relation names in a formula are denoted by uppercase characters.

3.1 Exception Representation

From the above logical semantics of an incomplete-rule, we define its intuitive interpretation as "**excluding exceptions from a result of evaluating it as a complete-rule**"[OBK95a]. Then as the relationship "exceptions of exceptions" can be regarded as a kind of hierarchy among exceptions, we call it *exception-hierarchy*. To obtain a set of exceptions, acquisition of its exceptions need to precede. It means that the exception-hierarchy represents an evaluation order for incomplete-rules.

To translate incomplete-rules into relational algebraic formulas and to evaluate them in proper order, we need to know their exceptions and the exception-hierarchy. However from previous knowledge representation we can get no information about them.

To solve these problems, we proposed a new representation, the *exc-representation*[OBK95a], in which a symbol "*exc*" means the exception, and the exception-hierarchy for a predicate r is recursively represented as $exc\text{-}r$, $exc^2\text{-}r$, $\cdots$; the predicate $exc^n\text{-}r(1 \leq n)$ denotes the exception of the predicate $exc^{n-1}\text{-}r$[3]. The superscript such as 2 in exc^2 is called the *degree*, and represents the exception-hierarchy explicitly. Accordingly the evaluation from an exc-predicate with a larger degree in order allows us to exclude all exceptions from an answer set. For a predicate $exc^n\text{-}r$, we call it the *exc-predicate*, and call r the *core* of exc-predicate.

Labeling the exc-predicates to the rules, which derive the exceptions of an incomplete-rules, allows us to solve the above problems. For example, let us consider the following set of simple rules about chemical property of some organic compounds.

---

[3]The predicates $exc^0\text{-}r$ and $exc^1\text{-}r$ are equal to the predicates r and $exc\text{-}r$, respectively.

$$\overline{electrolyte}(x) \Leftarrow organic\_compound(x) \qquad (6)$$
$$electrolyte(x) \Leftarrow carboxylic\_acid(x) \qquad (7)$$
$$\overline{electrolyte}(x) \leftarrow higher\_fatty\_acid(x) \qquad (8)$$
$$organic\_compound(x) \leftarrow carboxylic\_acid(x) \qquad (9)$$
$$carboxylic\_acid(x) \leftarrow higher\_fatty\_acid(x) \qquad (10)$$

The predicate symbol $\overline{electrolyte}$ represents "not electrolyte". These rules contain no information about both the exceptions of equations(6), (7) and the exception-hierarchy. However, representing equations(7), (8) as follows allows us to get the information about them.

$$[exc\text{-}\overline{electrolyte}(x)]electrolyte(x) \Leftarrow$$
$$carboxylic\_acid(x), \qquad (11)$$
$$[exc^2\text{-}\overline{electrolyte}(x)]\overline{electrolyte}(x) \leftarrow$$
$$higher\_fatty\_acid(x). \qquad (12)$$

Although we developed a method of acquiring exc-representation[OBK95b], we do not discuss it here, since it is irrelevant to the main subject.

3.2 Translation and Evaluation of an Incomplete-rule

From the above discussion, immediately we define the following translation of an incomplete-rule according to its intuitive interpretation. We assume that all rules which derive exceptions have already been labeled exc-predicate before the translation.

Definition 3 An incomplete-rule,

$$r \Leftarrow s_1, \cdots, s_m \qquad (13)$$

is translated into the following relational algebraic formula,

$$R = \pi_{\alpha}(\sigma_{\theta_1}(S_1) \bowtie \cdots \bowtie \sigma_{\theta_m}(S_m)) - exc\text{-}R \qquad (14)$$
$$\square$$

Similarly, an incomplete-rule with an exc-predicate is translated as follows.

Definition 4 An incomplete-rule with an exc-predicate,

$$[exc^n\text{-}p]r \Leftarrow s_1, \cdots, s_m \qquad (1 \leq n) \qquad (15)$$

is translated into the following relational algebraic formulas.

$$R = \pi_{\alpha}(\sigma_{\theta_1}(S_1) \bowtie \cdots \bowtie \sigma_{\theta_m}(S_m))$$
$$-(exc^{n+1}\text{-}P \cup exc\text{-}R) \qquad (16)$$
$$exc^n\text{-}P = \pi_{\alpha}(\sigma_{\theta_1}(S_1) \bowtie \cdots \bowtie \sigma_{\theta_m}(S_m))$$
$$-(exc^{n+1}\text{-}P \cup exc\text{-}R) \qquad (17)$$

If R is identified with P, exc-R in the above formulas is ignored. These formulas are selected with respect to a referenced predicate, that is, either r or exc^n-p.

$\square$

Also a complete-rule with an exc-predicate has two translations, which are selected similarly.

From the above discussion, equations (6)$\sim$(10) are translated as

$$\overline{EL}(x) = O\_CMP(x) - exc\text{-}\overline{EL}(x), \qquad (18)$$

$$exc\text{-}\overline{EL}(x) = C\_ACID(x) - $$
$$(exc^2\text{-}\overline{EL}(x) \cup exc\text{-}EL(x)), \qquad (19)$$

$$exc^2\text{-}\overline{EL}(x) = H\_F\_ACID(x), \qquad (20)$$

$$O\_CMP(x) = C\_ACID(x), \qquad (21)$$

$$C\_ACID(x) = H\_F\_ACID(x), \qquad (22)$$

where the relations, $EL(x)$, $O\_CMP(x)$, $C\_ACID(x)$ and $H\_F\_ACID(x)$ correspond to the predicates, $electrolyte(x)$, $organic\_compound(x)$, $carboxylic\_acid(x)$ and $higher\_fatty\_acid(x)$, respectively.

Note that equation(14) can be interpreted as the following rule.

$$r \leftarrow s_1, \cdots, s_m, \neg exc\text{-}r \qquad (23)$$

This rule is just identical to the rule in SDB, i.e. the rule with negative literal in its body. Accordingly we can evaluate the above formulas by the same method as SDB[Ull88,89]. Although we leave its details to the bibliography [Ull88,89], its basic idea is the stratification of predicates: predicates are stratified so that the strata of all predicates in the body of a rule are equal to or lower than the stratum of its head, and especially the strata of negative literals in its body should be lower than the stratum of its head.

The stratified predicates are evaluated in order of stratum: from lower to higher. In ALDB, the strata imply the exc-hierarchy; the larger the degree of exc-predicate is, the lower its stratum is[OBK95a]. Thus the evaluation in order of stratum reflects the evaluation order based on the degree mentioned in section3.1.

The result obtained by this ordinary method is certainly correct as an answer. Unfortunately all rule sets, or predicates are not always stratifiable. If unstratifiable, since the evaluation order is undecidable, we can make no answer. In ALDB, this corresponds to a problem called *multiple extensions* in nonmonotonic reasoning. In the next section, we discuss an answer set in that case and how to acquire it.

4. MULTIPLE ANSWER SET

4.1 Nixon Diamond Problem

Here we consider a famous problem called "Nixon Diamond"[Lif88] as a typical multiple extensions. We assume two pieces of knowledge, "Quakers are normally pacifists", "Republicans are normally not pacifists", represented as

$$pacifist(x) \Leftarrow quaker(x), \qquad (24)$$

$$\overline{pacifist}(x) \Leftarrow republican(x). \qquad (25)$$

These rules are unstratifiable, if exc-predicates are labeled to them as follows.

$$[exc\text{-}\overline{pacifist}(x)]pacifist(x) \Leftarrow quaker(x) \qquad (26)$$

$$[exc\text{-}pacifist(x)]\overline{pacifist}(x) \Leftarrow republican(x) \qquad (27)$$

The stratum of the head in equation(26), $s1$, has to be higher than the stratum of exc-$pacifist(x)$, $s2$, and similarly the stratum of the head in equation(27), $s3$, has to be higher than the stratum of exc-$pacifist(x)$, $s4$. In fact, $s2$ is equal to $s3$ and $s4$ is equal to $s1$. As a result, the number of strata is infinite.

We assume that Nixon is a Quaker and a Republican. Now we would like to know whether he is a pacifist or not. If we evaluate the equation(24) before the equation(25), we derive the fact "$pacifist(Nixon)$" from both it and the fact "$quaker(Nixon)$". Because of the conclusion, we can not derive the fact "$\overline{pacifist}(Nixon)$" from both the equation(25) and the fact "$republican(Nixon)$". If we reverse the evaluation order, we obtain the fact "$\overline{pacifist}(Nixon)$"and can not obtain the fact "$pacifist(Nixon)$". Namely there are two inconsistent extensions depending on each evaluation order. This is multiple extensions.

To decide whether he is a pacifist or not, we need to give priority among the above rules. For example, if we prioritize equation(24), then Nixon is a pacifist; otherwise not a pacifist. However since it is of great difficulty to decide such priority appropriately, we have no means to decide it uniquely. In fact, **we must consider all possible extensions**.

4.2 Definition of Multiple Answer Set

It is unreasonable to acquire all extensions in the case of multiple extensions, but ALDB should be able to make an answer appropriately also in that case. Now we consider an appropriate answer set in that case. First, as matter of course, the answer set should include the elements which are certainly correct as an answer; they are included in all extensions. For example, for a query "pacifist(x)?", these elements correspond to those who are not Republicans but Quakers:

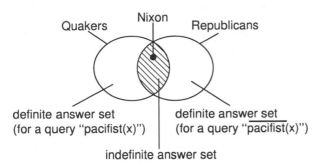

FIGURE 2   Definite answer set and indefinite answer set for Nixon Diamond

they are pacifists in any extension. We call a set of them the *definite answer set*.

Next, the answer set should include also the elements which may be correct as an answer. These elements have at least one extension including them. For the previous query, these elements correspond to those who are Quakers and Republicans like Nixon. We call a set of them the *indefinite answer set*. These two kinds of answer sets for Nixon Diamond are shown in Figure2. We call the answer set consisting of them the *multiple answer set*, which include all elements that are correct or may be correct as an answer. On the contrary, we call the answer set obtained by the ordinary method the *single answer set*.

For the elements of the indefinite answer set, we should add to each element a condition which make it correct as an answer. For the general discussion, let us consider the following rules.

$$\perp \leftarrow p_1(\boldsymbol{x}_1), \cdots, p_n(\boldsymbol{x}_n) \tag{28}$$

$$p_1(\boldsymbol{x}_1) \Leftarrow q_1 \tag{29}$$

$$\cdots$$

$$p_n(\boldsymbol{x}_n) \Leftarrow q_n \tag{30}$$

where each $\boldsymbol{x}_i(1 \leq i \leq n)$ is a set of variables, and each $\boldsymbol{q}_i$ is a conjunction of literals. We represent the set (or relation) defined by each conjunction $\boldsymbol{q}_i$ as S_i. It is noted that the intersection of sets $S_1, \cdots, S_n$ represents the indefinite answer set if there is no priority among equations(29)~(30). Because it is impossible to decide uniquely the truth-values of the predicates $p_1(\boldsymbol{x}_1), \cdots, p_n(\boldsymbol{x}_n)$ for an arbitrary element of the intersection. It means that multiple extensions exist.

For a query "$p_i(\boldsymbol{x}_i)$?", the number of extensions including a fact $p_i(\boldsymbol{a})$, where $\boldsymbol{a}$ is an instance of the indefinite answer set, is not always one. Because there are $2^{n-1} - 1$ combinations of the truth-values of the other predicates, and each combination corresponds to

one extension. Therefore the required condition need to be satisfied in all such extensions.

Moreover the form of the condition should be decided according to a given query uniquely. We should notice that in order to maintain database consistency, all of the facts $p_1(\boldsymbol{a}), \cdots, p_{i-1}(\boldsymbol{a}), p_{i+1}(\boldsymbol{a}), \cdots, p_n(\boldsymbol{a})$ should not be included in an extension including a fact $p_i(\boldsymbol{a})$; **at least one of them must be false.** This hold in all extensions including the fact $p_i(\boldsymbol{a})$. Fortunately the set of facts $p_1(\boldsymbol{a}), \cdots, p_{i-1}(\boldsymbol{a}), p_{i+1}(\boldsymbol{a}), \cdots, p_n(\boldsymbol{a})$ is decided uniquely according to both the query "$p_i(\boldsymbol{x}_i)$?" and the element of the indefinite answer set $\boldsymbol{a}$. Accordingly we represent an indefinite answer as

$$\boldsymbol{a}\langle p_1(\boldsymbol{a}), \cdots, p_{i-1}(\boldsymbol{a}), p_{i+1}(\boldsymbol{a}), \cdots, p_n(\boldsymbol{a})\rangle, \tag{31}$$

where $\langle p_1(\boldsymbol{a}), \cdots, p_n(\boldsymbol{a})\rangle$ represents a condition "if at least one of $p_1(\boldsymbol{a}), \cdots, p_n(\boldsymbol{a})$ is false". For the previous query "pacifist(x)?", we represent an indefinite answer for Nixon as

$$Nixon\langle \overline{pacifist}(Nixon)\rangle. \tag{32}$$

4.3   Acquisition of the Multiple Answer Set

In this section, we describe a relational algebraic formula to acquire the multiple answer set. Focusing on figure2, we can see that in fact neither quakers nor republicans are the exceptions of each other, and the exceptions of them are those who are quaker and republican. Since it is undecidable whether those who are quaker and republican are pacifist as mentioned above, we regard them as exceptions of both quakers and republicans. Generally, the elements of the intersection of the sets $S_1, \cdots, S_n$ are regarded as the exceptions of all rules, equations (29)~(30). Hence we introduce the following rule to define such exceptions.

$$[exc\text{-}p_1(\boldsymbol{x}_1), \cdots, exc\text{-}p_n(\boldsymbol{x}_n)] \leftarrow q_1, \cdots, q_n \tag{33}$$

Note that the rules of equations(29)~(30), and(33) are stratifiable; they can be evaluated by the ordinary way mentioned in the previous section.

For Nixon Diamond, we introduce the following rule.

$$[exc\text{-}pacifist(x), exc\text{-}\overline{pacifist}(x)] \leftarrow$$
$$quaker(x), republican(x). \tag{34}$$

The result of evaluating equations(24), (33) by the ordinary way is the definite answer set. However it is noted that the elements derived from the equation(33) are not only exceptions of equation(24) but the elements of the indefinite answer set. Now the predicates in the condition in equation(31) are the cores of exc-predicates in the label of the equation(33) without the predicate of the query. Accordingly we can make an indefinite answer set when acquiring the set of exceptions, if only add conditions to them, and can define the translation of the incomplete-rule to do it as follows.

Definition 5 The multiple answer set, P_i, for a query $p_i(\boldsymbol{x}_i)$ to equations(29)~(30), and(33) is acquired by evaluating the following relational algebraic formula.

$$P_i = (Q_i - exc\text{-}P_i) \cup \langle P_i \rangle \tag{35}$$

where Q_i is a result of evaluation of q_i, and $\langle P_i \rangle$ is a set as

$$\langle P_i \rangle = \{\langle a \rangle | a \in exc\text{-}P_i, \langle a \rangle = a \langle p_1(a), \cdots, p_{i-1}(a),$$
$$p_{i+1}(a), \cdots, p_n(a) \rangle \}. \tag{36}$$

□

In this definition, the set "$(Q_i - exc\text{-}P_i)$" corresponds to the definite answer set, and the set "$\langle P_i \rangle$" corresponds to the indefinite answer set. In fact, we need the operation different from relational algebra to make the elements of the indefinite answer set. However it is not important as it is easily realized.

In the translation of the incomplete-rule, if the rule with exc-predicate expressing its exceptions has the head like the equation(11), the translation is according to Definition3 or Definition4; otherwise like equation(34) according to Definition5.

In the result, we use the following formulas to acquire the multiple answer set for query "$pacifist(x)?$".

$$PAC(x) = (QUAK(x) - exc\text{-}PAC(x))$$
$$\cup \langle PAC(x) \rangle, \tag{37}$$
$$exc\text{-}PAC(x) = QUAK(x) \bowtie REPUB(x) \tag{38}$$

where the relations, $PAC(x)$, $QUAK(x)$ and $REPUB(X)$ correspond to the predicates, $pacifist(x)$, $quaker(x)$ and $republican(x)$, respectively. The equations(37), (38) are obtained by means of translating the equations(24), (34), respectively. The elements of the relation $PAC(x)$ are those such as equation(32).

5. CONCLUSIONS

In this paper, we have discussed a query procedure for allowing exceptions using relational algebra in ALDB. In the proposed method, by introducing both the multiple answer set and the rule to acquire it, we realized the query procedure which is able to derive answers from the unstratifiable predicate set. We believe that our approach is a novel attempt to multiple extensions. The result of this paper will be applied to the various knowledge base systems including rules containing exceptions independent of the domain.

We have two future problems: one is to introduce into our procedure some optimization like Magic sets [Ull88,89], and the other is to produce a hybrid query procedure based on both relational operations and SLD-refutation.

REFERENCES

[IB93]   G. Ido and N. Babaguchi. A Reasoning Procedure for Complete/Incomplete Rules, IPSJ, SIG-AI93-AI-86-3, pp.17-24(1993).

[IBK95] G. Ido, N. Babaguchi, and T. Kitahashi. Self Organization of Advanced Logical Database Handling Complete/Incomplete Knowledge, Journal of JSAI, Vol.10, No.4, pp.564-571(1995).

[Lif88] V. Lifschitz. Benchmark Problems for Non-monotonic Reasoning, Proc.2nd Workshop on Non-monotonic Reasoning, pp.202-219(1988).

[Min87] J. Minker. Foundations of Deductive Database and Logic Programming, Morgan Kaufmann (1987).

[Moo85] Robert C. Moore. Semantical Considerations on Non-Monotonic Logic, Artif. Intell., Vol.25, pp.75-94(1985).

[Nis93] S. Nishio. Knowledge Discovery in Very Large Database, J. IPS Japan, Vol.34, No.3, pp.343-350 (1993).

[OBK95a] K. Ohara, N. Babaguchi, and T. Kitahashi. Relational Algebraic Representation for Query Procedure in Advanced Logical DataBase Handling Incomplete Knowledge, Trans. IPS. Japan, Vol.36, No.6, pp.1433-1440(1995).

[OBK95b] K. Ohara, N. Babaguchi, and T. Kitahashi. Acquiring Hierarchic Structure from Knowledge Including Exceptions, Proc. 9th Annual Conference of JSAI, pp.61-64(1995).

[Rei80] R. Reiter. A Logic for Default Reasoning, Artif. Intell., Vol.13, No.1/2, pp.81-132(1980).

[Ull88,89] Jeffrey D. Ullman. Principles of Database and Knowledge-base Systems, two volumes, Computer Science Press(1988 and 1989).

Decision Support

DECISION SUPPORT SYSTEM FOR WATER-SUPPLY RISK MANAGEMENT

Xu Zongxue(*), Kawamura Akira(**), Kenji Jinno(**), Kazumasa Ito(*)
(*) AI Research Division, CTI Engineering Co., Ltd., Tokyo 103, Japan.
Email: xu@ctie.co.jp
(**) Department of Civil Engineering, Kyushu University, Fukuoka 812, Japan.
Email: kawamura@civil.kyushu-u.ac.jp

ABSTRACT

During the last two decades, a few of risk models have been developed and adopted. Recently, the technique of decision support system (DSS) has emerged as a potential approach for risk analysis. This paper describes how the risk models are integrated within a DSS and used to support the risk management in a practical water supply system. A decision support system (DSS) for risk management of water supply system is presented. An application is made in which the existing water supply system of Fukuoka region is analyzed. Given daily reservoir inflows, water demand data, and river flows, the potential water deficits and corresponding risk for different periods is analyzed. The conclusion is that the DSS techniques can improve the decision-making process by placing information in the hands of users, providing flexibility in the choice of operating policies and the presentation of results.

vances. To combine the powerful data management, interactive user interface and sophisticated simulation models into a single and convenient decision support system (DSS) has been a challenging appeal [KS93, BS94]. Because so many of problems that water management profession-als face require not only procedural mathematical simulation but also specialized knowledge and judgment, the potential appears high for the DSS technique to become a useful tool [SS89, NW+93, Wal93]. In this paper, a DSS for water-supply risk management is presented. The reliability, resiliency, vulnerability, and drought risk index (DRI) are used for identifying operational policies of water supply system. The preliminary conclusions may give some useful references for the operation of the water supply system in Fukuoka region.

1. INTRODUCTION

The progress in computer hardware and software technologies has been significant over the last three decades. The water management has received exceptional benefits from these ad-

2. SIMULATION MODEL FORMULATION

The most important component of a decision support system is the model base. The usefulness of the DSS always depends on the practical value of the mathematical model. Therefore, the first step in DSS development is to formulate and

FIGURE 1   Location of Fukuoka region

develop a suitable mathematical model.

Water Supply System of Fukuoka Region

Fukuoka region is a major political, economic and cultural center in Kyushu area, as shown in figure 1. The growing economics and urbanization over the last 20 years has drastically increased the water demand in this region. Due to inadequate underground water storage, both present and future water demands in this region will be primarily dependent on surface water sources. Due to the high seasonal variability of the precipitation and the steep channel slopes of most of the rivers and streams, over 10 reservoirs of various sizes have been constructed in Fukuoka region. They receive an average annual precipitation of about 1,800 mm. Over 60% of the precipitation falls during the period from April to August. The area of the water supply system is 1,156 km^2 and the total population is 2,02 million.

Simulation Model of Reservoir Operation

The simulation model of reservoir operation is a kind of technique to represent all the characteristics of the reservoir system by a mathematical description. A typical simulation model is simply a model that simulates the interval-by-interval operation of reservoirs with specified system characteristics, specified operating rules, and specified inflows at all locations during each interval.

The basic equation for reservoir simulation is the continuity equation, which records the reservoir storage at the end of each period and insures the conservation of mass among inflow, releases, spill and losses within each period,

$$\frac{dS(t)}{dt} = I(t) - Q(t) - R(t) - L(t) \qquad (1)$$

or in following discrete form,

$$S_t - S_{t-1} = (I_t - Q_t - R_t - L_t) * \Delta t, \\ t = 1,2,...,n \qquad (2)$$

Furthermore, the constraints on storage and release hold as well,

$$S_{t,\min} \le S_t \le S_{t,\max} \qquad (3)$$

$$Q_{t,\min} \le Q_t \le Q_{t,\max} \qquad (4)$$

in which $S_{t,\min}$ and $S_{t,\max}$ are the minimum storage and the capacity of reservoirs, respectively, $Q_{t,\min}$ and $Q_{t,\max}$ are the minimum and maximum flow which may be taken from reservoirs on the permitted water right.

Water-Supply Risk Model

Recently, risk has been studied intensively in water resources engineering field and many significant results have been achieved. Especially flood risk has been investigated in many literature [RR91, Xu93]. In this study, the analysis about the performance of water supply system focuses on system failure, i.e., the risk analysis for a water supply system. In which, the system performance is described from three different aspects: reliability, resiliency and vulnerability [JX+95].

Denote the state of system by random variable X_t at time t, which may be partitioned into two sets: S, the set of all satisfactory outputs, and F, the set of all unsatisfactory outputs. Assuming that the numerical indicator of the severity for the i-th failure is Se_i and the corresponding occurrence probability is p_i. Then the reliability, resiliency, and vulnerability may be expressed as follows,

$$\alpha = P\{X_t \in S\} \qquad (5)$$

$$\beta = P\{X_t \in S \, / \, X_{t-1} \in F\} \qquad (6)$$

$$\gamma = E\{Se\} \qquad (7)$$

Moreover, in order to easily diagnose different subsystems it is necessary to introduce an integrated index. In this study, the fourth criterion, drought risk index (DRI), as a linear weighted function of reliability, resiliency, and vulnerability, is introduced as,

$$\mu = w_1(1-\alpha) + w_2(1-\beta) + w_3\gamma \qquad (8)$$

where w_1, w_2 and w_3 are weights which need to be determined beforehand.

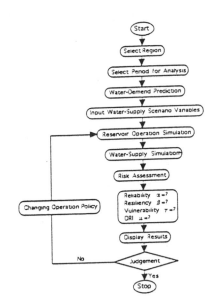

FIGURE 2   Flowchart of risk analysis

3. SYSTEM DEVELOPMENT

This DSS is an interactive, menu-driven consultation system designed as an advisory for the risk management of water supply systems. The flowchart of risk analysis is shown in figure 2. The process begins with the selection of region and period of interest for analysis. Then, a prediction model of water demand for the period of interest is retrieved from the model base. The water deficit in the period is calculated next. Finally, the system calculates the risk for different water supply scenarios in the region, corresponding to different water transfer ratios. For each alternative, the water deficit process can be calculated, thus the reliability, resiliency, vulnerability, and DRI are obtained. Once this has been done the risk curve is displayed, and the procedure terminates.

At the initial phase of development, this DSS is used to examine several advantages of this technique over the classical simulation approach. It mainly includes: (1) capabilities of the C language for numerical simulation; (2) merits of DSS techniques; (3) communication convenience of the DSS with managers. This DSS has been developed with Quick C and mainly consists of three modules: a system manager, a model base, and a data base. The system manager integrates other modules and controls the execution process of the system. It provides a friendly user interface, and aids in the visual comprehension of output used for water-supply risk management. The model base includes the simulation models related to the operation of water supply system. At the present level of development, only four kinds of models are included in this DSS: (1) "demd": a prediction model of water demand based on daily data; (2) "proc": a deterministic simulation model for calculating the reservoir storage and water deficit process on day-by-day basis; (3) "stat": a model to perform statistical analyses for water supply, water demand and water deficit process; (4) "risk": a model to calculate reliability, resiliency, vulnerability, and DRI based on the water deficit processes obtained by proc. The data base handles data storage, manipulation and extraction. It includes four kinds of data: water demand data, water supply data from three rivers, inflow data to eight reservoirs, and water taken from Chikugo river. The DSS is currently run on a NEC-9800 series computers and the compatible under MS-DOS 3.3 or later with a hard disk drive, VGA display device and laser printer. The system is developed by using Microsoft Quick C 2.0.

By providing decision support for water-supply risk management, the DSS is useful with some advantages, which may be summarized as follows: (1) the interactive menu-driven user interface guides users in carrying out risk analysis for water supply systems; (2) the color graphics for the model outputs supports an intuitive understanding of complex systems behavior; (3) the system may be used to risk management for complex water supply systems to determine optimal water-supply scenarios or assess consequences likely to occur due to the development of alternative water resources projects or the implementation of economic, environmental, institutional, or political con-

straints on a region's water resource development.

4. DSS APPLICATION: CASE STUDY

In this section, an application of the DSS in risk analysis for the water supply system of Fukuoka region is presented. As an example, eight reservoirs, Egawa, Terauchi, Minamihata, Sefuri, Magaribuchi, Zuibaiji, Kubaru, and Nagatani, together with Chikugo river and other three small rivers, currently supply water to Fukuoka city. The supply of water from these sources may fall shortage of demand in future, due to the increasing economy and population, especially if a drought occurs or the water demand target increases. How large will the water deficit and the corresponding risk be? Moreover, if the water demand increase because of population increasing or economics development, how much will the drought risk increase to? In these problems, the DSS developed was used to help the analysis. Because of the simplifications and assumptions made, the problems described here probably do not accurately reflect the whole situation. However, it may demonstrate what can be done by this approach and supply an important reference for practical operation of the water supply system.

This system has functions that with one simulation the following information can be obtained: (1) reservoir storage at the end of each period; (2) last storage for Fukuoka region at the end of each period; (3) reliability, resiliency, vulnerability, and DRI under different water-supply scenarios; (4) if a more serious drought occurs, how much will the risk of water shortage for Fukuoka region be?

In order to simulate the possible water supply and demand processes in future, the actual supply and demand processes should firstly be simulated. By changing water transfer ratio from Chikugo river, different relations between supply and demand, and different water deficit processes can

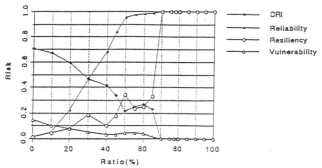

FIGURE 3   Water-supply risk in Fukuka city

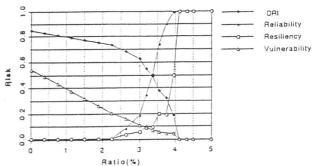

FIGURE 5 Risk analysis on population increase

be obtained. Then the reliability, resiliency, vulnerability, and DRI may be calculated for different cities or communities, which are shown in figure 3 and figure 4. With the development of economics and business in Fukuoka region, the population in the small communities increases quickly. In this paper, the population in Dazaifu city is assumed to increase by 5%, the risk curve is shown in figure 5. The serious drought occurred in 1994 at Fukuoka region, also has been analyzed by using this system and some significant conclusions were made about the operation of Fukuoka water-supply system. Figure 6 is the practical water-supply risk of Fukuoka city, in which it is assumed that Fukuoka city can take water from eight reservoirs as much as possible under water right. On the other hand, although the desalinization cost of sea water is higher than other measures the desalinization amount of sea water should increase gradually with the development of

economics in Fukuoka region because of its stability as a kind of water resource. If the daily desalinization amount of sea water is increased about 50,000 m^3 per day from January in 1994, the risk could be decreased significantly, which is shown in Figure 7.

5. CONCLUSIONS

The integration of DSS technique and simulation models is an efficient way for risk management in the operation of water supply systems. Not only does the DSS facilitate the examination of a series of scenarios in a short period that would be impossible by using traditional methods, but it also provides a dynamic output display and simulation which could be modified or updated by users at any time. This study demonstrates that the integration of

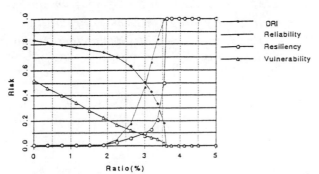

FIGURE 4 Water-supply risk in Dazaifu city

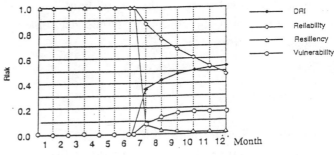

FIGURE 6 Water-supply risk of Fukuoka in 1994

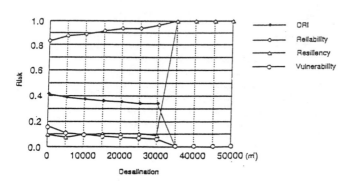

FIGURE 7 Risk analysis on desalization increase

DSS technique and risk analysis in the operation of a water supply system is useful and significant. Some preliminary conclusions about the planning and operation of the water resources system in Fukuoka region may be summarized as follows: (1) the water supply system of Fukuoka region is very vulnerable; (2) if the storage of reservoirs in Fukuoka city is sufficient it is feasible for other small cities and communities to share more water with Fukuoka city. If a drought with return period more than 10 years occurs in Fukuoka catchment, the water in Fukuoka city is not enough as well; (3) in order to increase the reliability of water supply in Fukuoka region, the development of new water sources with stable flow is of great urgency; (4) in order to increase the robustness of the entire water supply system in Fukuoka region, the unified operation for different subsystems is necessary, especially when some serious drought occurs as ones happened in 1994; (5) on the basis of unification for different subsystems, the "water bank" may be a efficient way to conquer serious drought in Fukuoka region.

REFERENCES

[BS89] M. Bender and S. Simonovic. Decision-Support System for Long-range Stream Flow Forecasting. *Journal of Computing in Civil Engineering*, ASCE, 8: 20-34, 1994.

[KS93] T. Kojiri and S. Sakakima. Decision Support System of reservoir Operation Considering Weather Forecast and Hydrography Similarity. In *Proceedings of the Symposium on Extreme Hydrological Events: Precipitation, Floods and Droughts*. Yokohama, Japan, July 1993.

[JX+95] K. Jinno, Z. Xu, K. Kawamura, and K. Tajiri. Risk Assessment of Water Supply System during Drought. *Water Resource Development*, 11: 185-204, 1995.

[NW+93] J. Napiorkowski, F. Wolbring, and G. Schultz. Expert System Application for Real-Time Risk Management during Drought. In *Proceedings of the Symposium on Extreme Hydrological Events: Precipitation, Floods and Droughts*, Yokohama, Japan. July 1993.

[RR91] P. Rasmussen and D. Rosbjerg. Evaluation of Risk Concepts in Partial Duration Series. *Stochastic Hydrology and Hydraulics*. 5: 1-16, 1991.

[SS89] S. Simonovic and D. Savic. Intelligent Decision Support and Reservoir Management and Operations. *Journal of Computing in Civil Engineering*, ASCE. 3: 367-385, 1989.

[Wal93] M. Walsh. Toward Spatial Decision Support Systems in Water Resources. *Journal of Water Resource Planning and Management*. 119: 158-169, 1993.

[Xu93] Z. Xu. Homogeneous Stochastic Point Process Model for Flood Risk Analysis. In *Proceedings of the Symposium on Extreme Hydological Events: Precipitation, floods and Droughts*, Yokohama, Japan. July 1993.

SUGGESTIONS ON DEVELOPING
"CREATIVE DECISION STIMULATION SYSTEMS (CDSS)"

Ronggui Ding and Wenxiu Han

Graduate School of Science and Technology, Okayama University

Okayama City 700, Japan

Email: ding@kanri.mech.okayama-u.ac.jp

Institute of Systems Engineering, Tianjin University

Tianjin City, 300072, China

ABSTRACT

This paper points out the necessity of developing "Creative Decision Stimulation Systems (CDSS)", proposes that the basic difference between CDSS and the existing DSS and ES is that CDSS aim to inspire and qualify the user's creative thinking by asking him/her meaningful questions while DSS and ES try to decrease the user's tedious work by providing him/her with data or solutions. Some methods and suggestions on the three most important activities in creative decision making are provided in this paper.

1. INTRODUCTION

The history of humanity will face a creative era in the near future. With the rapid development of our society, the amount of information, which has to be processed systematically, is becoming larger and larger, and the knowledge interchange between different fields is also becoming more and more necessary and complicated. Under such circumstances, we are recognizing that it will be more and more difficult for any individual to solve his problems based on his own creative ability, and we are in urgent need for some creativity support tools. On the other hand, though human beings are creatures with some extent of creative ability, there exist some obstacles to creativity, such as fixed concepts, e.g. "it is impossible ⋯ ", self-restrains, e.g. "it is silly to ⋯ ", and self-limitation, e.g. "I am not good at ⋯ ", and so on. We need some creativity support tools to remove these obstacles, too.

There is a lot of literature on creative thinking, much of it deals with the field of creative problem solving [KUN93, VAN87], and the results of this research work have certified the success in creativity support with the aid of AI. The meaning of "problem" is that an undesirable situation that is significant to and may be solvable by some agent although probably with some difficulty. In the real world, however, it is not necessary for a man, such as a manager, to make his decisions only when he is facing some kinds of problems, especially in a creative situation. He usually makes decisions not because he has some problems that need solving, but rather he has recognized some opportunities that he can take advantage of and these opportunities stimulate him to make decisions for grasping a larger value. We take this as a difference between creative decision making (CDM) and creative problem solving (CPS). Furthermore, even when decisions are made in a problem situation, there is also much room for us to generate creative objectives, alternatives, etc.. Based on this kind of thought, it is valuable for us to pay some attention to CDM, and it might show a different prospect beyond CPS.

We can divide the CDM process into three kinds of activities: opportunity-finding, idea-generation and decision-setting. We can hardly take that CDM could be divided into several phases like Simon's intelligence, design and choice [SIM73], and in fact, "the CDM process" can hardly be called a process, because each of the activities in CDM shares a common part with another.

In this paper, rather than using the word "support" as in DSS, we pay our most attention to "stimulation", for it is more suitable when helping arouse the decision maker's (DM's) motivation and inspire his creative thinking in all creative decision making activities, and propose "Creative Decision Stimulation Systems (CDSS)" as effective tools to help the decision maker (DM) make creative decisions. The next section states the difference between problem and opportunity, the third section proposes a brief view of CDSS, the fourth section provides some suggestions for the three activities in CDM, and the last section gives brief conclusions.

2. THE DIFFERENCE BETWEEN PROBLEM AND OPPORTUNITY

There is something odd about referring to a situation in which a decision has to be made as a "decision problem". Is it really a problem? Furthermore, is any decision made in a problem situation? We say in many situations, it is not. We can quote an example proposed by R. L. Keeney [KEE92]: A professional couple, the Lees, are considering a two-week vacation in Hawaii. After investigating several options through their travel agent, they have tentatively selected a vacation package that includes a week each on Oahu and Maui. But now, just before they sign up, the agent mentions another possibility. The Lees could extend their stay by one week and visit a third island, Kauai, at a good price. Even though their vacation from work are for two weeks, the Lees believe they could get the additional week off without pay. They decide to do it. With the range of vacation ideas somewhat expanded, the Lees are now dreaming about enjoying the tranquility of the islands. An image comes to mind of the South Seas, which they feel might be more exotic than Hawaii. They have often talked about a big South Seas trip sometime in the future. Maybe the future is now. They call their travel agent and ask about changing their plans to include two weeks in Hawaii and two weeks in Micronesia. And they decide that as long as they are taking their once-in-a lifetime trip, they should do it first class and not worry about the cost. The travel agent custom-designs such a vacation package, and now the Lees are again ready to sign up. This is a typical example of CDM. The decision maker (DM), i.e. the Lees, do not face a "decision problem", but owing to the travel agent's stimulation, they find an opportunity, and it results in a creative decision.

The DM usually thinks of decisions as problems to be solved, not as opportunities to be taken advantage of. There are at least two reasons why decision opportunities are not routinely identified. One reason is that many decision makers are not aware that identifying decision opportunities is an activity to which time should be allocated. Decision opportunities are not something thrust upon you, and in order to find them, you must first realize that you should be looking for them. The other reason is that there are few, if any, guidelines for identifying them. You must explore them and it is an unstructured task. One of the big differences between decision problems and decision opportunities is that you always know when you face a decision problem, but you don't necessarily know when a significant decision opportunity exists. Actions suggested by others will not inform you of a decision opportunity, identifying it requires your own effort, and it is mostly based on your creative thinking.

3. BRIEF VIEW OF CDSS

What are CDSS, and what are their objectives? We can propose a brief definition for CDSS, i.e. CDSS are computer-based stimuli that aim to inspire and qualify the DM's thinking when they are making creative decisions. While ES try to provide the user with some answers based on the experts' knowledge and DSS direct to support the DM to solve his decision problems with the aid of models and data, CDSS aim to provide the DM with some fundamental tools for his thinking, which can be selected by the DM himself, to help him find meaningful hypotheses rather than to provide him with complete, optimal solutions, and to stimulate him to do creative thinking with high quality by using suitable questions built into these tools.

More and more people now are finding themselves facing up to situations they have never dealt with before. In a creative environment, old recipes don't seem to fit. A creative decision maker who can no longer depend on referring to "how it was done before" need to be independent of his experience or "the cases" and to find new ways to finish his task. In order to grasp the profitable opportunities, discover wonderful alternatives and make creative decisions, the DM can usually rely on his own thinking, and only this mental operation, rather than data, can help him to pre-select in advance, to recognize what will be relevant even when they are apparently remote. The more senior the task is, the more it relies on the mental operation and the less it relies on detailed knowledge or models.

It is possible that a small number of mental operations could be responsible for all the thinking anyone ever does [RHO91]. On the other hand, questions are the driving forces of our thinking, and they form the very structure of our knowledge. Questions first help us identify useful information and second put the answers into patterns that make them useful. The mental operations are with questions built into them, and these questions are process questions, open rather than closed. The mental operations in the CDM activities act as effective tools that ensure the DM to think up all what is needed to handle his CDM. This is possible because they are pure process, and at a high level truly generic so they can be used whatever the situation is. Although these operations cannot guarantee that the right answers will be forthcoming, they can warn that some questions have not yet been answered. They have no sequence, the DM need to plan the sequence of his way through them.

There are many hypotheses concerning creative thinking [KUN93, VAN87]. Although they are all with some difference about it, a general opinion is that there are two kinds of actions, divergent and convergent actions in it. We can suggest that the mental operations in the CDM activities should involve enough divergent and convergent ones. One primary reason why these two

kinds of actions should be involved is that: if the DM uses only divergent actions, he probably would not achieve any sense of closure, and if he uses only convergent ones, the overemphasis upon judgement would lead to few, if any, creative results. But we think it is not suitable to take them as a process from divergent ones to convergent ones as they are usually taken [VAN87, KUN93], no matter whether it is an iterative process or not. The reason is that the creative thinking is basically unstructured, its fundamental motivation is to avoid any restraints. That is also the reason why we suggest that it is better to use mental operations as tools for CDM, because one tool can be used in many kinds of tasks, and if we could find out some mental operations that are suitable for any thinking activities, they would be sufficient for the CDM activities. In fact, some researchers have proposed such mental operations [RHO91]. The mental operations can usually be divided into three groups: mental operations relating to information, mental operations relating to judgement, and mental operations relating to ideas. And they can also be divided into hard ones and soft ones: hard mental operations deal with the more tangible side of thinking, they help the DM to be more objective than subjective with his thinking, and require to be used explicitly and quantitatively rather than implicitly and qualitatively; the others are soft ones. Hard mental operations act as analyzing, structuring and organizing, while soft ones go on inspiring and stimulating. Hard mental operations aim to be more efficient, soft mental operations are used for more effective. It is obvious that we cannot say that tools should be set according to a routine, even though we can use tools one by one according to our different tasks. The mental operations used in the CDM activities act as stimuli, and in what way to make them powerful is up to the DM himself. That is why we suggest CDSS as stimulation systems rather than generation systems.

Unlike that the models and model management systems are designed in DSS, and the knowledge-base, rules and reasoning mechanism are built in ES, CDSS just propose the basic tools for creative thinking, how to organize them into practical systems is up to the DM. Of course, CDSS also provide other necessary tools to help the DM do this job fast and easily. The guidelines and methods about how to use the mental operations to construct practical systems, i.e. CDSS, can be found in Ding and Osaki's paper "Research on 'Creative Decision Stimulation Systems (CDSS)'" [DIN96].

4. THREE ACTIVITIES IN CDM

. Opportunity-finding activity

We can use the method of stimulating the DM to think out a list of outcomes and obstacles to let him do the divergent actions in opportunity-finding activity. Outcomes are concerned with the things the DM would like to achieve, while obstacles are what might prevent him from achieving them. In this activity, CDSS can provide the DM with stimulating questions such as "wouldn't it be nice if ··· ? (WIBNI)" for outcomes and "wouldn't it be awful if ··· ? (WIBAI)" for obstacles, and furthermore, based on the answers to these questions, CDSS help the DM develop several tentative opportunities and state these opportunities using the format: "in what way might I ··· ? (IWWMI)" like means-ends method [KEE92]. The principles of Osborn's brain-storming method [OSB63] are also suitable in these actions. In order to defer all judgement and let the DM's thought flow and try for as many as he can, alternate words can be used as activators. This method can help stimulate a variety of perspectives by using synonyms for key words or phases in the original statements, and using different combinations of these words and redefinition statements such as "IWWMI ··· ? ". It would be apparent that these redefinitions are pretty abstract, but they could be stimuli for more concrete statements. On the other hand, convergent actions in opportunity-finding activity can be aided by examining the variable of ownership and outlook. Ownership is used to determine if the opportunity area is the one that the DM can influence and if he is motivated to deal with it. Outlook is used to access how familiar the area to the DM, how soon it needs to be dealt with, and how likely it is to change over time.

Opportunity-finding activity is the most important part in CDM and it is the very distinguishing feature

with which CDM differs from CPS. We want to suggest some useful mental operations as basic tools for this activity such as: the idea-aimed ones, e.g. Feel (listen to intuition), Escape (escape from musts), Purse (never stop exploring), Re-describe (change what is there), Pretend (pretend any play), and Challenge (question assumptions); the information- aimed ones, e.g. Observe (observe with all senses), Look in/out (any system relations?), Set level (what level and scope?), Categorize (under what label /category?), Code (the medium for the message); and the judgement-aimed ones, e.g. Distinguish (discern the pattern), Commit (commit to what is worthwhile), Test (test the rationale), Predict (assuming this we predict ···), and Compare (compare and contrast), etc..

. Idea-generation activity

There are a lot of approaches can be adopted for the idea-generation activity in CDM [MAC93, KUN93], and it is no need for us to repeat them in this paper. Here we just want to point out an aspect which should be pay more attention to: most of those methods generate ideas by using aspects of the domain that the DM cares for in an associated way. Most people are more comfortable starting with related aspects, especially the logical connection is apparent, but the process using unrelated stimuli can often help gain access to many of the "hidden" ideas that float around in our minds, i.e. ideas we may not be fully aware of until they are prompted by something else. It is regret that there are few methods having been developed. Here we want to suggest that natural language understanding based on key words might be helpful. As we know, there might be some mistakes in computer-based natural language understanding from some time to time. Even though it is a disadvantage to automatic translation systems, it would be stimulus for creative thinking. For example, usually shoppers pay for the goods they buy, but the result of computer-based natural language understanding might provide us with that the store pays the customers according to some kinds of grammar, and this could lead to the trading stamp idea which, in effect, paid shoppers a tiny amount for each purchase, etc..

We could suggest some useful mental operations for

this activity such as: the idea-aimed ones, e.g. Feel, Escape, Purse, Re-describe, Unform (go all flexible); the information-aimed ones, e.g. Look in/out, Set level, Specify (specify accuracy); and the judgement-aimed ones, e.g. Distinguish, Commit, Predict, and Interpret (make senses of data), etc..

. Decision-setting activity

With the experience, the intuition of himself or with the help of the opportunity-finding activity or the idea-generation activity, the DM could find the raw materials that form some alternatives for setting his decisions. Although some of his ideas may be well-developed, most of them are probably still in the rough. In the decision-setting activity, the DM faces with the task of deciding which ideas have the potential possibility for realizing his opportunities and which need additional work. Instead of trying to reject ideas, this activity also stimulate the DM's creative thinking.

According to VanGundy's opinion [VAN87], divergent actions in this activity consist of two kinds of sub-actions. The first ones involve generating criteria for screening, selecting and supporting ideas. These criteria are also creative ones which can help the DM reduce and refine the total pool of ideas. They are also used to develop ideas into more workable decisions. The second ones involve generating more ideas again. Convergent actions in this activity also involve two primary kinds of sub-actions. The DM diverges to screen, select and support his evaluation criteria, and converges to screen, select and support decisions. In this activity, the importance of affirmative judgement should be emphasized. Usually, our conditioning has taught us to make decisions using negative criticism. However, it can be more productive to also consider the positive value of ideas before rejecting them, which could be the "putting on your yellow hat" action as De Bond said [DEB85]. In this activity, we should try to stimulate the DM to combine ideas even the ideas appear workable at the outset. Sometimes deliberate analysis and toying around with ideas can suggest new approaches.

We'd like to suggest some useful mental operations for this activity such as: the idea-aimed ones, e.g. Escape, Re-describe, Pretend, and Challenge; the information-aimed ones, e.g. Observe, Look in/out, Set

level, and Specify; and the judgement-aimed ones, e.g. Distinguish, Commit, Test, Predict, Value (know and feel what matters), and Compare, etc..

5. CONCLUSIONS

The fundamental meaning of developing CDSS is that: CDSS propose a different direction for AI research, i.e. AI should not only aim to make machine work for human beings but also aim to let machine make human beings work more qualitatively. CDSS are future-directed systems for they try to help the DM find and realize his decision opportunities, while ES, DSS and most of the present decision making and problem solving methods are present-directed for they try to clarify and improve the present problem situations.

REFERENCES

[DEB85] De Mono, E., *Six Thinking Hats*, Little, Brown and Company, 1985

[DIN96] Ding R., H. Osaki, *Research on "Creative Decision Stimulation Systems (CDSS)"*, for publishing, 1996

[KEE92] Keeney, R. L., *Value-focused Thinking*, Harvard Univ. Press, 1992

[KUN93] Kunifuji, S., A Survey on Creative Thinking Support Systems and the Issues for Developing Them, *J. of Japanese Soc. of AI*, Vol. 18, No. 5, pp552-559, 1993

[MAC93] MacCrimmon, K., R.& C.Wagner, Stimulating Ideas Through Creativity Software, *Management Sci.*, Vol. 40, No. 11, pp 1514-1532, 1993

[OSB63] Osborn, A. F., *Applied Imagination*, 3rd ed., Charles Scribuer's Sons, 1963.

[RHO91] Rhodes, J., *Conceptual Toolmaking*, Basil Blackwell, 1991

[SIM73] Simon, H., The Structure of Illstructured Problems, *Artificial Intelligence*, pp181-201,1973

[VAN87] VanGundy, A. B., *Creative Problem Solving*, Greenwood Press Inc., 1987

Diagnosis

A METHOD OF DIAGNOSIS USING POSSIBILITY THEORY

Koichi YAMADA, Mitsuhiro HONDA
Advanced Technology Center, Yamatake-Honeywell Co., Ltd.
YBP West Tower, 134 Gohdo-cho, Hodogaya-ku, Yokohama 240, JAPAN
Email: yamada@atc.yamatake.co.jp

ABSTRACT

The paper proposes a diagnostic method using possibility theory. According to possibility theory, fuzzy set A on a set of possible causes U, B on possible symptoms V, and a fuzzy relation R on U×V could be understood as possibility distributions $\pi_U(u_i)$, $\pi_V(v_j)$ and a conditional possibility distribution $\pi_{V|U}(v_j|u_i)$, respectively. In this paper, we define diagnosis as a problem to obtain possibility that a given crisp subset of U is the causes of a crisp subset of V observed as symptoms. Possibility distribution $\pi_U(u_i)$ on U and conditional possibility distribution $\pi_{V|U}(v_j|u_i)$ are given as a priori knowledge. Then, a way to solve the problem is discussed and proposed. We also discuss a conventional approach using probability, and designate the difficulties that it suffers from, but our approach does not.
Keyword : Diagnosis, Possibility theory, Conditional possibility distribution

1. INTRODUCTION

There has been a way of diagnosis to solve an inverse problem of fuzzy relational equations, which express fuzzy causal relations between causes and symptoms [Terano78, Mo93]. The process of diagnosis in this approach is summarized as follows; A and B are fuzzy sets on all possible causes $U=\{u_i\}$, (i=1,...,N) and all possible symptoms $V=\{v_j\}$, (j=1,...,M), respectively. R is a fuzzy relation on U×V expressing fuzzy causalities between causes and symptoms. Then, fuzzy relational equations are given in the next Max-Min composition.

$$B(v_j) = \bigvee_{u_i}(A(u_i) \wedge R(u_i, v_j)), \qquad (1)$$

where $A(u_i)$, $B(v_j)$ and $R(u_i, v_j)$ are membership functions of A, B and R, respectively. $\vee$ and $\wedge$ express Max and Min operations. Then, the diagnosis is defined as deriving fuzzy set A given fuzzy set B and fuzzy relation R. This shows that the process of diagnosis is exactly the same as an inverse problem of fuzzy relational equations.

However, this approach has a problem for interpretation of the membership values of A and B [Yamada95]. In most of studies using this approach [Terano78, Mo93], $B(v_j)$ is interpreted as intensity of symptom v_j, while $A(u_i)$ as possibility or certainty that cause u_i is arising. The interpretation, however, confusingly mixes different kinds of measures - intensity and possibility - in the equations. Another interpretation is to regard $A(u_i)$ and $B(v_j)$ as intensities of occurrence. However in this case, Max-Min composition is not adequate, because the equation cannot express the natural relation that the stronger (weaker) the intensity of a cause is, the stronger (weaker) those of its symptoms tend to be. The third one is to consider both of $A(u_i)$ and $B(v_j)$ as possibilities. This apparently seems to be adequate, because fuzzy relational equations are considered to be based on possibility theory [Zadeh78, Nguyen78, Hisdal78]. However, again, this should be abandoned, because B should be a crisp set, since the symptoms are those actually observed.

In this paper, we deal with a diagnostic problem in which possibility of a given crisp set of causes is derived, when a crisp set of observed symptoms is given. As suggested in the above discussion, we do not employ fuzzy relational equations directly. Instead, we develop a new theory of diagnosis following the possibilistic interpretation of fuzzy relational equations. This approach could be considered as a kind of abduction [Peng90] based on possibility theory, because derived sets of causes are regarded as explanations of a given set of events (symptoms).

2. DEFINITION OF DIAGNOSIS

Let X and Y be variables that take values in U and V, respectively. Then, $A(u_i)$ and $B(v_j)$ in eq.(1) are considered as possibility distributions $\pi_X(u_i)$ and $\pi_Y(v_j)$, and $R(u_i, v_j)$ as a conditional possibility distribution $\pi_{Y|X}(v_j|u_i)$ [Hisdal78]. $\pi_X(u_i)$ and $\pi_Y(v_j)$ are given as marginal possibility distributions of joint possibility distribution $\pi_{(X,Y)}(u_i,v_j)$.

$$\pi_X(u_i) = \bigvee_{v_j \in V} \pi_{(X,Y)}(u_i, v_j). \qquad (2)$$

$$\pi_Y(v_j) = \bigvee_{u_i \in U} \pi_{(X,Y)}(u_i, v_j). \qquad (3)$$

In the rest of this paper, we describe possibility distributions simply as $\pi(u_i)$, $\pi(v_j)$, $\pi(v_j|u_i)$ and $\pi(u_i,v_j)$ instead of $\pi_X(u_i)$, $\pi_Y(v_j)$, $\pi_{Y|X}(v_j|u_i)$ and $\pi_{(X,Y)}(u_i,v_j)$, respectively. Furthermore, we assume that occurrence of multiple u_is and v_js is allowed, although possibility distributions are defined on variables taking a single value in principle.

Now, we are ready to define diagnosis based on possibility theory.

[Definition]

Assume that $\pi(u_i)$ and $\pi(v_j|u_i)$ are given as a priori knowledge representing possibility of occurrence of u_i and conditional possibility of v_j when u_i is present, respectively. Also assume that $Q^+ = \{v_1,...,v_m\}$, $(m \leq M, V \supseteq Q^+)$ is a set of observed symptoms and that $Q^- = \{v_{m+1},...,v_L\}$, $(m < L \leq M, V \supseteq Q^-)$ is a set of symptoms confirmed that they are not arising. The rest of symptoms $V - (Q^+ \cup Q^-)$ are uncertain if they are arising or not. Then, the diagnosis is defined as obtaining possibility $Poss(P | Q^+, Q^-)$ that $P=\{u_1,...,u_n\}$, $(n \leq N, U \supseteq P)$ is the causes of Q^+ and Q^-. $Poss(P | Q^+, Q^-)$ is defined as follows:

$$Poss(P | Q^+, Q^-) \equiv \pi(p_1^+,...,p_n^+,p_{n+1}^-,...,p_N^- | \\ q_1^+,...,q_m^+,q_{m+1}^-,...,q_L^-). \qquad (4)$$

In the above equation, p_i^+ and q_j^+ show that u_i and v_j are arising, and p_i^- and q_j^- means they are not. The right side of eq. (4) is a conditional possibility distribution that gives possibility of a value in $X_1 \times ... \times X_N$, when a value in $Y_1 \times ... \times Y_L$ is given. X_i is a variable which takes either p_i^+ or p_i^-, and Y_j takes either q_j^+ or q_j^-. By replacing values of X_i and Y_j by general values x_i and y_j, eq. (4) can be rewritten as follows:

$$Poss(P | Q^+, Q^-) = \pi(x_1,...,x_N | y_1,...,y_L), \qquad (5)$$

where $x_i = p_i^+$ when $u_i \in P$, $x_i = p_i^-$ when $u_i \notin P$, $v_j = q_j^+$ when $v_j \in Q^+$, and $v_j = q_j^-$ when $v_j \in Q^-$.

3. SOLVING DIAGNOSTIC PROBLEM

In this section, we discuss how to solve the diagnosis defined in the previous section.

3.1 Possibility distributions on X_i and Y_j

First, we examine possibility distributions of newly introduced variables X_i and Y_j. We assign $\pi(u_i)$ to possibility of $\pi(p_i^+) \equiv \pi_{Xi}(p_i^+)$, $(i \leq n)$, because $\pi(p_i^+)$ expresses possibility of u_i.

$$\pi(p_i^+) = \pi(u_i). \qquad (6)$$

Since variable X_i takes either p_i^+ or p_i^- as its value, we can obtain the next equation, if we assume that possibility distribution $\pi(x_i)$ is normal.

$$\bigvee_{x_i \in X_i} \pi(x_i) = \pi(p_i^+) \vee \pi(p_i^-) = 1.0. \qquad (7)$$

Therefore, $\pi(p_i^-)$ must be 1.0 except that $\pi(p_i^+) = \pi(u_i) = 1.0$. In this paper, we assume that $\pi(p_i^-)$ is always 1.0. This assumption might be adequate, since malfunction is considered to be a special case in general. In the result, we get the following:

$$\pi(x_i) = \begin{cases} \pi(u_i), & \text{if } x_i = p_i^+. \\ 1.0, & \text{if } x_i = p_i^-. \end{cases} \qquad (8)$$

Similarly, possibility distribution $\pi(y_j)$ is obtained as follows:

$$\pi(y_j) = \begin{cases} \pi(v_j), & \text{if } y_j = q_j^+. \\ 1.0, & \text{if } y_j = q_j^-. \end{cases} \qquad (9)$$

Since membership values $A(u_i)$, $B(v_j)$, and $R(u_i, v_j)$ in eq. (1) could be interpreted as $\pi(u_i)$, $\pi(v_j)$, and $\pi(v_j|u_i)$, respectively, $\pi(v_j)$ is derived from the following equation, if we have no information about u_i and v_j except $\pi(u_i)$ and $\pi(v_j|u_i)$.

$$\pi(v_j) = \bigvee_{u_i} (\pi(u_i) \wedge \pi(v_j \mid u_i)). \tag{10}$$

Notice that $\pi(u_i) \neq 0$ and $\pi(v_j) \neq 0$, because u_i and v_j are both possible causes and symptoms.

3.2 Decomposition of joint possibility distribution

In general, joint possibility distribution $\pi(x_i, x_{i'})$, $(i \neq i')$ is expressed as follows, using conditional possibility distribution $\pi(x_{i'} \mid x_i)$ [Hisdal78].

$$\pi(x_i, x_{i'}) = \pi(x_i) \wedge \pi(x_{i'} \mid x_i). \tag{11}$$

If causes x_i (u_i) are possibilistically independent of each other, the next equation holds [Hisdal78].

$$\pi(x_{i'} \mid x_i) = \pi(x_{i'}). \tag{12}$$

We can derive the following from the above two.

$$\pi(x_i, x_{i'}) = \pi(x_i) \wedge \pi(x_{i'}). \tag{13}$$

Eq. (13) shows that X_i and $X_{i'}$ are non-interactive [Zadeh78]. From the above discussion, when causes x_i are possibilistically independent of each other, eq. (5) can be rewritten as follows:

$$\text{Poss}(P \mid Q^+, Q^-) = \pi(x_1 \mid y_1, ..., y_L) \wedge ... \\ \wedge \pi(x_N \mid y_1, ..., y_L). \tag{14}$$

3.3 Relation between $\pi(y_j \mid x_i)$ and $\pi(x_i \mid y_j)$

As shown in eq. (11), joint possibility distribution $\pi(x_i, y_j)$ is given as follows:

$$\pi(x_i, y_j) = \pi(x_i) \wedge \pi(y_j \mid x_i). \tag{15}$$

Since conditional possibility distribution $\pi(y_j \mid x_i)$ is considered as expressing an implication, the next equation holds.

$$\pi(y_j \mid x_i) = \pi(x_i) \to \pi(y_j), \tag{16}$$

where $\to$ shows an implication.

By the way, since $\pi(x_i, y_j)$ shows an possibility of x_i and y_j in this paper, joint possibility distribution $\pi(x_i, y_j)$ should satisfy the commutative law, namely, $\pi(x_i, y_j) = \pi(y_j, x_i)$. However, when $\pi(y_j \mid x_i)$ in eq. (15) is replaced by

"$\pi(x_i) \to \pi(y_j)$", there are only a few definitions of implication that satisfy the law, though there have been proposed many definitions in various multi-valued logic [Togai85]. One of them is Gödelian logic. If we adopt it, eq. (16) can be rewritten as follows:

$$\pi(y_j \mid x_i) = \begin{cases} 1.0, & \text{if } \pi(x_i) \leq \pi(y_j). \\ \pi(y_j), & \text{if } \pi(x_i) > \pi(y_j). \end{cases} \tag{17}$$

If we substitute $\pi(y_j \mid x_i)$ in eq. (15) by eq. (17), eq. (15) is rewritten as follows:

$$\pi(x_i, y_j) = \pi(x_i) \wedge \pi(y_j). \tag{18}$$

This shows that x_i and y_j are non-interactive and T-independent [Nguyen78], which means $\pi(x_i, y_j) = T(\pi(x_i), \pi(y_j))$, where T is a function on $[0,1] \times [0,1] \to [0,1]$. Nguyen has proved that conditional possibility distribution is given in the next equation, when x_i and y_j are T-independent.

$$\pi(x_i \mid y_j) = \begin{cases} \pi(x_i, y_j), & \text{if } \pi(x_i) \leq \pi(y_j). \\ \pi(x_i, y_j) \cdot \dfrac{\pi(x_i)}{\pi(y_j)}, & \text{if } \pi(x_i) > \pi(y_j). \end{cases} \tag{19}$$

Since $\pi(u_i) \neq 0$ and $\pi(v_j) \neq 0$, the above equation gives the followings:

$$\pi(x_i, y_j) = \begin{cases} \pi(x_i \mid y_j), & \text{if } \pi(x_i) \leq \pi(y_j). \\ \pi(x_i \mid y_j) \cdot \dfrac{\pi(y_j)}{\pi(x_i)}, & \text{if } \pi(x_i) > \pi(y_j). \end{cases} \tag{20}$$

$$\pi(y_j, x_i) = \begin{cases} \pi(y_j \mid x_i), & \text{if } \pi(x_i) \geq \pi(y_j). \\ \pi(y_j \mid x_i) \cdot \dfrac{\pi(x_i)}{\pi(y_j)}, & \text{if } \pi(x_i) < \pi(y_j). \end{cases} \tag{21}$$

Since $\pi(x_i, y_j) = \pi(y_j, x_i)$, we finally get the next relation between $\pi(y_j \mid x_i)$ and $\pi(x_i \mid y_j)$ from eq. (20) and (21).

$$\pi(y_j) \cdot \pi(x_i \mid y_j) = \pi(x_i) \cdot \pi(y_j \mid x_i). \tag{22}$$

3.4 Decomposition of $\pi(x_i \mid y_1, ..., y_L)$

In this section, each term in eq. (14) is decomposed. From eq. (22), we can get the next equation.

$$\pi(y_1,...,y_L) \bullet \pi(x_i \mid y_1,...,y_L)$$
$$= \pi(x_i) \bullet \pi(y_1,...,y_L \mid x_i) \qquad (23)$$

Through the same examination as that in **3.3**, we can conclude that Y_j and $Y_{j'}$ are non-interactive, and obtain the following:

$$\pi(y_1,...,y_L \mid x_i) = \pi(y_1 \mid x_i) \wedge ... \wedge \pi(y_L \mid x_i). \qquad (24)$$

Considering that $\pi(y_1,...,y_L)$ is constant in the situation where Q^+ and Q^- are given, the next equation is derived from eq. (23) and (24).

$$\pi(x_i \mid y_1,...,y_L)$$
$$= \alpha \bullet \pi(x_i) \bullet (\pi(y_1 \mid x_i) \wedge ... \wedge \pi(y_L \mid x_i)), \qquad (25)$$

where α is a constant value. Finally, by substituting each term in eq.(14) with the above equation, we obtain the following:

$$\text{Poss}(P \mid Q^+,Q^-) \propto$$
$$\bigwedge_{i=1,N} \pi(x_i) \bullet (\pi(y_1 \mid x_i) \wedge ... \wedge \pi(y_L \mid x_i)). \qquad (26)$$

Since $\pi(x_i)$ is given by eq.(8), we can obtain the value of $\text{Poss}(P \mid Q^+,Q^-)$, if $\pi(y_j \mid x_i)$ is derived.

3.5 Conditional possibility distribution $\pi(y_j \mid x_i)$

This section examines a way to obtain $\pi(y_j \mid x_i)$ in eq. (26). From Gödelian implication given by eq.(17), the followings are derived.

$$\pi(q_j^+ \mid p_i^+) = \begin{cases} 1.0, & \text{if } \pi(u_i) \le \pi(v_j). \\ \pi(v_j), & \text{if } \pi(u_i) > \pi(v_j). \end{cases} \qquad (27)$$

$$\pi(q_j^- \mid p_i^+) = 1.0. \qquad (28)$$

$$\pi(q_j^+ \mid p_i^-) = \pi(v_j). \qquad (29)$$

$$\pi(q_j^- \mid p_i^-) = 1.0 \qquad (30)$$

Therefore, if we can determine the value of $\pi(v_j)$, values of $\pi(q_j^+ \mid p_i^+)$, $\pi(q_j^+ \mid p_i^-)$ and $\text{Poss}(P \mid Q^+, Q^-)$ can be obtained. When we have no information about u_i and v_j except $\pi(u_i)$ and $\pi(v_j \mid u_i)$, $\pi(v_j)$ should be derived from eq.(10). However in this case, we are trying to determine the values under the assumption that P is a set of causes in eq.(26). Therefore, eq.(10) should be applied with values of $\pi(u_i)=1.0$ for $i \le n$ and of $\pi(u_i)=0.0$ for $n<i \le N$. Then, $\pi(v_j)$ is given by the next equation.

$$\pi(v_j) = \bigvee_{u_k \in P} \pi(v_j \mid u_k). \qquad (31)$$

Summing up the above results, $\pi(y_j \mid x_i)$ is obtained as follows:

$$\pi(y_j \mid x_i) = \begin{cases} 1.0, & \text{if } x_i = p_i^+, y_j = q_j^+, \\ & \qquad \pi(u_i) \le \bigvee_{u_k \in P} \pi(v_j \mid u_k) \\ \bigvee_{u_k \in P} \pi(v_j \mid u_k), & \text{if } x_i = p_i^+, y_j = q_j^+, \\ & \qquad \pi(u_i) > \bigvee_{u_k \in P} \pi(v_j \mid u_k) \\ 1.0, & \text{if } x_i = p_i^+, y_j = q_j^- \\ \bigvee_{u_k \in P} \pi(v_j \mid u_k), & \text{if } x_i = p_i^-, y_j = q_j^+ \\ 1.0, & \text{if } x_i = p_i^-, y_j = q_j^- \end{cases} \qquad (32)$$

Looking at the cases where $y_j = q_j^-$, $\pi(y_j \mid x_i)$ is always 1.0 in the above equation. Therefore, terms with "i" where $m<i \le L$ in eq. (26) is unnecessary. This means that we do not have to consider Q^-, which is a set of symptoms confirmed that they are not arising. The diagnosis is reduced to a problem to obtain possibility $\text{Poss}(P \mid Q^+)$, which is given in the next equation.

$$\text{Poss}(P \mid Q^+) \propto$$
$$\bigwedge_{i=1,N} \pi(x_i) \bullet (\pi(y_1 \mid x_i) \wedge ... \wedge \pi(y_m \mid x_i)). \qquad (33)$$

The result shown above indicates that confirmation of no symptoms has no influence on the diagnosis. This is natural, however, because we have assumed $\pi(q_j^-) = 1.0$, which leads to eq.(28) and (30). If we want to utilize the information of Q^- for diagnosis, we should not assume that $\pi(q_j^-)$ are always 1.0, but should give some additional knowledge that gives $\pi(q_j^-) \ne 1.0$ or $\eta(q_j^+) \equiv 1 - \pi(q_j^-) \ne 0.0$ for all "j", where $\eta(q_j^+)$ is necessity of q_j^+.

The fact that a cause is not denied even if its symptoms are not observed indicates that possibility of a set of causes including ones that have nothing to do with the observed symptoms cannot be 0.0, if the set includes real causes. This means that too many sets of causes are obtained with a certain possibility, because there are many causes that have no relation with the observed symptoms in general. Therefore, it is appropriate that only sets of causes that are *parsimonious* [Peng90] are chosen as candidates of real causes. Here, a parsimonious set of causes is defined as one whose any proper set has possibility 0.0.

4. NUMERICAL EXAMPLE

(1) Case 1

Suppose that $\pi(u_i)$ and $\pi(v_j | u_i)$ are given as shown in table 1 for $U=\{u_1, u_2, u_3\}$ and $V=\{v_1, v_2, v_3\}$. In this case, $\pi(v_j)$ is calculated as shown in table 2 using eq.(31).

Now, we assume that symptoms v_2 and v_3 are observed. Using eq.(33), possibilities of all power sets of V are calculated as follows:

(a) When we assume $P = \{u_1, u_2, u_3\}$,
$$Poss(P | Q^+) \propto 0.4$$
(b) When we assume $P = \{u_1, u_2\}$,
$$Poss(P | Q^+) \propto 0.35$$
(c) When we assume $P = \{u_1, u_3\}$,
$$Poss(P | Q^+) \propto 0.7$$
(d) When we assume $P = \{u_2, u_3\}$,
$$Poss(P | Q^+) \propto 0.4$$
(e) When we assume $P = \{u_1\}$,
$$Poss(P | Q^+) \propto 0.35$$
(f) When we assume $P = \{u_2\}$,
$$Poss(P | Q^+) \propto 0.0$$
(g) When we assume $P = \{u_3\}$,
$$Poss(P | Q^+) \propto 0.6$$
(h) When we assume $P = \varnothing$,
$$Poss(P | Q^+) \propto 0.0$$

If we sort the cause sets with non-zero possibilities in the descending order, we get the following:

$\{u_1, u_3\}$, $\{u_3\}$, $\{u_2, u_3\}$, $\{u_1, u_2, u_3\}$, $\{u_1\}$, $\{u_1, u_2\}$

Since only parsimonious sets are chosen, candidates of the causes are $\{u_3\}$ and $\{u_1\}$ in this case.

(2) Case 2

If we use the same knowledge as that in case 1, and assume that $Q^+ = \{v_1, v_2\}$, possibility of each set of causes is derived shown below.

(a) $Poss(\{u_1, u_2, u_3\} | Q^+) \propto 0.4$
(b) $Poss(\{u_1, u_2\} | Q^+) \propto 0.4$
(c) $Poss(\{u_1, u_3\} | Q^+) \propto 0.0$
(d) $Poss(\{u_2, u_3\} | Q^+) \propto 0.4$
(e) $Poss(\{u_1\} | Q^+) \propto 0.0$
(f) $Poss(\{u_2\} | Q^+) \propto 0.0$
(g) $Poss(\{u_3\} | Q^+) \propto 0.0$
(h) $Poss(\{\varnothing\} | Q^+) \propto 0.0$

From the above, $\{u_1, u_2\}$ and $\{u_2, u_3\}$ are picked up as parsimonious sets of causes. The results say that plural causes are arising; one is u_2, and the other is u_1 or u_3. From table 1, we can see that symptom v_3 may arise, if any of the three causes happens, while it is not included in the observed symptoms. Therefore, this case demonstrates the capability that the proposed method can diagnose successfully, even if we fail to find some symptoms.

5. COMPARISON WITH PROBABILISTIC APPROACH

There is a well-known probabilistic way of diagnosis similar to the proposed approach. According to the Bayesian theorem, conditional probability $Pr(P | Q^+)$ is given in the next equation.

$$Pr(P | Q^+) = \frac{Pr(Q^+ | P) \cdot Pr(P)}{Pr(Q^+)}, \qquad (34)$$

where $Q^- = V - Q^+$ is assumed. In the equation, $Pr(Q^+)$ could be regarded as a constant value in the given situation. $Pr(P)$ is calculated from probability distribution on U, if all cases

Table 1   Possibility distributions given as a priori knowledge

ui \ vj	$\pi(v_j | u_i)$			$\pi(u_i) = \pi(pi^+)$
	v1	v2	v3	
u1	0.0	1.0	0.5	0.7
u2	0.9	0.0	0.4	0.4
u3	0.0	0.6	0.8	1.0

Table 2   Values of $\pi(v_j)$

P	$\pi(v1)$	$\pi(v2)$	$\pi(v3)$
{u1, u2, u3}	0.9	1.0	0.8
{u1, u2}	0.9	1.0	0.5
{u1, u3}	0.0	1.0	0.8
{u2, u3}	0.9	0.6	0.8
{u1}	0.0	1.0	0.5
{u2}	0.9	0.0	0.4
{u3}	0.0	0.6	0.8
$\varnothing$	0.0	0.0	0.0

are considered to be independent of each other. [Peng90]

$$Pr(P) = K \prod_{u_i \in P} \frac{Pr(u_i)}{1 - Pr(u_i)}, \qquad (35)$$

where K is constant. However, $Pr(Q^+ | P)$ cannot be calculated like $Pr(P)$, because all symptoms are NOT independent of each other. They must be given as a priori knowledge, and the quantity of knowledge is very huge (in the order of $2^{|U|} \times 2^{|V|}$). This means that it is almost impossible to use eq.(34) directly for diagnosis.

To cope with the problem, Peng[Peng90] has introduced *Conditional Causal Probability* instead of conditional probability. It is a probability that "a symptom v_j is caused by a cause u_i, when the cause arises", while conventional conditional probability is one that "a symptoms arises when a cause does". Using the Conditional Causal Probability c_{ij}, Peng showed that $Pr(Q^+ | P)$ is given in the next equation.

$$Pr(Q^+ | P) = \prod_{v_j \in Q^+} (1 - \prod_{u_i \in P}(1 - c_{ij})) \cdot$$
$$\prod_{u_i \in P} \prod_{v_k \in effect(u_i) - Q^+}(1 - c_{ik}) \qquad (36)$$

where $effect(u_i)$ is all possible symptoms that u_i may cause.

However, this approach still has a problem that the probability distribution on U must be very accurate. For example, suppose there are two subsets of causes P_1 and P_2, both of which include u_1 and u_2. In this case, our typical concern is a question such as which is bigger, $Pr(P_1 | Q^+)$ or $Pr(P_2 | Q^+)$. Unfortunately, however, the answer is sensitive to the absolute values of $Pr(u_1)$ and $Pr(u_2)$. The order in case that $Pr(u_1)=0.01$ and $Pr(u_2)=0.03$ may be different from that in another case that $Pr(u_1)=0.03$ and $Pr(u_2)=0.09$. This forces us to give very accurate probability distribution based on a vast amount of data.

In our approach, such inversion does not occur, if values of possibility distributions are relatively correct. Therefore, it might not be so difficult to construct a priori knowledge in the real world applications.

6. CONCLUSION

An approach of diagnosis based on possibility theory is proposed. This approach presupposes that causal relations between causes and symptoms are expressed in fuzzy relational equations. Differently from the conventional ones, however, it does not use the equations directly such as solving the inverse problem. Instead, it calculates possibilities of crisp sets of causes based on possibility theory, when a crisp set of symptoms is given. Then, a simple example is shown to illustrate how this approach works.

REFERENCES

[Hisdal78] E. Hisdal: Conditional possibilities independence and non-interaction, Fuzzy Sets and Systems, Vol.1, pp. 283-297 (1978)

[Mo93] H. Mo, et al.: Improvement of Reasoning Model for Fuzzy Diagnosis System, Journal of Japan Society for Fuzzy Theory and Systems, Vol. 5, No. 3, pp. 565-576 (1993) (in Japanese)

[Nguyen78] H. T. Nguyen: On conditional possibility distributions, Fuzzy Sets and Systems, Vol.1, pp. 299-309 (1978)

[Peng90] Y. Peng, J. A. Reggia: Abductive Inference Models for Diagnostic Problem-Solving, Springer-Verlag (1990)

[Terano78] T. Terano, et al.: Diagnosis of Engine Trouble by Fuzzy Logic, Proc. IFAC 7th World Congress, pp. 1621-1628 (1978)

[Togai85] M. Togai: A fuzzy inverse relation based on Gödelian logic and its applications, Fuzzy Sets and Systems, Vol.17, pp. 211-219 (1985)

[Yamada95] K. Yamada: Fuzzy Abductive Reasoning for Diagnostic Problems, IFSA World Congress '95, Vol. 1, pp.649-652 (1995)

[Zadeh78] L.A.Zadeh: Fuzzy sets as a basis for a theory of possibility, Fuzzy Sets and Systems, Vol.1, pp. 3-28 (1978)

Model-Based Program Debugging and Repair*

Markus Stumptner and Franz Wotawa
Christian Doppler Laboratory for Expert Systems
Institut für Informationssysteme, Technische Universität Wien
Paniglgasse 16, 1040 Wien, Austria, Europe
Email: {mst,wotawa}@dbai.tuwien.ac.at

Abstract

The current state of the art in integrated circuit design is based on the use of special hardware design languages such as VHDL. In the context of the development of an intelligent, knowledge-based debugging aid for VHDL programs, we are dealing with analysis and diagnosis of a subset of VHDL (which is similar to conventional concurrent programming languages). We present an adaptation of conventional model-based diagnosis methods to the debugging of VHDL expressions and signal assignments. The examination of possible faults in VHDL expressions leads to the use of fault models (i.e., a representation of typical errors) as an aid to focusing, and as a basis for proposing repair actions for small errors in these programs.

1 Introduction

Over the last decade, the model-based approach has achieved widespread use in hardware diagnosis [Rei87, dKW87]. While model-based techniques have also found recognition in the field of design in general [Wil90], their application to software design in particular has remained very limited. In more recent years, different authors have examined the use of model-based diagnosis for debugging logic programs [CFD93], but their efforts were limited to small examples. The work described in this paper attempts to extend this basis towards a more generally usable approach for diagnosing software, which is not limited to the particularly benevolent environment of pure logic programming languages, and also scales up to meaningful error search in programs of a more realistic size.

In the DDV (Design Diagnosis for VHDL) project, the authors have applied model-based diagnosis to the debugging of hardware designs written in the VHDL language [FSW95]. Hardware specification languages such as VHDL (an IEEE standard and the most widespread such language in current use) combine the syntax and expressiveness of a full conventional programming language with special parallel constructs (to simulate the concurrent internal workings of integrated circuits), data types (e.g., the multivalued logics used by circuit designers to describe different states of signal lines). VHDL development environments are delivered together with large libraries containing predefined standard circuits, such as the standard logical gate types. The main reason for the use of design languages is the attempt to avoid, by whatever means possible, the surfacing of design errors after a lot of money has been spent on production of the first prototypes of the board or chip. Circuit designs are successively refined to a degree where they can be automatically transformed ("synthesized") into the final gate-level representation. Large hardware designs (comprising multiple ASIC's and microprocessors) can reach dimensions of several 100.000 lines of VHDL code and thousands of components and signals at the top level. For such designs, typically written by a large team of designers, testing, fault detection and localization become very time-consuming activities, because they involve simulations of an actual circuit that can take hours or days, scanning of the resulting signal traces by hand, and subsequent identification of the error in a large program parts of which may have been produced by outside subcontractors. In effect, these designs present all the problems of software debugging with the additional problem that test runs and actual finding of misbehavior take a lot of manpower and real time. An automated assistant for fault-finding thus

*This work was co-sponsored by the Christian Doppler Laboratory for Expert Systems and Siemens Austria

presents the possibility for major cost savings in the design process.

The tool developed in the DDV project is used for testing and diagnosing purposes. The trace files produced by a simulation run are compared with so-called specification traces, i.e., traces produced by the simulation of a less detailed, already tested version of the program. The output of the simulation run is a set of discrepancies, i.e., signals whose values differ between the two trace files. The set of discrepancies is passed to the diagnosis generator, which attempts to find the source of the error, with a model that is based on static analysis of the data dependencies inside the program. This approach has the significant advantage that the design can remain unaltered. No special code annotations are used, and separate axiomatizations (as commonly used by design verification systems) or complex test benches are unnecessary.

The current version of the tool (which is being tested by actual developers at the moment) uses a very abstract domain model which can optimally be used on the top level of VHDL programs but is not as good while diagnosing the lowest levels of the decomposition hierarchy making up the system. The model is, in effect, based on the topology of the data dependencies between different parts of the program (different components in the finished design). Detailed time values and signal values are ignored. One reason for this approach was the size of the systems involved. Diagnosing a 100,000 line program by reasoning about the detailed semantics of each individual expression in the program is effectively impossible for performance reasons. The abstract representation delivers an answer in minutes (which is a fraction of the time used for executing the simulation of the test run). However, in circuits with complex structure (i.e., with large feedback loops), the ability to discriminate between different possible diagnoses is limited.

The work described in this paper attempts to improve diagnosis quality in these cases, by introducing a more sophisticated model which uses detail information on signal values and the structure of assignments to achieve stronger focusing and discard more incorrect explanations for the fault. This requires analysis of the expressions that compute the values assigned to signals. Such expressions can contain arbitrary user-defined functions that are defined using the Ada-like sequential part of the VHDL language. Therefore, techniques for diagnosing faults not only in functional expressions, but also in imperative programs, may be ultimately required. Interestingly, the fact that the structure of a design may change over time (whereas the structure of a finished circuit is usually assumed to be unchanging), implies that considering repair actions may be a useful heuristic for focusing on the actual errors.

The paper is structured as follows: We first present the basic definitions of Model-Based Diagnosis (MBD) as they pertain to our particular domain. We then give (very small) example of the basic VHDL code, and discuss the various error types that could occur, as well as their representation by fault models in the MBD paradigm and the use of repair in focusing the search. Finally, we provide an overview of related research.

2 Basic Definitions

For the standard definition of model-based diagnosis, see [Rei87].

Definition 2.1 (Programming Language) *A programming language $\mathcal{L}$ is a tuple (L, S) consisting of a syntax L and a semantic description S.*

We assume that the description of the $\mathcal{L}$ semantics is given in first order logic. We use the term $COMP$ to denote the diagnosis components (statements, expressions, blocks) of a program.

Definition 2.2 (Program) *A program Π is an element of L^* where L^* denotes all syntactically correct programs of $\mathcal{L}$.*

Given a set of observations OBS describing the actual behavior of the program and a system description SD representing the semantics of the program, a debugging problem basically consists of fixing the program so the SD of the new fixed program will produce the predicted observations.

Definition 2.3 (Program Diagnosis) *Let Π be a program written in $\mathcal{L}$ and OBS be observations on values of variables (signals) used in Π. A diagnosis of Π is a function Δ associating a single mode with every (diagnosis) component such that the following proposition holds:*

$$OBS \cup S(\Pi) \cup \{\Delta(C) | \Delta(C) \in m(C) \wedge C \in COMP\}$$

The diagnosis task can be described as the search for an assignment of fault models to diagnosis components that explains the behavior given by the observations. It can be easily shown that the number of fault models for a program is infinite in general. However, use of a finite set of fault models will usually be sufficient. The usage of fault models can be guided by introducing additional information.

3 Problem Description

Consider the following simple VHDL program, implementing a three input and-gate with delay.

```
entity AND3_GATE is
    generic( delay:  time := 0 fs );
    port( signal I1, I2, I3:  in std_logic;
        signal O: out std_logic );
end AND3_GATE ;

architecture BEHAVIOR of AND3_GATE is
begin
    O <= I1 and I2 after delay;
-- Correct: O <= I1 and I2
--              and I3 after delay;
end BEHAVIOR;
```

Assume further that we have detected a discrepancy for the test case: $I1 = 1$, $I2 = 1$, $I3 = 0$.

In this case the fault in the program can be found by simply looking at the functional dependencies, i.e., which signals are involved in the computation of the incorrect signals. However, such a representation is limited in its expressibility because it gives no hint how the bug in the program should be corrected, although a number of simple alterations would produce the desired effect. For example, the correction given above is minimal in terms of the number of expressions changed in the program. A second possible (but not minimal) repair step consists of making two small program changes, resulting in

```
...     signal DUMMY: std_logic;
begin
    DUMMY <= I1 and I2;
    O <= I3 and DUMMY after delay;
    ...
```

Note that if an expression were syntactically restricted to contain one top-level operator (i.e., no multiple and's), the second suggestion would actually be preferable. In general, it can be said that program repair is limited by

- The syntax and semantics of the programming language.
- The original program and a metric related to the complexity of repair actions.
- The given test cases.

The task of a diagnosis-repair algorithm must be to compute a repair plan using the restrictions above. These repair plans can be ordered using a metric which assigns a high cost if many changes are necessary and a low cost if this is not the case. By generating only plans with a bounded cost, the search space can be restricted. Note that for repairing a program, this original search space is infinite, since arbitrarily complex alterations can in principle be made to any program without detrimental effect on its correctness (e.g., adding variables that are never used in computing the result, and computations on them).

A restriction of the search space can also be achieved through assumptions about the possible types of faults. If a finite number of such assumptions, called *fault modes* in model-based diagnosis, is used, the search space is again restricted to a subset of the original one. Depending on the fault modes allowed, that subset can be finite or not. The assumption of finiteness can be made because we assume that the originally intended program can be produced by a small variation of the incorrect program. Fault modes for parts of a program can be viewed as repair proposals – a fault mode, by specifying a particular kind of error, also offers possible choices of how to correct it. Assuming a typical block structure as is usually the case with concurrent or imperative languages, fault modes can be associated with statements, expressions inside statements and statement blocks. We call the description of a fault mode a fault model.

For example, take a VHDL signal assignment (as shown above) that produces incorrect system behavior even though the values for the input signals are correct for that particular point in execution. The target signal on the left can be wrong, the expression can be incorrect (in which case one has to look inside the expression to find the error), or the statement itself should be replaced by another statement. The last fault mode can be expressed by a replacement of the statement with a null statement that has no effect on signal values in conjunction with the addition of another statement.

The addition of a new statement can be viewed as fault mode of a statement block. Since VHDL is a concurrent language, a statement can be added at the end of a block without changing the meaning of the other statements for a given point in time. This is not possible for sequential languages where the location of the statement has normally an immediate influence on the behavior, i.e., the execution of the subsequent code. Obviously, adding a statement is a complex change, due to the many degrees of freedom involved: statement type, argument signals, target signal, and the functions used to compute the output value must be determined. The only information available are the test case and the context, e.g., all other statements in the program. It would therefore be unrealistic to assume that adding statements results in a meaningful limitation on the search space except in very simple or restricted cases such as the one depicted above.

4   System Description

This subsection shows how behavior and fault models can be described using first order logic. We demonstrate the system description using only the VHDL signal assignment statement with a reduced syntax.

$SignalAssignment ::=$
$\qquad Signal <= Expression \text{ \bf after } TimeExpression$

$Expression ::=$
$\qquad Signal \mid \text{Constant} \mid$
$\qquad (Expression \, BinOp \, Expression)$

The semantics of this statement can be expressed as first order rules. In the system description, the correct behavior of a statement is associated with the *ok* mode of the statement. In other words, a statement S is assumed to work correctly if $ok(S)$ holds.

We introduce a predicate *type/2* indicating the type of a concurrent statement, for example $type(SA, signalAssignment)$ holds if SA denotes a signal assignment. VHDL differentiates between five such types: *signalAssignment, processStatement, generateStatement, assertionStatement, componentInstantiation.*

The system description that is used for diagnosis is the union of the individual statement descriptions as given below. Note that this presentation is simplified and does not take the full VHDL behavior into account. A more sophisticated description is used in practice.

Every signal defined in a VHDL program has an associated initialization value (at time 0). This behavior is described by facts of the form

$value(S, V, 0), \; event(S, V, 0)$

where V is either a value which is related to the type of S, or an explicitly given statement. This statement is part of the signal definition. The *event* predicate indicates that a signal has changed its value.

To define program execution we first must define what expressions evaluate to. This is specified via the predicate *evaluate* as follows.

$evaluate(E_1 \, Op \, E_2, V, T) \leftrightarrow (evaluate(E_1, V_1, T) \wedge$
$\qquad evaluate(E_2, V_2, T) \wedge Op(V_1, V_2, V))$
$evaluate(C, C, T) \leftrightarrow constant(C)$
$evaluate(S, V, T) \leftrightarrow signal(S) \wedge value(S, V, T)$

The following rule describes the semantics of a correctly working concurrent statement of the form

$SA = S <= E \text{ \bf after } TE$, where V is the assigned value, and T' the time the value appears on S (as opposed to the time when the statement is executed):

$mode(ok(SA)) \rightarrow$
$\qquad (active(SA, T) \wedge evaluate(E, V, T) \wedge$
$\qquad evaluate(TE, D, T) \wedge T' = T + D$
$\qquad\qquad \leftrightarrow value(SA, S, V, T'))$

The *active* predicate is true if a statement can be executed at time T. This is the case when at least one input signal changes its value.

$value(S, V, T) \wedge value(S, V', T') \wedge$
$\nexists T'' : T' < T'' < T \wedge value(S, V'', T'') \wedge V \neq V'$
$\qquad \rightarrow event(S, V, T)$
$event(S, V, T) \wedge input(S, St) \rightarrow active(St, T)$

If a statement St is active is not true at a time point T, we assume that the value of the signals changed by St is equal to the value it had before T.

$\neg active(St, T) \wedge value(St, S, V, T') \wedge$
$\nexists T'' : T' < T'' < T \wedge value(St, S, V'', T'') \rightarrow$
$\qquad value(St, S, V, T)$

If multiple concurrent statements attempt to assign values to the same signal, the signal needs to be *resolved* (see [VHD88] for more details). A concurrent statement that change a signal's value is called a driver for this signal. We introduce a predicate *drivers* defining all drivers for a signal.

$drivers(S, \{St_1, \ldots, St_n\})$

The value of the signal is now determined by the values of the signal drivers. Therefore add the following rule to the system description:

$drivers(S, D) \wedge valueD(S, V, T, D) \rightarrow value(S, V, T)$
$value(St, S, V, T) \rightarrow valueD(S, V, T, \{St\})$
$(value(St_1, S, V_1, T) \wedge valueD(S, V_2, T, \{St_2, \ldots\}) \wedge$
$\qquad resolved(V_1, V_2, V)) \rightarrow valueD(S, V, T, \{St_1, St_2\})$

5   Fault modes

We now come to the description of the various possible fault modes. First, consider an assignment SA that assigns to the wrong target signal, with S' being the correct signal:

$mode(wrongTarget(SA, S')) \rightarrow$
$\qquad (active(SA, T) \wedge evaluate(E, V, T) \wedge$
$\qquad evaluate(TE, D, T) \wedge T' = T + D$
$\qquad\qquad \leftrightarrow value(SA, S', V, T'))$

We can restrict the use of fault modes applicable to an incorrect signal assignment to those that deal with signals not already known to be correct due to observations. If we assume that the expression returning the value is incorrect, this is formalized as follows:

$$mode(wrongExpression(SA, E')) \rightarrow$$
$$(active(SA, T) \land evaluate(E', V, T) \land$$
$$evaluate(TE, D, T) \land T' = T + D$$
$$\leftrightarrow value(SA, S, V, T'))$$

where $SA' \equiv (S <= E' \text{ after } TE)$ is an alternative statement, and $timeExpression(SA, TE)$ holds. A fault model describing that a statement should be excised can be formalized by:

$$mode(wrongStatement(SA)) \rightarrow \top$$

Next, we describe the case where the behavior of the statement is not known, e.g. the ab mode usually used in model-based diagnosis. We denote this mode by $undef$.

$$mode(undef(SA)) \rightarrow$$
$$(active(SA, T) \land evaluate(TE, D, T) \land$$
$$T' = T + D$$
$$\rightarrow value(SA, S, \epsilon, T'))$$

ϵ is a value not used in any signal type indicating that the actual value is unknown. Operations on this value always cause as result the value ϵ.

Finally, a program could also be modified by adding complete statements to it and leave the rest unmodified. We do not discuss this (rather complex) operation due to space reasons.

Basically, the model-based diagnosis process works by attempting to find fault mode assignments that are consistent with the observations. Observations are represented by two facts for each observation.

$$value(S, V, T), event(S, V, T)$$

Contradictions occur if different values are assigned to the same signal:

$$(value(S, V, T) \land value(S, V', T) \land V' \neq V \land$$
$$V' \neq \epsilon \land V \neq \epsilon) \rightarrow \bot$$

The aim of the repair task is to determine a new expression or statement explaining the faulty behavior, i.e., replacing the incorrect part of the program with the correct one. The search space for this task can be limited by the knowledge of the correct behavior, e.g. by knowing which signals depend functionally on other signals, and by the assumption that the new expression should be a small variant of the old one. This assumption explains the difference between the automatic programming community and our notion of correcting faults. While automatic programming deals with the question of automatically writing programs

out of (a series of) examples or (logic) descriptions, we use an existing program and search for a small variant.

We do not discuss in this paper how temporal misbehavior can be found and corrected, but we claim this can also be achieved using appropriate fault models.

Note that the fault models as used above effectively induce a repair approach. If a diagnosis has been found, the associated fault models specify how to repair the program with regard to the test cases used. This is a significant difference to the standard hardware diagnosis approach, which is made possible by the malleability of software. In effect, finding the repair means that it can be immediately executed by altering the source code, whereas the opportunity for arbitrary modifications is usually very limited in hardware devices.

6 Related Research

The notion of using knowledge-based systems as a support for software and related design tasks is not new. Previous research efforts usually aimed at working with languages that had strictly limited expressiveness, such as telecommunications protocols [Rie93] or logic programming [CFD93]. The latter paper also presents arguments for the superiority of model-based reasoning over the algorithmic program debugging approach originating in [Sha83].

Much work has also been done in knowledge-based debugging support for software. We cite two examples. The PROUST system [Joh86] requires a separate description of the program to be manually developed and entered. Similarly, in [AC91], an axiomatization of the intended functionality is necessary. The **Talus** system [Mur88] attempts to match student solutions to example programs to find the source of divergent outputs. Both for PROUST and **Talus**, the goal is to aid students, i.e., novice programmers, in finding as many bugs as possible in small programs. In contrast, in our case, the goal is to help experts (at least, experienced users) orient themselves in very large programs of different structure and functional architecture.

Finally, we claim that our approach avoids many of the problems arising in automatic programming [WL69]. We do not deal with the idea of finding a program out of examples or a formal description. We only require that correct functioning of the program can be attained by small variations of the original, incorrect one. The search space for finding such a variant is much smaller and can be further restricted by making heuristic assumptions about fault models. Of course restricting the number of fault models may exclude diagnoses. However, in this case, the user or the program itself still has the choice of relaxing the assumptions at the cost of a larger search space.

7 Conclusion

In this paper, we have introduced a model representing the semantics of the subset of the VHDL language (that deals with concurrent statements and expression evaluation) and its usage for diagnosing real-world VHDL programs and making limited repair proposals.

We plan to integrate the more sophisticated system description presented in this paper into the existing VHDL diagnosis tool. This can be achieved by moderate alterations and use of the new system description. The existing diagnosis tool consists of a theorem prover which is used to generate contradictions from SD and OBS, and a hitting set generator that chooses fault modes and uses the theorem prover for computing diagnoses. See [Rei87, GSW89] for an introduction about model-based diagnosis and the hitting set algorithm. VHDL programs are parsed and automatically converted into the SD representation.

The first version of our diagnosis tool, including a tool for automatic detection of misbehaviors by comparing waveform traces of VHDL programs, is currently being used in practice. Experience has shown that the current tool is well suited for automization of regression tests and additionally help the VHDL programmer to focus on the right diagnoses. However, more information about possible diagnoses and hints about how to repair a misbehavior will increase the effectiveness of the tools in many cases.

The fact that fault models (usually used in model-based diagnosis to improve the diagnosis result) are equal to possible repair actions in the software domain is crucial to this approach. We have introduced a number of ways for limiting the infinite space of possible repair actions. In particular, we assume that the corrected program is only a variant of the actual one that can be found with limited alterations. This is a very useful heuristic for ordering the search space, although more involved methods must be used to deal with fundamental structural defects in the program [Liv94]. Unfortunately, such methods have also never been used with regard to larger systems. This is an ongoing research effort.

References

[AC91]   Dean Allemang and B. Chandrasekaran. Maintaining knowledge about temporal intervals. In *Proc. KBSE*, pages 136 – 143. IEEE, 1991.

[CFD93]   Luca Console, Gerhard Friedrich, and Daniele Theseider Dupré. Model-based diagnosis meets error diagnosis in logic programs. In *Proc. IJCAI*, pages 1494–1499, Chambery, August 1993. Morgan Kaufmann Publishers, Inc.

[dKW87]   Johan de Kleer and Brian C. Williams. Diagnosing multiple faults. *Artificial Intelligence*, 32:97–130, 1987.

[FSW95]   Gerhard Friedrich, Markus Stumptner, and Franz Wotawa. Model-based diagnosis of hardware designs. In *Proceedings on the Sixth International Workshop on Principles of Diagnosis*, Goslar, October 1995.

[GSW89]   Russell Greiner, Barbara A. Smith, and Ralph W. Wilkerson. A correction to the algorithm in Reiter's theory of diagnosis. *Artificial Intelligence*, 41(1):79–88, 1989.

[Joh86]   W. Lewis Johnson. *Intention-Based Diagnosis of Novice Programming Errors*. Pitman Publishing, 1986.

[Liv94]   Beat Liver. Modeling software systems for diagnosis. In *Proceedings on the Fifth International Workshop on Principles of Diagnosis*, pages 179–184, New Paltz, NY, October 1994.

[Mur88]   William R. Murray. *Automatic Program Debugging for Intelligent Tutoring Systems*. Pitman Publishing, 1988.

[Rei87]   Raymond Reiter. A theory of diagnosis from first principles. *Artificial Intelligence*, 32:57–95, 1987.

[Rie93]   Marc Riese. *Model-based diagnosis of communication protocols*. PhD thesis, EPFL Lausanne, 1993.

[Sha83]   Ehud Shapiro. *Algorithmic Program Debugging*. MIT Press, Cambridge, Massachusetts, 1983.

[VHD88]   IEEE Standard VHDL Language Reference Manual LRM Std 1076-1987, 1988.

[Wil90]   Brian C. Williams. Interaction-based invention: Constructing novel devices from first principles. In *Proceedings AAAI*, Boston, August 1990. The MIT Press.

[WL69]   R. J. Waldinger and R. C. T. Lee. Prow: A step toward automatic program writing. In *Proc. IJCAI*, Washington, DC, May 1969.

AN OBJECT-ORIENTED METHOD FOR CONTROL LOOP DIAGNOSIS AND MONITORING

P.A. Collier[1] and M.G. Casey[2]

[1]Department of Computer Science, University of Tasmania, GPO Box 252C, Hobart 7001, Australia
[2]Pasminco-Metals–EZ, GPO Box 377D, Hobart, 7001, Australia

ABSTRACT

We present a control loop diagnosis and monitoring method (CLDMM). This is a symbolic method which has been designed to be reusable. Reusability is facilitated by the object-oriented framework of CLDMM, and by generic messages, which can be adapted or used directly with CLDMM. CLDMM has been used in a prototype expert system at Pasminco Metals-EZ in Hobart, Australia.

1 INTRODUCTION

There has been much work recently on computer systems for diagnosis of malfunctions in process control. Conventional control techniques are only of limited use in such situations. Instead a more flexible symbolic representation of rules and procedures may be used to perform diagnosis. The rules and procedures may include heuristics derived from experience of the best operators over time, sometimes called shallow reasoning, or a more principled model of the process, sometimes called deep reasoning, see for example [DK91].

Symbolic representations, particularly when using production rules, are very flexible and unstructured. This makes verification of the represented knowledge very difficult. [RSB93] is an example of an attempt to overcome this problem in the context of process control and monitoring. [SWD+94] describe the CommonKADS framework, a method for specifying a conceptual model separately from its implementation.

A class-object hierarchy is a convenient and popular formalism in expert system shells, for example NEXPERT OBJECT™, G2 and LEVEL-5 OBJECT™. Classes are used to represent abstract properties of various concepts, like an instrument, or flow meter. If classes are arranged in a hierarchy then the properties of more general classes may be inherited by more specific classes, for example a flow meter is a type of instrument, and should inherit all properties of an instrument. A particular physical object will be represented by an instance of the relevant class. The key idea is to attach fault finding rules to the most general concept to which it applies. It will then apply to all the more specific classes and their instances. [RSB93] provide a good introduction to the use of rules in the context of a class object hierarchy.

The idea of attaching rules to classes is significant because it provides structure to the knowledge base. No longer are a collection of fragmented production rules required, with likely inconsistent treatment of different instruments, and complete omission of others. They are all covered by a uniform method. Special cases suggest that the underlying class structure may need modification.

The major contribution in this paper is a reusable method for monitoring and diagnosing problems in control loops. We call this the control loop diagnosis and monitoring method (CLDMM). CLDMM is object-oriented as it applies to a general control loop represented as a class. It combines an appropriate mixture of shallow reasoning and deep reasoning and is easily applied in practice. We also provide outline messages to offer advice to operators for any diagnoses produced by CLDMM. Of course, some work is still required to provide certain information about individual control loops (the instances) for the method to be useful in practice.

CLDMM is a traditional symbolic method, which does not attempt any form of mathematical modelling of a control loop. It is easily read and comprehended. In this sense, it serves as a practical illustration of the way that symbolic rules may be used in any expert system.

CLDMM is designed for use in a relatively slow process. It may be used every minute in response to a synchronous snap shot of process measurements and features extracted from those measurements. A different approach would be required for a more dynamic real-time process with asynchronous data available. [IGR92] describe techniques for handling this situation.

2 BACKGROUND

The work in this paper is based on a prototype expert system developed for the Strong Acid Leach (SAL) section of the Pasminco Metals–EZ (PMEZ) Zinc Refinery in Hobart, Tasmania. The system identifies several classes of faults similar to those outlined by [BL93]: "operator error, malfunction or failure of process equipment, or undesired changes in the conditions of the process environment".

The purpose of the SAL plant is to dissolve zinc and iron from a slurry, leaving a silver/lead residue which is a valuable by-product of the zinc process.

The dissolution of zinc and iron is achieved by the controlled addition of acid to the process slurry in two tanks in series. The residual solids are then separated from the solution in a series of three countercurrent thickeners. The underflow from the third thickener is filtered, the filtrate being returned to the thickeners' feed.

A process *control loop* consists of a *set point* (SP) and a *measured variable* (MV) together with some mechanism to change the value of the measured variable if it is not equal to the set point. Physically the mechanism might be a valve or a variable speed drive (VSD) for a pump. The physical setting of the valve position or drive speed is often determined by numerical algorithm, for example a classical PID (proportional/integral/derivative) controller. Such a controller is able to control the flow in a line given minor variations in pressure, for example.

The set point of a control loop may be set *locally* by manual input of a value or *remotely* by automatic calculation. For example, the flow of a process fluid may be determined by a local set point, but acid flow may be determined remotely depending on process acidity.

A remote set point may be determined by another PID controller called a *supervisory-control-loop*. The output of a supervisory-control-loop is not connected to a physical device such as a valve or variable speed drive, instead it is connected to the set point of its *slave-control-loop*. For example, a temperature controller may have available a supply of cold process fluid to add to a tank. The temperature controller has a set point and its measured variable is the process temperature. The output of the control loop is a set point for the flow of the cold process fluid.

A supervisory-control-loop can be related in one of two ways to its slave. Sometimes an increase in the set point of the supervisory control loop will correspond to an increase in the set point of the slave-control-loop. For example, if acidity is low as measured by the supervisory-control-loop, then increasing the set point of the acidity will increase the set point in the acid flow control loop. For other supervisory-control-loops an increase in the set point will correspond to a decrease in the set point of the slave-control-loop. For example, in the temperature controller described above, an increase in temperature set point should result in a decrease in the cold flow set point. We define the notion of polarity-up, which belongs to any control loop. If *polarity-up* is *same*, then an increase corresponds to an increase in the sense above. Conversely

if polarity-up is *opposite* then an increase corresponds to a decrease.

In some situations a *triple-cascade-control-loop* is used. A triple cascade control loop has two supervisory-control-loops. One supervisory-control-loop, in the middle, receives a set point from its supervisor and provides a set point for the slave-control-loop. For example, acidity may be measured periodically by an on-stream analyser. This is used as measured variable in a supervisory-control-loop which provides a conductivity set point. Conductivity is measured continuously by a supervisory control loop for acid flow. In this example, both of the polarity-ups are same.

We define the notion of a *controller*, which consists of a slave-control-loop attached to physical device and any supervisory-control-loops which provide its set point, or a set point of one of its supervisors. For example a triple-cascade-control-loop is one controller.

We follow the advice given by [Lee84] when discussing a computer system to assist with diagnosis and analysis of alarms: "It is not practical to rely on an unsystematic manual analysis and to write a separate program for each plant since the time and effort required are excessive. A systematic approach is necessary. In principle, this may be achieved by dividing the program into two parts, a basic analysis program of general applicability and a data structure for a particular plant."

Object-oriented design offers a method for implementing the advice from [Lee84]. Such objects can be implemented directly in many expert system shells. We have used NEXPERT OBJECT™ from Neuron Data.

[YKK+93] used an object-oriented representation for process equipment selection. This was also implemented in NEXPERT OBJECT. It is not clear that they have used object-oriented facilities in the same way as we have here, however ([YKK+93], figure 1) shows an object-oriented representation of different types of heat exchanger.

In CLDMM, relationships between the different types of control loop may be represented as a *class hierarchy* shown in figure 1. In this figure we show the classes of different control loops. The classes higher in the hierarchy represent general concepts, those lower down are more specific. Particular control loops will be objects which are instances of the appropriate class. For example, the SATH3OFTH2-F is an instance of a slave-control-loop class in the SAL plant. This controls the overflow flow from thickener 3 to thickener 2.

Properties of classes are inherited downwards through the class hierarchy. The class *VSD-control-loop* has the properties shown in figure 2.

Properties have values. Some of these values will be constants, such as the value for *Controlled-property* in each instance of the class *VSD-control-loop*, while others will be computed from the current state of the process.

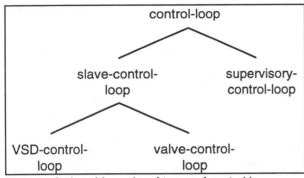

Figure 1 A class hierarchy of types of control loops.

We have argued previously [Col95] that models of reasoning for expert systems should be presented in KADS, Knowledge Analysis and Documentation System [SWD+94]. This is one of the best developed amongst several different modelling formalisms, see for example [FB93] and [Bar93]

A major contribution from the KADS methodology is that knowledge should be represented in layers. The layers encourage a principled representation of knowledge, where the domain itself is modelled directly. Frequently in the domain layer there is an object-oriented model. The KADS methodology discourages ad hoc production rules.

In this paper we have chosen not used KADS to present CLDMM. This choice reflects the fact that CLDMM is essentially a decision tree, which is a relatively simple structure. Dressing this up in KADS may serve only to obscure this simplicity.

3 DIAGNOSING AND MONITORING CONTROL LOOPS

For each control loop there are two basic attributes which may be of interest. Firstly, the control loop may have unsatisfactory performance. Secondly, there may be reasons which can cause unsatisfactory performance. Generally, for diagnosis and monitoring, we are not interested in the reasons unless unsatisfactory performance is detected.

We identify two overall categories for when a control loop has unsatisfactory performance: *desired-MV-lower* or *desired-MV-higher*. These concepts represent a situation where the measured variable of the slave control loop should be lower or should be higher respectively. For example:

- desired-MV-lower may be present if a set point is 20 m^3/hr and the corresponding measured variable is 10 m^3/hr.
- there may be a tank upstream which is overflowing and a plant model suggests a desired-MV-higher downstream to help overcome the problem.
- a tank temperature may be close to the high alarm value, suggesting a desired-MV-lower upstream at a heater exchanger.

In general a demand for desired-MV-lower or desired-MV-higher can be determined from one of the following classes of problems:

- a significant deviation between the set point and measured variable of the slave control loop (or any of its supervisors).
- a measured variable close to its high or low alarm

Alarm_HI	Measured_slope	Output_oscillating
Alarm_HIHI	Measured_variable	*Output_tracking_condition*
Alarm_LO	MV_close_to_zero	*Parent_controller*
Alarm_LOLO	MV_erratic	*Polarity_up*
Blocked_line	MV_high	Possible_blockage_5_mins
Changes_rapidly	MV_less_than_set_point	Set_point
Compensating_reason	MV_low	*Set_point_high*
Controlled_property	MV_more_than_set_point	*Set_point_low*
Controller_description	*MV_more_than_set_point_description*	Set_point_mode
Desired_MV_higher	MV_oscillating	Set_point_range
Desired_MV_lower	*MV_tolerance*	Set_point_slope
Full_range	Output	Supply_feed_high
In_use	Output_HI	Supply_feed_low
Master	Output_LO	
Measured_average	Output_mode	

Figure 2 The properties of the class *VSD-control-loop*. Properties in italics are statically defined, some of these are text messages for output, some define tolerances for the control loop. Properties in plain type are dynamical and their value depends on the current state of the process. (An instance of the class *VSD-control-loop* is generally also an instance of a *Pumper* class, which introduces extra properties concerning duplicated pumps; and an instance of the class *Flow-meter*, or some other instrument.)

values.

- an excess or depleted capacity upstream of the control loop.
- an excess or depleted capacity downstream of the control loop.

For each control loop a selection of these will apply, and will have to be tailored to the specific aspects of that control loop. The upstream and downstream capacity problems can be identified by local events (see the examples above), but deeper modelling of the plant would make these more reliable.

The reasons for a control loop to have unsatisfactory performance can be simple things such as output in manual or set point in local. More significant things such as a line blockage, pump turned off, or pump faulty are also possible causes. These causes may be handled by specific rules for each control loop, or by further object-oriented rules. For example, it is possible to set up an object-oriented decision procedure to count the number of pumps running in an object-oriented framework. The details of this are outside the scope of this paper.

Figures 3 and 4 contain CLDMM for diagnosing and monitoring control loops with the problem *desired-MV-higher*. CLDMM for *desired-MV-lower* are similar, but lack the check for a line blockage and analogues of *Output-HI—MV-LO—desired-MV-higher—VSD* and *Output-HI—MV-LO—desired-MV-higher—valve*.

For each controller, use of the decision procedures in figures 3 and 4 results in a problem diagnosis such as *Supply-feed-low—desired-MV-higher*. When such a diagnosis is generated, a message should be presented to the plant operators giving information on how to fix the problem. An example of the message format we use is shown in figure 5. Generic justifications are available for all of the diagnoses. These need to be augmented with specific reasons where available, and may be further specialised to plants other than SAL at Pasminco Metals-EZ.

4 USING THE METHOD FOR DIAGNOSIS

A prototype expert system has been installed at Pasminco Metals-EZ using CLDMM for analysing control loop problems. CLDMM as shown in figures 3 and 4 has been seen to provide reasonable advice. More work is needed to find the signatures of blocked lines, for example. This is a symptom of the more general problem of how to represent qualitatively, various features of the process trends which are readily apparent to experienced operators. This issue is now receiving serious attention, for example [CS90], the first in a series of papers, and [WD92].

The generic check lists shown in figure 5 were also seen as being appropriate. They do need to be carefully tailored for each individual control loop. With experience,

operators know the common faults which occur and any special peculiarities of each control loop. This should be reflected in the messages.

The major limitation of CLDMM is that it has no representation of time nor of likely fault propagation. This could be addressed by a supervisory procedure which filters messages produced. We have implemented a simplistic rule which only reports messages which are new since the method was last executed, one minute previously. A more sophisticated version would have a known dead band period for each control loop, or an acknowledgment from the operator that a problem has been fixed. There could also be some means of predicting that an upset in some control loop (or other problem detected by rules not described in this paper) would affect another control loop, and disabling messages for that control loop for an appropriate period. In dynamic real time systems this issue is much more significant with temporal reasoning and fault propagation modelling having received considerable attention in the literature, for example [PKB+92].

5 SUMMARY AND CONCLUSION

This work has two main aims. One is to combine a collection of potentially unstructured production rules for diagnosis and monitoring into a single coherent method, CLDMM. The second is to make CLDMM reusable. In these two aims we have succeeded. There still exist several outstanding problems such as qualitative feature extraction, temporal reasoning and fault propagation modelling.

6 ACKNOWLEDGEMENTS

This work has been supported by a CSIRO Collaborative Programme in Information Technology and an Australian Research Fellowship (Industry).

7 REFERENCES

[Bar93] Barbuceanu, M. 1993, Models: toward integrated knowledge modelling environments, *Knowledge Acquisition*, 5, 245–304.

[BL93] Becraft W. R. and P. L. Lee 1993, An Integrated Neural Network/Expert System Approach for Fault Diagnosis, *Computers chem. Engng.*, 17, 1001–1014.

[CS90] Cheung, J. T.-Y. and G. Stephanopoulos 1990, Representation of process trends—Part I. A formal representation framework, *Computers chem. Engng.*, 14, 495–510.

[Col95] Collier, P. A. 1995, A Conceptual Model for Monitoring and Fault Identification in Industrial Instruments using Data from a Distributed Control Computer, in G. F. Forsyth & M. Ali

For any *slave_control_loop*:

1   **if**   desired-MV-higher is true
    and   measured-slope < 0.2
    then   go to 2
    else   No-problem
If the measured variable is changing in the correct direction, then there is no need to look for reasons for the problem.

2   **if**   MV is within 5% of the maximum value
When the MV is already high there is little scope for increasing it further.
    then   Desired-MV-higher—controller
If we get to here then we have been unable to find a cause for the problem.
    else   go to 3

3   **if**   output is at its maximum value
    and   output-mode is "AUTO"
When the output of a control loop has ramped up automatically to its maximum value this indicates that there may be a fault in the process.
    then   go to 4
    else   go to 9

4   **if**   supply-feed-low is true
For example, a pump which provides supply feed may be turned off.
    then   Supply-feed-low—desired-MV-higher
    else   go to 5

5   **if**   slave-control-loop is a VSD-control-loop
If not a VSD-control-loop then it will be a valve-control-loop.
    then   go to 6
    else   go to 8

6   **if**   blocked-line is true
It is possible to find the signature of a blocked line for different types of pumps. For example, a a centrifugal pump will show a small decrease in pump amps.

    then   Blockage—desired-MV-higher—VSD
    else   go to 7

7   **if**   MV-close-to-zero is true
    then   Output-HI—MV-LO—desired- MV-higher—VSD
    else   Output-HI—desired-MV-higher—VSD

8   **if**   MV-close-to-zero is true
    then   Output-HI—MV-LO—desired-MV-higher—valve
    else   Output-HI—desired-MV-higher—valve

9   **if**   output-mode is manual
    then   Output-manual—desired-MV-higher
    else   go to 10

10   **if**   set-point-mode is local
    and   MV-less-than-set-point is false
    and   set-point is not within 5% of the maximum value of the set point
We should not recommend a change to the set point if the reason for the desired-MV-higher is that it is significantly less than the set point, or if the set point is near the top of its range. It may be worth recommending a change to the set point if there is a tank overflowing upstream, for example.
    then   Set-point-local—desired-MV-higher
    else   go to 11

11   **if**   there is a supervisory control loop
    then   go to 12
    else   Desired-MV-higher—controller

12   **if**   polarity-up of the supervisory control loop is opposite
    then   Use decision graph for desired-MV-lower with the supervisory control loop
    else   Use decision graph for desired-MV-higher with the supervisory control loop

Figure 3 CLDMM for the problem *desired-MV-higher*.

(eds) Industrial and Engineering Applications of Artificial Intelligence and Expert Systems, 273–280, Gordon and Breach.

[DK91]   Dvorak, D. and B. Kuipers 1991, Process Monitoring and Diagnosis: A Model-Based Approach, *IEEE Expert*, 6, 3, 67–74.

[FB93]   Ford, K. M. and J. M. Bradshaw 1993, *Knowledge Acquisition as Modelling*, Wiley, New York.

[IGR92]   Ingrand, F. F., M. P. Georgeff and A. S. Rao 1992, An Architecture for Real-Time Reasoning and System Control, *IEEE Expert*, 7, 6, 34–44.

[Lee84]   Lees F. P. 1984, Process computer alarm and disturbance analysis: outline of methods for systematic synthesis of the fault propagation structure, *Computers chem. Engng.*, 8, 91–103.

[PKB+92]   Padalkar S., G. Karsai, C. Biegl, J. Sztipanovits, K. Okuda and N. Miyasaka 1992, Real-Time Fault Diagnosis, *IEEE Expert*, 6, 3, 75–85

For any Supervisory control loop with the problem *desired-MV-higher*:

1   **if**   output is at its maximum value
     and   output_mode is auto
     then   Output-HI—desired-MV-higher—supervisor
     else   go to 2

2   **if**   output_mode is manual
     then   Output-manual—desired-MV-higher—
         supervisor
     else   go to 3

3   **if**   set-point-mode is local
     and   MV-less-than-set-point is false
     and   slave-MV-less-than-set-point is false
     and   set-point is not within 5% of the maximum
         value of the set point
We should not recommend a change to the set point if the reason for the desired MV higher is that it is significantly less than the set point, or the set point of the slave control loop is significantly less than the set point, or if the set point is near the top of its range. It may be worth recommending a change to the set point if there is a tank overflowing upstream, for example.
     then   Set-point-local—desired-MV-higher—
         supervisor
     else   go to 4

4   **if**   there is a supervisory control loop
The name of a supervisor of a Slave_control_loop is attached to the slot Supervisor.
     then   go to 5
     else   go to 6

5   **if**   polarity-up of the supervisory control loop is
         opposite
     then   Use decision graph for desired-MV-lower with
         the
         supervisory control loop
     else   go to 1 with the supervisory control loop
Repeat this decision procedure for the supervisory_control_loop's supervisor.

6   **if**   polarity-up of the supervisory control loop is
         opposite
If we get to here then we have been unable to find a cause for the problem.
The polarity of the controller is the same as that of its slave control loop. For example there may be a controller with slave and opposite polarity supervisor. In this case if no cause can be determined for a desired-MV-lower in the slave and then desired-MV-higher in the supervisor, we have to report a Desired-MV-lower—controller for the controller.
     then   Desired-MV-lower—controller
     else   Desired-MV-higher—controller

Figure 4 CLDMM for a *supervisory-control-loops* with the problem *desired-MV-higher*.

[RSB93]   Renard, F.-X., L. Sterling and C. Brosilow 1993, Knowledge Verification in Expert Systems combining Declarative and Procedural Representations, *Computers chem. Engng.*, 17, 1067–1090.

[SWD⁺94] Schreiber, G., B. Weilinga, R de Hoog, H Akkermanns & W. Van de Velde 1994, CommonKADS: A Comprehensive Methodology for KBS Development, *IEEE Expert*, 9, 28-37.

[WD92]   Whiteley, J. R. and J. Davis 1992, Knowledge-Based Interpretation of Sensor Patterns, *Computers chem. Engng.*, 16, 329–346.

[YKK⁺93] Yang, J., T. Koiranen, A. Kraslawski and L. Nistrom 1993, Obeject-Oriented Knowledge Based Systems for Process Equipment Selection, *Computers chem. Engng.*, 17, 1181–1189.

```
Increase flow on the following loop.
DCS tag          Equipment tag  Description
SATH3OFTH2-F FIC0052 Thickener 3 flow to thickener 2
    Reason: SATH3OFTH2-F: MV is 100% below set-point
    The flow is low because: One of the faulty pumps is required
```

Figure 5 An example of the message which is produced for the diagnosis *Supply feed low—desired MV higher*. The components of this message are included in slots of the instance SATH3OFTH2-F. *DCS tag, Equipmen tag* and *Description* are constant valued slots. *Reason* is generated at run time to describe the unsatisfactory performance and *The flow is low* is also generated at run time to describe a possible cause for unsatisfactory performance. In this case there are duplicate pumps available to drive the flow through SATH3OFTH2, both are out on overload.

A SELF-LEARNING FUZZY SYSTEM FOR AUTOMOTIVE FAULT DIAGNOSIS

Yi Lu

Department of Electrical and Computer Engineering
The University of Michigan-Dearborn
Dearborn, MI 48128-1491
USA
yilu@umich.edu

ABSTRACT

This paper describes a self-learning fuzzy system for automotive fault diagnosis. We first present a formal model for fuzzy engineering diagnosis including problem modeling, selecting optimal parameters, and a learning algorithm for adjusting fuzzy membership functions. A fuzzy diagnostic system developed under this model is expected to produce better diagnostic results, requires minimum process time, and can be easily adapted to different vehicle models. A fuzzy system for detecting Lambse fails in an Electronic Engine Controller(EEC) in a vehicle is developed and implemented. The system is designed to be a component of an End-of-Line test system at automobile assembly plant.

1. INTRODUCTION

The success of U. S. motor vehicle industry very much depends on the quality of product. As automotive electronic control systems have become more advanced and sophisticated in recent years, malfunction phenomena have also become increasingly more complicated. It has been well recognized in the automotive industry that effective vehicle diagnostic systems will play a key role in the competitive market of the 2000. In order to meet this challenge, the major US automotive companies are in the process of launching an end-of-line test system at every North American assembly plant. In order to accomplish this task, a system is designed to collect and analyze Electronic Engine Controller(EEC) data while the vehicle is dynamically tested. Operators drive the vehicles through a preset profile and the vehicles are either passed or failed according to the data collected during the tests. The decision is made based on two information sources, the EEC on-board tests and off-board test performed by the vehicle test system on EEC generated data. This project focuses on solving the second problem. The End-of-Line calibration engineer has two primary responsibilities. The first is to ensure that vehicles conform to engineering standards. The second is guard against falsely failing vehicles that do conform to Ford engineering standards.

Diagnosis of impending or actual faults through the interpretation of patterns in operating data or observed characteristics of a current malfunction are generally performed by specialists who are trained in the interpretation of such patterns. The existing techniques used in assembly plants for the off-board test are mainly on a trial and error basis and largely dependent on individual engineers' experience. The mechanics in the assembly plants have varying degrees of vehicle diagnostic expertise. They are in general very weak in fundamentally EEC operations, signals, and etc. Therefore, there is a strong need for developing an intelligent system to automatically diagnose EEC failures. The nature of the application requires such an intelligent system to provide **prompt and reliable diagnosis.** The work described in this paper is an attempt to develop such a system.

Fault diagnosis has been a classic engineering problem. Techniques for diagnosing faults in dynamic systems ranging from expert diagnosis with the assistance of various instruments to statistic models[RSP93] . In more recent years, artificial intelligence(AI) has played an important role in engineering diagnosis. Diagnostic knowledge usually involves several stages of reasoning and contain a mixture of knowledge gained by training, from maintenance and repair manual, and through accumulated experience. The AI based diagnostic techniques include model-based, Bayesian methods, cased-based methods, and rule-based expert system[CHO94, FMO84]. The major challenge for automotive diagnosis is that the electronic control components in modern vehicles are more like a blackbox. It is difficult for mechanics to diagnose faults accurately unless they are thoroughly familiar with the system specifications and functions. As a result, it is extremely difficult to develop a complete diagnostic model that can fully answer all the questions in terms of faults.

The theory of fuzzy is aimed at the development of a set of concepts and techniques for dealing with sources of uncertainty or imprecision and incomplete. It has been successful in many applications including control theory when gradual adjustments are necessary, air and spacecraft control, business and even the stock exchange. Fuzzy systems are particularly well suited to modeling nonlinear systems. The nature of fuzzy rules and the relationship between fuzzy sets of differing shapes provides a powerful capability for incrementally modeling a system whose complexity makes traditional expert system, mathematical, and statistical approaches very difficult. The following sections will describe a fuzzy model for automotive fault diagnosis.

2. A SELF-LEARNING FUZZY DIAGNOSTIC MODEL

The fuzzy diagnostic model is designed based on the following two major design criteria:
- Generic: The model should provides a systematic approach to build and test a diagnostic system applicable to any one typical problem within the scope of automotive fault diagnosis. For example, the model can be applied to detect vacuum leak in EEC as well as transmission shifts.
- Adaptive: When a diagnostic system is developed under the proposed fuzzy model to diagnose one particular type of faults, the system should be easily adapted to diagnose the same type of faults on different vehicle models. For example, a vacuum leak diagnostic system developed based on Ford Thunderbirds should be easily adapted to detect vacuum leak for Lincoln Town Car.

In order to satisfy these two criteria, the entire process in building the components in the fuzzy diagnostic model should be heavily automated.

Because of the nature of application, the fuzzy diagnostic model must satisfy the following constraints:
- no extra sensory devices can be added to the test sites to acquire extra data. The system should use exclusively the data currently available at the test site.
- The system should be efficient. The proposed system will be integrated as a component into a real-time diagnostic system in assembly plants for end-of-line test. According to the assembly plant operation, the speed of the system is an important factor to be considered during the design process.
- The system should be implemented on PC platform. Since PC or other desktop computers are the major computer platforms used in assembly plants.

The last two constraints put restriction on the development of algorithms: the computations used in the fuzzy diagnostic system should be efficient and the results should be reliable. The algorithms in the fuzzy model cannot be too complicated to require a large amount of execution time or demand the computing power beyond the desktop computers. A diagnostic system hinders any of these primary constraints, it will not likely to be used in the automobile assembly plant. These application constraints provide us guidance through the design of the fuzzy model. For example, when we examine several design options, if one option is more expensive to implement but does not warrant significant improvement of performance, the option will not be selected.

The fuzzy diagnostic model is illustrated in Figure 1. As shown in Figure 1, a fuzzy system consists of a fuzzy rule base, membership functions and an inference engine. The fuzzy rules can be completely characterized by a set of **control variables,**

$$X = \{x_1, x_2, ..., x_n \}.$$

Each control variable x_i is associated with fuzzy terms $\Sigma_i = \{ \alpha_i^1, ... \alpha_i^{p_i} \}$. In the domain fuzzy diagnosis, we have one unique solution variable y, which is the particular type of the fault we want to detect. Assume y is associated with fuzzy terms $\Gamma = \{ \tau_1, ..., \tau_q \}$. Each fuzzy rule has the format:

if $(x_{k1}$ is $\alpha_i^{k1})$AND$(x_{k2}$ is $\alpha_i^{k2})$AND...$(x_{km}$ is $\alpha_i^{km})$...**then** y is τ_j

where $m \leq n$, $\{x_{k1}, x_{k2}, ..., x_{km}\} \subset X$, $\{ \alpha_i^{k1}, \alpha_i^{k2}, ..., \alpha_i^{km} \} \subset \Sigma_i$, and $\tau_j \in \Gamma$.

The degree to which the fuzzy action is taken depends on the degree of truth in the antecedent proposition. Within this application domain, unconditional rules are not used.

A fuzzy system can be defined in terms of an input-process-output Model:
- what information flows into the system
- what basic or fundamental transformations are performed on the data, and
- what data elements are eventually output from the system.

In this application domain, the input information is the data that is useful for diagnosing the particular vehicle fault, or in other words the data of the parameters that

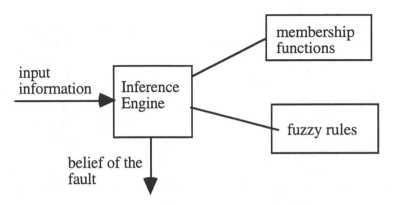

FIGURE 1 A fuzzy diagnostic model.

	ITOT_MAX	ITOT_MIN	TIME1_2	N2_SHIFT	LAG2	TP_REL2	N2_RISE
ITOT_MIN	**0.948**						
TIME1_2	-0.281	-0.293					
N2_SHIFT	0.158	0.174	0.070				
LAG2	-0.347	-0.355	0.547	-0.011			
TP_REL2	-0.072	-0.087	-0.038	0.663	-0.137		
N2_RISE	0.318	0.348	0.032	**0.889**	0.101	0.384	
N2_START	-0.018	-0.019	0.091	**0.908**	-0.111	0.789	0.616

FIGURE 2. An example of correlation matrix.

control the fault. The output is the degree of belief in the vehicle having the fault

The first step in the design is to isolate the critical performance variables. The performance variables consists of two exclusive sets of variables, CP(control parameters) and SP(solution parameters). In the diagnostic system, CP consists of all the variables that give influence to the particular type of fault which the system is designed to detect, and SP consists all the variables that describes the fault phenomenon. For example, at the vacuum leak problem[HaL95],

$$CP = \{\text{throttle position, Lambse\_1, Lambse\_2,} \\ \text{Idle\_speed\_DC, Mass air flow}\} \text{ and} \\ SP = \{\text{vacuum\_leak}\}.$$

However, in the automotive industry, the diagnosis of one particular type of fault can involve hundreds of variables. The model describing one particular type of fault is represented by a sequence of control variable and a single solution variable representing the fault. Therefore the determination of performance variable only involves the control variable. The decision process involves the following steps:

• identify variables that actually affect the model performance out of a population of possible variables,
• consolidate similar variables, and
• prune unimportant variables.

We propose two different methods: heuristic knowledge from engineers and statistic analysis. Assume the set of total parameters acquired at a test set be P. The knowledge engineers can generate an initial set of parameters, $CP^0 \subseteq P$, which contains variables that actually affect the fault model based on their knowledge expertise. If there is no such knowledge available, we can always set $CP^0 = P$.

The number of variables in CP^0 can be very large for a particular type of fault, and many can be redundant. The process of variable consolidation is accomplished by statistical analysis. We apply a correlation analysis across all variables in CP^0. The parameters highly correlated with each other are consolidated by using the following computational steps:
1. generate a correlation matrix based on data set X_i for the control parameters of car model i. The correlation matrix, COR_CP has dimensions d by d, where d is the number of control parameters for car model i.

2. set a correlation threshold.

3. For a parameter p_i, if COR_CP[i][j] is less than the correlation threshold, then keep p_i.

4. For any p_k and p_j, if COR_CP[k][j] is greater than the correlation threshold and p_k has more correlated parameters than p_j does, then prune off p_j

Figure 2 shows the correlation matrix of 8 variables useful for transmission shift detection. If we have a threshold being 0.85, the entries in the bold face indicate correlation. For example, ITOT_MAX and ITOT_MIN has a high correlation, and since they both have the same number of correlated parameters, so either one can be pruned. We also can easily see that N2_RISE, N2_SHIFT and N2_START are highly correlated. Since N2_SHIFT has two correlated parameters and each of the other two has only one correlated parameters, therefore N2_START and N2_RISE can be pruned. Note this process has only one design parameter, correlation threshold, cor_th. This parameter controls the number of control variables retained from this process, and should be determined by the domain engineering experts. Assume at the end of the consolidation process, we have a set variable CP^1, such that $CP^1 \subseteq CP^0$.

The process of pruning unimportant parameters was implemented by using stepwise regression. In this process, variables are added one at a time to the regression and the regression statistics are calculated. If a variable does not significantly affect the regression model, then the variable is discarded and another variable is tested. If the variable adds significant contribution to the regression, then the variable is kept in the model. At the end of this process, we obtain $CP^2 \subseteq CP^1$, that contain variables most pertinent to this particular type of fault.

The result at the end of this stage is a compact CP = $\{x_1, ..., x_m\}$ that gives efficient and accurate diagnosis, in another words, CP contains the minimum number of variables necessary to model the fault.

However, we need to point out that in the automotive industry, different car models may have different control variables influencing the same type of fault. In order to deal with such variation, CP should be the minimum super set of the control variables that give influence over the type of fault for all the car models under consideration. For example, a stereo HO2S model, both Lambse_1 and Lambse_2 give influence on vacuum leak, however a single HO2S car has only Lambse_1.

3. A SELF-LEARNING PROCESS

Fuzziness describes the degree of membership in a fuzzy set. It is also a measure of how well an instance (value) conforms to a semantic ideal or concept. The membership functions of a control variable is a control surface that responds correctly to a set of expected data points. The degree of membership can be viewed as the level of compatibility between an instance from the set's domain and the concept overlying the set. The membership function of a fuzzy set can assume almost any shape, but the closer the set surface maps to the behavior of a physical or conceptual phenomenon, the better our model will reflect the real world. A self-learning process for a fuzzy system involves two steps, elicit the shape of the fuzzy set (its surface morphology) and automatic tuning of the selected membership functions. Automatic tuning of membership function is to make the control surface (the system response) to react correctly to data by changing either the shape of the underlying fuzzy sets and/or the meaning of the rules.

Eliciting the membership function shape is important since it determines the correspondence between data and the underlying concept. It has been shown that triangular functions can be used to approximate system behaviors to nearly any degree of precision by careful placement of triangles with intensive overlaps and widths[Cox94]. Therefore we employ triangle fuzzy membership functions for all control variables in our fuzzy system(see Figure 3).

The desired characteristics of membership functions for this application are as follows:

• The domain of the fuzzy region should be elastic rather than restrictive.

• Every base fuzzy set must be normal, namely, it must have at least one membership value at 1.0 and at least one membership value at 0.0.

In the triangular membership function representation, the fuzzy terms of a variable are represented by a series of triangular functions each of which has the unity measure at its the apex, and the locations of the apices form the critical parameters of the membership functions of x. Figure 3 shows the critical parameters of fuzzy membership functions for variable x_1.

Formally, we define the membership function for the jth fuzzy term of a variable x_i as:

$$MBF_i^{j_i} = \begin{cases} 1 - \dfrac{a_i^{j_i+1} - x}{a_i^{j_i+1} - a_i^{j_i}} & a_i^{j_i} < x \leq a_i^{j_i+1} \\[2ex] 1 - \dfrac{a_i^{j_i} - x}{a_i^{j_i} - a_i^{j_i-1}} & a_i^{j_i-1} < x \leq a_i^{j_i} \\[2ex] 0 & x \leq a_i^{j_i-1} \ or \ x > a_i^{j_i+1} \end{cases}$$

for j = 2, 3, ..., k_i - 1,

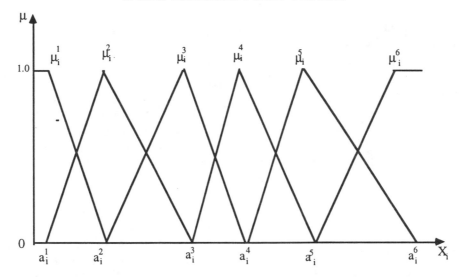

FIGURE 3 Fuzzy membership functions of a control variable.

$$MBF_i^1 = \begin{cases} 1 - \dfrac{a_i^{j_i+1} - x}{a_i^{j_i+1} - a_i^{j_i}} & a_i^{j_i} < x \le a_i^{j_i+1} \\ 1 & a_i^0 < x \le a_i^1 \quad \text{and} \\ 0 & x > a_i^1 \end{cases}$$

$$MBF_i^{k_i} = \begin{cases} 1 & a_i^{k_i} < x \le a_i^{k_i+1} \\ 1 - \dfrac{a_i^{j_i} - x}{a_i^{j_i} - a_i^{j_i-1}} & a_i^{k_i-1} < x \le a_i^{k_i} \\ 0 & x \le a_i^{j_i-1} \end{cases}$$

The regions in the middle of a variable are represented by triangles since their associated concept increases and then decreases, and the shouldered fuzzy sets are used to bracket the endpoints of a variable's fuzzy region. The overlapping point of triangles are at the apices of the triangles.

In the rule base, we have **q** rules associated with the solution variable y. Without losing generality, we assume rule s has the form:

IF $(x_{s1}$ is $\alpha_i^{s1})$AND$(x_{s2}$ is $\alpha_i^{s2})$AND...$(x_{sm}$ is $\alpha_i^{sm})$,

THEN y is τ_s

where $\{x_{s1}, ..., x_{sm}\} \subseteq CP$, and variables x_{si} and y have triangle membership functions. Each control variable x_i corresponds to a vector of critical parameters, $\overline{a}_i = (a_i^0 ... a_i^{k_i})$ that defines the triangular membership functions of the fuzzy terms associated with x_i. Similarly, the membership functions of the fuzzy terms associated with the solution variable y is a vector $\overline{b}$

$$\overline{b} = (b_1, ..., b_q)$$

and $\overline{b} \in R^q$ For rule s, the membership function of τ_s, μ_s, is specified by b_{s-1}, b_s and b_{s+1}.

The goal of the learning algorithm is to determine the critical parameters of the membership functions of control and solution variables using a supervised machine learning technique on a training data file. Assume the training data file contains H samples and each sample contains data acquired from one vehicle at the test site and the fault identity to indicate whether the vehicle is suffering the type s fault. This fault identity value is inserted by mechanics at the test site and the decision is made based on the mechanics' engineering experience.

The proposed machine learning algorithm is based on gradient descent optimal search algorithm. We assume the range of a solution y is normalized between 0 to 1. The machine learning algorithm is to determine the optimal values of parameter vectors for all control variables and the solution variable so the resulting fuzzy system will give the best performance in fault diagnosis. Let $\overline{a}_i = \{a_i^1, ..., a_i^{ki}\} \in R^m$ be the parameter vector of variable x_i. The constraints in this problem are:

$$\overline{a}_i^1 < \overline{a}_i^2 < ... < \overline{a}_i^{ki}.$$

Assume the H samples in the training data file has the form, $(\overline{x}^1, y^1), (\overline{x}^2, y^2), ...(\overline{x}^H, y^H)$, where $\overline{x}^i$ is the values of parameters acquired from vehicle j and y^j is the fault identity which represents the possibility of the vehicle having the fault. For example, if the fault under current diagnosis is vacuum leak, the numerical value of the fault identity represent a number of fuzzy terms "no vacuum leak", "highly possible vacuum leak", "possibility of vacuum leak is low," etc. The criterion function used in the learning process is:

$$\min_{z(\overline{a},\overline{b})} E = \frac{1}{2} \sum_{j=1}^{H} (y^i - y^{*i})$$

where y^{*i} is the output value of fuzzy reasoning system with ith input $\overline{x}^i$, $i = 1, ..., H$.

The constraint functions for the control variables are defined as:

$$g_i = (a_i^1 - a_i^2, \ a_i^2 - a_i^3, \ ..., \ a_i^{k_i-1} - a_i^{k_i}) < 0,$$

for $i = 1, ..., m$, and for the solution variable we have

$$h = (b^1 - b^2, \ b^2 - b^3, \ ..., \ b^{q-1} - b^q) < 0.$$

Let $z = (\overline{a}, \overline{b})$, then $\overline{g}(z) = (g_1, g_2, ..., g_m, h)$. By combining the criterion function with the set of constraints, we have the following objective function,

$$J(\overline{a}, \overline{b}) = E + r \phi(\overline{g}(z)),$$

where $r > 0$, and the penalty function is defined as:

$$\phi(\overline{g}(z)) = \sum_{i=1}^{m} \sum_{j=1}^{k_i} \frac{-1}{a_i^j - a_i^{j+1}}.$$

Using the gradient method, at the jth iteration we have

$$\tau_s(j) = \tau_s(j-1) - \lambda \frac{\partial J(z^j, r)}{\partial \tau_s}, \text{ for } s=1, ..., q, \text{ and}$$

$$\overline{a}_i(j) = \overline{a}_i(j-1) - \lambda \frac{\partial J(z^j, r)}{\partial \overline{a}_i}, \ i = 1, ..., m,$$

where $\dfrac{\partial J(z^j, r)}{\partial \tau_s} = -\sum_{i=1}^{N} (y^i - y^{*i}) \mu^s$ and

$$\frac{\partial J(z^j, r)}{\partial \overline{a}_i^j} = -\sum_{i=1}^{N} (y^i - y^{*i}) \sum_{s=1}^{q} \tau_s \frac{\partial \mu^s}{\partial \overline{a}_i^j} +$$

$$r[\frac{-1}{(a_i^{j-1} - a_i^j)^2} + \frac{1}{(a_i^j - a_i^{j+1})^2}]$$

for $1 \le j \le k_i$, $i = 1, ..., m$.

The initial values for $\overline{a}_i$ will be evenly located between a^1 and a^m, i.e. membership functions all have the same shape, which is the symmetric triangular functions evenly located in the interval of $[a^1 ... a^m]$.

The iteration stops when the difference between the two consecutive iterations is smaller than a threshold δ, namely,

$$| J(z^j, r) - J(z^{j-1}, r) | < \delta.$$

The self-learning model has been implemented to solve Vacuum leak diagnosis problem.

4. CONCLUSION

We have presented a self-learning fuzzy model for automotive fault diagnosis. In order for a system to be quickly adapted to different vehicle models in a short time period, a self-learning system is a critical for the end-of-line tests, since large quantity of vehicles are being tested and vehicle models change from time to time. The fuzzy diagnostic model we described above is innovative in vehicle diagnosis and it is generic with respect to off-line diagnosis of vehicle faults. In particular we described the two important components in the self-learning model is automatic pruning of control parameters and automatic generation of membership functions. The model is implemented to diagnose vacuum leak in EEC in vehicles and are tested on data collected from assembly plant test sites in Dearborn, Michigan, USA.

5. ACKNOWLEDGMENT

This project is supported by the CEEP grant from the University of Michigan-Dearborn

6. REFERENCES

[ChO93] Byeong-Mook Chung and Jun-Ho Oh, "Autotuning Method of Membership Function in a Fuzzy Learning Controller," Journal of Intelligent & fuzzy systems, Vol. 1, Issue 1, 1993.

[Cox94] Earl Cox, "The Fuzzy Systems Handbook," AP Professsional, 1994.

[FMO88] Filljov, M. Marinov, S. Ovcharov, "Engine Diagnostic Expert System", 18th International Symposium on Automotive Technology & Automation,30th May - 3rd June 1988.

[HaL95] Brennan T. Hamilton and Yi Lu, "Diagnosis of Automobile Failures using Fuzzy Logic," *The eightth IEA/AIE*, Melbourne, Autralia, June 5 ~ 9, 1995

[RSP93] Giorgio Rizzoni, Ahmed Soliman, and Kevin Passino, "A Survey of Automotive Diagnostic Equipment and Procedures," Society of Automotive Engineers, paper number #930769.

[ZYZ88] Zheng Xiaojun, Yang Shuzi, Zhou Anfa, Shi Hanmin, "A Knowledge-Based Diagnosis System for Automobile Engines", The International Journal of Advanced Manufacturing Technology, 1988.

Distributed AI

AN AUTOMATED REASONING SYSTEM FOR MULTI-AGENT COOPERATION

Lifeng He*   Hirohisa Seki   Hidenori Itoh
Department of AI and Computer Science
Nagoya Institute of Technology
Gokiso, Showa-ku, Nagoya 466, Japan
E-mail{helifeng, seki, itoh}@juno.ics.nitech.ac.jp

ABSTRACT

We present an automated reasoning system for multi-agent knowledge and time, which is a kind of multi modal logics, incorporating common knowledge and implicit knowledge. Such knowledge is essential when considering reasoning, planning and cooperating problem solving in distributed and dynamically changing environments. We first consider how to represent such common knowledge and implicit knowledge based on modal logic. We then describe how to use the framework of the possible-world semantics to capture the notations of common knowledge and implicit knowledge. The reasoning procedure of our system is based on a so-called semantic method, that is, a formula expressing multi-agent knowledge and time is first translated, according to the possible world semantics, into the set of corresponding first-order clauses. We then check the satisfiability of the set of translated first-order clauses by a general purpose first-order theorem proof procedure ME (the model elimination) ([10]), augmented with the capabilities of handling inequality predicates introduced by the translation. We apply the idea of theory resolution for reasoning about inequalities efficiently. We also show some experimental results of our automated reasoning system.

1   INTRODUCTION

Reasoning, planning and cooperating problem solving under multi-agent environments have recently attracted much attention in artificial intelligence (e.g., [1], [2], [8]). Thus research on automated reasoning about multi-agent knowledge and time becomes a very important topic in dynamically changing environments, distributed artificial intelligence (DAI) and the application of intelligent systems.

In multi-agent environments, it is often needed to reason about the states of knowledge of all agents considered, not simply those of an individual agent. Therefore, *common knowledge* and *implicit knowledge* ([5]) is necessary for reasoning in multi-agent environments. Common knowledge is such a knowledge that not only everyone knows some facts, but everyone knows that everyone knows them, and everyone knows that everyone knows that everyone knows them, and so on. Common knowledge is very important for reasoning in multi-agent environments. For example, if p is a common knowledge and agent a knows that agent b knows that $p \rightarrow q$, then agent a can conclude that agent b knows q.

Moreover, it is also often desirable to be able to reason about the knowledge which is implicit in a group. For example, in cooperating action, if agent a knows p and agent b knows $p \rightarrow r$, then together with them the group to which agent a and agent b belong would get the knowledge of r by means of communication, even though neither agent a nor agent b could individually know r. Intuitively, a group has implicit knowledge of ϕ if and only if the knowledge ϕ is distributed among the members of the group, and a group of cooperating agents cannot get ϕ unless ϕ is already implicit knowledge. It is obvious that implicit knowledge is a very useful notion in describing and solving reasoning, planning and cooperating problems in distributed environments ([4]).

In this paper, we present an automated reasoning system for multi-agent knowledge and time, a kind of multi modal logics, with incorporating common knowledge and implicit knowledge (hereafter abbreviated to *MAKTCI*) under the conditions of linear, discrete and synchronous time and unbounded memory. Actually, this system is an extended system of WEIHE ([6]).

Our reasoning procedure is based on the so-called semantic method ([12]). That is, the negation of a given

---
*This research is supported in part by the Japanese Ministry of Education and the Artificial Intelligence Research Promotion Foundation.

multi-agent knowledge and time formula is first translated by translation procedure, which is based on the possible world semantics, into a semantically equivalent first-order formula, from which we then derive a set of clauses. Then, we use a general-purpose first-order theorem proof procedure such as ME (the Model Elimination) ([10]) to check whether the derived set of clauses is unsatisfiable or not.

We should, however, deal with inequality predicates occurring in translated first-order formulas. We therefore consider an extension of ME, called *MEI* (ME with Inequalities), to deal with those inequality predicates appropriately.

A special reasoning mechanism, i.e., so-called *theory resolution* ([15]), is incorporated into our system to reason about inequalities efficiently.

The organization of the rest of the paper is as follows. In the next section, we introduce the syntax and the possible world semantics of MAKTCI. In section 3, we introduce our translation procedure. We then describe the implementation of our system in section 4, and show some experimental results in section 5. We give concluding remarks in section 6.

2   THE SYNTAX AND THE POSSIBLE WORLD SEMANTICS OF MAKTCI

2.1   THE SYNTAX OF MAKTCI

Our *language of MAKTCI* for m agent consists of a countable collection of symbols and all formulas defined over it. The countable collection of symbols is classified into the following groups:

   (i) *time variables* denoted by $t, t_1, t_2, \cdots$;
  (ii) *time constants* denoted by $0, 1, \cdots, T, T_i, \cdots$, where $T, T_i \in \{0, 1, 2, \cdots\}$;
 (iii) *agent constants* denoted by $a_1, \cdots, a_m$ $(m \geq 1)$;
  (iv) *proposition symbols* denoted by $p, q, r, \cdots$;
   (v) *modal operators* denoted by K, E, C, I;
  (vi) *time order relations*, which are $\leq, <$;
 (vii) *logical connectives* , which are: $\neg$ (negation) and $\wedge$ (conjunction);
(viii) *quantifier*, which is: $\forall$ (for all);
  (ix) *parentheses*, which are: (and).

The *modal operators* K, E, C and I come from *Know, Everyone, Common* and *Implicit* respectively.

Suppose that τ is a time constant or a time variable and a_i is an agent. Then formation rules for MAKTCI formulas are as follows:

- if p is a proposition symbol, then p^τ is a formula;
- if ϕ is a formula, then $K^\tau_{a_i}\phi$, $E^\tau\phi$, $C^\tau\phi$ and $I^\tau\phi$ are formulas;
- if ϕ is a formula, then $\forall(T \leq t)\phi$ is a formula;

- if ϕ, ψ are formulas, then $\neg\phi$, $\phi \wedge \psi$ are formulas.

For the sake of convenience, we also use logic connectives $\vee$, $\rightarrow$ and an existential quantifier $\exists$ in our language.

Intuitively, $K^\tau_{a_i}\phi$ $(E^\tau\phi)$ means that at time τ, agent a_i (everyone) knows ϕ, and $C^\tau\phi$ $(I^\tau\phi)$ means that at time τ, ϕ is a common (implicit) knowledge.

Example 2.1 Let F_{ex} be the following MAKTCI formula:

$$C^2p^3 \wedge K^1_{a_1}(\forall(2 \leq t_1)\neg p^{t_1} \vee q^4) \rightarrow K^7_{a_2}E^4p^3 \wedge I^5q^4$$

Suppose that we interpret p (q) as "The bus starts" ("The conference begins"), respectively. Then, the formula F_{ex} can be interpreted as follows: "If at two o'clock the bus starts at three o'clock is common knowledge and at one o'clock agent a_1 knows that 'either there is not any bus after two o'clock or the conference begins at four o'clock then at seven o'clock', agent a_2 must know that 'at four o'clock everybody knows that the bus starts at three o'clock', and at five o'clock the conference begins is an implicit knowledge.

2.2   THE POSSIBLE WORLD SEMANTICS OF MAKTCI

Following Halpern and Vardi ([7]), we consider a possible world semantics for our MAKTCI. A possible world model of MAKTCI for m agents can be defined as a tuple:

$$(\mathcal{R}, \pi, \mathcal{X}_{a_1}, \cdots, \mathcal{X}_{a_m})$$

where $\mathcal{R}$ is a set of *runs*, each of which indicates a possible history of a given multi-agents' system over time. We assume that each run proceeds along in discrete time points, thus for each r in $\mathcal{R}$ and n in $\mathcal{N}$, (r, n) represents the n-th time point of run r. On the other hand, π assigns a truth value to each proposition formula at every point (r, t), and $\mathcal{X}_l$ $(l \in \{a_1, \cdots, a_m\})$ is an equivalence relation on $\mathcal{R} \times \mathcal{N}$.

As a model of MAKTCI, we assume in this paper that each agent has unbounded memory and time is synchronous. Then, for all $l \in \{a_1, \cdots, a_m\}$, $\mathcal{X}_l$ has following four properties:

$$\forall rt \; (((r, t), (r, t)) \in \mathcal{X}_l) \qquad (1)$$

$$\forall rtr' \; (((r, t), (r', t)) \in \mathcal{X}_l \rightarrow ((r', t), (r, t)) \in \mathcal{X}_l) \qquad (2)$$

$$\forall rtr'r'' \; ((((r, t), (r', t)) \in \mathcal{X}_l) \wedge (((r', t), (r'', t)) \in \mathcal{X}_l) \\ \rightarrow ((r, t), (r'', t)) \in \mathcal{X}_l) \qquad (3)$$

$$\forall rr'T \; (((r, T), (r', T)) \notin \mathcal{X}_l \\ \rightarrow \forall(T \leq t)((r, t), (r', t)) \notin \mathcal{X}_l) \qquad (4)$$

The first three properties indicate that for each fixed t, $\mathcal{X}_l$ is reflexive, symmetric and transitive respectively ([3]), while the last one means that each agent keeps track of his whole history.

We denote by $M, (r_i, t_j) \models \phi$ that an MAKTCI formula ϕ is true at the point (r_i, t_j) of model M, truth being defined inductively as follows:

- $M, (r_i, t_j) \models p^\tau$ iff $\pi(r_i, \tau, p) = true$;
- $M, (r_i, t_j) \models K_{a_k}^\tau \phi$ iff $M, (r, \tau) \models \phi$ for all (r, τ) such that $((r_i, \tau), (r, \tau)) \in \mathcal{X}_{a_k}$;
- $M, (r_i, t_j) \models E^\tau \phi$ iff $M, (r, \tau) \models \phi$, for all (r, τ) such that $((r_i, \tau), (r, \tau)) \in \mathcal{X}_l$, where $l = a_1, \cdots, a_m$;
- $M, (r_i, t_j) \models C^\tau \phi$ iff $M, (r_i, t_j) \models (E^\tau)^N \phi$, for $N = 1, 2, \cdots$, where $(E^\tau)^1 \phi = E^\tau \phi$ and $(E^\tau)^N = E^\tau((E^\tau)^{(N-1)})$;
- $M, (r_i, t_j) \models I^\tau \phi$ iff $M, (r, \tau) \models \phi$, for all (r, τ) such that $((r_i, \tau), (r, \tau)) \in \mathcal{X}_{a_1}$ or $\cdots$ or $((r_i, \tau), (r, \tau)) \in \mathcal{X}_{a_m}$;
- $M, (r_i, t_j) \models \forall(T \leq t)\phi$ iff $M, (r_i, t) \models \phi$, for all (r_i, t) such that $t \geq T$;
- $M, (r_i, t_j) \models \phi \wedge \psi$ iff $M, (r_i, t_j) \models \phi$ and $M, (r_i, t_j) \models \psi$;
- $M, (r_i, t_j) \models \neg\phi$ iff $M, (r_i, t_j) \not\models \phi$.

From the above possible world semantics of MAKTCI and the property (4), we can conclude $K_l^T \phi \equiv \forall(T \leq t)K_l^t \phi$ for all $l \in \{a_1, \ldots, a_m\}$, $E^T \phi \equiv \forall(T \leq t)E^t \phi$, $C^T \phi \equiv \forall(T \leq t)C^t \phi$ and $I^T \phi \equiv \forall(T \leq t)I^t \phi$. That is, any kind of knowledge is *persistent* ([9]).

3 TRANSLATION PROCEDURE

The reasoning process in our automated reasoning system is as follows. The negation of the MAKTCI formula to be proved is first translated, according to the possible world semantics, into its equivalent first-order formula, from which we then derive a set of clauses. We use an extended general-purpose first-order proof procedure, say, MEI (ME with Inequalities) to check the satisfiability of the set of clauses. If the derived set of clauses is unsatisfiable, it means that the original MAKTCI formula is valid. Otherwise, it means that the original MAKTCI formula is not valid.

In order to translate an arbitrary MAKTCI formula into its equivalent first-order formula, we introduce predicate $X(r, r', t, l)$ to represent the possible world relation $((r, t), (r', t)) \in \mathcal{X}_l$ ($l \in \{a_1, \ldots, a_m\}$). We also use the same inequality predicate symbol $\leq$ to represent the translated version of the time order relation. Moreover, we use predicate P_i to represent each corresponding propositional symbol p_i in an MAKTCI formula. We denote our translation procedure by tr.

If a formula ϕ does not contain any modal operator like E, C and I, from the possible world semantics of

MAKTCI, the translation procedure is defined inductively as follows ([6]), where $1 \leq k \leq m$.

$$tr(p^\tau, (r_i, t_j)) = P(r_i, \tau)$$
$$tr(K_{a_k}^\tau \phi, (r_i, t_j)) = \forall r(\neg X(r_i, r, \tau, a_k) \vee tr(\phi, (r, \tau)))$$
$$tr(\forall(T \leq t)\phi, (r_i, t_j)) = \forall t(\neg(T \leq t) \vee tr(\phi, (r_i, t)))$$
$$tr(\phi \wedge \psi, (r_i, t_j)) = tr(\phi, (r_i, t_j)) \wedge tr(\psi, (r_i, t_j))$$
$$tr(\neg\phi, (r_i, t_j)) = \neg tr(\phi, (r_i, t_j))$$

The four properties of $\mathcal{X}_l$ ($l \in \{a_1, \ldots, a_m\}$), i.e., (1), (2), (3) and (4) can be translated as follows:

$$\forall rt\, X(r, r, t, l) \quad (5)$$
$$\forall rr't(X(r, r', t, l) \rightarrow X(r', r, t, l)) \quad (6)$$
$$\forall rr'tr''(X(r, r', t, l) \wedge X(r', r'', t, l) \rightarrow X(r, r'', t, l)) \quad (7)$$
$$\forall rr'Tt(\neg X(r, r', T, l) \rightarrow \neg(T \leq t) \vee \neg X(r, r', t, l)) \quad (8)$$

Next, we consider the translation procedure for $E^\tau \phi$, $C^\tau \phi$ and $I^\tau \phi$, where ϕ is an arbitrary MAKTCI formula. According to the possible-world semantics, translation procedure for $E^\tau \phi$ at the point (r_i, t_j) can be described as:

$$tr(E^\tau \phi, (r_i, t_j)) = tr(K_{a_1}^\tau \phi \wedge \cdots \wedge K_{a_m}^\tau \phi, (r_i, t_j))$$
$$= \forall r_1((\neg X(r_i, r_1, \tau, a_1) \vee tr(\phi, (r_1, \tau))) \wedge \cdots \wedge$$
$$\forall r_m(\neg X(r_i, r_m, \tau, a_m) \vee tr(\phi, (r_m, \tau)))$$

For the simplicity, we introduce a new variable l, called *agent variable* ($l \in \{a_1, \ldots, a_m\}$), into our language. Thus, that at time τ everyone knows ϕ, i.e., $E^\tau \phi$, can be expressed as $\forall l K_{a_l}^\tau \phi$, and the translation procedure for the formula $E^\tau \phi$ in point (r_i, t_j) is simplified as follows:

$$tr(E^\tau \phi, (r_i, t_j)) = tr(\forall l K_l^\tau \phi, (r_i, t_j))$$
$$= \forall l \forall r(\neg X(r_i, r, \tau, l) \vee tr(\phi, (r, \tau)))$$

On the other hand, the first-order formula translated directly according to the possible world semantics of common knowledge $C^\tau \phi$ would be a infinite form, so it is not very useful. In fact, $C^\tau \phi$ philosophically means that at time τ, ϕ is true at any possible world. Since our system is closed, i.e., the number of agents is finite and fixed, say, $a_1, \cdots, a_m$, the translation procedure for $C^\tau \phi$ at point (r_i, t_j) is given by the following formula:

$$tr(C^\tau \phi, (r_i, t_j))$$
$$= \forall l \forall r_1 \forall r_2(\neg X(r_1, r_2, \tau, l) \vee tr(\phi, (r_2, \tau))$$

Lastly, according to the possible world semantics, the translation procedure for implicit knowledge $I^\tau \phi$ at point(r_i, t_j) can be expressed as:

$$tr(I^\tau \phi, (r_i, t_j)) = \forall r(\neg X(r_i, r, \tau, a_1) \vee \cdots \vee$$
$$\neg X(r_i, r, \tau, a_m) \vee tr(\phi, (r, \tau)))$$

Using the agent variable l, the above formula can simply be rewritten as:

$$tr(I^\tau \phi, (r_i, t_j)) \equiv \exists l \forall r(\neg X(r_i, r, \tau, l) \vee tr(\phi, (r, \tau)))$$

Concluding the above discussions, the translation procedure for $E^\tau\phi$, $C^\tau\phi$ and $I^\tau\phi$ is defined inductively as follows:

$$tr(E^\tau\phi, (r_i, t_j)) = \forall l \forall r (\neg X(r_i, r, \tau, l) \vee tr(\phi, (r, \tau)))$$
$$tr(C^\tau\phi, (r_i, t_j))$$
$$= \forall l \forall r_1 \forall r_2 (\neg X(r_1, r_2, \tau, l) \vee tr(\phi, (r, \tau)))$$
$$tr(I^\tau\phi, (r_i, t_j)) = \exists l \forall r (\neg X(r_i, r, \tau, l) \vee tr(\phi, (r, \tau)))$$

Example 3.1 *(Continued from Example 2.1)*
By the translation procedure, the negation of F_{ex} in Example 2.1, is translated into the following corresponding first-order formula, where we assume $(0, 0)$ as the initial point.

$$tr(\neg F_{ex}, (0, 0)) = \forall r1 \forall r2 \forall l1 (\neg X(r1, r2, 2, l1) \vee$$
$$P(r2, 3)) \wedge \forall r3(\neg X(0, r3, 1, a_1) \vee \forall t1(\neg(2 \leq t1) \vee$$
$$\neg P(r3, t1)) \vee Q(r3, 4)) \wedge (\exists r4(X(0, r4, 7, a_2) \wedge$$
$$\exists l2 \exists r5(X(r4, r5, 4, l2) \wedge \neg P(r5, 3))) \vee$$
$$\forall l3 \forall r6(X(0, r6, 5, l3) \wedge Q(r6, 4)))$$

The soundness and completeness of our reasoning method can be shown by the following proposition, whose proof is omitted due to the lack of space.

Proposition 3.1 *Suppose that F is an arbitrary MAKTCI formula and F' is its translated first-order form. Then, F is valid iff F' is valid in the first-order sense.*

4   IMPLEMENTATION

Our system is based on ME (model elimination procedure)([10]), which is a general-purpose first-order proof procedure. As there exist inequality predicates in translated first-order formulas, ME can not be directly utilized to check their satisfiability. In this section, we discuss how to extend ME to MEI (ME with Inequalities), which can appropriately handle those inequality predicates introduced by the translation procedure. We also explain how to use theory resolution ([15]) in MEI.

4.1   THE REVIEW OF ME (MODEL ELIMINATION) PROCEDURE

Suppose that S is the set of clauses to be refuted, and C, called centre chain, is one clause in S. Generally, a clause has a form of $L_1 \vee \ldots \vee L_m$, where L_i $(1 \leq i \leq m)$ is a literal. ME procedure is defined as follows:

1. The extension operation: like the standard resolution operation applied to the leftmost literal, L, of C and a matching literal of some side clause in S, including C itself[1]. Instead of discarding

---

[1] Of course, before matching, the variables of the elected side clause must renamed so that it has no variable in common with C.

L, it is converted to a marked literal L^*. For example, if $C = Q(x) \vee P(y)$ and a side clause $B = R(u) \vee \neg Q(a)$, then the result of the extension operation is $R(u) \vee Q^*(a) \vee P(y)$.

2. The reduction operation: if the leftmost unmarked literal, L, of C is a matching literal of some marked literal K^* in C, i.e., there exists a substitution θ such that $L\theta = K^*\theta$, then the result is $C\theta$ with the leftmost literal deleted. For example, if $C = \neg Q(y) \vee P(y) \vee Q^*(a)$, then the result of the reduction operation is $P(a) \vee Q^*(a)$.

3. The contraction operation: deleting all of the the marked literals to the left of the leftmost unmarked literal. It is performed whenever possible after the extension and reduction operations.

4. The procedure terminates successfully if the empty chain is obtained.

4.2   HANDLING INEQUALITY CONSTRAINTS

We have to deal with inequality predicates occurring in a set of clauses. We assume the inequality axioms *INEQ* on the natural numbers, which include, among others, the following transitivity axioms:

$$(t_1 \leq t_2), (t_2 \leq t_3) \quad \longrightarrow \quad t_1 \leq t_3 \qquad (9)$$
$$(t_1 \leq t_2), (t_2 < t_3) \quad \longrightarrow \quad t_1 < t_3 \qquad (10)$$
$$(t_1 < t_2), (t_2 \leq t_3) \quad \longrightarrow \quad t_1 < t_3 \qquad (11)$$
$$(t_1 < t_2), (t_2 < t_3) \quad \longrightarrow \quad t_1 < t_3 \qquad (12)$$

We should make some extensions of ME procedure in order to use it in our case where inequality predicates exist. For example, the set of clauses $t_1 < 5 \vee P(t_1)$ and $t_2 < 2 \vee \neg P(t_2)$ is unsatisfiable (e.g., let $t_1 = t_2 = 7$), but we can not use ME procedure to prove it. Moreover, ordinary resolution in this case is quite inefficient. For example, it is obvious that there is no refutation based on $3 \leq T$ and the transitivity axiom $t_1 \leq t_2, t_2 \leq t_3 \longrightarrow t_1 \leq t_3$, but it is possible to derive an infinite number of consequences from them. Therefore we use theory resolution ([15]) for reasoning about inequalities in MEI. Theory resolution is an effective method for resolving these problems. Our inequality theory includes, for example, the transitivity axioms, i.e., the axioms of (9) $\sim$ (12). With theory resolution, the use of a transitivity axiom is restricted to occasions where there are two inequalities which can match the transitivity axiom.

4.3   How MEI Works

Suppose that the MAKTCI formula to be proved is $\mathcal{A} \rightarrow \mathcal{B}$, S is the set of clauses from the first order

formula translated from $\mathcal{A} \wedge \neg \mathcal{B}$. Further, suppose that the initial center chain, C, is a clause obtained from the first order formula translated from $\neg \mathcal{B}^2$. We denote a clause by $\mathcal{L} \vee \mathcal{I}$, where $\mathcal{L} = L_1 \vee \ldots \vee L_m$ is the disjunction of non-inequality literals, $\mathcal{I} = I_1 \vee \ldots \vee I_n$ is the disjunction of inequality literals.

Definition 4.1 A set of inequality clauses $\mathcal{S}_\mathcal{I} = \mathcal{I}_1 \wedge \ldots \wedge \mathcal{I}_m$ is unsatisfiable iff $\mathcal{S}_\mathcal{I} \cup INEQ \models \perp$.

The extension of ME, called *MEI* (ME with Inequalities), contains four operations: the extension operation, the reduction operation, the contraction operation and the $\leq$-reasoning operation.

The extension operation, the reduction operation and the contraction operation are basically the same as in ME, except that the extension operation and the reduction operation are only applied to members of $\mathcal{L}$.

The $\leq$-reasoning operation is based on theory resolution with the theory of inequality. We also consider the following operations applied to members of $\mathcal{I}$ for simplification:

(1) If some $I \in \mathcal{I}$ is false, it is deleted from $\mathcal{I}$;

(2) If a variable occurring in some $I \in \mathcal{I}$ does not occur in any literal of $\mathcal{L}$, I is deleted from $\mathcal{I}$;

(3) Suppose that $I_j, I_k \in \mathcal{I}$. If $\neg I_j$ implies $\neg I_k$, I_k is deleted from $\mathcal{I}$.

For example, apply the $\leq$-reasoning operation to $P(3) \vee \neg Q^*(T) \vee T < 1 \vee T < 4 \vee 3 < 1 \vee t_3 < 7$, the result is $P(3) \vee \neg Q^*(T) \vee T < 4$.

The contraction operation and the $\leq$-reasoning operation are performed whenever possible after the extension and reduction operations.

The procedure terminates successfully iff when $\mathcal{L}$ becomes empty, $\mathcal{I}$ is also empty or $\mathcal{I}_1 \wedge \ldots \wedge \mathcal{I}_n \wedge \mathcal{I}$ is unsatisfiable, i.e., $\mathcal{I}_1 \wedge \ldots \wedge \mathcal{I}_n \wedge \mathcal{I} \cup INEQ \models \perp$, where $\mathcal{I}_i$ $(1 \leq i \leq n)$ is a clause only containing inequality literals in $\mathcal{S}$ or earlier obtained by MEI. Otherwise, a new centre chain is selected and the procedure is repeated.

The completeness of MEI can be shown by the following lemma, whose proof is omitted because lack of space.

Lemma 4.1 *Let S be an unsatisfiable set of clauses. Then, $\square$ is derivable by MEI.*

We explain how MEI works, using the previous example.

Example 4.1 *(Continued from Example 3.1)* The following set of clauses is obtained from $tr(\neg F_{ex}, (0,0))$ in Example 3.1, where R_4, R_5, R_6 and L_3 are Skolem constants.

---

[^2]: This makes the proof goal-directed.

1. $\neg X(r1, r_2, 2, l_1) \vee P(r_2, 3)$
2. $\neg X(0, r_3, 1, a_1) \vee \neg P(r_3, t_1) \vee Q(r_3, 4) \vee 2 \leq t_1$
3. $X(0, R_4, 7, a_2) \vee X(0, R_6, 5, l_2)$
4. $\neg Q(R_6, 4) \vee X(0, R_4, 7, a_2)$
5. $X(R_4, R_5, 4, L_3) \vee X(0, R_6, 5, l_4)$
6. $\neg P(R_5, 3) \vee X(0, R_6, 5, l_5)$
7. $\neg P(R_5, 3) \vee \neg Q(R_6, 4)$
8. $\neg Q(R_6, 4) \vee X(R_4, R_5, 4, L_3)$

The satisfiability checking of the above set of clauses proceeds as follows, where $\mathcal{E}$ $(\mathcal{R}, \mathcal{C}, \leq)$ denotes to the extension operation (the reduction operation, the contraction operation, the $\leq$-reasoning operation), respectively. The clause 8 is selected as the center chain. Notice that by the property of $\mathcal{X}_l$ given in (8), $\neg X(r_1, r_2, t_1, l)$ and $X(r_1', r_2', t_2, l)$ are matching literals for each other if there is a substitution θ such that $r_1 \theta = r_1' \theta, r_2 \theta = r_2' \theta$ and $t_1 \theta \leq t_2 \theta$. The executing result is shown in Table 1.

9. $\neg X(0, R_6, 1, a_1) \vee \neg P(R_6, t_1) \vee \neg Q^*(R_6, 4)$
   $\vee X(R_4, R_5, 4, L_3) \vee 2 \leq t_1 \qquad \mathcal{E}$ with 2
10. $X(R_4, R_5, 4, L_3) \vee \neg X^*(0, R_6, 1, a_1) \vee$
    $\neg P(R_6, t_1) \vee \neg Q^*(R_6, 4) \vee X(R_4, R_5, 4, L_3)$
    $\vee 2 \leq t_1 \qquad \mathcal{E}$ with 5
11. $P(R_5, 3) \vee X^*(R_4, R_5, 4, L_3) \vee$
    $\neg X^*(0, R_6, 1, a_1) \vee \neg P(R_6, t_1) \vee \neg Q^*(R_6, 4)$
    $\vee 2 \leq t_1 \qquad \mathcal{E}$ with 1 & $\mathcal{C}$
12. $X(0, R_6, 5, l_5) \vee P^*(R_5, 3) \vee X^*(R_4, R_5, 4, L_3)$
    $\vee \neg X^*(0, R_6, 1, a_1) \vee \neg P(R_6, t_1) \vee \neg Q^*(R_6, 4)$
    $\vee 2 \leq t_1 \qquad \mathcal{E}$ with 6
13. $\neg P(R_6, t_1) \vee \neg Q^*(R_6, 4) \vee 2 \leq t_1 \qquad \mathcal{R}$ & $\mathcal{C}$
14. $\neg X(r_1, R_6, 4, l_3) \vee \neg P^*(R_6, 3) \vee \neg Q^*(R_6, 4)$
    $\qquad \mathcal{E}$ with 1 & $\leq$
15. $X(R_4, R_5, 4, L_3) \vee \neg X^*(0, R_6, 4, l_3) \vee$
    $\neg P^*(R_6, 3) \vee \neg Q^*(R_6, 4) \qquad \mathcal{E}$ with 5
16. $P(R_5, 3) \vee X^*(R_4, R_5, 4, L_3) \vee \neg X^*(0, R_6, 4, l_3)$
    $\vee \neg P^*(R_6, 3) \vee \neg Q^*(R_6, 4) \qquad \mathcal{E}$ with 1
17. $X(0, R_6, 5, l_3) \vee P^*(R_5, 3) \vee X^*(R_4, R_5, 4, L_3)$
    $\vee \neg X^*(0, R_6, 4, l_3) \vee \neg P^*(R_6, 3) \vee \neg Q^*(R_6, 4)$
    $\qquad \mathcal{E}$ with 6
18. $\square \qquad \mathcal{R}$ & $\mathcal{C}$

5   EXPERIMENTAL RESULTS

Our automated reasoning procedure for MAKTCI has been currently implemented in SICStus Prolog on Sun SPARCstation2/80MHz. Many MAKTCI formulas have been checked their satisfiability and some of those results are shown in Table 1. The results show that MEI is much more efficient than SATCHMOI ([6]) (SATCHMO [11] with Inequalities).

	MAKTCI formulas	Results	Runtime (*msec*)	
			SATCHMOI	MEI
1	$C^2 p^3 \wedge K_{a_1}^1 (\forall (2 \le t_1) \neg p^{t_1} \vee q^4) \to K_{a_2}^7 E^4 p^3 \wedge I^5 q^4$	valid	650	17
2	$K_{a_1}^3 E^1 p^2 \to K_{a_3}^2 (\forall (1 \le t) p^t \vee q^4)$	not valid	70	13
3	$\forall (0 \le t_1) E^{t_1} (\forall (1 \le t_2) p^{t_2} \wedge q^4) \to K_{a_1}^5 p^3 \wedge K_{a_2}^7 q^4$	valid	899	23
4	$E^2 (\forall (0 \le t_1) p^{t_1} \wedge q^4) \to K_{a_1}^5 p^3 \wedge K_{a_2}^7 q^4$	valid	550	23
5	$E^2 (p^3 \wedge q^4) \to K_{a_1}^5 p^3 \wedge K_{a_2}^7 q^4$	valid	260	19
6	$\neg K_{a_1}^4 q^4 \to \neg I^3 q^4$	not valid	55	7
7	$\neg K_{a_1}^4 p^3 \to \neg C^2 p^3 \wedge \neg E^4 p^3$	valid	60	3
8	$K_{a_1}^2 K_{a_2}^3 (p^{10} \vee E^5 q^2) \to K_{a_2}^6 (p^{10} \vee K_{a_3}^7 q^2)$	valid	350	10

Table 1: Some Experimental Results

6   CONCLUDING REMARKS

We have proposed a reasoning system for multi-agent cooperation, which can deal with common knowledge and implicit knowledge. We have shown how to capture the notions about common knowledge and implicit knowledge in our language and given the translation procedure for translating MAKTCI formulas into their corresponding first-order formulas. We have used an example to show our automated reasoning system how to check the validity of a multi-agent knowledge and time formula. Some experimental results were also shown.

There are some other reasoning methods for modal logics ([14], [13]), but the extensions of these methods to multi modal logics and how to handle common knowledge and implicit knowledge are not made yet. The possible world semantic method is, as we have shown in this paper, is useful for reasoning in multi modal logics and can handle common knowledge and implicit knowledge smoothly. Moreover, our method is easy to be extended to first-order MAKTCI.

For future work, we will apply our system to solve practical application problems. We will also try to incorporate belief into our system.

REFERENCES

[1] E. Ephrati, M.E. Pollack and S. Ur, 'Deriving Multi-Agent Coordination through Filtering Strategies', *Proceedings of IJCAI*, pp.679-685, 1995.

[2] E. Ephrati and J.S. Rosenschein, 'Multi-Agent Planning as a Dynamic Search for Social Consensus', *IJCAI*, pp.423-429, 1993.

[3] G.E. Hughes and M.J. Cresswell, 'An Introduction to Modal Logic', Methuen, London, 1968.

[4] J.Y. Halpern and Y. Moses, 'Knowledge and Common Knowledge in a Distributed Environment', *Proceedings of the 3rd ACM Conference on Principles of Distributed Computing*, pp.50-61, 1984.

[5] J.Y. Halpern and Y. Moses, 'A Guide to the Modal Logics of Knowledge and Belief: Preliminary Draft', *IJCAI*, pp.480-490, 1985.

[6] L. He, H. Seki and H. Itoh, 'WEIHE: An Automated Reasoning System for Multi-Agent Knowledge and Time', *Proceedings of PRICAI*, pp14-19, 1994.

[7] J.Y. Halpern and M.Y. Vardi, 'The Complexity of Reasoning About Knowledge and Time: Extended Abstract', *Proceedings of the Eighteenth Annual ACM Symposium on Theory of Computing*, pp.304-315, 1986.

[8] T. Hogg and C.P. Williams, 'Solving the Really Hard Problems with Cooperative Search', pp.231-236, *Proceedings of AAAI*, 1993.

[9] F. Lin and Y. Shoham, 'On the Persistence of Knowledge and Ignorance: A Preliminary report', In *the fourth Int. Workshop on Nonmonotonic Reasoning*, 1992.

[10] D.W. Loveland, 'Mechanical Theorem Proving by Model Elimination', *J. of the ACM 15*, pp.236-251, 1968.

[11] R. Manthey and F. Bry 'SATCHMO: a theorem prover implemented in Prolog', *Proceedings of 9th Conference on Automated Deduction*, 1988.

[12] C. Morgan, 'Methods for Automated Theorem Proving in Nonclassical Logics', *IEEE, Transactions on Computers*, vol. c-25, No.8, August 1976.

[13] A. Nonmengart, 'First-Order Modal Logic Theorem Proving and Functional Simulation', *IJCAI'93*, pp.80-85, 1993.

[14] H.J. Ohlbach, 'A Resolution Calculus for Modal Logics', *CADE'88 (LNCS 310)*, pp.500-516, 1986.

[15] M. Stickel, 'Automated Deduction by Theory Reasoning', *International Joint Conference on AI*, pp.1181-1186, 1985.

GUARDNET: A DISTRIBUTED AND CONCURRENT PROGRAMMING ENVIRONMENT FOR MULTI-AGENT SYSTEMS

Motoyuki Takaai, Hideaki Takeda and Toyoaki Nishida
Graduate School of Information Science,
Nara Institute of Science and Technology
8916-5 Takayama, Ikoma, Nara, 630-01 Japan
Phone:+81-7437-9-9211 ext.5316, FAX:+81-7437-2-5269
Email:{motoyu-t,takeda,nishida}@is.aist-nara.ac.jp

ABSTRACT

GuardNet is a programming environment that supports distributed and concurrent development of multi-agent systems. Since agents in multi-agent systems are distributed and work concurrently, they are often developed distributively and concurrently. Because GuardNet itself is a multi-agent system, GuardNet provides the following three services to support programmers developing such multi-agent systems; (1)exchanging agent specifications instead of agent programmers, (2)generation of agent templates from exchanged specifications, and (3)proxing agents for incomplete or under-programming agents. In this paper, we describe GuardNet's architecture, mechanism of these services, and practice of agent programming with GuardNet.

1. INTRODUCTION

In recent years, researchers are interested in problem solving and knowledge information processing by multi-agent systems. In multi-agent systems, a number of computer programs (so called *agents*) are connected to each other, so that larger and more complex problems can be solved. For example, we have proposed Knowledgeable Community as a framework of knowledge sharing and reuse, which is based on multi-agent systems [Ni94][Ta95]. Through out experience, we have realized that programming of multi-agent systems is not easy task.

It is because agents in multi-agent systems are distributed and work concurrently. In this paper, we show a new programming environment called GuardNet that supports distributed and concurrent development of multi-agent systems.

GuardNet generates a special support agent called Guardant for each agent to be programmed, and each Guardant supports the programmer for programming the original agent by communicating her/him and other Guardants. Guardants provide the following three services to support programmers of multi-agent systems;

(1) Exchanging agents specifications instead of agent programmers, (2) Generation of agent templates from exchanged specifications, and (3) Proxing agents for incomplete or under-programming agents.

The following sections are organized as follows. In Section 2, we show our basic approach to model agents, i.e., agent as virtual knowledge base. In Section 3, we show the architecture of GuardNet and Guardant. In next three sections, we show Guardant's three functions. In Section 7, we show GuardNet implementation and examples of GuardNet programming. In Section 8, we show how GuardNet supports programmers by using a test programming to build a simulation system of a multiple robot system. Section 9 summarizes the paper.

2. AGENT AS VIRTUAL KNOWLEDGE BASE

In this paper, we employ KQML(Knowledge Query and Manipulation Language)[Fi92] as communication protocol among agents. In KQML, agents are interpreted as virtual knowledge base (VKB), that is, each agent is expected to behave as knowledge base, even if they are not knowledge bases actually. It means that they accept messages of knowledge operation (query, deleting, addition etc.).

A KQML message includes message intention, sender's and receiver's names, name of knowledge representation language and so on. Intention is represented as *performative*, that is defined as the operations of knowledge base (i.e. query, reply, insert and delete information etc.).

Figure 1 shows the examples of KQML messages. `ask-if` means that the message is a query, `:content` means that the next is knowledge content, `:language` means that the next is name of knowledge representation language.

```
(ask-if :content (and (temple ?t todaiji)
                      (tel-number ?t ?tel))
        :language kif)
```

(a) Question Message

```
(reply  :content (and (temple t1 todaiji)
                      (tel-number t1 ''12-3456''))
        :language kif)
```

(b) Answer Message

Figure 1: Examples of question and answer message

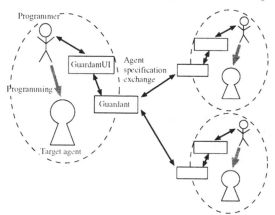

Figure 2: Connections among developers by GuardNet

3. SUPPORTS OF MULTI-AGENT SYSTEM DEVELOPMENT BY GUARDNET

We adopt a multi-agent system as the architecture of GuardNet. Since multi-agent system is used to build flexible and extensive systems, programming environments of multi-agent systems should be also flexible and extensive. Then it is natural to build the programming environment by a multi-agent system.

GuardNet provides a special support agent called Guardant and an editor-embedded agent for each programming agent (see Figure 2). We call the programming agent the *target* agent for the Guardant. We also call the other programming agents *outer* agents for the Guardant.

A Guardant support programmers for programming the target agent by communicating programmers via editor-embedded agents and programmers of outer agents via other Guardants.

The network of GuardNet is independent from the network of the developing agent system, and specifications of programming agents are exchanged agents on this network.

Figure 3 shows the architecture of Guardant. Guardant is not only the agent of GuardNet but also the agent of the developing agent system.

Specification manager supports cooperation among programmers during the decision processes of message specifications. The message specifications which are decided or under consideration are recorded to the target agent specification database. Agent proxing is realized by this database and outer agent specification

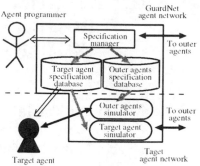

Figure 3: The architecture of Guardant

```
(ask-if :content (and (temple ?t $name)
                      (tel-number ?t ?tel))
        :language kif)
```

(a) Message Class

```
($name = todaiji)  => (?t = t1, ?tel = "12-3456")
($name = saidaiji) => (?t = t2, ?tel = "23-4567")
```

(b) Message Instance

Figure 4: Examples of Message specifications

database.

4. DECISION OF MESSAGE SPECIFICATIONS BY GUARDANT

Since one of the important features of agents is to manage messages, that is exchanging various messages with other agents, GuardNet supports the decision of *message specification* as specification of agents.

A programmer of an agent must discuss with other agent programmers, because the agent exchanges various messages with other agents. GuardNet supports this process of decision by exchanging requests of message specifications among Guardants instead of these programmers. This exchanging of requests are controlled by state transition diagram of conversation for action.

The benefits of this approach are as follows;

Message specifications are clearly defined:
Decision by human communication is often vague and uncertain. In this approach, each specification is explicitly represented, and certainty is verified by state transition diagram for action.

Message specifications are re-usable:
Message specifications are recorded in the database, so that we can use them in generating agent templates and proxing agents.

4.1 Message specification
Message Specifications are either message classes or message instances which are defined as follows;

Message Instance: message itself represented by message class name and substitution for variables

Message Class: abstract representation of message instance using variables

Message instance can be made from message class substituting constants for the variables. Guardant exchanges message classes for the decision of message specifications. Figure 4 shows examples of the message specifications which use KQML as message protocol and KIF [Ge90] and knowledge representation language.

4.2 Databases of message specifications in Guardant

Guardant have two databases. One is target agent specification database where the message specifications of the target agent are stored. The other is outer agent specification database where the message specifications that the programmer requests to other programmers are stored.

4.3 The process of manage specifications of decision

We show processes of the decision of message specifications as follows;

Here we assume that programmer U of agent A uses Guardant G, and U' of A' does G' too.

(1) U registers a request with the outline texts to G.
(2) The process of decision of message class.
 (2-a) U inputs a draft of the message class to G.
 (2-b) G sends the draft to G'.
 (2-c) G' shows the draft to U'.
 (2-d) U' inputs the reply (acceptance or counter-offer) for it to G'.
 (2-e) G' sends the reply to G.
 (2-f) G shows the reply to U.
(3) G and G' record draft and accepted (or promised) message class to their database.

In the process from (2-a) to (2-f), specification manager in Guardant runs according to state transition diagram of conversation for action [Wi88] until U and U' agree the specification. During this process, the conversations are recorded to their database by Guardants. Since Guardant prepares state transition diagram of conversation for action for each message specification, programmer can handle a number of processes of decision of some message classes at once.

4.4 Addition of message instances

Programmers can add message instances to decided message specifications of target agent. They are stored in target agent specification database with the message class in the Guardant.

Programmers can also do to the message specifications of outer agents, and they are stored in outer agent specification database.

Guardant can import the message instances related to the target agent from other Guardants.

5. SUPPORT OF THE AGENT CODING BY GUARDNET

GuardNet supports agent coding by using database of the message specifications.

```
(defagentkc foo
 :mbus-port 2392
 :message-handler 'foo-handler)
(setq foo-handler
 '(((((:performative . (?or ask-if))
     (:kif-content . (and (temple ?t $name)
                          (tel-number ?t ?tel)))
    (:language . kif)
    ) . do-foo-telnumber)
  (((:performative . (?or ask-if))
    (:kif-content . (and (temple ?t $name)
                         (open-time ?t ?o-time)))
   (:language . kif)
   ) . do-foo-open-time)))
(defun do-foo-telnumber (env)
"Title: An phone number of a temple
Comment: phone number is local number)"
  (let ((binding *kif-binding*))
  ))
(defun do-foo-open-time (env)
"Title: Admission time for visiting
Comment: $name is a temple name."
  (let ((binding *kif-binding*))
  ))
```

Figure 5: The interface part of an agent generated by Guardant

We developed an agent programming language called KC-Basic which runs on Common-Lisp. In KC-Basic, the message receiving parts of agent programs consist of the following elements;

(1) agent registration
(2) registration of message classes and functions to execute for them
(3) the bodies of the functions

When an agent receives a message of which pattern matches with one of the message classes in (2), the associated function for it is called and executed along with its body written in (3).

Guardant makes the template of this message receiving parts of agents the agent automatically from the massage specifications. Figure 5 shows the template from the message specifications in Figure 4. When the programmer modifies the function body of *do-foo-telnumber* and *do-foo-open-time*, this agent will be executable.

6. AGENT PROXING

Agent proxing by GuardNet is that Guardants works as proxies of programming agents in case that there does not exist necessary agents yet or exist only incomplete agents which are not good for use yet. Agent proxing can make the simulation of such incomplete agent systems. We realized the agent proxing by using specifications from message exchanging explained in Section 4.

6.1 Types of agent proxing

We provide two types of agent proxing.

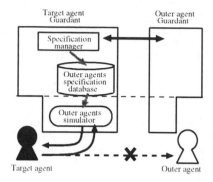

Figure 6: Outer Agent Imitating by Guardant

Agent proxing for outer agents: For verifying execution of the target agent, Guardant becomes proxy agents for outer agents to which the target agent wants to communicate. If some of the outer agents which the target agent want to communicate are not at work, we cannot verify the execution of target agent. By proxing such agents, it becomes possible to check the execution of target agent.

Agent proxing for the target agent: When the target agent is not good for use (for example in debugging), Guardant becomes a proxy agent for the target agent. If other agents want to communicate to the target agent, the Guardant responds to it instead of the target agent.

6.2 Type of message actions to simulate

We can identify the following four types of message actions that proxy agents should simulate as the normal agents do.

Ask: Send the query message and expect to get the answer

Reply: Send the answer for the query message

Tell: Send the message to tell information

Listen: Receive message to get information from outer agents

For example, simulation of reply action is as follows;

(1) Find specifications which match the incoming messages in the specification database.

(2) Compose reply messages by using the query message and the found specifications.

6.3 Agent proxing mechanism

In agent proxing for outer agents, Guardant prepares a simulator for each of them. The simlulators react as knowledge database which have the message instances of outer agents' specification database in Guardant(see Figure 6).

In agent proxing for target agent, the simulator react as knowledge database which have the message instances of target agent's specification database in Guardant.

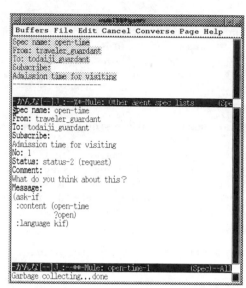

Figure 7: The developer of the traveler agent exchanges agent specifications with the other developer by Guradant

6.4 Control of the message flow

To proxy agents, it is necessary to control the message flow so that exchange of message with agent's proxy agents becomes possible.

When outer agents are already proxied, all messages from the target agent to the outer agents are received by Guardant and the real outer agents never receive these messages. Guardant plays the outer agents' roles and uses this agents' names for answer.

In the multi-agent system that uses "federation architecture"[Pa92], all messages are sent to a special agent called *facilitator* first. We use the facilitator to realize control of these message flows.

7. IMPLEMENTATION OF GUARDNET

7.1 GuardNet architecture

GuardNet consists of two types of agents, i.e., Guardant and GuardantUI.

Guardant continues working whether programmer is using Guardant or not, to wait for the messages from other Guardants always. Guardant manages all the data of message specifications. Guardant is implemented on GNU Common Lisp.

GuardantUI interacts with a programmer for Guardant. GuardantUI is implemented on Emacs Lisp (ver 19) and is embedded to the editor that programmer uses. A programmer can start and exit GuardantUI anytime.

7.2 Examples of system's execution

Here we show an example to illustrate how GuardNet works. In this example, there are an agent to represent a traveler, and some agents to represent temples.

Figure 7 shows a snapshot of GuardantUI, where the programmer of the traveler agent is asking to the

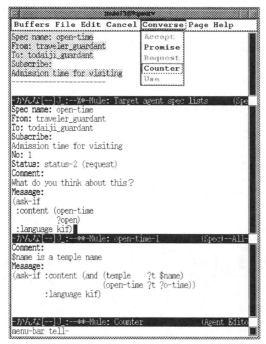

Figure 8: The developer of Todaiji-temple agent exchanges agent specifications with the other developer by Guradant

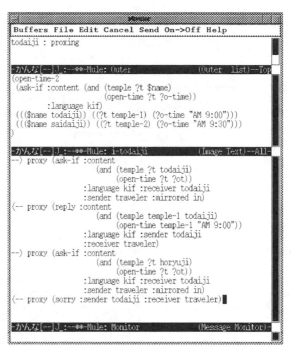

Figure 9: The developer of the traveler agent makes imitations of other agents by Guradant

programmer of Todaiji temple agent about possible time to visit the temple. The upper window shows the list of message specifications which are registered up to now. The specification about the information of the possible time to visit the temple is chosen and the contents of exchanging message specifications is shown in the lower window. The message specifications that the traveler agent programmer requests to Todaiji temple agent programmer are the same to those shown in Figure 4. The requested content for it is sent to the Guardant of Todaiji temple agent.

Figure 8 shows a snapshot when Guardant of Todaiji temple agent received the request, and replied a counter-offer. The upper window shows the list of message specifications which are registered up to now. The middle window shows the contents of the received requirement.

Figure 9 is a snapshot when the traveler agent programmer is exchanging the message with the todaiji agent which is currently proxied by the traveler agent's Guardant. The upper window shows that Todaiji temple agent is now in agent proxing. In the middle window, there is data of Todaiji temple agent's message specifications (see Figure 4), and the admission time for visiting Todaiji temple and Saidaiji temple are given as message instances. The lower window shows how execution of the traveler agent works with the proxy agent of Todaiji temple agent. Instead of Todaiji temple agent, Guardant of the traveler agent receives the message which was sent from the traveler agent to Todaiji temple agent and it replies a message to the traveler agent.

8. EXPERIMENTS ON AGENT PROGRAMMING BY GUARDNET

We used GuardNet to build a simulation system of a multiple robot system (see Figure 10) to show how it can support programmers of multi-agent systems.

There are the following three tasks to simulate.

- Robots bring the object to the user which he/she wants to get.
- Robots put objects in an area to the rack in order.
- Robots deliver objects.

We have two autonomous mobile robots (one has hands), an autonomous rack and an automatic door. In this test, we asked two programmers to build the system.

8.1 Meeting among programmers

Before decision of the agent programming by Guard-Net, the programmers had a meeting. In the meeting, they talked what agents they would make, and what information the agents exchange. In this case, they decided the architecture of the agent system, i.e., four agents for robots, a user-interface agent and a planning agent. One programmer took charge of four agents, the other took charge of two (see Figure 10).

8.2 The experiments on the decision of message specifications by Guardant

They made 25 messages. 8 messages in them are just re-used of the other messages, so that they have no decision processes. In the decision processes of 9 messages, there are at least one counter-offer. The other messages are accepted without any counter-offers.

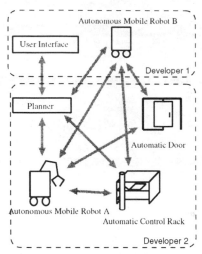

Figure 10: An example of agent system on physical world

The later the experiments proceeded, less counter-offers were made. It seems that the programmers formed representation patterns of message specifications.

The counter-offers can be classified into 6 groups as follows;

- Point out mistakes
- Modification of representation
- Addition of representation and information
- Fundamental change of representation
- Emphasis on his own opinion again
- Query

We can conclude the following advantages of Guard-Net from this experiment.

- Programmers can communicate and decide the agent specifications, regardless their physical distances.
- Each programmer can manage agent specifications concurrently.
- The decision process of the agent specifications are done in a relatively short time.
- Even if a programmer is not at work, Guardant can provide the agent specifications to other programmers instead of the programmer.

But we also obtained the following disadvantages;

- Many (more than about 5) parallel decision processes of message specifications often make the programmer confuse.
- Programmers may want to unify expressions of different messages, but GuardNet don't support it.

9. CONCLUSION

By management of message specification, Guard-Net supports decision of message specification, agent coding and debugging. To share the expression of message expression, GuardNet needs the function to manage ontology. In the KSE they study collaborative ontology construction on world wide web [Fa94].

Parman [?] is intelligent parametric design tool supporting collaborative engineering. The agent of Parman consists a design tool and a human user, and the relation of them is similar to our Guardant and developer.

10. SUMMARY

In this paper we described a multi-agent system GuardNet which supports cooperative development of multi-agent systems. It provides a new environment for cooperative, distributed, and asynchronous development of multi-agent systems.

REFERENCES

[Ni94] T. Nishida and H. Takeda: Towards the Knowledgeable Community, In K. Fuchi and T. Yokoi, editors, Knowledge Building and Knowledge Sharing, Ohmsha, IOS Press, 1994, pp.155–164.

[Fi92] T. Finin, J. Weber, G. Wiederhold, M. Genesereth, R. Fritzson, D. McKay, J. McGuire, P. Pelavin, S. Shapiro, and C. Beck: Specification of the KQML agent-communication language, Technical Report EIT TR 92-04, Enterprise Integration Technologies, 1992.

[Wi88] T Winograd: A language / action perspective on the design of cooperative work, In Irene Greif, editor, Computer-Supported Cooperative Work: A Books of Reedings, Morgan Kaufmann, 1988.

[Ta95] H. Takeda and K. Iino and T. Nishida: Agent Organization and Communication with Multiple Ontologies, the International Journal of Cooperative Information Systems (IJCIS),1995.

[Ge90] M. R. Genesereth and R. E. Fikes: Knowledge Interchange Format version 3.0 reference manual, Technical Report Logic-90-4, Computer Science Department, Stanford University,1990.

[Pa92] R. S. Patil, R. E. Fikes, P. F. Patel-Schneider, D. McKay, T. Finin, T. R. Gruber, and R. Neches: The DARPA knowledge sharing effort: Progress report, In Charles Rich, Bernhard Nebel, and William Swartout, editors, Principles of Knowledge Representation and Reasoning: Proceedings of the Third International Conference. Morgan Kaufmann, 1992.

[Ku94] Daniel Kuokka and Brian Livezy: A Collaborative Parametric Design Agent, Proceedings of the AAAI-94,pp.387-393

[Fa94] A. Farquhar, R. Fikes, W. Pratt, and J. Rice: Collaborative Ontology Construction for Information Integration, http://www-ksl.stanford.edu/KSL_Abstracts/KSL-95-63.html

Fuzzy Logic and Control

ASSIGNING WEIGHTS TO RULES OF AN EXPERT SYSTEM BASED ON FUZZY LOGIC

Mehdi R. Zargham and Leishi Hu
Department of Computer Science
Southern Illinois University
Carbondale, Illinois 62901
USA
Email: mehdi@cs.siu.edu

ABSTRACT

A new weight assignment method based on Dempster theory is proposed. The new method is applied to a fuzzy stock prediction system based on a proposed stock evaluation model. The performance of the system is tested by conducting a simulation on true stock market data. The average annual return of a portfolio selected by our system is compared to the return of the Standard & Poor's 500 Index. From the comparison, we determine that our system significantly outperforms the Standard & Poor's 500 Index return.

Key Words: Fuzzy Logic, Stock Market, Weight Assignment, Securities Analysis, Inductive Learning.

INTRODUCTION

Two kinds of systems exist in the real world: one for which we can find a mathematical model, the other for which we cannot find a mathematical model, or no mathematical model exists at all. Often, the latter model is implemented by using expert systems.

Expert systems are further divided into two categories: (a) classical systems using two-value logic and (b) fuzzy systems based on fuzzy logic, introduced by Zadeh in 1965 [Zad65]. For some applications, the fuzzy systems are considered more flexible and economical than the classical expert systems since fewer rules or combinations of rules are needed to cover more possible outcomes. Unlike a classical expert system, fuzzy systems can handle overlap or ambiguity between rules.

The rules are the true spirit of the expert systems. We can generally obtain rules in two ways: either from the experts directly or generated from the data indirectly [WM92]. In either case, the rules cannot be equally treated. So the problem arises: How can we assign some kinds of measurements, or weights, to the rules in order to represent their difference in importance? Given a set of data and a rule, we would like to determine the compatibility between the data

set and the rule. In other words, we want to find out the degree of evidence that the data set supports the rule. The first part of this paper proposes a solution to this problem.

In the second part of this paper, some fuzzy rules are generated based on Ben Graham's stock selection criteria [Blu77]. Simulation is done on the data from 1988 to 1993; a portfolio is selected for each year and the annual return of the portfolio is calculated using these fuzzy rules. The technique of proposed weight assignment is used in the simulation, and very good results are obtained.

A NEW WEIGHT ASSIGNMENT METHOD

The Problem is to find out the degree of evidence that a given data set supports a given rule. Since this problem involves calculating the degree of evidence, Dempster theory seems to be a good tool to solve this problem. In fact, Delgado and Gonzalez [DG93] have used Dempster theory to associate to a given rule some kind of belief or uncertainty measure about the statement "this rule is a true rule for the system." Their approach has given us some insight on our derivation of weight assignment method which is introduced below.

To describe the vagueness or imprecision in the assignment of an element to a crisp set, a measurement, known as *fuzzy measure,* is often used. The fuzzy measure assigns a value to each crisp set of the universal set signifying the degree of evidence or belief that a particular element belongs in the set. (Details about the fuzzy set theory and fuzzy measures can be found in [KF88].) There are two special important types of fuzzy measures: *belief measure* and *plausibility measure.* Basically, a belief measure is a quantity, denoted as Bel(A), which expresses the degree of support, or evidence, for a collection of elements defined by one or more of the crisp sets existing on the power set of a universe. Plausibility measure, denoted as Pl(A), is defined as the "complement of the belief of the complement of A."

$$Pl(A) = 1 - Bel(\overline{A}). \qquad (1)$$

So 1-Pl(A) will express the degree of support of the complement of the crisp sets existing on the power set of a universe. In [Dem67], it is shown that the plausibility measure is always greater than or equal to the belief measure, that is:

$$Pl(A) \geq Bel(A)$$

So the belief measure and plausibility measure are also called lower probability and upper probability, respectively. In this paper, we use all these terms while referring to them.

In the theory of evidence, there is also a direct generalization of the conditioning in probability theory given by Dempster [Dem67]. According to it, conditional plausibility and belief measures are calculated as:

$$Pl(B|A) = \frac{Pl(A,B)}{Pl(A)} \qquad (2)$$

$$Bel(B|A) = \frac{Bel(A,B)}{Bel(A)} \qquad (3)$$

where $Pl(A,B)$ and $Bel(A,B)$ indicate the joint plausibility and belief measures of A and B.

The conditional plausibility and belief measures can be used to derive a lower and an upper probability of evidence that are representative of how much a data set supports a given rule. Let's say each rule, R_i, in our system has the following form:

Rule R_i: IF A THEN B.

In the above rule, A, called antecedent part, is in the form of "X_1 is A_{1i} ... and X_n is A_{ni}" and B, called consequent part, is in the form of "Y_1 is B_{1i} ... and Y_m is B_{mi}," where X_j and Y_j are linguistic variables with A_{ji} and B_{ji} as their corresponding linguistic values. Also, let's assume that the data set contains a total of h data pairs; each data pair represented as (x_{1k},..., x_{nk},y_{1k},...,y_{mk}), where k=1,...,h. Furthermore, let's define the following notations that will be used in calculating the belief measure and plausibility measure of a rule.

AC_k = the compatibility degree of the antecedent and the k-th data pair. It is defined as:
$AC_k = Bel_k(A) = \mu_{A_{1i}}(x_{1k}) * \dots * \mu_{A_{ni}}(x_{nk})$

CC_k = the compatibility degree of the consequence and the k-th data pair. It is defined as:
$CC_k = Bel_k(B) = \mu_{B_{1i}}(y_{1k}) * \dots * \mu_{B_{mi}}(y_{mk})$

$\overline{CC}_k$ = the incompatibility degree of the consequence and the k-th data pair. It is defined as: $\overline{CC}_k = Bel_k(\overline{B}) = (1 - \mu_{B_{1i}}(y_{1k})) * \dots * (1 - \mu_{B_{mi}}(y_{mk}))$

The data set might contain more than one data pair. So the compatibility degree between a set of data and a statement can be calculated as the arithmetic average of the compatibility degree between the data set and the statement, which is the sum of the each compatibility degree divided by the number of data. Now we can derive the formula for calculating the belief and plausibility measures as:

$$Bel(R_i) = Bel(B|A) = \frac{Bel(A,B)}{Bel(A)}$$

$$= \frac{\sum_k \dfrac{Bel_k(A,B)}{h}}{\sum_k \dfrac{Bel_k(A)}{h}}$$

$$= \frac{\sum_k Bel_k(A) * Bel_k(B)}{\sum_k Bel_k(A)}$$

$$= \frac{\sum_k AC_k * CC_k}{\sum_k AC_k}$$

$$Pl(R_i) = Pl(B|A) = 1 - Bel(\overline{B}|A) = 1 - \frac{Bel(A,\overline{B})}{Bel(A)}$$

$$= 1 - \frac{\sum_k \dfrac{Bel_k(A,\overline{B})}{h}}{\sum_k \dfrac{Bel_k(A)}{h}}$$

$$= 1 - \frac{\sum_k Bel_k(A) * Bel_k(\overline{B})}{\sum_k Bel_k(A)}$$

$$= 1 - \frac{\sum_k AC_k * \overline{CC}_k}{\sum_k AC_k}$$

A direct interpretation of Bel(R_i) and Pl(R_i) is that they are the lower and upper probability of the evidence contained in the set of data which support the corresponding rule. That is, each rule is supported by the data to the degree Bel(R_i), and against by the data to the degree (1-pl(R_i)).

For example, if we have a rule with Bel(R_i)=0.5 and Pl(R_i)=0.6, the interpretation is that the data set will support the rule to the degree 0.5 and against the rule to the degree 0.4 (which means the data set will support the rule with the same antecedent and different consequence to the degree 0.4). If the interval is [1,1], the examples are 100% positive and 0% negative, which means the rule perfectly represents the data set.

If the interval is [0,0], 100% of the examples are negative. So all the data support the rule with the same antecedent and a different consequence.

To obtain a weight for each rule, we assume that each rule has just one linguistic variable in the consequent part; that is m=1. Thus each rule is represented as:

Rule R_i: if X_1 is A_{1i} ... and X_n is A_{ni} then Y is B.

Based on this assumption, we can show that the lower and upper probability of the evidence become equal, since

$$pl(R_i) = 1 - \frac{\sum_k AC_k * \overline{CC_k}}{\sum_k AC_k}$$

$$= 1 - \frac{\sum_k AC_k * (1 - CC_k)}{\sum_k AC_k}$$

$$= 1 - \frac{\sum_k AC_k - \sum_k AC_k * CC_k}{\sum_k AC_k}$$

$$= 1 - 1 + \frac{\sum_k AC_k * CC_k}{\sum_k AC_k}$$

$$= \frac{\sum_k AC_k * CC_k}{\sum_k AC_k} = Bel(R_i)$$

So we derive a simple formula to calculate the weight $W(R_i)$ by using $Bel(R_i)$ or $pl(R_i)$, that is:

$$W(R_i) = \frac{\sum_k AC_k * CC_k}{\sum_k AC_k} \qquad (4)$$

STOCK MARKET PREDICTION

One of the real-world problems without a mathematical model is stock market prediction. Selection of the right stocks to beat the markets is the main idea of investing in the stock markets. To design a profitable portfolio is nevertheless a difficult and complex problem. It needs to apply and combine a lot of experts' knowledge to analyze and evaluate stocks.

To develop our expert system for stock selection, Graham and Rea's 10 stock selection criteria were used [Blu77] as basis for generating rules. Benjamin Graham, the father of modern security analysis, believed in the overall efficiency of securities markets, but he also believed that pockets of inefficiency still existed. If any investor could pay close attention to the investment fundamentals and take advantage of under valuation and mispricing of individual securities, a high return could obtained in the future. In an article published shortly after Graham's death [Blu77], he listed 10 criteria which an investor could use to identify undervalued stocks. The 10 criteria for selecting stock are:

(1) An earnings-to-price yield at least twice the AAA bond yield,
(2) A price-earning ratio less than 40 percent of the highest average price-earnings ratio the stock had over the past five years,
(3) A dividend yield of at least two-thirds the AAA bond yield,
(4) Stock price below two-thirds of tangible book value per share,
(5) Stock price below two-thirds "net current asset value,"
(6) Total debt less than book value,
(7) Current ratio greater than two,
(8) Total debt less than twice "net current asset value,"
(9) Earning growth of prior 10 years at least at a 7 percent annual (compound) rate,
(10) Stability of growth of earnings in that no more than two declines of 5 percent or more in year-end earnings in the prior 10 years are permissible.

The definitions of the financial items in the above rules can be found in [CMB88] or other books about investment. The above criteria are mainly for screening out those stocks whose values are under-estimated, thus, with a large potential to grow in price. We can see in the 10 criteria that Graham has set some sharp boundaries to the factors, such as "at least twice" in rule (1), "40 percent" in rule (2), and so on. Actually in the real world, nobody will consider that a stock with earning-to-price yield 1.99 times the AAA bond yield is really worse than a stock with earning-to-price yield 2.01 times the AAA bond yield.

To implement Graham's rules in fuzzy logic, some changes must be made, though the spirit of the Graham's rule will always be kept. The variables that will be used in the fuzzy rules are defined as follows:

X_1 = (Earning Yield)/(AAA Bond Yield)

X_2 = (Highest average Price Earning Ratio over past 5 years)/(Current Price Earning Ratio)

X_3 = (Dividend Yield)/(AAA Bond Yield)

X_4 = (Book Value Per Share)/(Price)

X_5 = (Net Current Asset Value Per Share)/(Price)

X_6 = (Book Value)/(Total Debt)

X_7 = Current Ratio

X_8 = (Net Current Asset Value)/(Total Debt)

X_9 = Annual Earning Growth rate (Compound) of prior 10 years

X_{10} = Number of declines of 5% or more in year - End
        Earnings
while X_i is the input variable and Y is the output
variable.

Now, each Graham's criteria i (for i=1 to 10) can be
represented as the following five rules:
Rule (i.1) If X_i is VERY BIG   then
        Y is VERY GOOD,
Rule (i.2) If X_i is BIG   then
        Y is GOOD,
Rule (i.3) If X_i is MEDIUM   then
        Y is OK,
Rule (i.4) If X_i is SMALL   then
        Y is BAD,
Rule (i.5) If X_i is VERY SMALL   then
        Y is VERY BAD.

So based on Graham's criteria, 50 fuzzy rules are
generated since each criterion produces 5 rules,
respectively.

The definitions of the membership functions are
derived from the intention of Graham's rules and
observation of data. The membership functions of the
ten input variables are shown in Figure 1. The output
variable Y is defined as "the rating of the stock." It
represents the degree to which a stock matches one
criterion. The closer a stock matches the criterion, the
higher it is rated. The higher a stock is rated, the better
stock it is considered. The membership functions of the
output variable Y are shown in Figure 2.

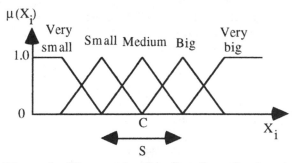

**Figure 1. The membership functions for input
variables X_i.**

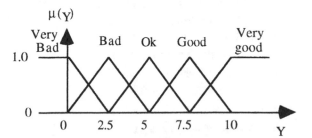

**Figure 2. The membership functions for output
variable Y.**

To select the best rules among the 50 rules, we used
equation 4 to assign a weight to each rule. James B.
Rea, Graham's partner in creating the 10 criteria, wrote
in the article   "Remembering Benjamin Graham -
Teacher and Friend" [Rea77] that "Each one of our ten
criteria had been considered of equal weight . . . And
we knew that wasn't quite right, but we had no
evidence of what else to do. We intuitively knew that
one criterion could had a higher weight that another . .
.". Also as mentioned before in the approach suggested
by Oppenheimer[Opp84], increasing the number of
criteria to select the stock may cause a decrease on
the annual return.
Both of the above arguments support our belief on
the importance of assigning proper weight to the rules.
So if we train the rules on the past data, we can
identify the comparatively more important rules to
enhance the performance of the system. The proposed
weight assignment method is used to train the rules.
Since the fuzzy rules we used have only one
antecedent and one consequence, we may further
simplify our weight assignment formula (which is given
in the previous section) as:

$$W(R_{ij}) = \frac{\sum_k \mu_{A_j}(X) * \mu_{B_j}(Y)}{\sum_k \mu_{A_j}(X)} ,$$

where k is the number of data.
Now the weights of rules can be used to calculate the
rating of the stocks.

SIMULATION

This section shows the results of a simulation running
on the stock market from 1988 to 1993. The historical
data are downloaded from Standard & Poor's
Compustat Database and involve 6,017 companies. In
our system, we simulate picking the top 20 highest-
rating stocks to form a portfolio, and buying these
stocks at the last month of 1988, 1989, 1990, 1991,
1992, then selling them exactly one year later with the
average return of each portfolio being calculated.
When invested with the portfolio, equal amounts of
money are assumed to be put into each stock.
The benchmark used for comparison is the return on
the S&P 500 Index, which is the standard market
return. The returns of the S&P 500 Index during these
years are as follows:
The results of the simulation are shown in Table 2.
The first row is the average return of the S&P 500
Index. The second row is the result of the simulation
using the rules with equal weights, which means that
each fuzzy rule is treated equally. The third row is the
result of a simulation using our weight assignment
method; the rules are trained on all the previous years'
data, and the average of the trained weights is used for
the selection of stocks.

Table 2 represents that all the Portfolios selected by our set of fuzzy rules (based on Graham's rules) outperform the return of the S&P 500 Index significantly. In addition, the weight assignment method produces much better performance than the approach using equally weighted rules for the 1988, 1989 and 1993 portfolios, especially for 1989.

CONCLUSION

In this paper, first, a new weight assignment method based on Dempster theory was proposed. Next, the new method was applied to a fuzzy stock prediction system based on a proposed stock evaluation model. The performance of the system was tested by conducting simulations on true stock market data. The average annual return of portfolio selected by our system was compared to the return of the Standard & Poor's 500 Index. From the comparison, we determined that our system outperformed the Standard & Poor's 500 Index return significantly.

Although our study mainly demonstrates the potential application of our weight assignment method to the area of securities analysis, we believe the theory can be applied as well to the other areas of decision making, especially those of rule-based evaluation and analysis, such as short-term electric load forecasting, credit scoring of loan application, and so on.

REFERENCES

[Blu77] P. Blustein, Ben Graham's last will and testament, Forbes, August 1, 1977, p. 43-45.

[CMB88] S. Cottle, R.F. Murray, and F.E. Block, Graham and Dodd's Security Analysis, 5th Ed., New York: McGraw-Hill Book Company, 1988.

[Dem67] A. P. Dempster, Upper and Lower Probabilities Induced by a Multivalued Mapping, Annual Mathematics Statistics 38 (1967), p. 325-339.

[DG93] M. Delgado and A. Gonzalez, An Inductive Learning Procedure to identify Fuzzy Systems, Fuzzy Sets and Systems 55, 1993, p. 121-132

[KF88] G. Klir and T. Folger, Fuzzy Sets, Uncertainty and Information, Englewood Cliffs, Prentice-Hall, 1988.

[Opp84] H. R. Oppenheimer, A Test of Ben Graham's Stock Selection Criteria, Financial Analysts Journal, September-October, 1984, p. 68-74.

[Rea77] J. B. Rea, Remember Benjamin Graham - teacher and friend, The journal of Portfolio management, Summer, 1977, p. 66-72.

[WM92] L-X Wang and M. Mendel, Generating Fuzzy Rules by Learning from Examples, IEEE Transactions on Systems, Man, and Cybernetics, v. 22, n. 6, November/December, 1992, pp. 1414-1427.

[Zad65] L. A. Zadeh, Fuzzy sets. Information and Control, 8:338-353, 1965.

Table 1: Return of S&P 500 Index during 1988-1993

Time Period	Return for S & P 500 Index
Dec. '88 ~ Dec. '89	+25%
Dec. '89 ~ Dec. '90	-7%
Dec. '90 ~ Dec. '91	+23.6%
Dec. '91 ~ Dec. '92	+4.6%
Dec. '92 ~ Dec. '93	+7.2%

Table 2: Average return for S&P 500 Index and different portfolios

	12/88-12/89	12/89-12/90	12/90-12/91	12/91-12/92	12/92-12/93
S&P 500 Index	25.0%	-7.0%	23.6%	4.6%	7.2%
Equal weights	21.04%	-2.85%	178.47%	43.98%	230.00%
Different weights	46.26%	81.40%	86.89%	23.88%	270.48%

DETECTION OF CHAOS IN PROCESS CONTROL SYSTEMS USING A JULIA SET METHODOLOGY

David W. Russell and James J. Alpigini *
Department of Electrical Engineering, Penn State Great Valley,
Malvern PA 19355, USA .
email: RZN@PSUGV.PSU.EDU
* Lockheed Martin Corporation, Camden, NJ, USA

ABSTRACT

The paper defines and demonstrates the occurrence of chaos in the iteration of mathematical formulae and shows that such systems can be exposed by their dependence on the initial values of variables and parametric coefficients. Filled Julia sets, although computationally intensive, give visual explanation of the fate of an iteration and indicate regions of crossover between stable and unstable operation. The paper then demonstrates that even the simplest process controllers may be subject to chaos in their component parts while still maintaining control of the primary outputs of the system. A variation on the Julia set is then described and offered as a possible "chaotic function analyzer" for such systems. The paper closes with some illustrations of transients taken from the stable and chaotic regions for a second order plant with a non-linear parameter estimator that is examined by the proposed method.

INTRODUCTION

The term "chaos" has taken on new meaning in the last decade. It is now not so much considered a synonym for random, erratic and unrepeatably confused behavior, but rather as a commonly seen and somewhat predictable phenomenon. Furthermore, in the past few years, it has been reputed that the trajectories of state variables in dynamic systems may not only demonstrate unstable, but truly, chaotic motion. As the parameters that describe a process model and/or its controller are computed, it is apparent that certain combinations of values have been shown to drive the system into chaotic motion. What is more surprising is the fact that the system outputs may remain stable, while the feedback signal is exercising chaotic orbits. In other words, a control mechanism may be successfully completing its assigned task while producing wildly varying values of control parameters in

order to handle the chaos. Chaotic transients have also been detected in real process control systems (e.g. [SSR84], [CS92].) It is the purpose of this paper to study the ease with which chaos can be induced into a system and to describe a methodology that may prove useful in the detection of the possibility of occurrence of chaotic motion. The paper includes the following sections:

- the occurrence of mathematical chaos
- the familiar Julia set
- chaos and process controllers
- computer simulations that describe the Julia-like method for an example system
- concluding comments and future research

MATHEMATICAL CHAOS

Chaotic behavior can occur in very simple mathematical iterations. Consider the rule: $y = y^2 + c$. In order to perform iterations on this, values for c and y_0 (the initial "seed value" of y) must be known. Figures 1 (a)-(f) show the first sixty iterations of $y = y^2 + c$ for various y_0 and c values. The data points are connected purely for clarity of observation. What is immediately observable is that the value of y_0 (*Figures 1(a) & (c)*) and the value of c (*Figures 1(a) & (b)*) are highly significant in the fate of the iteration. Furthermore (*Figure 1(e) &(f)*) even a 0.0001 change in c completely alters the chaotic waveform.

The importance of initial conditions in transient behavior is one of several indicators that chaotic processes may be present [MOO92] Another feature is that of period doubling, which can often be detected in real systems as unstable oscillation and "non harmonic ringing". The presence of chaotic elements in a control algorithm has obvious detrimental effects on the dynamic performance of a system. Chaotic values in variables occur by successive entry and redefinition of terms within a module, which is exactly the case in an in-line process control system. [MB86] present some very disturbing and well substantiated findings in this area and conclude:

"..under different conditions the adaptive controller gain behaves chaotically but still regulates the plant output..."

In other words, the plant may be controllable even though chaotic values permeate the parameters. Consequently, the adaptive control algorithm attempts to mathematically filter out such non-linearities which perhaps explains the irregular forms and trajectories of computed control output values that are required in practice in order to maintain stable operations.

JULIA SETS

The Julia Set provides a graphical representation of the of stable iterative orbits for a given range of some function f. More specifically [STR86], the *"filled-in Julia set* of f contains the points Z_0 for which Z_n stays bounded as [iteration] n $\rightarrow \infty$." The Julia set gives an immediate, graphical map of the stability of an iterated function across a range of Z. By coloring black the stable orbits of function f, and the unstable or chaotic orbits a different color, it is easy to recognize the stable and chaotic orbits. The process is numerically intensive and requires as many as 480 million calculations per graph.

The Julia set evaluation of an iterative mathematical function $f(Z, c)$ begins with a complex seed c (c_{real} + i $c_{imaginary}$) and a range of complex Z (Z_{real} + i $Z_{imaginary}$). For example, to find the Julia set of the complex function $Z^2 + c$, a single seed c is used, and Z is mapped to a computer display by each pixel being programmed to represent the complex plane in Z. The function is then iterated a predefined number of times. The trajectory of the function either remains within bounds, in which case the associated pixel is colored black, or the orbit escapes, in which case the pixel is set to some other color based on the rapidity with which it escapes.
A key component of the Julia set paradigm is the determination of the boundary value of the function that constitutes an escape. For example, the function $Z^2 + c$, may be split into its real and imaginary components as follows:

$$Z = x + y.i \qquad (1)$$
$$Z^2 + c = x^2 + 2.x.y.i - y^2 + c_{real} + i.c_{imaginary} \qquad (2)$$

In order to iterate this function beginning with an initial $Z_0 = x_0 + y_0$, the function is expressed as follows:

$$x = x_0^2 - y_0^2 + c_{real} \qquad (3)$$

$$y = 2.x_0. y_0 + c_{imaginary} \qquad (4)$$

Where x_0 and y_0 are either the seed values of Z or the result of the previous iteration.

Devaney [DEV90], shows that, for this function, the value of Z approaches infinity if ever the magnitude of either x or y becomes larger than 2.0. A boundary value of 2 may therefore be set for x and y, with the orbit of the function declared as escaped, or tending toward infinity, if this boundary is exceeded for either value. Figure 2 illustrates the Julia set of $Z^2 + c$, with a boundary of 2.0 used to determine orbit escape. The c_{seed} used was 0.1 + 0.8i and Z was ranged from -1.4 - 1.4i to 1.4 + 1.4i. The areas that are black represent the values of Z for which the function is bounded and therefore stable. The varying shades of gray and white represent orbits that escaped at different speeds.

CHAOS and PROCESS CONTROLLERS

It is proposed that the Julia set methodology can be adapted for use in the detection of chaos in a process control system. A chaotic control system will, by definition, demonstrate extreme sensitivity to the initial conditions, the values of the plant parameters, or both. It is desirable either to find the plant parameters that provide stability for a given set of initial conditions, or to find the initial conditions that provide stability for a set of plant parameters. Unlike a simple function such as $Z^2 + c$, the values in a process control system, such as the output or feedback, may oscillate or have an offset at a large magnitude while still remaining stable. For this reason, magnitude was not used to determine orbit escape, as would have been in a strict Julia Set.

Moon [MOO92] identifies *intermittent chaos* as "..periods of regular motion with transient bursts of chaotic motion; where the duration of the regular motion interval is unpredictable." The discreet parameter estimator feedback control plant presented by [MBR86] evidences chaos in the form of transient chaotic spikes, meeting Moon's criteria. Experimentation with the plant simulation described below demonstrated that the magnitude of the derivative of a variable, i.e. the velocity of the orbit, appeared a better determinant of orbit escape than the variable itself.

It was found that the Julia set methodology could be adapted to give a visual analysis of the motion of the plant values over time, or in this case, over periods of numeric integration. In the proposed "Julia-like" method, the color map is updated if the any plant variable derivative changes by more than some fixed amount during a time period. In this manner it is possible to record and color-code the transient behavior

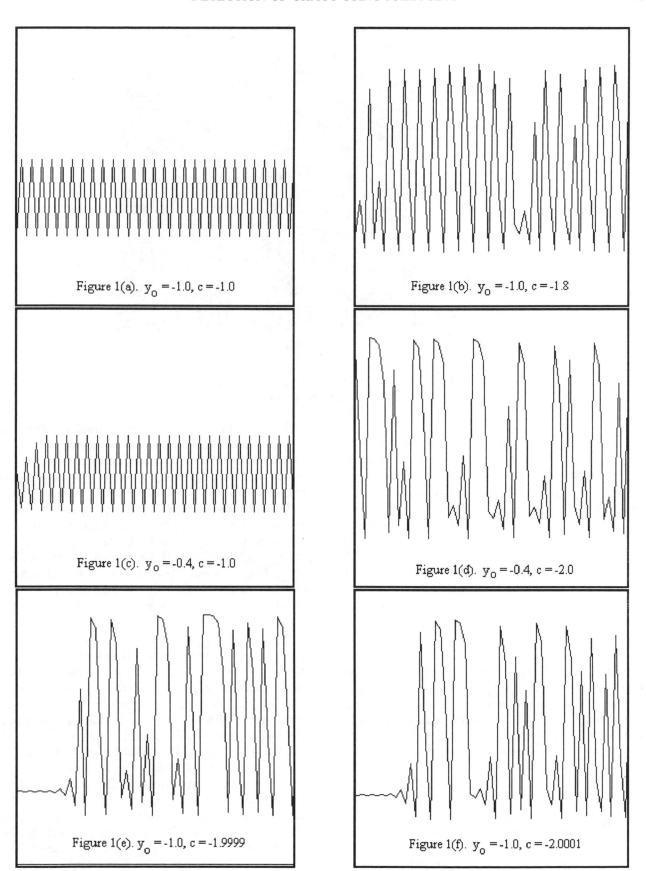

Figure 1(a). $y_0 = -1.0$, $c = -1.0$

Figure 1(b). $y_0 = -1.0$, $c = -1.8$

Figure 1(c). $y_0 = -0.4$, $c = -1.0$

Figure 1(d). $y_0 = -0.4$, $c = -2.0$

Figure 1(e). $y_0 = -1.0$, $c = -1.9999$

Figure 1(f). $y_0 = -1.0$, $c = -2.0001$

FIGURE 1 Various Iterations for $Y = Z^2 + c$

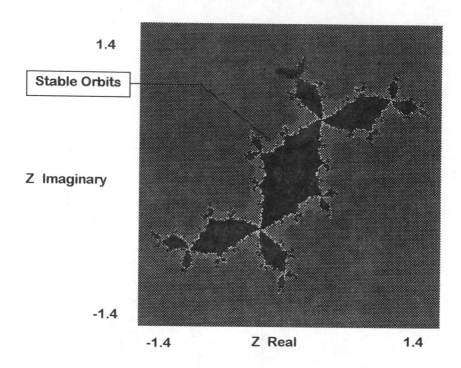

**FIGURE 2   A Julia Set of the complex function $Z^2 + c$, with a seed c of 0.1 + 0.8i,
iterated across a Z range of -1.4 - 1.4i to 1.4 + 1.4i.**

of the system.

Time graphs of signal amplitude, derived from control parameters selected from the non-black zones in the Julia set, do indeed show chaotic motion in the form of nonperiodic, transient spikes, whereas parameters in the black zone give stable performance.   The method is demonstrated in the computer simulations that follow.

COMPUTER SIMULATIONS

To demonstrate the use of a Julia set methodology in a control environment, computer simulations were performed for a discreet plant that demonstrates extreme sensitivity to initial conditions [MB86.] The model used is a second order linear plant with a non-linear parameter estimator feedback. The plant is modeled with no external perturbations, deriving the plant input solely from plant values with a deadbeat control mechanism. The discreet model used in this simulation is:

$$y_k = A.y_{k-1} + B.y_{k-2} + u_k \qquad (5)$$
$$\hat{a}_k = A + B.y_{k-2} / y_{k-1} \qquad (6)$$
$$u_k = -\hat{a}_k.y_k \qquad (7)$$

where: u_k = input, $\hat{a}_k$ = feedback, y_k = output, and A and B are parametric coefficients

Figure 3 illustrates a block diagram of the plant.

While holding the plant initial conditions of y_k, y_{k-1} and y_{k-2} constant, the plant simulation is exercised over a range of Parameter A and Parameter B values. To prevent skewed results, the simulation was programmed to halt in the event of a divide by zero condition. Each A and B value used is mapped to a pixel coordinate on the computer screen. For each set of Parameter values, or pixel, the plant algorithm is iterated beginning at the initial (seed) conditions. If the plant orbits remain stable after a fixed number of iterations have occurred, then the pixel is colored black. If the amplitudes of the plant velocity orbits change beyond a fixed magnitude in a single time unit, a transient spike is declared and the orbit is considered "escaped". The pixel is then colored according to how many iterations this escape took. The plant values used to determine orbit escape are the rates of change of the input u_k, feedback $\hat{a}_k$ and output y_k plant values. A black pixel indicates that all three values remain stable, while any other color indicates that at least one plant value evidences transient spikes.

Figure 4 is an image of a Julia like set calculated from this model. The band of black is a region of stability, where the output damps quickly to zero, and the feedback and derived input are stable. Above the black region is a band that has gray and black spotting. A color print would show that many of the black dots are actually blue, indicating that they do indeed escape, albeit slowly. This

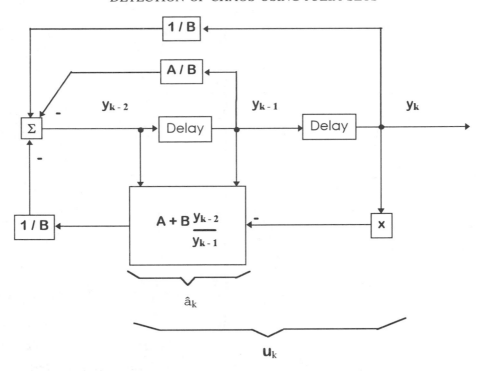

FIGURE 3 Block diagram of the discreet plant used for computer simulation

region is chaotic, where the output is controlled, but the feedback shows chaos, in the form of random spikes.

The upper and lower gray areas of the Julia like set are regions where the orbits escape very quickly and do not necessarily control the output. These regions are completely unstable and appear red on the screen.

Figure 5 is a normalized graph of the plant values over time, using values selected from a point in the black, stable region. It shows a very well behaved plant. Figure 6 on the other hand, is a normalized graph of the plant

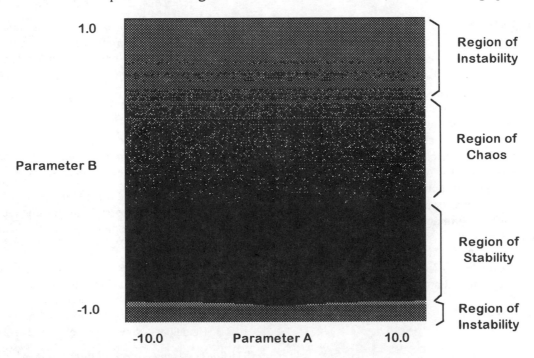

FIGURE 4 Julia-like Set for Discreet Plant with initial values of $Y_k = 1$, $Y_{k-1} = 0.5$ and $Y_{k-2} = 1.5$, with A and B ranging from -10 to 10

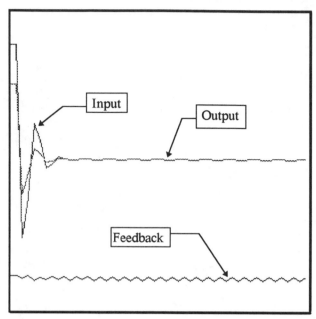

t = 0   **Parameter A = -3.7293723964**   t = 50

Parameter B = -0.1190525216

FIGURE 5 A graph over time for A and B selected in the Black (stable) region of the Julia like set

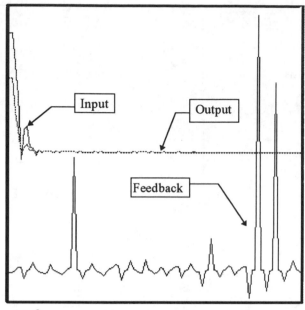

t = 0   **Parameter A = -3.7293723964**   t = 100

Parameter B = 0.1502142425

FIGURE 6 A graph over time for A and B selected in the spotted (chaotic) region of the Julia like set.

values over time selected from a point in the spotted area of the Julia set. It shows that the output and input are quickly damped to zero, whereas the feedback exhibits chaos in the form of irregular transient spikes.

CONCLUSION

This paper is an alert to process control engineers as to the possible occurrence of chaos in seemingly controlled plants. The chaos illustrated in this paper appears to be an artifact of the iterative nature of the parameter estimator paradigm. However, any methodology that involves models and simulation techniques has latent iterations within the numerical integration. With this in mind, it is proposed that induced chaos may occur without detection in a system and that the net effect is detrimental to the control system's mean-time-to-failure and to its dynamic performance. If the chaotic behavior is intrinsic to the adaptive schema, other non-iterative, possibly AI-based (e.g. [RUS91]) control methodologies may be more desirable. Research is underway at Penn State Great Valley into "chaotic function analyzers."

REFERENCES

[CS92]   S.G.Corcoran and Sieradzki., Chaos During the Growth of and Artificial Pit", *Journal of the Electrochem. Soc.*, 1992; 139 (6): 1568-73.

[DEV90] R. L. Devaney.*Chaos, Fractals, and Dynamics, Computer Experiments in Mathematics*, 1990; Addison-Wesley: 86-87.

[MB86] I. M. Y. Mareels and R.R. Bitmead. Non-linear Dynamics in Adaptive Control: Chaotic and Periodic Stabilization. *Automatica* 1986; 22 (6): 641-655.

[MOO92] F.C. Moon. *Chaotic and Fractal Dynamics*, 1992; John Wiley & Sons: 47-53.

[RUS91] D. W. Russell. Further Studies in A.I. Augmented Process Control using the BOXES Methodology, *A.I. in Real Time Control*, 1991; Pergamon Press: 75-80.

[SSR84] H.L. Swinney, R.H. Simoyi and J.C. Roux. Instabilities and Chaos in a Chemical Reaction. in *Chaos and Statistical Methods (Ed: Y. Kuramoto)* 1984; Springer-Verlag. Berlin, Heidelberg New York Toronto: 244-248.

[STR86] G. Strang. *Introduction to Applied Mathematics* 1986; Wellesley-Cambridge Press: 507-510.

FUZZY JOHNSON'S ALGORITHM FOR TWO-MACHINE FLOW SHOP

Tzung-Pei Hong[†] and Tzung-Nan Chuang[‡]

[†]Department of Information Management, Kaohsiung Polytechnic Institute
Kaohsiung, 84008, Taiwan, R.O.C.
Email: tphong@nas05.kpi.edu.tw

[‡]Institute of Management Science, Kaohsiung Polytechnic Institute
Kaohsiung, 84008, Taiwan, R.O.C.
Email: m831017@nas05.kpi.edu.tw

ABSTRACT

Scheduling mainly concerns allocating resources to jobs over time, under necessary constraints. In the past, the processing time for each job was usually assigned or estimated as a fixed value. In many real-world applications, however, the processing time for each job may vary dynamically with the situation. In this paper, fuzzy concepts are utilized in Johnson's algorithm for managing uncertain scheduling. Given a set of jobs, each with two tasks respectively executed on two machines and their membership functions for the processing time, the fuzzy Johnson's algorithm can get a scheduling result with a membership function for the final completion time.

1. INTRODUCTION

One kind of scheduling problems usually occurring in real-word applications are flow shop problems [MP93]. For jobs to be scheduled on two-machine flow shop, Johnson proposed an algorithm to get a minimum makespan. In the past, the processing time for each job was usually assigned or estimated as a fixed value. In many real-world applications, however, the processing time for each job may vary dynamically with the situation. If the processing time for each job is uncertain, then the finish time of a scheduling schema is apparently also uncertain.

Several theories such as *fuzzy set theory* [Zad88, Zim87], *probability theory*, *D-S theory* [Nas94], and approaches based on *certainty factors* [BS84], have been developed to manage uncertainty. Among them, fuzzy set theory is more and more frequently used in intelligent control, because of its simplicity and similarity to human reasoning [HIF94]. The theory has been applied to many fields such as manufacturing, engineering, diagnosis, economics, and others [GJ88, Kan92, KG84, Zim87]. In this paper, the fuzzy concept is utilized in the Johnson's algorithm to manage uncertain scheduling. Given a set of jobs, each with its membership function for the processing time, the fuzzy Johnson's algorithm proposed can get a scheduling result with a membership function for the final completion time (or the makespan).

The remaining parts of this paper are organized as follows. In Section 2, the conventional Johnson's algorithm is first reviewed. In Section 3, the concept of fuzzy set theory is introduced. In Section 4, the fuzzy Johnson's algorithm is proposed to schedule jobs with uncertain processing time on two-machine flow shop. In Section 5, an example is given to illustrate the new scheduling algorithm. Finally, conclusions are given in Section 6.

2. REVIEW OF JOHNSON'S ALGORITHM FOR TWO-MACHINE FLOW SHOP

A two-machine flow shop problem with makespan criterion is stated as follows. Given a set of n independent jobs, each with two tasks $(T_{11}, T_{21}, T_{12}, T_{22}, ..., T_{1n}, T_{2n})$ to be executed in the same sequence on two machines $(P_1$ and $P_2)$, the scheduling requests the minimum completion time of the last job. Johnson proposed a scheduling algorithm to solve the above problem. The algorithm is stated as follows [MP93]:

Johnson's Algorithm:
Input: A set of n jobs, each with two tasks to be respectively executed on two machines;
Output: A schedule with the minimum completion time of the last job;

Step 1: Form the group of jobs U which are shorter on the first machine than on the second; restated, $U = \{ i \mid t_{1i} < t_{2i} \}$ as the first priority group;

Step 2: Form the group of jobs V which are shorter on the second machine than on the first; restated, $V = \{ j \mid t_{2j} \leq t_{1j} \}$ as the second priority group;

Step 3: Sort the jobs within U in an ascending order of t_{1i}'s;

Step 4: Sort the jobs within V in a descending order of t_{2j}'s;

Step 5: Schedule the jobs on the machines in the sorted order of U and then in the sorted order of V.

Below, an example is given to illustrate Johnson's scheduling algorithm.

Example 1: Assume eight jobs J_1 to J_8 are to be scheduled. Each job (Ji) has two tasks (T_{1i}, T_{2i}) respectively executed on two machines. Assume the execution time of each task is shown in Table 1.

By Johnson's algorithm, the execution process is shown as follows.

Step 1: Form the group of U as $\{J_2, J_3, J_6\}$.

Step 2: Form the group of V as $\{J_1, J_4, J_5, J_7, J_8\}$.

Step 3: Sort the jobs in U as $\{J_3, J_2, J_6\}$.

Step 4: Sort the jobs in V as $\{J_5, J_4, J_7, J_1, J_8\}$.

Step 5: Schedule the tasks on the two machines. The complete optimal sequence is then $\{J_3, J_2, J_6, J_5, J_4, J_7, J_1, J_8\}$. The finial scheduling result is shown in Figure 1. The final completion time is then 37.

3. REVIEW OF RELATED FUZZY SET OPERATIONS

In this section, Fuzzy set operations related to this paper are reviewed. There are a variety of fuzzy set operations. Among them, three basic and commonly used operations are *complementation*, *union* and *intersection*. Let X be the universal set. The definitions of complementation, union, and intersection proposed by Zadeh are as follows [KF92][Zad88]:

(1) The *complementation* of a fuzzy set A is denoted by $\neg A$ and the membership function of $\neg A$ is given by:

$$\mu_{\neg A}(x) = 1 - \mu_A(x) \qquad \forall\, x \in X.$$

(2) The *intersection* of fuzzy sets A and B is denoted by $A \cap B$ and the membership function of $A \cap B$ is given by:

$$\mu_{A \cap B}(x) = \min\left\{ \mu_A(x), \mu_B(x) \right\} \qquad \forall\, x \in X.$$

(3) The *union* of fuzzy sets A and B is denoted by $A \cup B$ and the membership function of $A \cup B$ is given by:

$$\mu_{A \cup B}(x) = \max\left\{ \mu_A(x), \mu_B(x) \right\} \qquad \forall\, x \in X.$$

Although fuzzy set concepts are mainly used in linguistic domains, they are also used in numerical domains, where each number is assigned a membership value. Examples are Gazdik's fuzzy network planning [Gaz83], Klein's fuzzy shortest path [Kle89], Nasution's fuzzy critical path [Nas94], and so on. A

Table 1: The eight jobs for Example 1

Job	J_i	J_1	J_2	J_3	J_4	J_5	J_6	J_7	J_8
The first task	t_{1i}	5	2	1	7	6	3	7	5
The second task	t_{2i}	2	6	2	5	6	7	2	1

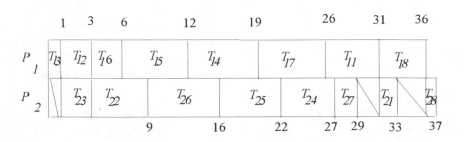

Figure 1: Scheduling results for Example 1

fuzzy addition operation is usually defined for fuzzy numerical domains as follows.

Let A and B be two fuzzy sets, where each element is a number. A and B can then be represented as follows:
$$A = [\ \mu_{a1}/x_{a1},\ \mu_{a2}/x_{a2},\ \bullet\ \bullet\ \bullet\ ,\ \mu_{an}/x_{an}],$$
$$B = [\ \mu_{b1}/x_{b1},\ \mu_{b2}/x_{b2},\ \bullet\ \bullet\ \bullet\ ,\ \mu_{bm}/x_{bm}].$$

The *addition* of fuzzy sets A and B is denoted as follows:
$$A + B = [\ \min(\mu_{a1}, \mu_{b1})/(x_{a1}+x_{b1}),\ \min(\mu_{a1},$$
$$\mu_{b2})/(x_{a1}+x_{b2}), ..., \min(\mu_{an}, \mu_{bm})/(x_{an}+x_{bm})\ ].$$

If the sums of two or more $(x_{ai}+x_{bj})$'s are the same, then only the one with the largest membership value is kept.

4. FUZZY JOHNSON'S ALGORITHM

Notations and assumptions used in this paper are first described as follows.

4.1. ASSUMPTIONS AND NOTATION

Assumptions:
- The jobs are not preemptive.
- Each job has two tasks to be executed in series.
- The execution-time membership function of each task is known.
- Each job is processed by two machines in exactly the same order.

Notation:
- n: the number of jobs.
- T_{ij}: the i-th task for the j-th job, i=1 or 2, and j= 1, 2, ..., n.
- t_{ij}: the fuzzy execution time (a fuzzy set) of T_{ij}, i=1 or 2, and j = 1, 2, ..., n.
- t_{ijk}: the k-th possible execution time of t_{ij}, $1 \leq k \leq$ |supp(t_{ij})|, where supp(t_{ij}) is the support of the fuzzy set t_{ij}
- $\mu(t_{ijk})$: the membership value of t_{ijk}
- P_i: the i-th machine, i=1 or 2.
- p_i: the fuzzy execution time (a fuzzy set) of the i-th machine task (P_i), i= 1 or 2.
- f: the final completion time of the whole schedule.

4.2. RANKING THE FUZZY SETS

Ranking two numbers is quite important in the original Johnson's algorithm. In this paper, fuzzy ranking methods have to be applied since each operand is a fuzzy set. Several fuzzy ranking methods have been proposed in the literature. Below, the averaging method is used an example to show the fuzzy Johnson's algorithm. Note that other ranking methods can also be used.

4.3. THE AVERAGING METHOD

The ranking by the averaging method is defined as follows.

Let A and B be two fuzzy sets, where each element is a number. A and B can then be represented as follows:
$$A = [\ \mu_{a1}/x_{a1},\ \mu_{a2}/x_{a2},\ \bullet\ \bullet\ \bullet\ ,\ \mu_{an}/x_{an}],$$
$$B = [\ \mu_{b1}/x_{b1},\ \mu_{b2}/x_{b2},\ \bullet\ \bullet\ \bullet\ ,\ \mu_{bm}/x_{bm}].$$

$$\text{Let } x_a^{ave} = \frac{\sum\limits_{i=1}^{|supp(A)|} (\mu_{ai} \times x_{ai})}{\sum\limits_{i=1}^{|supp(A)|} (\mu_{ai})}, \text{ and}$$

$$x_b^{ave} = \frac{\sum\limits_{i=1}^{|supp(B)|} (\mu_{bi} \times x_{bi})}{\sum\limits_{i=1}^{|supp(B)|} (\mu_{bi})}.$$

By the averaging method, we say $A > B$ if $x_a^{ave} > x_b^{ave}$.

Example 2 : Assume $A = \{0.4/3, 1.0/8, 0.8/10\}$ and $B = \{0.7/5, 1.0/7\}$. The ranking by the averaging method for A and B is calculated as follows:

$$x_a^{ave} = \frac{\sum\limits_{i=1}^{3} (\mu_{ai} \times x_{ai})}{\sum\limits_{i=1}^{3} (\mu_{ai})} = \frac{(0.4*3+1.0*8+0.8*10)}{0.4+1.0+0.8} = 7.8$$

$$x_b^{ave} = \frac{\sum\limits_{i=1}^{2} (\mu_{bi} \times x_{bi})}{\sum\limits_{i=1}^{2} (\mu_{bi})} = \frac{(0.7*5+1.0*7)}{0.7+1.0} = 6.2.$$

Since $7.8 > 6.2$, we then have $A > B$.

By using the averaging method, the fuzzy Johnson's algorithm is shown below.

Fuzzy Johnson's Algorithm:

Input: A set of n jobs, each with two tasks to be respectively executed on two machines; each task has a membership function for its processing time;

Output: A fuzzy schedule with a membership function for its completion time;

Step 1: For each task T_{ij}, find its average processing time t_{ij}^{ave} by the following formula:

$$t_{ij}^{ave} = \frac{\sum_{k=1}^{|supp(t_{ij})|} (\mu(t_{ijk}) \times t_{ijk})}{\sum_{k=1}^{|supp(t_{ij})|} \mu(t_{ijk})}.$$

Step 2: Form the group of jobs U which are fuzzily shorter on the first machine than on the second; restated $U = \{i \mid t_{1i}^{ave} < t_{2i}^{ave}\}$ as the first priority group.

Step 3: Form the group of jobs V that are fuzzily shorter on the second machine than on the first; restated, $V = \{j \mid t_{2j}^{ave} \le t_{1j}^{ave}\}$ as the second priority group.

Step 4: Sort the jobs within U in an ascending order of t_{ij}^{ave}'s.

Step 5: Sort the jobs within V in an descending order of t_{ij}^{ave}'s.

Step 6: Set the initial finishing time p_1 and p_2 of machines P_1 and P_2 as zero with fuzzy value 1.

Step 7: Schedule the first job J_i at the front of sorted U on the machines; restated, T_{1i} is assigned on P_1 and T_{2i} is assigned on P_2.

Step 8: Set $p_1 = p_1 + t_{1i}$ using the fuzzy addition operation..

Step 9: Set $p_2 = Max(p_1, p_2) + t_{2i}$, where Max is done by fuzzy ranking.

Step 10: Remove task J_i from U.

Step 11: Repeat Steps 7 to 10 until U is empty.

Step 12: Schedule jobs in V in a similar way (Steps 7 to 11).

Step 13: Set the final completion time $f = p_2$.

After Step 13, the scheduling is finished and the completion time with a membership function f has been found.

5. AN EXAMPLE

The following example shows how the fuzzy Johnson's algorithm works.

Example 3: Assume a group of five jobs are to be processed through a two-step operation. The first operation involves degreasing, and the second involves painting . Also assume the fuzzy execution time of each job for each step is listed in Table 2:

By the fuzzy Johnson's algorithm, the process proceeds as follows.

Step 1: For each task, find its average processing time as in Table 3.

Step 2: Form the group of jobs $U = \{J_2, J_4, J_5\}$.

Step 3: Form the group of jobs $V = \{J_1, J_3\}$.

Step 4: Sort the jobs in U as $\{J_2, J_4, J_5\}$.

Table 2: The processing time (hours) of each task for Example 3

Job j	degreasing (t_{1j})	painting (t_{2j})
1	{ 1.0/4 }	{ 1.0/3, 0.9/4 }
2	{ 1.0/1 }	{ 1.0/2, 0.9/3 }
3	{ 1.0/5, 0.9/6 }	{ 1.0/4 }
4	{ 1.0/2, 0.2/4 }	{ 0.7/2, 1.0/3 }
5	{ 0.5/4, 1.0/5 }	{ 1.0/6 }

Table 3: The average processing time (hours) of each task for Example 3

Job j	t_{1j}^{ave}	t_{2j}^{ave}
1	4	3.47
2	1	2.47
3	5.47	4
4	2.33	2.59
5	4.67	6

Step 5: Sort the jobs in V as $\{J_3, J_1\}$.

Step 6: Set $p_1 = p_2 = \{1.0/0\}$.

Step 7: Assign J_2 (the first job on U) on machines to execute.

Step 8: Set $p_1 = p_1 + t_{12} = \{1.0/0\} + \{1.0/1\} = \{1.0/1\}$

Step 9: Set $p_2 = \max(p_1, p_2) + t_{22}$

$$= \{1.0/1\} + \{1.0/2, 0.9/3\}$$

$$= \{1.0/3, 0.9/4\}$$

Step 10: Remove J_2.

Steps 11 and 12: Repeat the above steps. The results are shown in Table 4.

Step 13: The final completion is then:

$$f = \{0.5/20, 1.0/21, 0.9/22, 0.2/23, 0.2/24\}.$$

In the fuzzy Johnson's scheduling algorithm, many possible data items compose the fuzzy sets p_1 and p_2 after the fuzzy addition operation. As an alternative, an appropriate α-cut can be used to remove some unimportant items, thus making the fuzzy sets more concise. The α-cut method limits the number of possible results of fuzzy additions, and reduces the time complexity.

6. CONCLUSION

In this paper, fuzzy set concepts were used in Johnson's algorithm to schedule uncertain jobs for a two-machine flow shop. Given a set of jobs, each with two tasks respectively executed on two machines and their processing time membership functions, the fuzzy Johnson's algorithm can get a scheduling result with a membership function for the final completion time. The results can then help a manager have a wide overview of scheduling. In the future, we will try to apply other characteristics of fuzzy sets to the scheduling field. For example, we can apply other kinds of membership functions to other complicated scheduling problems.

REFERENCES

[BS84] B. G. Buchanan and E. H. Shortliffe, *Rule-Based Expert System: The MYCIN Experiments of the Standford Heuristic Programming Projects* (Addison-Wesley, MA., 1984).

Table 4: The scheduling result for Example 3

Job j	t_{1j}	t_{2j}	p_1 (after t_{1j} is added)	p_2 (after t_{2j} is added)
2	$\{1.0/1\}$	$\{1.0/2, 0.9/3\}$	$\{1.0/1\}$	$\{1.0/3, 0.9/4\}$
4	$\{1.0/2, 0.2/4\}$	$\{0.7/2, 1.0/3\}$	$\{1.0/3, 0.2/5\}$	$\{0.7/5, 1.0/6, 0.9/7\}$
5	$\{0.5/4, 1.0/5\}$	$\{1.0/6\}$	$\{0.5/7, 1.0/8, 0.2/9, 0.2/10\}$	$\{0.5/13, 1.0/14, 0.2/15, 0.2/16\}$
3	$\{1.0/5, 0.9/6\}$	$\{1.0/4\}$	$\{0.5/12, 1.0/13, 0.9/14, 0.2/15, 0.2/16\}$	$\{0.5/17, 1.0/18, 0.2/19, 0.2/20\}$
1	$\{1.0/4\}$	$\{1.0/3, 0.9/4\}$	$\{0.5/16, 1.0/17, 0.9/18, 0.2/19, 0.2/20\}$	$\{0.5/20, 1.0/21, 0.9/22, 0.2/23, 0.2/24\}$

[Gaz83] I. Gazdik, Fuzzy-network planning, *IEEE Transactions on Reliability* **32-3** (1983).

[GJ88] I. Graham and P. L. Jones, *Expert Systems - Knowledge, Uncertainty and Decision* (Chapman and Computing, Boston, 1988) 117-158.

[HIF94] S. Han, H. Ishii and S. Fujii, One machine scheduling problem with fuzzy duedates, *European Journal of Operational Research* **79** (1994) 1-12.

[Kan92] A. Kandel, *Fuzzy Expert Systems* (CRC Press, Boca Raton, 1992) 8-19.

[KG84] A. Kaufmann and M. M. Gupta, *Introduction to Fuzzy Arithmetic Theory and Applications* (Van Nostrand Reinhold, New York, 1984).

[Kle89] C. M. Klein, Fuzzy shortest paths, *Fuzzy Sets and Systems* (1989) 27-41.

[KF92] G. J. Klir and T. A. Folger, *Fuzzy Sets, Uncertainty, and Information* (Prentice Hall, New Jersey, 1992) 4-14.

[MP93] T. E. Morton and D. W. Pentico, *Heuristic Scheduling Systems with Applications to Production Systems and Project Management* (John Wiley & Sons Inc., New York, 1993).

[Nas94] S. H. Nasution, Fuzzy critical path method, *IEEE Transactions on Systems, Man, and Cybernetics* **24-1** (1994).

[Zad88] L. A. Zadeh, Fuzzy logic, *IEEE Computer* (1988) 83-93.

[Zim87] H. J. Zimmermann, *Fuzzy Sets, Decision Making, and Expert Systems* (Kluwer Academic Publishers, Boston, 1987).

PREDICTION OF PROPERTIES OF MIXTURE BY INTEGRATION OF PATTERN RECOGNITION AND LOGIC PROCEDURE BY USING FUZZY CLUSTERING

Hiroyuki Tanaka and Yuji Naka

Research Laboratory of Resources Utilization, Tokyo Institute of technology

4259 Nagatsuta, Midori-ku, Yokohama 226, Japan

Email:htanaka@pse.res.titech.ac.jp

ABSTRACT

New methodology is presented here to overcome the difficulty in predicting   non-linear cooperative effects caused by mixing components. One example in which such difficulties are found is the prediction of properties of mixed solvents. This work integrates two methodologies, fuzzy clustering as pattern recognition and logical procedure. Virtual meta-probability space and Lebesgue integral are adopted to develop mathematical formulation. This work provides an answer for the calculation of properties of mixed components, such as solubility from the data of prototypes in binary system. Development of applications to other objects is expected.

1. INTRODUCTION

Prediction of properties such as distribution coefficients of solvents in binary system has been carried out by analyses based on physico-chemical knowledge of thermodynamics and chemistry.   The problems that are not to be solved by such methodologies should be worked with other methodologies.   A target of this work is to give a framework for achieving the integration of pattern recognition(induction   inference)   and   logical procedure(deduction inference).

The following two guidelines are considered as the most important to promote integration.

1) Bottlenecks exist in integrating existing technologies such as neural network, KBS, and fuzzy system.   At present, there is neither obvious methodology nor tool to break these.

(2) Integration claims inextricably detailed exploration of individual methods in both induction and deduction phases. This situation gives serious complexity to integration as a result.

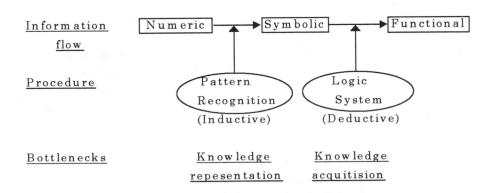

FIGURE 1   Bottlenecks in information flow

2. GENERAL ISSUES ON THE INTEGRATION OF PATTERN RECOGNITION AND LOGIC SYSTEM.

2.1. Bottlenecks for the integration of pattern recognition and logic system

FUGURE 1 shows general information flow of problem solving to extract functional information from numerical data. A bottleneck of knowledge representation arises in the interpretation process of numeric data into symbolic data. A concept of QI was proposed by Whiteley et al. to compromise this bottleneck[WD92]. Another bottleneck of knowledge acquisition exists in the transformation of symbolic data into functional information, which is given as rules, remains unsolved in their work. This bottleneck should be coped with for effective implementation of knowledge base system.

2.2. Limitations of existing methodologies for the integration of pattern recognition and logic system.

In case of applying an existing neural network such as back propagation network(BPN) as logic system, it is impossible to give theoretical explanation to calculated results. As for existing logic system, knowledge acquisition procedure depends on human expertise and the difficulty of implementing updates of rules are left unresolved. As for fuzzy set theory, the calculation of a membership function of a sum of sets is based on the following groundless equation for example.

$$\mu_k(S_m \cup S_n) = \max(\mu_k(S_m), \mu_k(S_n))$$

Here, $\mu_k(S_m)$ and $\mu_k(S_n)$ are membership functions that express the degrees of membership of k to sets S_m and S_n respectively. $\mu_k(S_m \cup S_n)$ is a membership function showing the degrees of membership of k to a sum of sets, $S_m \cup S_n$.

3. ORIGINALITIES OF THIS WORK

The first original feature is represented as follows. This work proposes simple domain-independent methodology to estimate quantitatively the cooperative effects of mixing. Probability membership functions, which are calculated by fuzzy clustering, are utilized for the integration of pattern recognition and logic procedure.

The second original feature is shown as follows. This work eliminates the dependency on human experience, for only two parameters, a cluster number(c) and a weight parameter(m), are left undecided in this system as described later. The former can be evaluated by physico-chemical knowledge and the latter can be estimated as a fitting parameter.

4. MATHEMATICAL FORMALISM.

4.1. Definition of σ-family of sets

A family $\mathfrak{I}$ of subsets of a set X, is defined as a σ-family

4.2. Introduction of meta-probability space

A set X, and a σ-family $\mathfrak{I}$, which is a family of subsets of X, are given in advance. We define P(A) as measure. We get complete additivity shown by

$$P(\sum_{n=1}^{\infty} A_n) = \sum_{n=1}^{\infty} P(A_n)$$

When X is a set , F is a σ-family composed of subsets of X and P is probability measure, probability measure space can be defined. Probability measure space is represented by $(X, \mathfrak{I}, P)$.

4.2.1. Definition of measurable function

Real function f(x) is defined as measurable on X.

4.2.2. Sum and product of measurable functions

If f(x) and g(x) are measurable functions, α f(x)+β g(x) becomes measurable(α, β are real).

4.2.3. Definition of a simple function

$\psi(x ; A_i)$, which is a characteristic function of a set A_i, is defined as follows.

$$\psi(x; A_i) = \begin{cases} 1, & x \in A_i \\ 0, & x \notin A_i \end{cases}$$

A partition of X by finite numbers of measurable sets is shown as

$$X = A_1 \cup A_2 \cup \cdots \cup A_n$$

A simple function $\psi(x)$ is defined as follows using X and real numbers $r_1, r_2, ..., r_n$.

$$\psi(x) = \sum_{i=1}^{n} r_i \psi(x; A_i)$$

4.2.4. Integral of a simple function and expectation

As a characteristic function is measurable, a simple function becomes measurable using the theorem described in (2). A measurable simple function defined by

$$\psi(x) = \sum_{i=1}^{n} r_i \psi(x ; A_i)$$

is a random variable and $E(\psi)$, expectation of ψ is represented as

$$E(\psi) = \sum_{i=1}^{n} r_i P(A_i)$$

5. THEORY FOR PREDICTING SUBSTANCE PROPERTIES

5.1. Definition of probability measure and introduction of fuzzy clustering

Here again, X and $\Im$ are defined as a set of objective substances and its σ-family respectively. We choose a substance which has several attributes as an example. A letter, $\mathbf{x}_k$ represents a vector of multivariate attributes. S_i represents a subset. A membership function is given as a two-variable characteristic function as follows.

$$\psi : X \times \Im \longrightarrow [0,1],$$
$$\psi(\mathbf{x}_k ; S_i) = P_{ik}$$

This characteristic function ψ becomes probability measure when represented as a function of a set S_i. Fuzzy clustering methodology is applied as pattern recognition to the calculation of membership functions of each substance to clusters. In the previous paper[TN94], we proposed the procedure to calculate membership functions of common solvents by adopting Snyder's parameters[Sny78] as vectors of representation space.

5.2. Algorithm of fuzzy clustering[Dun74]

The number of substances is denoted by n and the number of the attributes of each substance p. A variable x_{kj} indicates the jth numerical value of kth substance. An attribute vector of kth substance is represented by $\mathbf{x}_k = (x_{k1},...,x_{kp})$. $P_k(S_i)$ is defined as the degree of membership of $\mathbf{x}_k$ to a subset S_i. Dissimilarity of $\mathbf{x}_k$ is defined by Euclidean metric $\| \mathbf{x}_k - \mathbf{v}_i \|^2$. Here,

$\mathbf{v}_i$ is a prototype vector of ith cluster. With these definitions, the values of $P_k(S_i)$ and $\mathbf{v}_i$ are obtained by minimizing the following J.

$$J = \sum_{k=1}^{n} \sum_{i=1}^{c} (P_k(S_i))^m \| \mathbf{x}_k - \mathbf{v}_i \|^2$$

The sum of probabilities being 1 leads to the next.

$$\sum_{i=1}^{c} P_k(S_i) = 1$$

By applying Lagrange's method of undetermined multipliers to two equations above, we get.

$$\mathbf{v}_i = (1/\sum_{k=1}^{n} (P_k(S_i)^m) \sum_{k=1}^{n} (P_k(S_i))^m \mathbf{x}_k$$
$$P_k(S_i) = 1/\sum_{j=1}^{c} (\| \mathbf{x}_k - \mathbf{v}_i \| / \| \mathbf{x}_j - \mathbf{v}_i \|)^{2/(m-1)}$$
$$(1 \le i \le c, 1 \le k \le n)$$

By using these equations recursively, $P_k(S_i)$ and $\mathbf{v}_i$ are obtained as converged values. Here, c is a number of clusters and $2 \le c \le n$ is satisfied.

5.3. Structure of meta-probability space.

5.3.1. Definition of S_l, S_{mn} and $P_k(S_i)$

Membership functions in binary system are defined as follow. First, it is assumed that different substances, k and k' are mixed at the ratio of $\alpha : (1-\alpha)$. Secondly, the prototypes of clusters of S_m and S_n, which membership functions give the maximum values, are given by $\mathbf{x}_m'$ and $\mathbf{x}_n'$ respectively. With these premises, S_{mn}, which is a virtual fuzzy cluster in meta-probability space, is created by mixing k and k' at the ratio of $\alpha P_k(S_m):(1-\alpha)P_{k'}(S_n)$. A

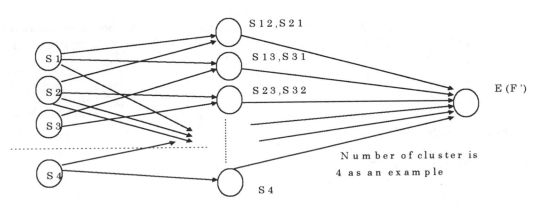

FIGURE.2   Flow of information in this system

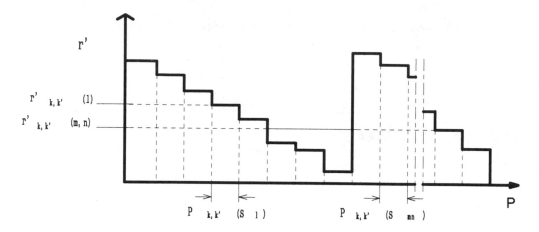

FIGURE.3   Lebesgue integral representing expectation of substance function

membership function of this mixture to a newly created cluster S_{mn} inherits the original membership function of k to the cluster S_m and that of k' to the cluster S_n. The other type of membership function of mixture to an original cluster S_l is also introduced. Theoretical background will be described clearly later. By the procedure above, we obtain

$$P_{k,k'}(S_l)$$
$$= (1/c)\{\alpha P_k(S_l) + (1-\alpha)P_{k'}(S_l)\} \quad (1)$$
$$(1 \le l \le c)$$

$$P_{k,k'}(S_{mn})$$
$$= (1/c)\{\alpha P_k(S_m) + (1-\alpha)P_{k'}(S_n)\} \quad (2)$$
$$(m \ne n, 1 \le m \le c, 1 \le n \le c)$$

5.3.2. Formal discussion of deductive function

When mixture contributes to a cluster S_l, the modification of equations are done as follows.

$$P(S_l^{kk'}) = P_{k,k'}(S_l) \qquad (3)$$
$$P(S_l^k) = (1/c)P_k(S_l) \qquad (4)$$
$$P(S_l^{k'}) = (1/c)P_{k'}(S_l) \qquad (5)$$
$$P(S_l^{kk'}|S_l^k) = \alpha \qquad (6)$$
$$P(S_l^{kk'}|S_l^{k'}) = 1-\alpha \qquad (7)$$

We follow the assumption that the probabilities that contribute to a cluster l are only $P_k(S_l)$ and $P_{k'}(S_l)$ and that these are probabilities of exclusive events. Substitution of equations from Eq.(3) to Eq.(7) into Eq.(1) makes

$$P(S_l^{kk'})$$
$$= P(S_l^{kk'}|S_l^k)P(S_l^k) + P(S_l^{kk'}|S_l^{k'})P(S_l^{k'}) \quad (8)$$

When mixture contributes to a cluster S_{mn}, the transformation of equations is done by

$$P(S_{mn}^{kk'}) = P_{k,k'}(S_{mn}) \qquad (9)$$
$$P(S_m^k) = (1/c)P_k(S_m) \qquad (10)$$
$$P(S_n^{k'}) = (1/c)P_{k'}(S_n) \qquad (11)$$
$$P(S_{mn}^{kk'}|S_m^k) = \alpha \qquad (12)$$
$$P(S_{mn}^{kk'}|S_n^{k'}) = 1-\alpha \qquad (13)$$

The probabilities that contribute to substances m and n are only $P_k(S_m)$ and $P_{k'}(S_n)$ respectively. Substitution of equations from Eq.(9) to Eq.(13) into Eq.(2) makes

$$P(S_{mn}^{kk'})$$
$$= P(S_{mn}^{kk'}|S_m^k)P(S_m^k) + P(S_{mn}^{kk'})P(S_n^{k'}) \quad (14)$$

Probabilities are denoted as nodes and information flows as arrows in FIGURE 2. The information flow is similar with that of a simplified form of GMDH(Group Method of Data Handling)[Hec90], a neural network which exhibits deductive function.

5.3.3. Definition of meta- probability measure space

The following additivity holds true for all l, l', m, m', n and n' ($l \ne l'$, $m \ne m'$, $n \ne n'$).

$$P_{k,k'}(S_l) + P_{k,k'}(S_{l'}) = P_{k,k'}(S_l \cup S_{l'}) \qquad (15)$$
$$P_{k,k'}(S_l) + P_{k,k'}(S_{mn}) = P_{k,k'}(S_l \cup S_{mn}) \qquad (16)$$
$$P_{k,k'}(S_{mn}) + P_{k,k'}(S_{mn'}) = P_{k,k'}(S_{mn} \cup S_{mn'}) \quad (17)$$

When $i > c^2$, if $P(S_i)=0$,
we have

$$\sum_{l=1}^{c} P_{k,k'}(S_l) + \sum_{m=1}^{c} \sum_{n=1(m \neq n)}^{c} P_{k,k'}(S_{mn})$$

$$= (1/c) \sum_{m=1}^{c} \sum_{n=1}^{c} \{\alpha P_k(S_m) + (1-\alpha)P_{k'}(S_n)\}$$

$$= \alpha \sum_{m=1}^{c} P_k(S_m) + (1-\alpha) \sum_{m=1}^{c} P_{k'}(S_n)$$

$$= \alpha + (1-\alpha)$$

$$= 1 \qquad \qquad (18)$$

We get complete additivity shown as

$$P(\bigcup_{i=1}^{\infty} S_i) = \sum_{i=1}^{\infty} P(S_i) = \sum_{i=1}^{c} P(S_i) = 1 \quad (19)$$

Whenever α, k and k' are changed, new binary clusters are created. As a number of the clusters is at most countable, the characteristic of arbitrariness of mixture is reflected on this complete additivity.

6. CALCULATION OF EXPECTATION

The following theorem of Bayes is derived from Eq.(14).

$$P(S_m^*|S_{mn}^{kk'})$$

$$= P(S_m^*)P(S_{mn}^{kk'}|S_m^*) \Big/ \{P(S_{mn}^{kk'}|S_m^*)P(S_m^*) + P(S_{mn}^{kk'}|S_n^{k'})P(S_n^{k'})\} \quad (20)$$

$$= P(S_m^*)\alpha \Big/ \{\alpha P(S_m^*) + (1-\alpha)P(S_n^{k'})\}$$

This probability shows the extent of contribution of a single substance k to a binary cluster S_{mn}. We make a following simple function defined by Eq.(21).

$$F' = \sum_{l=1}^{c} \psi(x; S_l) + \sum_{m=1}^{c} \sum_{n=1(m \neq n)}^{c} \psi(x; S_{mn}) \quad (21)$$

Area surrounded by a simple function in FIGURE 3 is the expectation of properties of mixed substances, that is calculated by Lebesgue integral defined by

$$E(F') = \int_X F' dP$$

$$= \sum_{l=1}^{c} r'_{k,k'}(l) P_{k,k'}(S_l) + \sum_{m=1}^{c} \sum_{n=1(m \neq n)}^{c} r'_{k,k'}(m,n) P_{k,k'}(S_{mn})$$

$$(22)$$

Newly added area shows the residual effects by mixing two substances.

7. EXAMPLE OF CALCULATION

While solubility parameter has been regareded as poptential for evaluationg functions of common solvents among many parameters, it involves a great difficulty to evaluate polar and donor-acceptor interactions. Snyder

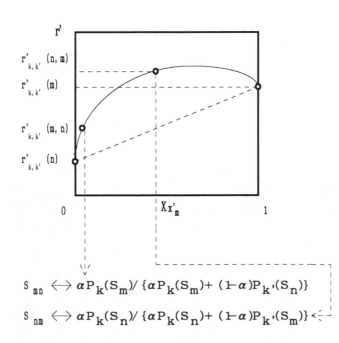

$$S_{mn} \longleftrightarrow \alpha P_k(S_m) / \{\alpha P_k(S_m) + (1-\alpha)P_{k'}(S_n)\}$$

$$S_{nm} \longleftrightarrow \alpha P_k(S_n) / \{\alpha P_k(S_n) + (1-\alpha)P_{k'}(S_m)\}$$

FIGURE 4 Nonlinear dependence of r' on X_m'

Table 1 Examples of membership functions of solvents

Number of clusters	triethyl amine	quinoline
I	0.5475	0.0000
II	0.4064	0.0000
III	0.0317	0.0000
IV	0.0081	1.0000
V	0.0032	0.0001
VI	0.0013	0.0000
VII	0.0018	0.0000

introduced three parameters x_e(denotes proton donorproperty), x_d(denotes proton acceptor property) and x_n(denotes strong polarity)[Sny78]. In this work, 62 solvents were adopted and classified into 7 clusters. Examples of membership functions of solvents, which were calculated by fuzzy clustering, are shown in Table 1. Here, 1.5 was adopted for the value of m. The membership function of quinoline to cluster IV is almost 1. This indicates that the solvent can be the the prototype of the cluster. The maximum value among membership functions of triethyl amine is much less than 1. This indicates that triethyl amime is not a prototype of any cluster. The prediction of properties of mixed solvents at arbitrary mixing ratio is completed by the procedure descrbed in Section 6. The correspondence between experimental data and membership functions of binary system, which is shown in FIGURE 4, is given by Eq.(20). In FIGURE 4, r' is a property of mixed prototypes and $X_{X'm}$ is a ratio of mixing of these prototypes.

CONCLUSIONS

1) Integration of both inductive and deductive actvities is shown.
2) Hierarchy structure is introduced by making meta-probability space to predict properties of arbitrarily mixed substances
3) Experimental procedure is shown in order, based on Bayes' theorem.
4)By introducing Lebesgue integral, robust mathematical formalism is given for an arbitrary ratio of mixing.

REFERENCES

[Dun74]   Dunn, J. C. , A Fuzzy Relative of the ISODATA Process and Its Use in Detecting Compact Well-Separated Clusters, *J.Cybern.*,**3**,3,32(1974)

[Hec90]   Hecht-Nielsen R., Neurocomputing, *Addison-Wesley Pub.* 155-162(1990)

[Sny78]   Snyder L.R.,Classification of Solvent Properties of Common Liquids, *J.of Chromatogr. Sci.*, **16**,223(1978)1)

[TN94]   Tanaka.H. and Y. Naka, Classi-fication of Solvents by Fuzzy Clustering, *Journal of Chemical Society of Japan*, No.1, 74-80(1994)

[WD92]   Whiteley, J. R. and J. F. Davis, Knowledge-based interpretation of sensor patterns, *Computers chem. Engng* ,**16**, 329-346(1992)

SELF-LEARNING FUZZY LOGIC CONTROL OF NON-LINEAR PROCESSES

S. H. Ghwanmeh, K. O. Jones and D. Williams
School of Electrical Engineering, Electronics and Physics,
Liverpool John Moores University,
Byrom Street, Liverpool, L3 3AF, United Kingdom.
Email: eeesghwa@livjm.ac.uk

ABSTRACT

The performance of a SLFLC has been evaluated in this paper. A robustness test is presented to demonstrate the performance of the SLFLC by examining its ability to produce a consistent set of fuzzy rules based on a predefined criteria and analysing its transient performance over a variety of on-line tests. The SLFLC has been applied to two laboratory liquid-level processes: an interacting process and a non-linear process. The on-line results presented show that even with very limited knowledge of the process, the self-learning algorithm is capable of producing a satisfactory performance with some degree of robustness and repeatability.

1. INTRODUCTION

Fuzzy Logic allows imprecise and qualitative information to be presented in a quantitative form. It also offers a rigorous and practical technique for manipulating qualitative information originally expressed in a linguistic form. The lack of knowledge to develop a mathematical model of a system can be compensated by human experience of the behaviour of the input/output characteristics [Zim91]. If such system experience can be encompassed in a rule-base, it becomes a potential candidate for Fuzzy Logic Control (FLC). Since the introduction of fuzzy set theory and approximate reasoning [Zad65, Zad73], FLC has been successfully implemented on a number of processes [Lee90].

The major difficulty in fuzzy logic control is in the development of an algorithmic rule-base that is both complete and consistent. Completeness means that at least one rule has a contribution to the controller output for every possible input state of the process. A fuzzy rule set is inconsistent if there are two rules with the same rule-antecedents but different rule-consequence. A possible solution to this rule generation problem is to use a Self-Learning Fuzzy Logic Controller (SLFLC) which is capable of generating and modifying the rule-base by learning the process based on performance measurement. SLFLC combines the functions of identification and control of the process, because it performs two tasks: firstly it observes the environment while providing an appropriate control action and secondly it uses the process response to the applied control signal to improve the rule base [PM79]. Other studies on the implementation of SLFLC have been reported [LA92] and [Spi92].

To make the SLFLC suitable for direct implementation on real-time industrial processes, there is a need to show its capability of generating adequate fuzzy rules for the controlled process with high repeatability. To achieve this, the generated fuzzy rules need to be extracted from the rule-base and tested for their capability to control the process as a fixed rule-base controller (FRBC). Little has been reported to demonstrate this characteristic. This paper deals with the development of a SLFLC which can self-learn fuzzy rules using knowledge from the controlled process during real-time control, and automatically produce the new fuzzy control rules based on its learning capability from past control signals. Evaluation of the SLFLC performance, robustness and repeatability have been shown via a series of on-line experiments on two liquid-level processes: an interacting process and non-linear process. Results are presented showing satisfactory performance in controlling processes which have non-linearity and dynamic complexity.

2. DESCRIPTION OF THE SLFLC

The SLFLC structure consists of two levels (Fig. 1): the lower level is a simple fuzzy logic controller which has been successfully implemented [Ton77] while the second level is a Self-Learning level in which a set of a fixed performance rules are used to modify a second set of control rules. The performance rules represent the required performance of the controller.

2.1. The Basic Level

The inputs to the controller are the conventional error (e) and change in error (ce), while the control signal (u) is a positional output. The two input signals are mapped from physical values to normalised values by using input scaling factors (GE and GC). The output fuzzy set of the controller is defuzzified, scaled by the output scaling factors (GU), and sent to the process. The fuzzy sets of the controller are formed upon a discrete universe of discourse of 14 elements for the error defined in the range (±6), 13 elements for the change in error defined in the range (±6), and 15 elements for the output defined in the range (±7) [DG86]. The membership function is triangular and each two adjacent membership functions have a cross-point level of 0.7. The linguistic labels were chosen as follows:

PB : Positive Big	NB : Negative Big
PM : Positive Medium	NM : Negative Medium
PS : Positive Small	NS : Negative Small
PO : Positive Zero	NO : Negative Zero.

The terms PO and NO are introduced to obtain finer control about the equilibrium state, where NO defines values slightly below zero and PO defines values slightly above zero. The Centre of Gravity (COG) defuzzification method has been employed because it provides a smoother control signal [BR78]. Additionally, the less computationally demanding 'Max-Min' method [MA74] was used for implication and inference.

2.2. The Fuzzy Self-Learning Level

The Self-Learning Level employs a process performance feedback, where the performance index issues a correction value to individual control actions that contributed to the present process state. It consists of a performance index and a rule modification mechanism.

Performance Index

The performance of the fuzzy controller can be described by the deviation of the actual response from that required. This can be expressed as a variable which is an approximate indication of the magnitude of the required correction. The output response can be monitored by the current error, e(nT), given by the difference between the process output and the setpoint; and the current change in error, ce(nT), given by the difference between the present error and the previous error. The performance index can take the form of a correction maker which issues the output corrections required from a knowledge of e(nT) and ce(nT). The linguistic performance rules (Table 1) can be transformed into a look-up table using the standard techniques of fuzzy calculus. The correction of the control action for a single-input single-output (SISO) process can be given as:

$$P_o(nT) = K P_i(nT) \qquad (1)$$

where P_i is the performance index reinforcement, P_o is the required output change, and K is the process static gain. If the values of the process input and output are scaled and then normalised to their maximum values, then K becomes unity.

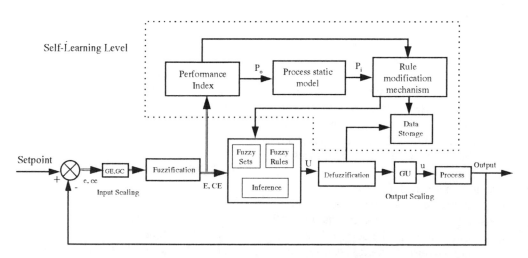

FIGURE 1 Block diagram of the SLFLC.

Table 1 The Performance Index.

Error	Change in Error						
	NB	NM	NS	ZO	PS	PM	PB
NB	NB	NB	NB	NM	NM	NS	ZO
NM	NB	NB	NM	NM	NS	ZO	PS
NS	NB	NB	NS	NS	ZO	PS	PM
NO	NB	NM	NS	ZO	ZO	PM	PM
PO	NB	NM	ZO	ZO	PS	PM	PB
PS	NM	NS	ZO	PS	PS	PB	PB
PM	NS	ZO	PS	PM	PM	PB	PB
PB	ZO	PS	PM	PM	PB	PB	PB

Rule Modification Mechanism

The rule modification mechanism is based on reconstructing the set of rules in the rule-base and can be described as follows: "if a process has a dead time of d samples, then the process output at the current sample time, nT, *has* primarily been produced by the controller output at sample time (nT-dT)". The basic structure of the original relation matrix is given by:

$$R'(nT) = E(nT\text{-}dT) \to CE(nT\text{-}dT) \to U(\text{nT-dT}) \quad (2)$$

and according to the correction value issued by the performance index, $P_i(nT)$, this relation should be modified to:

$$R''(nT) = E(nT\text{-}dT) \to CE(nT\text{-}dT) \to U(\text{nT-dT}) + P_i(\text{nT}) \quad (3)$$

where

$$E(\text{nT-dT}) \;\; = F \{ e(\text{nT-dT}) \}$$
$$CE(\text{nT-dT}) = F \{ ce(\text{nT-dT}) \}$$
$$U(\text{nT-dT}) \;\; = F \{ u(\text{nT-dT}) \}$$
$$P_i(\text{nT}) \qquad = \text{The correction of the control}$$
$$\text{action at time t = nT.}$$
$$F\{ \} \qquad = \text{Fuzzification procedure.}$$

If $R(nT)$ is the relation matrix at the present instant and $R(nT+T)$ is the modified one; then

$$R(nT+T) = \{ R(nT\text{-}dT) \cap \overline{R'(nT)} \} \cup R''(\text{nT}) \quad (4)$$

To keep the rule-base consistent, only rules with a new antecedent should be added to the rule-base while rules with different consequent and similar antecedent are modified. The rules removal approach adopted in this work can be expressed linguistically as *"Delete all the rules that are about the same as the one to be added"* [DG86].

3. PROCESS DESCRIPTION

Two systems were chosen to test the SLFLC; an interacting liquid-level process and a non-linear liquid-level process.

3.1. Interacting Liquid-Level Process

This process consists of five-interacting tanks and pneumatically operated control valve (Fig. 2). The position of the control valve can be adjusted by a voltage signal (0-6 V, corresponding to 0-100% open) fed through a voltage-to-current converter. The liquid level in each tank is measured using a pressure sensor producing 0-10 V, corresponding to 0-60 cm height. The level in the second tank (h_2) is controlled using the water feed into the first tank (q_i). It was found that for the controlled process there was a delay of 3 seconds from applying a step input until a change in the process output occurred. The sampling period was constrained by the analogue/digital converter card to 1 second.

3.2. Non-Linear Liquid-Level Process

This system consists of a water holding tank and a pneumatically operated control valve (Fig. 3). The shape of the tank is such that the cross-sectional area of the tank varies considerably with height. The height of the water in the tank is measured by a pressure transducer attached to the bottom of the tank (0-10 V, corresponding to 0-70 cm height). The inlet flow rate to the main tank can be adjusted by changing the position of the control valve by using a voltage signal (0-10 V, corresponding to 0 -100% open) fed through a voltage-to-current converter. The object was to control the height of water in the tank (h) using water feed in the top of the tank (q_i).

4. ON-LINE RESULTS

A series of tests have been performed on both processes to investigate the performance of the proposed SLFLC.

4.1. Interacting Process

Performance Evaluation

Three setpoints were chosen to test the convergence of the controller: 15 cm, 20 cm and 25 cm. The time response of the 'learning phase' is shown in Fig. 4 for setpoint 20 cm. The total number of rules generated at the end of this run was 32, with the controller having started with a zero rule-base. The SLFLC produced an acceptable steady-state accuracy (within 2%) without excessive overshoot (< 5%).

To test the rule base robustness, the rules generated at the end of the 'learning phase' for a setpoint of 20 cm were taken and used as a FRBC (i.e. learning stopped). The transient response for this test is shown in Fig. 5, for the same setpoint and operating conditions.

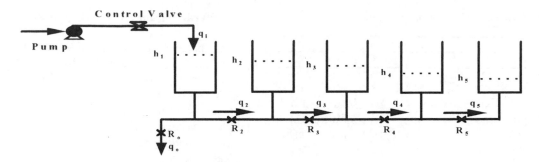

FIGURE 2 Multi-column interacting liquid level process.

It can be seen that the controller produced a similar response compared with that of the learning phase. The control signal for both cases (Fig. 4 and Fig. 5) is smooth without large deterioration, which is important for a real-time implementation since it prevents excessive excursions of the control valve [GJW95]. A significant change in the control action can produce a controller which may become sensitive and hence determine a relatively large rule-base which would need a significant increase in computation time.

To demonstrate the learning capability of the controller, the setpoint was changed from 20 cm to 25 cm and then to 15 cm, with the control task being to track the changing setpoint. Figure 6 shows the response for this test, as well as the rule-base growth. The initial control performance around the 20 cm setpoint is slightly underdamped but as time proceeds the control performance gradually improves, especially when the first setpoint change occurs (20 cm to 25 cm), because the rule-base has enough knowledge to cope with the new condition. Only 8 rules were added to the rule-base for this setpoint change, while for the second setpoint change (25 cm to 15 cm) 16 rules were added to the rule-base. For this case, the rule-base does not have enough knowledge about the new setpoint since the direction of the error and change of error have altered. The first change can be regarded as a continuation of the previous knowledge in the rule-base, while the second change requires new knowledge be added to the rule-base. Again, it can be seen that the control performance is acceptable. The total of 55 rules had been generated by the end the experiment.

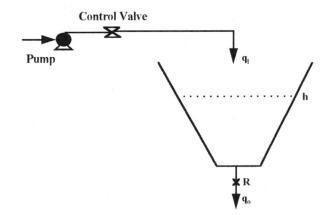

FIGURE 3 Non-linear liquid level process.

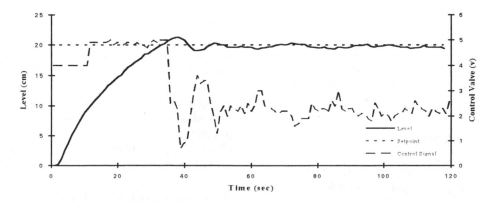

FIGURE 4 Response of the process in learning phase.

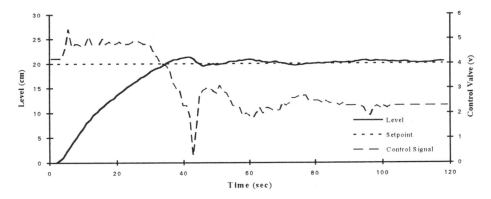

FIGURE 5 Response of the process with no-learning (FRBC).

Repeatability

In order to investigate the repeatability of the controller, five runs were carried out for a setpoint of 25 cm, for each run the controller started with an empty rule-base and had the same operating conditions. The total number of rules generated at the end of each run were 34, 37, 34, 27, and 30 respectively. The difference in the number of rules generated for each run is primarily caused by the noise content in the measured signals. Figure 7 shows the transient response of all the runs. It can be seen that the controller is highly repeatable except for run #4 which produced the minimum number of rules. Although the response has a large rise time, this is an acceptable controller response in terms of the steady-state accuracy and overshoot. Figure 8a and Fig. 8b show the rule-base contents for run #2 which gave the maximum number of rules, and run #4 which gave the minimum number of rules. The difference between the two rule-bases can be seen by looking at the values of the controller output corresponding to each pair of `e` and `ce` (Fig. 8c). Both rule-bases generated the same rules during the first 20 samples (i.e. large and medium errors). For the time between sample 20 and sample 80 (i.e. small and zero errors) the generated control rules were different. In run #2 the rule-base had effective rules which produced fast rise time, while in run #4 the knowledge in the rule-base was not enough to prevent the large rise time. It can be seen (Fig. 8c) that the difference between the rule-bases is concentrated in the centre region (i.e. small and zero errors). The large bars represent the rules that were in rule-base of run #2 and not in rule-base of run #4, while the small-bar regions illustrate that both rule-bases had the same rules in this region but with slightly different consequences.

In the second part of the test, the rules generated at the end of run #3 (which gave the average number of rules) was used as a FRBC using the same controller parameters as run #3. Again, five runs were carried out for the same setpoint (25 cm), with Fig. 9 illustrating the transient responses. Although, the learning mechanism was off, the controller gave satisfactory control results for each run.

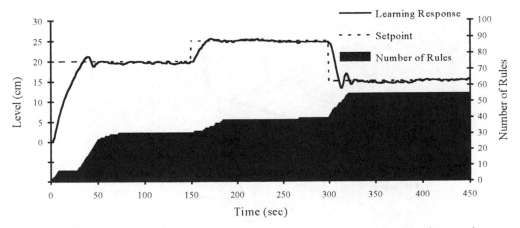

FIGURE 6 Response of the process for a varying setpoint with learning phase only.

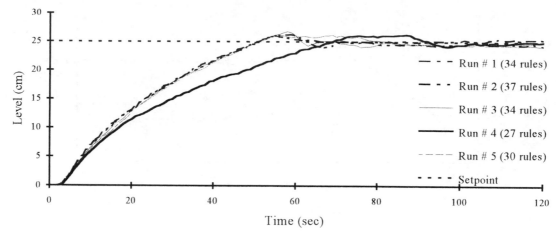

FIGURE 7 Response of the process for the repeatability test (learning phase) .

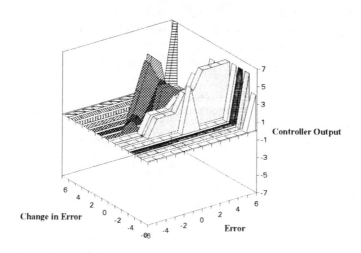

FIGURE 8a Final rule-base for run #2 (maximum number of rules).

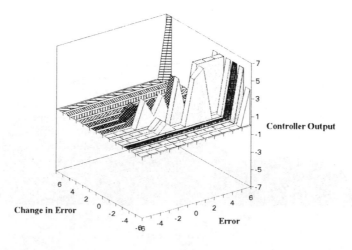

FIGURE 8b Final rule-base for run #4 (minimum number of rules).

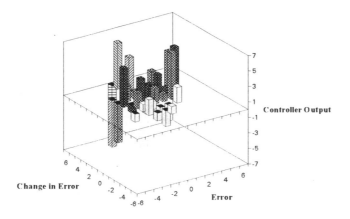

FIGURE 8c The difference between the rule-bases of run #2 and run #4.

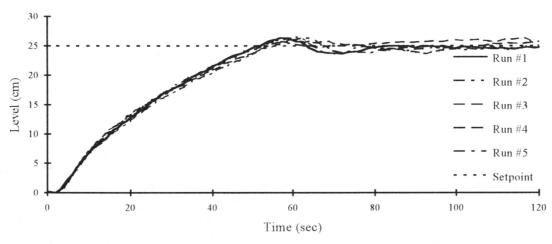

FIGURE 9 Response of the process for the repeatability test (FRBC).

4.2. Non-Linear Process

Performance Evaluation

On-line experiments were carried out on the non-linear process to illustrate the generic control performance of the proposed SLFLC. The setpoints were chosen: 20 cm (bottom region) and 50 cm (top region) to test the convergence of the controller in the operational regions of the process, with Fig. 10 showing the response of the 'learning phase' for setpoint 50 cm. With the controller having started with empty rule-base, a total of 37 rules were generated by the end of this run. It can be seen from the control signal that the control valve is 70% open during the first 15 seconds, while for the next 65 seconds it is fully opened. The reason behind this is due to the cross-sectional area variations. In the first region the cross-sectional variants is relatively small compared with the middle and the top regions. Accordingly, the rate of change of the level is fast in the first region, while it is slow in the middle and top regions. This produced, according to the performance index, a medium control signal during the first interval (15 secs.) and a large control signal during the next interval (65 secs.) to overcome the large variations in the cross-sectional area. The resultant response had an acceptable steady-state accuracy (within 1.5%) without excessive overshoot (< 4%).

To test the SLFLC robustness, the rules generated at the end of the 'learning phase' for a setpoint 50 cm were extracted and used as a FRBC. Figure 11 shows the transient response for this test, for the same setpoint (50 cm) and operating conditions. The controller produced a similar response compared to that of the learning phase. Both cases produced a smooth control signal (Fig. 10 and Fig. 11).

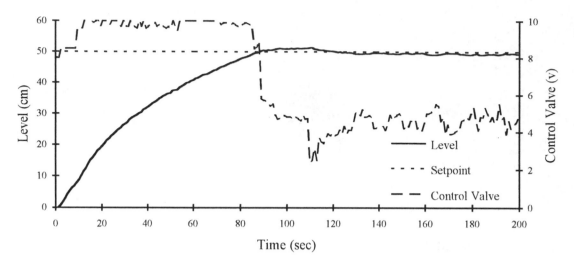

FIGURE 10 Response of the non-linear process in learning phase (top region).

Figure 12 shows the transient response of the 'learning phase' for the setpoint 20 cm (bottom region). In this region, the rate of change of height is relatively high. Therefore, to reduce the overshoot, the control valve is 70% open during the first 30 seconds producing a satisfactory rise time with small overshoot (< 6%). The total number of rules generated at the end of this run was 22.

Repeatability

Five runs were performed for the 20 cm setpoint, with each starting with an empty rule-base and using the same operating conditions. The total number of fuzzy rules produced for the runs were 30, 28, 30, 28, and 31

respectively. The difference in the number of rules generated is due to the non-linearity of the process. As Fig. 13 illustrates, the SLFLC is highly repeatable and produced an acceptable control response.

In the second part of the test, the rules at the end of run #1 were extracted and used as a FRBC using the same parameters as run #1. Once again, five experiments were performed for the same setpoint (20 cm) to investigate the repeatability of the controller. Figure 14 shows the controlled variable response for these runs. It can be seen that the controller produced satisfactory control results with the learning algorithm switched off.

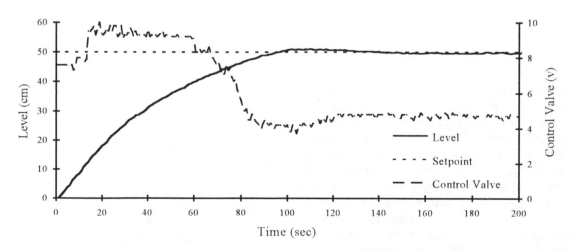

FIGURE 11 Response of the non-linear process with no-learning (FRBC).

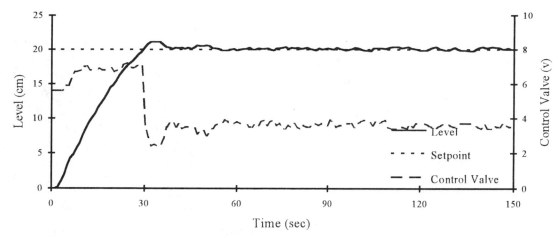

FIGURE 12 Response of the non-linear process in learning phase (bottom region).

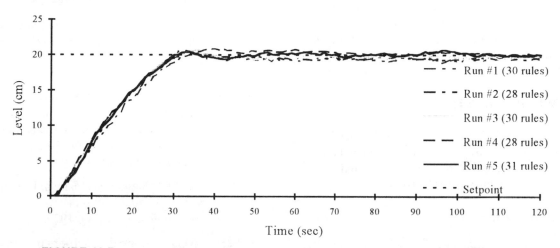

FIGURE 13 Response of the non-linear process for the repeatability test (learning phase).

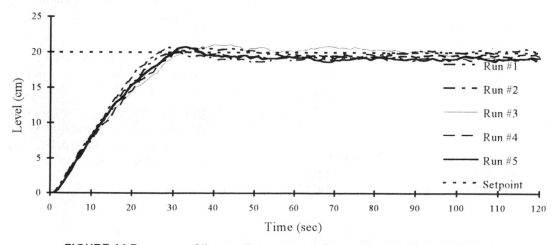

FIGURE 14 Response of the non-linear process for the repeatability test (FRBC).

5. CONCLUSIONS

A SLFLC has been presented which can perform relatively well on non-linear processes, where limited process knowledge is available. The proposed SLFLC can modify its current set of fuzzy control rules by self-learning new changes in process setpoints or process load disturbances. Despite the complexity and non-linearity of the processes, the on-line results show that the controller is able to yield a satisfactory performance with robustness and repeatability.

In conclusion, the proposed fuzzy system has the capability of generating appropriate rules for process control, under different operating conditions, by using knowledge gained from the process response during an on-line application. Additionally, the SLFLC can be used as an alternative approach to rule-base development. The robustness tests reveal that the rule-base generated by the SLFLC was complete and suitable for microcontroller implementation. Furthermore, the repeatability tests illustrate the capability of the controller to learn the process dynamics and provide an effective rule-base.

ACKNOWLEDGEMENTS

The authors gratefully acknowledge the financial support of Yarmouk University, Irbid, Jordan and the School of Electrical Engineering, Electronics and Physics, Liverpool John Moores University.

NOMENCLATURE

F{ } = Fuzzification procedure.
GC = Change of error scaling factors.
GE = Error scaling factors.
GU = Output scaling factors.
h_n = Liquid height.
q_n = Flow rate.
R_n = Restrictance valve.

REFERENCES

[BR78] M. Braae and D. Rutherford. Fuzzy relations in a control setting. *Kybernetes*, 7: 185-199, 1978.

[DG86] S. Daley and K. Gill . A design study of a self-organising fuzzy logic controller. *Proceedings of the Institute of Mechanical Engineering*, 200: 59-69, 1986.

[GJW95] S.H. Ghwanmeh, K.O. Jones and D. Williams. Comparison of different fuzzy implication approaches under self-learning fuzzy logic control. *LAAS International Conference on Computer Simulation*, Lebanon. 221-228, 1995.

[LA92] D.A. Linkens and M.F. Abbod. Self-organising fuzzy logic controller and the selection of its scaling factors. *Trans. of Inst. Mech. Eng.*, 14: 114-125, 1992.

[Lee90] C.C. Lee. Fuzzy logic in control systems: fuzzy logic controller. *IEEE Trans on Systems, Man and Cybernetics*, 20: 404-435, 1990.

[MA74] E.H. Mamdani and S. Assilian. A case study on the application of fuzzy set theory to automatic control. *Proceedings IFAC Symposium on Stochastic Control*, Budapest, Hungary, 1974.

[PM79] T.J. Procyk and E.H. Mamdani. A linguistic self-organising process controller. *Automatica*, 15: 15-30, 1979.

[Spi92] M.D. Spinrad. Self-organising fuzzy control. *ISA*, 2: 1161-1171, 1992.

[Ton77] R.M. Tong. A control engineering review of fuzzy systems. *Automatica*, 13: 559-569, 1977.

[Zim91] H.J. Zimmerman. Fuzzy set theory and its applications. *Kluwar Academic Publisher*, Boston, USA, 1991.

[Zad73] L.A. Zadeh. Outline of a new approach to the analysis of complex systems and decision process. *IEEE Trans. Syst. Man. Sybern.*, SMC-3: 28-44, 1973.

[Zad65] L.A. Zadeh. Fuzzy sets. *Information and Control*, 8: 338-353, 1965.

USING FUZZY LOGIC IN FEEDRATE CONTROL FOR FREE-FORM SURFACE MACHINING

Ralph W.L. IP[*] and Felix T.S. CHAN[+]

department of Manufacturing Engineering,
City University of Hong Kong,
Tat Chee Avenue, Kowloon, Hong Kong
Email: mewlraip@cityu.edu.hk

+School of Manufacturing and Mechanical Engineering, University of South Australia,
Warrendi Road, The Levels, South Australia 5905, Australia
Email: METSC@LEVELS.unisa.edu.au

ABSTRACT

Using constant feedrate cutting method to manufacture a free-form surface cannot produce an idea result, because the actual cutting speed on the cutting edges of a ball-nosed end mill changes with the surface gradient. A large group of machining factors will also affect the machining quality in the machining process, such as cutting edge sharpness. Although analytical approaches are currently used to determine the feedrate in NC code programming, they may not produce solutions when the machining factors are all considered. A fuzzy based feedrate control approach is reported in this paper, which can synthesize the machining factors and offers an optimum feedrate for each cutting point on the machined surface. The machining trial as conducted in the study showed that the quality of machining in term of surface roughness can be improved when the fuzzy approach is implemented.

1. INTRODUCTION

Since 1947, John Parsons investigated the practically use of numerical control (NC) techniques to manufacture aircraft components in Michigan, the NC technology has become an intelligent automation strategy in manufacturing industry over the half century. Research reports and literature have shown the merits and advantages of employing the NC concept for manufacturing tasks, especially on material removal process [Lea86, BB88, GG91, Lin94]. Milling and turning are two major traditional material removal process, using NC techniques to monitor the process can give a more consistent and higher accuracy output results. NC milling is used commonly in mould and die cavity machining in plastic and press work industry, e.g. toy and automobile manufacturing. In these manufacturing process, the milling cutter follows the pre-programmed contour paths to remove the excess material from a billet. The path control in NC milling machines is normally divided into four modes, i.e. 2-, 2 1/2-, 3- and 5-axis machining. In 3-axis machining model, the three Cartesian type machine axes, x, y and z, are simultaneously controlled to produce a specified contouring path, e.g. a straight line and a circular arc. The driven mechanism of the three axes is independent to each other, therefore, the speed in each axis can be individually controlled. Producing a true 3-dimensional mould cavity need to be performed on a 3-axis NC milling machine. However, the 5-axis machining, two extra tilting axes provided by the machine spindle, gives a larger kinematics range for the milling cutter in complex shape mould cavity crafting, e.g. free-form surface machining.

Research in the domain of free-form surface machining has emphasized on producing accurate geometrical features and economic machining tasks, some theories and approaches on geometrical surface description using parametric surface models [FP81, RA92], gouging avoidance and elimination [CLH+88, CJ89] and optimum tool path planning [Guy90] have been developed. However, the consideration of qualifying a machined free-form surface in terms of roughness and geometrical texture is little. In fact, the machined surface quality will depend upon a large group of machining factors, such as cutting speed, feedrate, depth of cut, cutting edge conditions, workpiece machinability, cutting edge coolig methods, and workpiece holding device rigidity. The classical material removal theory on static and dynamic cutting force analysis has shown the inter-relationship among the machining factors, research results in this area have been well documented [Boo75, AY95], and the currently used machining data can be found from engineering handbooks [Car80]. Unfortunately, the developed theories have considered slightly on free-form surface machining, little research has paid attention on quality control in such machining process. The aim of the paper as presented here is to report a new strategy which employs fuzzy logic approaches to control the feedrate for each cutter movement step in free-form surface machining. A better quality machined surface would be achieved because the feedrate is determined upon each cutting point rather than an average value over the surface.

2. ANALYTICAL FEEDRATE CALCULATION

The feedrate determination for milling process was originally from analytical material removal theory. The material removal rate (MRR) was formulated as:

$$MRR = f_m \, w \, t \qquad (1)$$

where f_m is the machining feedrate in mm/min, w and t are the width and depth of cut in mm, and the unit of MRR is mm^3/min. The feedrate in equation (1) gives the speed of the milling cutter crosses over the surface, however, counting the cutter advancing distance for each revolution would be an alternative method to determine the feedrate, see equation (2).

$$MRR = f \, t \, V_c \qquad (2)$$

The machining factor, cutting speed V_c, will be included in the material removal equation when the feedrate f is defined as mm/rev of the milling cutter. This expression gives useful information in cutter rotation speed calculation and cutter diameter selection in programming of NC codes. Choosing the cutter rotation speed and cutter diameter are normally limited by the mechanical design of the machine spindle. The cutting speed in equation (2) for milling can be expression as:

$$V_c = \frac{\pi \, D_n \, N}{1000} \qquad (3)$$

where D_n is the cutter nominal diameter in mm, N is the cutter rotating speed in rev/min, and V_c is the cutting speed in m/min.

In free-form surface machining, ball-nosed end mills are normally employed in finishing cut because they can easily match the curvature of the required part shape [VQ89]. Some research studies have reported various cutter path design methods for ball-nosed end mills, but most of them based upon the theory of rolling a sphere on the surface [CLH+88, CJ89]. Broomhead and Edkins [BE86] have formulated a cutting path equation for a ball-nosed end mill, and which used vector algebra to define the location of the cutter at each cutting point. The equation, known as cutter location (CL) equation, is given in equation (4), and the associated vectors of the ball-nosed end mill is illustrated in Figure 1.

$$CL = CC + \frac{D_e}{2} \left[n - A \right] \qquad (4)$$

In equation (3), the cutting speed V_c is in terms of the cutter nominal diameter and its rotating speed. In fact, the effective diameter D_e at the cutter contact point (CC) may be equal or smaller than D_n, it will depend upon the surface gradient at the point, see Figure 2.

The effective diameter D_e at the CC point can be given as the product of D_n to the surface gradient α, i.e. $D_e = D_n \sin \alpha$. Therefore, the actual cutting speed at the CC point of a ball-nosed end mill can be obtained by substituting the D_n by D_e in equation (3), it is given as:

$$V_c = \frac{\pi \, N \, D_n \, \sin \alpha}{1000} \qquad (5)$$

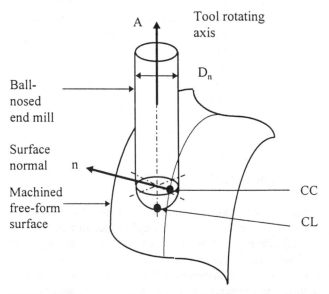

FIGURE 1 The associated vectors of a ball-nosed end mill in free-form surface machining.

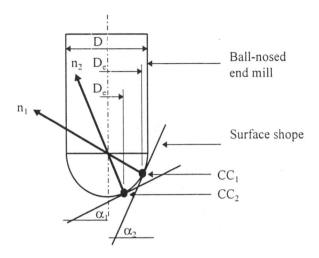

FIGURE 2 The variation of cutter effective diameter at different surface gradients.

The MRR in equation (2) for free-form surface machining will be in terms of the four dependent machining factors: f, t, N and D_n, and the independent part shape parameter: surface gradient α. The investigation on economic machining has emphasized on the determination of maximum MRR under the available machining conditions, such as workpiece machinability, tool material and the machine spindle power. In 1907, Taylor [Tay07] formulated an empirical tool life equation, as given in equation (6), which shows the relationship between the available tool life T and the cutting speed.

$$V_c T^m = C \qquad (6)$$

In Taylor tool life equation, C is basically a constant for the workpiece material, whereas the exponent m is the tool material characteristics. The data of C and m for a wide class of tool and workpiece materials are available from engineering handbooks [Car80]. The MRR in equation (2) in free-form surface machining can also be expressed in terms of the two dependent machining factors: f and t, the two independent material parameters: C and m, and the dependent tool parameter: T.

In designing an NC part programme for free-form surface machining, the cutter rotating speed N and tool nominal diameter D_n need to be initially specified. The common practice in choosing the rotating speed is to use equation (3) with the recommend cutting speed from engineering handbooks [Car80]. Selecting the nominal diameter for machining will directly relate to the surface part geometry and the NC machining model, e.g. 3- or 5-

axis. Smith and Stephenson [SS91] have recommended an approach which used the surface curvature theory to calculate the maximum ball-nosed end mill diameter for 3-axis surface machining. Gouging avoidance algorithms, such as the method proposed by Choi and Jun [CJ89], need to be considered in determining the tool nominal diameter for such machining process. When the cutter nominal diameter and rotating speed are chosen, the actual cutting speed at each cutting point on the free-form surface can be calculated from equation (5); and the remained tool life T is given from equation (6) for the specified tool and workpiece material.

The depth of cut, t in equation (2) is normally fixed at each machining level in the cutter path planning process, thus, a new set of CL data will be given from equation (4) when the cutter increases its penetration, see Figure 3.

The last depend machining factor in equation (2) for the calculation of optimum MRR is the feedrate f. In NC part programming process, a new feedrate needs to be determined and assigned to each new cutter movement programming block to compensate the variation of cutting speed due to the surface gradient alteration. Unfortunately, few research study has been found to report the compensated feedrate determination algorithm, and none application type NC code post-processors which can adjust the machining feedrate for a free-form surface are available on the market.

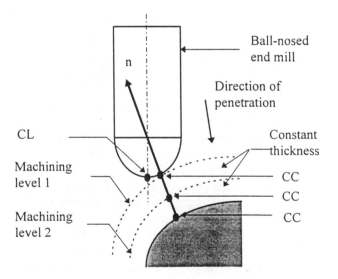

FIGURE 3 A constant depth of cut machining strategy for ball-nosed end mills.

3. FUZZY BASED FEEDRATE DETERMINATION IN FREE-FORM SURFACE MACHINING

Although the analytical approach, as discussed in Section 2, can be a possible method to calculate the feedrate for free-form surface machining, some important tooling factors which may affect significantly the machined surface quality have not been included in the approach, such as cutting edge conditions, cooling methods and tool holding rigidity. In fact, the roughness of a machined surface still depend highly upon the conditions of the cutter at a correct cutting speed and feedrate [Boo75, Lea86, GG91].

Using analytical methods to formulate an optimum feedrate approach for free-form surface machining may be difficult when the tooling factors are considered together. The authors suggested that employing fuzzy logic theory may be an effective way to resolve the problem. Some research studies have also reported the advantages of using fuzzy approach to resolve the engineering problems with a massive input database [Wil91, LDV94]. A fuzzy based feedrate control approach is addressed as follows.

The part shape parameters, surface gradient α and the dependent machining factor, depth of cut t, are treated as the input variables in the fuzzy approach, their membership functions can be simplified and described as triangular shape as illustrated in Figure 4.

The membership values μ, between 0 to 1, are governed by the shape of the membership functions, e.g. triangular, and the fuzzy sets are labelled in linguistic terms. The amplitude of the surface gradient on a free-form surface is characterized into a fuzzy set which contains three membership functions, they are labelled as Small (S), Medium (M) and Large (L) within the range of 0 to $\pi/2$, see Figure 4 (a). The membership functions of the input variable depth of cut, which are defined in term of the cutter nominal diameter D_n, are also characterized into three linguistic terms, i.e. Shallow (S), Middle (MID) and Deep (D), see Figure 4 (b).

A fuzzy inference system can be set up when the membership function of the input variables are identified. In the system, a group of fuzzy results are collected by the operation of linguistic production rules, and the rules are built upon logical rules, e.g. If α = S and t = D, then f = MID. Figure 5 shows the complete feedrate control inference system.

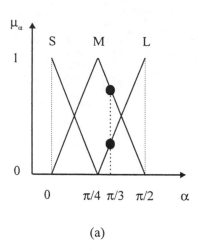

(a)

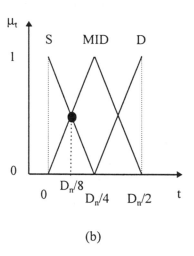

(b)

FIGURE 4 The membership functions of : (a) surface gradient; (b) depth of cut.

Depth of cut (t)

		S	MID	D
Surface gradient (α)	**S**	H	RH	MID
	M	RH	MID	RL
	L	MID	RL	L

FIGURE 5 Feedrate control inference system.

The feedrate f in the inference system is characterized into five membership functions' fuzzy set, they are labelled as High (H), Rather High (RH), Middle (MID), Rather Low (RL) and Low (L), see Figure 6.

Determining the feedrate for each cutter movement is given by executing the fuzzy inference system. For example, the surface gradient at a particular cutting point is measured as $\pi/3$ and the depth of cut is given as $D_n/8$. The membership values for μ_α and μ_t are measured and listed in Table 1.

Fuzzy reasoning process is then carried out to determined an appropriate feedrate from the fuzzy set. Among the various fuzzy reasoning methods, the most commonly used in industrial process control is the MAX-MIN fuzzy reasoning method. In the method, Mamdani's minimum operation rule [MG81] is used for fuzzy implication. The feedrate membership value μ_f is point-wise and the mathematical expression is given as:

$$\mu_f\,(\%\,f_{ref}) = (R_1 \wedge \mu_{f1}\,(\%\,f_{ref})) \vee (R_2 \wedge \mu_{f2}\,(\%\,f_{ref})) \qquad (7)$$

and

$$R_1 = \mu_{\alpha 1}\,(\alpha) \wedge \mu_{t1}\,(t),$$
$$R_2 = \mu_{\alpha 2}\,(\alpha) \wedge \mu_{t2}\,(t).$$

Figure 7 illustrates the MAX-MIN reasoning process for the two input variables, α and t.

Defuzzification is the final process of mapping the inferred fuzzy outputs, i.e. four shaded trapeziums as shown in Figure 7, to a non-fuzzy control space, e.g. feedrate control action to the NC milling machine. In mathematical expression, the process is given as: $\%\,f_{ref} =$ Defuzzifier (μ_f). The centre-of-area (COA) method is widely used in the defuzzification process to produce a non-fuzzy control action that best represents the possibility distribution of the inferred fuzzy variables. For n numbers of fuzzy inferred rules are used; let M is the moment of each fuzzy output membership function, and A is the area of the membership function. The defuzzified control variable, $\%\,f_{ref}$ calculated from the COA method is given as:

$$\%\,f_{ref} = \frac{\sum\limits_{i=1}^{n} \mu_{fi}\,M_i}{\sum\limits_{i=1}^{n} \mu_{fi}\,A_i} \qquad (8)$$

From the measured quantities in Table 1 and the results in Figure 7, the feedrate calculation results are listed in Table 2.

The calculated feedrate value, 0.5845, is only a percentage of the feedrate as recommended in engineering handbooks [Car80]. The required feedrate quantity for the surface machining needs to multiply the calculated percentage to the recommended feedrate value.

Table 1 Membership values of μ_α and μ_t at $\alpha = \pi/3$ and $t = D_n/8$.

$\alpha = \pi/3$		$t = D_n/8$	
MF = M,	$\mu_\alpha = 0.8$	MF = S,	$\mu_t = 0.5$
MF = L,	$\mu_\alpha = 0.2$	MF = MID,	$\mu_t = 0.5$

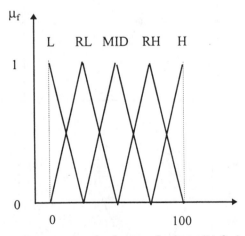

FIGURE 6 Membership functions of the feedrate.

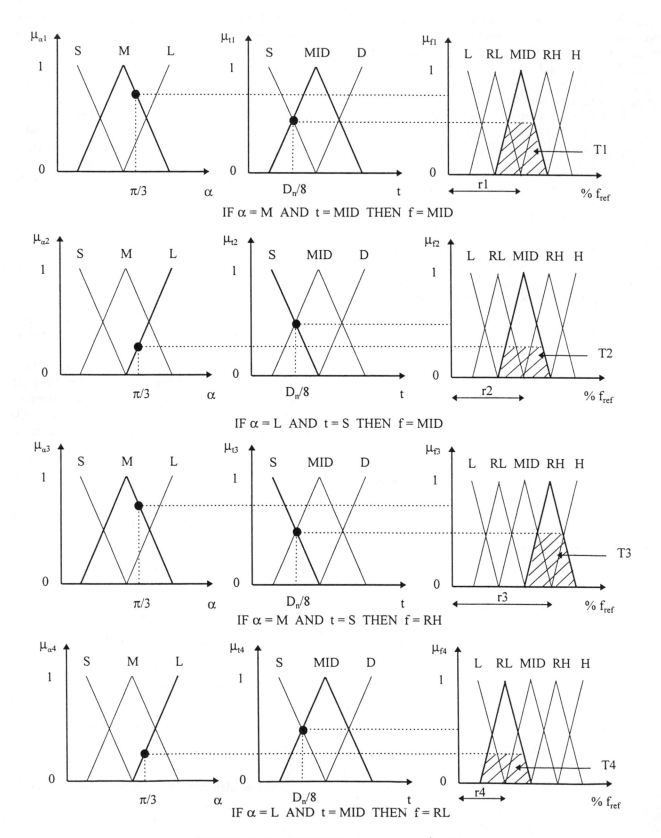

FIGURE 7 The MAX-MIN fuzzy reasoning method.

Table 2 The results of the feedrate defuzzification using the COA method.

T	μ_{fi}	A_i	Moment arm (r_i)	M_i	$\mu_{fi} M_i$	$\mu_{fi} A_i$
1	0.5	0.1875	0.5	0.09375	0.0469	0.0938
2	0.2	0.090	0.5	0.0450	0.0090	0.0180
3	0.5	0.1875	0.75	0.1406	0.0703	0.0938
4	0.2	0.090	0.25	0.0225	0.0045	0.0180
				Σ	0.1307	0.2236

$$\% f_{ref} = \frac{0.1307}{0.2236} = 0.5845$$

4. A MACHINING EVALUATION TRIAL

Although the principle of the fuzzy based feedrate control strategy for free-form surface machining has been developed, its performance needs to be validated. A machining trial was designed to compare the roughness of two machined free-form surfaces; one has and one has not used the feedrate control strategy. A 10 mm diameter HSS ball-nosed end mill was used on a 3-axis CNC machining centre to machine two half-lunar grooves on a 50 mm diameter steel rod. Two sets of NC codes were designed for the machining:
The first code set:
A constant feedrate was inserted at the first tool movement block.

The second code set:
Used the investigated fuzzy approach to calculate the feedrate and assigned them to each tool movement block

The roughness of the two machined surfaces was measured on the Taylor-Hobson surface texture assessment machine. The measured surface texture profile and the average roughness values for the two surfaces are illustrated in Figure 8. The measured results of Trace 1 were given from the surface produced by the first code set, and Trace 2 results were the surface machined by the second code set.

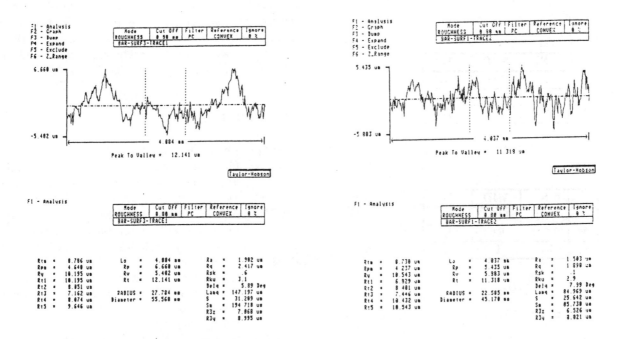

FIGURE 8 Surface roughness measurement results of the two machined surfaces.

5. DISCUSSION, CONCLUSION AND FUTURE WORK

The machining trial results indicate that the quality of the machined surface is improved when the proposed feedrate control strategy is applied. The arithmetic mean deviation of roughness R_a of the surface machined under constant feedrate is about 20 % larger than the surface produced by the fuzzy based feedrate control method. The machining trial considered only two machining parameters, i.e. surface gradient and depth of cut. However, the other machining factors, e.g. cutting edge conditions, e can be collected and treated as a group of input variables for the fuzzy inference system. The membership functions of the cutting edge condition will be similar to the membership functions of the surface gradient and depth of cut as shown in Figure 4, however, the linguistic terms used to describe the cutting edge conditions are Sharp (S), Medium (M) and Dull (D). The fuzzy inference system of three variables is given in Figure 9.

t

	S	MID	D
S	VH	H	RH
M	H	RH	MID
L	RH	MID	RL

α

e = S

t

	S	MID	D
S	H	RH	MID
M	RH	MID	RL
L	MID	RL	L

α

e = M

t

	S	MID	D
S	RH	MID	RL
M	MID	RL	L
L	RL	L	VL

α

e = D

Linguistic terms: VH: Very High, VL: Very Low

Figure 9 Feedrate control inference system for three fuzzy variables: α, t and e.

Although the process of creating the fuzzy inference system and finding the feedrate control value in defuzzification process will be proportionally longer for the addition of input variables, the calculation procedures and their complexity remain constant. When a large number of machining variables need to be considered simultaneous, the fuzzy method would offer the solution in simpler and quicker way than employing analytical approaches.

The research in fuzzy based feedrate control for free-form surface machining is moving on adaptive type membership function design. The authors found that the membership function of some machining factors will change with time, and temperature, such as the tool life and cutting edge hardness. An accurate result could be obtained when only if the membership functions are correspondingly corrected when the dynamic machining factors are included in the fuzzy variable set. More research could be done to study how the machining factor dynamic behaviour is described by the membership functions.

In conclusion, a new feedrate control method in free-form surface machining was studied. The incorrect feedrate assignment problem in NC coding process due to the alternation of surface gradient can be resolved using the investigated fuzzy based feedrate control approach. The quality of the machined surface in term of roughness can be improved when the approach is applied. A better result would be achieved if a large group of machining factors are considered in the fuzzy inference system. Using adaptive type membership functions to describe the dynamic behaviour of the machining factors need to be intensively studied.

REFERENCES

[AY95]   Abdou, G. and Yien, Y., 1995, Analysis of force patterns and tool life in milling operations, *International Journal of Advanced Manufacturing Technology*, Vol. 10, pp 11-18.

[BB88]   Bryan, L.A. and Bryan, E.A., 1988, *Programmable Controllers: Theory and Implementation*, Industrial Text: Chicago.

[BE86]   Broomhead, P. and Edkins, M., 1986, Generating NC data at the machine tool for the manufacture of free-form surfaces, *International Journal of Production Research*, Vol. 24, No. 1, pp 1-14.

[Boo75]   Boothroyd, G., 1975, *Fundamentals of Metal Machining and Machine Tools*, Scripta/McGraw Hill: Washington D.C.

[Car80] Carbaloy Systems Department, 1980, *Milling Handbook of High-Efficiency Metal Cutting*, G.E.: Michigan.

[CJ89] Choi, B.K. and Jun, C.S., 1989, Ball-end cutter interference avoidance in NC machining of sculptured surfaces, *Computer-Aided Design*, Vol. 21, No. 6, pp 371-378.

[CLH⁺88] Choi, B.K., Lee, C.S., Hwang, J.S. and Jun, C.S., 1988, Compound surface modelling and machining, *Computer-Aided Design*, Vol. 20, No. 3, pp 127-136.

[FP81] Faux, I.D. and Pratt, M.J., 1981, *Computation Geometry for Design and Manufacture*, Ellis Horwood: U.K Leatham, J.B., 1986, *Introduction to computer Numerical Control*, Pitman/Wiley.

[GG91] Gibbs, D. and Grandell, T.M., 1991, *An Introduction to CNC Technology*, Industrial Press: New York.

[Guy90] Guyder, M.K., 1990, Automating the optimization of 2 1/2 D axis milling, *Computing Industry*, Vol. 15, pp 163 - 168.

[Lea86] Leatham, J.B., 1986, *Introduction to computer Numerical Control*, Pitman/Wiley.

[LDV94] Li, H., Dong, Z. and Vickers, G.W., 1994, Optimal toolpath pattern identification for single island, sculptured part rough machining using fuzzy pattern analysis, *Computer-Aided Design*, Vol. 26, No. 11, pp 787-795.

[Lin94] Lin, S.C., 1994, *Computer Numerical Control: Essentials in Programming and Networking*, Delmar: Albany.

[MG81] Mamdani, E.H. and Gaines, B.R., 1981, *Fuzzy Reasoning and its Applications*, Academic: London.

[RA92] Rogers, D.F. and Adams, J.A., 1992, *Mathematical Elements for Computer Graphics*, 2 nd., McGraw-Hill: New York.

[SS91] Smith, G. and Stephenson, H., 1991, The selection of optimum tool size for three-axis and multi-axis machining of highly curved features, *Proceedings of the 7 th National Conference on Production Research*, pp 317-325.

[Tay07] Taylor, F.W., 1907, Tool life predication, *Transactions of SAME*, Vol. 28, pp 31-279.

[VQ89] Vickers, G.W. and Quan, K.W., 1989, Ball-mills versus end-mills for curved surface machining, *Transactions of the ASME*, Vol. 111, pp 22-26.

[Wil91] Williams, T., 1991, Fuzzy logic simplifies complex control problems, *Computer Design*, March, pp 90-102.

Genetic Algorithm

GA-BASED RECONSTRUCTION OF PLANE IMAGES FROM PROJECTIONS

Zensho Nakao, Yen-Wei Chen, and Fath El Alem Fadlallah Ali
Department of Electrical & Electronics Engineering,
Faculty of Engineering, University of the Ryukyus,
Okinawa 903-01, JAPAN
nakao@augusta.ie.u-ryukyu.ac.jp

ABSTRACT

A GA-based technique for reconstructing plane binary images from a minimal number of projections is presented; its effectiveness is demonstrated by reconstructing two-dimensional objects from their one-dimensional coded images. The algorithm gets projection data from *three* different viewing angles. An initial population of strings, each of which contains encodings of an image, is generated randomly. Typical genetic operations are performed on every generation of population to produce a new generation, where a *Laplacian* type constraint term is included in the fitness (*cost*) function, and two-dimensional genetic operators are developed. Results obtained so far are fairly good although there is still some room for modification and improvement.

INTRODUCTION

The problem of digitally reconstructing an image from its projections has become important during the past few decades. There are many fields where the practical applications of this problem exist. One of the important applications is computerized tomography in the medical diagnostic radiology. X rays are used to generate the projection data for a cross section of the human body. The problem is to manipulate the projection data in order to reconstruct a cross-sectional image depicting with very high resolution the morphological details of the body in that cross section. In nuclear medicine application the problem is to map the distribution of the concentration of a gamma-ray emitting isotope in a given cross section of the body. Nonmedical areas of application where images may be reconstructed from projections include radioastronomy, optical interferometry, electronmicroscopy, and geophysical exploration.

Many techniques have been employed to manipulate the projection data in order to reconstruct the original image. Algebraic Reconstruction Technique[RK76] is one of the well known techniques in this regard. The early eighties noticed several neural network applications in the filed of image processing and pattern recognition. Smith, Barret and Paxman presented a procedure for retrieval of objects from their coded images[SBP83]. Their reconstructed process is modeled as an optimization problem whose cost function is reduced by a simulated annealing process.

THE GENETIC APPROACH

In order to apply the genetic algorithm to the problem of image reconstruction from projections, an image has to be converted into a genetic string (*chromosome*). To achieve this, each pixel in the plane of the image is represented by an *allele* in a two-dimensional chromosme. A simple model example is shown in FIGURE 1.

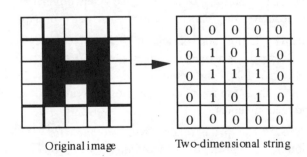

Original image   Two-dimensional string

FIGURE 1 Image encoding

235

FITNESS MEASURE

To each string is attached a *cost* value upon comparing the calculated projections of the image encoded in the matrix with the projections of the original image. An additional *apriori* knowledge about the smoothness of the original image is incorporated. This is done by a *Laplacian Constraint*. As the reconstructed image becomes more similar to the original one, the smaller becomes the mean square difference between the original and the reconstructed image projection data, and the smaller the Laplacian value and, hence, the smaller the cost of the chromosome. The following formula is used to return a chromosome cost:

$$Cost = E + \lambda \times \|LapCon\|^2,$$

where

$$E = \frac{\sum\limits_{\theta=1}^{c} \frac{1}{Width} \sum\limits_{S=1}^{Width} [P(\theta, S) - R(\theta, S)]^2}{c},$$

and

$$LapCon = [f(i+1, j) + f(i-1, j) + f(i, j+1) + f(i, j-1)] - 4f(i, j)$$

$P(\theta, S)$:   projections of the original image,

$R(\theta, S)$:   projections of the reconstructed image,

θ :   index of the projection angle,

S :   index of the projection subdata,

$f(i, j)$:   pixel weight,

λ :   real coefficient

GENERATION OF AN INITIAL POPULATION

Strings for the initial population are generated randomly with bias to provide as much as possible consistency among available projections. This is achieved with a guided randomness, using the three given original image projection data. This is accomplished in two steps:

Step1   Using one projection data ($\theta = 0^\circ$), dots are located randomly in each vertical section (to be called a *unit*) so that the number of filled pixels agrees with the number in the corresponding original projection subdata.

Step2   The grid is rotated with an angle ($\theta = 90^\circ$) to obtain a projection equivalent with the second projection. Following the step, a refining operation is performed: each unit is visited once. Comparing each unit

projection data with the original one, we may face one of three cases:

Case I: $P(\theta, S) = R(\theta, S)$:

All filled pixels are copied into a refined population pool.

Case II: $P(\theta, S) < R(\theta, S)$:

Among the filled pixels, a number equivalent to $P(\theta, S)$ is randomly selected and copied in the refined population pool.

Case III: $P(\theta, S) > R(\theta, S)$:

Dots enough to bring the two projections alike are drawn from other units (which are randomly selected).

For the third direction ($\theta = 135^\circ$), projection data is calculated and then compared to the original original in this direction. The three cases in **Step2** are considered and refinement is made.

THE GA OPERATORS

In the reproduction operation, an elitist selection scheme is used for selection of parents instead of the normal known schemes. In the elitist scheme, fittest strings are given chance to contribute in the next generation without undergoing any further GA operation, and strings with low fitness are replaced by offsprings mated from the more fitted strings. Thus, strings with high fitness values are copied directly in the new generation. From the elitist portion, parents are selected randomly, crossed over, mutated and the produced offsprings are copied into the new generation.

CROSSOVER

A regional crossover is used. At first, two positions P1 and P2 in the chromosomes are selected randomly. Then two blocks in the parents, which are determined by P1 and P2 are exchanged. As a result, two offsprings are generated as shown in the FIGURE 2 below.

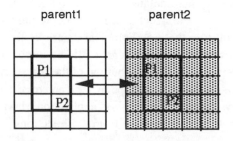

offspring1   offspring2

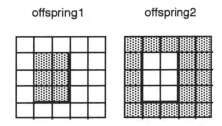

FIGURE 2 Crossover

MUTATION

Three mutation schemes are used: an ordinary *uniform* mutation whose application depends on the rate of convergence of the population average fitness, another mutation to be called *chameleon*, which is used with regular frequency, and a third *projection-weighted* mutation.

Uniform Mutation The allele selected for mutation takes the complementary value of the current value. In order to use the uniform mutation, the program calculates the percentage increase in the average fitness. If no increase is recorded then this mutation is injected into the program. Once a positive increase is recorded, the mutation is applied no more.

Chameleon Mutation This is a new mutation scheme. It does not simulate a natural phenomenon in natural genetics behavior, but its design is based on the characheristics of the concerned problem. It performs a small-scale survey of the area surrounding the allele selected for mutation; the adjacent neighboring eight pixels and the pixel itself are checked and the pixel is mutated toward the prevailing value (FIGURE 3). Chameleon mutation effectively provides local genesis of consistent and integrated subsegments. It repairs the side effects of the global pixel-wise replacement of the uniform mutation and crossover, and hence restores the feasiblity of the strings.

Projection-Weighted Mutation This operator makes use of the given original projection data: with a large-scale survey, a refining process, based on comparison of projection data, is performed and aims to provide most possible consistency between two related projection subdata and those of the corresponding original ones. The mutation process operates on both deterministic and probabilistic rules, using two projection data at a time. Distances between the two projections are calculated for each unit. Then application of mutation for pixels in each unit is based on the distance between the two projections. Every unit is visited exactly once. The unit projection weight is compared with that of the corresponding original one. If they are not the same, two cases emerge:

Case1: $P(\theta, S) > R(\theta, S)$ (FIGURE 4)

The number of dots equal to the difference between the two values is added randomly to the unit in location where the addition will improve the other angle profile.

Case2: $P(\theta, S) < R(\theta, S)$

The opposite of **Case1** is done, namely, the number of dots equal to the difference between the two values is removed randomly from the unit at locations where the removal will improve the other angle profile.

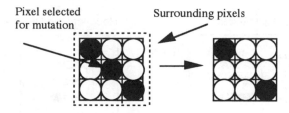

Pixel selected for mutation   Surrounding pixels

FIGURE 3 Chameleon mutation

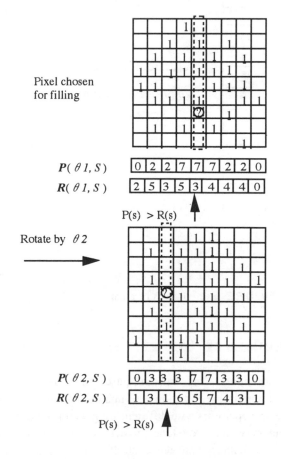

Pixel chosen for filling

$P(\theta 1, S)$ | 0 | 2 | 2 | 7 | 7 | 7 | 2 | 2 | 0 |
$R(\theta 1, S)$ | 2 | 5 | 3 | 5 | 3 | 4 | 4 | 4 | 0 |

$P(s) > R(s)$

Rotate by $\theta 2$

$P(\theta 2, S)$ | 0 | 3 | 3 | 3 | 7 | 7 | 3 | 3 | 0 |
$R(\theta 2, S)$ | 1 | 3 | 1 | 6 | 5 | 7 | 4 | 3 | 1 |

$P(s) > R(s)$

FIGURE 4 Projection-Weighted Mutation- Case1

To summarize, FIGURE 5 describes the process of the production of a new generation, applying the above mentioned operators.

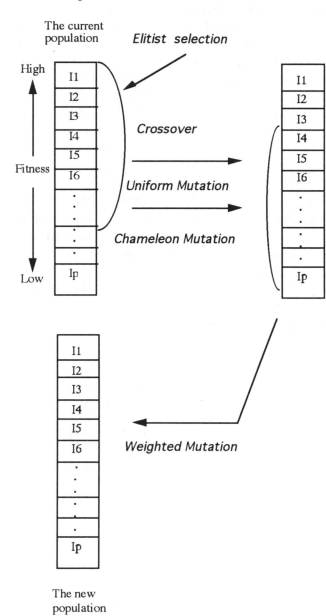

FIGURE 5 Production of a new population

EXPERIMENTS AND RESULTS

The experimental work, which required long and intensive computational time, was done through mutual interactions among the operators with different weights for the related parameters, which were tuned to achieve the elusive goal of a robust balance between controlled convergence and consistency in exploration and exploitation of information manipulated by the GA.

For the GA operations, many schemes were tested, and finally the appropriate schemes were adopted. The most elusive problem faced was the proliferation of highly fit and similar population members which usually results in the vanishing of significant features needed for new exploration areas which in turn results in a premature convergence in the fitness graph.

The method was applied to several different types of target images with the aim of provision of wide diversity of the target images. Results presented here are averages of five testing runs for each image. Through more than twenty testing runs for each, optimal values for the different parameters were obtained. Here are detailed parameter values:

Table 1 Main parameters used

Parameter	optimal value
Elitist selection rate (S_Rate)	0.9
Selection rate between parents for crossover	0.5
Mutation rate for uniform mutation	0.6
Mutation rate for chameleon mutation	0.5
Portion of population subjected to weighted mutation	60%
Population size	200
Number of angles of projections	3
First projection angle	0°
Second projection angle	90°
Third projection angle	135°
Balancing coefficient (λ)	0.001
Size of the image	45x45

FIGURE 6 below shows the development of a reconstructed image through progression of generations by the evolutionary technique.

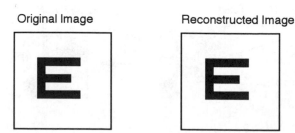

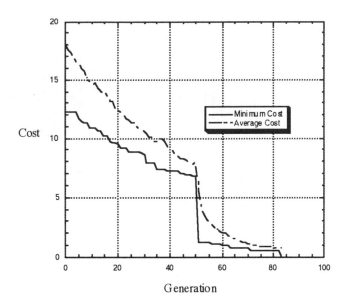

FIGURE 6 Reconstruction over generations

FIGURE 7 shows some images being reconstructed using the genetic agorithm.

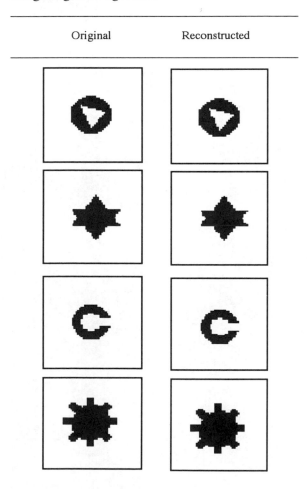

FIGURE 7 Original and reconstructed images

COMMENTS AND FUTURE WORK

A satisfactory agreement between the reconstructed image and the original one was obtained in most cases. Comparing the GA results with those of ART consolidated the feasibility of the evolutionary method. Extenstion of this algorithm to manipulate multi-density and three-dimensional images would mark out a valuable technique that would be of some use in the medical application of CT scanners, laser fusion, and plasma radiation CT.

REFERENCES

[GG84] S. Geman, and D. Geman: " Stochastic relaxation, Gibbs distributions, and the Bayesian restoration of images, " *IEEE Trans. on Patt. Anal. Mach. Int.*, *Vol. PAMI-6*, pp. 614-634, November 1984.

[G89] D. Goldberg: *Genetic Algorithms in Search, Optimization, and Machine Learning*, Addison-Wesley, 1989.

[M94] Z. Michalewicz: *Genetic Algorithms + Data Structures = Evolution Programs*, Springer-Verlag, 1994.

[RK76] A. Rosenfeld and A.C. Kak: *Digital Picture Processing*, Academic Press, 1976.

[SBP83] W.E. Smith, H.H. Barrett, and R.G. Paxman: "Reconstruction of objects from coded images by simulated annealing," *Optics Letters*, Vol. 8, No. 4, pp. 199-201, April 1983.

GENETIC ALGORITHM PROCESSOR FOR THE FREQUENCY ASSIGNMENT PROBLEM

Mehrdad Salami and Greg Cain
Department of Electrical Engineering,
Victoria University of Technology,
PO Box 14428 MCMC, Melbourne, VIC 8001 Australia.
Email: salami@cabsav.vut.edu.au

ABSTRACT

This paper presents a method of frequency assignment in cellular radio networks. We used a Genetic Algorithm implemented in hardware for setting the frequencies for the cells in the network to minimize the interference in the system. A new configuration of the system based on multiple genetic algorithm processors has been applied to the problem. The method has been successfully tested and some results are presented. Test results show the Genetic Algorithm Processor is capable of optimizing with high reliability.

1. INTRODUCTION

Genetic algorithms represent a class of adaptive search techniques which can be applied to optimization problems or in artificial intelligence. A genetic algorithm (GA) is a robust global optimization technique based on natural selection. The basic goal of GAs is to optimize functions called fitness functions. GA-based approaches differ from conventional problem-solving methods in several ways. First, GAs work with a coding of the parameter set rather than the parameters themselves. Second, GAs search from a population of points rather than a single point. Third, GAs use payoff (objective function) information, rather than auxiliary knowledge. Finally, GAs use probabilistic transition rules, not deterministic rules. These properties make GAs robust, powerful, and data-independent.

Due to its evolutionary nature, a GA will search for solutions without regard to the specific inner workings of the problem. Much of the interest in genetic algorithms is due to the fact that they provide a set of efficient domain-independent search heuristics without the need for incorporating highly domain-specific knowledge [Hol92]. There is now considerable evidence that genetic algorithms are useful for global function optimization and NP-Hard problems [Dav91].

In this paper we start with an explanation of how a Genetic Algorithm Processor (GAP) can be applied to an optimization task. Then the Multiple Genetic Algorithm Processor is introduced. Finally we apply our model to see how a GAP in a multiple configuration can be applied to the Frequency Assignment Problem (FAP).

2. THE GENETIC ALGORITHM PROCESSOR

A hardware genetic algorithm can be constructed to directly execute the operation of a genetic algorithm as shown in the Figure 1 [SC95a]. Such a processor can be used in situations where high throughput is required and where the logic of the genetic algorithm is expressible in simple units which can be synthesised in hardware. This is generally the case as genetic algorithms are inherently simple and contain only a few logic operations. Such speed enhancement is important for many applications such as adaptive filters.

The GAP maintains a population of bit strings in its memory. Each of these member strings represents a tentative solution to the problem at hand. In execution the GAP delivers a member string to the Problem Module (PM) which evaluates its performance and returns a response known as a *fitness value* which is stored in memory. The fitness value is used by the Selection Module (SM) to select suitable members for mating and reproduction to obtain new member strings for the next generation.

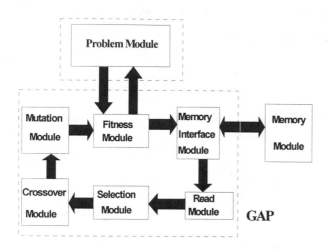

Figure 1 Internal architecture of the GAP

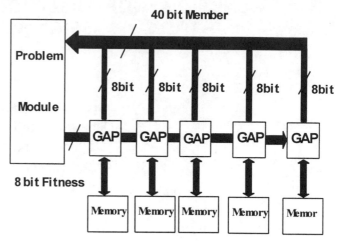

Figure 2 Splitting each member between five GAPs

In the GAP, mating involves switching substrings from two selected parent strings in simple operation which takes place in the Crossover Module (CM). New child strings formed in this fashion may then be subjected to a simple random bit-flipping process in the Mutation Module (MM). After mutation the newly formed strings are delivered to the Problem Module for evaluation. The PM returns a fitness value which is written to memory with the new member string.

The above steps continue until the FM determines that the current GAP run is finished. The whole GAP section was written in VHDL (VHSIC Hardware Descriptive Language) and simulated and synthesised by Mentor Graphics tools. After synthesizing the GAP Modules, the NeoCAD FPGA Foundry™ tools was used to convert the GAP modules to the FPGA (Field Programming Gate Array) technology.

3. A MULTIPLE GENETIC ALGORITHM PROCESSOR

In real applications the number of variables is often high and the member string length becomes long. This means that in a hardware implementation of a GAP, a wide data path would normally be required. However it is very important to keep the data path as small as possible (preferably 8 bits) for an FPGA implementation of the GAP. This problem can be handled by splitting the member string among several small GAP processors as shown in Figure 2 [SC95b].

Each 8-bit slice of the member string is assigned to a separate GAP which can then be implemented in a single FPGA chip. Thus each GAP maintains its own population of 8-bit strings and associated fitness values and conducts the crossover and mutation operations only on its own members.

In each evaluation cycle a 40-bit member string is delivered to the Problem Module (PM) which replies with an 8-bit response. All GAPs execute their internal operations concurrently but only one GAP unit can alter it's 8-bit slice of the 40-bit member string in each cycle and it is this unit which waits and receives the response through its fitness module. The remaining 32 bits of the 40 bit member string are made up from the existing strings delivered by the other four GAPs.

4. THE FREQUENCY ASSIGNMENT PROBLEM

Frequency assignment is an important problem in cellular radio networks. Algorithms for frequency assignment have been inspired by the classical graph colouring problem [Box78]. Early methods used algorithms based on regular hexagonal arrays but as real cellular networks are far from regular, these algorithms are not suitable.

In the classical graph colouring problem an algorithm has to colour the nodes of a graph in such a way, that nodes which are joined by an edge get different colours, and that the total number of used colours is minimal. The frequency assignment methods inspired by those algorithms asked for an input graph where base stations were nodes of the graph and an edge between two nodes essentially meant that those two base stations were not allowed to share a common frequency. The same information could be given in form of a matrix, the so called compatibility matrix.

The global technique comes down to finding a good frequency assignment to favour the same frequency reuse by sufficiently distant cells which allows the number of communications over the network to be maximized with a limited number of frequencies. Different optimization versions of FAP could be developed such as maximizing the total traffic, minimizing the number of frequencies used and minimizing the interferences over the network.

The satisfiability of this problem can be shown to be NP-complete because it is reduced to the graph k-coloring problem [GR82]. Many methods have been proposed to solve the FAP, including classic methods: graph coloring algorithms [GR82] and integer programming; and heuristic methods: neural networks [Kun91] and genetic algorithms [DH95].

Given the number of cells N, the number of available frequencies in the spectrum $Nfreq$, the total number of interference constraints defined for each pair of adjacent cells $Ninter$, the FAP can be modelled with a quadruple $<X,D,C,F>$ representing a constrained optimization problem with:

X = {Ci | Ci is a cell of the network, i ∈ [1..N]}.
D = {Fi | Fi is an available frequency of the spectrum, i ∈ [1..$Nfreq$]}.
C = T ∪ I
T= {Ti | Ti minimal number of frequencies necessary for Ci, i ∈ [1..N]}.
I = {Ii | Ii interference constraints between two cells, i ∈ [1..$Ninter$]}.
F = Fitness function of a frequency assignment of the network.

The traffic constraint of each cell is represented by Ti which is the minimal number of frequencies necessary for Ci to cover its maximum traffic communications. In reality, this maximum traffic value is defined by an estimation of the maximum number of mobiles which can move at the same time within this cell.

The interference constraints are represented by a matrix M of $N*N$ and with f(i,n) corresponding to the n-th frequencies of Ci, each element of M defines the set of constraints as follows:

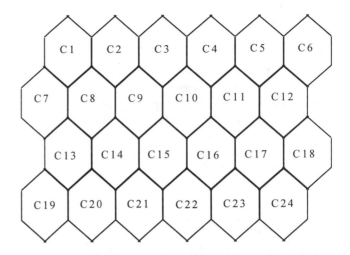

Figure 3 Assignment of frequency to cells. Each Ci consists of Ti frequencies

• M[i,j] with i ≠j represents the minimal number of channel separations required to satisfy the adjacent channel constraints between the cells Ci and Cj.

$$\forall\ n \in [1..Ti],\ \forall\ m \in [1..Tj],\ |\ f(i,n) - f(j,m)\ | \geq M[i,j]$$

• M[i,i] represents the channel separations necessary to satisfy the co-cell constraints:

$$\forall\ n,m \in [1..Ti],\ n \neq m,\ |\ f(i,n) - f(i,m)\ | \geq M[i,i]$$

• M[i,j]=0 means there is no constraint between the cells Ci and Cj.

A solution for the problem can be defined as S=<C1, C2, C3, ... CN> (Figure 3) and the fitness function may be defined as F: S→ {The number of interference constraints}. Each Ci is composed of Ti frequencies assigned to the Ci (Ci = <f(i,1), f(i,2), ... f(i,Ti) >) and F(S) is:

$$\sum_{i=1}^{N} \sum_{j=i+1}^{N} \sum_{k=1}^{T_i} \sum_{m=1}^{T_j} CA(i,j,k,m) + \sum_{i=1}^{N} \sum_{k=1}^{T_i} \sum_{m=k+1}^{T_j} CO(i,k,m)$$

where

CA(i,j,k,m) = 1   if |f(i,k) - f(i,p)| < M[i,j]
       = 0   Otherwise

and

CO(i,k,m) = 1   if |f(i,k) - f(i,p)| < M[i,i]
       = 0   Otherwise.

5. RESULTS

This section describes the performance of the GAP model to the FAP. The configuration is similar to Figure 2 with five GAPs except the memory length for each GAP is varied according to experimental requirements. Each experiment can be defined with three parameters:

nf : Number of frequencies available
nc : Number of cells
d : Density of matrix M.

The parameter d represents the interference from adjacent cells. It is proportional to the number of non-zero values in the matrix M. For all experiments the number of frequencies needed for each cell is limited to one $T_i=1$ ($i \in [1..N]$) which means $M(i,i)=0$ ($i \in [1..N]$). Furthermore all elements of matrix M are less than or equal to 1. This means that for two adjacent cells C_i and C_j the assigned frequency cannot be the same:

$$|f(i) - f(j)| \geq M(i,j)=1 \ (i,j \in [1..N]).$$

Table 1 shows the result of simulation for the FAP applied to the GAP. The first column shows the problem parameters. All simulations are averaged over 10 individual runs with the following parameters:

Fixed Parameters:

Number of generations =255.
Representation of the fitness value = 8 bits.
Population size = 50.
Probability of crossover = 80%.
Probability of mutation = 1%.

Variable Parameters:

nf = 4 or 8 or 16   and
nc = 50 or 75   and
d= 10% or 20%

Table 1 shows the difficulty of using the GAP to solve the problem if the number of available frequencies is equal to 4. On the other hand if the number of frequencies set to 16 then the GAP always can find the answer in less that 60 ms. For 8 frequencies it is better to use 50 cells only to get good results but the speed to find answer is still quite fast.

Table 1 The results of simulation for several configurations

nf	nc	d(%)	Succes	Time
4	50	10	84	135.1
4	50	20	78	131.1
4	75	10	76	163.8
4	75	20	65	129.4
8	50	10	99	97.6
8	50	20	94	85.6
8	75	10	93	66.9
8	75	20	85	74.9
16	50	10	100	2.7
16	50	20	100	21.7
16	75	10	100	60.9
16	75	20	97	59.1

6. CONCLUSIONS

In this research we have tested a model of a hardware genetic algorithms on the frequency assignment problem in cellular radio networks. the genetic algorithm processor can be constructed from a configuration of parallel processing units which enables it to handle many simultaneous variables. The simulation studies show that this configuration can find optimal frequencies with good reliability.

REFERENCES

[Hol92] Holland, J.H., "Adaptation in Natural and Artificial Systems", 2nd Edition, The MIT Press, Cambridge, Massachusetts, 1992.

[Dav91] Davis, L., "Handbook of Genetic Algorithms", International Thomson Publishing, New York 1991.

[SC95a] Salami, M., Cain, G., "Adaptive Hardware Optimization Based on Genetic Algorithms", Proceedings of The Eighth International Conference on Industrial Application of Artificial Intelligence & Expert Systems (IEA95AIE) , Melbourne, Australia , June

1995, pp. 363-371.

[SC95b]   Salami, M., Cain, G., "A PID Controller Based on a Multiple Genetic Algorithm Processor", The Proceedings of Control 95 Conference (Control'95), University of Melbourne, Melbourne, Australia , October 1995, pp. 359-362.

[Box78]   Box, F., "A Heuristic Technique for Assigning Frequencies to Mobile Radio Nets", IEEE Transaction on Vehicular Technology, VOL. VT-27, pp. 57-64, 1978

[GR82]   Gamst, A., and Rave, W., "On the Frequency Assignment in Mobile Automatic Telephone Systems", Proceedings of the IEEE Global Telecommunication Conference, GLOBECOM 1982, pp. 309-315, 1982.

[Kun91]   Kunz, D., "Channel Assignment for Cellular Radio Using Neural Networks", IEEE Transaction on Vehicular Technology, Vol. 40, pp. 188-193, 1991

[DH95]   Dorne, R., Hao, J.K., "An Evolutionary Approach for Frequency Assignment in Cellular Radio Networks", Proceedings of The IEEE International Conference on Evolutionary Computing (ICEC'95), The University of Western Australia, Perth, Australia, November 1995, pp. 539-545.

SELECTION OF REPRESENTATIVES USING GENETIC ALGORITHMS

Y. Hamamoto, Y. Tsuneta, C. Kanayama and S. Tomita

Faculty of Engineering, Yamaguchi University,

Ube, 755, Japan.

Email: hamamoto@csse.yamaguchi-u.ac.jp

ABSTRACT

A representative selection technique based on genetic algorithms is proposed. The performance of the nearest neighbor classifiers based on the technique is demonstrated in the small-sample, high-dimensional setting. The proposed technique offers significant advantages over conventional ones in terms of both the classification accuracy and data reduction rate.

1. INTRODUCTION

In practice, one often must deal with problems of non-normal distributions. In such situations, the use of nonparametric classifiers is recommended because they provide much high classification accuracy than those achieved by parametric classifiers. However, nonparametric classifiers have two serious disadvantages: one is their computational complexity and the other is the requirement for a very large amount of computer storage to retain the training set [Das 91]. One way to overcome these disadvantages is to select the small number of representatives from the training set.

The main problem when selecting representatives is computational complexity. For each class, consider a problem of selecting m representatives out of N samples. The best subset of m representatives out of N may be found by evaluating a criterion for all possible combinations of m representatives. However, the number of all possible combinations, i.e., $\binom{N}{m}$, becomes prohibitive even for modest values of N and m. Therefore, we use the Genetic Algorithms (GAs) [Gol 89] to avoid the exhaustive search. GAs are a class of search methods deeply inspired by the natural process of evolution. The proposed technique offers significant advantages over conventional ones in terms of both the classification accuracy and data reduction rate.

2. BOOTSTRAPPING

In our method, a new bootstrapping technique is first applied to the original training set of each class. Bootstrapping is similar to other resampling schemes such as cross-validation and jackknifing. There could be many possible ways to generate bootstrap samples. We will focus on the generation of the bootstrap samples. Assume that the priori probabilities of all classes are equal. Let $S_{N_i} = \{x_1^i, x_2^i, \cdots, x_{N_i}^i\}$ be an original training set from class ω_i where N_i denotes the training sample size of class ω_i. A bootstrap set $S_{N_i}^B$ of size N_i is generated as follows:

1. Select a sample $x_{k_0}^i$ from S_{N_i} randomly.

2. Find the r nearest neighbor samples $x_{k_1}^i, x_{k_2}^i, \cdots, x_{k_r}^i$ of $x_{k_0}^i$, using the Euclidean distance measure.

3. Compute a bootstrap sample $x_{i1}^b = \sum_{j=0}^r w_j x_{k_j}^i$, where w_j is a weight. The weight w_j is given by

$$w_j = \frac{\Delta_j}{\sum_{k=0}^r \Delta_k} \qquad 0 \le j \le r,$$

where Δ_j is chosen from a uniform distribution on $[0, 1]$.

4. Repeat steps 1., 2. and 3. N_i times to get $S_{N_i}^B = \{x_{i1}^b, x_{i2}^b, \cdots, x_{iN_i}^b\}$.

The most important difference between our bootstrapping and the ordinary bootstrapping [Efr 79] lies in the manner in which bootstrap samples are generated. In our bootstrapping, the bootstrap samples are created (not selected) by linearly combining the original training samples.

A 1-NN classifier with bootstrap samples is defined as follows:

Classify x into class ω_i if

$$\min_j[d(x, x_{ij}^b)] \leq \min_j[d(x, x_{kj}^b)] \qquad \text{for all } k \neq i,$$

where $d(\cdot, \cdot)$ denotes the Euclidean distance measure. In this bootstrapping, the selection of r is crucial, which depends on the training sample size N_i and on the distribution structures. From preliminary experiments, we recommend the use of the small values of r for non-normal distributions.

3. APPLYING GA TO REPRESENTATIVE SELECTION

In this section, we propose a technique of selecting representatives for the 1-NN classifier with bootstrap samples by using GAs. To simplify the notation, we omit the index representing class label. We describe the algorithm below.

1. Select the value of r.

2. Generate bootstrap samples.

3. Using GAs, find a subset of m representatives (bootstrap samples) for each class so that the estimated error rate is minimized.

4. Repeat step 3. five times.

5. Repeat steps 2.-4. changing the value of r.

6. Out of solutions obtained by GAs, find one which minimizes the error rate.

Next, we describe GAs in details. One needs to define the fitness function. In the genetic algorithm, a solution is represented by a chromosome. Here, each chromosome, which represents a subset of representatives for a class, is given by

$$G = \{g_1, g_2, \cdots, g_m\},$$

where g_i denotes the presence of the g_i-th bootstrap samples $x_{g_i}^b$. This coding, which comes from [Gre 85], was applied to the traveling salesman problem. For each class, a chromosome is given. This means that for the L-class problem, each individual has L chromosomes. By encoding the chromosomes corresponding an individual X_i, we can obtain m representatives for each class.

Note that representatives are selected from the bootstrap sample set of each class. First, the number of training samples misclassified by the 1-NN classifier

with the selected representatives is counted. Next, the error rate $Er(X_i)$ is estimated by dividing the number of the misclassified training samples by the total number of training samples. Note that bootstrap samples are independent of training samples. It should be stressed that we are not using an apparent error rate, but rather the bootstrap-based error rate. It is well known in the pattern recognition field that the apparent error rate is heavily biased [Fuk 90]. On the other hand, the bootstrap estimators can provide much smaller biases than the apparent error rate [Efr 86]. Therefore, the bootstrap-based error rate can be used as reasonable one of the true error rate. Using the bootstrap-based error rate $Er(X_i)$, the fitness function to be maximized is defined as follows:

$$J(X_i) = \{100 - Er(X_i)\}^2.$$

Moreover, the averaged fitness function is given by

$$\hat{J} = \frac{1}{T} \sum_{i=1}^{T} J(X_i),$$

where T denotes the population size. Using $\hat{J}$, the following ratio is defined:

$$\alpha = \frac{\hat{J} \text{ in the current generation}}{\hat{J} \text{ in the previous generation}}.$$

Now, we present a GA algorithm for selecting representatives.

1. Construct an initial population set $Q = \{X_i\}_{i=1,\cdots,T}$. Compute both J and $\hat{J}$ for Q.

2. Construct a population set M by selecting T individuals from Q according to the following probability $P(X_i)$:

$$P(X_i) = \frac{J(X_i)}{\sum_{k=1}^{T} J(X_k)}.$$

3. Select a pair X_j and X_k from M and do one point crossover with probability P_c.

4. Mutate M with probability P_m.

5. Compute J for M.

6. Update Q by selecting the best T individuals from $Q \cup M$ according to the fitness value J.

7. Compute $\hat{J}$ for Q.

8. Terminate if the number of generations exceeds the maximum number of generations or if $1.000 \leq \alpha \leq 1.005$. Otherwise step 2.

4. EXPERIMENTAL RESULTS

In general, when a fixed number of training samples is used to design a classifier, the error of the classifier tends to increase as the dimensionality of the data gets large. It is widely known that when the dimensionality of the data is high, the error rate of the 1-NN classifier seems to be biased severely [Fuk 90].

In order to study the performance of the GA-based NN classifier in high-dimensional spaces, we adopted the Ness data set [Nes 80]. This data set was used in [Nes 80] to study the performances of the linear, quadratic and Parzen classifiers in high-dimensional spaces. The available samples were independently generated from n-dimensional Gaussian distribution $N(\mu_i, \Sigma_i)$ with the following parameters:

$$\mu_1 = [0, \cdots, 0]^t, \qquad \mu_2 = [\Delta/2, 0. \cdots, 0, \Delta/2]^t$$
$$\Sigma_1 = I_n, \qquad \Sigma_2 = \begin{pmatrix} I_{n/2} & \mathbf{0} \\ \mathbf{0} & \frac{1}{2}I_{n/2} \end{pmatrix}$$

where Δ is the Mahalanobis distance between class ω_1 and class ω_2, and I_n denotes the $n \times n$ identity matrix. Note that in this data set, both the mean vectors and covariance matrices differ. The Bayes error varies depending on the value of Δ as well as n.

The purpose of the first experiment is to study the influence of both population size T and mutation rate P_m on the error rate. The following experiment was performed.

No. of classes:	2
Dimensionality n:	50
No. of training samples N:	100 per class
No. of test samples:	1000 per class
Values of Δ:	2, 6
Values of r:	$N/4, N/2, N-1$
Population size T:	10, 20, 30, 40, 50
Mutation rate P_m:	0.05, 0.10, 0.15
Crossover rate P_c:	1.0
No. of representatives:	5 per class
No. of trials:	100

Note that the training sample size is small for the dimensionality. It is recommended in general that the number of training samples per class should be at least five to ten times the dimensionality [JC 82]. All experimental results, which are averaged over 100 trials, are shown in Table 1. The resulting error rates were much less sensitive to the values of T and P_m. In all our experiments, GAs converge to a solution in the small number of iterations. The population size T mainly determines the computational time for the GA. As T increases, the computational time becomes large. As

shown in FIGURE 1, the fitness value obtained for $T = 30$ was comparable to one for $T = 50$. Hence, we selected $T = 30$ and $P_m = 0.1$ for this problem.

Next, in order to compare the proposed technique with other editing techniques such as the condensed nearest neighbor rule [Har 68] and reduced nearest neighbor rule [Gat 72] in terms of the error rate, the following experiment was performed.

No. of classes:	2
Dimensionality n:	50
No. of training samples N:	100 per class
No. of test samples:	1000 per class
Values of Δ:	2, 6
Values of r:	$N/4, N/2, N-1$
Classifiers:	the GA-based NN classifier ($T = 30$, $P_c = 1.0, P_m = 0.1$), CNN classifier [Har 68], RNN classifier [Gat 72], 1-NN classifier
No. of trials:	100

Results are shown in FIGURE 2. For comparison, the error rate of the 1-NN classifier is also presented. The GA-based NN classifier significantly outperforms the conventional 1-NN classifier as well as both CNN and RNN classifiers in terms of the classification accuracy. This advantage comes from the use of the bootstrap samples. It is widely believed that classifier performance can be improved by removing outliers. Our bootstrap technique is an effective means of removing outliers. In addition to the classification accuracy, unlike CNN and RNN techniques, this technique can provide high data reduction rates. Note that in CNN and RNN techniques, the number of representatives is itself a random variable, because it is a function of the training samples. On the other hand, in this technique, the data reduction rate is completely under the control of the algorithm.

5. CONCLUSIONS

We have proposed a technique of selecting representatives for the 1-NN classifier with bootstrap samples using GAs. The performance of the resulting classifier is demonstrated in the small-sample, high-dimensional setting. Experimental results show that the proposed technique offers significant advantages over conventional techniques such as CNN and RNN techniques in terms of both the classification accuracy and data reduction rate.

Table 1   Influence of the population size T and mutation rate P_m on the error rate[%].
The first line is the mean of the error rate, and the second line is the 95% confidence interval.

(a) $\Delta = 2$

P_m	10	20	30	40	50
0.05	28.7	28.8	29.0	29.3	29.1
	(28.3,29.1)	(28.5,29.2)	(28.6,29.3)	(28.9,29.8)	(28.8,29.5)
0.10	28.8	28.9	29.0	28.9	28.8
	(28.5,29.1)	(28.5,29.3)	(28.6,29.3)	(28.5,29.3)	(28.5,29.2)
0.15	28.9	28.6	28.9	29.0	28.9
	(28.6,29.2)	(28.3,29.0)	(28.6,29.3)	(28.7,29.4)	(28.5,29.2)

(b) $\Delta = 6$

P_m	10	20	30	40	50
0.05	1.99	1.95	2.07	1.94	1.91
	(1.91,2.07)	(1.89,2.02)	(1.98,2.15)	(1.88,2.00)	(1.83,1.99)
0.10	1.92	2.00	1.98	1.96	2.04
	(1.84,2.00)	(1.90,2.09)	(1.91,2.06)	(1.89,2.03)	(1.96,2.12)
0.15	1.97	2.00	1.94	1.97	1.96
	(1.89,2.05)	(1.91,2.08)	(1.85,2.02)	(1.90,2.05)	(1.88,2.04)

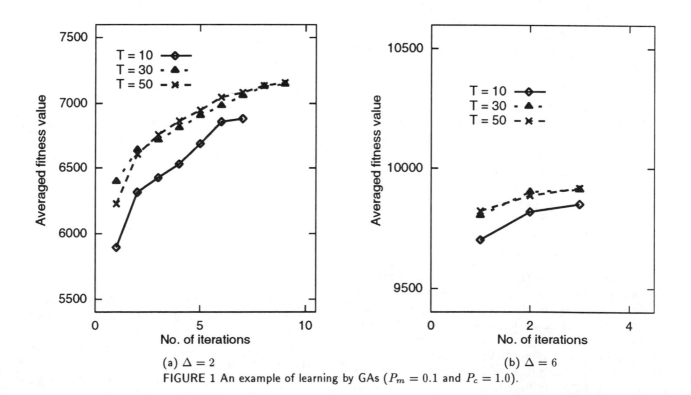

(a) $\Delta = 2$ (b) $\Delta = 6$
FIGURE 1 An example of learning by GAs ($P_m = 0.1$ and $P_c = 1.0$).

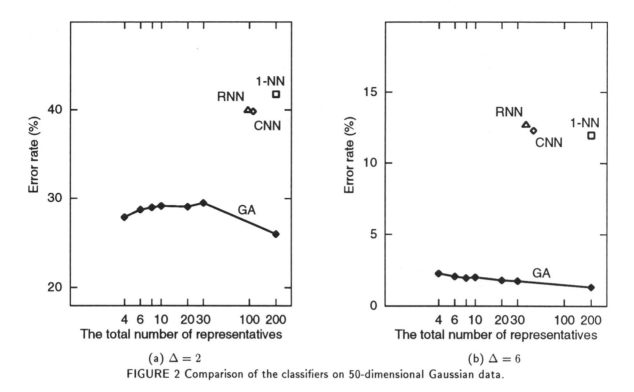

(a) Δ = 2   (b) Δ = 6

FIGURE 2 Comparison of the classifiers on 50-dimensional Gaussian data.

REFERENCES

[Das 91]   B. V. Dasarathy. Nearest neighbor pattern classification. the IEEE Computer Society, 1991.

[Efr 79]   B. Efron. Bootstrap methods: Another look at the jackknife. Ann. Statist., 7, pp. 1-26, 1979.

[Efr 86]   B. Efron. How biased is the apparent error rate of a prediction rule?. J. Amer. Statist. Assoc. 81, pp. 461-470, 1986.

[Fuk 90]   K. Fukunaga. Introduction statistical pattern recognition. 2nd Edition, Academic Press, 1990.

[Gat 72]   G. W. Gates. The reduced nearest neighbor rule. IEEE Trans. Inform. Theory, IT-18, pp. 431-433, 1972.

[Gol 89]   D. Goldberg. Genetic Algorithm in Search Optimization and Machine Learning. Addison-Wesley, Reading, MA, 1989.

[Gre 85]   J. J. Grefenstette et al.. Genetic algorithm for the traveling salesman problem. Proc. 1st. Int. Conf. Genetic Algorithms, 1985.

[Har 68]   P. E. Hart. The condensed nearest neighbor rule. IEEE Trans. Inform. Theory, IT-14, pp. 515-516, 1968.

[JC 82]   A. K. Jain and B. Chandrasekaran. Dimensionality and sample size considerations in pattern recognition practice. In P. R. Krishnaiah and L. N. Kanal, editors, Classification Pattern Recognition and Reduction of Dimensionality, pp. 835-855, North-Holland Publishing Company, 1982.

[Nes 80]   J. Van Ness. On the dominance of nonparametric Bayes rule discriminant algorithms in high dimensions. Pattern Recognition, 12, pp. 355-368, 1980.

USING THE BALDWIN EFFECT TO ACCELERATE A GENETIC ALGORITHM

John R. Podlena and Tim Hendtlass
Centre for Intelligent Systems,
School of Biophysical Sciences and Electrical Engineering,
Swinburne University of Technology,
P.O. Box 218 Hawthorn 3122.
Email : jrp@bsee.swin.edu.au & tim@bsee.swin.edu.au

ABSTRACT

The standard Genetic Algorithm, originally inspired by natural evolution, has displayed its effectiveness in solving a wide variety of complex problems. This paper describes the use of the natural phenomenon known as the "Baldwin Effect" (or cross-generational learning) as an enhancement to the standard Genetic Algorithm. This is facilitated via the use of artificial neural networks to store aspects of the population's history, and through the use of elitism in conjunction with a local heuristic. It also describes a method by which the negative side effects of a large elite sub-population can be counter-balanced by using an aging coefficient in the fitness calculation. The new algorithm is then tested on three standard genetic algorithm test functions, and on one commercial application of the genetic algorithm.

1. INTRODUCTION

The genetic algorithm (GA) is a paradigm which is based on both gene recombination and the Darwinian "survival of the fittest" theory. A group of individuals in some environment have a higher probability to reproduce if their fitness (their ability to thrive in this environment) is high. Offspring are created via a crossover operation (as in gene recombination) and mutation. The algorithm, described by Holland [Hol92], aims to increase the average fitness of the individuals over a number of generations.

In nature, evolution in higher forms of life also includes an accelerating feature known as the "Baldwin effect" [Bal]. This is the effect of passing down the learning within a generation's current population to the next generation. This collective learning is then used for greater chances of survival and for the acceleration of the population's genetic material into aquiring a generally more "fit" classification. One example of this is human development, which in a relatively very short time period has accelerated extremely fast in evolutionary terms, largely due to our ability to pass on learning to future generations.

The Baldwin effect works by effectively smoothing the surface of the genetic variety versus fitness landscape. An individual which is highly fit due to its genetic make-up in comparison to its neighbours forms a very high and thin plateau in this landscape, but with lifetime learning, and successive generations having "knowledge" of the environment, the plateau is transformed into a sloping hill. This also works in reverse, sudden pitfalls in terms of fitness transform into valleys. Multi-generational learning can also act as a kind of mapping, memory of good and bad regions of the environment/search space used and updated by successive generations to act as points of attraction or repulsion.

2. THE ADAPTED GENETIC ALGORITHM

The basic GA was modified in a number of ways to permit the creation of the Baldwin effect. Each evaluation of the cost function is used in the update of the **history network**, which is subsequently used in the breeding function as a breeding filter.

For all individuals:
- initialise population and history networks

Until error of best individual is below some minimal value Loop:
- use local heuristic on population
- calculate fitness based on cost and age
- update history network

- calculate probability for parenting new generation
- copy 'm' of the best individuals into new population (elitism)
- increment the age of replicated individuals
- breed individuals using history filter to fill next generation pool

EndLoop

3. THE HISTORY NETWORK

The history network consists of an artificial neural network called a "Cluster Network" [HW93], which is derived from the Linear Vector Quantisation (LVQ) network [Koh88] and the Probabilistic Neural Network [Spr88,Spr90]. However, the network differs in that there are continuously varying numbers of examples of each "class" and the procedure for updating the reference examples differs from either of these networks. Unlike the LVQ and PN networks, this network continuously learns. In this application, the network must classify a set number of "regions" within the problem search space, which are labelled either "good" or "bad" regions. The number of regions was set in all our experiments equal to the population size (this parameter has not yet been investigated). A single region is stored as a vector into the search space. The algorithm for training proceeds as follows:

for 'n' individual's per generation with the worst or best cost evaluation:

- find the closest vector end point in the history network to the point of this individual's genetic coordinate.

if distance between the two is < UpdateThreshold **and** winner is of same type (good/bad) **then**

- update the closest vector using n% of the old network vector and (100-n)% of the current individual's coordinate,

ie. $\bar{v}_{new} = (100 - n) \times \bar{v}_{win} + n \times \bar{v}_{old}$

else

- replace the vector least used in the breeding function with the new individual's coordinate as the vector's endpoint.

The update percentages used were found empirically, and relate directly to the decay of memory in the network over generations. In this way, as the number of generational iterations increases, the network will operate as a "mapping" of the good and bad regions in the search space, updated as new information becomes available. This mapping can then be used to filter potential new individuals depending on their genetic coordinate location in respect to the different good/bad regions on the map.

It is worth noting that research into the combinational use of genetic algorithms and neural networks has almost exclusively been restricted to the "evolution" of neural networks [IWC92]. Our review uncovered no previous use of a neural network in the classification of features within a GA's problem space.

4. THE LOCAL HEURISTIC

The Local Heuristic used for each test problem was the Directed Random Search (DRS), which alters the coordinates of an individual by the sum of a directed component and a random component. The directed component, the magnitude of which is controlled by a variance, is in the direction of the last successful change. The DRS then tests the individual's performance using the cost function, and if an improvement is found the individual takes on this new coordinate and the variance is increased, else the same magnitude in the opposite direction is checked with the same consequences. If neither change is successful variance is decreased. A fixed number of iterations of this local heuristic was used on each of the individuals in a given generation, thus the next generation would benefit not only from the information stored in the history network, but also that stored in the locally converged parent genes (a concept known as lamarckianism[AL93]).

5. BREEDING STRATEGY

The breeding strategy used in all trials of the enhanced algorithm contains two of the fundamental genetic algorithm operators, crossover and mutation. In all cases, real as opposed to binary numbers were used to form the chromosomal strings. The crossover method used was typical for real numbers, using either average or uniform crossover over single numbers, or uniform crossover across floating point numbers [JM91] for any given new individual's production. Mutation was restricted to a "creep" mutation [Dav91].

After each single individual was produced using the crossover and/or mutation operators, its genetic coordinates are run through the history network to discover which vector/classification it is closest to (the winning node). This individual's survival is then dependant on the following rules:

if the winning node classifies a bad region **and** (distance is < BadThreshold **or** RejectProbability < 0.5) **then**

- do not accept individual

elseif the winning node classifies a good region **and** (distance is < GoodThreshold **or** RejectProbability < 0.9) **then**

- do not accept individual

else

- accept individual

The algorithm can be summarised as the following rule; if an individual is too close to a known bad or good region, then reject it. The thresholds effectively control how close future generations can get to a known position without being removed (ie. the rejection probabilities). Thus genetic diversity is increased and subsequent generations are less likely to "repeat history" by re-exploring regions. The BadThreshold was found to be best set to 25% of the creep mutation, with the GoodThreshold value set to twice the desired solution precision (this effectively ensured that new individuals could be much closer to the good references than to the bad). These initial threshold values could then be decayed as the genetic algorithm converged in on a solution (ie. variance of the population $\propto$ tThresholds and mutation). The threshold values do not inhibit the local heuristic moving the individual towards a history network node. The disturbance of the history filter on the genetic algorithm's use of schema to perform hyper-plane sampling [Dav91] is therefore inherently limited by the thresholds, removing all but a few of the evaluations for those chromosomal combinations which are within the radius of the thresholds only.

It could be argued that the breeding filter (or selection procedure) as described above, may have a disruptive effect on the genetic algorithm's ability to converge due to crossover operating on a generally more diverse set of parents. The benefits for always copying 'm' of the most fit parents directly into the new generation (known as elitism, this is usually restricted to one or two individuals) are therefore two fold. Through elitism, information from the best parents is not lost via unfortunate crossover combinations. Secondly, the best individuals have a chance to make progress based on their local heuristic over a number of generations. However one of the dangers of such an extended elitist strategy is the possibility of highly trained parents that are stuck in local minima forever being copied directly into the next generation. To encourage diversity, each parent in this algorithm has an associated "age penalty" coefficient which effectively reduces the parents fitness to breed with time. This age penalty is implemented as a fitness modifier, where the resulting fitness is equal to:

$$fitness' = fitness - \left(fitness - \frac{\sum pop\_fitness_i}{pop\_size} \right) \times \frac{Age}{100}$$

where **fitness'** is the modified fitness, **fitness** is the individual's original fitness, and $\sum \frac{pop\_fitness_i}{pop\_size}$ is the average (unmodified) fitness of the population.

The *Age* value is incremented for all replicated individuals each generation. *Age* therefore effectively controls the number of iterations an elite individual survives. The strategy described above for parent retention differs from the more general forms of the genetic algorithm. In one method, called the Generational Replacement Genetic Algorithm (GRGA), the new generation replaces all the previous individuals (except those parents not crossed over). Another common method, Steady State Genetic Algorithm (SSGA), replaces only one or two members of the population. Results from the variation of the number of 'm' individuals retained from the previous generation are outlined in previous work [PH95].

Research into the disruptive effects of other simpler selection procedures with elitism, specifically on the implicit parallelism of the genetic algorithm [Esh89], has shown that no negative side effects were noticed on the tests performed [Dav91,SCE⁺89]. It was with this in mind that we proceeded to use three of the same functions to those used in the above research to test our algorithm.

6. THE TEST PROBLEMS

Four optimisation test problems were used on the modified algorithm. The first three are the Rozenbrock equation, F6 and F7 [14], which prove very difficult for hill climbing algorithms, and also difficult for a genetic algorithm with a restricted mutation radius . To successfully find the optimal solution for the test functions F6 and F7, the standard genetic algorithm becomes very reliant on the mutation operator. To further handicap the initial population, the "random" selection of individuals was permitted only to be outside a certain radius of the optimal solution point.

The functions tested were:

Rozenbrock: $100\left(x - y^2\right)^2 + (y - 1)^2$

F6: $0.5 + \dfrac{\sin^2 \sqrt{x^2 + y^2} - 0.5}{\left[1.0 + 0.001\left(x^2 + y^2\right)\right]^2}$

F7: $\left(x^2 + y^2\right)^{0.25}\left[\sin^2\left(50\left(x^2 + y^2\right)^{0.1}\right) + 1.0\right]$

These are shown in figures 1,2,3 and 4.

figure 1: *Rozenbrock's Saddle*

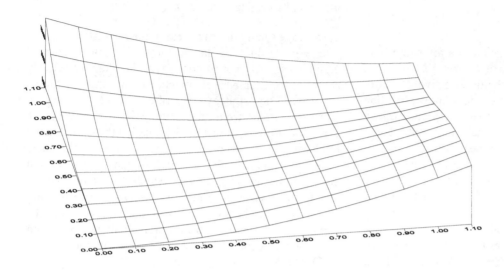

figure 2: *X values when Y is at optimum for F6*

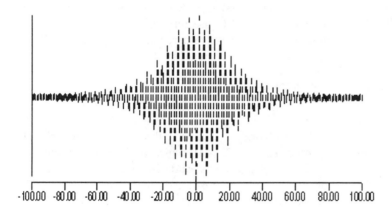

figure 3: *X values when Y is at optimum for F7*

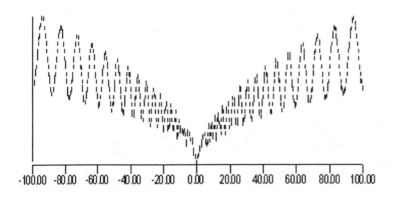

figure 4: *F7 Contour (centre is optimal)*

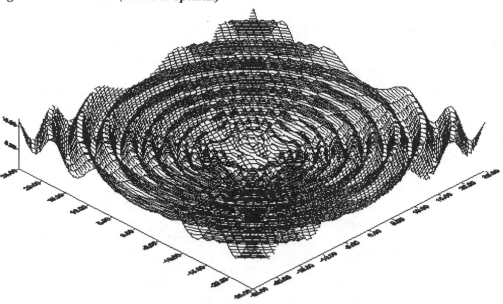

The three functions shown above map into three dimensions, and are therefore easier to understand, since the problem surface can be displayed. With this in mind we chose for the final test problem a commercial application of the GA which uses a complex cost function with five input parameters to solve an optimisation problem. The existing implementation was a hybrid implementation whereby an existing rule based scheduling algorithm became the cost function evaluator, with the GA population consisting of individual chromosomes each with the five input parameters set as genes. Thresholds were set based on the creep mutation radius and the precision of the answer required. The problem was an interesting test function due to its quantised search space, created by the rule based algorithm it consists of, and due to the limits on population size and number of iterations imposed by its "real world" use.

7. RESULTS

Table 1: Results	Rozenbrock	F6	F7	Optimiser
Number of History wins	2664	2040	2312	13
Number of History losses	325	925	668	8
Draw	11	35	20	6
average GA generations for optimal solution without History	113	96	165	6
average GA generations for optimal solution with History.	65	82	104	5.4

The test results are displayed in table 1. The "Number of History Wins" describes the number of times the history GA reached the optimal solution in less generations than the standard GA given the same initial population. In a similar way "Number of History Losses" and "Draw" were calculated. The average number of GA iterations for convergence for both the runs using History and those without are also displayed.

On the Rozenbrock saddle function, the algorithm quickly mapped out the valley displayed in figure 1. By filtering out most of the individuals closest to the worst positions on the slopes, and retaining those closer to the good solutions (but not too close), the algorithm quickly converged. On the F6 and F7 functions, the algorithm quickly mapped out the smaller outer valleys (figure 4), history filtering enabling more diverse individuals to be accepted, and thus enabling the algorithm to quickly jump the maximum cost bands to reach the optimum. Another interesting feature was when the GA was close to the optimum, but the mutation radius was too large for it to converge quickly. This problem was quickly overcome by our algorithm due to the history network reducing the search space significantly after a few generations in the same region.

The results on the final problem, the commercial application of the GA, displayed a slightly quicker convergence when using our enhanced GA in comparison to the standard GA runs. Since the number of iterations is restricted to 12, at least 25% of the run time is spent setting up the history mapping of the function, at which time the effects on the population begin to separate from the standard GA given the same starting populations.

8. CONCLUSION

The enhanced algorithm shows promise in its ability to effectively accelerate the convergence of the typical GA onto an optimal solution. On average, our algorithm produced convergence times 31% faster than the conventional GA. The amount of iterations our algorithm won by were found to be much higher than the amount of iterations it lost by in competition with the standard GA. The algorithm also performs reasonably well on a commercial application of the standard GA, on average speeding up convergence by 10%. The Baldwin effect reproduced via the history network and Lamarckian learning is the driving force behind these improvements. The algorithm shows its maximum usefulness when the cost function evaluations are taking a large amount of computer time (such as with the last test problem), as the filtering process need not call the cost function. Future testing and enhancements will include use of the algorithm on binary representations (using hamming distances) and on other higher dimensional cost functions. A distributed Baldwin enhanced GA is also envisaged which will have sub-populations each for a different cost factor, the combined mapping then being used to find the best individual for the combination of all the cost factors.

[Hol92]   John H. Holland, *Adaption in Natural and Artificial Systems*, second edition, MIT Press, 1992.

[Bal]   J.M, Baldwin, A New Factor in Evolution, *American Naturalist*, 30, pp441-451.

[HW93]   Tim Hendtlass and Janice Wells, The Discrimination of Short Acoustic Events by Artificial Neural Networks, *Proceedings of ANNES, New Zealand*, IEEE Press, 1993.

[Kow88]   T. Kohonen, Statistical Pattern Recognition with Neural Networks: Benchmark Studies, *Proceedings of the Second Annual IEEE International Conference on Neural Networks*, Volume 1, 1988.

[Spr88]   D. Sprecht, Probabilistic Neural Networks for Classification, Mapping and Associative Memory, *ICNN-88 Conference Proceedings*, 1988.

[Spr90]   D. Sprecht, Probabilistic Neural Networks, *Neural Networks*, November 1990.

[IWC92]   *International Workshop on Combinations of Genetic Algorithms and Neural Networks,* COGANN-92, IEEE Computer Society Press, California, 1992.

[AL93]   D. H. Ackley & M. L. Littman, *A case for Lamarkian evolution*. In C. G. Langton (Ed.), Artificial Life III, Reading, MA: Addison-Wesley, 1993.

[JM91]   Cezary Z. Janikow & Zbigniew Michalewicz, An Experimental Comparison of Binary and Floating Point Representations in Genetic Algorithms, *Proceedings of the Fourth International Conference on Genetic Algorithms*, pp 31-36, California, July 1991.

[Dav91]   Lawrence Davis, *Handbook of Genetic Algorithms*, Van Nostrand Rienhold, New York, 1991.

[PH95]   John R. Podlena and Tim Hendtlass, Evolving Complex Neural Networks That Age, *Proceedings of the IEEE International Conference on Evolutionary Computing*, 1995.

[Esh89]   Larry J. Eshelman, The CHC Adaptive Search Algorithm: How to Have Safe Search When Engaging in Nontraditional Genetic Recombination, in D.E. Goldberg (ed.), *Genetic Algorithms in Search, Optimisation, and Machine Learning*, Addison-Wesley Publishing Company, Reading, Mass. 1989.

[Dav91]   Lawrence Davis, Bit-Climbing, Representational Bias, and Test Suite Design, *Proceedings of the Fourth International Conference on Genetic Algorithms*, pp 18-23, California, July 1991.

[SCE⁺89] J. David Schaffer, Richard A. Caruana, Larry J. Eshelman and Rajarshi Das, A Study of Control Parameters Affecting Online Performance of Genetic Algorithms for Function Optimization, *Proceedings of the Third International Conference on Genetic Algorithms*, pp 51-60, California, June 1989.

Intelligent Tutoring

GENERATING COURSES IN AN INTELLIGENT TUTORING SYSTEM

Roger Nkambou, Marie-Claude Frasson, Claude Frasson
Département d'Informatique et de Recherche Opérationnelle, Université de Montréal
Montréal (Québec), H3C 3J7 CANADA
nkambou@iro.umontreal.ca

ABSTRACT

In this paper, we present a knowledge-based system intended to support automatic course generation for a particular target group. This generation is realized by a reasoning process on the curriculum and is guided by the set of knowledge which should be acquired by the learner. The idea is to produce a modifiable course graph that would allow to control the evolution of a learner's cognitive structure. The course produced is directly exploitable by an intelligent tutoring system (ITS) or by a human professor.

1. INTRODUCTION

Industrial training is time consuming and expensive. For instance, IBM US spends $2 billion a year on training including $1 billion for trainers' salaries; Each year, the US government spends $20 billion on military training [MW93]. Part of this training cost is relied to the instructional design and to the teaching process. Thus, a system that permits to rapidly build a course according to some needs could contribute to decrease these costs. Researches have been made to ease this process. Systems such as ISD Expert [Mer87], ISD Expert [Ten93], the GAIDA project [Gag93] have been specified to that effect. These systems offer a set of tools that are used in the instructional design process. However, the course building process still remains an explicit task of the instructional designer.

The system we are presenting in this paper uses design expertise to automatically generate a course from a set of parameters specified by the designer. The main ones are the expected knowledge to be acquired by the course and the target public which will be taking it. The course can be adapted by the designer who can also test it with a simulated student. The final course is an object that can be directly used by an intelligent tutoring system or by a human teacher.

In order to build such a system, we use a subject-matter (curriculum) model called CREAM[1] [NG96]. In this model, a curriculum is represented and organized according to three points of view: the domain knowledge (through the capability model (CREAM-C)), the peda-

gogical aspect (through the objective model (CREAM-O)) and the didactic point of view (through the resource model (CREAM-R)). These three models are combined in a transition network structure called CKTN (Curriculum Knowledge Transition Network).

After a brief presentation of the subject-matter model, we present the problematic of course generation, the generation process we developed and how CREAM sustains it.

2. CREAM: A SUBJECT MATTER REPRESENTATION MODEL FOR ITS

We consider the curriculum in ITS as a structured representation of the subject matter in terms of capabilities [GBW92], instructional objectives and pedagogical resources (learning materials). Achievement of instructional objectives contributes to the acquisition of capabilities. This achievement is supported by pedagogical resources through learning activities (exercises, demonstrations, problems, simulations...). CREAM implements domain, pedagogical and didactic aspects of a subject matter through a network organization of capabilities, of instructional objectives defined on these capabilities and of pedagogical resources supporting the accomplishment of instructional objectives. Using these three knowledge structures, we construct a curriculum knowledge transition model which contains particulars links between their elements.

2.1. THE CAPABILITY MODEL

A capability is a knowledge (or cognitive) unit stored in a person long term memory that allows him to succeed in the realization of physical, intellectual or professional activity. When we want to represent this kind of curricular knowledge, we ask ourselves what must be taught, that is the content we want the student to acquire. Three categories of capabilities are set up in CREAM: verbal information, intellectual skills (discriminations, concepts and rules) and cognitive strategies. The capability model (CREAM-C) is a multi-graph where nodes are capabilities (each type of node denotes a capability category) and links among them can be of several types: analogy, generalization-specialization, abstraction, aggregation, deviation.

---

[1] Curriculum REpresentation and Acquisition Model

2.2. INSTRUCTIONAL-OBJECTIVE MODEL

An instructional objective is the description of the behavior (or performance) that the student must demonstrate following a learning process. Several studies have shown the necessity of objective specification in teaching systems ([Les88]; [Web94]). We have taken this into account in our representation approach by introducing an instructional objectives model (CREAM-O) in which instructional objectives are represented and organized together by links of type *prerequisite, complementary or pretext*. These links can be defined by the instructional designer or generated by the curricular system by reasoning on its knowledge structures [NG96].

2.3. LEARNING RESOURCE MODELLING

Pedagogical resources are the means used by the teaching system to support the teaching/learning process. Several categories of resources are considered in CREAM: Those that support the teaching/learning process for the acquisition of knowledge (problems, demonstrations, exercises, hypermedia document...), *the expert type resources* that act as experts that can intervene in the teaching/learning process to help or critic the student in a specific activity (advisor, criticising systems, coach) and *physical and media resources* that represent basic teaching material (simulators, video, sound, pictures...). We have identified several types of links between resources: equivalence, abstraction, case of, use of, and auxiliary. The network defined by these relations represents the resource model (CREAM-R).

2.4. CKTN: COUPLING THE PRECEDING MODELS

The coupling of the preceding models creates special links between their elements. In the space resulting from this coupling, we group each objective together with its associated resources in an element we call a *transition*. The *input capabilities* of a transition are those that are prerequisites to the achievement of the objective involved in this transition. The *output capabilities* of a transition are those that are produced by the achievement of the objective involved in the transition. Thus, they produce two types of links between capabilities and objectives: *prerequisites* and *contribution* links. The resulting network is called CKTN (Curriculum Knowledge Transition Network). Figure 1 shows part of CKTN in the Baxter pump manipulation domain. Note that, for the generation process, we will only consider the objective included in the transition. The transition itself will be taken into account at the time of the teaching process. A *prerequisite* link from a capability C to an objective O expresses the fact that C is a precondition to the realization of O. A prerequisite link is caracterized by its nature (*mandatory, optional*) and by the minimum mastery level required on the source capability to be able to consider it sufficient

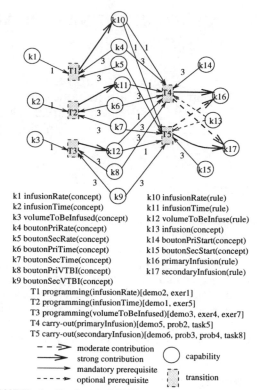

k1 infusionRate(concept)
k2 infusionTime(concept)
k3 volumeToBeInfused(concept)
k4 boutonPriRate(concept)
k5 boutonSecRate(concept)
k6 boutonPriTime(concept)
k7 boutonSecTime(concept)
k8 boutonPriVTBI(concept)
k9 boutonSecVTBI(concept)

k10 infusionRate(rule)
k11 infusionTime(rule)
k12 volumeToBeInfuse(rule)
k13 infusion(concept)
k14 boutonPriStart(concept)
k15 boutonSecStart(concept)
k16 primaryInfusion(rule)
k17 secondaryInfusion(rule)

T1 programming(infusionRate)[demo2, exer1]
T2 programming(infusionTime)[demo1, exer5]
T3 programming(volumeToBeInfused)[demo3, exer4, exer7]
T4 carry-out(primaryInfusion)[demo5, prob2, task5]
T5 carry-out(secondaryInfusion)[demo6, prob3, prob4, task8]

- - -> moderate contribution   ◯ capability
———> strong contribution
———> mandatory prerequisite
- - - -> optional prerequisite   ▦ transition

FIGURE 1: Part of CKTN on the Baxter Pump

for overstepping the link (*entry level*) and thus be able to eventually realize the objective. This entry level is specified by using a qualifying value taken from an evaluation (or acquisition) vocabulary. For example, in Klausmeier's vocabulary [Kla90], a capability of type "concept" can be recognized, identified, classified or generalized (4 acquisition levels) whereas a capability "rule" can be applied or transferred (only 2). Evaluation vocabularies are denoted in our system by an ordered set of integers representing different levels of acquisition. Since several vocabularies exists for describing the same type of capability, the designer must choose a vocabulary before proceeding to the construction.

A *contribution* link qualifies the way in which the realization of an objective contributes to the acquisition of a capability: It can be a *strong, moderate* or *weak* contribution. Therefore, several objectives can contribute to the acquisition of one capability.

In the following parts, we will show how CREAM approach supports the course generation.

3. GENERATING COURSE GRAPHS BY REASONING ON THE CREAM MODEL

The CREAM model can serve many educational purposes and can be exploited using a set of tools we have developed. In this paper, we describe the course generation part: the process that starts with a specification of objectives or capabilities and performs a traversal of the corresponding curriculum-knowledge structures to produce a course graph which will permit to reach the desired goals.

3.1. THE COURSE CONCEPT

In the teaching field, a course is a sequence of instruction periods dealing with a particular subject and aimed at the evolution of a student knowledge. In our model, we defined a course as a structured set containing three categories of objectives: global, specific and terminal objectives. A *global objective* being a statement expressed by the teacher to globally describe all the lasting changes (cognitive, affective and psychomotor) that he wishes to induce in his students' behavior during a course; A *specific objective* describes a set of behaviors that the learner should be able to demonstrate (specified in terms of the capabilities the learner should acquire); A *terminal* (or operational) *objective* is, in our context, an objective which describes a precise performance the student should achieve. In general, a specific objective is composed of several terminal objectives. This set of objectives is centered on a well-defined educational purpose and can tackle various themes around it.

If we group some parts that became apparent after our study of the different representations of the course concept ([Koe94], [Web94]), we state that a course is composed of three main parts:

- *descriptive part*: its title, its description and the set of general objectives. A general objective being a very abstract description of the course goals, most of the time a text resuming them.
- *course graph* constructed from the curriculum and containing a flat organization of the model objectives and appropriate links.
- *structural part* (described above): the structure of the different kinds of objectives and course themes.

In our system, a course doesn't consist of only one type of activity to accomplish but of a variety of them going from the teaching of verbal informations through a content presentation to complete problem solving activities. Each course comprises several kinds of pedagogical resources, connected to the course objectives and essential for their achievement. Their aim is to permit the learner to eventually master the notions involved in these objectives.

3.2. COURSE GENERATION PROBLEMATIC

Up until now, course creation, even on a model such as CREAM, has always been done manually. The instructional designer had to determine and choose his objectives himself, according to what he wanted to instil in his students. He then had to structure them, decide how he could make his students realize them, etc... All this could be very long and tedious and could result in badly-designed course if carried out too fast or carelessly.

The generative aspect of our method transforms completely the role of the designer. He now only has to specify the knowledge he wants to teach, supervise the process and approve or not the generated course. If one aspect of the course doesn't satisfy him, our toolkit offers him all the means to help him modify the original curriculum, change the initial knowledge specification for his course and even edit the course. This last thing is done by changing either the course structure or some elements of it (for example, adding, removing or modifying objectives or resources). Therefore, he is free to accept or not the generated course: he's the final judge in the process. Automatic course generation reduces the volume of his task and might then create more accurate and refined instructional material in addition to be less time-consuming and more cost-effective.

To generate a course is thus to allow an instructional designer to orient his efforts towards a different aspect of a course construction than the one concerned with the objectives to be chosen for the course. He can now concentrate on several aspects as: the course structure, the pedagogical resources which will help realize the objectives, the ways to teach the different kinds of knowledge involved (tutorial strategies), the definition of the various course themes and subthemes and the objectives to be associated with them, the relevance of the course or some of its objectives, etc... In fact, on the more declarative aspects of the course. In addition, he can, with our system, evaluate his courses by using the student interface at his disposal to simulate the evolution of a course.

In this paper, we precisely want to bring forth a solution to the problem of reducing the amount of time devoted to the conception of a course, by introducing an automatic course generation process and by proposing a system that performs this generation. This diminution will thus lead to lower training costs, an appreciable factor in industry.

Our approach is thus a key point in educational and industrial matter and is actually part of a large system which permits the specification, generation, edition, development and supervision of an entire course.

3.3. THE GENERATION PROCESS

We put forward two main approaches for a course generation: Generation of a course from a set of objectives to be achieved, or from a specification of the knowledge (capabilities) to be acquired.

First approach The first approach is relatively straightforward: a set of objectives is specified and sent to CREAM course generator. This one builds a course graph from the curriculum structures using this set of objectives. This graph is then sent to the structuring process which determines the global, specific and terminal objectives and creates automatically a course structure (which can be further edited with a course editor). A structuration algorithm has been developed for this purpose and will also be used for the second generation approach. It analyzes the way the objectives are linked in the course graph in order to classify them in either of the

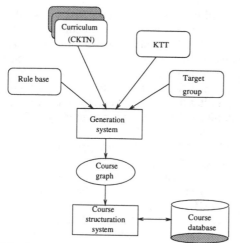

FIGURE 2: Course generation process

For a capability C prerequisite to an objective O,
IF TargetGroupeLevel(C) >= EntryLevel(C, O)
   THEN link (C,O) is acquired
IF TargetGroupeLevel(C) = NIL
   THEN link (C,O) is not acquired
ELSE link (C,O) is partially acquired

Marking of a capability

For a capability C prerequisite to objectives O1, O2, ..., On,
IF links(C,O1),(C, O2),...,(C,On) are acquired
   THEN C is possessed
IF links(C,O1),(C, O2),...,(C,On) are not acquired
   THEN C is not possessed
ELSE C is partially possessed

FIGURE 3: Some marking rules

objectives'categories. For example, all objectives without prerequisites will be considered as global. After the structuration, we can visualize the course structure and observe that it is made up of several learning hierarchies.

Second approach   This second approach is the one we will concentrate the most on. It consists in generating a course from a domain knowledge specification, i.e. a set of capabilities which we will call KTT (Knowledge To Transfer). This set represents the knowledge the student should master after taking the course we want to generate. Obviously, all its capabilities must belong to the curriculum which is to be concerned by our course. It is from this set that our generator will go through the CKTN in search of the objectives which will permit to acquire the set capabilities. Figure 2 presents all the elements that will allow us to generate a course with a CREAM model.

◆ *Student target group*   We define a target group as a student group state of knowledge on various capabilities which can be part of several subject matters (curriculums). For instance, the knowledge of a beginning nurse on the handling of the Baxter pump will not be the same as the one of an advanced nurse; So a course on this topic should not include the same objectives for the former as for the latter; the advanced nurse will waste her time learning things that she already knows. Thus, these two groups of nurses constitute two different student target groups and the generator should build a course well-suited to each one.

To determine these target publics and which capabilities they must include, the instructional designer must do a good cognitive task analysis in order to differentiate and classify the different categories of students and their hypothetical knowledge. If the designer has no idea about the state of knowledge of his students, he can generate a course with no target group input to his algorithm. He will then be free to readjust it for its needs and purposes.

In our model, the state of knowledge of a given target group is represented as a vector of couples (capability,

mastery level on this capability). For example, *[infusionRate(concept), identify]* or *[infusionRate (concept), 2]* is one element of a vector representing the knowledge of intermediate nurses on a concept belonging to a curriculum on the use of the Baxter pump (using KlausMeier vocabulary to describe the qcquisition level). We call this vector a TargetGroup.

◆ *DynCKTN*   Since the generation is performed by going through the CKTN graph defined on the chosen curriculum, we thus have to assign the target group state of knowledge to the capabilities and the links which constitute the CKTN to make it **dynamic**. The resulting graph is called a **DynCKTN** and we name this operation the **marking** of the CKTN. More precisely, it consists in attributing to each prerequisite link a value in {*acquired, partially acquired, not acquired*} indicating whether the minimum acquisition level on this link has been reached according to the target group, and to each capability a value in {*possessed, partially possessed, not possessed*} representing the acquisition standard of our target public on this knowledge and calculated from the levels assigned to the links. Figure 3 shows **some** of the rules used to calculate this marking.

Consequently, a generation without target public leads automatically to the marking of all the CKTN-prerequisites links as *not acquired* and therefore of all capabilities involved in the current CKTN as *not possessed*.

Example: Considering the CKTN in figure 1, we want to generate a course on the manipulation of the Baxter pump for intermediate nurses. Thus, a part of the TargetGroup can be described as: [(k1,2), (k2,1), (k3,2), (k4-k9,0), (k10,1), (k11,0), (k12,1), (k13,3), (k14-k17,0)]. The marking process produces the DynCKTN of figure 4.

◆ *Heuristics*   We now have a dynamic CKTN on which we have to reason to generate a course; We traverse it to determine which objectives have to be included in the course in order to permit the acquisition of the knowledge

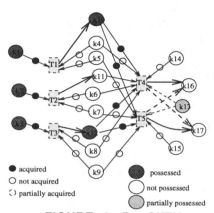

acquired
not acquired
partially acquired

possessed
not possessed
partially possessed

FIGURE 4: DynCKTN

specified in the KTT. To do this, we perform from each capability to acquire, a backward chaining traversal of the subgraph rooted at the capability in order to choose the objectives we judge necessary for the acquisition of the capability. We first evaluate the immediate prerequisite objectives of the capability and then, since some objectives possess mandatory prerequisite knowledge which in turn has contributing objectives, we have to trace the subgraph back until we reach an objective without any prerequisite or a capability already mastered by the student (as specified in the target group or seen by the marking).

The choice of objectives is carried out by applying heuristic rules introduced into the system and which consider several parameters: the links between capabilities and objectives (prerequisite or contribution), the knowledge in the KTT and also the DynCKTN. We defined (and actually implemented) three possible heuristics:

A **general inclusion** heuristic which consists in including all the sub-graph objectives in the course. This simple heuristic can lead to unnecessarily overloaded courses while it is possible that a course with fewer objectives could bring a learner to the acquisition of the same knowledge. On the other hand, it could leave more freedom to the designer for his personal objectives' choices among those selected by the generator and thus lead to a course with several possible plans. For instance, he could find some objectives unnecessary and take them out of the generated course (while ensuring that the entire KTT could still be acquired).

A second heuristic consists in choosing a certain number of objectives according to their contribution value to the capability we want to obtain. Thus, if an objective O contributes strongly to a capability C, it is sufficient to its eventual acquisition so we only select it and continue the generation in the subgraph rooted at O. If there is no objective of strong contribution, we have to pick, for instance, two medium-valued contributing objectives or three weak ones. In addition, we must also consider the marking of the capability because a partially possessed capability require a lower objectives' contribution for its acquisition than a non possessed one. So we take this last

parameter into account when setting the *contribution quota* needed for the acquisition of a capability.

The third heuristic developed consists in evaluating the complexity of the subgraphs rooted at each of the objectives we are considering. It permits to choose the objective with the highest probability of acquiring the considered capability. This complexity is expressed in terms of weights assigned to each capability of the subgraph. To calculate this weighted subgraph, we take into account the capabilities' marking, the value and marking of the prerequisite links and the contribution links value. For instance, a possessed mandatory capability will permit to realize an objective more likely than a non possessed optional one, and thus will weigh more. By adding up all the weights in each subgraph, we can evaluate each of their "global acquisition level" and make a more "intelligent" objective choice.

We therefore see that the chosen heuristic is a key element in the course generation. Totally different courses can result from the use of one or the other heuristics.

4. THE COURSE GENERATION WITH HEURISTIC REASONING

In the second approach, after the choice of the curriculum (from a list of availables curriculum on the subject matter), the optional specification of the target public, the specification of the KTT (pertaining to the CKTN and from which we remove capabilities already acquired according to the marking) and the choice of the heuristic, the task of the designer is finished until the delivery of the course. The automatic generation begins...

Each capability of the KTT receives a message of objectives' generation and the chosen heuristic determines which objectives will be used for its acquisition during a learning session. The three heuristics have been implemented and have led to different courses, as expected.

During the progress of the algorithm, we keep in a list the capabilities which could not be acquired with the CKTN actual objectives, according to the employed heuristic. These capabilities will bring about the introduction of a *course prerequisite* i.e. some additional knowledge the student should master prior to the beginning of the course.

••• general algorithm •••
in: CKTN, KTT, TargetPublic (optional), heuristic H
out: a Course C: structured set of objectives

- DynCKTN = Marking(TargetPublic, CKTN)
- newKTT = Reduce(KTT, DynCKTN)
- ∀ capability k in newKTT,
    CourseGraph = Generate(k, H)
    Put in a list all the capabilities which couldn't
    be acquired with the present CKTN.
- C = Structure(CourseGraph)
- Open(CourseEditor(C)) and show statistics.

The generated set of objectives is used to construct the course graph which is then passed to the structuring process. Then, the system opens a course editor which will allow the designer to modify his course if he wishes. A window of statistics also appears to give some information about the generation: the KTT specified, the number of objectives chosen and which ones, the KTT capabilities already possessed by the student (according to the target group) and the capabilities which require the course to have a prerequisite.

Example: Suppose we have the following KTT (scond approach): [k16, k4, k6, k8, k14] (knowledge to make a primary infusion). Our algorithm, using the first and second heuristics (H1 and H2) will produce the following course after reasoning on the DynCKTN of figure 4:

• **H1:** Course = {T1, T2, T4, T5} and capabilities {k5, k6, k7, k8, k9, k14, k15} need to have prerequisite objectives to be introduced
• **H2:** Course = {T1, T2, T4} and capabilities {k6, k8, k14} need to have prerequisite objectives to be introduced.

It is worth noting that the system permits an easy implementation of new heuristics since the heuristic is part of the input given to the generation algorithm. The designer can also describe his own heuristic and ask it to be implemented, so as to generate his courses in a more personal way; The constraints on an heuristic description being only that it must receive a capability as input, reason on a CKTN structure and output a set of objectives.

5. CONCLUSIONS

We presented a system for automatic course generation from a specification of knowledge to be taught, or from objectives that the generated course will permit to realize. The generation process consists in reasoning on the subject matter knowledge structures taking into account the knowledge of an optional target public. The generated course is directly usable by an ITS or a human teacher. The generation process uses heuristics in order to choose the content of the course. We have implemented three of these heuristics.

A functional framework of this generating system has already been implemented in Smalltalk and actually generates courses on a variety of domains. It is part of a complete authoring and ITS system that makes use of the course produced for learning purposes. It has been used for the construction of several courses in the SAFARI project [FG94]. In particular, a course for beginners' nurses on the handling of an intravenous infusion pump, a course for first year medicine students on the clinic exam in the intensive care unit and also a course on the Quebec highway code. This experimentation has permitted to test and validate our different heuristics. However, our system stays open to new heuristics.

We are now working on the implementation and testing of a third approach in which the instructional designer could specify objectives as well as capabilities for the system input. This could give him more flexibility in his course design.

Aknowledgements: We would like to acknowledge the MISCT (Quebec) for supporting this work under the SYNERGY program and also our two industrial partners, Novasys and Virtual Prototypes Inc.

REFERENCES

[FG94] Frasson, C. and Gauthier, G. A gradual software environment for developing tutoring systems. In *Advances in Artificial Intelligence - Theory and Application II*, 73-78. Windsor (Canada): IIAS.

[GBW92] Gagné, R.M., Briggs, L. and Wager, W. *Principles of Instructional Design*. Orlando, FL: Harcourt Brace Jovanovich, 4th edition.

[Gag93] Gagné, R. M. Computer-Based Instructional Guidance. In Spector, M.; Polson P. ; and Muraida, D., eds., *Automating Instructional Design: Concept and Issues.*; pp. 133-146, ETD. Englewood Cliffs, N. J.

[Kla90] Klausmeier, H. Conceptualizing. In *Dimensions of Thinking and Cognitive Instruction*, NJ:LEA: Hillsdale. 93-138

[Koe94] Koehorst, A.M. Curriculum graphs as a backbone for computer supported design and delivery of education. In proceedings of CSRIC,pp.31-36.

[Les88] Lesgold, A. Toward a theory of curiculum for use in designing intelligent instructional systems. In Mandl, H., and Lesgold, A., eds., *Learning Issues for Intelligent Tutoring* Systems. Berlin: Springer-Verlag.

[Mer87] Merrill, M. An expert system for instructional design. *IEEE-Expert* 2 (2) 25-37.

[MW93] Murray, T., and Woolf, B. Design and implementation of an intelligent multimedia tutor. In *AAAI'93 tutorials*.

[NG96] Nkambou, R. and Gauthier, G. Un modèle de représentation du curriculum dans un système tutoriel intelligent. In *Proceedings of ITS '96*; Montreal, Canada.

[Ten93] Tennyson, R. A Framework for Automating Instructional Design. In Spector, M.; Polson P. ; and Muraida, D., eds., *Automating Instructional Design: Concept and Issues*. ETD. Englewood Cliffs, N.J., pp. 191-217.

[Web94] Webster, J. Instructional Objectives and Bench Examinations in Circuits Laboratories. In *IEEE Transactions On Education* 37 (1), pp. 111-113.

OBJECT-ORIENTED MODEL OF KNOWLEDGE
FOR THE INTELLIGENT TUTORING SHELL

Ljubomir Jerinić

Institute of Mathematics, University of Novi Sad

Department of Computer Science

Trg Dositeja Obradovića 4, 21000 Novi Sad, Yugoslavia

Email: Jerinic@uns.ns.ac.yu

Vladan Devedžić

FON - School of Business Administration, University of Belgrade

Department of Information Systems

Jove Ilica 154, 11000 Belgrade, Yugoslavia

Email: edevedzi@ubbg.etf.bg.ac.yu

ABSTRACT

Rapid development of computer technologies and Artificial Intelligence (AI) methods, and introduction of computers into schools have yielded an ever-increasing use of AI methods in design of education software. The teachers in schools have become interested in taking an active role in designing educational software, and not being only passive users of these tools. However, the advancement of AI methods and techniques makes understanding of Intelligent Tutoring Systems (ITS) more difficult, so that the teachers are less and less prepared to accept these systems. As a result, the gap between the researchers in the field of ITS and the educational staff is constantly widening. This paper describes our efforts toward developing uniform data and control structures that can be used by a wide circle of authors, e.g., domain experts, teachers, curriculum developers, etc., who are involved in the building of ITS.

1. INTRODUCTION

The endeavour to design Intelligent Tutoring Systems and Intelligent Learning Environment [Fra88, BOP93, Gre95] has opened a new research direction in the application computers for educational purposes. These systems are capable to help solving difficult problems in the process of knowledge transfer, and enable an almost full individualization of the teaching process, as well as the student's advancement in the stages of acquiring knowledge and application of the new knowledge according to their capabilities, aspirations, previous knowledge and the like.

Also, ITSs stimulate the use of computers in education as a new education tool and controller of the teaching process, and opened new approaches for the use of computers in education.

On the other hand, it is generally accepted view within computer community that the 1990's is the decade of successful application of Artificial Intelligence (AI) expertise as real world systems. High on the list of the new software technologies that are going to be applied in the marketplace are Intelligent Systems (IS) or knowledge-based systems. Certainly, what distinguishes IS from the more traditional computer programs are not only their different names. Every computer program makes use of some kind of specialized knowledge to solve the problems of its application domain and is, therefore, in some sense knowledge based. The main difference lies not in use of knowledge but above all in the form of organization within them and in their general architecture.

In IS, the paradigm of problem solving in the application domain is explicitly expressed of as a separate entity, called knowledge base, instead of being hidden in the program's code. The knowledge base is managed by the separate, clearly identifiable, control component, usually a general purpose inference engine that is capable of making decision from the knowledge stored in knowledge base, answering questions about this knowledge, and determining the consequences implied by it. This means that the knowledge incorporated into intelligent systems is independent of the programs that manipulates it. That knowledge could be easily used by the users and potentially more easily modified or extended.

The intensive development of the methods of knowledge engineering and reasoning enables that education by the computers could be possible. The main goal of that approach is the complete individualization of the educational process. We assume, that the tutoring is a task of helping a student or pupil to learn particulars in a certain domain. This requires a special concern about the cognitive load that the student or pupil will bear in the learning process. The research on intelligent tutoring systems has produced the technology to do this task within the computer program. But, as the computers have entered teaching practice, teachers show an increasing interest in educational software systems. For these teachers, however, ITS is the most difficult one in terms of grasping the total aspects [MW90]. Of course, they have a chance to read articles about ITS but what they can understand through the articles is confined to the abstract level of comprehension. Experience in using ITS directly through a system that transperentizes the inner mechanisms of ITS would be a powerful tool to help them in comprehending it.

As research in ITS continues to produce more refined systems, the gap between the research community and educational community continues to become wider. The educators' understanding, and usage of these research results have been much slower than the research progress. As this theory-application gap becomes wider, it becomes more difficult for educators to participate in ITS research and application, and the result of the research becomes increasingly academic and unconnected to the pragmatic aspects of teaching and learning.

In designing and testing tutoring strategies several problems are encountered. First, the experience of learning via ITS is such a novelty that neither teachers nor theorists can anticipate many important issues. Second, teachers generally do not have well-articulated and mathematically or algorithmcally defined theories of learning. Third, relevant instructional and cognitive theories are not operationalized to a level easily realized as the computer programs.

Also, the educational programs are different among the distinct countries. Then, the required knowledge for learning some lesson in some domain is not the same even in one country itself (the north - south problems, country - town, etc.). The teachers want to have the active role in design and eventually update and improvements of intelligent tutors.

In view of the above trends and research issues, the goal of designing the EduSof [Jer94, JD95] system was to construct a conceptual framework for representing the objects, events, responses, reactions, and relationships involved in tutoring. Also, the aim was to build a highly-usable knowledge acquisition interface for rapid prototyping and easy creation, modification, deletion, and testing of both teaching concepts and tutorial strategies. Finally, our intention was to produce a system that enables teachers to build their own lessons without any computer or AI specialist.

2. OBJECT-ORIENTED KNOWLEDGE BASES

First, we describe a unified abstraction of different knowledge representation techniques and different models of expert knowledge in Expert System (ES) knowledge bases in general. That techniques (of a unified abstraction) are used while designing the shell for design intelligent tutoring lessons. The abstraction is derived and realized by applying object-oriented approach. The motivation for defining such an abstraction was to provide:

- a unified description and representation of different elements in ES knowledge bases;
- more general knowledge access methods for use in interactions between an ES knowledge base and the other ES modules at runtime, as well as in interactions between different modules of integrated ES building tools (ESBTs) at development time.

As a result, an object-oriented model of knowledge bases is developed and is called **OBOA** (OBject-Oriented Abstraction). It covers both design of knowledge bases and communication between the knowledge base and other ES and ESBT modules. However, the **OBOA** model sets only general guidelines for developing and using of object-oriented knowledge bases. It is open for fine-tuning and adaptation to particular applications. The ultimate practical goal of developing the **OBOA** model was to use it as the basis for building a class library of knowledge representation and knowledge access tools and techniques. At the time of writing, the implementation of the library is incomplete, but it is sufficient for exploring the ideas of the **OBOA** model.

The background for developing the **OBOA** model was several different ideas coming from the fields of knowledge and data modeling, representation and management. Gruber and Cohen [GC87] have described a hierarchy of knowledge representation techniques and tools, starting from most primitive ones to quite complex knowledge structures. Techniques for representing knowledge about external signals in real-time ESs, described in [Dev95], were developed as a class-like representation of properties of I/O data that real time ESs exchange with external processes they monitor and/or control. Finally, some ideas from the field of object-oriented databases were also adopted in the **OBOA** model.

On the other hand, we also investigate the possible application in design of intelligent tutors. A tutor can be defined as a provider of additional, specialized, or individualized instruction. We are concerned with the construction of a general framework for intelligent tutoring design, i.e., the intelligent tutoring shell. This shell can be used to create and maintain a curriculum for an individual domain. In this paper we introduce knowledge primitives for an interactive environment in which the teacher can manipulate graphical objects at different levels of abstraction, based on the **OBOA** model.

Building of an intelligent tutoring system requires the ability to model and reason domain knowledge, human thinking and learning processes, and the teaching process. Acquiring and encoding this large amount of knowledge is difficult and time consuming. We have been searching for efficient ways to do these knowledge engineering tasks. This paper describes our efforts toward developing uniform data and control structures that can be used by wide circle of authors, e.g., domain experts, teachers, curriculum developers, etc. who are involved during building the system. One of our goals is to build tools that will enable experts to use the computer directly, without any help of knowledge engineering.

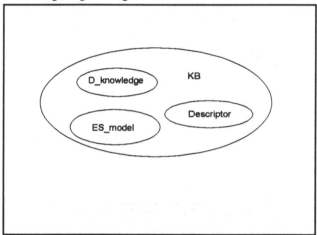

FIGURE 1   Knowledge base as a composite object

The field of AI in education is concerned with development of artificial intelligence techniques for the study of human teaching and for the engineering of systems that facilitate human learning. That field tries to answer to the questions that are a long term in nature: How can computer systems help learning and is it possible the measurement of learning progress? The term intelligent tutoring system is frequently used concerning the engineering side of the usage of artificial intelligence techniques in making the educational programs. In ITS the computational methods are used in support of AI activities such as planning, control, knowledge representation and acquisition, explanation, cognitive modeling and dialogue management. Also, the computational models are used to explore and evaluate alternate theories about learning.

It is possible to think of an ES knowledge base as a logical entity composed of three related parts (components), Fig. 1.: the domain knowledge, control knowledge (i.e., a model of the ES), and the third part that we call explanatory knowledge (see the discussion below). Domain knowledge is represented by using one or more knowledge representation techniques. In the most general case, domain knowledge is a structured record of many interrelated knowledge elements. These elements describe relevant domain models and heuristic, and can vary a lot in their nature, complexity, and representation. They can be everything from simple data to instantiated knowledge primitives such as frames, rules, logical expressions, procedures, etc., and even more complex knowledge elements represented by using either simple aggregations of the knowledge primitives or

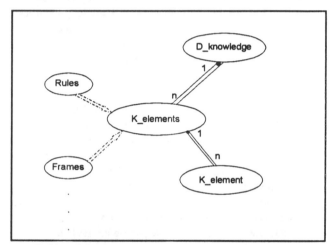

FIGURE 2   Organization of domain knowledge

conceptually different techniques based on the knowledge primitives and other (simpler) knowledge elements.

Knowledge elements of the domain knowledge are usually grouped to create meaningful collections[1]. One possible criterion of defining such a collection (the one used in the OBOAmodel) is the type of knowledge element that will be included in the collection. Therefore we can speak of collections of rules, frames, etc., or collections of more complex knowledge elements. Each such collection is homogeneous regarding the type of knowledge elements it contains. Generally, domain knowledge can be composed of n collections of knowledge elements, and there can be zero or more collections of elements of a particular type.

---

[1]The word "collections" is used rather than sets, lists, tables, etc., in order to avoid suggesting a particular implementation technique at the analysis phase.

The class diagram in Fig. 2. illustrates the organization of the elements of the domain knowledge as homogeneous collections, K_elements. A K_elements object is a collection of objects of the K_element class. The K_element class describes a single knowledge element of the domain knowledge and is elaborated below. The double line from the D_knowledge class icon to the K_elements class icon, with the point close to the D_knowledge icon, means that the D_knowledge class uses the K_elements class. The 1 and n at the ends of the double line denote the cardinalities of the corresponding classes: the D_knowledge class can have only one instantiation for a particular knowledge base, and there can be n instantiations of the K_elements class. Similarly, for each instantiation of the K_elements class, there can be n instantiations of the K_element class. K_elements is a generic (parameterized) class, which can be instantiated to generate classes denoting collections of knowledge elements of a particular type. The instantiation relationship between the corresponding classes is denoted by a dashed line, with an arrow pointing to the generic class.

For the sake of simplicity and basing this description on commonly known concepts, it is assumed that some knowledge primitives belong to the second highest level in the organizational hierarchy of knowledge elements. Therefore the classes' Frames, Rules, etc. are instantiations of the K_elements class, and denote homogeneous collections of frames, rules, etc., respectively. In the actual implementation, there can be several intermediate levels between the level of knowledge primitives and the highest (and conceptually the most abstract) level of knowledge elements, i.e., the level of K_elements and K_element classes.

K_element is an abstract base class with several subclasses inheriting its properties. In the Booch [Bo91] notation, the inheritance relationship among two classes is denoted by a solid line with an arrow pointing to the base class. Abstract base classes cannot have instantiations, only the objects of the Rule, Frame, Slot, If-clause, Signal, and other subclasses can exist in the knowledge base only. Again, we assume that the classes designating the knowledge primitives mentioned are derived directly from the K_element class, skipping more complex classes of knowledge elements. The OBOA model is opened for extensions by less common (or even new) knowledge element types. The K_element class defines common properties inherited by all types of knowledge elements and adding another knowledge element type requires only type-specific properties to be defined. Second, all the elements of a knowledge base are treated the same way, regardless of what their type is and how common (standard) or specific they are. This means that all the common operations on elements of the knowledge base (like updating, searching, deleting, etc.) use the interface of the K_element class, and element-specific operations are

provided through polymorphism in the corresponding subclasses.

Another interesting design detail in **OBOA** is the fact that certain types (classes) of knowledge elements, denoting part of more complex elements, are at the same level of hierarchy as the corresponding aggregate elements. For example If-clauses can be treated as "stand alone" knowledge elements, and rules. Such a design allows a particular If-clause object to be shared by several Rule objects, which is more flexible and more efficient, and can also simplify the rule compilation process [Dev95]. The classes in **OBOA** are designed by filling-in a standard template for class properties [Boo91].

Most elements of the K_element class have obvious meanings - a K_element object lets other objects (communicating with it) use its public interface functions to set and get its name, get its size, update it, etc. It should be

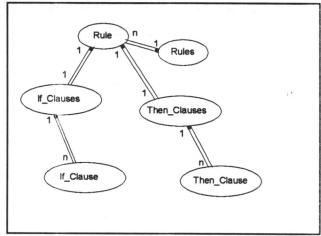

FIGURE 3   The Rule class diagram

stressed that the **OBOA** model strictly applies encapsulation, as a general and extremely important principle of object-oriented design: most attributes (fields) of classes and objects used in this design cannot be accessed directly. Client objects must use the server object's public interface functions. Public operations of the K_element class are only declared here, and are actually implemented in the derived classes using polymorphism. Thus it is possible to, for example, retrieve an arbitrary knowledge element from the knowledge base referring to the element as a K_element object and using the Get_K_El function. If the element is a rule, due to polymorphism, the code executed then will be the code of the Get_K_El function implemented in the Rule class. Each K_element object belongs to a collection of knowledge elements, which is specified by using the Collection_Ptr field, and uses public interface functions of the corresponding K_elements object.

To illustrate the design of a particular knowledge element type in the **OBOA** model, Fig. 3. shows the class diagram of the Rule class. It is assumed that only AND operator is used to connect the clauses in both the If-part and the Then-part of a rule. The main parts of such a rule are collections of If-clauses and Then-clauses, represented by the If-clauses and Then-clauses classes. An object of an If-clauses (Then-clauses) class is a collection of one or more If-clause (Then-clause) objects. Since in, many rule-based ESs rules are often divided in sets of rules or groups of rules to provide more efficient focusing of the inference process. Each Rule objects in the **OBOA** model can belong to one or more rule-collection objects, which are instantiations of the Rules class.

3. OBOA REPRESENTATION OF EduSof

In this section the usage of the **OBOA** model in the knowledge representation of the **Expert** module of **EduSof** is briefly presented. That knowledge representation is concerning some particular and concrete type of knowledge elements that are necessary for the realization of the **Expert** module in any ITS shell. One of the most important elements of knowledge for designing the intelligent tutoring lesson is the model of **Lesson**, and this is basic class needed for modeling the learning process. Any lesson is consisted of one or more issues, modeled by the class **Topic** in the **OBOA** model. We assume that the student must learn these issues, during mastering that lesson. The basic attributes of the lesson are: the name (**Name**), current topic (**CurrentTopic**, the issue or the problem introduced, defined and/or explained in the present moment), the task that is currently solved, the level of prerequisites for the student (**StudentLevel**), and so on.

The class **Lesson**, used for describing that elementary knowledge is inherited from the more general class **Frame**, and in the **OBOA** model that class is defined with:

Name: **Lesson**
Visibility: **Exported**
Cardinality: **n**
Base classes: **Frame**
Derived classes: **FirstLesson, LastLesson, HardLesson, EasyLesson, ...**
Generic parameters: -; only for generic class
Public Interface
  Operations: **SetTopic, GetTopic, UpdateTopic, DeleteTopic, CreateTopicCollection, GetTopicCollection, ...**
Implementation
  Uses: **Topic, Goal, ...**
  Fields: **Title, CurrentTopic, CurrentGoal, StudentLevel, TopicCollection_Ptr [], ...**
Persistence: **Static**; disk files
The issues that student must learn (class **Topic**) are realized with separate type of knowledge elements in the

EduSof system. Any topic could be introduced with text information (**Information**), graphically (the pictures and/or the diagram - **Image**), and/or with the simulation of some event (**Simulation**). Also, the additional or further explanation (**Explanation**) for that theme, or some suggestions (**Hints**) could be defined. The class **Topic** in **EduSof** is made for specifying and defining the issues or notions needed for lesson creation. That class in OBOA model is:

Name: **Topic**
Visibility: **Exported**
Cardinality: **n**
Base classes: **Frame**
Derived classes: **BasicTopic, DerivedTopic, MultidisciplinaryTopic, ...**
Generic parameters: -
Public Interface
  Operations: **SetInfo, GetInfo, UpdateInfo, ...**
Implementation
  Uses: **Topic, Goal, ... ;**
  Fields: **Name, InformationPtr, ImagePtr, LessonPtr, SimulationPtr, Explanation, Hints, QuestionPtr [ ], TaskPtr [ ], ...**
Persistence: **Static**

Abstract class **TQ** served for description of the common elements for the two comparable and connected classes, one for the definition of tasks or assignments (class **Task**) and the other class for the realization of questions or problems (class **Question**). The instances of that class are given to the student during the learning process. The definition of the class **TQ** is:

Name: **TQ**
Visibility: **Exported**
Cardinality: **0;**
Base Classes: **Frame**
Derived classes: **Task, Question, ...**
Generic parameters: **Answer, SolutionType, ...**
Public Interface
  Operations: **SetInfo, GetInfo, UpdateInfo, GiveFeedBack, ...**
Implementation
  Uses: **Topic, Goal, ...**
  Fields: **Name, InformationPtr, ImagePtr, LessonPtr, SimulationPtr, Solution, NextAction, ...**
Persistence: **Dynamic**

Method **GiveFeedBack** which handled the students response on some instance of **TQ** class, i.e., some task, hints, exercise, etc., is defined:

```
Method GiveFeedBack;
InheritFrom Rules, Frame, Student, Psychology
Begin
   AcceptAnswer;
   ElaborateRespons;
   Find(Psychology.Type, CurrentStudent.Level);
      (* find the model of student and he/her psychology
      type *)
   If Congratulate = "ON" Then
      GiveCongratulate(Psychology.Type,
                              CurrentStudent.Level);
   If Elaborate = "ON" Then
      GiveElaborate(Psychology.Type, CurrentStudent.Level);
   If Explain = "ON" Then
      GiveExplain(Psychology.Type, CurrentStudent.Level);
   If MoreInfo = "ON" Then
      GiveMoreInfo(Psychology.Type, CurrentStudent.Level);
   ....

      (* according to current student, he/her psychology
      type, parameterizing semantic net of frames and
      inherit from frame and rule class *)

End.
```

On the similar way the other classes and methods [Jer94, JD95] are defined in the OBOA model, which are needed in the creation of intelligent tutoring lessons.

6. DISCUSSION AND CONCLUSIONS

The purpose of the proposed OBOA model of knowledge bases and knowledge base management is to make a basis for applying the ideas of object-oriented software design methodology to ES knowledge organization, representation, and access. It covers all important aspects of knowledge bases, like their contents, knowledge representation techniques, using, updating, extending and maintaining the knowledge, etc. However, it is important to stress again that the OBOA model should be regarded as an open framework for developing ES knowledge bases, rather than as a closed set of design rules and organizational hierarchies.

The presented method for building a knowledge base suitable for the educational purpose is currently under development and the realization in the present form. In the current development state we test usage of EduSof in some summer schools in various subjects (chemistry, history, geography etc.). With introduced approach in building the intelligent tutors we realize the one of the goals we set down before the start of the realizing of the above concepts. The teacher's independents in design of the intelligent tutoring lessons were almost 90% after an hour of explanation how to use the EduSof shell. It is supposed that they have some knowledge of using the computers, a little knowledge in computer graphics, and some help of professional programmers' in making the simulation's procedures.

We will also try to realize the EduSof concept in the object-oriented approach in design the knowledge base. Also, the above principles will be incorporated in the other parts of intelligent tutoring systems, i.e., in the **Student**, **Tutor**, **Diagnostic** and **Interface** modules.

REFERENCES

[BOP93] Barna, P., Ohlsson, S., Pain, H., (Eds.) *Proceedings of AI-ED 93, Wold Conference on Artificial Intelligence in Education*, AACE Press, Charlottesville, VA, USA, 1990.

[Boo90] Booch, G. *Object-Oriented Design with Applications*. Redwood City, CA, Benjamin Cummings, 1990.

[Gre95] Greer, J. (Ed.) *Proceedings of AI-ED 95, Wold Conference on Artificial Intelligence in Education*, AACE Press, Charlottesville, VA, USA, 1995.

[Dev95] Devedžić, V. Knowledge-Based Control of Rotary Kiln. *Proceedings of The 1995 IEEE International Conference on Industrial Automation and Control: Emerging Technologies*, pp. 452-458, 1995.

[Fra88] Frasson, C. (Ed.) *Proceedings of Intelligent Tutoring Systems 1988*, ACM Press, USA, 1988.

[GC87] Gruber, T. and Cohen, P. Knowledge Engineering Tools at the Architecture Level. *Proceedings of The 10th IJCAI*, Vol.1, 100-103, 1987.

[JD95] Jerinić, Lj. and Devedžić, V. The EduSof Intelligent Tutoring System. *Proceedings of the Second International Conference on Graphics Education, EDUGRAPHICS'95*, 1995. (in print)

[Jer94] Jerinić Lj. Frames Technique, Inference Mechanisms and Knowledge representation in Design Educational Software. Proc. SYM-OP-IS '94, 107-110, 1994.

[MW90] Murray, T. and Woolf B. P. A Knowledge Acquisition Tool for Intelligent Computer Tutors. *SIGART Bulletin* Vol. 2, No. 2, 1-13, 1990.

Knowledge Acquisition

A DATA MINING TECHNIQUE FOR TIME SERIES DATA

Yoshinori Satoh and Akira Maeda

Systems Development Laboratory, Hitachi, Ltd.

1099 Ohzenji, Asao-ku, Kawasaki 215 JAPAN

e-mails: y-satou@sdl.hitachi.co.jp, maeda@sdl.hitachi.co.jp

ABSTRACT

Data Mining is a technique to extract nontrivial and previously unknown regularities in databases. Data mining is one of the key technologies in realizing intelligent plant diagnosis system and quality control systems. In this paper, we propose a new data mining algorithm that uses time series data as input, and extracts meaningful relationships between trend patterns in time series data. In this algorithm, time series data is transformed to a set of trend pattern symbols. After that, the algorithm searches for interesting combinations of the symbols that may reveal a hidden causal relationship of the target system. Experimental results are also described to present the effectiveness of the proposed method.

1. INTRODUCTION

Diagnosis systems have been introduced to various kind of plants. In these systems, more accurate and flexible system is required to reduce production costs and to improve product quality. Techniques of AI have been used in this field, however, knowledge acquisition is a bottleneck in constructing and maintaining these systems.

On the other hand, data mining has been extensively studied these days. Data mining is a technique for knowledge discovery from large data in real domain. Though applications of data mining have been developed in the field of database such as RDB(Relational DataBase), its application to time series data has hardly been studied. In this paper, we propose a new method of data mining for time series data.

Figure 1 shows an example of a diagnosis system using time series data mining. The data mining module works for finding symptoms that predict system faults from sensor data. These symptoms can be used for constructing expert systems and providing users useful information that is necessary to predict faults.

2. A FRAMEWORK FOR DATA MINING METHOD

Data mining is a technique to extract implicit, previously unknown, and potentially useful information from data. It has been recognized as a promising new field in the intersection of databases, AI, and machine learning. The

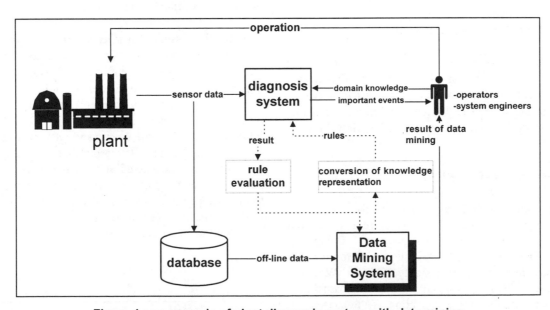

Figure 1 : an example of plant diagnosis system with data mining

goal is to find interesting, meaningful patterns from a large amount of data [SF91].

Compared to other techniques in AI, data mining deals with data in real domain and has features as following:

- **a large collection of input data:** Generally it is hard to process very large data on memory and such data must be processed using external storage. Therefore a fast algorithm that can process data on external storage is necessary.

- **multidimensional data structure**: In addition to the problem above, the real data may have many variables. For example, over 100 variables may be given to a system. This also requires a fast algorithm in the term of the number of variables.

- **input data may include missing, irrelevant, noisy values:** In most cases, real-world data cannot be complete and may include incorrect values. Statistical methods is useful to process such data.

- **redundancy information exists in input data:** If data structure is difficult to understand for a user, the user may use redundant variable or values for analysis. Data mining system must decide which information is necessary.

- **explicit result:** The knowledge extracted by data mining must be easy to understand because this is used by a user rather than a computer. Implicit and complex knowledge representation like that in neural networks is not suitable for data mining.

3. TIME SERIES DATA MINING

In this section we consider how to apply data mining method to time series data. Our goal is to find typical regularities extracted from multiple time series. A typical example of knowledge is: "When temperature decreases while humidity is constant, then shortly later it will rain."

3.1. ASSOCIATION RULES

Usually, an observed value of time series data from sensors is a pair of a time-stamp and a current value. So it is natural to save data order by time-stamp. Computing *association rules* is known as the data mining for this kind of data structure on external storage[AS94]. When all records are divided into disjoint subsets , an association rule gives regularity over a collection of the subsets. It

consists of conditional clauses X and single conclusion clause Y, $X \Rightarrow Y$. Each clause is an attribute value in the database. We say that some records are in a "case"[i], when these records have an equivalent value on an attribute. Regularity is computed over sets of cases. Notice that it gives statistical hypothesis of data. Hence association rule induction method is robust in dealing with irrelevant and noisy data. Detail is shown in [SF91].

3.2. PROBLEMS

As described above, association rule induction method is suitable for data structure of time series on external storage. Association rules ,however, have no ability to deal with the property of the time series. There exist problems as following:

- Patterns are important rather than instantaneous values, considered from the nature of the matter. Association rule induction method has no measure for this point.
- Time axis should be handled explicitly. Association rule induction method lacks the aspect like this.

Therefore, it is necessary to modify the framework for extracting rules.

3.3. APPROACH

We introduce SSTM(Signal-to-Symbol Transformation Method of Time series Data)[Nis93] for pattern matching. The basic concept of this method is to expand time series data into polynomial representation and the time series data are transformed to a series of line segments as shown in figure 2.

Extracting association rules is basically unsupervised leaning. The purpose of this method is to find not only "what kind of pattern is important", but to find "what is important data". This is why association rule induction algorithm is carried out without a supervisor. However, in plant diagnosis system for example, it is possible to get information about some events and we can regard it as a supervisor for mining. If the purpose of diagnosis is to find something connected with machine troubles, it is natural to analyze data observed at the time around the trouble happens. Supervisor also contains a measure to deal with time axis as shown figure 2.

The outline of our method is:

---

[i] Agrawal et al. use the term "transaction"[SF91].

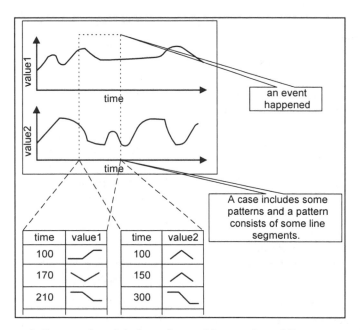

figure 2: time series data transformed to a series of line segments

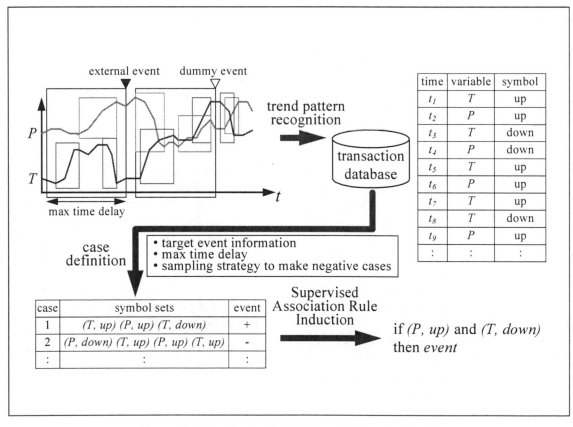

Figure 3: Time Series Data Mining System Architecture

1. **symbolization of time series data:** This provides a method to deal with patterns.

2. **case definition:** Training set for supervised mining consists of positive cases and negative cases. A positive case is a set of data that observed before an event(e.g. operational fault) happens, and a negative one is a set of data that observed under normal operation. Typically time span of cases could be max dead time(time delay) between observed values and an event. All cases have same time span for comparison.

3. **target of mining:** Some combinations of symbols are contained by many of positive cases and appear in positive cases rather than negative ones. This means we can find the characteristic patterns in positive sets.

Now we are ready to find knowledge from time series data. Figure 3 shows the flow of this method. In the next section, we will present the algorithm in detail.

3.4. ALGORITHM

First we introduce some basic concepts of association rules, using the formalism presented in [AS94]. Let $I = \{i_1, i_2, ..., i_m\}$ be a set of literals, called items. Let D be a set of transactions, where each transaction T is a set of items such that $T \subseteq I$. Associated with each transaction is a unique identifier, called its TID. We say that a transaction T contains X, a set of some items in I, if $X \subseteq T$. An association rule is an implication of the form $X \Rightarrow Y$, where $X \subset I, Y \subset I$, and $X \cap Y = 0$. The rule $X \Rightarrow Y$ holds in the transaction set D with confidence c if $c\%$ of transactions in D that contain X also contain Y. In short, confidence c is given by $P(Y|X) = P(X \cup Y) / P(X)$. The rule $X \Rightarrow Y$ has support s in the transaction set D if $s\%$ of transactions in D contain $X \cup Y$. In short, support s is given by $P(X \cup Y)$. Usually minimum threshold of support and confidence, called *minsup* and *minconf* are given by users.

The problem of discovering all association rules can be decomposed into two subproblems:

1. Find all sets of items(*itemsets*) that have transaction support larger than *minsup*. We say

that a set of item found here is a *large* itemset.

2. Use the large itemsets to generate the desired rules. A rule consists of condition part X and conclusion part Y, which has larger confidence value than *minconf*.

We have developed an algorithm for computing association rules in a supervised manner. In our method, the conclusion part of a rule is a class name such that the rule represents a feature of positive cases. Hence, it becomes possible to calculate *support* and *confidence* at the same time and integrate two subproblems above.

1) $C_1 := \{\{X\} | X \in D\}$

2) $s := 1$

3)   **while** $C_s \neq 0$ **begin**

4)   database pass: let L_s be the elements of C_s that have *support* over *minsup*

5) $L_s = \{c \in C_s | support \geq minsup\}$

6) $R_s = \{c \in R_s | confidence \geq minconf\}$

7)   candidate generation: compute C_{s+1} from L_s

8) $s := s+1$

9)   **end;**

10)   output result: $\bigcup_s R_s$

In addition to *support* and *confidence*, we define *accuracy* of describing positive class as a measure for rules.

$$accuracy := P(Y|X) \log \frac{P(Y|X)}{P(Y)}$$

Notice that *minconf* should larger than be $P(Y)$, because a rule is meaningless if *accuracy* is negative. We redefine the accuracy as $P(X)$ instead of $P(X \cup Y)$, because our method is supervised and Y is considered to be fixed.

Finally, we introduce the utility measure(*u*-measure) to represent the "goodness" of a rule[AMT94].

$$u = P(X)^\beta P(Y|X) \log \frac{P(Y|X)}{P(Y)}$$

where β is a parameter to define the weight between support and accuracy. If $\beta = 0$, *u*-measure doesn't concern about *support*. By using *u*-measure we can choose only important rules from many rules.

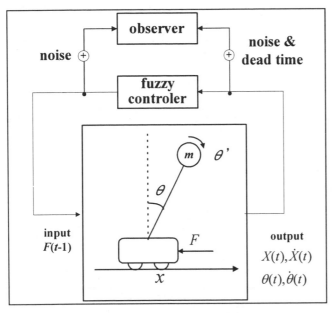

figure 4: inverted pendulum under control of fuzzy logic

sampling interval	10ms
dead time	50ms
noise	$\pm 10\%$

table 1 : observation conditions

intersection for 1 case	100ms (fixed)
minimum support	0.1
β (for u-measure)	0.8

table 2 : parameters for rule extraction

input variables	angle	Θ
	angular velocity	Θ'
	position	X
	velocity	X'
	force	F
	unnecessary variable * 4	random values

table 3: observed variables

No.	type
1	a ⟋ b
2	a ⟍ b

table 4 : type of membership functions

condition	conclusion
$\Theta>0$	F>0
$\Theta<0$	F<0

table 7 : fuzzy rules

variable	predicate	type of membership function	parameter	weight
Θ	positive	1	a=0,b=10	1
Θ	negative	2	a=-10,b=0	1
F	positive	numerical value	F=5	1
F	negative	numerical value	F=-5	1

table 5 : membership functions

	negative	a little negative	nearly zero	a little positive	positive
Θ	[-0.799, -0.479)	[-0.479, -0.160)	[-0.160, 0.159)	[0.159, 0.479)	[0.479, 0.799]
X	[-0.113, -5.120)	[-5.120, 0.011)	[0.011, 0.072)	[0.072, 0.134)	[0.134, 0.196]
X'	[-0.726, -0.432)	[-0.432,-0.138)	[-0.138, 0.155)	[0.155,0.449)	[0.449, 0.742]
Θ'	[-5.099, -2.020)	[-2.020, 1.058)	[1.058, 4.138)	[4.138, 7.217)	[7.217, 10.29]

	negative	positive
F	[-5,0)	[0,-5]

table 6 : boundaries for symbolization

4. EXPERIMENT

4.1. DESCRIPTION OF THE EXPERIMENT

In this section, we try to reproduce knowledge about controlling an inverted pendulum from time series data. A model of inverted pendulum under control of fuzzy logic is shown in figure 4. Two simple fuzzy rules used are shown in table 4, 5, and 7. All observed variables are shown in table 3. Initial status of the system is that $\theta(0) = 0.5$ degrees and $\dot{\theta}(0)$, $x(0)$, $\dot{x}(0) = 0$. The pendulum is going to vibrate periodically after the system starts. Observer can see input values $F(t\text{-}1)$ and output values $\theta(t)$, $\dot{\theta}(t)$, $x(t)$, $\dot{x}(t)$, where t means time of observation. Observation conditions are shown in table 1. All parameters for rule extraction are shown in table 2.

Now consider to find rules that exist in the behavior of the inverted pendulum system. Though symbolized values are given to the system for simplicity, these data contain almost the same information of data generated by SSTM. All numerical values are symbolized as shown in table 6. We assume that a user has no knowledge about the system. Four random variables in table 3 are also given as input to the system, in order to verify the robustness of the algorithm.

4.2. RESULTS

Results are shown in table 8 and table 9. Rule 3 and 5 in the both of the tables show correct fuzzy logic. Rule 6 about position of the supporting point in the both of the tables describes that "if the supporting point is off from the origin, move it to the opposite side". Rule 6 is not given as the fuzzy logic, but it is correct. It is also possible to describe the control logic by using rule 6. Other rules are also explainable in the same manner. Rule 0 and 4 about velocity of the supporting point describe that "the object moves forward to the direction of the force". Rule 1 and 2 about angular velocity in the both of tables are similar to rule 0 and 4. These rules describe a state of the pendulum while the direction of the force is constant.

The influence of dead time, noise and all unnecessary variables doesn't appear in the extracted rules. It is the effect of using statistical method and u-measure. Notice that data mining system tends to generate a great number of rules under two constraint conditions about *minsup* and *minconf*. As a result, most of rules are correct but unnecessary or

No.	condition	P(X)	P(X|Y)	u-measure
0	X'=a little negative	0.486	1	0.0183
1	Θ'=a little positive	0.408	1	0.0159
2	Θ'=positive	0.338	1	0.0137
3	Θ=negative	0.29	1	0.0121
4	X'=negative	0.256	1	0.0109
5	Θ=a little negative	0.252	0.992	0.00809
6	X=positive	0.148	0.986	0.00405

table 8: positive class is "F<0"

No.	condition	P(X)	P(X|Y)	u-measure
0	X'=a little positive	0.471	1	0.0162
1	Θ'=a little negative	0.415	1	0.0146
2	Θ'=negative	0.348	1	0.0127
3	Θ=positive	0.304	1	0.0114
4	X'=positive	0.254	1	0.00989
5	Θ=a little positive	0.213	1	0.00857
6	X=negative	0.127	1	0.00568

table 9: positive class is "$F \geq 0$"

already known. The u-measure, however, works well to select only good rules. It is also effective in excluding unnecessary variables. Hence, it is possible to discover important and useful knowledge.

5. CONCLUSIONS

We proposed a new data mining technique to discover knowledge from time series data. It is possible to deal with large, unreliable data that contain missing, irrelevant, and noisy values even if data contains unnecessary variable/values. We presented that the data mining method can be applied to time series data and the result of the simple experimentation shows our method is effective. As we described in section 3, it could be more powerful and general diagnostic technique for various kind of plants by introducing Signal-to-Symbol Transformation Method. Knowledge discovered by our method is readable and easy to integrate with domain knowledge of users, such as operators and plant engineers. This means that it can support users on various field of plants, such as quality control, system diagnosis, and equipment management.

It is also possible to integrate other kind of AI technique such as expert systems. In the field of plants, many AI techniques have been used. Knowledge acquisition, however, come to be a bottleneck to keep the facilities in good working order. We believe that our method gives one of the solutions on this problem.

6. REFERENCES

[AMT94] Hitoshi Ashida, Akira Maeda, *Yori Takahashi, Characteristic Rule Extraction Algorithm for Data Mining*, Proc. of Information Processing Society of Japan, pp.19-20, Vol.3, Tokyo, Japan, March 15-17,1994(in Japanese)

[AS94]Rakesh Agrawal, Ramakrishnan Srikant, *Fast Algorithms for Mining Association Rules*, Proceeding of the 20th VLDB Conference Santiago, Chile, 1994

[MCS93] Christopher J. Matheus, Philip K. Chan, and Gregory Piatetsky-Shapiro, *System for Knowledge Discovery in Databases*, IEEE Trans. on Knowledge and Data Engineering , Vol. 5, No. 6, December 1993.

[MTV94] Heikki Mannila, Hannu Toivonen, A.Inkeri Verkamo, *Efficient Algorithms for Discovering Association Rules*, AAAI Workshop on Knowledge Discovery in Databases, Eds. Usama M. Fayyad and Ramasamy Uthurusamy, pp.181-192, Seattle, Washington, July 1994

[Nis93]Takeshi Nishiya, *A Signal-to-Symbol* Transformation *Method of Time series Data for Detecting Signs of a Process Change*, Proc. of IEEE International Workshop on NEURO-FUZZY CONTROL, pp.345-351, Murolan, Japan, March 22-23,1993

[SF91]G. Pietetsky-Shapiro and W. J. Frawley, editors. *Knowledge Discovery in Databases*. AAAI press / The MIT press, Menlo Park, CA, 1991

DEVELOPMENT OF A KNOWLEDGE BASED SYSTEM USING A COMBINED KNOWLEDGE ACQUISITION AND KNOWLEDGE MODELLING APPROACH

Glenn D. Aguilar*, Hiroyuki Yamato** and Takeo Koyama**
* Research into Artifacts, Center for Engineering
University of Tokyo
4-6-1 Komaba, Meguro-ku Tokyo, Japan 153
** Department of Naval Architecture and Ocean Engineering
University of Tokyo
Email: glenn@race.u-tokyo.ac.jp

ABSTRACT

A knowledge based system for the definition of ship hull forms was developed. The knowledge base was constructed through a combination of knowledge acquisition and knowledge modelling approaches. Hull forms from published sources were scanned and a hull database constructed using an on-screen digitizing system. A knowledge acquisition tool for making selection rules and domain dictionary was written. A hull definition tool with an advisory system was implemented after an analysis and design process that made use of a knowledge modelling approach. Pieces of knowledge gathered during knowledge acquisition were integrated into the domain and expertise components of the model. Tools for the rapid generation of hull forms were made and showed the importance of using both approaches in building and using knowledge bases.

SHIP HULL FORM DESIGN AND KNOWLEDGE BASE CONSTRUCTION

The major inspiration behind this research was the task of developing a system that could be applied in a setting where expertise and design experience are considerably deficient. Such conditions exist in developing nations; technical resources are scarce and even non-existing. Expertise or experience in ship design and construction are quite limited. What is taken as standard design procedure in an engineering task is sometimes unheard of in the existing small and medium sized boat building industries. Most design and construction practices are empirical in nature with boat builders very reluctant to change or modify existing designs that have been used for generations. The development of a knowledge based system (KBS) was seen as a means of devising solutions to the transfer of technology and the institution of practical tools for hull form design.

A major task in the design of ships is the representation of the hull form. As early as the 1960s, the importance of mathematical modeling of hull forms has been recognized. The works of [RB59, TS61, Hay63, MK63, Tay63] are some of the pioneering efforts in ship hull representation. Polynomial curves, splines and other mathematical representations were used.

Geometric modeling systems have become more accurate with a wider range of splines and surfaces that model different shapes. Early curve representation used cubic splines that corresponds to the draftsman's spline. From the early cubic splines, parametric representations were developed which include Hermitic splines such as the Cardinal Spline, the popular Bezier curve, Bspline, Beta spline, Non-uniform rational B-Spline (NURBS). Surface representation includes the extension of curve definition into surfaces, generation of patches, tessellation and other such techniques.

Computer Aided Design (CAD) systems require geometric modelling approaches that meet accuracy, performance and aesthetics requirements. Adding functionality to a CAD system entails adding more knowledge available to the user when the system is employed. The Knowledge Based System (KBS) CAD must include rules and objects that automate some of the modeling procedures such as curve and surface fitting, constraint evaluation and relationship determination. Algorithms and calculations for analysis and evaluation of design results should be included for determining the viability of the design and aid in decision making. Ease of use and user-friendly interface with explanation capabilities were also considerations for an effective system. Related work involving CAD and knowledge bases in ship design include that of [CM91, SPS91, WBH90].

The approach in this research for the construction of a knowledge base can be divided into two: a bottom-up or gradual knowledge acquisition approach and a top-down or modelling approach. Gradual knowledge acquisition refers to the gathering of basic rules and inferences and later grouping them into categories. Modelling involved working from the top of the hierarchy or from categories towards the details. Subcategories can contain other levels of hierarchy; rules, frames, facts and other knowledge pieces consist the basic elements. When this research started, gathering the knowledge bits and pieces was the major activity but as the knowledge base was growing, modelling became the major activity as the need to structure the knowledge base and systems became obvious. At this stage, modelling covered not only the knowledge included in the system but everything else including user activities, database requirements and other non-knowledge base related entities and activities.

SYSTEM DEVELOPMENT APPROACHES

Two methods were used in developing different components of the system: OMT and KADS (Figure 1).

During the early phases of this research, focus was on the acquisition of knowledge; an activity considered to be the most difficult task in the development of a KBS system. Before system development, some time was devoted in the selection and application of an object oriented analysis and design approach mainly because of availability, maturity and the promised advantages of object orientation. The Object Modeling Technique by [RBP+91] was selected; it was also the approach employed by other researchers in a shipbuilding Computer Integrated Manufacturing (CIM) project [Ito94].

Two programs were developed using OMT; the first is named the Geometry Acquisition Tool and was developed to capture the mathematical representation of hull forms from two-dimensional lines drawings. The Selection Tool is the second program, a knowledge structuring and construction rule derivation program based on the Repertory Grid Technique [Kel55].

A hull definition system called the Hull Design Tool and a related knowledge based advisory system (Advise System) were developed based on a modelling approach involving the Knowledge Based System Analysis and Design Support (KADS). KADS can be considered as a framework for KBS design. It is an attempt to formalize the methodology of KBS development and involves the development of several models resulting from analysis and design phases. The primary function of KADS is to model knowledge explicitly in the system considering the system components and external agents or processes such as the user, knowledge sources and types, databases and related programs or systems [WSB92, BWV+87, TH93].

KNOWLEDGE ACQUISITION TOOLS: THE GEOMETRY ACQUISITION TOOL

A variety of data and references exist for the design of fishing vessels. Most are in the form of plans such as the lines drawing, general arrangement sketch, reports on performance including resistance, stability, seakeeping characteristics and others. Reference material from which these data come from include textbooks, technical literature, experts, design plans from shipyards and existing databases.

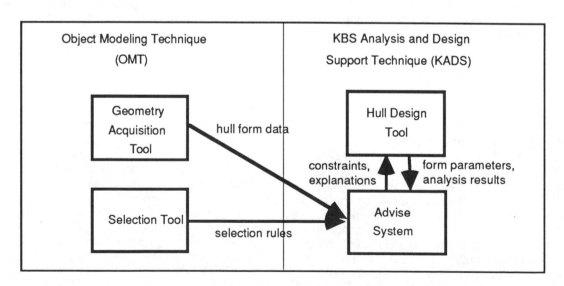

FIGURE 1. System development approaches and relationships between the resulting tools.

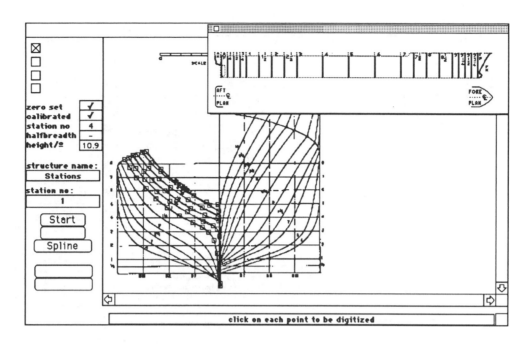

Figure 2. Main window of the Geometry Acquisition Tool.

Established textbooks and publications of Classification Societies provided excellent sources of constraints and requirements for hull forms. Such data defines most of the limits imposed on the desired hull form. The usual method of developing a hull representation for computer representation involves using the offsets table of a particular design. The approach used in this case was to scan lines drawings and digitize the image directly on the screen without the use of external digitizers. Most of the figures used as sources were from books and the technical literature. [AKY93]

Digitizing is normally done with the use of a digitizer where the position of a pointer and user actions such as clicking a button determines the position of a point within a coordinate system. Scanned images have some advantages over a digitizer including the ability to directly manipulate the image without using another peripheral.

Depending on the input figure, the process of scanning, image editing and digitization can take up to 30 minutes for a fairly complex hull form. It involves calibrating and clicking on the different points representing the sections. Each digitized structure or curve is labeled and relationships added. BSplines, Cardinal splines or Bezier curves are available for geometric modelling. Also included are hydrostatic calculations, database functions, symbolic/geometric representation of structures, forms and ship types (Figure 2).

The database provided hull forms for reference during the initial stages of design. Such data was also be used for related systems such as hydrodynamics calculations, structural analysis, general arrangement and other tasks in the stages of the design spiral. Besides the numerical representation of the curves, the stored, bit-mapped image itself is available for other purposes.

It is possible to automate the digitization process using image processing methods but a typical lines drawing has a lot of noise such as the grids and notations. Also, most scanned images had sufficient information required at this stage of design making this attractive option unnecessary.

KNOWLEDGE ACQUISITION TOOLS: THE SELECTION TOOL

It has been argued that design activities range from pure formation to pure selection [DL90]. Pure formation involves the creation of ideas by the designer while at the other end, pure selection involves choosing from several elements. Since most of the design elements in the later stages of ship design are mostly manufactured products, synthesis becomes more of a selection task. The requirement for selection is knowledge of all the alternatives and the process of search. Knowledge acquisition for such design alternatives involves the ranking of alternatives and definition of attributes that form the basis for choosing the best element.

The main approach that served as the basis for a knowledge acquisition tool called the Selection Tool was the Grid Repertory Technique, one of the more common methods used in knowledge acquisition. This technology was based on the Personal Construct Psychology (PCP) approach of [Kel55]. Several existing commercial and industrial software for knowledge acquisition are based on the Grid Technique but its use in design or other synthesis problems have been limited. The more well known products include Nextra, Aquinas and ETS. [Boo86, Boo89].

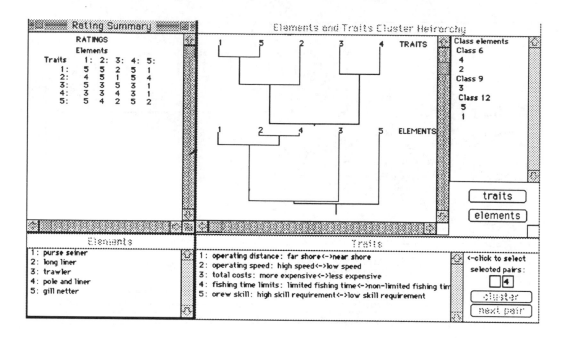

Figure 3. Main window of the Selection Tool.

Grid rating technology has been used successfully in the field of diagnostics, classification, selection and other analysis type problems. Its use in finding solutions to synthesis problem has been quite limited and some authors express doubts about its applicability in this domain. However, some activities in design involve the selection of elements without any formation or creation activities. Examples include selecting equipment from several brands or the initial size selection of parent ships based on mission requirements.

The primary function of the Selection Tool is to provide rules that would help in decision making when selecting the best from several choices. Decision elements are identified; traits that determine the ranking of each element are solicited and a ranking process is undertaken. Several viewpoints such as clusters, classification trees and outlines aid in finalizing the ranking of elements as shown in Figure 3. Rules for selecting the best element from a ranking grid based on traits is the final output of the system. Using clustering techniques also helped in determining built-in hierarchies among similar attributes describing the element set as well as relationships between the decision elements.

HULL DESIGN TOOL AND ADVISE SYSTEM

Modeling the process of hull definition was done using the KADS technique. Different design modules were identified and considered as tasks and subtasks in an hierarchical structure. Several models were produced including a task model, cooperation model, expertise model and domain model. The model of expertise consists of 4 "layers" as defined by [BWV+87]. These are the domain layer, inference layer, task knowledge and the strategy layer. This research did not include the strategy layer.

The task layer is on top of the inference layer and describes how the various inference types are combined to achieve a particular goal. The inference layer lies on top of the domain knowledge and includes *inference types*. Inference types describe the way domain concepts, relations or structures can be used to make inferences. It consists of an input, a knowledge source and an output. The knowledge source carries out an action on the input data and produces new information as output.

The Hull Design Tool was designed using the KADS approach. At first, an integrated, tightly coupled knowledge based system was the objective but as the system was evolving, the knowledge component was separated to add flexibility, make maintenance and implementation more convenient.

The purpose of the Hull Design Tool was the rapid design of hulls mainly through a *selection* and parameter *variation* process. Predetermined shapes, limits and constraints are used to produce curves for hull definition. Default shapes of specific hull sections or surfaces were stored and could be easily modified by changing a few simple parameters (Figure 4).

The Advise System stored the knowledge necessary to guide the user through all stages of creating a ship hull form. Hydrostatic calculations were built into the system for hull evaluation and several commonly used design charts included for the rough estimate of resistance.

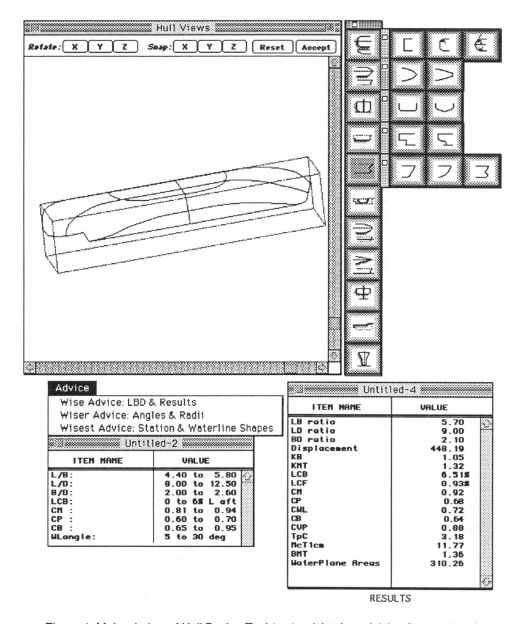

Figure 4. Main window of Hull Design Tool (top); advise from Advise System (bottom).

The main function of the Advise System was to provide the necessary information during the design process. Most of the knowledge contained resulted from using the Geometry Acquisition Tool and the Selection Tool. The Advise System could be activated separately from the Hull Design Tool and is not linked to the program. Flexibility in terms of usage resulted because the KBS did not control the design process; it only offered advise that the user use or might want to avail of. Well-experienced designers who use Hull Design Tool may or may not use the system for reference. Non-expert users are given a variety of recommendations on what parameters to change and the limits of each design variable (Figure 4).

Implemented in Nexpert Object, the Advise System included as the knowledge base classes and object descriptions of major components of fishing vessels, a database of hull forms, performance databases and processed parametric data. Rules normally applied during the initial stages in design, classification rules for various vessel types, selection rules on load items and inference control rules were also included. Explanation facilities include views of different objects existing at a particular state and information on how a particular decision or recommendation was determined.

Test cases showed accurate reproductions of designs from scanned drawings with the Advise System offering useful, necessary and required information. The combination of the Hull Design Tool and Advisory System resulted in rapid hull form generation within the range of the knowledge base. [Agu95]

CONCLUSIONS

The development of systems with KBS functions required not only expert knowledge, different reference materials and algorithms but also the capability to explicitly model and structure all relevant objects, procedures and knowledge structures. Knowledge base construction required the use of approaches that combined both modelling and acquisition methods. Knowledge acquisition methods were used to build the pieces of knowledge in the form of rules. Modelling was used to structure such knowledge pieces into a practical and easily maintained knowledge base. This dichotomy was established as components of the system were being implemented and proved practical in the construction of the knowledge base in the advisory tool. The different tools developed can also be integrated or converted into more comprehensive and powerful systems not only for hull form or ship design but also for similar problems in other domains.

REFERENCES

[Agu95]   Aguilar, G. D. *A knowledge based system for the preliminary design of hull forms.* Doctor of Engineering Dissertation. University of Tokyo Libraries, 1995.

[AKY93]   Aguilar, G. D., T. Koyama, and H. Yamato. An approach to knowledge acquisition for the hull form design of fishing crafts 2nd Report: Object oriented methodology for the rapid development of procedural tools and interactive elicitation for hull selection. *Proceedings Spring meeting of the Japan Society of Naval Architects. Tokyo.* No. 174. pages 401-406, 1993.

[Boo86]   Boose, J. H. *Expertise transfer for expert system design.* Amsterdam, 1986. Elsevier.

[Boo89]   Boose, J. H. Using AQUINAS as a knowledge acquisition workbench for knowledge based systems. *International Journal of Man-Machine Studies* 26 : (1) pages 3-28. January 1989.

[BWV+87]   Breuker, J., B. Wielinga, M. Van Someren, R. De Hoog, G. Schreiber, P. De Greef, B. Bredeweg, J.Wielemaker, J-P Billaut, M. Davoodi and S. Hayward. Model driven knowledge acquisition: interpretation models, *ESPRIT P1098, Deliverable D1 (Task A1),* University of Amsterdam & STL Ltd., 1987.

[CM91]   Calisal, S.M. and D. McGreer. An expert system for fishing vessel design. *Proceedings of the 4th International Marine Systems Design Conference*, Kobe, Japan May 26-30. SNAJ, KSNAJ, Science Council of Japan, 1991.

[DL90]   Dym, C. L. and K.E. Levitt. *AI approaches to engineering design: Taxonomies and search prescriptions.* Working Paper No. 28, Center for Integrated Facility Engineering, Stanford University, CA, 1990.

[Hay63]   Hayes, J. G. The optimal hull form parameters. *Symposium on Numerical Methods Applied to Shipbuilding*, Oslo, 1963.

[Ito94]   Ito, K. Product model for ship structure from the viewpoint of structural design. *8th International Conference on Computer Applications in Shipbuilding.* Ed. by J. Brodda and K. Johansson. Bremen, Germany, 5-9th Sept. 1994. North Holland Pub Co.

[Kel55]   Kelly, G. *The Psychology of Personal Constructs.* New York. 1955. Norton.

[MK63]   Miller, N. S. and Kuo, C. The mathematical fairing of ship lines. *European Shipbuilding* Vol. 12, No 4, 1963.

[RB59]   Rosingh, W.H.C.E. and J. Berghius. Mathematical Design of Hull Lines. International Shipbuilding Progress No 6, Jan. 1959.

[RBP+91]   Rumbaugh, J., M. Blaha, W. Premerlani, F. Eddy and W. Lorensen. *Object Oriented Analysis and Design.* New York, 1991. Prentice- Hall International.

[SPS91]   Stearns, H., P. Payne and G. Smith. Designing a Knowledge Based Ship Design System. *Proceedings of the 4th International Marine Systems Design Conference*, Kobe, Japan May 26-30. SNAJ, KSNAJ, Science Council of Japan, 1991.

[Tay63]   Taylor, F. Computer applications to shipbuilding, *Transactions of the R.I.N.A*, Part A. Volume 105, 1963.

[TH93]   Tansley, D.S.W and C.C. Hayball. *Knowledge-Based Systems Analysis and Design.* A KADS Developer's Handbook. 1993. Prentice Hall.

[TS61]   Theilheimer. F. and W. Starkweather. The fairing of ship lines on a high speed electronic computer, *David Taylor Model Basin report 1474*, 1961.

[WBH90]   Welsh, M., I.L. Buxton and W. Hills. The application of an expert system to ship concept design investigations. *Transactions of the R.I.N.A*, Part A. Volume 133, page 123, 1991.

[WSB92]   Wielinga, B.J., A.Th. Schreiber and J.A. Breuker. KADS: A modelling approach to knowledge engineering. Knowledge Acquisition, 4, March. pages 5-53, 1992.

A HUMAN-INTELLIGENCE-IMITATING QUESTION-ASKING STRATEGY FOR PROPOSITIONAL KNOWLEDGE-BASED SYSTEMS

Jinchang Wang, Ph.D[1]
Department of Business and Economics
Missouri Western State College
St. Joseph, MO 64507-2294, USA
EMail: wang@griffon.mwsc.edu

ABSTRACT

When initial data are not sufficient to accomplish inference in a knowledge-based system, more information may be needed. A question-asking strategy is to select questions ask about. We present a heuristic question-asking strategy for propositional knowledge-based systems. Our computational experiments showed that the new strategy is significantly better than the strategy which is currently used in many knowledge-based systems.

1. INTRODUCTION

Inference is to obtain logically implied facts from given data. Human beings make inference everyday, which can be as simple as "if it is Sunday then no postman comes to pick up mails" and "if one does not sleep enough then he may fall in sleep when driving in the highway", or as complicated as in lawsuits, disease diagnosing, and decision making in financial investment, management and military operations. Many computer knowledge-based systems have been created and developed in the past ten years, which mimic the human way of thinking, for the purpose of knowledge processing efficiency, knowledge sharing, knowledge dissemination, and knowledge portability.

In an application of a knowledge-based system, inference is carried out towards proving some goals which are called top-level-conclusions (TLC). For example, possible faults in an engine are TLCs in a diagnosing system for maintenance of an aircraft; required operating adjustments are TLCs of a real-time piloting control system.

It often happens that initial data are not sufficient for a knowledge-based system to reach any TLC, and more information is needed. The systems for diagnosing and consulting typically work in that circumstance. As in a disease diagnosing expert system, the system is usually unable to tell what disease a patient has based on the superficial symptoms, and would ask for more information and tests, such as family disease history, blood test, X-ray. A question-asking (QA) strategy is to identify the needed data which would guide inference to a TLC.

A good QA strategy would get a TLC proved after asking a few key questions. When costs of answering questions are taken into account, a good QA strategy would prove a TLC at a low cost. A poor QA strategy, on the other hand, would ask irrelevant, silly, and costly questions and retard the whole inference process.

Question-asking is a part of inference process. The target of inference is to prove a TLC. In many applications, an inference process can be divided into two parts, inference-running and inference-guiding. Inference-running is to see, based on the information on hand, whether a TLC is proved, and if so, what it is. Inference-guiding is to figure out proper missing information if no TLC can be proved by using the available data. For the inference-running algorithms, as SAT algorithms, one focuses on efficiency which is reflected in size of the search tree and/or running time. For an inference-guiding algorithm, which is to guide inference into a right direction, one focuses on effectiveness which is reflected in number (cost) of

---

[1] This research is partially supported by ONR grant #N00014-94-1-0355.

questions asked before a TLC is proved. If the objective of the algorithms for inference-running is "thinking quickly", then the objective of the algorithms for inference-guiding is "thinking on the right track". To obtain additional information an intelligent system has to contact its environment through, for example, running more sensors and equipment, doing more analysis and calculations, and talking to the users. Contacting environment usually takes longer time than running inference inside an intelligent system. A good QA algorithm would pick up a few key questions so that their confirmation would lead to a TLC. For the purpose of developing an efficient inference engine for a knowledge-based system, the algorithm for inference-guiding is as important as, if not more important than, the one for inference-running.

The question-asking problem is computationally hard. Inference in a propositional system is a notorious NP-complete problem. Question-asking is even harder. Even in a propositional Horn knowledge base where inference is linear, question-asking has been proved to be NP-hard [WV90].

The issue of developing an effective and efficient QA strategy has been considered by many researchers. The expert system EXPERT [HWL83] uses pre-ordered lists of rules and questions. PROSPECT [DGH79] uses a scoring function for selecting questions. The "Alpha-Beta" pruning strategy was introduced by Mellish [M85] for acyclic inference nets. De Kleer and Williams [DW87] developed an entropy-based technique for minimizing the number of measurements which are needed to detect the faulty components. The heuristic question-selection algorithms developed by Wang and Vande Vate [WV90] was experimentally proved effective for Horn clause systems [TW94]. Some variations of the Wang-Vande Vate QA algorithm were developed afterwards, taking questioning costs and uncertainties into account [W94] [WT96]. However, there has been no systematic research on the QA strategy for general propositional systems. Most of current knowledge-based systems are using a contingent QA algorithm, which is on a random base, with the backward-chaining inference process.

2. THE NEW QA STRATEGY

For convenience of conceptually describing our QA algorithm, we introduce some more terms. An assertion is observable if its value can be obtained directly from an information provider (a sensor or a user, for example) out of the system. If an observable assertion's value is not yet asked, it is called an unconfirmed observable assertion

(UOA). Most of the current knowledge-based systems are using a "contingent" question-asking strategy to deal with the situation that given data are not sufficient. It works with backward chaining inference procedure, as follows.

Contingent QA Strategy in Backward Chaining:
Step 0: Place a TLC in a stack.
Step 1: Pop out an literal from the top of the stack. If the literal is a UOA, ask about it, go to Step 2; otherwise select a rule which can prove it and put the premise assertions of the rule into the stack, go to Step 1.
Step 2: Obtain the value of UOA. Testing whether a TLC is proved. If so, stop; If not, go to Step 1.

That contingent QA strategy follows a way of thinking by humans. To prove a conclusion, certain logical rules and facts are needed. The values of observable facts are obtained from the user or some sources, while the values of inferred facts are determined by inference.

An advantage of the contingent QA strategy is that all the questions asked are relevant towards proving some possible TLC and no irrelevant questions are asked. Another advantage is that it is easier to answer the questions such as "why ask that question". A disadvantage, on the other side, is that it is kind of blind in picking up possible TLC and related rules. If there are more than one possible TLC, the strategy is not able to select a "more" possible TLC which may need to ask less questions. If there are more than one rules to prove an assertion, the strategy is not able to select a rule which is associated with less missing data.

People have another way of thinking when initial data are not sufficient. Initial data are reviewed and tentative inference is carried out to see which TLC is more likely. Among the rules which may prove a TLC and/or an assertion, identify a rule which has fewest missing facts. Questions are selected so that they may help prove the TLC by using the identified rules with minimum number of missing data.

The above way of guiding inference motivated us in developing our new method, pseudo-UOA-set algorithm, for question-asking in propositional systems. For a TLC A_i, an unconfirmed observable assertion set (UOA-set) of A_i is a set of UOAs such that if they were had certain values then A_i would be proved. The pseudo-UOA-set algorithm identifies a TLC associated with a small UOA-set and a set of rules so that if the UOAs in the UOA-set are of certain values, the TLC is proved. We selects questions from the UOA-set to ask, until: (a) an answer is obtained and it is not in favor of proving the TLC through that UOA-set, then another UOA-set has to be found; or (b) all UOAs in that UOA-set are asked, and answers are all in favor of proving the TLC, then the TLC is proved. A TLC may have many UOA-sets. Identifying a smallest UOA-set is harder than inference. We used a heuristic labeling algorithm to find a small UOA-set.

The following is the procedure of inference with the pseudo-UOA-set algorithm as guide. Without loss of generality, we assume that the knowledge base is in "if (premises) then (conclusion)" rules. This assumption does not exclude any particular cases in propositional logic since a proposition in any format can be converted into the format of "if...then..." rules. A fact can be expressed as a rule without premises. Let us represent an assertion as A_i, $i = 1, 2, ..., n$. Let $R(i,d)$ denote the d-th rule whose conclusion is A_i. Let $D(i)$ be the index set of rules that have A_i as the conclusion, $I(i,d)$ be the index set of the premises of rule $R(i,d)$.

Inference with Guide of Pseudo-UOA-set Algorithm.

BEGIN {Main} ;
Step 1, Inference-running.
Run an inference procedure. If a TLC is proved, stop;
Step 2, Inference-guiding.
Ignoring all the negation signs of assertions:
   Label the unconfirmed assertions and rules by using the LABELING procedure ;
   Select a potential top level conclusion A_c with a smallest label;
   Trace and find a UOA-set, say U, which is associated with A_c and has a smallest label;
   Select a question A_k to ask from the UOA set U;

Get a response from the user about the value of A_k;
Go to Step 1.
END {Main} ;

Procedure LABELING ;
BEGIN { LABELING } ;
Step 1:
Label each unconfirmed observable assertion A_i as $C_i = 1$.
Step 2:
Alternately use PROPAGATE and ASSIGN to label the assertions and clauses until no new labels are applied in either subroutine.
END { LABELING } ;

Procedure PROPAGATE ;
BEGIN { PROPAGATE } ;
While there is an unlabeled element t, in which all immediate causes have been labeled, label t according to the following rule:
If the element t is a rule $R(i,d)$ whose unconfirmed premises are all labeled, then label it as:

$$C_{id} = \sum_{j \in I(i, d)} C_j \; ;$$

If the element t is an unconfirmed assertion A_i, such that the rules with conclusion A_i are all labeled, then label it as:

$$C_i = C_{i\delta(i)},$$
$$where \; \delta(i) \in D(i) \; and$$
$$C_{i\delta(i)} = \min_{d \in D(i)} \{ C_{id} \}.$$

END { PROPAGATE } ;

Table 1. Number of questions asked by using three QA methods

KB Group	KB parameters	wild-random	contingent	pseudo-UOA-set
1	40,40,4,25	19.5	4.5	4.0
2	80,60,4,40	27.4	15.0	4.8
3	80,80,4,50	40.5	22.5	12.0
4	100,100,4,70	40.8	17.4	8.0
5	100,100,5,70	44.2	24.6	10.4
6	100,120,5,85	53.4	23.4	10.2
7	100,150,4,100	54.2	24.7	11.0

Procedure ASSIGN ;
BEGIN { ASSIGN } ;
   Among all labeled rules whose conclusions are not
      labeled, choose the one, say (k,d*), with a
      smallest label. That is:

$$C_{kd*} = \min_{R(i,d)\ labeled,\ A_i\ unlabeled} \{C_{id}\}.$$

   Let: $C_k = C_{kd*}$.
END { ASSIGN } ;

   Step 1 is a regular inference-running process. Any
inference procedure for propositional logic can be used
here. If a TLC cannot be reached, Step 2, the pseudo-
UOA-set question-asking algorithm, is needed to guide the
inference. The pseudo-UOA-set algorithm is a labeling
heuristic and works on the un-signed propositions. A
rationale for ignoring the symbol of negation is that the
values of assertions are unknown so we do not care
whether they occur in the proposition as negations or not
when we select questions to ask. The UOA-set selected
by this heuristic, however, is neither sound (i.e., it may
not lead to a TLC) nor complete (i.e., the algorithm may
not find all UOA-sets). Running time of the pseudo-
UOA-set algorithm is linear to the size of the knowledge
base if there are no cycles in the knowledge base, and
log-linear otherwise.

3. COMPUTATIONAL EXPERIMENTS

   We have carried out some computational experiments
comparing QA algorithms. In those experiments, the
inference-running procedure is Jeroslow-Wang method
[JW90], the inference-guiding procedure is one of the
three QA algorithms: our new pseudo-UOA-set QA
algorithm, the contingent QA strategy, and the wild-
random QA strategy. The contingent QA strategy, as we
described before, is used with the backward chaining
inference-running procedure, so that if an UOA is
encountered during backward chaining then it is asked for
confirmation. The wild-random QA strategy is to select
a UOA to ask from the pool of all UOAs. Table 1 shows
average numbers of questions needed to reach TLC's in
seven groups of knowledge bases by the three algorithms,
"wild-random", "contingent", and "pseudo-UOA-set".
The knowledge bases are randomly generated. Each group
has a set of parameters which are listed in the second
column. The four parameters of knowledge bases are
number of clauses, number of assertions, number of
assertions in a clause, and number of UOA's. We tested
twenty knowledge bases for each group of parameters.
One can see that pseudo-UOA-set QA algorithm is better
than the other two algorithms in all groups. Number of
questions asked before a TLC is reached by using pseudo-
UOA-set algorithm is only 18% - 30% of that by using
"wild-random", and 27% - 90% of that by using

Figure 1. Average Numbers of Questions Asked by Three QA Methods

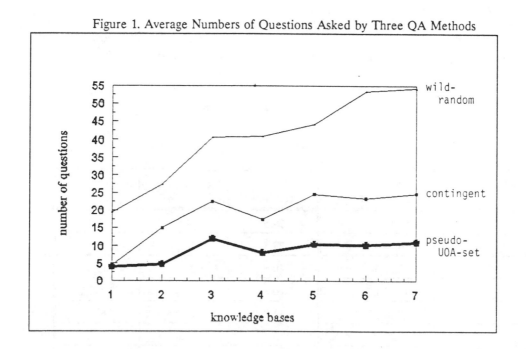

"contingent". Figure 1 compares the three algorithms graphically. It shows also a tendency that the larger the knowledge base the more significant "pseudo-UOA-set" outperforms the other two.

4. CONCLUSION

Our QA algorithm can better guide inference when given data are not complete. Our newly developed QA algorithm is two or three times better than the currently widely used algorithm. That suggests a great potential impact on efficiency and intelligence of knowledge-based systems, especially for problem-diagnosing systems, real-time control systems, intelligent emergency system, automatic personnel training systems and decision/command support systems where interacting with environment is typical during inference.

Developing question-asking strategies in propositional systems is a topic which has not yet been well explored. Most of the current knowledge-based systems are using "contingent" QA strategy which is to a large extent a fumbling method. Implementing a "smarter" QA strategy, as our QA algorithm, can shorten the whole inference process, save un-necessary answering costs, and make the system look more intelligent.

REFERENCES

[DGH79] R.Duda, J.Gasching and P.Hart. Model design in the PROSPECT Consultant System for Mineral Exploration, *Expert Systems in the Micro-electronic Age*, Edinburgh Univ. Press, Great Britain, 1979

[DW87] J.De Kleer and B.C.Williams. Diagnosing Multiple Faults, *Artificial Intelligence*, 32, 1987, pp.97-130.

[HWL83] F. Hates-Roth, D.A.Waterman, D.B.Lenat. Building Expert Systems, Reading, MA: Addison-Wesley, 1983

[JW90] R.Jeroslow and J.Wang. Solving Propositional Satisfiability Problems, *Annals of Mathematics and Artificial Intelligence*, 1, 1990, pp.167-187

[M85] C.S.Mellish. Generalized Alpha-Beta Pruning as a Guide to Expert System Question Selection, *Expert System 85, Proceedings of the Fifth Technical Conference of the British Computer Society Specialist Group on Expert Systems*, Univ. of Warwick, Cambridge Univ. Press, Cambridge CB2 1RP, pp.31-41

[TW94] E.Triantaphyllou and J.Wang. The Problem of Asking the Minimum Number of Questions in Horn Clause Systems, *Mathematical and Computer Modeling*, Vol.20, No.9, 1994, pp.75-87

[W94] J.Wang. Identifying Key Missing Data for Inference under Uncertainty, *International Journal of Approximate Reasoning*, Vol.10, 1994, pp.287-309

[WT96] J.Wang and E.Triantaphyllou. "A cost-effective question-asking strategy for propositional knowledge-based systems," to appear in *Annals of Mathematics and Artificial Intelligence*, 1996

[WV90] J. Wang and J. Vande Vate. "Question-asking strategies for Horn clause systems," *Annals of Mathematics and Artificial Intelligence*, 1, 359-370, 1990

INCREMENTAL RULE PRODUCTION : TOWARDS A UNIFORM APPROACH FOR KNOWLEDGE ORGANISATION

Mondher Maddouri and **Ali Jaoua**

Department of Computer Science,
Campus Universitaire, Le Belvédère, 1060, Tunis, Tunisia
E-mail: maddouri@ensi.rnrt.tn, alijaoua@avicene2.rnrt.tn

ABSTRACT:

Production rule based systems constitute well interesting knowledge based systems, since they generate non exclusive IF-THEN rules; whereas many of them are exponential in complexity. In this work, we propose a polynomial and incremental rule production method (the IRP method), that extracts IF-THEN rules from a set of data. Out of the standard, the IRP system provides tools for extracting "meta-knowledge", detecting poor attributes, ... A complexity and a precision analysis of the system is presented.

1. INTRODUCTION:

In rule extraction applications, the knowledge is represented by a set of objects stored in a relational data base [ZO94]. Each object constitutes a record of values of one "exogene" attribute (the attribute that will be explained) and some "endogene" attributes (the attributes used to explain the "exogene" attribute). The "endogene" attributes are continuous, whereas the "exogene" attribute is nominal and partitions the whole data base into some categories (natural classes) [ZAD92]. If the category of the object is mentioned in the data base, the object is called "resolved object"; otherwise it is called "unresolved object". The problem consists of extracting a set of IF-THEN rules from the resolved objects (knowledge acquisition), and using these rules in order to predict the exogene attribute value of an unresolved object (decision making). Many knowledge acquisition methods exist and can be classified in two categories: the decision tree methods and the production rule based methods. The decision tree methods use heuristic entropy measures to generate mutually exclusive IF-THEN rules [MCMK93]. The production rule based methods generate much more interesting non mutually exclusive IF-THEN rules but, generally, they are exponential in complexity [WK91].

In this paper, we present a production rule based method : the incremental rule production (IRP) based method. This method consists in transforming a relational data base to a binary relation (binary matrix) that will be covered by a minimal number of optimal rectangles (complete sub-matrix). Then, each rectangle is assumed to be an IF-THEN rule, since it can make an association between a set of objects which verify, simultaneously, a set of properties. The IRP method is based on a heuristic that makes the rule extraction incremental and polynomial in complexity. By incremental, we mean that, as the data base is dynamically modified by adding or deleting data, the knowledge base is also dynamically modified by adding or deleting rules. This is match more interesting than the fully regeneration of the knowledge base after each modification of the data base. The IRP method is applied to predict the category of a flower (Setosa, Versicolor or Verginica) using the values of four attributes (petal length, petal width, sepal length and sepal width). It is applied, also, to detect the Wisconsin Breast Cancer desease, when the values of nine clinical tests are given [WK91].

In section 2, we present a method to transform a relational data base to a binary relation, we specify the mathematical and the heuristic concepts to search for an optimal rectangle and we explain the rule extraction process of the IRP approach. In section 3, we use the concept of optimal rectangle to deal with some actual research futures out of the standard problem such that the incremental maintenance of the knowledge base and the "meta-knowledge" acquisition. Then, we propose an automatic method to detect poor predictive attributes. In section 4, we present the precision measures of the method and we specify a comparative complexity analysis.

2. THE INCREMENTAL RULE PRODUCTION METHOD (I.R.P.):

2.1. REPRESENTATION OF DATA:

In Human Science, we suggest that Human reasoning is based on decision rules made of vague notions [Zad77]. In fact, to decide if a flower is "Setosa" or not, it is interesting to know if it has a low petal length but it is not interesting to know if her petal length is 4 mm or if it is 4.5 mm. Consequently, a process of transforming a continuous attribute to some modalities is useful to simulate Human reasoning. This designates a well known problem in pattern recognition labelled "Modal representation problem".

The modal representation of a non binary attribute consists in its decomposition into k modalities which represent the different vague (linguistic) values of the attribute [Zad77]. Each linguistic value is modelled by an interval. For example, the attribute "Petal Length" will be decomposed into three modalities: "short", "medium" and "high" (Figure 1). The choice of the optimal number of modalities and the intervals (or the thresholds) associated constitutes the major difficulty of the existing approaches.

In our work, we present an automatic approach for modal representation. First, for every continuous attribute, draw the histogram representing the distribution of the attribute values. Each occurrence is coloured by the natural class of the object associated to it. Thus, each modality of the histogram represents a linguistic value of the attribute and it is associated to a natural class (the most frequent one in the modality).

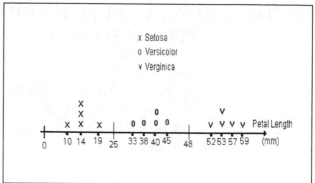

FIGURE 1: Histogram for modal representation of Petal Length attribute (IRIS data)

Second, detect the different modalities, and determine their natural class and the thresholds that separate them. Finally, if two disjoint modalities have the same natural class, we join them in one modality.

Figure 1 illustrates the modal representation of the Petal Length attribute obtained by the previous approach. The number of modalities and the choice of thresholds depend on the data and are dynamically modified after each addition of new objects to the data base.

Using this, we can transform an n-ary relation (that represents a relational data base) to a binary relation R (figure 2). R is defined between a set of objects O and a set of properties P, as a set of the Cartesian product OxP, by :

$R(o,p)=1$ if the object 'o' has the property 'p',
$R(o,p)=0$ otherwise.

O	petal length	petal width
o1	24	5
o2	11	4
o3	19	7
o4	33	5
o5	45	5

P / O	p1	p2	p3	C
o1	1	1	0	1
o2	1	1	0	1
o3	1	1	0	1
o4	0	1	1	2
o5	0	1	1	2

O: set of objects
C: flower category
   C=1: Setosa flower
   C=2: Versicolor flower

P: set of properties
   p1: low petal length=[0, 25] (mm)
   p2: low petal width=[0, 9] (mm)
   p3: medium petal length=[25, 48] (mm)

FIGURE 2: Modal representation of continuous attributes related to 5 flowers (IRIS Data base).

2.2. BASIC CONCEPTS:

Let O be a set of objects (Patterns) and P be a set of properties.

Definition 2.1: A binary relation between O and P is a subset of the Cartesian product OxP. An element of R is denoted by (x,y). ♦

For any binary relation R, we associate the following definitions:

* The domain of R is defined by
$$Dom(R) = \{e \mid \exists e' : (e,e') \subseteq R\},$$
* The codomain of R is defined by
$$Cod(R) = \{e' \mid \exists e : (e,e') \subseteq R\},$$
* The cardinality of R is defined by
Card(R)= number of pairs in R,

Definition 2.2: Let R be a binary relation defined between O and P, a rectangle of R is a Cartesian product of two sets (A,B) such as $A \subseteq O$, $B \subseteq P$ and $AxB \subseteq R$, where A is the domain of the rectangle and B is the codomain ♦

Definition 2.3: A binary relation R is elementary if and only if for all element $e \in Dom(R)$ and $e' \in Cod(R)$, there is a path between e and e'. ♦

Notation: PR=ϕ_R(a,b), Pseudo Rectangle, is used to denote the elementary relation containing the couple (a,b), RE to denote a rectangle and RE(a,b) to express that (a,b)∈ R.

Definition 2.4: Let R be a binary relation defined between O and P, a rectangle (A,B) of R is Maximal if $AxB \subseteq A'xB' \subseteq R \Rightarrow A=A'$ and B=B'. ♦

Definition 2.5: A rectangle containing (a,b) of a relation R is Optimal if it realises the maximal gain among all the maximal rectangles containing (a,b). The gain realised by a rectangle RE = (A,B) is measured by :
$$g(RE)= card(A)*card(B)-[card(A)+card(B)] ♦$$

Definition 2.6: A coverage of a relation R is a set of rectangles CV={RE1,........REn} of R such that any element (a,b) of R is contained in at least one rectangle of CV. ♦

Definition 2.7: A coverage CV defined by {RE1, ... , REn} of a relation R is minimal if it is formed by a minimal number of optimal rectangles. ♦

Example: Figure 3.a illustrates the example of elementary relation, R, which contains the item (o4, p2) of the initial relation representing the IRIS data base. We note that the optimal rectangle containing (o4, p2) is RE2(o4, p2) [figure 3.c] having the gain zero. The minimal coverage of R contains only the rectangles RE1 [figure 3.b] and RE2.

Heuristic for selecting an optimal rectangle : The problem of searching for an optimal rectangle containing (a,b) from a binary relation R is NP_hard [BJDEOG94]. An heuristic for searching a minimal coverage of optimal rectangles into a binary relation were used in [MJ95] and were successfully applied in many fields such that Data base Analysis, signature files [AJ96], documentary data bases and information systems [KJ96]. Recently, it is used to extract consensual expert rules in multi expert systems [GL94].

2.3. RULE EXTRACTION:

In paragraph 2.1, we proposed a modal representation approach that transforms a relational data base presented by an n-ary relation with non binary attributes to a binary relation R in OxP, where O is a set of objects and P is a set of properties. Each property is either a binary attribute or a modality of a non binary attribute such that : "low Petal length". The resultant relation R is a particular binary relation like that presented in figure 2.

With the modal representation of pattern data base, an optimal rectangle RE=(A,B) is a Cartesian product of the subset A of objects (domain of RE) by the subset B of properties (codomain of RE).

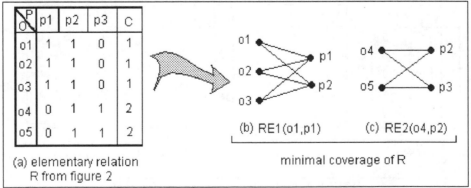

(a) elementary relation
   R from figure 2

(b) RE1(o1,p1)   (c) RE2(o4,p2)

minimal coverage of R

FIGURE 3: Example of elementary relation of figure 2

The optimal rectangle	The rule associated	Weight
TABLE 1 : Example of Production Rules extracted from the relation R of figure 3.		
E1={o1,o2,o3}x{p1,p2}/C=1	IF Low Petal Length AND Low Petal Width THEN 'SETOSA' flower.	3
RE2={o4,o5}x{p2,p3}/C=2	IF Low Petal Width AND Medium Petal Length THEN 'VERSICOLOR' flower.	2

The Cartesian product reflects the fact that all the objects of the domain verify simultaneously, all the properties of the codomain of RE. Then, we cannot find an other property p (p∉B) that is verified by all the objects of A. Similarly, we cannot find an other object o (o∉A) that verify all the properties of B. This is due to the definition of optimal rectangle. We have a high form of similarity between the objects of A based on the properties of B.

Each optimal rectangle RE of R can be associated to a natural class of objects, since the objects constituting the domain of RE may have the same value of categorical attribute (endogene attribute). In general cases, fewer objects of the domain of RE can have different natural class labels from the others, whereas they verify all the properties of the codomain of RE. We consider such fact as an error due to noisy data and we assign the value of the most frequent natural class to the optimal rectangle RE. In conclusion, for each optimal rectangle of R, we can assign a natural class, using the categorical attribute. From the definition of optimal rectangle and its semantic characteristics that we defined previously, we assume that the properties of the codomain of an optimal rectangle RE constitute the conditions of the rule that associates the rectangles category (the most frequently natural class) to the objects of the rectangle domain. So, each optimal rectangle RE of the binary relation R is an equivalent representation of a production rule in the set-theory [BJDEOG94]. The codomain of RE constitutes the action part of the rule and the domain cardinality of RE will be used to ponder the rule (conflict rule problem).

The problem of rule extraction from a relational data base is equivalent to the problem of searching for an optimal rectangle from a binary relation R. In the whole paper, the words "optimal rectangle" and "Production rule" mean the same thing. Since the problem of searching for an optimal rectangle is NP_hard [KJ96], we use the heuristic presented in (section 2.2) to generate a minimal coverage of optimal rectangles of R. The knowledge base is then deduced from the minimal coverage as presented in table 1.

3- OUT OF THE STANDARD:

3.1. KNOWLEDGE BASE MAINTENANCE :

In real world applications, the data base is usually modified by adding new objects to it or modifying old measurements [Can93]. Most learning methods require the fully reorganisation of the knowledge extracted from the data base every once the data base is modified. This is frequent in decision trees, statistical and neural net methods [WK91]. In our work, we use an algorithm that, once the data base is modified, it reproduces the new knowledge base, using only the last knowledge base and the information added to (or deleted from) the data base with a minimal cost.

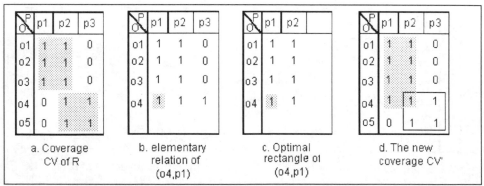

FIGURE 4: Addition of (o4,p1) to R (figure 3).

Since "production rule" is equivalent to "optimal rectangle", the "knowledge base" is equivalent to "minimal coverage" of the binary relation R. Then, the problem of knowledge maintenance deals with the modification of rectangular structures due to the addition of new couples or deletion of old couples from the binary relation (or both). In the literature there are very few maintenance algorithms. In general, these algorithms are developed for growing databases; only some of them, however, can be used for couple deletion. Here, we present two algorithms (taken from [KJ96]) : an algorithm for data addition and an algorithm for data deletion. Each algorithm is illustrated by an example.

Data addition :

In a dynamic environment, data additions occur frequently. So it is very useful to add data and maintain the data base organisation with minimum overhead. Our task is to construct a new minimal coverage of the base after each update using only the old coverage and the package of data to add.

> Algorithm Data-addition
> Begin
> 1. Let CV be the minimal coverage of R and Pqt be the entry to add.
> 2. Sort the couples of Pqt by the value of the gain function in decreasing order
> 3. While (Pqt$\neq\varnothing$) Do
>   * Select a couple (a,b) in Pqt by the sorted order
>   * Search PR : the pseudo-rectangle containing (a,b)
>   * Search RE: the optimal rectangle containing (a,b)
>   * CV = (CV - {r$\in$CV / r$\subseteq$RE })$\cup${RE}

> * Pqt= Pqt - { (X,Y)$\in$Pqt / (X,Y)$\in$RE}
> End While
> 4. Delete all the redundant pairs in the different rectangles in CV
> End

Example: Let the relation R be the relation of the figure 3. The package to add is Pqt={(o4,p1)}. Figure 4 presents the steps of incrementation. In the knowledge base, only the weight of the first rule (table 1) is updated from 3 to 4.

Data deletion :

The case of data deletion is ignored by most incremental data clustering algorithms. Here we propose a solution for this case. Let CV be the initial coverage of R and Pqt the package to delete, the algorithm proceeds as follows:

> Algorithm data-deletion
> Begin
>   1. Compute the union, RU, of all optimal rectangles containing a pair (a,b) to delete,
>   2. Delete from CV all rectangles containing a pair (a,b) to delete,
>   3. Delete from RU the pairs contained in Pqt,
>   4. Execute algorithm data-addition from the initial relation CV and the package to add RU.
> End

Example: Let the relation R be the new relation of figure 4. The package to delete is {(o4,p1)}. Figure 5 presents all deletion steps.

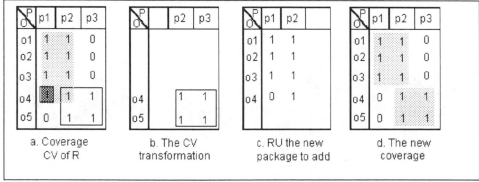

FIGURE 5 : Deletion of (o4, p1) from the relation R of figure 4.

3.2- KNOWLEDGE ORGANISATION:

The knowledge base is very large in usual knowledge systems that manage well developed and variant expert fields. In expert systems, we proceed by substituting the knowledge base into sub-knowledge bases dealing with defined subjects to deviate the research for the interesting rule to a small knowledge base [Dom88]. In our work, we propose an automatic manner to organise the knowledge base (substitute).

Definition 3.1 : Let R be a binary relation between O and P, and CV={RE1, RE2, ..., REn} its "minimal coverage". We define a super-relation SR between CV and P by :

SR(i,j)=1 if the property 'j' exists in the codomain of the rectangle REi of CV;
SR (i, j) = 0 otherwise. ♦

Definition 3.2 : A super-rectangle (SA, SB) of R is a rectangle defined on the super-relation SR. Similarly, we define a maximal and optimal super-rectangle and a minimal super-coverage of a binary relation R. ♦

In our proposition, we will substitute the whole knowledge base into a set of sub-knowledge bases {SKB_1, SKB_2, ...,SKB_k} such that each sub-knowledge base SKB_i is associated to a super-rectangle SR_i of the minimal super-coverage of R. The rules in SKB_i will be the synonymous of the rectangles constituting the domain of SR_i. When a user query is presented, it will be deviated to the sub-knowledge base that the super-rectangle associated contains the properties specified by the query in its domain.

Here, the same incremental algorithm used to extract the production rules, is also used to extract the meta-rules. Generally, we can consider the minimal coverage as a first level of data abstraction, the minimal super coverage as a second level, the minimal super super coverage as a third level, ... The last n-super coverage will contain only one rectangle (stopping condition).

Then, we build a hierarchy of abstraction levels (or knowledge levels) for the data base; that can be incrementally updated using the same incremental algorithm (a uniform approach for knowledge organisation).

Figure 6 shows the impact of using a "meta-rule" base on the number of rules tested in the IRIS knowledge base, when a complete query is presented. Over of 30% of the rules is not tested. Consequently, the CPU time required by the system will be greatly reduced. We note here that the number of rules tested by incomplete queries (when the user specify only one, two or three attribute values) is very less than the number of rules tested by complete queries. So, the saving time and the saving in number of rules tested by incomplete queries will be more important than those of the figure 6.

3.3- POOR FEATURES DETECTION:

The data requirements of the samples are quite simple. All we need is a set of observations and their correct associated classifications. The set of features for which we are given recorded observations are preselected. Thus we cannot expect miracles. If our selector does a poor job, and the features have poor predictive capabilities, the results will be poor. For example, if we are trying to predict the development of cancer in currently normal subjects, we may not be able to find any features that have a very high predictive potential.

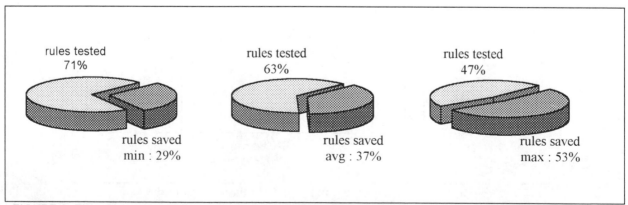

FIGURE 6: The percentage (min, avg, max) of reducing the rule research space when using a "meta-rule" base [the IRIS knowledge base].

R23: IF low sepal width AND large petal length
THEN Versicolor flower.
R24: IF low sepal length AND medium sepal width
THEN Versicolor flower.
R32: IF large sepal width AND large petal length AND
large petal width THEN Versicolor flower.

**FIGURE 7: Some rules for 'Versicolor' flower category
[IRIS knowledge base].**

In such case, an automatic method to detect the poor features is useful for any pattern recognition system. In the IRP system, we propose an automatic method to detect the poor features based on the rectangular representation. First, we define a poor feature for a natural class C, as an attribute that all its modalities are used in a condition part of some rules associated to the natural class C. In the figure 7, the conditions : Low Sepal width, Medium Sepal width and Large Sepal width appear in the condition part of the rules associated to the 'Versicolor' class. Second, we define a poor feature as a feature that is identified as poor feature for many natural classes. Then, like the Sepal length is identified as a poor feature for 'Versicolor' and 'Verginica' categories, the poor features identified will be removed from the data base.

Figure 8 presents the effect of removing poor features on the performance of the IRP system. The basic impact is that the rapidity of the system has been improved after removing the poor features; while the precision has not been modified.

The improvement of the time requirements of the method is due to the fact that the relational data base, and by consequence, the binary relation associated becomes more reduced and the number of operations made by the program is deminished.

4- EVALUATION MEASURES :

4.1. FLOWER CATEGORY RECOGNITION :

Here, the precision measures of the method are done on the IRIS data base [WK91]. It is a one hundred fifty patterns data base with three flower categories: 'Setosa', 'Verginica' and 'Versicolor'. Each category is represented by fifty patterns. Each pattern is represented by four continuous attributes: Petal length, Petal width, Sepal length and Sepal width. The problem consists to extract from the data base, the rules with which we can assign a certain flower category to a particular flower. These rules will be used, later, to predict a new coming flower category using only the measures of its Petal length, petal width, Sepal length and Sepal width.

The precision of the system is defined as the true error rate [MJ95]. The true error rate is the error rate, tested on the true distribution of cases in the population - which can be empirically approximated by a very large number of new cases gathered independently from the cases used to design the system [SLBGWM95]. In our work, we estimate the true error rate by the leaving-one-out error rate. The results are shown in Figure 9.

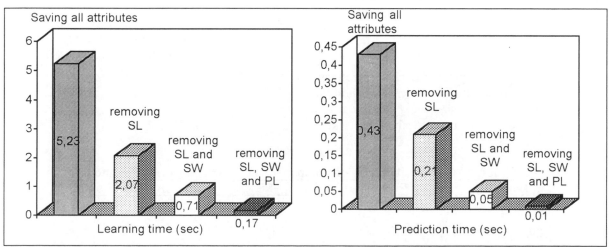

**FIGURE 8: Impact of removing poor features on the learning time and the prediction time, of the
IRP system**

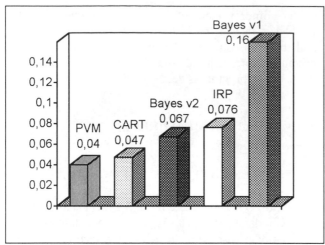

FIGURE 9: « Leaving-one-out » error estimation of five learning methods on the IRIS data base. We note that the IRP precision is acceptable.

4.2. COMPLEXITY ANALYSIS:

In this paragraph, we are interested only to the machine learning methods. The precision measure of the IRP method is very reliable relatively to other machine learning methods. Another important parameter of comparison is the complexity analysis. In general, the CPU time requirement of a method is proportional to its complexity. Here, we will study the complexity of a typical rule based method : the PVM (Predictive Value Maximisation) method, since the IRP method is a rule based method. The CART tree method is not a dynamically rule based method, but it will be studied. Let d be the number of objects (Objects), c the number of features of the data base and n the number of couples in the binary relation representing the data base (n=c*d).

In PVM rule method [WK91], first we choose an arithmetic operator for every attribute using the median method [$O(c*d)$]. Second, for each value of each attribute, we calculate the predictive value percentage, and search for its local maxima [$O(c*d^2)$]. Third, we establish the table of rules of size one. Each rule will be evaluated on the total data base. A threshold will be

chosen with which some one-sized rules will be removed [$O(c*d^2)$]. The remaining rules will be combined to produce rules of size two. Each rule of size two will be evaluated, a threshold will be chosen and some two-sized rules will be removed [$O((c*d)^2*d^2)$]. The complexity of the kth iteration is $O\left[(c*d)^{2^{k-1}} * d^{2^{k-1}}\right]$. Since we have c attributes (features), the total complexity is about : $O\left[\sum_{k=1}^{c} (n*d)^{2^k}\right]$. The PVM rule method is an exponential complexity method which cannot be applied to large data bases..

In the IRP method, the complexity analysis is that of the minimal coverage generation algorithm. The loop which calls Rect-Optimal routine is the most complex part of the algorithm. The cost analysis of this part gives the following results. First, to look for pseudo rectangle containing (a,b), when we use a matrix as a basic structure, the cost of this function is about $O(n)$. Second, to look for the optimal rectangle containing (a,b), the cost is about $O[n*(c+d)^2]$. Finally, to look for a minimal coverage, the number of iterations of the algorithm is equal to the number of rectangles of the relation. the cost of this part is majored by $O[n^2*(c+d)^2]$.

In the CART tree method [WK91], the number of nodes in the tree is about $O(d^2)$. To build any node, we will try the c attributes at most d times. Each time, we will evaluate the resulting tree [$O(c*d^2)$]. Then, to construct a complete tree, we require $O(c*d^4)$ operations. The pruning part of the algorithm consists of three actions. First, relatively to the leaving-one-out evaluation method, we will construct d trees. For each tree, we will test the k (k=d^2) nodes. For each node pruned, we will make $O(d^2)$ evaluations with $O(d^2)$ comparisons. This makes $O(c*d^{11})$. Second, we evaluate the k*d obtained trees on the test samples to choose one optimal tree [$O(d^5)$]. Finally, from the complete tree, we will derive the tree solution, by making the same k prunings (made on the chosen tree) [$O(c*d^{10})$].

TABLE 2 : Complexity/precision comparisons of machine learning methods.

Measures	PVM rule	IRP	CART tree
Error (Err|oo)	.040	.076	.047
Complexity	$\sum_{k=1}^{c} (n*d)^{2^k}$	$n^{2*}(c+d)^2$	$c*d^{11}$
Case c=d	$\sum_{k=1}^{d} d^{3*2^k}$	d^6	d^{12}

In this paragraph, we are interesting to the machine learning methods. In table 2, the error (Err_{100}: Leaving-one-out estimator error) measures of the PVM rule and CART tree methods are taken from [WK91]. Here, we accentuate the fact that the PVM method is exponential in complexity. This is due to the part of the algorithm that search for the combinations of simple conditions to build complexes ones. In the IRP system, we use a heuristic that enables us to identify a complexes condition rule without testing a combinatory number of possibilities. The CART tree method is not a production rule based method, since the rules deduced from the tree are mutually exclusive. Nevertheless, the complexity analysis proves that the IRP method is well faster than the CART tree method.

4.3- WISCONSIN B. CANCER DIAGNOSTIC:

The samples consist of 683 objects (patients), nine "endogene" attributes and two classes. We created ten randomly sampled data sets with 90% train and 10% test partitions; each method was tried on each of the ten data sets, and the results were averaged (Error Err_{10-CV}).

Table 3 summarises the results of two methods : the IRP method and the SIPINA method. The error rate is the error rate on the test cascs (the 10-fold cross-validation error rate).

SIPINA (Interactive System for Interrogation Process in a non tree form) [ZAD92] is a decision tree method that gives to the Human expert the possibility of joining or devising the tree leaves (in an interactive manner) in order to perform its precision. The IRP system has the advantage that it is a fully automatic system and can replace all the Human expert actions. Figure 4.3 proves that the IRP system performs well on this data set, and its precision is comparable to the SIPINA system.

5- CONCLUSION:

In this paper, we presented a knowledge based method: the IRP method. This method differs from the decision tree methods because it generates non mutually exclusive IF-THEN rules.

TABLE 3 : Comparative precision measures of the IRP and the SIPINA systems.

Methods	Error (Err_{10-CV})
IRP System	.0034
SIPINA	.0030

It defers also from many production rule based methods by its polynomial and incremental aspects. The complexity analysis proved that the conception of polynomial production rule based methods is possible. Out of the standard problem, the IRP system gives tools to deal with many research futures such that extracting "meta-knowledge" rules to reduce the rule research space and eliminating poor attributes to perform the system time cost. It has been also used to generate multidecisional rules and to reformulate incomplete user's queries [MJ95].

The IRP system handles precise and certain rules. It was applied to classification tasks such as flower class recognition from Fisher's IRIS data. It was applied to Wisconsin Breast cancer diagnostic to identify cancerous patients, ... It was proved that it has a precision value comparable to existing systems (PVM method, CART method, ...). Actually, the IRP system is being tested on other diagnostic applications, and we hope achieve better results compared to recent methods [HCK95]. Also, we are studying the possibility of using the IRP system in Arabic speech recognition fields in order to classify the CEPSTRE vectors.

In Human reasoning fields, it was proved that the linguistic representation of data is more appropriate to Human expert usage than the modal representation [Zad77]. In the IRP method, since a rectangle can be associated to many object categories, the rules generated are uncertain rules. Actually, we are studying a fuzzy version of the IRP method in order to deal with imprecision data representation and imprecision decision making in production rule based systems. This study includes a comparative evaluation work of fuzzy knowledge based systems.

ACKNOWLEDGEMENT:

Financial assistance for this work has been founded by SERST (Sécrétariat d'Etat à la Recherche Scientifique et à la Technologie de Tunisie), (PNM93: Organisation incrémentale de l'information). We thank also Mss Y. Hmaid and Mme I. Jedidi, students at the University of Tunisia, for their effort in implementing and evaluating this method.

REFERENCES :

[AJ96]   K. Arour, A. Jaoua, Signature extraction method in rectangular databases, will appear in *the journal of Information science*, 1996.

[BJDEOG94] N. Belkhiter, A. Jaoua, J. Deshernais, G. Ennis, H. Ounelli and M. M. Gammoudi, Formal properties of rectangular relations, *In Proceeding of the 9th International Symposium on computer and information sciences (ISCIS IX)*, p.310-318., Antalya Turkey, 7-9 November 1994.

[Can93] F. Can, Incremental clustering for dynamic information processing, *ACM transactions on information systems*, vol.11 N°2, p.143-164, April 1993.

[Dom88] C.H. Dominé, *Techniques de l'intelligence artificielle, un guide structuré*, DUNOD Informatique, 1988.

[GL94] M. M. Gammoudi & S. Labidi, An automatic generation of consensual rules between experts using rectangular decomposition of a binary relation, *In Proceedings of the XI Brazilian Symposium on Artificial Intelligence (SBIA 94)*, 1994.

[HCK95] HSU, Chun-Nan and Knoblock, Estimating the robustness of discovered knowledge, *In Proceeding of the First International Conference on Knowledge Discovery and Data Mining (KDD-95)*, Montreal, Published by the AAAI Press, 1995.

[KJ96] R. Khcherif, A. Jaoua, Incremental rectangular decomposition of documentary data base, *will appear in the journal of Information science*, 1996.

[MJ95] M. Maddouri & A. Jaoua, Incremental learning: proposition and evaluation of methods, *In Proceeding of the Fourth annual International conference (FT&T)*, Durham (USA), September 28, 1995.

[MCMK93] R. S. Michalski, J. G. Carbonell, T.M. Michell, Y. Kodratoff, *Apprentissage symbolique, une approache de l'Intelligence Artificielle*, CEPADUES Edition, 1993.

[SLBGWM95] N. Shahsavar, U. Ludwigs, H. Blomqvist, H. Gill, O. Wigertz and G. Matell, Evaluation of a knowledge-based decision-support system for ventilator therapy management, *journal of Artificial Intelligence in Medicine*, Volume 7 n°1, , p.37-52, February 1995.

[WK91] S. M. Weiss & C. A. Kulikowski, *Computer Systems that learn*, Morgan Kaufmann Publishers, 1991.

[ZAD92] A. Zighed, J. P. Auray, G. Duru, *SIPINA: méthode et logiciel*, Edition Alexandre Lacassagne-Lyon, 1992.

[Zad77] L. A. Zadeh, *Fuzzy sets and their applications to object classification and clustering analysis*, classification and clustering, New York Academic, 1977.

[ZO94] N. Zhong & S. Oshuga, Discovering concept clusters by decomposing data bases, *Data & Knowledge engineering*, 12, p.223-244, 1994.

ONTOLOGY-BASED INFORMATION GATHERING AND CATEGORIZATION FROM THE INTERNET

Michiaki Iwazume, Hideaki Takeda and Toyoaki Nishida
Graduate School of Information Science
Nara Institute of Science and Technology
8916-5 Takayama, Ikoma, Nara, 630-01 Japan
Email: {mitiak-i, takeda, nishida}@is.aist-nara.ac.jp

Abstract

In this paper, we propose a new method to develop more intelligent navigation system for the Internet using ontologies. We implemented a system called IICA (Intelligent Information Collector and Analyzer)which helps people to acquire knowledge from information resources on the Internet by gathering and categorizing information. We tested IICA for tasks on the World Wide Web (WWW) and the network news. The results of the experiments indicated that the ontology-based approach enable us to use heterogeneous information resources on the wide-area such as the WWW and the network news.

Keywords: Ontology, Information gathering, Text categorization, the WWW, Network news, Knowledge media

1 Introduction

Since the number and diversity of information sources on the Internet is increasing rapidly, it becomes increasingly difficult to acquire information we need. A number of tools are available to help people search for the information (for example [McBryan 94], [Maes 93]). However, these tools are unable to interpret the result of their search due to lack of knowledge. We need more intelligent systems which facilitate personal activities of producing information such as surveying, writing papers and so on.

In this paper, we propose an ontology-based approach to gathering and classifying information in order to realize intelligent agents to help personal activities of information production.

We implemented a system called "IICA" which helps people to acquire knowledge from the information resources on the Internet by gathering and categorizing information . Figure 1 shows the outline of IICA.

The function of IICA is twofold. (1) Information Gathering: IICA gathers WWW pages and USENET network news articles on the Internet in response to user's requests. IICA uses ontologies to compute the

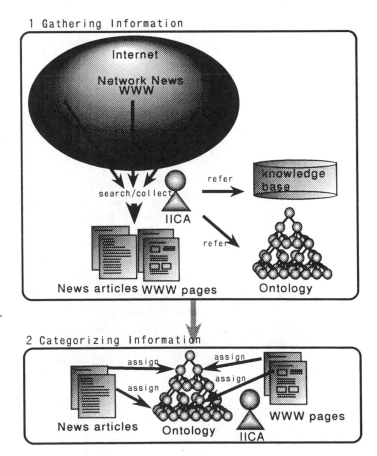

Figure 1: IICA: Intelligent Information Collector and Analyzer

similarity between the keywords given by the user and those extracted from candidate texts. In case there is no texts which contains the given keywords, IICA infers significant terms related to the given keywords and gathers texts concerned with these terms. (2) Information Categorizing: Furthermore, IICA categorizes the gathered texts by linking them with an ontology. The system helps nonprofessional users to search for information and understand the result of categorizing them by visualizing the ontological structures.

We tested IICA for tasks on the World Wide Web (WWW) and the network news. The results of the experiments indicated that the ontology-based approach enable us to use heterogeneous information resources on the Internet.

In Section 2, we will first explain the role of ontologies in gathering and categorizing text-like information. We also propose and describe weakly structured ontology which is developed from existing terminologies and thesauruses. In Section 3, we will show how IICA uses ontologies and heuristics to gather information intelligently, and we will explain a new method of text categorization using ontologies in Section4. In Section 5, we will describe the evaluation of the above two methods. In Section 6, we will explain the implementation of the prototype system briefly. In section 7, we will compare our work with related work. In section 8, we will discuss the advantage of our approach and summarize this paper.

2   Ontology

2.1   The Role of Ontology

An Ontology is specification of conceptualization which consists of a vocabulary and a theory [Gruber 91]. The role of ontologies in our approach is fourfold: (a) providing knowledge for agents to infer information which is relevant to user's requests, (b) filtering and classifying information (c) indexing information gathered and classified for browsing , and (d) providing a pre-defined set of terms for exchanging information between human and agents.

2.2   Weakly Structured Ontology

Unfortunately, development of ontologies is often a quite painstaking and time consuming task. Ontologies are often described in frame languages such as Ontolingua [Gruber 92] and knowledge representation

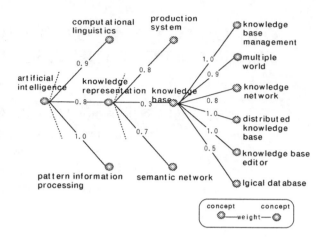

Figure 2: An Example of a Weakly Structured Ontology

languages based on first-order predicate logic. We believe that the difficulty comes from the fact the these languages is computer oriented media and not human-oriented media. Since most of our knowledge is in human media such as natural language documents, we have to somehow translate human-oriented media into computer-oriented media. As human-oriented media is often ill-structured, i.e., ambiguous, indefinite, vague, unstructured, unorganized and inconsistent, we need a tremendous amount of efforts on translating ill-formed information into well-formed information.

We decided to make use of weakly structured ontologies which is developed from existing terminologies, thesauruses [Iwazume 94], and technical books [Nishiki 94]. Weakly structured ontologies have only one type of associative relation between terms. Conceptual relations such as concept-value, class-instance, superclass-subclass, part-whole are not explicitly distinguished in the weakly structured ontologies.

In the following experiments, we use the ontology built from the information science terminology which has about 4,500 terms.

3   Ontology-based Intelligent Information Gathering

This chapter describes how IICA uses ontologies to gather information intelligently.

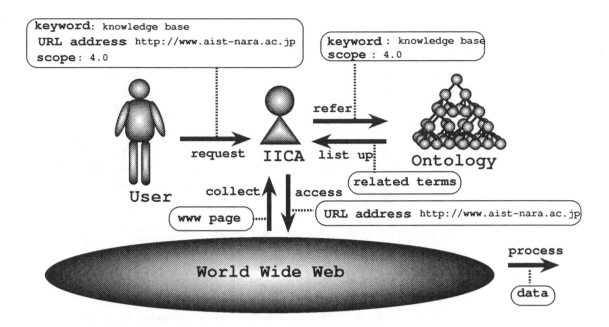

Figure 3: Outline of Information Gathering on the WWW

3.1   Inference of Related Terms to User Inputs

IICA uses ontologies to compute the similarity between the keywords given by the user and those extracted from candidate texts.

For example, suppose that a user wants to know information about "knowledge base". In case there is no texts which contains a term "knowledge base", IICA infers that significant terms related to "knowledge base" are not only terms containing the same string like "knowledge base management", "distributed knowledge base", and "knowledge base editor", but also terms related ontologically like "knowledge network" and "multiple world". IICA can also reason about the context from user's query. For example, when the input keywords are "semantic network" and "logical database", IICA interprets that the context is "knowledge representation". Inference about the context depends on the level of the terms and the weight values between terms. Users can control the scope of reasoning the context by specifying the threshold parameter.

3.2   Information Gathering on the WWW

In this section, we describe the search algorithm [1] on the WWW.

Figure 3 is outline of the information gathering on the WWW. IICA collects pages by (1)accessing HTTP or (2)searching the archive of WWW pages. In the former case, IICA gets the specified page by sending a URL address to its socket modules and accessing the specified host. The gathered page is added to the archive. All pages in the archive are managed by IICA with its file table . In the latter case, IICA searches the archives using the file table.

3.2.1   Algorithm

The algorithm is basically breadth-first searching. The difference is that IICA evaluates gathered pages and decides which anchor to access next. we show the algorithm as follows.

step1

Receive a set of keywords, starting URL address, scope of reasoning context and number of pages to gathered from the user.

---

[1]We should take notice that an automatic search on the WWW often bring about heavy loads on the network. In practical use, some heuristics such as restricting time and frequency to access to the network and avoiding concentrative access to particular hosts is necessary.

step2

> Match the keywords with terms in the ontology and list up terms relevant to the within the scope.

step3

> If the specified URL address exists in the close-list, search the page from the archive. Otherwise, retrieve the page by accessing HTTP.

step4

> If the number of pages is greater than the limit, exit the procedure. Otherwise, go to step5.

step5

> Parse the gathered page to extract URL addresses and labels in anchors and titles. If the addresses already exist in the open-list and close-list, discard them. Otherwise, add them to the open-list.

step6

> IF the terms listed up at step2 are included in the labels, score the labels using ontology. Otherwise, remove the label and the addresses from the open-list. Then Sort the open-list.

step7

> If there is no anchor in the page, pick up a URL address from the open-list. Then Go to step3.

3.2.2 Example

We describe an example of gathering pages on the WWW using above algorithm. Suppose that the user's keyword is "knowledge base", the starting URL address is "http://www.aist-nara.ac.jp/home-en.html", and the scope is 4.0 at step1 (see Figure 4). IICA generate a set of related terms to the keyword using the ontology (step2). As the specified URL address is not in the open-list or close-list, IICA retrieves the page (step3). Moreover, IICA extract 27 anchor labels and URL addresses from the page (step5), and score and sort them (step6). In this case, two anchors, "Graduate School of Information Science"(URL address: "http://www.aist-nara.ac.jp/IS/home-en.html", score : 4.0) and "Home Page of Department of Information Science in Kouchi University" (URL address: "http://www.is.aist-nara.ac.jp/", scope: 4.0) are added to the open-list.

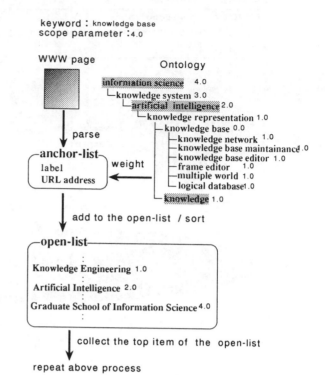

Figure 4: An Example of Information Gathering on the WWW

3.2.3 Heuristics

We often use various heuristics such as empirical knowledge and common sense, when we search for information on the WWW. For example, it is better for us to search for information about AI(artificial intelligence) using a heuristic that

> "the WWW page of institutes, laboratory often contains information about AI".

We decided to make use of heuristics. For instance, the heuristic that "if search for information on AI, go pages of laboratory" is described as follows:

    ''artificial intelligence'' → ''laboratory''

IICA gives priority over the pages which contain term "laboratory" and access them by using the heuristic.

    ''artificial intelligence'' → ''laboratory''
    ''artificial intelligence'' → ''institute''

3.3 Information Gathering on the Network News

Gathering articles on the network new is easier than gathering pages on the WWW, because the structure

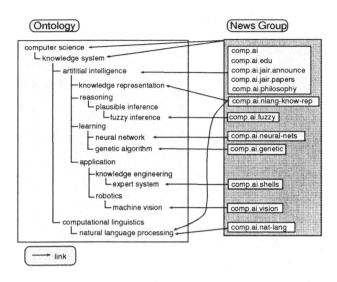

Figure 5: Classification of Newsgroups

of newsgroups is not as complex as that of the WWW. IICA classifies the newsgroups to gather articles efficiency. Figure 5 shows an example of classification of newsgroups.

Since newsgroups are often described in peculiar abbreviations (for example "comp"), it is difficult to match the description of the newsgroups for terms in the ontologies. IICA uses heuristics to classify the newsgroup. There are 511 abbreviations in 711 newsgroups of the domain "comp". Figure 6 shows the heuristics which is described in a list of pairs called an *association list*. The *car* of a pair is an abbreviation, and the *cdr* is an ontological term. It is not necessary to give heuristics to every newsgroup, because abbreviations are used in newsgroups of other domain such as "alt".

```
("comp" . "computer science")
("ai" ."artifical intelligence")
("edu" . "education")
("fuzzy" . "fuzzy inference")
("genetic" . "genetic algorithm")
("nat-lang" . "natural language processing")
("neural-nets" . "neural network")
("nlang-know-rep" . "natural language processing")
("nlang-know-rep" . "knowledge representation")
("shells" . "expert system")
("vision" . "machine vision")
("infosystems" . "information system")
```

Figure 6: Heuristics for Classification of Newsgroups

4 Ontology-Based Text Categorization

Ontology-based text categorization is the classification of documents by using ontologies as category definition. Conventional approaches focused only on the accuracy of categorization and left the easiness of human understanding out of consideration. Our purpose is extending the conventional methods using ontologies. Ontologies help people to interpret the result of categorizing texts by showing the ontological relations between texts.

In our approach, the process of text categorization is twofold: (1)Calculating similarity between a feature vector and a category vector, (2)Calculating similarity between category vectors (see Figure 7).

A feature vector is a vector which represents feature of a document. The feature vector is calculated from the term frequency and the inverse document frequency *A category vector* is a vector which represents the characteristic of a category. The category vector is calculated from the feature vectors of the document assigned to the category.

4.1 Algorithm

Category vectors and weights are calculated as follow procedures.

step1

Calculate the feature vectors of the gathered text.

step2

Classify gathered texts by calculated the feature vector.

step3

Calculate the category vectors from the classified texts.

step4

Repeat step2 and step3 until the category vectors converge.

step5

Calculate distance between the categories and renew weight between terms in the ontology.

The each initial category vectors is calculated from the feature vector of the texts which is assigned to the category by matching keywords.

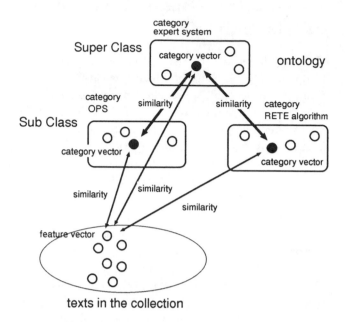

Figure 7: Text Categorization Using an Ontology

4.2   Vector Space Model

We use vector space model commonly used in the information retrieval studies to weight terms and calculate feature vectors [Salton 83].

The weight of term is a product of its term frequency (tf) and its inverse document frequency (idf).

The tf is the occurrence frequency of the term in the text. It is normally reflective of term importance.

The idf is a factor which enhances the terms which appears in fewer documents, while downgrading the terms occurring in many documents. It means that the document-specific feature are highlighted, while the collection-wide feature are diminished in importance. The weight of the term is given as

$$w_{ik} = tf_{ik} \times idf_k,$$

where tf_{ik} is the number of occurrences of term t_k in document i, and idf_k is the inverse document frequency of the term t_k in the collection of documents. A commonly used measure for the inverse document frequency is

$$idf_k = log(N/n_k),$$

where N is the total number of documents in the collection, and n_k is the number of document which contains a given term t_k. The collection of documents is the context within which the inverse documents frequencies are evaluated.

5   Evaluation

This chapter describes evaluation of our method.

5.1   Evaluation of Information Gathering

We tested an ontology-based method for information gathering tasks on the WWW. We evaluated our system by accuracy and efficiency.

5.1.1   Test of Accuracy

In order to evaluate its accuracy, we restricted 100 pages, and chose the 5 queries related to AI in English and the 5 queries related to sightseeing in Japanese. Then we ran IICA on the WWW in the following ways.

1. Breadth first search: IICA does't use ontologies. It traces hyperlinks on the WWW using breadth first algorithm.

2. Ontology based search: IICA use ontology-based search algorithm.

3. Ontology based search and heuristics: IICA use ontology-based search algorithm and heuristics.

We evaluated the result of the experiment according to the standard as follows.

○:   The collected page is directly related to user's queries.

△:   The collected page is not directly related to user's queries, but it is related to user's interests.

×:   The collected page is neither directly related to the user's queries nor related to user's interests.

Table 1 and Table 2 shows the results.

Table 1: Evaluation of Gathering Pages Relevant to Artificial Intelligence

search	○ (%)	△ (%)	× (%)
1. breadth first search	64.6	7.4	28.0
2. ontology	66.6	11.6	21.8
3. ontology and heuristics	67.8	10.6	21.6

Table 2: Evaluation of Gathering Pages Relevant to Sightseeing

search method	○ (%)	△ (%)	× (%)
1. breadth first search	57.4	8.4	34.2
2. ontology	59.5	15.6	24.9
3. ontology and heuristics	59.5	15.6	24.9

Table 4: Evaluation of Information Gathering — 2 keywords ("semantic network" and "production system")

search method	○	△	×
1. breadth first search	0	0	0
2. ontology	10	12	11
3. ontology and heuristics	18	23	15

Table 3: Evaluation of Efficiency of Information Gathering — 1 keyword ("knowledge base")

search method	○	△	×
1. bread first search	3	3	3
2. ontology	21	8	12
3. ontology and heuristics	44	13	25

5.1.2 Test of Efficiency

We tested search efficiency of our method. We restricted 500 search steps and chose the 2 queries related to AI in English. Then we ran IICA on the WWW in the above three ways.

Table 3 shows the search result to the query consists of one keyword *"knowledge base"*. Table 4 shows the search result to the query which consists of two keywords *"semantic network"* and *"production system"*. Here, the numbers in this Table indicate number of pages.

5.2 Evaluation of Text Categorization

5.2.1 Experiment on the Network News

We tested our method by categorizing 400 articles about "artificial intelligence" on the USENET network news. We chose newsgroups "comp". IICA classified the articles to 75 categories. Table 5 shows a part of the results. The table in the left-hand side shows the highest 20 categories and the number of articles and the table in the right-hand side is the lowest 20 categories and the number of articles.

In order to evaluate the result, we calculated Accuracy(A), Recall(R) and Precision(P) using the following equations:

$$A = \frac{No.\ of\ texts\ assigned\ to\ the\ correct\ category}{No.\ of\ total\ texts\quad in\ the\ collection},$$

$$R = \frac{No.\ of\ texts\ assigned\ to\ the\ correct\ category}{No.\ of\ total\ texts\quad in\ the\ category},$$

$$P = \frac{No.\ of\ texts\ assigned\ to\ the\ correct\ category}{No.\ of\ total\ texts\ assigned\ to\ the\ category}.$$

Table 5: Result of Classifying Articles

the top 20 categories and the number of texts		the low 20 categories and the number of texts	
program	48	VLSI	1
planning	31	statistics	1
aritificial intelligence	25	SQL	1
prolog	17	signal	1
software	16	psycology	1
inference engine	14	PC	1
classification	13	lisp	1
	12		
cognitive science	10	interface	1
expert system	9	informatics	1
C	8	DOS	1
Turing	8	device	1
neural network	7	design	1
TSP	7	connectionism	1
information	7	computer security	1
concept	7	compiler	1
communication	6	chess machine	1
search	6	brain	1
fuzzy	6	bag	1
IEEE	6	backpropagation	1
backtracking	6	analog computer	1

The result of calculation is shown in Table 6, where values of "Recall" and "Precision" are the average of all categories. We also analyzed misclassifications and discriminated them to 3 groups. The first group contains cases in which texts are assigned to the subclass of the correct category. The second group contains cases in which texts are assigned to the superclass of the correct category. And, the third group contains cases in which texts are assigned to the unrelated classes with the correct category. The result of the analysis is shown in Table 7.

Misclassification of the first and second groups is not serious, because the user can access the misclassi-

Table 6: Evaluation of Classifying Articles

Accuracy(%)	Recall(%)	Precision(%)
77.0	76.2	76.0

Table 7: Groups of Misclassifications

result type	No. of texts
texts assigned to the subclass category	26
texts assigned to the superclass category	5
texts assigned to the other category	51

Table 8: Revaluation of the Experiment

Accuracy(%)	Recall(%)	Precision(%)
85.3	85.1	85.1

fied items by tracing ontological relation between categories. Table 8 shows revaluation of the experiment regarding the two groups as correct. In conventional approaches, the misclassified items are not accessible. In contrast, IICA allows the user to search and reach the items by using ontological relations.

5.2.2   Experiment on the WWW

We also made an experiment of categorizing the about 500 pages concerned with AI in English and the about 800 page concerned with sightseeing in Japanese. In order to evaluate our method, we calculated recall and precision. The result is shown in Table 9.

Table 9: Evaluation of Categorization of WWW pages

	AI (English)	Sightseeing (Japanese)
Precision	81.9	79.0
Recall	80.5	70.0

6   IICA: Intelligent Information Collector and Analyzer

We implemented a prototype system of "IICA". IICA consists of fourfold modules: (1) user interface modules, (2) network modules, (3) inference modules, (4) database modules. User interface modules is described in Tck/Tk and perl scripts. Network modules is socket programs in C. Inference modules and database modules is implemented in Common Lisp. Figure 8 shows

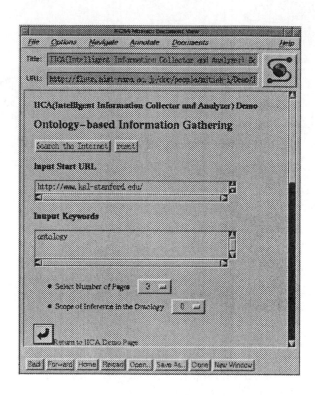

Figure 8: Interface of IICA

the interface the system using NCSA Mosaic.

IICA also has an ontological browsing tool realized as a *knowledge medium* which unifys the human oriented media and the computer oriented media (See Figure 9). It helps nonprofessional users to search for information and understand the result of categorizing them by visualizing the ontological structures.

7   Related Work

7.1   Internet Robots and Agents

There has been much recent work whixh is concerned with building an index of information on the Internet [Mauldin 94], [Balabanovic 95]. Another related area work attempts to automatically filter incoming information [Lashkari 94]. However, these tools are unable to interpret the result of their search due to lack of dmain knowledge.

In our approach, ontology-based interface helps nonprofessional users to search for information on the Internet and understand the result of categorizing them by visualizing. Ontology also make accurcy and efficiency better in information gathering.

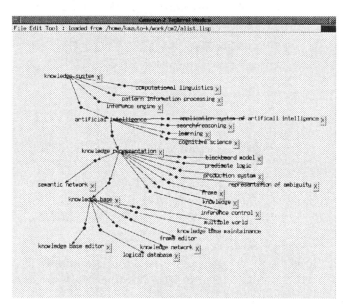

Figure 9: Visualization of the Ontological Structures

7.2 Information Retrieval and Text Categorization

There are studies on text categorization using structured knowledge such as thesaurus [Kawai92e] [Yamamoto95]. However, in these approaches, it is difficult to deal with changeable information resources, because a link between terms in the thesaurus is fixed, and category vectors are strongly depended on the initial learning data. Moreover, it is impossible to retrieve texts in categories similar to the current category, because there is no consideration of similarity between categories.

In the field of information retrieval, the approach based on using Kohonen's self-organizing map have been received attention [Kohonen 90], [Niki 95].

The merit of this approach is providing a collection ambiguous-query interface or an incremental browser for text databases by mapping a collection of documents into a two-dimensional map without teaching. However, it is often difficult for users to understand the meaning of the structure of the map which is organized from raw data.

In out approach, it is possible to use actual information resources by modifying not only category vectors dynamically but also weight between categories from gathered data. Furthermore, it is impossible to retrieve texts in categories similar to the current category by calculating similarity between categories.

8 Conclusion

In this paper, we proposed a new method of information gathering and text categorization using ontologies.

We implemented a system called "IICA (Intelligent Information Collector and Analyzer)" which helps people to acquire knowledge from the information resources on the wide-area network gathering and categorizing information.

IICA can deal with various types of text-like information, because most of our knowledge we can use are described as text form.

We tested our approach for two experiments: (1)gathering pages on the WWW, (2)categorization news articles and WWW pages .

We can conclude the following advantages of our approach from the results .

- Ontology and heuristics make accuracy and efficiency better in information gathering.

- Agent can understand which information is related to user's request using ontologies.

- IICA allows the user to search and reach the the misclassified items by tracing ontological relations.

- It is easier to develop weakly structured ontologies from terminologies and thesauruses than conventional methods.

- The ontology-based approach enables us to use heterogeneous information resources on the Internet.

The problem of the current system is that ontologies it uses are given and therefore not flexible both to users and information. We should consider learning of new terms from gathered texts and customizing of ontologies to user's interest and purposes.

References

[McBryan 94]   O. McBryan, "GENVL and WWWW:Tools for taming the Web", *In Proceedings of the 1th International WWW Conference*, 1994.

[Maes 93]   P. Maes and R. Kozierok, "Learning Interface Agents", *In Proceedings of AAAI*, 1993.

[Gruber 91]   T. R. Gruber, "The Role of Common Ontology in Achieving Sharable, Reusable Knowledge Bases", *Principles of Knowledge Representation and Reasoning – Proceedings of the 2nd International Conference*, pp. 601–602, 1991.

[Gruber 92]   T. R. Gruber, "Ontolingua: A mechanism to support portable ontologies", *Technical Report of Stanford University, Knowledge Systems Laboratory*, col. KSL 91-66, 1992.

[Iwazume 94]   , Michiaki Iwazume and Hideaki Takeda and Toyoaki Nishida, "Automatic classification of articles in network news and visualization of discussions – Intelligent News Reader", *Proceedings of the 8th Annual Conference of JSAI*, pp. 497–500, 1994.

[Iwazume 95]   Michiaki Iwazume, "Classification and organization of infomation by ontology", *Master's Thesis, Department of Infomation Processing, Graguate School of Infomation Science, Nara Institute of Science and Technology*, vol. NAIST-IS-MT351014, 1995.

[Nishiki 94]   Masanobu Nishiki and Hideaki Takeda and Toyoaki Nishida, "Extraction, Unification and Presentation ok Knowledge by Multi Agent System", *Proceedings of the 8th Annual Conference of JSAI*, pp. 505–508, 1994.

[Salton 83]   G. Salton, *Intoroduction to Modern Infomation Retrieval*, MacGraw-Hill, 1983.

[Mauldin 94]   M. L. Mauldinand and J. R. Leavitt, "Web-agent related research at the CMT", *In Proceedings of the ACM Special Interest Group on Networked Information Discovery and Retrieval*, 1994.

[Balabanovic 95]   , M. Balabanovic, "Leaning Information Retrieval Agents: Experiments with Automated Web Browsing", *Proceedings of the AAAI Spring Symposium*, pp. 13–18, 1995.

[Lashkari 94]   Y. Lashkari and M. Metral and P. Maes, "Collaborative interface agents", *In Proc of the 12th National Conference on Artificial Inteligence*, 1994.

[Yamamoto95]   Kazuhide Yamamoto and Shigeru Masuyama and Shuzo Maito, "An Automatic Classification Method for Japanese Texts using Mutual Category Relations", *IPSJ SIG Notes*, vol. 95, No. 27, pp. 7–12, 1995.

[Kawai92e]   Atsuo Kawai, "An Automatic Document Classification Method Based on a Semantic Category Frequency Analysis", *Transactions of Information Processing Society of Japan*, vol. 33, No. 9, pp. 1114–1122 1992.

[Kohonen 90]   T. Kohonen, "The Self-Organizing Map", *Proceedings of the IEEE*, vol. 78, No. 9, pp. 1464–1480, 1990.

[Niki 95]   Niki Kazuhisa and Katsumi Tanaka, "Information Retrieval Using Neural Networks", *Journal of Japanese Society for Artificial Intelligence*, vol. 10, No. 1, pp. 45–51, 1995.

Knowledge-based Systems

A KNOWLEDGE-BASED MESH GENERATION SYSTEM
FOR FORGING SIMULATION

Osamu Takata[1], Koukichi Nakanishi[2], Nariaki Horinouchi[1],

Hiroshi Yano[3], Tadao Akashi[3] and Toyohide Watanabe[4]

{System Engineering Div. I[1], Materials Div. I[2]} Toyota Central R&D Labs., Inc.,

Production Engineering Div. No. 5, TOYOTA MOTOR CORPORATOIN[3],

Department of Information Engineering, Nagoya University[4],

Nagakute, Aichi, 480-11, JAPAN[1]

Email: takata@kis.tytlabs.co.jp[1]

ABSTRACT

We have developed a knowledge-based system GENMAI (Artificial Intelligence Mesh GENerator) for automatically generating two-dimensional meshes, which is easily applicable to various fields of computational dynamics with structured meshes. We analyzed and modeled experts' mesh generation procedures in the simulation processes of forging deformations, vehicle aerodynamics and engine in-cylinder flows. This paper describes the outline of GENMAI focusing on the implementation techniques of search and inference, and then shows the functional ability of GENMAI applying to the forging simulation. The characteristics of GENMAI are as follows. (1) Plural solutions can be efficiently obtained at the same time in the search phase, using the global dependency and local dependency. (2) The meta-level inference method and its knowledge representation method are very applicable to various fields of analyses. We have applied GENMAI to forging simulation, and found it to have a prospect for practical use.

1. INTRODUCTION

Recently, improvement of computational performances has expanded the coverage of computational dynamics, for example, a structural analysis, a plasticity deformation analysis, a crash analysis and a fluid analysis. The process of computational dynamics consists of mesh generation (as the preprocessing), numerical computation, and evaluation of computational results (as the post-processing). So, mesh generation is one of the major tasks confronting computational dynamics so that the computational accuracy and the computational time depend directly on the quality of generated meshes.

Today, generating high quality meshes requires for us to composition of application-specific programs for each analytic field individually or to utilization of a commercial software product of mesh generation. However, in the former case, experts who have the domain-oriented knowledge and expertise associated with their application fields must write down programs by themselves, using the procedural language like Fortran. This work is very time consuming. In addition, such a developed program is not always applied to newly requested targets easily. This is because the computational methods and applicable fields are too strongly dependent on applications. For instance, the numerical simulation of the flow around a three-dimensional vehicle with real shape needs a computational time of about 50 hours on a super computer. It also takes 1-3 months to generate appropriate meshes, including the development of mesh generation programs.

In the latter case, even if we utilize a commercial software product of mesh generation, a large amount of domain knowledge and expertise attended to their application fields are required to make the mesh generation software usable instead of writing down mesh generation programs. In addition, even experts need a number of trials and errors to obtain appropriate meshes. Therefore, it is highly important to develop automatic mesh generation systems.

Generally, the approaches are classified into two different classes according to the simulating objects or simulation objectives: unstructured mesh class and structured mesh class. The unstructured meshes applicable to a structural analysis, a crash analysis, etc. were traditionally generated, using the Finite-Element Method (FEM), and the automatic mesh generation systems have been developed and used practically [Cav74, Lo85, PVM+87]. On the other hand, the structured meshes have been constructed analytically on the basis of the Finite-Difference Method (FDM), but the automatic systems has not yet been successfully provided since various kinds of domain knowledge must be managed in accordance with individual applications [BMP+88, And88, Dan91, DHS+92].

At least, it is difficult to develop automatic systems for generating structured meshes in comparison with the unstructured-mesh-oriented systems. However, the structured-mesh-oriented methods are more useful than the unstructured-mesh-oriented methods because the former is superior to the latter in point of the memory size and computational time when the same accuracy is attained.

This paper addresses our two-dimensional automatic system for generating structured meshes: Artificial Intelligence Mesh GENerator(GENMAI) which is easily applicable to various fields of computational dynamics. We analyzed and modeled experts' mesh generation procedures in the analysis processes of forging deformations, vehicle aerodynamics and engine in-cylinder flows. This paper describes the outline of GENMAI focusing on the implementation techniques of search and inference, and then shows the functional ability of applying GENMAI in the forging simulation.

2. OUTLINE OF GENMAI

2.1 OBJECTIVE

Our objective to develop a structured-mesh-generation system GENMAI is as follows:
(1) Users can generate high quality meshes in the same manner as experts do;
(2) The system is widely applicable to various fields

of computational dynamics, because the knowledge representation is flexible enough to expand the system to new analytic fields.

We model experts' trial and error processes first, and then implement the system with the knowledge-base approach for attaining to the issue (1). Moreover, we classify the knowledge and expertise of analytic fields into domain-dependent knowledge and procedural knowledge so as to be independent of analytic fields as for the issue (2). The procedural knowledge is represented as meta-rules and also the domain-dependent knowledge is organized as rules. GENMAI provides plural conflict resolution strategies so that meta-rules can control the application of rules and management of working memories case by case.

2.2 MODEL

First of all, we analyzed and modeled experts' mesh generation procedures in the analysis processes of forging deformations, vehicle aerodynamics and engine in-cylinder flows. As a result, the experts' trial and error processes are formulated as follows. Particularly, with a view to generating a high quality of meshes, number of trials and errors are repeated in the step (1).
(1) A given geometry is divided into plural small areas (called sub-regions) which are of comparatively simple shapes ((C) in FIGURE 1).
(2) Each sub-region is transformed into a structured mesh, using the transfinite interpolation method or Delaunay triangulation, etc. ((D) in FIGURE 1).
(3) The correspondences between the adjacent sub-regions are established structurally.
In fluid dynamics, a given geometry is often divided

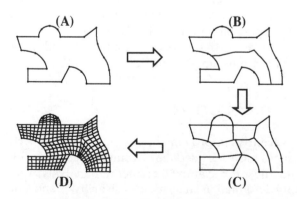

FIGURE 1 Mesh generation process

roughly into plural areas first so that the number of inflow is equal to that of outflow for these pre-divided areas ((B) in FIGURE 1).

The method of dividing a given geometry into sub-regions is commonly used in various kinds of analytic fields. Additionally, in the method of dividing, the usage and order of knowledge about groups to be applied are similar even if analytic fields are different. That is, a mesh generation system which does not depend on each analytic field can be developed by classifying the knowledge into analytic field-independent knowledge and analytic field-dependent knowledge.

2.3 SYSTEM CONFIGURATION

GENMAI consists of the inference part and search control part. The inference part is composed of the meta-knowledge base (where the procedural knowledge is stored), meta-inference engine, knowledge base (where the analytic field-dependent knowledge is stored), its inference engine and working memory. The search control part maintains and controls the conflict sets, using the concept of dependencies so that the working memory is inferred by the inference part effectively.

In GENMAI, the processing flow is controlled by meta-rules which are independent of individual analytic fields, and the following steps (1)-(5) are achieved by rules which depend on analytic fields. The processing flow in GENMAI is as follows. The steps (3)-(5) are repeated until all basic regions are transformed into the corresponding sub-regions.

(1) **Conversion of input data into the internal expressions**: The initial geometry data (such as basic regions, segments and points) which are input data are converted to objects.

(2) **Initialization of basic region**: In each basic region, candidates of both start points for division lines and direction vectors for each start point are calculated.

(3) **Decomposition of basic region**: This step is the main processing procedure in GENMAI. In each division line, a basic region is decomposed into two basic regions. As shown in FIGURE 2, GENMAI decomposes one basic region into plural patterns, provided that each pattern satisfies some predefined restrictions concerning the suitability between basic regions and division lines. The basic region which has no candidates of start points for the division line and satisfies such re-

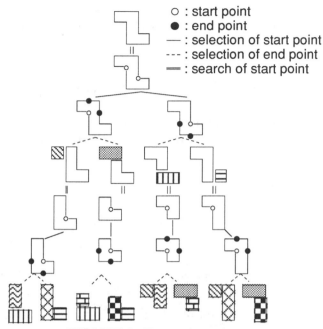

○ : start point
● : end point
— : selection of start point
--- : selection of end point
═ : search of start point

FIGURE 2 Examples of pattern decomposition into sub-regions

strictions becomes a sub-region.

(4) **Evaluation of decomposed patterns**: Feasible patterns obtained in the step (3) are evaluated for suitability (see Section 4.2) and then ordered. GENMAI selects the best N patterns and transfers the execution control to the step (2). At this point, the other patterns may be discarded in general. However, GENMAI keeps these temporarily discarded patterns with dependencies in order to make backtracking control possible when the best N selected patterns are failure.

(5) **Combination of decomposed sub-regions**: The decomposed sub-regions obtained at recursive steps (2)-(4) are composed and plural mesh patterns are generated.

(6) **Determination of the number of distributed points on each segment**: First, the number of distributed points on each segment is determined so as to be equal to the number of confronted sides in each sub-region. Second, in order to satisfy the above restrictions and minimize the summation of the distributed number, the number of distributed points on each segment is calculated, using the integer linear programming method.

(7) **Distribution of structured meshes**: The points on each segment are distributed into each sub-

region, using the transfinite interpolation method or Delaunay triangulation. Finally, mesh smoothing and adjustment procedures are executed.

3. IMPLEMENTATION OF GENMAI

This section explains the implementation of GENMAI focusing on the search method and inference method.

3.1 Search Method

A basic region is transformed into plural patterns which are composed of a pair of basic regions. This is because the basic region has plural start points (white-painted circles in FIGURE 2) for each division line and each start point may possess plural end points (black-painted circles in FIGURE 2). Repetition of the same operations may cause a problem of combinatorial explosion [BF81]. To cope with this problem, GENMAI introduces the following two dependencies: local and global. In this paper, we call plural patterns "multiple worlds", which are similar conceptually to contexts in ATMS [Kle86a].

As shown in FIGURE 2, the shape of the basic region transformed from a certain basic region may become the same as that of the one transformed from

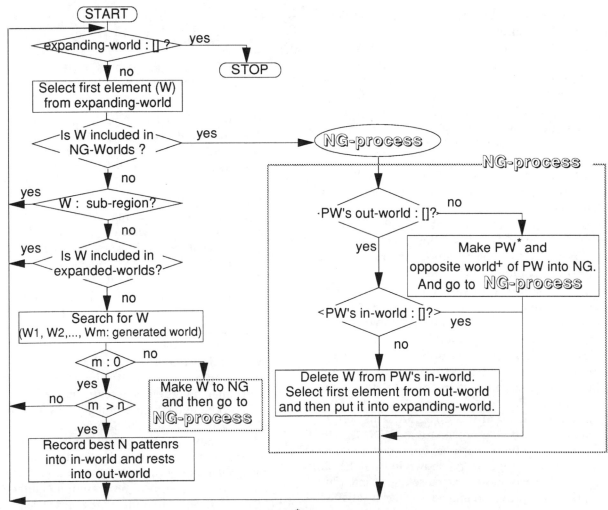

PW*: Parent World of W

opposite world+: world except of pairs of worlds

FIGURE 3 Search method

another basic region. In FIGURE 2, basic regions with the same hatching marks are the same shape. When basic regions of the same shape are generated, GENMAI combines such basic regions and treats them as one basic region. That is, when a new world is generated, GENMAI tests whether the new world is the same as one of the existing worlds or not. If the same world is found generated, GENMAI combines the newly generated world with the existing worlds and treats them as one world. Using the following simplified ATMS dependencies, GENMAI reduces useless worlds. These dependencies are stored at whole worlds globally. Moreover, when one of a pair of basic regions with the same world cannot be transformed into any sub-region finally, GENMAI discards the other basic region and makes the world "NG" [Kle86b].

- expanding-world: The world which will have to be expanded in the later search.
- expanded-world: The world which has been expanded in the past search.
- NG-world: The world which failed to be expanded in the past search.

FIGURE 3 shows the flowchart of search method to be effective on the basis of these local and global dependencies.

On the other hand, when the number of generated worlds exceeds N which is specified as an input data, GENMAI selects the best N worlds. If all of the best N selected worlds are transformed into a pair of sub-regions, the search finishes. However, if some world fails to be transformed into a pair of sub-regions, the backtracking is caused and the (N+1)th pattern is proceeded instead of the failure world. In GENMAI , the use of simplified TMS dependencies reduces useless searches, like a chronological backtracking [Doy79]. That is, the best N patterns and the (N+1)th pattern are related to TMS's Supported-List. These dependencies are stored at each world locally.

```
- Meta-rule(e.g. Determine start points of division line)
  meta_rule(nominate_points, BD, W):-
      get_data(basic_domain, BD, W, composed_segments, L),
      get_data_all(segment, L, W, start_point, M),
      -->
      inference_engine(wms_rule, division_nominate_rule, W, M).


- Rule(e.g. Restrict of adjacent lines with straight lines)
  rule(division_nominate_rule, rule1, W, POINT):-
      get_data(point, POINT, W, division_nominate_point, yes),
      get_data(point,  POINT, W, previous_segment,  P_SEG),
      get_data(point,  POINT, W, next_segment,  N_SEG),
      get_data(segment, P_SEG,  W, line_type,straight),
      get_data(segment,  N_SEG,  W, line_type,straight),
      intersect_angle(in, straight, straight,  P_SEG, N_SEG, W, ANGLE),
      put_data(point, POINT, W, test_angle, ANGLE),
      data_base([straight], test_intersect_angle, INT_ANGLE),
      INT_ANGLE < ANGLE,
      -->
      put_data(point, POINT,W, division_point, yes).
```

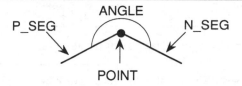

FIGURE 4 Examples of meta-rule and rule

3.2 INFERENCE METHOD

3.2.1 INFERENCE ENGINE

All the data available in GENMAI such as user-prepared data and system-defined data are managed uniformly as objects. GENMAI transforms basic regions into sub-regions on the knowledge base by the manipulations of instance attributes such as "get", "put" and "delete" operations. These operations are organized as the main tasks:

(1) Collection of instances : A meta-rule collects instances which satisfy the domain-independent restrictions;

(2) Application of the knowledge about analytic fields to instances: The collected instances are checked up by the restrictions about analytic fields. If an instance satisfies them, it is put with a new attribute. Otherwise, it is deleted from the working memory.

FIGURE 4 shows an example of a meta-rule and a rule to determine the start point of division line. First, the meta-rule collects point instances where attributes are start points of segments. Second, the rule checks up each instance, corresponding to adjacent lines which satisfy a restriction about straight line. If an instance satisfies the restriction, it is put into a division-point as "OK" and is selected as a start point for a division line.

In GENMAI, meta-rules control the procedural flow of operation, and rules represent individual knowledge about analytic fields. Therefore, it is possible to deal with different procedural flows by changing meta-rules, which are independent of analytic fields. Moreover, it is easy to maintain and acquire rules because they are classified by applied meta-rules.

3.2.2 CONFLICT RESOLUTION

In the knowledge-based system (production system), the conflict resolution strategy is an important

Table 1   Conflict resolution strategy

wm: working memory

Strategy name	Conflict resolution strategy
(1)wm-rule	Select the first element from instances. Then, test the condition part of the first element of rule-sets with it. If the condition of selected rule is satisfied, the action part is fired. Otherwise, proceed to the next element of rule-sets. If the first element of instances fails to match with any rule-sets, proceed to the next element of instances.
(2)wm-rules	Select the first element from instances. Then, test the condition part of all elements of rule-sets. If the condition of rule is satisfied, the action part is fired. Otherwise, proceed the next element of rule-sets. If the first element of instances fails to match with any rule-set, proceed to the next element of instances.
(3)wms-rule	In (1), test the condition with all instances, and if the condition of the selected rule is satisfied, the action part is fired. Otherwise, proceed to the next element of rule-sets.
(4)wms-rules	In (2), test the condition with all instances, and if the condition of selected rule is satisfied, the action part is fired. Otherwise, proceed to the next element of rule-sets.
(5)rule-wm	In (1), exchange instances and rule-sets.
(6)rule-wms	In (2), exchange instances and rule-sets.
(7)rules-wm	In (3), exchange instances and rule-sets.

task. It is necessary to apply a rule to working memories at the same time and also to prefer the best N ones which are specified in rules, because GENMAI generates the best N patterns in each basic region at the same time. It is difficult for a simple conflict resolution strategy such as MEA and LEX to control these complicated ones [Ric83]. Therefore, GENMAI provides 7 conflict resolution strategies shown in Table 1. These strategies are specified in meta-rule by the predicate "inference_engine" as shown below.

- inference_engine(strategy, rule-set, world, instances)

In the meta-rule shown in FIGURE 4, the rule-set which is named "division_nominate_rule" and points instances which are the start points of division lines

are interpreted by the conflict resolution strategy (3) (i.e. wms-rule). That is, if the restriction about straight line is satisfied for each start point, the action part of the rule is fired. These conflict resolution strategies are used for various kinds of synthesis problems.

4. APPLICATION IN FORGING SIMULATION

Forging is one of metal forming processes to put a workpiece between upper die and lower die and press it. This process is a kind of geometrical transformation from a simple shape to a complex shape. Thus, the shape of a workpiece is deformed in accordance with the progress of forging simulation with finite element method (FEM). Remeshing is very important in this large deformation problem because it is sensi-

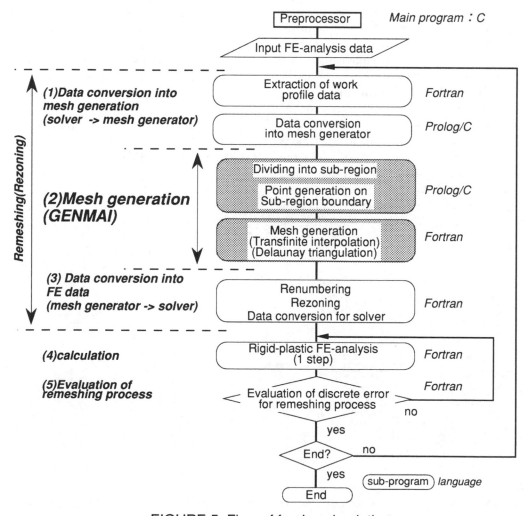

FIGURE 5 Flow of forging simulation

tive to computational accuracy. This remeshing process is time-consuming due to user's trial operations. In order to obtain high quality meshes, much domain knowledge derived especially from experimental heuristics is necessary.

Therefore, there is a strong demand for an automatic remeshing system which can assist users with numerical analysis in forging. The new forging simulation system would be developed so as to be combined with the rigid plastic deformation analysis program (solver) and GENMAI.

4.1 FLOWCHART OF FORGING SIMULATION

Our forging simulation system consists of five processing modules: data conversion for mesh generation, mesh generation, data conversion into finite element data, calculation using rigid plastic FEM and evaluation of remeshing process. The processing flow is illustrated in FIGURE 5. The global loop signifies the deformation progress: that is, movement of upper die. (In forging, the amount of movement is called "stroke".)

(1) **Data conversion for mesh generation**: Rigid plastic finite element data is converted for the mesh generator GENMAI. In particular, to make the workpiece as geometrical data at one time is very important. These data are extracted from the node-location data and the die-geometry data in the previous FEM calculation step.

(2) **Mesh generation**: Mesh data are generated by GENMAI. Here, more knowledge about forging

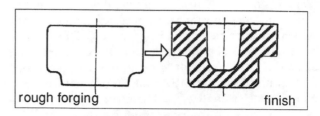

FIGURE 6 Typical forging process of idling gear

analysis is added to GENMAI.

(3) **Data conversion into finite element data**: The data obtained in the mesh generation module is converted as the rigid plastic analysis data such as boundary conditions and mesh connectivity.

(4) **Calculation using rigid plastic FEM**: Rigid plastic calculation is executed with FEM in one computational step. In this calculation, the updated-Lagrangian method is used. Therefore, in one step calculation the workpiece deformation is shown in the mesh distortion.

(5) **Evaluation of remeshing process**: Necessity of remeshing is evaluated by computational error such as mesh distortion and workpiece profile error due to overlapped elements with die profile. If necessary, the remeshing process is executed for the workpiece of the time before the next step FEM calculation.

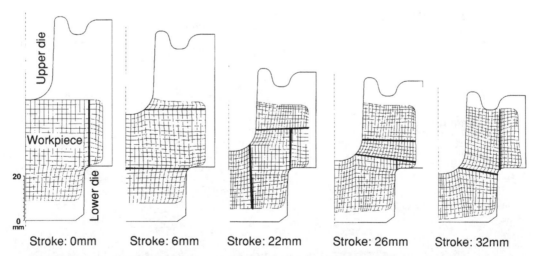

FIGURE 7 Sub-regions and structured meshes in each remeshing process

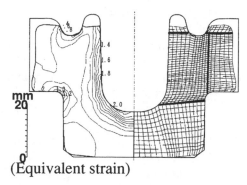

(Equivalent strain)

FIGURE 8 Computational result in final step

4.2 EXAMPLES

To check the validity of our system, we applied to a typically closed die forging process. This process is a hot forming process for idling gear and consists of four stages. The third stage is taken here, in which finish is formed from rough forging (see FIGURE 6). FIGURE 7 shows results of remeshing processes obtained as various strokes. In each figure, structured mesh and sub-regions which are divided by thick lines are indicated. It is clear that the sub-region shapes and meshes in the deformed workpiece generated at each stroke are appropriate. FIGURE 8 shows a calculated result in the final step. The right side shows the transformed sub-regions and distorted finite element. This result demonstrates the applicability to a complex shape. The left side indicates the equivalent strain distribution. A reasonable result has been obtained for this deformation analysis. The total execution time for

this example was five hours on an engineering workstation.

As mentioned in the previous section, it is necessary to select the best pattern in remeshing processes, because plural mesh patterns are generated at once. Our system can select the best generated mesh pattern, using the following five evaluation criteria concerned with sub-regions.

(1) Number of division lines composed of four-sides
(2) Difference between right-angle and four-side's angles
(3) Ratio of opposite sides
(4) Angle change in segments composed of four-sides
(5) Inclination of division line

The best pattern is selected from feasible ones by evaluating the summation of these criteria. Examples of evaluation results are explained in FIGURE 9. FIGURE 9 indicates plural sub-region patterns at one time generated from these in FIGURE 6. Our system selected the best pattern (pattern 1) from four patterns by using the above criterion, because other patterns include bad angle parts and bad ratio of opposite sides.

5. CONCLUSIONS

We have developed an automatic system GENMAI (Artificial Intelligence Mesh GENerator) for generating two-dimensional structured meshes, which can be easily applicable to various fields of computational dynamics.

The characteristics of GENMAI are as follows:
(1) Plural patterns can be efficiently obtained at

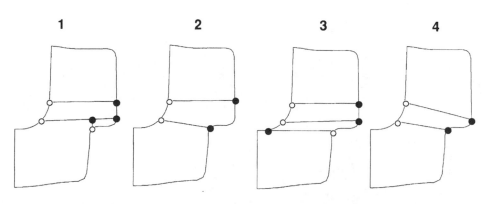

FIGURE 9 Evaluation of generated plural patterns

the same time by the search technique based on the global dependency (like ATMS method) and local dependency (like TMS method);

(2) The meta-level inference method and its knowledge representation method are highly applicable to various fields of computational dynamics.

We have applied GENMAI to forging simulation and found it to have a prospect of practical use, because our system makes it possible to do remeshing automatically in the simulation of plasticity deformation without expertise of mesh generation and plasticity deformation analyses.

In the future, it is necessary to establish the methodology by which the experts can write down their new domain knowledge about a new field. That is, a knowledge acquisition tool which is implemented with the methodology helps experts extracting a basic task in the mesh generation process, classifying them into the meta-knowledge and knowledge, and acquiring knowledge easily.

REFERENCES

[And88]   A. E. Andrews. Knowledge-Based Zonal Grid Generation for Computational Fluid Dynamics. In *Proc of NASA Conf. Publ.*, NASA-CP-3019: 73-80,1988.

[BF81]   A. Barr and E. A. Feigenbaum: *The Handbook of Artificial Intelligence*, William Kaufmann, Inc., 1981.

[BMP+88]   T. D. Blacker, J. L. Mitchiner, L. R. Phillips and Y. T. Lin. Knowledge System Approach to Automated Two- Dimensional Quadrilateral Mesh Generation. In *Proc. of ASME Int. Comput. Eng. Conf.*, 153-162, 1988.

[Cav74]   J. C. Cavendish. Automatic Triangulation of Arbitrary Planar Domains for the Finite Element Method. In *Int. J. Numer. Methods Eng.*, 8: 679-697, 1974.

[Dan91]   J. F. Dannenhoffer III. Computer-Aided Block Structuring Through the Use of Optimization and Expert System Techniques. *AIAA-91-1585-CP*, 654-661,1991.

[DHS+92]   P. D. Dabke, I. Haque, M. Srikrishna and J. E. Jackson. A Knowledge-Based System for Generation and Control of Finite-Element Meshes in Forging Simulation. In *J. of Materials Eng. and Perf.*, 1, 3: 415-428,1992.

[Doy79]   J. Doyle. A Truth Maintenance System. In *Artificial Intelligence*, 12: 231-272, 1979.

[Kle86a]   J. de Kleer: An Assumption-base TMS. In *Artificial Intelligence*, 28: 127-162, 1986.

[Kle86b]   J. de Kleer. Extending the ATMS. In *Artificial Intelligence,* 28: 163-196, 1986.

[Lo85]   S. H. Lo. A New Mesh Generation Scheme for Arbitrary Planar Domains. In *Int. J. Numer. Methods Eng.*, 21: 1403-1426, 1985.

[PVM+87]   J. Peraire, M. Vahdati, M. Morgan and O. C. Zienkiewicz. Adaptive Remeshing for Compressible Flow Computations. In *J. Comp. Phys.*, 7: 449-466, 1987.

[Ric83]   E. Rich. *Artificial Intelligence*. McGRAW-Hill, Inc., 1983.

A KNOWLEDGE BASED TOOL FOR CHECKING LARGE KNOWLEDGE BASES

Rym Nouira, Jean-Marc Fouet
Laboratoire d'Ingénierie des Systèmes d'Information,
Université Claude Bernard - Lyon1
43 Boulevard du 11 Novembre 1918 - 69622 Villeurbanne Cedex
email: nouira@lisisun.univ-lyon1.fr - fouet@lisisun.univ-lyon1.fr

ABSTRACT

We present a tool for detecting conflicts and redundancies among large knowledge bases.
First it allows the reduction of the combinatory by selecting among all the rules only those liable to generate conflicts ; it does this using properties of the rules and heuristics.
Then some attributes linking chunks of knowledge constitute access paths allowing to reduce the search duration for a particular information. They allow also to keep trace of the results of previously accomplished treatments that will be useful in future process, in order to avoid needlessly repeating treatments.
Finally as the knowledge used to accomplish these treatments and checks is itself expressed in inference rule formalism, the system may be applied to itself.

INTRODUCTION

To respond to the increasing importance of knowledge based systems, many methodologies and tools have been developed in order to assist building these systems. Aiming at modelling experts' knowledge and reasoning, the knowledge contained in the system is heuristic and not well defined. So traditional methods for validating software are no longer applicable.
The primary objective is to obtain a system that is on one hand valid, that solves problems correctly the way a human expert would. And on the other hand, it has to be complete, so as to solve all the problems that may be faced by the user. These are the objectives of the verification and validation tasks.
Verification [Gupta 91] consists of checking that a system is constructed correctly according to the specifications : "build the system right". It tries to detect programming errors. Errors are tightly related to the knowledge representation formalism, and in the case of rule based systems to the implicit control used by the inference mechanism.
Validation consists of verifying that the constructed system satisfies the needs of the users : "build the right system". For systems using inference rules formalism to represent knowledge, verification becomes a combinatorial problem. So a systematic and deep approach in detecting anomalies is difficult and becomes impossible when the base grows in size. It is therefore necessary to find ways to reduce the combinatory.
After a brief description of the actual best known tools and methodologies in validating knowledge bases, we shall begin by describing the environment in which the system has been developed. This will bring in evidence all the hypotheses considered in this work. Then we will define the various concepts used and the problematic, and proceed to a more detailed description of the treatments applied on rules.
A comparison of our system with some others will follow before concluding by a discussion of our approach stating its advantages and disadvantages.

1. STATE OF THE ART

Many tools have been developed to address the verification of knowledge based systems [Ayel and Rousset 90] [Coenen and Bench-cap 93] [Gupta 91] [Lopez et al 90] [O'Keefe and O'Leary 93]. They can be divided into two categories [O'Keefe and O'Leary 93]

1.1 - DOMAIN INDEPENDENT TOOLS

Domain independent tools try to detect anomalies which consist of an abuse or unusual use of the knowledge representation scheme.
Some are based on decision table methods (ESC [Cragen and Steudel 87] RCP [Suwa et al 82] CHECK [Nguyen et al 87]). They separate rules' conditions and actions parameters. Conditions are given along the X-axis and actions along the Y-axis. Algorithms then examine the existence of relationships among rows and columns.
The drawback of this method is that it is useful only for small rule bases, because the table grows to an unmanageable proportion. COVER [Preece and Shingal 92] constructs a graph representation of the rule base. This allows it to detect anomalies across numerous rules rather than between pairs of rules as is common with the table based approaches.

KB-reducer [Ginsberg 88] is a system for checking all potential inconsistencies and redundancies in rule bases. It transforms rules into a logical form, then it computes for each hypothesis its labels following the ATMS terminology [De Kleer 86]. Anomalies are detected during this labelling process. Some systems like INDE [Pipard 88] and [Agarwal and Tanniru 91], use Petri nets to represent the rule base which can then be tested for detecting inconsistencies, incompleteness and non fireable rules, using existing methods.

1.2 - DOMAIN DEPENDENT TOOLS

Domain dependent tools use meta-knowledge from the domain to verify the validity of the knowledge.
One of the best known examples in this category is EVA [Chang et al 90]. It consists of an integrated set of generic tools that enable the user to check the validity of the knowledge base. It also incorporates an extended structure checker which tests for synonymy between terms and inheritance information using meta-facts.
SACCO [Ayel 87] defines the coherence model as the set of all the properties and semantic constraints. The verification of the dynamic coherence is not systematic. In fact the system gives the expert means to limit them to some parts of the coherence model. This will be done by using meta-informations like the constraints relevance coefficient or the concerned concepts.
In TIBRE [Lalo 89] dynamic checks are entirely guided by the expert. He specifies the coherence constraints that he wants to verify. A solver, that is a Prolog program, checks for each of these constraints whether it can be obtained from a coherent fact base.

2. THE DEVELOPMENT ENVIRONMENT

Our tool is developed in the Gosseyn Machine Environment [Fouet 87] [Fouet 95] . The Gosseyn Machine enables the expert to create his knowledge about a domain. It is a non monotonic system, based upon first order and three valued logic. It uses the object and rule formalisms to represent knowledge.
The tool constitutes a component of a more complete system whose function is to provide an assistance to the expert throughout the construction of the knowledge base [Fouet & Nouira 95]. Each new created knowledge will be studied, confronted with some other chunks of knowledge contained in the base, looking for anomalies or deficiencies.
These checks will be of two types. Local checks, verifying that the description of the new knowledge satisfies the structural and semantic constraints. And global checks to make sure that its integration into the base will not generate new anomalies or imperfections. The results of these checks consist in a set of messages, sent to the user, indicating the chunks of knowledge that have to be corrected, completed or even created.
Once the new knowledge is judged valid, it will be integrated into the knowledge base, and in consequence the system will update a certain number of meta-attributes in order to take account of this new knowledge.

3. DEFINITIONS
3.1 - DEDUCTIVE KNOWLEDGE

Knowledge consists of rules, describing the expert's "know how", and constraints, stating how to maintain the coherence of the knowledge base, giving attributes properties and relations between the concepts in order to describe their semantic.
With the same syntax as rules, constraints describe in the left hand side (lhs) a situation that represents their application conditions: the description of the concerned objects, their type (the class(es) of which they will be instances), properties (the attributes with which they have to be described), and the relations between them (one is a value of an attribute of the other). Their right hand side (rhs) gives the conditions that have or have not to be satisfied.
Negative constraints will be used in searching for conflictual rules. Their rhs has only one term, using the specific meta-predicate "incompatible" with two literals representing the incompatible facts, such as, for instance :

is-instance-of (?p1 person) -> incompatible (retired (?p1)
exist (?z, grand-father (?p1 ?z))

In addition to these constraints, and in order to alleviate the expert's task of formulating the constraints, we have defined meta-attributes for some frequently used attributes. So the expert is asked directly when creating a new attribute to supply their values. Examples of such properties include monovalued or multivalued, arity, inverse, exclusive attributes (that cannot take the same value inside a given object), incompatible attributes (that cannot coexist inside a given object). Example :

C1: .. --> incompatible (father (?x ?y) son (?x ?y))
C2: .. --> incompatible (husband (?x ?z) wife(?x ?t))

3.2 - COHERENCE OF A RULE

Validating a rule locally consists in checking first that it is fireable : the situation described in the lhs (called SPA by [Ayel 90]) must consist of a coherent set of premises (according to the syntactic and semantic constraints) ; so does the situation obtained after the application of the rule: the union of the premises of the lhs and the rhs of the rule (called SPQ) .
Note that if the rule is non monotonic we will have to remove from the SPQ the premises of the lhs that will be negated once the rule is fired.

3.3 - CONFLICTUAL RULES

We assert a conflict between two rules when they cause the generation, during the resolution of a problem, of two incompatible facts.
Hence, searching conflicts consists in extracting couples of rules (R1, R2) that generate two conflictual facts according

to a negative constraint C. One must prove that they can be applied during the resolution of at least one problem.

Example. R1: ... -> att1(?x ?y), ..

R2: ... -> att2 (?z ?t), ..

C : ... -> incompatible (att1 (?x1 ?y1) att2 (?z1 ?t1))

Four conditions, listed below, have to be satisfied.

3.3.1 - Conflictual facts

We have to make sure that the two facts are effectively conflictual.

Therefore we have to verify for each of these two facts that it is concerned by the constraint. Since the lhs of the constraint specifies its application conditions, we have to make sure that the type and properties describing x and y in the first rule R1 are compatible with those concerning x1 and y1 in the constraint. The same verification will be made for z and t according to z1 and t1.

3.3.2 - Applicability during problem resolution

There is a potential conflict if :

- the lhs of R1 is compatible with the lhs of R2, or
- the incompatibility between these two left hand sides may be removed after R1 has been fired.

Example : R1 : A -> att1(x y)

R2 : not A -> att2(z t)

R3 : C -> not A

one may fire successively R1, R3 and R2.

3.3.3 - Satisfaction of the conditions of a constraint

If the two lhs-s are compatible, R1 and R2 may be fired in parallel. To have a conflict, we need the conditions of C to be satisfiable afterwards, in other words we need those conditions to be compatible with the SPQs (see above 3.2) of R1 and R2.

If, on the other hand, R1 and R2 cannot be fired simultaneously, then all we need (to have a conflict) is for the conditions of C to be compatible with either SPQ, unless we can prove that R1 has to be fired before R2, in which case all we need is the SPQ of R2.

3.3.4 - Existence of both conflictual facts

The rules have to be applied in such an order that even if one of them deletes the fact generated by the other one, we still obtain the existence of both facts at a given time. Here we distinguish between three situations :

1) If none of the rules deletes the fact generated by the other one, we get a conflict.

2) If R1 deletes the fact generated by R2 then at least one of the two following situations has to exist :

* either they may be fired simultaneously,
* or R1 may be applied before R2.

3) If both rules delete the fact generated by the other one, then they have to be able to be fired simultaneously.

3.4 - REDUNDANCY

We define a redundancy as the generation of the same fact more than once during the same problem resolution (the same value for the same attribute of the same object).

R1: .. -> att1(x y), ..

R2: .. -> att1 (z t), ..

Hence searching for redundancy consists of extracting couples of rules that are able to generate the same fact and verify (in analogy with the conflict conditions) that :

1) The two facts can be identical : x and z as treated in the rules may be instanciated with the same objects, as well as y and t.

2) Both rules may be applied during a problem resolution.

Unless the specification of the contrary (example: the use of certainty factors), redundancy is a general anomaly that doesn't depend on specific application conditions.

No matter whether the redundant facts exist simultaneously (the attribute is multivalued), or not (monovalued and non monotonic). We assume a redundancy in both these cases.

4 - TREATMENTS APPLIED TO RULES
4.1 - DETERMINATION OF VARIABLES TYPE

In order to enhance the efficiency and rapidity of the treatments and checks applied to rules, or their exploitation, it is important to identify as precisely as possible the type of each variable. This will be done during the creation of the new rule.

Starting by the variables, the type of which has been specified by the expert, the system proceeds incrementally. It determines the type of the other variables using the attribute type information.

If we obtain many candidates then the system will try to reduce the search space by confronting the properties describing the variables and treatments applied on them with the constraints defined on each of these classes. When an incompatibility is detected, the class will be discarded.

This is done during the validation of the rule.

4.2 - LOCAL CHECKS USING SATURATION

As stated above, the system will check whether both the application conditions of the rule and the situation obtained once applied are coherent sets of premises.

In both cases it will try to identify contradictory premises. The premises set is considered as a fact base (variables are assimilated to objects described by attributes and having relations between them). The system tries to complete it by generating the maximal number of properties about these objects that may be deduced from this initial set of facts. This will allow to fire the maximal number of constraints and hence, in the case of existence of indirectly contradictory facts, to augment the chances of identifying them.

To do that the system will do in turn forward chaining and backward chaining.

4.2.1 - Forward chaining

Attempting to complete the description of the objects (variables) used in the premises set, it tries to fire for each of them all the positive and negative constraints that concern their type.

The system starts by making a direct access to the concerned constraints by the index attributes. In fact when creating a new inference rule or a constraint and once the validation checks succeed, it adds a reference to it in the classes of the objects concerned.

The rules, the evaluation of which fails due to incompatible monotonic premises, will be eliminated from the list.

Positive constraints will generate true facts that have to be deduced in a real execution. And negative constraints will generate forbidden facts that must not exist in the fact base, For example, if we know for ?x (a person) that he is retired, we will add the property, exist (?z grand-father (?x ?z)) to the set of forbidden facts (according to the constraint C1 mentioned above in §3.1).

Incompatibility will be detected when we obtain that a given fact is simultaneously true and forbidden.

Once no constraint is fireable any more, the system switches to backward chaining.

4.2.2 - Backward chaining

During backward chaining the system tries to identify the source (generating rule) of each of the initial facts belonging to the current set of premises. Initial facts are those coming from the initial set of premises or obtained by backward chaining. Are not concerned those that may be supplied in an initial fact base nor those that do not express a real fact, but a situation, such as *false exist (?z att2 (?z ?x))* (there is no object having ?x as a value to its attribute att2)

Here again the system makes direct access to the rules concerned. In fact each inference rule is referenced in both the object attribute and the class of the first argument of each of its generated facts.

If for a given initial fact no rule is obtained, the system signals missing rules. If many rules are obtained we will benefit from the saturation already done over each of them by confronting their results to the current results of the saturation of the new rule. If they are incompatible, this means that the rule cannot be the origin of the initial fact, so it will be eliminated. If at the end we obtain only one candidate rule, it will be fired. And so we will delete the treated fact from the set of initial facts, and in exchange, we will add to this set, among the premises composing the lhs of the fired rule those that may be generated by other rules.

If many rules are obtained, they are kept, hoping that later, new facts may be generated allowing the system to discard some others, to finally obtain only one rule.

During backward chaining, we will again obtain true and forbidden facts (if in the lhs of a rule there is a premise specifying that a certain fact has to be false).

Example: is-instance-of (?p person) and
             false exist (?r resources (?p ?r)))   -> poor (?p)

If this is the only candidate rule concluding "poor", we may add "exist (?r resources (?p ?r))" as a forbidden fact.

As unknown facts may also be obtained when appearing in the lhs of a fired rule, incompatibility becomes the result of finding one fact in both lists.

For each of the generated facts we will keep trace of their origin by one of these two types of lists.

(premise$_i$ "Backward Ch." rule_name the_initial_fact)
(consequent$_i$ "Forward Ch." rule_name
                              the_premisses_instanciation_list)

Heuristics are used to decide of the moment to return to forward chaining. In fact a minimal number of new facts has to be obtained to augment the chances of obtaining incompatibility during forward chaining, and so avoiding treating needlessly all the remaining constraints.

Three factors will be considered. In addition to their number, we examine the concentration of the new facts: the more numerous the facts dealing with a given object, the more its description is complete so we have better chances to fire more rules. And finally, we will take into account the number of rules fired. In fact as each rule has been validated locally, this eliminates any risk of finding anomalies among the premises of one rule. Consequently, the more rules we have, the higher the interaction between them rises, so we may have better chances in detecting incompatibility.

The saturation stops when no rule is fireable anymore in forward nor backward chaining.

If the saturation succeeds, the rule is declared valid, and as a result we obtain the most complete description of the objects treated that the system was able to determine. This yields three lists, of true, forbidden and unknown facts.

And then the system proceeds to search for conflicts.

4.3 - DETECTION OF CONFLICTS

4.3.1 - Determination of the set of rules to compare with a newly created rule

For R2 to be conflictual with R1, at least one of the facts generated by R2 has to be incompatible with one of the facts generated by R1.

This will be done in two steps.

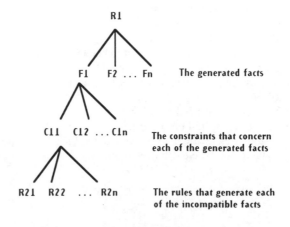

Steps in determining the set of rules to compare to the new one

4.3.1.1 - Determine the set of facts incompatible with those generated by a new rule

a) The system begins by extracting, for each fact generated by R1, the constraints that may concern it. Here again, it uses the index attributes : each constraint is referenced in the attribute and in the class of the first argument of each of the incompatible facts.

b) Then it checks, for each of the constraints obtained, whether it may effectively concern the objects treated in R1. It starts by unifying the two sets of variables and then verifies that the union of the two sets of premises constitutes a coherent fact base. This will be done again by the saturation, starting from the results obtained previously by the saturation of both the rule R1 and the constraint.

If the saturation succeeds, this means that effectively the objects treated in rule R1 may in addition satisfy the conditions necessary to the application of the constraint C, so C will be retained.

4.3.1.2 - Extract rules able to generate each of these incompatible facts

For each constraint retained, the system extracts the rules R2 that may generate the second incompatible fact. Here again, the system uses the index attributes and the variables type to make a direct access to each of the rules concerned.

4.3.2 - Determination of conflictual rules

For each of these rules R2, the system tries to verify each of the four previously stated conflict conditions (3.3.1-4).

4.3.2.1 - Verification of the first three conflict conditions

This is achieved by unifying the two sets of variables R1UC (union of the variables of R1 and C) on one side and R2 on the other side, and then verifying the compatibility of the union of the two corresponding sets of premises, again by the saturation which will start from the results previously obtained by the saturation of R1UC on one side and R2 on the other side.

Both the saturations (R1 with C, or R1UC with R2) may impose -in case an incompatibility due to non monotonic attributes is detected- restrictions on the order of application of the rules.

4.3.2.2 - Verification of the fourth conflict condition

As explained above according to the fact that each of the rules deletes or not the incompatible fact generated by the other, the system defines the order in which the rules should be fired so as to yield simultaneously both incompatible facts.

4.3.2.3 - Combination of the previous tests

Finally, the system confronts the two orders obtained in the previous verifications. If the orders are compatible (identical, or one is included in the other), then a conflict may happen. In the other case there is no conflict because the first three conditions on one side and the fourth on the other side cannot be simultaneously satisfied.

5. EXAMPLE

Given the rules

R1: age (?p ?age) and > (?age 70) -> old-person (?p)
R2: practise-sport (?p ?s) and age(?p ?age) -> < (?age 50)
R3: old-person (?p1) -> fragile(?p1)

and the constraint

C1: person (?p) -> incompatible (fragile(?p),
                                high-level-sportive (?p))

The expert creates the new rule:

R4: practise-sport (?p2 ?spo) and hard (?spo)
                                -> high-level-sportive (?p2)

The system processes in the following steps:

1) Searching for the constraints that may concern R4, it starts by selecting C1, since one of its incompatible facts may be unified with the fact generated by R4.

2) Then, in order to verify whether C1 concerns effectively the objects treated in R4, it unifies the variables of C1 with those of R4 starting by the arguments of the incompatible fact: ?p2 with ?p. Then it tries to fire over the obtained premises set: (person (?p2), practise-sport (?p2 ?spo), hard (?spo)) the constraints and rules available in forward and backward chaining. As no rule is fireable, the premises set is declared compatible, and the system deduces that R4 is concerned by C1.

3). The system goes on searching for the rules that may generate the contradictory fact fragile(?p), so it selects R3.

4). In order to check the existence of a conflict between R3 and R4, the system unifies the two sets of variables of R3UC1 and R4 starting by unifying ?p2 with ?p1. Then to check the compatibility of the obtained premises set: (person (?p2), practisee-sport (?p2 ?spo), hard (?spo)) old-person (?p2)), the system -while saturating- fires in forward chaining the rule R2 generating the facts age(?p ?age) and < (?age 50) (the fact age(?p ?age) is automatically deduced true because the attribute "age" is mandatory). And while looking for the source of the fact old-person (?p1) fires the only candidate rule R1, generating the fact > (?age 70).

Hence obtaining two contradictory facts < (?age 50) and > (?age 70), the system deduces that the rules may not be fired during problem resolution as they have contradictory premises. Consequently the rules R3 and R4 are declared not conflictual.

6. COMPARISON WITH SIMILAR SYSTEMS

Contrary to systems SACCO and TIBRE, we believe that if the expert has defined some constraints, he is expecting that they will always be satisfied. Therefore, we consider all the constraints in the checks, but as the checks are launched by the creation of a new rule we determine the set of constraints to be verified, based on the facts generated by the rule.

Another difference is that in our system, global checks are not dynamic, but we qualify them as semi-dynamic because we simulate a real execution by assimilating the premises available to a fact base from which we try to generate all the possible facts. This limits the number of backward chainings, but has the disadvantage of risking to be unable to settle between all candidate rules.

7. DISCUSSION AND CONCLUSION

We have presented a system that relies on meta-knowledge to reduce the combinatory and enhance the treatments. It has many advantages. First, the flexibility of the constraints formalism facilitates their expression. It allows even the translation of the semantic properties that are usually represented using specific meta-attributes. This allows avoiding programming specific treatments for them.

The determination of the type of the variables of the rules and the indexation of deductive knowledge allows limiting the combinatory and accelerating the search time.

Moreover there is a continuity in the treatments, each step re-uses the results obtained during preceding steps. In fact, the results obtained during the saturation of a new rule R1, will be exploited as a starting point for each of the constraints C. And the results of the saturation of the premises of R1UC, will also be used for each of the possibly conflictual rules R2. This will also be the case for the saturation of the premises of each rule, which will be used during backward chaining, when testing whether it is fireable. Finally as these heuristics are themselves expressed using inference rules formalism, the system may be applied on itself.

Nevertheless the system presents disadvantages.

First, concerning the constraints, the system allows the representation and treatment of only binary constraints, so constraints of a higher degree (n) will be transformed into $n*(n+1)/2$ binary constraints. But we suppose that constraints of a very high degree would be seldom.

A more tricky problem concerns updating knowledge. In fact if the addition of a new rule or attribute does not cause great damage and only requires additional tests, the modification of a rule, a constraint or an attribute type, will invalid many of the treatments and checks previously done and necessitate to do them again. In this case the continuity of the treatments that was considered as an advantage becomes an inconvenient.

Finally, this characteristic has also the consequence that the quality of many of the treatments done by the system will have repercussions on future processing. This makes each task delicate and therefore has to be done perfectly.

REFERENCES

[Agarwal and Tanniru 91] Agarwal R., Tanniru M. A Petri-Net based approach for verifying the integrity of Production systems. Knowledge-Based systems verification, Validation and Testing. Workshop notes from the 9th National Conference on Artificial Intelligence, AAAI-91 Anaheim CA, 17 July.

[Ayel 87] Ayel, M. Détection d'incohérences dans les bases de connaissances: SACCO. thèse d'état, université de Chambery, September 1987.

[Ayel 90] Ayel, M., Rousset, M.-C., La cohérence dans les bases de connaissances, CEPADUES, 1990.

[Chang et al 90] Chang C. L., Combs J. B., and Stachowitz R.A. A report on the expert system validation associate (EVA). Experts systems with applications. Vol 1, N° 3, p219-230.

[Coenen and Bench-cap 93] Coenen F. and Bench-cap T., Maintenance of knowledge based systems Academic Press 1993.

[Cragen and Steudel 87] Cragen B.J. and H.J. A decision table based processor for checking completeness and consistency in rule based expert systems. International journal of Man -Machine studies. Vol 26, p633-648, 1987.

[De Kleer 86] De Kleer J. An assumption based TMS. Artificial intelligence 28, p127 -162.

[Fouet 87] Fouet J.-M., Utilisation de connaissances pour améliorer l'utilisation de connaissances: La Machine GOSSEYN, Thèse d'état, Université Paris 6, September 1987.

[Fouet 95] Fouet J.-M., Utilisation de méta-connaissances pour l'acquisition, la transformation et la découverte de connaissances. Revue Ingénierie des systèmes d'information , Volume 3, N°2-3 1995.

[Fouet and Nouira 95] Fouet J. -M., Nouira R. Acquisition and structuration of knowledge using meta-knowledge. IJCAI , Workshop on Machine learning. Montreal, August 1995.

[Ginsberg 88]   Ginsberg A. Knowledge base system reduction: A new approach to checking knowledge bases for consistency and redundancy. In Proc. Seventh National Conference on Artificial Intelligence (AAAI-88), pages 585-589. AAAI-Press, Menlo Park, CA, 1988.

[Gupta 91]   Gupta U. Validating and verifying Knowledge-Based Systems. IEEE Computer Society Press, Los Alamitos, CA, 1991.

[Lalo 89] Lalo. A. La détection d'incohérences dans les systèmes d'ordre un. Thèse Université de Paris VI, 1989.

[Lopez et al 90] Lopez B., Mesguer P., Plaza E. Knowledge-based systems validation: A state of the art. AI Communications, 3 (2): 58- 72, 1990.

[Nguyen et al 87] Nguyen T. A., Perkins W. A., Laffey T. J., and Pecora D. Knowledge base verification. AI Magazine 8 (2): 69-75, Summer 1987.

[O'Keefe and O'Leary 93] O'Keefe R. M., O'Leary D. E. Expert systems verification and validation: A survey and tutorial. Artificial intelligence Review 7, p3-42 1993.

[Pipard 88]   Pipard E. Détection d'incohérences et d'incomplétudes dans les bases de règles- Le système INDE. Proceedings   Les systèmes experts et leurs application, Avignon June 88, p13-33.

[Preece and Shingal 92] Preece A.D. and Shingal R. Foundation and application of knowledge base verification. International journal of intelligent Systems, 9 (8) 683-701, 1994.

[Suwa et al 82] Suwa M., Scott A.C, Shortliffe E.H. An approach to verifying completeness and consistency in a rule-based expert system. AI Magazine 3 (4) 16-2, Fall 1982.

OKCFTR: TRANSLATORS FOR KNOWLEDGE REUSE

Kunhuang Huarng
Department of Finance
ChaoYang Institute of Technology
Taiwan, ROC
huarng@dec8.cyit.edu.tw

Ming-Cheng Chen
Department of Information Management
Ling Tung College
Taiwan, ROC
yoyo@imsun1.ltc.edu.tw

ABSTRACT

Reuse is considered as an applicable way to improve productivity, which is especially applicable for knowledge based systems. The reuse of knowledge can avoid the bottleneck of developing knowledge based systems and hence greatly improve productivity. Translation is considered as an approach to facilitating knowledge reuse. The number of translators can be reduced by using a canonical form in translation. In this article, it is shown how the OKCFTr can be used to translate facts among Insight 2+, IBM ESE, CLIPS, OPS5, NEXPERT, and KAPPA. Translation schemes and examples are listed to demonstrate the feasibility. The OKCFTr can be used to facilitate knowledge reuse and collaboration. Integrating with a user interface, the OKCFTr becomes an Expert System Prototyping System to support expert system prototyping. In addition, the OKCF can serve as a communication convention in networks.

1. INTRODUCTION

Reuse is considered as an applicable way to improve productivity, which is more applicable for knowledge based systems. The reuse of knowledge can avoid the bottleneck of developing knowledge based systems and hence greatly improve productivity. Translation is considered as an approach to facilitating knowledge reuse and collaboration. The use of a canonical form in translation has been proven effective: the number of translators for translation among N expert system building tools (ESBTs) has been reduced from $N \times (N-1)$ to $2 \times N$ [Esc91, HS96]. An Object Knowledge Canonical Form (OKCF) was applied to translate object knowledge successfully [HS96].

In this article, it is shown how the OKCFTr can be used to translate facts among six ESBTs: Insight 2+, IBM ESE, CLIPS, OPS5, NEXPERT, and KAPPA. These ESBTs are divided into two groups based on the support of uncertainty management systems. Translation schemes and examples are listed to demonstrate the feasibility. The OKCFTr can be used to facilitate knowledge reuse and collaboration among these ESBTs. Integrating with a user interface, the OKCFTr becomes an Expert System Prototyping System to support expert system prototyping. In addition, the OKCF can serve as a communication convention to reuse knowledge in heterogeneous ESBTs.

The OKCF and OKCFTr are described in sections 2 and 3 respectively. The translation schemes and examples of the OKCFTr are listed in section 4. This article concludes by discussing the contributions of the OKCFTr and future research direction.

2. OBJECT KNOWLEDGE CANONICAL FORM

An Object Knowledge Canonical Form (OKCF) was proposed in Backus Naur Form [Pag81] to facilitate knowledge reuse. The OKCF contains a data fact, a simple fact, a simple fact with uncertainty, a tuple fact, a class of facts and a structured fact [Hua93]. The *data fact* is a single item; for example a data fact in CLIPS is

```
(LARGE)
```

The *simple fact* contains two items: attribute and value. For example, a simple fact in CLIPS is

```
(SIZE LARGE)
```

The *simple fact with uncertainty* is a quadruple. For example, a simple fact with uncertainty in Insight 2+ is

```
SHAPE IS ROUND CONFIDENCE 80
```

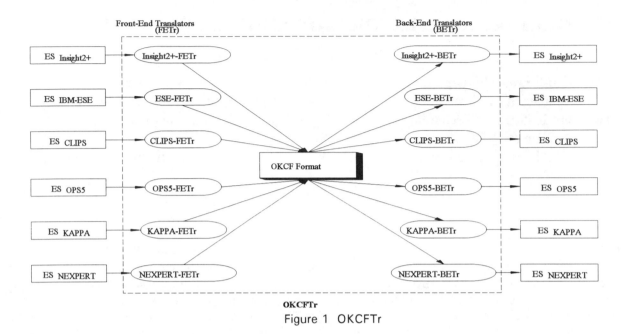

OKCFTr
Figure 1 OKCFTr

Table 1 A Comparison of Facts in ESBTs

ESBTs	Simple Fact	Simple Fact with Uncertainty	Tuple Fact	Group
Insight 2+		√		1
IBM ESE		√		1
CLIPS	√		√	2
OPS5	√		√	2
NEXPERT	√			2
KAPPA	√			2

where 'IS' and 'CONFIDENCE' are reserved words indicating that the following items are the value for the attribute and the uncertainty for the fact, respectively. Different notations and terms may be used in various ESBTs.

The *tuple fact* contains two or more items; for example, a tuple in CLIPS is

```
(SIZE LARGE MEDIUM SMALL)
```

Classes of facts and structured facts in the OKCF correspond to classes and instances in conventional object-oriented programming languages. In other words, structured facts are instances of classes of facts. The use of the OKCF in the translation of classes of facts and structured facts was successfully implemented [HS96], which is beyond the scope of this article.

The Interlingua Working Group developed a format called Knowledge Interchange Format (KIF) for knowledge reuse [GF92]. However, the OKCF supersedes the KIF in the support of object knowledge, uncertainty management systems, and translation schemes, as well as in the commitment to commercial ESBTs [Hua93].

3. OBJECT KNOWLEDGE CANONICAL FORM BASED TRANSLATORS

Object Knowledge Canonical Form based Translators (OKCFTr) have been built to translate facts among six ESBTs: Insight 2+, IBM ESE, CLIPS, OPS5, NEXPERT, and KAPPA. The OKCFTr contains front- and back-end translators, which translate knowledge from source ESBTs to the OKCF and from the OKCF to target ESBTs, respectively. These ESBTs are divided into two groups: Group 1 supports simple fact with uncertainty; Group 2 supports simple facts. The characteristics of these ESBTs

Table 2 Translation from Group 1 to Group 1

Insight 2+	OKCF	IBM ESE
SHAPE IS ROUND CONFIDENCE 80	SHAPE = ROUND UNCERTAINTY 80	THERE IS 0.6 EVIDENCE THAT SHAPE IS 'ROUND'
IBM ESE	OKCF	Insight 2+
THERE IS 0.8 EVIDENCE THAT SHAPE IS 'ROUND'	SHAPE = ROUND UNCERTAINTY 0.8	SHAPE IS ROUND CONFIDENCE 90

Table 3 Translation from Group 1 to Group 2

Insight 2+	OKCF	CLIPS
SIZE IS LARGE CONFIDENCE 80	SIZE = LARGE UNCERTAINTY 80	(SIZE LARGE 80)
IBM ESE	OKCF	OPS5
THERE IS 0.8 EVIDENCE THAT SIZE IS 'LARGE'	SIZE = LARGE UNCERTAINTY 0.8	(SIZE LARGE 0.8)

are analyzed. The command of executing the OKCFTr is also listed.

3.1 OKCFTr

In OKCFTr, front-end translators (FETr) translate knowledge from source ESBTs to the OKCF; back-end translators (BETr) translate knowledge from the OKCF to target ESBTs. For example, in Figure 1, $ES_{Insight\ 2+}$ can be translated to the OKCF via Insight 2+-FETr (a front-end translator for Insight 2+); an ES in the OKCF can be translated to ES_{CLIPS} via CLIPS-BETr (a back-end translator for CLIPS).

3.2 ESBTs Characteristics

According to the support of UMS, a comparison of these ESBTs is summarized in Table 1. Insight 2+ and IBM ESE support simple fact with uncertainty; hence, both are in Group 1. CLIPS and OPS5 support simple fact and tuple fact. NEXPERT and KAPPA support only simple fact. CLIPS, OPS5, NEXPERT, and KAPPA are in Group 2.

3.3 Command

OKCFTr was written in C and can be run on various platforms. OKCFTr can be issued by

```
OKCFTr -a -b input_file output_file
```

where a specifies a source ESBT;
      b specifies a target ESBT;
      a, b =   o,   OPS5;

    k,   KAPPA;
    c,   CLIPS;
    n,   NEXPERT;
    i,   INSIGHT 2+;
    b,   IBM ESE;
    m,   OKCF;

The knowledge in a source ESBT can be translated to the OKCF; the knowledge in the OKCF can be translated to a target ESBT. Meanwhile, the OKCF can also be either the source or target ESBT.

The format of the OKCF file is listed as follows.

```
ESBT: name

FACTS:
Attribute Value UNCERTAINTY Uncertainty
Attribute Value UNCERTAINTY Uncertainty
Attribute Value UNCERTAINTY Uncertainty
...
```

where ESBT records the name of the source ESBT which is necessary when the knowledge is translated to target ESBTs; FACTS record the facts translated.

4. TRANSLATION

Translation schemes between Groups 1 and 2 are quite different and are discussed separately: from Group 1 to Group 1, from Group 1 to Group 2, from Group 2 to Group 1, and from Group 2 to Group 2. Various translation techniques may be needed in different situations.

Table 4 Translation from Group 1 to Group 2

Insight 2+	OKCF	KAPPA
SIZE IS LARGE CONFIDENCE 80	SIZE = LARGE UNCERTAINTY 80	GLOBAL:SIZE_LARGE= 80)
IBM ESE	OKCF	NEXPERT
THERE IS 0.8 EVIDENCE THAT SIZE IS 'LARGE'	SIZE = LARGE UNCERTAINTY 0.8	(SIZE_LARGE) ("0.8")

Table 5 From Group 2 to Group 1

CLIPS	OKCF	Insight 2+
(SIZE LARGE)	SIZE = LARGE UNCERTAINTY 100	SIZE IS LARGE CONFIDENCE 100
NEXPERT	OKCF	Insight 2+
(SIZE) ("LARGE")	SIZE = LARGE UNCERTAINTY 100	SIZE IS LARGE CONFIDENCE 100
KAPPA	OKCF	IBM ESE
GLOBAL:SIZE = LARGE	SIZE = LARGE UNCERTAINTY 100	THERE IS 1.0 EVIDENCE THAT SIZE IS 'LARGE'

4.1 From Group 1 to Group 1

To perform translation in Group 1, two issues need to be noticed. First, ESBTs must provide similar uncertainty management systems (UMSs). For example, Insight 2+ and IBM ESE both provide Mycin type of UMSs [Sho85]. Therefore, the uncertainties in both ESBTs are equivalent semantically. Second, the ranges in the UMSs should be the same, otherwise a *mapping* is necessary. For example, the range in Insight 2+ is 0 to 100, while IBM ESE is -1.0 to 1.0. In this case, an uncertainty 80 in Insight 2+ is mapped to 0.6 in IBM ESE. The OKCFTr perform the mapping during the translation from the OKCF to target ESBTs when necessary. Translation examples between Insight 2+ and IBM ESE through the OKCFTr are listed in Table 2.

4.2 From Group 1 to Group 2

When the translation is performed from Group 1 to Group 2, two situations may happen. First, if target ESBTs support tuple facts such as CLIPS and OPS5, simple fact with uncertainty s in the OKCF can be translated to tuple facts (as attribute value uncertainty) in target ESBTs. For example, the knowledge in Insight 2+ and IBM ESE can be translated to CLIPS and OPS5, as listed in Table 3. Since target ESBTs do not support

UMSs, UMSs in source ESBTs should also be simulated and included in target ESBTs to maintain uncertainties. During inference, the uncertainties in the tuple facts should be extracted and processed properly by the simulated UMSs.

Second, if target ESBTs do not support tuple facts, a "concatenation" technique must be applied. Concatenation technique concatenates "attribute" and "value" in the OKCF into "attribute_value." The translated facts in target ESBTs appear as

   attribute_value uncertainty

By the concatenation technique, the knowledge in Insight 2+ and IBM ESE can be translated to KAPPA and NEXPERT, as listed in Table 4.

4.3 From Group 2 to Group 1

This type of translation is similar to the translation from Group 1 to Group 1 except there is no uncertainty information in source ESBTs. The simple facts without uncertainties are assumed to be complete certain. Complete certainties (100's) are added to the knowledge in the OKCF. Then, the uncertainties need to be mapped to the ranges of the target ESBTs if necessary. In Table 5, the knowledge in CLIPS and NEXPERT are translated to Insight 2+; KAPPA is translated to IBM ESE.

Table 6 From Group 2 to Group 2

OPS5	OKCF	KAPPA
(COLOR RED)	COLOR = RED UNCERTAINTY 100	GLOBAL:COLOR = RED
KAPPA	**OKCF**	**CLIPS**
GLOBAL:COLOR = RED	COLOR = RED UNCERTAINTY 100	(COLOR RED)
CLIPS	**OKCF**	**NEXPERT**
(COLOR RED)	COLOR = RED UNCERTAINTY 100	(COLOR) ("RED")

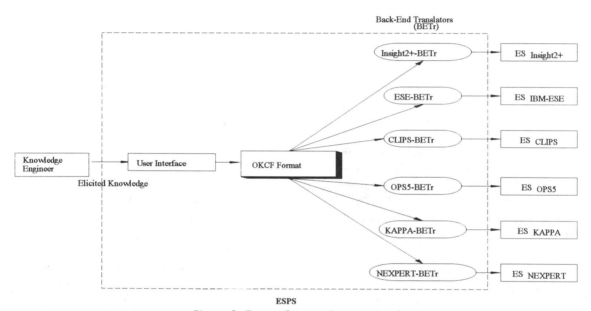

Figure 2 Expert System Prototyping System

4.4 From Group 2 to Group 2

The translation from Group 2 to the OKCF is to add complete certainties to the OKCF, as mentioned previously. If target ESBTs are in Group 2, the uncertainties are dropped. The translation examples among ESBTs in Group 2 are shown in Table 6.

5. CONCLUSION

The OKCFTr contribute to the field of knowledge based systems in several aspects:

• Knowledge translation and collaboration. Through the translation examples in this article, the OKCFTr have been demonstrated to translate knowledge among six ESBTs successfully. Translated knowledge can facilitate knowledge collaboration among heterogeneous ESBTs. More ESBTs will be studied and more translators will be included in the OKCFTr to provide wider coverage.

• Knowledge reuse. The OKCF is especially useful to translate knowledge in ESBTs no longer supported. Through the OKCF, the knowledge in "aged" ESBTs can be reserved and translated to active ESBTs for reuse. Hence, knowledge in "aged" ESBTs will not phase out simply because their ESBTs are not supported.

• Expert System Prototyping System. An Expert System Prototyping System (ESPS) is under development to integrate a user interface with the OKCFTr to generate prototype expert systems, as listed in Figure 2. Knowledge engineers can input the elicited knowledge to the user interface; the user interface formulates the knowledge to the OKCF format; the back-end translators can translate the knowledge into various ESBTs. For the same knowledge is represented in various ESBTs, the performance and feasibility of these ESBTs for particular problem domains can be tested.

• Communication convention. The OKCF can serve as a communication convention for heterogeneous ESBTs, as in Figure 3. When knowledge is to be distributed, the

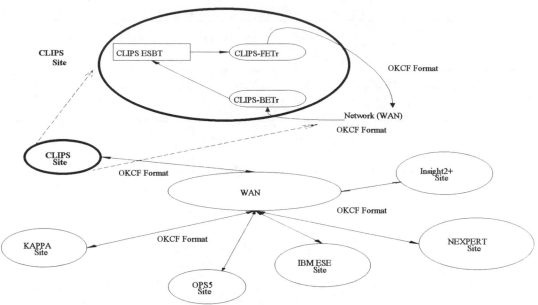

Figure 3 OKCF as Communication Convention

knowledge can be translated to the OKCF by the front-end translators. When knowledge is to be obtained by remote sites, the OKCF knowledge can be translated by the back-end translators into target ESBTs. In other words, all the knowledge in networks is in the OKCF format. Hence, the OKCF can serve as a communication convention among ESBTs.

REFERENCES

[Esc91] T. D. Escamilla, *An Object Oriented Canonical Form for Reusable Knowledge Bases*, Ph.D. Dissertation, Texas A&M University, 1991.

[GF92] M. R., Genesereth and R. E. Fikes, *Knowledge Interchange Format Reference Manual*, Ver. 3.0 Logic Group Report Logic-92-1, Stanford University, 1992.

[HS96] K. Huarng and D. B. Simmons, An Object Knowledge Canonical Form for Knowledge Reuse, *Expert Systems with Applications*, vol. 10(1), 1996.

[Hua93] K. Huarng, *An Object Knowledge Canonical Form*, Ph.D. Dissertation, Texas A&M University, 1993.

[Pag81] F. G., Pagan, *Formal Specification of Programming Languages: A Panoramic Primer*, 1981.

[Sho85] E. H. Shortliffe, Details of the Consultation System, In *Rule-Based Expert Systems: The MYCIN Experiments of the Stanford Heuristic Programming Project*, eds, B. G. Buchanan and E. H. Shortliffe, 1985.

INTEGRATION OF FORWARD AND BACKWARD INFERENCES USING EXTENDED RETE NETWORKS

Yong H. Lee and Suk I. Yoo

Department of Computer Sciences, Seoul National University, Seoul, Korea, 151-742

E-mail: lyh@plaza.snu.ac.kr

ABSTRACT

In this paper, it is explained that the forward and backward chaining can be naturally integrated on an extended Rete network via the generation of hypotheses from forward chaining inference. We introduce the concept of the hypothesis objects and augment the production rule format by adding the hypothesis condition elements in the LHS part of a rule, which control the forward and backward chaining invocation of rules. A rule can be used in forward directions, backward directions, or mixed directions. To be applied to the backward chaining, a modified version of Rete has been formulated by adding the hypothesis-and-nodes and backward direction edges to the node network. We then develop a backward chaining algorithm which takes advantage of already existing partial match results obtained by the previous forward chaining in the ERMI. The whole mixed inference process has the potential advantages over purely forward and backward chaining strategy.

1. INTRODUCTION

Most current knowledge engineering tools have been designed to provide the structures for representing domain knowledge by using rule structure (IF-THEN), and an inference mechanism for controlling the use of domain knowledge. In production rule systems, two very important reasoning control strategies are forward (data-directed) chaining and backward (goal-directed) chaining [HWB83]. Although forward and backward chainings are useful inference schemes, forward chaining approaches sometimes have the disadvantage of generating many facts not directly related to the problem under consideration, while backward chaining approaches have the disadvantage of perhaps becoming fixed on an initial set of goals and having difficulty shifting focus when the data available do not support them. An interesting possibility is to inference in both directions simultaneously or opportunistically. We call this inference mechanism as *mixed inference*. To remedy control limitations within the basic architecture, several works have been tried. Some knowledge engineering tools (ART, TIRS, ESE, KEE, etc.) can process both forward and backward chaining [JC94]. However, their chaining schemes are not be naturally integrated tightly into a single uniform control framework.

We presume the reader's familiarity with production systems [HWB83, BFK85] and Rete match algorithm [FOR81, FOR82, HUG86]. This paper describes how forward and backward chaining control can be naturally integrated into a single uniform control framework, Rete network, via the generation of hypotheses from forward chaining inference. We present an augmented version of the Rete, *Extended Rete for Mixed Inference (ERMI)*, that integrates the forward and backward chaining controls.

The overall configuration of mixed inference methodology is depicted in Fig. 1, which presents a high-level schematic control for inference based on ERMI. It consists of six modules: Rule System, Object System, Forward Chaining Module, Backward Chaining Module, Extended Rete network for Mixed Inference, and Rule Compiler. Backward pattern matching process in ERMI is motivated by the desire to make use of precomputed partial match results that already exist in the network, as well as to make use of shared Rete structures. When a Rete has been used in forward pattern matching for a while, it typically has nonempty partial result memories associated with many of the joins. All instantiations that have been matched a set of LHSs but not been fired are still stored in CS(Conflict Set). When a backward chaining is processed, we make use of those partial match results stored in memory nodes and those instantiations stored in CS, as will be shown in Section 4. This speeds up the backward matching process as well as forward matching process.

2. KNOWLEDGE REPRESENTATION

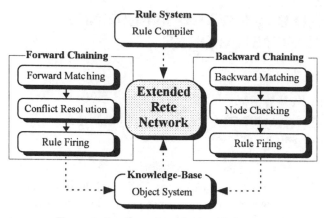

Figure 1. Configuration of mixed inference

2.1 OBJECT AND HYPOTHESIS

We describe the object representation briefly. An example of object class which represents a part of MPI engine of automobile is given in Fig. 2. The *hypotheses*, which denote some WMEs whose truth or falseness is not known on the basis of some evidence, are represented as formal objects. The *hypothesis class* is introduced to exploit an explicit representation of goals, to previous solution is applicable again, and to merge goals that can be achieved together. A hypothesis class is declared with the hypothesis declaration format in Fig. 3. The status slot denotes the current state of the hypothesis instance and can hold one value from { *active*, *inactive*, *suggested* }. The value slot represents the value of a hypothesis instance and can hold one value from { *true*, *false*, *unknown*, *notknown* }. A hypothesis object with *active* status means that it can be used in inference process. A hypothesis object with *inactive* status will not be used. A hypothesis object with *suggested* status and *unknown* value is considered to be a backward chaining goal. Such a hypothesis object is inserted into the *backward conflict set* (*BCS*), which denotes currently unattained goals. When a hypothesis object is verified (refuted), its value is set to *true* (*false*). A hypothesis object with *unknown* value means that it has not been evaluated. A hypothesis object with *notknown* value means that it has been evaluated but has not been verified or refuted. The further details can be found in [MY90, SK92, YKP90].

2.2 RULE REPRESENTATION

Rules are composed of four parts as follows.
    rule *<rule-name>* {
        if *<primary-condition>*
        when *<hypothesis-condition>*

```
Object-Class AFS {
    relations: part-of(ENGINE)
    attributes:
        float voltage prompt="What is the voltage of AFS?"
        float connector_voltage prompt="What is the voltage of connector?"
        string state  "good", "broken"
        string control_relay_afs_wire  "good", "broken"
        string earth  "good", "broken"
        int power
        string power_state  "good", "broken"
}
```

Figure 2. Example of object class definition

        do *<action>*
        }
The *<hypothesis-condition>* part is additionally introduced into ordinary if-then rule format and plays a crucial role in coordinating inference activities. The *<primary-condition>* which is the collection of patterns of object-attribute-value format, called *primary condition elements* (*PCEs*), indicates when the rule can potentially be applied. The *<hypothesis-condition>* also specifies a sequence of patterns, called *hypothesis condition elements* (*HCEs*), which represent hypotheses to be tested. A condition element, called *lhs-pattern*, consists of an object instance and a sequence of terms. A *term* can be the operator ↑ followed by an attribute and a specification of a value. If a rule is involved in forward matching process but its HCEs are not satisfied with current WM, they can be verified by backward rule chaining. An *<action>* represents the procedures to be executed when the LHS is successfully matched with the WM. The *suggest* action changes the status of the hypothesis object corresponding to the specified argument into *suggested* and makes a new top-level backward chaining goal to be inserted to BCS.

3. MIXED INFERENCE

3.1 EXTENDED RETE NETWORK

The matching algorithm developed for mixed inference is based on the Rete algorithm, which is well known to be an efficient matching algorithm. A large portion of the ERMI is organized conventionally [4, 5, 6] except for three augmentations: (1) we link nodes bidirectionally. (2) we add a new kind of two-input node: *hypothesis-And-node* (*Hand-node*). (3) we compile rhs-variables appearing in both LHS and RHS of a rule into the structure that is used for variable binding during backward rule matching.

    (1) *Connecting nodes bidirectionally*: In ordinary Rete networks, there is a kind of outgoing edge, called *forward edge* (*F-edge*), connecting pairs of node from top node (*F-predecessor*) to bottom node (*F-successor*) as in Fig.

```
Hypothesis-Class <hypothesis-name> {
    status: true, false, unknown, notknown ;
    value: active, inactive, suggested, asserted ;
    attributes: <attribute₁>, <attribute₂>, ...;
}
    <attributeᵢ> = <type> <name> prompt="prompt?" <certainty>
    <type> = int float char string
```

Figure 3. Hypothesis class definition

5(left). In ERMI, nodes are connected by two kinds of outgoing edges as shown in Fig. 5(right): F-edge and *backward edge* (*B-edge*) being directed from bottom node (*B-predecessor*) to top node (*B-successor*). F-edges are used for forward chaining as ordinary Rete algorithm while B-edges are used for the backward verification process of nodes corresponding to condition elements of rules.

(2) *Hypothesis-And-node*: When rules are compiled into network, all PCEs and HCEs are represented by one-input nodes denoting intracondition tests. A chain of one-input nodes representing a positive HCE is joined to the network structure for the preceding conditions of the rule by a Hand-node (as ① or ② in Fig. 6). Hand-node represents that it may be a starting point for backward chaining. When a token arrives at right input (as ③ or ④ in Fig. 6) of the two inputs of a Hand-node, it joins its left and right memory, and passes down the result of join. When tokens are passed into the left side (as ⑤ or ⑥ in Fig. 6) of a Hand-node, the Hand-node looks for WMEs stored in right memory to be consistent with the newly arrived token. For each consistent WME, the Hand-node sends out join result. If, however, there is no such WME and its right input memory is empty, a chain of one-input nodes (as ⑦ or ⑧ in Fig. 6) connected to its right input are combined. The combined result corresponds to all intracondition tests of a HCE. Then a goal for backward chaining is generated by applying the intercondition tests and the newly arrived token to the combined result, and then is inserted into the BCS. For example, suppose that tokens, $+(c_1 \uparrow a_1 \ 1 \ \uparrow a_2 \ 2 \ \uparrow a_3 \ 3)$ and $+(c_2 \uparrow a_1 \ 2 \ \uparrow a_2 \ 2)$, are passed to the network in Fig. 7, and α-mem (①) is empty. Then the token $+((c_1 \uparrow a_1 \ 1 \ \uparrow a_2 \ 2 \ \uparrow a_3 \ 3) (c_2 \uparrow a_1 \ 2 \ \uparrow a_2 \ 2))$ is generated and passed into the left-input of the Hand-node (③). The intracondition test $(c_3 \uparrow a_1 \ 3)$ is generated by combining a chain of one-input nodes (④) connected to the right input of ③. A new subgoal $(c_3 \uparrow a_1 \ 3 \ \uparrow a_2 \ 2)$ is inserted into the BCS by applying the intercondition $(c_1.a_2 = c_2.a_2 = c_3.a_2)$ and the token $+((c_1 \uparrow a_1 \ 1 \ \uparrow a_2 \ 2 \ \uparrow a_3 \ 3) (c_2 \uparrow a_1 \ 2 \ \uparrow a_2 \ 2))$ to the intracondition test, $(c_3 \uparrow a_1 \ 3)$.

(3) *Compiling rhs-pattern variables*: If the RHS of a rule has pattern variables, backward matching is done by using the unification algorithm. If a matching rule is

```
rule inj_15_1 (INJECTOR) {
    if      (ENGINE ↑diagNo 12)
            (CAR ↑start off)
            (INJECTOR ↑time <> normal)
    when    (APS ↑voltage < 4.5) <aps>
            (hApsBroken ↑value true)
    do      modify(<aps> ↑state broken)
            print("The state of APS is broken.")
}
rule aps_5 (APS) {
    if      (APS ↑voltage < 4.5)
            (AFS ↑power <> 12)
            (BATTERY ↑voltage < 11 ↑rpmVolt < 12)
    do      make(hApsBroken ↑value true)
}
```

Figure 4. Sample Rules for diagnosing MPI engine

found, the backward chaining module will instantiate the condition elements of the matching rule by replacing variables with the values they bound to. Each instantiation, then, is treated as a new subgoal. For such an instantiation, variables appearing in both LHS and RHS are compiled into the (*name*/<*variable*>) structures, called *binding constraints*, where *name* is either attribute name or class name. In Fig. 7, the binding constraint for rhs-pattern $(<c> \uparrow a_5 <m>)$ is $(c_1.a_3/<m>, c_2/<c>)$, which the variable $<m>$ denotes attribute $\uparrow a_3$ of the class c_1 in the first CE, and the variable $<c>$ is the element variable associated with the second CE that will be bound to the instance of class c_2. In order to find matching rules (i.e. corresponding P-nodes such as ⑤ in Fig. 7) for a backward goal efficiently, when a rule is compiled into the ERMI, a hash table is constructed according to the hypothesis class name as shown in Fig. 8. We make slots of hash table contain pointers to linked lists of items, <rule-name, P-node>.

3.2 MIXED INFERENCE IN ERMI

The mixed inference in ERMI (Algorithm: *Mixed-Inference*) consists of forward chaining phase and backward chaining phase, where forward chaining phase is the basic strategy. In general, it first applies the forward chaining (similar to ordinary Rete algorithm) and then activates the backward chaining (Algorithm: *Backward-Chaining*) based on the partial information obtained by the forward chaining. The forward *match-rule* and *execute-rule* steps are actually interleaved with the *backward-rule* steps. During backward chaining, all rules paticipating in the chaining are push into the *backward solution path stack* (*BSPS*).

Algorithm: *Mixed-Inference*
Step 1. If there is no change in WM, goto Step 4. Otherwise, for each change to WM, do the Step 2 and

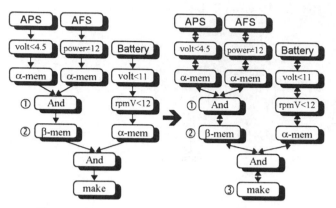

Figure 5. Linking nodes by bidirectional edges

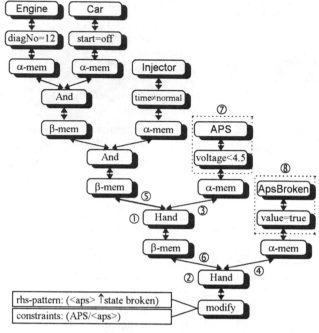

Figure 6. ERMI for rule inj_15_1 in Fig. 4

Step 3.

Step 2. Generate a token corresponding to the change to WM.

Step 3. (Match-rule) Propagate token through the ERMI network. Follow ordinary Rete matching algorithm in the nodes except for Hand-node. Update the CS and BCS if possible.

Step 4. (Select-rule) If a change to BCS occurred, then goto Step 6. If CS is empty, then goto Step 7. Otherwise, select one rule instantiation from the CS.

Step 5. (Execute-rule) Execute the selected rule and goto Step 1.

Step 6. (Backward-rule) If BCS is empty, then terminate with FAILURE. If BCS contains elements newly generated by a action *suggest* in the last execute-step, then select one of those elements. Otherwise, if BCS contains elements newly generated by a Hand-node, select one of those elements. Let the selected element be BGoal and goto Step 8.

Step 7. If BCS is empty, then terminate with FAILURE. Otherwise, select a element (BGoal) from BCS.

Step 8. Make the backward rule stack empty. Call Backward-Chaining(BGoal). Pop up rules in BSPS and execute them sequentially. Goto Step 1.

3.3 BACKWARD CHAINING IN ERMI

The backward chaining procedure consists of (1) backward matching and (2) checking of CS or nodes in ERMI (Algorithm: *Check-One-Input* and *Check-Two-Input*). In general, a backward chaining algorithm performs a blind search for facts and rules. However, backward chaining in ERMI makes use of the partial matches stored in the network.

(1) *Backward matching*: A backward chaining triggered for a goal means that it is not satisfied with current WM directly so that it should be solved by backward rule chaining. The first step for the backward rule chaining in the ERMI is to select an element, BGoal, from BCS by resolving conflicts. An element of BCS matches a RHS when the followings are satisfied: the RHS contains *make* or *modify* actions such that its rhs-pattern matches the pattern of the element of BCS. The CS consists of rule instantiations where an instantiation is a rule name and an ordered set of WMEs that satisfy that rule. Therefore, next step is to find an instantiation in CS such that its rule matches BGoal and its WMEs satisfy the binding constraints of the rule. If such an instantiation exists, backward matching can be terminated successfully by executing the rule instantiation immediately and propagating changes to WM to ERMI network. For example, consider a situation in Fig. 8. Suppose that the RHS of rule injector_5_1 is satisfied with WM, but not executed (its rule instantiation is stored in CS as ①). If the current goal for backward chaining is *hInjectorProb* (③), and rule symp_43 is selected as a matching rule by hashing *hInjectorProb*, node checking for verifying rule symp_43 will start from ④. If ⑥ is checked successfully and α-memory (⑤) is empty, another backward chaining for the goal combined from ⑦ (ECU ↑state broken) will be triggered. In this case, however, the new goal can be satisfied immediately by executing an instantiation (①) in CS.

If BGoal are not satisfied in CS, next step is to match the RHSs of rules against the BGoal. When matching rules are found, every matching rule is enumerated in the form of list. The list of matching rules is stored in stack

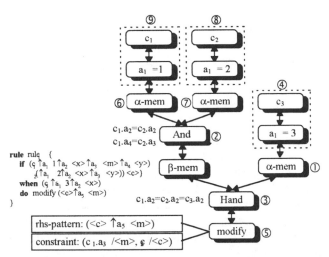

Figure 7. Backward matching and AND condition

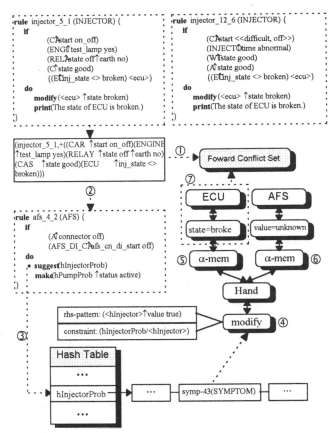

Figure 8. Reference to CS during backward matching

called *backward rule stack* (*BRS*). It proceeds by selecting and deleting a rule from the rule list on the top of BRS, and checking nodes associated with it. If a rhs-pattern contains variables that occurred in the LHS of the rule, substitutions, which are unifiers of the rhs-pattern and the BGoal, are generated by using binding constraints. In Fig. 7, if the current backward goal is ($c_2 \uparrow a_5$ 5), the substitution, {(5, <m>), (c_2, <c>)}, is produced by unifying the goal with the rhs-pattern. By applying the substitution to the binding constraint ($c_1.a_3$/<m>, c_2/<c>), the rhs-pattern of $rule_1$ matches ($c_2 \uparrow a_5$ 5) and $c_1.a_3$ is bound to 5. Then, the first CE of the $rule_1$, ($c_1 \uparrow a_1$ 1 $\uparrow a_3$ 5), is instantiated when combining nodes (⑨) through the left B-edges of And node (②).

Algorithm: *Backward-Chaining*(BGoal)
Step 1. If there is no instantiation in CS whose rule matches BGoal, then goto Step 2. Otherwise, push such an instantiation represented by (rule, tokens) into BSPS and return SUCCESS.
Step 2. If BGoal can not be achieved by asking user, then goto Step 3. Otherwise, ask user about BGoal. If the answer matches Bgoal, return SUCCESS. Otherwise, return FAILURE.
Step 3. Find all matching rules whose RHS matches BGoal by hasing BGoal, and create a list that consists of the matching rules. Call this list as RList. Reorder RList according to some arbitrary scheme, and push RList into BRS.
Step 4. If RList is empty, return FAILURE. Otherwise, select the first matching rule on the RList on top of BRS, remove it from the list, and call this rule CRule.
Step 5. Find a production node, Pnode, corresponding to CRule, and generate a substitution called BToken by applying its binding constraints to BGoal. Let B-

successor of Pnode be BSuccessor. If BSuccessor is the two-input node, then call Check-Two-Input(BToken, BSuccessor). Otherwise, call Check-One-Input(BToken, Bsuccessor). Let the return value be NewTokens.
Step 6. If the node checking in Step 5 fails (i.e. NewTokens is empty), then goto Step 4. Otherwise, Push CRule and NewTokens into the BSPS, and return SUCCESS.

(2) *Checking nodes in ERMI:* If one of the matching rules is selected, all nodes corresponding to condition elements of the selected matching rule are then checked to see if they will be verified or denied (Algorithm: Check-One-Input and Check-Two-Input). If it cannot be determined that a condition is true or false, the condition is added as a new goal to be explored by a backward matching. This process continues until a complete solution is found or no additional expansion is possible. There is no necessity for looking for an assertion in the WM that matches a current goal or subgoal. If there are matching assertions in WM, they are already stored in memory node, and can be verified by node checking through B-edges.

Algorithm: *Check-Two-Input*(BToken, TwoNode)

Step 1. If TwoNode is Not-node, then goto Step 4.

Step 2. Make LTokens, RTokens, and NewBToken empty. If the left input memory of TwoNode is not empty and a memory element matches BToken, insert such all elements to LTokens. Copy LTokens to LMemory. If the right input memory of TwoNode is not empty and a memory element matches BToken, insert such all elements to RTokens. Copy RTokens to RMemory. If RMemory is not empty, goto step 4.

Step 3. If LMemory is not empty, copy BToken to NewToken and extract a token (LToken) from LMemory. Replace variables in join condition of TwoNode with the corresponding values of LToken. Then insert the join condition into NewBToken. Let RBSuccessor be the B-Successor of the right input memory of TwoNode. Call Algorithm:Check-One-Input(NewBToken, RBSuccessor) and let RTokens be the return value. If RTokens is not empty, goto step 4. If LMemory is empty, return NULL. Otherwise, goto step 3.

Step 4. If LTokens is not empty, goto step 6. Copy RTokens to RMemory. While RMemory is not empty, repeat step 5.

Step 5. Copy BToken to NewBToken. Extract a token (RToken) from RMemory. Find a substitution instance by substituting values of RToken for corresponding variables in the joins condition of TwoNode, update the binding constraints according to the substitution instance, and then insert the join condition into NewBToken. Let LBSuccessor be the B-Successor of the left input memory of TwoNode. If the type of LBSuccessor is two input node, call Algorithm:Check-Two-Input(NewBToken, LBSuccessor). Otherwise, call Algorithm:Check-One-Input(NewBToken, LBSuccessor). Then, let LTokens be the return value. If LTokens is not empty, goto step 6.

Step 6. Compare each element of LTokens with the RTokens memory elements and join them into bigger tokens if they are satisfied with the join condition of TwoNode (The join operation is performed as in ordinary Rete algorithm). If any new bigger token is generated, then insert it into the β-memory of TwoNode as well as NewTokens. Return NewTokens.

Algorithm: *Check-One-Input*(BToken, OneNode)

Step 1. If the α-memory node of the OneNode is not empty, then return FAILURE.

Step 2. Generate a new subgoal by combining a sequence of B-successors of OneNode. Let the subgoal be IntraCond.

Step 3. Apply the binding constraints of BToken to IntraCond. Let the result be BNewGoal.

Step 4. Call Algorithm:Backward-Chaining(BNewGoal).

Then, let NewTokens be the return value. If NewTokens is empty, return NULL.

Step 5. Insert NewTokens into the α-memory of OneNode, then return NewTokens.

4. CONCLUSIONS

We have shown how forward and backward chaining control can be naturally integrated into Rete network. We have suggested a mechanism that integrates forward and backward chaining control via the generation of hypotheses from forward chaining inference. Within this augmented architecture, a wide range of inference paradigms can be implemented efficiently according to the specific features of the current problem to be solved: from those based on a forward chaining approach to those based on backward chaining activity. Through mixed chaining, the proposed mechanism generates exactly the sequence of facts needed to solve the problem with no wasted motion.

REFERENCES

[BFK+85] Brownston, R. Farrell, E. Kant, and N. Martin, *Programming Expert System in OPS5: An Introduction to Rule-Based Programming*. Reading, MA: Addison-Wesley, 1985.

[FOR81] C. L. Forgy, OPS5 user's manual, Tech. Rep. CMU-CS-81-135, Carnegie-Mellon Univ., 1981.

[FOR82] C. L. Forgy, "RETE-A fast algorithm for the many pattern/many object pattern matching problem," *Artificial Intelligence*, vol. 19, pp. 17-37, 1982.

[HUG86] Kenneth J. Hughes, An Introduction to the Rete Match Algorithm, 1986.

[HWB83] F. Hayes-Roth, D. A. Waterman, and D. B. Lenat (eds.), *Building Expert Systems*, Addison-Wesley, Reading, MA, 1983.

[JC94] Verlyn M. Johnson and John V. Carlis, Sharing and Reusing Rules: A Feature Comparison of Five Expert System Shells, *IEEE Expert: Intelligent Systems & Their Applications*, pp. 3-17, June 1994.

[MY90] Min, M. and Yoo. S. SAILOR: Design and Implementation of an Expert System Shell. *Korea Information Science Society*, 17(1), 163-172, 1990.

[SK92] Suk I. Yoo and Il Kon Kim, DIAS1: An Expert System for Diagnosing Automobiles With Electronic Control Units, *Expert Systems With Applications*, Vol. 4, pp. 69-78, 1992.

[YKP+90] I. Yoo, I. K. Kim, C. H. Park, H. J. Chang, T. G. Kim and M. K. Min, "HEXPERT: An expert system building tool," *Proc. 3rd IEEE Int. Conf. on Tools for Artificial Intelligence*, pp. 510-511, 1990.

VAREX: AN ENVIRONMENT FOR VALIDATING AND REFINING RULE BASES

Heribert Schimpe and Martin Staudt

RWTH Aachen, Wirtschaftsinformatik/Informatik V, D-52056 Aachen, Germany

Email: schimpe@wi.rwth-aachen.de

ABSTRACT

Quality of knowledge bases is one of the most important requirements for acceptance and reliability of expert systems. We discuss methods performing statical and dynamical analyses of rule bases and present a deductive and an inductive approach to justify refinement operations on the rule set. In addition we describe the implementation of both approaches and their integration into a general development environment for expert systems which is supported and controlled by an intelligent repository component. Our test application is an expert system for judging credit agreements which was designed in cooperation with experts of two leading German banks.

1 INTRODUCTION

Knowledge based systems, in particular expert systems have become successful within the last years in a wide range of application areas from classical diagnosis support systems to intelligent control components for industrial processes. The responsible employment and the acceptance of these systems as a substitute for human experts substantially depends on the quality of their knowledge bases and their basic inference capabilities. Hence, it is indispensable to have methods for systematical verification, validation and test of the system and its behavior.

In this paper we focus on validating knowledge bases of rule based expert systems, i.e. sets of rules, and neglect the problem of verifying the correctness of their inference machines or other system components.

An important question is at which stage of the development process of a rule base validation should take place. Following the general IEEE definition (No. 729-1983) validation is the process of evaluating software *at the end* of the development process to ensure compliance with software requirements. The most important requirement for expert systems is a behavior that is as close as possible to the one of a real human expert in the same domain. As a consequence only these experts respectively their decisions and actions for solving the same problem as the system can be recruited

in order to judge the knowledge base quality. Due to the complexity of knowledge engineering and acquisition it is usually impossible to directly map the experts knowledge into a formal framework of rules. The whole process is very time consuming. On the other hand the availability of human experts is restricted especially for revision and modification work that takes place after the real knowledge engineering phase. The implementation of a first rule base prototype within a concrete expert system shell takes along additional ballast which impedes a fast knowledge base revision together with the expert.

Thus validation and refinement steps should be performed as early as possible in the knowledge base design process. Following the idea of rapid prototyping in a first step a basic rule base can be acquired with expert support which is then statically and dynamically analysed in order to discover inconsistencies and deviations from the intended inference behavior. Dynamical analyses try to apply the rule base to real cases with known results. Based on these analyses several proposals for refining the rule base - i.e. modification of rule premises or conclusions, deletion and insertion of complete rules - are generated automatically. In an interactive step which involves the expert, the rule base then actually is being refined by selecting the appropriate measures. This validation-refinement sequence is repeated in several cycles until no further improvements can be achieved. The success of this procedure of course heavily depends on the number and variety of cases employed for dynamical analyses.

The rest of this paper is organized as follows: Section 2 introduces the technique of learning by knowledge base refinement and sketches results from literature and related work. Section 3 explains our two implemented approaches for refining rule bases. Section 4 presents the learning and refinement results we achieved. An overview of an integrated development environment for rule bases called VAREX is given in Section 5. Section 6 concludes with a summary.

2 LEARNING AND KNOWLEDGE BASE REFINEMENT

Knowledge Acquisition for expert systems is the process of extracting and formalizing expert knowledge in order to simulate the experts inference capabilities on a machine. Three different types of aquisition methods are usually distinguished: direct (the expert himself performs the knowledge transfer), indirect (support through a knowledge engineer) and automatic (machine learning esp. induction algorithms on existing cases and problem solutions) acquisiton. Whereas the third category obviously is applicable only to very limited problems the other two types are expensive and often referred to as the classical bottleneck of expert systems development.

Knowledge base refinement is an approach which tries to optimize the acquisition phase by a three step process: building a prototypical knowledge base (set of rules), applying analyses to the prototype and automatically generating refinement suggestions which are then rated and selected by the expert. This procedure has to be repeated in several cycles until no more refinement operation is possible or needed. The analyses can be grouped (following e.g.[MKL89]) into static and dynamic methods. *Static analyses* only look at the syntactic structure of the rule base without testing its inference behavior.The most common static tests perform consistency and completeness checks. They try to find redundant, subsumed and inconsistent rules as well as superfluous premises and cycles in and between rules. Completeness tests check e.g. unreachable conclusions, intermediate goals and premises of rules. Existing systems which offer static analysers are e.g. KB-Reducer [Gin88a], CHECK [NPL+85] and MELODIA [CD91]. Beside these pure static analyses various extended and more specific approaches exist which respect uncertainty or being domain sensitive, i.e. work with semantic information (e.g. [KNM85]). A simple example for the latter are value restrictions and mutually exclusive values. Statical defects of a rule base directly lead to possible repair measures, e.g. eliminating subsumed or redundant rules. Several fault situations require interactive operations by the knowledge engineer or expert, e.g. when breaking up rule cycles.

The generation of further refinement proposals is based on the *dynamic analyses* phase. Dynamic analyses require the execution of the prototype rules on test cases and compare the results with the expected results given by the expert [And92]. Occuring discrepancies are caused by shortcomings in the formalization of the experts knowledge. Following Valtorta [Val91] knowledge base refinement methods based on dynamic analyses fall into two categories: empirical and knowledge-based approaches.

Empirical approaches test the rule base behavior with all existing test cases with respect to input and output data and build up case statistics, i.e. record information which cases were classified correctly or wrongly, which rules were actually used to infer these results, which rule premises seem to be the most frequent obstacles to reach a correct decision, etc.

Based on these statistics and a set of heuristic rules the rule base can be modified mostly by rule *generalizations* and *specializations*. The modification proposals are usually rated corresponding to specific criteria such as minimality of change. The decision which of them actually should be carried out is usually undertaken interactively. Probably the best known empirical refinement system is SEEK and its fully automated successor SEEK2 [Gin88b].

Knowledge-based approaches also consider cases and the systems inference results but in addition often employ explicit domain theories which are used to guide the refinement process. Hence, they have a more deductive character. The process of identifying rule defects which are responsible for unintended system behavior may be nearly completly interactive as in TEIRESIAS [Dav79] as a kind of algorithmic debugging. Other methods are based upon error support and explanation structures, diagnosis/meta rules, or a standard for rule chains, as in LAS [SWM+85], ODYSSEUS [WCB96], and CONCRET [Lop91].

3 REFINING RULE BASES WITH TIM AND LEAF

In this section we present two approaches for knowledge base refinement, a deductive and an inductive one. Both methods presume the existence of a set of classified test cases. In the field of machine learning a method of this kind is called *supervised learning* [Anz92].

In contrast to pure inductive methods where past experiences are being *compiled* into new general heuristics [Alt93] we try to integrate the test cases into the existing rule base in a way that the refined rule base is expected to correctly decide even those cases which were not processed properly before.

This strictly differs from case-based *reasoning* where an analogy between a current working case and a reference case is looked for in order to apply, at least partially, the known solution of the reference case to reach a solution for the working case [Kol80] but not to modify a given set of rules.

3.1   Static Analyses

As already stated above the field of refining a knowledge base by static analyses is well understood. Many algorithms are described in the literature [Ign91]. Therefore in this paper we do not focus on static methods but just itemize those tests our implemented algorithms are able to perform:

- **Tests on inconsistencies:** redundant rules, subsumed rules, contradictory rules, unnecessary conditions, cyclic rules.

- **Tests on incompleteness:** unreachable goals, conclusions and conditions, unreferenced or illegal values, unreferenced or illegal scopes.

The importance of static analyses lies in their applicability as a preprocess to dynamic analyses: the less redundant the knowledge base is, the better are the rule refinement suggestions.

3.2   Dynamic Analyses

Dynamic analyses try to detect and repair discrepancies (suspected faults) between classified test cases and the system behavior. The goal is to improve the knowledge base by first detecting discrepancies between test case and system behavior (*localisation step*) and by secondly modifying the rule base (*refinement step*) by operations like inserting a rule condition or changing the conclusion. The refinement step is interactive, i.e., the system suggests an operation, which has to be verified by a human expert and will be carried through only in case of acceptance.

Note that in our approach the test cases play a different role than in the methods of theory revision (e.g. CLINT [deR92] or KRT/MOBAL [Wro94]). Whereas we use test cases for localizing faults automatically, in theory revision the user has to mark derived facts as incorrect (i.e., undertake the localisation step himself) and alternative inference paths are not analysed at all.

3.2.1   Definitions

The aim of comparing one or more test cases to the system behavior is to localize discrepancies which can be interpreted as modification hints. Together with strategic knowledge how to remove discrepancies refinement operations can be computed.

In most approaches for knowledge base refinement the localisation step is restricted to the comparison of the final system solution to the given classification of the test case. Therefore only different results yield refinement suggestions (*result-oriented methods*). But even when the final system solution and a given classification are identical, useful refinements can be obtained (*precondition-oriented methods*). Therefore it is necessary to include all possible system solutions and their underlying proofs respectively. Because a proof in an expert system is nothing else than a chain of rules derived by the inference engine we deal with such rule chains and call them *complete chains*. Modelling a rule chain as a graph-theoretic tree where each path describes a way from facts (leafs) to a diagnosis (root), a complete chain can be seen as a path in the tree.

Another advantage of the precondition-oriented approach is the possibility to ensure refinements which are *minimal* in the sense that the modifications are as little as possible. This yields to a refined knowledge base which is as close to the old one as possible. Minimal refinements are useful if the knowledge base is working well in general. An alternative is the *radical* modification of the system which is in some sense the opposite of minimalism. We define both criteria later on. Let us first define the terms *test case* and *complete chain* in more detail.

- A **test case** $C=(COND, \varphi^{exp})$ is a tuple containing a set $COND$ of attribute-value-pairs which represent observational facts and a correct classification φ^{exp}.

- A **complete chain** is a single path in the complete inference tree computed via depth-search represented as a recursive list.

We give a small example in table 1, reducing the different levels of inference steps in a complete chain to a single one.

3.2.2   The Deductive Approach TIM

The first approach for knowledge base refinement is a deductive one called TIM (Test and Improve Method).

The localisation step

The goal of the comparison between test case and complete chain is to get a statement of how similar they are. This (global) similarity depends on the (local) similarities of the separate comparisons between single case facts and parts of the complete chain. Therefore we suggest a deterministic finite automaton [HU79] as a suitable representation. The transitions are derived out of the following possible situations:

- result-oriented comparisons:

  1. The final results are identical.

  2. The final results are not identical.

Table 1: An Example: Test case and complete chain

	test case TC		complete chain CC
$COND$	financial-situation(negative) promoter-profile(very bad) guarantees(good)	List of Leaves	financial-situation(negative) promoter-profile(bad)
φ^{exp}	creditAgreement(yes)	φ	creditAgreement(no)

- precondition-oriented comparisons:

  1. Test case and complete chain are identical.

  2. All attribute-value-pairs of the case description are represented in the complete chain, but there are further facts in the chain (*subset relationship of attribute-value-pairs*).

  3. The test case contains all attribute-value-pairs of the complete chain and further more (*superset relationship of attribute-value-pairs*).

  4. Test case and complete chain correspond to their attributes, but not to all of the values.

  5. All attributes of the test case are represented in the chain, possibly with different values, but there are further attributes in the chain (*subset relation of attributes*).

  6. All attributes of the case facts correspond to facts in the complete chain, but there are further attributes in the case description (*superset relation of attributes*).

  7. None of the above possibilities holds.

Due to the independence of the comparisons of the final results (first two items) on the one hand and those of the preconditions (all other items) on the other hand, the combination yields to $2 * 7 = 14$ *classes of discrepancies*. We use a 32-state automaton called *fault-analysis-automaton*, including 14 final states (classes of discrepancies) and one declared initial state. The automaton uses informations about local comparisons as transitions. Those local comparisons are: attribute-value-existing (a-v-e), attribute-existing (a-e), not-existing (n-e), end-of-case (e-o-c), further-leaf (f-l), no-further-leaf (n-f-l), conclusion-identical (c-i), and conclusion-not-identical (c-n-i). The fault-analysis-automaton is defined as follows. A **fault-analysis-automaton** is a quintupel $(Q, \Sigma, \delta, q_0, F)$, where

- $Q = q_0, q_1, ..., q_{14}, q_a, q_b, ..., q_r$ is the finite set of states.

- $\Sigma = \{$a-v-e, a-e, n-e, e-o-c, f-l, n-f-l, c-i, c-n-i$\}$ is a finite input alphabet. Each sign represents a local comparison (see above).

- $\delta : Q \times \Sigma \Rightarrow Q$ is a transition function, defined as follows (for the purpose of readability represented as a diagram in Fig. 1)

- $F = q_1, ..., q_{14} \subseteq Q$ is the set of final states, corresponding to the classes of discrepancies.

Each comparison is a deterministic walk through the automaton (see figure 1) from the initial state to one of the 14 final states. Each of those can be interpreted as a fuzzy kind of similarity between test case and complete chain. It is important to note that such a comparison will be carried out for each possible complete chain with regard to each test case. Therefore, the localisation step yields one class of discrepancy for each comparison between a complete chain and a test case. In our deductive approach we treat the computed classes of discrepancies as a strong logical foundation for improvements.

The refinement step

The detection of discrepancies is followed by an interactive phase of modifications to the rule base. This could be done by using each computed class of discrepancy but this would trigger an enormous amount of modifications because the localisation step uses all possible complete chains for each comparison to a test case. Depending on the number of rules the number of complete chains can be very high. Therefore a filter of minimally necessary operations is needed.

The treatment of a discovered discrepancy is realised in TIM by a set of predefined procedures called *basic refinement operations*. These are: change-value, enlarge-scope, reduce-scope, change-conclusion, insert-literal, delete-literal, delete-rule, insert-rule, and insert-case-as-rule. The meanings of the procedures are intuitive.

The basic refinement operations are independent of the underlying classes of discrepancies in a modular sense. The presented ones are proven to be useful in

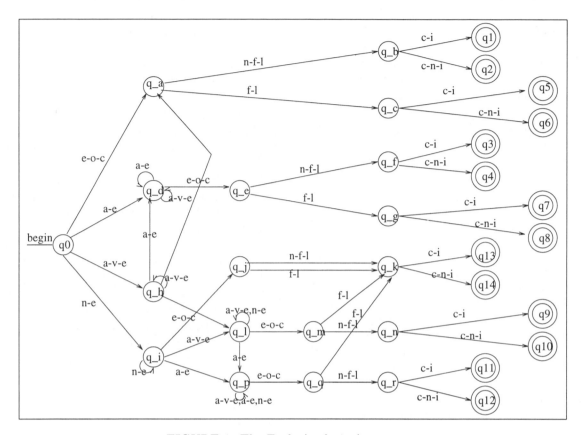

FIGURE 1: The Fault Analysis Automaton

the credit agreement application (see section 4). The relation of basic refinement operations to classes of discrepancies is an unambiguous one. Each class of discrepancy is assigned one or more basic refinement operations.

Consider, e.g. *TC* and *CC* in table 1: The comparison leads to class q_{12} (cf. figure 1) stating much similarity between case and chain with respect to the conditions but contradictory conclusions. There are two basic refinement operations assigned to q_{12} which lead to the following refinement suggestions:

insert-rule:

```
(IF financial-situation(negative) AND
    promoter-profile(bad) AND
    guarantees(good)
 THEN creditAgreement(yes))
```

insert-literal:

```
not(guarantees(good)) ==>
    (IF financial-situation(negative) AND
        promoter-profile(bad)
     THEN creditAgreement(no))
```

insert-rule:

```
(IF promoter-profile(very bad)
 THEN promoter-profile(bad))
```

Note that the instantiated basic refinement operations are not carried out automatically but must be verified by an expert.

The amount of complete chains may be huge, which could lead to a huge amount of possible suggestions for refinements. Although each suggestion is useful in some sense a human expert would not accept too much of them. Therefore, we propose a minimality measure in order to sort the suggestions and to enable filtering redundant ones.

The *minimality measure* is defined as follows. Let each atomic rule condition and each conclusion be a *part* of a rule. Then the minimality γ_{min} of a basic refinement operation *op* with respect to a knowledge base *kb* is defined as

$$\gamma_{min}(op, kb) = \sum_{\text{modified\_rules}(op,kb)} \frac{|\text{modified parts}|}{|\text{number of parts}|}$$

This measure regards all rules of *kb* which are modified by the operation *op*. Modifications of a rule condition and a rule conclusion have the same weight. The minimality of the basic refinement operation is defined as a sum over all modified rules.

Since one or more basic refinement operations are assigned to one class of discrepancy, the minimalities of all operations have to be added to get the minimality of a certain class of discrepancy. We propose to assign weights ω_{min}^{op} to the basic refinement operations leading to a definition for the *minimality of a class* with respect to a knowledge base *kb*:

$$\gamma_{min}(cl, kb) = \sum_{bas\_ref\_op} \omega_{min}^{op} \gamma_{min}(op, kb)$$

Many revision systems use metrics of this kind for defining minimal refinements, for example AUDREY II [WP93] or KRUST [CS95] Our definition is similar to KRUST, where the number of modified rule parts is taken into account. In addition we use user-definable weights and define *radicality* of a basic refinement operation or a discrepancy class in an analogous way [Kau92]. In the credit agreement application the minimality measure was proven to be useful.

The minimality measure allows rating the possible refinement operations: the minimal ones are presented to the user first. If a set of such modifications of the rules in the knowledge base, which are necessary for the correct treatment of the underlying test case, are accepted and carried through, no more possible refinement operations are presented. Since we take into account all complete chains, the sorting is essential for the acceptance of the system.

3.2.3 The Inductive Approach LEAF

Inductive approaches to knowledge base refinement treat the results of a single comparison between test case and system behavior as a weak logical foundation and just mark the results in suitable tabular form. The underlying idea in LEAF is to compare all given test cases quasi-simultaneously just counting the computed classes of discrepancy. Only the resulting frequency ranks are strong logical foundations for the refinement step.

Given a test case $C=(COND, \varphi^{exp})$ and applying the case to the expert system, we distinguish between four situations:

- **True-Positive-Decision (TP)**
  $XPS(C) \Rightarrow \varphi^{exp}$: φ is a inferable (*positive*) system solution and $\varphi = \varphi^{exp}$ holds (*true*).

- **False-Positive-Decision (FP)**
  $XPS(C) \Rightarrow \varphi, \varphi \neq \varphi^{exp}$: the inferable system solution differs from the given expert solution.

- **False-Negative-Decision (FN)**
  $XPS(C) \not\Rightarrow \varphi^{exp}$: the given expert solution φ^{exp} is not inferable by the expert system.

- **True-Negative-Decision (TN)**
  $XPS(C) \not\Rightarrow \varphi, \varphi \neq \varphi^{exp}$: the solution φ which differs from the given expert solution is not inferable by the system.

TPs and TNs are correct decisions which require no modification of the rule base. But they are important for marking those rules (called *significant rules*) which are enclosed in a complete chain infering the desired solution. The underlying problem is that refinements based on fault localisations of a single test case can possibly lead to a new rule base which does not handle the old correctly classified cases properly any longer. If we do not mark rules from successful chains the modification would just be a local improvement but the overall performance (number of correctly classified cases of the test set) would be much worse. Therefore we use all four kinds of decisions.

Unlike a deductive approach the local results of the comparison between test case and system behavior are stored in certain tables marking the corresponding rules as *not general enough* or *not specific enough*. The goal is to ensure that only those refinement suggestions τ are derived which yield an *empirical success* σ, defined as

$$\sigma(\tau) = \Delta gain - \Delta loss$$

where $\Delta gain$ is the number of cases in which the refinement operation τ changes a False-Positive- or False-Negative-Decision into a True-Positive one and $\Delta loss$ is defined vice versa.

We use 15 different tables for marking possible rule modifications. For example if a complete chain contains all facts of a single test case but one additional, this latter one is unnecessary assuming the test case is correct in the sense of supervised learning. Therefore the rules in the complete chain are potential candidates for generalization and are marked in a table named GEN/3. This table contains the names of the rules, a variable counting the number of cases in which the rule is a generalization candidate and a list of the corresponding test cases.

In a similar manner the SPEC/3 table is used for marking those rules being candidates for specialisation. This is the case, e.g., if the scope of an intervall has to be reduced.

Each entry increases the probability that the corresponding refinement operations will be carried out. The decision which of the possible modifications are the right ones for the best overall performance, will be based on heuristics.

The refinement step

The evaluation of the tables in our inductive approach LEAF is carried out by a set of heuristic rules which use the statistical informations of the tables to establish instantiated basic refinement operations. These operations are essentially equivalent to those of the deductive approach TIM. Let us look at an example of a heuristic refinement rule:

```
IF SPEC(rule, spec_potential, _)
AND ( SIGNIF(rule, anti-potential, _)
AND spec_potential > anti_potential
AND SPEC_PART(rule, newLiteral, number, cases)
THEN insert_literal(rule, newAttributeValuePair)
WITH ESTIMATED SUCCESS = number
```

The rule is to be read as follows: if the potential for specialization (spec-potential) is greater than the anti-specialization-potential (which is the number of cases in which the corresponding rule is marked as successfully applied) and the considered part for specialization is a new attribute then the basic refinement operation should be the insertion of the attribute-value-pair. The success is just estimated by the number of analysed cases because it cannot be guaranteed that the single refinement operation sufficiently improves the rule base for changing *all* corresponding cases into True-Positive-Decisions. The complete set of heuristics is contained in [Spe94].

3.2.4 Implementation

Both approaches were implemented in Prolog. An important feature of the tools is the ability to simulate suggested refinement operations. That means, the user can process a what-if-analysis as a help for his decision. Clearly, this feature is of high practical use for the experts, because the information about the system behavior before and after the modification can be better derived by a computer than be estimated by a human expert.

4 AN APPLICATION TO CREDIT AGREEMENT

The stiff competition in the banking field stresses the importance of qualified consultation of clients. This however requires a high degree of expert knowledge multiplication and high valuation and analysis capacity. Expert systems surely can make a contribution to fulfill these requirements.

During the past five years this led to a growing interest of banks in expert system technology in particular in the employment of expert systems in two areas: investment consulting and credit business.

Our example rule base which served for validating our refinement approaches sketched above was concerned with the problem of credit scoring, i.e., the analysis and decision whether clients of a bank should get a credit for carrying out a specific investment or not. The credit decision problem can be split into 5 main phases that are performed consecutively:

1. checking formal credit qualification (e.g. legal and disposing capacity),

2. checking personal credit worthness,

3. checking material credit worthness in the past, presence and future,

4. checking future prospects of a company with respect to business cycle and branch,

5. judging profitability of intended investments.

The most complex problem is phase 3 if the credit borrower is a company and its balance has to be analysed. Banks usually hold rules and regulations how they do this specific analysis strictly under lock and key in order to prevent respective adaptions of balances by the companies.

Together with an expert from the Dresdner Bank AG we acquired a first prototypical rule base, containing 61 rules, 74 metafacts and only 8 cases. The rules were formalized in a simple rule specification language called RSL [Kau92] whereas the case descriptions were reduced to the basic criteria used for the decision and the result itself and put onto (virtual) filing cards. The evaluation was restricted to our deductive approach TIM and the most interesting question was if the presented amount of refinement suggestions would be acceptable by the expert. It became clear that even though we had to deal with nearly 1000 complete chains, just a few suggestions (2-6) were presented. The number of presented suggestions depends also on the answer of the expert. Only if the expert rejects the first proposal(s), additional ones are presented by the system as already mentioned above. In our case nearly all suggestions were accepted, but the 8 test cases were of course not representative.

In an extended version the rule base contained about 200 rules. We could complete the case base with support from an expert of the Deutsche Bank AG to 114 test cases (44 positive and 70 negative examples). The larger amount of cases was neccessary for the inductive approach LEAF. In several simulations convincing results could be obtained. The number of correct classified cases could be increased from 82 (71.93 %) to 93 (81.58%) apart from the fact that most of the resulting refinements were accepted by the expert.

The dynamical approaches are comparable only in certain aspects. To give a global statement, the suggestions derived by the fault-analysis-automaton are easier to justify and often more concrete, on the other hand the inductive approach is more independent from the quality of the given test cases and more stable against exceptions. We plan to analyze if a combination of both approaches is superior.

5   AN INTEGRATED ENVIRONMENT FOR ASSURING KNOWLEDGE BASE QUALITY

Additional field studies in more comprehensive settings for validation and extension of the techniques described above require a comfortable environment for TIM and LEAF.

In our first prototype the rule base specification (using the language RSL) took place with a simple text editor on files. Case databases were represented in text files, too. The runnings, sequences and cycles of refining the initial rule base by TIM and LEAF were initiated manually. The interaction of both systems with the knowledge engineer was very poor. Result of the refinement process was again a text file containing RSL rules which had to be transformed manually into a concrete XPS language.

A definitely more advanced solution is to build an integrated environment coupling all involved subsystems with a semi-automatic control flow. Besides TIM and LEAF the following additional components of such an environment can be identified:

- a graphical *rule-base editor*,

- a *graphical interface for refinement decisions* which provides for visual preparation of refinement proposals, justifications on demand (explicit cases, statistics),

- an *XPS compiler* which maps the abstract rule language into an XPS shell data format,

- a *relational database management system* for storing the case databases, and finally

- a *repository* for storing rules and intermediate results, for documenting and tracing the refinement process, for providing basic process control and notification services as well as support for the design of other XPS components, e.g., the explanation component. To perform this task the repositiory needs a meta model for the complete development resp. refinement process and the data structures and actors. In addition it maintains meta knowledge of the applications.

Fig. 2 presents the architecture of our knowledge base design environment VAREX which provides a set of tools as mentioned above connected via IPC in a UNIX based setting. The environment is coordinated by the intelligent repository system ConceptBase [JGJ+95]. ConceptBase implements the object data model O-Telos which in particular has metamodelling facilities and provides for powerful deductive inference capabilities. Its suitability for the coordination of design environments has been proven in a variety of reasearch projects.

As suggested above we have developed a specific meta-model called *XPSModel* describing the typical structure and components of rule-based expert systems. This meta model also contains process information for the complete validation and refinement process, e.g. that the rule base editor has to give a commit notification before starting the statical analyses by TIM may take place or that depending on the experts judgement of refinement suggestions further validation cycles should be initiated. XPSModel constitutes the top layer of the repository.

The second layer is instantiated by a given application, i.e. contains information about its type of rules and the objects allowed together with characterizations such as single valued, multivalued, obligatory, main goal etc. Several value restrictions (e.g., enumeration of allowed values) for objects are stored at the third layer, which is the the actual working area, i.e. contains the concrete rule sets, information about refinement decisions, their justifications etc.

Besides checking consistency on and across the three modelling layers additional possibilities of employing the reasoning facilities of ConceptBase [SNJ94] which we currently don't consider are to perform parts of the statical analyses directly in the repository by designing specific integrity constraints or integrity views on the rule model, and secondly, to support simulation runs with the rule base prototypes.

6   CONCLUSIONS

We presented two approaches for knowledge base refinement of rule bases covering analysis and repair of most relevant types of static defects and inconsistencies. Dynamic analyses are performed with respect to collections of cases with known results. The resulting suggestions of refinement measures based on a fault analysis automaton and on specific statistics in combination with heuristics, respectively. Our techniques were implemented by two refinement tools called TIM and LEAF and were validated with a rule base intended to solve a real-world problem, namely credit

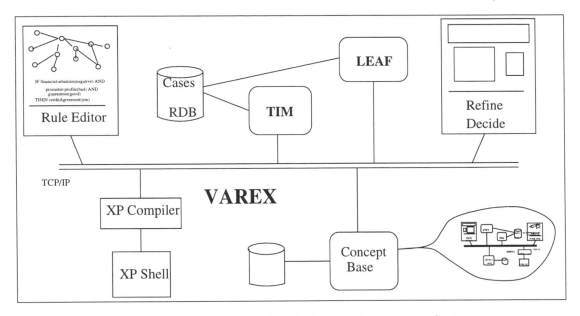

IF financial-situation(negative) AND
promoter-profile(bad) AND
guarantee(good)
THEN creditAgreement(yes)

Rule Editor

Cases
RDB

TIM

LEAF

Refine
Decide

TCP/IP

VAREX

XP Compiler

XP Shell

Concept
Base

FIGURE 2: VAREX: A Design Environment for Expert Systems

assignment for private clients of a bank. Although our validation results are satisfying as stated in section 4 it became obvious that in fact the initial rule base already was too good: Statical defects were avoided through unconscious consolidations by the knowledge engineer and led at least to better results than under real time pressure. In fact the knowledge engineering phase for our example application was very time consuming last but not least because we had too less methodological experience with this task.

With respect to the dynamical aspect the rule extraction process by the expert and the knowledge engineer took place more or less in an inductive way starting from the concrete case rather than just formalizing a present set of given general heuristics and rules.

Our approaches to knowledge base refinement have been proven to be a powerful method for changing a non-perfect rule base to fit a set of preclassified examples.

The next steps to be done are to continue the implementation of the protoype environment VAREX, especially the mapping of final rule sets to a commercial XPS shell. Second, we will try to acquire a much larger set of cases for our prototype KB from different sources, i.e. additional other banks, than before. This should give us more detailed hints for the quality of the empirical approach and its heuristics. Especially we are interested in including information about the success of assignment decisions, i.e. whether the borrowers turned out to be in fact as creditworthy as expected.

In addition we plan to work out another application

domain again in the banking field and based on that to develop a third refinement approach which explicitly employs a domain theory.

REFERENCES

[Alt93]   K.-D. Althoff et al. Induction and Case-Based Reasoning for Classification Tasks. In: Bock, Lenski & Richter (Eds.) *Information Systems and Data Analysis*, 1993, Springer, pp.3–16.

[And92]   E.P. Andert. Integrated Knowledge-based Systems Design and Validation for Solving Problems in Uncertain Environments. *Int. Journal of Man-Machine Studies*, 36, 357–373, 1992.

[Anz92]   Y. Anzai. *Pattern Recognition and Machine Learning*. Academic Press, 1992.

[CD91]   E. Charles and O. Dubois. MELODIA: Logical Methods for Checking Knowledge Bases. In: M. Ayel and J.P. Laurent (eds.): *Validation, Verification and Test of Knowledge-based Systems*, John Wiley & Sons, 95–106, 1991.

[CS95]   S. Craw and D. Sleeman. *Knowledge-based Refinement of Knowledge Based Systems*. Technical Report, University of Aberdeen, 1995.

[Dav79]   R. Davis. Interactive Transfer of Expertise: Acquisition of new Inference Rules. *Artificial Intelligence*, 12, 121–157, 1979.

[deR92]   L. de Raedt. *Interactive Theory Revision*. Academic Press, London, 1992.

[Gin88a]   A. Ginsberg. Knowledge Base Reduction: A New Approach to Checking Knowledge Bases for Inconsistencies and Redundancies. *Proc. AAAI*, 585–589, 1988.

[Gin88b]   A. Ginsberg. *Automatic Refinement of Expert System Knowledge Bases*. Pitman, London, 1988.

[HU79]   J.E. Hopcroft and J.D. Ullman. *Introduction to Automata Theory, Languages and Computation*. Addison-Wesley, 1979.

[Ign91]   J.P. Ignizio. *Introduction to Expert systems*. McGraw-Hill, 1991.

[JGJ+95]   M. Jarke, R. Gallersdörfer, M. Jeusfeld, M. Staudt and S. Eherer. ConceptBase - A Deductive Object Base for Meta Data Management. *Journal of Intelligent Information Systems*, 4(2), 167–192, 1995.

[Kau92]   B. Kauert. *Static and Dynamic Analyses of Rule Sets*. Diploma thesis (in German), RWTH Aachen, Germany, 1992.

[KNM85]   G. Kahn, S. Nowlan and J. McDermott. MORE: An Intelligent Knowledge Acquisition Tool. *Proc. IJCAI-85*, Los Angeles, 581–584, 1985.

[Kol80]   J.L. Kolodner. *Retrieval and Organisational Strategies in Conceptual Memory: A Computer Model. Ph.D. Thesis*, Yale University, 1980.

[Lop91]   B. López. CONCRET: A Control Knowledge Refinement Tool. In: M. Ayel and J.P. Laurent (eds.): *Validation, Verification and Test of Knowledge-based Systems*, John Wiley & Sons, 191–206, 1991.

[MKL89]   R.L. de Mántaras, P.F. Kerdiles and H. Larsen. *Validation Methods and Tools for Knowledge-based Systems*. Informatik-Fachberichte 227, Springer, 1989.

[NPL+85]   T.A. Nguyen, W.A. Perkins, T.J. Laffey and D. Pecora. Checking an Expert Systems Knowledge Base for Consistency and Completeness. *Proc. IJCAI-85*, Los Angeles, 375–378, 1985.

[SNJ94]   M. Staudt, H. Nissen and M. Jeusfeld. Query by Class, Rule and Concept. *Journal of Applied Intelligence*, 4(2), 133–157, 1994.

[Spe94]   A. Sperber. *Case-based Learning in Expert Systems by Empirical Analyses*. Diploma thesis (in German), RWTH Aachen, Germany, 1994.

[SWM+85]   R.G. Smith, H.W. Winston, T.M. Mitchell and B.G. Buchanan. Representation and Use of Explicit Justifications for Knowledge Base Refinement. *Proc. IJCAI-85*, Los Angeles, 673–680, 1985.

[Val91]   M. Valtorta. Knowledge Base Refinement: A Bibliography. *Journal of Applied Intelligence*, 1(1), 87–94, 1991.

[WCB96]   D.C. Wilkins, W.J. Clancey and B.G. Buchanan. Overview of the ODYSSEUS Learning Apprentice. In: T. Mitchell (ed.): *Machine Learning: a guide to current research*, Kluwer Academic Press, Boston, 369–373, 1986.

[WP93]   J. Wogulis and M.J. Pazzani. A Methodology for Evaluating Theory Revision Systems: Results with Audrey II. *Proceedings of the Thirteenth IJCAI Conference*, Chambry, France, 1993.

[Wro94]   S. Wrobel. *Concept Formation and Knowledge Revision*. Kluwer Academic Publishers, Dordrecht, 1994.

Knowledge Representation

A Formalism for Defeasible Inheritance

Nadim Obeid
Department of Computer Science, University of Essex,
Wivenhoe Park, Colchester CO4 3SQ, England.
Email: obein@essex.ac.uk

ABSTRACT

In this paper, we present a formalism for defeasible inheritance reasoning which we can show that it is computationally tractable. Links are interpreted as sentences in the language, $L_\Rightarrow$, of a conditional logic **D** which is suitable for such type of reasoning. Defeasible inheritance reasoning is simply capturable by employing simple and intuitive principle and strategies. We compare our approach to other approaches such as the default based-based approach by [POO 85, POO 88] and the path-based approach by [TOU 86] and show that it performs better representationally and/or in terms of capturing the required reasoning.

1. Introduction

The practical need for ways of representing and handling taxonomic information led to the development of inheritance systems within Artificial Intelligence (AI). Network representations in general were presented without any formal semantic analysis and inheritance networks were no exception. However, following many ocriticisms, a great deal of effort began to address the problem of providing an account of meaning of networks which is independent of the domain of application and the implementation. The work by [HAY 79] and [CES 75] along this line proceeded by following an approach where the meaning of a network is specified via a mapping into the language of classical First Order Predicate Calculus (FOPC). This, in fact, explains why networks were considered for some time to be just notational variants of syntactically restricted first order theories. The only distinction was, perhaps, that a network may support specialized inference algorithms. However, the attempt to incorporate incomplete (defeasible) information into inheritance systems such as FRL by [ROG 77], KRL by [BOW 77], and NETL by [FAH 79] showed that the question of precise meaning of networks is far from settled. It has been shown by [FTR 81] that the inference algorithms built into these systems were naive and could lead to unintuitive results. It was then concluded that the the translation into FOPC could not be natural as the informal interpretation required a nonmonotonic consequence relation, and thus networks could be translated to nonmonotonic logics. Attempts in this direction by [ETH 88] interpret networks in terms of default

logic. These attempts take the view that inheritance networks could be viewed as syntactically restricted default theories and the default logic. However, they seem to have failed to capture the natural patterns of nonmonotonic inheritance reasoning.

In this paper, we present a *computationally tractable* formalism for defeasible inheritance reasoning. Links are interpreted as sentences in the language, $L_\Rightarrow$, of a conditional logic **D** which is suitable for such type of reasoning, and not as (default) rules as it is the case in some of the approaches mentioned above. It can be shown that the formalism presented here permit us to extend the expressive power of inheritance networks. Nonmonotonic inheritance reasoning is simply capturable by employing simple and intuitive principles and strategies. We compare our approach to other approaches such as the default-based approach by [POO 85] and [POO 88] and the path-based approach by [TOU 88] and show that it performs better representationally and/or in terms of capturing the required reasoning.

2. Inheritance Networks and Classical Logic

Inheritance networks are mainly employed to achieve tractable and fast inference mechanisms that are comparable to and even perform better than first-order theorem provers. However, there are other features such as modularity and usability, which, though very useful in practical applications, have proved to be very difficult to capture formally. Modularity, on the one hand, concerns the fact that semantic networks localize information in a way that first-order logic does not and it does not seem very obvious that it can do so without some additional logical constraints. Usability, on the other hand, has more to do with the ease with which semantic networks can be altered to capture the changes required as more information becomes available.

The approaches which map networks into the language of FOPC have overlooked and/or restricted at least one of the above two features. For instance, Cercone and Schubert in [CES 75] claim that if there are no exceptions, inheritance networks are just notational variants of First Order Predicate Calculus (FOPC). This can easily be proved to be imprecise. Even if it were possible to argue otherwise, the impact of reducing *simple semantic networks* to first order theories will be at the expense of modularity and usability.

3. Defeasible Inheritance and Default Logic

An Attempt by [ETR 83] which is further followed by [ETH 88], has been made to present a formal reconstruction of nonmonotonic inheritance networks (inheritance networks *with exceptions*) in terms of semi-normal defaults.

This approach can best be shown in terms of the following example.

Example 3.1. Let S = {Penguins are birds, birds fly, penguins do not fly}.

Etherington suggests that S can be represented by the following default theory $\Delta = \langle D, W \rangle$ where

$$W = \{(\forall x)(penguin(x) \rightarrow bird(x))\}$$
$$D = \{bird(x) :fly(x)\&\neg penguin(x)/fly(x)\}$$

So, given that only *bird(j)* then, based on the available information, $fly(j)\&\neg Penguin(j)$ is consistent; hence $fly(j)$ can be inferred. Given only *penguin(j)* we may conclude *bird(j)* using FOPC knowledge, but *penguin(j)* blocks the application of the default, preventing the derivation of *fly(j)*.

Touretzky in [TOU 86], however, *correctly* argues that the above reconstruction is inappropriate for the following reasons: (1) adding new information to the knowledge base may require some modification of the defaults which are already in the knowledge base; (2) the translation of a link depends on other links in the network; and (3) the translation ignores the essential "hierarchical" nature of inheritance networks. (cf. [ETH 88] for a reply to Touretzky's criticism and in favour of default logic).

4. Representational Issues

Whichever representation is chosen for the links of an inheritance network, it has to support an additional form of reasoning which is different from *logical entailment*, especially for defeasible links (i.e., links with exceptions). We partially agree with Etherington that a defeasible link should be captured by a default. However, default logic allows us to only reason with defaults but never about defaults. This is because defaults are meta-language rules which can neither be negated nor nested, thus preventing us from reasoning about them in useful ways so as to capture some natural patterns of inference.

The use of conditionals allows us some flexibility in that we could now reason about links as well as reasoning with links in an inheritance network. In particular, it allows us formally, with some additional axioms, to capture the principle of defeasible inheritance: if X is a *subclass* of Y then *members of X inherit the properties which members of Y have unless stated otherwise*.

4.1. Language

The language $L_{\Rightarrow}$ is that of First Order Predicate Calculus (FOPC) (with V, ¬, & and → as connectives) together with an intensional operator (defeasible implication) $\Rightarrow$.

More explicitly, $L_{\Rightarrow}$ has the following vocabulary: a set of individual variables x, y, z, ..; a set of individual constants a, b, c, ...; a set of n-ary predicates symbols P1, P2, P3, ...; commas and parentheses; and the symbols T (true) and F (false), the connectives ¬, →, &, V, $\Rightarrow$, and the universal quantifiers ∀ and ∃. The individual variables and constants comprises the set of terms of $L_{\Rightarrow}$. The meta-variables A, B, C, ... represent arbitrary formulas in $L_{\Rightarrow}$.

The well-formed formulas (wff) of $L_{\Rightarrow}$ consists of the least set containing T and F and such that:

(1) if P is an n-ary predicates symbol and $t_1,...,t_n$ are terms then $P(t_1,...,t_n)$ is a wff.

(2) If A and B are wff then ¬A, A&B, AVB, A → B and A $\Rightarrow$ B are wff.

(3) if x is an individual variable and A is a formula, then so is $(\forall x)A$.

The definition of ∃ is the usual one, $(\exists x)A \equiv \neg(\forall x)\neg A$. The standard notions of scope and of free and bound variables are assumed. The notation of A[t/x] indicates the result of taking a bound alphabetic variant A' of A such that no bound variable in A' occurs free in the term t, and then substituting t for all free occurrences of x in A'.

We define L and M in terms of $\Rightarrow$ as follows:

$$MA =_{def} \neg(T \Rightarrow \neg A) \text{ and } LA =_{def} (T \Rightarrow A).$$

4.2. Representation of defeasible inheritance

We distinguish between two different types of links in inheritance networks:

(1) A *strict* link (one with no exceptions) such as penguin are birds will be expressed as a first order sentence $(\forall x)(P(x) \rightarrow B(x))$.

(2) A *normal defeasible* link (one with exceptions) such as "birds fly" which is represented by a normal default B(x) :C(x)/C(x) will be expressed in $L_{\Rightarrow}$ as MC(x) → (B(x) $\Rightarrow$ C(x)) where the quantifier ∀ is omitted. Suppose that we have the fact P(Tweety), then we will have the following sentences: P(Tweety) → B(Tweety) and MC(Tweety) → (B(Tweety) → C(Tweety)) where the reading of the second sentence is as follows: if it is consistent that Tweety flies (If we cannot infer that it does not fly) then if Tweety is a bird then it flies.

5. The Conditional Logic D

In this section we present the conditional logic **D** which, as we shall show in the next section, is adequate to capture defeasible inheritance reasoning. It should be noted that our main interest is in the the principles which govern the operator "M".

5.1. Model theory

Definition 5.1. A selection function model (f-model) is a structure $\mathbf{M} = <W,f,g>$ where W is a non-empty set, g is a truth assignment function for atomic wff and elements of W and f is a function from W x P(W) to P(W), where P(W) is the power set of W and W x P(W) is the Cartesian product of W and P(W). Informally, W is a set of worlds. For wff A, the notation $[A]_M$ is used to stand for the set of worlds in **M** in which A is true. For convenience, reference to **M** will be omitted. If propositions are identified with sets of possible worlds, then [A] stands for the proposition expressed by A. The function f then picks out a set of possible worlds f(w,[A]) for every world w and proposition A.

We employ the notation $\mathbf{M},w \models_g A$ to mean that A is *true in the model* **M** at world w. For convenience, reference to g will be omitted except when a confusion may arise.

Definition 5.2. *Truth at a world w in a model* $\mathbf{M} = <W, f, g>$ is defined by

$\mathbf{M},w \models p$	iff	$g(w,p) = $ true for p atomic
$\mathbf{M},w \models A\&B$	iff	$\mathbf{M},w \models A$ and $\mathbf{M},w \models B$
$\mathbf{M},w \models \neg A$	iff	not$(\mathbf{M},w \models A)$
$\mathbf{M},w \models A \Rightarrow B$	iff	$f(w, [A]) \subseteq [B]$

5.2. The system D

The system **D** is the smallest conditional logic closed under the rules of FOPC, together with the inference rules (RCM), (RCE) and the axiom schema (D), (ID), (CC) and (INH) which are as follows:

Inference Rules:

(RCM) From $B \rightarrow C$ infer $(A \Rightarrow B) \rightarrow (A \Rightarrow C)$.

(RCE) From $A \rightarrow B$ infer $A \Rightarrow B$

Axiom Schema:

(D) $LA \rightarrow MA$

(ID) $A \Rightarrow A$

(CC) $(A \Rightarrow B)\&(A \Rightarrow C) \rightarrow (A \Rightarrow B\&C)$

(INH) $(MB \rightarrow (A \rightarrow B))\&(MC \rightarrow (B \rightarrow C)) \rightarrow (MC \rightarrow (A \rightarrow C))$

Since our main concern is with the logic of "M"/"L", it is not so difficult to motivate the rules (RCM) and (RCE). It is enough to notice that if we substitute T for A in these rules then we obtain:

(RCM') From $B \rightarrow C$ infer $LB \rightarrow LC$

(RCE') From B infer LB

These are reminiscent of rules of modal logic. The inheritance axiom schemata simply states that if *defeasibly* A's are B's and *defeasibly* B's are C's then *defeasibly* A's are C's. In particular, it follows that if A's are B's and *defeasibly* B's are C's then *defeasibly* A's are C's. This is exactly the principle which we want to employ when we are in a situation similar to the following:

Example 5.3.

Joe is a bird and *defeasibly* birds fly

In this case we want to be able too infer that Joe flies if this is the only information that is available to us. However, if we add

Joe's wings are broken and birds with broken wings don't fly

then we are no longer able to infer that Joe flies. There will be more discussion of how reasoning is carried out in section 6.

5.3. Soundness and completeness

D is sound and complete. Soundness is straightforward. To show the completeness of **D**, we first consider the conditional logic **B** where **B** is the the smallest conditional logic closed under (Mp) and (RCM).

B is complete. The details of this proof are omitted here due to lack of space. We shall present the proof in an extended version of this paper.

For the completeness of **D**, further conditions must be imposed on f in the f-models.

(id) $f(w,[A]) \subseteq [A]$

(cc) if $f(w,[A]) \subseteq [B]$ and $f(w,[A]) \subseteq [C]$ then $f(w,[A] \subseteq [B] \cap [C]$.

(in) if $(f(w,[B]) \neq \emptyset$ implies $[A]\subseteq[B])$ and $(f(w,[C]) \neq \emptyset$ implies $[B]\subseteq[C])$ then $(f(w,[C]) \neq \emptyset$ implies $[A]\subseteq[C])$

6. Strategies for Inheritance Reasoning

In this section we discuss some of the principles which ought to be followed in order to capture defeasible inheritance reasoning.

Principle 1. Defeasible inheritance inference is based on both the presence and absence of information. A representation in the language of **D** of a defeasible inheritance link is :

$MP \rightarrow (A \rightarrow B)$

where A represents the positive premise(s), P represents the exception(s) and B the conclusion of the inference. What distinguish the above sentence from a classical implication (a strict link) is the presence of MP which tells us that the absence of $\neg P$ brings forth the conclusion. The reasoning is carried out in the logical system **D** by allowing MP to be assumed (without having to be discharged) as long as there is no contradiction.

Example 6.1.

(1) Penguin(Tweety),

(2) Penguin(Tweety) $\rightarrow$ Bird(Tweety)

(3) MFly(Tweety) $\rightarrow$ (Bird(Tweety) $\rightarrow$ Fly(Tweety))

Intuitively we should infer "Fly(Tweety)". The idea is to assume MFly(Tweety) and use it without having to discharge it as long as no contradiction occurs. Then we have "Bird(Tweety) $\rightarrow$

Fly(Tweety)". Form this and the other facts we may, employing the rules of FOPC, infer Fly(Tweety).

Principle 2. Defeasible inheritance inference is drawn taking into account only what is being represented.

Example 6.2. If to the facts and links represented in Example 6.1, we add

(4)M¬Fly(Tweety)   →   (Penguin(Tweety)   →   ¬Fly(Tweety)).

then intuitively in this case we should infer ¬Fly(Tweety). However, using the inference machinery of the system **D** and the strategy that we could assume MA as long as there is no contradiction, we could go on to assume both MFly(Tweety) and ¬MFly(Tweety). But in doing so we obtain inconsistency as we will then be able to infer both Fly(Tweety)&¬Fly(Tweety). To conform to the requirement of logic, the only way out is to throw away either (3) or (4). Without (3) we can infer ¬Fly(Tweety) and without (4) we can infer Fly(Tweety).

What we have here is a typical example of a situation where we fail to apply the *exception-first* principle which is in fact our next principle.

Principle 3. Make use first of those sentences whose positive premises are about more specific (exception to other) concepts.

If we look carefully at the situation we notice that "Penguin" could be considered as an *exception* to (or more specific than) "Bird". Therefore, the sentences directly relevant (or more relevant) to "penguin" should be attempted first. That is, the *exception-first principle* imposes an order on which sentences should be first used in deriving further information.

However, logical systems in general do not comply with the *exception-first* principle. Hans and McDermott in [HMC 87] argue that circumscription and default logic are not suitable for representing commonsense reasoning because there are situations where they fail to capture commonsense intuitions. In fact, it could be shown that non-compliance with the *exception-first* principle is the reason why they fail to capture commonsense reasoning in many situations including the famous Yale shooting problem and the frame problem.

In Example 6.2 we should first apply the sentence whose positive premise is about "Penguin". Thus, we make the assumption M¬Fly(Tweety). Therefore, we can infer ¬Fly(Tweety).

7. Our Approach Versus Preferential Approaches

In this section, we review Poole's and Touretzky's approaches to reasoning with inheritance networks and discuss their performance in comparison with ours.

7.1. Poole's theory

Poole in [POO 85] and [POO 88] presents an alternative motivation, justification, semantics and syntax for default reasoning. He argues that default reasoning can be compared with scientific reasoning. Scientific theories are sets of postulates (defaults) which are used to explain certain phenomena. These theories should be the simplest sets of defaults which can explain the phenomena and are consistent with the world. Consistency means that there is at least one (logical) model of the system (the scientific theory and the facts).

The language of Poole's logical system is an extension of the language of the first-order predicate calculus. It consists of *facts* and *defaults*.

     \<fact> ::== \<wff>

     \<default>   ::==   \<Universal   quantifier> ASSUME \<wff>

Facts, which include the logical axioms of the form A → B, are taken to be true. A default can be taken to mean that that for each of the universally quantified variables before 'ASSUME', an instance of the \<wff> can be employed if it is consistent with the facts and the other defaults used. A default, to which we refer as A(x) ⇒ B(x) where ∀x is omitted, should be interpreted as "can be used to explain" or "is an acceptable form for a hypothesis", rather than "typical" or "most".

The semantics is given as follows. If H is the set of facts and Δ is the set of defaults. We say G is *explainable* (provable from the total system) if there exists some δ, a set of instances of the elements of Δ, such that

     H ∪ δ is consistent, and H ∪ δ ⊨ G

δ is said to be the theory that explains G. δ is like a "scientific" theory, H ∪ δ is the corresponding "logical" theory, with the corresponding model-theoretic semantics.

To account for the problems which occur when reasoning with defaults when more than one answer can be produced and one is to be preferred, Poole in [POO 85] introduces the notion of one theory being applicable whenever another is. He divides the facts into two classes:

H_n: the facts which are necessarily true.

H_c: the facts which happen to be true in the case under consideration.

Definition 7.1. Let δ and δ' be two explanations of a theory with different (possibly conflicting) conclusions A and A' respectively. I.e. let (δ,A) and (δ',A') be two solutions from H = H_c (∪ H_n): then (δ,A) is more specific then (δ',A'), i.e. (δ,A) ≤ (δ',A') if,

for every set H_1, we have:

(i)   if H_n ∪ F_1 ∪ δ ⊨ A

(ii)   and H_n ∪ H_1 ∪ δ' ⊭ A

(iii)   then $H_n \cup H_1 \cup \delta' \models A'$.

(δ,A) is strictly more specific than (δ',A'), i.e. $(\delta,A) < (\delta',A')$ if $(\delta,A) \leq (\delta',A')$ and $not((\delta',A') \leq (\delta,A))$.

When $(\delta,A) < (\delta',A')$ then we choose (δ,A).

Consider the following example

Example 7.2.

$$\Delta = \{P(x) \Rightarrow \neg Fly(x), B(x) \Rightarrow Fly(x)\}$$
$$H = H_n \cup H_c \text{ where}$$
$$H_n = \{P(x) \rightarrow B(x)\} \text{ and } H_c = \{P(a)\}.$$

There are two competing solutions: (δ,A) where $\delta = \{B(a) \Rightarrow Fly(a)\}$ and $A = Fly(a)$, and (δ',A') where $\delta' = \{P(a) \Rightarrow \neg Fly(a)\}$ and $A' = \neg Fly(a)$.

It can easily be shown that here we have $(\delta',A') \leq (\delta,A)$ and $not((\delta',A') \leq (\delta,A))$.

Let us, in the example above, change the status of "$P(x) \rightarrow B(x)$" from being an IS-a link without exception to a default rule as in the following:

$$\Delta = \{P(x) \Rightarrow B(x), B(x) \Rightarrow Fly(x), P(x) \Rightarrow \neg Fly(x)\}$$
$$H_c = \{P(a)\}, H_n = \phi.$$

In this situation, it is not clear whether "$P(a)$" is more specific/informative than "$B(a)$". But, Poole's definition would still produce the same answer as above.

So far, we have only shown examples where one solution is preferred to another. Here is an example where Poole's definition, while admitting the presence of ambiguity, does not help in resolving it.

Example 7.3. $H_n = \phi, H_c = \{R(Nixon), Q(Nixon)\}$

$$\Delta = \{R(x) \Rightarrow \neg P(x), Q(x) \Rightarrow P(x)\}$$

According to Poole there are two solutions:

(δ,A) where $\delta = \{R(Nixon) \Rightarrow \neg P(Nixon)\}$ and $A = \neg P(Nixon)$ and

(δ',A') where $\delta' = \{Q(Nixon) \Rightarrow P(Nixon)\}$ and $A' = P(Nixon)$.

Poole's definition here does not prefer any of the solutions.

7.2. Touretzky's approach

Tourtesky in [TOU 86] presents a syntactic definition of a notion of preference. The analysis is based on inheritance systems such as FRL where links between nodes are considered as defaults. As such, every rule is a default rule. Touretzky gives a definition of a notion of "inferential distance" which he distinguishes from the notion of "shortest path" and he only considers normal defaults, i.e. rules of the form: $P(x): Q(x)/Q(x)$.

Definition 7.4. Let d and d' be two normal defaults. $d < d'$ iff there exist a default with the same prerequisite as d and with the prerequisite of d' as its consequent or there exist a default d" such that $d < d''$ and $d'' < d'$.

Touretzky employs the above definition in order to compare proof sequences. The definition of a proof sequence is adopted from Reiter's definition of a default proof. A proof sequence is a sequence of defaults which allow the inference of a result from a set of closed literals. Then Touretzky sets up the following principles:

(a)   proof sequences are ordered according to their maximal elements.

(b)   two proof sequences are contradictory if they result in contradictory conclusions.

(c)   an extension in which a conclusion depends on a proof sequence S such that there is a contradictory sequence S' where S '< S is rejected.

Example 7.5.

$H_c = \{P(a)\}$.

$\Delta = \{d_1(x) = P(x) \Rightarrow B(x), d_2(x) = B(x) \Rightarrow Fly(x)\}$
$d_3(x) = P(x) \Rightarrow \neg Fly(x)\}$

For $(1 \leq i \leq 3)$, let d_i stands for $d_i(a)$.

There are two contradictory proof sequences.

$S = (d_1,d_2)$ which yields $Fly(a)$ and $S' = (d_3)$ which yields $\neg Fly(a)$.

From the above definition, we obtain: $d_1 < d_2$ and $d_3 < d_2$. Thus, according to principle (a), we have $S' < S$. Therefore, we accept S' and reject S by applying principle (c).

Example 7.6. $H_c = \{R(Nixon), Q(Nixon)\}$

$\Delta = \{d_1(x) = R(x) \Rightarrow \neg P(x), d_2(x) = Q(x) \Rightarrow P(x)\}$

As in Poole's approach, there are two contradictory proof sequences:

$S = (d_1)$ which yields $\neg P(Nixon)$ and

$S' = (d_2)$ which yields $P(Nixon)$.

d_1 and d_2 are not comparable, therefore S and S' are not comparable: there is no choice to be made here.

7.3. Our approach versus Poole's and Touretzky's approaches

One of the advantages of our approach over all the others is representational: in our approach we may allow links to be nested and dependent on each others. For instance, we may allow expressions of the form: if penguins inherit the ability to fly from birds then they have wings. It is impossible to represent such expressions in a default based (e.g., Poole's) or path-based (e.g., Touretzky's) approach.

Furthermore, an important aspect which characterizes Poole's approach is that it does not conform to Reiter's default logic. Consider the following example which illustrate how relevant the preference notion suggested by Poole is in a situation where no contradiction arises.

Example 7.7.

$$\Delta = \{P(x) \Rightarrow HW(x), B(x) \Rightarrow Fly(x)\}$$

H = $H_n \cup H_c$ where

$H_n = \{P(x) \to B(x)\}$ and $H_c = \{P(a)\}$.

According to Poole there are two solutions:

(δ, A) where $\delta = \{B(a) \Rightarrow Fly(a)\}$ and $A = Fly(a)$, and (δ', A') where $\delta' = \{P(a) \Rightarrow HW(a)\}$ and $A' = HW(a)$, and (δ', A') is preferred over (δ, A).

However, in our approach and in Reiter's default logic, we would infer that Fly(a) and HW(a), as the two conclusions do not contradict each others. It should be noted that other definitions have been proposed in an attempt to bring together Reiter's defaults and Poole's defaults.

Poole's theory is simple as the preference criterion which he presents is completely domain-independent. However, contrary to our own approach, there are problems as McDermott in [MCD 87] points out. Proving (i) above would require finding an arbitrary subset of facts that entail A when combined with δ but do not entail $\neg$A when combined with δ'. The situation with (ii) is even worse. (ii) requires us to show that *every* set of formulae that entails $\neg$A when combined with δ' does not entail A when combined with δ.

Touretzky's approach, however, considers preference only in one type of situation which can be summarized as follows: there is some fact H, and some rule H $\Rightarrow$ A which is concurrent to some proof sequence from H to A' and where A and A' are contradictory. Then, A is preferred to A'.

In contrast to our approach, (1) Touretzky's approach does not distinguish between rules without exceptions and rules with exceptions; (2) it does not even allow for default rules like A&B :C/C. However, it has been shown that Touretzky's principles open to improvement (cf. [FRG 87] for some extension of this approach).

7. Concluding Remark

We have present a formalism for defeasible inheritance reasoning. Links are interpreted as sentences in the language, $L_\Rightarrow$, of a conditional logic **D** which is suitable for such type of reasoning, and not as (default) rules as it is the case in most of the *translational* approaches. It can be shown that the formalism presented here permit us to extend the expressive power of inheritance networks. However, this is outside the scope of this paper and we shall elaborate on it in a forthcoming publication. We have shown that the defeasible inheritance principle is simply capturable as an axiom schemata **D**. However, to properly capture defeasible inheritance reasoning there was a need for additional principles such as the ones proposed in section 6. We have also shown that both representationally and in terms of capturing defeasible inheritance reasoning that our system perform better than Poole's ([POO 85] and [POO 88]) and the path-based ([TOU 86]) approaches.

References

[BOW 77] Borbow D. and Winograd T., (1977), An Overview of KRL, A Knowledge Representation Language, *Cognitive Science 1 (1)*, 3-46.

[CES 75] Cercone N. and Schubert L. K., (1975), Towards a State Based Conceptual Representation, *Proc. IJCAI, 83-90*.

[ETH 88] Etherington D., (1988), *Reasoning with Incomplete Information*, Research Notes in Artificial Intelligence, Pitman, London.

[ETR 83] Etherington D. and Reiter R., (1983), On Inheritance Hierarchies With Exceptions, *Proceedings of AAAI-83*, Washington, D. C..

[FAH 79] Fahlman S., (1979), NETL: A System for Representing and Using Real-World Knowledge, cambridge, MA, The MIT Press.

[FTR 81] Fahlman S., Touretzky D. and Van Roggen W., (1981), Cancellation in a Parallel Semantic Network, *Proccedings of Seven IJCAI*, Vancouver, B. C.

[FRG 87] Froidevaux C. and Grumbach S., (1987), Priority on Knowledge in Non-Monotonic Reasoning, *Proc. COCGNITIVA*, Paris, 189-197.

[HAY 79] Hayes P., (1979), The Naive Physics Manifesto, in *Expert Systems in The Electronic Age*, (Ed. D. Mitchie), Edinburgh University Press, 242-270.

[MCD 87] McDermott D., (1987), AI, Logic, and the Frame Problem, in: The Frame Problem in AI, F. M. Brown (eds.), Morgan Kaufmann, Los Altos, 1987.

[MOI 87] Moinard Y., (1987), La Specificite en Logic des Defauts, *Report INRIA, No. 642*, Rennes. NY, 373-384.

[POO 85] Poole D. L., (1985), On the comparison of Theories: Preferring the Most Specific Explanation, *Proc. 9th International Joint Conference on Artificial Intelligence*, 144-147.

[POO 88] Poole D. L., (1988), A Logical Framework for Default Reasoning, *Artificial Intelligence* 36, 27-47.

[ROG 77] Roberts R., and Goldstein I., (1977), The FRL Primer, *AI Memo No. 408*, Artificial Intelligence Laboratory, MIT, Cambridge, MA.

[TOU 86] Touretzky D., (1986), The Mathematics of Inheritance Systems, Pitman, London.

[WOO 75] Woods W., (1975), What's in a Link: Foundations for Semantic Networks, in: *The Psychology of Computer Vision*, P. H. Winston (eds.), McGraw-Hill Book Company, 157-209.

KNOWLEDGE COMPILATION FOR INTERACTIVE DESIGN OF SEQUENCE CONTROL PROGRAMS

Yasuo Namioka,   **Toshikazu Tanaka**
Systems and Software Engineering Laboratory,
Research & Development Center, TOSHIBA Corporation
Kawasaki, Kanagawa, Japan, 210.
Email: namioka@ssel.toshiba.co.jp

ABSTRACT

This paper presents a *knowledge compilation* method which improves the performance of planning-based interactive systems. The method automatically generates *state space models* from *domain knowledge*. It is generally difficult to adopt planning to aid interactive design because designers require quick response time in their work, and because design problems have more than several million states. We have been developing a planning-based VIsual Programming System for Sequence control program SpeciFiCation Design (VIPS/S-SFCD). The System addresses the difficulties by (i) dividing the state space models into the machine state space and the equipment state space, and by (ii) automatically generating the machine-level state space models from domain knowledge in short time without errors. The knowledge compilation also reduces search space in planning and the number of planning steps. Systems adopting our approach are capable of designing interactively, improving the reliability and quality of control specifications, and rapidly adapting with modification of a complex plant.

1 INTRODUCTION

Many AI researchers have focused on improving the performance of "knowledge-based" software systems such as for diagnosis [Chandra83], [Keller91], [Selman91], design, and planning [Kautz94], [Mitchell90], [Levi92]. There are many knowledge compilation approaches [Goel91] for the performance improvement [Chandra83], [Keller91], [Selman91], [Kautz94], [Levi92].

We have been researching/developing a planning-based VIsual Programming System for Sequence control program SpeciFiCation Design (VIPS/S-SFCD) [Nam95a]. In this system, designers make control specifications interactively. The planning method of VIPS/S-SFCD is a kind of method that searches on a state space model [Whitney69]. The control specifications are formed into SFC (Sequential Function Chart), and are input to an automatic programming system CAD-PC/AI [Nak90], [Sada90], [Mizu92].

It is generally difficult to adopt planning to aid interactive design because designers require quick response time in their work, and because complex plants have more than several million states. Therefore, three problems stand out: (i) representing the model, (ii) constructing and maintaining the model, and (iii) searching on the model. In order to solve the problems, we take the following approaches: (i) divide the state space model into machine state spaces and equipment state spaces, (ii) automatically generate the machine-level state space models from domain knowledge using knowledge compilation in short time without errors, (iii) reduce search space in planning and the number of planning steps with a concept of scope [Nam95a] and the knowledge compilation method.

This paper presents a *knowledge compilation* method which improves planning performance in planning-based interactive systems for sequence control program design, and representations of the state space models and the domain knowledge.

2 INTERACTIVE DESIGN

In the interactive design process of VIPS/S-SFCD [Nam95a] (FIGURE 1), designers visually indicate an initial state and goal conditions, and VIPS/S-SFCD generates a control specification segment.

In VIPS/S-SFCD, example-based programming and planning are adopted to realize interactive design. VIPS/S-SFCD consists of a knowledge base and a management part as shown in FIGURE 2. The knowledge base contains domain knowledge, state space models, and generated control specifications. The domain knowledge consists of general knowledge for plants (generic models) and specific knowledge for a plant (specific models). For simplicity, this paper describes domain knowledge without the classification into two models. The management part consists of seven subparts such as the Generators and Editors of Domain Knowledge [Sada90], the State Space Model Generator for knowledge compilation, the Icon Operator and Visual Instruction Interpreter for interaction between designers and

VIPS/S-SFCD [Nam95b], the Control Specification Generator (Planner) [Nam95a].

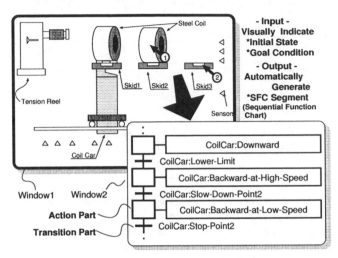

FIGURE 1: Planning-based Interactive Design

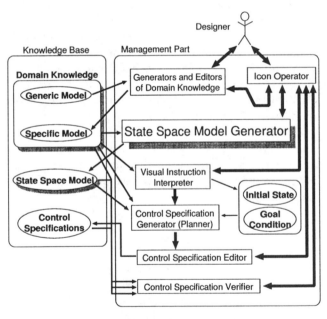

FIGURE 2: VIPS/S-SFCD

3   STATE SPACE MODEL

To deduce optimal and valid solutions, the planning, based on state space search [Wilkins88], should be reduced to a shortest path problem on a weighted directed graph representing the state space of the equipment. However, it is not practical to represent a complicated and enormous state space of the equipment with a usual weighted (directed) graph. It takes too enormous calculation time to generate equipment state nodes and arcs of actions/side effects from domain knowledge in search.

Therefore, this paper proposes a state space model for equipment, in which separated state space models for individual machines have interaction among machines, and in which equipment state nodes refer to machine state nodes in the separated state space models. The model allows (1) to reflect state transition effects of particular machines upon state transitions of other machines in search, (2) to generate valid machine-level state space models before search, and (3) to dynamically create new valid equipment state nodes in search. The state space models for machines are automatically generated from domain knowledge (states, actions, side effects, constraints on states and actions) by knowledge compilation. The generated machine state nodes and arcs linking the machine state nodes are appropriately restricted to satisfy the constraints on states and actions in domain knowledge.

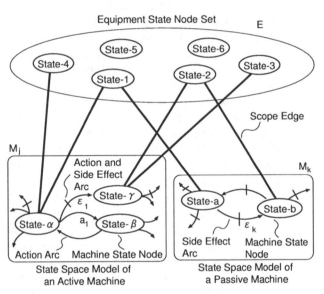

FIGURE 3: State Space Model for Equipment

The state space model for the equipment (FIGURE 3), consists of state space models for machines, scope edges, and equipment state nodes set (State-1, ...). There are two types of state space model of machines, because machines can be classified into active machines and passive machines. The active machines (Coil Car, Tension Reel, ...) have actions and side effects to change the machine state. The passive machines (Skid1, Skid2, ...) have only side effects to change the machine state. So, state space models for active machines consist of action arcs (ex. a_1), action and side effect arcs (ex. ε_1), and machine state nodes (ex. State-α). State space models for passive machines consist of side effect arcs and machine state nodes. In State-α (FIGURE 4), action arcs are represented by attributes (Backward at High Speed, Upward). Values of the attributes are the resulting machine state nodes after the actions. In Arcs-α (FIGURE 5), side effect arcs represented by attributes like "Side Effect1". Values of the attributes are the resulting machine state nodes after the side effect.

The interactions among machines are represented by side effects arcs and side effects of domain knowledge. A side effect consists of pre-condition and effect. When the pre-condition is satisfied, the states of the machines in the effect are changed in same time.

Scope edges are pointers to link machine state nodes and equipment state nodes. Equipment state nodes, such as State-1 in FIGURE 3, represent equipment state and consist of the scope edges. Each equipment state has no direct link to other equipment state nodes.

Name:	Coil Car State-α
Vertical Position:	Lower Limit
Horizontal Position:	Stop Point2
Coil Loading State:	Not Loading
Action State:	Stopping
Arcs:	GOM Arcs of Coil Car State-α,
	AOM Arcs of Coil Car State-α,

FIGURE 4: Machine State Node

Name:	AOM Arcs of Coil Car State-α
Effective Actions:	Backward at High Speed, Upward
Effective Side Effect:	Side Effect with Upward
Backward at High Speed:	Coil Car State-δ
Upward:	Coil Car State-β
Side Effect with Upward:	Coil Car State-γ

FIGURE 5: Arcs

Name:	Equipment State-1
Coil Car:	Coil Car State-α
Skid1:	Skid1 State-a
Skid2:	Skid2 State-a
Skid3:	Skid3 State-b
Tension Reel:	Tension Reel State-e
Table:	Table State-h
Def. Pinch Roll:	Def. Pinch Roll State-k
T. Pinch Roll:	T. Pinch Roll State-s
B. Pinch Roll:	B. Pinch Roll State-s
Belt Wrapper:	Belt Wrapper State-e
Shear:	Shear State-i

FIGURE 6: Equipment State Node

To change the equipment state (to create a new equipment state node), the state space models of machines are used. The following steps represent a transition from one equipment state node to another equipment state node **(1)** Select a machine to make active (Coil Car) and look up the machine state node (State-α). **(2)** Select an action of the machine (a_1: Upward) and check whether the action caused any side effect (ε_1: Side Effect 1)on the equipment states. **(a)** If the action caused no side effects, look up the machine state node (State-β) to which the action goes. **(b)** If the action caused a side effect, look up the machine state node and related machine state nodes (State-γ) to which the side effect goes, if any. **(3)** Create a new equipment state node (State-2) if the equipment state is valid. The validity is judged with constraints on states and actions in domain knowledge especially with the layer of the constraints: state combination denial layer.

If the knowledge compilation is not adopted, in planning, the search step **(2)** should be changed to procedures which is equal calculation time to the procedures in section 4.5 Method of Knowledge Compilation.

4 KNOWLEDGE COMPILATION

Knowledge compilation to generate machine-level state space models use domain knowledge about states, actions, side effects, and constraints on states and actions. It consists of two procedures, as shown in FIGURE 7. First, primitive state sets are used to generate machine state nodes, which are valid combinations of primitive states. Then actions and side effects are used to add arcs representing valid state transition of machines to machine state nodes. Constraints on states and actions are used to check validity of machine state nodes and arcs.

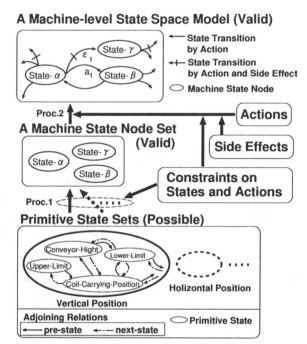

FIGURE 7: Outline of Knowledge Compile Process

4.1 STATES

The structure of states is used to generate valid machine state nodes. Adjoining relations among primitives are used to generate valid arcs in machine-level state space models. The State Space Model Generator refers to the constraints on states and actions from each primitive state to validate the nodes and arcs.

The states have three levels: primitive state, machine state, and equipment state. A primitive state, such as "Lower Limit", directly corresponds to a sensor signal, a logical connection of sensor signals, or other signals on the PC. A state item is a set of primitive states. Each primitive state is classified into a state item. For example, state item "Vertical Position" consists of primitive states such as "Lower Limit", "Coil Carrying Position", and "Upper Limit". We also use the state item name for the attribute names of machine state nodes in the state space model. Each machine has one or more state items. A machine state is a combination of primitive states of each state item of the machine. Thus a set of machine states is a Cartesian product of state items. For example, Coil Car has a machine state <Lower Limit, Stop Point2, Not Loading, Stopping>. A machine state node, as shown in FIGURE 4, is an implementation of the machine state. An equipment state is a combination of the machine states of each machine. Thus a set of equipment states is a Cartesian product of machine state sets. For example, Delivery equipment has an equipment state <Coil Car State-α, Skid1 State-a, ... >. An equipment state node, as shown in FIGURE 6, is an implementation of the equipment state.

```
Name:         Lower Limit
State-of:     Coil Car
State Item:   Vertical Position
Pre-State:    nil
Next-State:   Coil Carrying Position, Conveyor Height
Constraint:   Coil Car GOM Lower Limit,
              Coil Car AOM Lower Limit, ...
        ...          ...
```

FIGURE 8: Lower Limit

A primitive state has attributes about "state item", "adjoining relations", "constraints on states and actions", "relations between primitive states and PC signals", and figures. FIGURE 8 represents a primitive state, e.g. "Coil Car Lower Limit". This primitive state is a state of Coil Car, is an element of Vertical Position, has no pre-state adjoining relation, has next-state adjoining relations to other primitive states in the same state item such as Coil Car Coil Carrying Position and Coil Car Conveyor Height, and has constraints on states and actions for each operation mode.

4.2   ACTIONS

State transitions (effects) in the world are classified according to whether the state transitions depend on the particular situation (context) in which the action is executed. In this paper, the state transitions, independent of the situation, are represented by actions. The context-dependent state transitions are represented by side effects. In the knowledge compilation, the actions are used to generate valid arcs in machine-level state space models.

```
Name:          Upward
Action-of:     Coil Car
Causes:        Side Effect 1, ...
State Item:    Vertical Position, Action State
Change Method: (Vertical Position,   Change along Next-State),
               (Action State,   Change Pointing to States), ...
Pointed State: (Action State,   Rising)
Weight:        10
          :
```

FIGURE 9: Upward

FIGURE 9 shows an action "Upward" which is an "Action-of" Coil Car, which "Causes" Side Effect 1, which change "State Item" Vertical Position and Action State,

which change the State Items with "Change Method" Change along Next-State and Change Pointing to States. A change method "Change Pointing to States" changes a primitive state of machine state to another primitive state shown by the attribute "Pointed States". Each state transition by the actions is given the value of the "Weight" in our planning algorithm. A weight value is defined from the action's efficiency in operations.

4.3   SIDE EFFECTS

In the knowledge compilation, side effects are used to generate arcs which represent the context-dependent state transitions in machine-level state space models.

FIGURE 10 shows a side effect called "Side Effect 1". A side effect consists of attributes: conditions and effects. If the conditions are satisfied, machine states are changed according to the effects.

```
Name                :   Side Effect 1
Conditions
 · Caused by        :   Upward
 · Initial States   :   Not Loading, Lower Limit,
                        Stop Point, Supporting
 · Changed to       :   Coil Carrying Position
 · Common State Item :  Horizontal Position
Effects             :   Loading, Not Supporting
```

FIGURE 10: Side Effect 1

4.4   CONSTRAINTS

Constraints on states and actions are used to validate machine state nodes and arcs in machine-level state space models. Constraints on states and actions consist of four layers: Operation Mode Layer, Machine Layer, Primitive State Layer, State Combination Denial Layer.

```
Name:             Coil Car GOM Lower Limit
Operation Mode:   General Operation Mode
Machine Layer:    Coil Car GOM
State Combination
Denial Layer:     Improper State 1, ...
Horizontal Position:  Line Center, ... , Stop Point2, ...
Coil Loading State:   Not Loading
Action State:     Stopping, Moving for Backward, ...
Actions:          Upward, Forward, Backward, ...
```

FIGURE 11: Coil Car GOM Lower Limit

On the Operation Mode Layer and the Machine Layer, constraints are classified with each operation mode, e.g. General Operation Mode (GOM) and Automatic Operation Mode (AOM), and each machine.

On the Primitive State Layer, internal constraints of each machine are represented. These constraints show all possible pairs of primitive states and are used to check validity of machine states and arcs in knowledge compilation. FIGURE 11 shows constraints (Coil Car GOM Lower Limit) on this layer. The attributes "Horizontal Position", "Coil Loading State", and "Action State" show all possible pairs of a primitive state "Lower Limit" and primitive states of other state items. The attribute "Actions" shows which action can

Proc.1 : Generate Valid Machine States
1   **begin** for each machine M in equipment **do**
2   **begin** for each primitive state combination $PSC1$
            in the Cartesian product of state items of M **do**
3   **begin** for each primitive state $P1$ of $PSC1$ **do**
4   Look up the constraint $C1$ for general operation mode from $P1$;
5   Check whether all primitive states of $PSC1$ except for $P1$ satisfy
                $C1$
6   **end**;
7   If all of constraints looked up from primitive states of $PSC1$ are
                satisfied by $PSC1$ **then**
                Create a machine state node $MSN1$
8   **end**;
9   **end**

Proc.2 : Generate Valid Machine-level State Transition Arcs
1 **begin** for each machine M in equipment **do**
2   **begin** for each operation mode OM **do**
3   Create a list $MSNL$ of machine state nodes valid on OM;
4   **begin** for each machine state node $MSN1$ of $MSNL$ **do**
          /*Create a candidate list of actions valid at $MSN1$ on OM */
5   Look up constraints on OM from each primitive state of $MSN1$;
6   Create the intersection of action lists AL, which are listed
            in the attribute "Actions" of each of the constraints;
7   **begin** to find machine state nodes to which $MSN1$ is changed
            to by each action $AC1$ of AL **do**
8   Deduce primitive state combinations, e.g. $PSC2$, $PSC3$,
            which correspond to machine states to which $MSN1$ is
            changed by $AC1$;
9   Evaluate conditions of each side effect SE, related to $AC1$
            with regard to M;
10   If any SE is effective **then**
11   Deduce primitive state combinations, e.g. $PSC4$,
              which correspond to machine state to which $MSN1$
              is changed by $AC1$ and SE;
12   Check validity of $PSC2$, $PSC3$, $PSC4$ ... on OM;
13   Look up machine state nodes, e.g. $MSN2$,
              which correspond to the valid primitive state combinations
14   **end**;
15   **end**;
16   Create arcs based on correspondence between actions,
          side effects, and machine state nodes deduced by
          the actions and side effects
17   **end**;
18 **end**

FIGURE 12: Procedures of Knowledge Compilation

be carried out at "Lower Limit". All primitive states have the constraints for each operation mode.

It is possible for the representation of the Primitive State Layer to easily deal with complex constraints, because designers focus on rules for each primitive state, and because the State Space Model Generator synthesizes the rules for each machine state.

4.5 PROCEDURES

State space models for machines are generated by the two procedures Generate Valid Machine States (Proc.1) and Generate Valid Machine-level State Transition Arcs (Proc.2), as shown in FIGURE 12.

The following are examples of executing Proc.1. **1.** Select Coil Car as M. **2.** Select <Lower Limit, Stop Point2, Not Loading, Stopping> as $PSC1$. **3.** Select Lower Limit as $P1$. **4.** Look up Coil Car GOM Lower Limit as $C1$. **5.** In this example, Stop Point2, Not Loading and Stopping are satisfied because the primitive states are included in the attribute value of Horizontal Position, Coil Loading State and Action State. **7.** Create a machine state node as shown in

FIGURE 4, when all of constraints looked up from primitive states of $PSC1$: <Lower Limit, Stop Point2, Not Loading, Stopping> are satisfied by $PSC1$.

The following are examples of executing Proc.2. **1.** Select Coil Car as M. **2.** Select Automatic Operation Mode AOM as OM. **3.** Create a valid machine state node list $MSNL$ (Coil Car State-α, ...) which are valid on AOM. **4.** Select Coil Car State-α as $MSN1$. **5.** In steps **5** and **6**, create a candidate list of actions which are valid at $MSN1$: Coil Car State-α on AOM. Look up Coil Car AOM Lower Limit, Coil Car AOM Stop Point2, Coil Car AOM Not Loading and Coil Car AOM Stopping. **6.** Create the intersection of action lists (Upward, Backward, ...) as AL. **7.** Select Upward as $AC1$ to change $MSN1$: Coil Car State-α. **8.** At PSC1 <Lower Limit, Stop Point2, Not Loading, Stopping>, Lower Limit is changed by "Change along Next-State" of Upward with Coil Carrying Position or Conveyor Height, and Stopping is changed by "Change Pointing to States" of Upward with Rising. Therefore PSC1 was changed to PSC2<Coil Carrying Position, Stop Point2, Not Loading, Rising> or PSC3<Conveyor Height, Stop Point2, Not Loading, Rising>. **9.,10.** At PSC2<Coil Carrying Position, Stop Point2, Not Loading, Rising>, with regard to Coil Car, the conditions (Caused by, Initial State, Changed to) of SE: Side Effect 1 are satisfied by AC1 (Upward), MSN1 (Coil Car State-α), Coil Carrying Position of PSC2. Primitive state "Supporting" of a machine "Skid" and condition "Common State Item" are evaluated in Planning. **11.** Therefore PSC4<Coil Carrying Position, Stop Point2, Loading, Rising> is created from PSC2 by Side Effect 1. **12.** Check validity of $PSC2$, $PSC3$, $PSC4$, ... on AOM. In this example, PSC2 and PSC4 was valid. **13.** Therefore Coil Car State-β and Coil Car State-γ were looked up. **16.** Create arcs as shown in FIGURE 5 based on correspondence between actions, side effects and machine state nodes.

4.6 RESULTS OF KNOWLEDGE COMPILATION

To evaluate the knowledge compilation method, practical state space models for machines have been generated from domain knowledge. The domain knowledge contains most of states, actions, side effects and constrains of the major machines in delivery equipment of a steel plant.

Table 1 shows the number of machine state nodes and arcs of state space models for machines in VIPS/S-SFCD. The generated machine state nodes and arcs linking the machine state nodes are restricted to satisfy the constraints on states and actions in domain knowledge. For example, Coil Car has four state items and the state items are sets of primitive states. There are 2304 primitive state combinations in a Cartesian product of the state items. VIPS/S-SFCD selected 352 (GOM) and 92 (AOM) valid combinations from the a Cartesian product and generated 352 machine state

Table 1: Generated machine state nodes and arcs

machines	primitive state combs.	nodes GOM/AOM	arcs AOM
Coil Car	2304	352 / 92	125
Tension Reel	450	17 / 13	14
Skid1	2	2 / 2	2
Skid2	2	2 / 2	2
Skid3	2	2 / 2	2
Table	2	2 / 2	2
Def. Pinch Roll	18	4 / 4	4
T. Pinch Roll	2	2 / 2	2
B. Pinch Roll	2	2 / 2	2
Belt Wrapper	16	8 / 6	8
Shear	2	2 / 2	2

nodes. For example, Coil Car has 17 actions and side effects. VIPS/S-SFCD selected 125 valid arcs (AOM) from the set of all possible arcs (possible: 39168, GOM: 5984, AOM: 1564). VIPS/S-SFCD took at most 20 minutes to generate machine state nodes and arcs which table 1 show. It is impossible for designers to generate such results in this short time. There is much possibility that designers make some mistakes carelessly.

To evaluate our approach, we have been experimenting with control specification generation using the state space models for machines. For the experiments, we used 354 test cases that were equivalent to 1 - 30 machine actions. VIPS/S-SFCD generated control specifications equivalent to 2865 actions in total. In interactive design, the specifications equivalent to 1 - 7 actions were frequently generated. VIPS/S-SFCD took 1 - 7 seconds to generate the specifications, proportional to the number of actions. VIPS/S-SFCD took 17, 29, or 47 seconds to generate control specifications equivalent to 14, 21, or 29 actions respectively. It is impossible to generate such short times without knowledge compilation before planning.

These results show that systems that adopt our approach, are able to design interactively , to improve the reliability and quality of control specifications, and to deal rapidly with reconstruction of complex plants.

5   CONCLUSION

This paper describes a knowledge compilation method for interactive design of sequence control programs. In our approach, VIPS/S-SFCD derives the machine-level state space networks (state space models) from domain knowledge (states, actions, side effects and constraints on states and actions). This is done quickly and without errors before planning, which improve planning performance.

It is possible for the systems adopting the knowledge compilation to realize the following:

- Design of the control programs interactively since planning efficiency has been improved.

- Improvement of the reliability and quality of control specifications since state space modes have been generated without human error.

- Rapid adaption of the reconstruction of complex plants since state space models have been generated automatically and quickly.

REFERENCES

[Bylander91] Bylander, T.: A Simple Model of Knowledge Compilation, *IEEE EXPERT*, pp. 73-93, April, 1991.

[Chandra83] Chandrasekaran, B., et al.: Deep versus compiled knowledge approaches to diagnostic problem solving, *International Journal of Man-Machine Studies* 19(5), pp. 425-436, 1983.

[Goel91] Goel, A. K.: Knowledge Compilation(A Symposium), *IEEE EXPERT*, pp. 71-73, April, 1991.

[Kautz94] Kautz, H., et al.: An Empirical Evaluation of Knowledge Compilation by Theory Approximation, *Proceedings of AAAI*, pp. 155-161, 1994.

[Keller91] Keller, R. M.: Applying Knowledge Compilation Techniques to Model-Based Reasoning , IEEE EXPERT, pp. 82-87, April, 1991.

[Levi92] Levi, K. R.: An Explanation-Based-Learning Approach to Knowledge Compilation, *IEEE EXPERT*, pp. 44-51, Jun., 1992.

[Mitchell90] Mitchell, T. M.: Becoming Increasingly Reactive, *Proceedings of AAAI Vol. 2* pp. 1051-1058, July, 1990.

[Mizu92] Mizutani,H., et al.: Automatic Programming for Sequence Control, *Proceedings of the IAAI-92*, pp. 315-331, 1992.

[Nak90] Nakayama,Y.,et al.: Model-based Automatic Programming for Plant Control, *Proceedings of IEEE CAIA90* , pp. 281-287,   1990.

[Nam95a] Namioka,Y.,et al.: Planning-based Visual Programming System for Sequence Control, *Proceedings of IEEE CAIA95*, pp. 186-194, Feb., 1995.

[Nam95b] Namioka,Y.,et al.: "Knowledge-Based Visual Programming for Sequence Control,"(in Japanese) *Human Interface News and Report*, Vol. 10 No. 2, pp. 161-168, Apr., 1995.

[Newell82] Newell, A.: "The Knowledge Level," *Artificial Intelligence*, Vol.18, No.1, 1982, pp. 87-127.

[Sada90] Sadashige,K.,et al.: A construction method of the plant model used for the program generation, (in Japanese) *Proceedings of the 8th Annual Conference of JSAI*, pp. 575-578, 1990.

[Selman91] Selman, B., et al.: Knowledge Compilation Using Horn Approximations, *Proceedings of AAAI*, pp. 904-909, 1991.

[Whitney69] Whitney, D. E.: State Space Models of Remote Manipulation Tasks, *IEEE Trans. on AC.*, Vol. AC-14, No.6, pp. 617-623, 1969.

[Wilkins88] Wilkins, D. E.: *"Practical Planning: Extending the Classical AI Planning Paradigm,"* Morgan Kaufmann Publishers, 1988.

SHAPE REASONING IN CAD SYSTEMS

José C. Damski   and   **John S. Gero**
Key Centre of Design Computing
Department of Architectural and Design Science
University of Sydney, NSW, 2006, Australia
email : {jose,john}@arch.su.edu.au

ABSTRACT

This paper presents a logic-based representation for shapes and objects which allows topological reasoning, among other features. This representation is suitable to be used in current CAD systems as it can be mapped from a shape's numerical representation.

1 INTRODUCTION

Current CAD systems allow designers to represent shapes and objects largely as graphic objects in drawings only, without any semantic information attached to them. This is equivalent to a system that knows words and the rules for sentence formation (the syntax) in a language. Such a system does not understand the meaning of each sentence and the relation among them. Currently there is a need for CAD systems that could be able to interact with designers, give feedback, and reason at a more abstract level.

The basic level of shape reasoning in a CAD system concerns the relations among shapes (or objects in 3D) in a drawing. The system should be able to recognize if one shape touches another, for example, or even describe such a relation, analysing and understanding the drawing. Other relations such as inside, overlap and cover give much richer information about the drawing then endpoints and properties of lines.

Beyond the topological and spatial information some other types of reasoning are also desirable, such as shape emergence and shape semantics (GDJ95). Shape emergence deals with shapes that are not in the drawing but can be "seen" by humans as derived shapes. Shape semantics is the relation among shapes which gives meaning in a particular context.

In this article we present a logic-based representation of shapes that allows a uniform representation for straight and non-straight boundaries that can also be extended to 3D objects bounded by planes and surfaces. From this representation we present topological and spatial reasoning as examples of symbolic reasoning over CAD drawings.

2 REPRESENTATION

In order to represent shapes symbolically we map each line in the drawing as the edge of a halfplane. The halfplane, which is a subset of an idealized plane around the drawing, is the basic unit of our system. In the following subsections we define this concept more precisely. In this paper we do not address how a line segment in a CAD system can be extended from the edge of the plane, and assume it is trivial for straight lines.

2.1 BACKGROUND

We define a halfplane slightly differently to its definition in geometry. In geometry two halfplanes are divided by a line, and the points on that line do not belong to either of the halfplanes. In our concept there is no line, only a conceptual border that divides two sets of points. Each set of points defines one halfplane. More formally:

- U is a region defined by a set of points $p(x, y)$.
  $U = \{p(x, y)\}$

- U always can be divided into exactly two subsets A and B, defined by:
  $A = \{p(x, y) : f(x, y) > 0\}$, and
  $B = \{p(x, y) : f(x, y) \leq 0\}$

- A and B are non-empty, closed sets.

The set B is the *complement* of set A, denoted by $\overline{A}$, where we only consider elements in U. Arbitrarily we can assign the truth value True to A and therefore False to $\overline{A}$. For a given halfplane x, the predicate $hp(x)$ has a truth value in the same way. For instance, in Figure 1(a) the halfplane a is shown by the shaded area. Its complement, $\overline{a}$ is the unshaded area bounded by U and can be represented by $\neg hp(a)$, regardless of the specific truth value assigned to it.

By convention, we assign with truth value True to the halfplane where its name lies, as $hp(a)$ shows in Figure 1(a).

Given the halfplanes a and j, if $a \subset j$, the logical representation equivalent to this condition is: $hp(a) \to hp(j)$.

According to these principles, all shapes expressed by halfplanes can be rewritten as a Wff (Well Formed Formula) in first order logic. The aim of this representation is to map shapes into halfplanes, and from them, into predicates to be manipulated according to first order logic principles.

2.2 REGIONS AND SHAPES

We define regions and shapes as:

Region Given n halfplanes, a *region* R is defined by a conjunctive formula of n $hp(x)$, as

$$R \text{ is } hp(x_1) \wedge hp(x_2) \wedge ... \wedge hp(x_n)$$

Since each halfplane can have the truth value True or False, each *region* is an *interpretation* [1] of this formula. This means, for a given n halfplanes we have 2^n different regions. Some of them may not be *visible* in the drawing, as demonstrated in (DG96).

Shape A shape is a disjunctive formula of regions, represented by $R_1 \vee R_2 ... R_n$.

In order to illustrate these definitions, Figure 1(b) shows three halfplanes a, b and c.

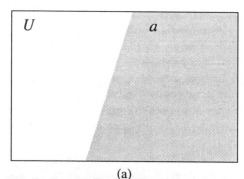

(a)

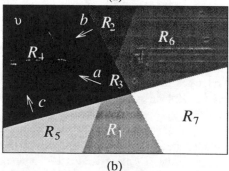

(b)

Figure 1: **(a) Halfplane** a **in** U. **(b) Halfplanes** a, b **and** c.

Where the region R_1 is described by the formula:

$$\neg hp(a) \wedge hp(b) \wedge \neg hp(c)$$

A shape S_1 might be $R_1 \vee R_3$, which is represented by:

$$(\neg hp(a) \wedge hp(b) \wedge \neg hp(c)) \vee (\neg hp(a) \wedge hp(b) \wedge hp(c))$$

and can be simplified further to: $\neg hp(a) \wedge hp(b)$

In order to describe the relation among all halfplanes, we can write a unique topological constraint using the formula: $hp(a) \wedge \neg hp(b) \to hp(c)$ which logically means the region R_2 is inside the halfplane c. This formula holds the topological relation among all regions.

The predicate $hp(a)$ makes no strong assumptions about the boundaries of the halfplane. As a consequence this same representation is equally applicable to non-straight boundaries of halfplanes.

3 REASONING

In this section we address two types of spatial relationships of interest in CAD systems: **topological** relationships and **directional** relationships. The first one deals with relations, such as *overlap*, *touch*, *contains*, etc, that are invariant under geometrical transformations. Those relations do not change if the set of shapes is rotated, scaled or translated. The second one deals with the relative position of shapes according to a pre-defined cardinal system. This results in relationship such as *left*, *right*, *above*, *below*. The halfplane representation can deal with these two types of relationship in an unified way.

The combination of these two types of relationships allows semantic queries about a drawing, such as "which room is located next to the kitchen ?" or "which room is located next to and left of the hall ?" in an architectural domain.

3.1 TOPOLOGY

Egenhofer (Ege91) identifies eight fundamental relations between two shapes: **overlap**, **disjoint**, **inside** (and the opposite: **contains**), **touch** (meets at the border), **cover** (and the opposite: **coveredby**) and **equal**, as shown in Figure 2.

Using the halfplane representation the six basic relations are defined as:

touch There are two ways in which two regions can touch each other:

border when two regions share at least one edge. Given a region R_1 expressed by:

$$hp(x_1) \wedge hp(x_2) \wedge ... \wedge hp(x_n)$$

a region R_2 expressed by:

$$hp(y_1) \wedge hp(y_2) \wedge ... \wedge hp(y_n)$$

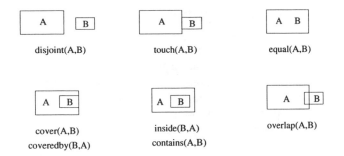

Figure 2: Possible relations between two shapes (after (GPP95)).

they are **border adjacent** iff they differ in only one $hp(x_i)$, such as $hp(x_i)$ in R_1 is $\neg hp(x_i)$ in R_2. The predicate $touch\_border(R_i, R_j)$ is used for this relation. For example, in Figure 1(b), the region R_3 is represented by the formula $\neg hp(a) \wedge hp(b) \wedge hp(c)$ and the border adjacent regions are:

$$R_1 \text{ is } \neg hp(a) \wedge hp(b) \wedge \neg hp(c)$$
$$R_4 \text{ is } hp(a) \wedge hp(b) \wedge hp(c)$$
$$R_6 \text{ is } \neg hp(a) \wedge \neg hp(b) \wedge hp(c)$$

Once a shape is a collection of regions, the regions adjacent at the border of a shape is the set of regions border adjacent to each region of that shape.

corner when two regions share one vertex. Given a region R_1 expressed by:

$$hp(x_1) \wedge hp(x_2) \wedge ... \wedge hp(x_n)$$

a region R_2 expressed by:

$$hp(y_1) \wedge hp(y_2) \wedge ... \wedge hp(y_n)$$

they are **corner adjacent** iff they differ exactly in two literals $hp(x_i)$, $hp(x_j)$ such as $hp(x_i)$, $hp(x_j)$ in one region is $\neg hp(x_i)$, $\neg hp(x_j)$ in the other. The predicate $touch\_corner(R_i, R_j)$ is used for this relation. For example, in Figure 1(b), the region R_3 is represented by the formula $\neg hp(a) \wedge hp(b) \wedge hp(c)$ and the corner adjacent regions are:

$$R_2 \text{ is } hp(a) \wedge \neg hp(b) \wedge \neg hp(c)$$
$$R_5 \text{ is } hp(a) \wedge hp(b) \wedge \neg hp(c)$$
$$R_7 \text{ is } \neg hp(a) \wedge \neg hp(b) \wedge \neg hp(c)$$

The corners for a shape are the collection of the corner adjacencies of all the regions of it.

overlap Given two shapes S_1 and S_2 described as: S_1 is $R_1 \vee R_2 \vee ... \vee R_n$ and S_2 is $T_1 \vee T_2 \vee ... \vee T_m$ where R_i and T_j are regions. S_1 overlaps S_2 if:

$$\exists i \; \exists j \mid R_i = T_j$$

where $R_i = T_j$ means these regions are defined by the same halfplanes, and it is represented by: $overlap(S_i, S_j)$.

disjoint Given the two shapes S_1 and S_2 described earlier, S_1 is *disjoint* to S_2 if:

$$\forall i \; \not\exists j \mid R_i = T_j$$

represented by $disjoint(S_i, S_j)$.

inside Given the two shapes S_1 and S_2 described earlier, S_1 is said to be *inside* S_2 ($S_1 \subset S_2$) if the following expression is true:

$$\forall i \; \exists j \mid R_i = T_j$$

represented by the predicate $inside(S_i, S_j)$.

cover Given the two shapes S_1 and S_2 described earlier, S_1 *covers* S_2 if the following formula is true:

$$inside(S_1, S_2) \wedge \exists R_i \mid inside(R_i, S_1) \wedge$$
$$touch\_border(R_i, R_{i_{border}}) \wedge \neg inside(R_{i_{border}}, S_2)$$

represented by the predicate $cover(S_i, S_j)$.

equal Given the two shapes S_1 and S_2 described earlier, S_1 is said *equal* S_2 if the following expression is true:

$$(\forall i \; \exists j \mid R_i = T_j) \wedge (\forall i \; \exists j \mid T_i = R_j)$$

represented by the predicate $equal(S_i, S_j)$.

From these relations we can infer logical consequences. A complete table of such results can be found in (Ege91).

3.2 DIRECTION

As the topological relation does not carry any information about spatial relation among shapes, it is necessary to have a labelling system to provide directional information, such as above, below, left and right. In order to implement this in our representation it is necessary to give a semantic denotation for each halfplane. This denotation can be giving by the following declaration

- The universe of discourse U is represented by a rectangle, whose boundaries are numbered from 1 to 4 in a clockwise direction, as shown in Figure 3.

- Each halfplane has its symbol and denotation according to the endpoints of its border. Figure 3 shows the possible combinations among sides, where the border of each halfplane is represented by a dotted line. Halfplane labels are consistently placed above or to the left of the boundary of the halfplane. Arbitrarily name **above** as the upwards direction and **left** as westwards direction.

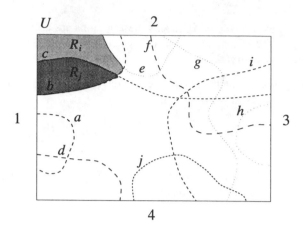

Figure 3: Halfplanes $a - j$.

- The denotation for each halfplane is shown in Table 1, assuming Left as opposite direction of Right and Above as opposite direction of Below.
- The name of each halfplane is given on the side its denotation is, as shown in the example Figure 3.

With this systematic way of labeling halfplanes and denotations, we can infer the relative position of one region to another, in relation to a given halfplane. A one-step inference from given regions R_1, R_2 and R_3 is shown in Table 2.

In order to compare the relative position between two regions it is necessary to eliminate the halfplanes in common and analyze the differences between the resulting formulas.

Where the predicate components are the same, the two regions lie in the same halfplane and therefore lie one the same side of the boundaries of the halfplane. Differences imply that the regions lie on opposite side of the boundary of a halfplane. Take as an example the following two formulas representing two arbritary regions R_i and R_j shown in Figure 3:

$$R_i \; hp(a) \wedge hp(b) \wedge hp(c) \wedge hp(d) \wedge \neg hp(e) \wedge hp(f) \wedge hp(g) \wedge hp(h) \wedge hp(i) \wedge hp(j)$$
$$R_j \; hp(a) \wedge hp(b) \wedge \neg hp(c) \wedge hp(d) \wedge \neg hp(e) \wedge hp(f) \wedge hp(g) \wedge hp(h) \wedge hp(i) \wedge hp(j)$$

The difference is:

$$R_i \; hp(c)$$
$$R_j \; \neg hp(c)$$

i.e. R_i lies on the other side of the boundary of halfplane c to R_j. The specific interpretation depends on the denotation of the specific case, and in this case means R_i is *above* R_j.

4   APPLICATION

In this section we present a simple example in order to illustrate the concepts shown in this paper. An example of an architectural plan of a two bedroom house is shown in the Figure 4.

Table 1: **Halfplane denotation.**

Halfplane	Begin-end	Denotation
a	1-1	Left
b	1-2	Above-Left
c	1-3	Above
d	1-4	Above-Right
e	2-2	Above
f	2-3	Above-Right
g	2-4	Left
h	3-3	Left
i	3-4	Above-Left
j	4-4	Above

Table 2: **Basic directional inferences among regions.**

Given	Given	Result
$above(R_1,R_2)$	$above(R_2,R_3)$	$above(R_1,R_3)$
$below(R_1,R_2)$	$below(R_2,R_3)$	$below(R_1,R_3)$
$left(R_1,R_2)$	$left(R_2,R_3)$	$left(R_1,R_3)$
$right(R_1,R_2)$	$right(R_2,R_3)$	$right(R_1,R_3)$

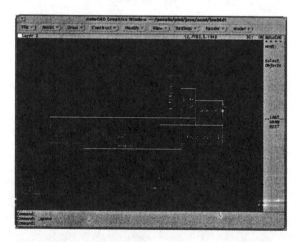

Figure 4: **Simple architectural plan of a two-bedroom house (after Frank Lloyd Wright).**

We can extend each straight line segment up to the border of an imaginary plane that embeds the whole plan, as shown in Figure 5(a). Each line now represents a border that divides this plane into two halfplanes. They are labelled from a to o.

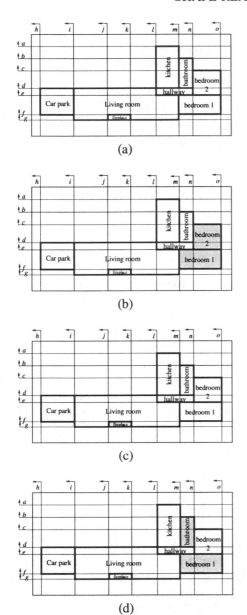

Figure 5: **Some topological and directional relations.**

We will examine three examples of possible inferences from this plan. Some details are not presented in order to give a better understanding of the whole process.

inside - The predicate $inside(S_i, S_j)$ is satisfied if $S_i \rightarrow S_j$ is true. The living room (S_3) and the fireplace (S_4), shown in the Figure 5(c). are represented by:

S_3 is $R_5 \vee R_6 \vee R_7 \vee R_8 \vee R_9 \vee R_{10} \vee R_{11} \vee R_{12} \vee R_{13} \vee R_{14} \vee R_{15}$

S_4 is R_{13}

so $inside(S_4, S_3)$ is represented by $S_4 \rightarrow S_3$, or the logical equivalent $\neg S_4 \vee S_3$, which replacing the shapes by regions is:

$\neg R_{13} \vee R_5 \vee R_6 \vee R_7 \vee R_8 \vee R_9 \vee R_{10} \vee R_{11} \vee R_{12} \vee R_{13} \vee R_{14} \vee R_{15}$

where $\neg R_{13} \vee R_{13}$ makes this formula true for every interpretation. Therefore S_4 is inside S_3.

touch - The area for bedroom_1 (S_1) is defined as: S_1 is $R_1 \vee R_2$, where the regions R_1 and R_2 are defined as:

R_1 is $\neg hp(a) \wedge \neg hp(b) \wedge \neg hp(c) \wedge \neg hp(d) \wedge \neg hp(e) \wedge hp(f) \wedge hp(g) \wedge \neg hp(h) \wedge \neg hp(i) \wedge \neg hp(j) \wedge \neg hp(k) \wedge \neg hp(l) \wedge \neg hp(m) \wedge hp(n) \wedge hp(o)$

R_2 is $\neg hp(a) \wedge \neg hp(b) \wedge \neg hp(c) \wedge \neg hp(d) \wedge \neg hp(e) \wedge hp(f) \wedge hp(g) \wedge \neg hp(h) \wedge \neg hp(i) \wedge \neg hp(j) \wedge \neg hp(k) \wedge \neg hp(l) \wedge \neg hp(m) \wedge \neg hp(n) \wedge hp(o)$

The area for bedroom_2 (S_2) is defined as: S_2 is $R_3 \vee R_4$, where the regions R_3 and R_4 are defined as:

R_3 is $\neg hp(a) \wedge \neg hp(b) \wedge \neg hp(c) \wedge hp(d) \wedge hp(e) \wedge hp(f) \wedge hp(g) \wedge \neg hp(h) \wedge \neg hp(i) \wedge \neg hp(j) \wedge \neg hp(k) \wedge \neg hp(l) \wedge \neg hp(m) \wedge \neg hp(n) \wedge hp(o)$

R_4 is $\neg hp(a) \wedge \neg hp(b) \wedge \neg hp(c) \wedge \neg hp(d) \wedge hp(e) \wedge hp(f) \wedge hp(g) \wedge \neg hp(h) \wedge \neg hp(i) \wedge \neg hp(j) \wedge \neg hp(k) \wedge \neg hp(l) \wedge \neg hp(m) \wedge \neg hp(n) \wedge hp(o)$

The predicate $touch(S_i, S_j)$ tests all regions of S_i against S_j verifying if they satisfy the conditions for touching (at corner or border) as presented in the section 3.1, for regions. In the example for shapes S_1 and S_2 shown in the Figure 5(b), there are 4 possible combinations:

$$(R_1, R_3), (R_1, R_4), (R_2, R_3) \text{ and } (R_2, R_4).$$

One of them (R_2, R_4) satisfies the condition of touch at the border because they differ in only one halfplane $hp(e)$ where it is negated in R_2 and it is not in R_4. Therefore, the regions S_1 and S_2 touch at some part of their borders.

direction - Bedroom_1 (S_1) as represented above and the bathroom is represented by S_5 as $R_{16} \vee R_{17}$. The regions are represented by:

R_{16} is $\neg hp(a) \wedge \neg hp(b) \wedge hp(c) \wedge hp(d) \wedge hp(e) \wedge hp(f) \wedge hp(g) \wedge \neg hp(h) \wedge \neg hp(i) \wedge \neg hp(j) \wedge \neg hp(k) \wedge \neg hp(l) \wedge \neg hp(m) \wedge hp(n) \wedge hp(o)$

R_{17} is $\neg hp(a) \wedge \neg hp(b) \wedge \neg hp(c) \wedge hp(d) \wedge hp(e) \wedge hp(f) \wedge hp(g) \wedge \neg hp(h) \wedge \neg hp(i) \wedge \neg hp(j) \wedge \neg hp(k) \wedge \neg hp(l) \wedge \neg hp(m) \wedge hp(n) \wedge hp(o)$

The denotation of halfplanes, based on Table 1, is :

Halfplane	Begin-End	Denotation
a, b, c, d, e, f, g	1-3	Above
h, i, j, k, l, m, n, o	2-4	Left

In order to compare S_1 and S_5, it is necessary to compare each region of S_1 against each region of S_5. Comparing R_1 (from S_1) against R_{16} (from S_5), the difference is:

R_1 is $\neg hp(c), \neg hp(d), \neg hp(e)$

R_{16} is $hp(c), hp(d), hp(e)$

From the denotation table all these halfplane (c, d, e) imply that R_{16} *is above* R_1. In the same way, the comparision between all pairs of regions gives the following result:

R_{16} *is above* R_1, R_{16} *is above* R_2, R_{16} *is left* R_2, R_{17} *is above* R_1, R_{17} *is above* R_2, R_{17} *is left* R_2.

The relation between S_1 and S_5 is a statistical summary of the above relations, therefore it is clear that S_5 is *above* S_1 and S_1 is partially to the *right* of S_5. These shapes are shown in Figure 5(d).

5   CONCLUSION

Numerical representations of shapes and objects are the traditional modes of representation in CAD systems. They have well known and obvious properties. They provide a ready access to the geometric manipulations of CAD systems which are required of graphical presentations. However, they fail to provide ready access to any other form of shape reasoning. As CAD systems become more sophisticated so the demands on them to carry out shape reasoning increase. It is at this stage that representations capable of directly supporting such reasoning are required. This paper has presented one such representation, a representation based on first order logic. The grounding of the logic differs from the traditional grounding of geometry in that the atom of the grounding is not an infinitesimal point as is the case with geometry but is rather a halfplane.

The use of halfplanes as the grounding of the logic carries with it a number of advantages, three of the most important of them are:

Objects in Three Dimensions - With very minor changes this logic-based representation may be extended to describe objects in three dimensions. Halfplanes become halfspaces which divide the space (or volume) into two halfspaces (or halfvolumes), and the logic representation remains the same.

U becomes a volume defined by a set of points $p(x, y, z)$ which is divided into exactly two subsets A and B, defined by:

$A = \{p(x, y, z) : f(x, y, z) > 0\}$, and

$B = \{p(x, y, z) : f(x, y, z) \leq 0\}$.

The definition of regions is directly extensible to a definition of volumes and the definition of shapes is directly extensible to a definition of objects. The reasoning concerning topology and direction is directly extensible to objects.

Classes of Shapes Which May be Represented - Numerical representations vary quite dramatically between those capable of representing shapes with straight-line boundaries and those with curved-line boundaries. Shapes with curved-line boundaries are significantly more difficult to represent than those with straight-line boundaries. The effect of this variation in representation is that all operations on shapes need two versions to cover both representations. The use of a logic-based representation grounded on halfplanes makes no direct assumptions concerning the straightness or otherwise of the boundary of the halfplane. Therefore, the definitions of regions and shapes and the reasoning concerning topology and direction are applicable to both classes of shapes.

Classes of Operations - In addition to the two classes of reasoning operations presented in this paper other operations which normally require a re-representation of the numerical representation are directly applicable to the logic-based representation. Of particular interest in graphical systems is shape emergence, embedded shape emergence is available as a direct consequence of this representation of regions as is a subset of illusory shape emergence.

6   ACKNOWLEDGEMENTS

José Damski thanks CNPq/Brazil for his scholarship. Part of this work is funded by a grant from the Australian Research Council.

REFERENCES

[DG96]   J. C. B. Damski and J. S. Gero. A logic-based framework for shape representation. *CAD Journal*, 1996. (to appear).

[Ege91]   Max J. Egenhofer. Reasoning about binary topological relations. In O. Günter and H.-J. Schek, editors, *Advances in Spatial Databases*, pages 143–160, Berlim, August 1991. Springer-Verlag.

[GDJ95]   J. S. Gero, J. C. B. Damski, and H. J. Jun. Emergence in CAAD systems. In *Preprints of CAAD Futures'95*, volume 2, pages 49.1–15, Singapore, 1995.

[GPP95]   Michelangelo Grigni, Dimitris Papadias, and Christos Papadimitriou. Topological inference. In *Proceedings of the Fourteenth International Joint Conference on Artificial Intelligence*, volume 1, pages 901–906, August 1995.

Logic Programming

APPLICATIONS OF LOGIC PROGRAMS WITH FUNCTOR SET TO AUTOMATED PROBLEM SOLVING UNDER UNCERTAINTY

Hiroshi SAKAI and Akimichi OKUMA

Department of Computer Engineering, Kyushu Institute of Technology,

Tobata, Kitakyushu, Japan, 804

Email: {sakai,okuma}@comp.kyutech.ac.jp

ABSTRACT

We sometimes have uncertain knowledge like 'We do not know the real thing but we know a set which the real thing exists in.', which we name knowledge by rule of thumb. In this paper, we think automated problem solving with uncertain attribute values by rule of thumb. We first show that there are several problems related to this uncertainty and it is necessary to discuss the automated problem solving in such situation. Then, we propose a framework of logic programs with functor *set* and refer to the interpreter for programs. Finally, we apply this framework to solve some problems. We also show the application areas of our framework.

1. INTRODUCTION

It is very important problem that how we handle the incompleteness and the uncertainty in problem solving, information systems and knowledge based systems. There are several approaches for this problem, for example, hypothetical reasoning[PGA87], abduction[Ko90], default reasoning[Rei80], disjunctive logic program[LMR92], near-horn prolog[Lo87], fuzzy prolog[MBP87, MSD87], null value[Co70, Za84], incomplete relational database [Li79, Li81], rough set theory[Pa91], etc. Here, we mainly discuss the area related to null value problem, uncertain attribute values and logic program.

We sometimes have uncertain knowledge like 'We do not know the real thing but we know a set which the real thing exists in.', which we name *knowledge by rule of thumb*. For example, we do not know the precise number of persons, but we know the maximum number is 8. We have to take persons to sightseeing. There are two cars, the one is 5-seater and the other is 10-seater. In such case, we probably select the 10-seater car. Because, 10-seater car makes no trouble in any case. Namely, the rule of thumb is a way to escape from the uncertainty, even if the precise information is uncertain. This concept is clearly a variation of null value and it is also related to the lower and upper approximation in rough set theory. In incomplete relational database[Li81], we see the concept of rule

by thumb. We think that we need to develop this concept on logic programming, and we have been discussing such issues[Sa94A,94B,94C]. In [Lak93], we can see a discussion on deductive databases with null values.

In this paper, we refer to the logic programs with functor *set* for handling knowledge by rule of thumb and we discuss the applications of these programs by examples.

2. EXAMPLES

Now in this section, let's consider the following examples which are expressing knowledge with uncertain attribute values by rule of thumb.

Example 1. Let's consider the following knowledge for influenza, which may not be real: *If a person's temperature is slightly high or high, his headache is slightly bad or bad and his cough is bad, then he is with influenza. There are two persons, Mike and Suzan. As for Mike, temperature is normal or slightly high, headache is slightly bad and his cough is bad. As for Suzan, temperature is high, headache is slightly bad or bad and her cough is bad. In this situation, how do we diagnose them?* The answers are probably as follows:

(1) As for Mike, he may be with influenza. Because if his temperature is slightly high, then his symptoms satisfy the condition of influenza.

(2) As for Suzan, she is surely with influenza. Because, her symptoms satisfy condition in every case.

In Example 1, every knowledge has uncertainty, but the answer (2) is not affected by the uncertainty. The answer (1) is also helpful for us, because it shows us the condition which is necessary for holding query. We can see such question answering in [Lak93]. Lakshmanan named it *hypothetical query answering* and insisted on the usefulness of this hypothetical query answering.

Now, we go to the second example, which is a revised version in [Sa94B].

Example 2. Let's consider the next situation: *The color of criminal's shirt was about black, navy or gray and his height was about 5.5, 5.6 or 5.7 feet. As for the person a,* *the color was about black or gray and the height was about 5.6 or 5.7. As for b the color was about navy or gray and the height was about 5.7 or 5.8. In this situation, who is the criminal ?*

In this example there may be several interpretations for the 'about', however if we see such knowledge as knowledge by rule of thumb, then the answers are as follows:

(1) The person *a* is a criminal, because *a* satisfies every condition in spite of uncertain knowledge.

(2) The person *b* may be a criminal. Because, if height of *b* is 5.7 feet then *b* satisfies the condition.

Namely, the concept of knowledge by rule of thumb will be applicable to such cases.

We have shown two examples which is discussing knowledge with uncertain attribute values. We think there are many cases and we think it is important to give a framework which can explicitly handle such knowledge. For handling above two examples and others by logic programming, it is natural that we introduce a functor *set* into prolog. We intuitively would like to denote the above Example 2 as follows:

```
(A) criminal(X):-
       wore(X,set([black,navy,gray])),
       height(X,set([5.5,5.6,5.7])).
(B) wore(a,set([black,gray])).
(C) wore(b,set([navy,gray])).
(D) height(a,set([5.6,5.7])).
(E) height(b,set([5.7,5.8])).
```

In real, the above (A) to (E) is a logic program with functor *set*. For this program, the interpreter which responds the above answers (1) and (2) in Example 1 and Example 2 will be helpful for us. Such interpreter will be a good tool for automated problem solving. However, standard prolog interpreter does not know the meaning of the functor *set*. We need to develop a new interpreter which can understand the meaning of the functor *set*.

3. LOGIC PROGRAMS WITH FUNCTOR *SET*

Now in this section we discuss a framework of logic programs with functor *set*.

3.1. Syntax of Programs

We introduce a new functor *set* to prolog and develop a new interpreter which recognizes the meaning of *set*. More formally, we handle *n*-ary predicates $p(x_1,..., x_{n-1}, x_n)$, where the value of x_n depends upon every value of $x_1,..., x_{n-1}$ and the domain for the value x_n is a finite set. Such constraint may be called *functional dependency* [GM78]. Namely, if every value of $x_1,..., x_{n-1}$ is fixed, then the value of x_n is uniquely fixed. However, we do not know the real value of x_n for some ground terms

$x_1,..., x_{n-1}$. For such ground terms we know a set of ground terms $\{y_1,..., y_k\}(k \geq 2)$ which a real value exists in. This comes from the concept of the rule of thumb. Namely, we use predicates $p(x_1,...,x_{n-1},set([y_1,...,y_k])$ instead of $p(x_1,...,x_{n-1},x_n)$. For terms $x_1,...,x_{n-1}$ with variables, we see x_n is $set(\cup_k LIST_k)$ such that $p(x'_1,..., x'_{n-1},set(LIST_k))$ holds where $x'_1,..., x'_{n-1}$ are ground instanciated terms for $x_1,..., x_{n-1}$. We call the atom with a functor *set FS-atom* from now on.

According to the definition of FS-atoms, we define the following *program clauses*.

$$\psi \leftarrow \phi_1,..., \phi_m \ (m \geq 0),$$

where the body implies the conjunction of atoms $\phi_1,..., \phi_m$ and $\psi, \phi_1,..., \phi_m$ are either an atom or an FS-atom. A *logic program with functor set* is a finite set of program clauses. As for goal, we deal with a conjunction of atoms.

In the definition of FS-atom $p(x_1,...,x_{n-1}, set(LIST))$, every element in *LIST* is only ground term. Namely we are now handling the uncertainty on attribute values, however there may be other cases for using the functor *set*. For example,

(1) A case some elements in the *LIST* have variables.

(2) A case some elements in the *LIST* have functor *set*.

(3) A case some elements in the *LIST* are atoms.

The above cases cause not only the uncertainty on attribute values but also another uncertainty. The (3) will be a problem in the higher order logic. We think these cases are next problems and we do not handle such cases here.

3.2. Two Evaluation Systems for A Program

First, we define an *extension* of program P. For every term *set(LIST)*, we call every element in the *LIST* an *extension of set(LIST)*. In every FS-atom $p(x_1,...,x_{n-1}, set(LIST))$, if we replace the *set(LIST)* with its extension then we get an atom, which we call an *extension of the FS-atom*. For example, an FS-atom *height(a,set([5.6, 5.7]))* has two extensions *height(a,5.6)* and *height(a,5.7)*. The extension for every atom which does not have *set(LIST)* is atom itself. For a program clause $\psi \leftarrow \phi_1, ..., \phi_m$, if ψ' is an extension of the head ψ, then

$$\{ \psi' \leftarrow \phi'_1,..., \phi'_m \mid \phi'_j \text{ is an extension of } \phi_j \}$$

is an *extension of the program clause*. For a program P, if we replace every program clause with its extension then we get a logic program without functor *set*, which we call an *extension of the program P*. We denote the set of all extensions of P by *EXT(P)*. We know that there exists a real program in *EXT(P)*, but we do not know which is the real program. In Example 2, the clause (A) has an extension, (B), (C), (D) and (E) has two extensions, respectively. Therefore, *EXT(P)* consists of 16 extensions. By

using *EXT(P)*, we define the two concepts '*may hold*' and '*surely hold*' as follows;
- *A ground formula F may hold, if F holds in some extensions in EXT(P).*
- *A ground formula F surely holds, if F holds in all extensions in EXT(P).*

For a formula *F* with variables, we have to consider substitutions θ such that $F\theta$ holds in some extensions. From now on we call two evaluation systems by *May-system* and *Sure-system*, respectively.

3.3. Evaluation Systems and Problem Solving

Let's consider the two evaluation systems and automated problem solving in Example 1 and Example 2. If a ground formula *F* may hold, then there are some extensions which satisfy the formula *F*. Namely it implies that formula *F* holds if one of the extensions holds, which corresponds to the hypothetical query answering. If a formula *F* has some variables and $F\theta$ holds for a substitution θ, then we get the θ is an answer. Furthermore, if a ground formula *F* surely holds, then we get the answer that formula *F* holds in spite of the uncertain knowledge. By introducing the two definitions, we can formally handle the problem solving like Example 1 and Example 2.

3.4. An Interpreter for Two Systems

We briefly show the interpreter for two systems. Our interpreter consists of the following two programs.
- *Program translator*
- *Prover for problem solving*

They are realized by K-prolog and the size of the both programs is about 25 KB. We first initialize every program by program translator, which is realized as a predicate *init*. Then, we use problem solving systems, which are realized as predicates *may* and *sure*. When we evaluate a query $\leftarrow G(G$ is a conjunction of atoms) in May-system, we execute the query $\leftarrow may(G)$. In Sure-system, we do $\leftarrow sure(G)$. We will show the real execution in the subsequent section.

Now we refer to the overview of the program translation. This translation is necessary for the prover. Therefore, we add an atom *headpred(LIST)*, where *LIST* is a list of head in every program clause with functor *set*, to the program. In Example 2, we add the following atom to the program.

```
(F) headpred([criminal(X),wore(Y,Z),
             height(R,S)]).
```

The translator *init* picks up every head in the list and sequentially asserts the following meta-facts.
- *mfact(mfact_num,sub_num,extension_of_clause)*,
- *ic(list_of_mfact_num)*.

The *mfact_num* is the total number for every extension of program clauses, *sub_num* is the subnumber of every extension. We identify every extension as a hypothesis. For every program clause, we have to remark that we can use only one extension in a resolution. For example, we can use either *height(a,5.6)* or *height(a,5.7)* for the FS-atom *height(a,set([5.6,5.7]))*. We can not use both *height(a,5.6)* and *height(a,5.7)* in a resolution. In order to avoid this problem, we use the second atom *ic(list_of_mfact_num)*. If the element in *list_of_mfact_num* is only one, then the program clause has no functor *set* and there exists no uncertainty. According to these translated meta-facts, we can handle every extension of program *P*. The original program and the translator *init* are automatically deleted for efficient query processing after translation.

Now, we go to the proof procedures in two evaluation systems. As we have mentioned, we identify every extension as a hypothesis and use a framework of hypothetical reasoning. In May-system, prover first tries to solve a problem by using clause without functor *set*. If impossible, then it uses a set of hypotheses which satisfy the constraints *ic(list_of_mfact_num)*. In Sure-system, the prover tries to find all possible solutions in May-system, then it checks all possible solutions correspond to all extensions of program or not. The preliminary implementation of the above procedures is in [Sa94B]. In [Sa94A], we are touching the semantics for May and Sure-systems.

4. AN APPLICATION TO PROBLEM SOLVING UNDER UNCERTAINTY

Now in this section, we discuss an issue such that to find a sequence of actions that will achieve some goal. This type of problem solving is the so called '*state and state transformation*' method. The monkey-banana problem also belongs to this area. We cope with the state and state transformation method under uncertainty and we apply logic programs with functor *set* to it.

4.1. An Example Related to State and State Transformation Method

In order to clarify issues, we show the third example. **Example 3.** Let's consider the next situation: *We have a set of states ST={a,b,c,d,e,f,g} and we can select either s1 or s2 in every state. For selection of s1 or s2, we know the following state transitions $\delta : ST \times \{s1,s2\} \rightarrow ST$.*
$\delta(a,s1)=b, \delta(a,s2)=c, \delta(b,s1)=d, \delta(b,s2)=e, \delta(c,s1)=c, \delta(c,s2)=c, \delta(d,s1)=f, \delta(d,s2)=g, \delta(e,s1)=d, \delta(e,s2)=g, \delta(f,s1)=g, \delta(f,s2)=g, \delta(g,s1)=f, \delta(g,s2)=g.$
However as a constraint for the selection, once we make a selection in every state we always have to take the same

*selection in every state. In this situation, how can we go
to the goal state g from initial state a ?*

The Example 3 seems to be a standard problem related to
state and state transformation method. We can express
this example by the following program.

```
(A)  p(a,set([delta(1,b),delta(2,c)])).
(B)  p(b,set([delta(1,d),delta(2,e)])).
(C)  p(c,set([delta(1,c),delta(2,c)])).
(D)  p(d,set([delta(1,f),delta(2,g)])).
(E)  p(e,set([delta(1,d),delta(2,g)])).
(F)  p(f,set([delta(1,g),delta(2,g)])).
(G)  p(g,set([delta(1,f),delta(2,g)])).
(H)  path(X,Y):-p(X,delta(M,Y)).
(I)  path(X,Y):-p(X,delta(M,Z)),path(Z,Y).
(J)  headpred([p(X,Y),path(P,Q)]).
```

The clauses (A) to (G) imply state transitions, (H) and (I)
do the rule of connection. The (J) implies all heads in
clauses, which is necessary for program translation.
However, when we deal with such programs, we often
have infinite loop problem. In the above program, our
prover also caused infinite loops in states c, f and g.
Because, $\delta(c,s1)=c$ and $\delta(c,s2)=c$ hold. Namely, every
selection in state c transfers c to c.

4.2. An Improvement of Prover for Handling Infinite Loops

Once our preliminary prover falls in infinite loops, the
prover continues the same search until the stack overflow. Namely, our prover could not find another solutions. Therefore we improved the prover not to fall in any
infinite loops. In real, we realized two variations. Every
strategy is as follows:

(1) *Prover restricts the available counts of every clause
in a refutation. Namely, if a clause is called more
than a threshold in a resolution, then we identify it
as infinite loop.*

(2) *Prover keeps the sequence of goals which are unified
to the head of rules and it checks the current goal is
not in the sequence.*

In both strategies, if the prover finds an infinite loop then
the prover forces not to continue the resolution. The strategy (2) comes from [Ge87]. The prover by strategy (1)
whose available count of every clause is 10 responded 101
answers for a query ←*mayall(path(a,g))*. On the other
hand, the prover by strategy (2) responded 20 answers for
the same query. The 101 answers have redundancy for
loops and the 20 answers have no redundancy for loops,
but these answers are the same. Of course, we adopted the
prover by strategy (2) for loop detection. The following is
a part of the improved prover.

```
mayall(X):-
    stime,abolish(num,1),assert(num(0)),
    solve(X,[],RES,[],G,LO,GO),
```

```
    sorteq(RES,RES1),gonum(COUNT),
    disp(RES1,X,LO,GO),nl,fail.
mayall(X):-etime,!.
solve(X,XX,RES,G1,G2,LO,GO):-
    functor(X,Y,NUM),Y/==(,),!,
    solvesub(X,XX,RES,G1,G2,LO,GO).
solve(X,XX,RES,G1,G2,LO,GO):-
    functor(X,(,),2),arg(1,X,X1),
    solvesub(X1,XX,R1,G1,G3,LO1,GO1),
    (LO1==1->(LO=1,RES=R1,G2=G3,GO=GO1,!);
    (arg(2,X,X2),solve(X2,R1,RES,G3,G2,LO,GO)).
solvesub(X,XX,RES,G1,G2,LO,GO):-
    loopcheck(X,G1,NUM),
    (NUM==1->(LO=1,RES=XX,G2=G1,GO=X,!);
    solveatom(X,XX,RES,G1,G2,LO,GO)).
solveatom(X,XX,RES,G1,G2,0,GO):-
    clause(X,true),RES=XX,G2=G1.
solveatom(X,XX,RES,G1,G2,LO,GO):-
    clause(X,Y),Y/==true,golist(X,XC),
    solve(Y,XX,RES,[XC|G1],G2,LO,GO).
solveatom(X,XX,RES,G1,G2,0,GO):-
    mfact(M1,0,X),constraint(hp(M1,0),XX),
    add(hp(M1,0),XX,RES),G2=G1.
solveatom(X,XX,RES,G1,G2,LO,GO):-
    mfact(M1,M2,(X:-Y)),M2/==0,
    constraint(hp(M1,M2),XX),
    add(hp(M1,M2),XX,RES1),golist(X,XC),
    solve(Y,RES1,RES,[XC|G1],G2,LO,GO).
```

In the above program, predicate *solve*, *solvesub* and
solveatom take the important role. The atom *solve(X,XX,
RES,G1,G2,LO,GO)* for goal X keeps a list of currently
used hypothesis numbers in *XX*, a list of currently solved
goals in *G1*. If there exists a refutation, then a list of
totally used hypothesis numbers and solved goals are
assigned to *RES* and *G2*, respectively. If there exists an
infinite loop for solving an atom q, then the *solve* assigns
LO=1 and *GO*=q. Namely, predicate *solve* dynamically
checks the constraints for hypotheses and infinite loops in
search. As for predicate *solvesub*, it mainly takes the role
of loop detection and it forces not to continue the resolution in case of infinite loop. The predicate *solveatom*
tries to solve the first argument X which is a subgoal and
an atom. In this case, it first tries to solve X without
using hypothesis. If there exists an input clause, then it
continues the resolution and calls *solve*. Otherwise, it
uses hypothesis to solve X. At the same time, *solveatom*
checks candidate hypothesis satisfies constraints or not.
Furthermore in *solveatom*, if the current subgoal is unified by using a rule, then the subgoal is added to the list of
solved goals. As for the *loopcheck(X,G1,NUM)*, X is a
current subgoal and *G1* is a list of solved subgoals. If
there exists a loop, then 1 is assigned to *NUM* else 0 is
assigned to *NUM*. The predicate *member* seems to be
useful for checking X is an element of *G1*. However, the

unification algorithm assigns values to variables in *X*, so we can not use *member* in this case. The *loopcheck* picks up a list of arguments in the same predicate atom in *G1*, and then it checks every corresponding arguments are both variables or both the same constant. If so, the *X* is an element of *G1* and the *loopcheck* knows an infinite loop occurs.

4.3. Real Execution in Example 3

Now, we show the real execution in Example 3.

```
yes
?-consult(prover.pl).
yes
?-consult(examp3.pl).
yes
?-init.
Executing...
Initialization complete!!
EXEC_time = 0.017(sec)
yes
?-listing(hypo).
mfact(1,0,p(a,delta(1,b))).
mfact(2,0,p(a,delta(2,c))).
    :   :   :
mfact(14,0,p(g,delta(2,g))).
mfact(15,1,(path(_408,_406):-p(_408,delta(_41
6,_406)))).
mfact(16,1,(path(_412,_410):-p(_412,delta(_42
0,_418)),path(_418,_410))).
yes
?-listing(ic).
ic([1,2]).
ic([3,4]).
    :   :   :
ic([13,14]).
yes
?-mayall(path(a,g)).
[1] Answer = path(a,g)
Hp 1:0 = p(a,delta(1,b))
Hp 3:0 = p(b,delta(1,d))
Hp 8:0 = p(d,delta(2,g))
[2] Answer = path(a,g)
    :   :   :
[6] Loop Detected at path(g,g)
Hp 1:0 = p(a,delta(1,b))
Hp 3:0 = p(b,delta(1,d))
Hp 7:0 = p(d,delta(1,f))
Hp 11:0 = p(f,delta(1,g))
Hp 14:0 = p(g,delta(2,g))
[7] Answer = path(a,g)
    :   :   :
[39] Loop Detected at path(c,g)
Hp 2:0 = p(a,delta(2,c))
Hp 6:0 = p(c,delta(2,c))
EXEC_time = 7.360(sec)
yes
```

```
?-sure(path(a,g)).
EXEC_time = 0.756(sec)
yes
?-sure(path(X,Y)).
[1] Answer = path(c,c)
[2] Answer = path(d,g)
[3] Answer = path(e,g)
[4] Answer = path(f,g)
[5] Answer = path(g,g)
[6] Answer = path(b,g)
EXEC_time = 6.587(sec)
X = X,
Y = Y
```

Now we refer to the above execution. We first consult prover and program expressing Example 3. Then, we initialize the program by *init* subcommand. After the initialization, the program examp3.pl and *init* is deleted and *mfact(1,0,p(a,delta(1,b)))* and others are asserted. Constraints *ic([1,2])* and others are also asserted. However, every user does not have to know these internal data structure. Then, we execute the query ←*mayall(path(a,g))*. The system responds 39 answers including 19 loop detections for this goal. The almost execution time is used to display answers. In every answer, we can see the selection of *s1* or *s2* as hypothesis. Namely, if we select these hypotheses, then we can go to place *g* from *a*. The response to the next query ←*sure(path(a,g))* is nothing, which implies the refutation failed. Finally, we execute ←*sure(path(X,Y))* which implies 'Find all two states where the selection of *s1* and *s2* does not affect the total path.' The system responds 6 answers for the most difficult query. The last answer *path(b,g)* implies that every path from place *b* surely goes to place *g*. We think that the above discussion will affect the areas like geometrical information systems, flight route planning systems, car navigation systems, etc. We think our interpreter will be a good tool for handling such problems with uncertainty.

4.4. Theoretical Aspects of Loop Detection Procedures

Now, we briefly discuss the theoretical aspects of loop detection procedures. We implemented above two procedures for escaping from the local stack overflow by infinite loops. Because, we could not use May and Sure-systems in case of infinite loop. The strategy (1) is clearly not proper. We can not show how we decide the available counts of every clause. On the other hand, the strategy (2) seems to be proper, namely this strategy seems to be complete for two logical consequences in May and Sure-system. However, we have not given the theoretical results in case of infinite loops. We need to refer the discussion in [TS86].

5. SOME APPLICATION AREAS AND PERSPECTIVE

Now in this section, we refer to the application areas of our interpreter and our perspective. We are now touching the following two applications.

(1) *Application to the inference engine for incomplete and fuzzy relational databases.*

(2) *Application to the consistent labeling problem on combinatorial analysis.*

As for the fuzzy relational databases, we can see every relation as a set of facts in logic programs. Therefore, we will easily be able to apply our framework to fuzzy relational databases. We can explicitly express the uncertain knowledge by using functor *set* in relational databases. In real, our interpreter is providing the external interpretation in [Li79]. We are going to realize some programs which simulate the relational algebra for fuzzy relational databases. We have another problem for fuzzy relational databases. We are now handling the functor *set* whose element is a discrete set. Therefore, the number of extension is absolutely finite. However, we may have a continuous set as an argument of functor *set*. In such cases, the number of extension is not finite and our prover can not deal with such problems. Namely, we have to improve our interpreter with functor *interval(minimum, maximum)*. As for the consistent labeling problem, the May-system finds the possible solutions and the Sure-system finds a special case in the May-system.

6. CONCLUSIONS

We first clarified the necessity for the rule of thumb. When we do not know the precise knowledge, we often use the rule of thumb for uncertain knowledge. Namely, we need a framework which can explicitly handle such uncertainty. For this issue, we introduced a functor *set* to the logic program and developed an interpreter which can understand the meaning of functor *set*. Then, we added infinite loop detection function to this interpreter. Without this function, our interpreter caused local stack overflow and we could not use May and Sure-systems. This function forces not to continue the resolution in case of infinite loop. We think this framework will be useful tool for making application systems and expert systems.

This work is partially supported by the Grant-in-Aid (No.07780338) in 1995, The Ministry of Education in Japan.

REFERENCES

[Co70] E.F.Codd. A Relational Model of Data for Large Shared Data Banks. *Communication of the ACM* 13:377-387, 1970.

[Ge87] A.V.Gelder. Efficient Loop Detection in Prolog Using the Tortoise-Hare Technique. Journal of the Logic Programming, 4:23-31, 1987.

[GM78] H.Gallaire and J.Minker editors. *Logic and Databases*, Plenum Press, 1978.

[Ko90] R.Kowalski. Problems and Promises of Computational Logic. *Computational Logic* (Ed. J.W.Lloyd), pp.1-34, Springer-Verlag, 1990.

[Lak93] L.Lakshmanan. Evolution of Intelligent Database Systems: A Personal Perspective. *Incompleteness and Uncertainty in Information Systems*, pp.189-208, Springer-Verlag, 1993.

[Li79] W.Lipski. On Semantic Issues Connected with Incomplete Information Databases. *ACT Transaction of Data Base Systems*, 4:262-296, 1979.

[Li81] W.Lipski. On Databases with Incomplete Information. *Journal of the ACM*, 28:41-70, 1981.

[LMR92] J.Lobo, J.Minker and A.Rajasekar. *Foundations of Disjunctive Logic Programming*, MIT press,1992.

[Lo87] D.Loveland. Near-Horn Prolog. In *Proceedings of Fourth International Conference on Logic Programming*, pp.456-469, MIT Press, 1987.

[MBP87] T.P.Martin, J.F.Baldwin and B.W.Pilsworth. The Implementation of FPROLOG. *Fuzzy Sets and Systems*, 23:119-129, 1987.

[MSD87] M.Mukaidono, Z.Shen and L.Ding. Fuzzy Prolog. In *Proceedings of Second IFSA Congress*, pp.844-847, 1987.

[Pa91] Z.Pawlak. *Rough Sets*. Kluwer Publishing,1991.

[PGA87] D.Poole, R.Goebel and R.Aleliunas. Theorist: A Logical Reasoning System for Defaults and Diagnosis. *The Knowledge Frontier: Essays in the Representation of Knowledge*, pp.331-352, Springer-Verlag, 1987.

[Rei80] R.Reiter. A Logic for Default Reasoning. *Artificial Intelligence*, 13:81-132, 1980.

[Sa94A] H.Sakai. On Semantics for the LPII(Logic Programming with Incomplete Information). *Transaction of Information Processing Society of Japan*, 35:706-713, 1994(in Japanese).

[Sa94B] H.Sakai. Prolog Extension Based on Another Fuzziness. In *Proceedings of the third International Conference on Fuzzy Logic, Neural Nets and Soft Computing*, pp.561-562, Fuzzy Logic Systems Institute, 1994.

[Sa94C] H.Sakai. Some Proof Procedures and Their Application for Realizing Another Fuzzy Prolog. *AI'94*, pp.283-290, World Scientific Publishing, 1994.

[TS86] H.Tamaki and T.Sato. OLD-resolution with Tabulation. In *Proceedings of the third International Conference on Logic Programming*, pp.84-98, Springer Verlag, 1986.

[Za84] C.Zaniolo. Database Relations with Null Values. *Journal of Computer and System Sciences*, 28:142-166, 1984.

CHANNEL ROUTING WITH CONSTRAINT LOGIC PROGRAMMING AND DELAY

Neng-Fa Zhou

Faculty of Computer Science and System Engineering
Kyushu Institute of Technology
Iizuka, Fukuoka, Japan, 820.
E-mail: zhou@mse.kyutech.ac.jp

ABSTRACT

Channel routing is a well-known NP-complete problem in VLSI design. The problem is to find routing paths among a group of terminals that satisfy a given connection requirement without overlapping each other. This problem can be regarded as a constraint satisfaction problem. For a HV channel where there is only one horizontal layer and one vertical layer, the problem can be described easily in finite-domain constraint logic programming languages. However, for a nHV channel where $n > 1$, the modelization is not so straightforward because some entailment constraints are involved. We use delay to implement the entailment constraints. The resulting program is very simple, but demonstrates good performance that is comparable to that of previous programs.

1. INTRODUCTION

VLSI layout design consists of two phases: the first phase, called *placement*, determines the positions of the modules on the VLSI chip, and the second phase, called *routing*, connects the modules with wiring. *Channel routing* is a kind of routing where the routing area is restricted to a rectangular channel. A channel consists of two parallel rows with terminals on them. A connection requirement, called a *net*, specifies the terminals that must be interconnected through a routing path. A channel routing problem is to find routing paths for a given set of nets in a given channel such that no paths overlap each other [Bur86]. There are a lot of different definitions of the problem that impose different restrictions on the channel and routing paths. Figure 1 shows an example. There are two nets in the problem, N_1 requiring t_1 (the first terminal of the top row) and b_3 (the third terminal of the bottom row) be connected, and N_2 requiring b_1 and

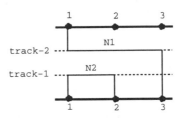

FIGURE 1   Single-layer channel.

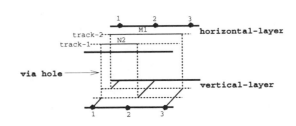

FIGURE 2   Two-layer channel.

b_2 be connected. The routing path for each net consists of one horizontal segment placed on some track and several vertical segments. In a single-layer channel, only planer problems are routable. For example, if N_2 requires b_1 and t_2 be connected, then the problem is unsolvable. In practice, there are multiple layers available in a channel for routing. Figure 2 shows a solution for the example in a two-layer channel.

The problem has been studied extensively in the VLSI design community since Hashimoto and Stevens [HS71] proposed it in 1971. Lapaugh has proved that the problem is NP complete [Lap80]. Many algorithms have been proposed for the problem [Deu76, FFL92, LSS94, Tak92, YK82]. Most of the traditional algorithms are based on graph algorithms. Recently, several modern heuristic search algorithms, for example, neural networks [Tak92], simulated annealing [BB88] and genetic algorithms [LSS94], have been proposed.

The channel routing problem is a finite-domain constraint satisfaction problem (CSP) [Mac86]. We concentrate on the *dogleg-free multi-layer* channel routing problem where the routing path for every net consists of only one horizontal line segment parallel to the two rows of the channel and several vertical line segments perpendicular to the two rows, and the routing area in the channel is divided into several pairs each of which consists of a horizontal layer for horizontal segments and a vertical layer for vertical segments. For each net, we need to determine the horizontal layer and the track on which the horizontal segment lays. The vertical layer for the net is determined automatically to be the one in the same pair as the horizontal layer. For this problem, each net to be routed can be treated as a variable whose domain is a set of all pairs of layers and tracks. To minimize the routing area means to to minimize the number of tracks.

Simonis [Sim90] has applied CHIP [DVS+88], a constraint logic programming language, to two-layer and three-layer channel routing problems where there is only one vertical layer involved. In [Zho94], we have implemented a forward checking algorithm that uses a special data structure called state tables. In this paper, we give a program for solving multilayer channel routing problems. The program uses the finite-domain constraint solving and delay facilities. It is very simple, but demonstrates good performance comparable to previous programs for the the Deutsch's difficult problem. Duchier and Huitouze have written a program [DH96] in CLP(FD) that is based on the same idea and demonstrates similar performance.

This paper is organized as follows: In Section 2, we define the channel routing problem in detail. In Section 3, we describe the program. In Section 4, we give the experimental results. In section 5, we compare our approach with other approaches and discuss the directions for improving the program.

2.   CHANNEL ROUTING

A *channel* consists of two parallel horizontal rows with terminals on them. The terminals are numbered 1, 2, and so on from left to right. A *net* is a set of terminals that must be interconnected through a *routing path*. The *channel routing problem* is to find routing paths for a given set of nets in a given channel such that no segments overlap each other, and the routing area and the total length of routing paths are mini-

$$N_1 = \{t_2, t_5\}$$
$$N_2 = \{b_1, b_6\}$$
$$N_3 = \{b_2, b_4\}$$
$$N_4 = \{t_3, t_9\}$$
$$N_5 = \{b_3, t_4, b_5\}$$
$$N_6 = \{t_6, b_7\}$$
$$N_7 = \{t_7, b_{11}\}$$
$$N_8 = \{b_8, b_{10}\}$$
$$N_9 = \{b_9, t_{10}, b_{12}\}$$
$$N_{10} = \{t_{11}, t_{12}\}$$

FIGURE 3   The set of nets in the example problem.

mized.

There are a lot of different definitions of the problem that impose different restrictions on the channel and routing paths. We consider the *dogleg-free multilayer* channel routing problem which impose the following three restrictions: First, the routing path for every net consists of only one horizontal segment that is parallel to the two rows of the channel, and several vertical segments that are perpendicular to the two rows. This type of routing paths is said to be *dogleg-free*. Second, the routing area in a channel is divided into several pairs of layers, one called a *horizontal layer* and the other called a *vertical layer*. Horizontal segments are placed in only horizontal layers and vertical segments are placed in only vertical layers. The ends of segments in a routing path are connected through *via holes*. There are several tracks in each horizontal layer. Minimizing the routing area means minimizing the number of tracks. Third, no routing path can stretch over more than one pair of layers. Thus, for each net, we only need to determine the horizontal layer and the track for the horizontal segment. The positions for the vertical segments are determined directly after the horizontal segment is fixed. In the following, we use nHV to denote a $2 \times n$-layer channel that has n pairs of horizontal and vertical layers.

For example, Figure 3 shows a set of nets. The terminals on the ith column of the top and bottom rows are denoted as t_i and b_i respectively. Figure 4 depicts a 2HV channel and the routing paths for the nets.

Two constraint graphs are created based on the given set of nets: one directed graph called a *vertical constraint graph* G_v and one indirected graph called a *horizontal constraint graph* G_h. In G_v, each vertex corresponds to a net and each arc from vertex u to vertex v means that net u must be placed *above* net v if they are placed in the same horizontal layer. The

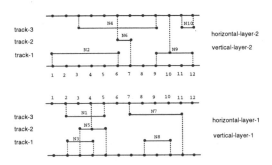

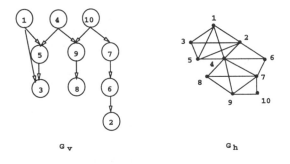

FIGURE 4   One solution for the example.

G_v   G_h

FIGURE 5   Constraint graphs.

relation above does not necessarily reflect the physical configuration of tracks. A track with number t_1th is said to be above another track with number t_2 in the same layer if t_1 is greater than t_2. In G_h, each vertex corresponds to a net and there is an edge between two vertices u and v if net u and net v cannot be placed on the same track. Figure 5 depicts the constraint graphs for the set of nets shown in Figure 3. There is an arc from vertex 1 to vertex 5 in G_v because t_5 is included in N_1 and b_5 is included in N_5. If N_5 is placed above N_1 or on the same track as N_1 in the same horizontal layer, then the vertical segments on the fifth column will overlap. There is an edge between vertex 1 and vertex 2 in G_h because the segment $(2,5)$ connecting the two farthest terminals in N_1 and the segment $(1,6)$ connecting the two farthest terminals in N_2 overlap each other. Notice that the relation above is not transitive unless there is only one vertical layer in the given channel. For example, in a 2HV channel, N_4 can even be placed below N_8 in the same horizontal layer if N_9 is placed in a different horizontal layer.

The *depth* of a net u in G_v is computed as follows: If u lies at the top of G_v, then u's depth is 0; otherwise, suppose u has n predecessors $v_1, v_2, \ldots, v_n$, then u's depth is $\max(\{d_{v1}, \ldots, d_{vn}\})$ where d_v denotes the

depth of v. There may exist cycles in G_v. In this case, all the vertices in a cycle have the same depth. The *length* of a routing path is the sum of the lengths of the horizontal and vertical segments in the path. For a horizontal segment whose left-most terminal number is l and whose right-most terminal number is r, the length of the segment is $r - l + 1$. Let t be the number of tracks in each horizontal layer. The length of a vertical segment between the ith track and the top row is $t - i + 1$ and the length of a vertical segment between the ith track and the bottom row is i.

3. PROGRAM

Channel routing problem is a CSP: Each net is treated as a variable whose domain is the set of all pairs of layers and tracks. The constraints are represented by the two constraint graphs G_v and G_h. Simonis [Sim90] has applied CHIP to the two-layer and three-layer channel routing problems where there is only one vertical layer involved. The relation *above* is represented as disequalities ($>$). However, as we have described in Section 2, the relation *above* is not transitive for general multi-layer channel routing problems and thus cannot be represented as disequalities. We present now a program in CLP(FD) for the problem that uses the corouting facility.

The formulation described above does not directly suit CLP(FD) because the domains in CLP(FD) are restricted to sets of atomic values. To make the formulation suitable to CLP(FD), we concentrate on the tracks and number them uniquely as 0, 1, and so on. In this way, we can still associate each net with a domain variable that indicates the global track number. After the global track number is determined, the layer and the track in the layer can be easily computed. Let L be the number of horizontal layers and T be the number of tracks in each horizontal layer. The domains of all variables are $0..L \times T - 1$. Let t be the global track number for a net. The layer is $V//T + 1$ and the track number in the layer is $t - (t//T) \times T + 1$, where the operator $//$ denotes integer division.

Figure   6   shows the program. The call `generate_vars(N,L,T,Vars)` generates N variables whose domains are $0..L \times T - 1$ and each of which corresponds to a net. The call `generate_constraints(Vars,T)` generates constraints among the domain variables. The call `label(Vars)` assigns values to variables.

It is very straightforward to generate horizontal constraints. For each pair of variables X and Y, if they cannot lie on the same track, then emit the inequality constraint $X \neq Y$.

```
route(N,L,T):-
    generate_vars(N,L,T,Vars),
    generate_constraints(Vars,T),
    label(Vars).

label([]):-!.
label(Vars):-
    choose(Vars,Var,Rest),
    indomain(Var),
    label(Rest).
```

FIGURE 6   Program in CLP(FD).

different_track(X,Y):-X≠Y.

However, generating vertical constraints is not so straightforward. Suppose there is an arc between two variables X and Y in Gv. The following entailment constraint declaratively specifies the above relation: "if X and Y lie in the same layer, then X must be greater than Y". Procedurally, the constraint can be described as follows:

```
delay above(X,Y,T) if var(X), var(Y).
above(X,Y,T):-
         nonvar(X),!
         Y ∉ X..(X//T)*T+T-1.
above(X,Y,T):-
         X ∉ (Y//T)*T..Y.
```

Figure 7 illustrates the relation. If both X and Y are variables, then delay the constraint; if X has gotten a value, then Y cannot be routed in the shadow area $X..(X//T) * T - 1$; if Y has gotten a value, then X cannot be routed n the shadow area $(Y//T) * T..Y$.

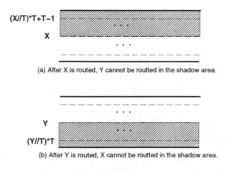

(a) After X is routed, Y cannot be routted in the shadow area.

(b) After Y is routed, X cannot be routed in the shadow area.

FIGURE 7   The constraint X *is above* Y

4.   EXPERIMENTAL RESULTS

In this subsection, we present the results for the Deutch's problem obtained with B-Prolog and compare them with the best results known now.

4.1   B-PROLOG

B-Prolog is an emulator-based Prolog system. Its performance is comparable in general to that of emulated SICStus-Prolog (version 2.1), a commercial system developed at Swidish Institute of Computer Science. For a group of search programs that do a lot of backtracking, B-Prolog is about forty percent faster than emulated SICStus-Prolog.

The finite-domain constraint solver is mostly written in canonical-form Prolog where input and output unifications are separated and determinisms of clauses are denoted explicitly. The performance of the constraint solver is better than that of clp(FD) (version 2.2), a system developed at INRIA, and Eclipse (version 3.5.1), a system developed at ECRC. For the 64-queens problem, B-Prolog takes 0.8 second on a SPARC-10, whereas Eclipse takes 1.6 seconds and clp(FD) takes 2.8 seconds on the same computer.

4.2   PROGRAMS AND BENCHMARKS

The program can minimize the number of tracks and the total length of routing paths by using branch & bound. It is only around 350 lines long excluding comments, blanks, the data for the nets, and the code for displaying solutions.

The Deutsch's difficult problem is used as the benchmark. The benchmark suite given in [YK82] are well used in the VLSI design community, among which the Deutsch's difficult problem is a representative one. The problem is to route a set of 72 nets on a channel where there are 174 terminals on each row. There are 117 arcs in the constraint graph G_v and 846 edges in G_h.

4.3   HEURISTICS

The order in which variables are instantiated can affect the efficiency of the algorithm dramatically. We use the following rules to choose a variable.

1. Choose first a variable whose corresponding net lies at the bottom in G_v.

2. Choose first a variable with the smallest domain.

3. Choose first a variable whose corresponding net has the greatest degree in G_v.

	HV	2HV	3HV	4HV
Initial bound	40	20	14	10
Best solution	28	11	7	5

Table 1   The best solutions found in five minutes.

4. Choose first a variable whose corresponding net has the greatest degree in G_h.

5. Choose first a variable whose corresponding net lies at the bottom of G_v.

The first rule ensures that the nets at the bottom of G_v are routed before those above them. All the other rules are consistent with the *first fail principle* [Hen889]. Choosing first a variable that has the smallest domain and participates in the largest number of constraints can usually make a failure occur earlier. The second rule is only used in the forward checking algorithm.

4.4   SOLUTIONS

We have run the programs on a SPARC-10 many times by asking them different questions.

Question 1:

What solutions for HV, 2HV, 3HV and 4HV channels can be found in five minutes that require the minimum numbers of tracks.?

Table 1 shows the answers. The row *Initial bound* depicts the initial bound on the number of tacks, and the row *Best solution* gives the best solution obtained in five minutes.

We have tried several combinations of rules for choosing variables. The combination 1-5-3-4 demonstrates the best performance for all the programs and all the types of channels. The best HV solution obtained is known to be optimal in terms of the number of tracks [KSP73]. The optimal solutions for the remaining types of channels have not yet been reported.

The CHIP program described in [Sim90] found an optimal HV solution for the same problem in less than 30 seconds. Takefuji's programs found the same best solutions. The router described in [FFL92] found in less than one second a solution for 2HV that requires 10 tracks, but it does not require segments in a routing path to be in only one pair of layers.

Question 2:

Are the best solutions optimal? The program failed to prove the optimality of the solutions in 12 hours.

HV	2HV	3HV	4HV
1.9	1.8	2.0	26.1

Table 2   The times (in seconds) required to find the first best solutions.

HV	2HV	3HV	4HV
5954	4102	3567	3364

Table 3   The lengths of the best solutions found in one hour.

Question 3:

How many seconds does it take to find the first best solution for each type of channel? Table 2 shows the answers. The heuristics 1-5-3-4 is used.

Question 4:

What shortest solutions can be found in one hour? Table 3 shows the lengths of the solutions. Figures 8 to 9 show the solutions for HV and 4HV channels.

FIGURE 8   Number of tracks = 28, Length=5954

5.   CONCLUSION

This this paper, we described a program for the channel routing problem in a finite-domain constraint programming language. There have been a huge number of algorithms proposed to solve the channel routing problem. Recent algorithms tend to be very complicated and thus are very difficult to implement. Furthermore, when the restrictions on the channel or routing paths change, the algorithms must be redesigned. Compared with these traditional algorithmic approaches, our approach is declarative and very simple. The program can be easily adapted to other types of routing problems by modifying the definitions of domains, constraints and heuristics.

The program can be improved in several directions. Firstly, the current program only use general heuristics for ordering variables. It would be more efficient

FIGURE 9   Number of tracks = 5, Length = 3364.

to use some problem specific heuristics used in traditional algorithms. For example, such information about nets concerning the lengths of nets, types of nets (two-terminal nets, multi-terminal nets, nets connecting only terminals at the top, nets connecting only terminals at the bottom, etc.) can be used to choose variables. Secondly, the program can be improved by introducing heuristics for choosing appropriate values for selected variables. These two improvements should be justified by experiments. For this purpose, a large number of benchmark problems must be tested. Thirdly, the program can be executed in parallel on a multi-processor computer or a network of computers. Parallel search is a promising technique that can be used to find good solutions and prove optimality of solutions.

ACKNOWLEDGEMENTS

The idea originated from my early work described in [Zho94] and discussion with Dr. S.L. Huitoze and Dr. D. Dechier who contributed the key idea of using delay.

REFERENCES

[BB88]   Brouwer, R.J. and Banerjee, P., A Parallel Simulated Annealing Algorithm for Channel Routing on a Hypercube Multiprocessor, in: *IEEE Int. Conf. Comput. Design*, 1988, 4-7.

[Bur86]   Burstein, M, Channel Routing in:*Layout Design and Verification*, North-Holland, 1986, 133-167.

[DH96]   Duchier, D. and Huitouze, S.L.: Channel Routing with CLP(FD), submitted to PACT'96, 1996.

[Deu76]   Deutsch, D.N., A Dogleg Channel Router, in: *Proc. 13th Design Automation Conference*, 1976, 425-433.

[DVS+88] Dincbas, M., Van Hentenryck, H., Simonis, H., Aggoun, A., Graf, T., and Berthier, F., The Constraint Logic Programming Language CHIP, in: *Proc. FGCS'88*, 1988, 693-702.

[FFL92]   Fang, S.C., Feng, W.S., and Lee, S.L., A New Efficient Approach to Multilayer Channel Routing Problem, in: *Proc. of the 29th ACM/IEEE Design Automation Conference*, 1992, 579-584.

[HS71]   Hashimoto, A and Stevens, S., Wire Routing by Optimizing Channel Assignment within Large Apertures, in: *Proc. 8th Design Automation Workshop*, 1971, 155-169.

[KSP73]   Kernighan, B.W., Schweikert, D.G., and Persky, G., An Optimum Channel-routing Algorithm for Polycell Layouts of Integrated Circuits, in: *Proc. 10th Design Automation Workshop*, 1973, 50-59.

[Kum92]   Kumar, V., Algorithms for Constraint Satisfaction Problems: A Survey, in: *AI Magazine*, 1992,32-44.

[Lap80]   LaPaugh, A.S., *Algorithms for Integrated Circuits Layout: An Analytic Approach*, PhD Dissertation, MIT Lab. of Computer Science, 1980.

[LSS94]   Liu, X., Sakamoto, A. and Shimamoto, T., Genetic Channel Router, in: *IEICE Trans. Fundamentals*, 1994, E77-A:492-501.

[Mac86]   Mackworth, A., Constraint Satisfaction, in: *Encyclopedia of Artificial Intelligence*, John Wiley & Sons, 1986, 205-211.

[Sim90]   Simonis, H., *Channel Routing Seen as a Constraint Problem, Tech. Rep.*, TR-LP-51, ECRC, Munich, July, 1990.

[Tak92]   Takefuji, , *Neural Network Parallel Computing*, Kluwer Academic Publishers, 1992.

[Hen889]   Van Hentenryck, P., *Constraint Satisfaction in Logic Programming*, MIT Press, 1989.

[YK82]   Yoshimura, T. and Kuh, E.S., Efficient Algorithms for Channel Routing, in: *IEEE Trans. CAD*, 1:25-35 (1982).

[Zho94]   Zhou, N.F., A Logic Programming Approach to Channel Routing, in: *Proc. 12th International Conference on Logic Programming*, MIT Press, 159-173, 1994.

DESIGN OF DATABASE INTERFACE TO ILP FOR BUILDING KNOWLEDGED BASE

Keiko Shimazu and Koichi Furukawa,
Graduate School of Media and Governance, Keio University,
Endo 5322, Fujisawa, Kanagawa, Japan
kshimazu@mag.keio.ac.jp, furukawa@sfc.keio.ac.jp

ABSTRACT

This paper presents a framework of a machine learning system based on an experimental study using Progol, an inductive logic programming (ILP) system developed by Muggleton et al. The main purpose of our framework is to establish technologies for applying ILP to knowledge discovery in databases; i.e. to generate rules for describing the databases. This paper reports our successful result in characterizing specific shapes which are hidden in international three dimensional geometric data. We focussed on three important issues:

(1) how to extract appropriate positive examples from raw data,

(2) how to generate appropriate negative examples,and

(3) how to generate appropriate background knowledge.

In particular, we found that both near miss data in the negative examples and super concepts in background knowledge are critical for making obtaining appropriate hypotheses in a reasonable computation time. We also found a way to define a finite search space by raw input data.

1. INTRODUCTION

One of the most important issues in machine learning is to apply its techniques to knowledge acquisition. A new school of machine learning called inductive logic programming (ILP) has been recently growing and its new techniques have been applied many to problems in various fields such as medicine design, genome analysis, medical diagnosis and so on [Lav 93, Fay93]. Further more, because of employing background knowledge, ILP may represent a breaking of the "Feigenbaum" bottleneck of knowledge

acquisition or knowledge discovery.

However, it is not easy for naive users, to apply the system to their own problems.

In this paper, we consider an example of finding certain shapes from geometric data using one of the most powerful inductive logic programming systems, Progol[Mug 91]. We discuss various issues concerning the preparation of appropriate input data to Progol. The issues include data representation, the selection of a relevant data set, the generation of good negative examples, and the choice of appropriate background knowledge.

We performed two experiments. Both of them were for characterizing particular shapes international three dimensional geometric data. In the first experiment, negative examples representing non target shape were generated from the raw data. In the second experiment, negative examples were automatically generated from positive examples. We should mention that the latter is much more relevant for obtaining the target hypothesis than the former.

2 A BRIEF INTRODUCTION OF PROGOL

Progol employs PROLOG language interpreter for writing and testing logic programs. In addition, PROLOG programs can be automatically generated from examples using the inductive algorithm built into Progol [Mug 92].

In the following we show how to represent input data to Progol, using a family relationships example. This example learns the grandfather concept. In a mode declaration, the target concept predicates and the predicates for defining the target concepts, are declared differently. The former predicate(s) will appear in the head of a

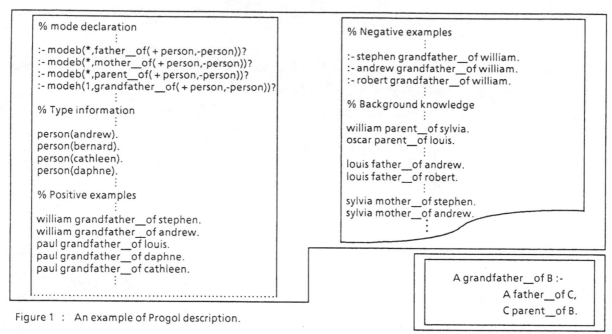

Figure 1 : An example of Progol description.

Figure 2 :   Example of target hypothesis description in Progol.

target hypothesis, whereas the latter will appear in the body. In this example, the predicate "grandfather__of" is declared as a head-predicate and the three predicates, "father__of", "mother__of" and "parent__of", are declared as body predicates. These declarations are designated respectively by "modeh" and "modeb". Each predicate can have both input and output variables, "+" or "-". Type information is used to define a domain of each variable type. If an instance of a clause belongs to the corresponding type, it can be used either as input data or output data. A positive example is given by a statement such as

william grandfather__of sylvia.
A negative example is given by a statement such as

:- stephen grandfather__of william.
Both of them are statements about head predicates.
Background knowledge is given by a set of statements using body predicates;

sylvia mother__of stephen (see Figure 1).
Then the target hypothesis is inferred in PROLOG language description (see Figure 2).

3   EXPERIMENTAL STUDY ON IGES DATA

Initial Graphic Exchange Specification (IGES)

format data were used as real data. The format is an international 3D format and it is adopted by such application software as Intergraph [5]. In the IGES database, it is easy to identify primitive shapes, (100 for circle, 110 for line, and so on), by their numbers (as shown in Figure 3).

However, if any composite shapes are constructed by a set of lines, there is no way to identify what kind of shape the data represent, unless we compute the geometry of the data. Especially, in the three dimensional representation, it is hard to identify shapes by manual operation with eye inspection without a precise rotation operation.

Machine learning systems, especially ILP, have the potential to automatically identify shapes from 3D raw data. In our experiment the shape of diamond was inferred successfully.

4   INPUT DATA DESIGN

Progol can not handle with raw data in any database. This was our motivation for developing an Progol interface to database. Artificial raw data clause, positive examples and negative examples are generated automatically. Mode declarations , type information and components descriptions, which are used for describing the target hypothesis, are input by manual. In particular, the direct handling with raw data is

```
Entities          110    1   3   1   0           LINE    9D  44
                  110   25   0   1   1   0   0   001010000D  45
                  110    1   3   1   0           LINE   10D  46
                  110   26   0   1   1   0   0   001010000D  47
                  110    1   3   1   0           LINE   11D  48
                  102   27   0   1   1   0   0   000000001D  49
                  102    1   3   1   0          COMP CRV  3D  50
                  100   28   0   1   1   0  53   000000000D  51
                  100    1   3   1   0          CIRCLE   1D  52


3D representation  110, -28.27776, 0., 49.14188, -28.27776, 0., 17.36911;      33P  19
and                110, -28.27776, 0., 17.36911, 35.90323, 0., 17.36911;       35P  20
Decimal fractions  110, 35.90323, 0., 17.36911, 35.90323, 0., 49.14188;        37P  21
                   110, 35.90323, 0., 49.14188, -28.27776, 0., 49.14188;       39P  22
                   102, 4, 33, 35, 37, 39;                           41P  23
                   110, 28.84261, 0., -15.97039, 5.49617, 0., -56.40762;       43P  24
                   110, 5.49617, 0., -56.40762, 52.18906, 0., -56.40762;       45P  25
                   110, 52.18906, 0., -56.40762, 28.84261, 0., -15.97039;      47P  26
                   102, 3, 43, 45, 47;                             49P  27
                   100, 0., 0., 0., 15., 0., 15., -0.;                  51P  28
```

Figure 3 :   IGES format data example

4.1 PREPARATION OF ARTIFICIAL RAW DATA

In our experiment, three dimensional data in IGES were used. Ideally, it is desirable to extract "real" examples from a given IGES database. Unfortunately, we did not have any IGES database having "oracle" information specifying a target shape.

An obvious solution to this problem is to ask IGES database creators to add oracle type information to sets of points, as comments. An alternative solution is to develop a user interface program displaying 3D objects with suitable rotation operations to make it possible for users identify target objects by visual inspection.

In this experimental study, we assumed oracle type information and made an artificial IGES database, which we called as an intermediate file. This is an "artificial raw database" with three parts: artificial raw data clauses to be used in background knowledge, mode declaration clauses, and type information clauses (see Figure4).

Line clauses were generated automatically from a real database (see 1st step in Figure 5).

realized by using artificial raw data, mode declaration and type information.

4.2 GENERATING POSITIVE EXAMPLES

In our artificial raw data, through the "oracle" information some line data are identified as "diamond". To be more concrete, positive examples were generated automatically by extracting four line clauses following the "diamond" tag (see the 2nd step in Figure5).

```
% Mode declarations
:- modeh(1, diamond ( + node, + node, + node, + node))?
:- modeb(*, quadrangle ( + node, + node, + node, + node))?
:- modeb(*, distance ( + node, + node, -number))?

% Type information
node([X,Y,Z]) :- line([X,Y,Z,__,__,__]).
node([X,Y,Z]) :- line([__,__,__,X,Y,Z])
% BK – artificial raw data clauses
line([x1, y1, z1, x2, y2, z2]).
line([x2, y2, z2, x3, y3, z3]).
line([x3, y3, z3, x4, y4, z4]).
line([x4, y4, z4, x1, y1, z1]).

line([x11, y11, z11, x22, y22, z22]).
line([x22, y22, z22, x33, y33, z33]).
line([x33, y33, z33, x44, y44, z44]).
line([x44, y44, z44, x11, y11, z11]).
```

Figute 4 : Mode declarations , Type information & raw data

4.3 GENERATING APPROPRIATE NEGATIVE EXAMPLES

Since it is empirically known that near miss data have high relevancy as negative examples, the most critical issue is how to generate near miss data from positive examples [Lav 91]. In our real data, there are a lot of negative data which were not diamond but were still made of four lines. These could be used as negative examples, but it would be easy to have too many inappropriate negative clauses.

To obtain such near miss data, our negative examples were made of positives by modifying only one of the four elements (see the 3rd step in Figure 5). Our procedure of generating appropriate negative examples is as follows. Note that this is presented assuming two dimensional data for better understanding. First, negative examples must have the same structures as positives. As our positive examples in two dimensional presentation have four arguments:

diamond([10,5],[8,8],[11,10],[13,7]).,

negative examples must have exactly the same structure:

:- diamond([x1,y1],[x2,y2],[x3,y3],[x4,y4]).

Secondly, we change exactly one argument of a positive example so as to generate a negative example. (see Figure 6).

4.4 BACKGROUND KNOWLEDGE AS INTERFACE TO DATABASE

Background knowledge defines components predicates in inductive logic programming.

All clauses for representing components were generated by manual at the time of preparation of the intermediate file. In our interface, type information says that all arguments, which will be used for each predicates described in mode

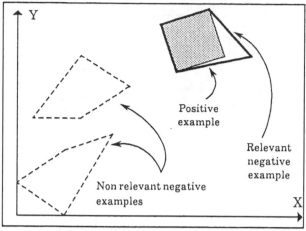

Figure 6: Generating relevant negative example

declaration, refer to an artificial raw data. Therefore, it provides a mechanism for defining a finite search space in terms of database (see Figure 4).

5 A RESULT OF OUR EXPERIMENTAL STUDY

We inferred the target concept :
if a quadrangle has nodes , A, B, C and D,
and four square distances of A to B, B to C,
C to D and D to A are same,
then it must be a diamond,
as in Figure 7.

6 RELEVANCY OF NEGATIVE EXAMPLES

We performed a comparative study to judge relevancy of negative examples for two cases : negative examples extracted randomly from raw data and these generated by modifying positives.

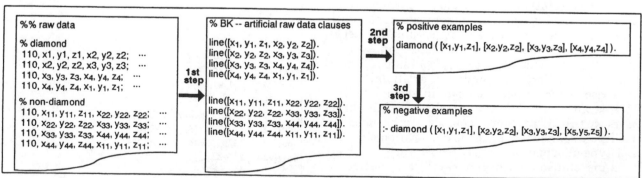

Figure 5 : How to generate intermediate file and input data to ILP.

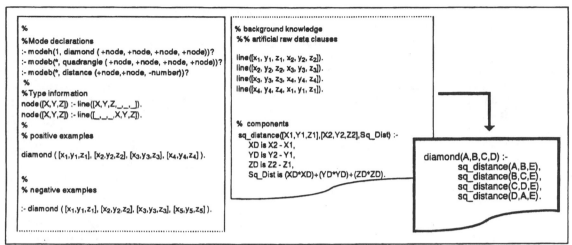

Figure 7 : Inferring target hypothesis.

6.1 NEGATIVE EXAMPLES EXTRACTED RANDOMLY FROM RAW DATA

Our intermediate file involved many non-diamond examples, even though they were made by four lines, as we mentioned earlier. We tried to use these data as negative examples to perform a comparative experimental study. It means that all negative examples were not created from positive examples, but extracted from raw data.

In other words, negated formulas were generated from intermediate file directly as negative examples. This means that all non-diamond shapes, which were made by 4 lines, were transformed to negative example clauses.

As a result, in order to set the target hypothesis, we needed over 50 negative examples (see Figure 8).

6.2 NEGATIVE EXAMPLES GENERATED BY MODIFYING POSITIVE EXAMPLES

One of our interests was the relevancy of negative examples. In this experimental study, all our negative examples were generated from positive clauses, by changing only one argument .

As a result, in order to get the target hypothesis we needed to generate around 25 negative examples (see Figure 8 again). W e note that if we ant o reduce computational complexity, then data relevancy is very important.

7   CONCLUSION

In our experiment, we tried to establish a design principle for supplying appropriate input data to an ILP system, Progol, from a database.

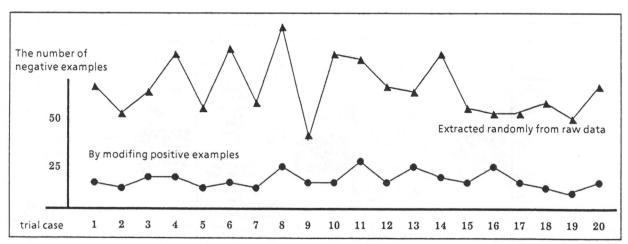

Figure 8   :   How many negative examples are needed

We noted the importance of finding an appropriate method of generating positive examples from the database. Usually, a database has no "oracle" for a machine learning system, so it is important to know how to add an interface to generate the intermediate file. Our interface generated the intermediate file as an artificial raw data. By extracting clauses as related raw data to background knowledge, and by setting type information clause and mode declaration, we realized a filter from the raw data to positive examples.

Generating negative examples is another critical issue. We were able to show the importance of generating near miss negative examples from the given positive examples.

Finally, we demonstrated the importance of defining a finite domain in terms of all the constants appearing in a given database. We successfully defined it by using the type information and mode declaration of Progol. This report concerns the first step of constructing a general interface between a database and an ILP system. Such an interface has a big possibility of generating a knowledge base cheaply from a huge database (see Figure9).

ACKNOWLEDGEMENTS

We would like to express special thanks to Alan Robinson. This study owes a great deal to his valuable comments. Also, we would also like to thank Yoshiko Katoh for her helpful work.

REFERENCES

[Mug 91]   Mggleton, S. : Inductive Logic Programming. New generation Computing, 8(4), pp. 295-317 (1991). Programming. New generation Computing, 8(4), pp. 295-317 (1991).

[Mug 92]   Muggleton, S., Srinivasan, A. and Brain, M. : Compression, significance and accuracy. In proceedings of the 9th International Workshop on Leanings, pp. 338-347, Morgan Kaufmann (1992).

[Lav 91]   Lavrac, N., Dzeroski, S. and Grobelnik, M. : LEARNING NONRECURSIVE DEFINITIONS OF RELATIONS WITH LINUS, Proceedings of Machine Learning - EWSL-91, European Working Session on Learning, Lecture Notes in Artificial Intelligence, Subseries of Lecture Notes in Computer Science, Edited by J. Siekmann, Spring - Verlag (1991).

[Lav 93]   Lavrac, N., Dzeroski, S., Pirnat, V. and Krizman, V. : The use of background knowledge in learning medical diagnostic rules. Applied Artificial Intelligence 7, pp. 273-293 (1993).

[Fay 93]   Fayyad, U., and Uthurusamy, R. : Progress Challenges in Knowledge Discovery in Databases, Proc. KDD-95

[Ester 95]   Ester, M., Kriegel, H. and Xu, Xiaowei : A Databese interface for Clustering in Large Spatial Databases. In proceedings of the !st international Conference on Knowledge Discovery and Data mining, pp. 94-99 (1995).

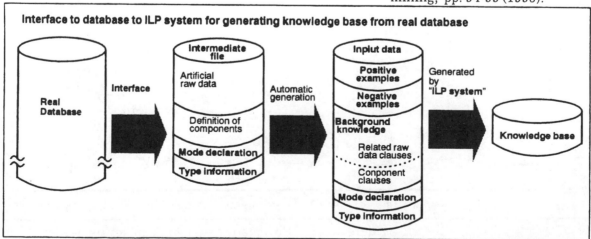

Figure 9 :   Interface to ILP .

AN INCREMENTAL SOLVER FOR CONSTRAINT HIERARCHIES OVER REAL INTERVALS

Duong Tuan Anh
Regional Computer Center

Kanchana Kanchanasut
Computer Science Program,

Asian Institute of Technology
GPO Box 2754 Bangkok 10501, Thailand
Email: anh@rccsun.ait.ac.th

ABSTRACT

This paper presents an incremental solver for constraint hierarchies over real intervals. This solver makes use of a technique to deal with constraint deletion in Constraint Satisfaction Problems (CSPs) over real intervals in which we avoid resolving the remaining system from scratch. The solver also imports some techniques from dependency-directed backtracking to determine which constraints should be relaxed from an over-determined constraint hierarchy so that the remaining constraint system can be solvable.

1. INTRODUCTION

The well-known shortcomings of exact value constraint reasoning in dealing with inequalities, non-linear constraints, inexact data or in solving under-constrained or over-constrained problems lead to the advent of *interval constraint reasoning*. Interval constraint satisfaction (ICS) provides a more general framework for CSPs. ICS has been applied in many fields, such as temporal, physical, and spatial reasoning (see the survey [Dav87], financial planning [Hyv91], constraint-based design [HFH90]. Some authors, such as Davis [Dav87], Hyvonen [Hyv89], Sidebottom & Havens [SiH92], Benhamou & Older[BeO93], and Lhomme [Lho93] developed arc-consistency (AC) algorithms for CSPs where the variable domains are real intervals. These algorithms make use of Interval Arithmetic [Moo66] to compute the refined domains.

Many real life problems involve reasoning in dynamic environments. For example, in an constraint-based design process, the designer may add more constraints to specify further the problem, or relax some constraints when there is no solution. Every time a constraint has been added or removed we need to check if there still exists solutions in the CSP. We can modify the traditional AC-3 algorithm for static CSPs over real intervals to become the algorithm to handle the addition of a constraint [BeO93]. However, the main difficulty in developing AC algorithm for dynamic CSPs is the problem involving the deletion of a constraint: how to avoid resolving the remaining system from scratch when a constraint is removed from the CSP.

In [AnK95] we proposed an efficient algorithm for constraint deletion where re-execution from scratch can be avoided. The method we use in the algorithm is based on the observation that deletion of a constraint from a CSP has only local effects on the constraint network.

Real dynamic CSPs are usually *over-constrained*, and a best answer a constraint solver could give is a partial solution satisfying all but the least important constraints. Preferences among constraints may be declared explicitly by associating a strength denoting the degree of requirement to non-required constraints, in contrast with required constraints that are mandatory. A theory of *constraint hierarchy* is given in [BoM88]. *Constraint relaxation* is a commonly used method to manage over-determined constraint hierarchies. The technique for constraint deletion in [AnK95] is intended to be applied in an incremental algorithm to solve constraint hierarchies over real intervals. This solver also needs a strategy to determine which constraints should be relaxed from an over-determined constraint hierarchy so that the remaining constraint system can become solvable. The strategy we use here is somewhat similar to *dependency-directed-backtracking* [StS77].

The paper is organized as follows. Section 2 reviews some basic definitions and previous results. Section 3 describes an incremental algorithm to solve constraint hierarchies over real intervals. Related research is briefly discussed in section 4. Section 5 contains a summary and some final remarks.

2. BASIC DEFINITIONS AND PREVIOUS RESULTS

In this section we recall some basic definitions and previous results on arc-consistency for Dynamic CSPs over real intervals.

2.1 Real Intervals

A *real interval* is specified by a *lower* and an *upper* bound. A bound consists of a real numeral and a bracket symbol. A square bracket indicates that the bound is closed and a round bracket indicates the bound is open. For instance, $[0,1]$ denotes the set $\{ x \mid 0 \leq x \leq 1 \}$ and $[0,1)$ denotes $\{ x \mid 0 \leq x < 1 \}$.

2.2 Constraint Satisfaction Problems over Real Intervals

A *static* CSP involves a set of *variables* V and a set of *constraints* S. Each variable X in V takes value in its respective real *domain* D_X. A constraint specifies which values from the domains of its variables are compatible. Here constraints can be in form of equations and inequalities. A solution of a CSP will be the assignment of the form: $x_1 \leftarrow I_1, x_2 \leftarrow I_2, ...x_i \leftarrow I_i, ...$where each I_i is a real interval.

Let v(C) denote the set of variables in constraint C. A useful function for describing AC algorithms over real domains is *projection*, denote $\prod$, which maps a constraint C and a variable $X \in v(C)$ to a subset of D_X, the domain of X. $\prod_X(C)$ is the set of values for X which is consistent with the constraint C and with all the domains of the other variables. Given a constraint C and a related variable X, we can compute the projection of C on X using *Interval Arithmetic* with some generalizations [SiH92].

In a *dynamic* CSP, we can add a new constraint to the constraint network (*a restriction*) or delete an old constraint from it (*a relaxation*) at any time. DACR, a new arc-consistency algorithm we have proposed for solving Dynamic CSPs over real intervals is given in [AnK95]. The algorithm DACR consists of two procedures: one for adding a new constraint to and another for deleting an old constraint from a constraint system.

3. AN INCREMENTAL SOLVER FOR CONSTRAINT HIERARCHIES OVER REAL INTERVALS

A constraint hierarchy H is a set of constraints which is divided into partitions $H_0,...,H_n$ such that all constraints in H_i are preferred over those in H_{i+1} and the constraints in H_0 are required to be satisfied. The preference strength of each partition is called *preference level*.

A solution to a constraint hierarchy H, is a mapping of the variables in H to their respective real domains, that satisfies all the required constraints (H_0). Given two solutions θ and σ, θ is *locally-predicate-better(lpb)* than σ if a) θ and σ both satisfy exactly the same number of constraints in each level until level k, and b) in level $k+1$ θ satisfies more constraints than σ.

3.1 An Incremental Algorithm for Solving Constraint Hierarchies over Real Intervals

An incremental constraint solver has to check arc-consistency of a constraint hierarchy which often evolves gradually. On the insertion of a new constraint, the solver uses procedure *Add* (in [AnK95]) to maintain arc-consistency of the system. If the system is consistent, the solver can go forward. Otherwise it has to deal with a conflict before it can go forward. There are three steps in dealing with a conflict on the insertion of a new constraint into a constraint hierarchy:

1. Identify the group of non-required constraints which are pertinent causes of the conflict.

2. Relax the least important constraint in this group using the technique of constraint deletion in [AnK95]. If the relaxation still can not remove the conflict, relax the second least important constraint

and so on until the system (including c) becomes arc-consistent. If all constraint relaxations do not solve the conflict, the newcomer constraint should be removed.

3. For each relaxed constraint c_r, any constraints which were previously relaxed due to adding c_r can become the next constraints to be inserted again to the system.

In order to identify the possible causes of a failure in the process of propagation we have to develop a technique which is somewhat similar to *dependency-directed-backtracking* developed by Stallman & Sussman [SuS77]. During execution we have to record *dependencies* among active constraints in the system. When the real interval of a certain variable becomes empty, using the recorded dependencies we can detect all the activated constraints which directly or indirectly contribute to that failure. These constraints form a group of possible causes of the conflict.

Definition 1 (Narrower). The active constraint c is a *narrower* of T ($T \in v(c)$) if it actually caused the reduction of D_T.

Definition 2 (Child /Father constraint). c_k is *a child constraint of* c_j (conversely c_j is an *father constraint* of c_k), written $c_j \rightarrow c_k$, iff $\exists T \in v(c_k)$ s.t. c_j is a narrower of T.

Property 1. $c \rightarrow c$ *iff c is a narrower.*

We can keep track of the relation $\rightarrow$ among the constraints since this relation is induced by the local propagation. A directed graph can be constructed to represent this relation among the constraints.

Definition 3 (Descendant/Antecedent constraint). c_k is *a descendant constraint of* c_j (conversely c_j is a *antecedent constraint of* c_k), *iff* $c_j \rightarrow^* c_k$, $\rightarrow^*$ *is a transitive closure of* $\rightarrow$.

Given a system S and a constraint $c \in S$, we can compute the set of all antecedent constraints of c:

$$FT(c) = \{c_j \in S \mid c_j \rightarrow^* c \}$$

The set FT(c) represents the constraints that directly or indirectly narrowed the domains of the variables of c.

In a process of arc-consistency, if some variable of constraint c gets an empty domain, the pertinent causes of this failure is the set of all antecedent constraints of c.

The algorithm for inserting a new constraint to a constraint hierarchy over real domains is given in Algorithm 3.1. It can deal with conflicts that may occur on the insertion of a new constraint c to the constraint hierarchy H. The variables *Satisfied*, *Unsatisfied* and *Untried* denote respectively the set of satisfied constraints, the set of relaxed constraints and the set of untried constraints. And *OS(c)* denotes the set of all constraints which had been relaxed to make way for inserting c to the constraint system.

In Algorithm 3.1, *Relax_Set* denotes the set of candidate constraints for relaxation. The set contains only the non-required constraints belonging to $FT(c_j)$ and less preferred than the newcomer constraint c because relaxing a constraint of the same strength would lead to a different solution but not better (*lpb*) solution, and relaxing a stronger constraint would lead to a worse solution. It's obvious that any solution generated by Algorithm 3.1 is a *lpb* solution to the current constraint hierarchy.

Notice that if we have more than one candidate constraints with the same strength for relaxation, we must choose any one of them, say c1, to be relaxed and establish a *choice point* where another candidates can be chosen later to derive alternative best solutions. That is the responsibility of three procedure calls *check-alternative(c1)* in the above algorithm. The established choice points will help to search alternative solutions later. (In the cases we concern ourselves to find only one best solution for the constraint hierarchy, we should remove the three procedure calls *check-alternative(c1)* out of the above algorithm.)

If all the constraints in *Relax_Set* were removed but we still can not insert the non-required constraint c to H, c itself should be relaxed. In this rare case, the algorithm transfers all the relaxed constraints due to c to the *Untried* set so that they can be re-inserted afterwards.

The computation overhead for dealing a conflict in Algorithm 3.1 depends on the cardinality of *Relax_Set*. Notice that this set is empty in the cases that the constraints in a hierarchy are processed in the

Algorithm 3.1

procedure Insert_to_hierarchy(c,H,Satisfied,Unsatisfied,Untried)
/* c is a constraint from Untried Set */
begin
  if Add(c,H) is not successful **then**   /* the addition of c leads to conflict */
    Let c_j be the constraint
       whose narrowing process makes D_v of some $v \in v(c_j)$ empty.
    Compute the set $FT(c_j)$, the set of all antecedent constraints of c_j.

    $Conf\_Set := \{c_k \in FT(c_j) \mid level(c_k) > 0\}$
    $Relax\_Set := \{c_k \in FT(c_j) \mid level(c_k) > 0 \wedge level(c_k) > level(c)\}$

    if Conf_set is not empty and Relax_set is empty **then**
      Unsatisfied := Unsatisfied $\cup$ { c };
      check-alternative(c);
    else
      Forward := false; Removed = $\emptyset$;
      while Relax_Set is not empty $\wedge$ not Forward **do**
      begin
        Let c_r be the least important constraint in Relax_Set.
        Relax_Set := Relax_Set\$\{c_r\}$; Satisfied = Satisfied\$\{c_r\}$
        Unsatisfied := Unsatisfied $\cup$ { c_r };
          Removed := Removed $\cup$ { c_r };
          Delete(c,H);
          if Add(c,H) is successful **then** Forward := true;
      end;
      if not Forward **then**   /* c is unsatisfied */
        if c is required **then halt** /* inform "no admissible solution" */
        else
          Unsatisfied := Unsatisfied $\cup$ { c };
          check-alternative(c);
          Untried := Untried $\cup$ Removed; /* insert all constraints in Removed to Untried Set */
        endif
      else
        OS(c) := Removed;
        forall $c_r \in$ Removed **do**
        begin
          Untried := Untried $\cup$ OS(c_r);
          check-alternative(c_r);
        end
      endif
    endif
  endif.
end.

descending order of importance (i.e. stronger constraints come to the system first). This observation can help to keep the overhead low by controlling the order in which the constraints are inserted to the hierarchy. The more we can maintain the order of inserting constraints close to the ideal order, the less computation overhead we will have for the Algorithm 3.1.

3.2 Obtaining Alternative solutions

The established choice points build up *a tree* representing a solving process. There are two steps for finding one alternative solution.

(a) Go back to the most recent choice point, choose another candidate constraint to be relaxed and insert the previously chosen candidate to Untried set.

(b) Go forward until the *Untried* set become empty.

When all candidates at a choice point have been tried, we can remove this choice point out of the tree. Then go up to the next choice point to do the same two steps to obtain other alternative solutions. When all the choice points in the tree are removed, we obtain all the solutions of a hierarchy.

A prototype implementation of the solver has been written in C about 1200 lines and experimented with example hierarchies consisting of linear constraints on real intervals. The experimental results are quite promising with respect to its performance efficiency.

4. RELATED WORK

There have been so few efficient algorithms for solving constraint hierarchies incrementally. The Delta-Blue algorithm developed by Freeman-Benson, Maloney & Borning [FMB90] is the first incremental solver for constraint hierarchies. It directs a dataflow graph by which we can know for each variable which constraint determines its value, and deduces a straightforward sequence of function applications which will satisfy as many constraints as possible. But this algorithm achieves exact-value constraint solving, limits constraints only in form of equations and it can not solve the under-constrained hierarchies -the number of constraints is less than that of variables. To our knowledge, extending

Delta-Blue algorithm to be a solver for hierarchies over real intervals is a very hard problem.

Hua et al.[HFH90] suggested a Dynamic constraint satisfaction algorithm over real domains as the central part of a constraint-based design system. This paper presents a special case of constraint solving over real domains: *directed constraint network*. There exists a partial ordering on the variables of the system which makes each constraint to be a causal relationship among its variables and enables interval propagation to flow only in one direction. This algorithm also uses an ordering on constraints similar to constraint hierarchy and adopts a scheme of constraint relaxation similar to the scheme in our paper.

Menezes & Barahona [MeB93] also use the same *dependency-directed-backtracking* technique in their incremental method to solve hierarchies of constraints over finite domains. But apart from this, there are some major differences between the solver in [MeB93] and the solver in this paper:

(1) Our solver uses the newcomer constraint as a factor to determine the set of candidate constraints to be relaxed to make way for it, whereas their solver does not.

(2) Their solver removes all the constraints which are possible causes of a conflict in a batch at every backward phase while our solver tries to minimize the number of relaxed constraints. Our solver does not relax more constraints than necessary.

(3) In their solver, when a constraint c is relaxed, all the constraints inserted in the active store after c and *related to* c will be removed from the active store to *untried store* so that they can be re-inserted to the system afterwards. This reset step makes their solver more expensive in space and time. Due to the special feature of our constraint deletion technique, our solver does not require this constraint reset step.

(4) In fact there is *no backward phase* in our solver when dealing with a conflict. Whenever there is a conflict, our solver just deletes some constraints, if necessary, and then go ahead.

(5) Due to the constraint reset step in backward phase, much more trailing and untrailing work is required

in their solver than in ours.

5. CONCLUSIONS

We have presented our constraint deletion algorithm and its application in an incremental solver for constraint hierarchies over real intervals with locally-predicate-better comparator. This incremental solver uses dependency-directed backtracking technique to deal with conflicts and adopts a simple scheme to search multiple solutions. We have also discussed some implementation issues of the solver. A prototype implementation of the solver has been written in C and experimented. The experimental results show that the solver is sound and quite promising with respect to its performance. However, the algorithm complexity still needs to be fully assessed.

The proposed solver can be used as the kernel of a Hierarchical Constraint Logic Programming (HCLP) system over real intervals and with the comparator *locally-predicate-better*. Our further research will also involves developing an HCLP interpreter over real intervals.

REFERENCES

[AnK95] D.T.Anh & K. Kanchanasut. *Arc-consistency for Dynamic Constraint Satisfaction Problems over Real Intervals.* Proc. IEA 95 AIE, Melbourne, Australia, June 5-9, 1995, pp247-254.

[BeO93] F. Benhamou & W.J. Older. *Applying Interval Arithmetic to Real, Integer and Boolean Constraints.* Technical Report, Bell Northern Research, Computing Research Laboratory, Ottawa, Ontario, Canada.

[BoM88] A. Borning & M. Maher. *Constraint Hierarchies and Logic Programming.* Tech. Report 88-11-10, Comp. Sc. Dept., Univ. of Washington, November 1988.

[Dav87] E. Davis. *Constraint Propagation with Interval Labels.* Artificial Intelligence, Vol 32, pp 281-331.

[FMB90] B.N. Freeman-Benson, M. Maloney, & A. Borning. *An Incremental Constraint Solver.* Com.

ACM, Vol. 23, No.1, pp 54-63.

[HFH90] K.Hua, B. Faltings & D. Haroud. *Dynamic Constraint Satisfaction in a Bridge Design System.* Proc. Expert Systems in Engineering: Principle and Applications, Int. Workshop, Vienna, Austria, Sept. 1990.

[Hyv89] E. Hyvonen. *Constraint Reasoning based on Interval Arithmetic.* Proc. IJCAI'89, pp 1193-1198.

[Hyv91] E. Hyvonen. *Interval Constraint Spreadsheets for Financial Planning.* Proceedings of First International Conference on Artificial Intelligence Applications on Wall Street IEEE Press, New York.

[Lho93] O. Lhome. *Consistency Techniques for Numeric CSPs.* Proc. IJCAI'93, pp 232-238.

[Mac77] A.K. Mackworth. *Consistency in Networks of Relations.* Artificial Intelligence, Vol. 8, pp 99-118.

[MeB93] F. Menezes & P. Barahona. *Preliminary Formalization of an Incremental Hierarchical Constraint Solver.* EPIA'93, Porto, Portugal, Oct.

[Moo66] R.E. MOORE. *Interval Analysis.* Englewood Cliffs, New Jersey, Prentice-Hall.

[SiH92] G. Sidebottom & W.S.Havens, *Hierarchical Arc Consistency for Disjoint Real Intervals in Constraint Logic Programming.* Proc. Workshop on Constraint Logic Programming Systems: Design and Applications, Washington D.C., USA, Nov. 1992.

[StS77] J.M. Stallman & G.J. Sussman. *Forward Reasoning and Dependency-Directed Backtracking in a System for Computer-Added Circuit Analysis.* Artificial Intelligence, Vol. 9, No.2, pp 135-196.

[WiB93] M. Wilson and A. Borning. *Hierarchical Constraint Logic Programming.* J.Logic Programming, Vol. 16, pp 227-318.

Machine Learning

COIN CLASSIFICATION USING A NOVEL TECHNIQUE FOR LEARNING CHARACTERISTIC DECISION TREES BY CONTROLLING THE DEGREE OF GENERALIZATION

Paul Davidsson

Department of Computer Science, Lund University
Lund, Sweden, S–221 00
Email: Paul.Davidsson@dna.lth.se

ABSTRACT

A novel method for learning characteristic decision trees is applied to the problem of learning the decision mechanism of coin-sorting machines. Decision trees constructed by ID3-like algorithms are unable to detect instances of categories not present in the set of training examples. Instead of being rejected, such instances are assigned to one of the classes actually present in the training set. To solve this problem the algorithm must learn characteristic, rather than discriminative, category descriptions. In addition, the ability to control the degree of generalization is identified as an essential property of such algorithms. A novel method using the information about the statistical distribution of the feature values that can be extracted from the training examples is developed to meet these requirements. The central idea is to augment each leaf of the decision tree with a subtree that imposes further restrictions on the values of each feature in that leaf.

1 INTRODUCTION

One often ignored problem for a learning system is how to know when it encounters an instance of an unknown category. In many practical applications it cannot be assumed that every category is represented in the set of training examples (i.e., they are open domains [Hut94]) and sometimes the cost of a misclassification is too high. What is needed in such situations is the ability to reject instances of categories that the system has not been trained on.

In this article we will concentrate on an application concerning a coin-sorting machine of the kind often used in bank offices.[1] Its task is to accept and sort (and count) a limited number of different types of coins (for instance, a particular country's), and to reject all other coins. The vital part of the machine is a sophisticated sensor that the coins pass one by one. The sensor measures electronically five properties (diameter, thickness, permeability, and two kinds of conductivity) of each coin, which all are given a numerical value. Based on these measurements the machine decides of which type the current coin is: if it is of a known type of coin, it is sorted, otherwise it is regarded as an unknown type and is rejected.

The present procedure for constructing the decision mechanism is carried out mostly by hand. A number of coins of each type are passed through a sensor and the measurements are recorded. The measurements are then analyzed manually by an engineer, who chose a minimum and a maximum limit for each property of each type of coin. Finally, these limits are loaded into the memory of the machine. When the machine is about to sort a new coin, it uses the limits in the following way: if the measurement is higher than the minimum limit and lower than the maximum limit for all properties of some type of coin, the coin is classified as a coin of this type. If this is not true for any known type of coin, the coin is rejected.

Thus, in the present method, which is both complicated and time consuming, it is the skill of the engineer that decides the classification performance of the coin-sorting machine. Moreover, this procedure must be carried out for every new set of machines (e.g., for each country's). In addition, there are updating problems when a new kind of coin is introduced. This is not only applicable when a new denomination is introduced, or when the appearance of an old denomination is changed. It is, for instance, not unusual that the composition of the alloy is changed. In fact, this happens often undeliberately as it is difficult to get exactly the same composition every time and, moreover, there are sometimes trace elements of other metals in the cauldron. Another kind of problems comes from the fact that it is difficult to make all the sensors exactly alike. As a consequence of this and the fact that all machines used for the same set of coins use the same limits, each sensor must be calibrated. Moreover, this calibration is not always sufficient and ser-

---

[1] The work presented in this paper has in part been carried out in collaboration with Scan Coin AB (Malmö, Sweden), a manufacturer of such machines.

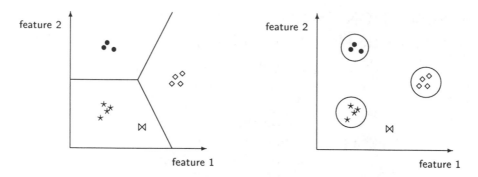

Figure 1: Discriminative (left) versus characteristic (right) category descriptions.

vice agents must sometimes be sent out to adjust the limits on particular machines.

However, if we construct a method which automatically learns the decision mechanism, it would be possible to bring down the effects of these problems. In short, the task to be solved can be described as follows: develop a system that computes the decision mechanism of a coin-sorting machine automatically from a number of examples of coins.

2 CHARACTERISTICS OF THE APPLICATION

To begin with, this is a case of learning from examples in that we can safely assume that there is a number of coins with known identities which can be used to train the system. Moreover, as the working of the machine has been described above, we can clearly divide it into two separate phases, the initial training phase and the classification phase. Thus, there is no need to make demands on incrementality upon the learning algorithm.

There are also some demands on performance. The method must be time-efficient both in the learning phase, since the long-term goal is to let the end users themselves adapt the machine to a set of coins of their choice and we cannot expect the bank clerks to be willing to spend much time training the machine, and in the classification phase, since the machine must be able sort approximately 800 coins per minute. Moreover, as the current hardware is rather limited the method also must be memory-efficient. However, as the hardware probably will be upgraded in the near future, we will not bother much about the space problem. The demands on classification performance that must be met by the learning system before it can be used in a real world application are rather tough: not more than 5% of known types of coins are allowed to be rejected and very few misclassifications (less than 0.5%) of any type of coins are accepted. Consequently, more than 99.5% of unknown types of coins should be rejected.

Another aspect of our learning task concerns a problem that often has been ignored in the machine learning research. Namely, how the learning system should "know" when it encounters an instance of an unknown category. It is for practical reasons impossible to train the learning system on every possible kind of coin (genuine or faked). Thus, we must assume that the system can be trained only on the types of coins it is supposed to accept. As been pointed out by Smyth and Mellstrom [SM92], the only way of solving this problem is to learn *generative*, or *characteristic*, category descriptions that try to capture the similarities between the members of the category. This, in contrast to learning *discriminative* descriptions that can be seen as representations of the boundaries between categories. The difference between these kinds of descriptions is illustrated in Figure 1. It shows some instances of three known categories (⋆, •, and ◇), and examples of possible category boundaries of the concepts learned by a system using discriminative descriptions (to the left) and by a system using characteristic descriptions (to the right). In this case, a member of an unknown category (⋈) will be categorized wrongly by a system using discriminative descriptions, whereas it will be regarded as a member of a novel category by a system using characteristic descriptions.

3 RESULTS FROM A PRELIMINARY STUDY

A preliminary study was carried out by a Master's student (Mårtensson [Mar94]). Three methods were evaluated: induction of decision trees using the ID3 algorithm [Qui86], learning neural networks by the backpropagation algorithm [RHW86], and computing Bayesian classifiers [TG74]. However, since all these methods in their original versions learn discriminative descriptions, they must be modified in order to learn characteristic descriptions. For instance, in the ID3 algorithm each leaf of the decision tree was augmented with a subtree in order to impose further restrictions on the feature values. A lower and an upper limit are computed for every feature. These will serve as tests:

if the feature value of the instance to be classified is below the lower limit or above the upper limit for one or more of the features, the instance will be rejected, i.e., regarded as belonging to a novel class, otherwise it will be classified according to the original decision tree. Thus, when a new instance is to be classified, the decision tree is first applied as usual, and then, when a leaf would have been reached, every feature of the instance is checked to see if it belongs to the interval defined by the lower and the upper limits. If all features of the new instance are inside their interval the classification is still valid, otherwise the instance will be rejected. In this method the lower and upper limits were assigned the minimum and maximum feature value of the training instances of the leaf respectively. Since this approach will yield a *maximum specific description* (cf. Holte et al. [HAP89]), we will refer to it as ID3-Max.

The main result of Mårtensson's study was that the performance in terms of classification accuracy of three methods was almost equivalent. However, as it became clear that the company wanted explicit classification rules (e.g., for service and maintenance reasons), the neural network approach was given up. It also became apparent that the method must be able to learn disjunctive concepts (i.e., classes that correspond to more than one cluster of training instances in the feature space). In addition to the fact that the actual appearance of a denomination change now and then (and often both are valid as means of payment), angular coins will be in different positions in the sensor resulting in different values in the diameter, and, as mentioned earlier, the alloy is different in different batches. These irregularities will tend to divide the measurements of one type of coin into two or more separate clusters. This was one of the reasons why the Bayesian classifier approach was given up. Another reason was that it does not provide explicit minimum and maximum limits for each parameter, which is what should be programmed into the decision mechanism of the machine.

Since the decision tree approach (1) by far was the most time-efficient in the classification phase, (2) was reasonable fast in the learning phase, (3) learned explicit rules (and provides explicit min and max limits for each parameter), and (4) was good at learning disjunctive concepts, it was selected as the most promising candidate for solving the problem.

Although the ID3-Max algorithm showed promising results, an improvement was desired for mainly two reasons: Firstly, the ID3-Max algorithm is not sufficiently robust, e.g., an extremely low (or high) value of one parameter of one coin in the training set could ruin the whole decision mechanism. This problem is related to the problem of noisy instances which will be discussed in the last section of this paper. Secondly, and perhaps more important, it is not possible to control the degree of generalization (i.e., the po-

sition of the limits). For example, too many coins were rejected as the ID3-Max algorithm did not generalize sufficiently. For these reasons, a more dynamic and robust method for computing the limits has been developed.

4 THE NOVEL LEARNING TECHNIQUE

The novel technique developed to solve these problems is also based on the ID3 algorithm and constructs subtrees out of lower and upper limits in a way similar to ID3-Max. The central idea of the method is to make use of statistical information concerning the distribution of the feature values of the instances in the leaf. For every feature we compute the lower and the upper limits so that the probability that a particular feature value (of an instance belonging to this leaf) belongs to the interval between these limits is $1 - \alpha$.

In this way we can control the degree of generalization by choosing an appropriate α-value. The lesser the α-value is, the more will the algorithm generalize. For instance, if it is important not to misclassify instances and a high number of rejected (not classified) instances are acceptable, a high α-value should be selected.

If X is normally distributed stochastic variable, we have that:

$$P(m - \lambda_{\frac{\alpha}{2}}\sigma < x < m + \lambda_{\frac{\alpha}{2}}\sigma) = 1 - \alpha$$

where m is the mean, σ is the standard deviation, and some common values of λ are:

$$\lambda_{0.05} = 1.6449, \quad \lambda_{0.025} = 1.9600, \quad \lambda_{0.005} = 2.5758$$

Thus, we have, for instance, that the probability of an observation being larger than $m - 1.96\sigma$ and smaller than $m + 1.96\sigma$ is 95%.

In order to follow this line of argument we have to assume that the feature values of each category (or each leaf if it is a disjunctive concept) are normally distributed. This assumption seems not too strong for most applications. However, as we typically cannot assume that the actual values of m and σ are known, they have to be estimated. A simple way of doing this is to compute the mean and the standard deviation of the training instances (belonging to the current leaf):

$$m^* = \frac{\Sigma x_i}{n}, \quad \sigma^* = \sqrt{\frac{\Sigma(x_i - \overline{x})^2}{n-1}}$$

To get a nice interpretation of the interval between the upper and lower limit, we have to assume that these estimates are equal to the actual values of m and σ. This is, of course, too optimistic, but it seems reasonable to believe (and will be shown in Section 6) that the method is of practical value also without this interpretation. Anyway, the intended statistical interpretation suggests that the probability of a feature of an instance of a category being larger than lower limit and smaller than upper limit for $\alpha = 0.01$ is 99%.

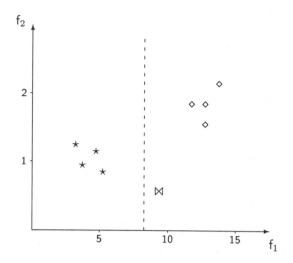

Figure 2: The feature space of the example. The boundary between the two categories ($\star$ and $\diamond$) induced by the ID3 algorithm is represented by the vertical dashed line.

5 A SIMPLE EXAMPLE

We will now present a very simple example to illustrate the method. All instances are described by two numerical features, and the training instances belong to either of two categories: the $\star$-category or the $\diamond$-category. The system is given four training instances of each category.

The feature values of the training instances of the $\star$-category are: (3.0, 1.2), (3.5, 0.9), (4.5, 1.1), (5.0, 0.8) and the $\diamond$-category training instances are: (11.5, 1.8), (12.5, 1.5), (12.5, 1.8), (13.5, 2.1). Figure 2 shows the positions of the instances in the feature space.

If these training-instances are given to the ID3 algorithm, the output will be the decision tree shown in Figure 3. (Or a similar one, depending on the cut-point selection strategy. In all examples presented here the cut-point is chosen by first sorting all values of the training instances belonging to the current node. The cut-point is then defined as the average of two consecutive values of the sorted list if they belong to instances of different classes.) This tree represents the decision rule: if $f_1 \leq 8.25$ then the instance belongs to the $\star$-category, else it belongs to the $\diamond$-category. The classification boundary that follows from this rule is illustrated in Figure 2 by a vertical dashed line. If we now apply the deci-

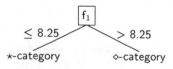

Figure 3: The decision tree induced by the ID3 algorithm.

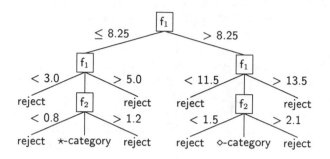

Figure 4: The decision tree induced by ID3-Max.

sion tree to an instance of another category ($\bowtie$) with the feature values (9.0,0.5), it will be (mis)classified as an instance of the $\diamond$-category.

Let us apply the method based on the maximum specific description to this problem. Given the training instances it will produce the decision tree in Figure 4 which will reject all instances outside the dotted boxes in Figure 5.

If we now apply the augmented decision tree to the $\bowtie$-category instance, we first use the decision tree as before resulting in a preliminary classification which, still as before, suggests that it belongs to the $\diamond$-category. However, as we proceed further down the tree into the appended subtree, we will eventually encounter a test that brings us to a reject-leaf (i.e., we check whether the new instance is inside the dotted box, and find out that it is not). As a consequence, the instance is rejected and treated as an instance of a novel, or unknown, category.

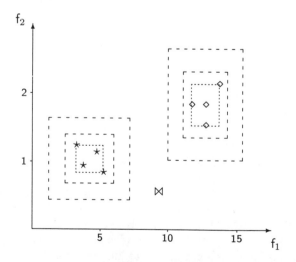

Figure 5: The acceptance regions of the maximum specific description (dotted boxes) and those resulting from the method based on statistical distribution with $\alpha = 0.05$ (inner dashed boxes), and with $\alpha = 0.001$ (outer dashed boxes).

	Canadian Coins			Foreign Coins		
	correct	miss	reject	correct	miss	reject
ID3	99.7%	0.3%	0.0%	0.0%	100.0%	0.0%
ID3-Max	83.7%	0.0%	16.3%	0.0%	0.0%	100.0%
ID3-SD 0.1	62.1%	0.0%	37.9%	0.0%	0.0%	100.0%
ID3-SD 0.05	77.5%	0.0%	22.5%	0.0%	0.0%	100.0%
ID3-SD 0.01	92.2%	0.0%	7.8%	0.0%	0.0%	100.0%
ID3-SD 0.001	97.9%	0.0%	2.1%	0.0%	0.0%	100.0%
ID3-SD 0.0001	98.9%	0.0%	1.1%	0.0%	0.0%	100.0%
desired	100.0%	0.0%	0.0%	0.0%	0.0%	100.0%

Table 1: Result from training set containing Canadian coins (averages over 10 runs).

If we apply the method based on statistical distribution with $\alpha = 0.05$, the lower and upper limits will have the following values: for the $\star$-category 2.2 (f_1 lower), 5.8 (f_1 upper), 0.6 (f_2 lower), and 1.4 (f_2 upper), and for the $\diamond$-category 10.9, 14.1, 1.3, and 2.3 respectively. These limits will yield a decision tree with the same structure as that of the maximum specific method but with different values on the rejection branches, and will cover the areas marked by the inner dashed boxes in Figure 5. Such an area, which we will call *acceptance region*, can be interpreted as meaning that, if the assumptions mentioned above were correct and if the features are independent, 90.2% (0.95×0.95) of the instances of the category are inside the box. Just as with the maximum specific tree this tree will reject the $\bowtie$-category instance. We can also see that the lesser α-value that is chosen, the more will the algorithm generalize. The outer dashed boxes correspond to the acceptance region for $\alpha = 0.001$, i.e., 99.8% of all instances of the category are inside the region.

6 TEST RESULTS

In our experiments two databases were used, one describing Canadian coins contains 7 categories (1, 5, 10, 25, 50 cent, 1 and 2 dollar), and one describing Hong Kong coins that also contains 7 categories (5, 10, 20, 50 cent, 1, 2, and 5 dollar). All of the 5 attributes (diameter, thickness, conductivity1, conductivity2, and permeability) are numerical. The Canada and Hong Kong databases were chosen because when using the company's current method for creating the rules of the decision mechanism, these coins have been causing problems.

In each experiment 140 (7×20) instances were randomly chosen for training and 700 ($2 \times 7 \times 50$) instances for testing. This scenario is quite similar to the actual situation where you in the training phase expose the system only to the coins

of one country, but in the classification phase also confront it with coins of other countries. Each experiment was performed with the original ID3 algorithm, the maximum specific tree algorithm (ID3-Max), and the algorithm based on statistical distribution (ID3-SD) for the α-values: 0.1, 0.05, 0.01, 0.001, and 0.0001.

Table 1 shows the classification results when training on the Canadian coin database. We can see that all foreign coins (i.e., Hong Kong coins) are rejected, except of course for the ID3 algorithm. Neither were there any problems with misclassifications. In fact, the Canadian coins that are misclassified by the ID3 algorithm are rejected by the algorithms learning characteristic descriptions. However, the requirements of classification accuracy (less than 5% rejects of known types of coins and very few misclassifications (less than 0.5%)) are met only by the ID3-SD algorithm with $\alpha = 0.001$ and 0.0001, which illustrates the advantage of being able to control the degree of generalization.

In Table 2 the results when training on the Hong Kong coin database are shown. As indicated by the percentages of misclassifications of known types of coins, this is a more difficult problem. Although there are two α-values (0.001 and 0.0001) that meet the requirements, they are very close to the acceptable number of rejects (0.001) and misclassifications (0.0001) respectively. Results from experiments using other databases can be found in Davidsson [Dav95].

A potential problem for the ID3-SD algorithm is when the training set consists of only a few training instances of one or more of the categories (or clusters). One would think that when the number of training examples of a category decreases there is a risk that the estimates of the mean value and the standard deviation (which are fundamental for computing the acceptance regions) will not be sufficiently good. However, preliminary experiments indicate that the classification performance decreases only slowly when the training

	Hong Kong Coins			Foreign Coins		
	correct	miss	reject	correct	miss	reject
ID3	98.3%	1.7%	0.0%	0.0%	100.0%	0.0%
ID3-Max	79.7%	0.0%	20.3%	0.0%	0.0%	100.0%
ID3-SD 0.1	60.0%	0.0%	40.0%	0.0%	0.0%	100.0%
ID3-SD 0.05	74.8%	0.0%	25.2%	0.0%	0.0%	100.0%
ID3-SD 0.01	88.9%	0.0%	11.1%	0.0%	0.0%	100.0%
ID3-SD 0.001	95.1%	0.3%	4.6%	0.0%	0.0%	100.0%
ID3-SD 0.0001	96.3%	0.5%	3.2%	0.0%	0.0%	100.0%
desired	100.0%	0.0%	0.0%	0.0%	0.0%	100.0%

Table 2: Result from training set containing Hong Kong coins (averages over 10 runs).

examples get fewer. As can be seen in figure 6, it handles the problem of few training instances much better than the maximum specific description approach which, in fact, has been suggested as a solution to the related problem of small disjuncts (cf. Holte et al. [HAP89]). In the coin classification domain small disjuncts could, for instance, be caused by irregularities in the composition of the alloys.

7   NOISY AND IRRELEVANT FEATURES

The ID3-SD algorithm is better at handling noisy data than the ID3-Max algorithm in the sense that an extreme feature value for one (or a few) instance(s) will not influ-

ence the positions of the limits of that feature in ID3-SD as much as it will in ID3-Max. This is illustrated in Figure 7 where a single instance with a single noisy feature value corrupts the acceptance region of the Max-algorithm whereas the acceptance region of the SD-algorithm is affected to a lesser extent.

A method for further reducing the problem of noisy instances, would be to use the acceptance regions to remove instances that are (far) outside their acceptance region and then recalculate the region. For instance, if we remove the noisy instance in the figure and recalculate the acceptance region, we get the region shown in the right picture in the figure. This method for *outlier detection* is currently under evaluation.

However, in this paper we have used the traditional ID3 algorithm as a basis, an algorithm that is not very good at handling noisy data in the first place. In fact, there is a triv-

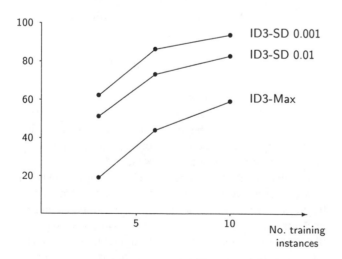

Figure 6: The percentage of correctly classified instances of known categories (Canadian coins) as a function of the number of instances of each category in small training sets (averages over 10 runs). The remaining instances were rejected.

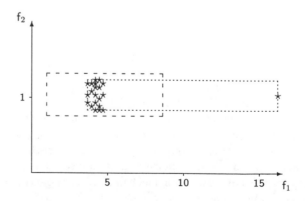

Figure 7: A category with twenty good and one noisy instance. The maximum specific description correspond to the dotted box. The dashed box correspond to the description resulting from the ID3-SD algorithm with $\alpha = 0.1$.

ial solution to the problem with noisy data: Use a pruning method (cf. Mingers [Min89]) to cut off the undesired branches, or use any other noise tolerant algorithm (e.g., C4.5 [Qui93]) for inducing decision trees, and then compute the subtrees as before for the remaining leaves.

Moreover, the original ID3-algorithm is quite good at handling the problem of irrelevant features (only features that are useful for discriminating between the categories in the training set are selected). But since the suggested method computes upper and lower limits for every feature and use these in the classification phase, also the irrelevant features will be subject for consideration. However, this potential problem will typically disappear since an irrelevant feature often is defined as a feature which value is randomly selected according to a uniform distribution on the feature's value range (cf. Aha [Aha92]). That is, the feature values have a large standard deviation, which will lead to a large gap between the lower and the upper limit. Thus, as most values of the feature will be inside the acceptance region with regard to this feature, the feature will still be irrelevant for the classification.

8   THE GENERALITY OF THE SD APPROACH

The main limitation of the SD approach seems to be that it is only applicable to numerical attributes. The maximum specific description method, on the other hand, requires only that the features can be ordered. Thus, one way of making the former method more general is to combine it with the latter method to form a hybrid approach that is able to handle all kinds of ordered features. We would then use the statistical method for numerical attributes and the maximum specific description method for the rest of the attributes. Moreover, nominal attributes could be handled by accepting those values present among the instances of the leaf and reject those that are not. In this way we get a method that learns characteristic descriptions using all kinds of attributes. However, the degree of generalization can, of course, only be controlled for numeric features.

The statistically based approach for creating characteristic descriptions is a general method in the sense that we can take the output from any decision tree induction algorithm, compute a subtree for every leaf, and append them to their leaf. In fact, the approach can, in principle, be applied to any empirical learning method, supervised or unsupervised, using the hybrid approach. However, if the instances of a category corresponds to more than one cluster in the feature space (cf. disjunctive concepts), the method will probably work better for algorithms that explicitly separates the clusters, i.e., where it is possible to find out which cluster a particular instance belongs to. If this is the case, the acceptance regions can be computed separately for each cluster. Otherwise, we must compute only one acceptance region for the

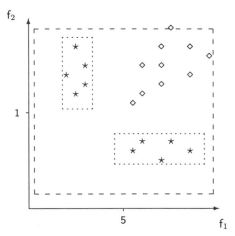

Figure 8: The acceptance regions for the $\star$-category computed by ID3-SD (dotted boxes) and by an algorithm that does not explicitly separate clusters of category instances (dashed box). ($\alpha = 0.05$).

whole category, which probably will result in a too large acceptance regions (see Figure 8).

The problem with disjunctive categories is also a partial answer to the question: Why bother building a decision tree in the first place? Could we not just compute the lower and the upper limits for every category and test unknown instances against these? The main problem with such an approach would be that when a new instance is to be classified, it might be assigned to two or more classes. The reason for this ambiguity is, of course, that the acceptance regions of two or more categories may overlap. Moreover, even if the regions do not overlap, there will be problems dealing with disjunctive concepts for the reason mentioned above. Thus, in either case, we must have an algorithm that is able to find suitable disjuncts of the concept (which, in fact, is an unsupervised learning problem), a task that ID3-like algorithms normally are quite good at. However, Van de Merckt [dM93] has suggested that for numerical attributes a similarity-based selection measure is more appropriate for finding the correct disjuncts than the original entropy-based measure used in the empirical evaluations presented here.

The procedure for augmenting an arbitrary empirical learning algorithm X is as follows: train X as usual, then compute the limits for every category (i.e., cluster) in the training set as described earlier. When a new instance is to be classified, first apply X's classification mechanism in the same way as usual, then check that all features values of the new instance are larger than the lower limit and smaller than the upper limit. Thus, it is not necessary to represent the limits in the form of decision trees, the main point is that there should be a method for comparing the feature values of the instance to be classified with the limits.

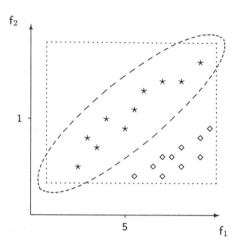

Figure 9: The acceptance regions for the ⋆-category computed by IB1-SD (dotted boxes) and by an algorithm that takes into account covariation among features (dashed ellipse). ($\alpha = 0.05$).

To illustrate the generality of the SD approach some experiments with its application to a nearest neighbour algorithm very similar to IB1 [AKA91] are described in Davidsson [Dav96]. Moreover, if we are not restricted to representing concepts by decision trees, we are not forced to have separate and explicit limits for each feature. As a consequence, we do not have to assume that features are independent. If we are able to capture covariation among two or more features we would be able to create acceptance regions that closer match the distribution of feature values (i.e., regions that are smaller but still cover as many instances). This is illustrated by the example shown in Figure 9. The acceptance region for the ⋆-category computed by IB1-SD does not fit the training instances very well and does actually also cover all training instances of the ◇-category despite that the members of the two categories forms clearly separated clusters. The acceptance region computed by the algorithm able to capture covariances between features, on the other hand, do not cover any of the ◇-instances.

To implement the latter algorithm it is necessary to apply multivariate statistical methods. In particular the following theorem is useful: assuming that feature values are normally distributed within categories/clusters we have that the solid ellipsoid of x values satisfying

$$(x - \mu)^T \Sigma^{-1} (x - \mu) \leq \chi_p^2(\alpha)$$

has probability $1 - \alpha$, where μ is the mean vector, Σ is the covariance matrix, χ^2 is the chi-square distribution and p is the number features. (See for instance [JW92].) This theorem can be used to compute a weighted distance from the instance to be classified to the "center" of the category/cluster.

If this distance is larger than a critical value (dependent of α) the instance is rejected. Thus, instead of estimating m and σ, we have to estimate μ and Σ for each category/cluster from the training set. An algorithm based on this idea together with some very promising experimental results are presented in Davidsson [Dav96].

Unfortunately, such an algorithm is not applicable to the coin sorting problem described above. There are two main reasons for this: (i) the algorithm does not provide explicit min and max limits for each feature, and (ii) the algorithm is too slow in the classification phase; first, the instance to be classified has to be compared to a large number of the training instances, and then, a number of time consuming matrix calculations are necessary in order to compute the weighted distance.

9   CONCLUSIONS

We have applied a novel method for learning characteristic decision trees, the ID3-SD algorithm, to the problem of learning the decision mechanism of coin-sorting machines. The main reason for the success of this algorithm in this application was its ability to control the degree of generalization (an ability which, of course, requires the learning of characteristic descriptions). To author's knowledge, ID3-SD is the first algorithm in which this can be done both explicitly (i.e., by specifying a parameter, the α-value) and without making any ad-hoc assumptions (i.e., it is based on sound statistical reasoning).

In our experiments α-values about 0.0001 has given the best results, but we do not know whether this holds generally. Development of methods to determine automatically the appropriate degree of generalization belongs to future research. Future work will also include an evaluation of how other empirical learning methods can be improved by the SD approach. In this perspective, we have in this paper only described an application of the general method to the ID-3 algorithm (i.e., it can be regarded a case study).

ACKNOWLEDGEMENTS

I wish to thank Eric Astor, Robert M. Colomb, Paul Compton, Len Hamey, Måns Holgersson, and Anders Holtsberg for fruitful discussions and helpful suggestions.

REFERENCES

[Aha92]   D.W. Aha. Generalizing from case studies: A case study. In *Ninth International Workshop on Machine Learning*, pages 1–10. Morgan Kaufmann, 1992.

[AKA91]   D.W. Aha, D. Kibler, and M.K. Albert. Instance-based learning algorithms. *Machine Learning*, 6(1):37–66, 1991.

[Dav95] P. Davidsson. ID3-SD: An algorithm for learning characteristic decision trees by controlling the degree of generalization. Technical Report LU–CS–TR: 95–145, Dept. of Computer Science, Lund University, Lund, Sweden, 1995.

[Dav96] P. Davidsson. *Autonomous Agents and the Concept of Concepts*. PhD thesis, Department of Computer Science, Lund University, Sweden, 1996. (In preparation.).

[dM93] T. Van de Merckt. Decision trees in numerical attribute spaces. In *IJCAI-93*, pages 1016–1021. Morgan Kaufmann, 1993.

[HAP89] R.C. Holte, L.E. Acker, and B.W. Porter. Concept learning and the problem of small disjuncts. In *IJCAI-89*, pages 813–818. Morgan Kaufmann, 1989.

[Hut94] A. Hutchinson. *Algorithmic Learning*. Clarendon Press, 1994.

[JW92] R.A. Johnson and D.W. Wichern. *Applied Multivariate Statistical Analysis*. Prentice-Hall, 1992.

[Mar94] E. Mårtensson. Improved coin classification, (Master's thesis). LU–CS–EX: 94–2, Dept. of Computer Science, Lund University, Sweden, 1994. (In Swedish).

[Min89] J. Mingers. An empirical comparision of pruning methods for decision tree induction. *Machine Learning*, 4(2):227–243, 1989.

[Qui86] J.R. Quinlan. Induction of decision trees. *Machine Learning*, 1(1):81–106, 1986.

[Qui93] J.R. Quinlan. *C4.5: Programs for Machine Learning*. Morgan Kaufmann, 1993.

[RHW86] D.E. Rumelhart, G.E. Hinton, and R.J. Williams. Learning internal representations by error propagation. In D.E. Rumelhart and J.L. McClelland, editors, *Parallel Distributed Processing: Explorations in the Microstructure of Cognition. Vol.1: Foundations*. MIT Press, 1986.

[SM92] P. Smyth and J. Mellstrom. Detecting novel classes with applications to fault diagnosis. In *Ninth International Workshop on Machine Learning*, pages 416–425. Morgan Kaufmann, 1992.

[TG74] J.T. Tou and R.C. Gonzales. *Pattern Recognition Principles*. Addison-Wesley, 1974.

EVALUATION OF ATTRIBUTE SELECTION MEASURES IN DECISION TREE INDUCTION

TuBao Ho, TrongDung Nguyen

Japan Advanced Institute of Science and Technology – Hokuriku

Tatsunokuchi, Ishikawa, 923-12 JAPAN

Email: bao, nguyen@jaist.ac.jp

ABSTRACT: Many attribute selection measures have been proposed for decision tree induction, but little was known regarding their experimental comparative evaluation. This paper aimed at two following objectives. One was the introduction of a new attribute selection measure (R-measure) for decision tree induction. The second was an experimental evaluation of the performance of inductive systems using R-measure and eight different attribute selection measures used by machine learning community: gain-ratio, gini-index, gini'-index, Relief, J-measure, d_N distance, relevance, χ^2.

Key words: machine learning, decision tree induction, attribute selection measure, rough sets.

1 INTRODUCTION

Research in machine learning has grown rapidly in recent years as a promising way to support the (semi)-automatic construction of knowledge bases for artificial intelligence applications. A major focus of study has been supervised inductive learning in which decision tree induction is certainly the most active and applicable research area. The goal of decision tree induction is that from a set of labelled examples to induce a classifier which can be used to predict correctly classes of unseen instances. During the last decade, many decision tree induction methods have been developed, most notably CART [Breiman et al., 1984], ID3 [Quinlan, 1986] and its optimized version C4.5 [Quinlan, 1993].

Two crucial problems for decision tree induction are *attribute selection* (choosing the "best" attribute to split a node in the decision tree) and *pruning* (avoiding overfitting). Among attribute selection measures, the most popular may be informa-tion gain and gain-ratio in ID3 and C4.5, Gini-index of diversity in CART, relevance [Baim, 1988], d_N distance of partitions [López de Mantaras, 1991], χ^2 and G statistics [Liu and White, 1994], J-measure [Smyth and Goodman, 1993], Relief and Gini'-index [Kononenko, 1994].

In [Mingers, 1989], the author claimed that the attribute selection measures affect the size of a tree but not its accuracy, which remains the same even when attributes are selected randomly. Subsequently, many authors did not hold this opinion and tried to evaluate the importance of attribute selection measures or biases in estimating attributes, e.g., [Buntine and Niblett, 1992], [Kononenko, 1995], [Liu and White, 1994]. However, it has been always necessary to evaluate selection measures in more reliable situations (e.g., with more data sets and resampling techniques). This paper focused on two following objectives. One was to introduce a new attribute selection measure, called R-measure, inspired by the measure of dependency degree between attribute sets from the rough set theory [Pawlak, 1991]. The second was to evaluate and compare the performance of decision tree induction systems using different attribute selection measures which are gain-ratio, gini-index, Relief, gini'-index, J-measure, d_N distance, relevance, χ^2, and R-measure.

2 ATTRIBUTE SELECTION MEASURES

In order to facilitate a common understanding of different attribute selection measures, we use in this paper the definitions presented in [Liu and White, 1994] and [Kononenko, 1995]. Suppose that we are dealing with a problem of learning a classifier with k classes C_i ($i = 1, k$) from a set O of examples described by a set of attributes. We assume that all attributes are discrete each of which is with a finite number of possible values. Let $n_{..}$ denotes the total number of examples in O, $n_{i.}$ the number of objects from class C_i, $n_{.j}$ the number of objects with the j-th value of the given attribute A, and n_{ij} the number of objects from class C_i and with the j-th value of A. Let further

$$p_{ij} = \frac{n_{ij}}{n_{..}}, \quad p_{i.} = \frac{n_{i.}}{n_{..}}, \quad p_{.j} = \frac{n_{.j}}{n_{..}}, \quad p_{i|j} = \frac{n_{ij}}{n_{.j}}$$

denote the approximation of the probabilities from the training set O. Let

$$H_C = -\sum_i p_{i.} log p_{i.}, \quad H_A = -\sum_j p_{.j} log p_{.j},$$

$$H_{CA} = -\sum_i \sum_j p_{ij} log p_{ij}, \quad H_{C|A} = H_{CA} - H_A$$

be the entropy of the classes, of the values of the given attribute, of the joint example class–attribute value, and of the class given the value of the attribute, respectively (all logarithms introduced here are of the base two).

In our work eight attribute selection measures published in machine learning literature were considered, and also R-measure was introduced. All of them are either information-based or statistic-based measures. The most popular is gain information defined as the information about class membership which is conveyed by a given attribute A

$$Gain = H_C + H_A - H_{CA} \qquad (1)$$

The state-of-the-art decision tree algorithm ID3 [Quinlan, 1986] and its optimized version C4.5 use the gain-ratio [Quinlan, 1993]

$$GainR = \frac{Gain}{H_A} \qquad (2)$$

The measure d_N [López de Mantaras, 1991] is based on the definition of distance between two partitions and can be rewritten as [White and Liu, 1994]

$$d_N = 1 - \frac{Gain}{H_{CA}} \qquad (3)$$

The author has shown two experiments with the data sets "hepatitis" and "breast cancer". In [Smyth and Goodman, 1993] the authors introduced

J-measure with a claim that it possesses desirable properties as a rule information measure

$$J = \sum_j p_{.j} \sum_i p_{i|j} log \frac{p_{i|j}}{p_{i.}} \qquad (4)$$

Gini-index used in decision tree learning algorithm CART [Breiman et al., 1984] is another well-known measure

$$Gini = \sum_j p_{.j} \sum_i p_{i|j}^2 - \sum_i p_{i.}^2 \qquad (5)$$

In [Kononenko, 1994], Kononenko reformulated a measure called Relief from algorithm RELIEF [Kira and Rendell, 1992] and showed that Relief is a promising measure which is enable to efficiently deal with noisy data, missing values and multi-class problem

$$Relief = \frac{\sum_j p_{.j}^2 \times Gini'}{\sum_i p_{i.}^2 (1 - \sum_i p_{i.}^2)} \qquad (6)$$

where Gini'-index is highly correlated with the Gini-index

$$Gini' = \sum_j \left(\frac{p_{.j}^2}{\sum_j p_{.j}^2} \sum_i p_{i|j}^2\right) - \sum_i p_{i.}^2 \qquad (7)$$

Baim introduced a selection measure called *relevance* and showed an experiment in craniostenosis syndrome identification [Baim, 1988]

$$Relev = 1 - \frac{1}{1-k} \sum_j \sum_{i \neq i_m(j)} \frac{n_{ij}}{n_{i.}}, \qquad (8)$$

$$i_m(j) = arg \ max_i \left\{\frac{n_{ij}}{n_{i.}}\right\}$$

Two other statistics-based measures of interest are χ^2 and G statistics (however we will not consider G statistics as it is mere $Gain$ multiples with a constant)

$$\chi^2 = \sum_i \sum_j \frac{(e_{ij} - n_{ij})^2}{e_{ij}}, \quad e_{ij} = \frac{n_{.j} n_{i.}}{n_{..}} \qquad (9)$$

$$G = 2n_{..} Gain \times log_e 2, \quad e = 2.7182... \qquad (10)$$

3   R-MEASURE

We proposed a new selection measure, called R-measure, which is inspired from the degree of the dependency of a set of attributes P on a set of attributes

R in the rough set theory [Pawlak, 1991]. The starting point of the rough set theory is the assumption that our "view" on elements of the object set O depends on equivalence relations $E \subseteq O \times O$. Two objects $o_1, o_2 \in O$ are called to be *indiscernible* in E if $o_1 E o_2$. The *lower* and *upper* approximations of any $X \subseteq O$ consisting all objects which surely and possibly belong to X, respectively, regarding a relation E. These approximations are defined as

$$E_*(X) \triangleq \{o \in O : [o]_E \subseteq X\} \qquad (11)$$

$$E^*(X) \triangleq \{o \in O : [o]_E \cap X \neq \varnothing\} \qquad (12)$$

where $[o]_E$ denotes the equivalence class of objects indiscernible with o in the equivalence relation E. A subset P of attributes determines an equivalence relation that divides O into equivalence classes each contains objects with the same values on all attributes of P. A key concept in the rough set theory is the *degree of dependency* of a set of attributes P on a set of attributes R, denoted by $\mu_R(P)$ $(0 \leq \mu_R(P) \leq 1)$, defined as

$$\mu_R(P) = \frac{card(\bigcup_{[o]_P} R_*([o]_P))}{card(O)} \qquad (13)$$

Our analysis and experiments have shown that the measure $\mu_R(P)$ in the rough set theory is not robust with noisy data and not enough sensitive with "partial" dependency between R and P. ¿From the definitions in (11) and (13), we can write

$$\bigcup_{[o]_P} R_*([o]_P) = \{o \in O : [o]_R \subseteq [o]_P\} \qquad (14)$$

By Eq.(14) we see that $\mu_R(P)$ relies on one basic assumption: each equivalence class $[o]_R$ in (14) is completely included in some equivalence class $[o]_P$. However this condition is somewhat strict and $\mu_R(P)$ can not measure the dependency of P on R if the partition generated by R do not contain any cluster included in a cluster of the partition generated by P. Also, $\mu_R(P)$ does not work well in the case when two partitions generated by R and P are nearly identified (R and P are highly mutual dependent) but their equivalence classes do not satisfy the above assumption. Unfortunately, these situations occur quite often in real-world data. To avoid this limitation, we proposed a new measure of the dependency of P on R defined as follows

$$\tilde{\mu}_R(P) = \sum_{[o]_R} max_{[o]_P} \frac{card([o]_R \cap [o]_P)^2}{card(\mathcal{O}) card([o]_R)} \qquad (15)$$

The main difference between $\mu_R(P)$ and $\tilde{\mu}_R(P)$ is the latter measures the dependency of P on R in maximizing the predicted membership of an instance in the family of equivalence classes generated by P given its membership in the family of equivalence classes generated by R.

We illustrate the above remark about $\mu_R(P)$ and $\tilde{\mu}_R(P)$ by the information table presented in [Pawlak et al., 1995].

	Temperature	Headache	Flu
e_1	normal	yes	no
e_2	high	yes	yes
e_3	very_high	yes	yes
e_4	normal	no	no
e_5	high	no	no
e_6	very_high	no	yes
e_7	high	no	no
e_8	very_high	no	no

Table 1. Information table

Let $P = \{Flu\}, R = \{Headache\}$. The equivalence classes w.r.t. *Flu* are $\{e_1, e_4, e_5, e_7, e_8\}$, $\{e_2, e_3, e_6\}$ and the equivalence classes w.r.t. *Headache* are $\{e_1, e_2, e_3\}$, $\{e_4, e_5, e_6, e_7, e_8\}$. We can check that $\mu_{Headache}(Flu) = 0$, and it means that according to $\mu_R(P)$, *Flu* is absolutely independent of *Headache*. However, from Table 1 we can see intuitively that the fact *Headache = yes* probably implies *Flu = yes* and the fact *Headache = no* probably implies *Flu = no*. We can check that $\tilde{\mu}_{Headache}(Flu) = \frac{17}{30}$, and it means that according to $\tilde{\mu}_R(P)$, there is some degree of dependency of *Flu* on *Headache*.

Let describe $\tilde{\mu}_R(P)$ in the statistic notation used in [Liu and White, 1994]. Suppose that $R = \{A_1, A_2, ..., A_r\}$, and $P = \{B_1, B_2, ..., B_p\}$. Let $n_{.j_1 j_2 ... j_r}$ denotes the number of objects with the j_1-th, j_2-th,..., j_r-th values of $A_1, A_2, ..., A_r$, respectively, and $n_{i_1 i_2 ... i_p | j_1 j_2 ... j_r}$ the number of objects with the i_1-th, i_2-th,..., i_p-th values of $B_1, B_2, ..., B_p$ and with the j_1-th, j_2-th,..., j_r-th values of $A_1, A_2, ..., A_r$, simultaneously. We also denote by $p_{.j_1 j_2 ... j_r}$ and $p_{i_1 i_2 ... i_p | j_1 j_2 ... j_r}$ the approximations of these probabilities from the training set. We can rewrite

$$\tilde{\mu}_R(P) = \sum_{j_1, ..., j_r} p_{.j_1 ... j_r} max_{i_1, ..., i_p} p^2_{i_1 ... i_p | j_1 ... j_r} \qquad (16)$$

In a special case of (16) when P stands for the class attribute and R stands for the given attribute A, $\tilde{\mu}_R(P)$ can be used to measure the dependency of the class attribute on A. We call this R-measure which is written as

$$\tilde{\mu}_A = \sum_j p_{.j} \, max_i \{p_{i|j}^2\} \qquad (17)$$

4   EXPERIMENTAL METHODOLOGY

The studies of Minger, White and Liu mentioned in section 1 have shown how difficult it is to be thorough and fair when evaluating the accuracy of learning methods. It seems that it is not adequate to evaluate methods either with a small number of data sets or with only a single train-and-test experiment. In order to attain a reliable estimation of the mentioned attribute selection measures we designed the experiments as follows

- implement programs for different measures in an unique scheme;
- use a large number of public data sets;
- use the cross-validation technique for the evaluation.

4.1   IMPLEMENTATION OF SELECTION MEASURES

Nine mentioned attribute selection measures are implemented in the common scheme CLS (Concept Learning System [Hunt et al., 1966]) of decision tree induction. This scheme can be briefly described in the following steps

1. Choose the "best" attribute by given *selection measure*
2. Extend tree by each attribute value
3. Sort training examples to leaf nodes
4. If training examples unambiguously classified Then Stop Else Goto steps 1

Nine systems are implemented based on modifications of the program C4.5 written in C programming language [Quinlan, 1993]. In this paper, in order to evaluate the original attribute selection measures we considered only the results of these programs on unpruned trees but not on pruned trees. The pruning technique of cost complexity [Breiman et al., 1984] is used in these programs.

4.2   DATA SETS

In this subsection we briefly describe the data sets used in our empirical comparisons. Twelve data sets

are taken from the UCI repository of machine learning databases. They included Wisconsin Breast Cancer, Congressional Votes, Audiology, Hayes-Roth, Heart Disease, Image Segmentation, Ionosphere, Lung Cancer, Promoters and Splice-Junction of sequences in molecular-biology, Solar Flare, Soybean Disease, Satellite Image, Tic-Tac-Toe endgame, King-rook-vs-King-pawn. Their properties are listed in Table 2 with the following abbreviations: att, inst, and class stand for the number of attributes, instances, and classes; type stands for the type of attributes (symbolic or mixture of symbolic and continuous).

4.3   CROSS VALIDATION

Traditionally in supervised learning examples provided are usually divided into two sets of training and testing data. Training data are used to produce the classifier by a method and testing data are used to estimate the prediction accuracy of the method. A single train-and-test experiment is often used in machine learning for estimating performance of learning systems.

Data sets	att	inst	class	type
Vote	16	435	2	sym
Breast cancer	9	699	2	sym
Soybean-small	35	46	4	sym
Tic-tac-toe	9	862	9	sym
Lung cancer	56	32	3	sym
Hayes-roth	5	160	3	sym
Kr-vs-kp	36	3196	2	sym
Audiology	69	226	24	sym
Splice	60	3190	3	sym
Promoters	58	106	2	sym
Heart-disease	13	303	2	mix
Solar Flare	13	1389	7	mix

Table 2. Properties of chosen data sets

It is recognized that multiple train-and-test experiments can do better than single train-and-test experiments [Weiss and Kulikowski, 1991]. Recent works showed that cross validation is a suitable resampling technique for accuracy estimation, particularly the 10-fold cross validation, e.g. [Kohavi, 1995]. However, cross validation is still not widely used in machine learning as it is computationally expensive.

To obtain a fair estimation of the performance of mentioned selection measures, we carried out a 10-fold cross validation in our experiments. The data set O is randomly divided into 10 mutually exclusive subsets $O_1, O_2, ..., O_{10}$ of approximately equal size.

data set	GainR	Gini	Gini'	Relief	J-m	R-m	d_N	Relev	χ^2
Vote	**5.5**	6.2	6.2	7.2	6.4	6.4	*7.6*	6.4	6.2
Breast cancer	6.7	**5.9**	8.1	6.4	6.1	6.1	*7.7*	6.6	6.0
Soybean-small	2.2	2.2	**0.0**	**0.0**	2.2	2.2	2.2	2.2	**0.0**
Tic-tac-toe	12.0	13.8	16.4	*18.2*	12.9	14.1	13.7	**11.9**	13.6
Lung cancer	23.3	**10.0**	13.3	20.0	16.6	13.3	*36.6*	20.0	**10.0**
Hayes-roth	25.0	25.0	**20.6**	23.3	24.4	23.7	24.4	25.6	*26.3*
Kr-vs-kp	0.5	**0.3**	*1.2*	*1.2*	0.4	0.5	3.2	**0.3**	**0.3**
Audiology	23.4	22.5	*45.9*	22.1	24.8	**21.6**	31.7	44.1	21.7
Splice	8.0	7.9	14.4	*28.1*	**6.9**	8.4	10.4	10.3	8.4
Heart-disease	**26.7**	29.1	28.6	30.0	27.5	27.7	*31.7*	30.1	27.8
Flare	28.4	28.7	29.3	29.2	**27.3**	27.5	*29.8*	29.3	28.1
Promoters	**11.5**	14.5	17.1	20.3	14.9	14.5	*29.5*	17.2	16.0
Average	13.3	12.8	15.5	15.9	13.1	12.8	*17.6*	15.7	**12.7**

Table 3. Error rates estimated for different measures

Each attribute selection measure is tested 10 times. Each time for $k \in \{1, 2, ..., 10\}$, a decision tree is generated on $O \setminus O_k$ and tested on O_k. The error rate of each measure is the average of its error rates after 10 running times.

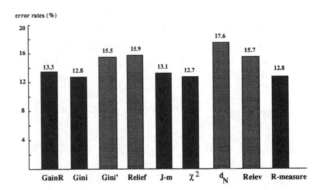

Figure 1: Graphical view of error rate average

5 EXPERIMENTAL RESULTS

Table 3 shows the estimated error rates of nine attribute selection measures obtained in our experiments. Each line presents error rates of nine measures for a data set in which the bold number indicates the measure with minimum error rate and the italic number indicates the measure with maximum error rate. The abbreviations J-m and R-m stand for J-measure and R-measure, respectively.

6 CONCLUSION

The results of our experiments on attribute selection measures can be summarized as follows:

- We reconfirm that the quality of the decision tree construction is affected by the choice of attribute selection measures.

- Different from conclusions of some works mentioned above, our experiments showed that the error rates of Gini'-index, Relief, Relevance and particularly d_N distance were relatively higher

than those of others, and they were somehow unstable (as showed the results with data sets Lung-cancer, Audiology, Splice, Promoters).

- Measures gain-ratio, Gini-index, J-measure, R-measure, and χ^2 are relatively similar as their error rates are not really much different.

- The overestimation phenomenon for multi-valued attributes (i.e., J-measure, Gini-index, relevance tend to linearly increase with the number of values of an attribute; and gain-ratio, d_N distance, Relief decrease for informative attributes and increase for irrelevant attributes) studied with artificial data in [White and Liu, 1994], [Kononenko, 1995] does not appear in our experiments with real-world data.

- In comparison with estimation of learning performance by a single train-and-test experiment, e.g., [Buntine and Niblett, 1992] or our own, the cross-validation often gives higher error rates. Thus, the cross validation technique can allow us to avoid an usual overoptimistic estimation of learning performance.

- R-measure seems to be a good measure for attribute selection in decision tree induction. Its error rates are always among the lowest and moreover R-measure is stable with different data sets. Another advantage of R-measure is that it is easy to be applied in multivariate decision tree induction for producing decision trees where several attributes are used to split a node as it measures directly the relationship between two attribute sets.

An extension of our experiments will be made with more data sets and with noisy data on pruned trees. Careful evaluation in these conditions will hopefully allow us to obtain more reliable results.

References

[Baim, 1988] Baim, P.W. (1988). A method for attribute selection in inductive learning systems. *IEEE Trans. on PAMI*, 10, 888-896.

[Breiman et al., 1984] Breiman, L., Friedman, J., Olshen, R., Stone, C., *Classification and regression trees*, Belmont, CA: Wadsworth, 1984.

[Buntine and Niblett, 1992] Buntine, W., Niblett, T., A further comparison of splitting rules for decision-tree induction. *Machine Learning*, 8, 75-85, 1992.

[Hunt et al., 1966] Hunt, E.B., Martin, J., Stone, P.T., *Experiments in induction*, Academic Press, 1966.

[Kira and Rendell, 1992] Kira, K., Rendell, L., A practical approach to feature selection. *Proc. Int. Conf. on Machine Learning*, 249–256, 1992.

[Kohavi, 1995] Kohavi., R, A study of cross-validation and bootstrap for accuracy estimation and model selection. *Proc. Int. Joint Conf. on Artificial Intelligence IJCAI'95*, 1137-1143, 1995.

[Kononenko, 1994] Kononenko, I., Estimating attributes: Analysis and extension of RELIEF. *Proc. European Conf. on Machine Learning ECML'94*, 171–182, 1994.

[Kononenko, 1995] Kononenko, I., On biases in estimating multi-valued attributes. *Proc. Int. Joint Conf. on Artificial Intelligence IJCAI'95*, 1034–1040, 1995.

[Liu and White, 1994] Liu, W.Z., White, A.P., The importance of attribute selection measures in decision tree induction. *Machine Learning*, 15, 25–41, 1994.

[López de Mantaras, 1991] López de Mantaras, R., A distance-based attribute selection measure for decision tree induction. *Machine Learning*, 6, 81–92, 1991.

[Mingers, 1989] Mingers, J., An empirical comparison of selection measures for decision-tree induction. *Machine Learning*, 3, 319–342, 1989.

[Pawlak, 1991] Pawlak, Z., *Rough sets: Theoretical aspects of reasoning about data*, Kluwer Academic Publishers, 1991.

[Pawlak et al., 1995] Pawlak, Z., Grzymala-Busse, J., Slowinski, R., Ziarko, W., Rough sets, *Communication of the ACM*, November 1995, Vol. 38, No. 11, pp. 89–95.

[Quinlan, 1986] Quinlan, J.R., Induction of decision trees. *Machine Learning*, 1, 81–96, 1986.

[Quinlan, 1993] Quinlan, J.R., *C4.5: Programs for machine learning*, Morgan Kaufmann, 1993.

[Smyth and Goodman, 1993] Smyth, P., Goodman, R.M., Rule induction using information theory. In G. Piatetsky-Shapiro and W. Frawley (Eds.) *Knowledge discovery in Database*, AAAI Press, 1993.

[White and Liu, 1994] White, A.P., Liu, W.Z., Bias in information-based measures in decision tree induction. *Machine Learning*, 15, 321–329, 1994.

[Weiss and Kulikowski, 1991] Weiss, S. M., Kulikowski, C. A., *Computer Systems That Learn*, Morgan Kaufmann, 1991.

FEATURE SELECTION AND CLASSIFICATION
– A PROBABILISTIC WRAPPER APPROACH

Huan Liu and Rudy Setiono
Department of Information Systems and Computer Science
National University of Singapore
Kent Ridge, Singapore 119260
Email: {liuh,rudys}@iscs.nus.sg

ABSTRACT

Feature selection is defined as a problem to find a minimum set of M features for an inductive algorithm to achieve the highest predictive accuracy from the data described by the original N features where $M \leq N$. A probabilistic wrapper model is proposed as another method besides the exhaustive search and the heuristic approach. The aim of this model is to avoid local minima and exhaustive search. The highest predictive accuracy is the criterion in search of the smallest M. Analysis and experiments show that this model can effectively find relevant features and remove irrelevant ones in the context of improving the predictive accuracy of an induction algorithm. It is simple, straightforward, and providing fast solutions while searching for the optimal. The applications of such a model, its future work and some related issues are also discussed.

1. INTRODUCTION

The problem of feature selection can be defined as finding relevant M features among the original N attributes, where $M \leq N$, to define the data in order to minimize the error probability or some other reasonable selection criteria. Feature selection has long been the focus of researchers of many fields - pattern recognition, statistics, machine learning (see Section 2). Many methods have been proposed. In general, they can be classified into two categories[1]: (1) the filter approach [AD94, KR92], i.e., the feature selector is independent of a learning algorithm and serves as a filter to sieve the irrelevant and/or redundant features; and (2) the wrapper approach [JKP94], i.e., the feature selector as a wrapper around a learning algorithm relying on which the relevant features are determined. The major advantage to the wrapper model is that it utilizes the induction algorithm it-self as a criterion in selecting features since in the context of learning classification rules, the purpose of feature selection is to improve the performance of an induction algorithm. However, incorporating an induction algorithm in the process of feature selection is not without a cost. (More discussion below).

In each category, methods can be further divided into two types: exhaustive or heuristic search. The difficulty about feature selection can be stated as follows: except in a few very special cases, the optimal selection can only be done by testing all possible sets of M features chosen from the N features, i.e., by applying the criterion $\binom{N}{M} = \frac{N!}{M!(N-M)!}$ times. If there are M relevant features, the total number of times is $\sum_{i=1}^{M} \binom{N}{i} = O(N^M)$. This is prohibitive when N and/or M is large. In practice, heuristic methods are the way out of this exponential computation. Heuristic methods in general make use of low order (first or second) information[2] to estimate relevance of features approximately. Although the heuristic methods work reasonably well [Qui93, LW93], it is certain that they miss out the features of high order relations, for example, the parity problem. On one hand, it is a problem of exponential explosion; on the other hand, it is very likely that some relevant features will be omitted if the heuristic approach is taken. Our goal is to have an algorithm that can explore the high order relations among the N features and find M relevant features, with high probability, but without resorting to exhaustive search. In this work, the feature selection problem is redefined in terms of predictive accuracy of an inductive algorithm: to find the smallest set of M features for an inductive algorithm to achieve the highest accuracy.

2. RELATED WORK

The problem of feature selection has long been an active research topic within statistics and pat-

---

[2] First order information contains only one feature, second order information two features, etc.

tern recognition [WDJ80, DK82], but most work in this area has dealt with linear regression [Lan94] and under assumptions that do not apply to most learning algorithms [JKP94]. They pointed out that the most common assumption is monotonicity, that increasing the number of features can only improve the performance of a learning algorithm[3].

In the past few years, feature selection has received considerable attention from machine learning and knowledge discovery researchers interested in improving the performance of their algorithms and cleaning data. There are many heuristic feature selection algorithms. The RELIEF algorithm [KR92] assigns a "relevance" weight to each feature, which is meant to denote the relevance of the feature to the target concept. RELIEF samples instances randomly from the training set and updates the relevance values based on the difference between the selected instance and the two nearest instances of the same and opposite classes. According to [KR92], RELIEF assumes two-classes classification problems and does not help with redundant features. If most of the given features are relevant to the concept, it would select most of them even though only a fraction are necessary for concept description. The PRESET algorithm [Mod93] is another heuristic feature selector that uses the theory of Rough Sets to heuristically rank the features, assuming a noise-free binary domain. In order to consider higher order information among the feature. It is suggested in [LW93] to use high order information gain in feature selection. Since the last two algorithms do not try to explore all the combinations of features, it is likely that they fail on the problems like Parity and Majority functions where the combinations of a small number of features do not help in locating the relevant features. Chi2 [LS95] is another heuristic feature selector, it automatically discretizes the continuous features and removes irrelevant continuous features based on the chi-square statistics and the inconsistency found in the data. However, it cannot handle nominal features. In [JKP94], forward selection and backward elimination wrapper models are studied. Nevertheless, no conclusion is given on which one is better and no guideline is offered on which problems which method should be used. It can also be observed from the results that, in general, the latter achieves lower error rates, and the former achieves smaller numbers of features. The classic exhaustive method can be found in [AD94]

which is called FOCUS, in which the authors proposed several heuristic versions to speed up the process, assuming a noise-free binary domain.

Another common understanding is that some learning algorithms have built-in feature selection, for example, ID3 [Qui86], FRINGE [PH90] and C4.5 [Qui93]. The results in [AD94] suggest that one should not rely on ID3 or FRINGE to filter out irrelevant features. Since C4.5 conducts test on each individual feature as well, it is not proper either to use C4.5 to find the minimum set of features. It is expected (to be shown in Section 4) that it will fail on the parity problems.

A summary is that the exhaustive search approach is infeasible in practice; and the heuristic search approach can reduce the computational time significantly, but cannot explore the combined effects among the features, will fail on hard problems (e.g., the parity problem) or cannot remove redundant features. It is right time for the third approach that relies on neither heuristics nor exhaustive search in producing solutions and with high probability, selects the optimal and/or near-optimal set(s) of relevant features.

3. PROBABILISTIC WRAPPER MODEL

This probabilistic approach is a modified version of Las Vegas Algorithms [BB96]. Las Vegas algorithms make probabilistic choices to help guide them more quickly to a correct solution. One kind of Las Vegas algorithms uses randomness to guide their search in such a way that a correct solution is guaranteed even if unfortunate choices are made. As we mentioned earlier, heuristic search methods are vulnerable to the datasets of high order relations. Las Vegas algorithms free us from worrying about such situations by evening out the time required on different situations. The time performance of a Las Vegas algorithm may not be better than that of some heuristic algorithms. With high probability, data that took a long time deterministically are now solved much faster, but data on which the heuristic algorithm was particularly good are slowed down to average by the Las Vegas algorithm. The following algorithm generates random subsets of N features; for each subset $S1$, a learning algorithm is applied to the training data in order to obtain its estimated error rate $err1$, the smallest subset with the lowest error rate is kept. In a Las Vegas algorithm, it is guaranteed that given sufficiently long time, it finds the optimal solution. In search of the mini-

---

[3]The monotonicity assumption is not valid for many induction algorithms used in machine learning. See dataset 1 in Section 4 which is reproduced from [JKP94].

mum set of M features, LVW outputs every current best. As shown in the algorithm, the computation of each $err1$ is based on $S1$ and D_{train}; D_{test} is only used in computing $err2$. That is, at the end, the learning algorithm is applied to both training and testing data and produces its estimated error rate on the testing data. This is the rate reported below in experiments. K is the specified number maximum runs, k is the number of runs, C is the smallest number of features at present, err is the current smallest error rate. Function randomSet produces set $S1$ of features at each run, $C1$ is the number of features in $S1$. LearnAlgo can be any induction algorithm (C4.5 and ID3 are chosen in our experiments).

LVW algorithm
    $err = 0; k = 0; C = 100;$
    repeat
      $S1 = \text{randomSet}();$
      $C1 = \text{numOfFeatures}(S1);$
      $err1 = \text{LearnAlgo}(S1, D_{train}, \text{NULL});$
      if ($err1 < err$ OR
        ($err1 = err$ AND $C1 < C$)) {
        output the current best;
        $k = 0; err = err1;$
        $C = C1; S = S1; \}$
      $k = k + 1;$
    until err is not updated for K times;
    $err2 = \text{LearnAlgo}(S, D_{train}, D_{test});$

Some analysis shows that LVW can give a good solution, or an optimal solution if K is sufficiently large. With a good pseudo random number generator [PTVF92], the selection of an optimal subset of M features can be considered non-replacement experiments. The probability of finding the optimal subset is $\frac{1}{2^N - k}$ at the kth experiment, although the probability of finding the subset after k experiments is still $\frac{2^N - 1}{2^N} \times \frac{2^N - 2}{2^N - 1} \times ... \times \frac{1}{2^N - k} = \frac{1}{2^N}$, where N is the number of original features. In our experiments, K (shown in the LVW algorithm) is approximated as $60 \times N$. The larger an N is, the more trials LVW will try. When N is large, this approximation of $K \ll 2^N$. This analysis assumes there is one optimum. If there exist l optima, at the kth tossing, the probability of finding one optimum would be $\frac{l}{2^N - k}$.

4. EMPIRICAL STUDY

Although the induction algorithm in the LVW algorithm can be any kind, it is important that the induction algorithm be fast since the time complex-

ity of LVW is bound by the number of runs and the time complexity of the induction algorithm. Among many choices, we chose C4.5 and ID3 in our experiments. The C4.5 program is the program that comes with Quinlan's book [Qui93]; the ID3 results were obtained by running C4.5 and using the unpruned trees.

Two types of datasets are chosen in experiments. One type is artificial data so that the relevant features are known before feature selection is conducted, which includes CorrAL [JKP94], Monks1-3 [TBB$^+$91], and Parity5+5. The other type is real-world data including Credit, Vote, and Labor [Qui93, MA94]. The choice of these datasets can simplify the comparison of this work with some published work. These datasets were used in [JKP94] in which comparisons with different methods were described. For the artificial datasets, no cross-validation is done. For the real-world datasets, 10-fold cross validation is used to obtain the estimated accuracy on the training data. On the choice of options of C4.5, following [JKP94], we use "-m1" C4.5 flag which indicates that splitting should continue until purity on the artificial datasets, the default setting on the real-world datasets.

Artificial Data:

1. **CorrAL** The data was designed in [JKP94]. There are six binary features, A_0, A_1, B_0, B_1, I, and C. Feature I is irrelevant, feature C is correlated to the class label 75% of the time. The Boolean target concept is $(A_0 \wedge A_1) \vee (B_0 \wedge B_1)$. Both ID3 and C4.5 chose feature C as the root. This is an example of datasets in which if a feature like C is removed, a more accurate tree will result.

2. **Monk1, Monk2, Monk3** The datasets were taken from [TBB$^+$91]. They have six features. The training datasets provided were used for feature selection. Monk1 and Monk3 only need three features to describe the target concepts, but Monk2 requires all the six. The training data of Monk3 contains some noise. These datasets can be used to show that a feature selector selects either only the relevant features or the relevant ones plus others.

3. **Parity5+5** The target concept is the parity of five bits. The dataset contains 10 features, of which 5 are uniformly random (irrelevant). The training set contains 100 instances randomly selected from all 1024 instances. Another independent 100 instances are drawn to form the testing set. Most heuristic feature selectors will

fail on this sort of problems since an individual feature does not mean anything.

Real-World Data:

4. **Vote** This dataset includes votes from the U.S. House of Representatives Congress-persons on the 16 key votes identified by the Congressional Quarterly Almanac Volume XL. The data set consists of 16 features, 300 training instances and 135 test instances.

5. **Credit** (or CRX) The dataset contains instances for credit card applications. There are 15 features and a Boolean label. The dataset was divided by Quinlan [Qui93] into 490 training instances and 200 test instances.

6. **Labor** The dataset contains instances for acceptable and unacceptable contracts. It is a small dataset with 16 features, a training set of 40 instances, and a testing set of 17 instances.

Results are shown in Tables 1 and 2. In the column of Err, $x(p\%)$ means that there are x instances misclassified, the percentage is p. For all the artificial datasets, LVW did find all the relevant features. For example, LVW rediscovered that features 1, 2 and 5 are relevant for Monk1, all six features for Monk2, features 2, 4 and 5 for Monk3, five features for Parity5+5, features A_0, A_1, B_0, B_1 for CorrAL. These results give us confidence on LVW that using accuracy as a criterion, relevant features can be selected even in the presence of noise (e.g., Monk3). For the real-world datasets, there is no knowledge about which features are relevant. However, the comparison between with and without LVW can still be performed along three dimensions: (1) tree size, (2) error rate (Err), and (3) number of features used (# Att). The results of ID3 and C4.5 with and without LVW are summarized in terms of the three dimensions. For ID3, all the figures improved after LVW is applied, i.e., except for Monk2, the number of features is reduced, tree size is smaller, and error rate is decreasing. For C4.5, the improvement is not clear-cut. Although all datasets but Monk2 have their number of features reduced, the significantly decreased error rates for CorrAL, Monk1, and Parity5+5 come with an increase in tree size. This is not without a reason (see discussion below).

The experimental results from [JKP94] are reproduced here in Table 3 for a reference purpose. See more details in the paper. Before (Bf) means *before feature selection*, Forward (Fw) means *forward stepwise selection*, Backward (Bw) means *backward stepwise selection*, Relieve (Rl) is a modified version of Relief [KR92], because of significant variance in the relevance rankings given by Relief [JKP94].

5. CONCLUSION

In a wrapper model, feature selection is closely linked to an induction algorithm, in other words, LVW is only constrained by the limitations of the induction algorithm. If the induction algorithm can handle noisy data, missing values, both continuous and discrete values, so can LVW. In this work, C4.5 is used and no special effort is needed to tailor the datasets in order to run LVW. To achieve the lowest possible error rate is the aim of both the feature selector and the induction algorithm. In addition, LVW produces intermediate solutions while working toward better ones.

A general belief is that the fewer features, the simpler the decision trees since irrelevant features are removed. However, Murphy and Pazzani [MP94] find that the smallest trees typically have lower predictive accuracy than slightly larger trees; exhaustive search for the simplest consistent theories does not necessarily lead to improvement. That means that using accuracy as a criterion may not lead to the simplest tree. This is truly reflected in the results of these datasets. The size of a decision tree is the number of leaves of the tree plus 1. The size measure does not show how many features are contained in the tree. Our experimental results show that in pursuit of high accuracy, LVW can reduce the number of features, improve the accuracy, but may not always reduce the tree size. A measure which combines the three dimensions may help in achieving high performance in all the dimensions.

Our experience with LVW is that it is slow in running many trials ($O(K)$) of different patterns. Since every random pattern should be tested by the induction algorithm, if its time cost is C_{Ind}, the minimum cost of LVW is $O(K * C_{Ind})$. For cross validations, this cost should be increased by another factor related to the number of folds of cross validations.

Due the slowness of LVW, it is not recommended to use it in applications where time is a critical factor. If it is used just once and for all for some period of time, the slowness does not do much harm. Researchers have been trying to speed up the wrapper approach [Lan94]. LVW can play a role of a bench mark for comparisons with other heuristic methods. All the heuristic FS algorithms are deterministic. Heuristic algorithms designed for particular applications can run very fast. LVW can be used to check if a fast solution is also a good one when designing

Table 1: Results of tree size, error rate (Err) and number of features (# Att.) for ID3 with/without LVW.

Learner	ID3					
Measure	Size		Err		# Att.	
LVW	w/o	with	w/o	with	w/o	with
CorrAL	13	13	0(0.0%)	0(0.0%)	6	4
Monk 1	43	12	101(23.4%)	12(2.8%)	6	3
Monk 2	174	174	131(30.3%)	131(30.3%)	6	6
Monk 3	42	19	42(9.7%)	0(0.0%)	6	3
P5+5	87	63	0(0.0%)	0(0.0%)	10	5
Vote	25	7	7(5.2%)	4(3.0%)	16	4
Credit	117	76	39(19.5%)	31(15.5%)	15	6
Labor	12	7	3(17.6%)	3(17.6%)	16	4

Table 2: Results of tree size, error rate (Err) and number of features (# Att.) for C4.5 with/without LVW.

Learner	C4.5					
Measure	Size		Err		# Att.	
LVW	w/o	with	w/o	with	w/o	with
CorrAL	7	9	2(12.5%)	1(6.2%)	6	4
Monk 1	18	38	105(24.3%)	24(5.6%)	6	3
Monk 2	46	46	148(34.3%)	148(34.3%)	6	6
Monk 3	12	12	12(2.8%)	12(2.8%)	6	3
P5+5	19	63	22(22.0%)	0(0.0%)	10	5
Vote	7	7	4(3.0%)	4(3.0%)	16	4
Credit	44	30	40(20.5%)	30(15.0%)	15	6
Labor	7	7	3(17.6%)	3(17.6%)	16	4

a heuristic algorithm.

It is noticed that the slowness is caused by the learning algorithm. This significantly limits the application of this probabilistic model. It is our current interests to find a criterion that does not rely on the accuracy of a learning algorithm. If checking the satisfaction of the new criterion can be made faster, the probabilistic model can be applied to more applications. Hence, one line of our current research is to get rid of the wrapper approach and go for the filter one (there are other reasons in addition to the speed issue). We have been trying to find such a criterion whereby the speed of the probabilistic method can be improved significantly without sacrificing the end results.

ACKNOWLEDGEMENTS

Thanks to the two anonymous referees for their comments on an earlier version of this paper and Y.C. Chew for transforming the graphs in [JKP94] into Table 3.

REFERENCES

[AD94] H. Almuallim and T.G. Dietterich. Learning boolean concepts in the presence of many irrelevant features. *Artificial Intelligence*, 69(1-2):279–305, November 1994.

[BB96] G. Brassard and P. Bratley. *Fundamentals of Algorithms*. Prentice Hall, New Jersey, 1996.

[DK82] P.A. Devijver and J. Kittler. *Pattern Recognition: A Statistical Approach*. Prentice Hall International, 1982.

[JKP94] G.H. John, R. Kohavi, and K. Pfleger. Irrelevant feature and the subset selection problem. In *Machine Learning: Proceedings of the Eleventh International Conference*, pages 121–129. Morgan Kaufmann Publisher, 1994.

[KR92] K. Kira and L.A. Rendell. The feature selection problem: Traditional methods and a new algorithm. In *AAAI-92, Proceedings Ninth National Conference on Artificial Intelligence*, pages 129–134. AAAI Press/The MIT Press, 1992.

[Lan94] P. Langley. Selection of relevant features in machine learning. In *Proceedings of the AAAI Fall Symposium on Relevance*. AAAI Press, 1994.

Table 3: Experimental results reported in John et al 94: Bf - Before, Fw - Forward, Bw - Backward, Rl - Relieve. X means the figure is not available in the original paper.

ID3 Algorithm									
	TreeSize			ErrorRate(%)			Attributes		
Dataset	Bf	Fw	Bw	Bf	Fw	Bw	Bf	Fw	Bw
CorrAL	X	X	X	X	X	X	X	X	X
Monk3*	X	X	X	X	X	X	X	X	X
Parity5+5	109	13	63	49	50	0	10	3	5
Vote	25	13	37	5	4	2	16	3	15
Credit	117	66	98	20	21	19	15	4	13
Labor	12	7	12	18	18	18	16	3	12

C4.5 Algorithm												
	TreeSize				ErrorRate(%)				Attributes			
Dataset	Bf	Fw	Bw	Rl	Bf	Fw	Bw	Rl	Bf	Fw	Bw	Rl
CorrAL	11	5	13	5	19	19	0	19	6	2	5	5
Monk3*	8	13	8	6	1	2	1	2	6	3	2	2
Parity5+5	X	X	X	X	X	X	X	X	X	X	X	X
Vote	7	4	7	7	3	3	3	3	16	1	15	15
Credit	44	16	44	41	21	19	21	18	15	3	14	14
Labor	X	X	X	X	X	X	X	X	X	X	X	X

[LS95] H. Liu and R. Setiono. Chi2: Feature selection and discretization of numeric attributes. In *Proceedings of the 7th IEEE International Conference on Tools with Artificial Intelligence*, 1995.

[LW93] H. Liu and W.X. Wen. Concept learning through feature selection. In *Proceedings of First Australian and New Zealand Conference on Intelligent Information Systems*, 1993.

[MA94] P.M. Murphy and D.W. Aha. UCI repository of machine learning databases. FTP from ics.uci.edu in the directory pub/machine-learning-databases, 1994.

[Mod93] M Modrzejewski. Feature selection using rough sets theory. In P.B. Brazdil, editor, *Proceedings of the European Conference on Machine Learning*, pages 213–226, 1993.

[MP94] P.M. Murphy and M.J. Pazzani. Exploring the decision forest: An empirical investigation of occam's razor in decision tree induction. *Journal of Art. Intel. Res.*, 1:257–319, March 1994.

[PH90] G. Pagallo and D. Haussler. Boolean feature discovery in empirical learning. *Machine Learning*, 5:71–99, 1990.

[PTVF92] W.H. Press, S.A. Teukolsky, W.T. Vetterling, and B.P. Flannery. *Numerical Recipes in C*. Cambridge University Press, Cambridge, 1992.

[Qui86] J.R. Quinlan. Induction of decision trees. *Machine Learning*, 1(1):81–106, 1986.

[Qui93] J.R. Quinlan. *C4.5: Programs for Machine Learning*. Morgan Kaufmann, 1993.

[TBB+91] S.B. Thrun, J. Bala, E. Bloedorn, I. Bratko, B. Cestnik, J. Cheng, K. De Jong, S. Dzeroski, S.E. Fahlman, D. Fisher, R. Hamann, K. Kaufman, S. Keller, I. Kononenko, J. Kreuziger, R.S. Michalski, T. Mitchell, P. Pachowicz, Reich Y., H. Vafaie, W. Van de Welde, W. Wenzel, J. Wnek, and J. Zhang. The monk's problems: A performance comparison of different learning algorithms. Technical Report CMU-CS-91-197, Carnegie Mellon University, December 1991.

[WDJ80] N. Wyse, R. Dubes, and A.K. Jain. A critical evaluation of intrinsic dimensionality algorithms. In E.S. Gelsema and Kanal L.N., editors, *Pattern Recognition in Practice*, pages 415–425. Morgan Kaufmann Publishers, Inc., 1980.

KNOWLEDGE DISCOVERY FROM NUMERICAL DATA

Chie Morita and Hiroshi Tsukimoto

Systems & Software Engineering Laboratory,
Research & Development Center, Toshiba Corporation.
70, Yanagi-cho, Saiwai-ku, Kawasaki 210, Japan.
E-mail: {chie,tukimoto}@ssel.toshiba.co.jp

ABSTRACT

Since many real data consist of numerical data, the discovery of understandable propositions from numerical data is important, and can be regarded as inductive learning with continuous classes. However, inductive learning in AI can be applied only to classification problems with discrete classes, and cannot be applied to the problems with continuous classes. One of the authors previously presented an algorithm for discovering understandable propositions from numerical data. The algorithm consists of normalization, multiple regression analysis and the approximation of multilinear functions by continuous Boolean functions. However, since the approximation algorithm is exponential in computational complexity, it can hardly be applied to real databases. This paper presents a polynomial approximation algorithm and applies the algorithm to a stock database to discover understandable propositions representing static relations and dynamic relations among stock prices.

1 INTRODUCTION

Since many real data consist of numerical data, the discovery of understandable propositions from numerical data is important. The following table shows an example of numerical data.

x_1	x_2	...	x_9	x_{10}
780	1280	...	456	845

In this table, $x_i (i = 1, ..., 9)$ are independent variables and x_{10} is a dependent variable, that is, x_{10} is represented as the function of $x_i (i = 1, ..., 9)$. If x_{10} is discrete, the problem is a classification. Therefore, the discovery of understandable propositions from numerical data can be regarded as inductive learning with continuous classes.

Inductive learning in AI can be applied to classification problems with discrete classes, but cannot be applied to problems with continuous classes. For example, C4.5 does not work well for the classification problem with continuous classes [Qui93].

One of the authors previously presented an algorithm for discovering understandable propositions from numerical data [Tsu94]. The basic idea is to obtain propositions from prediction models such as regression functions. The algorithm consists of the following steps.

1. The numerical data are normalized to [0,1].

2. Multiple regression analysis is performed.

3. The multilinear function obtained by multiple regression analysis is approximated by a continuous Boolean function.

The third step is based on the space of multilinear functions, which is made into a Euclidean space, when the domain is $\{0, 1\}$ or [0,1] [TM94] [TM96]. By extending the domain from $\{0, 1\}$ to [0,1], continuously valued logical functions satisfying all axioms of classical logic are obtained, which are called continuous Boolean functions in this paper [Tsu95]. A function obtained by multiple regression analysis in which data are normalized to [0,1] belongs to this Euclidean space. Therefore, the function can be approximated by a continuous Boolean function.

However, since the above approximation algorithm is exponential in computational complexity, it can hardly be applied to real databases. This paper presents a polynomial approximation algorithm when the regression function is linear. The basic idea is to generate terms from the lowest order up to a certain order. The algorithm is applied to a stock database. AI researchers have been interested in finance and have applied prediction methods to stock data[Wal91]. We discover understandable propositions representing static relations and dynamic relations among stock prices.

Section 2 explains the space of multilinear functions. Section 3 presents a polynomial approximation algorithm. Section 4 describes the discovery of propositions from a stock database.

The following notations are used. $x, y, \ldots$ stand for variables. $f, g, \ldots$ stand for functions.

2   THE SPACE OF MULTILINEAR FUNCTIONS

Definition 1 *Multilinear functions of n variables are as follows:*

$$\sum_{i=1}^{2^n} a_i x_1^{e_1} \cdots x_n^{e_n},$$

where a_i is real, x_i is variable, and e_i is 0 or 1.

In this paper, for simplification, $\sum$ stands for $\sum_{i=1}^{2^n}$, and n stands for the number of variables. For example, multilinear functions of 2 variables are as follows:

$axy + bx + cy + d.$

2.1   Multilinear functions of the domain $\{0, 1\}$

Theorem 2 *The space of multilinear functions ($\{0, 1\}^n \rightarrow \mathbf{R}$) is the Euclidean space spanned by the atoms of Boolean algebra of Boolean functions. Proof is omitted due to space limitations. Refer to [TM96].*

2.2   Multilinear functions of the domain $[0, 1]$

Multilinear functions of the [0,1] domain are briefly explained in this subsection [TM94].

Definition 3 *Definition of τ*

Let $f(x)$ be a real polynomial function. Consider the following formula:

$f(x) = p(x)(x - x^2) + q(x),$

where $q(x) = ax + b$, where a and b are real, that is, $q(x)$ is the remainder.

τ_x is defined as follows:

$\tau_x : f(x) \rightarrow q(x).$

The above definition implies the following property:

$\tau_x(x^k) = x$, where k is any natural number.

In the case of n variables, τ is defined as follows:

$\tau = \prod_{i=1}^{n} \tau_{x_i}.$

For example, $\tau(x^2 y^3 + y + 1) = xy + y + 1.$

Theorem 4 *The space of multilinear functions ($[0, 1]^n \rightarrow R$) is a Euclidean space with the following inner product: $\langle f, g \rangle = 2^n \int_0^1 \tau(fg) dx$. Proof can be found in [TM94].*

Definition 5 *Logical operations are defined as follows:*

AND	:	$\tau(fg)$,
OR	:	$\tau(f + g - fg)$,
NOT	:	$\tau(1 - f)$.

Theorem 6 *The functions obtained from Boolean functions by extending the domain from $\{0, 1\}$ to $[0, 1]$ can satisfy all axioms of classical logic with the logical operations defined above. Proof can be found in [Tsu95].*

Therefore, in this paper, the functions obtained from Boolean functions by extending the domain from $\{0, 1\}$ to [0,1] are called continuous Boolean functions.

For example, x, $1 - y(= \bar{y})$ and xy are continuous Boolean functions, where $x, y \in [0, 1]$. x means the direct proportion and $1 - y$ means the "inverse proportion".

The orthonormal functions are the functions which can be obtained from the atoms of Boolean algebra by extending the domain from $\{0, 1\}$ to [0,1]. For example, in the case of 2 variables, the orthonormal functions are as follows:

$xy, x(1 - y), (1 - x)y, (1 - x)(1 - y).$

Hereinafter, the domain is [0,1]. Since multilinear functions can be represented as vectors, multilinear functions are also described as $(f_i), (g_i) \ldots$, where f_i and g_i stand for the components.

3   APPROXIMATING A MULTILINEAR FUNCTION BY A CONTINUOUS BOOLEAN FUNCTION

3.1   Basic approximation algorithm

Theorem 7 *Consider that a multilinear function is approximated by the nearest continuous Boolean function. Let (f_i) be a multilinear function. Let $(g_i)(g_i = 0$ or $1)$ be a continuous Boolean function. The approximation algorithm is as follows:*

$$g_i = \begin{cases} 1 (f_i \geq 0.5), \\ 0 (f_i < 0.5). \end{cases}$$

Proof The nearest continuous Boolean function minimizes $\sum(f_i - g_i)^2$. Each term can be minimized independently and $g_i = 1$ or 0. Therefore, the above approximation algorithm is obtained.□

For example, let

$f = 0.6x - 1.1y + 0.3$

be a multilinear function. The function is transformed to

$f = -0.2xy + 0.9x(1 - y) - 0.8(1 - x)y + 0.3(1 - x)(1 - y).$

By the above approximation algorithm, the function is approximated to

$x(1-y)$.

The approximation algorithm can be regarded as a pseudo maximum likelihood method using the principle of indifference[Tsu94]. The principle of indifference states that a probability distribution is uniform when we have no information[Key21].

The above approximation algorithm is exponential in computational complexity, and therefore, unrealistic when variables are many.

3.2 The condition that x_i exists in the continuous Boolean function after approximation.

A polynomial approximation algorithm is presented when the function is linear. Hereinafter, multilinear functions are limited to linear functions.

Let a linear function be

$$p_1 x_1 + ... + p_n x_n + p_{n+1},$$

and let the orthonormal expansion be

$$a_1 x_1 \cdots x_n + a_2 x_1 \cdots \bar{x}_n + ... + a_{2^n} \bar{x}_1 \cdots \bar{x}_n.$$

Theorem 8 a_i's are as follows:

$$a_1 = p_1 + ... + p_n + p_{n+1},$$
$$a_2 = p_1 + ... + p_{n-1} + p_{n+1},$$
$$...$$
$$a_{2^{n-1}} = p_1 + p_{n+1},$$
$$a_{2^{n-1}+1} = p_2 + ... + p_n + p_{n+1},$$
$$...$$
$$a_{2^n-1} = p_n + p_{n+1},$$
$$a_{2^n} = p_{n+1}.$$

Proof a_1 can be obtained by substituting $x_1 = \cdots = x_n = 1$ in the following formula.

$$p_1 x_1 + ... + p_n x_n + p_{n+1} = a_1 x_1 \cdots x_n + a_2 x_1 \cdots \bar{x}_n ... + a_{2^n} \bar{x}_1 \cdots \bar{x}_n$$

The other coefficients can be obtained in a similar manner. □

Theorem 9 The condition that $x_{i_1} \cdots x_{i_k} \bar{x}_{i_{k+1}} \cdots \bar{x}_{i_l}$ exists in the continuous Boolean function after approximation is as follows:

$$\sum_{j=i_1}^{i_k} p_j + p_{n+1} + \sum_{1 \le j \le n, j \ne i_1,...,i_l, p_j < 0} p_j \ge 0.5.$$

Proof Consider the existence condition of x_1 in the continuous Boolean function after approximation. (For simplification, this condition is called the existence condition). Because $x_1 = x_1 x_2 \cdots x_n \vee x_1 x_2 \cdots \bar{x}_n \vee ... \vee x_1 \bar{x}_2 \cdots \bar{x}_n$, the existence of x_1 equals the existence of the following terms:

$$x_1 x_2 \cdots x_n, x_1 x_2 \cdots \bar{x}_n, ..., x_1 \bar{x}_2 \cdots \bar{x}_n.$$

The existence of the above terms means that all coefficients of these terms $a_1, a_2, ..., a_{2^{n-1}}$ are greater than or equal to 0.5 (See 3.1). That is,

Min$\{a_i\} \ge 0.5 (1 \le i \le 2^{n-1})$,
Because a_i's $(1 \le i \le 2^{n-1})$ are

$$a_1 = p_1 + ... + p_n + p_{n+1},$$
$$a_2 = p_1 + ... + p_{n-1} + p_{n+1},$$
$$...$$
$$a_{2^{n-1}} = p_1 + p_{n+1},$$

each $a_i (1 \le i \le 2^{n-1})$ contains p_1. If each p_j is non-negative, $a_{2^{n-1}} (= p_1 + p_{n+1})$ is the minimum because the other a_i's contain other p_j's, and therefore the other a_i's are greater than or equal to $a_{2^{n-1}} (= p_1 + p_{n+1})$. Generally, since each p_j is not necessarily non-negative, the Min$\{a_i\}$ is a_i which contains all negative p_j. That is,

Min$\{a_i\} = p_1 + p_{n+1} + \sum_{1 \le j \le n, j \ne 1, p_j < 0} p_j$,

which necessarily exists in $a_i (1 \le i \le 2^{n-1})$, because $a_i (1 \le i \le 2^{n-1})$ is

$p_1 + p_{n+1} + ($arbitrary sum of $p_j (2 \le j \le n))$.

From the above arguments, the existence condition of x_1, Min$\{a_i\} \ge 0.5$, is as follows:

$$p_1 + p_{n+1} + \sum_{1 \le j \le n, j \ne 1, p_j < 0} p_j \ge 0.5.$$

Since $p_1 x_1 + ... + p_n x_n + p_{n+1}$ is symmetric for x_i, the above formula holds for other variables; that is, the existence condition of x_i is

$$p_i + p_{n+1} + \sum_{1 \le j \le n, j \ne i, p_j < 0} p_j \ge 0.5.$$

Similar discussions hold for $\bar{x}_i$, and therefore we have the following formula:

$$p_{n+1} + \sum_{1 \le j \le n, j \ne i, p_j < 0} p_j \ge 0.5.$$

Similar discussions hold for higher order terms $x_{i_1} \cdots x_{i_k} \bar{x}_{i_{k+1}} \cdots \bar{x}_{i_l}$, and therefore we have the following formula:

$$\sum_{j=i_1}^{i_k} p_j + p_{n+1} + \sum_{1 \le j \le n, j \ne i_1,...,i_l, p_j < 0} p_j \ge 0.5. \quad \square$$

3.3 Generation of DNF formulas

The approximation algorithm generates terms using the above formula from the lowest order up to a certain order. A DNF formula can be generated by taking the disjunction of the terms generated by the above formula. A term whose existence has been confirmed does not need to be re-checked in higher order terms. For example, if the existence of x is confirmed, then it also implies the existence of xy, xz ,..., because $x = x \vee xy \vee xz$; hence, it is unnecessary to check the existence of xy, xz,.... As can be seen from the above discussion, the generation method of DNF formulas includes reductions such as $xy \vee xz = x$.

Let

$$f = 0.65 x_1 + 0.23 x_2 + 0.15 x_3 + 0.20 x_4 + 0.02 x_5$$

be the linear function. The existence condition of $x_{i_1} \cdots x_{i_k} \bar{x}_{i_{k+1}} \cdots \bar{x}_{i_l}$ is

$$\sum_{j=i_1}^{i_k} p_j + p_{n+1} + \sum_{1 \leq j \leq n, j \neq i_1, \ldots, i_l, p_j < 0} p_j \geq 0.5.$$

In this case, each p_i is positive; therefore the above formula can be simplified to

$$\sum_{j=i_1}^{i_k} p_j \geq 0.5.$$

For x_i, the existence condition is $p_i \geq 0.5$. For $i = 1, 2, 3, 4, 5, p_1 \geq 0.5$, and therefore x_1 exists. For $x_i x_j$, the existence condition is $p_i + p_j \geq 0.5$. For $i, j = 2, 3, 4, 5$, $p_i + p_j < 0.5$, and therefore no $x_i x_j$ exists. For $x_i x_j x_k$, the existence condition is $p_i + p_j + p_k \geq 0.5$. For $i, j, k = 2, 3, 4, 5$, $p_2 + p_3 + p_4 \geq 0.5$, and therefore $x_2 x_3 x_4$ exists. Because higher order terms cannot be generated from x_5, the approximation algorithm stops. Therefore, x_1 and $x_2 x_3 x_4$ exist and the DNF formula is the disjunction of these terms, that is,

$$x_1 \vee x_2 x_3 x_4.$$

3.4 Computational complexity of the approximation algorithm and error analysis

The computational complexity of generating the mth order terms is a polynomial of $\binom{n}{m}$, that is, a polynomial of n. Thus, the computational complexity of generating up to the kth order terms is $\sum_{m=1}^{m=k} \binom{n}{m}$, which is a polynomial of n. Therefore, the computational complexity of generating DNF formulas from the linear functions is a polynomial of n. However, the sum of the number of up to the highest (=nth) order terms is $\sum_{m=1}^{m=n} \binom{n}{m} = 2^n$; therefore, the computational complexity of generating up to the highest order terms is exponential. If it is desired that the computational complexity be reduced to a polynomial, the terms are generated only up to a certain order.

When the domain is $\{0, 1\}$, Linial showed the following theorem[LMN93]:

Assume that the probability distribution is the uniform distribution.

$$\sum_{|S| > k} \hat{f}(S)^2 \leq 2M2^{-k^{1/2}/20},$$

where f is a Boolean function, S is a term, $|S|$ is the order of S, k is any integer, $\hat{f}(S)$ denotes the Fourier Transform of f at S and M is the circuit size of the function.

The above formula shows the high order terms have very little power; that is, low order terms are informative. Therefore, a good approximation can be obtained by generating terms up to a certain order; that is, the computational complexity can be reduced to a polynomial by adding small

errors. When the domain is [0,1], a similar theorem holds, which will be presented in another paper.

4 DISCOVERING UNDERSTANDABLE PROPOSITIONS FROM NUMERICAL DATA

First, the procedures are described. Second, we confirm that the algorithm presented in this paper works well compared with another algorithm. Since no other inductive learning algorithm can work well for continuous classes, *iris* data is used, whose attributes are continuous and whose classes are discrete. Finally, we present two kinds of experiments with stock data. One is devoted to static relations among stock prices. The other is devoted to dynamic relations among stock prices.

4.1 Procedures

The procedures are as follows.

1. Normalize data to [0,1].

   There are a few methods for normalization. In this paper, we use the following method:

   Let a stand for the maximum data and let b stand for the minimum data.

   $$y = (x - b)/(a - b),$$

   where x is a given data and y is the normalized data.

2. Execute multiple regression analysis.

   A linear function is obtained by the usual multiple regression analysis.

3. Approximate it by a continuous Boolean function.

   A continuous Boolean function is obtained by the approximation algorithm in the preceding section. In this paper, generating terms are terminated when terms are firstly generated.

4.2 *iris* data

iris data consists of four numerical attributes: sepal length(=a), sepal width(=b), petal length(=c) and petal width(=d); and three classes: *setosa*, *versicolour* and *virginica*. A linear function is obtained for each class by multiple regression analysis. As a result, three linear functions are obtained as follows:

setosa	:	$\bar{c}$
versicolour	:	$\bar{b}\bar{d} \vee bcd \vee bc\bar{d}$
virginica	:	d

The accuracy of the results of our algorithm is 92.0%, and

the accuracy of the results of C4.5 is 93.6%. Therefore, our algorithm is a little worse than C4.5. However, the results of our algorithm will be more accurate as a result of improvements such as variable selection.

4.3 Stock data -static relations

4.3.1 Static relations among industries

The data consists of 200-days' stock prices of 200 companies listed on the Tokyo stock market in 1990. The 200 companies were selected from all industries. The 200 companies are classified into 24 industries as follows.

> fishery, construction, food, textile, paper, chemical, oil, petrochemical, steel, nonferrous metals, machinery, electrical products, shipbuilding, motor vehicle, precision machinery, trading, finance, real estate, insurance, transportation, electric power, gas, communication, film

Consider the relationships among the stock prices of industries, where the stock price of an industry is the average stock price of the companies in the industry. The database is transformed to the following one. I_i stands for the average stock price of the companies in the industry. I_k stands for the average stock price of the companies in an industry.

I_1	I_2	...	I_{23}	I_k
780	1280	...	456	845

We adopted electrical products as the class. First, we have obtained a good linear function whose average error is 0.038. Second, we have obtained the following proposition from the linear function:

$$e = tp\bar{t_r}(m \vee \bar{f}),$$

where e, t, p, m, t_r and f stand for electrical products, textile, petrochemical, machinery, transportation and finance. This proposition means that electrical products is high(low), when textile is high(low), petrochemical is high (low), transportation is low(high) and (machinery is high(low) or finance is low(high)). The accuracy of the proposition is 86.3%, where a stock price is high when the normalized datum is greater than 0.95 and a stock price is low when the normalized datum is less than 0.05. Thus we have obtained a good proposition.

Fig.1 shows the stock prices of electrical products, transportation and machinery. Textile, petrochemical and finance are omitted for simplification. X axis stands for time. The unit is day. Y axis stands for stock prices. Since the stock prices in Fig.1 are biased in order to increase the visibility, the absolute values are meaningless (Fig.2 is

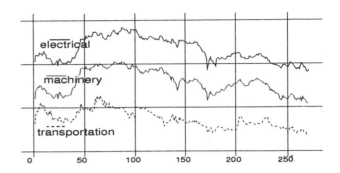

FIGURE 1: A static relation among industries

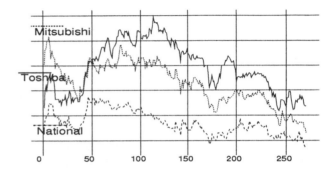

FIGURE 2: A static relation in the electrical industry

the same). From Fig.1, we can intuitively confirm that the proposition obtained is reasonable.

4.3.2 Static relations among the companies in an industry

There may be some relationships among the companies in the same industry. We checked the electrical industry, where there are 16 companies. We adopted Toshiba as the class. First, we have obtained a good linear function whose average error is 0.042. Second, we have obtained the following proposition from the linear function:

$$t = m_i\bar{n}(hm_e \vee hf \vee m_ef),$$

where t, m_i, n, h, m_e and f mean Toshiba, Mitsubishi, National, Hitachi, Meidensha and Fujitsu, respectively. This proposition means that Toshiba is high(low) when Mitsubishi is high(low), National is low(high) and so on. The accuracy of the proposition is 85.2%, where a stock price is high when the normalized datum is greater than 0.95 and a stock price is low when the normalized datum is less than 0.05. Thus we have obtained a good proposition. Fig.2 shows the stock prices of Toshiba, Mitsubishi and National. From Fig.2, we can intuitively confirm that the proposition obtained is reasonable.

4.4 Stock data -dynamic relations

It is interesting to discover the dynamic relations among the stock data. Let's represent future stock prices by the past stock prices. Let x_k stand for a stock price. Let x_is stand for the other stock prices. Let x_{in} stand for the nth day's stock prices. The problem is formalized as follows:

$$x_{k(n+l)} - x_{kn} = f((x_{1n} - x_{1(n-m)}), ..., (x_{in} - x_{i(n-m)}), ..., (x_{200n} - x_{200(n-m)})),$$

where $f(x)$ stands for a linear function. In this paper, $l = 3$ and $m = 5$.

4.4.1 Dynamic relations among industries

Consider the relationships among the stock prices of industries. We adopted electrical products as the class. The database is transformed to the following one.

I_1	I_2	...	I_{24}	I_k
-80	128	...	-47	8

The data are filtered by the moving average of 8 days before multiple regression analysis. First, we have obtained a good linear function whose average error is 0.068. Second, we have obtained the following proposition from the linear function:

$$e = \bar{t}ef(\bar{c} \vee g),$$

where e, t, f, c and g mean electrical products, textile, finance, chemical and gas. The above proposition means that electrical products increases when textile has decreased, electrical products has increased, finance has increased and so on. The accuracy of the proposition is 81.5%. Thus we have obtained a good proposition. Note that the propositions obtained depend on l and m.

4.4.2 Dynamic relations among the companies in an industry

There may be some causal relationships among the companies in the same industry. We checked the electrical industry. We adopted Toshiba as the class. First, we have obtained a good linear function whose average error is 0.12. Second, we have obtained the following proposition from the linear function:

$$t = sy(\bar{h} \vee t \vee \bar{m}),$$

where t, s, y, h and m mean Toshiba, Sharp, Yokogawa, Hitachi and Mitsubishi. This proposition means that Toshiba increases when Sharp has increased and Yokogawa has increased and so on. The accuracy of the proposition is 84.2%. Thus we have obtained a good proposition. Note that the propositions obtained depend on l and m.

5 CONCLUSIONS

One of the authors previously presented an algorithm for discovering understandable propositions from numerical data. However, since the approximation algorithm is exponential in computational complexity, it can hardly be applied to real databases. This paper has presented a polynomial approximation algorithm and applied the algorithm to a stock database to discover understandable propositions representing static relations and dynamic relations among stock prices. Future work will include the extension of the polynomial approximation algorithm to non-linear multiple regression analysis.

References

[Key21]   J.M. Keynes: *A Treatise on Probability*, Macmillan, London, 1921.

[LMN93]   N. Linial, Y. Mansour and N. Nisan: Constant depth circuits, Fourier Transform, and Learnability, *Journal of the ACM*, Vol.40, No.3, pp.607-620, 1993.

[Qui93]   J. R. Quinlan: *Programs for Machine Learning*, Morgan Kaufmann, 1993.

[TM94]   H. Tsukimoto and C. Morita: The Discovery of Propositions in Noisy Data, *Machine Intelligence 13*, Oxford University Press, 1994.

[TM96]   H. Tsukimoto and C. Morita: Efficient algorithms for inductive learning-An application of multi-linear functions to inductive learning, *Machine Intelligence 14*, Oxford University Press, 1996 (To appear).

[Tsu94]   H. Tsukimoto: The discovery of logical propositions in numerical data, *AAAI'94 Workshop on Knowledge Discovery in Databases*, pp.205-216, 1994.

[Tsu95]   H. Tsukimoto: On continuously valued logical functions satisfying all axioms of classical logic, *Systems and Computers in Japan* Vol.25, No.12, pp.33-41, SCRIPTA TECHNICA, INC., 1994.

[Wal91]   *Proceedings of the First International Conference on Artificial Intelligence Applications on Wall Street*, 1991.

A KNOWLEDGE REVISION LEARNER USING ARTIFICIALLY GENERATED EXAMPLES

Megumi Ishii†, Hussein Almuallim‡, and Shigeo Kaneda†

†NTT Communication Science Laboratories
†1−2356, Take, Yokosuka−shi, 238−03, Japan
†Email: {megumi, kaneda}@nttkb.ntt.jp
‡King Fahd University of Petroleum and Minerals
‡POBOX 801, Dhahran, 31261, Saudi Arabia
‡Email: hussein@ccse.kfupm.edu.sa

ABSTRACT

Attribute-type-learners which express examples as attribute vectors cover many application fields. When it is difficult to collect a sufficient number of examples, from which to acquire knowledge, the most appropriate learners are revision learners. Such learners can acquire knowledge from a small number of examples because they also employ knowledge provided by human experts. There are, however, only a few revision learners of the attribute-type. Thus, a framework is required that would allow non-revision learners, which cannot learn from both given knowledge and examples, to be used as revision-learners. We propose a revision learner that generates artificial examples from knowledge given by humans, inputs these artificial examples together with collected examples, and acquires knowledge by using non-revision-learners. Evaluating the proposed revision learner, which uses the Irvine-Database and the ID3 learner, reveals that knowledge acquisition can be performed more accurately than that based on either given knowledge or collected examples alone. Thus its effectiveness as a knowledge acquisition tool is demonstrated.

1. INTRODUCTION

Attribute-type-learners[MCM83], [Qui86], [CN89], which express examples as attribute vectors, have many application fields[AWG91], [BP91], [AAYK94], [HS93], [DWGT94]. In many application domains, however, it is difficult to collect a number of examples that is sufficient to acquire knowledge of acceptable quality level[KNT91]. Learners that learn solely from examples are not helpful, in such a situation. Instead, one has to employ a *revision learner*, which learns from collected examples (*real-examples*), combined with prior knowledge ascribed by humans (*human-made-knowledge*).

In the real world, it is difficult to forecast which learner would be optimal for each task since optimality depends upon the task. As a result, various types of revision learners should be applied to a task and the most accurate acquired knowledge should be selected. There are, however, only a few revision learners of the attribute-type[TN94]. Accordingly, a framework is required in which *non-revision-learners*, which can not learn from both human-made-knowledge and examples, can be turned into revision learners.

To achieve this framework, we propose a revision learner that generates examples from human-made-knowledge (*artificial-examples*) and inputs them, along with real-examples, into non-revision-learners. As a result, the non-revision-learners can use the information contained in human-made-knowledge. In this paper, the non-revision-learner employed in the proposed learner is called an *internal-learner*. A problem that is encountered when combining artificial-examples with real-examples is determining the weights (significance) of the artificial-examples originating from human-made-knowledge, relative to the real-examples. In this paper, these weights are determined through parameter tuning using the cross-validation method.

Section 2 formally proposes the revision learner and Section 3 evaluates it using ID3[Qui86] as the internal-learner. Related work is described in Section 4. Finally, a conclusion is given in Section 5.

2. REVISION LEARNER USING ARTIFICIALLY GENERATED EXAMPLES

This section proposes a revision learner that satisfies the requirements described in the previous section. This learner can, by converting human-made-knowledge into examples, "inject" human-made-knowledge along with real-examples into arbitrary attribute-type-learners.

2.1 Knowledge expression format

2.1.1 Examples

The proposed learner is meant for classification tasks. Each example is expressed as an attribute value vector and a class name:

$$(v_1, v_2, \cdots, v_i, \cdots, v_n : Class\_value).$$

Here, n is the number of attributes, v_i is some value of the i-th attribute, and $Class\_value$ is the name of the class as determined by the attribute values.

The number of attribute values for each attribute is finite, and an unobserved attribute value (missing value) is represented as a question mark "?".

2.1.2 Expression of human-made-knowledge

The proposed learner employs Internal Disjunctive Expression (*IDE*) format rules[Hau88] to express human-made-knowledge. Under the IDE syntax, only one attribute is referred to in each disjunct which makes up a Conjunctive Normal Form (CNF) formula. Human-made-knowledge expressed as rules, decision trees, or decision lists can be easily expressed using IDE format rules (the conversion details are omitted in this paper).

IDE format rules

Consider the following disjunct:

$$d_i \equiv (N_i = v_{i1}) \vee (N_i = v_{i2}) \vee, \cdots, (N_i = v_{ik})$$

where $v_{i1}, v_{i2}, v_{i3}, \cdots, v_{ik}$ are values of attribute N_i. An IDE formula consists of disjuncts of the above form connected by the $\wedge$ operator as follows:

$$IDE \equiv d_1 \wedge d_2 \wedge d_3 \wedge, \cdots.$$

Rules can then be expressed as follows using IDE:

$$if \ \ IDE \ \ then \ \ Class = Class\_value.$$

Negation is not allowed in IDE format rules. However, since the number of attribute values for each attribute is finite, the negation

$$N_i = not \ v_j$$

means that all attribute values other than v_j for N_i can be expressed by listing them within the disjunctive expression.

2.2 Revision learner using artificially generated examples

This subsection describes the flow of the proposed revision learner technique. A block diagram of the learner is shown in Figure 1.

Flow of the revision learner

Step 1: Generate artificial examples conforming to the given human-made-knowledge.

Step 2: Merge the human-made-knowledge with the real-examples. This is done by assigning to the real and artificially generated examples proper weights that reflect their soundness[1].

Step 3: Acquire knowledge. The above examples are input into the internal-learner to obtain final knowledge.

The following describes Step 1 and Step 2 in more detail.

2.2.1 Generating artificial examples

We assume that the human-made-knowledge is expressed as a *set* of IDE format rules. The following three rules listed below, a pre-determined fixed number of artificial examples (called the *number-of-rule-examples*) are generated from each IDE format rule.

Rules for generating an artificial example

Let R be an IDE format rule.

1) An attribute not appearing in any rule of a given IDE format rule set is taken as a missing value, "?".

2) For an attribute appearing in R, an attribute value is selected at random from those values appearing within the conditions of that attribute.

---

[1]In the evaluation described later, the weight of real-examples is taken as 1.

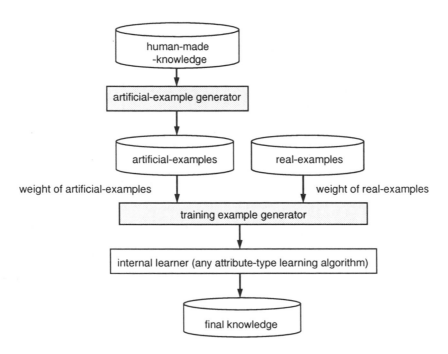

FIGURE 1: Block Diagram of the Proposed Learner.

3) For an attribute not appearing in R but appearing in other rules, an attribute value is selected at random from the set of values of that attribute.

An example of generating artificial-examples is described below. Assume that there are four attributes, $N_1, N_2, N_3,$ and N_4, with N_1 taking on the values of $\{0, 1\}$, N_2 the values of $\{0, 1, 2, 3\}$, N_3 the values $\{0, 1, 2\}$, and N_4 the values of $\{0, 1\}$. Assume that the human-made-knowledge includes the rule

$$if \quad (N_1 = 1) \wedge (N_2 = 1 \vee N_2 = 2) \quad then \quad Class = 0,$$

and that N_3 is used in other IDE format rules while N_4 does not appear in any of the IDE format rules.

The above rule states that the value of N_1 is 1, so N_1 has value 1 in the artificial examples produced from this rule. The value of N_2 in the rule is 1 or 2, and thus, the value allocated to N_2 in the artificial examples is selected randomly from 1 and 2. Since N_3 does not appear in the rule (but does in other IDE format rules), its value in the artificial examples is selected randomly from 0, 1 and 2. The value for N_4 is "?" in the artificial examples.

The following is a set of artificial examples generated for the above case when the number-of-rule-examples is 8.

$$(1, 2, 1, ?; 0) \qquad (1, 2, 0, ?; 0)$$
$$(1, 1, 0, ?; 0) \qquad (1, 2, 2, ?; 0)$$
$$(1, 1, 1, ?; 0) \qquad (1, 1, 0, ?; 0)$$
$$(1, 2, 2, ?; 0) \qquad (1, 1, 2, ?; 0)$$

When learning is performed by an internal-learner using only artificial examples generated as described above, it is desirable to completely reproduce the information in the original rules. In other words, it is required that the knowledge generated from artificial examples have the same level of accuracy as that of the human-made-knowledge.

In this generation method, the values for the attributes in the rules considered *don't care*[2] are selected at random from all values possessed by the attribute. In general, then, the greater the number-of-rule-examples, the more accurately the original human-made-knowledge should be reflected. Nevertheless, as described below, it has been verified experimentally that human-made-knowledge can actually be reproduced with a relatively small number-of-rule-examples.

---

[2] These are the attributes to which rule 3) is applied and have no influence on the classification (according to the human-made-knowledge). An attribute not appearing in R but appearing in other rules does influence the classification. For such attributes, rather than performing a random value selection, it would be smarter to use a prior distribution that is either inferred or given by a user.

2.2.2 Combining human-made-knowledge and examples

On combining human-made-knowledge and real-examples, a problem arises as to how much significance should be attached to human-made-knowledge and real-examples respectively. In the proposed learner, real-valued weights are assigned to the artificial and real examples. These weights measure the soundness of these two types of examples. Determining the best weights is a sort of parameter tuning which can be handled using the cross-validation method, based on the accuracy on the test examples.

Specifically, while tuning the weights of artificial examples and real-examples, cross validation is performed on each weight combination. The most accurate weight combination is then selected and the weights are assigned to the artificial and real examples. In the descriptions below, accuracy with respect to test examples refers to accuracy obtained by cross validation. A simple method for implementing the weight assignments is to appropriately duplicate the examples so that the number of copies of an example is proportional to its weight. An advantage of this method is that it can be applied to any internal learner. In this paper, however, we have chosen to modify slightly Quinlan's C4.5 program, which we use as our internal learner, in order to handle the real-valued weights explicitly instead of duplicating the examples. Modifying Quinlan's package to do this is quite easy because the package already associates a weight with each example for the purpose of handling unobserved attribute values (missing values)[3].

3. EVALUATION

Some experiments were performed using natural data to clarify the effectiveness of the proposed learner.

3.1 Evaluation method

In the following evaluations, ID3 (strictly speaking, its upgraded version, C4.5[Qui92]) is used as the internal-learner. The C4.5 pre-pruning function was turned off, but the post-pruning function was left on under default conditions so that example information could be reflected as much as possible in the decision trees generated by C4.5.

The evaluation data sets were the Soybean-Large,

TABLE 1: Data Sets for Experiments.

	# of classes	# of attributes	# of examples
Soybean-Large	19	35	683
Tic-Tac-Toe	2	9	958
Monks-3	2	6	432

TABLE 2: Number of Examples for Human-made-Knowledge and Test.

	# of examples for human-made-knowledge	# of real-examples	# of test examples
Soybean-Large	60	60	563
Tic-Tac-Toe	70	70	818
Monks-3	20	20	392

Tic-Tac-Toe , and Monks-3 which are all included in the Irvine-Database of the University of California at Irvine. The contents of each data set are shown in Table 1.

Due to the fact that the Irvine-Database does not contain human-made-knowledge, such knowledge was simulated by running C4.5 on a subset of the examples of each evaluation data set for this experiment. Human-made-knowledge was obtained by converting each path running from the generated decision tree root to a leaf into one IDE format rule since C4.5 expresses knowledge as a decision tree.

Table 2 shows the number of examples used to mimic human-made-knowledge, the number of real-examples, and the number of test examples for Soybean-Large, Tic-Tac-Toe and Monks-3. The number of test examples represents the number of examples used to predict the accuracy under cross validation. The number of cross validation was five. Five sets of human-made-knowledge were prepared for each data set.

This experiment evaluated (1) the reproducibility of human-made-knowledge information by the number-of-rule-examples, (2) the increase in accuracy achieved by the proposed learner and (3) the increase in accuracy of the final knowledge as determined by the quality of the human-made-knowledge.

---

[3] As found from our evaluation results, accuracy changes little even if the weight is changed past the decimal point. The example-copy method is therefore more realistic for learning algorithms other than C4.5.

TABLE 3: Reproducibility by the number-of-rule-examples.

	accuracy(%)	accuracy decrease(%)
human-made-knowledge	65.6	
number-of-rule-examples = 2	61.7	3.9
number-of-rule-examples = 4	64.9	0.7
number-of-rule-examples = 16	65.0	0.6

3.2 Reproducibility by the number-of-rule-examples

The reproducibility of human-made-knowledge information was evaluated by comparing the accuracy of human-made-knowledge and regenerated knowledge with respect to test examples. The latter knowledge is acquired from artificial examples generated from the human-made-knowledge. Five sets of human-made-knowledge were generated in this experiment, and for each of these sets, five data sets were prepared for the test examples taken from the examples not used for human-made-knowledge. Artificial examples were then generated from human-made-knowledge, and using the above five sets of test examples, the average accuracy was determined for 1) the knowledge generated from artificial examples and 2) human-made-knowledge.

Table 3 shows the results for Soybean-Large. The accuracy shown for each number-of-rule-examples is the average accuracy of the above 25 trials. When the value of the number-of-rule-examples is four or more, the difference in accuracy is within 1%. This shows that human-made-knowledge information can be reproduced even with a small number of values per attribute in the Soybean-Large domain (the average is three).

To ensure proper reproducibility, we have chosen to fix the number-of-rule-examples to 16 in the following evaluations. Due to the fact that the number of attribute values for each attribute of Tic-Tac-Toe and Monks-3 is about the same as that of Soybean-Large, it is considered that human-made-knowledge information can be reproduced with a small number-of-rule-examples in these cases as well.

3.3 Improvement in accuracy

In this experiment, the weight of the real-examples is taken as 1. Figure 2 shows the improvement of accuracy for Soybean-Large, Tic-Tac-Toe, and Monks-3. The vertical axis represents the average accuracy on test examples and the horizontal axis represents the weight given to artificial examples. This weight is shown here as the weight for the all artificial examples generated from one IDE format rule, or in other words, the weight of an IDE format rule. The solid lines represent the obtained accuracy with the proposed learner, with Soybean-Large indicated by ●, Tic-Tac-Toe by ▲ and Monks-3 by ■. A value of zero for the weight of an IDE format rule corresponds to the accuracy of the real-examples only, and the dashed lines labeled "human-made-knowledge" correspond to the human-made-knowledge itself (before generating artificial examples).

From Figure 2, it can be seen that the obtained accuracy is changed using the weights and that tuning by cross validation can find the best weight for IDE format rules. We found that the proposed learner provided more accurate knowledge than human-made-knowledge or than the knowledge generated from the given real examples alone. In addition, accuracy is better when the real-example side is weighted more strongly. This is considered to be due to the fact that real-examples have much more information concerning the target tasks than artificial examples in three evaluation domains. Note that the best point here does not appear as a sharp peak; high accuracy is obtained over a relatively wide region.

3.4 Change in accuracy with quality of human-made-knowledge

The influence of the quality of human-made-knowledge on the accuracy of the final knowledge is evaluated for (1) human-made-knowledge conforming to real-examples[4] and (2) human-made-knowledge contradicting real-examples. The data set is Soybean-Large.

Human-made-knowledge conforming to real-examples

Figure 3 shows how the accuracy of the final knowledge varies with the quality of human-made-knowledge conforming to real-examples. Sixty examples are allotted beforehand for generating human-

---

[4] Human-made-knowledge was generated based on the same probability distribution as the learning concept.

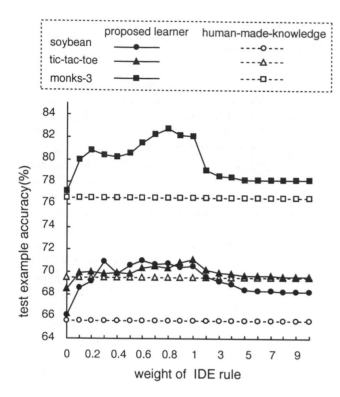

FIGURE 2: Accuracy Improvement (Irvine-Database).

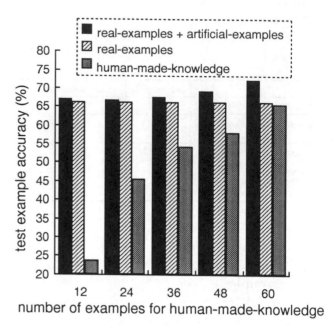

FIGURE 3: Rise in Accuracy by Quality of Human-made-Knowledge (Consistent Knowledge.)

made-knowledge, and the remaining examples are divided between 60 real-examples and 563 test examples.

In this experiment, the amount of information is equated to the quality of knowledge. The amount of information is defined as the number of examples used for generating human-made-knowledge[5]. The horizontal axis represents the number of examples used for generating human-made-knowledge; the quality of human-made-knowledge rises as the graph moves toward the right.

The vertical axis represents the accuracy of the final knowledge. The column labeled "real-examples + artificial-examples" indicates the accuracy of the final knowledge acquired by the proposed learner; the one labeled "real-examples" indicates the accuracy of the knowledge acquired using only real-examples; and the one labeled "human-made-knowledge" simply indicates the accuracy of human-made-knowledge. The chart shows that accuracy increases as the quality of

human-made-knowledge increases. This demonstrates that accuracy rises as the amount of knowledge conforming to real-examples increases. This is a reasonable result.

Human-made-knowledge contradicting real-examples

Figure 4 shows the accuracy achieved when human-made-knowledge is generated which contradicts real-examples. This contradiction is introduced by adding noise (class errors) to the artificial examples used for generating human-made-knowledge. The horizontal axis represents the number of noisy examples within the 60 used for generating human-made-knowledge (those whose class has been changed). In other words, the quality of human-made-knowledge decreases as one moves toward the right.

When using human-made-knowledge that contradicts real-examples, accuracy starts to drop due to the addition of noise, as shown by Figure 4. The weight of artificial examples is adjusted automatically to eliminate inappropriate human-made-knowledge by parameter tuning using cross validation. As a result, performance is prevented from becoming worse than that with real-examples only. This is an important

---

[5] The increase in the number of examples by the increase in the weight of human-made-knowledge does not mean the amount of information in human-made-knowledge increases. The reason for this is that the weight does not change the number of examples for generating human-made-knowledge.

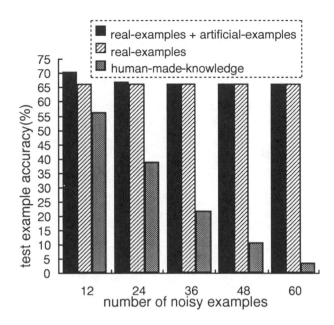

FIGURE 4: Decline in Accuracy with Corruption of Human-made-Knowledge (Contradictory Knowledge).

feature of this system. On the other hand, a limitation of the system is that when real-examples are too noisy to predict the accuracy using cross validation, it may provide undesirable knowledge based on the noisy real-examples.

4. RELATED RESEARCH

Much research has been conducted on Theory Revision for combining knowledge and examples [Coh91], [Ham91], [Rae92], [TS93]. Most of this research, however, considers only learners that use the first order predicate logic format for knowledge expression. There are also techniques that enable an attribute-type-learner to learn both knowledge and examples. One technique is [TN94] for ID3. AQ[MCM83] is one learner that handles knowledge and examples[6]. With these conventional techniques, non-revision-learners are restructured individually to become a revision learner. While [Nag89] proposed a revision learner that used artificially generated examples, it can not, however, guarantee the elimination of poor human-made-knowledge. The greatest difference between the proposed learner and other conventional techniques is

---

[6] The AQ learner from the beginning employed the same expression format for examples and rules, and is therefore considered to be one type of revision learner.

that it provides any attribute-type-learners with the ability to combine real-examples with human-made-knowledge without degrading in accuracy. Moreover, the proposed method can be applied to not only attribute-type-learners but also to the Nearest Neighbor method, Neural Nets and statistical classification methods (e.g., multiple regression analysis).

5. CONCLUSION

In this paper we have proposed a revision learner that can acquire more accurate knowledge by combining existing human-made-knowledge and examples. It generates examples from human-made-knowledge (artificial-examples) and inputs them together with collected examples (real-examples) into an non-revision-learner. In this process, the weights of artificial and real-examples is a problem that must be addressed, and here we applied parameter tuning using the cross-validation method to determine optimal weights. The most significant feature of the proposed revision learner is that it can employ any learning algorithm as a revision learner as long as it expresses examples as attribute vectors.

An evaluation was performed using C4.5 as the non-revision-learner and the Irvine-Database provided by the University of California at Irvine. The results showed that more accurate knowledge could be obtained compared to using either existing knowledge or examples alone.

In future work, we plan to confirm the effectiveness of our method by applying the proposed revision learner in real world domains using real human-made-knowledge.

References

[AAYK94] H. Almuallim, Y. Akiba, T. Yamazaki, and S. Kaneda, "Induction of Japanese-English Translation Rules From Ambiguous Examples and a Large Semantic Hierarchy," *Journal of Japan Society of Artificial Intelligence*, Vol. 9, No. 5, pp.730-740, Sept. 1994.

[AWG91] C. Apte, S. Weiss, and G. Grout. , "Predicting Defects in Disk Drive Manufacturing, A Case Study in High-Dimensional Classification," in Proc. of the 7th Conf. on Artificial Intelligence for Applications, pp.212-217, 1991.

[BP91] J. Bala and P. Pachowicz, "Application of Symbolic Machine Learning to the Recognition of Texture Concepts," in Proc. of the 7th Conf. on

Artificial Intelligence for Applications, pp. 224-230, 1991.

[CN89] P. Clark and T. Niblett, "The CN2 Induction Algorithm," *Machine Learning*, Vol. 3, No. 4, pp.261-284, 1989.

[Coh91] W. Cohen, "The Generality of Overgenerality," in Proc. of the 8th Int. Workshop on Machine Learning, pp.490-494, 1991.

[DWGT94] M. Dyne, L. Woolery, J. Gryzmala-Busse, and C. Tsatsuulis, "Using Machine Learning and Expert Systems to Predict Preterm Delivery in Pregnant Women," in Proc. of the 4th Conf. on Artificial Intelligence for Applications, pp. 344-350, 1994.

[Ham91] R. Hamakawa, "Revision Cost for Theory Refinement," in Proc. of the 8th Int. Workshop on Machine Learning, pp.514-518, 1991.

[Hau88] D. Haussler, "Quantifying Inductive Bias: AI Learning Algorithms and Valiant's Learning Framework," Artificial Intelligence, Vol. 26, No. 2, pp.177-211, 1988.

[HS93] L. A. Hermens and J. C. Schlimmer, "A Machine-learning Apprentice for the Completion of Repetitive Forms," in Proc. of the 9th Conf. on Artificial Intelligence for Applications, pp.164-170, 1993.

[KNT91] Y. Koseki, Y. Nakakuki, and M. Tanaka, "DT: A Classification Problem Solver with Tabular-Knowledge Acquisition," in Proc. of the 3rd Int. Conf. on Tools for AI, pp.156-163, 1991.

[MCM83] R. S. Michalski, J. G. Carbonell, and T. M. Mitchell (Eds.), Machine Learning, Tioga, Palo Alto, Calif, 1983.

[Nag89] Y. Nagano, "Reserch on Knowledge Aquisition for Constructing Expert Systems," Waseda University master thesis, 1989 (in Japanese).

[Qui86] J. R. Quinlan, "Induction of Decision Trees," Machine Learning, 1(1), 81-106, 1986.

[Qui92] J. R. Quinlan, "C4.5 Programming for Machine Learning," Morgan Kaufman, San Mateo, California, 1992.

[Rae92] L. Raedt, "Interactive Theory Revision," Academic Press, London, 1992.

[TN94] K. Tsujino and S. Nishida, "Inductive Generation of Classification Knowledge from Examples and Correlation Tables," in Proc. of Japanese Knowledge Aquisition for Knowledge-Based Systems Workshop, pp.183-195, 1994.

[TS93] S. Tangkitvanich and M. Simura, "Learning From an Approximate Theory and Noisy Examples," in Proc. of the 11th National Conference on Artificial Intelligence, pp.466-471, 1993.

LEARNING PROBABILISTIC NETWORKS FROM DATA

Bozena Stewart
Department of Computing and Information Systems
University of Western Sydney, Macarthur
Campbelltown, NSW, 2560, Australia
Email: b.stewart@uws.edu.au

ABSTRACT

Probabilistic networks provide a theoretically sound mechanism for handling uncertainty in expert systems but their use in practical applications has been limited by the difficulty of their construction. In the last few years an active area of research has been the development of machine learning methods for automated learning of probabilistic networks from empirical data. This paper presents a heuristic scheme that recovers from data a subclass of probabilistic networks — decomposable networks. Two different implementations of the method are proposed and their performance is illustrated on two data sets of real cases.

1. INTRODUCTION

An important problem in Artificial Intelligence (AI) is machine learning of probabilistic networks from empirical data. Probabilistic networks provide a reliable and coherent mechanism for representing and reasoning with uncertain knowledge in expert systems but their use in practical applications has been limited by the difficulty of their construction. So far, they have been constructed by a manual process based on interviewing domain experts, which is very time consuming and requires a close cooperation between the domain expert and a knowledge engineer. In many application domains, large databases of examples are now becoming available and it would be more convenient to use machine learning techniques to derive probabilistic networks automatically from data. During the last few years researchers have developed a variety of different schemes to perform this task. To date, however, none of these schemes have been successfully applied to real world problems. In the majority of cases, only the results obtained from simulated data have been reported in the literature. The main reason for the difficulty of recovering probabilistic networks from real data is the nature of the problem itself. It has been shown by Bouckaert [Bou94] that the problem of learning probabilistic networks which have the minimum number of edges (minimal I-maps) is NP hard. This implies that only exponential-time algorithms can be devised to learn optimal networks. Polynomial-time algorithms can be designed to learn only approximate probabilistic networks which have non-minimal graphical structures containing some redundant edges. But if the derived structures are reasonably close to optimal, such networks may be adequate for many practical applications.

This paper presents a heuristic method, based on stochastic simulation, that recovers from data a subclass of probabilistic networks — *decomposable networks*, in polynomial time. Although the method is approximate, it has some important optimality properties described in Section 4.5. Two different implementations are proposed and their performances are illustrated using two databases of real examples. Both algorithms, DN1 and DN2, are extensions of our earlier algorithm TCN [Ste94]. The results obtained show that the proposed approach is capable of learning decomposable networks from moderately large databases of real cases. The recovered networks provide a close approximation of the probability distribution contained in the data and can be used for performing probabilistic inference. Furthermore, their structure provides a valuable insight into the dependency relationships that exist among the attributes in the data.

The paper is organized as follows. Section 2 presents a brief overview of probabilistic networks, Section 3 gives a short summary of the previous work on learning of probabilistic networks, Section 4 presents our algorithms for learning decomposable networks from data, Section 5 presents experimental results, and Section 6 concludes the paper and suggests possible future work.

2. AN OVERVIEW OF PROBABILISTIC NETWORKS

A probabilistic network consists of two components: 1) a graphical representation of domain structure, and 2) numerical parameters quantifying the structure. There are three classes of probabilistic networks – *Bayesian networks*, *Markov networks*, and *decomposable networks*. Figure 1 illustrates an example of each type of network.

In a Bayesian network, the structure of a domain is represented in terms of a directed acyclic graph in which the vertices correspond to the variables of the domain, and the edges represent the relationships between the variables. Each vertex in the graph is quantified with a conditional probability distribution specifying conditional probability of each value of the corresponding random variable given every possible combination of values of its parents. In a Markov network, the domain structure is represented by an undirected graph, where the vertices represent the random variables and the edges represent symmetric associations between the variables. The graph is quantified with numerical quantities called evidence potentials. In a decomposable network, the domain structure is represented in terms of a triangulated undirected graph which is quantified with the marginal probability distributions of the cliques in the graph. A triangulated graph is a graph containing no cycles of length ≥ 4 without a chord. A clique in a graph is a maximal subset of vertices in which each pair of vertices is connected by an edge.

A Bayesian network can be transformed into an equivalent Markov network by including so called *moral edges* and dropping directions. A moral edge is an edge obtained by connecting two unconnected parent nodes that have a common child node. For example, by connecting the vertices X_2 and X_4 in Figure 1(a) we obtain the moral edge (X_2, X_4). The moral graph can then be transformed into a triangulated graph by including fill-in edges so that the condition for a triangulated graph is satisfied. The key property of a probabilistic network is that the missing links can be interpreted in terms of conditional independence. We say that two variables X and Y are conditionally independent given the variable Z if $P(x|y, z) = P(x|z)$, where x, y, z denote all possible values of X, Y, Z, respectively. In a probabilistic network, a missing link between two vertices X_i and X_j implies that X_i and X_j are conditionally independent given some subset of the remaining variables.

A probabilistic network represents the joint probability distribution of the domain variables and this distribution can be computed using an appropriate product form of the corresponding numerical parameters. In a decomposable network, the joint distribution is computed as the product of the distributions of the cliques divided by the product of the distributions of their intersections. For theoretical details refer to Pearl [Pea88] or Neapolitan [Nea90].

3. A BRIEF OVERVIEW OF METHODS FOR LEARNING PROBABILISTIC NETWORKS

The earliest work related to learning of probabilistic networks was done by Chow and Liu [CL68] in the late 1960's. They investigated the problem of approximating a discrete joint probability distribution of n random variables by a product of $n - 1$ second order distributions or the distribution of tree dependence. They discovered that the tree distribution which minimizes the Kullback-Leibler cross-entropy measure is the distribution corresponding to the maximum weight spanning tree (MWST), where the weight of the branch (X_i, X_j) is defined by the mutual information measure

$$I(X_i, X_j) = \sum_{x_i, x_j} P(x_i, x_j) log \frac{P(x_i, x_j)}{P(x_i)P(x_j)} \geq 0 \quad (3.1)$$

Chow and Liu's algorithm first computes all $n(n-1)/2$ information weights $I(X_i, X_j), i = 1, 2, ..., n - 1, j = 2, 3, ..., n, i < j$, and then constructs a maximum weight spanning tree. The algorithm is efficient in time and space requirements $(O(n^2)$ in both) but produces reliable results only when the underlying distribution is tree structured.

Chow and Liu's algorithm provided a basis for Rebane and Pearl algorithm [Pea88] for learning polytrees. A polytree is a directed graph with no directed or undirected cycles (a tree with multiple roots). Rebane and Pearl discovered that if a joint distribution is of polytree dependence then its underlying undirected graph is the MWST of Chow and Liu. Their algorithm first recovers a MWST and then assigns the directionality to the branches of the tree to the maximum extent possible. The algorithm is efficient in time and space requirements $(O(n^2))$ but is reliable only for polytree distributions. Rebane and Pearl derived an important relationship for mutual information weights, which states that if two variables X and Y are conditionally independent given a variable Z then the following inequality holds:

$$I(X, Y) < min[I(X, Z), I(Y, Z)] \quad (3.2)$$

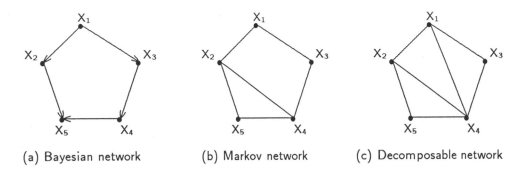

Figure 1: Probabilistic networks.

This relationship is used in Section 4.5 to provide a theoretical justification for our method.

Learning of general multiply connected networks is a difficult problem. Several algorithms have been developed that use a two stage strategy to derive a probabilistic network. In the first stage an underlying undirected graph, called a skeleton, is recovered from data, and in the second stage, edges of the skeleton are assigned orientations consistent with the data. These methods aim to identify optimal (minimal) network structures by using exact statistical tests to discover the conditional independence relations contained in the data. The worst case complexity of these algorithms is exponential in the number of vertices in the graph. The best known algorithms in this class of methods are the IC algorithm of Pearl and Verma [PV91] which recovers from data a hybrid graph containing three types of edges: undirected, directed and bi-directed, and the PC algorithm of Spirtes and Glymour [SG91] which recovers Bayesian networks and has polynomial worst case complexity for sparse networks.

Since the exact algorithms are too inefficient for practical applications, researchers have been investigating heuristic schemes that could learn good approximate networks. This class of methods includes the K2 algorithm of Cooper and Herskovits [CH91] that uses a Bayesian method to learn Bayesian networks; the CB algorithm of Singh and Valtorta [SV93] which combines the approaches of the PC algorithm and the K2 algorithm to generate Bayesian networks; the algorithm of Lam and Bacchus [LB94] which uses the minimal description length principle to learn Bayesian networks; and the Constructor system of Fung and Crawford [FC90], which learns Markov networks. Of these heuristic schemes, the K2 algorithm, and the minimal description length algorithm of Lam and Bacchus have polynomial-time worst case complexity and thus show most potential for practical applications.

4. LEARNING DECOMPOSABLE NETWORKS FROM DATA

This section presents our heuristic method, which employs stochastic simulation to recover from a database of empirical examples a decomposable network consistent with the data. Two different strategies for choosing the edges to be added to the graph are proposed and their performance is compared. The first strategy is implemented in the algorithm DN1 which is a modified version of our earlier algorithm, TCN [Ste94]. The modification involves the use of a different heuristic for determining moral edges, described below. The DN1 algorithm begins with an empty graph and then repeatedly determines the worst fitting edge and adds it to the graph until the resulting decomposable network fits the data or the maximum clique size exceeds a predetermined limit. The second strategy is implemented in the algorithm DN2 and differs from the first strategy in that it repeatedly determines the worst fitting node and the worst fitting edge of that node and adds that worst fitting edge to the graph. In most cases the two algorithms derive almost identical graphs, but sometimes some of the weaker edges may be different. Both algorithms use the same goodness of fit procedure (GFT) to determine the worst fitting edges and to test goodness of fit of the resulting network. Both algorithms converge to optimal (minimal) graphs for three classes of networks: trees, polytrees, and networks in which the longest undirected cycle is of length three. Experiments with simulated data have shown that if an underlying graph is sparse, both algorithms recover a sparse triangulated graph. Experiments with real data have produced sparse structures for databases with well-chosen attributes, and dense structures for databases with highly correlated attributes. It will be shown later that the worst case complexity of both algorithms is polynomial.

In this section we first describe the goodness of fit

procedure and then present the DN1 and DN2 algorithms.

4.1. The GFT Algorithm for Testing Goodness of Fit of a Decomposable Network

The goodness of fit procedure used by the algorithms DN1 and DN2 to determine the worst fitting edges and to test goodness of fit of the derived networks is based on stochastic simulation and the χ^2 test. The basic idea of the procedure is this: if a probabilistic network represents the joint distribution contained in a given data, then stochastic simulation of the network should produce simulated data that contains the same joint distribution as the original data. A comparison of the two distributions using a suitable statistical technique should reveal if they are equivalent or not.

The algorithm measures closeness of the joint distributions contained in the original and simulated data by comparing pairwise probabilities computed from the original data with those computed from the simulated data. If two distributions are equivalent, then the corresponding pairwise marginal probabilities must be equal. Hence it is a necessary condition. The χ^2 test for contingency tables is used to test whether the two sets of pairwise distributions are equivalent. The GFT algorithm computes the conditional probabilities of a current learnt network from the original data. Thus stochastic simulation of the network will produce data in which mutual information weights of the edges included in the graph are the same (or nearly so) as the corresponding weights in the original data. However, for the genuine edges omitted from the graph, the corresponding weights will be different.

For each pair of vertices X_i, X_j the GFT procedure determines the difference between the mutual information weight $I(X_i, X_j)$ computed from the original data and the corresponding weight computed from the simulated data. It then selects as the worst fitting edge the edge for which the weight difference is the largest. In addition, for each node X_i in the graph the GFT algorithm computes the sum of the positive weight differences of all edges adjacent to X_i, that is, $(X_i, X_j), j = 1, ..., n, i \neq j$, and also determines the node's worst fitting edge as the adjacent edge with the largest weight difference. It then selects as the worst fitting node the node with the largest sum of the positive weight differences.

We now show that the GFT algorithm is able to detect edges omitted from the graph. If two variables X_i and X_j are dependent in the original data, and the edge (X_i, X_j) has been omitted from a recovered network, then in the data simulated from this network X_i

and X_j will be conditionally independent given some subset of variables. This difference will be reflected in the values of the marginal probabilities $P(x_i, x_j)$ computed from the original and simulated data, and will result in different values of the corresponding information weights $I(X_i, X_j)$ and the χ^2 terms. Similarly, if a genuine moral edge has been left out from a recovered network then the corresponding conditional independence in the simulated data will be different from the conditional independence in the original data, and this will also result in different values of the corresponding information weights and the χ^2 terms.

4.2. Complexity of the Goodness of Fit Algorithm

The time required to generate a single example using stochastic simulation algorithm of Henrion [Hen88] is of the order $O(n)$, since all n vertices have to be visited. The time required to process one example, that is, to update the probabilities $P(x_i, x_j)$, is of the order $O(n^2)$, since $n(n-1)/2$ probabilities have to be updated. Thus the time required to generate and process a single example is of the order $O(n^2)$. If C examples are simulated, the time required will be of the order $CO(n^2)$. The computation of model χ^2 value is also performed in $O(n^2)$ steps. Hence the complexity of the GFT algorithm is of the order $CO(n^2)$.

In addition to execution time, it is necessary to consider storage requirements. With each node of the directed triangulated graph, there is stored a table of conditional probabilities of the values of the node conditional on every possible combination of the values of its parents. The storage required for each table is an exponential function of the size of the parent set. For example, if a variable has k parents and if all variables are binary, the table will require 2^k storage locations. Hence the storage requirements of the GFT algorithm are of the order $O(2^k)$, where k is the size of the largest parent set.

4.3. The DN1 Algorithm

The DN1 algorithm begins with the most inconsistent network, an empty graph having all the vertices but no edges, and gradually builds up the network by adding one edge at a time. It maintains a moralized and triangulated graph so that the conditional independence relations in the evolving graph correspond to conditional independence relations in the unknown underlying probability distribution. During each pass it adds the worst fitting edge, computes additional moral edges, computes new fill-in edges, directs the edges,

procedure GFT

begin

1. From a given triangulated DAG and the associated conditional probability distributions generate examples using stochastic simulation (Henrion [Hen88]) and compute pairwise probabilities $P(X_i, X_j)$ for all combinations of values of variables X_i and X_j. Also compute corresponding mutual information weights $I(X_i, X_j)$.

2. Compute an edge with the largest difference between the mutual information weights computed from the original and simulated data.

3. Compute a node X_k which has the largest sum of the positive weight differences of its adjacent edges $(X_k, X_j), j = 1, ..., n, j \neq k$. (A positive weight difference is the difference between the mutual information weight computed from the original and simulated data that is > 0.) Also determine the worst fitting edge of the worst fitting node. (That is, the adjacent edge with the largest positive weight difference.)

4. Compare the probability distributions $P(X_i, X_j)$ computed from the original and simulated data by computing χ^2 value for contingency table test.

5. If overall χ^2 value is greater than the threshold for a given level of significance, return 0 else return 1.

end

Figure 2: The GFT algorithm.

and performs goodness of fit test. This process is repeated until the network fits the data, or the maximum clique exceeds a given threshold. During each pass through the while loop the DN1 algorithm applies the GFT algorithm described above to test goodness of fit of a current network and determine the "worst fitting" edge to be added to the graph. Before adding a new edge to the graph, the DN1 algorithm removes all fill-in edges. But unlike our earlier algorithm, TCN, it leaves all the moral edges in the graph. After a new edge has been added to the skeleton, the algorithm computes additional moral edges, if any, using the heuristic *MoralEdges* described below, and then computes a new fill-in using the fill-in algorithm of Tarjan and Yannakakis [TY84]. It is necessary to use a heuristic to determine moral edges because the general problem of finding a moral graph is NP complete. A maximum clique in the resulting triangulated graph is computed using the algorithm of Gavril [Gol80] and the undirected graph is transformed into an equivalent directed acyclic graph using maximum cardinality search algorithm (MCS) of Tarjan and Yannakakis [TY84]. The justification for the heuristic *MoralEdges* is the fact that the mutual information weight $I(X_i, X_j)$ of two marginally independent variables X_i, X_j is equal to zero, and an empirical observation that, in general, moral edges have a much lower information weight than the genuine dependencies and also lower than other conditionally independent pairs of variables. The reason for requiring that the weight difference be negative is that a negative weight difference indicates that in the original data the value of the weight is lower than in the simulated data. That suggests that the missing edge in the current graph represents an incorrect conditional independence. By requiring that the ratio be as small as possible, we ensure that only the most significant moral edges will be included. The heuristic has been tested on a variety of simulated data and has produced good results.

4.4. Complexity of the DN1 Algorithm

We show that the DN1 algorithm has polynomial-time worst case complexity. The worst case occurs when an underlying graph is completely connected. Then all $n(n-1)/2$ edges will have to be added to the graph, where n is the number of vertices. The complexity of adding one edge to the graph is determined by the complexity of the GFT algorithm, which is of the order $CO(n^2)$, where C is the number of cases simulated, and n is the number of vertices. Thus the time required to add one edge to the graph is of the order $CO(n^2)$, and hence the worst case complexity of the DN1 algorithm is of the order $CO(n^4)$.

As explained in Section 4.2, storage requirements for a directed triangulated graph are an exponential function of the size of the largest parent set. The algorithm

procedure *DN1*

begin

1. Start with an empty graph $G = (V, E)$.
2. Test goodness of fit of the current graph and determine the worst fitting edge.
3. **while** the graph does not fit the data **and** maximum clique $< MAXCLIQUE$ **do**
    4. Remove all fill-in edges, if any.
    5. Add the worst fitting edge to the graph.
    6. Compute new moral edges using the heuristic *MoralEdges*.
    7. Triangulate the graph.
    8. Compute maximum clique.
    9. **if** maximum clique $< MAXCLIQUE$ **then**
        10. Direct the graph using MCS algorithm.
        11. Compute conditional probability distributions.
        12. Test goodness of fit of the current graph and determine the worst fitting edge.
    endif
endwhile

end

Figure 3: The DN1 algorithm.

will be efficient for sparse networks but will be slow for large dense networks.

Our empirical studies have shown that, in general, for small training data sets (< 1000 cases), the number of examples generated in the GFT procedure, C, should be at least half of the size of the training set. For large data sets (more than $10,000$ cases) the fraction can be much smaller.

4.5. Performance of the DN1 Algorithm

Given sufficient data, the DN1 algorithm converges to optimal structures for three classes of networks: trees, polytrees, and networks with the longest undirected cycle of length three. If an underlying graph is sparse, the algorithm converges to a sparse network.

First we show that the DN1 algorithm recovers a maximum weight spanning tree of Chow and Liu [CL68] when an underlying distribution is of tree dependence. Since the algorithm begins with an empty graph, containing all the vertices but no edges, initially there will be no relationships between the variables in the network. That is, in the data simulated from the empty network the variables will be mutually independent. Thus the mutual information weights computed from the simulated data will be equal to zero

(or close to zero). The algorithm uses the GFT procedure to determine the next edge to be added to the graph, which selects the edge for which the difference between the mutual information weight computed from the original data and the corresponding weight computed from the simulated data is the greatest. When applied to a tree dependent distribution, the GFT algorithm selects the strongest edge of MWST first, then it chooses the next strongest edge, and so on. The property (3.2) ensures that the DN1 algorithm will add edges of MWST to the network, and not the edges corresponding to conditionally independent variables. The DN1 algorithm recovers an optimal structure also for distributions of polytree dependence. It has been shown by Rebane and Pearl [Pea88] that an underlying undirected skeleton of a polytree dependent distribution is the MWST of Chow and Liu. Similarly as for tree distributions, the algorithm recovers an underlying MWST for polytree distributions. Unlike a tree, however, a polytree contains converging pairs of arrows with unconnected tail nodes, known as vee structures. A polytree has to be transformed into a moral graph to become a decomposable network. Since in a polytree the tail nodes of vee structures are marginally independent, and thus have zero (or very small) mutual information weight, the heuristic *MoralEdges* will

procedure *MoralEdges*

begin

repeat

Find an edge e_i satisfying the following conditions:

1. e_i has a negative weight difference, $wd_i < 0$, and $|wd_i|$ is maximum, where wd_i is the difference between the information weights computed from the original and simulated data;

2. e_i does not already occur in the current graph (excluding the fill-in edges);

3. e_i is adjacent to two edges e_j, e_k which occur in the current skeleton;

4. e_i has the smallest value of the following ratio:

$$ratio = max\{I_i/I_j, I_i/I_k\}$$

where *ratio* is less than some prespecified threshold, and I_i, I_j, I_k denote the mutual information weights (3.1) of the edges e_i, e_j and e_k respectively.

until no suitable edge can be found

end

Figure 4: The algorithm for determining moral edges.

determine such moral edges quite reliably. Due to the property (3.2) the algorithm converges to an optimal structure for any distribution that has an underlying undirected graph in which the longest cycle is of length ≤ 3. In such distributions only first-order conditional independence relations occur and thus satisfy the property (3.2). When a joint probability distribution has a sparse underlying graph, many of the conditional independencies will be first-order and satisfy the property (3.2). Hence the algorithm will recover a sparse graph.

4.6. The DN2 Algorithm

The algorithm DN2 is derived from DN1 by replacing step 5 with the following statement:

5. Add the worst fitting edge of the worst fitting node to the graph.

The DN2 algorithm repeatedly adds the worst fitting edge of the worst fitting node to the graph until the model fits the data or a maximum clique exceeds a specified maximum size. Since this operation is of the order $O(n^2)$, the worst case complexity of the DN2 algorithm is the same as that of the DN1 algorithm. For the same reasons as in the case of the DN1 algorithm, the DN2 algorithm converges to optimal structures for the same three classes of networks as the DN1 algorithm. The only difference is that the edges are added in a different order.

In general, the DN2 algorithm produces almost identical graphs as the DN1 algorithm. But differences may occur in some of the weaker edges. The DN2 algorithm focuses on a total node fit and thus tends to add more edges to nodes with many weaker adjacent edges and may omit some genuine weak edges adjacent to nodes with only a few adjacent edges.

5. EXPERIMENTAL RESULTS

This section presents the results of experiments with two data sets obtained from the machine learning repository at the University of California, Irvine [MA94] — the image segmentation data, and the letter image recognition data. One problem with learning probabilistic networks from real data is that a resulting structure is not known in advance and thus it is difficult to assess whether a derived structure is correct or not. The GFT algorithm computes an overall χ^2 value of the model and the learning process is halted when the value exceeds a pre-specified threshold or the size of a maximum clique reaches a specified limit. In some cases it may be necessary to stop learning earlier to avoid a large maximum clique and a possible storage overflow. Frequently, the recovered network represents a good approximation of the joint distribution even if the χ^2 threshold has not yet been reached. Alternatively, to check how well a derived structure corresponds

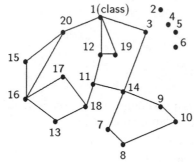

Skeleton of image segmentation data
(2000 simulations in the GFT procedure).

(a)

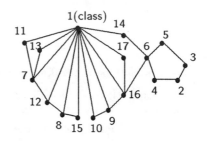

Skeleton of letter image recognition data
(2000 simulations in the GFT procedure).

(b)

Figure 5: Derived skeletal structures.

Table 1: Experiments with Image Segmentation Data

Image Segmentation Data — 20 attributes, 2100 training cases, 210 test cases						
Simulations in GFT	Max Clique	# Edges incl. moral & fill-in	# Edges in Skeleton	Learning Time (sec)	Accuracy %	
					Training	Testing
(DN1) 1000	4	22	15	293	97.5	98.6
(DN1) 2000	4	30	21	1247	98.5	99.5
(DN2) 1000	4	26	19	382	98.2	99.05
(DN2) 2000	4	26	19	729	98.2	99.05
MWST	2	19	19	1	92.1	91.0
C4.5 Classification Accuracy				Unpruned	98.7	98.1
				Pruned	96.7	97.6

to the data, we could perform probabilistic inference on the derived network and compare the inferred results to the known values. One way to do this is to use the recovered network as a probabilistic classifier and classify examples in the database from which it was derived (training data) as well as unseen examples in a test data set. If the network corresponds to the data, that is, if both the structure and the associated conditional probabilities are consistent with the data, we would expect a high classification accuracy on the training set. The accuracy on the test set will depend on how representative of the total population the training data were. If the training set was too small and did not include many value combinations found in the test data, the accuracy may be low. In the experiments described below, the classification accuracy of the derived networks is also compared to the accuracy of MWST of Chow and Liu, and to the accuracy obtained using the machine learning system C4.5 [Qui93]. The experiments were performed on Pentium 100 PC using Linux operating system.

Experiment 1: *Image Segmentation Data*

The data used in this experiment contains examples of image segmentation into seven classes: sky, cement, window, brick, grass, foliage and path. We used the larger data set of 2100 cases as the training set and the small set of 210 cases as the test set. The data contains 19 continuous attributes and the class attribute with 7 values. Since our program works only with discrete attributes, the continuous attributes had to be discretized prior to using the system. We have developed a heuristic algorithm that computes an approximate maximum entropy discretization. That is, for a given number of intervals it computes the interval boundaries so that the frequencies of observations are approximately equal in all the intervals. In both data sets the attribute number 4 has only one value, 9, and the attributes 5 and 6 have 2 and 3 values respectively. The remaining 16 attributes were discretized into 15 intervals. Figure 5(a) shows the graph derived by the DN1 algorithm using 2000 simulated examples in the GFT procedure. The results obtained using both the DN1 and DN2 algorithms are presented in Table 1.

Table 2: Experiments with Letter Image Recognition Data

Letter Image Recognition — 17 attributes, 16,000 training cases, 4000 test cases						
Simulations in GFT	Max Clique	# Edges incl. moral & fill-in	# Edges in Skeleton	Learning Time (sec)	Accuracy %	
					Training	Testing
2000	3	29	26	3415	91.4	82.2
MWST	2	16	16	17	76.4	74.2
C4.5 Classification Accuracy				Unpruned	94.8	75.9
				Pruned	91.6	75.6

The graphs produced by the two algorithms differed in some of the weak edges but the classification results were very similar.

Experiment 2: *Letter Image Recognition Data*

The letter image recognition data set contains 20000 examples representing images of the 26 upper case letters generated from 20 different commercial fonts. Each example has 16 discrete attributes which assume integer values between 0 and 15, representing various primitive numerical features. In this experiment, the first 16000 cases were used as the training set and the remaining 4000 cases as the test set. Since both learning algorithms produced the same network structure, only one result is presented. Figure 5(b) illustrates the structure derived from the training set using 2000 simulated examples in the GFT procedure and Table 2 shows the classification results.

6. CONCLUSIONS

This paper presented a method based on stochastic simulation that recovers from a database of examples a decomposable network consistent with the data. Two implementations were proposed, the DN1 and DN2 algorithms, which differ in the order in which the edges are added to the graph. The results of the experiments presented in this paper (and other experiments not reported here) show that this approach to learning probabilistic networks from data is capable of learning good approximate networks from real data. Such networks closely approximate the probability distribution contained in the data, provide a visual representation of the relationships among the attributes in the data, and can be used for performing probabilistic inference. However, some difficulties must still be resolved before the method can be applied to very large databases. The most critical problem is the size of the maximum clique in the derived graphs. As the maximum clique increases, the storage requirements of the network increase exponentially with its size. Likewise, the efficiency of probabilistic inference is an exponential function of the maximum clique size. Thus our major effort will concentrate on designing strategies for reducing the maximum clique size as much as possible. Our future research will also address alternative stopping criteria for halting the learning process, and the development of strategies for adapting the structure and associated conditional probabilities in the light of new cases, so that the network can improve its performance over time.

References

[CL68] C. K. Chow and C. N. Liu, "Approximating discrete probability distributions with dependence trees", *IEEE Trans. Inf. Theory* **14** (1968) 462-467.

[Bou94] R. R. Bouckaert, "Properties of Bayesian belief network learning algorithms", In: *Proceedings of the Tenth Conference on Uncertainty in Artificial Intelligence* (Morgan Kaufmann, San Francisco, California, 1994) 102-109.

[CH91] G.F. Cooper and E. Herskovits, "A Bayesian method for constructing Bayesian belief networks from databases", In: *Proceedings of the Seventh Conference on Uncertainty in Artificial Intelligence*, UCLA (1991) (eds. B. D. D'Ambrosio, P. Smets and P. P. Bonissone) (Morgan Kaufmann, 1991) 86-94.

[FC90] R. M. Fung and S. L. Crawford, "Constructor: A System for the Induction of Probabilistic Models", In: *Proceedings of AAAI*, Boston, Massachusetts (1990) 762-769.

[Gol80] M. C. Golumbic, *Algorithmic Graph Theory and Perfect Graphs* (Academic Press , London, 1980).

[Hen88] M. Henrion, "Propagating uncertainty in Bayesian networks by probabilistic logic sampling", In: *Uncertainty in Artificial Intelligence 2* (eds. J. F. Lemmer and L. N. Kanal)(North-Holland, 1988) 149-164.

[LB94] W. Lam and F. Bacchus, "Using new data to refine a Bayesian network", In: *Proceedings of the Tenth Conference on Uncertainty in Artificial Intelligence* (Morgan Kaufmann, San Francisco, California, 1994) 383-390.

[LS88] S. L. Lauritzen and D. J. Spiegelhalter, "Local computations with probabilities on graphical structures and their applications to expert systems (with discussion)", *J. Roy. Statist. Soc. Ser. B* **50** (1988) 157-224.

[MA94] P. M. Murphy and D. W. Aha, *UCI Repository of machine learning databases* (Irvine, CA: University of California, Department of Information and Computer Science, 1994).

[Nea90] R. Neapolitan, *Probabilistic Reasoning in Expert Systems.* (John Wiley and Sons, New York, 1990).

[Pea88] J. Pearl, *Probabilistic Reasoning in Intelligent Systems: Networks of Plausible Inference* (Morgan Kaufmann, San Mateo, 1988).

[PV91] J. Pearl and T. S. Verma, "A theory of inferred causation", In: *Principles of Knowledge Representation and Reasoning : Proceedings of the Second International Conference* (eds. J. A. Allen, R. Fikes and E. Sandewall) (Morgan Kaufmann, San Mateo, 1991).

[Qui93] J. R. Quinlan, *C4.5:Programs for Machine Learning* (Morgan Kaufmann, San Mateo, 1993).

[SV93] M. Singh and M. Valtorta, "An algorithm for the construction of Bayesian network structures from data", In: *Uncertainty in Artificial Intelligence: Proceedings of the Ninth Conference* (Morgan Kaufmann, San Mateo, California, 1993) 259-265.

[SG91] P. Spirtes and C. Glymour, "An algorithm for fast recovery of sparse causal graphs", *Social Science Computer Review* **9:1** (Duke University Press, 1991) 62-73.

[Ste94] B. Stewart, "Learning triangulated causal networks from data", In: *AI'94: Proceedings of the 7th Australian Joint Conference on Artificial Intelligence* (eds. C. Zhang, J. Debenham, and D. Lukose) (World Scientific Publishing Co, 1994) 76-83.

[TY84] R.E. Tarjan and M. Yannakakis, "Simple linear-time algorithms to test chordality of graphs, test acyclicity of hypergraphs, and selectively reduce acyclic hypergraphs", *SIAM J. Comput.* **13** (1984) 566-579.

A PRACTICAL OBJECT-ORIENTED CONCEPT LEARNING SYSTEM IN CLINICAL MEDICINE

Gou Masuda*, Norihiro Sakamoto** and Kazuo Ushijima*

* Department of Computer Science and Communication Engineering, Kyushu University.
6-10-1 Hakozaki, Higashi-ku, Fukuoka 812-81, Japan
** Department of Medical Informatics, Kyushu University Hospital.
3-1-1 Maidashi, Higashi-ku, Fukuoka 812-82, Japan
{masuda,ushijima}@csce.kyushu-u.ac.jp,nori@med.kyushu-u.ac.jp

ABSTRACT

Concept learning is one of the most central areas in machine learning. This paper describes our new concept learning system. It is based on existing learning systems, but is designed with object-oriented technology to improve their performance in real application area. This system can offer the user more extensibility and a more useful interface than existing systems. Because object-oriented technology provides more powerful modeling capability for real world objects, the object-oriented concept learning system can deal with the complex objects in application field, especially clinical medicine.

1   INTRODUCTION

Concept learning aims to induce generalized descriptions of a concept from instances of it. A concept learning system examines examples whose class or category for a concept is already known, constructs descriptions for categories, and classifies unseen cases into corresponding categories according to their properties. In general a property of an example is represented by its attributes. This classification procedure by a concept learning system bears great analogy to the diagnostic process in clinical medicine. A medical doctor examines the symptoms of and laboratory data for a patient which represent the patient's medical condition. The doctor then determines which clinical condition the case belongs to. There have thus been many reports of concept learning systems being applied to clinical medicine[Hudson84, Mangasarian94, Peter94]. However they are not sufficiently practical from the viewpoint of users or clinical doctors.

It is usually impossible to find the most suitable description for a concept by the simple application of one method. Therefore, a number of heuristics and hybrid strategies have been proposed for efficient classification. However most of these techniques have been experimental and have not been implemented in a practical system.

Our system aims to provide end-users in clinical medicine with the ability to integrate many powerful concept learning methods and to apply a concept learning system of their own construction to analyzing their own data. We designed and implemented the 'object' used in concept learning systems in completely object-oriented parts. Our new concept learning system was assembled from these parts. This allows end-users easily to integrate new heuristic methods into the system when those methods are implemented as object-oriented parts. Moreover, an excellent graphical user interface(GUI) used for selecting heuristic methods and analyzing input data and result will be easily developed through object-oriented programming.

In Section 2 we describe the concept learning system C4.5 on which our model is based. Section 3 presents our object-oriented learning system model. In Section 4, we evaluate our new concept learning system through the result of a preliminary experiment. In Section 5, we discuss the prospective merits of object-oriented models and related works, and Section 6 concludes this paper.

2   CONCEPT LEARNING SYSTEM C4.5

This paper focuses on the extension of C4.5[Quinlan93] which is a widely known concept learning system using a decision tree approach[Quinlan86].

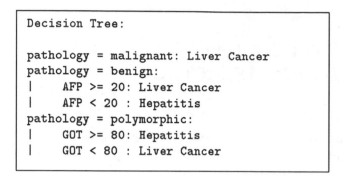

FIGURE 1: An example of a decision tree

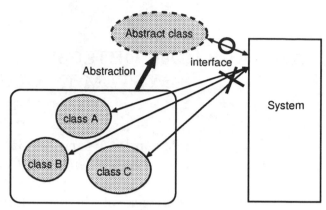

FIGURE 2: Abstract class: the system can deal with the class A, B and C in the same way through the interface defined in the abstract class.

2.1 DECISION TREE APPROACH

C4.5 represents a learning result as a decision tree, and induces classification rules from it. A decision tree of C4.5 consists of leaf nodes and branch nodes. A branch node specifies some test to be carried out on a single attribute value. A leaf node indicates a category. A case travels from the root node to a leaf node, being tested on an attribute value at a branch node. When the case reaches a leaf node, a category of the case is decided to be that recorded at the leaf node.

The learning algorithm of C4.5 is based on the divide-and-conquer method. It is summarized as follows.

(1) If all cases belong to a single category, the tree is a leaf labeled with that category.

(2) Otherwise, select a test based on one attribute.

(3) Divide the training set into subsets, each corresponding to one of the possible outcomes of the test.

(4) Apply the same procedure to each subset.

On constructing a decision tree, C4.5 employs an information-based heuristic called gain ratio criterion. This heuristic method is intuitively clear and often produces a simple tree successfully. To construct a simpler and more predictive tree, C4.5 uses pruning. Figure 1 shows an example of a decision tree constructed by our system.

2.2 ADVANTAGES AND DISADVANTAGES OF C4.5

Concept learning systems such as neural networks and genetic algorithm classifier systems repre-

sent what is learned in their own special forms while C4.5 represents it in the forms of a tree and rules. These are more easily understood by humans. Therefore, in C4.5 a user can compare a concept learned by the system with the existing knowledge of the user or expert. Another advantage of C4.5 is that it doesn't require the coding of input data or the decoding of output data.

The limitation of C4.5 on the expression power of a decision tree has been pointed out [Quinlan93]. To solve this problem, Brodley and Utgoff proposed the Linear Machine Decision Tree[Brodley92], and Murthy *et al.* proposed an oblique decision tree[Murthy94]. To determine which heuristic method is the best for inducing the most appropriate decision tree in a target domain remains another problem, because it depends on the properties of the domain.

3   OBJECT-ORIENTED CONCEPT LEARNING SYSTEM MODEL

We propose a new concept learning system model which is based on object-oriented technology in order to meet the needs of application fields, especially the needs of clinical medicine. In this model a heuristic method for a concept learning system such as a decision tree approach in C4.5 is regarded as a complex object. We decompose a complex object into element objects such as nodes and a tree in a decision tree. This decomposition makes it possible to modify, exchange and add heuristic methods, and to build up the kind of concept learning systems that are really required in an application domain.

Table 1: Overview of the class design

Class name	Roles
Manager	Controlling central behavior of the system.
Attribute[†]	Defining the fundamental interface such as returning the attribute name.
Attributevalue	Holding a value of an attribute.
Category	Holding a name of a category for a concept.
Categories	Working as a container of **Category**.
Case	Storing and managing the whole information on a case.
Cases	Working as a container for **Case**.
Node[†]	Holding the basic information on nodes.
Tree	Defining the operation on a tree such as counting its nodes, displaying and pruning it.
TreeGenerator	Generating an unpruned and pruned tree in cooperation with **AttributeTester**.
AttributeTester[†]	Determining the best test for dividing cases.
ErrorEstimator[†]	Estimating errors in the pruning tree.
TreeEvaluater[†]	Evaluating unpruned and pruned trees for a set of cases.

[†] Abstract class

3.1 CLASS DESIGN

First, we determine the object model of the concept learning system. Table 1 shows a list of classes and their roles. This model is based on C4.5, but is designed to enable a user to modify and exchange objects by decreasing and clarifying dependences among them. This is achieved through the use of abstract classes[Timothy91] in object-oriented technology.

The abstract class, which doesn't have its own objects, works as a base class of other classes. It is used to define the interface of the classes which we want to manage in common with other classes. All classes which have this interface are implemented as subclasses of the abstract class. When those classes are used in the program, we can deal with them in the same way using the interface (Figure 2). This makes the program more extensible as well as more abstract. We are using abstract classes where the system needs to be capable of extension.

The **Manager** class controls the central behavior of the system. It generates instances of **TreeGenerator**, **Cases**, and **TreeEvaluater** classes, and makes the instance of **TreeGenerator** construct a tree (including pruning). The **Attribute** class is an abstract class for attributes of a case. Since attributes are grouped into discrete attributes, continuous attributes and others, this class has at least two subclasses **DiscreteAttribute** and **ContinuousAttribute**. The **Attribute** defines the methods which are common to such attributes.

The **Case** class holds a list of **AttributeValues**, an instance of the **category** which the case is classified into, and the weight of the case. Every case knows the list of **Attributes** in use through a class variable.

When it is asked about its attribute value by an instance of another class in constructing a decision tree, it answers. The **Cases** class corresponds to a set of cases. From the point of view of the control flow of the system, it reads names of attributes, categories, and attribute values from a name file and reads cases from a data file. Then it produces instances of **Attribute**, **Category**, **Categories** and **Case**.

There are several kinds of nodes in a decision tree. The **Node** class defines the interfaces of such nodes. In the current system, the **Node** has the following three subclasses: **LeafNode**, **BranchNode** and **ContinuousNode**. The **LeafNode** class corresponds to a leaf node. The **BranchNode** class corresponds to a branch node and knows the attribute which is tested at the node. The **ContinuousNode** class inherits the properties of **BranchNode**. It holds a threshold value on this node in addition to what a **BranchNode** holds.

The **AttributeTester** is an abstract class. It defines only interfaces to other classes. The actual method which decides the best test is defined in its subclasses. In the current system, the **Gaincalculator** class is used. It tests the attributes with the gain ratio criterion used in C4.5.

The **ErrorEstimator** is also an abstract class. The way it estimates errors is substantially described in its subclasses. The **PessimisticErrorCalculator** class is one of the methods used to estimate errors in pruning. For C4.5, pessimistic pruning is used.

The **TreeEvaluater** class has two subclasses: **TrainingEvaluater** and **TestingEvaluater**. The former corresponds to the training set and the latter to the testing set of cases.

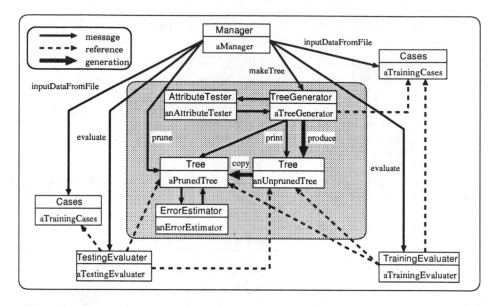

FIGURE 3: The structure of the system: the inner mesh box represents the entire **TreeGenerator** class. This complicated object includes some other objects.

3.2 THE STRUCTURE OF THE SYSTEM

We implemented a concept learning system using the object-oriented class design in Section 3.1. The system structure is shown in Figure 3. The learning process is performed as follows:

(1) First, *aManager*, an instance of **Manager**, is generated.

(2) *aManager* generates *aTrainingCases*, an instance of **Cases**, and then it reads each attribute name and training cases from files. At this time, instances of **Attribute**, **Category**, **Categories** and **Case** are generated and initialized.

(3) *aManager* generates *aTreeGenerator*, an instance of **TreeGenerator**, and gives it the pointer to *aTrainingCases* and lets it construct a decision tree.

(4) *aTreeGenerator* generates an instance of **AttributeTester**, and makes the instance test the attributes. Following this result, the *aTreeGenerator* divides the set of cases, and constructs *anUnprunedTree*, an instance of **Tree**.

(5) *anUnprunedTree* is copied to *aPrunedTree*, an instance of **Tree**, *aManager* makes the instance prune the tree.

(6) *aPrunedTree* generates an instance of **ErrorEstimator**, which co-operates to prune the tree.

(7) *aManager* generates *aTrainingEvaluater*, an instance of **TreeEvaluater**, and gives it the pointers to both trees, and makes the instance evaluate them for training cases.

(8) If a testing set exists, *aManager* generates *aTestingCases*, an instance of **Cases**, lets it read the testing cases from the file, and makes the instance evaluate both trees for them too.

As discussed in Section 3.1, this system is designed to enable a user to modify and exchange objects easily. For example, when a new sort of a node is introduced, all that a user has to do is to add it as a subclass of **Node**. Because the **Node** is defined as an abstract class, the system can deal with this new class in the same way to deal with other subclasses of **Node**. It is thus not necessary for a user to consider other objects in the system. In addition, if a new method is necessary in testing attributes instead of gain ratio criterion, a user has only to implement it as a new method of a subclass of **AttributeTester** by the same reason. Furthermore, if a user wants to modify the gain ratio criterion, he has only to modify the **Gaincalculator**. If another method is necessary in pruning, a user has only to add a new method to **Tree** and **Node** (Figure 4).

We have just finished implementing this system in C++. Our application domain is clinical medicine where a graphical user interface that not only graphically displays the decision tree but also enables a

Table 2: The result of the experiment on HCC data set

	Tree size	Training set(213 cases) Errors [%]	Testing set(100 cases) Errors [%]
Unpruned tree	42	2.8	38.0
Pruned tree	33	4.2	34.0

user to exchange heuristic methods easily is a necessary function. Because the system is developed on an object-oriented model, it is easy for us to add such an interface.

4 EVALUATION

We carried out a preliminary experiment in order to evaluate the ability of our system in clinical medicine. In the experiment, we used a clinical data set of Hepatic Cell Cancer(HCC). The HCC data set had 313 cases with 202 attributes, 56 discrete and 146 continuous. Each case is classified as either "Yes" or "No" depending on whether the cancer recurred within one year after the surgical operation. The data set includes some noise data.

The procedure for the experiment was as follows.

- Divide the data set into training and testing sets.

- Construct a decision tree from the training set and prune the tree.

- Evaluate two trees (unpruned and pruned) on training and testing sets.

In the machine learning area, the data set is divided into training and testing sets randomly and experiments are performed many times in oder to eliminate the deviation caused by selection of data sets. However in clinical medicine, because the diagnosis of a particular case requires the knowledge which medical doctors obtained from previous cases, we should divide the data set into training and testing sets by time, i.e. the day of a surgical operation. We divided 313 cases into 213 older cases for a training set and 100 newer cases for a testing set. Table 2 shows the result. The system created this result in 4.20 CPU seconds on IBM Personal Computer 850 with 64MB RAM. This result shows that our system is practical enough to be applied in real clinical medicine, which has more cases with more attributes.

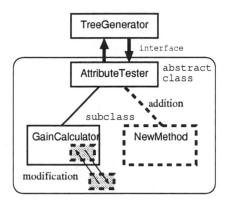

FIGURE 4: Addition and modification of the heuristic methods

5 DISCUSSION

In the object-oriented concept learning system model described herein, a case is expected to behave not as a tuple in C4.5 but as a complex object. Object-oriented modeling of a concept learning system leads to object-oriented data modeling of cases or examples. Data in clinical medicine have a complex structure, which it is impossible to express in a tuple, requiring object-oriented technology. A case in an object-oriented model hides its internal data, and makes only the interface public to the system. Therefore, on testing an attribute, the **AttributeTester** can test not only a value of an attribute, but also a combination of attributes. Object-oriented technology provides a powerful modeling capability of real world objects, which is very important in clinical medicine. For example, a patient object has many attributes and most of them are neither numerical data nor plain text. The patient object changes the data it contains over time but object-oriented concept learning systems can handle such data. This increases the power with which we can express concepts and widens the application fields.

A related work was reported in $\mathcal{MLC}$++[Kohavi94] by Ronny Kohavi *et al.* The main goal of $\mathcal{MLC}$++ is to provide researchers and experts with a wide variety of tools to experiment with and develop learning algorithms. In contrast, the model we propose aims to change and add heuristic methods in smaller object-oriented parts. $\mathcal{MLC}++$ does not consider changing object parts interactively through the use of a GUI as we propose.

6 CONCLUSION

In this paper, we have proposed a new concept learning system. It is based on an existing system, but is extended to be more flexible and more useful in clinical medicine. The principal idea is to separate a concept learning system into smaller objects or parts using object-oriented technology and modeling. Object-oriented modeling of cases will extend attribute-value learning system to message-value learning system. We carried out a preliminary experiment and showed that our system is useful with a larger database.

In the future work, we will model objects in clinical medicine, extend the system to a message-value learning system and develop the interactive GUI we described in this paper. The performance and effectiveness of the system to real clinical database then will be examined.

ACKNOWLEDGEMENTS

We would like to thank Masayuki Kamachi for his advice on programming and for his encouragement. We would also like to thank Nick May for his advice on English.

REFERENCES

[Brodley92] Brodley, C. E., and Utgoff, P. E.: Multivariate versus Univariate Decision Trees, COINS Technical Report 92-8, 1992.

[Hudson84] Hudson, D. L.: Derivation of rule-based knowledge from established medical outlines, Computers in Biology and Medicine, Volume 14, 1984.

[Kohavi94] Kohavi, R., John, G., Long, R., Manley, D., and Pfleger, K.: $\mathcal{MLC}++$: A Machine Learning Library in C++, Tools with AI '94, 1994.

[Mangasarian94] Mangasarian, O. L., Street, W. N., and Wolberg, W. H.: Breast Cancer Diagnosis and Prognosis via Linear Programming, Mathematical Programming Technical Report 94-10, 1994.

[Murthy94] Murthy, S. K., Kasif, S., and Salzberg S.: A System for Induction of Oblique Decision Trees, Journal of Artificial Intelligence Research 2 1-32, 1994.

[Peter94] Peter, K., Mrajan M., Jernej Z.,Kurt K., and Ivan M.: Decision Trees Based on Automatic Learning and Their Use in Cardiology, Journal of Medical Systems 8, 1994.

[Quinlan86] Quinlan, J. R.: Induction of Decision Trees, *Machine Learning,1*, pp. 81–106, 1986.

[Quinlan93] Quinlan, J. R.: *C4.5: Programs for Machine Learning*, Morgan Kaufmann Publishers, San Mateo, CA, 1993.

[Timothy91] Timothy, A. B.: *An Introduction to Object-Oriented Programming*, Addison-Wesley Publishing Company, Inc. 1991.

SAMPLING EFFECTIVENESS IN DISCOVERING FUNCTIONAL RELATIONSHIPS IN DATABASES

Atsuhiro Takasu[1], Tatsuya Akutsu[2] and Moonis Ali[3]

[1]Research and Development Department
National Center for Science Information Systems
3-29-1 Otsuka, Bunkyo-ku, Tokyo 112, Japan
E-mail: takasu@rd.nacsis.ac.jp

[2]Computer Science Department, Gunma University
1-5-1 Tenjin-cho, Kiryu, Gunma 376, Japan
E-mail: akutsu@keim.cs.gunma-u.ac.jp

[3]Department of Computer Science, Southwest Texas State University
San Marcos, Texas 78666-4616, U.S.A.
E-mail: MA04@academia.swt.edu

ABSTRACT

Functional relationship among attributes in relational databases has wide a variety of applications, as a typical rule, appearing in databases. It can be also regarded as a constraint for database design, query optimization, intelligent query processing, and so on. This paper discusses the sampling effectiveness for discovering both exact and approximate functional relationships through several experiments using a very large real bibliographic database and compares the previously derived theoretical result on sample complexities. It also showns that the sample size required for discovering approximate functional relationships is very small.

1   INTRODUCTION

Knowledge discovery and data mining in databases (KDD) is a very active database research area which aims at summarizing and extracting useful regularities hidden in databases. The progress of information processing and computer networking technologies enables data to be generated and distributed more quickly and widely. As a result, there is such a flood of data that the amount is now more than human beings can process with conventional data management tools. Formerly, this problem was limited to special fields such as the analysis of observation data sent by sattelite; however, it has now become general problem. Consequently, database systems are required to perform more intelligent functions such as filtering and summarizing data, and KDD technology is expected to provide a promising solution to this problem.

KDD research was started by applying machine learning techniques to database analysis. Since then, research has clarified the problems specific to this area and addressed them. There are two important research directions in KDD: enlargement of the variety of discovered rules and improvement of computational efficiency. As for the enlargement of rules, KDD handled two kinds of rules at the early stage. One is tuple oriented rules which are generally obtained by abstracting the data. For example, Han [Ca91] presented a relation abstraction mechanism using concept hierarchy. The other is attribute oriented rules such as correlative and functional relationships among attributes. For example, Żyktow [ŻB91] proposed a KDD method for correlative relationships using χ^2 test, while Cohen described a method for discovering statistical inter-attribute relationships based on regressive analysis [Co93]. The variety of rules has been gradually enlarged, and nowadays more complicated rules such as Bayesian networks [SM95] are handled.

As for the computational efficiency, it becomes more important as the amount of data increases rapidly and as more complicated rule discovery is required. Sampling is one of the possible way to overcome this problem. Valiant [Va84] proposed an approximate learning model in which the size of samples required to induce approximately correct concepts from examples are mathematically obtained [Bl90, Na91]. In the KDD field, some studies have been done on sampling (e.g., [Sh93]) to improve the efficiency of the computation. However, more extensive research is needed to make the KDD technology applicable to real databases.

We have been studying sampling effectiveness to discover functional relationships in databases. This relationship is a fundamental regularity and it is generally used as a constraint that databases should obey. As a constraint, it is called functional dependency (FD) and is used to design and maintain the consistency of relational databases [Ul88]. Hereafter the terms functional dependency and functional relationship are used interchangeably and abbreviated to FD. As a constraint, the exact FD is used. The FD is useful as a rule too. It can be used, for example, in query processing and guessing unknown values in distributed databases. As a rule, the exact FD is too strict since it does not allow any exceptions, but the approximate FD has broader applications. In our previous work [AT92, AT93], we analyzed the sampling effectiveness theoretically and showed the necessary size of a sample in order to extract FDs from samples. In this paper, we confirm the theoretical result through several experiments using a very large database containing bibliographic information about books and journals stored in Japanese universities. Furthermore, we discuss the effectiveness in finding FDs that are approximately correct in the original database.

The remainder of this paper is organized as follows. Section 2 gives basic notations and definitions about the FD and then describes the previous work briefly. In Section 3, we discuss the effectiveness of sampling through experiments. Conclusions are given in Section 4.

2   PREVIOUS WORKS

In this paper, we consider how to extract FDs from a n-ary relation R. Let the schema U of R be $(A_1, \cdots, A_n)$. For a tuple t in R, $t[A_i]$ denotes the i'th attribute value of tuple t. For a subset $A_{i_1} A_{i_2} \cdots A_{i_k}$ of U, $t[A_{i_1} A_{i_2} \cdots A_{i_k}]$ denotes the subtuple $(t[A_{i_1}], t[A_{i_2}], \cdots, t[A_{i_k}])$ of t.

An FD over U is denoted as $X \rightarrow A$, where $X \subseteq U$ and $A \in U$ holds, and means that for any relation R over schema U,

$$(\forall t, s \in R)(t[X] = s[X] \Rightarrow t[A] = s[A]) \qquad (1)$$

holds. For a relation R, an FD $X \rightarrow A$ is said to be *consistent* with R if the condition (1) holds for the relation R.

In order to handle the approximation of FD, let us define the error of an FD.

[**Definition 1**] [AT93] For a set R of tuples and an FD f, *error* of f for R is defined as $\dfrac{|v|}{|R|}$, where v satisfies the following conditions:

(1) f is consistent with $R - v$,

(2) $\dfrac{|v|}{|R|}$ is the minimum value among sets satisfying the condition (1).

In this definition, v represents minimum exceptions of tuples about f. Using the error ε, *approximation level* is defined as $1 - \varepsilon$.

[**Example 1**] Let us consider an FD *Language* $\rightarrow$ *Subject* over a relation in Table 1, then a tuple set $\{1, 2, 4, 6, 9\}$ satisfies the conditions in Definition 1 and both the error and the approximation level are 0.5. Note that there are other tuple sets satisfying the conditions (e.g., $\{1, 3, 4, 6, 9\}$) except for the tuple set $\{2, 3, 5, 7, 8, 10\}$ which does not satisfy the condition (2) of the Definition 1.

Approximately functional dependency over R denoted by $X \xrightarrow{a} A$ is an FD with the approximation level a for R. The approximation level of an FD f for a relation R is denoted by $al(f, R)$. $X \xrightarrow{1.0} A$ stands for an exact FD $X \rightarrow A$. For a relation R and a real number a, an approximate FD $X \xrightarrow{b} A$ is said to be *approximately consistent* if $b \geq a$ holds.

Approximate dependency inference problem is defined as:

Table 1: An Example Bibliographic Relation

Id	Author	Language	Title	Subject
1	Robert	English	Design	Complexity
2	John	English	Vector Space Model	Information Retrieval
3	Jane	English	Unix and Posix	Operating System
4	Mary	French	Sampling and Probability	Statistics
5	Michael	French	Skimming of Information	Information Retrieval
6	Jimmy	Spanish	Supervised Learning	Machine Learning
7	Kate	Spanish	Database Design	Database
8	Richard	Spanish	Object-Orientation	Database
9	Boris	German	Formal Language Learning	Machine Learning
10	Georg	German	Automatic Indexing	Information Retrieval

Instance :   A relation R over a schema U, an attribute $A \in U$, and real number a

Problem :   Find approximate FDs $\{f \mid al(f, R) \leq a\}$.

When a is 1.0, we call this problem an exact dependency inference problem.

In the approximate dependency inference, we evaluate the correctness of FDs derived from samples using *error parameter* ε and *confidence parameter* δ.

[Definition 2] The *sample complexity* for the exact dependency inference problem is the necessary size of a sample from which any exact dependency inference algorithm can enumerate FDs $\{f \mid (1 - al(f, R)) \geq \varepsilon\}$ with the probability of at least $1 - \delta$.

In order to derive the sample complexity of the exact dependency inference, let us consider the probability that an FD f is consistent with sample tuples although f is inconsistent with the original relation R. Let $lhs(f)$ and $rhs(f)$ respectively stand for the left hand and right hand sides of an FD f. For an FD f, if a sample contains at least one pair (s, t) of tuples such that $t[lhs(f)] = s[lhs(f)]$ and $t[rhs(f)] \neq s[rhs(f)]$ hold (termed as *violation pair*), then an exact dependency inference algorithm can find that f is inconsistent. Therefore, the sample complexity can be obtained by considering the probability that the sample contains at least one violation pair for every FD that is consistent with R. In order for this probability to be lower than the confidence parameter δ, the inequality

$$s > \frac{1}{1 - 3\varepsilon}\left(r\sqrt{2\left(1 - \left(\frac{\delta}{(n+1)S(n)}\right)^{\frac{2}{\varepsilon r}}\right)} + 2\right) \quad (2)$$

where r is $|R|$ and $S(n)$ is the number of FDs composed

of n attributes, that is 2^n.

When $r \to \infty$, the expression (2) is less than

$$\frac{1}{1 - 3\varepsilon}\left(2\sqrt{\frac{-\ln\left(\frac{\delta}{(n+1)S(n)}\right)}{\varepsilon}r} + 2\right). \quad (3)$$

See [AT93] for proofs of these inequalities.

By the inequality (3), the sample complexity for the exact dependency inference problem is $O(\sqrt{\frac{1}{\varepsilon}\ln\frac{1}{\delta}}n\ln n)$. For example, when r is one million and $\varepsilon = \delta = 0.1$, the sample complexity is about 30,000 (3% of the original relation).

3   EXPERIMENTAL RESULTS

In this section we discuss the accuracy of the dependency inference from samples through several experiments. In these experiments we use a union catalog database that contains bibliographic information about books and journals stored in Japanese universities. The data of this database is used for the online catalog. This database has been compiled by registering the bibliographic information of books and journals which the university libraries purchased. Currently about 350 universities participate in this union catalog database system and its size is growing by 66 Gbyte. In the experiments we used data of Japanese books in the database. This relation consists of about 100 attributes and contains about 1 million tuples. We selected 10 attributes from this relation whose values are less than 60 bytes because almost all values are different in long valued attributes and that does not

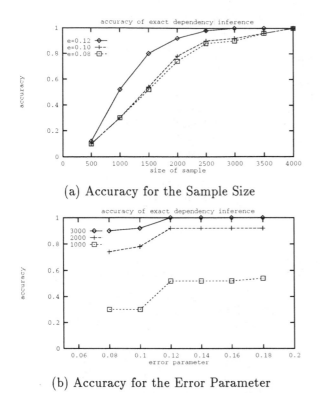

(a) Accuracy for the Sample Size

(b) Accuracy for the Error Parameter

Figure 1: Experimental Sample Complexity

make sense in FDs. The selected attributes include the author's name, publisher, registered date and so on. As a result, the relation used in the experiments consists of 10 attributes and contains about 1 million records whose size is about 50M bytes.

We generated samples from the relation described above randomly without duplication. The sample size ranged from 500 to 50000 and 50 sample sets were prepared for each sample size.

The first experiment is on the correctness of sampling in an exact dependency inference problem. We first defines the accuracy of exact dependency inference from samples. For a sample S and an error parameter ε, assume that exact dependency inference generates a set F of FDs. If for every FD f in F, the approximation level of f for the original relation R is greater than $1 - \varepsilon$, then the dependency inference is said to have succeeded. For a sample size s and an error parameter ε, the accuracy for s is defined as the proposition of succeeded inferences for ε over the total number of inferences for samples of size s. This accuracy corresponds to $1 - \delta$ for a confidence parameter δ described in the previous section.

For any combination of sample size ranging from 500 to 50000 and error parameters {0.08, 0.10, 0.12, 0.14, 0.16, 0.18}, we executed an exact dependency inference for the samples and calculated the accuracy. The graph in figure 1 (a) depicts the change of accuracy for the sample size when error parameters are {0.08, 0.10, 0.12}. By the expression (3), the confidence δ is $O(\frac{1}{e^{s^2}})$, that the accuracy $(1 - \delta)$ is rapidly converged. The graph shows the accuracy converges to 1.0 when the sample size is around 4000. As described in the previous section, the theoretical sample complexity for the error parameter of 0.1, the cofidence parameter of 0.1 (i.e., the accuracy of 0.9), and the relation size of 1 million is about 30000. However, the graph shows that the accuracy exceeds 0.9 when the sample size is about 3500 for the error parameter of 0.1, which is about 1/8.5 of the theoretical one. This experiment indicates that there is possibility to decrease the upper bound of the theoretical sample complexity.

The graph in figure 1 (b) depicts the change of accuracy for the error when the size of samples are {1000, 2000, 3000}. By the equation (3), the confidence δ is $O(\frac{1}{e^{\varepsilon}})$, and the accuracy $(1 - \delta)$ is also converging rapidly. The graph shows the accuracy is converged when the error parameter is around 0.12. The convergent value depends on the sample size, and in this experiment the value is 1.0 when the sample size is more than 3000. Note that this is related to the fact that the accuracy reaches 1.0 when the sample size is 3000 in figure 1 (a).

The second experiment is on the effectiveness of sampling in the approximate dependency inference. In the exact dependency inference the error is one-sided, that is, any dependency inference algorithm never eliminates an FD f when f is consistent with an original relation R. However, in the approximate dependency inference, we need to consider two types of errors. Let us consider to extract approximate FDs whose approximation levels are more than α, and assume to execute the approximate dependency inference with the approximation level β to a sample S. Note that the approximation level $al(f, R)$ of an FD f to the original relation is different from $al(f, S)$.

Assume that we obtain a set F of FDs. Then, one type of error is the inference of FD f such that f is approximately consistent with the original relation R, but approximately inconsistent with the sample rela-

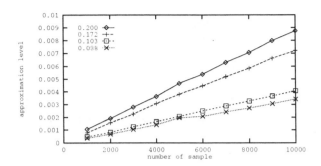

Figure 2: Errors in Samples

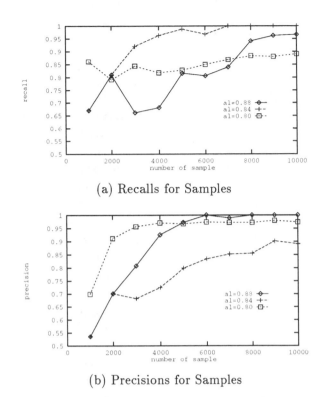

(a) Recalls for Samples

(b) Precisions for Samples

Figure 3: Recall and Precision Parameters in Samples

tion, that is, $al(f, R) \geq \alpha$ holds but $al(f, S) \geq \beta$ does not hold. The other is that f is approximately consistent with the sample relation, but inconsistent with the original relation, that is, $al(f, S) \geq \beta$ holds but $al(f, R) \geq \alpha$ does not hold. In order to evaluate these two kinds of errors, we use the recall parameter and the precision parameter [Sa89] defined as follows:

$$recall \equiv \frac{|\{f \mid al(f, R) \geq \alpha, al(f, S) \geq \beta\}|}{|\{f \mid al(f, R) \geq \alpha\}|}$$

$$precision \equiv \frac{|\{f \mid al(f, R) \geq \alpha, al(f, S) \geq \beta\}|}{|\{f \mid al(f, S) \geq \beta\}|}.$$

Firstly we show how the error in the original relation is changed in samples. For four FDs whose errors are respectively 0.200, 0.172, 0.103 and 0.088 in original relation, we examined their errors in various size of samples. Figure 2 shows the change of the error which is the average value of 50 samples for each sample size. This graph shows that the error is almost proportional to the size of sample. In the experiment, we also observed the linear relationship between the approximation levels of original relation and sample relation for the fixed sample size.

Next, we discuss the recall and the precision parameters through an experiment. For each pair of the approximation level α and the sample size s, we calculated the approximation level β using Figure 2 based on the linearity of β with α and the sample size s. Figure 3 depicts the average values of the recall and the precision parameters obtained from 50 sample relations for each pair of $\alpha = \{0.88, 0.84, 0.80\}$ and sample size ranging from 1000 to 10000. This graph shows that the recall parameter tends to increase as the size of sample and converges when the sample size is about 10000 in this experiment. Similar tendency is observed

for the precision parameter. These experiments indicate that we can infer approximate FDs with high accuracy from a small size of sample compared with original relation.

Two features are seen in the approximate dependency inference. Firstly, the recall and the precision paramters of approximate dependency inference converge at larger sample size than the accuracy of the exact dependency inference. This means that we need larger sample for approximate dependency inference. Secondly, the converged value differs according to α. In the approximate dependency inference, the precision and the recall parameters differ based on the value of β. We will obtain higher recall and lower precision for lower β, and vice versa. Therefore, we need to use appropriate value of β to obtain better recall and precision. This tradeoff makes the approximate dependency inference more difficult. Intuitively, this problem causes the larger sample size and smaller converged value, however, for more precise discussion we need to analyze the approximate dependency inference problem from theoretical point of view. We will report the theoretical analysis in future paper.

4   CONCLUSIONS

In this paper, we discussed the sampling effectiveness for both the exact and the approximate dependency inference problems through several experiments using a very large real bibliographic database. These experiments showed that the confidence parameter converges as the error parameter and the sample sizes increase, which is also derived from the theoretical result. Furthermore, the experimental sample complexity is far smaller than the theoretical one, which means that the sampling is more effective than the theoretical result indicates. In this paper, we also shows that the sampling is effective for the approximate dependency inference problem using the recall and the precision parameters.

Two problems are left. One is the gap between the theoretical and the experimental results of the sample complexity in the exact dependency inference. Current theoretical sample complexity is far more larger than the experimental one, and consequently, it causes the redundant calculation in applications. We will examine this problem from experimentally by making experiments for other databases as well as check the theoretical result in order to lower the upper bound of the sample complexity.

The other is theoretical analysis of the sampling effectiveness for the approximate dependency inference. The similar method to the exact dependency inference will be able to applied to this problem. This is an another future research direction.

References

[AT92]   T. Akutsu and A. Takasu. "On PAC learnability of functional dependencies". In *Proc. of the Third Workshop on Algorithmic Learning Theory (ALT'92)*, pages 229–239, 1992.

[AT93]   T. Akutsu and A. Takasu. "Inferring Approximate Functional Dependencies from Example Data". In *Proc. of AAAI workshop on Knowledge Discovery in Databases*, pages 138–152, 1993.

[Bl90]   A. Blumer, A. Ehrenfeucht, D. Haussler, and M. Warmuth. "Learnability and the Vapnik-Chervonenkis dimension". *Journal of the ACM*, 36:929–965, 1990.

[Co93]   P. Cohen, et al. "Automating Path Analysis for Building Causal Models from Data: First Results and Open Problems". In *Proc. of AAAI workshop on Knowledge Discovery in Databases*, pages 153–161, 1993.

[Ca91]   Y. Cai, N. Cercone, and J. Han. "Attribute-Oriented Induction in Relational Databases". In G. P. Shapiro and W. J. Frawley, editors, *Knowledge Discovery in Databases*, pages 213–228, Cambridge, MA, 1991. AAAI/MIT.

[Na91]   B. K. Natarajan. *"Machine Learning - A Theoretical Approach"*. Morgan Kaufman, 1991.

[Sh93]   G. Piatetsky-Shapiro and C. J. Matheus. "Measuring Data Dependencies in Large Databases". In *Proc. of AAAI workshop on Knowledge Discovery in Databases*, pages 162–173, 1993.

[Sa89]   G. Salton. *"Automatic Text Processing "*. Addison-Wesley Publishing Company, 1989.

[SM95]   P. Spirtes and C. Meek. "Learning Bayesian Networks with Discrete Variables from Data". In *Proc. of 1st Intl. Conf. on Knowledge Discovery and Data Mining*, pages 294–299, 1995.

[Ul88]   J. D. Ullman. *"Principles of Database and Knowledge-base Systems - volume 1 "*. Computer Science Press, 1988.

[Va84]   L. G. Valiant. "A Theory of the Learnable". *Communication of the ACM*, 27(11):1134–1142, 1984.

[ŻB91]   J. M. Żyktow and J. Baker. "Interactice Mining of Regularities in Databases". In G. P. Shapiro and W. J. Frawley, editors, *Knowledge Discovery in Databases*, pages 31–53. AAAI/MIT, Cambridge, MA, 1991.

SIGNAL INTERPRETATION IN TWO-PHASE FLUID DYNAMICS THROUGH MACHINE LEARNING AND EVOLUTIONARY COMPUTING

Bogdan Filipič[1,2], Iztok Žun[2] and Matjaž Perpar[2]

[1]Department of Intelligent Systems, Jožef Stefan Institute
Jamova 39, 61000 Ljubljana, Slovenia
E-mail: *bogdan.filipic@ijs.si*

[2]Laboratory for Fluid Dynamics and Thermodynamics
Faculty of Mechanical Engineering, University of Ljubljana
Aškerčeva 6, 61000 Ljubljana, Slovenia

ABSTRACT

The paper shows how techniques of machine learning and evolutionary computing can assist in making human expertise in sensor data interpretation explicit and suitable for computer execution. The study refers to a specific task in two-phase fluid dynamics, i.e. the interpretation of probe signals detected in gas–liquid flow. Given a raw probe signal, the corresponding two-state signal needs to be constructed which denotes the presence of the two phases. Due to the lack of knowledge about the processes on a micro scale, no exact procedure exists for accomplishing this task. However, operators are capable of interpreting visually presented probe signals through experience and intuition. To imitate their performance, a prototype signal interpretation procedure was designed manually, and its parameters tuned with genetic algorithms. In an alternative approach, skill acquisition was performed automatically, using inductive machine learning. The induced signal interpretation procedures were tested successfully on air–water pipe flow under laboratory conditions.

1. INTRODUCTION

Sensor data interpretation is a demanding task that requires skilled operators and is difficult to automate. In two-phase fluid dynamics, it is one of the key techniques needed to characterize the flow phenomena. Of great importance for the oil industry and chemical engineering is modelling and control of the gas–liquid flow. This is the interacting flow of two phases, gas and liquid. To obtain a particular flow characteristic, such as phase fraction and phase velocity, one should first be able to discriminate between the two phases in the flow. However, due to difficulties in interface structure identification, there is no ideal phase discrimination procedure. For practical purposes, various intrusive and non-intrusive phase detection techniques have been proposed over the last few decades. They are based on detecting the local phase change through some property, such as conductivity, which is different for the two phases. Most often, probes emitting two-state signals are used to detect the phase surrounding the probe tip. The probe signal then needs to be properly interpreted to obtain the flow characteristic of interest.

Existing signal interpretation methods reconstruct the two-state signal from a raw probe signal by static application of predefined threshold values. Although such a reconstruction is often sufficient to acquire particular flow characteristics, the signal interpretation itself may significantly disagree with expert understanding of the underlying processes. Experienced operators are capable of detecting inconsistencies in two-state signals generated by the existing methods. Moreover, given a visual representation of a probe signal, they are able to reconstruct the corresponding two-state signal.

These observations motivated us to develop a probe signal interpretation method for gas–liquid flow based on human expertise. To encode the procedure carried out by operators, the knowledge acquisition approach was undertaken. In a traditional consultation-based manner, a prototype procedure was designed manually, relying on dynamic application of thresholds to the probe signal. The threshold settings were tuned by means of evolutionary computing, where a two-state signal reconstructed from the training probe signal by an operator served as a reference in the optimization process. The resulting signal processing method was evaluated in gas-liquid flow measurements, where it proved highly accurate and capable of adaptation to specific flow conditions.

The study further focused on a question challenging from the knowledge engineering perspective: Is it possible to derive an appropriate signal interpretation procedure automatically? The issue was explored using the inductive machine learning approach. Examples of expert interpretation of a probe signal were submitted to a learning system which then generated a procedure for reconstructing the two-state signal. Although less general than the manually derived procedure, the machine-induced signal interpretation was of comparable quality and reflected operator performance.

In the rest of the paper, analytical and experimental aspects of signal interpretation in gas–liquid flow measurements are given, and the two approaches to the development of the interpretation procedure based on human expertise described. Results of the experimental evaluation are presented and compared. The paper concludes with a summary of findings of this study and questions remaining open for further investigation.

2. SIGNAL INTERPRETATION IN GAS–LIQUID FLOW MEASUREMENTS

Analytical treatment of gas-liquid flow [Ish75] assumes that an individual point x in the flow field can be in three possible states, i.e. in liquid phase, gas phase or phase interface, and defines the respective structural functions. Knowing the gas phase structural function $M_G(x,t)$ (which equals 1 when point x is in gas phase, and 0 otherwise), the local void fraction, i.e. the proportion of gas phase, $\alpha(x)$, is under stationary conditions defined by

$$\alpha(x) = \lim_{T \to \infty} \frac{1}{T} \lim_{\varepsilon \to 0} \int_t^{t+T} M_G(x,t)\, dt \qquad (1)$$

where T is the integration time which must be sufficient to ensure the convergence, and ε is the phase interface thickness. Limiting ε to zero and introducing the gas bubble residence time T_{Gi}, Equation (1) yields

$$\alpha(x) = \lim_{T \to \infty} \sum_i \frac{T_{Gi}}{T} \qquad (2)$$

which serves to practically determine the local void fraction value at a selected location in the gas-liquid flow.

In practice, phase detection is most frequently performed by electrical and optical probes. In our investigation a resistivity probe was used, mounted in a vertical pipe with upward air–water flow as outlined in Figure 1.

An example of the probe voltage signal and its corresponding two-state signal reconstructed by an operator are shown in Figure 2. Peaks in the probe signal

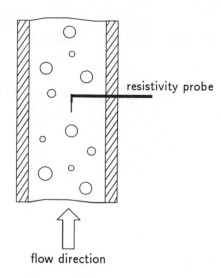

FIGURE 1. Gas–liquid flow in a vertical pipe: Phase detection with a resistivity probe

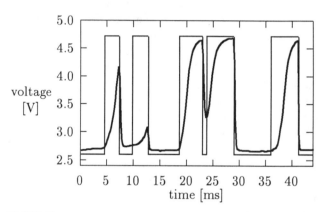

FIGURE 2. An example of the probe signal and the corresponding two-state signal denoting the presence of the gas and liquid phases

correspond to the interaction of the probe with gas phase (bubbles). For a particular event, the signal rise time is related to probe drying, and the fall time to probe wetting. The rise time is typically longer than the fall time. As the micro scale processes during the penetration of the probe through phase interface are not known, the transformation of a probe signal to the two-state signal necessarily involves *ad hoc* criteria. Existing techniques of probe signal interpretation range from elementary single-threshold to recently introduced multi-threshold methods.

In single-threshold phase discrimination, time intervals over which the probe signal exceeds a predefined threshold value are declared to denote the presence of gas, while others liquid. The method is simple and efficient, but in weak correspondence with underlying physical phenomena. It ignores the different nature of liquid-to-gas and gas-to-liquid transitions and is unable to detect peaks under the threshold level.

An improved method, two-threshold phase discrimination, assumes different threshold values to define liquid-to-gas and gas-to-liquid transitions during the reconstruction of the two-state signal. A weakness of the two-threshold method, when applied rigidly, is its inability to detect peaks appearing in the probe signal so close to each other that the liquid level is not reached in between.

Advanced digital signal processing allowed for multi-threshold techniques which take into consideration additional attributes of the probe signal. The CMHT discrimination method [Gel92], for example, involves a three-threshold procedure with an additional gas-to-liquid transition level for peaks not reaching the maximum signal amplitude. Moreover, the phase discrimination procedure of Liu and Bankoff [LB93] applies a compound criterion considering relations among neighbouring points and threshold values in the signal to recognize the liquid phase.

Although rather different, the aforementioned techniques share a common characteristic. They apply thresholds to the probe signal in a semi-static manner by setting their values fixed over intervals of given length and only adjusting them to possible changes of signal amplitude. While this is sufficient for ideal laboratory conditions, it may turn out to be inadequate in more realistic environments where noise in the signal or signal drift can appear, and under demanding flow conditions with a high proportion of gas phase, resulting in complex probe signals.

3. MANUAL ENCODING OF THE SIGNAL INTERPRETATION PROCEDURE AND ITS REFINEMENT

A new interpretation procedure for probe signals detected in gas-liquid flow has been proposed as an attempt to take into account human expertise [ŽFP+95]. The development consisted of two steps: the design of a prototype interpretation procedure and tuning of its threshold settings. Operator skills in probe signal interpretation were captured through both steps. While in the first step this was accomplished by manual encoding, the refinement in the second step was achieved through the evolutionary optimization process, where examples of signal interpretation demonstrated by an expert served as a reference.

In the first step, numerous examples of signal interpretation were analysed and discussed with operators. Important observations from this step were as follows:

- There are two key elements of probe signal interpretation: the liquid-to-gas and gas-to-liquid transition.

- The transitions are determined by characteristic points in the probe signal, where the gradient changes from steady or decreasing to increasing (liquid-to-gas transition), and from increasing or steady to decreasing (gas-to-liquid transition).

- Points of transitions roughly correspond to local minima and maxima in the signal.

- Additional attributes, such as the value of the slope, may help to successfully handle irregularities in the signal, such as noise and signal drift.

These were the guidelines for constructing the prototype signal processing procedure. The procedure operates as follows: digitized probe signal R_i, $i = 0 \ldots N$, is inspected point by point, and the two-state signal is generated simultaneously as output. For each point R_i, the values of current local minimum, R_L, and maximum, R_H, are possibly updated and a phase determined for R_i, depending on its relation to local extrema. During signal inspection, two basic thresholds, S_1 and S_2, are applied. They are introduced through procedure parameters p_1 and p_2, given as percentages of the signal amplitude, $R_{\max} - R_{\min}$. The following is a sketch of the computer implementation of the proposed procedure:

$$S_1 = p_1 * (R_{max} - R_{min});$$
$$S_2 = p_2 * (R_{max} - R_{min});$$

```
for i=1 to N do
    Update(R_L , R_H );
    if (R_i > R_{i-1}) and (R_i - R_L > S_1)
        or (R_i < R_{i-1}) and (R_H - R_i < S_2)
    then phase = gas
    else phase = liquid;
end_for;
```

Unlike in other procedures discussed in Section 2, the thresholds are here related to local extrema in the signal. Moreover, detecting local minima and maxima involves an additional threshold value to prevent small oscillations in the signal from being treated as local extrema. These features make the procedure less sensitive to noise and suitable for processing complex probe signals.

The idea behind introducing tunable thresholds was to make the method capable of adaptation to various flow conditions, signal-to-noise ratios and probe characteristics. To adjust the threshold values to particular conditions, a training probe signal was selected and the corresponding two-state signal constructed by an expert, using computer graphics aids. The expert two-state signal served as a reference during threshold optimization.

Optimization was carried out with a genetic algorithm [Gol89]. Possible solutions were vectors of threshold values represented as binary strings. The genetic algorithm evolved a population of procedure instances with different threshold settings. Each instance was evaluated with respect to its ability to reproduce the expert two-state signal from the training probe signal. The quality measure (fitness) of a particular procedure instance was the proportion of points in the generated two-state signal matching the related points in the expert two-state signal (see [ŽFP⁺95] for a detailed description of the genetic algorithm setup).

The optimization process resulted in a tuned signal interpretation procedure almost perfectly imitating the operator. In a detailed experimental verification including different probe geometries and flow conditions, it consistently outperformed the existing signal interpretation methods [ŽFP⁺95].

4. MACHINE LEARNING ACQUISITION OF OPERATOR SKILLS

Probe signal interpretation is an expert skill that is easier to demonstrate on concrete examples than to articulate explicitly. This calls naturally for the application of machine learning techniques, capable of automated knowledge acquisition from examples. Particularly in the area of attribute-based learning, a number of systems have been developed which are suitable for practical applications. With this in mind, we have explored the possibilities of automatic construction of a signal interpretation procedure from examples of operator performance. The machine learning system applied in this study was Magnus Assistant [Mla94].

Magnus Assistant belongs to a family of learning systems that, given training examples from a particular domain, induce the laws governing the domain in the form of a decision tree. Training examples need to be specified in terms of attribute values and classes. The resulting decision tree represents a classification procedure which can be applied to new, previously unencountered examples described only by attributes. The tree induction algorithm of Magnus Assistant is an extension of the well-known ID3 learning algorithm [Qui79]. Of particular importance for real-world applications are its tree pruning mechanisms that allow for learning from imperfect data.

Learning probe signal interpretation was performed on pairs of training probe signals and corresponding expert two-state signals, used to tune procedure

thresholds in previous experiments. Selected features of the probe signal were used as attributes for learning, and the class was the phase (liquid or gas) denoted by the two-state signal. Specifically, each point R_i in the probe signal represented a learning example and was described by the following attributes:

- VALUE – value of R_i as a fraction of the signal amplitude, i.e. $R_i/(R_{max} - R_{min})$,
- GRADIENT – direction of signal value change from R_{i-1} to R_i, i.e. *increasing* if $R_{i-1} < R_i$, *steady* if $R_{i-1} = R_i$, and *decreasing* if $R_{i-1} > R_i$,
- D_EXTR – distance of R_i from the last local extremum detected in the signal, expressed as a fraction of the signal amplitude, i.e. $(R_i - R_L)/(R_{max} - R_{min})$ for increasing and $(R_H - R_i)/(R_{max} - R_{min})$ for decreasing segments of a signal (R_L and R_H stand for local minimum and maximum, respectively).

While GRADIENT is a discrete attribute, VALUE and D_EXTR are continuous with values ranging from zero to one. For the purpose of machine learning, their values were also discretized. The discretization step used was 0.01.

Induction of signal interpretation procedures from examples with Magnus Assistant was not a single straightforward experiment but rather an iterative process. To construct accurate and robust procedures, a proper degree of decision tree pruning was to be determined. Tree pruning is aimed at eliminating inaccurate parts of a decision tree based upon small subsets of training examples. Such parts make the tree overspecialized with respect to the training set and may result in poor classification accuracy on new, unseen examples. In Magnus Assistant, several parameters affect the degree of tree pruning. Two of them were tuned in our experiments: *class frequency threshold*, which defines the minimum required relative frequency of the majority class examples occurring in a leaf of a decision tree, and *minimal weight threshold*, which defines the minimum proportion of training examples covered by a leaf. In addition, the so-called postpruning mechanism was employed which, after tree construction, checks for possible additional reductions of the tree. Optimal values of these parameters are domain dependent and difficult to determine. In our case, an empirical approach was undertaken, where trees were induced at various parameter settings and their classification accuracy tested on unseen examples. An additional criterion was the transparency and understandability of decision trees. Among several trees with similar classification accuracy, the one that better reflected operator performance was preferred.

5. EXPERIMENTAL EVALUATION

The manually constructed signal interpretation procedure, tuned via genetic algorithm, as well as the automatically induced procedures, were tested on air-water bubbly flow in a vertical pipe under laboratory conditions. Different flow conditions were created, and signals detected by resistivity probes at various radial positions in the pipe processed to obtain local void fraction values. From the resulting local values global void fraction was calculated for each experiment. On the other hand, the global value was, for the sake of comparison, measured with the quick closing valves (QCV) technique. Detailed results for the genetic-algorithm-tuned phase discrimination are given in [ŽFP+95], where it is shown that the new approach outperforms previous techniques both in quality of signal processing and measurement accuracy. In this section, results of the machine learning approach are presented and compared with the results of other techniques.

By means of machine learning, compact and transparent classification procedures were generated. To illustrate, for bubbly flow where liquid volumetric flux j_L was set to 0.45 m/s and gas volumetric flux j_G to 0.054 m/s, and probe signals were not contaminated with noise (further referred to as test case 1), the following procedure was obtained:

```
if VALUE <= 0.05 and
   GRADIENT = increasing and
   D_EXTR > 0.02
or VALUE > 0.05 and
   GRADIENT = (increasing or steady)
or VALUE > 0.05 and
   GRADIENT = decreasing and
   D_EXTR <= 0.01
then PHASE = gas
else PHASE = liquid
```

It was derived from 5,000 learning examples, i.e. subsequent points in the training signal, which correspond to one second recording time at the sampling frequency of 5 kHz. This is the maximum size of the learning set accepted by the employed version of the learning program. The class frequency threshold was set to 99%, the minimal weight threshold to 2%, and postpruning was applied. The procedure performs on a previously unseen probe signal as illustrated in Figure 3. By properly detecting small as well as unseparated peaks, it closely follows the expert way of phase discrimination.

In test case 2 (bubbly flow, $j_L = 0.45$ m/s, $j_G = 0.129$ m/s), a strong noise component was present in some probe signals, making discrimination harder.

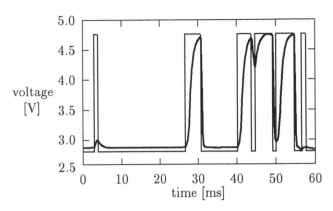

FIGURE 3. Performance of the automatically induced phase discrimination procedure on a probe signal in test case 1

The following discrimination procedure was induced for this case:

```
if VALUE > 0.09 and
   VALUE <= 0.12 and
   GRADIENT = increasing and
   D_EXTR > 0.01
or VALUE > 0.12 and
   GRADIENT = (increasing or steady)
or VALUE > 0.09 and
   GRADIENT = decreasing and
   D_EXTR <= 0.05
then PHASE = gas
else PHASE = liquid
```

Its performance on a noisy testing signal is illustrated in Figure 4. It can be seen that the procedure avoids possible critical errors of interpreting periodical noise waves as the presence of the gas phase.

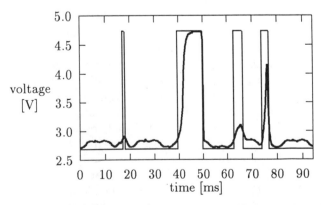

FIGURE 4. Performance of the automatically induced phase discrimination procedure on a noisy probe signal in test case 2

TABLE 1. Global void fraction values for air-water bubbly flow obtained in two test cases by different phase discrimination methods: ST – single-threshold discrimination (threshold set at 0.185), TT – two-threshold discrimination (thresholds set at 0.2 and 0.8), LB – discrimination by Liu & Bankoff, CMHT discrimination, GA – manually encoded procedure tuned by a genetic algorithm, ML – procedure induced by machine learning. QCV measurement yielded 7.6% for test case 1 and 16.5% for test case 2.

Signal process-	Global void fraction (%)	
ing method	Test case 1	Test case 2
ST	6.1	11.7
TT	5.6	10.7
LB	6.7	12.2
CMHT	7.0	14.7
GA	7.3	15.0
ML	7.5	13.8

For both test cases, void fraction profiles (i.e. local void fraction values along equidistant radial positions in the pipe) were calculated. The profiles were finally integrated to obtain global void fraction values. Table 1 compares these results with those of other techniques mentioned in Section 2.

6. CONCLUSIONS

In this interdisciplinary study, human understanding of gas–liquid flow phenomena was employed to develop new techniques of interpreting probe signals detected in the flow. The operator's ability to construct the target two-state signal from the raw probe signal was first acquired manually and refined through optimization on examples provided by an operator. Skill acquisition was later automated utilizing attribute-based inductive machine learning.

Several characteristics make the resulting procedures applicable to gas–liquid flows in industrial processes. First, signal interpretation is done in a natural manner, following expert performance as closely as possible. This greatly reduces the possibility of critical errors that are typical of traditional 'blind' discrimination procedures. Second, the new approach is capable of adaptation to specific flow conditions. This can be done either by tuning the proposed model procedure or by learning from examples. Finally, the resulting procedures require no complex data manipulations. Instead, they are based on elementary operations and attributes. Once adjusted to particular flow conditions, the method can therefore process probe signals on-line.

In view of the applied methodology, an important finding is that, at least for two-phase flows with well separated phases as used in the experiments, automated construction of signal interpretation procedures is possible. The automatically induced procedures are however less general than the manually encoded and optimized procedure. The reason for this is the presence of the VALUE attribute in current results which limits their applicability. We expect to overcome this weakness by exploring alternative signal features for learning and applying relational rather than attribute-based learning systems.

ACKNOWLEDGEMENTS

This work was supported by the Slovenian Ministry of Science and Technology under contracts No. P2–5095–0782 and J1–7255–0106. Thanks are due to Martin Bajc for computer implementation of the procedure to extract attribute values from digitized signals, and to Dunja Mladenić for providing a customized version of the Magnus Assistant learning program.

REFERENCES

[Gel92]   C. W. M. van der Geld. CMHT Phase discrimination computer program. Eindhoven University of Technology, Eindhoven, The Netherlands, 1992.

[Gol89]   D. E. Goldberg. *Genetic Algorithms in Search, Optimization and Machine Learning*. Addison-Wesley, Reading, MA, 1989.

[Ish75]   M. Ishii. *Thermo-Fluid Dynamic Theory of Two-Phase Flow*. Eyrolles, Paris, 1975.

[LB93]   T. J. Liu and S. G. Bankoff. Structure of air-water bubbly flow in a vertical pipe II – Void fraction, bubble velocity and bubble size distribution. *International Journal of Heat and Mass Transfer*, 36 (4): 1061–1072, 1993.

[Mla94]   D. Mladenić. Magnus Assistant. Technical report IJS DP-6938, Jožef Stefan Institute, Ljubljana, 1994. (In Slovenian).

[Qui79]   J. R. Quinlan. Discovering rules by induction from large collection of examples. In D. Michie, editor, *Expert Systems in the Microelectronic Age*, Edinburgh University Press, 1979, pp. 168–201.

[ŽFP+95]   I. Žun, B. Filipič, M. Perpar and A. Bombač. Phase discrimination in void fraction measurements via genetic algorithms. *Review of Scientific Instruments*, 66 (10): 5055–5064, October 1995.

Manufacturing

Hybrid AI solutions and their application in manufacturing

L. Monostori, Cs. Egresits, B. Kádár

Computer and Automation Research Institute, Hungarian Academy of Sciences
Kende u. 13-17, H-1518 Budapest, POB 63, Hungary
Phone: (36 1) 1665-644, Fax: (36 1) 1667-503, e-mail: laszlo.monostori@sztaki.hu

ABSTRACT

Artificial neural networks (ANNs) are successfully applied in different fields of manufacturing, mostly where multisensor integration, robustness, real-timeness, and learning abilities are needed. Since the higher levels of the control and the monitoring hierarchy require symbolic knowledge representation and processing techniques, the integrated use of the symbolic and subsymbolic approaches is straightforward. The paper describes and compares two hybrid AI solutions for supervision and control of manufacturing processes with different degrees of integration. The first experiences gained by their usage are outlined. Finally, further possible applications of these hybrid solutions in intelligent manufacturing environment are enumerated.

1. INTRODUCTION

For state classification and monitoring purposes the application of *pattern recognition techniques* is very straightforward. The first competent results were published in the early seventies by T. Sata et. al. [22] and W.H. Dornfeld, and J.G. Bollinger [6]. *Hierarchical structure* of intelligent machine tool controllers was suggested by K. Matsushima and T. Sata in 1980 [12]. In their scheme, the lower levels consist of adaptive controllers and process pattern recognizers, designed off-line. The higher levels are more global and provide data processing over a longer period of time. The results of higher levels are manifested as changes in the lower level parameters. The conclusion of the paper is that off-line and on-line *learning and self-organizing techniques* are crucial for these intelligent controllers to be able to operate the machine in an optimal manner.

There are many signals which correlate to the condition of the manufacturing processes, and they were subject of different monitoring algorithms. However all the efforts to find a unique signal, which is applicable from different (technical, economic) points of view, failed. Consequently, the necessity of sensor integration was relatively early recognized in this field. A number of *multipurpose monitoring systems* were developed on the basis of multisensor integration and parallel processing through multiprocessor systems [25,14]. The majority of these approaches used pattern recognition techniques for learning and classification [13].

In recent years, *ANNs* were successfully applied for supervision and control of manufacturing processes [16]. The main results of these investigations are the following:

- multisensor integration through ANNs,
- classification of wear states of cutting tools,
- estimation of flank wear,
- incorporation of cutting parameters into the learning and classification phases,
- inverse modelling of the cutting process through neural networks, and
- application of inverse models for tool monitoring [3,4,5,17,21].

As a logical consequence of these and similar investigations, some authors report on the development of *"neuro monitoring systems"*, using accelerator cards for neural network computations [5,16]. The next step is easily predicted: the application of special VLSI neural chips to further accelerate the computations and get solutions with minimal space and energy assumption.

Investigations confirmed that - similarly to our present conception of biological structures - adaptive ANN techniques seem to be a viable solution for the lower level of intelligent, hierarchical control and monitoring systems. Since the higher levels of the control and monitoring hierarchy require mostly symbolic knowledge representation and processing, the *integration of symbolic and subsymbolic methods* is straightforward [1,16].

Several techniques for integrating expert systems and neural networks have emerged over the past years [8]:
- *Stand-alone models* of combined expert system and neural network applications consist of independent software components not interacting in any way.
- *Transformational models.* In one direction, *neural networks* are often used to quickly adapt to a complex, data-intensive problem, to provide generalization, and to filter errors in the data. The trained networks are transformed into expert systems

for reasons such as knowledge documentation and verification, the desire of stepwise reasoning, and for explanation facilities. Less commonly, in the *expert system to neural network transformational model*, knowledge from the expert system is used to set the initial conditions and training set for the neural network, and the neural network evolves from there.
- In *loosely-coupled models* neural network and expert system components communicate via *data files*. Here, the ANN module can perform *pre-processing*, *post-processing* or *co-processing* tasks.
- *Tight coupling* passes information via *memory resident data structures* rather than external data files improving its interactive capabilities.
- *Fully-integrated expert system/neural network models* share data structures and knowledge representation. Communication between the different components is accomplished via the dual (symbolic and neural) nature of the structures. Reasoning is accomplished either co-operatively or through a component designated as the controller.

The fundamental aim of the paper is to describe and compare two solutions for supervision and control of manufacturing processes through the integration of symbolic and subsymbolic techniques:
- *hierarchically connected hybrid AI system* (tight-coupled model)
- *symbiotic type of hybrid AI system* by integrating *neural* and *fuzzy* approaches (fully integrated model).

The first experiences gained by these hybrid solutions are outlined. Finally, their further possible applications in intelligent manufacturing environment are enumerated.

2. *HYBEXP*: A PC BASED HYBRID CONTROL AND MONITORING SYSTEM

2.1 CONCEPT OF HIERARCHICALLY TRUCTURED HYBRID AI SYSTEMS FOR MANUFACTURING APPLICATIONS

A *hierarchical structure* of intelligent machine tool controllers was suggested as mentioned in [22]. This approach was generalized in the *concept of a hierarchical monitoring and diagnostic system for manufacturing cells* [15]. Model based and pattern recognition based algorithms characterized the lower machine tool level, which was connected to the cell level subsystem with symbolic knowledge representation and processing techniques.

A concept for coupled hybrid AI systems mostly for manufacturing and diagnostic applications was developed within the framework of different

cooperations between the *Institute of Electrical Measurement (EMT), University of Paderborn* and the *Computer and Automation Research Institute (CAI), Hungarian Academy of Sciences* as well as the *Institute of Manufacturing Technology, Technical University of Budapest*. In this concept, networks outputs are conveyed to an expert system which provides process control information. On the base of accumulated knowledge the hybrid systems influence the functioning of the subsymbolic levels, generate optimal process parameters and inform the user about the actual state of the process.

This *tight coupling* approach has some clear advantages:
- it fits to the monitoring-control hierarchy of manufacturing cells regarding both the form and speed of information processing,
- modular structure enabling and facilitating the use of commercial tools,
- faster development,
- clear interfaces,
- easier integration into existing manufacturing environment.

In the following the hierarchically coupled system developed in Hungary will be described.

2.2   ELEMENTS OF THE *HYBEXP* SYSTEM

In the realization developed in the Computer and Automation Research Institute, Budapest, an artificial neural network simulator called *NEURECA* constitutes the *lower, subsymbolic level*. It provides the following main functions in an integrated framework:
- definition of different statistical and spectral features for various channels,
- on-line feature computation,
- automatic feature selection,
- manipulation, visualization of pattern files,
- ANN learning with back propagation (BP) algorithm,
- classification, estimation of unknown patterns,
- standardized (DDE) interfaces to other programs, etc.

NEURECA was written in C++ using its object oriented nature enabling to dynamically vary the network structure during learning and to implement different ANN models including *neuro-fuzzy* approaches [19]. Figure 1 illustrates some functions of the system during learning, e.g. visualization of the network (strengths of weights and node activations are characterized by different colors), the normalized inputs and outputs, the error curve during learning, visualization of network's outputs and the corresponding target values, etc.

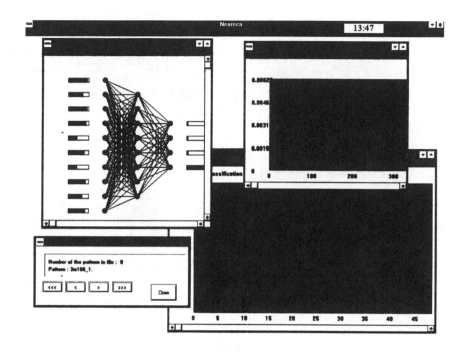

Figure 1
Some visualisation facilities of *NEURECA*

The system (and its predecessor) was successfully applied for monitoring and control of different manufacturing processes (turning, milling) [17].

Figure 1 refers to the four-class recognition problem of milling tools (sharp tools, tools with an average wear of teeth of 0.25 0.45 mm respectively, and tools with broken (missing) insert).

The *higher, symbolic level* is based on the commercially available *GoldWorks III* expert system shell [20]. *GoldWorks* provides frame-level and Lisp-level access to Dynamic Data Exchange files, so *GoldWorks* applications can directly access data in other applications through Microsoft's DDE interface.

Figure 2 illustrates the coupling and functioning of different submodules of the developed hybrid system. This solution incorporates the *NEURECA* neural network simulator (A) at the lower level and the symbolic part (B) at the higher level. The levels communicate with each other through the Microsoft's DDE interface (I).

Both the symbolic and neural subsystems are connected to the machine tool (the machine tool controller is incorporated). The symbolic part forwards (II) process parameter information (feed rate, depth of cut, cutting speed) to the machine tool (C). The

generated indirect signals (e.g. force components, vibration) are measured and conveyed (III) to the subsymbolic part (A) of the hybrid system.

In the figure the machine tool is substituted by a *simulator* of the manufacturing process (called *SIMURECA*), enlightening the test and demonstration of the system (2.4).

2.3 CONTROL AND MONITORING OF THE MILLING PROCESS BY THE *HYBEXP* SYSTEM

Configuration of the *SIMURECA* and *NEURECA* subsystems can be initiated from the symbolic level (Fig. 3). Type of manufacturing (e.g. turning, milling), the signals to be measured (e.g. force, vibration), the number of considered features, and the type of task (classification or estimation) to be fulfilled by *NEURECA* can be defined. All of the other tasks (e.g. activation of corresponding signal processing routines, neural networks, communication between the subsystems) proceed automatically. Behind this configuration process there are rules that govern this process.

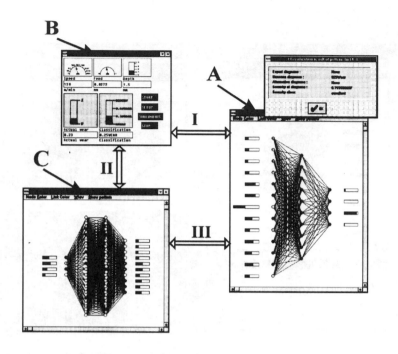

Figure 2
Components of the *HYBEXP* PC based hierarchically structured hybrid AI system

According to previous investigations [17] reliable ANN models for classification of cutting tools or for tool wear estimation using indirect signals can only be constructed if they handle process parameter (e.g. cutting speed, feed rate, depth of cut) information. process.

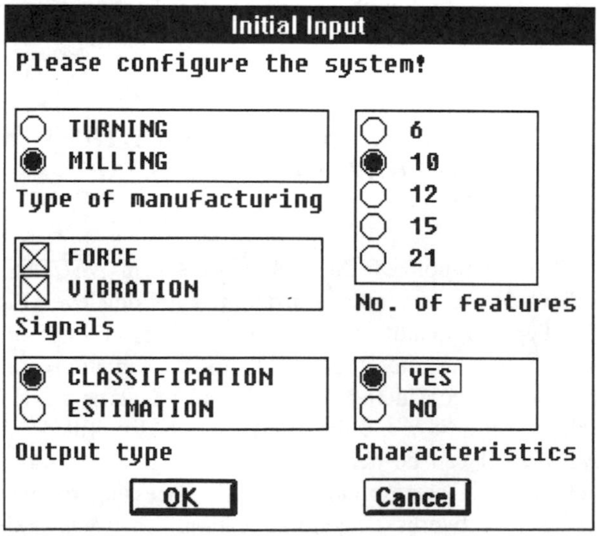

Therefore, the inputs of ANN models used in *NEURECA* additionally incorporate cutting parameters to indirect signal features. Models for both classification and estimation can be used. They are selected by the hybrid part of the *HYBEXP* during the described configuration

In both cases the results are conveyed to the hybrid part, where using additional stored knowledge (e.g. the type and number of cutting tools available, actual cutting parameters, the parts to be machined, etc.) different decisions can be made. *HYBEXP* can initiate e.g. machine stop, tool change, modification of cutting parameters (AC control) or change of parts to be machined.

Using high level explanation and visualization capabilities of *GoldWorks* different levels of information can be forwarded to the user (Fig. 2), or *HYBEXP* can work also as a *decision support system*.

Figure 3
Configuration of *NEURECA* and *SIMURECA* from the symbolic level of *HYBEXP*

2.4 SIMULATION OF MANUFACTURING ROCESSES BY ANN TECHNIQUE

Simulation of manufacturing processes is an open problem with high importance. There are a number of analytical methods but none of them proved to be reliable enough. Adaptability is required, i.e. systems which can adapt themselves to the varying manufacturing conditions. With the version of *NEURECA* (*SIMURECA*) an attempt was made to apply ANN techniques in simulation of manufacturing processes (Fig. 2). On the base of accumulated knowledge and actual process parameters (cutting speed, feed rate, depth of cut), *SIMURECA* estimates the selected features of the force and vibration signals. These features are forwarded to the *NEURECA* subsystem, which fulfils the estimation or classification assignment.

Using training patterns characterizing sufficiently broad range of cutting parameters, reliable estimates of force and vibration signal features have been generated for face milling. These estimates can be used also for process planning purposes. On the base of these results, the simulation of ultraprecise turning has been initiated.

3. NEURO-FUZZY SUPERVISION OF MANU-FACTURING PROCESSES

3.1 NEURAL, FUZZY, AND NEURO-FUZZY SYSTEMS

Fuzzy conditional statements or *fuzzy if-then rules* are expressions in the form of *IF A is A1 THEN B is B1*, where A and B are *linguistic variables*, labels of *fuzzy sets*, A1 and B1 are *linguistic values* characterised by *membership functions, (MBFs)* [9].

Fuzzy systems lie in some sense between traditional AI systems (knowledge based expert systems, KBESs) using structured knowledge representation in a symbolic manner and neural networks which cannot directly encode structured knowledge (Table 1).

Know./Encod.	Symbolic	Numerical
Structured	KBES	Fuzzy system
Unstructured	-	ANN

Table 1.
Taxonomy of model-free estimators [9]

Neural and fuzzy systems encode sampled information in a parallel-distributed framework. Both frameworks are numerical. Besides learning ability, model-free estimation is, after all, the central computational advantage of neural networks. The cost is system *inscrutability*. In contrast to the neural approach, knowledge representation in fuzzy systems is

comprehensive for users. Using fuzzy systems, the user can determine which rules contributed how much membership activation to a concluded output. Knowledge can be added or deleted without disturbing stored knowledge.

Fuzzy systems are considered to be a *natural link* between symbolic and subsymbolic approaches. On the one hand they can work in real-time circumstances and handle uncertainties as artificial neural networks, on the other hand they can manage both symbolic and numerical information. However, it is very hard to identify the fuzzy rules and tune the membership functions of the fuzzy reasoning.

It seems that best performance can prospectively be obtained by combining neural and fuzzy approaches, integrating their benefits. The resulting *neuro-fuzzy system*, is a *hybrid system*, where the system architecture remains fuzzy, but using neural learning techniques, it can be trained automatically.

There were significant previous approaches for using *fuzzy sets* and *fuzzy clustering* in monitoring of cutting tools [10,24].

The fundamental aim of this part of the paper is to demonstrate and compare the applicability of *neural* and *fuzzy systems* for monitoring and control of manufacturing processes. Combined use of the neural and fuzzy techniques is illustrated through a *neuro-fuzzy (NF)* system with structure and parameter learning. The applicability of the new solution is compared with the results gained by pure ANN techniques, and with previous results reached by a commercially available NF package. Some possible further developments are enumerated, and the role of hybrid learning approaches in realisation of intelligent manufacturing systems is highlighted.

3.2 A NEURO-FUZZY MODEL WITH STRUC-TURE AND PARAMETER LEARNING

In this section, the NF model used in the investigations described later in the paper, is introduced. It constitutes an integral part of *NEURECA* general purpose, object oriented ANN simulator. The implemented neuro-fuzzy model basically follows the main objectives of the solution described by Lin and Lee [11]. The system consists of five layers (Fig. 4).

The *linguistic nodes* in layer one and five represent the input (F_1-F_n.) and output (O_1-O_m) linguistic variables respectively.

Nodes in layers two and four are *term nodes* acting as membership functions (MBFs) to represent the terms of the given linguistic variable. Each neuron of the third layer represents one fuzzy rule (*rule nodes*). Layer three links define the *preconditions* of the rule nodes, and layer-four links incorporate the rules' *consequences*.

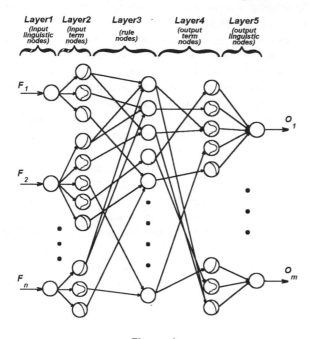

Figure 4.
5-layer structure of the implemented neuro-fuzzy model
(The arrows represent signal flow in forward direction)

Initially these layers are *fully connected* representing all possible rules.

Let node inputs be denoted as u, and node outputs as a! Using superscript and subscripts for indication of layer and node numbers respectively, the nodes in the different layers function as follows.

The nodes in layer 1 just transmit input values to the next layer. Therefore, $a_i^1 = u_i^1$, with unity link weights, $w_i^1 = 1$. A layer 2 node performs a membership function, e.g. a bell-shaped function, used in the investigations of section 4 , with a mean of $m_{i,j}$ and variance of $\sigma_{i,j}$ of the *j*th term of the *i*th input linguistic variable F_i. The link weights can be interpreted as $w_{i,j}^2 = m_{i,j}$.

Performing precondition matching of fuzzy rules, the rule nodes in layer 3 fulfil fuzzy AND (min) operation with $w_i^3 = 1$. The nodes in layer 4 integrate the fired rules having the same consequence by a fuzzy OR (sum) operator also with $w_i^4 = 1$. In forward pass, layer 5 performs *center of area* defuzzification. Here $w_{i,j}^5 = m_{i,j}^5 * \sigma_{i,j}^5$ [11].

The *learning algorithm* in the Lin-Lee model consists of four consecutive steps (Fig. 5).

- *Determination of MBFs by self organised clustering.* Kohonen's unsupervised feature-maps algorithm and *N-nearest-neighbours* heuristics are used for centering the mean values and for estimating the variance of the bell-shaped membership functions.
- *Selection of the most important fuzzy rules by competitive learning.*
- *Elimination and combination of rules.* After the second step, layer 4 weights represent the strength of the existence of the corresponding rule consequence. Using these weights, at most one consequence of a rule node will be kept. If all the links between a rule node and the layer-four nodes are deleted, then this rule node can be eliminated since it has no effect on the outputs. A rule combination is also performed to reduce the number of rules.
- *Adjustment of the parameters of the MBFs by supervised back propagation learning.*

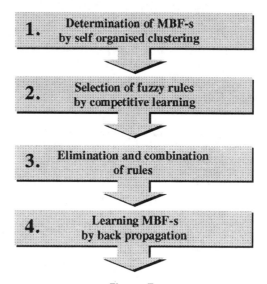

Figure 5.
Four steps of the learning algorithm in the Lin-Lee model

3.3 CUTTING EXPERIMENTS, PREPROCESSING OF SIGNALS

During the cutting (face milling) experiments carbon steel (Ck 45) was machined on a vertical milling machine with cutting speed (v = 151.6 m/min), tooth feed (s$_i$= 0.047 mm), axial depth of cut (a = 1.5 mm) and different stages of tool wear using a six-tooth 125 mm diameter WIDAX M40 cutter with ISG Böhlerit TPAN 1603 PPN R121 inserts. Cutting force components and mechanical vibration of the workpiece holder were measured with a sampling frequency of 5 kHz.

Similarly to previous ANN investigations, numerous (100 - 150) statistical and spectral features were computed from the measured signals, and the SFS (sequential forward search) feature selection procedure was accomplished with a statistical criterion using the S_b between class and S_w within class scatter matrices [17].

The investigations and comparisons described in this section refer to state classification of milling tools. Given 4 wear classes (sharp tools, tools with an average wear of teeth of 0.25 0.45 mm respectively, and tools with broken (missing) insert), the task was to generate and compare ANN and NF structures which are able to reliably classify unknown patterns characterising different wear states.

3.4 PERFORMANCE OF THE NEURO-FUZZY APPROACH

Six features (F_1-F_6) of the force components were selected using the SFS feature selection method (among these 6 features, 3-3 features came from the F_x and F_z force components). Similarly to previous investigations to be referred to later, where a commercial NF package was used [18], 4, 3, 2, 3, 4, 4 MBFs were respectively assigned to the linguistic variables. Corresponding to the described 4 class problem, the output linguistic variable (tool state) had 4 MBFs. Using the neuro-fuzzy model described in Sect. 3, the following network structure was determined (Fig. 6).

This figure also illustrates the normalised input features and the output of the network for a given pattern.

After initialisation, taking all the possible combinations, 4*3*2*3*4*4 = 1152 rule nodes were generated. During the competitive learning phase 1138 (!) of them were deleted, resulting in a network structure with 14 rules (Fig. 6). This *self organisation* is a very important feature of the chosen model. The number of eliminated links was 4594.

Figure 7 illustrates the result of parameter learning on the MBFs of feature 5 (F_5) with normalised x axis. Their mean values and variances were adjusted to reach a kind of optimum regarding the network error for training patterns. The generated 14 rules are enumerated in linguistic form in Table 2. The first rule in this Table can be interpreted as follows: *IF F_1 is high and F_2 is high and F_3 is high and F_4 is low and F_5 is high and F_6 is high THEN output is Wear025* (average wear of teeth is 0.25 mm). 97.9% of 47 patterns in the training set was properly classified after the self organised clustering and competitive learning phases of the 3 step learning algorithm. Having adjusted the parameters of the MBFs, 100% recognition ability was experienced within *some tens of learning epochs*. (After the very fast two first steps of the learning process, BP just fine tunes the parameters). The 32 test patterns were classified after these stages with 96.9 and 100% security.

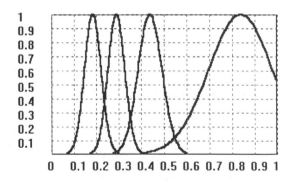

Figure 7.
Learned membership functions for F_5(quadratic mean value of F_z)

3.5 COMPARISON WITH ANN AND PREVIOUS NF INVESTIGATIONS

In the ANN investigations described here the *back propagation (BP)* technique was applied for three layer networks. All the four structures (with 2, 4, 6, 10 hidden elements respectively) converged to 100% recognition

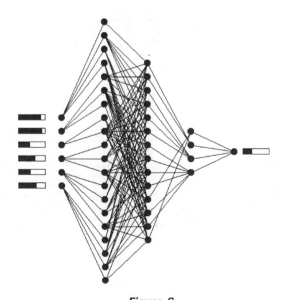

Figure 6.
NF structure for the four class problem, after learning, with normalised inputs and output

rate within the learning set. The recognition of test patterns varied from 93,75% to 96,88%, depending on the network structure.

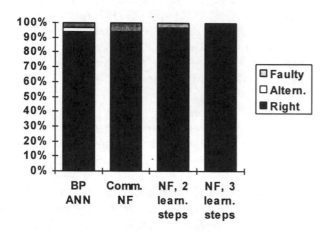

Figure 8.
Recognition performance of different approaches for test patterns

The best result achieved by a commercial NF package with 17 initial rules generated by an expert with the same 6 force features was 93.75% recognition of test patterns. Figure 8 illustrates the recognition abilities of trained ANNs and the two NF approaches for test patterns.

(The region "alternative" characterises the uncertainty of classifications caused by the different number of hidden elements in BP networks. Through self organisation, this problem does not come up in the NF approaches applied here.) The figure demonstrates the significance of structure and parameter learning.

4. CONCLUSIONS

The paper surveyed the most important steps of the process of applying *pattern recognition* (*PR*) *techniques*, *artificial neural networks* (*ANNs*), and *hybrid artificial intelligence* (*AI*) techniques in manufacturing.

Two hybrid AI solutions for supervision and control of manufacturing processes were described:

- a hierarchically connected coupled model, and
- a fully integrated model based on the integration of neural and fuzzy techniques.

The concept of *hierarchically structured hybrid AI systems* for manufacturing applications was introduced, the structure of the *HYBEXP* system developed in the Computer and Automation Institute was illustrated.

	F1	F2	F3	F4	F5	F6	Output
1	high	high	high	low	high	high	Wear025
2	high	high	high	low	high	m_high	Wear025
3	m_high	m_high	medium	low	high	m_high	Wear045
4	m_high	m_high	medium	low	medium	m_high	Wear045
5	m_high	m_low	medium	low	high	m_high	Wear045
6	m_high	m_low	medium	low	medium	m_high	Wear045
7	m_low	m_low	low	low	medium	m_low	Broken
8	m_low	low	low	low	medium	m_low	Broken
9	low	low	low	low	medium	m_low	Broken
10	m_high	m_low	medium	low	medium	low	Broken
11	m_low	low	low	high	medium	low	Sharp
12	m_low	low	low	low	medium	low	Broken
13	m_low	low	low	low	low	low	Broken
14	low	low	low	high	medium	low	Sharp

Table 2
The set of rules remained after learning

The first experiences gained by its applications were described. Further projects were initiated to investigate the applicability of the outlined hierarchical approach
- in production control and scheduling of manufacturing systems,
- in distributed control systems incorporating numerous ANN models,
- in holonic type systems.

Artificial neural networks have proven to be equal, or superior, to other pattern recognition learning systems over a wide range of domains, also in supervision of manufacturing processes.

The paper compared ANN and fuzzy solutions, and suggested their integrated use in supervision and control, with the aim to eliminate the above shortcomings. The first results seem to be promising. Having limited the investigations to the state classification of milling tools with constant operating parameters, the *NF technique* with structure and parameter learning illustrated in the paper showed superior performance to the BP solution and previous investigations with a commercial NF system.

It was shown that NF systems as hybrid learning systems can comply with the fundamental requirements of intelligent manufacturing, i.e. *real-time nature, learning ability, handling of uncertainties, managing and learning both symbolic and numerical information.* This approach has the additional benefits of *explicit knowledge representation* which is easily *modifiable* and incorporates a kind of *explanation facility*.

Further research was started to incorporate cutting parameters (cutting speed, feed per tooth, depth of cut) into the NF approach, and to apply this technique for tool wear estimation and process modelling.

It is expected that the above and similar hybrid AI solutions fitting into the process outlined in the title, can provide sufficient frameworks for the solution of numerous problems in manufacturing, and can prospectively contribute to the future realisation of intelligent manufacturing systems.

5. ACKNOWLEDGMENTS

This work was partially supported by the *National Research Foundation, Hungary*, Grants No. T 014514, and T016512 (Fundamental research for intelligent manufacturing). The milling experiments were accomplished at the Technical University of Budapest in 1993. The author expresses his gratitude to Prof. M. Horváth head of the Institute of Manufacturing Technology and his colleagues, especially to Dr. S. Markos.

The concept of hierarchically structured hybrid AI systems for manufacturing applications have been developed within a scientific cooperation with the Institute of Electrical Measurement, University of Paderborn, Institute of Manufacturing and Production Engineering, University of Kaiserslautern, as well as the Computer and Automation Research Institute, Hungarian Academy of Sciences. The Hungarian contribution was supported by the PHARE ACCORD Programme of the European Union, Grant No.: H 9112-0216 (Hybrid knowledge processing in CIM). Special thanks to the heads of the above institutes Prof. D. Barschdorff and Prof. G. Warnecke

REFERENCES

[1] Barschdorff, D., Monostori, L.: "Symbolicism and connectionism: rivals or allies in intelligent manufacturing?", Proc. of The 23rd CIRP International Seminar on Manufacturing Systems, 6-7 June, 1991, Nancy, France, Section 4, pp. 14-28.

[2] Barschdorff, D., Monostori, L., Wöstenkühler, G.W., Egresits, Cs., Kádár, B.: "Approaches to coupling connectionist and expert systems", Preprints of the Second International Workshop on Learning in IMSs, April 20-21, 1995, Budapest, Hungary, pp. 591-608.

[3] Chryssolouris, G., Domroese, M.: "An experimental study of strategies for integrating sensor information in machining", Annals of the CIRP, Vol. 38/1/1989, pp. 425-428.

[4] Dornfeld, D.A.: "Neural network sensor fusion for tool condition monitoring", Annals of the CIRP, Vol. 39/1/1990, pp. 101-105.

[5] Dornfeld, D.A.: "Unconventional sensors and signal conditioning for automatic supervision", Proc. of the AC'90, III. CIRP International Conference on Automatic Supervision, Monitoring and Adaptive Control in Manufacturing, 3-5. Sept, Rydzyna, Poland, 1990, pp. 197-233.

[6] Dornfeld, W.H., Bollinger, J.G.: "On line frequency domain detection of production machinery malfunctions", Proc. of the 18th. Int. Machine Tool Design and Res. Conf., London, 1977, pp. 837-844.

[7] Erdélyi, F.: "Control of machine tools and discrete manufacturing processes", Dissertation for the C.Sc. degree of the Hungarian Academy of Sciences, Miskolc, Hungary, 1993.

[8] Kandel, A.; Langholz, G.: "Hybrid architectures for intelligent systems", CRC Press, Boca Raton, 1992.

[9] Kosko, B.: Neural networks and fuzzy systems; Prentice Hall, New Jersey, 1992.

[10] Li, P.G., Wu, S.M.: "Monitoring drilling wear states by a fuzzy pattern recognition technique", J.l of Engineering for Industry, Vol. 110, 1988, pp. 297-300.

[11] Lin, C.H., Lee, C.S.G.: "Neural-network-based fuzzy logic control and decision system", IEEE Trans. on Comp., Vol. 40, Dec. 1991, pp. 1320-1336.

[12] Matsushima, K. and Sata, T.: "Development of intelligent machine tool", Journal Faculty Eng., Univ. Tokyo, vol. 35, No. 3, 1980, pp. 299-314.

[13] Monostori, L.: "Learning procedures in machine tool monitoring, Computers in Industry, North-Holland, Vol. 7, 1986, pp. 53-64.

[14] Monostori, L.; Bartal, P.; Hermann, Gy.; Horváth, L.; Nacsa, J.; Pasztirák, G.; Soós, J.; Zsoldos, L.: "Multipurpose, complex machine tool monitoring systems, their programming and application", Preprints of the 4th International Conference on the Manufacturing Science and Technology of The Future, MSTF'89, Stockholm, Sweden, 6-9 June 1989, pp. 345-356.

[15] Monostori, L., Bartal, P. and Zsoldos, L.: "Concept of a knowledge based diagnostic system for machine tools and manufacturing cells", Computers in Industry, Vol. 15, 1990, pp.95-102.

[16] Monostori, L.; Barschdorff, D.: "Artificial neural networks in intelligent manufacturing", Robotics and Computer-Integrated Manufacturing, Vol. 9, No. 6, 1992, Pergamon Press, pp. 421-437.

[17] Monostori, L.: "A step towards intelligent manufacturing: Modelling and monitoring of manufacturing processes through artificial neural networks", Annals of the CIRP, 42, No. 1, 1993, pp. 485-488.

[18] Monostori, L., Egresits, Cs.: "Modelling and monitoring of milling through neuro-fuzzy techniques", Preprints of the 2nd IFAC/IFIP/IFORS/ Workshop on Intelligent Manufacturing Systems, June 13-15, 1994, Vienna, Austria, pp. 381-386.

[19] Monostori, L.; Egresits, Cs.: "On hybrid learning and its application in intelligent manufacturing", Preprints of the Second International Workshop on Learning in Intelligent Manufacturing Systems, April 20-21, 1995, Budapest, Hungary, pp. 655-670.

[20] N.N.: "GoldWorks III, User's guide", Gold Hill, Inc., Cambridge, Massachusetts, 1993.

[21] Rangwala, S.S., Dornfeld, D.A.: "Learning and optimatization of machining operations using computing abilities of neural networks", IEEE Transactions on Systems, Man, and Cybernetics, 19, No. 2, March/April, 1989, pp. 299-314.

[22] Sata, T., Matsushima, T., Nagakura, T., Kono, E.: "Learning and recognition of the cutting states by the spectrum analysis", Annals of the CIRP, Vol. 22/1/1973, pp. 41-42.

[23] Tönshoff, H.K., Wulsberg, J.P., Kals, H.J.J., König, W., van Luttervelt, C.A.: "Developments and trends in monitoring and control of machining processes", Annals of the CIRP, Vol. 37/2/1988, pp. 611-622.

[24] Wang, M., Zhu, J.Y., Zhang, Y.Z.: "Fuzzy pattern recognition of the metal cutting states", Annals of the CIRP, Vol. 34/1/1985, pp. 133-136.

[25] Weck, M., Monostori, L., Kühne, L.: "Universelles system zur Prozeß- und Anlagenüberwachung", Vortrag und Berichstband VDI/VDE-GMR Tagung "Verfahren und Systeme zur technischen Fehlerdiagnose", Langen, FRG, 1984, pp. 139-154.

[26] Westkämper, E.: "Zero-defect manufacturing by means of a learning supervision of process chains", Annals of the CIRP, Vol. 43/1/1994, pp. 405-408.

PERFORMANCE OPTIMIZATION OF FLEXIBLE MANUFACTURING SYSTEMS USING ARTIFICIAL NEURAL NETWORKS

S.Cavalieri

University of Catania-Faculty of Engineering
Institute of Informatic and Telecommunications
Viale.A.Doria, 6 Catania 95125 (ITALY)
Email:cavalieri@iit.unict.it, Fax:+39 95 338280

ABSTRACT

The paper presents a strategy for performance evaluation of flexible manufacturing systems based on AI techniques. The author proposes a Hopfield-type neural model capable of optimizing the productivity of any flexible manufacturing system. This is one of the main goals in the automated factory environment in which productivity is often limited by poor use of available resources. The aim is therefore to exploit these resources to the full. The solution proposed is an original one and features several advantages, the most important of which is the possibility of optimal or close to optimal solutions in a polynomial time, proportional to the size of the flexible manufacturing system. In addition, the possibility of simple, economical hardware implementation of the Hopfield model favours integration of the model in the automated factory environment, allowing real-time supervision and optimization of productivity.

INTRODUCTION

A Flexible Manufacturing System (FMS) is an interconnected system of material processing stations capable of automatically processing a wide variety of part types simultaneously and under computer control [Pim90]. The system features flexibility in routing parts, part processing operations, coordination and control of part handling, and in using appropriate tooling. The required flexibility in virtually every aspect of the manufacturing process requires careful coordination of the different components of the manufacturing system. Furthermore, planning and controlling the movement of parts through the system can be a complex problem, above all when the main aim is that of maximizing the productivity of the manufacturing system. This can be achieved by scheduling resources (e.g. determining the sequence in which the machines operate) in such a way as to utilize them fully.

Evaluating a manufacturing system consists of providing information about its behaviour, the most important information being the productivity of the system. Literature provides well-known techniques for modelling and performance evaluation of FMS using a particular class of Petri nets, called Event Graphs. The theoretical conditions and strategies of performance evaluation are well documented [Hil89][Laf92] [Zur94]. Performance evaluation can be applied to an existing FMS, i.e. one in which the use of resources is already defined, to assess performance features such as productivity. This is a very simple task and does not entail great computational complexity. If, on the other hand, the aim is to schedule resources in such a way as to maximize productivity, performance evaluation by means of Event Graphs is a much more complex process and requires a great deal of time and calculating power. It is, in fact a combinatorics task of a non-polynomial dimension (i.e.an NP-hard problem). This obviously represents a limit to the applicability of performance optimization to a highly complex FMS (e.g. featuring a large number of machines). There are numerous contributions in literature presenting heuristic algorithms for performance optimization by Event Graph. Although these algorithms do not always guarantee an optimal solution to the optimization problem, they have the evident advantage of low calculating times One of the most important contributions in this field is provided by [Laf92], in which two heuristic algorithms are proposed. Although they feature high percentages of optimal solutions, some remarks need to be made. One concerns the definition of certain conditions surrounding the optimization problem to be solved. Some algorithmic solutions are extremely imprecise in the specific formulation of the surrounding conditions. In the Adjustment Heuristic Algorithm (AHA) **[Laf92]**, for instance, the conditions introduced to respect some of the constraints are quite rough ones. In addition, although the algorithms are heuristic and therefore reduce the set of solutions to be explored, the time required to reach a solution cannot be established a priori, and may range in an interval whose upper bound is the time required to calculate an exhaustive solution. The aim of this paper is to present an alternative to these heuristic algorithms, based on the use of artificial intelligence, which can overcome the limits outlined above. The author proposes a Hopfield-type neural model to implement the Event Graph modelling the FMS. Once the function to be optimized (e.g. productivity) and the relative surrounding conditions have been established, the neural network provides a solution to the optimization problem. The paper will illustrate the advantages of the proposed strategy in great detail, focusing on how it overcomes the above-mentioned limits of the algorithmic approach.

1.SOME REMARKS ON EVENT GRAPHS

Literature provides for several examples on the use of Petri Nets to the modelling and scheduling of Flexible Manufacturing Systems **[Aly90][Shi91] [Tsa92]**. In particular, the class of Petri Nets denoted as Timed Event Graphs is widely used to model FMSs, as it allows performance evaluation and performance optimization of these systems **[Hil89][Laf92][Zur94]**. The aim of this section is to give a brief description of Event Graph and the related performance optimization methodology.

An Event Graph is a Petri Net in which each place has one input transition and one output transition. Each transition can have an associated firing time, in which case the event graph is called timed. The presence of a token in a place enables firing of the output transition for that place. Firing may occur in null time, if the transition is immediate, or in the time associated with the transition if it is timed. An event graph is said to be strongly connected if there is a path joining any pair of nodes of the graph. An elementary circuit in a strongly connected event graph is a direct path that goes from one node back to the same node, while any other node is repeated. It can be

demonstrated that the number of tokens in any elementary circuit is not varied by transition firing **[Com71]**. If $M(\gamma)$ indicates the number of tokens in the elementary circuit γ, and $\mu(\gamma)$ is the sum of all the firing times of the transitions belonging to the elementary circuit γ, the relation:

$$C(\gamma) = \frac{\mu(\gamma)}{M(\gamma)}$$

is called the cycle time. If Γ indicates the set of elementary circuits, the critical circuit is defined as the one with the highest cycle time. If we call:

$$C^* = \max_{\gamma \in \Gamma} C(\gamma)$$

any $\gamma \in \Gamma$ such circuit for which $C(\gamma) = C^*$ is a critical circuit. In the case of Event Graphs with deterministic timed transitions which fire as soon as they are enabled, the firing rate of each transition in the steady state is given by $\lambda = 1/C^*$ **[Chr83]**. This result highlights that the critical circuit is the one that actually binds the speed (i.e. the throughput) of the FMS. As consequence, if we want to increase the speed of the system, we have to increase the number of tokens in each critical circuit. On the basis of this consideration, FMS performance optimization problem can be formulated as minimizing the quantity:

$$\sum_{i=1}^{n} u_i \cdot x_i \qquad (1)$$

where u_i are the p-invariants of the event graph **[Zur94]** and x_i represents the number of tokens in place P_i, under the condition :

$$M(\gamma) \ge \alpha \cdot \mu(\gamma) \qquad (2)$$

where α is the performance required. As can be seen, condition (2) forces the number of tokens in each elementary circuit to be increased, according to the performance index α. In this way, the firing rate of each transition in the steady state assumes higher value, improving the throughput of the system. Condition (2) may not be the only one. In the case of job-shop systems, for example, one machine may carry out several operations sequentially. The corresponding event graph will feature particular elementary circuits called command circuits for each machine. Each command circuit specifies the sequencing of the jobs on the corresponding machines. Indicating as Γ_c the set of command circuits, the total number of tokens, $M(\gamma^c)$, $\forall \gamma^c \in \Gamma_c$ has to be equal to 1. More than one token in a command circuit would, in fact, correspond to the impossible situation in which the relative machine is performing two jobs at the same time.

When there command circuits in an event graph, (2) therefore becomes:

$$\begin{cases} M(\gamma^*) \geq \alpha \cdot \mu(\gamma^*) \quad \forall \gamma^* \in \Gamma^* & (2') \\ M(\gamma^c) = 1 \qquad \quad \forall \gamma^c \in \Gamma_c & (2'') \end{cases}$$

where $\Gamma^* = \Gamma - \Gamma_c$.

On account of the conditions (1),(2') and (2''), optimization of the performance of a manufacturing system can be achieved by solving an integer linear problem where at least as many constraints as the elementary circuits in the graph must be considered. Actually, this represents a major drawback of such an approach, since it requires a great deal of time to find a solution. In [Laf92] two heuristic algorithms to reduce this computational burden are proposed. (The reader is referred to [Laf92] for a description of the heuristic techniques proposed).

In the following sections the author presents an alternative strategy for FMS performance optimization. The proposal is based on use of a particular neural model (the Hopfield type model, which has proved to be particularly suitable for solving optimization problems). The neural approach to the performance optimization of FMSs offers numerous advantages which will be illustrated in detail in Section 4.

2. SOME REMARKS ON THE HOPFIELD NEURAL NETWORK

The use of neural networks to solve optimization problems was first proposed by Hopfield and Tank [Hop82][Hop86][Tan86]. They devised a neural model, known as the Hopfield model, which is able to offer valid solutions to discrete combinatorics optimization problems. The computational capacity offered by the Hopfield model was first demonstrated by application to the Travelling Salesman Problem (TSP). Since then, several researchers have made contributions to both the neural model itself and its application to various optimization problems [Coh83][Hec90]. The Hopfield model is based on a single-layer architecture of neurons, the outputs of which are fed back towards the inputs. The output of the x-th neuron, O_x, which assumes values ranging between 0 and 1, is linked to the input of the same neuron, U_x, by the following sigmoidal monotonic increasing function:

$$O_x = g(U_x) = \frac{1 + \tanh(\frac{U_x}{u_0})}{2}$$

where the parameter u_0 controls the effective steepness of the function: the lower it is, the steeper the function. Each feedback between the output of the i-th neuron and the input of the x-th neuron has an associated weight, w_{xi}, which determines the influence of the i-th neuron on the x-th neuron. An external bias current, I_x, is then supplied to the x-th neuron. The dynamics of the Hopfield network are described by:

$$\frac{d U_x}{dt} = -\frac{U_x}{\tau} + \sum_i w_{xi} \cdot O_i + I_x \qquad (3)$$

where τ is a user-selected decay constant.

Hopfield [Hop84] showed that if the weights matrix $W = [w_{xi}]$ is symmetrical and if the function g is a steep-like curve (i.e. $u_0 \to 0$), the dynamics of the neurons described by (3) follow a gradient descendent of the quadratic energy function, also known as the Lyapunov function:

$$E = -\frac{1}{2} \cdot \sum_x \sum_i w_{xi} \cdot O_x \cdot O_i - \sum_x I_x \cdot O_x \qquad (4)$$

Under the same hypothesis, Hopfield [Hop84] showed that the minima of the energy function (4) coincide at the corners of the hypercube defined by $O_x \in \{0,1\}$.

These theoretical results allow a solution to a particular optimization problem to be obtained from the stabilized outputs of the Hopfield network, by the following method. First the surrounding conditions of the optimization problem are identified and expressed in the form of the energy function given by (4). Generally, a real coefficient is multiplied to each condition, in order to weight its influence over the solution to the optimization problem. By comparison of the energy function obtained with the function (4), the weights and bias currents are expressed in terms of these coefficients, and so linked to the surrounding conditions of the problem to be solved. If the weights matrix obtained is symmetrical and the function g is steep-like, the stable output of the Hopfield model obtained by imposing the weights and biases previously calculated corresponds to a minimum of (4) and thus to a solution to the problem. In the following section this methodology will be applied to solve the FMS performance optimization problem.

3. USING THE HOPFIELD NETWORK TO MODEL AND OPTIMIZE PERFORMANCE OF FMSs

The aim of this section is to show how the Hopfield network is used for performance optimization of a FMS. The strategy is divided into two steps: modelling the FMS through the neural model and performance optimization.

Modelling. The first step in the strategy proposed is representing the FMS using a Hopfield neural network. This is achieved by mapping the Event Graph modelling the FMS with the neural model. The Event Graph is mapped onto the Hopfield neural model by making each place P_i in the Event Graph correspond to the neuron O_i in the neural network. The value (1 or 0) of the output of each neuron O_i models the presence or absence of a token in the place P_i modelled: if the output is 1 the place corresponding to the neuron contains a token, if it is 0 the place contains no tokens. It is clear that the proposed coding of the neural output is based on the necessary assumption that each place in the Event Graph contains at most one token. In general there is no limit in an Event Graph to the maximum number of tokens in a place. Such a limit can only be imposed by the particular features of the FMS modelled (e.g. if a place models a buffer, the maximum number of tokens in the place is equal to the real capacity of the buffer). In [**Laf92**] it is demonstrated that it is always possible to modify a strongly connected graph into an equivalent one in which each place possesses at most one token. Fig.1 shows the transformation which makes place P (which was supposed containing at most two tokens) correspond to two places, P' e P'', containing at most one token each. In such a case the transition tnew has a null firing time.

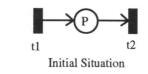

Initial Situation

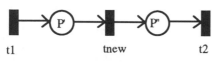

Extended Situation

FIGURE 1   Equivalence between Event Graphs

In the following an example of the mapping proposed will be shown. Let us consider the Event Graph shown in Fig.2, modelling a job-shop composed of four machines M1, M2, M3 e M4, which can manufacture three types of parts denoted by R1, R2 and R3. The production mix is 25%, 25% and 50% for R1, R2 and R3, respectively. As can be seen, in the model there are four command circuits relating to the four

machines: $\Gamma^c=\{\gamma^c_1,\gamma^c_2,\gamma^c_3,\gamma^c_4\}$ where γ^c_1=(P13, P14, P15, P16), γ^c_2=(P17, P18, P19), γ^c_3 = (P20, P21), and γ^c_4=(P22, P23, P24, P25).

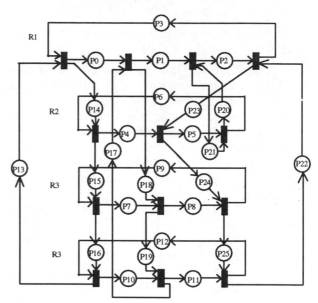

FIGURE 2   Event Graph Model of a Job-Shop

On account of what previously said, it can be assumed that the Event Graph of Fig.2 features each place containing one token at most. Fig.3 shows the Hopfield network corresponding to the Event Graph in Fig.2.

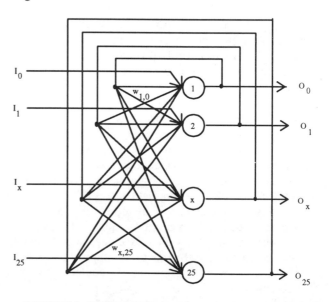

FIGURE 3   Hopfield Neural Model Equivalent to the Event Graph shown in Figure 2

As can be seen, the mapping between the Event Graph modelling the FMS and the Hopfield network allows the architecture, i.e. the number of neurons it contains, to be defined. The Hopfield network corresponding to the Event Graph in Fig.2 possesses, in fact, 26 neurons, one for each place in the event graph. The figure also shows the bias currents I_x and weights $w_{x,i}$ (x,i=0..25), the values of which are established using the method illustrated in the following sub-section.

FMS Performance Optimization. FMS performance optimization can be achieved by means of the Hopfield network through a number of steps. First, all the constraints of the optimization problem are expressed in terms of Lyapunov energy functions. Then, on the basis of the methodology outlined in Section 2, the weights and bias currents of the Hopfield network are obtained, so that it can provide a solution to the problem. These steps are examined in greater detail below, with reference to the surrounding conditions (1),(2') and (2") which, as mentioned previously, are always present in the problem if FMS performance optimization. At times condition (2") may not be present, i.e. if there are no command circuits.

Condition (1) can also be expressed in the following form:

$$\text{minimize } (\sum_{i=1}^{n} u_i \cdot x_i)^2$$

This condition can be expressed in the following term of the Lyapunov energy function:

$$\frac{A}{2} \cdot (\sum_x \sum_i u_x \cdot O_x \cdot u_i \cdot O_i)$$

Condition (2'), relating to performance, is represented by the following Lyapunov energy function term:

$$\frac{B}{2} \cdot \sum_{\gamma^*} (\sum_{x:P_x \in \gamma^*} O_x - M(\gamma^*))^2$$

where the first sum is extended to $\forall \gamma^* \in \Gamma^*$. For each $\gamma^* \in \Gamma^*$, the second sum is extended to $\forall O_x$ such that the corresponding place P_x in the Event Graph belongs to γ^*. As can be seen this condition imposes in each circuit $\gamma^* \in \Gamma^*$ a number of tokens given by $M(\gamma^*)$,

which, according to (2'), must be $\geq \lceil \alpha \cdot \mu(\gamma^*) \rceil$.

Finally, condition (2"), relating to imposing a single token in each of the command circuits, corresponds to the Lyapunov function term:

$$\frac{C}{2} \cdot \sum_{\gamma^c} (\sum_{x:P_x \in \gamma^c} O_x - 1)^2 \qquad (5)$$

where the first sum is extended to $\forall \gamma^c \in \Gamma_C$. For each $\gamma^c \in \Gamma_C$, the second sum is extended to $\forall O_x$ such that the corresponding place P_x in the Event Graph belongs to γ^c. As can be seen this condition imposes in each command circuit γ^c a single token (i.e. $M(\gamma^c)=1$, $\forall \gamma^c \in \Gamma_c$).

As said in Section 2, the real coefficients A,B and C weights the influence of each term on the others. The energy function relating to the surrounding conditions outlined above becomes:

$$E = \frac{A}{2} \cdot (\sum_x \sum_i u_x \cdot u_i \cdot O_x \cdot O_i)$$

$$+ \frac{B}{2} \cdot \sum_{\gamma^*} (\sum_{x:P_x \in \gamma^*} O_x - M(\gamma^*))^2 \qquad (6)$$

$$+ \frac{C}{2} \cdot \sum_{\gamma^c} (\sum_{x:P_x \in \gamma^c} O_x - 1)^2$$

It is very important to point out that the minimization of this energy function leads to the maximization of the FMS productivity. Infact, the minimization of the energy function (6) as whole is achieved when each of the three terms is minimized. The minimization of the term multiplied by the coefficient A, leads to the minimization of the quantity (1), the minimization of the terms multiplied by B and C leads to the respect of the surrounding conditions (2') and (2") respectively.

Comparison between (6) and (4) determines the following weights and bias currents for the neural network:

$$w_{xi} = -A \cdot u_x \cdot u_i - B \cdot n_{\gamma^*_{x,i}} - C \cdot n_{\gamma^c_{x,i}} \qquad (7)$$

$$I_x = +B \cdot \sum_{\gamma^*:P_x \in \gamma^*} M(\gamma^*) + C \cdot n_{\gamma^c_x} \qquad (8)$$

where $n_{\gamma^*_{x,i}}$ represents the number of circuits in the set Γ^* to which the places with index x and i simultaneously belong, $n_{\gamma^c_{x,i}}$ represents the number of circuits in the set Γ_c to which the places with index x and i simultaneously belong, and $n_{\gamma^c_x}$ represents the number of circuits in the set Γ_c to which the place with index x belongs. As said before, $M(\gamma^*)$ refers to the number of tokens in each circuit γ^*.

As can be seen the contribution of the term

$$+B \cdot \sum_{\gamma^*:P_x \in \gamma^*} M(\gamma^*)$$ to the bias currents I_x is not constant,

as $M(\gamma^*) \geq \left\lceil \alpha \cdot \mu(\gamma^*) \right\rceil$, according to (2'). This consideration highlights that the Hopfield-type neural model presented in Section 2 may be able to solve the problem being dealt with here, only if a variability in the bias currents is introduced. These currents can not a-priori be fixed, but must be varied during the neural iteration in order to cause the number of tokens in

each circuit $\gamma^* \in \Gamma^*$, to be no less than $\left\lceil \alpha \cdot \mu(\gamma^*) \right\rceil$. The modification of the bias currents may be performed according to the following algorithm. First the weights of the neural model are fixed according to (7). The bias currents are fixed according to:

$$I_x = +C \cdot n_{\gamma^c_x}$$

which was obtained from (8) removing the term

$$+B \cdot \sum_{\gamma^*:P_x \in \gamma^*} M(\gamma^*).$$ At each neural iteration all the non-activated neurons are considered. For each of them all the circuits to which the corresponding place belongs are considered. The number of tokens for each of these

circuits is counted. If it is less than $\left\lceil \alpha \cdot \mu(\gamma^*) \right\rceil$, i.e. condition (2') is not met for that circuit, the bias current of the neuron being considered is increased by

$$\delta B \cdot \sum_{\gamma^*:P_i \in \gamma^*} \left\lceil \alpha \cdot \mu(\gamma^*) \right\rceil.$$ The value δB represents a user-selected fraction of B. In this way, the increment of the bias current forces the network to satisfy the condition (2').

Table 1 shows an example of the solution provided by the neural approach here presented. The solution refers to the job-shop performance optimization problem, modelled by the Event Graph in

Fig.2. The performance index α was fixed to 1/6 and the following values of u_i in (1) were selected: $u_i=1$ $\forall i \in [0,12]$ and $u_i=0$ $\forall i \in [13,25]$. For each place of the Event Graph, the table shows the distribution of the tokens determined by the neural network. As can be verified, this solution produces the best performance for the whole system.

4. NEURAL NETWORK APPROACH VERSUS HEURISTIC ALGORITHMIC SOLUTIONS

The neural approach proposed for FMS performance optimization presents a number of advantages over heuristic algorithmic solutions like those shown in [Laf92], the most significant of which are illustrated below.

- Some algorithmic solutions are extremely imprecise in formally specifying the surrounding conditions. In the Adjustment Heuristic Algorithm (AHA), presented in [Laf92] for instance, the objective function used to maximize the performance of the event graph shown in Fig.2 is of the following kind:

$$f = \sum_{i=0}^{12} x_i + 1000 \cdot \sum_{i=13}^{25} x_i \qquad (8)$$

in which the coefficient 1000 has the aim of limiting the number of tokens in the places with an index between 13 and 25, i.e. places belonging to the command circuits. Although the aim of condition (8) is to highly penalize these places to avoid more than one token in the command circuits, it is quite a rough condition as it does not explicitly distinguish between the command circuits. The neural solution presented here allows more accurate definition of the surrounding conditions. As shown in Section 3, the surrounding condition relating to the total number of tokens in the control circuits, i.e. (5), can be specified exactly and also allows the values of the relative weights and bias currents to be determined by means of a mathematical equality.

- Another limit of heuristic solutions is represented by calculation times. Although the algorithms are heuristic and thus reduce the set of solutions to be explored, the time required to reach a solution

Table 1 Optimal Solution by the Neural Approach.

P	l	a		c	e	s																			
0	1	2	3	4	5	6	7	8	9	1	1	1	1	1	1	1	1	1	1	2	2	2	2	2	2
										0	1	2	3	4	5	6	7	8	9	0	1	2	3	4	5
0	0	1	1	1	1	0	0	1	0	0	1	0	1	0	0	0	1	0	0	1	0	0	1	0	0

cannot be foreseen a priori, and may range in an interval whose upper bound is the time required for an exhaustive solution. The heuristic Convex Optimization Algorithm for instance, again proposed in [Laf92], is based on a branch-and-bound search algorithm. Although it is theoretically capable of finding an optimal solution at the first step of the computation, the number of steps required to find the optimal solution cannot be determined. The time required for the neural solution proposed here, on the other hand, is foreseeable and easy to calculate. As shown in [Meh93][Tak86], the calculation time is proportional to the size of the weights matrix, more specifically to the number of neurons squared. In the FMS performance optimization solution by the neural approach, this time is polynomial with the number of places in the event graph, i.e. the size of the FMS to be optimized.

- Another advantage of the solution proposed lies in the quality of the solution provided. Literature provides examples of how Hopfield networks guarantee high percentages of optimal or close to optimal solutions [Aiy90]. This result was confirmed during the tests conducted by the author. Several FMS performance optimization problems were considered, and for each of them a very high number of optimal or very close to optimal solutions was obtained.

- A last advantage relates to the impact of the solution proposed on the automated factory environment. One of the main requirements of performance optimization is the possibility of easy integration into the production cycle of an FMS, so as to guarantee on-line monitoring and optimization of the industrial process. This can only be achieved if the performance optimization strategy can easily be engineered. Literature shows how simple and economical the hardware implementation of the Hopfield neural model is [Ver89], thus favouring integration in the automated factory environment.

FINAL REMARKS

The paper has presented an original approach to FMS performance optimization, based on the modelling of an FMS by a Hopfield neural network. The solution the network provides determines the FMS configuration which will maximize performance. From tests carried out on a large number of examples, it was found that the quality of the neural solution is always high, as the solution reached is always optimal or close

to optimal. The paper has illustrated the advantages of the solution proposed, the most significant of which is the possibility of hardware implementation of the neural model; this not only further reduces the time required to calculate the optimal solution but also and above all makes it possible to integrate the model proposed in the FMS production cycle. In this way on-line supervision and optimization can be obtained.

REFERENCES

[Aiy90] S.V.B.Aiyer, M.Niranjan, and F.Fallside, "A Theoretical Investigation into the Performance of the Hopfield Model", IEEE Transaction on Neural Networks, vol.1, n.2, pp.204-215, June 1990.

[Aly90] R.Y.Al-Yaar and A.A.Desrochers, "Performance Evaluation of Automated Manufacturing Systems using Generalized Stocharstic Petri Nets", IEEE Transaction on Robotics and Automat., vol.6, no.6, pp.621-639, 1990.

[Chr83] P.Chretienne, "Les Reseaux de Petri Temporisèes", Thèse d'Etata, University Paris VI, Paris, 1983.

[Coh83] M.A.Cohen, S.G.Grossberg, "Absolute Stability of Global Pattern Formation and Parallel Memory Storage by Competitive Neural Networks", IEEE Transaction on Systems, Man. and Cybernetics, vol.13, pp.915-826, 1983.

[Com71] F.Commoner, A.W.Holt, S.Even, A.Pneli, "Marked Directed Graphs", J.Comput. Syst. Sci. , vol.5, n.5, pp.511-523, 1971.

[Hec90] R.Hecht-Nielsen, "Neurocomputing", Reading, MA: Addison-Wesley, 1990.

[Hil89] H.P.Hillion, J.M.Proth, "Performance Evaluation of Job-Shop Systems using Timed Event Graphs", IEEE Transaction on Automatic Control, vol.34, no.1, January 1989, pp.3-9.

[Hop82] J.J.Hopfield, "Neural Networks and Physical Systems with Emergent Collective Computational Abilities", proceedings National Academy of Sciences, vol. 79, pp.2554 - 2558, April 1982.

[Hop84] J.J.Hopfield, "Neurons with Graded Response Have Collective Computational Properties Like those of two-state Neurons", proceedings National Academy of Sciences 81:3088-3092, May 1984.

[Hop86] J.J.Hopfield, D.W.Tank, "Neural Computations of Decisions in Optimization

Problems" Biol.Cybern., vol.52, pp. 141-152, 1986.

[Laf92] S.Laftit, J.M.Proth, X.L.Xie, "Optimization of Invariant Criteria for Event Graphs", IEEE Transaction on Automatic Control, vol.37, no.5, May 1992, pp.547-555.

[Meh93] M.K.Mehmet Ali, F.Kamoun, "Neural Networks for Shortest Path Computation and Routing in Computer Networks", IEEE Transaction on Neural Networks, vol.4, no.6, November 1993, pp.941-954.

[Pim90] J.R.Pimentel, "Communication Networks for Manufacturing", Prentice-Hall International Editors, 1990.

[Shi91] H.Shih and T.Sehiguchi, "A Time Petri Net and Beam Search based On-Line FMS Scheduling System with Routing Flexibility", Proc.IEEE Conf.on Robotics and Automat., Sacramento, CA, 1991, pp.2548-2553.

[Tak86] M.Takeda, J.W.Goodman, "Neural Networks for Computation: Number Representations and Programming Complexity", Appl. Opt., vol.25, no.18, pp.3033-3046, 1986.

[Tan86] D.W.Tank and J.J.Hopfield, "Simple 'Neural' Optimization Netwotks: An A/D Converter, Signal Decision Circuit, and a Linear Programming Circuit", IEEE Trans. Circuits. Syst. ,vol.CAS-33,no.5,pp.533-541, 1986.

[Tsa92] C.J.Tsai and L.C.Fu, "Modular Approach for Petri Net Modelling of Flexible Manufacturing Systems Adaptable to Various Task-Flow Requirements", Proc.of 1992 IEEE Conf.Robotics and Automat., Nice, France, 1992, pp.1043-1048.

[Ver89] M.Verleysen, P.G.A.Jespers, "An Analog VLSI Implementation of Hopfield's Neural Network", IEEE Micro, December 1989, pp.46-55.

[Zur94] R.Zurawski, "Petri Nets and Industrial Application: A Tutorial", IEEE Transaction on Industrial Electronics, December 1994, vol.41, n.6, pp.567-583.

REAL-TIME INTELLIGENT MONITORING SYSTEM IN MANUFACTURING SYSTEM

Sun Yu, Yu Zhigang, Zhang Shiqi
CIMS Research Center, Department of Manufacturing Engineering
Nanjing University of Science & Technology
Nanjing, Jiangsu, 210094, P.R.China
Tel: +86-25-4315615,   FAX: +86-25-4431622

ABSTRACT

On the basis of analyzing the causes of fault in manufacturing system, this paper introduces the concept of Real-time Intelligent Monitoring System (RIMS) in which control process is divorced from diagnosis. Then the architecture of RIMS is provided.

1. INTRODUCTION

Advanced Manufacturing System(AMS) is a state-of-the-art manufacturing system that is organically combined with modern information technology, synthetic automation technology, modern enterprise management technology and advanced manufacturing techniques. Its development is due to the requirement and challenge of modern industry. With the introduce of related new techniques and technologies, one of the developing tendencies of AMS is that its complexity and intelligence are higher and higher, its production scale and investment are larger and larger. As a highly integrated intelligent manufacturing system, it appears very important to provide an intelligent status monitoring system, which ensures the manufacturing system working safely, reliably, efficiently and high-quality. Therefore, real-time intelligent monitoring system is becoming one of the main components of AMS or Intelligent Manufacturing System.

In manufacturing system, factors of faults are diversified and changed with the alteration of the manufacturing pattern. During the production process of manufacturing workshop, the faults have such features as follows:

− Because the flexible production becomes the aim of manufacturing system, manufacturing process is geared to the batch production of mid-volume and mid-variety products. There are various kinds, procedures and variable status of manufacturing existing in manufacturing system. Accordingly, the kinds of faults are multiple.

− Because of the numerous signals and parameters related to the different complex manufacturing equipment, the features of faults are diversified.

− Because the functional elements of system are independent in structure but close-related in functions, the faults have properties of implicity, abruptness, association and concurrence.

− Because the complex manufacturing procedures are changeable, it is difficult to collect all the a priori failure symptoms and pattern samples.

According to these features, how to process the productive data of the manufacturing system efficiently is the kernel problem to realize the intelligence and real time of monitoring system. In the former integrated information processing, traditional methods are usually used, such as state table method, state equation method, fault tree method, mathematic model method, etc. They can only diagnose some specified faults by analyzing the characteristic signals obtained from the monitored objects. Once the system is developed, its capability of diagnosis is also primarily determined. So its flexibility is not high. Especially, the status information in status monitoring has less link with control decision in control system. There is no suitable solution for this problem. Mostly, the information is processed by those who use it, which results in low efficiency. Moreover, there are many factors such as the large amount of information, variety of variables, all kinds of disturbances existing in the acquisition and transmission of information, combinatory explosion during the computation of processing ⋯ . They bring about a series of problems, such as conflict results, uncertain status, etc. Though expert system can solve some of those problems, its running rate is too slow to meet the real-time need because of its symbol-based reasoning manner. So, new processing methods fitted to the status monitoring of AMS must be sought to overcome the drawbacks of the former information processing methods.

In the previous research, our research group has made deep study on the real-time intelligent monitoring of Flexible Manufacturing System(FMS). According to the structure characteristics of FMS, an architecture of Inspection and Monitoring System based on software platform was established. In this architecture, a rule-based analogous expert system was developed to execute real-time status monitoring on the basis of the real-time & multitask operating system, an off-line knowledge based expert system was provided to carry out the intelligent fault diagnosis. Good effects have been achieved. Recently, new techniques such as artificial neural networks and multi-sensor data fusion are combined into our study in addition to the previous research. Here, manufacturing shop is taken as research object and a new architecture of Real-time Intelligent Monitoring System of AMS is established. It provides a new approach for the real-time status monitoring and information integration of manufacturing system.

2. THE ARCHITECTURE OF RIMS

2.1 The Meaning of RIMS

Here RIMS we discussed includes three aspects:

① According to the principle of equipment grouping and region centralization, all kinds of status signals and parameters collected by multisensor are preprocessed and centralized. Finally, RIMS forms a "mapping" of the manufacturing system in operation for controllers.

② RIMS carries out the real-time intelligent monitoring by making use of the centralized information so as to inspect the disturbances and faults in an on-line manner. Then the influence on the manufacturing system caused by faults is analyzed. The related control decision is formed and provided to the corresponding controllers, which execute the dynamic schedule and emergency operation to ensure the safety and efficiency of the running system.

③ When a fault or irregularity occurs in the system, an on-line schedule is executed after processings as mentioned before. Then the diagnosis system is driven, an off-line fault diagnosis is accomplished, and appropriate guide of maintenance was provided. High real-time processing is not needed in this part.

The relationship between RIMS and other components of manufacturing system is demonstrated by Fig.1.

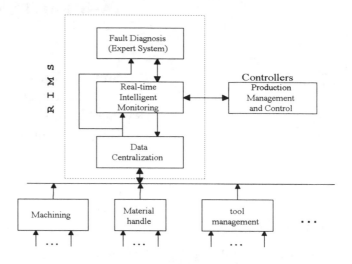

Fig.1 The Relationship of RIMS with other system

2.2 The Architecture of RIMS

The architecture of RIMS is illustrated by Fig.2.

RIMS is composed of status inspection and pre-fusion module, data fusion and feature extraction module, real-time monitoring module, fault diagnosis module, status and performance evaluation module. The functions of these modules are introduced in the following parts.

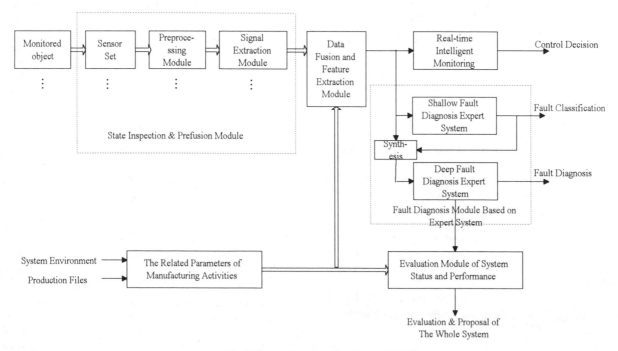

Fig.2 The system architecture of RIMS

2.3 Status Inspection and Pre-fusion Module

This module mainly finishes mapping from physical space to feature space, and then all kinds of features for the followed processes are provided. So it is the physical basis of real-time intelligent monitoring. As shown in Fig.2, while the system is running, all kinds of status signals of the monitored object (equipment, procedure, environment, etc.) are measured by distributed sensor network, collected by general data acquisition units and preprocessed by signal processing devices. The preprocess includes smoothing, filtering, adjustment and compensation, etc. The corresponding feature extraction and parameters' estimation are implemented according to the type of the signals, which turn original information into the form of feature space. In fact, this procedure is a pre-fusion process which lowers the dimension of feature space. Many switch signals adopted by manufacturing system are usually used directly. In addition, multiple parameters of each inspected object is measured by a sensor set, which can improve the reliability of status inspection and enlarge the space coverage and the time coverage due to the redundancy of the multisensor. On the other hand, such process enlarges the dimension of the measuring space. In order to ensure the real-time monitoring, simple time-domain characteristic parameters are computed and extracted to shorten the time of data processing, such as the mean value and variance of speed or acceleration, the crest value of vibration signals, etc. These extracted parameters are feeded into the fusion and feature extraction module according to the principle of equipment grouping and region centralization. Actually, this module belongs to the inspecting device itself.

2.4 Data Fusion and Feature Extraction Module

This module mainly accomplishes organization and management of feature space. Since the features obtained from different inspected objects by signal pre-fusion are dispersed, they need to be fused and organized to form an integrated characteristic information which can indicate current running status of the whole manufacturing system. AMS is a system with multitask and varied working status, even it engages in the same manufacturing activities, the corresponding feature spaces are different when their tasks are different. Therefore, these extracted features need to be further processed and converted to relative features which are not sensitive to the tasks to improve the flexibility of monitoring system. For example, if the rotational speed is set to v_0 before machining, but the real measured value is v_x, the relative error

$$r_v=(v_x-v_0)/v_0 \in [0,1]$$

is a speed feature which is not sensitive to the machining task, and it is changed with the degree of deviation of current status from the normal. It is obvious that $r_v \ll v_x$. This process of fusion can greatly improve the stability and converging rate of neural networks. The fused features are selected to form specific input pattern space for real-time monitoring module and intelligent fault diagnosis expert system, which can use the

characteristic variables as few as possible to express current running status of the whole system.

2.5 Real-time Intelligent Monitoring Module

According to the architecture of RIMS, real-time intelligent monitoring is its kernel. So we construct the real-time intelligent monitoring module through parallel processing of artificial neural network, in which RCE model is selected. RCE network consists of three processing layers, and its topology is illustrated by Fig.3.

Here we only briefly explain the input and output of RCE network. Its input pattern is composed of the fused features, which probably include the symptoms of faults or disturbances. The aim of real-time intelligent monitoring is not only fault diagnosis, but also it is more important to analyze the probable influence of the faults occurred on manufacturing system and to form control decisions, which are provided to controllers to intervene in the behaviors of system. Accordingly, coarse fault classification is accomplished in this module.

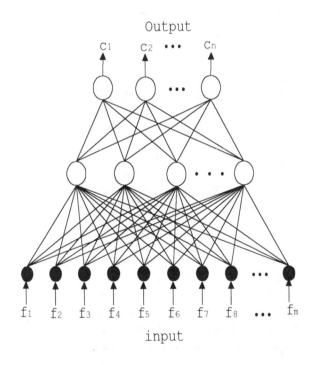

Fig.3 The Topology of RCE Network

2.6 Fault Diagnosis Module Based on Expert System

This module is made up of a shallow fault diagnosis expert system based on neural network and a deep knowledge-based fault diagnosis expert system. Its task is to diagnose and evaluate current status of system. Shallow fault diagnosis expert system obtains feature patterns from data fusion and feature extraction module, recognizes and classifies the faults with neural network. So the shallow knowledge inference is completed. Once a fault occurs, shallow expert system primarily locates it quickly. Then deep fault diagnosis expert system is awaken and its corresponding module is driven to make deep diagnosis and evaluation. So a detailed fault report is created. But deep fault diagnosis expert system doesn't work until a fault is recognized by shallow fault diagnosis expert system. Such architecture can greatly improve the diagnosis efficiency of expert system. In shallow fault diagnosis expert system, a three-layer BP network is selected. Different from control decision, fault diagnosis is a nonlinear mapping with multi-input and multi-output. It requires that the neural network can recognize concurrent multi-fault. Here BP network is just able to complete such nonlinear mapping.

2.7 The Evaluation Module of System Status and Performance

In this module the results and fault reports from other modules are collected. Then combined with the information and environment parameters of manufacturing activities, the suitable evaluation of the status and performance of manufacturing system in a period and some advice on system improvement are given. When new faults occur, this module provides an interface to the knowledge base of expert system. So RIMS can enlarge the capacity of the knowledge base through the evaluation module.

When many faults or deviations occur at the same time, further data fusion is made to give the degree of effects on manufacturing system. This evaluation is described by following equations:

$$P = p_{max} * K \qquad <1>$$

$$K = 1 + \sum_{j=1}^{n-1} p_j * c_j \qquad <2>$$

Where n is the number of multiple faults or deviations occurred, P is the probability of whole system stopping operation, p_{max} is the maximal value of probability among all faults which results in the system stopping, K is a coefficient related to all faults, c_j is a weight coefficient, and p_j is the stopping probability created by other faults.

3. CONCLUSIONS

RIMS is one of kernel components of AMS. In this paper the concept of RIMS which is described from three aspects is proposed. On the basis of this concept an architecture model of RIMS is established. Because of the complexity of RIMS, the research concentrates on manufacturing workshop, but its rationality have been verified in AMS through experiments.

REFERENCES

[SLZ95] Sun Yu, Lu Baochun, Zhang Shiqi, "The Study and Development of the Real-time Intelligent and Monitoring System in Manufacturing System", *ISTM/95 1st International Symposium on Test and Measurement*, Taiyuan, China, Aug.10-14, 1995, pp.505-508

[SLZ94] Sun Yu, Lei Bing, Zhang shiqi, "The Study and Development of Integrated Measuring and Monitoring System in FMS", *The Third International Conference on Automation, Robotics and Computer Vision (ICARCV'94) Proceedings (Vol. 1 of 3)*, Singapore, November 9-11, 1994, pp.299-301

[LSZ95] Lei Being, Sha Qinhua, Zhang Shiqi, "A Study of Intelligent Monitoring System in Intelligent Manufacturing", *Proceedings of the International Symposium on Intelligence, Knowledge and Integration for Manufacturing*, March 28-31, 1995, Southeast University, Nanjing, P.R.China, pp.98-103

[Pau81] L. F. Pau, "Fusion of multisensor Data in Pattern Recognition", *Pattern recognition Theory and Applications*, edited by J.Kittler, K.S.FU and L.F.Pau, Q327.N2, 1981, 189—201

[JSRH89] Don Johnson, Scott Shaw, Steven Reynolds, Nageen Hinayat, "Real-time blackboards for sensor fusion", *SPIE Vol.1100 Sensor Fusion II(1989)*, pp.61—72

[AB85] P.Allen and R.Bajcsy, "Two Sensor Are Better Than One: Example of Integration of Vision and Touch", in *Proc. Third Int. Symposium on Robotics Research*, Giralt, Eds. Cambridge, MA: MIT Press, 1985

[Hov88] S.A.Hovanessian, Introduction to Sensor Systems, NewYork: Addison-Wesley, 1988

[PGS88] John C.Pearson, Jack J.Gelfand, W.E.Sullivan, Richard M.Peterson and Clay D.Spence, "Neural network approach to sensory fusion", *SPIE Vol.931 Sensor Fusion(1988)*, pp.103—108

[HH89] J.Hoskins, D.Himmelblau, "Fault detection and diagnosis via neural networks", *Dechema-Monographs*, vol.116-VCH Verlagsgesellschaft, 1989

SIMULATION OF SCHEDULING RULES IN A FLEXIBLE MANUFACTURING SYSTEM USING FUZZY LOGIC

*Kazerooni, A., *Chan, F.T.S., *Abhary, K., and *Ip, R.W.L.
*School of Manufacturing and Mechanical Engineering, University of South Australia,
The Levels 5095, SA, Australia
Email: 9402014g@lv.levels.unisa.edu.au
*Dept. of Manufacturing Engineering, City University of Hong Kong, Hong Kong

ABSTRACT

Two decision rules for real-time dispatching of parts using fuzzy logic and fuzzy sets have been developed and tested in a simulated flexible manufacturing system (FMS). Routing flexibility is an important aspect of FMSs. Routing selection method proposed in this paper uses fuzzy sets to incorporate system status in decision-making. The next decision rule is concerned with real time machine loading. A conventional dispatching rule always blindly pursues a single objective, but in practice, however, more than one objective may often be perceived to be simultaneously considered. Regarding this fact, a method for combining of two dispatching rules using fuzzy logic has been suggested.

1. INTRODUCTION

A flexible manufacturing system (FMS) is designed to combine the efficiency of a mass-production line and the flexibility of a job shop to produce a variety of workpiece on a group of machines and other work stations connected by an automated material handling system. The overall system is under a computer control.

General FMS operation decisions can be divided into two phases: planning and scheduling. The planning phase considers the pre-arrangement of parts and tools before the FMS begins to process. At this stage it is necessary to evaluate the likely components to be produced by the FMS. This leads to a requirement for machine tool specification, material handling methods and many other consideration. The scheduling phase deals with routing parts while the system is in operation. It involves a set of tasks to be performed. There are trade-offs between early and completion of a task, and between holding inventory and frequent production changeover [Gra81]. Scheduling problems have been proven to belong to the family of NP-complete problems that are very difficult to solve [Bak74].

A serious constraint on the development of more sophisticated automated manufacturing systems is the lack of specific policies for the scheduling and control of the shop floor [BH86]. Since FMS is a relatively recent technology, it is not yet known if policies and procedures designed to schedule and control traditional manufacturing processes are appropriate for such advanced manufacturing technologies [Ste83, Kus86]. Thus, in order to enhance the performance of existing FMSs and to allow for further development of automated manufacturing systems, proper procedures for the scheduling and control of these automated systems must be developed and documented.

Dynamic nature of such systems demands for a scheduling procedure that is reactive and sensitive to the system's status instead of a predictive one. Because of the complexity of such systems, the analytical approach has proved to be extremely difficult, even with several limiting assumptions. In the face of the difficulties associated with analytical techniques, researches in this area have relied on computer simulation of both real and hypothetical FMSs [Ram90]. Its greatest strength lies in its ability to help users to analyse complex systems such as FMS where the complex decision-making logic makes other types of analysis difficult to apply and prone to error [OHV93].

It is claimed that simulation can not optimise the system's performance [Peg91]. It is essentially true if a decision-making strategy that is independent to the system's status is used. In other words the simulation technique can be used as an optimisation technique while the decision-making strategy is updated with the passage of the simulation time.

Routing selection method proposed in this paper uses fuzzy sets to incorporate system status in decision-making. The next decision rule is concerned with real time machine loading. A conventional dispatching rule always blindly pursues a single objective, but in practice, however, more than one objective may often be perceived to be simultaneously considered. Regarding this fact, a method for combining of two dispatching rules using fuzzy logic has been suggested.

2. REVIEW OF RELEVANT LITERATURE

In the past three decades scheduling of production systems has been dramatically studied. An exhaustive and well classified of early works has been done by Panwalker and Iskander [PI77]. Some recent works can be found in [GGB89] and [MW90]. Ramasesh focuses on simulation-based research of job shop scheduling problems in a dynamic and multiple machines environment in which processing times are stochastic. Montazeri and van wassenhove [MW90] briefly review the literature on scheduling rules for manufacturing systems. Then they apply various scheduling rules to a dedicated FMS. They finally conclude that the performance of scheduling rules depends very heavily on the criterion chosen as well as on the configuration of the production system.

As a matter of fact the scheduling problems in FMSs are more complicated compared with those problems in job shops. Depends on the degree of the flexibility of the FMS environment, the nature of the system is much more dynamic than a conventional production system. In such system scheduling of parts is associated with consideration of material handling devices (like AGVs), jigs and fixtures and alternate routing.

Like job shops, a wide range of scheduling rules are applied in FMSs. Stecke and Solberg [SS81] in their simulation study present that the heuristic rules known to be superior in convention job shops, perform poorly in FMS. Tang et al. [TYL93] use a random job shop FMS to study the effect of decision making at different decision points. They consider many decision points in their study and conclude that the effects of various dispatching rules at decision points have a different degree of influence on the various criteria such as lead time, tardiness, number of machine blocked, and WIP. They found that no single criterion can always dominate other. Denzler and Boe [DB87], Egbelu and Tanchoco [ET84], and Sabuncuoglu and Hommertzheim [SH89] reach to the same conclusion.

With extension of heuristic rules, some researchers come to the conclusion that ordinary heuristic rules can only include simple factors e.g. related to time, such as due dates, arrival times, operation times and so on. Watanabe [Wat90] in his work introduces a new fuzzy logic algorithm to determine the priority of jobs using fuzzy inference rules. He considers the quickness requirement and the profit of production indices.

The priority of a job in a queue is given by two fuzzy rules:

- if the slack time is short and the profit index is high, then the priority of the job is high
- if the slack time is long and the profit index is low, then the priority of the job is low

He shows that profit is significantly improved by adapting the proposed method in comparison with ordinary rules. Grabot and Geneste [GG94] state that workshop management is a multi-criteria problem and propose a way to use fuzzy logic in order to build aggregated rules allowing to obtain a compromise between the satisfaction of several criteria. They take the combination of the SPT rule and the slack time rule as an example using following rules:

Rule 1.   If (operation duration is low) and (slack time is low) then (priority is high)

Rule 2.   If (operation duration is high) and (slack time is high) then (priority is low)

The fuzzy propositions are defined by membership functions. The satisfaction degree of the antecedents of the rules is evaluated through a fuzzy pattern matching procedure, and a defuzzification method is applied to calculate a precise priority on the basis of this satisfaction.

The authors then compare the results with some other scheduling rules and conclude that the combination of the rules allows to define an aggregated rule with an intermediate performance on the different criteria selected.

Kazerooni et al. [KCA95] in their recent work, point out on some vagueness and uncertainties in scheduling rules and to improve the performance measure of average tardiness, propose a fuzzy logic approach. They focus on characteristic of system's status instead of jobs to select a scheduling rule.

3. THE FMS MODEL

The first step in any FMS simulation study is to establish the system configuration and then to develop a simulation model. The FMS studied in this paper includes six general purpose workstations, one loading/unloading station, and a vehicle staging area as shown in Figure 1.

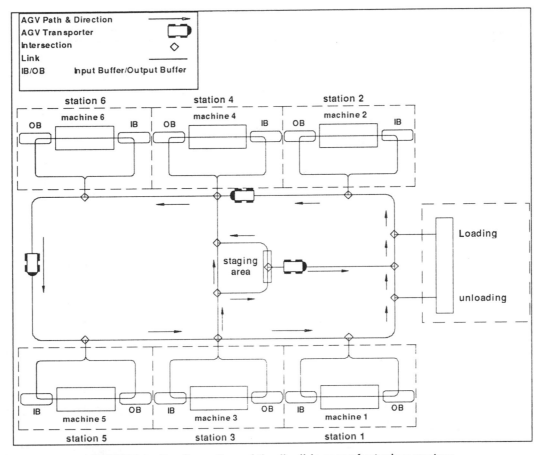

FIGURE 1 Configuration of the flexible manufacturing system

Each workstation has one input and one output buffer queue. Each queue has a limitation of 10 jobs as capacity. Three AGVs are assigned to transport the parts among stations. An automated device is employed to take jobs from AGVs and mount them to machines. The speed of AGVs is assumed to be constant during travel and is 14 m/min. The path of AGV is known and called network. When the AGV complete the transfer and is freed, it goes to the staging area. To reach the staging area, the AGV travels through a number of intersections. At each intersection any request for the free AGV is checked and if any request is detected it would be sent to the requested station. For the same production period, there are seven different type of jobs to be selected randomly from a discrete distribution. Incoming jobs are in batch of five. After arriving to the system, processing time are assigned to parts. Depend on the part type, there are number of alternative routings. This is the first decision point. How a part selects one of those alternatives is described in next section. The alternative routings and related processing times are shown in Table 1.

The following assumptions are made in the simulation model:

- All parts are processed according to their determined
- sequence of operations.
- Processing times, including set-up times are known deterministically.
- The inter arrival of parts is triangularly distributed.
- Raw materials, tools, jigs, fixtures, etc. are present and released immediately when required.
- The same pallets are used for different parts.
- The machines are continuously available for production.
- Local buffer capacity is limited and the number of AGVs is fixed.
- Once the operation is started, it can not be interrupted before completion.
- Distances between stations are known.
- Static due date is assumed, i.e., due date = arrival time + (processing time + travelling time) × constant. This is known as TWK rule.

Part type	Machine No. 1	2	3	4	5	6	Operation sequence			
1	——	30	——	70	——	40	2	4	6	
	40	——	55	——	50	——	1	3	5	
	——	——	45	60	——	75	4	3	6	
2	——	40	40	——	——	50	2	3	6	
	——	70	——	60	50	——	2	4	5	
	70	——	——	80	30	——	1	5	4	
3	40	——	50	30	——	60	1	3	4	6
	——	50	——	——	70	80	5	6	2	
	60	40	30	——	50	——	1	3	5	2
	——	50	——	30	40	——	2	4	5	
4	30	——	——	40	20	——	4	1	5	
	——	15	35	——	65	——	2	3	5	
	——	20	——	35	15	——	2	4	5	
5	——	40	——	70	——	——	2	4		
	——	——	25	——	45	15	3	5	6	
	15	25	35	——	——	——	1	2	3	
6	——	75	——	40	35	——	4	5	2	
	——	45	40	25	——	——	2	4	3	
	——	30	25	——	40	——	3	5	2	
7	——	——	——	——	65	50	6	5		
	——	65	45	——	25	——	5	3	2	
	——	25	——	15	——	40	2	4	6	

Table 1 The Alternative Routeings, Operation Times and Operation Sequences

4. DECISION POINTS AND ASSOCIATED RULES

For each part processed through the FMS there are five decision points as shown in Figure 2. As a part arrives the system, it should select a routing between several alternative routings (decision point 1). Then it request for a AGV for travelling (decision point 2). All parts transferred to input queues will be selected by machines for processing (decision point 3). After completion of process, jobs reside in the output buffer queues and compete to access idle AGVs (decision point 4). When more than one AGV try to occupy an intersection, a decision making will occur (decision point 5).

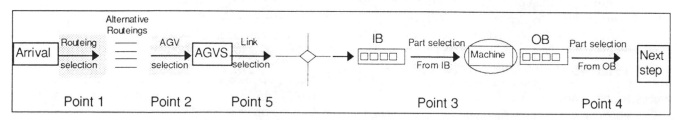

FIGURE 2 Decision points

4.1 Selection of a Routing

After being released from loading station, the part will select a routing as a sequence of operations. To select a routing, following factors is considered:

- minimising number of machines blocked
- minimising total processing time
- minimising travelling cost
- minimising number of processing steps

4.2 Parts Select AGVs

After selection of appropriate routing, the part will select an AGV to be transferred to the next station. The transporter selection rules commonly used here include: CYC (CYClic priority), RAN (RANdom priority), LDS (largest distance to requested station), SDS (Shortest Distance to the requested Station), and POR (Preferred Order Rule).

4.4 Machines Select Parts

When a machine becomes idle and more than one job is waiting in its preceding queue, it has to select which part to be processed next. The following scheduling rules are considered, FIFO (First In First Out), SPT (Shortest Processing Time), SRPT (Shortest Remaining Processing Time), SLRO (SLack per Remaining number of Operations), SIO (Shortest Imminent Operation time).

4.4 AGVs Select Parts

When more than one job are waiting in output buffer queue, they compete to obtain an Idle AGV to be transported to the next station. The following scheduling rules are considered, SPT (shortest processing time), EDD (earliest due date), FIFO (first in first out), SIO (Shortest Imminent Operation time), and LPT (Longest processing time).

4.5 Intersections select AGVs

When more than one AGV try to seize an intersection, the following rules (link selection rules) determine which one goes first, FCFS (First Come First Serve), LCFS (Last Come First Serve), CLSD (Closest to Destination), FRTD (Farthest from Destination).

5. OUTLINE OF THE SIMULATION MODEL

There are many commercially available simulation packages such as SIMAN & ARENA, SIMFACTORY and WITNESS. A set of important attributes while selecting simulation software for manufacturing is given by Law and McCourse [LM92]. The scheduling model based on the illustrated FMS was implemented in SIMAN & ARENA simulation program.

5.1 Enhancement of Scheduling Rules' Performance Using Fuzzy Logic and Fuzzy Sets

In this paper we propose use of fuzzy logic and fuzzy sets [Zad84] in two decision-making points. First we apply fuzzy sets when an incoming part should select one of defined routings. Since parts enter to the system over the time, selection of routing is a dynamic decision making.

Considering the objectives mentioned in section 4.1, it is hard to select a routing. To overcome this difficulty we apply fuzzy membership to find the contribution of a routing to an objective. For each job type the following judgments can be always made:

1) The possible minimum number of jobs in queues will be zero while the maximum number will not reach Q_B.
2) The possible minimum processing time will be T_A, while the maximum time will not reach T_B.
3) The possible minimum travelling cost will be C_A, while the maximum cost will not reach C_B.
4) The possible minimum number of processing steps will be P_A, while the maximum number of processing steps will not reach P_B.

Now we can build a fuzzy membership function to evaluate the contribution of a routing to an objective. When objective (1) is considered, for example for part type 3, four alternative routings are considered. Routings number 1 and 3 have four operations on four different machines. Since the queue of each machine has a capacity of 10 jobs, therefore Q_B for routing number 1 and 3 is 41 (i.e. 4×10+1) and for routing number 2 and 4 is 31 (i.e. 3×10+1). A simple function can be built to evaluate the contribution of a routing to objective (1):

$$\mu_1(x) = (Q_B - Q_x) / (Q_B - 0)$$

The membership of each routing can be calculated as shown in Figure 3.

For objectives (2), (3) and (4), a similar membership function can be built. Each objective can be weighted using Analytic Hierarchy Process (AHP) method that is beyond of this paper. To find the final membership of all alternative routings, the following formula is used:

$$\mu_A(j) = \sum_{k=1}^{N} \left(\frac{(W_k)}{\sum_{k=1}^{N} W_k} \times \mu_k(j) \right)$$

Where

$j = 1,2,...M$ alternative routings

$k = 1,2,...N$ objectives

$\mu_k(j)$ = membership of routing j to objective k

$\mu_A(j)$ = membership of routing j to all objectives

W_k = weight of objective k

The solution for the above equation is Max$\{\mu_A(j)\}$.

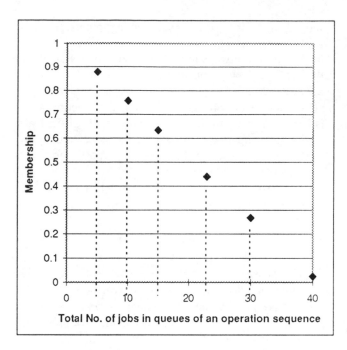

FIGURE 3 Membership function

When a machine becomes idle, a part from the preceding queue must be selected. When more than one part are waiting in the queue, a fuzzy logic is applied to combine two scheduling rules that are SLRO and SIO. Reader can find the reason for selecting these two rules in [KCA95].

At this section we propose a fuzzy logic approach to determine the tightness of the production system. Let us start with some definition of rules.

Rule 1   If multiplier K is low, and machine utilisation level is high then the due-date tightness of preceding queue is high

Rule 2   If multiplier K is high, and machine utilisation level is low then the due-date tightness of preceding queue is low

The fuzzy propositions of the antecedent of the rules are represented by membership functions as described in Figure 4 and Figure 5. Also fuzzy propositions of the concluding parts are presented in Figure 6.

Mamdani's min-max method is used to perform fuzzy inference rules [Mam74], refer to Figure 7 and Figure 8. In that method, the membership functions of the concluding parts of the rules are cut at min $\mu_i(v)$ where i=1 or 2 and v=K or L. The global conclusion is given by the maximum value of these shapes. Let us show the performance of the procedure with an example.

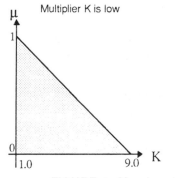
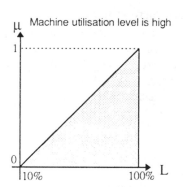

FIGURE 4 Membership functions for antecedent of rule 1

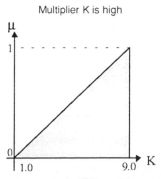
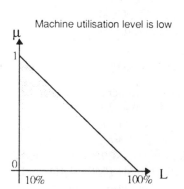

FIGURE 5 Membership functions for antecedent of rule 2

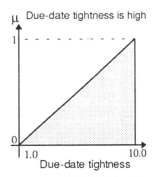
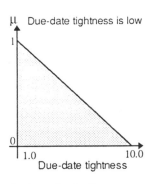

FIGURE 6 Conclusion modelling for rule 1 and rule 2

Consider at the moment that the rules are applying, multiplier K is 6.0 and shop load level is 90%. The due-date tightness can be found as follows:

1. Determine the fuzzy rules, rules 1 and 2.
2. Construct membership functions to represent the fuzzy declaration, Figures 4, 5 and 6.
3. Find the membership for each fuzzy proposition, Figures 7 and 8 (left sides).
4. Use min-max method to perform fuzzy inference rule, Figures 7 and 8 (right sides).

The final priority of jobs are determined by using the following formula:

$$Priority = LVF (S1 \times IOT + S2 \times SLRO)$$

where:
LVF = Low Value First, IOT = Imminent Operation Time, and SLRO = Slack / Remaining number of Operations

5.2 Simulation Experiments

The scheduling model based on the example FMS was implemented in ARENA™ simulation program and the routing selection and dispatching rule using fuzzy logic and fuzzy sets were coded in FORTARAN 77.

A pilot run shows no significant difference between link selection rules because of the few numbers of AGVs in the system. Therefore for the rest of the simulation, the FCFS link selection rule is selected. 125 possible combinations were tested for examining the effect of various decision points. Each set was simulated by three replications of 10580 minutes per run for one production period. A warm up length to pass transient portion of the simulation replication is set equal to 500 minutes. All results were tested at 0.975 percent confidence.

6. RESULTS AND DISCUSSION

For the given configuration and data, the FMS was simulated with different scheduling rules. Table 2 gives a detailed output. In total, 375 (i.e. 3×125) simulations were executed and 32 simulations are selected to compare their performance measures. Figure 9 is a bar chart of outstanding performances for easy comparison.

The results show that use of common scheduling rules satisfy only one of the criteria and may performs poorly regarding others. For example experiment number 15 gives a good average flow time but high average tardiness and lateness. Experiments 12, 13, and 14 are almost the same as experiment 15. Thanks to fuzzy logic, we can have a good performance regarding two or more criteria. For example experiments 8, 9, 20, 22, 30 and 31 give good average flow time, tardiness, and lateness all together.

7. CONCLUSION

A simulation model was developed to solve the real time production scheduling. We have showed that a fuzzy logic-based decision-making can be easily employed in a real-time simulation model. From the above discussion, it is clear that the combination of rules plays an important role while evaluating the system performance. Thus it must be emphasised that in an FMS environment, a single rule for a set of objectives is not desirable. In this paper we have also showed how to apply fuzzy logic and fuzzy sets to decision-making, and we have presented the excellent result of fuzzy rules in satisfying several objectives. It can be seen from Table 2 that those experiments that use fuzzy logic as scheduling rule at level 2, outperform all other rules while considering all objectives together. The results show that the fuzzy inferencing yields a good amount of improvement to the operation of the system, mainly because it allows simultaneous optimisation of SIO and SLRO scheduling rules. This research has shown a significant improvement in comparison with the previous studies, in that we have considered a fuzzy approach that can incorporate system's status in real time decision-making.

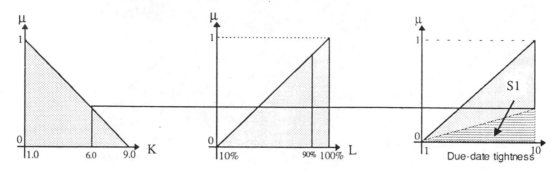

FIGURE 7   Conclusion modification for rule 1

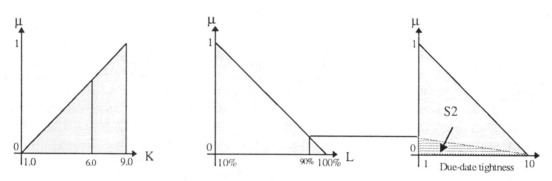

FIGURE 8   Conclusion modification for rule 2

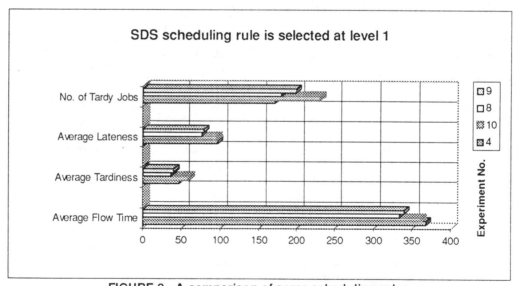

FIGURE 9   A comparison of some scheduling rules

Table 2 Simulation Results

Experimnet No.	Scheduling Rule Level			Performance Measure			
	1	2	3	Average Flow Time	Average Tardiness	Average Lateness	No. of Tardy Jobs
1	SDS	SIO	FIFO	375	51	100	**162**
2	SDS	SLRO	SPT	471	102	118	300
3	SDS	SIO	SIO	374	52	100	**160**
4	SDS	SLRO	LPT	367	46	95	170
5	SDS	SIO	LPT	384	57	104	180
6	SDS	SPT	SIO	415	88	137	157
7	SDS	SPT	FIFO	417	89	135	171
8	SDS	**FUZZY**	SIO	**332**	**36**	**74**	178
9	SDS	**FUZZY**	LPT	**338**	**39**	77	198
10	SDS	FIFO	SIO	360	58	94	230
11	SDS	FIFO	FIFO	358	59	98	224
12	LDS	SIO	SPT	**331**	47	97	155
13	LDS	SIO	LPT	**333**	48	97	162
14	LDS	SIO	SIO	**330**	47	97	155
15	LDS	SIO	FIFO	**327**	42	92	155
16	LDS	SPT	SPT	349	59	105	**144**
17	LDS	SPT	FIFO	348	58	106	**140**
18	LDS	SLRO	SIO	346	**37**	**64**	231
19	LDS	SLRO	LPT	**332**	47	96	159
20	**LDS**	**FUZZY**	**SIO**	**332**	**36**	**74**	178
21	**LDS**	**FUZZY**	**LPT**	**338**	**39**	77	198
22	**LDS**	**FUZZY**	**FIFO**	**332**	**37**	77	187
23	CYC	SIO	SIO	**334**	47	95	**155**
24	CYC	SIO	FIFO	**332**	46	95	**159**
25	CYC	SLRO	SIO	352	42	**67**	234
26	CYC	SLRO	LPT	**332**	46	95	166
27	CYC	SPT	SIO	346	57	104	**144**
28	CYC	SPT	LPT	351	59	103	**151**
29	**CYC**	**FUZZY**	**SIO**	342	**41**	76	200
30	**CYC**	**FUZZY**	**LPT**	337	**37**	73	193
31	**CYC**	**FUZZY**	**FIFO**	**332**	**37**	77	184
32	CYC	FIFO	SIO	381	73	106	246

Scheduling rule level 1 is for selection of AGV by part, level 2 is for selection of part by machine and level 3 is for selection of part by AGV.

REFERENCES

[Bak74] K.R. Baker. *Introduction to Sequencing and Scheduling.* Wiley, New York 1974.

[BH86] J. Bessant and B. Haywood. The Introduction of flexible Manufacturing Systems as an Example of Computer Integrated Manufacturing. *Operations Management Review.* **14** (6) 5-12, 1986.

[DB87] D.R. Denzler and W.J. Boe. Experimental Investigation of Flexible Manufacturing System Scheduling Rules. *International Journal of Production Research..* **25** (7) 979-994, 1987.

[ET84] P.J. Egbelu and J.M. Tanchoco. Characterization of Automated Guided Vehicle Dispatching Rules. *International*

Journal of Production Research. **22** (3) 359-374, 1984.

[GG94] B.Grabot and L. Geneste. Dispatching Rules in Scheduling: a Fuzzy Approach. *International Journal of Production Research.* **32** (4) 903-915, 1994.

[GGB89] Y.P. Gupta, M.C. Gupta and C.R. Bector. A review of scheduling rules in flexible manufacturing systems. *International Journal of Computer Integrated Manufacturing.* **2** (6) 356-377, 1989.

[Gra81] S.C. Graves. A review of production scheduling. *Decision Science* **13** (4) 646-675, 1981.

[KCA95] A. Kazerooni, F.T.S. Chan and K. Abhary. Simulation of Flexible Manufacturing Systems via Fuzzy Control Scheduling. In *6th International Conference on Manufacturing Engineering,* IEAust, Melbourne, Australia, 29 Nov-1 Dec. **1**, 95-100, 1995.

[Kus86] A. Kusiak. Application of Operational Research Models and Techniques in Flexible Manufacturing Systems. *European Journal of Operational Research.* **24**. 336-345, 1986.

[LM92] A. Law And M.G. McComas. How to select simulation software for Manufacturing Application. *Industrial Engineering,* **7** (7) 29-33, 1992.

[Mam74] E.H. Mamdani. Application of fuzzy algorithms for control of simple dynamic plant. *Proceeding of IEE,* **121**(12) 1585-1588, 1974.

[MW90] M. Montazeri and L.N. Wassenhove. Analysis of scheduling rules for an FMS. *International Journal of Production Research.* **28** (4) 785-802, 1990.

[OHV93] J.F. O'Kane, D.K. Harrison and V.I. Vitanov. An AI Approach To Scheduling in Flexible Manufacturing Systems. In *3rd International Conference on Factory 2000.* 24-28, 1993.

[Peg91] C.D. Pegden. *Introduction to Simulation With SIMAN.* Systems Modelling Co., 1991.

[PI77] S. Panwalker and W. Iskander. A survey of scheduling rules. *Operations research.* **25** (1) 45-61, 1977.

[Ram90] R. Ramasesh. Dynamic job shop scheduling: A survey of simulation research. *OMEGA International Journal of Management Science.* **18** (1) 43-57, 1990.

[SH89] I. Sabuncuoglu and D.L. Hommertzheim. An Investigation of Machine and AGV Scheduling Rules in an FMS. In *Proceeding of 3rd ORSA/TIMS Conference,* in Elsvier, Amsterdam. 261-266, 1989.

[SS81] K.E. Stecke and J. Solberg. Loading and Control Policies for Flexible Manufacturing Systems. *International Journal of Production Research.* **19** (5) 481-490, 1981.

[Ste83] K.E. Stecke. Design, Planning, Scheduling and Control Problems of Flexible Manufacturing Systems. *Reprint from J.C. Baltzer Scientific Publishing Co,* Basel, Switzerland, 1983.

[TYL93] L.L. Tang, Y. Yih and C.Y. Liu. A Study on Decision Rules of Scheduling Model in an FMS. *Computers in Industry.* **22** 1-13, 1993.

[Wat90] T. Watanabe. Job-shop sch using fuzzy logic in a computer integrated manufacturing environment. In *5th International conference on system research, information and cybernetics* in Baden-Baden, Germany, 150-58, 1990.

[Zad84] L.A. Zadeh. Making Computers Think Like People. *IEEE Spectrum,* 26-32, Aug. 1984.

Monitoring

AN EXPERT SYSTEM FOR THE MONITORING AND SIMULATION OF THE AXAF-I

Kai H. Chang[+], Mark Rogers[*], and Richard McElyea[*]

+ Department of Computer Science
and Engineering
Auburn University
Auburn, AL 36849-5347 USA
Email: kchang@eng.auburn.edu

* Payload Operations Laboratory
George C. Marshall Space Flight Center
National Aeronautics and Space Administration
MSFC, AL 35812   USA

ABSTRACT

Advanced X-ray Astrophysics Facility - Imaging (AXAF-I) has been scheduled for a space shuttle launch in late 1998 at the Kennedy Space Center. The AXAF-I will have an expected five year life time for the science mission phase. During the science mission phase, the operation and management of the flight and ground systems are personnel intensive. The purpose of the expert system presented in this paper is intended as a tool to reduce the cost for the mission operations. An expert system prototype that is capable of providing operation monitoring, simulation, and training for the AXAF-I has been developed. The hierarchical (or object-oriented) design has made the development, testing, and revision of the prototype efficient. It will be one of the pioneer mission payload operation systems at the Enhanced Huntsville Operations Support Center, Marshall Space Flight Center (MSFC), Huntsville, Alabama, which is scheduled to be operational in September 1998.

Key words: Expert systems, practical applications, knowledge-based systems, mission payload operations, object-oriented programming.

1. INTRODUCTION

Advanced X-ray Astrophysics Facility - Imaging (AXAF-I) is a spacecraft for X- ray emitting sources observation and has been scheduled for a space shuttle launch in late 1998 at the Kennedy Space Center. Its main objectives are "to determine the nature of astronomical objects ranging from normal stars to quasars, to understand the nature of the physical processes which take place in and between astronomical objects, and to add to our understanding of the history and evolution of the universe."

[AX94] The AXAF-I will have an expected five year life time for the science mission phase. During the science mission phase, the operation and management of the flight and ground systems are personnel intensive, requiring system experts on duty around the clock. The purpose of the expert system presented in this paper is intended as a tool to reduce the cost for the mission operation.

The AXAF-I will be operated at a nominal elliptical orbit of 10,000 by 100,000 kilometers (km), approximately 296 km above the earth. The spacecraft will be in contact with the Operation Control Center (OCC) through the Deep Space Network (DSN) every eight hours for a period of 45 minutes. During the 45-minute contact time, two types of telemetry data will be transmitted through the DSN. The first type of telemetry consists of the recorded data during the off-contact time. This data is temporarily stored for later forwarding to the OCC for analysis. The second type of telemetry consists of the real-time data from the spacecraft. This data will be processed in real-time to detect any anomaly and, if required, commands will sent to the spacecraft to provide corrective operations while in contact.

The telemetry data from the spacecraft can be divided into two categories: the science observation data and the engineering status data (telemetries). The science data contains the outputs from the X-ray sensing devices and will be forwarded to the AXAF-I Science Center for interpretation; while the telemetries will be monitored by the OCC for the operation diagnosis of the spacecraft. The expert system is designed to assist the training and the daily operation at the OCC.

Because of the evolving nature of the AXAF-I, its specifications will not be finalized until a later date. However, the AXAF-I operations system must be ready before that in order to provide the required personnel training and system integration. The system would require a simulator to provide operation data for system verification, testing, and training purposes. A high fidelity simulator will not be available until the specifications have been finalized. For this reason, a low fidelity simulator is built for the operations and management system development.

The functions of the system can be divided into three parts: monitoring, simulation, and training. The purpose of monitoring is to provide the operators the telemetry and status display and spacecraft control functions. The simulation and the training functions are tightly coupled, i.e., during the training mode, the simulator is activated to provide simulated flight telemetry. Since various personnel will be using the system, an appropriate operation mode must be assigned to each user to guarantee the safety of the spacecraft. At this moment, three modes have been established: administrator, instructor, and operator/student.

In the administrator mode, the user has all the editing and operation capabilities. Since program or system setup editing can lead to performance changes, only limited personnel are given this authority. The instructor mode allows a senior operator to activate and manipulate the simulator to provide desired operation scenarios for operators-in-training (or students). The operator/student mode provides the daily operation functions. In this prototype, the focus has been limited to the engineering data simulation and monitoring. The system is hosted in a Silicon Graphics Indigo-2 (SGI) workstation running the IRIX operating system. The expert system tool used is the G2 system from Gensym Corporation [G292].

Although expert systems have been applied through a wide spectrum of applications [SE94], due to the high reliability requirement of the mission payload operations, they have not been actually adopted in the daily operations. The autonomous monitoring system in the Extreme Ultrviolet Explorer (EUVE) science operations center (ESOC) [LG95] uses simple AI techniques to reduce the manpower requirement for its operations. Before the deployment of the system, the ESOC was staffed by 2 operators 24 hours per day, 7 days per week. Currently, it is staffed for one shift (8-hour) per day, 7 days per week. During the off shifts, the system would inform personnel about any anomaly through paging. If no action is taken within a certain time, the paging will be escalated to more personnel.

This expert system will be one of the pioneer mission payload operation systems at the Enhanced Huntsville Operations Support Center (Enhanced-HOSC), Marshall Space Flight Center (MSFC), Huntsville, Alabama, which is scheduled to be operational in September 1998. The AXAF-I program will use the Enhanced-HOSC where the SGI workstations will be standard. The current mission payload operation systems have been successfully deployed since the 70's and are still providing the needed services. However, it is becoming obvious that the in-

creasing performance demand and the high maintenance cost have promoted the need for more efficient systems. The distributed workstations environment is a possible solution which is planned to be implemented in the new MSFC control center - Enhanced-HOSC. Along with a few systems such as the science payload operation expert system for the International Space Station, Alpha, the system presented here is among the NASA's payload operations system pioneers.

2. AXAF-I DESCRIPTION

The AXAF-I includes the following six major components [AX94].

Structures and Mechanical: It provides the housing of all components of the AXAF-I and integrates them into one piece. No monitoring and management services are required for this subsystem.

Thermal Control System (TCS): It uses temperature sensors and heaters located at various places to maintain the temperatures of sensitive elements.

Electrical Power System (EPS): It generates, stores, and conditions the electrical power for the AXAF-I. It consists of a pair of solar arrays, three batteries, and a power control unit.

Communication, Command and Data Management (CCMD): This is the control center of the AXAF-I. It provides all the communication, command, and data management services. Its major components include a pair of low-gain antenna and transponders, an on-board computer, a solid state recorder, and an ultra stable oscillator.

Pointing Control and Aspect Determination (PCAD) and safing: Its purpose is to "provide all attitude control and determination functions necessary to point the AXAF-I at desired science targets, maneuver to new target, safe the spacecraft in response to detected failures, and to maintain solar position to the sun." [AX94]

Integrating Propulsion: It provides the control and propulsion to achieve the desired location and orientation specified by the PCAD.

For the monitoring and management purposes, sensors are installed throughout the spacecraft to provide the component status information. There are two types of sensors, analog (i.e., numeric) and bi-level (i.e., on/off). Since the number of sensors is in the order of hundreds and complex relationships exist among the sensor readings, the discussion of this paper will focus on the Electrical Power System (EPS) to provide a clear presentation. At the time of this expert system development, the EPS contains 17 analog telemetry outputs and 28 bi-level telemetry outputs. Most of the analog telemetry outputs concern

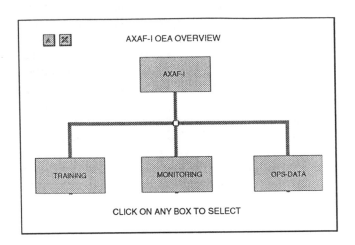

Figure 1 System functional diagram

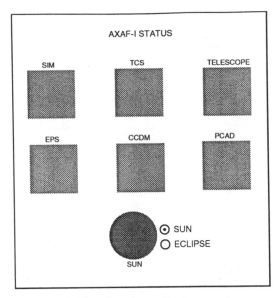

Figure 2 Top level monitoring workspace

with voltage and current; while most of the bi-level outputs concern with the status of components, like on/off and enabled/disabled. One major factor affects the functioning of the EPS is the sun/eclipse status. In the sun, the solar arrays provide all the power for the operation needs and maintain the full charge for the batteries. During eclipse, the batteries will provide the power. The number of eclipses is limited throughout the operation lifetime of the AXAF-I (<160) and most eclipses will be less than 2 hours long [EPS94].

The function of each system component is simulated by a simulator. This paper will discuss the simulator of the PCAD subsystem. A command to the PCAD consists of the desired pointing target and the time of execution. When the execution time arrives, the desired direction is first compared to the current pointing direction and, depending on the angle difference, the movement of the spacecraft will be divided into stages, i.e., each stage can only move a limited angle. The movement is achieved by braking the reaction wheels. A reaction wheel is assigned to each of the X, Y, and Z axes and can be control independently. The functionality of the PCAD is mainly expressed in terms of mathematics equations and the details can be found in [MC95].

3. STRUCTURE DESIGN OF THE EXPERT SYSTEM

This prototype has a hierarchical design and takes into consideration for future expansion and modification. The top-level workspace (or window) of the system is shown in Figure 1. Depending on the role of the user, different functionality will be available. An operator or student can choose MONITORING and OPS-DATA; while an instruc-

tor can also choose TRAINING. An administrator will have access to all functions, including the "AXAF-I" block. The user can choose a function by clicking on a desired block and a subsequent workspace will pop out. However, if the user clicks on a block that he/she does not have the privilege, nothing would take place. The MONITORING workspace provides the telemetry monitoring capability; while the TRAINING workspace allows an instructor to manipulate the simulator of the spacecraft. The OPS-DATA workspace will contain the operation procedure document and will be completed when the operation procedures become available. The "AXAF-I" workspace allows an administrator to modify the system (including programs and system setup).

3.1 MONITORING

The top-level monitoring workspace is shown in Figure 2. Under normal status, the icon color of each subsystem is green. When a subsystem is signaled for a warning, its icon color will be changed to red. In addition, a sun icon is shown in this workspace. When the AXAF-I is in the sun, the sun icon color is orange; while in eclipse, the color is gray. For simulation purpose, two buttons are provided to the user for sun or eclipse selection. When a subsystem icon is clicked upon, a subworkspace containing its components will be popped out (see Figure 3). The user may continue the process of traversing down the hierarchy by clicking on the desired components. However, when a primitive object is clicked upon, a subworkspace containing all of its real-time telemetry readings will be shown (see Figure 4).

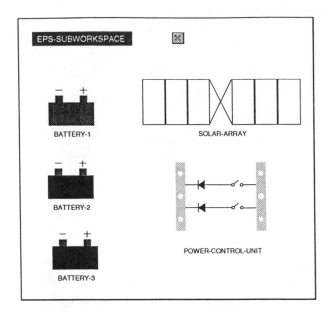

Figure 3 Workspace for the EPS subsystem

In any monitoring workspace, the color of an icon indicates the status of its corresponding component (or object). For example, under normal operation status, the original icon color is displayed. When a warning is signaled for an object, its icon color may be changed to orange or red (depending on the level of warning). The warning status of a compound object (e.g., EPS) is normally an "OR"ed result of its subcomponents, e.g., a warning for the BATTERY-1 or the SOLAR-ARRAY will trigger a warning for the EPS.

One important advantage of this hierarchical status display is that it allows an operator to monitor the top level display under normal operation. When a warning is signaled for any subsystem, the anomaly can be pin-pointed by traversing down the *colored* icons. Since multiple workspaces can be opened at one time, concurrent monitoring of different components can be done easily.

In addition to the icon color changes, a separate warning messages window is also provided on the top left corner of the screen. A typical message comes with a warning buzz and indicates the type, time, and location of the anomaly. Some messages also suggest the actions to take.

3.2 TRAINING

The training mode allows an instructor to set up the training scenarios and/or manually modify the simulated telemetry data. The top level training workspace is shown in Figure 5. At this time the training scenario editor has not been completed yet. In the simulator, most telemetries are simulated by formulas. The simulated value of each analog telemetry, T, is the sum of a varying number, N, and a user-controlled bias, B, i.e., $T = N + B$. Each N is designed in such a way that it will vary within the normal operation range of the simulated telemetry. Since each B is pre-set to 0 (or whatever proper offset), under normal operation, the simulated telemetries are always within the valid ranges. However, the user, i.e., instructor, can vary the bias B to make a simulated telemetry below or above its

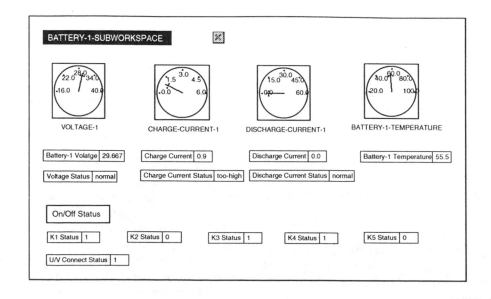

Figure 4 Workspace for primitive object BATTERY-1

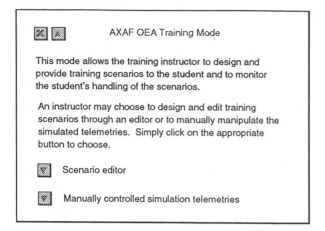

Figure 5 Training mode workspace

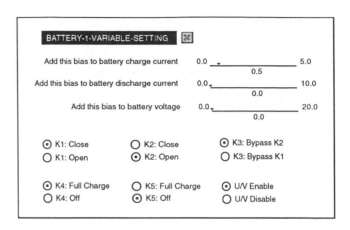

Figure 6 Simulation workspace for Battery-1

normal operation range. For example, the following formula simulates the DISCHARGE-CURRENT-BAT-1 (discharge current of BATTERY 1).

(the current second / 5 + 10.0) * (1 - the state of SUN) + discharge-current-bat-1-bias

While in the sun, i.e., the state of SUN = 1, the DISCHARGE-CURRENT-BAT-1 should be 0.0 ; while in eclipse, i.e., the state of SUN = 0, the value should range from 10.0 to 22.0. The term "the current second" is the *second* reading from the real-time clock of G2 which var-

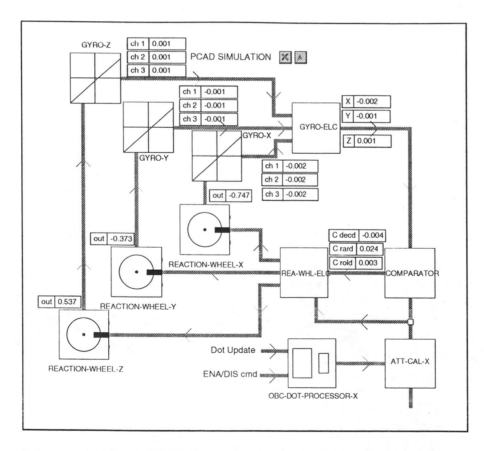

Figure 7 Simulation workspace for the PCAD subsystem

ies between 0 and 59. The *discharge-current-bat-1-bias* represents the bias, B, that the instructor controls. In the case of *discharge-current-bat-1-bias*, its value may range from 0.0 to 10.0 and is preset to 0.0 (see Figure 6).

The instructor can modify the simulated telemetry by adjusting the bias. In Figure 6, for example, the charge-current, discharge-current, and the voltage of BATTERY-1 can be modified by moving the slides. Certain bi-level telemetries can also be modified by selecting buttons. The example presented here is extremely simple, but it shows how the interdependency between components, e.g., SUN and BATTERY-1, can be simulated through formulas.

Although some telemetries can be simulated by simple formulas, most component functionality must be simulated by more complicated mathematics models. For example, the pointing computation process of the PCAD subsystem is passed from one component to the next. Procedural simulation is required to model these functions. In the implementation, rules are used to activate appropriate procedures by checking and setting flags. The simulation workspace for the PCAD is shown in Figure 7. When clicked upon, the workspace for each component will be popped out. For example, Figure 8 shows the workspace for the Detailed Operations Timeline (DOT) commands processor. The DOT commands are stored in a file and can be read one by one by the simulator. The workspace indicates the current simulation time, the next command execution time, and the desired pointing angles. If no ac-

tion is taken by the instructor, the simulator will execute the next command when the simulation time reaches or has passed the execution time. However, the instructor has the freedom to modify the data presented in any of the boxes. As another example, Figure 9 shows the simulation workspace for Gyro-Y. It allows the instructor to adjust the performance parameters of the gyro.

3.3 SYSTEM DESIGN

The G2 development environment provides the object-oriented programming capability for rules, objects, and workspaces. The design of the prototype tried to utilize this capability by first identifying the involved entities. The entities include (1) physical objects, (2) rules, and (3) telemetry variables.

3.3.1 PHYSICAL OBJECTS

The physical objects of the AXAF-I are classified into classes. The AXAF-OBJECT is the top level class and all subsequent subclasses and object instances inherit from

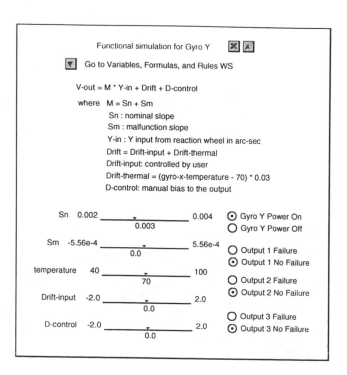

Figure 9 Simulation workspace for Gyro Y

Figure 8 Simulation workspace for the DOT Processor

Class name	battery
Superior class	eps-component
Attributes specific to class	voltage is given by a bat-voltage; charge-current is given by a bat-current; discharge-current is given by a bat-current; status has values normal or warning; voltage-status has values normal, too-high, or too-low; k5-status is given by an on-off-var;
Inherited attributes	component-temperature is given by a temperature

Figure 10 Portion of the definition table of object class BATTERY

this class. For example, BATTERY is a subclass of EPS-COMPONENT, which is a subclass of AXAF-OBJECT. A portion of the definition of BATTERY is shown in Figure 10. A physical object is an instance of an object class. It is used to represent a single or a collection of components. A single component object (called primitive object), e.g., BATTERY-1, contains all of its related telemetry readings and status. An object representing a collection of components (called compound object), e.g., EPS (an instance of AXAF-OBJECT), contains the integrated status of the compound object. A portion of the table of BATTERY-1 is given in Figure 11.

3.3.2 RULES

The rules determine when a telemetry reading is out of range and when to signal a warning, and suggest actions to take. A rule for the charge current checking for the BATTERY class is given in Figure 12. In the middle of the figure, it can be seen that G2 provides a natural-language like rule syntax [G292]. In addition, each rule is associated

Names	BATTERY-1
Component temperature	77.0
Voltage	25.5
Charge current	0.25
Discharge current	0.0
Voltage status	normal
K5 status	0

Figure 11 Portion of the definition table of object instance BATTERY-1

| Options | not invocable via backward chaining, not invocable via forward chaining, may cause data seeking, may cause forward chaining |

for any battery whenever the charge-current of the battery receives a value and when the value of the charge-current > the Batt-charge-current-sun-k5-off-max of battery-limits and the state of sun = 1 and the k5-status of the battery = 0 then conclude that the charge- current-status of the battery is too-high

Focal classes	battery
Focal objects	none
Rule priority	2

Figure 12 Partial table of the battery CHECKING-FOR-CHARGE-CURRENT-TOO-HIGH-WHEN-SUN-K5-OFF rule

with a list of attributes that may define the search and application scopes. For example, the figure indicates that the rule should not be invoked via backward or forward chaining. It should be invoked only when a new charge-current is received. The figure also indicates that the search space should be limited to the instances of BATTERY (see Focal classes).

3.3.3 TELEMETRY VARIABLES

Every telemetry reading sent to the OCC for monitoring must be defined in the G2 environment. The reason is that different telemetries may need different treatments. For example, some telemetries should cause forward chaining immediately after its reception; while other telemetries should be accessed only when a backward chain-

Batt charge current eclipse	0.01
Batt charge current sun k5 off min	0.2
Batt charge current sun k5 off max	0.4
Batt charge current sun k5 full charge min	2.5
Batt charge current sun k5 full charge max	4.2
Batt discharge current eclipse min	10
Batt discharge current eclipse max	22
Batt discharge current sun	0.01
Batt voltage min	24.5
Batt voltage max	35

Figure 13 Portion of the definition table for the BATTERY-LIMITS

ing has requested them. The telemetries of AXAF-I are defined as *variables* in G2. They are organized in a hierarchy as well. The top level class of the variable is AXAF-VAR which is a subclass of the system-defined QUANTITATIVE VARIABLE. Subclasses are then defined for voltage, current, temperature, etc.

Each telemetry reading has a valid range in a specific operating mode. Instead of specifying these ranges within the telemetry variable definitions, they are defined and specified as a separate entity. In this design, telemetry ranges associated with a physical object are collected and defined in a table. Figure 13 shows the range table for the BATTERY. This arrangement provides easiness to locating a desired range and performing modification - especially when multiple modifications are needed.

4. CONCLUSION AND FUTURE EXTENSION

The preliminary review of the prototype have been encouraging. In the past, any functional prototype for simulation or monitoring purpose would require a team of two to three developers at least ten months. However, the functional prototype presented here consisting of both simulation and monitoring capabilities were completed within a two-month period by one person. The readily available expertise of the NASA scientists and engineers was one of the major factors. Other factors include the expert system technology and the tool selected for the project, i.e., G2. Major findings are listed in the follows.

Hierarchical design: The hierarchical design of the objects and telemetries provides a conceptually clear framework for prototype. It not only provides a reasonable way to specify the entities and relationships, but also makes the system development and testing very efficient. This design also allows the user to choose the appropriate monitoring levels for different components which reduces the mental faculty burden of the user.

System flexibility: In addition to the demonstration of feasibility, the prototype has also shown the flexibility of accommodating the system changes. Specification changes can be easily updated by editing the rules or the attributes of the related objects. This feature is extremely important in this system because the detailed specifications of AXAF-I are still evolving.

Graphical user interface: The powerful graphics capabilities of the Silicon Graphics workstation and the G2 environment have made the mission operations monitoring more pleasant. By using the combination of rule-based reasoning, status display colors, and different icon designs, the user can easily identify the status of any system components. The labor intensive telemetry observation

task has now been accomplished by the expert system.

G2 environment: During the course of the design, development, and testing of this prototype, G2 has been confirmed to be a powerful expert system development tool. Most of the positive features mentioned above are directly related to the G2 capabilities. Its natural-language-like rule syntax is easy to follow; while the attribute tables of objects provide an efficient way to describe the domain and configurate the relationships among entities. Although G2 has proved itself to be a powerful tool, the learning curve to achieve proficient understanding of the tool can be long and costly.

At this moment, the prototype is still under constant revision and evaluation. On a later stage, an operational system will be re-developed based on the design and experience learned from the prototype. The new system will also have to consider the integration with other systems, e.g., the DSN. The complete report of this prototype can be found in [CR95].

5. REFERENCES

[AX94]   Advanced X-ray Astrophysics (AXAF) - Mission Operation Plan, AMO-1010, Mission Operations laboratory, George C. Marshall Space Flight Center, NASA, April 1994.

[CR95]   K. Chang, M. Rogers, and R. McElyea, "A Real-time Expert System for the Monitoring and Management of the AXAF-I Flight and Ground Systems," Technical report, Payload Operations Lab., George C. Marshall Space Flight Center, NASA, Hunstville, July, 1995.

[EPS94] Electrical Power Subsystem - AXAF-I Preliminary Design Audit Presentation Package, TRW Space & Electronics Group, September 9, 1994.

[G292]   G2 Reference Manual, Version 3.0, Gensym Corporation, Cambridge, MA, July 1992.

[LG95]   M. Lewis, F. Girouard, F. Kronberg, P. Ringrose, A. Abedini, D. Biroscak, T. Morgan, and R. F. Malina, "Lessons Learned from the Inttoduction of Autonomous Monitoring to the EUVE Science Operations Center," Publication 640, Center for EUV Astrophysics Technology Innovation Series, University of California, Berkely, CA, 1995.

[MC95]   Richard McElyea, The Mathematics Model for PCAD - a note, Payload Operations Lab., George C. Marshall Space Flight Center, NASA, Hunstville, AL, July, 1995.

[SE94]   P. Selffridge, "AI at Work, the Best of CAIA'94 - Editorial Comments", IEEE EXPERT, Vol. 9. No. 5, October, 1994, pp. 2.

KNOWLEDGE-ENHANCED CO-MONITORING IN COAL-MINES

Wolfram Burgard[1], Armin B. Cremers[1], Dieter Fox[1],
Martin Heidelbach[2], Angelica M. Kappel[1] and Stefan Lüttringhaus-Kappel[1]
[1]Institut für Informatik III, Universität Bonn, Römerstr. 164, D-53117 Bonn
[2]Ruhrkohle Bergbau AG, T 1.2, Shamrockring 1, D-44623 Herne
E-mail: {wolfram,abc,fox,heidelba,angelica,stefan}@cs.uni-bonn.de

ABSTRACT

Detection of underground fires is an important security task in hard-coal mining. Automated fire detection systems are usually based on the monitoring of carbon monoxide (CO). Systems using conventional technology based on threshold and tendency observations, however, generate a large number of false alarms. We show how CO-concentrations can be forecast by appropriate models of the physical and chemical processes. We furthermore describe a rule-based specification system utilizing forecasting for CO-monitoring. The improvement of this approach over the conventional is threefold. First, the number of false alarms is reduced by 50%, at least. Simultaneously, the thresholds for warnings and alarms can be reduced so that, second, the detection of real fires becomes both quicker and more reliable. Third, heuristic rules for fire detection and suppression of false alarms as well as the control of the forecasting can be described in a declarative way. While our system is still in a prototypic stage, the three major German hard-coal mining companies decided to use our approach in their CO-monitoring systems.

1 INTRODUCTION

In hard-coal mining preventive techniques against fires and explosions have been very successful in the 20th century, especially in avoiding explosions by intrinsically safe (electrical) equipment and reducing damage by strong mine rescue parties. Moreover, fire barriers are installed all over the mines. Sufficient supply of fresh air avoids dangerous concentrations of methane (CH_4), while it also reduces the heat and simultaneously improves the quality of the air. However, due to the self inflammability of the sulphureous hard-coal fires occur in every modern coal mine. Furthermore, defect machines or parts of machines, as for example defective rollers of band conveyors could cause a smoldering fire.

Enormous technical effort is taken to detect fires as early as possible and to extinguish them before miners are endangered or substantial damage takes place. In order to enable a rapid detection, ventilation velocity, carbon monoxide (CO) and CH_4 concentrations are constantly monitored using several hundreds of sensors in an average coal mine.

While the ventilation velocity and CH_4-concentrations can be monitored not to exceed fixed security limits, the interpretation of CO-concentrations and CO-curves is quite more complicated. This is due to the fact that various activities carried out underground by the miners such as blasting operations or driving Diesel-powered vehicles produce similar amounts of CO as open fires. Moreover, processes outside the mine produce CO which, if drawn into the mine, increases the CO-concentration in a similar way as a smoldering fire. Such processes are smog or combustion processes, like cars on nearby roads, factories, and power plants.

In the beginning, CO was monitored by specially trained staff. Since the end of the seventies, the Ruhrkohle Bergbau AG uses process computers to control the CO-concentrations measured by the CO-sensors [SSSS82]. Such a process computer installed on a mine monitors for each CO-sensor the mean CO-values averaged over a time period of one minute. In this system [Kar87], apart from absolute thresholds increasing CO-concentrations and special properties of CO-curves are used as indicators of open or smoldering fires. Furthermore, special filtering heuristics are applied to detect CO-curves coming from blasting operations which otherwise would produce a fire alarm.

Figure 1 shows a typical line plot of the minutely averaged CO-concentrations at several CO-sensors for about 16 hours without fire (the y-axis is scaled in ppm/100 CO). On one hand, large variations can be observed, and on the other hand, a slope of just 6 ppm CO per 24 hours could indicate a smoldering fire.

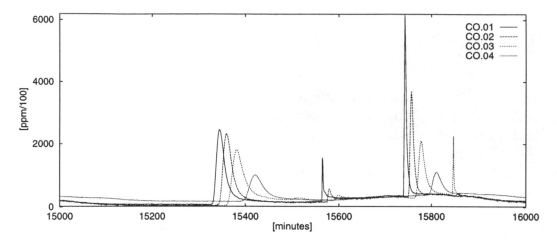

Figure 1: Typical CO-curves of different devices

To be on the safe side, thresholds for warnings and alarms in the current system have been set to very low values, causing a serious number of false alarms. The number of false alarms is further multiplied by the fact, that the clouds of CO (and other gases as well) are streaming with the air through the mine, causing series of similar (false) alarms at successive sensors. In all, there may be more than eighty warnings and alarms per day. The huge number of alarms are a critical factor burdening staff in the mine observation center. On one hand, a substantial amount of time is spent for checking CO-curves, or even for controlling the corresponding parts of the mine by the miners in order to verify the situation is really safe. On the other hand, the staff's sensitivity to actual fire events may be reduced.

In this paper we describe how the number of fire alarms can be reduced significantly by combining sub-symbolic and adaptive techniques with traditional knowledge representation methods. This is achieved by adaptive techniques forecasting CO-curves in the mine combined with a rule-based representation of heuristic rules for fire detection and the control of the forecasting. The next section describes the method to forecast CO-values. Section 3 illustrates the knowledge based part of the system. In Section 4 we describe how we integrate the forecasting and the rule-based part. Finally, in Section 5 we demonstrate the performance of the system by an application example.

2   FORECASTING OF CARBON MONOXIDE BY ADAPTIVE TECHNIQUES

In the fire detection system currently in use every CO-sensor is observed independently of the other CO-sensors. For this reason a warning or alarm has to be triggered whenever the concentration of CO reaches a fixed threshold. Because the air is streaming along the roadways, high concentrations of CO at a CO-sensor can be observed some minutes later at the following CO-sensor.

As an example consider the CO-curves in Figure 1. Near minute 15400, typical CO-curves of a single blasting operation measured at different devices located one after another in the air stream are shown. While the qualitative similarity between the different curves is obvious, the quantitative CO-curves do not match exactly. For the currently used system, much effort was spent to define heuristic classification rules to cope with this variety of CO-curves.

Our approach is to model the change of the CO-curves between two CO-sensors exploiting the similarity of the corresponding CO-curves. This allows us to predict the CO-curves at successor devices. For every time step we compute the expected CO-value of a CO-sensor based on the CO-curve of the preceding CO-sensor. For a reliable prediction it is extremely important to know the time delay of the air stream between such two devices as well as the mapping between the CO-curves. The time delay, unfortunately, is neither constant nor is it provided accurately enough by the monitoring system. We therefore developed an adaptive method that extracts this information from sample data given by the CO-curves of two adjacent devices (see Section 2.1). Concerning the mapping of CO-curves, it is obviously not possible just to shift the CO-curve of the first device by the time delay and to use these values as prediction for the second device. Our method to model this mapping based on sample data is described in Section 2.2.

For illustration we use the CO-curves of two adjacent

CO-sensors of the mine Blumenthal taken in March 1994. The devices were placed in an area of the mine with a lot of mining activities.

2.1 Determination of the time delay

To keep our method as adaptive as possible it only relies on the local topology of the mine, i.e. the predecessor/successor relation of the CO-sensors. All the information necessary to determine the time delay is contained in the sample data given by the CO-curves of the two corresponding CO-sensors. By repeated usage of this method on updated samples the time delay can be adapted to the changing environment in the mine. In the approach described in [Kar87] the determination of the time delay based on sample data depends on significant CO-values above 3.0 ppm. Especially on weekends or for devices far away from mining activities this method is not applicable. We therefore introduce a method extracting the time delay independently of such peaks.

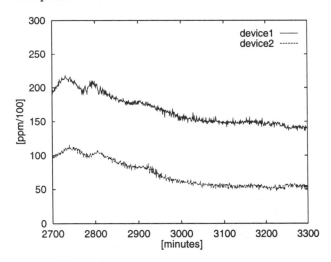

Figure 2: The noisy values are correlated

If we look at a sufficiently long time interval of ten hours as shown in Figure 2 a high similarity in the CO-curves becomes apparent even for values below 3.0 ppm. This similarity is due to the long term changes in the CO-curves. The high frequency noise is not correlated at all and has to be filtered out first. Assuming Gaussian noise in the high frequency we use a Gaussian filter to delete it. The result of this filter is given in Figure 3.

To extract the time delay from the data we only have to find the offset of the two time series leading to the best match of the values. For this task we have to explore a one dimensional search space to get the time shift with the best evaluation. To evaluate the quality of the match for each time shift we consider

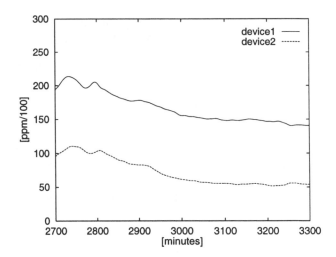

Figure 3: The values after Gaussian filter

the CO-values as n instantiations of two random variables. Thus we can use the cross-correlation between these values as evaluation function for the matches. The cross-correlation $r_{xy} \in [0; 1]$, see equation (1), is a quantitative measure of the linear dependency between n values of two random variables X and Y [HEK89]:

$$r_{xy} = \frac{\sum_{i=1}^{n} x_i y_i - n\bar{x}\bar{y}}{\sqrt{\left(\sum_{i=1}^{n} x_i^2 - n\bar{x}^2\right)\left(\sum_{i=1}^{n} y_i^2 - n\bar{y}^2\right)}} \quad (1)$$

2.2 Approximation of the Mapping

The prediction of the CO-concentration at a CO-sensor is of course most important for high values. We assume the change of the CO-curves in time to be caused by two types of processes. On one side there are chemical processes leading to a loss of CO until the concentration reaches equilibrium at about 2 ppm. On the other side we have the physical process of whirling which aims at the same concentration of a gas over the whole available space. A CO-peak changes under the influence of the first process in a way such that the integral under the peak becomes smaller but the shape of the CO-curve stays constant. The process of whirling leaves the overall amount of CO unchanged but aims at a flattening of the CO-curve.

A mathematical model of these processes depends on the exact knowledge of all facts influencing loss of CO and whirling of air. Among many other influences there are the temperature of the air, the composition of gases in the air, the concentration of pulverized coal in the air and the exact geometry of the mine. These facts are not available and because of the mining activities the environment is in permanent change. For these reasons we decided to extract the parameters of the model of these processes from examples. This leads to a model which constantly adapts itself to the specific facts given in a certain place at a certain time in the mine.

In our simple model we assume a loss of CO linear with the CO-values. The whirling, in contrast, can be modeled as a Gaussian process [Ari56, Lit70, Tay53]. This leads to the formula given in (2) computing the prediction value p_t for the succeeding CO-sensor at time t.

$$ p_t = a \left(\sum_{i=-n}^{n} \left(\frac{\omega}{\sqrt{2\pi}\sigma} e^{-\frac{1}{2}(\frac{i}{\sigma})^2} \right) c_{t-o+i} \right) + b \;\; (2) $$

Mathematically the whirling process is computed by a weighted sum of the value itself and n preceding and succeeding values. The coefficients of this sum are given by a Gaussian probability density function with mean at the actual value. To ensure the commutativity of the whirling process and the loss of CO, the coefficients are normalized by ω to sum up to 1. The standard deviation σ of the Gaussian defines the strength of the whirling process. $C_{t-o-n}, \ldots, c_{t-o+n}$ are the corresponding CO-values of the preceding CO-sensor and o is the time delay between both measuring devices. Finally the decomposition process is modeled by a linear function $y = ax + b$ where a describes the decomposition in time o and b the average minimum CO-concentration.

The task is to adapt the parameters o, σ, a and b of (2) to a given set of samples. The sample consists of the CO-values of two adjacent CO-sensors taken over a period of one week. This leads to about 10000 values for each device. In order to map each value of the preceding CO-sensor on the corresponding value of the other device we first have to compute the time delay o of the air stream between the devices. This is done with the method described in Section 2.1. Given this value we still have to determine the remaining parameters of the two processes described above.

To extract the parameter σ for the whirling we perform a one dimensional search over different standard deviations of the Gaussian. A deviation of zero does not change the values at all and the higher the deviation the more the flattening influences the peaks. Thus, the starting point for the search is a standard deviation $\sigma = 0$. Having the linear loss in mind we want to find the deviation which leads to the most linear coherence between the values. This is why we use the cross-correlation as a measure for the best standard

deviation. We regard a model of the whirling as the best one if it yields the highest cross-correlation value for the resulting curves.

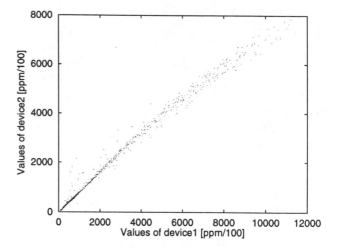

Figure 4: Mapping of shifted raw data

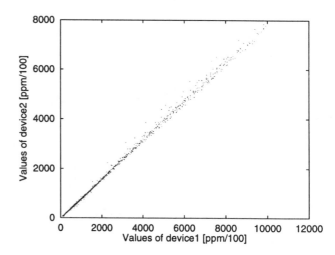

Figure 5: Mapping after Gaussian whirling

After having found the parameters of the whirling model for the two devices, we just have to apply this model to the CO-values of the first device and map the resulting values to the values of the second device, shifted by the time delay o. Such a mapping is shown in Figure 5. To demonstrate the influence of the Gaussian whirling we show in Figure 4 the input data before having computed the whirling. Obviously the Gaussian process leads to a significantly more linear dependency of the values.

The relation of the CO-values shown in Figure 5 is approximated by linear regression [HEK89]. As Figure 6 illustrates, the resulting mapping allows an accurate prediction of CO-curves. The figure contains

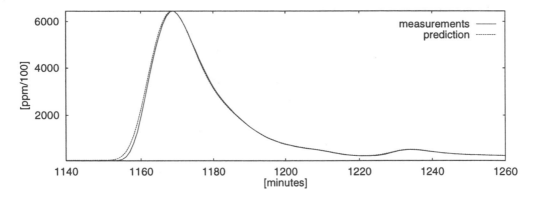

Figure 6: Predicted and measured CO-values of a blasting operation

the measured CO-curve of a CO-device and the predicted CO-curve based on the values of the preceeding device. Note that the shown time interval is different from the one used to generate the mapping.

Having provided a method for forecasting CO-values, we are next going to define a technique for the specification of the rule-based part of the fire-detection system.

3 EXECUTABLE RULE-BASED SYSTEM-SPECIFICATION

High-level specification tools are essential for an appropriate specification and integration of the various modules of our system. In contrast, the basic heuristics of the system currently in use are expressed by complex flow-charts, which are hard to survey and to maintain. Our aim was to cure this deficiency using an executable, descriptive, and rule-based specification language.

Recently, *Evolving Algebras* (EAs) [Gur91, Gur94] turned out to be a powerful rule-based specification tool. The semantics of EAs have been investigated intensively within the last years [Gur94], and various applications in different fields are known [Bör94a]. EAs allow problem-oriented specifications tailored very closely to the given subject at an appropriate level of abstraction, supporting easy system verification. We decided to use EAs as their properties lead to more transparency of the underlying algorithms and heuristics for fire detection and thus to easier maintenance of the system. In this way, EAs contribute to enhance security in underground coal-mining.

EAs are suitable for an efficient implementation using logic programming. A Prolog-based EA-implementation for sequential non-modular EAs emphasizing efficiency exists [Kap93]. But for our system for fire detection, it was necessary to have modularity and parallelism. Each part of the complex analyzing meth-

ods constitutes a module of its own. Each module is executed in parallel for all sensors in the mine. Different modules represent distributed processes communicating via streams. The streams also realize the synchronization of these asynchronously running processes. Figure 7 shows the architecture of our fire detection system. The boxes represent different modules, and streams are indicated by arrows.

For this purpose we developed SEA (**S**treams in **EA**s) which provides the required means to implement this high-level system-specification.

3.1 The Specification Language SEA

SEA is a modular language allowing parallel execution of modules. In addition to partial functions, SEA supports partially bound data structures, e.g. streams. An evolving algebra is an SEA-module consisting of a module head, function declarations, a start rule and a set of transition rules.

The module head consists of the name and the interface of the module. The interface defines the name and arity with which the module will be called. It may be called either as an n-ary predicate in Prolog-programs or as an $(n-1)$-ary function in SEA-modules.

Module parameters defined in the module interface are internal SEA functions by definition. They may optionally be annotated with modes.

External functions are defined by a predicate name and arity p/n. Inside the actual SEA-module, an external function is visible as an $(n-1)$-ary function.

Internal functions are local SEA-variables with optional mode annotations.

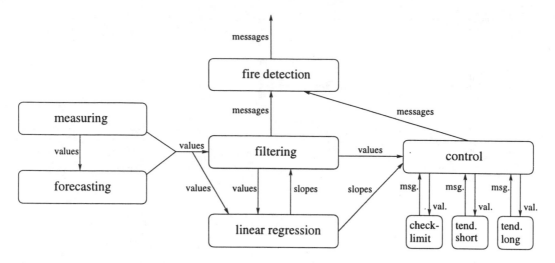

Figure 7: Module hierarchy of the fire detection system

A start rule is an action-formula, which is executed at the beginning of a computation.

A transition rule is a condition-formula followed by an action-formula, which is executed if the conditions are true.

Conditions compare terms by the use of built-in relations like =, <, >.

Actions are either function updates `f := t`, a `stop` operation terminating the computation, a binding of a partially bound data structure `t1 = t2` or a generation of a new temporary variable `var v` constituting an unbound part of a partial data structure.

Macros may occur either in condition or in action parts of a transition. Macros are defined outside the algebra by the command `macro Name => Replacements`, where `Replacements` are conditions or updates. Each occurrence of a macro name in a rule will be expanded at compile-time. Using macros rules become shorter and more declarative.

When a module is called at first the start rule is executed generating the initial state of the algebra. Then a loop is entered; in each round the update-part of the first rule whose conditions evaluate to true is carried out until a `stop` operation occurs. The `stop` operation terminates the module call.

SEA provides a graphical EA-visualization tool which allows a look at indicated functions of an algebra at any time. In this way, the system development became both quicker and safer.

```
algebra check_limit defines
        check_limit(device,value,message);

local warning_value,alarming_value;
external co_warning_value/2, co_alarming_value/2;

start warning_value := co_warning_value(device),
      alarming_value := co_alarming_value(device);

if   value < warning_value &
     value < alarming_value
then message = [], stop;

if   value < alarming_value &
     value >= warning_value
then message = ["Warning-level exceeded"], stop;

if   value >= alarming_value
then message = ["Alarming-level exceeded"], stop;
```

Figure 8: Fire detection module checking CO-limits

An example of an SEA-module for fire detection is given in Figure 8. The algebra `check_limit` defines the control of incoming CO-values with respect to given warning and alarm thresholds. As a more sophisticated example, we will subsequently show some rules for the filtering of blasting curves. The fire detection system identifies underground blastings and filters their high CO-values. According to [DMT93, SSSS82] a blasting curve is divided into three parts (see Figure 9): the first part starts at the beginning of the curve and ends at the first turning point `T`. The second part starts at `T` and ends at the maximum `M` of the curve. The third

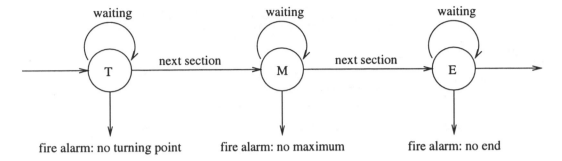

Figure 9: Graph representation of the algorithm filtering blasting curves

part starts at M and ends when the curve reaches its starting level again. T and M are detected by analyzing the slope of the curve. The end E of the curve, is found by comparing the current CO-value with the stored entrance level of the curve. Each part of the curve must be passed within a given time interval. Otherwise, an alarm is triggered and the filtering is stopped. Accordingly, the filtering algorithm has three SEA-rules for each section of a blasting curve: the first means waiting for T, M, or E, respectively, the second means going to the next section and the third gives a fire message if the time limit for this section has been exceeded. Figure 10 shows the rules for the M-section of a blasting.

4   INTEGRATION LEVEL OF THE SYSTEM

The overall system is realized by three permanently running Unix processes: the data acquisition interface, the monitoring module, and the graphical user-interface. The monitoring module receives its input as a set of streams containing CO-values, one per active CO-sensor. The monitoring module runs a considerably large set of parallel sub-processes under an and-parallel execution model. One of these parallel sub-processes is a controlling module which launches and supervises further processes monitoring a single CO-sensor each. These processes in turn fork to perform their subtasks. Synchronization between any pair of sub-processes is achieved through streams, which supply a (typically numerical) value per time unit.

The controlling module is also responsible for determining a monitoring method for each CO-sensor, i.e. whether forecasting can be applied or conventional techniques have to be used. The monitoring method is subject to change, when forecasting becomes no longer available due to the malfunction of the preceeding device. Furthermore, in case of modified forecasting circumstances like moving sensors or changed speed of air, the controlling module automatically launches temporary processes to determine new values for the whirl-

```
macro set_value =>
    head(new_values) = head(values),
    new_values := tail(new_values).

macro mark_value =>
    head(new_values) = void,
    new_values := tail(new_values).

macro next_values =>
    minutes := minutes + 1,
    slope := tail(slope),
    values := tail(values),
    flags := tail(flags).

if   filtering = on & turning_point = yes &
     maximum = no & minutes < maxminutes &
     head(slope) >= 0
then mark_value, next_values;

if   filtering = on & turning_point = yes &
     maximum = no & minutes < maxminutes &
     head(slope) < 0
then maximum := yes, mark_value, next_values;

if   filtering = on & turning_point = yes &
     maximum = no & minutes = maxminutes
then head(messages) =
         "Alarm: no maximum in blasting curve",
     messages := tail(messages),
     filtering := off, set_value, next_values;
```

Figure 10: Some rules of the module filtering blasting curves

ing parameters and the time delay. Before these values are used, they are checked by special SEA-rules.

The computation of predicted values is handled in an SEA-rule-based way, like any of the other monitoring methods. Given whirling parameters and the time delay from the sensor database, a time delay of n

Alarm type	Messages of system in operation		Messages of prototype with forecasting	
	Category A	Category B	Category A	Category B
Warning threshold exceeded	72	56	4	0
Alarm threshold exceeded	14	7	4	0
Fire tendency	20	20	4	4
Sum	106	83	12	4

Figure 11: Comparison of messages with and without forecasting

minutes is implemented by prepending a list of n void values to the predecessor's stream of CO-values. Due to the underlying Prolog's way of representing data structures in memory, no copying of the stream is necessary, it remains shared. Next, the shifted stream passes through a filter computing the Gaussian kernel corresponding to the whirling of the real CO concentration. The resulting stream of predicted values is subtracted from the actual CO-values of the current CO-sensor, and the difference stream is connected to the subsequent analysis modules.

5　Application Results

In order to document the enhancement of the current fire detection system in coal mining, we focused our interest on ten subsequent CO-sensors. The corresponding CO-values were measured during two days in march 1994, where several blasting operations were carried out. These blastings caused more than 100 alarms at the observed sensors. The alarms can be divided into two groups:

- Incoming values are compared with a warning and alarming threshold. A warning CO-level is announced by an optical signal, an alarming CO-level produces an optical and an acoustical signal.

- Over some given time intervals (10, 30, 60 and 180 minutes, 8, 16 and 24 hours) the slope of the CO-curve is computed by linear regression. If the slope exceeds a given threshold, a fire tendency message will be given. Computing the tendencies, it is possible to recognize even slowly but monotonous rising CO-values, which are indicating smoldering fires.

The current fire detection system filters high CO-values coming from blasting curves by some well evaluated heuristics, see 3.1. A blasting curve for example must have its turning point and maximum within fixed time intervals. All values measured by a CO-sensor

that are marked as filtered are not taken into account for the computing of tendencies.

Our improved fire detection system prototype relies on the forecasting technique. Instead of analyzing the absolute CO-values we only consider the difference between the measured and the predicted value. This difference is treated like a new, virtual CO-sensor and we use the same algorithm to compute the tendencies and to monitor thresholds like before. As mentioned above, we decreased the thresholds indicating warning and alarming levels to the half.

The table shown in Figure 11 compares the messages produced by the current system with the messages given by our enhanced fire detection system. While category A contains all ten CO-sensors, category B includes only the eight CO-sensors from category A which are not next to the two blasting areas. Notice that for CO-sensors next to mining activities the additional amount of CO cannot be predicted by the preceeding CO-sensors. The sensors in category B are most interesting for the forecasting method because their CO-curves can be predicted and they generally produce sequences of alarms.

Because of the blasting operations during the two observed days the current system produced 72 threshold warnings, 14 threshold alarms, and 20 fire tendency alarms for all ten sensors (category A). These numbers are reduced significantly by the forecasting method: the number of warnings decreased from 72 to 4 which is a reduction of 95 percent. The alarms were reduced by more than 70 percent such that only 4 alarm messages remained. Only 4 out of 20 fire tendencies were given by our system.

Considering only the sensors of category B the improvement becomes even better. 56 threshold warnings and 7 threshold alarms were produced by the current system. With the forecasting method all these messages were eliminated. As can be seen in the figure all fire tendency alarms were produced by the sensors of category B so that there is no further improvement compared to category A.

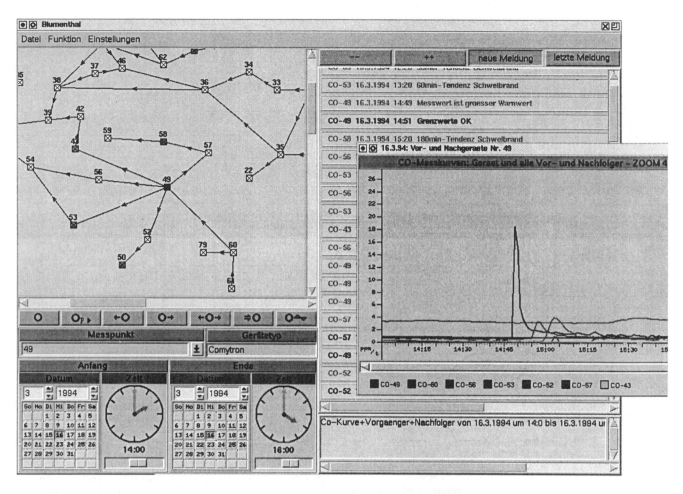

Figure 12: A snapshot of the user interface

The results of our analyzing method based on forecasting are very encouraging. By the use of our enhanced system, the staff in the mine observation center could be relieved from many false fire alarms. As a conclusion, our system based on forecasting reduces the number of false alarms to about 12%. It is safe to say, that our enhanced system for fire detection generally reduces the number of false alarms by more than 70%. This rate will be further increased, as improved classification methods for CO-curves, where no forecasting can be applied, are currently under development.

6 GRAPHICAL USER INTERFACE

The new CO-monitoring system comes along with a mouse-driven graphical user interface (see Figure 12). The user interface has been developed by the use of the modern and productive tool Tcl/Tk generating X11 applications. As the implementation of our system has mainly be done using the logic programming paradigm, we created a tool combining the languages Prolog and Tcl/Tk. More informations on our **PAT** tool can be found in the net[1].

On the right side of the system's main window actual warnings and fire alarms are displayed. The upper side of the left part shows a directed graph representing the CO-sensors of the mine and their predecessor/successor relations. This mine model is a base for the automatic computation of forecasting parameters. It may also be used to select a device for displaying its stored measuring or forecasting curves in a separate window.

7 CONCLUSIONS

The detection of underground fires is an important security task in hard-coal mining. The current fire detection system running on a process computer produces many false alarms caused by mining activities. In this paper we showed how CO-values can be forecast by appropriate models of the physical and chemical pro-

---

[1]http://www.informatik.uni-bonn.de/~angelica/PAT-WWW

cesses. This offers a great potential for the improvement of fire detection in hard-coal mines. First, the analysis of CO-curves can operate on the difference between the measured and forecast CO-concentrations. Second, heuristic thresholds for warnings or alarms can be adjusted more precisely, which allows a faster detection of fires. Practical applications of this new method show a reduction of false alarms by more than 70%.

Our specification language SEA extends the Evolving Algebra approach by stream parallelism. SEA has been used for the specification of the overall monitoring process. The high level of abstraction supports verification and improves transparency of the CO-monitoring system. At the same time, the SEA-compiler automatically compiles SEA-specifications into efficient Prolog code.

Thus, our system shows how the combination of knowledge based and sub-symbolic techniques leads to an improvement of transparency, reliability, and security.

REFERENCES

[Ari56]   R. Aris. On the dispersion of a solute in a fluid flowing through a tube. In *Proceedings of the Royal Society of London*, volume 253 of *A*, pages 67–77, 1956.

[BBC+90]   K.-H. Becks, W. Burgard, A.B. Cremers, A. Hemker, and A. Ultsch. Parallel process interfaces to knowledge systems. In *Proceedings of the International Conference on Parallel Processing in Neural Systems and Computers*, 1990.

[BCG+91]   Wolfram Burgard, Armin Bernd Cremers, Judith Grebe, Reinald Greve, Stefan Lüttringhaus, Frank Mücher, and Lutz Plümer. Knowledge based planning of underground lighting in hardcoal mining. In R. Vichnevetsky and J.J.H. Miller, editors, *Proceedings of the 13th IMACS on Computers and Applied Mathematics*, volume 4, pages 1751–1752, 1991.

[Bör94a]   Egon Börger. Annotated Bibliography on Evolving Algebras. [Bör94b].

[Bör94b]   Egon Börger, editor. *Specification and Validation Methods*. Oxford University Press, 1994.

[CTB94]   Armin B. Cremers, Sebastian Thrun, and Wolfram Burgard. From AI technology research to applications. In *Proceedings of the 13th World Computer Congress*, volume 3. Elsevier Science B.V. (North Holland), 1994.

[DMT93]   Bericht über die Untersuchung der Prozeßrechneranlage zum Erfassen und Verarbeiten wettertechnischer Meßdaten mit Schreiberersatz auf dem Bergwerk Ewald / Schlägel und Eisen. DMT-Gesellschaft für Forschung und Prüfung mbH, August 1993.

[Gur91]   Yuri Gurevich. Evolving Algebras. A tutorial introduction. *Bulletin of EATCS*, 43:264–284, 1991.

[Gur94]   Yuri Gurevich. Evolving Algebras 1993: Lipari Guide. In Börger [Bör94b].

[HEK89]   Joachim Hartung, Bärbel Elpelt, and Karl-Heinz Klösener. *Statistik: Lehr- und Handbuch der angewandten Statistik*. R. Oldenbourg Verlag München Wien, 9 edition, 1989.

[Kap93]   Angelica M. Kappel. Executable Specifications based on Dynamic Algebras. In A. Voronkov, editor, *Logic Programming and Automated Reasoning*, volume 698 of *Lecture Notes in Artificial Intelligence*, pages 229–240. Springer Verlag, 1993.

[Kar87]   Hans-Jürgen Kartenberg. Bessere Brandüberwachung in Strecken mit großen Wetterströmen. Glückauf-Forschungshefte 48, Nr. 2, 1987.

[Lit70]   A.B. Littlewood. *Gas Chromatography*. Academic Press, 2 edition, 1970.

[SSSS82]   Anton Stark, Horst Seyfarth, Hans Sredenscheck, and Joachim Steudel. Stand und Entwicklung der Brandfrüherkennung bei der Bergbau AG Lippe. *Glückauf*, 118(8):3–7, 1982.

[Tay53]   Sir Geoffrey Taylor. Dispersion of soluble matter in solvent flowing slowly through a tube. In *Proceedings of the Royal Society of London*, volume 219 of *A*, pages 186–203, 1953.

MEDI-VIEW
-- AN INTELLIGENT ICU MONITORING SYSTEM

Kevin Kennedy[1], Michael McKinney[1] and Yi Lu[2]

[1]Ford Diagnostic Service Center
1700 Fairlane Boulevard
Allen park, MI 48101, USA

[2]Department of Electrical and Computer Engineering
The University of Michigan-Dearborn
Dearborn, MI 48128-1491
USA
yilu@umich.edu

ABSTRACT

This paper describes an intelligent monitoring system, Medi-View, for the application to the Intensive Care Unit in hospitals. The system is served as a bedside monitor displaying the patient's vital signs from various devices and diagnosing arrhythmia's. One of the salient features of this program is that it takes the complexities of medicine and simplifies them into a user friendly interface. Medi-View provides the primary physician and nurses with easy access, immediate retrieval of accurate and current information, and on-line diagnosis.

1. INTRODUCTION

The hospital bedside monitors help process enormous amounts of patient data and display information in easy-to-understand formats. The data to be displayed on the monitors come from bedside medical devices, local area networks, hospital database systems and manual input. With the increasing data load in Intensive Care Unit(ICU), the need to improve acquisition, storage, integration and presentation of medical data has become of the utmost importance in present and future bedside monitors. The main problem of designing a data acquisition is the definition and configuration of the user interface that must allow the inexperienced user to interact with the computer intuitively. Emphasis must be put on the construction of a pleasant, logical and easy-to-handle graphical user interface[Das90].

The potential for a computer based system in medicine has already been realized in medical communities. In Spain bioengineers have designed *PATRICIA* for use in the ICU[Mor89]. *PATRICIA* monitors the gases in a patient's heart and uses this data to determine the need for mechanical ventilation. Two expert systems are currently being used in England. The first is *POEMS* (Post Operation Expert Medical System) which was commissioned in 1992 and used at St James University Hospital, Leeds UK. The *POEMS* detects irregularities in the patient's vital signs and then suggests candidate diagnoses for less experienced medical staff. The second assists in determining the need to admit persons suffering with chest pain into the Critical Care Unit(CCU), and is called *ACORN* (Admit to the Ccu OR Not). *ACORN* was commissioned in 1987 and is used by the Accident and Emergency Department of Westminster Hospital, UK.

However, most of the current monitoring systems used in the ICU are for storing information and has keyboard-entry interface[SHF92]. If a record of an electrocardiogram (EKG) is desired a hardcopy must be made. Test results, medication records, as well as the patient's medical history are kept in hardcopy form on a chart. Computers can be difficult to use for an individual not familiar with the computer's operation manual. For examples, the monitors at both Children's Hospital and St. Joseph's Hospital in Detroit both use function keys to interface with the medical staff. The medical staff must become quite familiar with these systems before they can comfortably and efficiently use these systems as an effective tool. In many urgent cases, while there are patients being taken care of, it is easy to forget the correct keys or punch wrong keys.

The greater demand to reduce the mortality and morbidity in ICU's has made intelligent monitoring systems the vanguard of bedside monitoring. This project is an attempt to incorporate an expert system into a

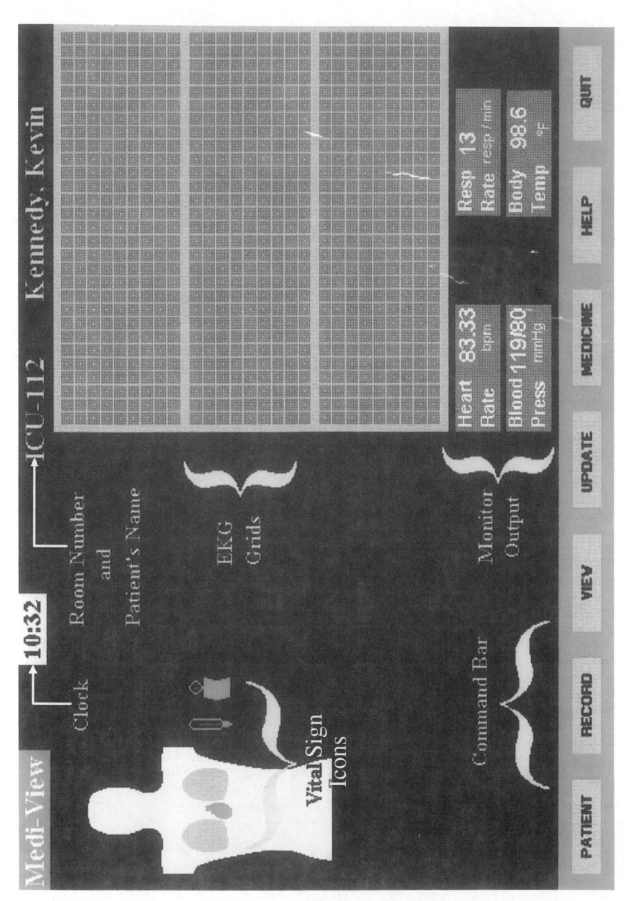

FIGURE 1 Main Screen of Medi-View

bedside monitoring system. Medi-View is a user friendly, graphics based monitoring system incorporating a "Rule Based Expert System" to diagnose arrhythmia in the Intensive Care Unit (ICU).

2. SYSTEM DESCRIPTION

The main screen of the Medi-View is shown in Figure 1. The overall system is shown in Figure 2. The system takes vital signs and hospital data base as input, and displays patient's personal information, vital signs, EKG and performs cardiac diagnosis by an expert system.

Medi-View's main screen displays patient's personal data, EKG and the vital signs including the heart rate, respiration rate, body temperature and blood pressure. In addition, the system stores drug tables for quick reference. As a complete system, Medi-View, while it performs the duties of a normal monitoring instrument, provides the advice of a Cardiologist by incorporating an expert system.

The main goals of Medi-View are to:
- design a user friendly diagnostic tool for the ICU staff using graphics interface, and
- provide the expertise of an onboard cardiologist.

2.1 Main Screen Configuration and graphics interface

The main screen is designed based on studies on human-computer interfaces. Many meetings took place between this design team and the medical professionals. Studies have shown that graphics interface is much easier to learn and to use[Das90, Imh91]. In particular for the bedside monitoring, graphics interface requires little training and causes fewer operator errors in urgent cases. Based on our meetings with the medical professionals, the graphics interface is much preferred to the computer function key driven interface. Therefore, Medi-View uses icons and pictures to interface with users. These icons allow the medical staff to point and click with the mouse to select the different commands, instead of using function keys. From interviews with ICU head nurses, insight was gained on the configuration of the main screen.

The graphics interface in Medi-View consists of mouse interaction with icon or pictures displayed on the main screen and the command bar. The screen shown in Figure 1 is displayed after the initial startup of the Medi-View program. This screen shows vital sign icons, the command bar, and the EKG grid. The torso of the human body as well as the vital sign icons are displayed to show. The EKG grid is created by plotting two sets of lines, one horizontal the other vertical. The command bar shown at the bottom of the main screen is a menu of icons on the bottom of the screen that allow the ICU staff to perform different tasks. From interviews with medical professionals, the menu has the following features:

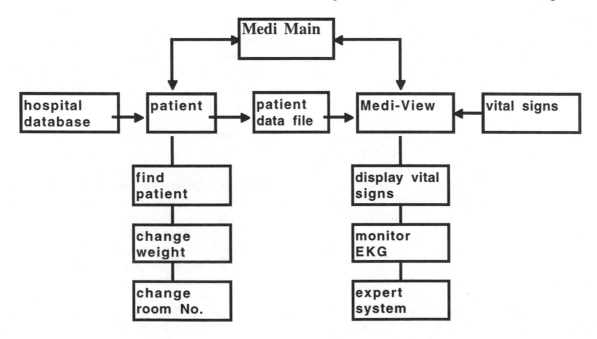

FIGURE 2. System block diagram of Medi-View

• "PATIENT" box allows the view of the candidate's personal information

• "MEDICINE" box allows a quick reference of Intravenous or "Code" drugs

• "HELP" box is an on-line manual to allow the staff to refresh on the meaning of each icon.

• "RECORD" box allows staff to record a patient's vital sign data.

• "UPDATE" box allows staff to update a patient's vital sign data and view this information

• The "QUIT" box allows the staff to exit from the diagnostic screen.

The information to be displayed under the command "PATIENT" is retrieved by implementing the *Patient Data Algorithm[Imh91]*. A data file includes the patient's name, medical ID number, weight, age, phone number, social security number, hospital ward, personal doctor, and that doctor's phone number. The patients name will be displayed on the main screen as well as their ID number. From this algorithm, the patient's weight is used as an output to the *Drug* algorithm to calculate the drip rate for intravenous drugs.

In the ICU, there are two categories of drugs, "Intravenous" drugs (IV) administered on a periodic basis and "Code" drugs administered during emergency procedures. By clicking the command bar on the "MEDICINE" box, information on these drugs can be viewed. The *Medicine Algorithm* [Fer91, Imh91] performs this function. For intravenous and "Code" medicine, a table of drugs is available on-line so the user can enter the drug name to retrieve the necessary information on that drug. Intravenous drugs need their drip rate (rate to administer the drug) calculated, therefor the *Medicine Algorithm* will gather the patient's weight from the *Patient Data Algorithm*, the concentration and dosage from a data file, and the name of the drug from the medical staff to complete this task. "Code" drugs have preset dosages at a set concentration, so these will appear for each drug viewed. Both tables will allow addition, change or deletion of drugs.

2.2 Intelligent cardiac system

The on-line intelligent cardiac system in Medi-View provides EKG display and the automatic diagnosis of abnormal cardiac cycles.

The cardiac cycle is monitored by up to three EKG "Leads". A "Lead" emulates the electric rhythm of the heart. Lead II is displayed on the top most EKG grid. If desired the middle EKG grid will display Lead I and the bottom grid will display Lead III (for simulation purposes). Lead II is the most important to monitor since the full cardiac cycle is derived from it. Thus for Medi-View, Lead II is automatically activated every time the heart is monitored. In order to ensure that only the current cardiac cycle is being displayed on the grid, a subroutine was created to overlay a single column of grid squares prior to plotting on that grid column. This method allows to maintain the continuity of the desired displaying without drastically slowing down other components of the system.

We have developed an expert system to assist the ICU staff to diagnose abnormal cardiac cycles. The expert system is driven by a *Cardiac Evaluation Algorithm*. The motivation for integrating such an expert system in the ICU is that it has been noted that considerable problems arise for medical staff who are relatively inexperienced. Deciding when to take a specific action based on monitoring signs and other available data, or distinguishing between major or minor changes of the monitoring signs and diagnostic expectations, are complex tasks. They are especially difficult when senior medical support is not readily available[Dub90]. This expert system will help alleviate the problem by detecting abnormal cardiac cycles (arrhythmia's) using the expert knowledge built in the system.

To help detect these abnormalities in the patient's EKG, Medi-View uses rule-based reasoning. The condition portion of the rules is used to evaluate the current cardiac cycle and the heart rate of the patient, and compare it with known arrhythmia conditions. Once a match occurs, the action portion can display the causes of the arrhythmia and the course of action for medical personnel to take. From the advise of the cardiologists, Medi-View uses Lead II of the EKG for diagnosing arrhythmia's.

To diagnose arrhythmia's, the "expert" system must detect and interpret each cardiac cycle with separate rules for detection and interpretation of each arrhythmia. Lead II is used because it provides the best depiction of the cardiac cycle. Medi-View begins evaluating a cardiac cycle from an initial "R" peak to a second "R" peak(see Figure 3). Medical personnel consider a cardiac cycle to begin with a "P" peak and continue to a second "P" peak. Medi-View redefines the cardiac cycle for two reasons, first is that the heart rate is measured by the periodicity of "R" peaks, therefor the heart rate calculation is simplified. Secondly, "P" peaks can be absent in certain arrhythmia's, making them a poor choice for defining the start of the cardiac

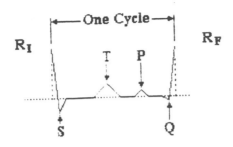

FIGURE 3. An example of cardiac cycle.

1.	A peak occurs when the sign of the slope changes. (i) P, R, and T have a <u>positive to negative</u> slope sign change. (ii) Q and S have a <u>negative to positive</u> slope sign change.
2	The magnitude of the peak defines the positive peaks (i) $\|R\| > 11$ mV (ii) $0 < \|P\| \le 6$ mV (iii) 6 mV $< \|T\| <= 11$ mV.
3.	Since the magnitudes of Q and S can be the same, the order of occurrence differentiates between them. (i) The first negative peak is S. (ii) The second negative peak is Q.

FIGURE 4. Peak detection rules.

1.	Premature Atrial Complex (i) No T peak (ii) P peak exists (iii) Heart Rate changes drastically (more than 10% of the normal rate).
2	Atrial Fibrillation (i) No P peak (ii) T peak exists (iii) Heart Rate > 100 Beats Per Minute.
3.	Ventricular Tachycardia (i) No T peak (ii) No P peak (iii) Width of "QRS" complex is wider than normal (iv) Heart Rate > 100 Beats Per Minute.

FIGURE 5. The Arrhythmia Rules

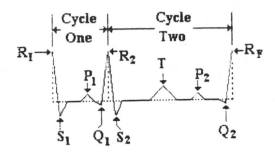

Premature Atrial Complex

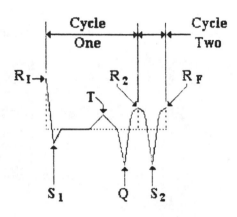

Atrial Fibrillation

Ventricular Tachycardia

FIGURE 6. Examples of three major arrhythmia's cycle

cycle. Since the cardiac cycle is a periodic wave, the beginning of each cycle (period) can be redefined with no elements of the cycle being lost. An example of a cardiac cycle is shown in Figure 3. The first stage of the expert system detects the peaks of the cardiac cycle. As each point is plotted on the EKG the magnitude of that point and the change in magnitude (slope) between the new point and the prior point is recorded. The rules of Peak Detection are shown in Figure 4.

This expert system uses the three most common arrhythmia's formulated from the interviews with the cardiologists. These include Ventricular Tachycardia, Atrial Fibrillation and Premature Atrial Complex. The "Arrhythmia Rules" for these are shown in Figure 5. Figure 6 shows examples of each arrhythmia's cycle.

When an arrhythmia is diagnosed, an audible beep is sounded to alert the medical staff to it's occurrence. Once a

key is depressed the beep stops and the therapy and possible causes are displayed. Again, the system is **not** meant to replace the cardiologist, instead it gives the staff the **knowledge of the "expert"** until a human expert can arrive on the scene.

3. CONCLUSION

We have presented an intelligent ICU system, Medi-View. The system provides a user-friendly graphics interface, intelligent data retrieval including drug references and the electrocardiogram (EKG) output, and incorporates an expert system in diagnosing cardiac arrhythmia in patients being treated in the ICU.

This system is implemented on DOS in the "C" programming language. The system has been tested on all.

the vitals. Specifically, Medi-View has gone through the following test and evaluation:

• the user interface was tested thoroughly for any defects.

• expert system test. The system was tested on seven data files. Four of these contained normal cardiac cycles with varying rates and varying vital signs. The other three contained the data for each type of arrhythmia. Each arrhythmia was correctly identified and diagnosed. A second test with the probabilities of each arrhythmia being 1 in 50 was used to evaluate the combined system. Again the arrhythmia's were correctly identified, and the audible beep sounded even if the user was calculating a drug dosage or any of the other command bar functions

• Medi-View was evaluated by a number of medical professionals at the Providence hospitals in Detroit, Michigan. As Jeannette Thai, an M.D., commented, "the system is really good. Most medical personnel are not familiar with the ins and outs of the computer, so the icons allow for a person to become very familiar with the system. The help file is an added tool in learning the system quickly. The expert system has a lot of potential."

In summary, Medi-View is a successful user-friendly intelligent system for ICU monitoring. Medi-View's user friendly graphics interface allow to reduce the time and cost in training staff members. The ability of intelligent data retrieval and cardiac diagnosis makes Medi-View more than a monitoring system. Lastly, Medi-View can be configured for different critical care units.

4. ACKNOWLEDGMENT

We are very grateful to our medical advisor Kathleen Kennedy, M.D., at the Providence hospitals in Detroit, Michigan. We would like to thank the nurses, bioengineers and cardiologists at the Providence hospitals with whom we had numerous meetings.

5. REFERENCES

[Das90] Dasta, JF, "Computers in Critical Care: Opportunities and Challenges," DICP, 24(11), (Nov.):1084-9, 1990.

[Dub90] Dubin MD, Dale. 1989. Rapid Interpretation of EKG's. 4th ed. Florida: COVER Publishing Co..

[Fer91] Ferri MD, Fred, Practical Guide to the Care of the Medical Patient. 2nd ed. New York: Mosby Year Book, 1991.

[Gil90] Gilpin, EA. and others. " Predicting 1-Year Outcome Following Acute Myocardial Infarction: Physicians versus Computers," Computers & Biomedical Research, 23(1), (Feb.):46-63, 1990.

[Imh92] Imhoff, M., "Acquisition of ICU Data: Concepts and Demas," International Journal of Clinical Monitoring & Computing, 9(4), (Dec.):229-37, 1992.

[Mor89] Moret-Bonitto, Vicente and others, The PATRICIA Project. IEE Engineering in Medicine and Biology, 12(4), (Dec.):59-68, 1989.

[SHF92] Schwaiger,J. and Haller, M. and Finstere, U., "A Framework for the Knowledge-Based Interpretation of Laboratory Data in Intensive Care Units Using Deductive Database Technology," *Proceedings - the Annual Symposium on Computer Applications in Medical Care*, 13-17, 1992.

Neural Network

COMPARISON OF TWO LEARNING NETWORKS FOR TIME SERIES PREDICTION

Daniel Nikovski

School of Computer Science
Carnegie Mellon University
Pittsburgh, Pennsylvania 15213, USA
Email: Daniel.Nikovski@cs.cmu.edu

Mehdi Zargham

Department of Computer Science
Southern Illinois University at Carbondale
Carbondale, Illinois 62901, USA
Email: mehdi@cs.siu.edu

ABSTRACT

Hierarchical mixtures of experts (HME) [JJ94] and radial basis function (RBF) networks [PG89] are two architectures that learn much faster than multilayer perceptrons. Their faster learning is due not to higher-order search mechanisms, but to restricting the hypothesis space of the learner by constraining some of the layers of the network to use linear processing units. It can be conjectured that since their hypothesis space is restricted in the same manner, the approximation abilities of the two networks should be similar, even though their computational mechanisms are different. An empirical verification of this conjecture is presented, based on the task of predicting a nonlinear chaotic time series generated by an infrared laser.

INTRODUCTION

The problem of predicting the continuation of a given time series is fundamentally a machine learning problem, since such a continuation requires the extraction by the prediction method of the rules that govern the dynamics of the time series. This extraction has to be done from examples of past values of the time series. The use of past values of the time series for reconstructing the dynamics of the underlying dynamic system is justified by Takens' theorem [WG94].

If the dynamic system that generates the time series is linear, two suitable forms for representing the inferred knowledge are regression coefficients and transfer functions (ARIMA models) [LS83]. However, when the underlying dynamic system is nonlinear, other representations are necessary. One possibility is to use locally linear models that store all past values of the time series and then produce a prediction based on locally linear interpolation [GW94]. While giving good predictions, such models are obviously useless when the time series is of infinite duration.

The other possibility is to use learning methods such as multilayer perceptrons (MLP) or other learning networks that concentrate the extracted knowledge in a small number of parameters, which are adjusted by on-line learning rules [GW94]. The requirements to such learning networks are that they should be universal function approximators, have good generalization power, be able to learn quickly from examples, and be implementable in hardware. The most popular learning networks, MLP, possess all of the above features except fast learning; convergence of MLP for nontrivial problems is in the order of hundreds of thousands of iterations and can be impractical for most real-time engineering applications even if it is accelerated by hardware.

The slow learning in MLP has motivated the development of alternative learning architectures that can learn faster. Two such learning models are radial basis functions [PG89] and hierarchical mixtures of experts [JJ94]. The two models are much faster than MLP because a significant part of their processing is linear, though in a different way. Both methods have been used for time series prediction — RBF by Casdagli [Cas89] and HME by Waterhouse and Robinson [WR95].

The approximation power and convergence time of each of the two methods have been compared with MLP so far [Cas89, JJ94], but on different test problems. It is thus difficult to see which of the two methods is better for time series prediction or any other nonlinear identification task, and it is of practical interest to compare them on a single test problem. Section 2 describes the chosen problem and the testing procedure. The architecture and performance on the test problem of RBF and HME are presented in sections 3 and 4 respectively. Section 5 provides a statistical comparison of the two methods, and section 6 discusses the results and concludes.

DESCRIPTION OF THE PROBLEM

Benchmarking of nonlinear time series prediction methods requires test problems that are representative of the problems commonly encountered in practice. Such nonlinear time series have been provided by the organizers of the Santa Fe Institute competition in time series prediction [GW94]. We used as test problem one of these time series, consisting of output of a CH_3 far-infrared laser. This nonlinear chaotic time series has been predicted successfully with a locally linear model by Sauer [Sau94] and with a time-delay neural network by Wan [Wan94]. Wan used an embedding dimension of 8; this value for the dimensionality of the embedding space was adopted in our experiments too, which means that our networks had 8 inputs each. The original data were measured as integers in the range from 0 to 255, which introduces quantization error of about 0.2%.

Based on the provided time series data, one training and nine testing data sets were prepared. The training set had 992 examples, and the testing sets had 1000 examples each. RBF and HME used the training set to estimate their parameters. The first of the testing sets was used for cross-validation - i.e., the estimated model that produced best out-of-sample error on the first testing set was assumed to be the optimal one for the respective architecture. The out-of-sample error of this model was tested on the remaining eight testing sets in order to compare the two learning methods. All errors reported in the paper are normalized root mean squared errors (NRMSE). For the purpose of comparison with linear prediction methods, a recursive least squares (RLS) autoregressive model was tested too [LS83].

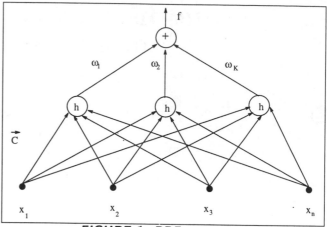

FIGURE 1. RBF network

PREDICTION WITH RADIAL BASIS FUNCTIONS

Radial Basis Functions (RBF) have been known in approximation theory as a powerful and computationally efficient method for interpolation and approximation of functions [PG89]. With the development of the connectionist approach to computation, it was noticed that the computation performed by these functions can be carried out by parallel and distributed processing elements similar to multilayer perceptrons [PG89].

An RBF network consists of a number of units whose output is a nonlinear function h of the unit's input (Fig. 1).

The output of the network y is given by

$$y = f(\mathbf{x}) = \sum_{\alpha=1}^{K} \omega_\alpha h(\| \mathbf{x} - C_\alpha \|)$$

where $\mathbf{x}$ is the vector input to the network, ω_α are the weights that should be estimated, C_α are vectors called knots, and $\| \cdot \|$ is the L^2 norm on the space of input vectors. In our experiments the knots C_α were chosen to be a subset of the training examples; other possible approaches include various unsupervised clustering schemes or supervised adjustment of the knots. The function h belongs to the class of radial basis functions [PG89]. For our experiments, the linear function $h(r) = r$ was used.

The estimation of the weights ω_α is a generalized linear squares problem and is solvable in time $O(K^3)$ in the number of knots K. A plot of the in- and out-of-sample NRMSE versus the number of knots is given in Fig. 2 for the problem of laser output prediction.

The in-sample error decreases gradually to zero, as expected. The out-of-sample error follows closely, but

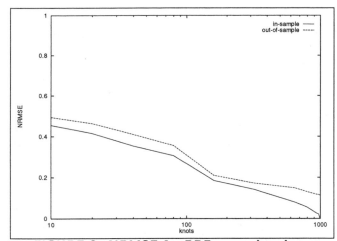

FIGURE 2. NRMSE for RBF approximation

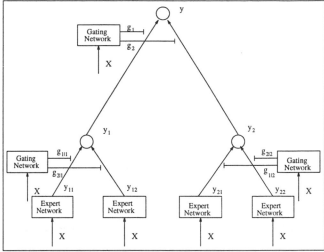

FIGURE 3. HME architecture

flattens to 0.10 when the number of knots approaches the number of training examples. Some overfitting occurs; the lowest out-of-sample error is attained for 990 knots, 2 less the number of training examples. Consequently, the optimal model from cross-validation with the first testing set is the model with 990 knots.

This model was tested on the remaining 8 testing sets, resulting in average out-of-sample error $\bar{E}_{RBF} = 0.10212$.

PREDICTION WITH HIERARCHICAL MIXTURES OF EXPERTS

The HME architecture is a general nonlinear function approximation model introduced by Jordan and Jacobs [JJ94]. In particular, it can be used for nonlinear regression in time series prediction. The architecture consists of a tree-like hierarchy of gating nodes and a set of experts at the leaves of the tree (Fig. 3).

If HME is to be used for nonlinear regression, each of the experts forms a linear prediction $\hat{y}_{ij}$ of the true output of the network y for a given input vector $\mathbf{x}$:

$$\hat{y}_{ij} = \mathbf{u}_{ij}^T \mathbf{x}$$

The purpose of the gating nodes is to assign weights to the predictions $\hat{y}_{ij}$ of the individual experts - these weights should be proportional to the precision with which an expert approximates the function at a particular location. For this purpose, the weights depend on the input $\mathbf{x}$ too. For the case of binary trees, shown in Fig. 3, two values g_1 and g_2 are output by the gate at the root of the tree, with their sum equal to 1:

$$g_i = \frac{e^{\xi_i}}{\sum_k e^{\xi_k}}, \ i = 1, 2, \ k = 1, 2$$

$$\xi_i = \mathbf{v}_i^T \mathbf{x}$$

Here the vector $\mathbf{v}_i$ parametrizes the gating expert. The predicted output $\hat{y}$ is then a weighted average of the predictions at the lower level $\hat{y}_i$:

$$\hat{y} = \sum_{i=1}^{2} g_i \hat{y}_i$$

The weights of the gating node at the root of the tree perform a partitioning of the input space into two parts, in each of which one of the weights is close to 1, while the other weight is close to 0. In this way a "soft split" is formed, whose steepness and orientation are determined by the vector $\mathbf{v}_1 - \mathbf{v}_2$. The greater the length of this vector, the sharper the split is. The splitting continues recursively on the lower levels of the tree.

The predictions at the lower level are formed in a similar way, this time using directly the predictions of the experts:

$$\hat{y}_i = \sum_{j=1}^{2} g_{j|i} \hat{y}_{ij}$$

Jordan and Jacobs [JJ94] derived learning rules for the experts and the gates on the basis of a probability model and a maximum-likelihood approach. The probability distribution of the output of a particular linear expert is given by

$$P(y|\mathbf{x}, \mathbf{u}_{ij}) = \frac{1}{\sqrt{2\pi}} e^{-\frac{(y-\hat{y}_{ij})^2}{2}}$$

By differentiating the likelihood that the observed output y is generated by the HME model with a partic-

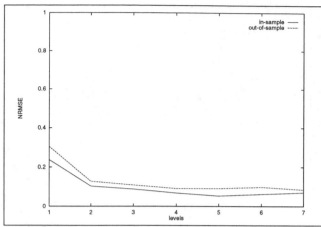

FIGURE 4. NRMSE of HME approximation

ular set of parameters θ, the following on-line gradient ascent learning rules result:

$$\Delta\mathbf{u}_{ij} = \rho h_i h_{j|i}(y - \hat{y}_{ij})\mathbf{x}$$

$$\Delta\mathbf{v}_i = \rho(h_i - g_i)\mathbf{x}$$

$$\Delta\mathbf{v}_{ij} = \rho h_i(h_{j|i} - g_{j|i})\mathbf{x}$$

where ρ is a learning rate and the quantities h_i, $h_{j|i}$ and h_{ij} are posterior probabilities defined as follows:

$$h_i = \frac{g_i \sum_j g_{j|i} P_{ij}(y|\mathbf{x},\theta)}{\sum_i g_i \sum_j g_{j|i} P_{ij}(y|\mathbf{x},\theta)}$$

$$h_{j|i} = \frac{g_{j|i} P_{ij}(y|\mathbf{x},\theta)}{\sum_j g_{j|i} P_{ij}(y|\mathbf{x},\theta)}$$

$$h_{ij} = \frac{g_i g_{j|i} P_{ij}(y|\mathbf{x},\theta)}{\sum_i g_i \sum_j g_{j|i} P_{ij}(y|\mathbf{x},\theta)}$$

Jordan and Jacobs have also derived another set of learning rules, based on the Expectation-Maximization (EM) developed in statistics [JJ94]. The EM learning rules learn faster than the ones based on the maximum-likelihood approach.

In the HME architecture, the number of estimated parameters and hence the achieved generalization depend on the number of levels in the hierarchy. For the laser test problem, the number of levels was varied and the in-sample and out-of-sample errors were plotted, similarly to the testing with RBF networks. The results are shown in Fig. 4. The tree was binary and the learning rate ρ was set to 0.01. For each model, 10000 iterations were performed.

The least out-of-sample error was attained for a seven-level network (128 experts). This network was used for comparison on the remaining 8 testing sets. Notice that this was not the network with smallest in-sample error.

COMPARISON OF THE TWO LEARNING MODELS

The relative performance of RLS and the best RBF and HME models is shown in Table 1.

While the superiority of HME and RBF over RLS is obvious, it is not clear how the RBF and HME networks compare with each other. To this end a paired two-tailed t-test was performed on the 8 testing sets to verify the hypothesis that one of the two networks is superior. The results are shown in Table 2.

Table 2 shows that there is no statistical significance to the hypothesis that RBF and HME have different performance on the learning task at hand.

CONCLUSIONS

The experiments presented in the previous sections demonstrate that the HME and RBF architectures possess comparable approximation abilities on a non-linear identification task. While it might be difficult to show analytically such similarity in the general case, the presented empirical results suggest that the two learning networks consider similar hypothesis spaces (or, in other words, impose similar smoothness constraints on their approximations.) This observation can be used to infer expectations about the performance on a given task of one of the networks, knowing the approximation error of the other. One application might be to train fast an RBF network and if

Table 1. NRMSE of RLS, RBF, and HME on the laser output time series.

	RLS	RBF	HME
In-sample	0.43309	0.00247	0.07028
Out-of-sample ($\bar{E}$)	0.44586	0.10212	0.10414

Table 2. Comparison of RBF, HME, and RLS: t-statistic and attained significance level p.

	t	p
RBF vs. HME	0.355286	0.440666
RBF vs. RLS	42.07407	$< 10^{-11}$
HME vs. RLS	42.73772	$< 10^{-11}$

the performance is satisfactory, proceed with the more time-consuming training of a HME network, whose approximation is piecewise linear and because of that can be more readily used for prediction and control applications.

REFERENCES

[Cas89] Casdagli, M. (1989). Nonlinear prediction of chaotic time series. *Physica D*, vol. 35, 335-356.

[GW94] Gershenfeld, N.A., and A.S. Weigend (1994). The future of time series: learning and understanding. In Weigend, A.S, and N.A. Gershenfeld (Eds.) (1994) *Time Series Prediction*. Reading, MA: Addison-Wesley, 1-70.

[JJ94] Jordan, M.I., and Jacobs, R.A. (1994). Hierarchical mixtures of experts and the EM algorithm. *Neural Computation*, vol. 6, 181-214.

[LS83] Ljung, L., and T. Söderström (1983). *Theory and Practice of Recursive Identification*. Cambridge, MA: MIT Press.

[PG89] Poggio, T., and F. Girosi (1989). *A Theory of Networks for Approximation and Learning*. AI Memo 1140, MIT AI Lab.

[Sau94] Sauer, T. (1994). Time series prediction by using delay coordinate embedding. In Weigend, A.S, and N.A. Gershenfeld (Eds.) (1994) *Time Series Prediction*. Reading, MA: Addison-Wesley, 175-193.

[Wan94] Wan, E.A.(1994). Time series prediction by using a connectionist network with internal delay lines. In Weigend, A.S, and N.A. Gershenfeld (Eds.) (1994) *Time Series Prediction*. Reading, MA: Addison-Wesley, 195-217.

[WR95] Waterhouse, S.R., and A.J. Robinson (1995). Nonlinear prediction of acoustic vectors using hierarchical mixtures of experts, in G. Tesauro, D.S. Touretzky, and T.K. Leen, (Eds.), *Neural Information Processing Systems 7*, Cambridge, MA: MIT Press.

[WG94] Weigend, A.S, and N.A. Gershenfeld (Eds.) (1994) *Time Series Prediction*. Reading, MA: Addison-Wesley.

AN EVOLUTIONARY ALGORITHM FOR THE GENERATION OF UNSUPERVISED SELF ORGANISING NEURAL NETS.

Tim Hendtlass,

Centre for Intelligent Systems, School of Biophysical Sciences and Electrical Engineering,Swinburne University of Technology,John Street, Hawthorn 3122, Victoria, Australia. Email: tim@bsee.swin.edu.au

ABSTRACT

An evolutionary algorithm is described that allows the speedy generation of unsupervised self organising maps. The maps produced are memory efficient in that they use almost the minimum number of nodes required to hold a specified level of detail about the training set. This level of detail can be explicitly controlled. Compared to conventional SOM techniques, the one described is faster and produces nets with far smaller numbers of nodes. The algorithm, although described here in terms of generating three dimensional nets, is applicable to the generation of nets of any desired dimensionality.

1 INTRODUCTION.

Evolutionary algorithms have been used for some time to assist in the selection of parameters that specify supervised artificial neural networks [ASP94, BBM92, CC93, Co92, PH95a]. These parameters can include the number of nodes, how these nodes are interconnected and the learning rates to be used.

Typically, a number of networks (a population) with randomly chosen parameters are generated. These are then trained by conventional techniques (for example back propagation) before being sorted into order of performance. A new generation is produced of new individual networks, each developed by combining parts from each of two networks from the current generation (the parents). The better a network performs in the current generation, the more it is involved in the breeding process. The number of networks in the new generation need not be the same as the number in the current generation. The new generation becomes the current generation and the individual networks are trained by conventional techniques, sorted into order and used to produce the next generation. This process continues until the best performing network in a generation meets or exceeds some performance criteria. It is the bias towards the better performing individual networks being parents that drives successive generations towards improved performance.

Refinements have been added to this basic technique, such as permitting the best performing few networks from one generation to be copied intact into the next generation with an aging penalty to combat elitism [PH95b] or adding a collective memory that minimises the tendency to re-explore old ground [PH96].

A traditional self organising map (Kohonen network) consists a number of nodes, usually in a regular array. [See for example De88, Ko88, WM76]. Each node is specified by its position and the information (exemplar) stored within it. As an input is presented, every node in the net calculates how similar its stored exemplar is to the input. The node whose exemplar is closest is called the winner and it, and its physical neighbours, are updated by making their exemplars more like the input. To try to spread information more uniformly across the net, a conscience is used. Nodes that are winning more frequently than average are penalised (have their similarity scores reduced), while nodes winning too infrequently are given slight assistance.

Application of evolutionary algorithms to developing unsupervised artificial networks is hampered by one basic problem. Since the network is unsupervised there is no way of knowing how well each node is doing over the whole data set and so no way to sort them into performance order. However, the basic concept of the evolutionary algorithm is still applicable and this paper describes how it has been implemented for a self organising map with a dynamically varying structure. The user is spared from having to decide how many nodes should be used in the self organising map and where these should be placed.

2 EVOLVING AN UNSUPERVISED NET.

A number of changes are required to the basic evolutionary algorithm described above to allow breeding unsupervised neural networks.

Firstly, one considers generations of nodes rather than generations of networks.

Secondly, the regular birth of whole new generations is replaced by the periodic birth of a new node. Similarly the wholesale death of the members of a generation when a new generation is born is replaced by the death of individual nodes as necessary. The number of nodes in a generation is not fixed.

Finally, the breeding process is changed. Unsupervised networks consist of nodes that are described by small packets of information, called chromosomes. In breeding two of these are crossed over; one (or more) parts from one parent's chromosomes are combined with one (or more) parts from the other parent's chromosome by a cut and paste process. Alternatively, the parent's chromosomes are averaged to produce a new chromosome. The new chromosome often has the same structure as each of the two parents but containing a mix of the information from the parents. Slight random changes to the values in the chromosome are permitted to occur (mutation) so as to introduce new material not present in either parent. This new chromosome describes a child node. In self organising map networks the chromosome contains two types of information, the positions of the node and the value of the node exemplar. Each child node has two types of parents rather than just two parents. The first type of parent has only one member, the current input to the SOM, which defines the exemplar of the child node. The second type of parent has one more member than the dimensionality of the self organising map being developed (four parents for a three dimensional self organising map such as that described in this paper). The position of the new node is derived from the positions of these parents.

With breeding no longer taking place at regular intervals (that is after each of the current generation has been scored), it is necessary to devise triggers that cause birth events and death events. These are derived from the frequency with which nodes win as the training examples are presented to the population of nodes. If a single node is winning far too frequently, a birth occurs. Similarly if a node never wins, it dies. Birth and death perform a similar function to that of a conscience in a conventional self organising map.

The net consists of a number of nodes in a three dimensional space. Each node contains one exemplar, a composite of the inputs that this node is currently representing. This exemplar has the same dimensionality as the input. During the evolution of the net the number of nodes will vary, the exemplars in the nodes will alter, and the nodes move in space so that their spatial position is related to their exemplar and therefore the group of inputs they are representing. Initially the size of the net will be small but will grow as the net becomes a more detailed representation of the input information. Once trained, information about an input example can be obtained by noting the positions of the nodes with the maximum responses.

3 THE EVOLUTIONARY CYCLE.

The basic net training cycle consists of presenting and scoring the current input, rewarding the best performing node, calculating the position of this input in the net space and moving all nodes so that their distance from the input's position is more compatible with their response to the input. It is also possible that a new node may be added to the net (birth), two very similar nodes merged together (marriage) or a node no longer performing any useful function in the net removed (death). Each of these steps is described in more detail below.

3.1 Scoring the net.

The responses of all the nodes in the network to the current input are calculated in turn. The response of the nth node will be referred to as out[n]. Considering the input and the exemplar stored in node K as two n dimensional vectors ($\underline{i}$ and $\underline{k}$) with an angle of theta between them, the score for a node is the product of the cosine of theta and the ratio of the shorter length vector to the longer length vector. This reduces to:

out[k] = ($\underline{i}.\underline{k}$) / (longer length)^2

The maximum possible response is one for a perfect match, when $\underline{i}$ and $\underline{k}$ are identical, and the minimum -1 when $\underline{i}$ and $\underline{k}$ are exactly opposites. After the responses of all the nodes has been calculated, the node with the highest response is the winner [win1]. The second best responding node is [win2] and so on.

3.2 Rewarding the winner.

The winning node is rewarded by having its activity increased by:

(1-out[win1])*abs((out[win1]-out[win2])/out[win1])

Note that the first factor ensures that the poorer the winning nodes response, the larger the reward. A perfect match results in no reward. The second factor especially rewards the winning node when it's response is far better than any other nodes or moderates the response if more than one node is almost equally representing this input. A high activity is taken as an indication that a node is trying to represent too diverse a range of inputs. A birth follows (see below), producing a new node to share this representation.

3.3 Updating the winner.

The winner's exemplar is updated by combining the existing exemplar with the current input. The amount of updating done is controlled by a learning parameter, L1:

$\underline{k}_{new}$ = L1 * $\underline{i}$ + (1-L1) * $\underline{k}_{old}$

In this way the exemplar in time becomes a composite of

the inputs that this node is representing (for which it has the highest output).

3.4   Estabishing the position of the current input in the net.

The position that the current input should occupy in 3 dimensional space is calculated from the responses of the four best responding nodes. The calculation assumes that the response of these nodes are linearly related to the distance between the input and the node in the three dimensional map.

The process of finding the ideal position of the current input is illustrated in figure 1 below. Let the desired distance from the winning node (win1) be D1, the desired distance from the second node (win2) be D2 and so on. In step 1 the desired distance from win1 is used to define a sphere. The position sought must lie somewhere on this surface. Step 2 shows the circle of points that are also the correct distance from the second node. Next two points on this circle are identified that most closely meet the desired distance to win3. Finally the point whose distance from win4 is closest to the D4 is selected. The point chosen will be at exactly the correct distance relative to the winner, whose output is presumably the most meaningful, but only the best approximation to the second, third and fourth placed nodes in turn — whose outputs are probably in descending order of meaning.

3.5   Moving the nodes.

Once the position of the input is established, the positions of all other nodes in the net are adjusted to be in better accord with their responses to this input. Let CD be the current distance between the input position in 3 dimensional space and the position of node N. The ideal target distance (TD) between them is linearly related to the difference in outputs of the input ($\underline{i}$ with an 'output' of unity) and of node N (exemplar $\underline{n}$). By trigonometry this is:

$$TD = \mathrm{Sqrt}(|\underline{i}|^2 + |\underline{n}|^2 - 2 * |\underline{i}| * |\underline{n}| * \cos \Theta)$$

where Θ is the angle between the input vector $\underline{i}$ and node N exemplar $\underline{n}$. The node N is moved along the line joining its present position and the position of the input to a new position:

$$ND = CD + (TD - CD) * L1 * \mathrm{out}[\mathrm{win1}] * (1 / (1 + CD)).$$

The actual amount it is moved depends:

- on the factor L1 used to update the winners exemplar
- on the output of the winning node (the higher this is the more faith we may have in the position we calculated for the input and so the more position adjustment we can reasonably make)

inversely on the actual current distance from the input to node N. The further this distance the less reliable the estimate of target position and thus the less corrective movement we should make.

3.6 Birth.

If the accumulated activity of the winning node is above a threshold (L2), a new node is placed at the input's position in the net with the current input values as its exemplar. The activity of this new node and the winning node are set to zero. Over the totality of inputs the placing of new nodes and the movement of existing ones produces an ordered position dependent response across the net to the range of inputs in the training set.

3.7 Death.

One further parameter is kept for each node, the number of input example since this node was last the

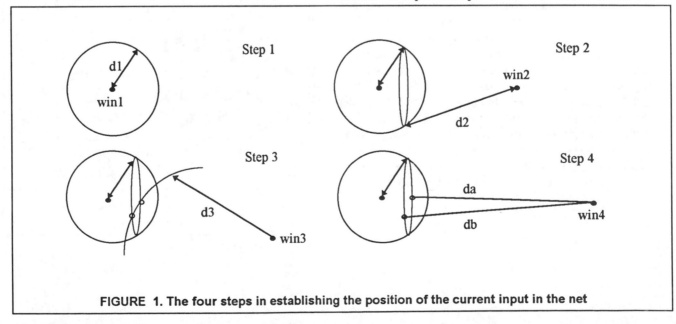

FIGURE 1. The four steps in establishing the position of the current input in the net

winner. This number, the idle time, is incremented every example this node doesn't win and zeroed when it wins. If the idle time ever exceeds the number of examples in the data set, this node has outlived its usefulness and it is removed from the net.

3.8 Marriage.

Finally it is possible that two nodes may try to represent the same subset of inputs. In this case they will move closer together. If two nodes are within a user specified capture range (L3) and have exemplars that are sufficiently similar (as measured by the response to the current input), they are married by being averaged. The idle time of the new node is set to the minimum of the idle times of the nodes married and the activity of the new node is set to zero. The requirement that the responses of the two nodes be very similar is to prevent marriage of two dissimilar nodes that happen to move close while passing each other on their way to their final (different) positions.

3.9 Stability.

Two mechanisms ensure the long term stability of a fully trained network. A node which exactly represents its input examples will receive no increase to its activity when these are presented to the network. As a result it will not parent any new nodes nor will it be moved. This ideal situation is rare when mapping real life data.

If a node is the best, but not exact, single representation of a number different inputs it will continue to receive small rewards and will in time parent a new node. This new node will be very close to the old node and move closer as inputs are presented. As a result the two nodes will soon (if not immediately) be married together and start the process of accumulating activity again. This type of dynamic equilibrium is normal with real life data and the marriage capture distance L3 sets the solution detail. The smaller L3, the more detailed the representation in 3 dimensional space. L1 controls how fast the nodes move to their final positions and the noisier the data, the lower its value should be. L2 controls how fast the net reaches its final size. If it is set too low an excess of nodes may be bred. Although these will eventually be removed, this may take time.

The description of the net above could readily be altered to provide mapping into a different dimensional space.

The main difference would be in the placing of the current input in the M dimensional space. This always requires the positions and response values of the M+1 best responding nodes.

4. RESULTS.

The approach described above has been used to map a number of data sets. Simple sets consisting of a number of equi- presented randomly chosen vectors are quickly mapped in one to two passes through the data set. Replicating and adding random noise to the vectors so as to produce sets of vectors each derived from one of the original vectors causes no problem until the noise is large enough to cause overlap in the derived sets. At this stage the action of the user variable L3 can be clearly seen, it controls the detail of the mapping, the lower the value of L3 the more detailed (the larger) the network produced. When vectors are unevenly replicated, so that one or more appear infrequently in the training set, it can take more passes through the training set before these infrequent examples are identified and represented in the network. However, for a training set with 20 vectors each replicated 20 times (with random noise added) and 5 vectors replicated only twice (also with random noise), all of the 25 vector groups were identified within ten passes. An increase in the marriage rate to match the birth rate is a good indication of a fully trained net.

FIGURE 2 A sample of the Frey and Slate character set

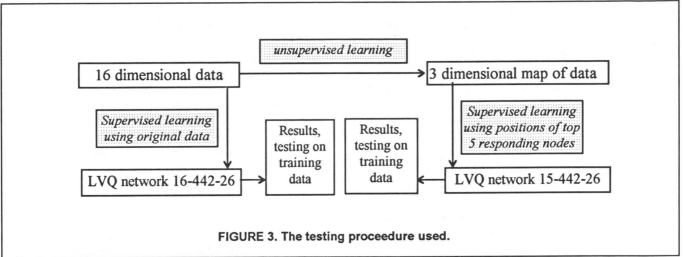

FIGURE 3. The testing proceedure used.

A more detailed test was conducted using Frey and Slate's [FS91] character recognition data, a small sample of which is shown in figure 2 above. This data set consists of the 26 uppercase letters of the English alphabet in each of 20 different fonts, providing a total of 520 classes. Each of these have been subject to distortions and noise and then preprocessed as described in [FS91] to reduce each to a set of 16 numbers. It is apparent that the noise level of some of these examples is high.

A Linear Vector Quantisation (LVQ) network [De88, Ko88b] with 16 inputs, 442 Kohonen nodes and 26 outputs is able to learn the last 4000 members of this set and correctly classify them to an accuracy of 91.7%. As a test of the performance of the 3 dimensional net development algorithm, this same data was used to produce a 3 dimensional net. After a given number of passes through the data set the positions of the top five maximum responses for each example was recorded. These were used to train an LVQ network with 15 inputs (the x, y and z coordinates of each of the top five responding nodes), 442 Kohonen nodes and 26 outputs. This testing proceedure is shown diagramatically in figure 3 above. The top five nodes were used so that the number of inputs to the two LVQ networks was as similar as possible.

Despite being mapped from 16 dimensions into three dimensions, the 3 dimensional net preserved the information in the original data well enough that the LVQ network was able to correctly identify the input in a substantial number of cases. The percentage correct after four passes of the network are shown in below, the network is an LVQ (15-442-26) with L1=0.1, L2=1, L3=0.01

Passes through data set	1	2	3	4
% correct	59.3%	62.7%	65.3%	71.2%
3D net size	132	206	270	321

Two different factors may explain the roughly 20% drop from 91.7% to 71.3% in recognition accuracy. Firstly, although there is in all probability some redundancy in the original 16 dimensional data, it must be expect that some information would be lost mapping into only three dimensions. Secondly, the net was not yet fully trained, the birth rate being higher than the combined marriage and death rates. Processing could not be continued as the net reached the maximum size able to be supported under the current version of the software.

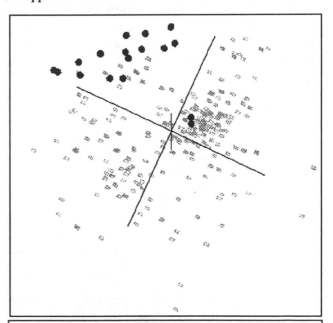

Figure 4. A plot of the network trained using the reduced data set. The seventeen nodes that respond to the letter V have been drawn as large spheres, all other nodes are smaller cubes. The predominant group of V nodes are clearly visible at the upper left. Less obvious is the subgroup of two nodes just to the right of the origin. The nodes towards the center of the large group respond to more examples of V than those at the edges of the group.

Further testing was therefore carried out on a limited data set, all examples of nine letters being removed from the original data set. The letters removed (A,F,I,L,O,P,T,U,W) were the letters the partially trained network had best identified. Using this reduced data set, now with 2604 examples, to generate a 3 dimension map, a stable net of 273 nodes was produced after 25 passes through the data set. The net size stayed basically constant from then on and is show in figure 4 above. An LVQ network with 15 inputs, 340 Kohonen nodes and 17 outputs was able to correctly classify 79% of the examples. For comparison, an LVQ network with 16 inputs, 340 Kohonen nodes and 17 outputs working from the reduced original data set, was correctly able to classify 92.8%.

5 CONCLUSION.

The modified evolutionary algorithm described has been shown to be able to generate unsupervised self organising maps. The process does not require many passes through the data and the resulting net has almost no passenger nodes. When tested on the training data set almost every node is the best responding node to one or more examples. Such a network is therefore quite a memory efficient way of holding information. Although described in terms of a three dimensional self organising map, the process is adaptable to produce a self organising map with any number of dimensions.

6 REFERENCES.

[ASP94] Peter J. Angeline, Gregory M. Sanders, Jordan B. Pollack, An Evolutionary Algorithm that Constructs Recurrent Neural Networks, *IEEE Transactions on Neural Networks*, vol. 5, no.1, January 1994.

[BBM92] Frank Z. Brill, Donald E. Brown, Worthy N. Martin, Fast Genetic Selection of Features for Neural Network Classifiers, *IEEE Transactions on Neural Networks*, Vol. 3, No. 2, March 1992.

[CC93] C.H. Chu, C.R. Chow, A Genetic Algorithm Approach to Supervised Learning for Multilayered Networks, *World Conference on Neural Networks*, vol. 4, pages 744-747, 1993.

[Co92] *International Workshop on Combinations of Genetic Algorithms and Neural Networks*, COGANN-92, IEEE Computer Society Press, California, 1992.

[De88] DeSieno D. Adding a conscience to competitive learning. *Proc. Int. Conference on Neural Networks*, Volume 1 pp117-124. IEEE Press New York. 1988.

[FS91] Peter W. Frey, David J. Slate, Letter Recognition Using Holland-Style Adaptive Classifiers, *Machine Learning*, 6, 161-182, 1991.

[Ko88a] Kohonen *T Self-Organization and Associative Memory*. Second Edition. Springer-Verlag New York 1988.

[Ko88b] Kohonen T et al Statistical Pattern Recognition with Neural Networks: Benchmark Studies. *Proceedings of the Second Annual IEEE Conference on Neural Networks*, Volume 1. IEEE Press New York 1988.

[PH95a] John R. Podlena, Tim Hendtlass, A Modified Genetic Algorithm Applied to Neural Network Design, Proc. *The Sixth Australian Conference on Neural Networks*, ACNN95, February 1995.

[PH95b] John R. Podlena, Tim Hendtlass, Evolving Complex Neural Networks That Age, *Proc. ICEC'95*, November 1995.

[PH96] John R Podlena, Tim Hendtlass, Using the Baldwin Effect to Accelerate a Genetic Algorithm, *Proc. IEA96AIE*, Japan, June 1996.

[WM76] Willshaw D J and von der Malsburg C. How Patterned Neural Connections Can be Set Up by Self-organization. *Proc. R. Soc. London B* Volume 194 pp 431-445 1976.

EXPLOITING DON'T-CARE INFORMATION IN NEURAL NETWORK LEARNING

Chung-Yao Wen

Department of Electronics Engineering

Hwa-Hsia College of Technology and Commerce

Taipei, Taiwan

Email: cywen@fuzzy.ee.hwh.edu.tw

ABSTRACT

In this paper, we present a novel neural network architecture called *M-net*, which exploits the don't-care information in training multilayer feedforward neural networks. Our method takes advantage of the user's prior knowledge as well as the neural network's ability to learn from examples. The user's prior knowledge is encoded in the form of don't-care inputs to reduce the number of training patterns required to represent a function. We derive the learning rule of *M-net* in the context of error backpropagation, and demonstrate its use on the priority decoding problem. Compared with conventional backpropagation networks, *M-net* drastically reduces the learning time while achieving superior quality.

1. INTRODUCTION

Neural networks are usually used as a "universal" learning device, where no prior knowledge is required and learning is based solely on examples. However, many of the real world problems exhibit structures that can be described concisely in certain forms. Taking advantage of such knowledge can greatly improve the efficiency and the quality of learning.

In this paper, we present a method for learning from both examples and rules in multilayer feedforward neural networks. Our rules take the form of **don't-care inputs** (a term borrowed from the computer aided design community). Although a training pattern containing don't-care inputs is functionally equivalent to a possibly large set of patterns without don't-care inputs, it takes only about 2 times as much computation to complete a learning cycle. We call the new network architecture *M-net*, where the letter *M* stands for "margins", implying intervals of tolerance or uncertainty.

The organization of the paper is as follows. Section 2 formally describes the framework of learning and the network architecture, and the learning algorithm. Section 3 gives the performance of *M-net* on the priority decoding problem. Section 4 describes related work on exploiting don't-care information. Section 5 concludes the paper.

2. THE LEARNING PROBLEM

In this section, we present *M-net*, our method of learning in the presence of don't care inputs. After a brief summary of notations, we give an overview of the conventional learning problem and the backpropagation algorithm. We then describe the new learning problem, and present the *M-net* learning algorithm, which is an adaptation of the backpropagation algorithm.

2.1. Notations

A training pattern is an ordered pair of input and output vectors. For convenience of discussion, we discriminate between training patterns with don't-care inputs, denoted by $[\pi^p, \rho^p]$ for the pattern p, and those without don't-care inputs, denoted by $[\xi^q, \zeta^q]$ for the pattern q. We associate with π^p the set Ψ^p, which contains all ξ such that ξ is the result of some instantiation of the don't-care inputs in π_p. For example, let x denote the don't-care input in the binary training pattern $\pi = 1x$, then the set Ψ for π is simply $\{10, 11\}$. All vector quantities are subscripted to denote a particular element in the vector.

Each neuron in the multilayer feedforward network is given a unique integer index as its name, and the output for neuron i is o_i. Define O as the set of neuron indices in the output layer, and I the set of neuron indices in the input layer. Let $w_{j,k}$ denote the weight of the connection from neuron k to neuron j, where neuron k is one layer below neuron j. Let o^q denote the output vector of the neurons given the input pattern ξ^q.

2.2. Conventional Learning Problem and Backpropagation

The conventional learning problem (i.e., one that does not allow don't-care inputs) uses the following error function:

$$E = 0.5 \times \sum_{\forall q, \{\forall i \in O\}} (\zeta_i^q - o_i^q)^2$$

The output o_i^q is

$$o_i^q = \tanh(h_i^q)$$

$$h_i^q = \sum_j w_{i,j} o_j^q$$

Using backpropagation, the weight update rule is

$$\Delta w_{i,j} = -\eta \frac{\partial E}{\partial w_{i,j}} = \eta \sum_q \delta_i^q o_j^q,$$

where

$$\delta_i^q = \tanh'(h_i^q)(\zeta_i^q - o_i^q)$$

if $i \in O$, otherwise

$$\delta_i^q = \tanh'(h_i^q) \left(\sum_k w_{k,i} \delta_k^q \right)$$

2.3. *M-net* Learning – Back-propagation With Don't-care Inputs

We now present the new learning problem and the *M-net* architecture. *M-net* is a multilayer feedforward network, with enhanced neurons and weight connections that propagate output and error "margins" (hence the letter *M*), as oppose to crisp outputs and errors in standard neural networks. The *M-net* learning algorithm is an extension of backpropagation for dealing with don't-care inputs.

For convenience of discussion, we assume all training patterns have some don't-care inputs. This does not affect the generality of our work, since all fully specified patterns can be represented by don't-care patterns with singleton Ψ sets. The training pattern $[\pi^p, \rho^p]$ is functionally equivalent to the set of patterns

$$\{ [\xi^r, \rho^p] \mid \xi^r \in \Psi^p \}$$

Under the standard frame work of learning, the error function translates to

$$E = \sum_{\forall p} E^p$$

$$E^p = 0.5 \times \sum_{\xi^r \in \Psi^p, i \in O} (\rho_i^p - o_i^r)^2 \qquad (1)$$

A global minimizer for (1) would map all patterns in Ψ^p to ρ^p, which is exactly what we want. However, the size of Ψ^p is exponential in the number of don't-care inputs in π^p, and computing (1) directly would be too costly. An alternative error function that achieves the same mapping is:

$$E^p = 0.5 \times \sum_{i \in O} \left(\left(\rho_i^p - \min_{\xi^r \in \Psi^p} o_i^r \right)^2 + \left(\rho_i^p - \max_{\xi^r \in \Psi^p} o_i^r \right)^2 \right) \qquad (2)$$

However, computing (2) directly is still computationally expensive. To eliminate the combinatorial explosion in the min and max terms, we settle for an approximation to (2) as follows. Let the interval $V_{min}...V_{max}$ be the range of all possible input and output values. For each input neuron i connected to a don't-care input of the pattern π^p, we have

$$\min_{\xi^r \in \Psi^p} o_i^r = V_{min}$$

$$\max_{\xi^r \in \Psi^p} o_i^r = V_{max}$$

For each input neuron i connected to the fully specified input k of the pattern π^p, we have

$$\min_{\xi^r \in \Psi^p} o_i^r = \max_{\xi^r \in \Psi^p} o_i^r = \pi_k^p$$

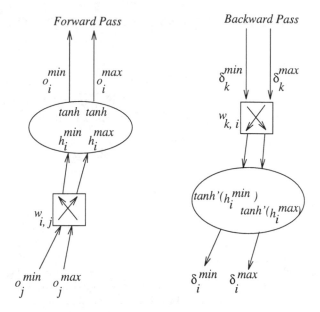

Figure 1. The information propagated during the forward and the backward passes. Note that each connection now acts as a binary switch controlled by the sign of its weight.

For each non-input neuron i, we bound its output value (conservatively) with respect to the pattern π^p as follows:

$$\min_{\xi^r \in \Psi^p} o_i^r \geq \tanh \left(\sum_{w_{i,j} \geq 0} w_{i,j} \min_{\xi^r \in \Psi^p} o_j^r + \sum_{w_{i,j} < 0} w_{i,j} \max_{\xi^r \in \Psi^p} o_j^r \right)$$

$$\max_{\xi^r \in \Psi^p} o_i^r \leq \tanh \left(\sum_{w_{i,j} \geq 0} w_{i,j} \max_{\xi^r \in \Psi^p} o_j^r + \sum_{w_{i,j} < 0} w_{i,j} \min_{\xi^r \in \Psi^p} o_j^r \right)$$

The above formulae can be applied recursively from the output layer to the input layer to obtain the concrete bounds for all neurons. The bounds are conservative because we consider each neuron output in isolation, while the output bounds of the neurons are in fact dependent. The conservative bounds eliminate the combinatorial explosion problem and allow the same type of learning as in standard backpropagation – the forward and backward passes can be performed locally on the individual neurons, even for patterns with don't-care inputs.

In summary, our learning method is different from standard backpropagation in the following aspects:

- During the forward pass, each neuron i propagates two values o_i^{min} and o_i^{max}, which denote the minimum and the maximum outputs of the neuron (conservative bounds) given a particular training pattern. The min and the max outputs are calculated by the input functions h_i^{min} and h_i^{max} respectively.

- During the backward pass, each neuron i propagates two values δ_i^{min} and δ_i^{max}, which denote the error derivatives with respect to o_i^{min} and o_i^{max} respectively.

The modified network architecture is shown in Figure 1. The weight update rule is derived from the error derivatives as in backpropagation. The only complication comes from the input functions, which may not have a derivative when the weight is 0. In this case we take the derivative as if the weight is positive [1]. Since the weight updates have greater impacts on the error derivatives than in standard backpropagation, we adopt the incremental learning approach, meaning that the weights are updated for each training pattern in a learning cycle. We observe significant performance improvements in our implementation if incremental update is used instead of batch update.

2.4. Summary of *M-net* Learning

The forward and the backward passes for *M-net* are summarized below.

For the input neuron i ($i \in I$) that is connected to a don't-care input of the pattern π^p,

$$o_i^{min,p} = V_{min}$$

$$o_i^{max,p} = V_{max}$$

For the input neuron i that is connected to the fully specified input k of the pattern π^p,

$$o_i^{min,p} = o_i^{max,p} = \pi_k^p$$

The error function is

$$E = \sum_{\forall p} E^p$$

$$E^p = 0.5 \times \sum_{i \in O} \left((\rho_i^p - o_i^{min,p})^2 + (\rho_i^p - o_i^{max,p})^2 \right)$$

The output of the neuron i (i not in I) for the pattern π^p is

$$o_i^{min,p} = \tanh\left(h_i^{min,p}\right)$$
$$= \tanh\left(\sum_{w_{i,j} \geq 0} w_{i,j} o_j^{min,p} + \sum_{w_{i,j} < 0} w_{i,j} o_j^{max,p} \right)$$

$$o_i^{max,p} = \tanh\left(h_i^{max,p}\right)$$
$$= \tanh\left(\sum_{w_{i,j} \geq 0} w_{i,j} o_j^{max,p} + \sum_{w_{i,j} < 0} w_{i,j} o_j^{min,p} \right)$$

Given the pattern π^p, the weight update rule is

$$\Delta w_{i,j}^p = \eta \left(\delta_i^{min,p} o_j^{min,p} + \delta_i^{max,p} o_j^{max,p} \right)$$

where

$$\delta_i^{min,p} = \tanh'(h_i^{min,p}) \left(\rho_i^p - o_i^{min,p} \right)$$

---

[1] Alternatively, we could use a continuous function that approximates the binary switching function

$$\delta_i^{max,p} = \tanh'(h_i^{max,p}) \left(\rho_i^p - o_i^{max,p} \right)$$

if $i \in O$, otherwise

$$\delta_i^{min,p} = \tanh'(h_i^{min,p}) \left(\sum_{w_{k,i} \geq 0} w_{k,i} \delta_k^{min,p} + \sum_{w_{k,i} < 0} w_{k,i} \delta_k^{max,p} \right)$$

$$\delta_i^{max,p} = \tanh'(h_i^{max,p}) \left(\sum_{w_{k,i} \geq 0} w_{k,i} \delta_k^{max,p} + \sum_{w_{k,i} < 0} w_{k,i} \delta_k^{min,p} \right)$$

After the neural network is trained, it can be used in a production environment where the input vectors are fully specified, in which case either the min or the max output value can be used (since they are identical). A salient feature of the *M-net* architecture is that it can be used even when the environment supplies **unknown** or **imprecise** inputs. In both cases, the range of the input (i.e., its minimum and maximum values) is fed to the network, and the average of the min and the max output values can be used. The performance of *M-net* in such environments is beyond the scope of this paper.

3. PERFORMANCE

In this section, we compare *M-net* against standard backpropagation. Although backpropagation is slow compared with more advanced learning methods, our performance study is still valid because *M-net* is not limited to backpropagation. We can use any error-driven learning method in *M-net* while keeping the same architecture and performance benefits.

The performance of the learning methods varies with the problem, and we do not claim the example we use is universal in any sense. However, it does show the advantages of our method. The example we use is a 16-bit priority decoder, which can be fully defined by a set of rules shown on Figure 2. [2]:

The priority decoder is learned by a 2-layer neural network with 32 hidden neurons, using at most 5000 cycles. We use incremental weight updates for *M-net* and batch weight updates for backpropagation (because we notice backprogagation performs better with batch updates on this problem). The learning rate, the momentum term and the cross-validation process are identical for both methods. *M-net* uses exactly the 17 training patterns shown above, while backpropagation uses randomly generated training patterns [3]. The quality of learning, or the error of the network over the input space, is estimated by the average error per pattern over 10000 random patterns. The learning statistics for *M-net* and backpropagation is summarized in Figure 3.

The error of backpropagation is much larger than *M-net*. The total time for learning, which is estimated by the number of learning cycles times the size of the training set, is

---

[2] The 1's and 0's shown are logical values. In our implementation, the floating point values 1.0 and -1.0 are used for the logical values 1 and 0, respectively.

[3] The randomization assumes that for each input, the binary values 1 and 0 are equally likely.

```
f(1,x,x,x,x,x,x,x,x,x,x,x,x,x,x,x)=(1,0,0,0,0,0,0,0,0,0,0,0,0,0,0,0)
f(0,1,x,x,x,x,x,x,x,x,x,x,x,x,x,x)=(0,1,0,0,0,0,0,0,0,0,0,0,0,0,0,0)
f(0,0,1,x,x,x,x,x,x,x,x,x,x,x,x,x)=(0,0,1,0,0,0,0,0,0,0,0,0,0,0,0,0)
f(0,0,0,1,x,x,x,x,x,x,x,x,x,x,x,x)=(0,0,0,1,0,0,0,0,0,0,0,0,0,0,0,0)
f(0,0,0,0,1,x,x,x,x,x,x,x,x,x,x,x)=(0,0,0,0,1,0,0,0,0,0,0,0,0,0,0,0)
f(0,0,0,0,0,1,x,x,x,x,x,x,x,x,x,x)=(0,0,0,0,0,1,0,0,0,0,0,0,0,0,0,0)
f(0,0,0,0,0,0,1,x,x,x,x,x,x,x,x,x)=(0,0,0,0,0,0,1,0,0,0,0,0,0,0,0,0)
f(0,0,0,0,0,0,0,1,x,x,x,x,x,x,x,x)=(0,0,0,0,0,0,0,1,0,0,0,0,0,0,0,0)
f(0,0,0,0,0,0,0,0,1,x,x,x,x,x,x,x)=(0,0,0,0,0,0,0,0,1,0,0,0,0,0,0,0)
f(0,0,0,0,0,0,0,0,0,1,x,x,x,x,x,x)=(0,0,0,0,0,0,0,0,0,1,0,0,0,0,0,0)
f(0,0,0,0,0,0,0,0,0,0,1,x,x,x,x,x)=(0,0,0,0,0,0,0,0,0,0,1,0,0,0,0,0)
f(0,0,0,0,0,0,0,0,0,0,0,1,x,x,x,x)=(0,0,0,0,0,0,0,0,0,0,0,1,0,0,0,0)
f(0,0,0,0,0,0,0,0,0,0,0,0,1,x,x,x)=(0,0,0,0,0,0,0,0,0,0,0,0,1,0,0,0)
f(0,0,0,0,0,0,0,0,0,0,0,0,0,1,x,x)=(0,0,0,0,0,0,0,0,0,0,0,0,0,1,0,0)
f(0,0,0,0,0,0,0,0,0,0,0,0,0,0,1,x)=(0,0,0,0,0,0,0,0,0,0,0,0,0,0,1,0)
f(0,0,0,0,0,0,0,0,0,0,0,0,0,0,0,1)=(0,0,0,0,0,0,0,0,0,0,0,0,0,0,0,1,0)
f(0,0,0,0,0,0,0,0,0,0,0,0,0,0,0,0)=(0,0,0,0,0,0,0,0,0,0,0,0,0,0,0,0,1)
```

Figure 2. Rules base of a 16-bit decoder

Method	M-17	BP-100	BP-1200	BP-2400	BP-4800	BP-9600
Error	0.0013	0.4159	0.2402	0.0988	0.0744	0.0653
Convergence	1.4118	0.0000	0.0000	0.0000	0.0002	0.0018
Cycles	5000	600	5000	5000	5000	5000

Figure 3. Summary of learning. M denotes *M-net*, and BP denotes backpropagation. The numbers following M or BP denote the sizes of the training sets. Convergence is estimated by the final error per pattern on the training set.

also much larger for backpropagation. For example, the time is 85000 for *M-net* , and 48 million for backpropagation using 9600 training patterns. Recall that a *M-net* learning cycle is only 2 times as expensive as a backpropagation learning cycle, which means that *M-net* reduces the learning time by a factor of 280, while achieving even higher quality.

However, we notice that *M-net* has difficulties converging to a small error on the training set. Two factors may contribute to the convergence problem – the non-smoothness of the neuron input functions (at 0 weights), and the conservative error approximation. The former is alleviated using incremental weight updates, while the latter is an intended trade-off of accuracy for speed. However, learning with *M-net* has much better predictive power than backpropagation given the same training error.

4.   RELATED WORK

Drucker [Dru90] proposes an ad hoc architecture for handling don't-care and don't-know attributes in ordnance identification. He uses a special weight setting method for learning, in which special input tokens are used to represent don't-care and don't-know attributes. Lee and Hsu [LH91] investigate the handling of don't-care inputs in the standard framework of learning by expanding the training sets in different ways. Based on the empirical result from a sample problem, they conclude that it is best to replace the don't-care inputs by the minimum and the maximum values, or by the full range of values. Both of these solutions suffer the combinatorial explosion problem.

5.   CONCLUSIONS

In this paper, we present a new network architecture called *M-net*, which exploits don't-care inputs in the training set. *M-net* uses a conventional multilayer feedforward network, and its learning proceeds in a distributed manner as in conventional backpropagation. We apply *M-net* on the priority decoding problem, and shows that it achieves higher quality with less learning time than standard backpropagation. Future work includes mapping *M-net* to learning rules other than back-propagation, and extending the idea to handling don't-care output terms.

REFERENCES

[Dru90] H. Drucker. Implementation of a neural net expert system in the presence of don't care and don't know features. In *International Symposium on Circuits and Systems*, New Orleans, 1990.

[LH91]   Mahn-Ming Lee and Ching-Chi Hsu. The handling of don't care attributes. In *International Joint Conference on Neural Networks*, Singapore, 1991.

Extended Tree based Regression Neural Networks for Multifeature Split

Sook Lim[*], **Sung Chun Kim**[**]

* Department of Computer Engineering, YoSu National Fisheries
University, YoSu, ChonNam 550-749, Korea
E-mail : limsook@yosu.ynfu.ac.kr
** Department of Computer Science, Sogang University Marpo,
Seoul 121-742, Korea

Abstract

In this paper, we propose the extended tree based regression neural networks to multifeature split that preserve the structure of regression trees. We also suggest a supervised learning algorithm that uses the information of the regression trees as the initial network state. The mapped neural network has the better performance than the regression trees. The competitive learning scheme in the learning algorithm has an effect to prune the over-grown tree without degrading the learning performance. We apply our system to the skeletal age prediction problem in radiology.

1. INTRODUCTION

The general hetero-association tasks are identified as classification problem in pattern recognition or regression problem in regression analysis[BFO+84]. In classification, the issue is to decide the class category of an observation input data, while the task of regression is to predict the value of response variable. There exist tree structured approaches for classification and regression. These tree structured approaches have been appealed many researchers since its white box nature gives an easy human understanding. One of the main tree design issues is to find an optimal size tree with the best generalization to unseen data. However, these tree approaches have a limitation that once a wrong path is taken at an internal node, there is no way of recovering from the mistake. The layered neural networks can avert this wrong choice because of adaptability beyond first hidden layer that allows some corrective actions.

In artificial neural networks, each application task requires its own network architecture, so that the architecture has to be designed with appropriate configuration topology. Therefore, it is necessary to find algorithms that optimize not only the weights for a given architecture, but also the number of layers and the number of neurons per layer.

Combining the faster learning of tree structured method with the adaptibility such as soft nonlinearity of layered neural networks can be founded in [Set90,SY90]. In early approaches, case of single feature split regression tree, it has been dealt with only parallel hyperplane on feature axis, after weight adjustment. There exists a limitation of small enhancement in performance, since connections in first hidden layer are considered with only one input variable and bias node.

In this paper, we construct the extended tree based regression neural networks for multifeature split, yielding the optimal boundaries with hyperpolyhedron shape, not having the only hyperbox parallel to each axis. In our method, it gives an effect of extension from single feature split regression to multifeature split regression through the full connection between all input variables and all the neurons in first hidden layer. The competitive learning scheme in the learning algorithm can effectively prune the over-grown tree without degrading the learning performance. We apply our system to the skeletal age prediction problem in radiology.

2. BACKGROUND

Given input-output mapping examples, we try to find the optimal number of disjoint subset with the homogeneous property that captures input space grouping and output space grouping. In pattern classification problem, the output domain is usually unordered, discrete set of labels of pattern categories. The output space grouping is represented by the majority class and the complexity measure such as entropy gives a numerical quality of homogeneity of the group[SG82]. In regression problem, the output domain is real-valued continuous subset of real numbers. The output grouping is represented by the mean of output values of patterns belonging to each subset and the complexity measure is the variance.

2.1 Regression trees

In regression problem, the aim of process is to predict the value of the dependent variable based on a given input data. The regression tree, the sibling of decision tree, is one of nonparametric approaches to regression problems. Instead of using partitioning measures such as the mutual information gain that are suitable for classification tasks, the least square regression error is used to build the regression tree[BFO+84]. In each stage of tree growing, threshold value for split is determined by minimizing the least square regression error due to a partitioning. This error reduction $r[t_j]$ is given as follows

$$r[t_j] = \frac{1}{N}\sum_i (y_0^i - \bar{y}_0)^2 - \left(\frac{1}{N_1}\sum_i (y_1^i - \bar{y}_1)^2 + \frac{1}{N_2}\sum_i (y_2^i - \bar{y}_2)^2\right)$$

where t_j is a threshold value on the j-th feature

axis, N, N_1 and N_2 are respectively the numbers of data in the current node, the left and the right child nodes. y_0^i, y_1^i and y_2^i are response variables with mean $\bar{y}_0$ taken over the N data in current node, with means $\bar{y}_1$ and $\bar{y}_2$ taken over the N_1, N_2, respectively.

The tree growing algorithm terminates when the reduction of the squared regression error reaches the specified amount. Each terminal node of the regression tree yields the mean of output values from training data. FIGURE 1(a) presents an example of a regression tree along with the corresponding regression surface for a two dimensional problem.

Drawback of regression tree is that the determined threshold value does not correspond to the optimal boundary for dividing of input space in each stage, another drawback is that a flat and rough regression surface equivalent to histogram, as shown FIGURE 1(b), is not appropriate to the continuous output domain.

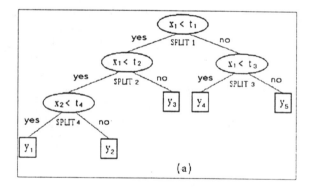

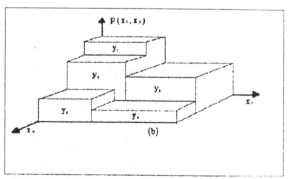

FIGURE 1. a) Example of regression tree for a two dimensional problem on (x_1, x_2).
b) The regression surface corresponding to a), where $P(x_1, x_2)$ is a approximator, and predicted output variable y_i is a constant.

2.2 Neural network mappings of tree structured methods

Feedforward neural network has been considered to be a universal function approximator. However, the accompanying backpropagation algorithm has the slow convergence behavior. Radial basis function network gives another way for the functional approximation problems[MD89,Spe91]. Usually, k-mean clustering algorithm gives the center and size of each radial basis function and output estimate for each cluster is learned by LMS learning rule.

Regression tree has been the basis for the construction of many neural networks and fuzzy inference systems. Sethi and Yu applied regression tree to entropy net. And they attached more neurons in the third layer of mapped entropy net in order to enhance the output prediction performance[SY90]. Prager combined CART(classification and regression trees) with CMAC(cerebellar model arithmetic computer)[Pra94,Alb75]. Parger enlarged the hyperboxes of terminal nodes to form overlapping regions so that multiple terminal nodes are activated to a given input pattern. Lim and Kim suggested fuzzy inference systems based on regression trees[LK95]. Lim and Kim recalculated the center and variance of each hyperbox to get more accurate information on training patterns belonging to each terminal node and mapped to triangular shape fuzzy associate memory bank.

The significant issue in the construction of layered neural networks is to decide of the appropriate network size for given problem. The network size is one of important parameters that can affect the learning time and the network performance. Pruning unimportant nodes and removing unnecessary connections can be one of solutions for this network configuration problem[SD88]. Also, there exists a method that begins with a small network and gradually grows to one of the appropriate sizes [Fre90]. Rather than growing a network by adding one neuron at a time, the entropy nets approach relies on a mapping that converts from decision tree to layered, partially connected neural networks[Set90]. Similarly, the entropy net approach is applied to build from regression tree to layered partially connected neural networks[SY90]. Mapping rules are as follows. The first hidden layer of the layered networks corresponds to the internal nodes of the tree. The second hidden layer neurons represent the terminal nodes of the tree. The final layer consists of distinct classes in entropy net model, while final layer in regression model consists of a set of neurons that represent the response variable values that are evenly distributed over output range. It is possible to map decision trees or regression trees automatically into a layered network of neurons in according to these rules. In these approaches, it defines exactly the number of neurons needed in each of the layers of neural network. Thus, mapped networks have fewer connections in comparison to standard layered neural networks that are fully connected in the first and the second layer.

3. EXTENDED REGRESSION NEURAL NETWORK FOR MULTI FEATURE SPLIT

3. 1 Network architecture

Looking at a regression tree, a threshold value t for a split function is midpoint of nearest points, n_1 and n_2, of adjacent partitions. Thus hyperbox does not represent the actual data distribution belonging to it except that sample data set is uniformly distributed. This is illustrated in FIGURE 2. Here, m_1 and m_2 are mean values of each partitions.

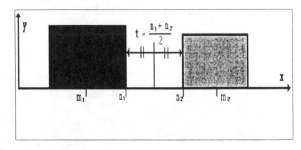

FIGURE 2. Property of threshold value t

For better accuracy, more elaborated adjustment scheme has been employed in layered neural network which is mapped from regression tree[Set90, SY90]. Considered components in learning phase on layered neural network are threshold value and corresponding input variable at each stage of tree growing. This approach has the limitation that adjusted result is the hyperbox which is parallel on feature axis, as before.

We suggest layered neural network, refer to MRNN(multifeature split regression neural network), which all input variables are fully connected with neurons of first hidden layer. The most important consequence of learning on MRNN is that obtained boundaries are hyperpolyhedron in shape which represents the boundaries of close to the actual distribution of data, not just parallel hyperbox on axis. FIGURE 3 presents the MRNN of previous FIGURE 1. For example, from the root node of FIGURE 1(a), input variable x_1 is connected to neuron assigned to $w_1 x_1 + w_0 \geq 0$ with weight value $w_1 = 1.0$ and bias input is linked with threshold value $w_0 = t_1$.

Compared to Sethi and Yu, our network architecture has more naturally extended tree structure[SY90]. Sethi and Yu store information in the third layer which is populated with multiple neurons and accompanying links, whereas we store information in the first layer by extending only connection links.

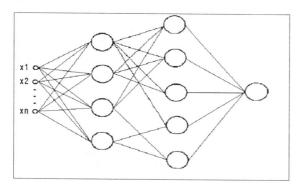

FIGURE 3. MRNN. First hidden layer neurons correspond to the internal nodes and second hidden layer corresponds to the terminal nodes of regression trees, in FIGURE 1(a). Third layer is attached to second hidden layer for predicted output variable.

3. 2 Network training

We train only the first layer of the networks to preserve the gray box nature of the multifeature regression tree. The initial setup of weights and the other configuration of the networks are given from the regression tree results using the training data. The resulting multifeature split tree structured neural networks are trained by the backpropagation algorithm. The winner – take – all strategy defines the errors of the second layer using hyperbolic tangent as follows.

error =   1.0 – activation for the winner neuron

error = – 1.0 – activation for the loser neurons.

Here, the winner neuron is decided to have the highest value in output space using Gaussian basis function.

The Gaussian basis function is defined in the output space as follows.

$$g_c(y_p) = \exp\left[-\frac{(y_p - \bar{y}_c)^2}{2\sigma^2_{y_c}}\right]$$

Here, y_p is the output value of given training datum, and $\bar{y}_c$ and $\sigma^2_{y_c}$ are the mean and the variance of output values of traing data belong to a neuron in the second layers. The mean and the variance of output values of winning neuron can be adapted as follows.

$$\sigma^2_{wn}(\text{new}) = \sigma^2_{wn}(\text{old})(1-\alpha) + (y_p - \bar{y}_{wn}(\text{old}))^2\alpha$$

$$\bar{y}_{wn}(\text{new}) = \bar{y}_{wn}(\text{old})(1-\alpha) + y_p\alpha$$

Here, α is $1/(n+1)$ and n is the ratio of number of training samples divided by power of 2 to the number of fan_in connection links to winning neuron.

Output of MRNN is a weighted average of mean output, $\bar{y}_c$ values of the second layer neurons as follows.

$$\hat{y} = \frac{\sum\limits_c w_c \, \bar{y}_c}{\sum\limits_c w_c}$$

Here, weights are the activations of the second layer neurons normalized to 0,1 range. Normalized activations are cut–offed at 0.1 level. If all activation are less than cut_off value, two highly activated neurons are considered in the weighted average.

3.3 Performance evaluation

By removing the constraint of single feature split regression tree that any separating hyperplane is to be orthogonal to the selected feature coordinate axis, the proposed networks can adapt to learn the more precise partitions of input space of given problem. Thus we expect the proposed networks to show a better performance than the original regression tree.

The performance of a regression system can be measured by the mean square error(MSE) between the output estimate of a regression system($\hat{y}$) and the given reference output(y_i) over all sample data.

$$MSE = \left(\frac{1}{n} \sum_{i=1}^{n} (\hat{y} - y_i)^2 \right)^{1/2}$$

Here, n is the number of samples.

The experiment was performed using the bone age data set. The problem is to predict the child skeletal age from the X ray image of hand[BG93]. The radiologist requires many years of experience to interpret X ray image of hand and to give a proper treatment to a child of abnormal growth. The radiologist uses the physical measurements obtained from the X ray image, and the sex and race information in order to assess the bone age. The bone age data set consists of four input features obtained from the measurement of hand X ray image and one output feature-the predicted bone age decided by the radiologist. There are 532 examples. We use the first 400 examples as the training set and the last 132 examples as the testing set.

The results of the experiment are shown in Table I and II. Here, RT stands for the single feature split regression tree, SRNN for single feature split tree structured regression neural networks and MRNN for multifeature split tree structured regression neural networks, proposed method. Table I shows performance comparision to original regression trees and its further learning schemes, SRNN and MRNN. Each entry of Table I denotes the MSE error as defined above, i.e, means disagreement between the radiologist and system prediction value. MRNN gives a better performance in all case. Each entry of Table II shows the number of clusters after learning. Entries for SRNN and MRNN show the survival neurons in the second layers that win at least one training data. The entries for the last two rows means that tree prunings are done for overgrowing trees. This implies that our neural net learning algorithm reconstructs the neural network topological

structure towards a reduced sized, optimal configuration without degrading the performance.

Table I. Regression Errors By Root Mean Square For Experiment.

% Variance reduction to (user specified)	RT	SRNN	MRNN
20 (training)	1.620	1.910	1.664
(testing)	1.855	1.657	1.521
10	1.195	1.833	1.780
	1.801	1.609	1.590
5	0.843	1.847	1.620
	2.023	1.806	1.657

Table II. Number of Clusters For Experiment.

% Variance reduction to (user specified)	RT	SRNN	MRNN
20	5	5	5
10	15	8	8
5	47	15	15

4. CONCLUSION

In this paper, we have presented the 3-layered feedforward neural networks that preserve the structure of the regression tree and the supervised learning algorithm that uses the information of the regression tree results as initial network configuration. Performance of multifeature based regression network is better than regression tree and single feature based regression network by yielding the optimal boundaries with hyperpolyhedron shape, not just having the hyperbox. Our competitive learning scheme has an effect pruning the over-grown tree without degrading the learning performance.

5. REFERENCE

[BFO+84] L. Brieman, J. Frieman, R. Olshen, and C.J. Stone, *Classification and Regression Trees*, Wadsworth

Books, Belmont Calif., 1984.

[Set90] I.K. Sethi, "Entropy nets: From decision tree to neural networks," proc. IEEE, Vol. 78, pp. 1605–1613, 1990.

[SY90] I.K. Sethi and G. Yu, "A Neural Network Approach to Robot localization using Ultrasonic Sensors," proc. of 5th IEEE International Symposium on Intelligent Control, pp. 513–517, 1990.

[SG82] I.K. Sethi and G.P.R. Sarvarayudu, "Hierarchical classifier design using mutual information," IEEE Trans., PAMI, Vol. 4, pp. 441–445, July, 1982.

[MD89] T.J. Moddy and C.J. Darken, "Fast learning in networks of locally tuned processing units," Neural Computation, Vol.1, pp. 151–160, 1989.

[Spe91] D.F. Specht, "A general regression neural network," IEEE Trans. Neural Networks, Vol. 2, No. 6, pp. 568–576, 1991.

[Pra94] R.W. Prager, "CART / CMAC Hybrid: Regression trees with Interpolation," proc. of 12th IAPR International Conference on Pattern Recognition, Vol. II, Conf. B: Pattern Recognition and Neural Networks, pp. 476–478, 1994.

[Alb75] J.S. Albus, "A new approach to manipulator control : cerebellar model articulation controller (CMAC)," ASME, Trans. G : Journal of Dynamic Systems, Measurements and Control, pp. 220–227, 1975.

[LK95] S. Lim and S.C. Kim, "Design of fuzzy inference systems using regression trees," proc. of IEEE International Conference on Neural Networks, Perth, Australia, vol.1, pp. 345–348, 1995.

[SD88] J. Sietsma and R.J.F.Dow, "Neural Net Pruning-Why and How," proc. of IEEE International Conference on Neural Networks, San Diego, vol. 1, pp. 325–333, 1988.

[Fre90] M. Frean, "The Upstart Algorithm: A Method for Constructing and Training Feedforward Neural Networks," Neural Computation 2, pp. 198–209, 1990.

[BG93] J. M. Boone and G. W. Gross, "Skeletal age evaluation from hand radiographs using neural networks," proc. of World Congress on Neural Networks '93, Vol. 1, pp. 190–193, 1993.

MIXING STOCHASTIC EXPLORATION AND PROPAGATION FOR PRUNING NEURAL NETWORKS.

Eric Fimbel
Departamento de Electricidad
UNEXPO, Puerto Ordaz
Calle China, Villa Asia, Altavista, Puerto Ordaz, Venezuela.
E-Mail efimbel@conicit.ve - Fax (58) 86 62 87 79

ABSTRACT.

Any neural network may contain **redundant** connections, whose contribution to the dynamics of the network is low. In the extreme case, a connection is **useless** when it can be removed without changing anything.

This paper gives a formal definition of useless connections and presents a new method for pruning neural networks based upon alternative steps of stochastic exploration and information propagation. The method works on any network with increasing, discrete activation functions.

It has been implemented and compared with an exploratory pruning algorithm. The stochastic-propagative method is robust and always converges quickly; on the other hand, the exploratory algorithm explodes in about 10% of the test cases.

The method has been applied to continuous activation functions, previously discretized. In this case, the error is equivalent to a noise in the states of the nodes. Experimentally, the convergence time grows up linearly with the precision of digitalization.

1. INTRODUCTION.

Any neural network may contain **useless** connections, as a result of learning algorithms or stochastic distribution of weights. Pruning such connections is convenient for several reasons. First of all, the pruned network is **more compact;** it uses less memory and run time algorithms are more efficient. Then, when pruned, the **hidden structure** of a network appears clearly.

Still, the detection of useless connections is a **highly combinatorial process**. In order to prove that a connection Wij is useless, we have to prove that for any combination of entries of node j, the elimination of Wij does not change the state of j.

Due to some interesting properties, proven in [FIM'95], the utility of connection Wij gives information about the utility of other connections Wkj. **Propagation** of the property can be done in order to reduce the size of the search space. Still, the combinatory remains very high.

This fact leads us to the use of **stochastic techniques** such as simulated annealing[CHU92] to complete the propagation, instead of classical exploratory methods.

Our stochastic-propagative method uses alternative steps of propagation and stochastic exploration to determine gradually all the useful connections. The remaining connections are useless.

The method can prune **any kind of network** with **increasing, discrete** activation functions**. Continuous** activation functions, previously discretized, can be processed, too. In this case, the effect of the pruning is equivalent to a digitalization noise on the entries of the nodes, which is mostly harmless [CHO90,HER90].

The pruning of a network is used to get **optimized versions** of a trained network, ready to run on low performance computers or controllers, for instance in distributed control systems.

Another application of pruning, which really motivated this work, is the simplification of a network in order to study its **possible limit cycles** [FIM95].

This paper is organized as follows:

-the **first part** gives an intuitive definition of useless connections. Important points are discussed (redundant vs. useless, effects of pruning, other techniques, etc.).

-In the **second part,** the concepts used by the pruning method are formally defined. Several theorems about utility are given.

-The **third part** gives an overview of the algorithm, and presents experimental results. A version of the algorithm adapted to the elimination of useless nodes (instead of connections) is given.

1.1 Useless connections.

A connection between nodes i and j is useless if a change in the state of i can never change the state of j. For instance:

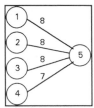

FIGURE 1. A network with useless connections.

The network of Fig.1 uses bipolar equations:

$$S(i,t+1) = \text{if } E(i,t) >= 0 \text{ , } 1 \text{ else } -1 \quad (1)$$

$$E(i,t) = \sum_j W_{ji} * S(j,t)$$

In this network, the connection 4-5 is useless: you can see that node 4 can never change the decision of nodes 1,2,3.

There is no straightforward relationship between the utility of a connection and its weight. For instance:

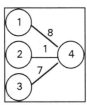

FIGURE 2. A network without useless connections.

The network of Fig.2 uses bipolar equations (1). Every connection is useful, even 2-4: node 2 can change the decision of nodes 1 and 3 in some cases (S1=-1, S3=1).

The following example is a simplified representation of a Purkinje cell of the cortex, inhibited by a single basket cell, with excitatory connections from other cortical columns:

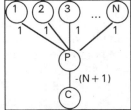

FIGURE 3. A network without useless connections.
The network of Fig.3 uses Pitts Mac Culloch equations:

$$S(i,t+1) = \text{if } (E(i,t) >= A), 1 \text{ else } 0. \quad (2)$$

$$E(i,t) = \sum_j W_{ji} * S(j,t)$$

Node P receives excitatory signals from nodes 1, ..,N. When the state of C is low(0), any coalition of A nodes in high state(1) switches P into high state. When the state of C is high, the state of P is low. Every connection is useful.

1.2 Heuristic definition vs. Strict definition.

A strict definition of utility may imply a **complex or non-effective proof-mechanism**. For instance, if the number of possible entries of a given node is infinite, our definition is only half decidable: only the useful connections can be determined in a finite time.

As a consequence, previous works use **estimators** of utility, which are easier to compute than a strict definition [CHO90,MOZ89].

Still, the inaccuracy of the criterion may bring unpredictable **side effects at run time**: errors may be amplified by the no linear activation functions; noisy entries increase the probability of such cases. We can see from [KAR90] and [MOZ89] that oversimplifying a network may lessen its abilities.

1.3 Useful nodes vs. useful connections.

Special care should be taken in the definition of useless nodes: the elimination of a node rests one dimension to the state vector, which is no little change. In previous works, the utility is defined from a limited, functional point of view. or instance in [BRO92], the utility of a node is estimated from the sensitivity of a set of output nodes to the variations of its state. A node pruning algorithm using a functional definition can be found in [FIM'95].

However, we prefer the following definition, which is altogether more general and more adapted to emergent computation: from an informational point of view, **a node is useless if its state can be computed by an effective algorithm from the rest of the state vector.** This condition grants that the Kolmogorov information of the state vector with and without the node is constant [DEL94].

1.4 Combinatory of the search space.

In order to check the utility of a connection i-j, we have to check the combinations of states of the entry nodes of j.

Suppose the network contains **nn** nodes, with an average of **na** entry connections, and the activation function has **nv** possible values. There are **nn*na** connections to check, and for every connection Wkj, there are **nv^na** combinations of the entry nodes of j to check. The size of the search space is:

$$\textbf{nummax = nn*na*nv}^{\textbf{na}} \qquad (3)$$

For instance, for a network of 100 nodes, an average of 30 entry connections, and binary activation function (2 possible states), we should perform:

$$100*30*2^{30} > 3*10^{13} \text{ steps.}$$

1.5 Classification of pruning methods.

We can now classify the pruning methods (of nodes or connections) as follows:

-Strict definition of utility, complete elimination. These exploratory algorithms remove every useless element. The pruning has no side effects at run time, but the computational cost may be high, even when heuristics are used.

-Heuristic definition of utility, complete elimination. An estimator of utility is given, and every element considered as useless is removed. Some useful elements may be removed by mistake. Hence, there are side effects at run time.

-Strict definition of utility, heuristic elimination. The pruning only removes useless elements, but some useless elements may remain. The pruning has no side effects at run time. The Stochastic-Propagative method belongs to this category.

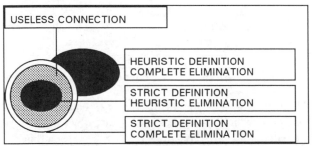

FIGURE 4. Summary of pruning methods.

1.6 Redundancy, Utility, and Learning abilities.

Redundant elements have three positive effects: they increase the resistance to noise, to damaged connections, and they improve learning abilities.

-Noise: redundant elements that are not exactly useless may change the dynamics in some marginal cases. However, if the entry is noisy, these cases have a high probability to appear. Hence, no element which is not absolutely useless should be removed.

-Damaged connections: in artificial neural networks, this property has no direct application.

-Learning abilities: when we remove connections, the network specializes itself but the learning abilities lessen, due to the elimination of free variables (each weight behaves as a free variable for learning).

Nevertheless, the nervous system provides an elegant solution to the conflict flexibility-specialization: in both central and peripheral nervous systems, several periods of expansion take place during the growth, when both axons and synapses are created. Each of these period is followed by a time of activity, when the redundant axons and synapses created during expansion periods are physically eliminated in order to specialize the nervous system [CHU92]. In [FIM'95], an experimental program based upon this idea is presented.

1.7 Discretization of continuous activation functions.

Continuous activation functions have to be discretized before using the stochastic-propagative method. When replacing the function Si by its discretized approximation, the pruning algorithm may remove "not-exactly-useless" connections. This, in turn, introduces a discretization noise in the entries of the nodes [OLL91]. The effects of such a noise on a network (feed forward or general), have been studied in previous works [CHO90,CHU92,HER91].

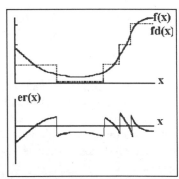

FIGURE 5. Discretization of a continuous function.
fd(x) is the discrete version of f(x).
er(x) is the discretization error fd(x)-f(x).

2 DEFINITIONS AND PROPERTIES.

We present here the formal definition of utility, and several general properties. We define some specific concepts (contribution range, entry boarder, etc.) and properties used by the Stochastic-propagative method. The proofs of all the theorems can be found in [FIM'95].

Let R be any network, **n** the number of nodes. Let $F_i()$ be the activation function of node i:
$S(i,t+1) = F_i(E(i,t))$, for every i, where:

$$E(i,t) = \sum_j W_{ji} * S(j,t) \text{ where:}$$

Wab is the weight of the connection between a and b.

Definition 1: the connection W_{ki} of R is **useless** iff:
for every possible state of the nodes of R,$(S_1,..,S_n)$,
for every possible value X of $F_k()$,

$$F_i(\sum_j W_{ji} * S_j) = F_i(\sum_{j \neq k} W_{ji} * S_j + W_{ki} * X) \qquad (4)$$

Definition 2: A connection is **useful** iff it is not useless.

The previous definitions apply to any kind of activation function, discrete or continuous. From now, we will limit ourselves to **increasing, bounded,** activation functions such as Pitts Mac Culloch, Bipolar, Sigmoïde, Atanh, etc.

Definition 3: $f()$ is **bounded** in the range [a,b] iff for every x, $f(x) \in [a,b]$, and if there exist values of $f(x)$ arbitrarily close to a and b.

When $f()$ is bounded, every term of the entry of a node $E(i,t)$, is bounded, hence, we can introduce the concept of **contribution range** of a node i to the entry of a node j.

Definition 4: The **contribution** of node j to node i in the given state of the network $(S_1,..S_n)$ is the term $W_{ji}*S_j$.

Definition 5: if $F_j()$ is increasing and bounded in $[V_1,V_m]$, the **contribution range** of node j to node i is defined as:

$$DELTA(j,i) = abs(W_{ji})*abs(V_1-V_m) \qquad (5)$$

Let $EMIN(j,i)$ and $EMAX(j,i)$ be the lowest and highest possible values of the contribution of j to i. An equivalent definition of the contribution range is:

$$DELTA(j,i)=EMAX(j,i)-EMIN(j,i) \qquad (6)$$

Theorem 1: if $F_k()$ is increasing and bounded in $[V_1,V_m]$, W_{ki} is useless in the network R iff
for every $(S_1,..,S_n) \in [V_1,V_m]^n$

$$f(\sum_{j \neq k} W_{ji} * S_j + EMIN(k,i)) = f(\sum_{j \neq k} W_{ji} * S_j + EMAX(k,i)) \qquad (7)$$

Interpretation: instead of checking all the possible values of Sk, we just have to check the extreme values of the contribution of k to i.

The following result allows us to deduce the utility of a connection by means of weight comparisons. We now suppose that the network R is homogeneous, i.e. that every node uses the same activation function: $F_i() = f()$. We will see later that this constraint can be easily released.

Theorem 2. If $f()$ is increasing and bounded in $[V_1,V_m]$, W_{ki} is useless in the network R, then every W_{gi} such that $abs(W_{gi}) <= abs(W_{ki})$ is useless.

Interpretation: if W_{ki} is useless, every connection W_{gi} with smaller absolute value than W_{ki} is useless; in reverse, if W_{ki} is useful, every connection W_{gi} with higher absolute value is useful.

The following results use homogeneous networks with increasing, bounded, **discrete** activation functions.:

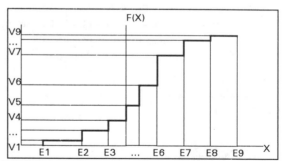

FIGURE 6. An increasing, bounded, discrete function.

Definition 6: Let $f()$ an increasing, bounded, discrete function which possible values are $(V_1,..,V_m)$. The **boarders** of $f()$ are the values $(E_1,..,E_{m-1})$ such that:
$x < E_1 \qquad => f(x) = V_1,$
$E_{i-1} < x < E_i => f(x) = V_i$ for every i en $[2,m-1]$
$E_{m-1} < x \qquad => f(x) = V_m$

When the entry of node j passes through a boarder, the state of j changes. The following result allows an incremental determination of useful connections based upon the boarders.

Theorem 3. Let f() be an increasing, bounded, discrete function whose possible values are (V1,..Vm) and whose boarders are (E1,..,E_{m-1}). Let C be a subset of the nodes of the network R, Let Eh be an arbitrary boarder and (S1,..Sn) an arbitrary state of the nodes of R. We define:

$$DIF(Eh,S1..Sn)=(Eh-\sum_{k \in C} Wki * Sk + \sum_{k \notin C} EMIN(k,i))$$
(8)

Let D be a set of nodes such that $D \cap C = \varnothing$. If:

$$\sum_{k \in D} DELTA(k,i) \geq dif(Eh,S1..Sn) \geq 0,$$
(9)

and if for every $g \in D$:

$$dif(Eh,S1..Sn) > \sum_{\substack{k \in D \\ k \neq g}} DELTA(k,i)$$
(10)

then for every $k \in D$, Wki is useful.

Interpretation: we consider a combination of states of the nodes of C which corresponds to a value of the entry of i, "close to and lower than" a boarder. If there exist nodes that do not belong to C, whose total contribution sets the entry of i higher than the boarder, if all these nodes participate (when removing one node, the entry of i stays under the boarder) then each of these nodes is useful.

3 THE PRUNING METHOD.

3.1 introduction.

For every node j, the useful connections Wij are detected and marked by a two-steps process: **propagation** of utility and **stochastic search of boarders.** The process stops when no new useful connection is detected. Unmarked connections are useless, and removed from the network.

The pruning uses a set **Uj** of nodes i such that Wij is useful, **Ij**, the complementary set of nodes (i.e. R-Uj), and **DISTMIN,** the lowest distance reached from an entry value of j, E(j,t), and a boarder. The process starts from Uj empty and DISTMIN infinite. At each step, the stochastic search decreases DISTMIN and the propagation increases Uj.

At each step:
-the **stochastic search** try to find a combination of states of the nodes of Uj that gets closer than DISTMIN from some boarder. The nodes of Ij are set to their lowest contribution.
-The **propagation of utility** looks for a subset S of Ij with a total contribution range higher than DISTMIN (theorem 3). If S exists, its elements are rested from Ij and added to Uj, and DISTMIN is decreased.
-The process **ends** when both Uj and DISTMIN are steady.

3.2 The stochastic search.

At each step of the algorithm, the stochastic search explores every boarder. For each boarder B:
-it **starts from a random combination** of contribution of the nodes of Uj, the nodes of Ij are set to their lowest possible contribution.
-it **chooses a random** node i in Uj.
-it **chooses a random contribution** for i:
-if the entry gets closer from the boarder, the change of value is accepted.
-if the distance from the boarder increases from Δ, the change of value is accepted with a probability given by the **Boltzmann distribution** [CHU92]:

$$\mathbf{proba}(\Delta) = e^{-\Delta/kTempe} , \text{where}$$
(11)

Tempe is a parameter called temperature, which defines the level of determinism of the system (T=0: deterministic system; high values of T: random system.).

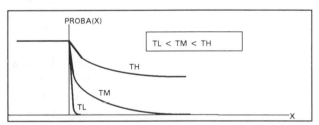

FIGURE 7. Probability of a change with Δ=X

The decision is taken by a **Monte Carlo algorithm** [VAS94]: a random p in [0,1] is chosen with uniform distribution; the change is accepted iff p <= proba(Δ).

Sometimes, the entry may become higher than the current boarder: such values are accepted, but the distance from the boarder is multiplied by a factor>>1 as shown in Fig.8.

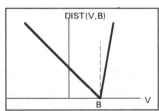

FIGURE 8. Distance between entry V and boarder B.

3.3 Experimental results.

We implemented the stochastic-propagative algorithm as well as a combinatorial algorithm for control. This algorithm uses an heuristic exploratory search to get close from the boarders [FIM'95].

12 typical configurations of networks (n entry nodes, 1 output node) have been designed and tested 10 times each.

TABLE 1. test cases.

Test	number of Wkj	values of f()	number of boarders	useless nodes
1	32	5	4	0
2	32	2	1	0
3	32	5	4	0
4	32	5	1	4
5	32	2	1	31
6	64	16	15	0
7	64	16	1	0
8	64	16	1	2
9	64	16	1	2
10	128	128	1	0
11	128	128	1	127
12	128	256	1	127

The results show an **absolute accuracy** of the stochastic algorithm. Moreover, in most of the cases, both algorithms get the same entry configuration to reach the lowest possible distance from boarders.

The stochastic algorithm is **very constant**, (between 50 and 1300 steps). The number of connections seems to increase the response time in a less-than-linear manner.

On the contrary, the exploratory algorithm is **very unstable** when there are useless connections: it explodes in 2 of the 12 cases (for each case, 10 times of 10).

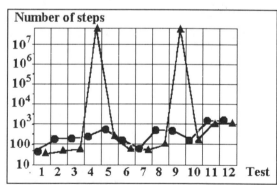

FIGURE 10. Number of steps for every test case. Circles=Stochastic, Triangles=Exploratory.

3.4 More results.

When working on **discretized activation functions**, the number of bits of digitalization does not change very much the performance of the stochastic algorithm. For instance tests number 11 and 12 are obtained from the same activation function, digitalized on 6 and 7 bits.

The stochastic algorithm **does not need annealing**: we start from a high temperature, we freeze the system at once, and we repeat the operation several times. This result was predicted in [VAS94]: when the neural network has a Curie Point, the convergence is fastened when the temperature is set alternately above and below this point.

4. CONCLUSIONS.

A **formal definition of useless connections** has been given, and a new pruning method which uses alternates steps of utility propagation and stochastic search has been presented. The method is designed to remove connections from networks with discrete activation functions, but it can be extended easily to the elimination of useless nodes, and to continuous activation functions.

The pruning can be used to **optimize a trained neural network,** to adapt it to slow machines such as micro controllers or computers of distributed control systems.

On the other hand, we are using the pruning to simplify neural networks in order to **study their dynamic properties** and their possible limit cycles [FIM95].

REFERENCES.

[BRO92] Brown E., *Tamaño, Generalización y Robustez en redes neurales multicapa,* tesis de maestria, Universidad Simón Bolívar, 92.

[CHO90] Choy J.Y., Choy C.H., *Sensitivity analysis of multilayer perceptron with differentiable activation functions* IEEE trans.,neural networks Vol.1 Num.2 06-90

[CHU92] Churchland P., Seknowski T.,*The computational brain,* MIT Press, 92.

[DEL94] Delahaye J.P., *Information, complexité et hasard,* Hermes editeur, 94.

[FIM95] Fimbel E., Brito F., *Orbits and quotient circuits in neural networks,* IEA95AIE Conference, 06 95.

[FIM'95] Fimbel E., *Eliminación de conexiones inútiles mediante un método propagativo-estocástico,* CNIASE95, 10 95.

[HER90] Herz J., Krogh A., Palmer R., *Introduction to the theory of neural computation,* Addison Wesley, 90.

[KAR90] Karnin E., *A simple procedure for pruning back propagation trained neural networks,* IEEE trans. on neural networks Vol.1, Num.2, 06-90.

[MOZ89] Mozer M., Smolensky P., *Skeletonization: a technique for trimming the fat from a network via relevance assessment,* advances in neural information processing I, Touretzki D.S., Ed. Morgan Kaufmann, 89.

[OLL91] Ollero A., *Control por computadora,* Alfaomega, 91

[VAS94] Vasquez J.M., *Búsqueda analítica del punto de Curie en cristales magnéticos y su interrelacion con redes neurales,* tesis de maestria, Univ. Simón. Bolívar, 94.

A NOTE ON THE GENERALIZATION ERROR
IN NEURAL NETWORKS

Yoshihiko Hamamoto, Toshinori Hase,
Yoshihiro Mitani and Shingo Tomita
Faculty of Engineering, Yamaguchi University, Ube, 755 Japan
Email: hamamoto@csse.yamaguchi-u.ac.jp

ABSTRACT

This paper discusses the training sample size N needed to keep the generalization error of multilayer ANN classifiers constant, while increasing the dimensionality n. Experimental results show that if one increases the training sample size N as $n^{3/4}$, the generalization error of multilayer ANN classifiers does not depend on the dimensionality. This is independent of the true Bayes error and hidden unit size.

1. INTRODUCTION

In neural networks, the problem of how well a classifier can perform on samples that it has not seen during training is very important [Bau90, Rau94]. In particular, a considerable amount of effort has been made to discuss the training sample size needed to achieve a given generalization error of ANN classifiers [CT92, AFS92, Rau93, RJ91]. Despite significant efforts in the past, however, only weak recommendations can be given to practitioners.

Now, consider that while increasing the dimensionality n, one has to increase the training sample size N as n^{α}, in order to keep the generalization error of ANN classifiers constant. Hamamoto et al.[HUT93] show experimentally that when $\alpha = 1$, no increase in the generalization error of multilayer ANN classifiers is observed as the dimensionality increases. Recently, Raudys [Rau95] points out that when $\alpha = 1/2$, the generalization error of single-layer ANN classifiers does not depend on the dimensionality. In this paper, we study the dependence of the generalization error of multilayer ANN classifiers on the dimensionality when $\alpha < 1$.

2. MULTILAYER ANN CLASSIFIER

We will study ANN classifiers with one hidden layer. The neurons in the input layer correspond to the components of the feature vector to be classified. The hidden layer has m neurons. The neurons in the output layer are usually associated with pattern class labels. The most popular approach to multilayer ANN classifiers is to fix the size of the network, i.e., the value of m, and to train the multilayer ANN classifier with m neurons in the hidden layer using the simple back propagation algorithm without a momentum term [RHW86]. According to a suggestion by Khotanzad and Lu [KL91], initial weights of the networks were randomly selected from [-0.5, 0.5] interval. Raudys [Rau95] notes that the selection of the initial weights has a significant influence on the generalization error. Learning was terminated after either a specified number of steps or when the mean-square output error, averaged over the training set, dropped below a specified threshold. The rate of convergence is significantly affected by the learning rate c. Unless the momentum term is used, Rumelhart et al.[RHW86] recommend the use of small values of c, in order for the networks to avoid getting trapped in a local minimum. From preliminary experiments, we used $c = 0.1$. It is well known that the generalization error of multilayer ANN classifiers is influenced by the value of m [RJ91]. Hence, the selection of m is always crucial in the design of multilayer ANN classifiers. Usually, heuristic approaches are used to select the value of m for a particular problem. In this paper, multilayer ANN classifiers were separately designed for 2 different values of m separately. That is, 2 multilayer ANN classifiers were evaluated.

3. EXPERIMENTAL RESULTS

The generalization error is the most effective measure of the performance of a classifier. In order for the estimated generalization error to be reliable in predicting the future classification performance of a classifier, the training and test samples must be statistically independent [DK82]. In error estimation literature, the holdout method [Fuk90] has been used, because it maintains the statistical independence between the training and test sets. Fukunaga [Fuk90] notes that the holdout method works well if the available samples are independently generated by a computer. Moreover, the Bayes error of artificial data is known. On the other hand, the true Bayes error of real data is never known. For the above reasons, we adopted the holdout method with artificial data.

It should be noted that the estimated generalization error is a function of the training and test sets. In order to reduce the influence of the test sample size, a large test sample should be used for error estimation. In this paper, the generalization error of a classifier was estimated by using 1000 test samples per class independently generated from distributions of the training samples.

We briefly describe an artificial data set [Van80], which we call the Ness data set. The performance of the Parzen classifier was compared with the linear and quadratic classifiers on this data set [Van80]. The available samples were independently generated from n-dimensional Gaussian distributions $N(\boldsymbol{\mu}_i, \Sigma_i)$ with the following parameters:

$$\boldsymbol{\mu}_1 = \mathbf{o}, \qquad \boldsymbol{\mu}_2 = [\Delta/2, 0, \cdots, 0, \Delta/2]^T$$

$$\Sigma_1 = I_n, \qquad \Sigma_2 = \left[\begin{array}{cc} I_{n/2} & 0 \\ 0 & \frac{1}{2}I_{n/2} \end{array} \right]$$

where Δ is the Mahalanobis distance between class ω_1 and class ω_2, and I_n denotes the $n \times n$ identity matrix. Note that in this data set, both the mean vectors and the covariance matrices differ. The Bayes error varies depending on the value of Δ as well as n.

In order to study the dependence of the generalization error on the dimensionality, the following experiment was conducted.

No. of classes	:	2
Dimensionality n	:	10, 20, 30, 40, 50
No. of training samples	:	$N_1 = N_2 = \lfloor n^\alpha \rfloor$
No. of test samples	:	1000 per class
Values of Δ	:	2, 4, 6
Values of α	:	1/2, 3/4
Hidden unit size m	:	8, 256
No. of trials	:	100

where $\lfloor x \rfloor$ denotes the greatest integer less than or equal to x.

The estimated generalization errors were averaged over 100 trials. Experimental results are shown in Figures 1-3. Experimental results show that when $\alpha = 3/4$, the generalization error of multilayer ANN classifiers does not depend on the dimensionality. When $\alpha = 1/2$, in contrast with single-layer ANN classifiers, the generalization error of multilayer ANN classifiers clearly increases with the dimensionality.

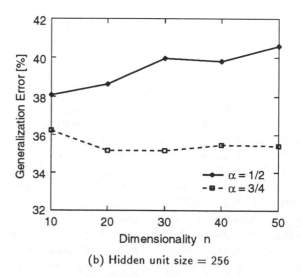

(a) Hidden unit size = 8   (b) Hidden unit size = 256

FIGURE 1   Dependence of the generalization error on the dimensionality ($\Delta = 2$)

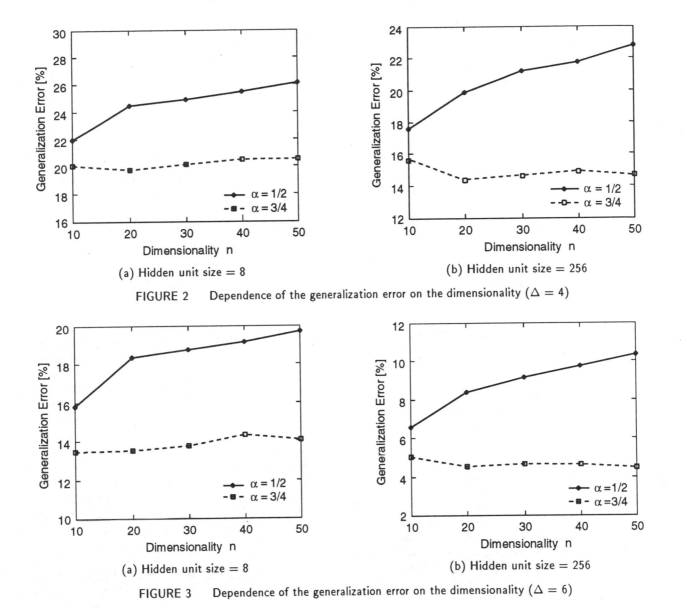

FIGURE 2 Dependence of the generalization error on the dimensionality ($\Delta = 4$)

FIGURE 3 Dependence of the generalization error on the dimensionality ($\Delta = 6$)

4. DISCUSSION

In the statistical pattern recognition field, Raudys [Rau76] points out that in order to keep a recognition accuracy constant, one must increase the training sample size N as n for linear classifiers and must increase that as n^2 for quadratic classifiers. On the other hand, multilayer ANN classifiers can be considered one of nonparametric classifiers. It is widely believed that many more samples are needed to design nonparametric classifiers than parametric classifiers such as linear and quadratic classifiers [Fuk90, DH73]. Nevertheless, multilayer ANN classifiers have better small sample

property than parametric classifiers. Our results suggest that while increasing the dimensionality, the training sample size needed to keep the generalization error constant can be directly determined from the dimensionality. We believe that this gives practical advice to designers and users of ANN classifiers.

ACKNOWLEDGMENTS

We would like to thank Prof. S. Raudys for giving us important references.

REFERENCES

[AFS92] S. Amari, N. Fujita and S. Shinomoto. Four types of learning curves. Neural Computation, 4, pp. 605-618, 1992.

[Bau90] E. B. Baum. When are k-nearest neighbor and back propagation accurate for feasible sized sets of examples?. Proc. of the EURASP Workshop on Neural Networks, L. B. Almeida and C. J. Wellekens Eds., Delft, pp. 2-25, Springer-Verlag, Berlin, 1990.

[CT92] D. Cohn and G. Tesauro. How tight are the Vapnik-Chervonenkis bounds?. Neural Computation, 4, pp. 249-269, 1992.

[DK82] P. A. Devijver and J. Kittler. Pattern recognition: A statistical approach. Prentice/Hall, 1982.

[DH73] R. O. Duda and P. E. Hart. Pattern classification and scene analysis. John Wiley & Sons, 1973.

[Fuk90] K. Fukunaga. Introduction to statistical pattern recognition. Second Edition, Academic Press, 1990.

[HUT93] Y. Hamamoto, S. Uchimura and S. Tomita. Evaluation of artificial neural network classifiers in small sample size situations. Proc. of 1993 Int. Joint Conf. Neural Networks, Nagoya, Vol.2, pp. 1731-1735, 1993.

[KL91] A. Khotanzad and J.-H. Lu. Shape and texture recognition by a neural network. in I. K. Sethi and A. K. Jain, Eds. Artificial Neural Networks and Statistical Pattern Recognition: Old and New Connections, Elsevier Science Publishers, pp. 109-131, 1991.

[Rau76] S. Raudys. On dimensionality, learning sample size and complexity of classification algorithms. Proc. Third Int. Conf. Pattern Recognition, pp. 166-169, 1976.

[Rau93] S. Raudys. On shape of pattern error function, initializations and intrinsic dimensionality in ANN classifier design. Informatica, 4, 3-4, pp. 360-383, 1993.

[Rau94] S. Raudys. Why do multilayer perceptrons have favorable small sample properties?. in E. S. Gelsema and L. N. Kanal, Eds. Pattern Recognition in Practice IV, pp. 287-298, Elsevier Science B. V., 1994.

[Rau95] S. Raudys. Unexpected small sample properties of the linear perceptrons. Technical report on research activities in LAFORIA-IBP, University PARIS VI, LAFORIA 95/17, 1995.

[RJ91] S. Raudys and A. K. Jain. Small sample size problems in designing artificial neural networks. in I. K. Sethi and A. K. Jain, Eds. Artificial Neural Networks and Statistical Pattern Recognition: Old and New Connections, Elsevier Science Publishers, pp. 33-50, 1991.

[RHW86] D. E. Rumelhart, G. E. Hinton and R. J. Williams. Learning internal representations by error propagation. in Parallel Distributed Processing: Explorations in the Microstructure of Cognition, ch. 8, MIT Press, 1986.

[Van80] J. Van Ness. On the dominance of nonparametric Bayes rule discriminant algorithms in high dimensions. Pattern Recognition, 12, pp. 355-368, 1980.

A Proposal of Emotional Processing Model

Kaori Yoshida, Masahiro Nagamatsu and Torao Yanaru

Dept. of Computer Engineering, Kyushu Institute of Technology

1-1 Sensui, Tobata, Kitakyushu city, Fukuoka pref. 804 JAPAN

E-mail address: kaori@human1.comp.kyutech.ac.jp

ABSTRACT

The purpose of this paper is to introduce a basic concept of the Emotional Processing System. The proposed model is realized by the two basic methods; (i) the one is to introduce a matrix into the model which is called frame as a set of arbitrary attributes extracting the basic features of emotional words, which are represented by any values considering how well fit the items of the frame, (ii) a new terminology including meaning of an artificial image of emotion, an *emoton* including emotional characteristics. The model discussed here, can be considered as a kind of emotional processing system, which has a possibility to be developed to human cognitive process model.

1 INTRODUCTION

Recently, there have been many discussions about more convenient and human friendly facilities. Developments of these facilities are well supported in several engineering fields like artificial intelligence and cognitive science. It is important to study human cognitive process, so as to put human sensibility into a computer, and we have to set up some mechanisms which treat emotional information to materialize human friendly facilities.

In connection with the thinking above, it has been suggested by Yanaru et. al. that an emotion-processing system based on fuzzy inference and its subjective observations [TYK94]. In that paper, they proposed Image codes as an expression of the mixed emotion word, taking into account Plutchik's theory [Plu60] and Young's theory [P.T67]. We wanted to expand them to more higher level, therefore the investigation applied to Frame Oriented Theory [YY95] has begun.

On the other hand, many scientists have been interested in our brain and mind, and are studying aggressibly toward the realization of a brain-like computer. We are also studying brain from psychological and physiological viewpoints for the purpose of realization of such the computer equipped with the function of emotional information processing.

First, we introduce the whole emotional system proposed by us. Next, we propose the emotional processing model based on Frame Oriented Theory as a part of the whole emotional system. And last, we could show the interesting simulated results, using the neural cell model, in which the movement of emotons takes dual emotional states.

2 BASIC CONCEPTS

2.1 Scientific Significance of Realization of the Emotional Processing System

When we think of the meaning of the realization of the emotional system from scientific viewpoint, the modeling methodology and mechanism sill give some benefit to human to compensate and enhance the understandings of real human activity and feeling reaction, almost of which, until now, have been done through psychology and philosophy. In this sense, the fundamental modeling study closely related physiological mechanism and reactions will be very important in the coming new science and engineering.

2.2 Image of the Whole Emotional System

An image of the whole emotional system proposed by us is shown in Figure 1. In above left, several kinds of signals from outer world toward internal world of brain are received on several kinds of sensors. These signals are processed and categorized in emotional receptors as a mapping of effectors corresponding to a kind of interface to human represented by frame items. A set of the emotional receptors is connected by artificial

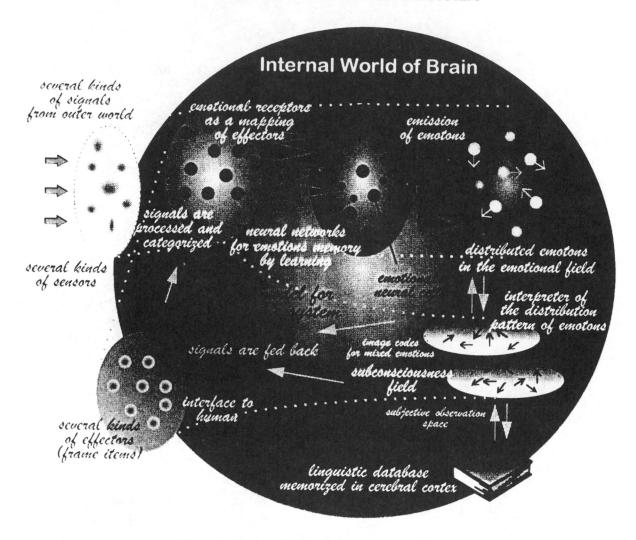

Figure 1: Image of the whole emotional system

neural network to a set of artificial emotional neural cells. The neural network corresponds some kinds of emotional memory, and several kinds of emotons as emotional particles are generated and emitted from the emotional neurons. Such generated emotons are distributed in the virtual emotional field. Applying a certain method for evaluation to the movement of the distributed emotons in the field, for example affine mapping method after mathematically normalizing or standardizing the distributed patterns of the particles, emotional interpretation by the emotional words are performed. On the contrary, the interpreted signals are fed back to emotional receptors through several kinds of effectors.

2.3 What kinds of Applications of the Emotional Processing System can be considered?

It must be considered that, is it really possible to realize so much the flexible emotional system enough to satisfy human feeling? before considering the application of the emotional system. My thinking is yes, under the fairly limited restriction. The reason is that when we think of the present situation of the study on artificial intelligence, the scientific approach and the development of the engineerings are so much advanced that a lot of the expert systems in several fields are available under the some restrictions. This makes us hope the possibility of realization of even the emotional system as long as being supported and compensated by some human handling, as well as the present available expert systems on artificial intelligence.

Then what is the restriction? Many restrictions are easily noticed, when we think of the difference of the construction materials between nervous circuits and electric circuits on logic method, the great difference of the number of constructing components and of the mechanism, and so on. Even if these kinds of differences exist between on real human and on artificial realized system, the realization of artificial emotional system may be possible and valuable in the sense that the emotional output from the system imitates well the human emotional reactions, as well as the artificial intelligent systems do well the human intelligence.

What kinds of application of the emotional processing system can be considered? In our laboratory, several applications have already been promoted. However, these applications are still partial realizations of the whole emotional processing system represented in Figure 1, supposed that the image codes of the several kinds of emotions, corresponding the mixed emotion proposed by R. Plutchik, are derived from the emotional field (or subconsciousness field) after interpretation of the movement of the generated and distributed *emotons*. Then, if we could realize the whole emotional system combined to the pure emotional processing part, represented in Figure 1, the more useful applications will be considered besides the applications mentioned above.

3 EMOTIONAL PROCESSING MODEL

3.1 Frame Oriented Theory

As is generally known, it can be considered that we may have some kinds of conceptual images in our intellectual activities. And, we are sensible to the conceptual image connected with the other ones. We express them by the patterns based on frame.

Generally, as we recognize a thing or concept, it is necessary to differentiate a particular one from others. We define a *frame* as a group of constructing items called *frame items*, and represent the frame by the following matrix :

$$\begin{pmatrix} f_{11} & \cdots & f_{1n} \\ \vdots & \ddots & \vdots \\ f_{m1} & \cdots & f_{mn} \end{pmatrix},$$

where $1 \leq i \leq m$, $1 \leq j \leq n$, $f_{ij} : frame\ item$. In short, a frame is a set of arbitrary attributes, and a frame item is an element of the set. Frame items are categorized by a column or a row.

In classifying a thing by its attributes, considering how well the attribute fits the item, we may use any value. Therefore in general, concept here, emotional words can be represented by the following pattern based on the frame.

$$\begin{pmatrix} p_{11} & \cdots & p_{1n} \\ \vdots & \ddots & \vdots \\ p_{m1} & \cdots & p_{mn} \end{pmatrix},$$

where $1 \leq i \leq m$, $1 \leq j \leq n$, $p_{ij} : any\ value\ [0,1]$.

When we consider that arbitrary concepts are analogous or not, it is needed in general to have some primitives. The primitive concepts are established obtaining the distinguishable condition by the computation along with the theory [YY95], and it can be revealed that an arbitrary concept belongs to a certain category.

3.2 Frame for Emotional Memory

Some concepts are sometimes constructed by several partial concepts. On the other hand, a frame can express only a partial concept, so if we want to express all the kind of concept, it is necessary as many frames as the number of partial concepts. Here, we prepare a certain frame which has several categories, the pattern based on the frame is regarded as the *image pattern* of the concept. Considering how well its attributes fits the frame item, the pattern can be represented by some values.

We introduce a certain frame for emotional memory in Figure 2. The frame items are categorized by a column.

A–1	A–2	A–3
B–1	B–2	B–3
C–1	C–2	C–3

A. **PHYSIOLOGICAL SYMPTOMS**
    (A–1) Lump in throat
    (A–2) Change in breathing
    (A–3) Heart beat faster
B. **NONVERBAL REACTIONS**
    (B–1) Changed facial expression
    (B–2) Changed in voice
    (B–3) Changed in gesturing
C. **VERBAL BEHAVIOR**
    (C–1) Silence
    (C–2) Short utterance
    (C–3) Speech melody change

Figure 2: A certain frame

joy

0.07	0.03	0.96
0.07	0.07	0.19
0.00	0.11	0.67

fear

0.63	0.96	1.00
0.78	0.30	0.11
1.00	0.37	0.04

anger

0.37	0.89	0.96
0.89	0.74	0.59
0.56	0.19	0.48

sadness

0.96	0.56	0.70
0.59	0.41	0.07
1.00	0.22	0.15

disgust

0.33	0.15	0.44
0.81	0.26	0.11
0.93	0.26	0.04

shame

0.56	0.26	0.85
0.78	0.19	0.15
1.00	0.33	0.04

guilt

0.63	0.11	0.70
0.59	0.11	0.04
1.00	0.19	0.07

Figure 3: An example image patterns

The frame items of the mentioned frame are selected from the subjects expressing symptom and reaction for a specific emotion based on the questionnaire table [PK89].

3.3   Frame Model

We introduce the frame model of emotional memory. A frame has a data-structure for representing a stereotyped situation of human activities. The system must know something about cause-effect, time sequences, state, process, and types of knowledge. And the concept is the definition of a category as the internal representation that enables the individual to determine the category membership of objects [TS94].

We constructed the frame of emotional state considering Plutchik's circular model and Rorschach measures [PK89]. In other words, frame is a set of arbitrary attributes, and emotional words are represented by some values considering how well fit the items of the frame.

We show some examples in Figure 3 based on the frame mentioned before (Figure 2). The patterns of each emotional words are decided by considering degree of fitness to the frame items. The values of the patterns here depend on the data from psychological point of view and expanded from Image code [TYK94] (, which were suggested in the paper of Yanaru et. al., taking the intensities of some words show by Plutchik

[Plu60] into account as standard intensities for normalization). It takes the value, unity if the attribute fit the item completely, and zero if it doesn't the item completely.

The introduced image patterns of several emotional words are static ones so far, we will report in the future on the studies concerning dynamic image patterns as an expansion of the static ones.

3.4   Proposal of Emoton

We introduce a new terminology, an *emoton* which has image of an artificial emotional particle. The emotons are organized to construct a small universe of the emotional memory, those of which are characterized by several features such as a location, a color, a movement and so on.

These several kinds of features can be derived from taking the several mathematical treatments into the matrices. The vectors in Figure 4, as an example of the mathematical treatments, are derived from the computation of eigen vectors corresponding the maximum eigen value. It may be considered that each of the vectors, corresponding to each of the matrices in Figure 3, locates in the universe (and moves if the dynamical data are introduced).

What feature is appropriate to correspond to what vector among the derived vectors is not clear. However, the locations of the emotons in Figure 4 strongly reflect on the multifactor analytic theory of emotion proposed by Robert Plutchik [Plu60].

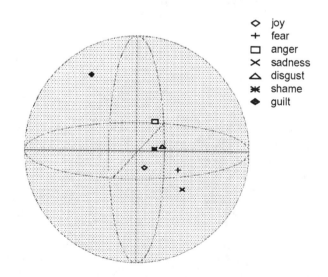

Figure 4: An image of *emoton*

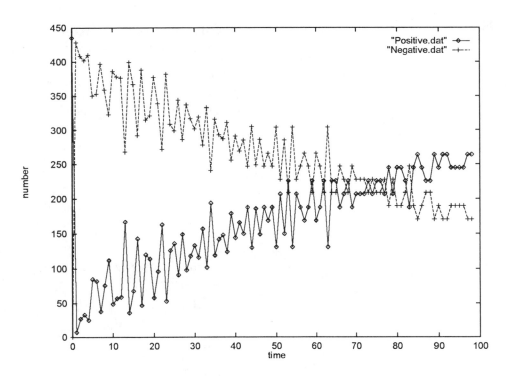

Figure 5: The transition of a number of P-Emoton and N-Emoton (initial state:joy)

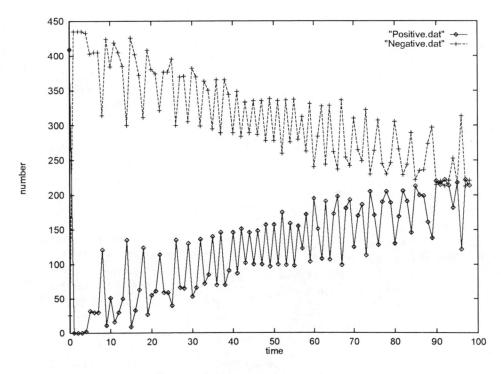

Figure 6: The transition of a number of P-Emoton and N-Emoton (initial state:sadness)

3.5 Dual System Using Neural Cell Model

We constructed the neural network model, which are output represents positive or negative state on the basis of current value criteria, regarding an Emoton as a neural cell. Here, Emoton represents current emotional state only positive or negative. We define an Emoton which represents positive state as a positive-Emoton(P-Emoton), and negative state as a Negative-Emoton(N-Emoton). This kind of Emoton, which represents only two state, is considered as an essential of emitting motivation in the next stage of the emotional processing system construction, in prospect.

The operation is mathematically described in the following:

$$U_i(t) = aV_i(t-1) + \sum_{j=1}^{N} W_{ji} X_j(t-1), \qquad (1)$$

$$\begin{aligned} &if \ U_i(t) \geq \theta \\ &\quad then \ X_i(t) = 1, \ V_i(t) = U_i(t) - p, \\ &\quad else \ X_i(t) = 0, \ V_i(t) = U_i(t), \end{aligned} \qquad (2)$$

where, $X_j(t-1)$ stands for input pulses, $X_i(t)$ output, $U_i(t)$ internal potential and $V_i(t)$ represents membrane potential. This approximates a real neuron better than the conventional nonlinear analogue neural cell do [SM92].

4 COMPUTER SIMULATION

We should take data out from the linguistic and the action model influenced by emotional information, but made the input data here supposing the results of the questionnaire based on the frame, because of total system construction uncompleted. The time sequence is represented by durable pulse input, and unit-time is considered from $t = 0$ to $t = 100$.

Figure 5 shows the result of the simulation at the initial state of joy, and Figure 6 at sadness. We can easily notice that a number of P-emoton increases and outrun N-Emoton around $t = 70$ in the first case(joy), but a number of P-Emoton does not outrun in the second case(sadness). It's considered that the state of Emoton, positive of negative, is under the influence of initial emotional state.

5 CONCLUSION

We introduced the emotional processing model, but we still know far too little about the contents and structure of knowledge of the emotional states.

We don't know how closely we must adjust brain model to foster appropriate mechanism and performance, but we believe that the construction of truly emotional model is sufficiently likely to justify beginning study and policy planning now.

References

[Gra88] Stephen R. Graubard, editor. *The Artificial Intelligence Debate*. The MIT Press, 1988.

[PK89] Robert Plutchik and Henry Kellerman, editors. *EMOTION: Theory, Research, and Experience*, Vol. 4. Academic Press, Inc., 1989.

[Plu60] Robert Plutchik. The multifactor/analytic theory of emotion. *Journal of Psychology*, Vol. 50, pp. 153–171, 1960.

[P.T67] P.T.Young. Affective arousal. *American Psychologist*, Vol. 22, pp. 32–40, 1967.

[SM92] Akiyama, Shigematu and Matsumoto. A dynamic neural cell model driven by pulse train. In *Proceeding of International Symposium (NIS'92)*, Iizuka, 1992.

[TS94] Esther Thelen and Linda B. Smith. *A Dynamic Systems Approach to the Development of Cognition and Action*. The MIT Press, 1994.

[TYK94] T.Yanaru, T.Hirota and N.Kimura. An emotion-processing system based on fuzzy inference and its subjective observations. *International Journal of Approximate Reasoning*, Vol. 10, pp. 99–122, 1994.

[YY95] K.Yoshida and T.Yanaru. An application of frame oriented theory to decision processing and evaluation of medical diagnosis. In *Proceedings of the International Joint Conference of the Fourth IEEE International Conference on Fuzzy Systems and the Second International Fuzzy Engineering Symposium (FUZZ-IEEE/IFES'95)*, Vol. 2, pp. 497–502, Pacifico YOKOHAMA, 1995.

RECURSIVE PREDICTION ERROR ALGORITHM FOR THE NUFZY SYSTEM TO IDENTIFY NONLINEAR SYSTEMS

B.T. Tien and G. van Straten

Dept. Agricultural Engineering and Physics, Wageningen Agricultural University
Bomenweg 4, 6703 HD Wageningen, The Netherlands
E-mail: Biing-Tsair.Tien@User.AenF.WAU.NL

ABSTRACT

In our early work, an integrated neural-fuzzy approach named NUFZY has been developed for nonlinear system identification. From the literature recursive methods have been applied with success to neural networks. Due to the resemblance of the NUFZY system and multi-layered neural networks recursive methods may also be suited for the neural-fuzzy approach. In this paper the recursive prediction error algorithm is adapted for on-line parameter tuning of the NUFZY system. Examples are presented to demonstrate the feasibility of this recursive identification to models of the NUFZY type.

1. INTRODUCTION

In the identification of systems which may contain time-variant properties, a recursive scheme for adjusting the system parameters is considered as a potential method, especially where on-line tuning is required to follow the varying characteristics of the system. The commonly used back-propagation algorithm for neural network training, where the steepest descent gradient serves as the search direction, is not suitable for recursive adaptation, because it encounters problems of slow convergence. In contrast, a recursive prediction error algorithm based on the alternative approximate Gauss-Newton search direction, was reported to have improved learning capabilities [CCB+90]. The recursive prediction error (RPE) algorithm was shown to have similar convergence properties as its off-line counter part in the case of linear systems [LS83]. More specifically, for a chosen criterion the estimated parameters obtained by the RPE method will converge with probability one either to a stationary point (local minimum) or get stuck at the boundary of that point as time goes to infinity. The asymptotic convergence property of the estimates made it attractive to adopt these ideas to the multi-layered networks and to extend to application of nonlinear

systems [CCB+90]. While comparing to on-line identification of neural networks with back-propagation training, provided the networks were previously trained well enough in an off-line way, the fast convergence of the RPE method is appealing in case where the system parameters are slowly time-variant.

Though lots of success have been done with neural networks identification, the information representation of the internal network structure still seems to be problematic, as little information can be extracted about the actual functioning of the system. On the other hand, fuzzy rule based models do have content but seem difficult to train. Like neural networks, also fuzzy systems are universal approximators that can approximate any real continuous function on a compact set to arbitrary accuracy [WM92]. Hence, based on the similar structure and functional equivalence between fuzzy systems and neural networks, an integrated neuro-fuzzy system, named NUFZY, has been developed for the purpose of system identification [TS95]. The developed NUFZY system is characterized by a triple-layered networks. Just like a neural networks it is capable of identifying nonlinear static systems off-line, but the internal structure can be interpreted in the form of rules. Due to the resemblance of the NUFZY system and multi-layered neural networks, it is attractive to try to adapt the recursive prediction error to the NUFZY system in order to identify nonlinear dynamical systems. In this paper, first a set of parameter values is determined by the orthogonal least squares method in an off-line manner. With these identified parameters as an initialization, the RPE method then is employed to proceed with on-line tuning of the NUFZY system. The main goal of this paper is to investigate the applicability of this RPE method for the NUFZY system, especially for application of on-line identification.

This paper is formulated as follows. The structure of the developed NUFZY system and its corresponding characteristics are briefly presented in section 2. The main theme of constructing the RPE algorithm and its implementation are described in section 3. Examples are demonstrated in section 4. In the end, conclusions are addressed.

2. STRUCTURE OF NUFZY

The developed NUFZY system, as shown in Fig. 1, is characterized by a triple-layered feedforward network. The first and second layer of NUFZY deal with the antecedent part of the fuzzy rule base and the third layer concerns the consequent part of the fuzzy rule base. The NUFZY system performs a Takagi and Sugeno type of fuzzy inference [TS85], i.e. the consequent part is formed as a linear combination of the premise variables. In the extreme case the output takes as crisp real values, which is also the method used in the NUFZY system. Given a system with ni input variables x_i, where each x_i has its own N_i membership functions, $i=1,..,ni$, and with *no* outputs $\hat{y}_j$, $j=1,...,no$, the fuzzy rules can be expressed in the form

R^r : **if** (x_1 is $A^r_{k1}(x_1)$ and .. and x_i is $A^r_{ki}(x_i)$ and .. and x_{ni} is $A^r_{Kni}(x_{ni})$)

   then ($\hat{y}^r_1 = g^r_1, .. , \hat{y}^r_j = g^r_j, , \hat{y}^r_{no} = g^r_{no}$)   (1)

where superscript r denotes the *rth* fuzzy rule and $A^r_{ki}(x_i)$ represents the *kith* linguistic label of x_i with respect to the fuzzy rule R^r. It is also noted that the membership function in the consequent part is expressed in the form of a singleton value denoted by g^r_j in the *rth* fuzzy rule. In this paper the fuzzy sets at the input have bell shaped Gaussian membership functions. The AND operation is implemented as the algebraic product and a simplified centroid of gravity (COG) defuzzification are used to construct the NUFZY reasoning functions. In the following an outline is given of the essential parts of the NUFZY system that will be used in the recursive estimation.

LAYER 1   The input node x_i, $i=1,2,...,ni$, just serves to distribute the input into the first layer nodes with fixed weights of unity. The node on the first layer, named *membership node* and denoted as $\Phi_{ki}(x_i)$, represents a membership function that performs fuzzification of the input variables. Each x_i has its own N_i linguistic labels characterized by membership function $\Phi_{ki}(x_i)$ that are numbered as $1,2,..,ki,..,Ni$, for $i=1,...,ni$. Inputs x_i's are used directly as arguments of the membership functions and the corresponding graded values, i.e. fuzzified values $\mu_{ki}(x_i)$ which represents the node output of this layer, are determined accordingly. The fuzzified value $\mu_{ki}(x_i)$ is obtained from the Gaussian membership function

$$\mu_{ki}(x_i) = \exp(-\frac{1}{2}\frac{(x_i - c_{i,ki})^2}{\sigma^2_{i,ki}})$$

with ki = 1,..,Ni, i = 1,..,ni   (2)

where $c_{i,ki}$ and $\sigma_{i,ki}$ are the *kith* center and bandwidth of $\Phi_{ki}(x_i)$ for the input x_i, respectively. For ease of notation, the $\mu_{ki}(x_i)$ are reordered sequentially and denoted as $\alpha_m(x_i)$ with subscript $m = 1,...,M$, where M is the total number of membership function of all input variables. M can be obtained by $M = N1 +... + Nni = \sum Ni$, $i=1,...,ni$. $\alpha_m(x_i)$ is expressed as

$$\alpha_m(x_i) = \exp(-\frac{1}{2}\frac{(x_i - c_m)^2}{\sigma^2_m})$$

with m = 1,.., M ; $M = \sum Ni$; i = 1,..,ni (3)

where $\alpha_m(x_i) = \mu_{ki}(x_i)$ with index $m = \sum Nj+ki$; $j=1,..,i-1$; $i=1,..,ni$. The center c_m and bandwidth σ_m corresponding to $c_{i,ki}$ and $\sigma_{i,ki}$ are defined as above. Hence, for a specific input vector $x = [x_1 \ x_2 \ ..x_i...x_{ni}]^T$, the corresponding graded membership values can be denoted in a vector form, $\alpha = [\alpha_1 \ \alpha_2 \ ... \ \alpha_m \ ... \ \alpha_M]^T$. The vector α just stacks the outputs of the membership functions for all inputs. In this layer the node parameters that are to be determined are $c_{i,ki}$ and $\sigma_{i,ki}$ (or c_m and σ_m). Initially, the centers are chosen as equally spaced on the range of x_i (from a set of training data) and the values of variance of x_i are taken as bandwidths. By the above definitions, it is easy to deduce the derivatives of the node outputs wrt parameters, i.e. $\partial\mu_{ki}(x_i)/\partial c_{ki}$ ($\partial\alpha_m(x_i)/\partial c_m$) and $\partial\mu_{ki}(x_i)/\partial\sigma_{ki}$ ($\partial\alpha_m(x_i)/\partial\sigma_m$).

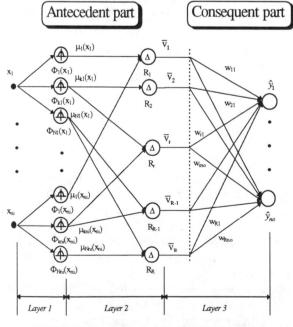

FIGURE 1 The structure of NUFZY system.

LAYER 2   In this layer the node, called a *rule node*, represents a fuzzy rule and is denoted as Rr, $r=1,2,..,R$; where R is the total number of all fuzzy rules given by $R=\Pi$ Ni ,$i=1,..,ni$. Inputs to the *rule node* are the graded membership values from the *membership node* of layer one. Each *rule node* performs a two-step operation as will be described later. According to Eq.(1) each *rule node* involves only one membership function for each input. Therefore, the existence of a connection between a *rule node* and a *membership node* is presented with a value of either 1 or 0. These connections can be defined by a relationship matrix **RM** with dimension $(R \times M)$, where M is the total number of all membership functions of all inputs as defined in Eq.(3). Each row r of **RM** represents the status of the antecedent part of a fuzzy rule. The element value of 1 represents a link between the *rth rule node* and the corresponding *membership node*, whilst element value of 0 indicates no connection. Hence, the relationship matrix constructs an initial fuzzy rule base. For instance, suppose we have three inputs, x_1, x_2, x_3, and each has 2, 2, 2 membership functions with values μ_{11}, μ_{12}, μ_{21}, μ_{22}, μ_{31}, μ_{32}, respectively. In this case, $R=2*2*2=8$ and $M=2+2+2=6$. A relationship matrix denoted as $RM_{(2,2,2)}$ constructs a 8 by 6 matrix as follows.

Relationship Matrix $RM_{(2,2,2)}$

	m =	1	2	3	4	5	6 (=M)
		μ_{11}	μ_{12}	μ_{21}	μ_{22}	μ_{31}	μ_{32}
r=1		1	0	1	0	1	0
r=2		0	1	1	0	1	0
r=3		1	0	0	1	1	0
r=4		0	1	0	1	1	0
r=5		1	0	1	0	0	1
r=6		0	1	1	0	0	1
r=7		1	0	0	1	0	1
r=8=R		0	1	0	1	0	1

There is a two-step operation to generate the node output in this layer.

(1) Step one : the transient firing strength v_r is obtained by

$$v_r = \hat{\Pi}_{m=1}^{M} \alpha_m = \hat{\Pi}_{i=1}^{ni} RM(r,a_i:b_i) \cdot \mu_i \qquad (4)$$

where $\hat{\Pi}$ means a product operation accompanied by the **RM** matrix. $RM(r,a_i:b_i)$ represents the partial elements from a_i to b_i of the *rth* row vector of the **RM** matrix, where $b_i = \Sigma Nk$, $k=1,..,i$, and $a_i = b_i - Ni +1$. μ_i is a vector of all membership value of input x_i as given by $\mu_i = [\mu_{i1} \ \mu_{i2} \ .. \ \mu_{iki} \ .. \ \mu_{iNi}]^T$. Eq.(4) just specifies that the output v_r of rule r is obtained by just multiplying the membership functions involved in rule r, according to connections shown in Fig. 1.

(2) Step two: the normalized firing strength $\bar{v}_r$ is calculated as

$$\bar{v}_r = \frac{v_r}{\sum_{r=1}^{R} v_r} \qquad (5)$$

In this layer there is no parameter to be determined. However, the derivative of rule node output wrt its node input, i.e. $\partial \bar{v}_r / \partial \alpha_m$, can be derived.

LAYER 3   The node, denoted as $\hat{y}_j$, $j=1,..,no$, stands for the NUFZY model output. The link in this layer represents a weight parameter denoted as w_{rj}, for $r=1,..,R$, $j=1,..,no$, and connects node $\hat{y}_j$ and Rr. The product $\bar{v}_r w_{rj}$ actually represents the unknown variable g'_j in the consequent part of the fuzzy rule as in Eq.(1). By applying the simplified centroid of gravity (COG) defuzzification method this node then just performs as a summation operation such that the model output is obtained by

$$\hat{y}_j = \frac{\sum_{r=1}^{R} v_r \times w_{rj}}{\sum_{r=1}^{R} v_r} = \sum_{r=1}^{R} w_{rj} \bar{v}_r = \mathbf{w}_j^T \bar{\mathbf{v}} \qquad (6)$$

where $\mathbf{w}_j$ is the weight parameter vector given by $\mathbf{w}_j = [w_{1j} \ .. \ w_{rj} \ .. \ w_{Rj}]^T$ and $\bar{\mathbf{v}}$ is a normalized firing strength vector given by $\bar{\mathbf{v}} = [\bar{v}_1 \ .. \ \bar{v}_r \ .. \ \bar{v}_R]^T$ with element $\bar{v}_r$ defined by Eq.(5). It is noted that the output is linear in the weight parameters. In the NUFZY system the linear property will be retained for other choices of *t-norm* operators for the AND connection in the fuzzy rules as well as for other choices of the membership functions. Also note that the $\bar{v}_r$ can be viewed as fuzzy basis functions (FBF), so that the NUFZY output forms a FBF expansion [WM92]. Thus, the NUFZY system has the property of a universal function approximator. The derivatives of the NUFZY output wrt weight parameter and normalized firing strength can be easily derived.

It is interesting to compare the relation of the NUFZY system to other neural networks, such as radial basis function networks (RBF). The main difference is that characteristic information from the system is embedded in the internal connection of the NUFZY system. This allows the NUFZY system to represent the system behavior in a rule form deduced from the data set. Further, in the NUFZY system the antecedent parts (layer 1 and 2) are only partially connected (defined by the **RM** matrix) rather than the full connection in RBF networks. However, the partially connected networks in the NUFZY system does not affect its function as a universal approximator.

3. PREDICTION ERROR ALGORITHM

The recursive prediction error algorithm can be summarized as follows [LS83],

$$\varepsilon(t) = y(t) - \hat{y}(t)$$

$$\hat{\Lambda}(t) = \hat{\Lambda}(t-1) + \gamma(t)[\varepsilon(t)\varepsilon(t)^T - \hat{\Lambda}(t-1)]$$

$$L(t) = P(t-1)\Psi_\theta(t)[\lambda(t)\hat{\Lambda}(t) + \Psi_\theta^T(t)P(t-1)\Psi_\theta(t)]^{-1} \qquad (7)$$

$$\hat{\theta}(t) = \hat{\theta}(t-1) + L(t)\varepsilon(t)$$

$$P(t) = \{P(t-1) - P(t-1)\Psi_\theta(t)[\lambda(t)\hat{\Lambda}(t) +$$
$$\Psi_\theta^T(t)P(t-1)\Psi_\theta(t)]^{-1}\Psi_\theta^T(t)P(t-1)\}/\lambda(t)$$

where $d = \dim \theta$. With no outputs, the sensitivity derivative $\Psi_\theta = [d\hat{y}/d\theta]^T$ being a d by no matrix is the transpose of the derivative of the model output $\hat{y}$ versus the model parameters θ. In the NUFZY system, the unknown parameter set, θ, can be regarded as either the output weights $\mathbf{w}$ in the consequent part of the NUFZY system or as the center $\mathbf{c}$ and bandwidth σ of the membership function in the antecedent part of the NUFZY system, or both. Depending on the choice of the parameter set of interest, once the sensitivity derivative Ψ_θ of the NUFZY system has been determined, one can apply the recursive prediction error algorithm to tune these parameters (Eq.(7))

In the following examples some initial values of the RPE are set as follows. The initial value of the estimated covariance matrix of prediction error, $\hat{\Lambda}(0)$, is set as $0.1*I$ (no by no matrix). The covariance matrix $\mathbf{P}(0)$ is initialized as a d by d diagonal matrix with diagonal element of 10000, where d depends on the choice of parameter set. It is desirable to set the forgetting factor $\lambda(t) < 1$ at the initial stage in order to require rapid adaptation and then let $\lambda(t) \to 1$ as $t \to \infty$. Hence $\lambda(t)$ and the gain sequence, $\gamma(t)$, are chosen as [LS83]

$$\lambda(t) = \lambda_0 \lambda(t-1) + (1-\lambda_0) \qquad (8)$$

$$\gamma(t) = 1 - \frac{\lambda(t)}{\lambda(t) + \gamma(t-1)} \qquad (9)$$

where $\lambda_0 = 0.99$ and $\lambda(0) = 0.95$.

It is noted that the purpose of real-time identification is to track the time-varying parameters, but in the presence of noise it is always impossible to accurately follow parameters that change fast. Obviously, a tradeoff exists between the tracking ability and the noise sensitivity and only the slow time variation of the parameters can be achieved by the recursive identification. Hence, where time variability is studied in the following examples, we have assumed that the systems have slow varying dynamics, as is particularly obvious in the example of plant growth in agriculture.

4. EXAMPLES

The following two examples demonstrate the implementation of the RPE method in the NUFZY system for identifying nonlinear MISO systems. The parameter sets of the NUFZY system that are desired to be tuned can be defined as either $\theta = \mathbf{w}$, or $\theta = [\mathbf{w}\ \mathbf{c}]^T$, or $\theta = [\mathbf{w}\ \mathbf{c}\ \sigma]^T$. Previous studies showed that good approximation could be achieved by merely tuning of output weights (i.e. $\theta = \mathbf{w}$) [Tie95]. Hence, the examples presented here are based on the RPE tuning of output weights only.

EXAMPLE 1 This example is equivalent to example 4 of [NP90]. The dynamical system is given as

$$y(k+1) = \frac{x_1 x_2 x_3 x_5 (x_3 - 1) + x_4}{1 + x_2^2 + x_3^2} \qquad (10)$$

where $[x_1, x_2, x_3, x_4, x_5] = [y(k), y(k-1), y(k-2), u(k), u(k-1)]$. In this example the NUFZY net is employed to generate an one-step-ahead prediction $\hat{y}(k+1)$. First, 500 training data points are generated by the plant Eq.(10) with a random input signal uniformly distributed in the interval of [-1,1]. These training data are used to train the NUFZY system with the orthogonal least squares method in order to get a trained output weight, denoted as $\mathbf{w}_{ols}$. These weights $\mathbf{w}_{ols}$ can be regarded to represent the optimal parameter values from this batch of training data. The number of membership functions is assigned as two to each input variable. Hence, initially the total number of fuzzy rules will be 32. After OLS training it is found that only 26 rules are significant. These weights are then used as initialization of the RPE method for the subsequent validation step (i.e. $\mathbf{w}_{ini} = \mathbf{w}_{ols}$). As a validation set, 1000 pairs of data are generated according to Eq.(10) based on an input sequence u(k) given by

$$u(k) = \begin{cases} \sin(\dfrac{2\pi k}{250}) & 0 \le k \le 500 \\ 0.8\sin(\dfrac{2\pi k}{250}) + 0.2\sin(\dfrac{2\pi k}{25}) & 500 < k \le 1000 \end{cases} \qquad (11)$$

Fig. 2(a) shows the result of the validation. It can be seen that the NUFZY system gives excellent prediction. Fig. 2(b) presents the variation of the output weights tuned by the RPE method during the validation. It is observed that little variation on the weights is made by the RPE tuning which is initialized with $\mathbf{w}_{ols}$. This reveals that the $\mathbf{w}_{ols}$ have already caught up the global behavior of the plant.

On the other hand, if the validation is initialized with initial weights set to zero (i.e. $\mathbf{w}_{ini} = \mathbf{0}$, rather than $\mathbf{w}_{ols}$), with RPE tuning a similar result as Fig. 2 (a) is also obtained but with a little less accuracy than the previous one. However, the variation of the tuned weights is quite different from the previous one as shown in Fig. 2(c). At

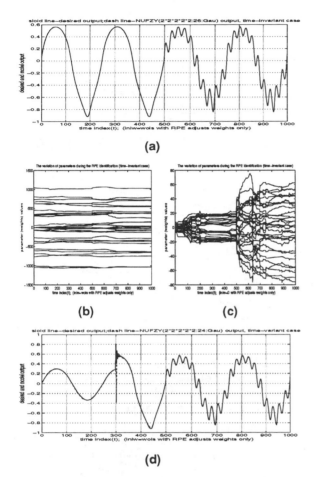

(a)

(b)   **(c)**

(d)

FIGURE 2 NUFZY predicted output superimposed on the desired output (they are hardly distinguishable): (a)time-invariant case; (d) time-varying case. The variation of the identified weight of the NUFZY system during the validation for time-invariant case : (b) $w_{ini}= w_{ols}$; (c) $w_{ini}= 0$.

the 500*th* time step the weights are clearly readjusted as the input signals with different frequency come in. This implies that the RPE initialized with **0** was trapped on some local minimum and the tuning proceeds only locally. When the change of frequency of the input signals takes place, the RPE readjusts these weights to another (local) minimum in order to get a good fit. This example demonstrates that it pays to initialize the RPE tuning with parameters obtained off-line from a good excitation signal. However, if not, it still fits the system well on that local excitation signal.

[NP90] also identified a parallel model (i.e. the past model predictions are components of input vector) requiring 100,000 steps of training. In order to compare with the aforementioned simulation, this parallel approach is also adopted for the NUFZY system. As a result a very good prediction is obtained as well (not

shown here) and our model accuracy of the NUFZY system is far superior than theirs. In [NP90[, they used multi-layer neural networks trained with back-propagation. In contrast the NUFZY system requires less training efforts and has a lower model complexity. For the training phase, 500 samples are used in one step using the OLS identification in contrast to 100,000 steps of back-propagation adaptation. With respect to the model complexity, only 32 weights has to be adjusted in the NUFZY networks while 320 weights were used by them. The key is that good performance can be achieved by just tuning the output weights of the NUFZY system. This makes it very appealing for fast identification since the problem then becomes linear-in-the-parameters.

In order to study time variability, the dynamical system is forced to change its status from Eq.(10) to the next status governed by

$$y(k+1) = \frac{x_1 x_2 x_3 x_5 (x_3 - \mathbf{2}) + x_4}{\mathbf{3} + x_2^2 + x_3^2} \qquad (12)$$

where, as before, $[x_1,x_2,x_3,x_4,x_5] = [y(k), y(k-1), y(k-2), u(k), u(k-1)]$. The following procedure was used to test this time-variant case. The training data are generated as before with a random input signal uniformly distributed in the interval [-1,1], but the first 250 training points were generated by following Eq.(10) and the remaining 250 by following Eq.(12), thus simulating a sudden change in parameters. Next, a validation data set was forced by first generating 300 points according to Eq.(12) and then another 700 according to Eq.(10), which means that system turns back to its original status. The arrangement of data in this way is desired to examine the adaptation ability of RPE method. The validation result of the RPE on-line tuning initialized with w_{ols} shows that good prediction is still attained as shown in Fig 2(d). It is also observed that the RPE tuning converges fast to a new working point when the system parameters switch to another values at the 300*th* time step.

EXAMPLE 2 In agriculture plant growth, which is related to its surroundings, inherently represents a system with a nonlinear character and some undetected time-variant parameters. In this example, the dry matter production of tomatoes as a function of the environmental factors, such as temperature, CO_2 concentration, and radiation, was considered. Data were available from the experiments in Wageningen [BH93]. Three experiments were done on three different growing seasons in 1988. During the experiment period, every 7 - 10 days the dry matter production of tomato was taken as destructive measurement for the utilization of simulation model.

With respect to the NUFZY identification, the goal of the NUFZY modeling is to identify the dynamical growing process of tomato and to predict the total dry weights (TDW) of tomato at the next sampling date. In this case the averaged radiation (RAD) and the averaged ambient CO_2 concentration of the greenhouse at the sampling interval from t(k-1) to t(k) are used as inputs. For TDW(k) and $\hat{\text{TDW}}$(k+1), the measured and predicted total dry weight of tomato at sampling date t(k) and t(k+1), the NUFZY is used to approximate the plant growth model as

$$\hat{\text{TDW}}(k+1) = f_{\text{NUFZY}}(\text{RAD}(k), CO_2(k), \text{TDW}(k)) \qquad (14)$$

In the initialization of the NUFZY model, data of experiment 1 and experiment 3, totally 31 tuples $(\text{RAD},CO_2,\text{TDW})$, were used for initialization. Data of experiment 2 (15 data pairs) were used for model on-line prediction. Owing to the limited data length the RPE tuning of the NUFZY model is done by repeatedly feeding these data to the NUFZY model up to 5 times to tune the output weights, which are initially set to zero. At the end of initialization process, it was found that the best identified results were obtained by NUFZY(3*4*2;Gau). This notation means that the input variables were assigned to RAD, CO_2, and TDW, respectively, 3, 4, and 2 Gaussian membership functions. In the on-line prediction, those identified parameter values as obtained from the initialization process were applied as initial values. The results are depicted in Fig. 3. The predicted TDW by the NUFZY model was located within 95% confidence interval of the measured TDW for both initialization and on-line prediction, showing that a good prediction of TDW has been achieved. This example exhibits the feasibility of the RPE tuning for the NUFZY system on a real world application.

5. CONCLUSIONS

In this paper, the sensitivity derivatives can be derived for the NUFZY system that allows to use the recursive prediction error method for tuning of the model and to identify the unknown dynamical nonlinear system. With RPE method, examples demonstrate that good model accuracy can be achieved by just tuning the consequent weights of the NUFZY system. Since this problem is linear-in-the-parameters it is also convenient to apply an orthogonal least squares method to the first batch of data, in order to obtain a set of optimal weight parameters as initialization for the RPE method that enables it to attain a faster adjustment for on-line tuning.

REFERENCES

[BH93]   Bertin N. and E. Heuvelink . Dry matter production on a tomato crop: comparison of two simulation models. *J. Horticultural Science*,**68**,995-1011,1993.

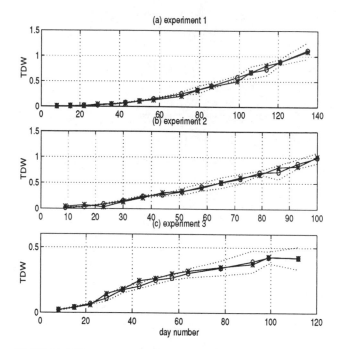

FIGURE 3 The measured TDW of tomatoes (circle-solid line) with 95% confidence interval (dotted line) and NUFZY(3*4*2:24;Gau) predicted TDW (star-solid line).

[CCB+90]   Chen S., C.F.N. Cowan, S.A. Billings, and P.M. Grant. Parallel Recursive Prediction Error Algorithm for Training Layered Neural Networks. *INT. J. Control*, **51**, 1215-1228, 1990.

[LS83]   Ljung L. and T. Stöderström. Theory and Practice of Recursive Identification. Cambridge, MA:MIT press, 1983.

[NP90]   Narendra K.S. and K. Parthasarathy. Identification and Control of Dynamical Systems Using Neural Networks. *IEEE Trans. Neural Networks*, **1**, 4-27, 1990.

[TS85]   Takagi T. and M. Sugeno. Fuzzy identification of systems and its applications to modeling and control. *IEEE Trans. Syst., Man, Cybern*, **15**, 116-132, 1985.

[TS95]   Tien B.T. and G. van Straten. Neural-Fuzzy systems for Non-linear System Identification -Orthogonal Least Squares Training Algorithm and Fuzzy Rule Reduction. *Preprints - 2nd IFAC/IFIP/EurAgEng Workshop on Artificial Intelligence in Agriculture*, pp 249-254, May 29-31, 1995, Wageningen, The Netherlands.

[Tie95]   Tien B.T. The Recursive Prediction Error Algorithm for the NUFZY system. Wageningen Agricultural University, Dept. Agri. Eng. and Physics, *MRS Report 95-10*, 1995.

[WM92]   Wang L.X. and J. M. Mendel. Fuzzy basis function, universal approximation, and orthogonal least-squares learning. *IEEE Trans. Neural Networks*, **3**, 807-814, 1992.

REFLECTIVE LEARNING OF NEURAL NETWORKS

Yoshiaki Tsukamoto and Akira Namatame
Dept. of Computer Science, National Defense Academy,
Yokosuka, 239, JAPAN
E-mail:{tsuka, nama}@cc.nda.ac.jp

ABSTRACT

This paper introduces the notion of an adaptive neural network model with reflection. We show how reflection can implement adaptive processes, and how adaptive mechanisms are actualized using the concept of reflection. Learning mechanisms must be understood in terms of their specific adaptive functions. We introduce an adaptive function so that a neural network is able to adjust its internal structure by itself to evolving environments by modifying its adaptive function and associated learning parameters. We also investigate reflective learning among multiple neural network modules. In multiple modules setting, two types of reflection are modeled: each neural network module learns on its own by adjusting its adaptive function, while at the same time, each neural network module mutually interacts and learns as a group to obtain the coordinated adaptive function and learning parameters.

1 INTRODUCTION

The ability to learn is the most important property of the living-systems. Evolution and learning are the two most fundamental adaptation processes and their relationship is very complex. Studying the evolution and development processes of biological systems can reveal how the structures are formed through interaction with the environment of the living systems in nature. These structural adaptation mechanisms in biological systems can suggest ways of building adaptable structure so that finally the network grows to a configuration suitable class of problems characterized by the training patterns. The initial step is to explore the aspects of this relationship by defining the process of evolution as the process of learning procedure to be adjusted [DM90]. Taking this approach, we view evolution as a kind of an adaptive process of learning mechanism. Here, the learning process itself is the object of evolution. In most of the current neural network models, learning is done through modification of the synaptic weights of neurons in the network [BB91][Lee91][Nad89]. This kind of learning is basically a parameter adaptation process. The structure of neural networks should be evolved and developed rather than pre-specified, and we need to develop a framework for an adaptable process. We introduce an adaptive neural network model with reflection as a framework for an adaptive process evolving in a dynamic environment. Adaptation is viewed either as a modification of one's behavior or as a modification of one's environment. Learning mechanisms must be understood in terms of their specific adaptive functions. We introduce an adaptive function of a neural network, and an self-reflective or an adaptive process of a neural network is modeled to adjust its internal structure to evolving environments by modifying its adaptive function and its associated learning parameter. We also investigate reflective learning among multiple neural network modules [GSL93]. In multiple modules setting, two types of reflection may occur: each network module learns on its own by adjusting its adaptive function and its associated learning parameter, while at the same time, each network module mutually interacts and learns as a group to obtain the coordinated adaptive functions and learning parameters. To adapt successfully to its conditions, an autonomous agent needs to learn from previous situations and apply the learnt knowledge to subsequent situations. We also introduce the concept of a reflective agency model with the neural network learning capability and provide the design model of large-scale neural networks using those multiple reflective neuro-agents. The large-scale and heterogeneous neural networks are made up of multiple and reflective neuro-agents which are individually trained in order to represent features defined over homogeneous feature spaces.

2 FORMULATION OF ADAPTIVE NEURAL NETWORK

The list of the feature values is represented as $d = (d_1, d_2, ..., d_n) \in \mathbf{D}$ where the variables $d_1, d_2, ..., d_n$ take the Boolean values. We denote training examples as the ordered pairs $<d_t, c_t>$ where $d_t \in \mathbf{D}$ and $c_t \in \{0, 1\}$. We define the adaptive function over the training set $<\mathbf{D}, \mathbf{C}> = [<d_t, c_t>: t = 1, 2, ..., T]$ as follows:

$$F(d) = \alpha\, G^{+}(d) - \beta\, G^{-}(d) + \theta \qquad (2.1)$$

where $G^{+}(d_p)$ and $G^{-}(d_p)$ are defined

$$G^{+}(d_p) = \sum_{d \in C^{-}} S(d, d_p)$$

$$G^{-}(d_p) = \sum_{d' \in C} S(d', d_p) \qquad (2.2)$$

and

$$S(d, d_p) = \sum_{i=1}^{n} d_i d_{pi} + \sum_{i=1}^{n} (1 - d_i)(1 - d_{pi}) \qquad (2.3)$$

We define the $T \times 2$ aggregate-matrix $T(\mathbf{D}, \mathbf{C})$ defined over the training set $<\mathbf{D}, \mathbf{C}>$ as

$$T(\mathbf{D}, \mathbf{C}) = \{ G^{+}(d_p), G^{-}(d_{p'}) \} \qquad (2.4)$$

The aggregate-matrix is obtained as

$$T(\mathbf{D}, \mathbf{C}) = \mathbf{D}\mathbf{D}^{T}[\mathbf{C}, 1 - \mathbf{C}] \qquad (2.5)$$

In the next theorem, we will show that the adaptive function is obtained from the aggregate-matrix $T(\mathbf{D}, \mathbf{C})$.

Theorem 2.1 [NT92] Suppose the aggregate-matrix $T(\mathbf{D}, \mathbf{C})$ is linearly separable, that is if there exists some learning parameters, α, β and θ satisfying

(i) For a positive training example $d_p \in \mathbf{C}^{+}$

$$\alpha\, G^{+}(d_p) - \beta\, G^{-}(d_p) + \theta > 0 \qquad (2.6)$$

(ii) For a positive training example $d_{p'} \in \mathbf{C}^{-}$

$$\alpha\, G^{+}(d_p) - \beta\, G^{-}(d_p) + \theta \leq 0 \qquad (2.7)$$

Then we can define the adaptive function

$$F(x) = \sum_{i=1}^{n} w_i x_i + \theta \qquad (2.8)$$

with the connection weights w_i, $i = 1, 2, ..., n$,

$$w_i = \alpha \sum_{d_p \in C^{-}} d_{p_i} - \beta \sum_{d_{p'} \in C} d_{p_i'} \qquad (2.9)$$

The weighing-evidence scheme is that it would not demand that every feature of some object be present, instead it would only weigh the evidence that object is present. That is, an object becomes active whenever the weighted summation of the present features proceed the threshold level. The number weights are assigned to each feature, and

this is based on the theme of weighing-evidence [Min85]. The adaptive function also serves as an activation function of the weighing-evidence learning scheme.

3 REFLECTIVE LEARNING AMONG MULTIPLE NEURAL NETWORK MODULE

In this section, we provide the reflective learning model with multiple neural network modules. The learning process, as a group, can be deployed in a manner parallel to the learning process of the individual network module. That is, each network module has a specialized representation of its characteristics in the form of the adaptive function. To apply cooperative learning, each individual network module needs to adjust, adapt, and learn from working with other network modules. The group of individual network modules can offer something not available in the individuals. In a multi-module learning environment, two types of learning may occur: the multi-modules can learn as a group, while at the same time, each network module can also learn its own by adjusting a learnt adaptive function. Cooperative learning would require the exchange of knowledge held by the individual network modules. At the cooperative stage, individual network modules can put forward their learnt knowledge for consideration by other network modules using their own local experiences [Bra91][Sia90]. Figure 1 illustrates the concept of cooperative learning. The overall system is constructed from several interacting learning modules. We then need to discuss how an assembly of modules cooperate to learn using their individual adaptive functions that represent their own internal states. Each adaptive function defined over an individual network module is encapsulating a specific set of adaptive function and its associated learning parameter as its knowledge obtained from a different training set. At the cooperative stage, each network module puts forward their learnt knowledge.

We consider a general learning problem that takes the learning space $\mathbf{W} = [\mathbf{D}, \mathbf{C}]$. A given network module can specialize in learning to discriminate certain subspace of the whole learning space $\mathbf{W} = [\mathbf{D}, \mathbf{C}]$. Learning system architecture composed of assemblies of such network modules need to specify the interactions among modules. We may consider the following two types of cooperation

Type 1: Cooperation of Homogeneous Neural Network Modules

Type 2: Cooperation of Heterogeneous Neural Network Module

If we know in advance that a set of training examples may be naturally divided into subsets that correspond to distinctive subtasks, modularization is a promising way of combating these problems. A central idea of the distributed

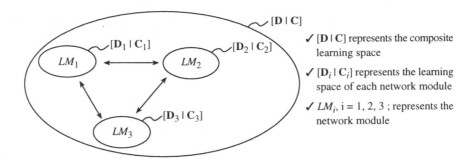

Figure 1 The illustration of cooperative learning among independent network modules.

✓ [**D** | **C**] represents the composite learning space

✓ [$\mathbf{D}_i$ | $\mathbf{C}_i$] represents the learning space of each network module

✓ LM_i, i = 1, 2, 3 ; represents the network module

learning is the conceptual decomposition of complex problems. A large-scale training set is decomposed into several subsets and each subset is used to train learning modules. The overall network is constructed from these pre-trained learning modules. The architecture is then characterized by a modular network architecture that facilitates functional network modules and network composition. Distributed learning corresponds to the decomposition of the training data set $\mathbf{W} = [\mathbf{D}, \mathbf{C}]$ into subsets $\mathbf{W}_i$, i = 1, 2, ..., n. Type 1 decomposes a learning space $\mathbf{W} = [\mathbf{D}, \mathbf{C}]$ into several subsets, $\mathbf{W}_i = [\mathbf{D}_i, \mathbf{C}_i]$, i = 1, 2, ..., n, where $U\mathbf{W}_i = \mathbf{W}$. After applying inductive learning program on each subset, it synthesizes those results. Each module has learned to be an expert to learn to handle a subset of the complete set of the training cases. However, traditional approaches require a complex coordinating mechanism of integrating those individually trained modules. Cooperative learning of Type 2 corresponds to the attribute-based decomposition. In the attribute-based decomposition scheme, each network module gets only the data corresponding to a subset of attributes. In this method of decomposition, however, the results of the individual network modules are under specified with respect to the whole problem. In these cases it is more efficient to use the inductive learners where the data resides and then synthesize the results, rather than collect the data at one place and then analyze it. On the other hand, the modularization scheme would be little use if there does not exit such an appropriate learning procedure to train high-level modules separately and to integrate those functionally pre-specified modules efficiently. Backpropagation learning (BP) is often used to train a system composed of several different modular networks, each of which perform different subtasks on a different training set, there will generally be strong interference effects that lead to slow learning. Several researches have attempted to reduce interference [Hry92][JJN+91][Sik92]. The central idea of these research

is to use the global error function, which is a weighted linear combination of the outputs of the local networks. Unfortunately, except the case that we know in advance that a set of training examples may be naturally divided into subsets that correspond to distinctive subtasks and there is no interference with the weights of local networks that specialize in different training case, these approach based on BP does not encourage localization.

We now discuss cooperative learning among multiple homogeneous neural network modules, and show the coordination mechanism. Without loss of generality, we can consider the case of two network modules. We denote the training set of each network module as $<\mathbf{D}, \mathbf{C}_i>$, i = 1, 2, ..., n. The integrated training set is described as $<\mathbf{D}, \mathbf{C}>$. The aggregate-matrix of each network module is represented by $T_i(\mathbf{D}, \mathbf{C})$, where $T_i(\mathbf{D}, \mathbf{C})$ i = 1, 2, ..., n are $T_i \times 2$ matrices respectively. At the cooperative phase, each network module trained with the initial training set $<\mathbf{D}, \mathbf{C}_i>$ i = 1, 2, ..., n modifies its adaptive function for coordination as follows: Each network module, i = 1, 2, ..., n modify its aggregate-matrix as follows:

$$T(\mathbf{D}, \mathbf{C}) = \begin{bmatrix} T_1(\mathbf{D}, \mathbf{C}) \\ T_2(\mathbf{D}, \mathbf{C}) \\ \vdots \\ T_n(\mathbf{D}, \mathbf{C}) \end{bmatrix} \tag{3.1}$$

where

$$T_i(\mathbf{D}, \mathbf{C}) = \left[\sum_{j=1}^{n} \mathbf{D}_i \mathbf{D}_j^T \mathbf{C}_j, \sum_{j=1}^{n} \mathbf{D}_i \mathbf{D}_j^T (1 - \mathbf{C}_j) \right] \tag{3.2}$$

Theorem 3.1 (*Composition of Type* 1): If the aggregate-matrices $T_j(\mathbf{D}, \mathbf{C})$ are linearly separable for some α, β and θ, then the whole training set $<\mathbf{D}, \mathbf{C}>$ is also linearly separable.

The connection weights w_i, $i = 1, 2, ..., n$, are given as the summation of the weights w_i^j, $j = 1, 2, ..., n$ obtained from each subset of each subset $<\mathbf{D}_i, \mathbf{C}_i>$, $j = 1, 2, ..., m$ of the training set $<\mathbf{D}, \mathbf{C}>$.

$$\begin{cases} w_i = \sum_{j=1}^{m} w_i^j \\ w_i^j = \alpha \sum_{d_i \in C_j} d_{t_i} - \beta \sum_{d_i \in C_j} d_{t_i'} \end{cases} \qquad (3.3)$$

The cooperative learning among multiple heterogeneous neural network modules and the coordination mechanism are described as follows: The aggregate-matrix of each network module by $T(\mathbf{D}_i, \mathbf{C})$, $i = 1, 2, ..., n$, where $T(\mathbf{D}_i, \mathbf{C})$ $T \times 2$ matrices respectively. Each network module trained with the initial training set $<\mathbf{D}_i, \mathbf{C}>$, $i = 1, 2, ..., n$ modifies its adaptive function for integration by simply adding the aggregate-matrix of the other network module as:

$$T(\mathbf{D}, \mathbf{C}) = T(\mathbf{D}_1, \mathbf{C}) + T(\mathbf{D}_2, \mathbf{C}) + \cdots + T(\mathbf{D}_n, \mathbf{C})$$

where

$$T(\mathbf{D}_i, \mathbf{C}) = \mathbf{D}_i \mathbf{D}_i^T (\mathbf{C}, 1 - \mathbf{C}) \qquad (3.4)$$

Theorem 3.2 (*Composition of Type* 2): If the aggregate-matrices $T(\mathbf{D}_i, \mathbf{C})$ $i = 1, 2, ..., n$, are linearly separable for some α, β then the whole training set $<\mathbf{D}, \mathbf{C}>$ is also linearly separable for α, β and $\theta = \sum_{i=1}^{n} \vartheta_i$. The connection weights w_j, $j = 1, 2, ..., m$, are given

$$w_j = \begin{cases} w_j^1 & for & d_j \in \mathbf{D}_1 \\ w_j^2 & for & d_j \in \mathbf{D}_2 \\ \quad \vdots \\ w_j^n & for & d_j \in \mathbf{D}_n \end{cases} \qquad (3.5)$$

where

$$w_i^j = \alpha \sum_{d_i \in C_j} d_{t_i} - \beta \sum_{d_i \in C_j} d_{t_i'}$$

4 AN AGENT MODEL WITH CONNECTIONIST LEARNING CAPABILITY

Most often, when people in AI use the term agent, they refer to an entity that functions continuously and

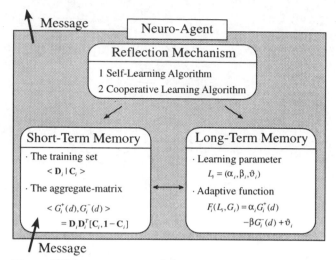

Figure 2 The components of a neuro-agent with reflective learning capability.

autonomously in an environment in which other processes take place and other agents exist [AG88][DM90]. In this section, we provide an integrated framework for realizing the reflective and self-autonomous agency model. Each agent, endowed with the neural network learning capability with reflection, is trained to represent a specific problem domain and is specialized to interact with the learning environment. The contents of an agent's memory defines its internal state. Each agent also has procedures called methods that specify his behavior. An agent is equipped with the neural network learning capability to adapt its behavior to his environment and automatically acquires knowledge from experience. The basic components of the agent model with the self-reflection is depicted in Figure 2.

It consists of the knowledge processing and the internal memory. The knowledge processing consists of the message processing component, the training set, and the learning mechanism. The internal memory is made of many smaller processes, termed as neuro-object. A neuro-object becomes active whenever some of its properties are active. Each of those neuro-objects must already have learned its own response to that same signal. Messages mean different things to different neuro-objects, and each neuro-object must learn its own, different way to react to that message. Each neuro-object learns its own specific and appropriate response. In order to do this, Each neuro-object must have its private knowledge to tell it how to respond to every message. Each neuro-object learns its way of the response to the message by the learning method. We are also concerned with adaptation in multi-agent domains in which the agents are actively learning from their experiences. A reflective agency

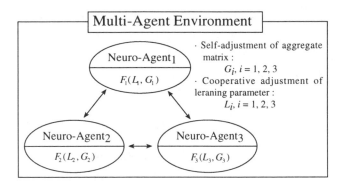

Figure 3 The reflective learning in a multi-agent environment.

model consists of individual agents that learn locally as a precursor to form a multi-agent system. The common knowledge with respect to the adaptive functions of the independent agents may induce cooperation. The overall system is constructed from several interacting learning modules. The concept of coreflective learning process with adjusting the adaptive function and the learning parameter of each individual is shown in Figure 3.

5 SELF-ORGANIZATION OF NEURO-AGENCY

When thinking about the nature of an agent, we should think about levels of analysis or levels of granularity. We might think about a particular actor, like a human being, as an agent. But we might think about an entire organization as an agent. So there needs some aspect of compositionality. That is, we need to be able to put things together and call them an agent and would like to have a recursive or comparable structure that would allow us to build agents at different levels of granularity. One approach of the composition is to model an agent based on an aggregation of hierarchy. In an aggregation hierarchy an agent could be made of a number of other agents with many levels. Such an organization has many roles and defines the behavior of agents in a particular context, how they relate to the environment as well as to each other. The overall system is constructed from several interacting learning modules, termed as multi neuro-agents.

In this section, we describe a way of organizing the set of multiple network modules, implemented as neuro-agents, into the structured networks, defined as the agency. The mechanism has a strong similarity to the nature self-organizing and growing process. The growth starts from the set of the unstructured neuro-agents, and the set of neuro-agents is let to self-organize into the whole neuro-agency with the nesting hierarchical structure. The growth process

is guided by the self-organizing mechanism of each neuro-agent, and the overall agency is constructed from interaction among those neuro-agents. We begin by reviewing the construction and operation of a single module, neuro-agemt. We then discuss how an assembly of neuro-agents cooperate to learn as a function of their individual internal states.

We formulate an assembly of neuro-agents as the sets of the training examples described as follows: A set of the training examples of each neuro-agent is formulated by n attributes, $X_1, X_2, ..., X_n$, and each of them has multi-valued attribute with the domain $Dom(X_i)$. The group of neuro-agents is denoted by $W = \{A_i : i = 1, 2, ..., n\}$.

Definition 5.1 A set of the training examples of each neuro-agent A_i is defined as

$$C_i = \{X_1 : Dom(X_1), X_2 : Dom(X_2), ..., X_n : Dom(X_n)\} \tag{5.1}$$

We define meta/sub relations in the group of neuro-agents. We define the group of

$$W = \{A_1, A_2, ..., A_k\}$$

with the meta-sub relations among neuro-agent such as

$$\text{for } C_i, C_j \in C \to C_i \succ C_j \text{ or } C_i \prec C_j \tag{5.2}$$

the society of neuro-agent with a generalization hierarchy, or simply neuro-agency.

The i-th attribute of each training instance is described by a vector ψ_i of length m_i, the j-th position of the vector ψ_i being either 1 or 0, indicating that the j-th value of the attribute X_i is or is not present, respectively. Each object, therefore, is indiced by a set of the feature vector, termed as the feature code, $(\psi_1, \psi_2, ..., \psi_n)$ of length Σm_i defined over the attribute space $X = X_1 \times X_2 \times ... \times X_n$ is denoted as

$$\psi_A(O_i) = \{\psi_{A1}(O_i), \psi_{A2}(O_i), ..., \psi_{An}(O_i)\} \tag{5.3}$$

Each instance is also indiced by its agent code, which is defined over the generalization hierarchy $C = C_1 \times C_2 \times ... \times C_k$ defined as

$$\varphi_C(O_i) = \{\varphi_{C1}(O_i), \varphi_{C2}(O_i), ..., \varphi_{Ck}(O_i)\} \tag{5.4}$$

where

$$\varphi_{Cj}(O_i) = \begin{cases} \varphi_{Cj}(O_j) & \text{if } O_i \in O_j \\ 0 & \text{otherwise} \end{cases} \tag{5.5}$$

The set of those training instances

$$[\{\psi_A(O_i), \varphi_C(O_i)\} : O_i : i = 1, 2, ..., k] \tag{5.6}$$

consists of the internal model of each neuro-agent.

We now show the self-organizing algorithm of the society of neuro-agent with a generalization hierarchy. Such a self-organizing algorithm involves the generation of the set of the training examples of the meta agent from those of the neuro-agents.

Generating a meta agent : With this method a new meta neuro-agent A is generated from several neuro-agents A_i. The generation of the meta agent is described as the process of generating the set of the training examples of the meta agent P from those of the neuro-agents, P_i, $i = 1,2,...,k$, as follows:

$$P_s = \bigwedge_{i=1}^{k} P_s = \bigwedge_{i=1}^{k} \bigwedge_{O_i \in C_i}^{k} \Psi_A(O_s) \qquad (5.7)$$

Generating subclasses : With this method some subclass C_i, with the class-attribute P_i, $i = 1, 2, ..., k$, is generated from its super classes C with the class-attribute P as follows:

$$P_i = \bigwedge_{O_i \in C_i} \Psi_A(O_i) \qquad (5.8)$$

A recursive model of generating a generalization hierarchy is shown in Figure 3.

6 CONCLUSION

This paper showed how an adaptive neural network model with reflection can implement adaptive processes. We introduced the concept of an adaptive function, and learning mechanisms were understood in terms of their specific adaptive functions. A neural network module was modeled to adjust its internal state to evolving training environments by modifying its adaptive function. We also investigate reflective learning among multiple network modules. In multiple modules setting, two types of reflection were considered: each network module mutually interact and can learn as a group, while at the same time, each network module can also learn on its own by adjusting its adaptive function. The overall system is constructed from several interacting learning modules. We then discussed how an assembly of modules cooperate to learn as a function of their individual internal states. Many separate network modules, each of which is combines their individually defined adaptive functions by coordinating learning parameters. We have also used the agent-oriented model as architecturally neutral metaphor for describing massively parallel distributed and cooperative neural network modules. An agent-oriented model for the specification of neural network models consists of cooperative distributed processing elements called autonomous neuro-agent with self-reflection. Each neuro-agent is encapsulating a specific set of knowledge obtained from a different training set. At the cooperative stage, neuro-agents put forward their learnt knowledge to obtain coordinated learning parameters.

REFERENCES

[AG88]   H. B. Alan and Gasser. *Readings in Distributed Artificial Intelligence*. Morgan Kaufmann (1988).

[BB91]   J. A. Barnden and J. B. Pollack(eds). *A High-Level connectionist Models*. Ablex Publishing (1991).

[Bra91]   P.B. Brazdil. *Learning in Multi-agent Environments. In Proceedings of the Second Workshop on Algorithmic Learning Theory*: 15-29, 1991.

[Chu88]   P. M. Churchland. *Matter and Consciousness*. The MIT Press, 1988.

[DM90]   Y. Demazeau and J. P. Müller(eds). *Decentralized AI*, Elsevier Science, 1990.

[GSL93]   S. Giroux, A. Senteni and G. Lapalme. Adaptation in Open Systems Reflection as a Backbone. *In Proceedings of the Int. Conf. on Intelligent and Cooperative Information Systems*, 114-123, 1993.

[Hry92]   T. Hrycej. *Modular Learning in Neural Networks*. Wiley-Interscience, 1992.

[JJN+91] R. A. Jacob, M. Jordan, S. Nowlan and G. Hinton: Adaptive Mixtures of Local Experts, *Neural Computation*, 79-87, 1991.

[Lee91]   T. C. Lee. *Structure Level Adaptation for Artificial Neural Networks*, Kluwer Academic Publishers , 1991.

[Mas94]   Maes: Learning Agents, *Communications of the ACM*, Vol. 37, No. 7, 30-40, 1994.

[Min85]   M. Minsky. *The Society of Mind*, Simons & Schster, 1985.

[Nad89]   J. Nadal. Study of a growth algorithm for a feedfoward network, *In Proceedings of the Int. J. of Neural Networks*, vol. 1, 55-59, 1989.

[NT92]   A. Namatame and Y. Tsukamoto: Structural Connectionist Learning with the Complementary Coding, *International Journal of Neural Systems*, Vol. 3, No. 1, 19-30, 1992.

[Sia90]   S. E. Sian. The Role of Cooperation in Multi-agent Learning Systems. in *Cooperative Knowledge-Based Systems*, Springer-Verlag, 67-84, 1990.

[Sik92]   R. Sikora. Learning Control Strategies for Chemical Processes : A Distributed Approach, *IEEE Expert*, June, 35-43, 1992.

Neural Network Applications

COLLISION AVOIDANCE USING NEURAL NETWORKS LEARNED BY GENETIC ALGORITHM

Nicolas Durand   **Jean-Marc Alliot**

Laboratoire d'Optimisation Globale ENAC/CENA
7 Avenue E Belin, 31055 Toulouse CEDEX, FRANCE
e-mail: durand@cena.dgac.fr, alliot@dgac.fr,

ABSTRACT

As Air Traffic keeps increasing, many research programs focus on collision avoidance techniques. In this paper, a neural network learned by genetic algorithm is introduced to solve conflicts between two aircraft. The learned NN is then tested on different conflicts and compared to the optimal solution. Results are very promising.

AIR TRAFFIC CONTROL AND COLLISION AVOIDANCE

As Air Traffic keeps increasing, the ATC[1] system overload becomes a serious concern. For the last twenty years, different approaches have been tried, and many solutions proposed. In a few words, all theses solutions are between the two following extreme positions:

On the one hand, we can imagine an ATC system where every trajectory would be planned and where each aircraft would follow its trajectory with a perfect accuracy. With such a system, no reactive system would be needed, as no conflict[2] between aircraft would ever occur. This solution is close to the ARC-2000 hypothesis, which has been investigated by the Eurocontrol Experimental Center [5].

On the other hand, there could be an ATC system where no trajectories are planned. Each aircraft would flight its own way, and all collisions would have to be avoided by reactive systems. Each aircraft would be in charge of its own safety. This could be called a completely free flight system. The free flight hypothesis is currently seriously considered for all aircraft flying "high enough".

Of course, no ATC system will ever totally rely on only one of these two hypothesis. It is quite easy to understand why. A completely planned ATC is impossible, as no one can guarantee that each and every trajectory would be perfectly followed; there are too many parameters that can not be perfectly forecasted: meteorological conditions (storms, winds, etc.), but also breakdowns in aircraft (motor, flaps, etc) or other problems (closing of landing runaway on airports, etc.). On the other hand, a completely reactive system looks difficult to handle; it would only perform local optimizations for trajectories. Moreover, in the vicinity of departing and landing areas, the density of aircraft is so high that trajectories generated by this system could soon look like Brownian movements.

An ATC system can be represented by an assembly of filters, or shells. A classical view of the shells in an ATC system could be:

1. Airspace design (airways, control sectors, ...), When joining two airports, an aircraft must follow routes and beacons; these beacons are necessary for pilots to know their position during navigation and help controllers visualize the traffic. As there are many aircraft simultaneously present in the sky, a single controller is not able to manage all of them. So, airspace is partitioned into sectors, each of them being assigned to a controller. This task aims at designing the air network and the associated sectoring.

2. Air Traffic Flow Management (ATFM) (strategic planning, a few hours ahead), With the increasing traffic, many pilots choose the same routes, generating many conflicts over the beacons inducing overloaded sectors. Traffic assignment aims at changing aircraft routes to reduce sector congestion, conflicts and coordinations.

3. Coordination planning (a few minutes before entering in the sector), This task guarantees that each new aircraft entering a control sector does not overload the sector.

4. Tactical control in ATC centers (up to 20 mn ahead), At this level, controllers solve conflicts between aircraft.

5. Collision avoidance systems (a few minutes before collision). This level is activated only when the previous one has failed. This level is not supposed to be activated in current situations.

---

[1] Air Traffic Control

[2] 2 aircraft are said to be in conflict if their altitude difference is less than 1000 feet (305 meters) and the horizontal distance between them is less than 8 nautical miles (14800 meters). These two distances are respectively called vertical and horizontal standard separation

Each level has to limit and organize the traffic it passes to the next level, so that this one will never be overloaded.

In this paper, we present a problem solver that can handle the collision avoidance problem (level 5 filter) with reactive techniques. This problem solver is based on a neural network, which was built by a genetic algorithm.

EXISTING REACTIVE TECHNIQUES

The most well known concept on reactive collision avoidance is certainly the ACAS/TCAS concept. It is already implemented in its two first versions (TCAS-I and TCAS-II). It is a very short term collision avoidance system (less than 60 seconds). It should only be thought as the last security filter of an ATC system. Using TCAS to control aircraft would probably end in serious problems. The TCAS algorithm is based on a the application of a sequence of filtering rules, which give the pilot a resolution advice.

Another simple technique to do reactive control has been investigated by [4]. The idea is to consider each aircraft as positive electric charges, while the destination of the aircraft is a negative charge. This way, each aircraft creates a repulsive force proportional to the inverse of the square of the distance, while the destination behaves like an attractor. This technique has a serious drawback. Symmetries can not be broken. This problem was solved by [9]. This system is slightly more complex, but the general idea is to add non symmetrical force: a force which has the direction of the repulsive force +90 degrees, and a module which is a small fraction of the module of the repulsive force is added to the repulsive force. This system solves the symmetrical problem. However, there are still some drawbacks: the different parameters of the attractive and repulsive forces are arbitrarily set, and it is unclear to define how to find optimal values. Moreover, the shape itself of the forces is also arbitrarily set. But the main problem of this system is that it forces aircraft to modify their headings, but also their speeds. Unfortunately, the range of available speeds is very limited for aircraft flying at their requested flight level. Moreover, it is technically very difficult to change aircraft speed with a continuous command, as aircraft engines are easily damaged by this kind of operations.

Our system only allows heading modification and solves very complex two aircraft conflict, with almost optimal trajectories. Moreover, the system is very fast, as soon as the neural network has been built. Building neural networks with GA has already been done. An application quite similar was the problem of car parking described in [8]. However, our problem is definitely more complex.

MODELING THE PROBLEM

The problem we want to solve is the following. An aircraft flying at a constant speed detects another aircraft fly-

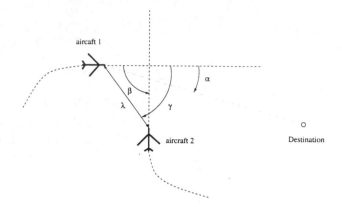

Figure 1: The neural network inputs of aircraft 1.

ing at the same altitude (more or less 1000 feet) in a 20 nautical miles diameter disk. We want to build a neural network that modifies the heading of this aircraft when there is a conflict (respecting operational constraint of 45 degrees maximum per 15 seconds). The other aircraft is supposed to have the same embarked system so that it also detects the first aircraft and reacts using the same neural network with different inputs.

The system uses an embarked radar to detect other aircraft. Consequently, all the inputs of the neural network must be given by the radar information.

USING A NEURAL NETWORK

In our problem, it seems clear that if no conflict occurs, no neural network is needed to solve it. Consequently, at each time step, we will first check if both aircraft can connect their destination without changing their heading and without generating conflicts. In that case, we do not modify aircraft headings. If we detect a conflict in less than twenty minutes, we compute a new heading for both aircraft with the NN.

The inputs

7 inputs are used by the neural network (see figure 1) :

- The heading of the destination α and its absolute value $|\alpha|$ (in degrees).

- The distance to the other aircraft λ and its gradient $\frac{d\lambda}{dt}$.

- The bearing of the other aircraft γ (in degrees)

- The converging angle of the trajectories β.

- A bias set to 1.

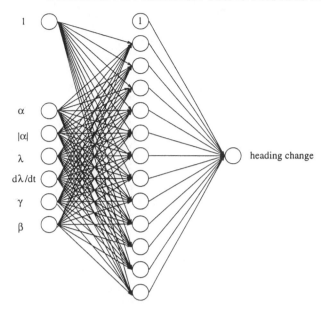

Figure 2: The neural network structure.

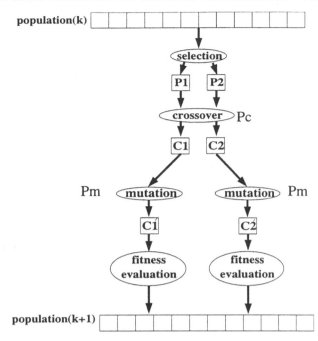

Figure 3: GA principle

The neural network structure

The neural network structure used is as simple as possible. A 3 layers network is used (see figure 2) and returns a heading change of 45 degrees maximum (for a time step of 15 seconds). The activation function used is the following :

$$act(s) \;=\; \frac{1}{1 + e^{-s}}$$

The first layer takes the 6 inputs described above plus the bias. The second layer holds 13 units, while the third layer holds the output unit.

Learning the neural network weights

Classical back propagation of gradient can not be used in our case because conflict free trajectories are not known in every configuration. They could be calculated for conflicts involving $n = 2$ aircraft, but the problem is not solvable for $n > 2$. As we plan to extend our system to more than two aircraft, we decided to use unsupervised learning with GA. However, we will compare the results of our network with optimal trajectories computed by LANCELOT[3] [2] to validate our hypothesis.

GENETIC ALGORITHMS

Figure 3 describes the main steps[4] of GAs that were used in this paper: first a population of points in the state space is randomly generated. Then, we compute for each population element the value of the function to optimize, which we will call *fitness*. Then the *selection process* reproduces elements according to their fitness. Afterwards, some elements of the population are picked at random by pairs. A *crossover operator* is applied to each pair and the two parents are replaced by the two children generated by the crossover. In the last step, some of the remaining elements are picked at random again, and a *mutation operator* is applied, to slightly modify their structure. At this step a new population has been created and we apply the process again in an iterative way. The different steps are detailed in the following.

Coding the problem

Here, each neural network is coded by a matrix of real numbers that contains the weights of the neural network.

Selection

A method called "Stochastic Remainder Without Replacement Selection" [3] was used. First, the fitness f_i of the n elements of the population is computed, and the average $a = \sum f_i/n$ of all the fitness is computed. Then each element is reproduced p times in the new population, with $p = truncate(n \times f_i/a)$. The population is then completed using probabilities proportional to $f_i - p\,a/n$ for each element.

Crossover

The crossover operator used is the barycentric crossover : 2 parents are recombined by choosing randomly $\alpha \in [-0.5, 1.5]$ and creating child 1 (resp child

---

[3] Large And Nonlinearly Constrained Extended Lagrangian Optimization Techniques
[4] We use classical Genetic Algorithms and Evolutionary Computation principles such as described in the literature [3, 7].

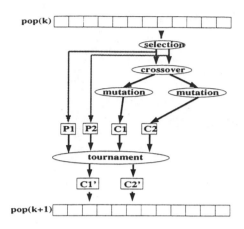

Figure 4: GA and SA mixed up

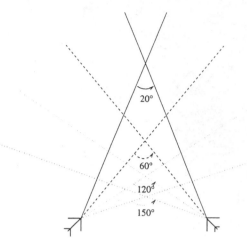

Figure 5: 4 configurations at the same speed.

2) as the barycentre of some randomly chosen weights of $(parent_1, \alpha)$ (resp $(parent_1, 1 - \alpha)$) and $(parent_2, 1 - \alpha)$ (resp $(parent_2, \alpha)$). In the further applications, the crossover probability used is 60%.

Mutation

The mutation operator adds a gaussian noise to one of the weights of the neural network. The mutation probability used here is 15%.

Simulated Annealing Tournament

GA can be improved by including a Simulated Annealing process after applying the operators [6]. For example, after applying the crossover operator, we have four individuals (two parents $P1, P2$ and two children $C1, C2$) with their respective fitness. Afterward, those four individuals compete in a tournament. The two winners are then inserted in the next generation. The selection process of the winners is the following: if $C1$ is better than $P1$ then $C1$ is selected. Else $C1$ will be selected according to a probability which decreases with the generation number (any cooling scheme used in simulated annealing can be used). At the beginning of the simulation, $C1$ has a probability of 0.5 to be selected even if its fitness is worse than the fitness of $P1$ and this probability decreases to 0.01 at the end of the process. A description of this algorithm is given on figure 4. Tournament selection brings some convergence theorems from the Simulated Annealing theory. On the other hand, as for Simulated Annealing, the (stochastic) convergence is ensured only when the fitness probability distribution law is stationary in each state point [1].

Other global data are required by the Genetic Algorithm such as the number of generations, the number of elements, the percentage of elements to cross and the percentage of elements to mutate.

Computing the fitness

One of the main issues is to know how to compute the *fitness* of a chromosome. The constrained problem to solve takes the following criteria into account :

- Aircraft trajectories must be conflict free.

- Delay due to deviation must be as low as possible.

To compute the fitness, a panel of different conflict configurations is created. The fitness is computed as follow :

$$F = \frac{1}{D} e^{-V}$$

D is the average delay due to deviations and V is the average number of conflict violations.

The learning examples

To learn the weights of the neural networks, 12 configurations were created. In each configuration, at $t = 0$ aircraft are 20 nautical miles distant.

- in 4 configurations, aircraft have the same speed and converge with different angles (20, 60, 120, 150 degrees, see figure 5).

- in 4 configurations, aircraft have different speed, their headings are calculated to generate a conflict (one aircraft speed is 500 knots and the other one is 300, 350, 400, and 450 see figure 6).

- in 2 configurations, aircraft have opposite headings and the same speed (see figure 7).

- in 2 configurations, aircraft have the same heading but different speeds (see figure 8).

Because of symmetries, these 12 configurations summarize all the situations that can happen. We will call "positive

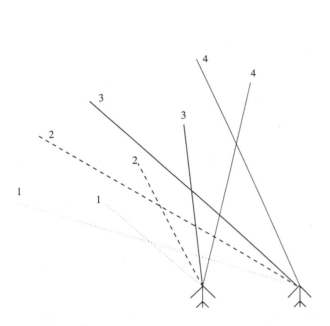

Figure 6: 4 configurations at the different speeds.

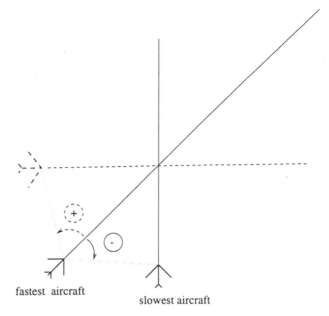

Figure 9: Symmetrical configurations.

configuration" (see figure 9) a configuration in which the angle between the slowest aircraft and the fastest is positive. When a "negative configuration" occurs, the symmetrical positive configuration is used in the neural network to calculate the deviation. Therefore, some of the inputs and the output are given the opposite sign.

NUMERICAL RESULTS

The neural network was learned using the following parameters :

number of generations : 500
population elements : 500
percentage of crossover : 60
percentage of mutation : 15
simulated annealing for crossover : yes

Optimal solutions to the different configurations were calculated using gradient method such as LANCELOT. LANCELOT has the great advantage to find the optimal solution to our problems but requires much more time (one hour on HP720). Controlling aircraft in real time with this technique is not possible. However, it is interesting to compare optimal solutions found by LANCELOT to solutions learned by the neural network. Learned solutions are obviously less optimal, but the loss of optimality is not significant (the delay induced by the neural network is never more than 4 times the minimal delay, which is generally very small).

The configurations used to compare the neural network to optimal solutions are not learned configurations. We

Figure 7: 2 configurations of facing aircraft.

Figure 8: 2 configurations of facing aircraft.

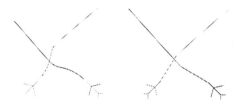

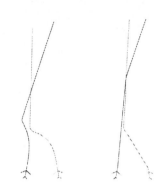

Figure 10: Neural network solution (left), optimal solution (right).

want also to validate the capacity of the NN to generalize to non-learned situations. For each solution, we give the mean lengthening of the trajectories in percentage :

Figure 11: Neural network solution (left), optimal solution (right).

- Figure 10 gives an example of conflict at 90 degrees in which aircraft have the same speed. Neural network (1.08%) and optimal solution (0.26%) are similar. The NN solution mean lengthening is quite worse than the optimal solution lengthening.

- Figure 11 gives an example of conflict at 15 degrees in which aircraft have the same speed. Such a conflict is particularly difficult to solve. Solutions are different, but for such a difficult conflict, the neural network (2.30%) gives a solution that is robust and quite as good as the optimal solution (2.23%). This conflict is the most difficult conflict to solve (in the 5 examples presented). It is interesting to see that the difference of lengthenings is the smallest.

- Figure 12 gives an example of aircraft at different speeds (400 and 500 knots) with crossing at a small angle (30 degrees). The neural network solution (1.32%) appears very similar to the optimal solution (0.28%). Its lengthening is quite worse.

- Figure 13 gives an example of aircraft crossing on the same route. This problem is easy to solve and solutions are similar. The NN solution (1.18%) is robust but worse than the optimal solution (0.25%).

- Figure 14 gives an example of aircraft flying on parallel routes at different speeds. This problem is easy to solve. Solutions are similar. The NN solution (1.02%) is robust but worse than the optimal solution (0.21%).

Figure 12: Neural network solution (left), optimal solution (right).

Thess 5 examples show that the principal advantage of th NN is to be very robust. It does not give optimal solu-

2

tions. However, we can see that it gives very good solutions for difficult conflicts. Tests done on non-learned situations gave as good results as tests done on learned situations.

CONCLUSION

Using a simple neural network to solve a conflict between 2 aircraft has given very good results. It was shown above that the neural network could be easily learned by a genetic algorithm without knowing the optimal solutions. The next step of this work will consist in extending the problem to conflicts involving more than 2 aircraft. As the problem becomes very combinatory, some hypothesis will probably have to be made to limit the size of the neural network. The third step will be to integrate climbing an descending aircraft in the model and to generate vertical manœuvres. The results presented above should be very soon used in a Test Bench to check their validity on real traffic.

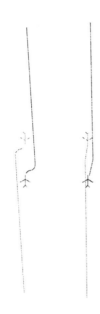

Figure 13: Neural network solution (left), optimal solution (right).

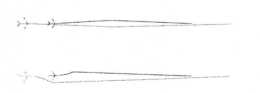

Figure 14: Neural network solution (down), optimal solution (up).

References

[1] Emile Aarts and Jan Korst. *Simulated annealing and Boltzmann machines.* Wiley and sons, 1989. ISBN: 0-471-92146-7.

[2] A.R. Conn, Nick Gould, and Ph. L. Toint. A comprehensive description of LANCELOT. Technical report, IBM T.J. Watson research center, 1992. Report 91/10.

[3] David Goldberg. *Genetic Algorithms.* Addison Wesley, 1989. ISBN: 0-201-15767-5.

[4] H. Gruber. Comparaison de diverses méthodes d'intelligence artificielle pour la résolution de conflit en contrôle de trafic aérien. Rapport de stage, Centre d'Etudes de la Navigation Aérienne, 1992.

[5] Fred Krella et al. Arc 2000 scenario (version 4.3). Technical report, Eurocontrol, April 1989.

[6] Samir W. Mahfoud and David E. Goldberg. Parallel recombinative simulated annealing: a genetic algorithm. IlliGAL Report 92002, University of Illinois at Urbana-Champaign, 104 South Mathews Avenue Urbana IL 61801, April 1992.

[7] Zbigniew Michalewicz. *Genetic algorithms+data structures=evolution programs.* Springer-Verlag, 1992. ISBN: 0-387-55387-.

[8] Marc Schoenauer, Edmund Ronald, and Sylvain Damour. Evolving nets for control. Technical report, Ecole Polytechnique, 1993.

[9] Karim Zeghal. *Vers une théorie de la coordination d'actions, application à la navigation aérienne.* PhD thesis, Universite Paris VI, 1994.

IDENTIFYING CHEMICAL SPECIES IN A PLASMA USING A NEURAL NETWORK

Phil D. Picton*, Adrian A. Hopgood, Nicholas St. J. Braithwaite and Heather J. Phillips
*Faculty of Applied Sciences, Nene College, Northampton, England, NN2 6JD.
Email: phil.picton@nene.ac.uk
Faculty of Technology, The Open University, Milton Keynes, England, MK7 6AA.

ABSTRACT

This paper addresses the problem of identifying the chemical species present in a plasma using optical emission spectra. Two neural networks are initially explored, a multi-layer perceptron and a linear associative memory which is trained using the Penrose pseudo-inverse matrix. However, examination of the spectra leads to a proposal of a more appropriate architecture in which individual neurons are trained on individual species.

1. INTRODUCTION

A plasma is an ionised gas, such as occurs in a fluorescent light. Plasma deposition is a technique that uses plasmas for depositing thin films (e.g. diamond-like carbon) on the surface of materials. One way of estimating the chemical composition of a plasma, and hence the resulting deposition process, is to analyse the optical emission spectrum of the light that is emitted from a plasma. However, the spectral data are often complex and noisy, and so conventional methods are unlikely to succeed. Neural networks [Pic94] are very good at classifying patterns, and have therefore become widely used in this sort of problem. As well as their ability to deal with noisy or incomplete data, they also have the potential to be very high speed. This makes them a good candidate technology for real-time analysis of spectra, which is essential for the control of a deposition process.

Usually neural networks are trained by showing examples of individual patterns, together with the appropriate output response. After training the network would be expected to classify similar patterns. However, some applications require the identification of individual patterns when they are presented as a mixture of patterns. Most neural networks which have been trained on patterns A and B (for instance), would classify a mixture of A and B as a different pattern, C. This paper is concerned with an application where the network is required to identify the presence of pattern A and B in the mixed pattern. Spectral data analysis is one such problem. The spectra of individual objects, such as the optical spectral emissions from specific chemicals [BRU93, MDO+94], contain peaks at known wavelengths and so could be used for training. The problem is then identifying individual species when presented with the spectrum of a mixture of chemicals.

For the purposes of this paper a mixed pattern will be considered to be a weighted sum of individual patterns.

Thus, if the individual patterns are X_1, X_2 to X_P, then a mixed input pattern is:

$$X = \sum_{i=1}^{P} k_i X_i$$

2. MULTI-LAYER PERCEPTRON (MLP)

Consider a simple case where a network has to learn to recognise two patterns. When presented with a mixture of the two patterns the ideal output would be at least the recognition of one or other of the patterns. If we take the simplest possible solution in which the problem is found to be linearly separable, then only one layer of neurons is required. So the network would consist of two neurons, one to recognise pattern X_1 and the other to recognise pattern X_2.

Assume that the function of the network is to produce 0 and 1 outputs, where 1,0 corresponds to pattern X_1, and 0,1 corresponds to pattern X_2, and that patterns closest to pattern X_1 would produce a 1,0 output and patterns closest to pattern X_2 will produce a 0,1 output.

Neuron 1 will have learned a set a weights W_1, and neuron 2 will have W_2. When the neurons are presented with pattern X_1, the weighted sums (net) are:

$$net_1(1) = X_1.W_1 \qquad net_2(1) = X_1.W_2$$

When presented with X_2:

$$net_1(2) = X_2.W_1 \qquad net_2(2) = X_2.W_2$$

If a sigmoid output function is used, when the network is shown X_1, the output of neuron 1 should be 1, which means that $net_1(1) > 0$, and $net_2(1) < 0$. Similarly for pattern X_2, $net_1(2) < 0$, $net_2(2) > 0$.

If a mixed pattern is presented, $X_3 = k_1X_1 + k_2X_2$, the weighted sums will be:

$$net_1(3) = (k_1X_1 + k_2X_2)W_1 = k_1net_1(1) + k_2net_2(1)$$

$$net_2(3) = (k_1X_1 + k_2X_2)W_2 = k_1net_1(2) + k_2net_2(2)$$

There is no way of knowing whether these weighted sums will produce 0s or 1s as outputs. Since $net_1(1) > 0$, so that when it is multiplied by k_1 (which is assumed to be less than 1) the value of the weighted sum will decrease but will still be greater than 0. However, the addition of $k_2net_2(1)$ which is negative, could cause the weighted sum to become negative also. So even in the simplest case of two patterns in a linearly separable problem, mixed patterns can produce unpredictable results.

3. LINEAR ASSOCIATIVE MEMORY

A well known linear associative memory uses the Penrose pseudo-inverse matrix to set the weights [Koh84]. This is applicable to problems where the stored patterns are linearly independent, a condition which can often be found in practical problems. Olmos *et al* [ODP+91] used this method to identify isotopes using γ-ray spectra of a mixture of the isotopes.

In terms of matrices, the network can be described as follows:

$$K = (X^tX)^{-1}X^tX_i$$

where X_i is the input pattern which equals KX i.e. the weighted sum of the training patterns X. Thus the proportion of each of the trained patterns, K, is produced as the output when the weights are:

$$W = (X^tX)^{-1}X^t$$

This is the formula for the Penrose pseudo-inverse matrix. So, under the condition that the stored patterns are linearly independent, the pseudo-inverse gives the values for the weights to be stored in a single-layer network. When a mixed pattern appears as the input, the output values are the proportions of each of the trained patterns in the mixed pattern.

4. OPTICAL EMISSION SPECTRA

Closer examination of typical optical emission spectra indicates that in the majority of cases the peaks from different species are not coincident. This leads to the conclusion that it should be possible to attach the inputs of an individual neuron to the wavelengths at which known peaks occur for a particular species. A layer of neurons is produced where each neuron identifies a specific chemical. Each neuron is trained using examples of plasmas which are known to contain that particular chemical.

The advantages of this method are that it is quick to train and can adapt to the particular plasma process that is being analysed. Training is achieved using Kohonen learning [Pic94, Koh84].

As an example, Figures 1, 2 and 3 show examples of the spectra from various plasmas. For this particular experiment the vacuum chamber which holds the plasmas was initially evacuated, then either a single gas introduced or a mixture of two. This process was repeated several times, and the spectra recorded. First Argon (Ar) was introduced and the spectrum shown in Figure 1 found, then Hydrogen (H) and the spectrum in Figure 2 found. Finally a plasma was generated with a mixture of Argon and Hydrogen (Ar/H) and the spectrum in Figure 3 recorded.

Since the absolute value of the peaks is of no great importance when it comes to identifying the presence of a species, the data are normalized. Normalization is carried out locally at each neuron, so that a neuron always receives a vector of unit length. Two neurons are used in this network - one to identify Argon and one to identify Hydrogen, as shown in Figure 4. The values that are used to train the neurons are taken from the examples of single gas plasmas. Appendix 1 shows the data used for training and testing, and indicates the wavelengths that were selected as the sampling points in the spectra. Note that in Appendix 1, the first 10 rows show the wavelengths that are used to detect Argon, and the last 3 rows show the wavelengths that were used to detect Hydrogen. The weights after training are shown in Table 1.

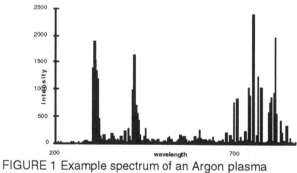
FIGURE 1 Example spectrum of an Argon plasma

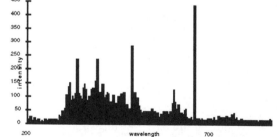
FIGURE 2 Example spectrum of a Hydrogen plasma

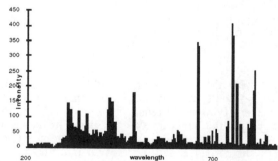
FIGURE 3 Example spectrum of a plasma containing Argon and Hydrogen

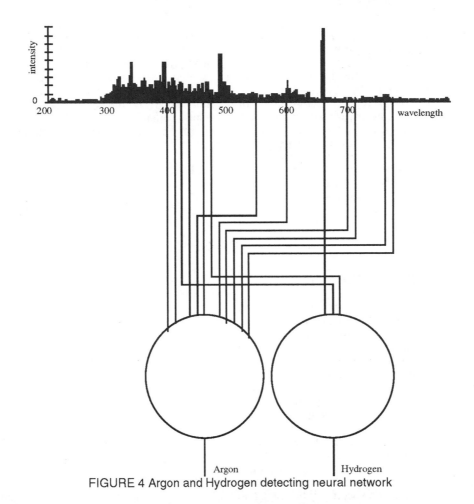

FIGURE 4 Argon and Hydrogen detecting neural network

Table 1a Trained weights for Ar detector

w1	w2	w3	w4	w5	w6	w7	w8	w9	w10
0.192	0.220	0.077	0.046	0.054	0.087	0.140	0.144	0.797	0.445

Table 1b Trained weights for H detector

w1	w2	w3
0.254	0.686	0.668

Table 2 Only Argon present

Chemical	Ar 1	Ar 2	Ar 3	Ar 4	Ar 5	Ar 6	Ar 7
Argon	0.964	0.964	0.966	0.9819	0.986	0.984	0.982
Hydrogen	0.490	0.875	0.869	0.934	0.9621	0.893	0.966

Table 3 Only Hydrogen present

Chemical	H 1	H 2	H 3	H 4	H 5	H 6	H 7
Argon	0.344	0.278	0.290	0.325	0.301	0.231	0.257
Hydrogen	0.970	0.990	0.988	0.975	0.987	0.991	0.969

Table 4 Output values for each mixture

Chemical	Ar/H 1	Ar/H 2	Ar/H 3	Ar/H 4	Ar/H 5
Argon	0.977	0.983	0.987	0.945	0.862
Hydrogen	0.918	0.926	0.944	0.916	0.853

The original plasmas and the ones containing mixtures of Argon and Hydrogen are then used to test the neurons using the network shown in Figure 4. The results are shown in Table 2, 3 and 4.

The results in Table 4 are as expected - high values for both Argon and Hydrogen. In Table 3 the results are also as expected - high values for Hydrogen and low values for Argon. The results in Table 2 however were unexpected. In these experiments, only Argon had been deliberately introduced into the chamber, so the spectra were expected to show peaks for Argon but not for Hydrogen. Closer examination of the data shows that there are small peaks present at the wavelengths associated with Hydrogen, so that there was a residual Hydrogen presence in the plasma. When the vacuum chamber was re-examined a leak was found which was the cause of this residual Hydrogen.

5. CONCLUSIONS AND FURTHER WORK

It would seem that as far as recognising individual patterns when presented with a mixture of patterns, the mlp can be ruled out. If the patterns are known to be linearly independent, then the Penrose pseudo-inverse matrix provides a set of weights for a single layer network that gives the relative amounts as its outputs.

For applications such as spectral analysis, the Penrose pseudo-inverse matrix is unnecessary and overly complicated. A network that consists of a single layer of neurons, where each neuron is trained on the normalized vectors produced by sampling the spectra at specific wavelengths, is sufficient to detect the presence of species from the optical emission spectra. It has the advantage that it can be trained for specific systems.

The method described has been specifically designed to be able to recognise species but not to measure the quantity of that species. However, it would be useful to know the relative quantities of a particular species over time. For example, in controlling a plasma deposition process, an action might be to increase the amount of Argon in the plasma. It would be useful to know at a later stage if the amount of Argon has increased, i.e. a qualitative measure rather than a quantitative measure is all that is required. One way that is currently being explored is to use the size of the input vector, which has to be calculated during normalization, as an indication of relative quantity. It is yet to be determined if this is sufficient, and is therefore an area of future research.

REFERENCES

[BRH93]   C. M. Bishop, C. M. Roach and M. G. Hellermann. Automatic Analysis of JET Charge Exchange Spectra Using Neural Networks. *Plasma Phys. Control Fusion*, 35: 765-773, 1993.

[Koh84]   T. Kohonen. *Self-Organization and Associative Memory*. Springer-Verlag, Berlin. 1984.

[MDO+94]   C. R. Mittermayr, A. C. J. H. Drouen, M. Otto and M. Grausserbauer. Neural Networks for Library Search of Ultraviolet Spectra. *Analytica Chimica Acta*, 294: 227-242, 1994.

[ODP+91]   P. Olmos, J. C. Diaz, Perez, J. M., P. Gomez, V. Rodellar, P. Aguayo, A. Bru, G. Garcia-Belmonte and J. L. de Pablos. A New Approach to Automatic Radiation Spectrum Analysis. *IEEE Transactions on Nuclear Science*. 38: 971-975, 1991.

[Pic94]   P. D. Picton. *Introduction to Neural Networks*. Macmillan Press Ltd, Basingstoke, Hampshire. 1994.

APPENDIX 1 TABLES OF DATA

Table 5 Data from seven spectra containing only Argon (Ar)

nm	Ar 1	Ar 2	Ar 3	Ar 4	Ar 5	Ar 6	Ar 7
415.8	952.69	102.22	72.802	162.99	150.24	176.71	154.22
420.1	1005.2	60.26	63.646	226.91	200.03	235.83	158.89
451.1	208.74	21.452	30.79	96.678	69.448	75.868	63.566
484.8	14.098	11.788	22.224	79.04	62.946	54.904	13.192
549.5	127.52	55.09	35.928	55.818	51.262	40.86	33.76
603.2	236.14	57.044	42.324	77.71	64.376	85.6	82.688
696.5	741.66	115.32	85.102	93.888	118.33	116	83.424
706.7	811.14	124.17	105.19	116.61	97.914	90.95	86.862
750.4	2387.6	809.47	607.46	597.79	660.87	630.3	574.14
763.5	1237.9	535.14	422.65	350.88	329.54	344.16	251.68
434	135.77	80.542	66.89	132.63	95.65	83.736	25.572
486.1	24.89	65.112	54.994	122.73	117.84	64.582	32.994
656.3	26.002	214.4	192.11	184.49	151.63	75.564	41.516

Table 6 Data from seven spectra containing only Hydrogen (H)

nm	H 1	H 2	H 3	H 4	H 5	H 6	H 7
415.8	999.57	49.302	76.724	123.81	100.47	81.118	79.78
420.1	1195.8	104.86	107.77	228.29	228.62	135.16	107.29
451.1	654.93	33.808	43.256	99.372	91.222	87.722	25.058
484.8	652.39	161.93	135.51	278.11	294.25	286.88	183.11
549.5	169	16.612	29.574	45.99	39.516	40.8	15.946
603.2	489.05	29.28	31.846	59.838	68.236	39.892	22.924
696.5	35.102	15.748	7.2517	4.8437	3.6677	2.7857	4.0057
706.7	34.516	0.6477	0	10.038	2.8217	9.7617	0
750.4	40.448	9.7177	1.5017	18.496	21.258	0	7.9997
763.5	45.934	2.5957	7.3157	26.354	12.982	14.868	0
434	762.09	98.698	91.938	177.82	160.33	156.17	75.048
486.1	1036.6	281.84	281.76	466.95	523.37	432.55	275.52
656.3	1277.4	255.9	234.46	652.67	584.83	427.59	172.7

Table 7 Data from five spectra of mixtures of Argon and Hydrogen (Ar/H)

nm	Ar/H 1	Ar/H 2	Ar/H 3	Ar/H 4	Ar/H 5
415.8	113.99	111.64	107.05	36.366	15.954
420.1	110.55	161.17	126.6	41.132	42.012
451.1	35.544	40.534	45.442	18.416	9.0417
484.8	75.264	70.696	28.1	81.43	27.054
549.5	25.654	30.968	42.144	19.522	14.38
603.2	44.464	64.61	64.078	36.67	10.304
696.5	43.334	73.552	78.916	29.822	4.9557
706.7	48.128	67.33	102.54	15.666	19.332
750.4	357.61	501.34	542.29	209.3	55.052
763.5	206.05	253.91	292.98	116.7	19.178
434	73.264	63.094	42.682	56.04	34.566
486.1	139.38	95.082	47.956	91.838	53.234
656.3	324.27	211.26	84.056	219.02	199.81

A METHOD FOR SELECTING LEARNING DATA IN THE PREDICTION OF TIME SERIES WITH EXPLANATORY VARIABLES USING NEURAL NETWORKS

Hisashi Shimodaira

Research and Development Department, Nihon MECCS Co., Ltd.

4-13-27 Midori-Cho, Koganei-Shi, Tokyo 184, Japan

ABSTRACT

In the prediction of time series using multi-layer feedforward neural networks, there are two practical methods for selecting learning data: the moving window data learning method and the similar data selective learning method with the correlation coefficient based similar data selection method which we proposed in a previous paper. In this paper, for time series data with explanatory variables, the predictive performance by the two methods was investigated by numerical simulations. With the time series whose nature is choppy, that by the latter was considerably better than that by the former. With the time series whose nature is smooth, that by the former was slightly better than that by the latter. According to these results, it is conjectured that the latter is effective for a time series whose nature is choppy.

1. INTRODUCTION

According to the simulation results of time series predictions by Lambert [LH91], the predictive accuracy by recurrent networks was better than that by multi-layer feedforward networks, whereas the former took much more learning time than the latter. Thus it is desirable to explore a method of obtaining more accurate prediction values using feedforward networks. The objective of this paper is to explore such a method.

The predictive performance by multi-layer feedforward neural networks depends on how to select learning samples from the historical database of a time series. In the practical application, two kinds of methods have been used: the moving window data learning method (MWDL method) [PH90], the similar data selective learning method (SDSL method) [PH92], [Sh95]. In the SDSL method, data which are similar to the data used to make a prediction, are selected from the historical database as learning samples. The similarity between the data group used to make a prediction and a candidate data group for learning samples, is measured with the distance between them.

In a previous paper [Sh95], we proposed a method of weighting the distance by the power function of correlation coefficients for the time series, which we call the correlation coefficient based similar data selection method (CSDS method). According to the results of simulations on single variable time series, it was found that with a time series whose nature is choppy, the predictive performance by the SDSL method with the CSDS method is considerably better than that by the MWDL method [Sh95]. In this paper, we extend the method to time series with explanatory variables and compare the predictive performance by the two methods through numerical experiments. As the object for the numerical experiments, we used measured data on warm-up and pull-down time in building air conditioning. In most buildings in Japan, the air conditioning equipment is stopped for the night and they are defined as the amount of time required to reach the desired room temperature in winter and the desired room discomfort index in summer, respectively, from the time when the air conditioning equipment is started.

2. PREDICTION PROCEDURES AND METHODS OF SELECTING LEARNING SAMPLES

We explain prediction procedures using a time series with a response variable y_i ($i=1, \cdots, t, \cdots, \infty$) and an explanatory variable x_i ($i=1, \cdots, t, \cdots, \infty$). As in the numerical experiments mentioned later, when the values of points for t and $(t-1)$ and previous to them for x_i and y_i, respectively, were observed, y_t is predicted. Using the group of d and $(d-1)$ data on an explanatory and a response variable, respectively, as input to the network, Y_t is output as the predicted value of y_t as follows:

$$(x_{t-d+1}, \ldots, x_{t-1}, x_t, y_{t-d+1}, \ldots, y_{t-2}, y_{t-1}) \longrightarrow Y_t \qquad (1)$$

At the training stage, using the group of d and $(d-1)$ data as input to the network, $Y_{t'}$ is produced as follows:

$$(x_{t'-d+1}, \ldots, x_{t'-1}, x_{t'}, y_{t'-d+1}, \ldots, y_{t'-2}, y_{t'-1}) \longrightarrow Y_{t'} \qquad (2)$$

The weight values of the network are estimated using the error backpropagation method [RM86] in which the sum of the errors between $y_{t'}$ and $Y_{t'}$ for n learning samples, is minimized. At the prediction stage, these estimated weight values are used. In the SDSL method, learning samples are selected from the range for selecting learning samples in the historical database in the order that the similarity between the data used to make a prediction and the candidate for learning samples, is greater. In order to quantify the degree of the similarity, both the data groups are regarded as two points in the d-dimensional and $(d-1)$-dimensional Euclidean space and the sum of the distance between them is calculated. It is assumed that the smaller the distance is, the greater the similarity is; and n learning samples are selected in the increasing order of the distance. To attach a sense to the distance using data with different units, the data in the range for selecting learning samples, are standardized by the following equations:

$$x_i^s = (x_i - m_x)/d_x \qquad (3)$$

$$y_i^s = (y_i - m_y)/d_y \qquad (4)$$

where the superscript s denotes the standardized value; m_x and m_y are the mean values for x_i and y_i, respectively; and d_x and d_y are the mean absolute deviations defined as follows:

$$d_x = \frac{1}{n_x} \sum_{i=1}^{n_x} |x_i - m_x| \qquad (5)$$

$$d_y = \frac{1}{n_y} \sum_{i=1}^{n_y} |y_i - m_y| \qquad (6)$$

where n_x and n_y are the total number of data. The measurements of the distance in the CSDS method [Sh95] are extended as follows. The weighted Manhattan distance is defined by

$$D = \sum_{i=0}^{d-1} |\rho_{xi}|^m |x_{t-i}^s - x_{t'-i}^s|$$
$$+ \sum_{i=1}^{d-1} |\rho_{yi}|^m |y_{t-i}^s - y_{t'-i}^s| \qquad (7)$$

The weighted Euclidean distance is defined by

$$D = \{\sum_{i=0}^{d-1} |\rho_{xi}|^m (x_{t-i}^s - x_{t'-i}^s)^2\}^{\frac{1}{2}}$$
$$+ \{\sum_{i=1}^{d-1} |\rho_{yi}|^m (y_{t-i}^s - y_{t'-i}^s)^2\}^{\frac{1}{2}} \qquad (8)$$

where ρ_{xi} $(0 \leq \rho_{xi} \leq 1)$ is the correlation coefficient between y_t and x_t and i is the time difference or lag. ρ_{yi} $(0 \leq \rho_{yi} \leq 1)$ is the autocorrelation coefficient for y_t and i is the time lag. In the above equations, ρ_{xi} and ρ_{yi} to the m-th power are used to adjust the influence of their magnitude on the distance calculation. The optimum value for m is a part of the investigation. Eqs. (7) and (8) are basically the weighted distances traditionally used to identify clusters of data in a multidimensional space [KR90]. In these equations, the greater the

correlation coefficients is, by the greater factor the coordinate is scaled. ρ_{xi} and ρ_{yi} represent the degree of dependences of y_t on x_{t-i} and y_{t-i}, respectively. Therefore, by Eqs. (7) and (8), distances in which we attach great importance to past data which have more influence on the value to be predicted, are calculated.

Let us define the first points of the ranges for selecting learning samples as x_s and y_s. Strictly, ρ_{xi} needs to be calculated by the data sets (x_s, y_{s+i}), (x_{s+1}, y_{s+1+i}), $\cdots$, (x_{t-i-1}, y_{t-1}), (x_{t-i}, y_t). However, because y_t is unknown, the approximate value is calculated by the data sets except (x_{t-i}, y_t). Strictly, ρ_{yi} needs to be calculated by the data sets (y_s, y_{s+i}), (y_{s+1}, y_{s+1+i}), $\cdots$, (y_{t-i-1}, y_{t-1}), (y_{t-i}, y_t). However, because y_t is unknown, the approximate value is calculated by the data sets except (y_{t-i}, y_t).

In Eqs. (7) and (8), it is assumed implicitly that the correlation coefficients between y_t and x_{t-i}, and $y_{t'}$ and $x_{t'-i}$ are the same value and the autocorrelation coefficients between y_t and y_{t-i}, and $y_{t'}$ and $y_{t'-i}$ are the same value. If this assumption holds for all the past data, we can fix x_s and y_s and enlarge the ranges for selecting learning samples, as observed values are added to the database, to utilize the past data effectively. If the assumption does not hold for all the past data, we need to determine the size of the range for selecting learning samples in which the assumption approximately holds and move them (sift x_s and y_s) forward according to a predicting point.

While in the Manhattan distance, the differences of coordinates themselves are considered, in the Euclidean distance, the squares of them are considered. Thus, the latter is more affected by the larger values of differences of coordinates than the former. The effect of the definition of distances used on the predictive performance, is investigated by the numerical experiments.

In the MWDL method, all the data in the moving window which is allocated in the immediately previous range of a predicting point, are grouped to form the learning samples according to Eq. (2). In this paper, we define the size of the moving window by d and n.

3. NUMERICAL EXPERIMENTS

A. Overview of Numerical Experiments

We used data on temperature and relative humidity measured at every hour in two buildings. We calculated the amount of time required to reach the desired room temperature and discomfort index by the linear interpolation using these data and used it as the measured warm-up and pull-down time. To investigate the nature of the time series in advance, the autocorrelation coefficients were calculated using all the data and the correlogram which is the plot of the autocorrelation coefficient versus the time lag (τ), was made. The autocorrelation coefficient tends to reflect the essential smoothness of the time series [Sh88]. Very smooth series exhibit autocorrelation coefficients which stay large even when τ is large, whereas very choppy series tend to have autocorrelation coefficients which are nearly zero for large τ's. In addition, the correlation coefficients between the response variable and a candidate for the explanatory variables, were calculated, and the explanatory variables used as input were selected based on them.

As mentioned in Section 2, the value of the response variable was predicted using (d–1) values for the response variable and d values for each explanatory variable. Holidays were not included in the prediction object, because the air conditioning equipment was not operated on holidays. The neural network used consists of one input layer with [(k+1)d–1] node (k: the number of explanatory variables), one hidden layer with h nodes, and one output layer with one node. The predictive performance depends on the range from which learning samples are selected, the number of data d, and the number of learning samples n. An optimum value of the node number (h) of the hidden layer depends on the characteristics of the time series as well as d and n. The parameters d, n, and h are not fixed; but determined as a part of the investigation in the numerical experiments. The optimum combination of the values of the parameters was sought by performing simulations in which these values were changed little by little.

As the activation function for the hidden and the output layer, the logistic function ($g(x)$=1/($1+e^{-x}$)) is used. The input data are scaled so that the values are within the range [0.2, 0.8]. The magnitude of the input data may vary according to each predicting point. This scaling allows the magnitude to be in the same range. In the backpropagation, we used the method in which the sum of errors for all the learning samples is propagated backward. We used 0.1 as the learning rate for the backpropagation method and 0.9 as the coefficient for the inertia term. The network was trained until the half of the sum of the errors in the output node becomes less than 0.001 and 0.0005 for the warm-up and pull-down time, respectively.

With the SDSL method, because according to preliminary tests, the predictive performance in the case where the top points of the range for selecting learning samples were shifted, was better than that in the case where those were fixed, the former was used. As its size, we used 40 days. However, when a predicting day is within 40 days from the top of the data, the learning data were selected from the range between the top of the data and the predicting day. The effect of m in Eqs. (7) and (8) was investigated by taking integer as its value.

To measure the predictive performance, we use correlation coefficients between the measured and the predicted values (CRC), mean values (mean) and standard deviations (σ) of the absolute errors (ABE), and mean values and standard deviations of the relative errors (RLE). The standardized absolute error is defined by

$$ABE = |Y_i - y_i| \times 100/y_{max} \qquad (9)$$

where y_{max} is the maximum value of the measured values in the range where the predictions were performed. The relative error is defined by

$$RLE = |Y_i - y_i| \times 100/y_i \qquad (10)$$

It should be noted that the RLE can become considerably large, because the numerator and the denominator can be in the same degree in magnitude in the case where the values to be predicted is very small.

B. Predictions about a Time Series on warm-up time

We used 76 days of data from December in 1993 to March in 1994. The air conditioning equipment was started at a time between 8 a.m. and 9 a.m. according to the cold of the morning. Assuming the desired room temperature to be 20℃, we calculated the amount of the warm-up time and used it as the measured value. Fig. 1 shows its time series data. For example, 94.1 on the horizontal axis shows that the data at the part belong to January in 1994. As the explanatory variables, we used the room temperature and the relative humidity at 8 a.m. on the day and the room temperature at 12 a.m. on the previous day. Usually, on next days of holidays, the warm-up time becomes longer, because the building is cooled and its temperature becomes lower. The last explanatory variable was used to consider automatically whether the air conditioning equipment was operated or not. Fig. 2 shows the correlogram. We can see from Fig. 1 and Fig. 2 that the time series has a rather choppy nature. As an example, Fig. 3 shows the correlation coefficient between the data on the warm-up time and the room temperature at 8 a.m. for τ. It has a large value for τ=0, whereas it has small values for τ's equal to and larger than 1. Thus it is found that a small range of the past data has the influence on the warm-up time on the predicting day. The prediction calculation was started from the 24th day from the top of the data.

Table 1 shows the optimum parameter values for each method by which the best prediction performance was obtained, where, for example, 7-11-1 denotes the network with 7 input nodes, 11 hidden nodes, and one output node. Table 2 shows the results by the SDSL method in which these parameter values were used. In the table, DST=1 denotes the Manhattan distance; and DST=2, the Euclidean distance. COR=0 denotes that the weighting is not used in Eqs. (7) and (8), and COR=1 denotes that the weighting is used. The optimum value of m for both distances was 2. These results show that the predictive performance by the SDSL method was considerab-

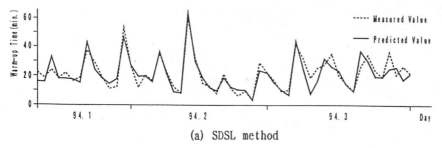

(a) SDSL method

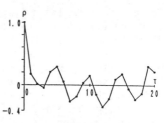

Fig. 2 Correlogram for warm-up time

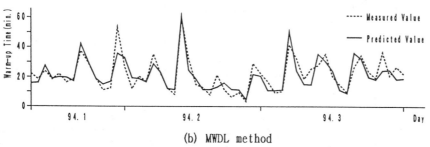

(b) MWDL method

Fig. 1 Measured and predicted values for time series on warm-up time

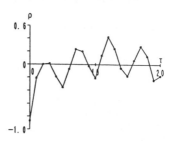

Fig. 3 Correlation coeff-
icients between warm-up
time and room temperature
at 8 *a.m.*

Table 1 Optimum parameter values for the prediction of warm-up time

Method	d	Network	n
SDSL	2	7-11-1	20
MWDL	2	7-8-1	22

Table 3 Prediction results for warm-up time with MWDL method using optimum parameter values

CRC	ABE mean	ABE σ	RLE mean	RLE σ
0.874	7.21	5.78	21.7	16.0

Table 2 Prediction results for warm-up time with SDSL method using optimum parameter values

DST	COR	m	CRC	ABE mean	ABE σ	RLE mean	RLE σ
1	0	—	0.823	8.12	7.74	24.0	16.9
	1	1	0.888	6.92	5.85	20.5	16.5
		2	0.907	6.31	5.13	18.8	15.2
		3	0.887	6.97	5.64	20.5	15.4
2	0	—	0.827	7.95	7.38	24.7	21.6
	1	1	0.873	7.68	6.16	22.5	16.3
		2	0.909	6.10	5.29	17.8	16.0
		3	0.910	6.26	4.95	19.4	16.0

ly improved by weighting the distance using autoc-orrelation coefficients in a proper degree. The best performance is obtained in the case of the Euclidean distance with $m=2$. The predicted values in this case are shown in Fig. 1 (a). Compared with the case where the weighting was not used, this case offers a 9.92% increase in the *CRC*; a 23.3% and a 28.3% reduction in the *mean* and σ of the *ABE*, respectively; and a 27.9% and a 25.9% reduction in the *mean* and σ of the *RLE*, respectively.

Table 3 shows the results by the MWDL methods in which the optimum values of the parameters were used. The predicted values in this case are shown in Fig. 1 (b). From Tables 2 and 3 and Fig. 1, it is found that the SDSL method with the best parameter values performs better than the MWDL method. Compared with the MWDL method, it offers a 4.0% increase in the *CRC*; a 15.4% and an 8.5% reduction in the mean value and the standard deviation of the *ABE*, respectively; and an 18.0% reduction in the mean value of the *RLE*.

C. Predictions about a Time Series on pull-down time

We used 82 days of data in the summer seasons in 1993 and 1994. The air conditioning equipment was started at 7:45 every day. In the air cooling, because the comfort depends on humidity as well as temperature, the discomfort index defined by the

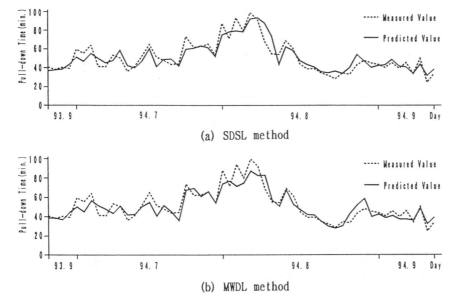

(a) SDSL method

(b) MWDL method

Fig. 4 Measured and predicted values for time series on pull-down time

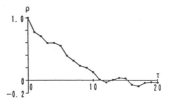

Fig. 5 Correlogram for pull-down time

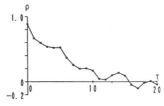

Fig. 6 Correlation coefficients between pull-down time and room discomfort index at 7 *a. m.*

Table 4 Optimum parameter values for the prediction of pull-down time

Method	d	Network	n
SDSL	6	23-3-1	22
MWDL	6	23-1-1	19

Table 5 Prediction results for pull-down time with SDSL method using optimum parameter values

DST	COR	m	CRC	ABE mean	ABE σ	RLE mean	RLE σ
1	0	—	0.886	6.02	5.16	11.5	8.93
	1	5	0.899	5.67	4.76	11.3	9.07
		6	0.908	5.50	4.48	10.8	8.18
		7	0.898	5.54	5.00	11.1	10.0
2	0	—	0.893	5.88	4.96	11.6	9.12
	1	2	0.885	5.89	5.29	11.6	10.1
		3	0.890	5.76	5.14	11.4	9.96
		4	0.886	6.01	5.05	12.0	9.88

Table 6 Prediction results for pull-down time with MWDL method using optimum parameter values

CRC	ABE mean	ABE σ	RLE mean	RLE σ
0.915	5.46	4.48	10.5	7.60

following equation, were used as variables:

$$DI = 0.72(t_a + t_w) + 40.6 \qquad (11)$$

where t_a and t_w are dry bulb and wet bulb temperature, respectively. Assuming the desired discomfort index to be 75, we calculated the amount of the pull-down time and used it as the measured value. Fig. 4 shows its time series data. As the explanatory variables, we used the room and outdoor discomfort index at 7 a.m. on the day and the room discomfort index at 12 a.m. on the previous day. Usually, on next days of holidays, the pull-down time becomes longer, because the building is warmed and its temperature becomes higher. The last explanatory variable was used to consider automatically whether the air conditioning equipment was operated or not. Fig. 5 shows the correlogram. The autoc-

orrelation coefficient value becomes gradually smaller with τ. We can see from Fig. 4 and Fig. 5 that the time series has a rather smooth nature. As an example, Fig. 6 shows the correlation coefficient between the data on the pull-down time and the room discomfort index at 8 a.m. for τ. Its value is large for $\tau=0$ and becomes gradually smaller with τ. Thus it is found that a considerably broad range of the past data has the influence on the pull-down time.

The prediction calculation was started from the 28th day from the top of the data. Table 4 shows the optimum parameter values for each method by which the best prediction performance was obtained. Table 5 shows the results by the SDSL method in which these parameter values were used. These results show that with the Manhattan distance the predictive performance was slightly improved by the CSDS method, whereas with the Euclidean distance it was slightly degraded. The best performance is obtained in the case of the Manhattan distance with $m=6$. The predicted values in this case are shown in Fig. 4 (a). Table 6 shows the results by the MWDL method in which the optimum values of the parameters were used. The predicted values in this case are shown in Fig. 4 (b). The MWDL method performs slightly better than the SDSL method with the best parameter values.

D. Summaries of the Results of the Numerical Experiments

The results of the numerical experiments show that the predictive performance by the two methods depends on the nature of the objective time series. With the time series on the warm-up time whose nature is choppy, the performance by the SDSL method is considerably improved by using the CSDS method. When the most appropriate degree was used for the autocorrelation function, the performance was considerably better than that by the MWDL method. With the time series for the pull-down time whose nature is smooth, the performance by the MWDL method was slightly better than that by the SDSL method.

In the case of the time series whose nature is choppy, that predictive performance by the SDSL method with the CSDS method is better than that for the MWDL method, means that to learn the nature and structure of the time series, data near the predicting day are not necessarily sufficient; and similar data which are selected from the past broad range, are required. In the case of the time series whose nature is smooth, that the predictive performance by the MWDL method is better than that by the SDSL method, means that the data near the predicting day are more required than the past data, because the influence of the former is stronger than that of the latter.

4. RELATED WORKS AND FEATURES OF THE PROPOSED METHOD

Peng [PH92] proposed to use weighted distances in which derivatives of the output variable with respect to the input variables were used as the weights. With the ordinary feedforward neural networks, the predictive performance in the case where the weighting was used, was rather inferior to that in the case where the weighting was not used. With the network in which the input is linked to the output by both nonlinear and linear terms, the performance of the former was better than that of the latter; and the former offers a 5.4% and a 17.1% reduction in the mean value and the standard deviation of the relative errors, respectively, over the latter. However, the derivative of the output variable with respect to the input variable

does not necessarily represent the dependence of the output variable on the input variable.

The originality of the CSDS method proposed by us consists in that the correlation coefficient which represents the dependence between two points in a time series, is employed as the weight for the distance calculation. It is so devised that using the correlation coefficient in its m-th power function allows its magnitude to adjust the degree of weighting.

5. CONCLUSIONS

The results of the numerical experiments show that the predictive performance by the two methods depends on the nature of the objective time series as in the case of single variable time series. With the time series whose nature is choppy, the performance by the SDSL method is considerably improved by using the CSDS method. When the most appropriate degree was used for autocorrelation function, the performance was considerably better than that by the MWDL method. With the time series whose nature is smooth, the performance by the MWDL method was slightly better than that by the SDSL method. In the previous paper [Sh95], the same results were obtained. According to these results, it is conjectured that the SDSL method with the CSDS method is effective for a time series of which nature is choppy.

REFERENCES

[KR90] L. Kaufman and P. J. Rousseeuw. *Finding groups in data.* John Wiley & Sons, NY, 1990.

[LH91] J. Lambert and R. Hecht-Nielsen. Application of feedforwad and recurrent neural networks to chemical plant predictive modeling. In *Proceedings of International Joint Conference on Neural Networks*, 1:373-378, 1991.

[PH90] T. M. Peng, N. F. Hubele and G. G. Karady. Conceptual Approach to the Application of Neural Network for Short-term Load Forecasting. In *Pros. of the 1990 IEEE International Symposium on Circuit and Systems*, 2942-2945, 1990.

[PH92] T. M. Peng, N. F. Hubele and G. G. Karady. Advancement in the Application of Neural Networks for Short-term Load Forecasting. *IEEE Transactions on Power Systems*, 7(1):250-257, 1992.

[RM86] D. E. Rumelhart and J. L. McClelland. *Parallel Distributed Processing.* MIT Press, Cambridge, 1986.

[Sh88] R. H. Shumway. *Applied Statistical Time Series Analysis.* Prentice-Hall, NJ, 1988.

[Sh95] H. Shimodaira. A Method for Selecting Similar Learning Data Based on Correlation Coefficients in the Prediction of Time Series Using Neural Networks. In *Proc. of the Eighth International Conference on IEA/AIE-95*, 109-117, 1995.

A NEW SECOND–ORDER ADAPTION RULE AND ITS APPLICATION TO ELECTRICAL MODEL SYNTHESIS

Jan Wilk, Eva Wilk[1] and Bodo Morgenstern
University of the Federal Armed Forces, D–22039 Hamburg
[1]Hochschule für Film und Fernsehen "Konrad Wolf", D–14482 Potsdam
email: jan.wilk@unibw-hamburg.de

ABSTRACT

Analog circuit simulation requires the knowledge of current and voltage behaviour and physical parameter dependencies of the devices used in the electrical circuit. Although a couple of nonlinear device models are implemented in the circuit simulator, the analog model of the hardware–realization of a standard–circuit device, as for example a digital Gate or an A/D converter from Texas Instruments, is seldom available.

In this case, the terminal behaviour and parameter dependencies of voltages and currents have to be measured, and the results must be made usable for the circuit simulator.

We use neural networks to approximate the terminal behaviour of electrical devices, maintaining the parameter dependencies. We have improved the adaption rule by an adaptive evaluation of the learning parameters to accelerate the approximation time. The network paradigm can be automatically transformed either into a netlist of an electrical subcircuit (for SPICE–simulation for example) or into a mathematical description language (for a behavioural simulator like SABER for example). Examples demonstrate the very accurate representation of nonlinear electrical devices for circuit simulation.

INTRODUCTION

In the design of microsystems, analog and digital components are often combined. To adapt digital devices to analog simulation, macromodels are developed. A macromodel is an analog circuit that approximates the current and voltage terminal behaviour and the parameter dependencies of the electrical device. We use neural network adaption for an accurate and fast development of macromodels.

A current or voltage terminal behaviour of an electrical circuit can be approximated very accurately by an appropriate network paradigm and adapted activation rules in a mutlilayer feedforward neural network with backpropagation approach. The exactness of the approximation to the original terminal behaviour is driven by the activation rule, the size of the neural network and the number of learning steps. We have improved the rate of convergence of the neural network adaption by an evaluation of the learning parameters with respect to the output error.

The size of the neural network determines the size of the resulting electrical subcircuit and the simulation time.

The input data are normalized and activation rules adjusted to the characteristic curves are chosen in order to decrease the size of the neural network which is necessary for yielding an accurate approximation.

The time dependency is modelled either by resistor–capacitor–combinations in the netlist or by additional delay neurons (time–delay networks) in the neural network.

The adapted neural network is then transformed into an electrical subcircuit consisting of controlled voltage or current sources and non–linear resistors. A transformation into a mathematical description language is possible, too. It leads to a behavioural model of the neural network. It can be used in a simulation program with behavioural interface.

We present macromodels of digital gates for analog circuit simulation. The accuracy of the circuit simulation with these models is very good, compared to the measured data. Furthermore, we show how macromodels can be generated automatically with neural network adaption.

APPROXIMATION BY NEURAL NET-WORKS

Multidimensional, non–linear, continuous functions can be approximated with a multilayer feed-forward neural network with backpropagation approach based on the classical gradient–descent method. It approximates any countinuous input–output mappings in a given tolerance range [Fun89]. In a feedforward neural network, the synaptic connection topology contains no closed synaptic loops. The network consists of a given number of neural cells in each layer. Each neuron is connected to all neurons of the neighbour layers. It is activated by a non–linear function.

The backpropagation network is a powerful tool for an approximate implementation of the function mapping $\Phi : \mathbf{x} \in R^n \xrightarrow{\Phi} \mathbf{y} \in R^m$, with $(x_k, y_k), k = \overline{1, N}$ where $\mathbf{y}_k = \Phi(\mathbf{x}_k)$ [Cyb89].

For macromodeling with neural networks, this can be reduced to two mappings:

- R $\longrightarrow$ R: This mapping is realized by a neural network with one hidden layer and sigmoidal activation functions.

- $R^n \longrightarrow$ R: This mapping is realized by a neural network with two hidden layers and sigmoidal activation functions.

The realization of the described mappings leads to the exactness needed for macromodeling. The size of the network depends on the size of the training set and the number of input variables. Using pre-processing functions like scaling and normalization and an adaption of the activation rules to the functional behaviour of the circuit results in a network reduction.

The convergence of the neural network is influenced by the network paradigm and by the learning law. It has to be taken into consideration that the neural network has to be transformed into a netlist of an electrical subcircuit. Feedback loops in the neural networks have to be avoided as they might cause convergence problems during analog circuit simulation.

We use backpropagation networks for synthesis of macromodels with neural networks. Doing this we are enabled to use higher order network paradigms like second order sigma–pi neurons. Their net input is determined not only by summation of the weighted output of single neural cells, but also by a multiplicative combination of the activation values of several cells. The advantage is that they are not bound to

be linear separable. Therefore, the resulting network paradigm is invariant to scaling, translation and rotation [Zell94].

A MODIFIED ADAPTION RULE

Generalized delta–rule is applied to use interpolative and associative properties of multi–layered feed-forward neural networks. Slow convergence is the major disadvantage of its direct application. Modification of the basic algorithm by using variable learning rate may lead to an improvement of the rate of convergence [Jac88], [Zhou93]. In some methods the rate of convergence depends on the initial value of the learning rate, and in other methods the values for the learning rates simply are estimated. The momentum of the adaption rule is either constant or not taken into consideration.

An improvement of the convergence properties of the basic algorithm can only be achieved by a formal description of the two constant learning parameters. Their adaption to the approximation error in each training step accelerates the gradient descent – and thus the rate of convergence.

Calculation of the Learning Parameters

Fig. 1 represents the analytical evaluation of the learning parameters. We apply the least square

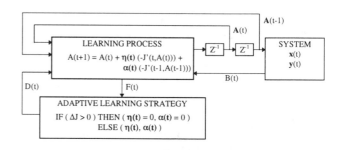

FIGURE 1: Strategy for the adaptive correction of the learning parameters

error criterion to the output layer and the generalized delta rule to the second order sigma–pi networks. This new adaption rule links the system information $B(t)$ (including the pattern information and the approximation error) to the adaptive correction of the learning parameters $D(t)$, and the state vector of the neural network $A(t)$. The result is a new state vector, $A(t+1)$.

The initial values are chosen randomly and the patterns are measured data. Therefore, an improvement of the convergence speed can only be achieved by an adaptive variation of the learning parameters. This is equivalent to a variation of the network information, $D(t)$. The learning parameters are determined by the approximation error of the last two training steps.

We apply a second order sigma–pi network assuming that the network parameters only change if this minimizes the approximation error

$$E(t, A(t+1)) < \gamma(t)E(t-1, A(t)), \quad (1)$$

with $E(t)$ approximation error, $A(t)$ state vector and $\gamma(t) \in]0;1[$ scaling parameter at step t, respectively.

First order and second order approximation errors can be considered separately, as we separate the network activity into first order activity (first order connections and threshold values), and second order activity (i.e. second order connections).

To set up a system of equations we exploit these separated parts of the network structure. It links the least square error criterion of first and second order connection to the learning parameters. The solution yields relations that enable us to determine the learning parameters, because they are now adapted to the approximation error.

Eqn. (2) describes the condition for the minimization of the approximation error:

$$E(t, A(t+1)) - \gamma(t)E(t-1, A(t)) =$$
$$\eta(t)\|E'_A(t, A(t))\|^2 +$$
$$\alpha(t)\mathrm{tr}\left\{E'_A(t, A(t))E'_A(t-1, A(t-1))\right\}. \quad (2)$$

It is set up both for first and for second order errors and solved for $\eta(t)$ and $\alpha(t)$. This results in

$$\alpha(t) = \frac{\Delta E_2 C_1 - \Delta E_1 C_2}{B_2 C_1 - B_1 C_2}$$
$$\eta(t) = \frac{\Delta E_1 - \alpha(t)B_1}{C_1} \quad (3)$$

with the abbreviations

$$\Delta E_i = E(t, A_i(t) - \gamma(t)E(t-1, A_i(t))$$
$$B_i = \mathrm{tr}\left\{E'_A(t, A_i(t))E'_A(t-1, A_i(t-1))\right\}$$
$$C_i = \|E'_A(t, A_i(t))\|^2;$$

$i = 1, 2$, indicating the first and second order connections, respectively.

The development of the analytical equations (3) for the learning parameters $\eta(t)$ and $\alpha(t)$ enable an optimal determination of the gradient descent in the error–plane in each training step. Thus the learning parameters are directly related to the approximation error. This accelerates the minimization of the total approximation error.

Convergence Behaviour

The efficiency of the presented algorithm is now demonstrated by the approximation of an XOR function. The selected neural network has two inputs and one output. It uses second–order sigma–pi neurons. The activation function is a sigmoidal function. A set of randomly chosen starting parameters is used for all networks in order to yield a base for a comparison of the new learning rule to other, similar algorithms. They all are trained with the backpropagation algorithm. The convergence behaviour of the "modified backpropagation algorithm with adapted learning parameters" can now be compared to vanilla backpropagation ($\eta = 0.5$, $\alpha = 0$) and extended backpropagation ($\eta = 0.5$, $\alpha = 0.1$). Figures 2 and 3 show the results of this comparison. It can be seen that the convergence of the modified backpropagation algorithm is achieved in 1/1000th of the steps that are required by the other algorithms.

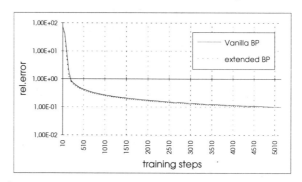

FIGURE 2: Convergence behaviour of backpropagation algorithms with constant learning parameters (approximation of XOR–function)

MODELING OF ELECTRICAL DEVICES FOR ANALOG CIRCUIT SIMULATION

The design of macromodels for electrical simulation can widely be done automatically owing to the application of neural networks: The terminal behaviour of the original circuit is measured. The measured data are scaled or normalized. They represent

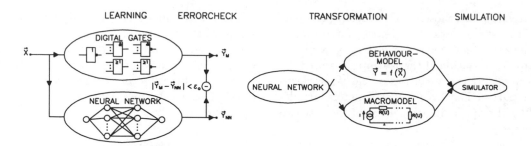

FIGURE 4: Synthesis of macromodels for digital gates

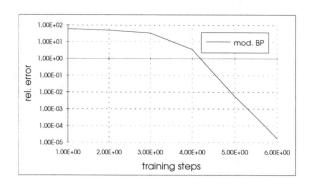

FIGURE 3: Convergence behaviour of the modified back-propagation algorithm with adaptive learning parameters (approximation of XOR–function)

the training set of the neural network. The adaption of the network is continued until a predefined, minimal error is reached.

Finally, the network paradigm can automatically be transformed either into a SPICE netlist or into a mathematical description language.

Fig. 5 shows the algorithm.

Partitioning

A modular approach is used for the development of macromodels. The electrical device is represented by three main modules which describe input behaviour, output behaviour and transfer behaviour. The transfer behaviour module includes the logical function in the modeling of digital gates. Each module is approximated separately by a neural network. Another neural network with an extended number of inputs and/or outputs is set up for the modeling of physical parameter dependencies.

Redundancy and Latency

We exploit redundancy in the set of input data and latency in the learning rule to reduce the learning time: A differential quotient approximates the gradient of the input function. Input data in the part of the function with constant gradient are used only sparsely. More data is used from this part if the error in the following learning step is large. To exploit latency, the connection strengths and the parameters of the activation rules are changed only when the output error is too large.

Time Dependency

Time dependency of the input pattern normally is not taken into consideration in neural networks. For the time–dependence representation of digital gates, a RC–element is added to the netlist. Different delays for rise and fall time are steered by a switching diode. This yields an exact modeling of delay time. We use a multiple delay of the input set as an additional input pattern of the neural network (time–delay neural network, [Zell94]) for analog electrical devices.

Temperature Dependency

Temperature dependency of the electrical device is considered by an extra input voltage of the neural network. It is represented by a temperature dependend resistor that is driven by a constant current source for the circiut simulation.

EXAMPLES

We have used the presented method for the development of macromodels for digital gates. Figure 6 represents the HSPICE–simulation of the input terminal behaviour of an AND gate macromodel and the corresponding measured data. Figure 7 shows

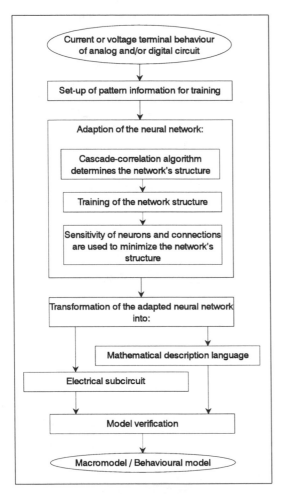

FIGURE 5: Algorithm for automatical generation of models for electrical circuit simulation

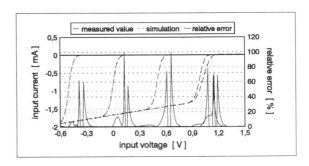

FIGURE 6: Input behaviour of an AND gate: Input current vs. two input voltages. Simulation results of the macromodel and measured data

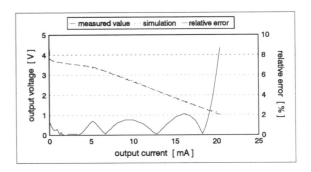

FIGURE 7: Output behaviour of an AND gate: Output voltage vs. output current. Simulation results of the macromodel and measured data

the terminal behaviour of the output of the same macromodel and the corresponding measured data. The mean error is less than 2% in both cases.

Figure 8 shows the macromodel representation of an ideal A/D converter for analog circuit simulation. The non–linear A/D converter behaviour (error and time delay) is described by additional modules.

CONCLUSIONS

The adapted representation of electrical devices for simulation with an analog circuit simulator is yielded accurately and with little development time by using second–order neural network adaption:
The input behaviour, the output behaviour and the transfer behaviour of a digital gate are separately approximated by neural networks.Each neural paradigm is then transformed into an analog subcircuit. The connection of these subcircuits results in an ana-

log circuit macromodel.

The proposed improvement of the generalized delta rule causes an acceleration of convergence speed. The learning parameters momentum and learning rate can now be *calculated* for each step. They depend on the approximation error.

Examples of macromodels of an digital gate and an A/D converter for simulation with an analog circuit simulator show very good accuracy. Adaption of the neural network and transformation of the network paradigm into an analog subcircuit can be done automatically for existing – measured – data of current and voltage terminal behaviour.

The proposed procedure is characterized by little development time and an accurate macromodel–representation of electrical devices.

REFERENCES

[Cyb89] G. Cybenko: "Approximation by superposition of a sigmoidal function", *Math. Control, Systems, Signals*, vol. 2, 1989, pp. 303–314.

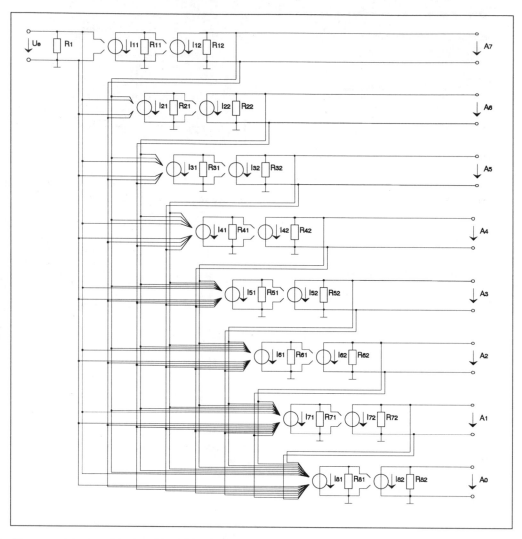

FIGURE 8: Macromodel of an ideal 8–Bit A/D converter for use in analog circuit simulation. The number of nodes is very small because of the development via neural network representation. This minimizes simulation time in analog circuit simulation

[Fun89] K. Funahashi: "On the approximate realization of continuous mappings by neural networks", *Neural Networks*, no. 2, 1989, pp. 192–193.

[Jac88] R.A. Jacobs: "Increased rates of convergence through learning rate adaption", *Neural Networks*, vol. 1, 1988, pp. 295–307.

[Spe93] A. Sperduti, A. Starita: "Speed up learning and network optimization with extended back propagation" *Neural Networks*, Vol. 6, 1993, pp. 365–383

[Zhou93] S. Zhou, D. Popovic, G. Schulz–Eckhoff: "An improved learning law for backpropa-

gation networks", *IEEE International Conference on Neural Networks*, San Francisco, March 1993

[Zell94] A. Zell: "Simulation Neuronaler Netze", Addison–Wesley, Bonn 1994.

REAL OPTION VALUATION WITH NEURAL NETWORKS

Alfred Taudes, Martin Natter, Michael Trcka
Department of Industrial Information Processing
Vienna University of Economics & Business Administration
Pappenheimgasse 35, A-1200 Vienna, Austria
Email: Alfred.Taudes@wu-wien.ac.at

ABSTRACT

We propose to use neural networks to value options when analytical solutions do not exist. The basic idea of this approach is to approximate the value function of a dynamic program by a neural net, where the selection of the networks weights is done via Simulated Annealing. The main benefits of this method as compared to traditional approximation techniques are that there are no restrictions on the type of the underlying stochastic process and no limitations on the set of possible actions. This makes our approach especially attractive for valuing real options in flexible investments. We, therefore, demonstrate this method by valuing flexibility to costly switch production between several products under various conditions.

KEYWORDS

Real Options, Neural Networks, Capital Budgeting, Simulated Annealing, Flexible Manufacturing Systems

1 MOTIVATION

While in finance the importance of option valuation is well established, option valuation techniques have recently gained significance in capital budgeting, too. In this field, there is a research direction called "Real Options" dealing with the use of option valuation techniques to valuate the possibilites of altering the mode of operation of an irreversible investment project in response to — a priori stochastic — changes of the economic environment. For projects that embed only one of several simple types of "flexibility" and in-

volve traded commodities, analytical option valuation techniques developed in finance can be used directly[1]. However, most real-world investments allow several modifications whose values are usually not independent (see e.g. [Kul95b]). Consider, for example, a production facility: normally here one can choose when to invest, to abandon the project for salvage value, to shut down production temporarily if the market price of the good produced is lower than its average costs and to change inputs and/or outputs in response to changes in the respective markets (see e.g. [SS90]). In addition to this extension of possible actions, the validity of the assumption that the value underlying the option follows a Geometric Brownian Motion allowing analytical solutions to financial options is much less established for real options and — as the example given above shows - often a multivariate stochastic process is necessary to describe the relevant environment.

In the case of such general real options valuation problems it therefore seems necessary to resort to approximate, numerical methods. In finance such methods are well-known. [GS85] group them into methods that approximate the underlying process directly (Monte Carlo Simulation and binomial process) and ones that solve the partial differential equation derived as partial equilibrium condition resulting from no-riskless-arbitrage (finite difference methods and numerical integration). Special numerical methods for valuing real options were developed by [Kul95a] and [Kam95]. Both are finite, discrete time dynamic programming formulations, where the latter considers only the case of a trinomial lattice approximation of Geometric Brownian Motion. All these methods assume that the type of underlying stochastic process is given and its parameters estimated. Furthermore, in the

---

[1]For a recent survey of the state of the art in the field of real options see [Tri95].

case of a multivariate stochastic process the state space grows combinatorially so that the backward recursion procedure used to solve the dynamic program quickly becomes computationally expensive.

We propose a method that is based on observed trajectories of the stochastic process only. There are no restrictions on the type of process used, no parameter estimates are needed. The basic idea of the approach is to approximate the value function of the dynamic program with a neural network that for each mode of operation has the current state as input and yields the mode to be chosen as output. The selection of weights is done via Simulated Annealing. A detailed description of the method is given in chapter 2. Chapter 3 is devoted to an application to flexible manufacturing systems: We study the value of a switching option that allows the change of output in the presence of switching costs[2]. Chapter 4 summarizes our findings and indicates directions for further refinement of the approach presented.

2   NEURAL NET APPROXIMATION FOR REAL OPTION VALUATION

Given that the owner of a project has the possibility to change the operation mode of a project in a persistent way, it is clear that the value of such a project and the optimal decision rule that specifies an optimal action for each possible time/environment combination have to be determined simultaneously via the solution of a dynamic program. Such a model has been developed in [Kul95a]. It assumes that in each period t, $t = 0, \ldots, T$[3], the project under consideration can be operated in one of several "modes of operation", $m = 1, \ldots, M$. π, the profit per period, depends on this operation mode and the uncertain state in which the environment is in t. The evolution of the environment is governed by a family of random variables Θ_t, $t = 0, \ldots, T$, i.e. $\pi = \pi(\theta_t, m)$ and θ_t being a realization of Θ_t. Modes can be changed in order to maximize the discounted sum of profits. However, there is a cost of switching between two modes j, k, denoted by c_{jk}, with $c_{ii} = 0\ \forall i$.

In [Kul95a], the optimal path of configuration changes in response to the development of the environment is determined via backward recursion using Bellman's equation of dynamic programming as follows:

lows:

- In the last period T the effect of the configuration chosen on future periods is 0[4] so that in response to an observed environment θ_t the configuration l is chosen for which

$$F(\theta_T, m) = \max_{l=1,\ldots,M} (\pi(\theta_T, l) - c_{ml}) \qquad (1)$$

where m is the configuration in period $T - 1$ and $F(\theta_T, m)$ is the value of the project at time T, a function of m, the mode in $T - 1$, and the observed environment θ_T.

- Bellman's equation of dynamic programming then states that the value of the project at time t is given as

$$F(\theta_t, m) = \max_{l=1,\ldots,M} ([\pi(\theta_t, l) - c_{ml}] + \mu E_t[F(\Theta_{t+1}, l)]) \qquad (2)$$

where $l, m = 1, \ldots, M$, $t = 0, \ldots, T - 1$, μ is the discount factor, defined by $\frac{1}{1+\rho}$, with ρ as the exogenously specified discount rate and $E_t[F(\Theta_{t+1}, l)]$ the expected value of the project in $t + 1$ conditional on the development up to period t as documented in the environments observed and modes chosen[5].

The optimal configuration policy for a particular development of the environment $(\theta_t, t = 0, \ldots, T)$ is the sequence of modes for which in each period the maximum on the right hand side of formula (1) is obtained for the particular value of θ_t. Thus, if the starting mode is mode m, the "value of flexibility" is given by

$$E_0[F(\Theta_0, m)] - E_0 \left[\sum_{t=0}^{T} \pi(\Theta_t, m) \frac{1}{(1+\rho)^t} \right] \qquad (3)$$

where the second term is the value of the project with a fixed mode m.

Instead of using this solution method, we propose to approximate $F(\theta_t, m)$ via a neural network whose input layer has nodes for every component of the current state, i.e. (t, θ_t, m). Thus, it is not necessary to make assumptions in order to be able to compute $E_t[F(\Theta_{t+1}, l)]$.

---

[2] An analytical model of similar type without switching costs is developed in [TH90].

[3] Thus, the model in [Kul95a] is a finite, discrete time model. Models in continous time and/or with infinite time horizon can be constructed in a similar, albeit mathematically more involved, way and are described in [DP93].

[4] If the salvage value is higher than 0, this has to be considered.

[5] For instance, if the stochastic process defined by the family of random variables Θ_t is a finite Markov Chain where the mode chosen does not influence future environments, $E_t[F(\Theta_{t+1}, l)] = \sum_{j=1}^{s} F(j, l) p_{\theta_t j}$ where $p_{\theta_t j}$ is the probability that the environment will change from θ_t to j.

Based on the payoffs of the different modes at time t, the period, and the given switching costs a feedforward neural network model[6] has to 'learn' to decide which of the given modes should be selected ("activated") to maximize the expected value of the sum of the discounted payoffs over all periods.

In a 'learning' (estimation) phase, the network model is parametrized on a large number of different payoff time series for the different modes (e.g. payoffs of different types of products which can be produced by a flexible manufacturing system (FMS)) to maximize the sum of the discounted payoffs. In this phase, the network is trained to map the inputs through a feed forward network onto the outputs in such a way that the output unit (every output unit representing a mode) having maximum activation makes the highest contribution to the sum of all discounted payoffs, i.e. the network learns the value function. So, in each period we choose the mode (product) for which the network has the highest output activation. In the validation phase the rules learnt by the network model are applied to a sample of new payoff series of the different modes and the expected value is calculated. For every mode (product) we have a seperate network model consisting of the following inputs:

- the payoffs of all modes in period t

- period t

- the mode of the previous period $t-1$ (this is important in order to know which switching costs apply)

We use the period as one of the inputs, because in an early period - ceteris paribus - it may be optimal to switch from mode i to mode j and accept relatively high switching costs c_{ij} while, in a later period, the same decision may not be optimal as there are only a few (or no) periods left in which j can possibly generate higher payoffs. The system must 'know' the mode of the previous period because of the fact that switching costs c_{ij} reduce the payoffs if a different mode is chosen in a new period.

It is assumed that the payoffs follow a Markov process [DP93, p. 62f]. Therefore, future payoffs only depend on the current payoffs (and not on any previous ones), and it is sufficient to use the payoffs in period t as inputs. Payoffs and period information are represented as real values. Information about the previous modes are dummy-coded ($m_i = 1$ representing the activ mode and $m_j = 0$ the inactiv modes). For

---

[6]For an introduction into neural network models see e.g. [HKP93] or [Rip93].

estimation purposes inputs for the payoffs and period are rescaled into the $[0; 1]$ interval.

For every mode we use one network with inputs as described above. Each of the networks has one output unit. The activations of the outputs represent the profitability of their "activation" given the inputs.

We propose to start with a 2 layer feedforward neural network with logistic activation functions because of its property as a universal approximator [Whi89]. The number of hidden units is varied between 0 (reducing the system to a logistic model without hidden units) and 10. With an increasing number of hidden units more complex functions can be mapped from inputs to outputs. To avoid overfitting of the network models different strategies are used. A very common practice is to use only part of the data for estimation and use the other part as a validation set. Performance of the model is then measured on this validation set.

SELECTION OF NETWORK WEIGHTS BY SIMULATED ANNEALING:

Network weights should be set in such a way that they produce optimal strategies for any series of payoffs generated.

In our approach, we use Simulated Annealing [KGV83, AK89], a global optimization method with a stochastic element, to estimate the network weights. Simulated Annealing is an iterative method that accepts improvements of the target function ($F(\theta_0, m)^{new} - F(\theta_0, m)^{old} > 0$) with probability 1. The objective value $F(\theta_0, m)$ is the sum of the discounted profits over all periods and series generated. To escape from local minima, a stochastic component (see equation 4) is introduced which permits - with diminishing probability - the acceptance of solutions that are worse than the present one. After initializing the networks weights with small (uniformly distributed) random numbers and calculation of the starting value of the objective function $F(\theta_0, m)^{old}$, the following steps are repeated until the target function stabilizes:

1. Randomly select a decision variable (network weight)

2. Generate a new random value for this variable

3. Determine the new strategy by the networks and evaluate the new strategy

4. Accept the new strategy if $F(\theta_0, m)^{new} - F(\theta_0, m)^{old} > 0$ or

$$\left(1 + e^{-\left(\frac{F(\theta_0, m)^{new} - F(\theta_0, m)^{old}}{c}\right)}\right)^{-1} > U[0; 1]$$

(4)

5. If a new weight was accepted, set $F(\theta_0, m)^{old} = F(\theta_0, m)^{new}$

6. Reduce control parameter c

where $U[0;1]$ represents a uniformly distributed random variable and c a control parameter regulating the stochastic influence. Usually, after every iteration c is multiplied by a constant smaller than but close to 1. When c gets closer to zero, decisions become more deterministic, which means that mainly improvements of the objective function are accepted. The starting value of c has to be high enough to allow for totally stochastic behavior of the system.

3 APPLICATION TO CAPITAL BUDGETING FOR FMS

In the following, we apply the neural network approach to determine the production program of an FMS for different switching costs, volatilities and payoff series. A flexible manufacturing system which can be used to produce 3 different types of products (modes) is evaluated. Only one product can be produced at a given time. Depending on the payoffs $Z_{i,t}$ of the 3 products $i = 1, .., 3$, the network has to select the mode of production in each period t. With "mode of production", we denote type of product that is produced by the FMS. The payoff of a product is given by its sales price minus its production costs.

Using a Monte Carlo Simulation, payoffs are maximized for 100 different payoff series generated by function 5 and validated on 10000 new series generated by the same function. The expected value of the discounted payoffs calculated over all different series represents the value of the investment (FMS). The sum of the maximized discounted payoffs over all planning periods may be used as a limit for the maximum price of the FMS. For our study, we use an interest rate of 10% per year. The life time of the FMS is assumed to be 50 periods (months).

The decision about which product will be produced in a given period is based on the payoffs generated in the case of its production. In our model, we assume that the payoffs Z of product i follow a given linear trend η_i and random disturbances of the following type:

$$Z_{i,t+1} = Z_{i,t} * (1 + \eta_i) + \xi_i * \gamma \qquad (5)$$

where ξ_i represents product i's dependency on random disturbances and γ a uniformly distributed random variable between -1 und 1. The valuation of the FMS is performed for different volatilities ξ_i of the

Table 1: Parameters for generating payoffs

Product	Z_0	η	ξ_A	ξ_B	ξ_C
1	30	0,02	20	40	100
2	40	0,15	30	60	150
3	50	0,1	40	80	200

payoffs with starting payoffs Z_0 and trend parameters η as given in table 1. Table 2 shows switching costs c_{ij} which have to be considered when in period t the FMS is used to produce a different product than in the previous period. We used 100 series for network

Table 2: Switching costs

	I	II	III	IV
$c_{0,1}$	27	135	270	540
$c_{0,2}$	13	65	130	260
$c_{0,3}$	18	90	180	360
$c_{1,2}$	24	120	240	480
$c_{1,3}$	16	80	160	320
$c_{2,1}$	25	125	250	500
$c_{2,3}$	29	145	290	580
$c_{3,1}$	27	135	270	540
$c_{3,2}$	39	195	390	780

training and 10000 series for validation, respectively. In a first step, we tested the generalization ability of the neural network model for a given constellation of parameters ξ_i and γ and different numbers of hidden units. The starting value for Simulated Annealing, c, was $c_0 = 200$. After every iteration c was multiplied by 0.99994 until the system stabilized. Table 3 presents the results for the different number of hidden units tested and the two phases of estimation and validation. We can see that a network without hidden units (a logit model) generalizes best. Therefore we select this architecture for the valuation of the FMS with different parameters (switching costs, etc.).

Table 3: Expected value of the discounted payoffs for 100 (estimation) and 10000 (validation) series generated

# hid. units	0	1	2	3	4	5	10
estim.	5194	5134	5147	5205	5175	5062	5200
valid.	4943	4782	4804	4882	4810	4766	4896

3.1 VALUE OF MACHINES WITH ONLY ONE PRODUCTION MODE

As a comparision for the value of the FMS, we calculate the expected value for a production system that is restricted to the production of only one of the three products (as given by the second part of formula 3). This yields the results given in table 4, row 1. Therfore, in case of a production system that has only one of the 3 production modes, we would choose product 3 because it has the highest expectation.

Table 4: Expectation values (row 1) and standard deviations (rows 2-4) of the discounted payoffs for the case of a unflexible production system

payoff	product 1	product 2	product 3
$E_0[\ldots]$	2042	2388	2630
A	2880	3910	4660
B	5750	7810	9320
C	14380	19530	23300

Table 5: Expected value of the discounted payoffs for the case that we produce the product that has the highest payoff in a given period (index a); expected values for the estimation phase of the network (index b); expected values and standard deviations of the network for the validation phase (index c),

	I	II	III	IV
A_a	5578	4992	4242	2743
B_a	8893	8321	7606	6176
C_a	18753	18214	17540	16192
A_b	5658	5412	5168	4832
B_b	9069	8738	8513	8036
C_b	19350	18992	18637	17848
A_c	5930	5653	5354	4937
	$\sigma=2920$	$\sigma=2960$	$\sigma=3041$	$\sigma=3057$
B_c	9555	9242	8862	8374
	$\sigma=5769$	$\sigma=5831$	$\sigma=5850$	$\sigma=5929$
C_c	20422	20047	19657	18460
	$\sigma=14298$	$\sigma=14394$	$\sigma=14457$	$\sigma=14678$

3.2 SIMULATION RESULTS

The results of the estimation and validation phase are displayed by table 5 (indices b and c) where, in addition to the expected values, standard deviations of the Monte Carlo Simulation are presented. Rows A, B, and C show the results for the different volatility parameters. The columns (I, II, III, and IV) represent the results for the different switching costs assumed.

As a comparison for the network solution, we also calculated the expected value of the discounted payoffs for the case that in each period we choose the production mode yielding the highest payoff of that period regardless of switching costs. The results of this rule are presented in table 5 (index a). From table 5 (indices a,c), we see that the decision rules learnt by the neural network provide significantly better results than that simple rule.

Because the expected value of the payoffs does not change for different volatilities of the payoffs, the maximum discounted payoffs for an (unflexible) manufacturing system that can only produce one type of products (as presented above) are the same (e.g. 2630 for product 3) for all cases treated here.

Table 5 (index c) shows that - especially for high payoff volatilities - an FMS has a much higher (up to 600 %) value than an unflexible manufacturing system that can produce only the product with the highest expectation. From table 5 (index c), we can also see that the value of the FMS is more dependent on the

volatility than on the switching costs. Even in the case of high switching costs and high volatilities the value of the FMS is 18460 (see table 5, cell C_c,IV), which is only about 10% less than for switching costs that are 20 times lower (cell C_c,I). On the other hand, the value of the FMS is only about 6000 for low volatilities. The standard deviations of the discounted payoffs of the FMS are similar to those of an unflexible manufacturing system which produces only product 1 (see tables 5, index c and 4, product 1). Because of the potential possibility to switch between different production modes, the use of an FMS reduces the variance of the sums of the discounted payoff values. In our example, the variances of the FMS are comparable to those of the product with the lowest variance, but much smaller than those of products 2 and 3. Hence, the switching option seems to avoid extremely poor results.

3.3 SWITCHING BEHAVIOUR

A look at table 5 shows that the difference between the expected values found by the network (index c) and those calculated based on the rule that the product with maximum payoff is produced (index a) is highest for low volatility and high switching costs (see cells A,IV). While in case of low switching costs the network tends to prefer the production of the product with the highest payoff, the analysis of the networks trained on the basis of high switching costs have a much lower tendency to change from one mode to another, if there

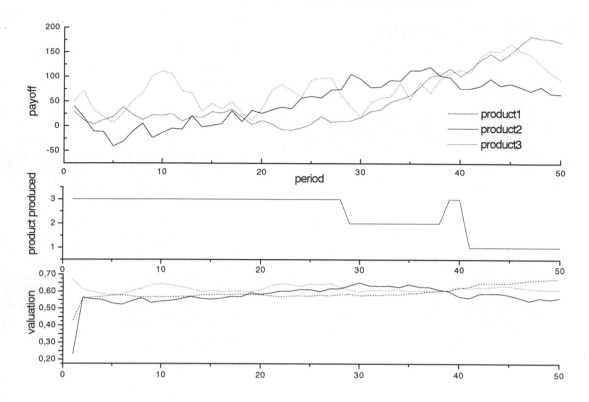

Figure 1: Payoffs of the 3 products, production mode chosen by the network, and valuation by the network

is only a slightly higher payoff for another product.

The analysis of one specific payoff series showed that in case of low switching costs the production mode is changed 10 times (by the network) over 50 periods but only once in case of high switching costs. Figure 1 shows one of the generated payoff series for the 3 production modes as well as the corresponding activation levels of the 3 output units and the networks' selection of the modes. For low switching costs (I), the (hysteresis) range in which the production mode is not changed even if the production of another product would yield higher payoffs is very small (see e.g. figure 1 between periods 5 and 6, where product 1 has higher payoffs than product 3, or periods 42 to 45 where production mode 1 is proposed by the network allthough product 3 has higher payoffs). Differences of up to 10 units are accepted. If the difference is higher, the production mode is changed. The width of the hysteresis is about 40% of the switching costs. In case of switching costs II, which are 5 times as high as switching costs I, the hysteresis lies between 30 and 50. For switching costs III, which are 10 times as high as switching costs I, the hysteresis is 70. For the highest switching costs (IV), which are 20 times the costs of I, the production modes are not changed even if there are differences up to 90 (which is about 20 % of the switching costs).

4   CONCLUSION

The traditional investment analysis leads to wrong results if the investment object can adapt to a changing environment. The possibility to react to new situations can be treated as a real option which increases the value of the investment object. In the literature, dynamic programming, portfolio theory and decision analysis are used to valuate such real options. All these methods are mathematically tentative and only work under rigorous assumptions. In cases where analytical methods cannot be used, valuation of the options has to be performed numerically. In our approach, we propose a method that is based on observed trajectories of the stochastic process only. No restrictions on the type of process used are made and no parameter estimates are needed. We approximate the value function with the aid of a neural network. As a pilot application, we valuated a flexible manufacturing system which had the option to change the production mode (3 different products). If the production mode is changed, switching costs are caused. It was shown that the value of the switching option can be very high - especially in case of high volatility. The rate of volatility has shown to have a higher influence on the value of the FMS than the switching costs. The results (variances of the discounted payoff rates) indicate that — as compared to a production system which is restricted to only one mode of production — the option of flexibility (as given with an FMS) reduces the risk of very poor results.

We would suggest three interesting directions for future research on the basis of the approach presented, i.e. its application to real-world data, the investigation of an additional 'shut-down' mode to limitate losses in cases where the profitability of all products becomes negative, and its application to a situation where several products are involved and only a subset of modes can be chosen.

REFERENCES

[AK89] Aarts, E.H.L. und Korst, J., *Simulated Annealing and Boltzmann Machines*. Chichester, Wiley, 1989.

[DP93] Dixit, A., Pindyck, R., *Investment Under Uncertainty*, Princeton, 1993.

[GS85] Geske R. and Shastri K., Valuation by Approximation: A Comparison of Alternative Valuation Techniques, *Journal of Financial and Quantitative Analysis*, 20 (1), 45–71, March 1985.

[HKP93] Hertz, J., Krogh, A., Palmer, R., *Introduction to the Theory of Neural Computation*, 6., Redwood City, 1993.

[HT90] Hodder, J., Triantis, A., Valuing Flexibility as a Complex Option, *The Journal of Finance*, 2, 1990.

[Kam95] Kamrad B., A Lattice Claims Model for Capital Budgeting, *IEEE Transactions on Engineering Management*, 42 (2), May 1995, 140–149, 1995.

[Kul95a] Kulatilaka N., The Value of Flexibility: A General Model of Real Options. In: L. Trigeorgis, editor, *Real Options in Capital Investment — Models, Strategies, and Applications*, Praeger, 87–107, 1995.

[Kul95b] Kulatilaka N., Operating Flexibilities in Capital Budgeting: Substitutability and Complementarity. In L. Trigeorgis, editor, *Real Options in Capital Investment — Models, Strategies, and Applications*, Praeger, 121–132, 1995.

[KGV83] Kirkpatrick, S., C.D. Gelatt, and M.P. Vecchi, Optimization by Simulated Annealing. *Science*, Vol. 220, No. 4598, 671–680, 1983.

[Rip93] Ripley, B.D., Statistical Aspects of Neural Networks. In: Barndorff-Nielsen, O.E., Jensen, J.L. und Kendall, W.S., editors, *Networks and Chaos — Statistical and Probabilistic Aspects*. London, Chapman & Hall, 1993.

[SS90] Sethi A.K. and Sethi S.P., Flexibility in Manufacturing: A Survey. *International Journal of Flexible Manufacturing Systems*, 2, 289–328, 1990.

[TH90] Triantis A.J. and Hodder J.E., Valuing Flexibility as a Complex Option. *Journal of Finance*, 45 (2), 549–565, June 1990.

[Tri95] Trigeorgis L., Real Options: An Overview. In: L . Trigeorgis, editor, *Real Options in Capital Investment — Models, Strategies, and Applications*, Praeger, 1–28, 1995.

[Whi89] White, H.: Connectionist Nonparametric Regression: Multilayer Feedforward Networks Can Learn Arbitrary Mappings. *Working Paper*, Department of Economics, University of California, San Diego, 1989.

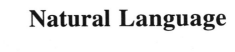

Natural Language

AUTOMATIC SEGMENTATION AND TAGGING OF HANZI TEXT USING A HYBRID ALGORITHM

An Qin and Wing Shing Wong
Department of Information Engineering
The Chinese University of Hong Kong, Shatin, N.T., Hong Kong
E-mail: {anq4, wswong}@ie.cuhk.edu.hk

ABSTRACT

In this paper we present a simple but effective approach to segment and to perform word tagging on Hanzi-based languages. The essence of the approach is based on a hybrid algorithm, which combines the learning capability with the flexibility of a digital processing system. Our ideas was implemented by a system with 72 tags and a corpus with 111,068 Chinese words (173,497 Chinese characters). Our closed test results reached 99.5% segmentation accuracy and 95.3% tagging accuracy. The opened test results for the specified domain is 91.6% segmentation accuracy and 89.6% tagging accuracy. Various applications of this system are also discussed in this paper.

1 INTRODUCTION

Chinese is a character-based language. However, the basic unit of meaning is a word [GB/T92] as in English, with some words containing only one character and others may contain several characters. In an alphabetic language, a word is defined explicitly by using space and word segmentation is not a problem. In the Chinese language, however, a sentence consists of a string of characters, with no delimiter to mark word boundaries. So, how to segment a sentence into words is a fundamental, non-trivial problem. Several automated word segmentation systems have been developed since 1980. The methods employed in these systems range from lexicon and occurrence probability based to rule-based [CLP91].

Word tagging is another fundamental issue in text analysis, for character-based as well as alphabetic languages, such as English. The problem is how to tag each word in the sentence with a proper part of speech, for example, noun, verb, and so on. It is natural to use the part of speech as the tag, but word tagging can provide even more information about a word than part of speech. This tagging provides the basic syntactic information of the word. For English text, this is a well-researched topic. Several systems with encouraging results have been put into use [Booth85][Greene71][Mar83]. In 1992, Bai [Bai92] developed an automatic tagging system of Chinese text that works on segmented texts. The result on closed tests (the definition of a closed test is given in section

3) showed that the accuracy rate is 95.2%; processing speed on a SUN/4 480 machine was about 250 Chinese characters per second.

A high performance system of automatic segmentation and tagging of Chinese text can find various applications, including:

1. Computational linguistic analysis of Chinese texts

The system can be used to analyze Chinese articles. It will provide linguists with various statistical data about characters, words, parts of speech, sentences, as well as syntactic patterns.

2. Information retrieval system

Before keywords extracting or keywords matching, the Chinese text has to be segmented into words.

3. Speech recognition systems and Chinese characters input systems.

So far, many systems have been developed to recognize consonants, vowels and tones of Chinese speech[Han93][Lee93]. The translation from syllables in the format of *Pinyin* into characters is not a trivial problem because of many homonyms in Chinese language. This is also the bottleneck of the *Pinyin* input method of Chinese characters. Information of the whole word and its part of speech is needed to help resolve the ambiguity.

4. Text-to-speech translation system

To synthesis a fluent speech from text, the information of word boundaries is needed[CLP91][XY93]. Furthermore, many characters have more than one pronunciations. The determination of the correct pronunciation can usually becomes unique once the word to which the character belongs is known.

5. Chinese character code translation system

There are several computer coding schemes in use for Chinese characters. The most common ones are GB, which is used for *Jianti Zi*, and Big5, which is used for *Fanti Zi*. The translations between Big5 to GB are not one-to-one translations. The information of the word and its part of speech is needed to help resolve the translation ambiguity.

In a human, the process of word segmentation and identifying the part of speech of the word is usually done subconsciously, in parallel, and interactively. Traditionally, however, automated word segmentation and word tagging systems usually carried out these tasks separately. To resolve ambiguities, many systems adopt rules. Interestingly, many rules are based on parts of speech [CLP91]. There are three main problems when more and more rules are added to the system. Firstly, since the interaction among the rules is sometimes unpredictable, adding rules to improve the result for certain scenarios may decrease the overall accuracy. Secondly, if the number of rules is increased, the processing speed slows down rapidly. Lastly, adding rules typically requires knowledge experts and system developers. The procedure is normally beyond the capability of a system user. The mechanism for updating or entering additional knowledge is complicated in general.

In this paper, we propose a computational procedure which combines word segmentation with word tagging, but the system remains simple. It is motivated by recent research on neural networks and systems with learning capability. These systems are more flexible for dynamic updating. A prototype system, Automatic Chinese proCESsing System (ACCESS), was implemented with a corpus containing 83 manually segmented and tagged text files, a total of 111,068 words, or 173,497 Chinese characters. A dictionary based on this corpus is created, including 14,400 entries with occurrence probability information specified. The tag set contains 72 tags, including 25 punctuation marks. Bi-gram model and dynamic programming algorithm are used. The results of the closed test are: segmentation accuracy is 99.7%, tagging accuracy is 95.7%. The results of opened tests (the definition of the opened test is given in section 3) in a specified domain are : segmentation accuracy is 91.6%, tagging accuracy is 89.6%. Processing speed on a SUN SPARC 10 machine is 4,300 characters per second.

The organization of the paper is as follows. The architecture of the system will be described in section 2. In section 3 the algorithm of ACCESS and its performance are presented. Some limitations and future work are discussed in section 4. Some conclusions are drawn in section 5.

2   ARCHITECTURE OF ACCESS

To overcome the disadvantages of rule-based systems mentioned in section 1, we propose an approach which is motivated by recent research on neural networks and systems with learning capability. These systems are more flexible for dynamic updating. However, the classic neural networks have rigid network structures which make them difficult to handle inputs with loosely defined formats, such as a sentence. In [WC94], a so-called *hybrid system* was proposed which combines the learning capability with the flexibility of a digital processing system. The concept is applied to the current issue of word segmentation and tagging. A prototype system, ACCESS, was developed and its architecture is shown in Figure 1. Like the classic neural networks, the hybrid network also has a feedback loop for learning, a forward propagation path, namely the scoring system, and a training set, namely the corpus. It contains two important data sets: the lexicon and the grammar table. These data sets, together with

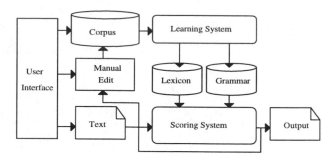

FIGURE 1: The Architecture of ACCESS

the set of system parameters used in the scoring system, form the core of the hybrid system. Learning is accomplished through updating the lexicon, the grammar table, and occasionally the system parameters.

From the user's viewpoint, the system operates in the following manner. The content of the lexicon and the grammar table are initially empty. The system operator fills in the data sets by training the system with manually prepared, segmented and tagged texts. After the system has been sufficiently trained, it is ready to process input queries. Inputs are entered in the form of Chinese text files represented in GB code format. Each character is represented by a code word with no special marking between words. The system performs the word segmentation and tagging function by calculating scores for different interpretations and selecting the highest scoring interpretation as the output of the query. The output path can be looped back to provide feedback for self-learning or supervised learning, depending on whether a human operator is involved or not. In the self-learning mode, the learning system compares the output of the scoring system with the output target threshold, the lexicon and the grammar table are updated only if the target threshold is reached or exceeded. System parameters can also be updated, but with a more complicated procedure as described in [WC94]. In the supervised learning mode, a human supervisor is prompted whenever the output scores are below a target threshold. The supervisor then performs the segmentation and tagging by hand. The results are then used to train the system. Notice that, in this mode, the supervisor only needs to be experienced in the segmentation and tagging functions as a user. No deep knowledge of the internals of the system is required.

In this system, the lexicon provides the information of words such as the occurrence frequency, the part of speech, pronunciation in *Pinyin* format, and the information needed to translate Chinese character code between different coding systems.

The grammar table is a two-dimensional matrix, $G = \{G(P_1, P_2)\}$, where $G(P_1, P_2)$ is the probability that the tag pair (P_1, P_2) occurs in the corpus. Table 1 shows some elements of G extracted from the grammar table of ACCESS.

In the linguistic terminology, the information in the lexicon is called unigram information, whereas

TABLE 1: Some Elements of the Grammar Table of ACCESS

(NN, NLN)	$9.209e^{-3}$	(NN, U)	$7.504e^{-5}$
(NN, BB)	$4.738e^{-3}$	(NLN, VT)	$2.090e^{-3}$
(VT, NN)	$2.212e^{-2}$	(VT, VT)	$3.420e^{-3}$
(VT, BB)	$5.253e^{-4}$	(U, VT)	$8.576e^{-4}$
(PP, DV)	$4.309e^{-3}$	(DV, NN)	0.0
(DV, DV)	$7.633e^{-3}$	(DV, I)	$5.489e^{-4}$
(I, NN)	$1.548e^{-2}$	(BB, PP)	$6.711e^{-3}$

NN - common noun, NLN - locative noun, VT - transitive verb, DV - verbal adverb, I - preposition, PP - personal pronoun, U - measure word, BB - full stop punctuation.

that of the grammar table is called bi-gram information[Nak89]. An n-gram is a string of n consecutive tags. Unigram and bi-gram statistics are well researched in English and Chinese texts. A number of techniques make use of these statistics for information processing [HZFRQ86][Joh89][Suen79].

3 ALGORITHM OF THE SCORING SYSTEM AND EXPERIMENTAL RESULTS

The maximum matching method [CLP91] is a well known basic algorithm to segment a Chinese sentence. Starting from the first character, we segment the sentence in the following way. Assume that the first $i-1$ characters have been segmented, and the remaining sentence is $A_i A_{i-1} \cdots A_{N-1}$. The algorithm tries to find the longest word in the lexicon of the form $A_i A_{i+1} \cdots A_{i+k-1}$ and map that part of the sentence to that word. If the word has more than one tag, it will be tagged using the one with the highest occurrence probability. If no word beginning with the character A_i is found in the lexicon, the A_i is tagged as an unknown word with word-length one. This routine is repeated recursively until all the characters in the sentence are processed.

The basic algorithm has two main weaknesses which can be improved on: the tagging and segmentation steps are done independently, and no bi-gram information is utilized. In our work, an approach is developed to segment and tag concurrently while keeping the system structure simple. The basic idea is to analyze all the possible character and tag combinations using the available unigram and bi-gram information, and to choose the highest scoring combination as the final answer. To illustrate the basic concept, consider an example where the basic algorithm fails.

他正在屋里写计划.
He is writing a plan in the room.

The output of the basic algorithm is listed in the upper part of Table 2, and the correct output is listed in the lower part.

The basic algorithm fails in two places. "正在"(be doing) is incorrectly segmented as a word by wrongly

TABLE 2: The Output of the Basic Algorithm and the Correct Output

Word	Tag	Tag Name	Meaning
他	PP	personal pronoun	He
正在	DV	verbal adverb	be (doing)
屋	NN	noun	room
里	NLN	locative noun	in...
写	VT	transitive verb	write
计划	VT	transitive verb	plan(verb)
.	BB	punctuation B	full stop
他	PP	personal pronoun	He
正	DV	verbal adverb	be (doing)
在	I	preposition	in or at
屋	NN	noun	room
里	NLN	locative noun	in...
写	VT	transitive verb	write
计划	NN	noun	plan(noun)
.	BB	punctuation B	full stop

using the maximum length is the best assumption. "计划" (plan) is tagged as verb because it has two tags and is used as a verb more often than as a noun in the corpus. Unfortunately, the correct interpretation in this case should be a noun.

When the bi-gram information is used in this example, a more accurate outcome can be obtained. In particular, according to Table 1, the probability that an adverb is followed by a noun is zero. This would rule out the first error. Also, after a transitive verb, the probability of seeing a noun is high. This implies the second error is not a good candidate. Hence, by judiciously using this information both errors can be avoided.

To exploit the bi-gram information systematically, we use a method that is similar to the approach used in [Church88] and [WC94].

In the following discussion, we assume all sentences are terminated with a punctuation mark. Define a *sentence fragment* be a sentence with only one punctuation mark or part of a sentence that begins from the head of the sentence or after a punctuation mark and ends with a punctuation mark. Hence, a sentence fragment has one and only one punctuation mark.

When a sentence is presented for analysis, it is firstly broken into sentence fragments. Each sentence fragment is subsequently segmented into a number of *tokens*. A token is defined here to be an ordered pair, (W, P), consisting of a word, W, followed by a part of speech tag, P. A token is said to be valid if the word, W, exists in the lexicon and has a valid part of speech usage P. Characters in a sentence fragment that do not correspond to any valid word in the lexicon form the *unknown tokens*. An unknown token is of the form (C, Z), where C is a single character and a label Z. Associated with each sentence fragment there is a *token list* consisting of all the valid tokens and unknown tokens that can be derived from the sentence fragment.

TABLE 3: The Token List of the Example

Word	Tag	Meaning	Frequency
他	PP	he	4.470e-3
正	DV	be(doing)	4.923e-4
正在	DV	be(doing)	2.591e-4
在	DV	be(doing)	2.332e-4
在	I	in or at	1.334e-2
屋	NN	room	1.036e-4
里	NLN	in...	1.866e-3
里	U	0.5km	5.182e-5
写	VT	write	2.850e-4
计	VT	count	2.591e-5
计划	VT	plan(verb)	1.036e-4
计划	NN	plan(noun)	9.069e-5
划	VT	paddle	2.591e-5
.	BB	full stop	1

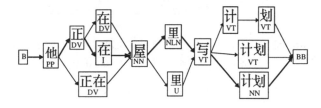

FIGURE 2: Partition Graph

For our previous example, the corresponding token list is contained in Table 3.

From this token list, a graph, called a *partition graph*, is constructed as shown in Figure 2. The graph essentially summarizes all the feasible ways of partitioning the given sentence fragment into the list of tokens. Each token in the token list forms a node in the partition graph. Hence, each node can be associated with a word, the word in its corresponding token. Nodes are connected if and only if the last character of one node immediately precedes the first character of the other node in the presented sentence fragment. The partition graph has a special node, the B node which symbolizes the beginning of a sentence fragment. It is connected to tokens whose associated words start with the first character in the sentence fragment. The partition graph also has a unique node which is associated with the unique punctuation mark in the sentence fragment. Notice that a path starting from the B node and ends with the punctuation mark node identifies a solution to the segmentation and tagging problem for the presented sentence fragment.

In the lexicon of ACCESS, the maximum length of a word is 8 characters, and the average length of the word is 2.07. The maximum number of the tags of a word is 8, and the average number of the tags of a word is 1.14. They are system parameters and independent of the length of the sentence fragment, N. So the partition graph can be generated in $O(N)$ time, and

the space complexity is also $O(N)$.

We define a scoring function to evaluate the possible paths from the node B to the punctuation mark node in the graph. From the viewpoint of segmentation, we hope the occurrence frequency of each word is as high as possible, so we should maximize the probability

$$P_{\text{seg}} = \prod_{i=1}^{M} p(w_i)$$

where M is the number of nodes in the path, and $p(w_i)$ is the probability that word w_i occurs in the corpus. Because $p(w_i) < 1$, maximizing P_{seg} implies that the smaller number of words is better, which is the idea of maximum matching method [CLP91] in a sense.

From the viewpoint of tagging, we hope the occurrence frequency of each tag pair is as high as possible, so we should maximize the probability

$$P_{\text{tag}} = \prod_{i=1}^{M} G(P_{i-1}, P_i)$$

where M is the number of nodes in the path; P_{i-1} and P_i are the tag of word w_{i-1} and w_i respectively; and $G(P_{i-1}, P_i)$ is the probability that the tag pair occurs in the corpus.

To maximize these two measures, we have

$$P = P_{\text{seg}} \times P_{\text{tag}} = \prod_{i=1}^{M} p(w_i) G(P_{i-1}, P_i)$$

Usually $p(w_i)$ and $G(P_{i-1}, P_i)$ are very small, to avoid the lower overflow during multiplications, the logarithm function is used and the final scoring function becomes

$$F = \ln P = \sum_{i=1}^{M} [\alpha \ln p_i + \ln G(P_{i-1}, P_i)] \qquad (1)$$

where α is a system parameter that should be fine-tuned to optimize the accuracy of the system. It is used to adjust the relative influence of the unigram and bi-gram information. When $\alpha > 1$, unigram information plays a more important role than bi-gram information. Some manual tuning of α was done to achieve good accuracy in the output. In ACCESS, α is set at 4.5. In [WC94], a more systematic approach to tune system parameters such as α is described. It was not followed here due to huge amount of training data needed for this application.

The optimal path is defined by the criterion

$$F_{\text{opt}} = \max_{k \in R} F_k \qquad (2)$$

where R is the set of all the paths from the node B to the punctuation mark node.

A number of algorithms can be used to find the optimal path in the graph, including Branch-and-bound, Shortest Path and Dynamic Programming

[Cormen90]. Among them, Dynamic Programming is the easiest one to implement and runs in linear time in terms of the number of nodes. Apply the Dynamic Programming Algorithm and the scoring function (1), we get the optimal path, which is drawn in thick lines in Figure 2.

To evaluate accuracy of the method, we define the segmentation accuracy as the percentage ratio:

$$\text{Segmentation accuracy} = \frac{N_S}{N_C} \times 100\%$$

where N_S is the number of characters in the correctly segmented words and N_C is the total number of characters in the input text.

Tagging accuracy is defined as the percentage ratio:

$$\text{Tagging accuracy} = \frac{N_T}{N_W} \times 100\%$$

where N_T is the number of words with the correct tags and N_W is the total number of words in the input text.

There are two testing experiments. One is called the closed test, which means using the corpus to train the system and testing the system using the same corpus. In this test, there is no new words and grammar patterns, so the accuracy is usually high. This test shows the consistence in the corpus and the correctness of the system. The other testing experiment is called the opened test, which means using the corpus to train the system and testing with other data. In this test, new words and new grammar patterns are usually found so that the accuracy of the output is decreased. This test shows the robustness and the usability of the system. It is the real situation when the system is put into use.

We did both closed test and opened test of our system. We did the opened test in the following way: for each article, A_i, in the corpus, using all the articles except A_i to train the system, and then testing article A_i. Finally, the results of the average accuracy of segmentation and tagging are obtained.

The system is designed for an unlimited domain, but a large corpus is required to obtain a high accuracy of the opened test. In the current stage of ACCESS, we have to limit the domain in the economic news in Hong Kong and China. However, to expand the lexicon, we also put some articles in other domain into the corpus. There are 83 articles in the corpus, and 40 articles are in the specified domain. The opened test results will be better after the corpus is expanded. To illustrate the performance of the system in the specified domain, we did another opened test only with the articles in the specified domain. We listed the results of all the three tests in Table 4.

4 LIMITATIONS AND THE FUTURE WORK

In ACCESS, bi-gram information is used in addition to unigram information. Since content words which include nouns, verbs, pronouns, etc. have a stronger correlation with preceding and succeeding words than

TABLE 4: The Results of the Closed Test and the Opened Test

Segmentation Accuracy of Closed Test	99.5%
Tagging Accuracy of Closed Test	95.3%
Segmentation Accuracy of Opened Test	87.0%
Tagging Accuracy of Opened Test	86.2%
Segmentation Accuracy of Opened Test in the Specified Domain	91.6%
Tagging Accuracy of Opened Test in the Specified Domain	89.6%
Processing Speed on SUN SPARC 10 (characters per second)	4,300

function words such as prepositions and conjunctions, the performance of ACCESS in tagging content words is better than its performance in tagging function words. For example, the word "在" can be tagged as "adverb" to mean "be doing", or as "preposition" to mean "in", depending semantically on the words that follow. In the sentence, "他在 很快地写."(*He is writing very quickly.*), "在" should be an adverb because it is followed by the verb "写". However, in the sentence, "他在 很大的屋里."(*He is in a very big room.*), "在" should be a preposition because it is followed by the noun "屋". Currently, ACCESS will fail in the first sentence because in the context, "他"(*he*) and "很"(*very*), "在" is tagged as "preposition" more often in the corpus. Bi-gram information cannot handle this type of long distance dependence. Function words poise a key problem in many natural language processing systems. Some methods have been presented to solve this problem [Huang93]. It was not incorperated current in ACCESS due to the issue of complexity.

There are many words like "计划"(*plan*) that can be tagged as either verbs or nouns. ACCESS may possibly fail in tagging these words because the semantic information is usually needed to clarify these cases. For example, in the sentence, "销售计划 写完 了." (*The sales plan has been written.*), for the first four characters, there are four combinations that are correct syntactically: "销售"(*sale*) and "计划"(*plan*) can both be tagged as either a "transitive verb" or a "noun". There are two combinations for "计划写": one is a noun followed by a verb, (this is a popular expression of the passive voice in Chinese); the other is a phrase of two verbs, which means *to plan to write something*. Currently, "计划" is wrongly tagged as "verb" in this case, which is the popular case of the pattern "*Somebody plans to write something.*". To resolve these ambiguities, the semantic information may be needed.

Unknown word is another key problem in a natural language processing system. As the corpus size increases, it is likely that less unknown words will be encountered. However, some unknown words, such as proper nouns and new words cannot be exhaustively listed. Errors coming from these words will affect the correct segmentation and tagging of neighboring words as well. ACCESS offers some limited capability in handling unknown words since the optimization algorithm may lead to a correct interpretation even when some of the nodes are unknown. However, a more powerful technique to handle unknown words is needed to boost the accuracy to higher than 99%.

5 CONCLUSION

In this paper, we presented a simple but effective method for the segmentation and tagging of Hanzi text. We developed a system ACCESS, which processes Chinese text. The system can be generalized to other Hanzi texts with suitable modification. ACCESS takes unigram and bi-gram information and contains two key data structures – the lexicon and the grammar table. The core procedure is based on a hybrid algorithm which enables the system to be updated easier than rule-based systems. The performance of ACCESS is satisfactory. Various applications of this system are also presented in this paper.

ACKNOWLEDGMENT

We would like to thank Dr. Yang Gu of the Chinese University of Hong Kong and Professor Changning Huang of the Tsinghua University of China for helpful discussions concerning this work and valuable comments on computational linguistics issues of the Chinese language.

REFERENCES

[Bai92]   Bai, S.H., Xia, Y. and Huang, C.N., "Automatic Part of Speech Tagging System of Chinese", *Technical Report*, Tsinghua University, Beijing, China, 1992.

[Booth85] Booth, B.M., "Revising CLAWS", *ICAME News*, vol. 9, pp. 29-35, 1985.

[Church88] Church, K.W., "A Stochastic Parts Program and Noun Phrase for Unrestricted Text", *Proceeding 2nd Conference on Applied Natural Language Processing*, pp. 695-698, Austin, Texas, 1988.

[CLP91]   *The Status and Progress of Chinese Language Processing*, Version 2.0, Technology, Association for Common Chinese Code, International, pp. 152-167, March, 1991.

[Cormen90] Cormen, T.H., C.E. Leiserson and R.L. Riverst, *Introduction to Algorithms*, MIT Press, 1990.

[GB/T92]   *Contemporary Chinese Language Word Segmentation Specification for Information Processing* , National Standard of the People's Republic of China, GB/T 13715-92.

[Greene71] Greene, B.B. and Rubin, G.M., "Automated Grammatical Tagging of English", Department of Linguistics, Brown University, Providence, Rhode Island, 1971.

[Han93]   Han, Di, Rodman, R.D. and Joost, M.G., "A Voice System for the Computer Entry of Chinese Characters", *Computer Speech and Language*, vol. 7, no. 1, Jan., 1993.

[HZFRQ86] *Words Frequency Dictionary of Contemporary Chinese Language*, Beijing Linguistics College Press, 1986.

[Huang93] Huang, Changning and Guo, Chengming, "Example-Based Sense Tagging of Running Chinese Text", *Proceedings of the Workshop on Very Large Corpora*, Ohio State University, Columbus, Ohio, USA, pp. 102-112, June 22, 1993.

[Joh89]   Johansson, S. and Hofland, K., "Frequency Analysis of English Vocabulary and Grammar Based on the LOB Corpus", Clarendon Press. Oxford, 1989.

[Lee93]   Lee, Lin-shan, et al., "Golden Mandarin (I) - a Real-Time Mandarin Speech Dictation Machine for Chinese Language with Very Large Vocabulary", *IEEE Transactions on Speech and Audio Processing*, vol. 1, no. 2, April, 1993.

[Mar83]   Marshall, I., "Choice of Grammatical Word-Class Corpus", *Computer in the Humanities*, vol. 17, pp. 139-150, 1983.

[Nak89]   Nakamura, M. and Shikano, K., "A Study of English Word Category Prediction Based On Neural Networks" , *ICASSP-89: 1989 IEEE International Conference on Acoustics, Speech and Signal Processing*, vol. 2, pp. 31-4, 1989

[Suen79]   Suen, C.Y., "N-gram Statistics for Natural Language Understanding and Text Processing", *IEEE Trans. Pattern Anal. Machine Intell.*, vol. PAMI-1, no. 2, pp 164-172, Apr. 1979.

[WC94]   Wong, W.S. and Chuah, M.C., "A Hybrid Neural Network for Address Normalization with Spelling Correction Capability", *IEEE Expert*, no.6 Dec. 1994.

[XY93]   Xu, Jun and Yuan, Baozong, "New Generation of Chinese Text-to-Speech System", *Proceedings TENCON'93, 1993 IEEE Region 10 Conference on "Computer, Communication, Control and Power Engineering"*, Beijing, Oct. 1993.

COMPUTER-AIDED NEWS ARTICLE SUMMARIZATION

Hisao Mase, Hiroshi Tsuji, and Hiroshi Kinukawa
Systems Development Laboratory, Hitachi Ltd.,
3-6-1 Bakuroh-machi, Chuo-ku, Osaka, 541 Japan.
Email: mase@sdl.hitachi.co.jp

ABSTRACT

This paper presents a parameter-settable text summarization system for news articles. The basic system design concept involves having the system generate various types of summaries that meet diverse user preferences. From a statistical viewpoint, we design 24 text skimming parameters for the system. These parameters are categorized into five groups: locations, sentence types, keywords, special expressions, and miscellanies. Experimental simulation employing 16 Japanese sample news articles shows that the parameters are powerful enough to generate various types of summaries for text summarization.

1. INTRODUCTION

It takes a lot of time and effort to set apart the information we really need from a variety of sources. Generating text summaries is one technical solution to reduce the load. Text summaries can be used to indicate whether the original text is of interest or not and to inform the reader of the main contents of the text [Sal89].

As for reports news displayed on electronic bulletin-boards, since it is impossible for readers to read original texts on the spot, a complete summary of the news is required instead. However, since it is difficult for a computer to present a "automatic summarization" for the wide range of news topics, summaries have to be made by expert editors. Since this work must be processed in real time, the editors are under extreme pressure.

Many researchers are engaged in developing a fully automatic text summarization system. The method of choice is to adopt key sentences extraction by computing a score for each sentence in the text. The scoring is based on rhetorical relations [Mii94], keyword distributions and frequencies [Luh59], and so on. Some systems combine several features to compute the scores [Kup95, Mas90, Pai90, Yas89]. These approaches are not useful in the news article summarization support systems because the quality of the summaries is not good enough and the features are not tunable by the user. We suppose that a news article summarization support system should be able to generate various types of summaries based on news categories or editors' preferences.

2. SUMMARIZATION WORK MODEL

To design a computer-aided system, we believe it is important to review the human text summarization process. In this process, An editor often repeats the steps shown in FIGURE 1(a). Consequently, similar steps should be employed when summarizing in collaboration with the text summarization system. Our proposition for a collaborative model that separates the work of the editor from that of the system is as follows (FIGURE 1(b)):

STEP 1: reading - We read the original text.
STEP 2: constructing - We specify the parameters that show how to summarize the text.
STEP 3: draft - The system makes a rough summary.
STEP 4: refining - We refine the draft.

In this proposition, one problem is to identify what kind of information we should specify for the system in STEP 2. Efficiency is needed in STEP 2 and STEP 3 because these steps may have to be repeated again and again until a relevant summary is generated. To solve this problem we adopt the text skimming parameter value definition approach. These parameters are defined from a statistical viewpoint. The editor can get various types of summary drafts simply by setting different combinations of parameter values. Since the editor's requests are reflected directly in the summaries, the system can generate more relevant summaries, and the work load for STEP 4 is reduced.

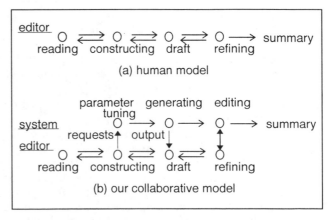

FIGURE 1   Text summarization work model.

3. PARAMETERS FOR GENERATING SUMMARIES

There are varying level of article structures as shown in FIGURE 2: titles, paragraphs, sentences, clauses, phrases, and words. As such, first, let us identify the categories of our text skimming parameters.

The parameters are categorized into five groups. "Locations" define the locations in texts to be checked. "Sentence types" define the types of sentences (fact sentences, the author's opinions, etc.) to be included. "Keywords" define the sentences to be included; determination based on particular words. "Special expressions" define what clauses, phrases or words in sentences should be included. Other parameters (summary length, etc.) are categorized into miscellanies.

The parameters in "locations" are as follows:

(1) *Titles (Headlines)*: The editor can select whether the title is to be included in the summary or not.

(2) *Paragraphs*: When we make a summary, we often limit the scope of the analysis. This parameter allows the editor to select the paragraphs to be analyzed.

(3) *Head sentences of the article*: Head sentences often give a feel of the contents of some news. The editor specifies the number of head sentences to be considered.

(4) *Head sentence of each paragraph*: The head sentence of each paragraph in some types of articles, such as editorials, often describes the content of the paragraph.

(5) *Sentences in the head paragraph*: The sentences in the head paragraph also give a feel of the contents of news.

"Sentence types" have the following six parameters for summarizing news articles. Note that one sentence may be classified into several types:

(1) *Opinion sentences*: Opinion sentences are identified by focusing on phrases. In Japanese, sentences are regarded as opinions if the last phrases include special auxiliary verbs such as "べき(should)".

(2) *Fact sentences*: It is difficult to identify fact sentences because there are greater variety of expression types in the ending phrases. Thus, we adopt a convenient rule; The sentence presents a fact if it does not present an opinion.

(3) *Cause/Reason sentences*: The cause or the motivation is essential in summaries covering accidents or criminal event articles in general. Cause/reason sentences are identified by focusing on special expressions such as conjunctive words "なぜなら(because)".

(4) *Result sentences*: Result sentences are also identified by focusing on special expressions such as conjunctive phrases "その結果(as a result)".

(5) *Example sentences*: These sentences are identified by focusing on conjunctive words.

(6) *Conversation sentences*: In Japanese, particular marks (e.g., 「」) in which direct narration is enclosed are also used to emphasize the different phrases; distinguishing them is done by focusing on the last word of the part enclosed in brackets and the word just after the bracket "」". If the last word of the part enclosed is a verb and the word just after the bracket "」" is the function word "と", then this is identified as a conversation.

All articles include a couple of words that designate the key sentences. "Keywords" have the following three parameters:

(1) *Title keywords*: The title is regarded as the shortest summary, and it includes a couple of keywords. We suppose that key sentences borrow several keywords from the title. For this parameter, if a sentence includes noun words existing in the corresponding title more than a fixed count N (in the experiments presented in chapter 5, N=2), then the sentence is identified as a key sentence.

(2) *Thematic keywords*: We also use keywords extracted from the body. Three features are used to distinguish keywords from noisy words: word frequency, word location and subject of each sentence.

The score $W(w_i)$ of each noun word w_i is computed based on the above three features as follows:

$$W(w_i) = \alpha * F_i + \beta * L_i + \gamma * S_i$$

where F_i is the frequency of w_i, L_i is the reciprocal of the earliest sentence w_i appears in (e.g., 1st: 1, 2nd: 1/2, etc.), S_i is the number of sentences in which w_i is the subject, and α, β, and γ are weights (in chapter 5, $\alpha = 1$, $\beta = 0$, $\gamma = 3$). The highest N words are extracted as the keywords (in chapter 5, N=5 or 3).

(3) *Pre-defined keywords*:

The summaries of certain articles, e.g., on the death of famous person and information on exchange rates, have fixed forms. As for articles of the former type, the summaries always include the name, the age, the occupation of the person, and the cause of his/her death.

In general, it is relatively easier to identify key sentences by searching for sentences that include pre-defined keywords. To identify the domain of an article, the editor must define "requisite keywords" and "arbitrary keywords". An article that includes all of the requisite keywords and at least one of the arbitrary words (if defined) pre-defined as keywords of a domain is identified as an article on that domain. Furthermore, in order to identify the key sentences in the article, the editor must pre-define keywords which should appear in the key sentences. In the case of exchange rates, for example, the editor may define the following three types of keywords:

requisite keywords: 円(yen), 為替相場(exchange rate)
arbitrary keywords: 銀行(bank), 暴落(fall)
keywords in key sentences:
　　　為替相場(exchange rate), 終値(closing price)

"Special expressions" have the following seven parameters:

(1) *Past information*: Some sentences describe the background or a past event that includes no new information. In general, such information is not important because the purpose of articles is to inform of things new. In Japanese news articles, the past progressive form is often used.

(2) *Parallel phrases*: Sometimes, there is a list of items, some of which are able to be omitted. The sentence "He went to the USA, Canada, Italy, and France." may be summarized into "He went to the USA, Canada, etc.". The word "など(etc.)" is inserted into the summary to

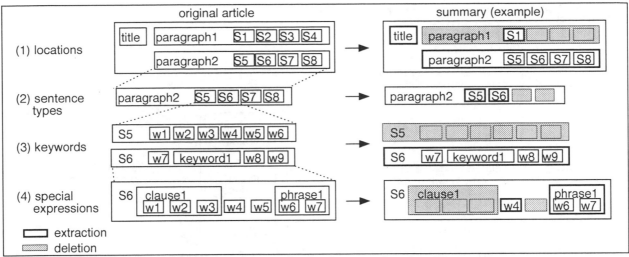

FIGURE 2   Summarization images of each category.

inform of the existence of other items.

(3) *Ordered phrases*: Some sentences include phrases that follow numbers and these numbers are in a sequence. If these ordered phrases are included, only the first phrase remains in the summary and the others are deleted.

(4) *Brackets*: Phrases enclosed in parentheses or brackets i.e., "(   )" or "[   ]", are often supplements.

(5) *Redundant phrases*: At times, redundant phrases can be omitted. For example, the word "西暦(the year)" in "西暦２０００年(in the year 2000)" is redundant.

Other parameters in this category are (6) *conjunctive phrases* and (7) *adverbial phrases*.

"Miscellanies" has three parameters which cannot be included in any of the above four categories:

(1) *Nominalization*: This involves the sentence styles. In Japanese, almost all sentences end with verbs, but some end with nominalized verbs to express conciseness.

(2) *Abbreviation*: We often use abbreviated words to express sentences concisely. For example, the word "television" is abbreviated to "TV" and the United States of America to the USA.

(3) *Summary length*: The editor can define the summary length either absolutely or relatively. Each value (alternative) of the parameter has a weight that is proportional to the importance of the parameter. The summary is generated based on the other parameters, and if the generated summary length is longer than the length the editor specifies, the score for each word is computed and ranked. Sentences are extracted from higher to lower until the point is reached where the summary length exceeds the length specified by the editor.

4. PROTOTYPE SYSTEM

Our prototype system has 24 parameters as shown in Table 1 and allows the editor to tune the value of each parameter, most of which have four alternatives.

"Extraction 1" is to extract only the parts that are relevant to the parameter. "Extraction 2" is to extract some of the parts that are relevant to the parameter. Whether the other parts are extracted depends on the other parameters. "Deletion" is to delete the parts that are relevant to the parameter. "Ignored" is to ignore the parameter.

If the editor sets parameter A and B to be "extraction 1", the system extracts only the parts that are relevant to both A and B. If the editor sets A and B to be "extraction 2", it extracts the parts that are relevant to either A or B. These alternatives have a priority order: deletion, extraction 1, extraction 2. If the two selected parameters contradict each other, the system generates summaries according to this priority. Some parameters (for the titles or the nominalization) have only two selections (apply or do not apply) and some have additional selections such as the addition or deletion of keywords by the editor.

As shown in FIGURE 3, the summarization process consists of the following three main modules:

(1) *Parameter analysis*: For each parameter, whether the parts in the article are relevant or not is analyzed. We collected about 70 rules from 120 sample news articles to analyze the parameters. These rules are registered in the parameter rules (see FIGURE 3). The results of this analysis are recorded into the analysis log.

(2) *Key parts extraction*: The key parts are extracted according to the parameter values the editor has set.

(3) *Summary Generation*: The parts are modified into a summary through the nominalization, the abbreviation, and the deletion of subject words which appear repeatedly in the summary.

FIGURE 4 shows the user interface for the parameter setting. The editor sets the value of each parameter by clicking a mouse button. The original news article and its summary are displayed side by side. The editor refines the summary, while referring to the original article.

The execution time needed to summarize the average article, containing 500 characters, is about 5 seconds. The

Table 1   List of parameters for summarization.

categories		parameters	
01	locations	01	titles (headlines)
		02	paragraphs
		03	head sent. of the article
		04	head sent. of each paragraph
		05	sent. in the head paragraph
02	sentence types	06	opinion
		07	fact
		08	cause/reason
		09	result
		10	example
		11	conversation
03	keywords	12	title keywords
		13	thematic keywords
		14	pre-defined keywords
04	special expressions	15	past information
		16	parallel phrases
		17	ordered phrases
		18	brackets
		19	redundant phrases
		20	conjunctive phrases
		21	adverbial phrases
05	miscellanies	22	nominalization
		23	abbreviation
		24	summary length

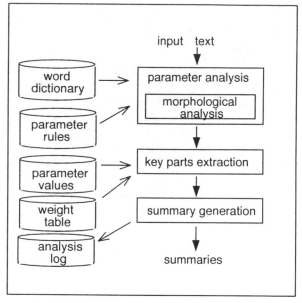

FIGURE 3   Flow of the summary generation process.

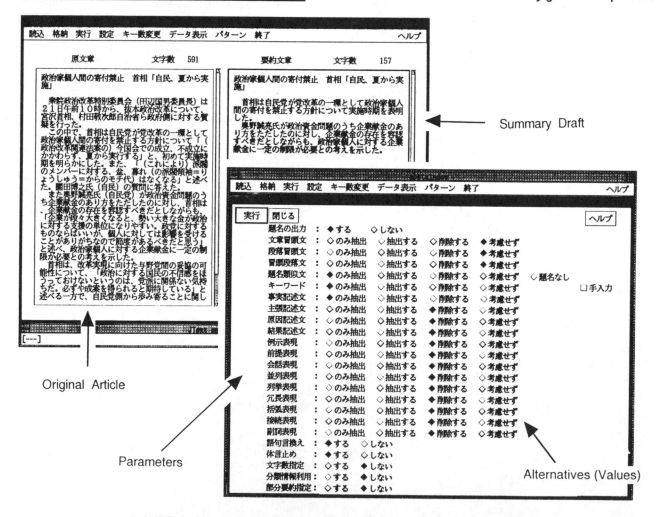

FIGURE 4   User interface for parameter setting in our prototype system.

Table 2 Data on sample articles and rate of the parts that are relevant to each parameter.

article no.	01	02	03	04	05	06	07	08	09	10	11	12	13	14	15	16	avg.
domain of the article*	tri	crm	soc	cnt	pol	cnt	cnt	eco	ent	pol	pol	crm	pol	crm	pol	acc	
article length(character)	1178	464	343	1104	310	349	116	499	420	567	655	662	356	282	343	274	495
number of paragraphs	8	4	3	9	2	4	1	5	3	4	3	4	2	1	3	3	3.7
number of sentences	15	6	7	14	4	4	2	8	6	6	8	10	4	3	3	5	6.6
parameters	rate of the parts that are relevant to each parameter																
head sent. of the article	13	30	16	10	37	34	48	18	27	13	22	18	24	51	38	41	28
head sent. of each para.	62	81	68	76	68	100	48	92	61	87	51	63	49	51	100	65	70
sent. in the head para.	29	49	36	10	55	34	100	18	44	13	42	21	52	100	38	41	43
opinion	0	13	0	0	0	0	0	0	0	0	0	0	0	0	0	23	2.3
fact	100	87	100	100	100	100	100	100	100	100	100	100	100	100	100	77	98
cause/reason	0	0	0	0	0	0	0	0	0	0	0	0	0	0	0	0	0
result	0	0	0	0	0	0	0	0	0	0	0	0	0	0	0	0	0
example	0	0	0	0	0	0	0	0	0	0	0	0	0	0	0	0	0
conversation	10	0	0	40	0	41	0	33	48	57	0	46	0	15	51	0	21
title keywords	69	66	53	32	37	0	100	35	70	50	41	55	52	51	38	76	52
thematic keywords(5 kinds)	81	100	100	94	100	100	100	100	100	87	83	100	100	100	72	100	95
thematic keywords(3 kinds)	61	85	93	81	100	85	100	100	92	87	83	92	100	100	72	100	89
pre-defined keywords	37	0	0	0	0	0	0	37	0	0	0	0	0	0	0	0	4.6
past information	2	4	4	0	32	0	0	11	16	17	6	15	10	3	0	6	7.9
parallel phrases	1	6	0	1	3	3	0	0	0	0	0	0	0	0	2	0	1
ordered phrases	0	0	0	0	0	0	0	0	0	0	0	5	0	0	0	0	0.3
brackets	5	9	4	4	0	2	10	2	2	10	11	2	2	11	0	0	4.6
redundant phrases	2	1	0	4	0	2	0	0	0	2	2	0	2	1	2	1	1.2
conjunctive phrases	1	1	0	0	0	0	0	0	0	1	0	0	0	0	0	0	0.2
adverbial phrases	1	1	4	3	1	1	0	4	1	2	1	3	5	1	1	6	2.2
nominalization	1	0	0	1	2	1	4	1	1	0	0	2	0	1	0	0	0.9
abbreviation	0	2	0	0	1	0	0	0	0	1	0	0	1	0	0	0	0.3

*Note: tri=trial, crm=crime, cnt=foreign countries, soc=society, pol=politics, eco=economy, ent=entertainment, acc=accident

Table 3 Examples of the setting of parameters.

no.	parameters	set1	set2	set3	set4
01	titles (headlines)	—	—	—	—
02	paragraphs	—	—	—	—
03	head sent. of the article	—	—	—	—
04	head sent. of each para.	—	—	◎	—
05	sent. in the head para.	—	—	—	—
06	opinion	×	—	—	—
07	fact	◎	—	—	—
08	cause/reason	×	—	—	—
09	result	×	—	—	—
10	example	×	×	×	×
11	conversation	×	—	—	—
12	title keywords	◎	—	—	—
13	thematic keywords	◎	—	◎	—
14	pre-defined keywords	—	—	—	◎
15	past information	×	×	×	×
16	parallel phrases	×	×	×	×
17	ordered phrases	×	×	×	×
18	brackets	×	×	×	×
19	redundant phrases	×	×	×	×
20	conjunctive phrases	×	×	×	×
21	adverbial phrases	×	×	×	×
22	nominalization	○	○	○	○
23	abbreviation	○	○	○	○
24	summary length	—	—	—	—

Notes:
◎ extracts only the parts that are relevant to that parameter.
× deletes the parts that are relevant to that parameter.
○ applies that parameter for summarization.

Table 4 Comparison of the correctness ratio (upper,%) and the relative length of summaries (lower,%).

no.	set1 A	B	C	set2 A	B	C	set3 A	B	C	set4 A	B	C
01	50	50	0	0	100	0	50	50	0	50	50	0
	34.4			54.8			54.2			25.9		
02	100	0	0	100	0	0	100	0	0			
	55.2			61.9			60.8					
03	33	0	67	33	33	33	33	33	33			
	48.7			61.8			84.3					
04	0	0	100	100	0	0	67	0	33			
	5.0			65.3			71.3					
05	50	0	50	50	0	50	100	0	0			
	36.5			46.5			62.9					
06	0	0	100	100	0	0	100	0	0			
	0.0			91.7			77.7					
07	100	0	0	0	100	0	100	0	0			
	85.3			35.3			85.3					
08	67	0	33	100	0	0	100	0	0	67	0	33
	31.5			75.8			83.2			34.9		
09	50	0	50	50	50	0	50	50	0			
	40.5			40.7			70.7					
10	50	50	0	100	0	0	100	0	0			
	23.5			61.6			61.6					
11	50	0	50	50	0	50	100	0	0			
	36.0			45.8			64.3					
12	50	0	50	50	0	50	50	50	0			
	15.9			58.3			71.0					
13	100	0	0	33	0	67	100	0	0			
	41.6			35.1			81.7					
14	50	0	50	50	0	50	100	0	0			
	39			39			82.6					
15	0	0	50	100	0	0	100	0	0			
	26.5			94.8			71.7					
16	50	0	50	50	50	0	50	50	0			
	47.4			55.8			65.3					
avg	51	6	43	63	17	20	80	14	6	60	20	20
	30.0			58.9			68.9			28.6		

▨ The best set of the four.   The best set: 83 14 3
58.4

execution time to summarize the same article with a different set of parameters is less than one second.

5. EXPERIMENTAL SIMULATION

We used 16 Japanese news articles to evaluate the effect of each parameter. Table 2 shows the rate of the parts that are relevant to each parameter and the data on the articles. The parameters belonging to "locations" effectively shortened the articles. The title keywords parameter and the conversation parameter are also effective, but half of the articles have no conversations (the rate is 0%). The past information parameter and the brackets parameter are relatively available for shortening the articles. The rates are lower compared with the parameters in "locations".

To evaluate our system, we defined four types of parameter sets as shown in Table 3. Set 1 generates the shortest possible summary; only fact sentences with title keywords and thematic keywords are extracted. Set 2 focuses on the importance of the sentence location; only the head sentences of each paragraph are extracted. Set 3 focuses on the sentences with thematic keywords; only sentences including thematic keywords are extracted. Set 4 considers the "pre-defined keywords" parameter.

We evaluated the correctness of each summary by comparing it with a summary written by human experts. All of the summaries made by the experts consisted of two or three sentences and the summary length was less than 90 characters. We divided the sentences in these summaries into three ranks. In rank A, the sentences were included almost wholly in the summary by our system. In rank B, the sentences were included only partly in the summary by our system. In rank C, the sentences were not included in the summary by our system. We calculated the correctness ratio with the following equation for each rank: $Correctness\ ratio\ (\%) = (X\ /\ Y) \times 100$ where X is the number of sentences in the rank and Y is the number of sentences in the expert summary.

Table 4 shows the correctness ratios and the relative summary length generated by our system. As for set 1, the average correctness ratio for rank A is 51%, which was not good enough for use as the input of summary refinement in STEP 4 mentioned in chapter 2. The average relative summary length was 30%, which is relevant. As for set 2, the average correctness ratio was 63%; better than the one of set 1. However, the average relative summary length was two-fold that of set 1. The average correctness ratio of set 3 was 80%; much better than that of set 1 or set 2, while the average relative summary length was 68.9%, which was the worst. The average correctness ratio of set 4 was better than that of set 1, and the summary length was relevant. However, since set 4 applies the "pre-defined keywords" parameter, which is limited to particular domains of articles, set 4 cannot be used for the summarization of most news articles.

These results shows that the editor cannot get relevant summaries if s/he has no means to set parameter values. On the other hand, any editor can get the best possible summary of the four by generating four types of summaries with our system. As shown in Table 4, if the editor can get the best summary from among the four types, the average correct ratio can 83%, which is good enough for the summary to continue on to refinement, while the average summary length is 58.4%, which is rather higher.

6. CONCLUSIONS

This paper presents a parameter-settable text summarization system for news articles. Our approach is applicable to other languages besides Japanese. Experimental results have shown that our system is powerful enough to generate summary drafts.

To improve our prototype system, we have several problems to overcome. First, we have to consider the possibility of automatic parameter setting for given articles. Secondly, we have to consider the automatic acquisition of parameter rules to be applied for the parameter analysis. Third, we have to consider the user interface to do the work more effectively.

ACKNOWLEDGMENT

We wish to thank the people working for the News Broadcasting Dept. and New Media Dept. of The Yomiuri Shimbun for giving us the chance to do this research, their thoughtful comments, and their helpful advice.

REFERENCES

[Kup95] J. Kupiec, et. al. A Trainable Document Summarizer. Proc. of 18th ACM SIGIR Conference, pp. 68-73, July, 1995.

[Luh59] H. P. Luhn. The automatic creation of literature abstracts. IBM J. Res. Develop., Vol. 2, 1959.

[Mas90] H. Mase, et. al. Key sentences extraction from the explanatory texts. Tech. Report of Inst. of Electronics, Information and Communication Engineers, NLC89-40 1990.

[Mii94] S. Miike, et. al. A full-text retrieval system with a dynamic abstract generation function. Proc. of 17th ACM SIGIR Conference, pp. 152-161, 1994.

[Pai90] C. D. Paice. Constructing literature abstracts by computer: Techniques and prospects. Information Processing and Management, Vol. 26, 1990.

[Sal89] G. Salton. Automatic Text Processing. pp. 439-448, Addison-Wesley Publishing Company, 1989.

[Yas89] H. Yasuhara, et. al. Summarizing Support System COGITO. Journal of Information Processing Society of Japan, Vol. 30, No. 10, October, 1989.

A CONNECTIONIST/SYMBOLIC DEPENDENCY PARSER
FOR FREE WORD-ORDER LANGUAGES

Jong-Hyeok Lee, Taeseung Lee, Geunbae Lee

Dept. of Computer Science and Engineering, Pohang University of Science and Technology

San 31 Hyoja-dong Nam-ku, Pohang 790-784, Republic of Korea

E-mail: jhlee@vision.postech.ac.kr

ABSTRACT

This paper proposes a dependency parser for free word-order languages on the basis of both connectionist and symbolic techniques. In the first connectionist phase, two neural nets are used to identify syntactic dependencies between individual words, specifying the grammatical role of dependency in each case. All the possible dependencies are kept in a syntactic dependency graph (SDG). Then, in the second symbolic phase, structural disambiguation is performed by a constraint satisfaction algorithm, in which various constraints on dependency structure are propagated through SDG to filter out inconsistent dependencies. For an efficient control, a predicate-driven constraint propagation as well as the forward checking algorithm is used. Experimental evaluation showed a quite promising result with parsing accuracy of above 90% for almost all Korean sentences of less than 10 words.

1. INTRODUCTION

Traditionally most syntactic parsers of a rigid word-order language like English have been developed on the basis of a phrase-structure grammar (PSG). However, it has been argued that PSG is not adequate for a relatively free word-order language like Korean or Japanese which have no syntactic constraints on the ordering of nominal arguments of verb. Furthermore, the variable word order often results in discontinuous constituents. A phrase-structure tree for such a scrambled sentence would have crossing branches, which PSG disallows. Unlike PSG, a dependency grammar (DG) does not divide a sentence up into constituents; instead, it identifies the grammatical relations that connect one word to another. This is known to be advantageous for dealing with discontinuity [Cov90]. Almost all known parsing algorithms are based on PSG, while efficient parsing strategies for DG are still unavailable.

In this paper, we propose a hybrid connectionist/symbolic dependency parser for free word-order languages, which consists of two phases. The first phase identifies all possible syntactic dependencies that connect one word to another while specifying the type of dependency in each case. All the identified dependencies are collectively recorded in a syntactic dependency graph (SDG). Then, the second phase removes inconsistent dependencies from SDG. The Yale system is used for Romanized Korean expressions.

2. KOREAN, FREE WORD-ORDER LANGUAGE

From the viewpoint of word order typology, Korean is a relatively free word-order language in which grammatical functions like *subject* and *object* are expressed by a variety of inflectional suffixes. On the contrary, English is a rigid word-order language, so grammatical functions can be usually determined by word ordering, and sometimes by prepositions. The following are major characteristics of Korean which should be considered when designing a parser:

Word Order Variations Korean has almost no constraints on word order, except that the main verb must appear in final position of a sentence. This may often results in discontinuous constituents that is hard to handle in PSG.

Governor's Postpositioning In verb-final languages like Korean, a governor should appear after its dependent in normal sentences, while English has no consistent rules of governor-dependent word order [Kod87].

Complex Word Structure A Korean word called *eojeol* usually consists of one lexical stem accompanied by one or more inflectional suffixes. In dependency analysis, the governing attribute of *eojeol* is determined by its lexical stem, while the dependent attribute by its

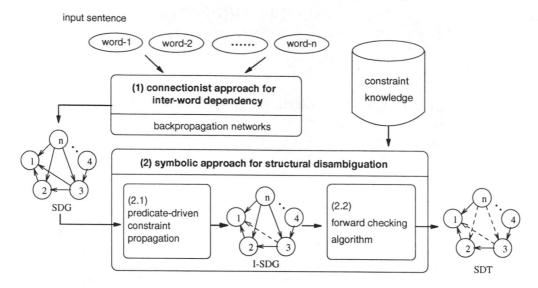

FIGURE 1: Overall architecture of two-phase dependency parser

inflectional suffixes such as case-markers, auxiliary particles, and verb endings.

Auxiliary Particles Instead of Case-markers Korean has a topic-marker *(n)un* as well as a subject-marker *ka/i*. The topic-marker can be used in place of some case-markers like *subject* and *object*. In that case, it becomes difficult to distinguish the type of case marking. Moreover, lots of other auxiliary particles can also take place of case-markers.

Frequently-Occurred Ellipses In Korean, many constituents may be omitted more frequently than in English. Although the missing parts need not be recovered completely, the type of their grammatical relations need be inferred to correctly interpret the other relations.

3. OVERVIEW OF TWO-PHASE DEPENDENCY PARSER

Since, unlike a phrase structure, a dependency structure relies on relationships between words, all the nodes are terminal nodes and need not be in linear order at all. This advantageous in free word-order languages. And also the syntactic dependency structure can be an ideal bridge between semantic structure and morpheme chain. In our dependency structure, a total of 11 grammatical roles are used for labeling the syntactic dependency: *subject, subject2, object, object2, indirect object, complement, adverbial, adjectival, subordinate, coordinate,* and *connective*.

In the proposed two-phase dependency parser (FIGURE 1), the first connectionist phase identifies all possible syntactic dependencies between individual words, specifying the grammatical role of dependency in each case. Here, we don't have to check all possible pairs of words in a sentence because of the *governor's postpositioning* constraint of Korean. All the identified dependencies are collectively recorded in a syntactic dependency graph (SDG).

Since SDG is built only from the viewpoint of inter-word level, it is natural that lots of structural ambiguities appear in it. So, the second symbolic phase is to resolve the structural ambiguities by removing all inconsistent (i.e. ungrammatical) dependencies from SDG, and thus to produce a correct syntactic dependency tree (SDT). The structural disambiguation is formalized as a constraint satisfaction problem, in which a predicate-driven constraint propagation as well as the well-known forward checking algorithm is used for an efficient control.

4. IDENTIFYING SYNTACTIC DEPENDENCIES BETWEEN WORDS

Among the inflectional suffixes, the case-markers are essential to identify the syntactic dependency between a predicate and its nominal arguments. In normal cases, a symbolic rule-based system can work well just relying on case-markers. But, if case-markers are omitted or replaced with auxiliary particles, it becomes very difficult or even impossible for the rule-based system to identify the syntactic dependency without some sort of semantic information (categories, features). What is worse, it is very hard to find regularities of *syntactic* dependency between *semantic* categories. This is why we adopt a connectionist

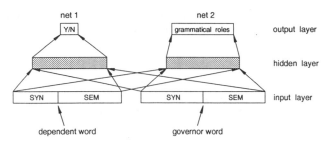

FIGURE 2 Two backpropagation nets for learning syntactic dependency

approach which makes it possible to learn the unclear underlying regularities from a training set.

4.1. LEARNING SYNTACTIC DEPENDENCIES

Two backpropagation nets are used: one for checking whether the syntactic dependency is possible, and the other for determining the type of its grammatical role. Each net consists of three layers as shown in FIGURE 2. The input layer, which is shared by the two nets, consists of 196 binary units (value 0 or 1). Among them, 56 binary units are for representing 14 syntactic features (4 nodes for each feature), and 140 units for encoding 2,000 semantic features of two words. SYN stands for the local representation of syntactic features, and SEM for the distributed coding of semantic features. For output layers, the first net has a single binary unit indicating Yes/No of dependency, while the second net has five real-valued units each of which produces the plausibility value of its corresponding grammatical role. One hundred real-valued hidden units encode the mapping from the input units to output units from a training set, and the hidden layers are fully connected between the input and output layers.

First, a total of 463 sentences were randomly selected from a large corpus, and then based on them two training sets, one for each net, were constructed. Each training example of two training sets consists of two feature vectors for two input words together with a feature vector for out dependency. The training set for the first net consists of 722 positive and 7,322 negative examples, while the second net uses only the positive ones. After the training phase, the trained nets were tested against both the training and test sets, which showed error rates of 9% and 16%, respectively.

4.2. SYNTACTIC DEPENDENCY GRAPH

A syntactic dependency graph (SDG) to keep all possible syntactic dependencies between words is a labeled weighted digraph as shown in FIGURE 3. A node stands for a syntactic/semantic category of word, and an arc for the syntactic dependency between them with an arrow directed from *governor* to *dependent* being labeled with its grammatical role. An arc weight, a plausibility value, is computed by the neural net on the basis of semantic as well as syntactic features. This means that semantic knowledge is reflected in arc weights to some extent. All words are numbered to indicate their linear order in a sentence.

5. STRUCTURAL DISAMBIGUATION

5.1. DISAMBIGUATION AS CSP

In SDG, any word with two or more incoming arcs indicates that its syntactic role is not determined yet. Such dependent word has to be assigned a single incoming arc (i.e. grammatical role) toward its unique governor. Since our goal is to find all such assignments to words such that the assignments satisfy all dependency constraints, structural disambiguation can be viewed as a constraint satisfaction problem (CSP) over a finite domain as discussed in [Mar90]:

- A set of variables V_i : *dependent* words
- A set of domain D_i : arcs (grammatical roles)
- A set of constraints C_i : dependency constraints

As illustrated in FIGURE 1, CSP for structural disambiguation is solved by two different control schemes. The first one, a predicate-driven constraint propagation, relies on the valency of predicates to eliminate inconsistent arcs which can not participate in any dependency interpretation. The second one is the well-

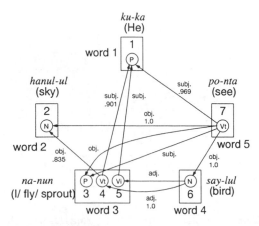

FIGURE 3 Initial SDG for sentence *ku-ka hanul-ul na-nun say-lul po-nta* (He sees a bird flying in the sky)

known forward checking algorithm by [Har80] to filter out the remaining inconsistent arcs.

5.2. DEPENDENCY CONSTRAINTS

The major constraints on syntactic dependency, which are so linguistically universal that they can be also applied to other languages, are summarized as follows:

Uniqueness of Governor Every word, except the root word, has just one syntactic governor [Mel88].

Postpositioning of Governor A governor appears after its dependent in a sentence. This holds true in most OV languages like Korean and Japanese [Kod87].

Uniqueness of Category Out of several nodes for a multi-category word, just one node can participate in a dependency interpretation.

Non-Crossing of Arcs In dependency structure, any arc cannot cross another. This constraint is sometimes called *adjacency condition* or *projectivity*. [Cov90].

Uniqueness of Grammatical Role No word can govern more than one word with the same grammatical role.

5.3. PREDICATE-DRIVEN CONSTRAINT PROPAGATION

Since a predicate expresses a central core of a sentence, the syntactic form and function of its arguments is predictable from the predicate. The number and nature of obligatory arguments which a predicate takes are given by its valency. Typically, verbs have a valency value between 1 and 3. To remove from SDG the inconsistent arcs which can not participate in any dependency interpretation, a predicate-driven constraint propagation is

performed in which all the obligatory arguments of a predicate can be predicted by its valency:

- For every predicate of SDG (in right-to-left word order), do the following:
  (1) Fix the obligatory argument which has a candidate with a single incoming arc, and then filter out inconsistent arcs through constraint propagation.
  (2) Try to fix all as-yet-undecided obligatory arguments using each of their candidates. If any try leads to an ungrammatical dependency structure after constraint propagation, then remove from SDG the incoming arc toward the candidate.
- Transfer the filtering result, an intermediate syntactic dependency graph (I-SDG), to the next FC algorithm.

5.4. FORWARD CHECKING ALGORITHM AND ITS HEURISTICS

Now, to further filter out inconsistent arcs that still remain in I-SDG, and thus to produce a correct syntactic dependency tree (SDT), the forward checking (FC) algorithm by [Har80] is performed, which is one of the well-known hybrid schemes of backtracking and constraint propagation. Whenever a new variable instantiation is made, the domains of all as-yet-uninstantiated variables are filtered to contain only those values that are consistent with this instantiation. If the domains of any of these uninstantiated variables become null, then failure is recognized, and backtracking occurs [Kum92].

In FC algorithm, the order in which variables are

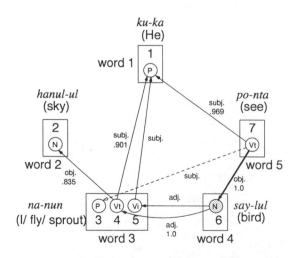

FIGURE 4 Predicate-driven constraint propagation using the valency of predicate *po-nta* (see) (1)

FIGURE 5 Predicate-driven constraint propagation using the valency of predicate *po-nta* (see) (2)

chosen for instantiation is very important to the efficiency of backtracking search. One powerful variable-ordering heuristic by [Bit75] is to select the variable with the fewest possible remaining alternatives. To realize it, variables (i.e. words) for instantiation are selected according to the following:

- Select preferentially the word with the fewest incoming arcs.
- If there exists more than one satisfying the above, then select the word with the fewest categories.

Once a variable is selected for instantiation, it can have several values available. Since the order in which these values are considered can have substantial impact on the time to find the first solution, we also need a good value-ordering heuristic. Now we prefer the incoming arc with the largest plausibility value at the selected word position.

6. STEP-BY-STEP EXAMPLE

In this section, we will give a step-by-step example of structural disambiguation. Consider an example sentence *ku-ka hanul-ul na-nun say-lul po-nta* (He sees a bird flying in the sky). FIGURE 3 shows its initial SDG, where some information is omitted to simplify the following explanation.

First, the predicate-driven constraint propagation will be performed in the right-to-left predicate order. The rightmost predicate is a transitive verb node 7, *po-nta* (see), and it has a valency value of 2, from which its two obligatory arguments *subject* and *object* can be predicted.

The transitive verb *po-nta* can fix its *object* argument because the noun node 6 *say-lul* (bird) is a unique candidate with a single incoming arc. In FIGURE 4, the fixed argument and its *object* grammatical role are indicated by the shaded node 6 and the bold arc (7,6), respectively. Here, as a notation, let (i,j) represent an arc with directionality from node i to node j. Since the other *object* arcs (7,2) and (7,3) come to violate the constraint of *uniqueness of grammatical role*, they are filtered by constraint propagation. What we have to do next is to try to fix the as-yet-undecided *subject* argument. Out of two *subject* candidates, node 1 and node 3, taking node 3 leads to parsing failure, because after constraint propagation word 1 and word 2 come to have no governor, which violates the constraint of *uniqueness of governor*. So, the arc (7,3) should be removed from SDG as illustrated in FIGURE 5. In a similar way, the valency of next two predicates, node 5 and node 4 makes it possible to filter out inconsistent arcs. FIGURE 6 shows the intermediate syntactic dependency graph (I-SDG) after all the predicate-driven constraint propagation.

Now, the FC algorithm is carried out to eliminate the remaining structural ambiguities. According to the variable-ordering heuristic, the word 3 *na-nun* (fly) is selected for variable instantiation because it has the fewest incoming arcs except the root word. Since the selected word 3 (i.e. node 4) has only one incoming arc, the value ordering is unnecessary. Once the word 3 is assigned the incoming arc (6,4) labeled *adjectival*, the domains of all as-yet-uninstantiated nodes should be filtered by constraint propagation. In this example, however nothing happens. For the next instantiation, the variable-ordering heuristic selects the word 1 (i.e. node 1) because it is the last as-yet-uninstantiated word. Out of its two incoming *subject* arcs

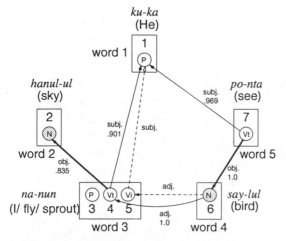

FIGURE 6 Intermediate SDG after all predicate-driven constraint propagation

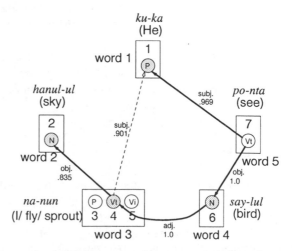

FIGURE 7 Syntactic dependency tree(SDT) after the forward checking algorithm

Table 1 Result of Dependency Parsing

Sentence size (ejeols)	Number of sentences	Average number of dependencies	Number of correctly parsed sentences	Parsing accuracy (%)
below 3	37	5.96	37	100
4-5	94	10.74	94	100
6-7	25	18.19	25	100
8-9	20	24.62	18	90
10-11	11	29.67	9	81.8
12-13	14	37.42	10	71.4
above 14	11	58.92	7	63.6
Total	212	17.40	200	94.3

(4,1) and (7,1), the value ordering heuristic prefers the arc (7,1) because of its larger plausibility value. Now, since no arcs violate the dependency constraints, FIGURE 7 shows a final syntactic dependency tree (SDT) as a parsing result. In case of ambiguous sentences, other SDTs can be also produced by backtracking.

7. EXPERIMENT

The proposed dependency parser has been implemented on a Sun 10 workstation in C language. In the experiment, two hundred and more sentences of various constructions were parsed, the length of which ranged from 2 to 24. The experimental results are summarized in Table 1. For almost all sentences of less than 10 *eojeols* (Korean words), the dependency parser shows a surprising result with high parsing accuracy of above 90%. This may be because the first connectionist phase for identifying inter-word dependency relies on semantic features as well as syntactic ones. On the other hand, for long sentences, the parsing accuracy dropped radically to about 70%. The longer a sentence is, the more rapidly the number of candidate dependencies increases. As a result, although the proposed parser still has some remaining problems, it shows a very promising possibility for dependency parsing.

8. CONCLUSIONS

We proposed a new dependency parser for free word-order languages like Korean based on a hybrid connectionist/symbolic technique, and its experimental evaluation showed a high parsing accuracy for almost all Korean sentences of less than 10 words. But, for its practical use there still remain some problems to solve: we need a smaller but more effective set of semantic features

for training the neural nets, and also we have to develop the method how to effectively handle the long-sentence problems.

REFERENCES

[Bit75]   Bitner, J., Reingold, E. M.: *Backtrack Programming Techniques* **18** (1975) 651--655

[Cov 90]   Covington, M. A.: *A Dependency Parser for Variable-Word-Order Languages*. Research Report AI-1990-01, Artificial Intelligence Programs, Univ. of Georgia, (1990)

[Har80]   Haralick, R., Elliot, G.: *Increasing Tree Search Efficiency for Constraint-Satisfaction Problems*. Artificial Intelligence **14(3)** (1980) 263--313

[Kod87].   Kodama, T.: *Studies in Dependency Grammar*. Tokyo: Kenkyusha (1987) (written in Japanese).

[Kum92]   Kumar, V.: *Algorithms for Constraint-Satisfaction Problems:* A Survey. AI Magazine **Spring** (1992) 32--44

[Mar90]   Maruyama, H.: *Structural Disambiguation with Constraint Propagation*. Proceedings of the 28th Annual Meeting of the ACL (1990) 31--38

[Mel88]   Mel'cuk, I.A.: *Dependency Syntax: Theory and Practice*. New York: State University of New York Press (1988)

INTELLIGENT SUPPORT FOR CONSTRUCTION AND EXPLORATION OF ADVANCED TECHNOLOGICAL INFORMATION SPACE

Toshiyuki Matsuo, Toyoaki Nishida
Graduate School of Information Science, Nara Institute of Science and Technology
8916-5 Takayama, Ikoma, Nara, 630-01 Japan
Email:{tosiy-m,nishida}@is.aist-nara.ac.jp

ABSTRACT

In this paper, we describe a practical method of extracting, structuring, summarizing and integrating technological information from technical papers in metallurgy. The heart of the method is a packet of domain specific knowledge called *KPs* (Knowledge Pieces) in which procedures for extracting and structuring technological information are embedded. We studied information structure of ten technical papers in metallurgy and constructed about a hundred *KPs*. We implemented a system called METIS(METallurgy papers Intelligent Surveyor) which interprets technical papers in metallurgy written in a mark-up language and produces a variety of summaries and surveys such as structured technical summary, visualization of similarities, differences of relevant papers, and cause-effect relations. We have undertaken qualitative and quantitative evaluation of METIS against 186 technical papers so far. The evaluation demonstrates reliability and robustness of our method.

1 INTRODUCTION

According to rapid progress of multimedia and network technology, technical information is spread more widely around researchers and engineers. Although what we want is systems that can classify, summarize, and produce technological information effectively from natural-language texts, few systems have been used in practice. Much more effort is needed to understand those papers which can be retrieved from conventional data base system in science, because those papers are left unclassified or unsummarized. Retrieving by keyword matching with full text or with an abstract often generates too many papers which make impossible to extract technological information.

In cooperative work with an expert in super alloy, which is a specific domain in metallurgy, at National Research Institute for Metals in Japan, we have been carrying out research into automatic extraction of technological information in metallurgy. We have three policies in our research as follows. (1)Content processing, (2)Reliability and robustness, and (3)Application of present artificial intelligence technology.

We studied information structure of ten technical papers in metallurgy. The analysis leads us to a conclusion that there is a generalized schema in papers, and that we get a bright prospect to extract and structure reliable information with a schema automatically.

The purpose of this paper is to develop and apply a practical method of extracting, structuring, summarizing and integrating technological information in metallurgy. We implemented a system called METIS (METallurgy papers Intelligent Surveyor) which interprets technical papers in metallurgy written in a mark-up language and produces a variety of summaries and surveys such as structured technical summary, visualization of similarities, differences of relevant papers, and cause-effect relations.

The heart of the method is a packet of domain specific knowledge called *KP* (Knowledge Pieces) in which procedures for extracting and structuring technological information from technical papers are embedded. The analysis of contents in metallurgy papers leads to construct a hundred *KPs*. METIS extracts technological information by naive natural language process specified by keywords or key phrases embedded in *KPs*, and produces summarized information specified by *KPs*. We have undertaken qualitative and quantitative evaluation of METIS against 186 technical papers so far. METIS produces reliable results especially with typical papers, but METIS produces unsatisfactory results because of the lacks of appropriate *KPs*. We get a prospect that METIS with addition of more relevant *KPs* would produce satisfactory results.

In section 2, we present our approach to extract, structure, summarize and integrate technological information from technical papers. In section 3, we describe a system called METIS which produces a variety of summaries and surveys. In section 4, we present qualitative and quantitative evaluation of METIS against

186 technical papers so far. Finally, we present a comparison with related work.

2   OUR APPROACH

We analyzed information structure of ten papers in metallurgy. The results of the analysis made it clear that we could implement a system with naive natural language process, that is, we developed an appropriate knowledge representation method.

We studied information structure of papers in metallurgy. The analysis of papers' structure leads us to a conclusion as follows. (1)The most papers in super alloy consists of a title, author, experimental procedure, experimental results and conclusions. (2)Researchers have an aim on experimental procedure and experimental result. A more detail analysis leads us to a conclusion that papers should be divided into detail sections such as material for experiments, chemical composition, preparation before experiments, methods of experiments, experimental results and discussions. (3)Figures and tables, which show microstructure analysis by a microscope and summarize experimental results, produce the most important technological information in a paper. But it does not seem that there is a practical technology that extracts information from figures and tables automatically.

The results of the analysis make it clear that we can implement a system with three policies as follows. (1)We represent a paper's knowledge piece in metallurgy as a schema which has a standardized structure. (2)We represent papers' contents as a combination of a hundred schema. (3)We represent technological information in metallurgy papers with schema and naive natural language process, which is mainly pattern matching with technical terms and domain specific expressions.

We show a knowledge representation method which is a packet of domain specific knowledge called KP (Knowledge Pieces) in which procedures for extracting and structuring technological information from technical papers in metallurgy are embedded. KPs have a parent-child relationship between relevant KPs. KP has four facilities as follows.

1. to select sentences
   S := a set of sentences; C := conditions to extract a sentence; E := conditions to exclude a sentence; are given.
   S' := a set of selected sentences; is generated.
   $S \times C \times E \rightarrow S' \subseteq S$

   Ex1.   S = {all sentences in a paper}

   C = {"ALLOY 5", "ALLOY 241"}
   E = {}
   S' = {a set of sentences which include "ALLOY 5" or "ALLOY 242"}
   Ex2.   S = {a set of sentences}
   C = {"addition"}
   E = {"in addition", "additional"}
   S' = {a set of sentences which include "addition" and do not include "in addition" and "additional"}

2. to extract features
   S := a set of sentences;
   Xf := conditions to extract features;
   Cf := a set of extracted concepts;
   Af := a set of co-occurrent words with Cf;
   are given.
   Vf := a summary of features;
   is generated.
   $S \times Xf \rightarrow Vf$
   provided that $Xf = \{\langle c_f, a_f \rangle \mid c_f \in Cf, a_f \in Af\}$

   Ex.   S = { "The strain-controlled low-cycle fatigue properties of CMSX-2 at intermediate temperatures are also greatly improved by using the high-gradient process."}
   Cf = {"achieve","affect","improve", ...}
   Af = { "is","are","not","strongly","greatly", "primary", ...} are given.
   Feature Vf of cause-effect relations is generated. Vf = { "are also greatly improved by"} is generated.

3. to structure information
   A KP is structured recursively by merging child of the KP.
   K := a KP; F := a feature;
   $2^{F \times K} \rightarrow K$
   Ex.   A KP for a summary of a paper is structured by merging the child of the KP, that are material, experimental method, discussion, figure information and table information.

4. to intersect relevant KPs
   A KP is structured by intersecting relevant KPs.
   S := a set of sentences
   $K_1 := \{K_{11}, K_{12}, \cdots, K_{1i}, \cdots, K_{1m}\}$
   $K_2 := \{K_{21}, K_{22}, \cdots, K_{2j}, \cdots, K_{2n}\}$ are given.
   provided that both K_1 and K_2 are a set of relevant KPs respectively.
   V:={a set of sentences } is generated.
   $K_1 \otimes K_2 \rightarrow V$
   V:= $\{V_{ij} \mid 1 \leq i \leq m, 1 \leq j \leq n \}$
   provided that V_{ij} is a set of sentences that satisfy extracting conditions of both K_{1i} and K_{2j}.

Ex. S={a set of sentences that include
cause-effect relations}
K_1= {KP for creep strain}
K_2= {KP for heat treatment} are given.
V={V_{11}}
V_{11}={a set of sentences that include
cause-effect relations, creep strain and
heat treatment} is generated.

A *KP* is defined by items as follows. A name of
KP, conditions to extract information, conditions to
exclude information, a set of child *KP* name, a set of
relevant *KP* name, method of information process, a
set of co-occurrent words, storage area for extracted
data, storage area for processed data, storage area for
merged data and storage area for intersected data.

3   OVERVIEW OF METIS

Table 1 summarizes functions of METIS. The architecture of METIS is applicable to other domains, because
definitions of *KP* and an algorithm for natural language process based on *KPs* are independent.

First, METIS applys *KPs* to papers, and produces
a *KP* instance which represents structured technical
summaries as itemized expressions.

[1]Structured technical summary   METIS produces
a *KP* instance which represents structured technical
summary through *information structuring* process of
KPs. There are five *KPs*, i.e., for metal materials, for
experimental results, for discussion, for tables and for
figures (see Figure 1). METIS produces a *KP* instance
which represents experimental result through *information structuring* process of *KP* for creep properties, fatigue properties, corrosion properties and other properties. METIS produces a *KP* instance which represents metal material through *extracting features* process.

**[2]Structured summary of experimental meth-
ods**   METIS produces a *KP* instance which represents
structured summary of experimental methods through
information structuring process of *KP* for morphology, heat treatments and so on(see Figure 2). METIS
produces a *KP* instance which represents morphology
through *information structuring* process of *KP* for orientation and preparation method. METIS produces a
KP instance which represents orientation through *extracting features* process.

Second, METIS extracts information about cause-
effect relations. Cause-effect relations are the most

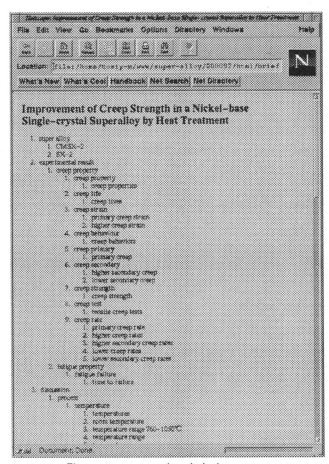

Figure 1: structured technical summary

important information in a paper. METIS *selects sentences* including cause-effect relations and produces
three expressions of cause-effect relations.

**[1]causes to give an influence on metal prop-
erty**   (see Figure 3) METIS expresses cause-effect relations from a viewpoint that a experimental process
or microstructure have some influence on some metal
properties. METIS *selects sentences* indicating experimental results, *intersects relevant KPs* for metal properties, experimental process and microstructure, and
summarizes them in a table form.

[2]contents of cause-effect relations   (see Figure
4) METIS expresses cause-effect relations by *extracting
features* about relations, experimental methods, structure and metal properties per a sentence, and summarizes them with a powerful and heuristic algorithm in
a table form.

[3]directional graph of cause-effect relations   (see
Figure 5) METIS expresses cause-effect relations by di-

Table 1: Functions of METIS

	function	
summary	(1)structured technical summary (2)structured summary of experimental methods (3)causes to effect metal property (4)directional graph of cause-effect relations (5)contents of cause-effect relations (6)sentences including cause-effect relations (7)information of figures and tables	(1)automatic (2)semi-automatic (3)automatic (4)automatic (5)automatic (6)automatic (7)automatic
survey	(1)plots of similarities between papers on two dimensional space (2)plots of similarities between papers on three dimensional space (3)similarities of relevant papers (4)distribution of papers from a viewpoint of cause-effect relations (5)merged cause-effect relations	(1)automatic (2)automatic (3)automatic (4)automatic (5)semi-automatic

Alloy: SRR99
Morphology: Single Crystal
   Orientation: [011]
   Preparation method: epitial crystallization
Heat treatment: solution heat treatment + annealing
   Solution heat treatment: 1553K for 1h + 1563K
   for 2h + 1573K for 0.5h + 1578K for 0.5h;
   cooling to 1273K at 0.7K s-1
   Annealing: 1353K for 4h + 1143K for 16h
   Precipitates: $\gamma\prime$
     Form: Cubic
     Geometrical between matrix and $\gamma\prime$: coherent
Experiment: creep test
   Test temperature: 1173K
   Test Load: 300Mpa
   Number of specimen: 2
     Gage length: 25mm
     Cooling of Specimen 2: under load
   Test condition: shown in Table 1
Investigation of microstructures: SEM observation
Investigation of microstructures: TEM observation
   Sample orientation: parallel to (011) planes
   Microscope: Phillips CM30

Figure 2: structured summary of experimental method

rectional graph that has a cause as starting point, an effect as ending point, and a relation as a link. METIS locates *extracted features* (see Figure 4) on three-dimensional space. The features are categorized into three , i.e., metal property, experimental process and microstructure. The location of the features are predetermined by METIS. METIS draws a line as a relation between causes and effects.

Third, METIS calculates similarities between papers in metallurgy from structured summaries(see Fig-

ure 1). A *KP* for structured summary consists of five *KPs*, i.e., metal material, experimental results, discussion, table information and figure information. METIS calculates similarities between relevant *KPs*, and sums up them into a similarity between papers by the following equations.

⟨Similarity between Paper A and Paper B⟩

$$= \sum_i \langle \text{weight} \rangle \times \langle \text{similarity between } KP_i \rangle$$

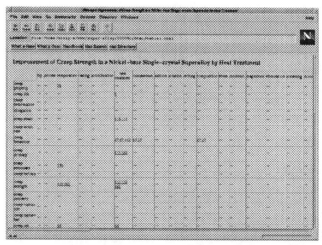

Figure 3: Causes to give an influence on metal property

This figure shows that there are some sentences including cause-effect relations between metal properties and experimental process/microstructure. A numeral(s) indicates a sentence(s) including cause-effect relations between relevant metal properties and experimental process/microstructure.

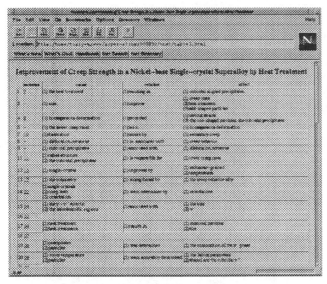

Figure 4: Contents of cause-effect relations

This figure indicates that some relations exists between causes and effects per sentence. When you click a numeral(s) that indicates ID of a sentence including cause-effect relations, METIS shows a relevant sentence.

$\langle$similarity between $KP_i\rangle$

$$= \left(1 - \frac{\text{NUM}_i(A \cap B)}{\text{NUM}_i(A)} \times \frac{\text{NUM}_i(A \cap B)}{\text{NUM}_i(B)}\right)^2$$

$\text{NUM}_i(X)$

$= \langle$the number of items extracted in KP_i of paper$X\rangle$

Matrix $S = (s_{ij})$ is defined as a similarity matrix between n papers, where s_{ij} is a similarity between i-th paper and j-th paper. s_{ij} satisfies the conditions $\{s_{ij} \geq 0, s_{ij} = s_{ji}, s_{ii} = 0\}$. $s_{ij} = 0$ means that i-th paper and j-th paper are the same in s summarized information level. METIS applies MDA(Multi Dimensional Analysis) method to similarity matrix S, and get a matrix $A = (a_{ij})$. This matrix means a location of each paper in a $n-$dimensional space. MDA calculates this matrix, for the distance between papers in this space of matrix $A = (a_{ij})$ to indicate the similarity between papers. METIS calculates matrix A and plots them on two or three dimension space(see Figure 6).

4 EVALUATION

METIS summarizes 87 papers published at 1992 national conference in super alloy, calculates similarities

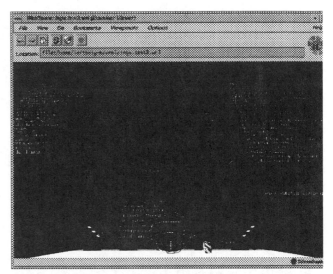

Figure 5: Directional graph of cause-effect relations

This figure shows cause-effect relations drawn on a three-dimensional space. When you click the line between a cause and an effect, METIS shows a relevant sentence including the cause-effect relation.

and plots them on three dimensional space. The results indicate usefulness of our method with KPs. We

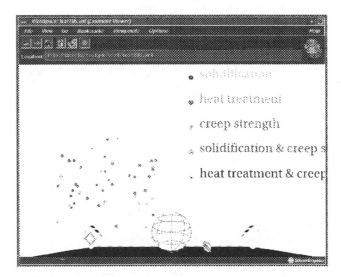

Figure 6: Plots similarities between papers on three dimensional space

This figure shows a similarity between papers by a length between spheres on three-dimensional space. Each sphere indicates a paper respectively.

Table 2: evaluated results

	A	A^-	B	C	D	E	avr
typical samples(3/99)	0	1	2	0	0	0	81
random samples(9/87)	0	1	3	2	3	0	55
12/186 papers	0	2	5	2	3	0	61

have a bright prospect of automatic papers' classification into some clusters.

Next, we have undertaken quantitative evaluation of METIS against 186 papers so far(see Table 2). The expert evaluates the technical summaries produced by METIS to five ranks, i.e., from A, B, C, D and E. We give five ranks to scores from 0 to 100, i.e., A, A^-, B, C, D, E are 100, 93, 75, 50, 25, 0 scores. Although METIS produces reliable results with typical papers, the expert in metallurgy points out that those summaries produced by METIS lack important keywords which must be essentially extracted. There are two features of papers in metallurgy which do not give rise to a satisfactory evaluation, (1)to discuss relation between production process and properties, (2)to discuss very general properties (e.g. tensile property). Lacks of important keywords are caused by lack of appropriate KPs. Although it seems that an addition of more relevant KPs would improve evaluation of METIS, we have to apply other artificial intelligence technology(e.g. mechanical learning technology) in order to (semi-) automatically construct appropriate KPs.

5   COMPARISON WITH RELATED WORK

This work is considered to be an effort of acquiring knowledge from texts and representing knowledge by naive natural language process. Previous and frontier work based on script theory in this direction involves: SAM(Script Applier MEchanism[Cu78] and FRUMP(Fast Reading Understanding and Memory Program) [De79]. Claire Cardie[Cl94] shows current work based on case-based approach. SAM and FRUMP could acquire knowledge from texts about short stories and newspapers, and there were not detail evaluations of them from users. The system of [Cl94] is limited to summarize domain-specific sentences. Niki and Tanaka[Niki95] shows work on automatic classification of papers using neural networks. This system produces functions of dynamic generation of schema, browser of contents with a variety of abstraction and vague information retrieval.

The evaluation of METIS demonstrates reliability and robustness, differently from SAM and FRUMP.

METIS integrates technological information, that is not achieved by the system of [Cl94]. While the system of [Niki95] does not use domain-specific knowledge to extract information, METIS uses domain-specific knowledge which enables it to investigate details of contents. Therefore METIS can perform structural extraction of information which is embedded in structure of contents.

6   SUMMARIES

We applied artificial intelligent technique to extract, structure, summarize and integrate information from papers in metallurgy. The heart of our method is a packet of domain specific knowledge called KP. We have undertaken qualitative and quantitative evaluation of METIS. The evaluation demonstrates reliability and robustness of our method. An interesting open problem is (1)to acquire and represent deep knowledge, so that METIS can produce the most important information from summarized results and (2)to research on an intelligent integration of extracted information.

References

[Cl94] Claire Cardie, "A Case-Based Approach to Knowledge Acquisition for Domain-Specific Sentence Analysis", Proceedings of the Eleventh National Conference on Artificial Intelligence, pp.798-803,1994.

[Cu78] Culling ford,R.E., "Script Application: Computer Understanding of Newspaper Stories. PhD thesis", Department of Computer Science,Yale University,New Haven,CT,Technical Report 116, 1978.

[De79] DeJong,G.F., "Skimming Stories in Real Time: An Experimental in Integrated Understanding. PhD thesis", Department of Computer Science,Yale University,New Haven,CT,Technical Report 158, CT,Technical Report 158,1979.

[Niki95] Kazuhisa Niki and Katsumi Tanaka, "Information Retrieval Using Neural Networks", Journal of Japanese Society for Artificial Intellingence, Vol.10, No.1, pp.45-51,1995.(In Japanese).

Planning and Scheduling

ADAPTATION OF A PRODUCTION SCHEDULING FRAMEWORK TO DISTRIBUTED WORK ENVIRONMENTS

Taketoshi Yoshida[†]   **Masahiro Hori**[‡]

[†] NCC Technology Center, IBM Japan Ltd.
19-21 Nihonbashi Hakozaki-cho, Chuo-ku, Tokyo 103, Japan
Email: yoshida@trl.ibm.co.jp

[‡] Tokyo Research Laboratory, IBM Japan Ltd.
1623-14 Shimotsuruma, Yamato-shi, Kanagawa-ken 242, Japan
Email: hori@trl.ibm.co.jp

ABSTRACT

This paper reports our experience in developing a scheduling system for the semiconductor fabrication process, and extending the design applying an object-oriented scheduling framework. First, we explain features of semiconductor fabrication processes, and scheduling methods including the one we have actually applied to a real fabrication line. We then introduce a production scheduling framework, which consists of three loosely coupled subsystems: a schedule model, a scheduling engine, and a graphical user interface. By elaborating the scheduling engine subsystem, we show how the framework can be adapted to the enhanced design of the fabrication line scheduler.

1 INTRODUCTION

The concept of enterprise integration, whose aim is facilitating coordination of work and information flow across organizational boundaries, now constitutes a direction for enterprise systems in the coming future [Petrie 92]. The basis of this concept in manufacturing is the realization of flexible systems, which is not only efficient but extensible and maintainable. In line with this trend, it is becoming more and more important to integrate applications participating in a manufacturing environment. Scheduling systems, as one of the constituent, must realize a high capability for dealing with evolution in floor environments through the introduction of various kinds of flexibility.

From the viewpoint of information technology, one of the primary issues in developing application systems is to improve the extensibility and productivity by using software and knowledge reuse technologies [Johnson 88, Musen 92]. Many scheduling expert systems have been developed individually to solve similar problems. Some issues that must be resolved are how to circumvent differences among application-specific scheduling methods and provide composable units for reuse. Issues related to the reusability as well as the extensibility of manufacturing systems have recently been attracting increased attentions both in the research community and in industry [Aarsten 95, SEMATECH 95].

We have worked for a component-oriented methodology for developing knowledge systems, and elicitation of reusable problem-solving components from existing scheduling expert systems [Hori 94, Hori 95]. On the basis of the experiences, we have designed an object-oriented framework for production scheduling systems, and implemented it as a C++ class library. The framework has been used to develop a scheduling system currently used in a production line for electromechanical devices, and is now being extended to reflect the design of a scheduling system being used in an actual semiconductor fabrication plant. In the remainder of this paper, we first explain features of semiconductor fabrication processes, and investigate variations of scheduling methods. We then introduce our production scheduling framework, and show how the framework can be adapted to the variations.

2 SCHEDULING OF SEMICONDUCTOR FABRICATION PROCESSES

The scheduling domain may be divided into a wafer fabrication specification and scheduling policies. The former describes the information about products, processes, and resources. The latter are ways of optimizing manufacturing operation relying on empirical knowledge.

2.1 Wafer fabrication line

The features of semiconductor fabrication processes are briefly described below, following the terminology given by Dayhoff et al. [Dayhoff 87]. A typical wafer fabrication line consists of mainly four *processes*: photo, etching, diffusion, and thin-film processes. The physical entities of these processes are called *production cells*, sometimes called *production bays* or *production aisles*. A production cell consists of *work stations*. A work station is a set of *equipments*. Equipments are scheduled for such as preventive maintenance.

An equipment executes an instance of a process, called a *process step*. A sequence of process steps is called a *process flow* or *process routing*, which is highly reentrant throughout equipment.

Multiple products are dealt with through a production line. Wafers are aggregated into *lots*, which are units of processing and transportation. A lot has the information such as a customer order and due date. Each work station deals with some lots that are on different process steps.

Characteristics of such production lines are the processing of large numbers of operations (for example, over 100 operations), and long manufacturing cycle times (typically 20 to 50 days). The status of a production floor is subject to changes due to unexpected equipment failures, which often disrupt the predetermined schedule.

2.2 Approaches to the Scheduling

Wafer fabrication lines (often abbreviated as FAB) are usually managed by two departments: a production and a manufacturing department. The production department is responsible for wafer-in and wafer-out, and the main objective is to minimize the tardiness of lots. The manufacturing department, on the other hand, fulfills the wafer-out plan and is responsible for improving equipment utilization. Therefore, the fabrication lines are managed on a day-to-day time scale by the production department and on a minute-to-minute (or second-to-second) scale by the manufacturing department.

The lot progress is reviewed each day by the production department and the results are used to revise the production schedule. In this phase, the main performance measure is due date satisfaction of the lots. Priorities are assigned to lots, for example, to minimize their tardiness.

An operator at a cell should decide the lot progress sequence on the basis of the current status of the cell such as the availability of lots and their priorities, the equipment status, and the amount of work in process in the cell. The scheduling activities at a cell are reduced to an assignment of lots to equipment.

The assignment method determines details of how equipment is used as well as when the assigned operations start and finish. However, this method is not necessarily practical in the wafer fabrication line, because combinatorial explosion will occur due to hundreds of equipment and thousands of lots involved. Furthermore, equipment failures may cause frequent modification of schedules, and real-time optimal rescheduling is hard to achieve.

The dispatching heuristics is practical for a single performance criterion such as due date satisfaction and equipment utilization maximization. We have developed a scheduling system that employs dispatching heuristics, taking account of the average tardiness of lots. Some of the dispatching rules are listed below.

- EDD (Earliest Due Date First)

The lot with the earliest due date has the highest priority.

- SST (Smallest Slack Time First)

This rule is a variant of EDD. The lot with the smallest slack time has the highest priority.

- FIFO (First-In First-Out)

This rule is deemed to be fair, in that lots are processed in the order of FAB-IN time.

- SRT (Shortest Remaining Time First)

The lot with the shortest remaining cycle time has the highest priority.

- LRT (Longest Remaining Time First)

The lot with the longest remaining cycle time has the highest priority.

- SRT/FIFO (SRT + FIFO)

This rule is a variant of SRT dispatching heuristics. The lots are prioritized by FIFO if they were released into FAB before the user-specified time. The other lots are prioritized by SRT. The lots by FIFO have the higher priorities than the ones by SRT.

- LRT/FIFO (LRT + FIFO)

This rule is a variant of LRT dispatching heuristics. The lots are prioritized by FIFO if they were released into FAB before the user-specified time. The other lots are prioritized by LRT. The lots by FIFO have the higher priorities than the ones by LRT.

- Opt_AT (Optimization with respect to tardiness)

This rule employs a simulated annealing algorithm [Kirkpatrick 83], which is a stochastic optimization technique, exploiting neighborhood search algorithms. This rule reduces the tardiness of lots. The initial solution of this rule is generated by EDD and hence the performance becomes equal to or better than the one by EDD. Since execution of this rule is time-consuming, the applicable number of lots should be restricted for practical use of this algorithm.

- Opt_NT (Optimization with respect to tardy lots)

This rule is also employs the simulated annealing algorithm, and tries to reduce the number of tardy lots. The applicable number of lots should be restricted for this rule, too.

The following simulation results show the effectiveness of the simulated annealing algorithm.

```
 50 lots;  63.4 % (  6.5 sec)
100 lots;  58.5 % ( 46.0 sec)
150 lots;  56.2 % (153.5 sec)
200 lots;  53.9 % (357.5 sec)
```

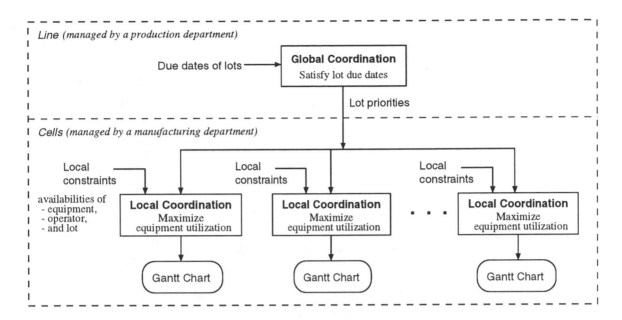

Figure 1 Overview of the distributed scheduling

The values in the brackets indicate the computational times required on IBM RS/6000 model 730. As mentioned above, we can observe the exponential increase of the computational time. This results indicate that we obtained about 36% improvement against EDD for 50 lots case, 41% for 100 lots, 44% for 150 lots, and 46% for 200 lots cases. The further details of the experimental results are found in [Yoshikawa 93].

The other dispatching heuristics employ a heap sort algorithm, which has the computational complexity of $O(n \log_2 n)$, where n is the number of lots to be considered.

It must be noted here that these scheduling heuristics are practical in a real situation, and the system has been working in an actual wafer fabrication plant. However, there exists a limitation of the total performance since the system focuses on a single criterion.

The primary objectives of the above mentioned wafer fabrication lines stems from the two departments: minimization of lot tardiness from the production department, and maximization of equipment utilization from the manufacturing department. Taking account of this work environment, we came up with a design of a distributed manufacturing scheduling system, which combines the dispatching and assignment method, so that conflicting objectives can be mediated. The essence of this approach is illustrated in Figure 1.

The local coordination is performed on arrival of a lot or on completion of an operation by an equipment. By doing reactive forward assignment, available equipment is occupied with incoming lots, and the results are presented in the form of a Gantt chart, so that the operators in the production cells can follow the instructions. The fundamental algorithm of the assign-

ment method is adopted from the scheduling system for steelmaking processes [Numao 91].

This design realizes a kind of hierarchical scheduling systems, and is not new in the literature in itself. For example, a two-layered scheduling system is proposed in [Lee 91]. Our central claim, however, is more on the aspects of reusability and extensibility in a software architecture throughout the entire life cycles of evolving scheduling systems embedded in the volatile manufacturing environment. In the next section, we will introduce an object-oriented framework for production scheduling systems, and explain how it can be adapted to the design of the distributed scheduling system.

3 SCHEDULING FRAMEWORK

An object-oriented framework is a set of cooperating classes that make up a reusable software architecture for a particular class of applications such as financial trading [Birrer 93] and plant process control [Pirklbauer 94]. The concept of a framework is clearly distinct from that of a collection of classes, which may provide useful functionality but does not prescribe the design information recurring in a family of applications [Johnson 88].

The scheduling framework that we have developed consists of three loosely coupled subsystems: a schedule model, a scheduling engine, and a graphical user interface. The schedule model plays the role of an application domain model, providing the assumptions in a manufacturing area for the purpose of a scheduling application. The other two subsystems are integrated with the model according to individual requirements. The rest of this section gives further details of

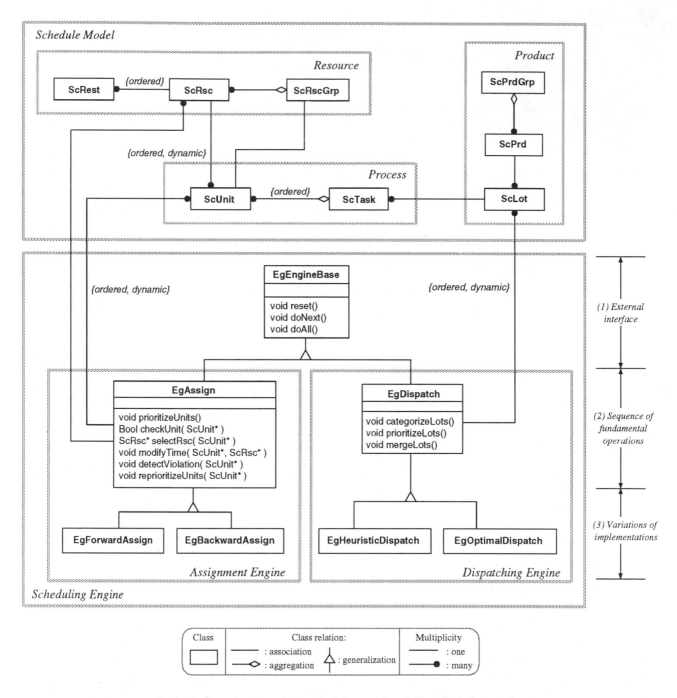

Figure 2 Organization of the schedule model and the scheduling engine

3.1 Schedule model

A schedule model includes entities and their relations, which form a model of an application domain, namely, a domain ontology. As depicted in the upper half of Figure 2, the schedule model further consists of three subsystems, which are relevant to the manufactur-

the schedule model and scheduling engine, which are depicted in Figure 2.

ing facility (*Resource*), production process (*Process*), and product family (*Product*). The diagrammatic notation in this paper follows the OMT methodology [Rumbaugh 91].

The resource subsystem includes a resource group (ScRscGrp), an individual manufacturing equipment (ScRsc), and the associated resting time (ScRest) with an equipment. A resource group is a collection of alternative equipments, each of which is associated with unavailable time intervals. The process subsystem includes a primitive unit of operation (ScUnit), and a sequence of operations (ScTask). Furthermore, a product type (ScPrd), a group of products (ScPrdGrp), and a unit of order acceptance and shipment (ScLot) are included in the product subsystem.

The semiconductor fabrication process described in the previous section is easily mapped onto this schedule model. That is, an *equipment* is regarded as ScRsc, a *work station* as ScRscGrp, a *process step* as ScUnit, a *process flow* as ScTask, a *product* as ScPrd, and a *lot* as ScLot.

Some concepts in the wafer fabrication description above do not appear in the schedule model, because the model dose not refer to entities that are not essential to activities in production scheduling. For example, a production cell is a characteristic entity in the wafer fabrication domain, but it is not included in the schedule model so that the model can be reusable in other domains other than semiconductor fabrication.

3.2 Scheduling engine

The scheduling engine subsystem, which is shown in the lower half of Figure 2, is stratified into three layers for extensibility and modifiability. The first layer provides external clients with simple, common interfaces that are defined in the abstract class EgEngineBase. The interfaces can be used to invoke a scheduling method in two ways. One is to go through all the problem-solving steps by calling doAll after using reset for initialization. The other is to initialize an engine by calling reset, and then repeat a single problem-solving step by calling doNext. The coverage of a single step depends on how doNext is specified in the second layer. Invocation and iteration of a scheduling engine is usually controlled by the end users. However, the user needs to know only these simple interfaces, regardless of which engine is being used.

The second layer provides declarations of fundamental operations that are given in subclasses of EgEngineBase, namely, EgAssign and EgDispatch. EgAssign assigns units (ScUnit) to resources (ScRsc), fixing the starting and finishing time of each unit. EgDispatch, on the other hand, determines the priority of each lot (ScLot) without regard to actual assignments of units to resources. This results in greater flexibility and easier extensibility of the scheduling engine subsystem. This subsystem was initially provided solely with the assignment engine, and the dispatching was integrated as the result of our observations in the semiconductor manufacturing domain. This extension, however, did not require any change in the rest of the framework.

It is also important to note here that every lot is indirectly associated with units via a production sequence (ScTask), as shown in Figure 2. This whole-part relation between a lot and units articulates the roles of the two scheduling methods. That is to say, the dispatching engine, which primarily deals with lot priorities, is concerned with the global manufacturing

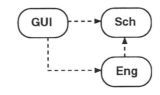

(a) Typical scheduler

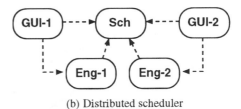

(b) Distributed scheduler

Sch : Schedule model   **GUI** : Graphical user interface
Eng : Scheduling engine   - - ➤ : subsystem dependency

Figure 3 Configurations of the subsystems

status. In contrast, the assignment engine is concerned with the status on a more detailed level. This explicitness is revealed by referring to the schedule model.

EgAssign defines a sequence of operations that are common to typical scheduling methods, namely, a forward and a backward scheduling procedures. The former tries to assign each unit to an appropriate resource as early as possible, up to the earliest starting time of a corresponding lot, while the latter does the same as late as possible, down to the due date of a corresponding lot. A sequence of operations common to the above two is defined as EgAssign's doNext operation, using fundamental operations such as checkUnit and selectRsc. An important point here is that scheduling methods are not fully executable yet in this second layer, but are specified with a control structure that is a sequence of fundamental operations to be performed. To put it in another way, this layer provides designs, or high-level specification, of various scheduling methods.

Individual implementations of the fundamental operations are given in the third layer. Therefore, the operations of a forward and a backward scheduling method are fully specified in EgForwardAssign and EgBackwardAssign, which are subclasses of EgAssign. According to this layered architecture of a scheduling engine subsystem, it is possible to customize a method, which requires modification of a fundamental operation.

4 ADAPTATION TO THE DISTRIBUTED SCHEDULER

The scheduling framework can be customized by modifying the configurations of subsystems and selecting appropriate classes, and often by subclassing these classes in each subsystem. The most typical configuration is obtained by elaborating the three subsystems in the original framework as they are (Figure 3(a)).

On the other hand, if the production cycle time is extremely long, it is possible to divide production lines into smaller regions, and then attempt to schedule with regard to two different criteria. The configuration depicted in Figure 3(b) is applicable to such a situation. This configuration is a sort of distributed schedulers, since the two scheduling engines and their human specialists cooperate via a common schedule model.

As mentioned earlier, entire production cycles of semiconductor fabrication processes are rather long, ranging from 20 to 50 days, and manufacturing equipment is managed on a scale of minutes. Typical evaluation criteria are due date satisfaction and rates of equipment utilization. These two criteria often conflict with each other. For example, if the due date criterion is favored, the batching in diffusion process and the set-up in the photo process are often performed inefficiently, resulting in decreased equipment utilization. In contrast, if the efficiency of batching is considered first, it is more likely to cause many due-date violations. This contention can be mediated by using two types of scheduling engines, EgDispatch and EgAssign, complementarily. The dispatching engine determines the priority of each lot, so that tardiness can be reduced. The assignment engine binds units to resources, taking account of resource utilization.

The scheduling framework can be easily customized to be a composite scheduler for the fabrication process, as shown in Figure 3(b). Such a composite scheduler includes a common schedule model shared by two different scheduling engines entailed with their own user interface subsystems. It must be noted here that each engine subsystem exploits a different portion of the schedule model as shown in Figure 2. The dispatching engine modifies the priority attributes associated with each lot object, while the assignment engine establishes bindings between units and resources.

5   CONCLUSION

In this paper, we have focused on the issues in flexibility of production scheduling systems, and shown the extensibility and reusability of the scheduling framework. We are currently working for enhancement of a scheduling system in a real manufacturing environment. Following the evolution of the system, we are going to pursue quantitative and qualitative metrics of reusable knowledge and software design, which are indispensable to flexible scheduling systems to be integrated with manufacturing enterprises.

Acknowledgment

The authors wish to express their gratitude to Mr. Hirofumi Yoshikawa of ITS Co. Ltd. for his help to implement the dispatching rules and their simulation studies.

REFERENCES

[Aarsten 95] A. Aarsten, G. Elia & G. Menga : G++: A pattern language for computer-integrated manufacturing. In J. Coplien & D. Schmidt, Eds., *Pattern Languages of Program Design*, pp. 91–118, Addison-Wesley, Reading, MA (1995).

[Birrer 93] A. Birrer & T. Eggenschwiler : Frameworks in the financial engineering domain: An experience report. In O. Nierstrasz, Ed., *ECOOP '93 – Object-Oriented Programming*, pp. 21–35, Lecture Notes in Computer Science 707, Berlin: Springer-Verlag (1993).

[Dayhoff 87] J. Dayhoff & R. Atherton : A model for wafer fabrication dynamics in integrated circuit manufacturing. *IEEE Trans. on SMC*, Vol.SMC-17, No.1, pp.91–100 (1987).

[Hori 94] M. Hori, Y. Nakamura & T. Hama : Configuring problem-solving methods: a CAKE perspective. *Knowledge Acquisition*, Vol.6, No.4, pp.461–488 (1994).

[Hori 95] M. Hori, Y. Nakamura, H. Satoh, K. Maruyama, T. Hama, S. Honda, T. Takenaka & F. Sekine : Knowledge-level analysis for eliciting composable scheduling knowledge. *Artificial Intelligence in Engineering*, Vol.9, No.4, pp.253–264 (1995).

[Johnson 88] R. Johnson & B. Foote : Designing reusable classes. *Journal of Object-Oriented Programming*, Vol.1, No.2, pp.22–35 (1988).

[Kirkpatrick 83] S. Kirkpatrick, C. Gelatt Jr & M. Vecchi : Optimization by simulated annealing. *Science*, **220**, pp.671–680 (1983).

[Lee 91] J. Lee, M. Shu & M. Fox : A hierarchical scheduling expert system: KAIS-3, *Proceedings of The World Congress on Expert Systems*, pp.3030–3039, Orlando, Florida (1991).

[Musen 92] M. Musen : Dimensions of knowledge sharing and reuse. *Computers and Biomedical Research*, Vol.25, pp.435–467 (1992).

[Numao 91] Numao, M. & Morishita, S. : Cooperative scheduling and its application to steelmaking processes. *IEEE Trans. on Industrial Electronics*, Vol.38, No.2, pp.150–155 (1991).

[Petrie 92] C. Petrie, Ed.: *Enterprise Integration Modeling: Proceedings of the First International Conference*, MIT Press, Cambridge, MA (1992).

[Pirklbauer 94] K. Pirklbauer, R. Plosch & R. Weinreich : Object-oriented process control software. *Journal of Object-Oriented Programming*, Vol.7, No.2, pp.30–35 (1994).

[Rumbaugh 91] J. Rumbaugh, M. Blaha, W. Premerlani, F. Eddy & W. Lorenson : *Object-Oriented Modeling and Design*. Prentice Hall, Englewood Cliffs, NJ (1991).

[SEMATECH 95] SEMATECH Inc. : *Computer Integrated Manufacturing (CIM) Application Framework Specification 1.2*, #93061697E-ENG (1995).

[Yoshikawa 93] H. Yoshikawa : An experimental performance analysis of neighborhood search algorithms for large-scale single machine tardiness problem. *IEEE Int. Workshop on Neuro-Fuzzy Control*, Muroran, Japan (1993).

APPLICATION OF GENETIC ALGORITHM TO ALLOY CASTING PROCESS SCHEDULING

Masuhiro Ishitobi

Knowledge Industry Department, Nuclear Energy Center, Mitsubishi Materials Corporation,
1-3-25, Koishikawa, Bunkyo-ku, Tokyo 112, Japan.
EMAIL: tobi@kid.enec.mmc.co.jp

ABSTRACT

A scheduling system has been developed for an alloy casting process. It is designed to automatically build draft schedules by using expert's heuristic rules and to provide Graphical User Interface with which user can manually improve the draft rapidly. However, the system is not designed to generate near optimal schedules as the calculation time for exhaustive search would be intolerably long. We applied Genetic Algorithm in order to have better schedules in a shorter time. Results of application and performance of GA are described.

1   INTRODUCTION

Expert System for factory scheduling is generally required to provide the following features:

- good quality of generated schedule
- tolerable calculation time
- excellent modifiability of generated schedule
- excellent operability
- extendibility

The system employs expert's knowledge to generate schedule. While shallow (simple) knowledge may be acquired from expert easily and used for automatic scheduling, deeper (complicated) knowledge must be obtained to build effective and practical schedules. However the latter is hardly acquired and maintained in most cases. Furthermore it may not be sufficient as there may be better solutions beyond the expert's knowledge.

We have built a knowledge-based scheduling system for an alloy casting process, which is later referred as "basic system." The system employs human expert's simple job assigning knowledge to build initial schedules. It is not designed to provide optimal schedules by using deeper knowledge or implementing exhaustive search. Because deriving such knowledge is usually time-consuming and a such system tends to end up with difficulties in maintaining knowledge-base as the process configuration and assigning rules change frequently. Instead the system provides Graphical User Interface with which the user can change an initial schedule to the final schedule easily.

However, it is still desirable to have a capability to build near optimal initial schedules. Thus we have studied several different techniques such as Genetic Algorithm (GA) [Gol89], Simulated Annealing [KGV83] in order to obtain better, near optimal initial schedules. The outline of the basic scheduling system and the application of GA among other search techniques are described.

2   PROCESS DESCRIPTION AND SCHEDULING

The alloy casting factory produces copper-based alloy product, whose attributes are alloy type, cross-sectional shape (pillar or slab), size and length. The process consists of three furnaces and two casting lines. Base metal is melt at the furnaces and fed to the casting lines continuously. Only two of three furnaces can be operated as each casting line holds molten metal only from one furnace. The casting line uses various casting units. The attributes of the casting unit are the size and the number of molds. Casting conditions are determined based on the product attributes. The maximum number of possible casting conditions is 12 (3 furnaces x 2 casting lines x 2 mold patterns). The number of jobs, alloy types and cross-sectional sizes for a month are about 120, 50 and 50 respectively.

A typical schedule (gantt chart) is schematically shown in

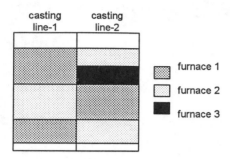

FIGURE 2-1 Schematic schedule

3.1 PROCESS MODELING

Modeling the process is straightforward. Two casting lines are considered as the machines which process jobs. The furnace to be used is considered as an job attribute. Thus generating an initial schedule is defined to assign about 120 jobs on the casting lines while satisfying the constraints.

The job attributes are shown below.

- product attributes (alloy type, size and length)
- furnace type
- machine type (casting line)
- the number of mold in casting unit
- setup time
- start time
- finish time
- due date
- processing time

The constraints are summarized below.

- restrictions on the use of the resources (furnace, casting line, casting unit) according to the product attributes.

- inter-machine constraints: no job can be assigned on one casting line when another job using the identical furnace in the same time period is assigned on another casting line.

- restrictions on the furnace 1 operation: the length of continuous operation of the furnace 1 is not allowed to exceed 10 days. And at least 4 days preparation period should be taken before next operation starts.

- resource availability: no job can be assigned when the corresponding furnaces or casting lines are in shutdown mode.

3.2 SYSTEM FUNCTIONS

The scheduling flow and the major functions are described.

1) The system initially loads job-related information, the heuristic rules for setting the job attributes, and the setup time matrix for deriving the setup time for a given pair of consecutive jobs.

2) The job attributes are set by using the rules.

3) The job order is determined according to their due dates. The job order for the furnace 1 is determined firstly and the order for other furnaces next. This heuristic assignment is introduced by the human expert.

4) Hill-climbing search is implemented.

5) The result is displayed on a scrollable gantt chart. Then the user modifies the result by grabbing, moving, dropping and

FIG. 2-1.

The schedule is built by human expert in order to reduce total setup time and total tardiness. If there are any differences in casting conditions such as alloy type, furnace and casting unit including the number of molds between two consecutive jobs, non-zero setup time up to 8 hours is required before processing second job. Therefore, the jobs having same casting conditions should be processed consecutively to minimize the setup time as far as the total tardiness is kept small.

In reality, as the number of jobs is relatively large compared to the maximum capacity of the process and due dates are not uniformly scattered throughout a month, satisfying two goals is usually very difficult.

The time length of cumulative continuous operation of a special furnace (referred as "furnace 1") is limited to 10 days. In addition, 4 days preparation period is necessary once the furnace 1 operation is terminated. Other limitations are imposed on the use of the resources such as furnaces and casting lines.

A schedule for a next month is built by the factory staff at the end of month. It usually takes half a day to build an initial schedule, which is to be modified at least every week to cope with due date changes. As the scheduling activity is time consuming, there's a strong reason to introduce such a system.

3   BASIC SCHEDULING SYSTEM

The system is designed to automatically generate initial schedules and provide user interface functions with which   the user can easily modify them to have ready-to-go schedules. The system is outlined below.

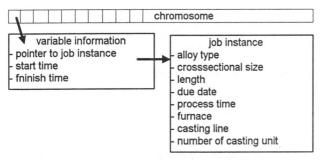

FIGURE 4-1 Chromosome Design

resizing the job by the mouse, and may start rescheduling. This function is necessary as the system is not designed to automatically provide complete schedules.

The search algorithm is described.

a) Calculate scores of all the jobs by the equation:

$$score = setup\ time + (-0.01) \times tardiness \qquad (1)$$

$$total\ score = \Sigma\ score$$

b) Create a list of jobs according to their scores. The first element of the list has the biggest score which means the worst. Only jobs having non-zero scores are included in the list.

c) Take a job from the list; choose the nearest two possible insert points in both upstream and downstream directions.

d) Move the job to the point. Rebuild the schedule.

e) Recalculate the total score.

f) If the new total score is smaller than the previous one, make the movement valid and go to a).

g) Abandon the movement. If the list is empty, then go to i).

h) Go to c)

i) Terminate the search.

3.3 IMPLEMENTATION

The system is implemented on SUN Sparc 20. The schedule engine is written by lisp-like language. Motif is used for building GUI.

4 APPLICATION OF GA

4.1 MOTIVATION

The job attributes are hardly changed during the search as no generic simple rules for changing them are found and a blind search in changing them is not expected to be practical in terms of calculation time. Although it may be possible to extract expert's knowledge about "when" and "how" to change them, this kind of knowledge tends to be obsolete in the near future.

We applied Genetic Algorithm (GA) in order to introduce the following merits:

- improvement in terms of total setup and total tardiness.
- ease of maintenance efforts by introducing less human knowledge.
- faster calculation time.

4.2 GENETIC ENCODING

A chromosome represents the job order, which is similar to the path representation for TSP. An element of the chromosome is a pointer to the structure having the job information shown in FIG. 4-1. Values in the variable information may be changed during the search, while the information of the job instances remains constant. Start and finish time are determined when the chromosome is decoded into the phenotype expression which is a casting schedule. The pointer to the job instance is set during the initialization process and may be changed when a mutation is occurred.

4.3 GENETIC DECODING AND SCORE CALCULATION

Genetic decoding procedure is described below. It is somewhat time-consuming, as all the constrains are satisfied during the process.

1) Take a job from the chromosome and append it to one of the following lists until the chromosome becomes empty.

  - list-1: using furnace 1 and casting line 1
  - list-2: using furnace 1 and casting line 2
  - list-3: using furnace 2 or 3 and casting line 1
  - list-4: using furnace 2 or 3 and casting line 2

2) Take jobs from list-1 and list-2 and assign the job which can be assigned earlier (see the left schedule in FIG. 4-2).

  constraints:

  - no assignment during furnace and casting line shutdown periods.

  - 4 days shutdown must be required for furnace 1 if the cumulative operation exceeds 10 days or if the operation is interrupted due to shutdown.

3) Take jobs from list-3 and list-4 and assign the job which can be assigned earlier (see the right schedule in FIG. 4-2).

  constraints:

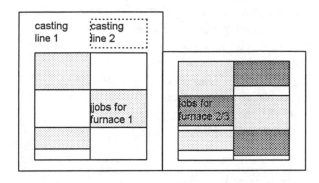

FIGURE 4-2 Decoded schedule

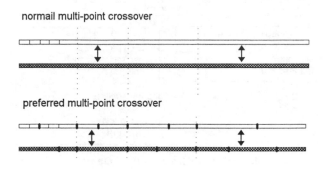

FIGURE 4-3 Corssover operation

- no assignment during furnace and casting line shutdown periods.

- no assignment overlapping the jobs previously assigned.

Then the fitness of the chromosome is calculated based on the decoded schedule by using the following equation:

$$fitness = (score) - 1 \qquad (2)$$

where score is calculated by the equation (1).

4.4   GENETIC OPERATION

CROSSOVER OPERATION

We applied simple two-point partially mapped crossover [GL85] and two-point preferred crossover operations as shown in FIG. 4-3.

The two-point preferred crossover method, which is based on the sub-tour exchange crossover, randomly selects crossover points from a set of points which never destroy "successful" jobs. This operation is considered to preserve job clusters having same casting conditions, which however may eliminate diversity of chromosomes. Unlike the sub-tour exchange crossover, the method does not explicitly identify matching jobs between two parents in order to speed up the calculation.

MUTATION OPERATION

Mutation operation is introduced with respect to the position and the attributes of a job. The former mutation procedure selects two genes randomly and swaps them within the same chromosome. The latter randomly selects a single gene and changes its attributes randomly, in other words, selects another job instance if exists.

4.5   REPRODUCTION

The conventional roulette model is used. Chromosomes in a next generation are reproduced based on the scaled which is calculated by the equations:

$$scaled\ fitness\ 1 = (score \times 1000)^3 \qquad (3)$$

scaled fitness 2 =

if score > avg. score in a generation then

0.2 + (score - avg. score) / (max. score - avg. score)

$$else\ 0.2 \qquad (4)$$

In addition, the worst chromosome is replaced with the best chromosome.

4.6   INITIALIZATION

Two initialization procedures are used. The former initializes the job order and the job attributes randomly. The latter initializes the job attributes randomly, but the job order according to their due dates. It is expected to have a smaller total setup time in the latter case because the jobs having same casting conditions tend to have close due dates.

5   RESULTS AND DISCUSSION

Two different methods were used to build initial schedules by using actual customer order and stock information.

5.1   BASIC SYSTEM

The result by the basic system is summarized below in terms of the total setup and the total tardiness. The calculation time was almost an hour.

Table 5-1 Result with Basic System

total setup	154
total tardiness	-3020

Table 5-3 Best Result with GA

total setup	131
total tardiness	-2971

5.2 GENETIC ALGORITHM

Preliminary survey was implemented in order to determine appropriate calculation conditions. Various options considered are summarized in Table 5-2.

Then the preferred calculation conditions were derived.

- crossover option: preferred
- crossover points: 4
- mutation rate: 0.5
- scaling option: scaled fitness 2
- initialization option: random

The final calculations with the above conditions were implemented with the population 100, 200, 300 and the generation 1000. The convergence performances are shown in FIG. 5-1 and the best result is shown below. The calculation time was less than 20 min.

5.3 DISCUSSION

It was found that GA could find better solutions faster than the basic system. This was expected because the basic system was designed to provide initial schedules to be modified by the user. However, GA has the obvious advantages:

- superior capability in searching.
- less case dependent and simpler algorithm; the basic system may generate an inferior initial schedule for a given month when the job assigning knowledge becomes obsolete.
- faster calculation time.

The following issues remain unsolved.

Whether the decoding procedure reflects the characteristics of survived chromosomes properly has not been validated.

Studies for finding better and faster decoding procedures have to be pursued.

Adjusting the processing time length of the job, dividing a single job into several smaller jobs may have to be considered to provide practical schedules.

GA is good at global search but not at microscopic one. For instance, when the total tardiness may be attributed to a few jobs, a conventional rule-based system can find them instantly and improve the total tardiness.

The equation used to calculate the fitness has not been validated. Further investigation have to be done.

GA program will be embedded in the basic system as a callable subroutine. Also hybridization between rule-base paradigm and GA will be studied.

REFERENCES

[Gol89] Goldberg, D.E.: Genetic Algorithms in Search, Optimization and Machine Learning, Addison-Wesley Publishing Company Inc., 1989.

[KGV83] Kirkpatrick, S., Gelatt Jr., C.D. and Vecchi M.P.: Optimization by Simulated Annealing, Science , Volume 220, pp.671-680, 1983.

[GL85] Goldberg, D.E. and Ligle, R.Jr.: Alleles, Loci the Traveling Salesman Problem, Proc. 1st ICGA, 1985.

Table 5-2 Calculation Options for Preliminary Survey

population	100
generation	100
crossover option	random(option 1)/preferred(option 2)
crossover points	2/4
crossover rate	0.8
mutation rate	0.1/0.5
scaling option	scaled fitness 1(option 1)/ scaled fitness 2 (option 2)
initialization option	random(option1)/due date(option2)

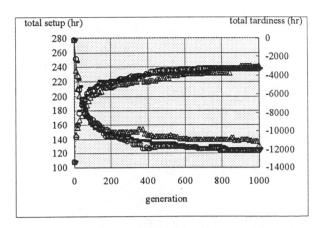

FIGURE 5-1 Convergence Performance

DESIGN AND DEVELOPMENT OF AN INTEGRATED INTELLIGENT PROCESS PLANNING SYSTEM TAKING CAD DATA FILES AS INPUT

Shu-Chu Liu
Department of Management Information Systems,
National Pingtung Polytechnic Institute, Pingtung 91207, Taiwan R.O.C.
Email: sliu@unix3cc.nppi.edu.tw

ABSTRACT

Automatic process plan generation is a major step for manufacturing automation because automatic process planning generation can directly extract design information from Computer Aided Design (CAD) files and translate this information into manufacturing information. This research develops an Integrated Intelligent Process Planning System (IIPPS) to simulate a human expert in a consistent and efficient way. The system includes the following modules: 1) A Feature Extraction Module for drawing out features designed in CAD, 2) A Machining Operation Expert System Module for feature and operation mapping and for generating various process sequence alternatives, and 3) A Sequencing Module to obtain a better process plan based on limited resources and constraints. System usability analysis is considered to verify IIPPS. The average usability value is 0.87 which is recognized as good. Thirty cases are used to validate the performance of IIPPS. The results show IIPPS can perform better in terms of accuracy than an expert who has ten years of experience.

1. INTRODUCTION

In the past, most process plans were generated manually in industry. Process planning involves translating the data defined in design into the data defined in manufacturing. The tedious information translation required for generating process plans is error-prone and time-consuming [CWW91]. Computers can substitute for human experts after the development of technical supports such as artificial intelligence (AI). This would result in process plans that are more consistent and the generation of process plans would be faster. In addition, process planners are in short supply, an intelligent process planning system would alleviate this. However, several drawbacks in existing process planning systems have been reported as follows: (1) lack of good representation and poor transfer of information between design and manufacturing [AZ88], (2) resources and constraints not considered in a global manner for sequencing problems [HL88], and (3) difficulty in maintaining knowledge bases. The primary purpose of this research is to design and build an integrated intelligent process planning system (IIPPS) to simulate human expertise and conquer the above drawbacks. The system which
(1) can link directly with and apply to general CAD models.
(2) can generate sufficient information regarding process planning.
(3) can assist designers to overcome their lack of manufacturing knowledge.

2. COMPONENTS IN IIPPS

IIPPS consists of three modules: (1) Feature Extraction Module, (2) Machining Operation Expert System Module, and (3) Sequencing Module. The relationship between these modules is depicted in FIGURE 1. In FIGURE 1, CAD input can be any CAD software with 3D IGES format as output. After the IGES file is generated from CAD part design, the feature extraction module is applied to extract and classify the features from CAD. Feature extraction entails recognizing and extracting features based on some algorithms proposed in another paper [Kyp83] and developed in this research. After a feature is extracted, the feature is classified as one of the predefined features which are stored in a database. After each feature in the drawing is recognized, the machining operation expert system module is applied for feature-operation mapping. This machining operation expert system includes process, machine, tool, fixture, and machining parameter databases and feature mapping, process mapping, and tool and machining parameter knowledge bases.

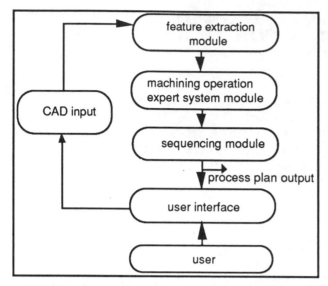

FIGURE 1 Module's relationships in IIPPS

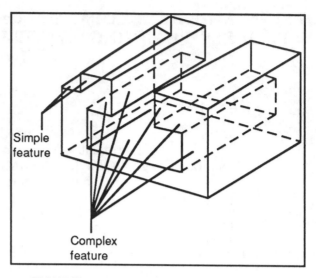

FIGURE 2 Simple and complex features

After the operations for features have been determined, the sequencing module is invoked for sequencing these operations. A sequencing algorithm is developed based on minimal times of machine, fixture, tool setup and shortest path method under constraints such as fixture, machining, etc. After the sequence is determined based on the heuristic algorithm proposed in sequencing module, better process plans can be generated which include operations, machine, tool, fixture, machining parameter, and sequence. The interface displays these outputs (e.g., process plans) to users and receives the input (e.g., material, quantity, etc.) from users.

3. A FEATURE EXTRACTION MODULE

The automatic feature extraction module presented in this section consists of three basic procedures - a data file translator, a part form feature classifier and a feature relationship classifier. The first procedure translates a CAD data file into a proposed object oriented data structure. The second procedure classifies different part geometric features obtained from the data file translator into different feature families. The third procedure determines the relationship among extracted features based on a proposed algorithm.

Most CAD system's information files are kept in IGES format. The proposed data structure translation procedure is designed to translate the IGES format into a new object-oriented data structure. Simple and complex features (FIGURE 2) are extracted using a proposed methodology that includes the concave edge test method as proposed by [Kyp83] and other feature extraction algorithms [Chi88, Sha90, KN93]. Features that can be merged or decomposed are recognized thereby completing the process of translating the IGES data file into the

proposed object oriented data structure. The algorithm for merging features is as follows:
(1) Two features have the same machining direction.
(2) At least one face in one feature is on the same plane as one face in the other feature.
(3) Another feature is adjacent to these two corresponding faces.
(4) Two features are not adjacent to each other.
The algorithm for decomposing features is as follows:
(1) Put the total faces in the feature into F*.
(2) Check whether F* is empty. If it is, stop. Otherwise, go to step 3.
(3) Recognize one sub-feature from this feature (face group).
(4) Delete the recognized feature's faces which are not shared with other sub-features. Go back to step 2.

The features defined previously are then classified into families, and the process ends by searching for relationships among the recognized features. For further details, please reference [Liu94].

4. A MACHINING OPERATION EXPERT SYSTEM MODULE

A machining operation is a basic element for Computer-Aided Process Planning (CAPP). A good machining operation knowledge base is the foundation for a good CAPP system. This section describes an expert system for machining operations. There are four basic steps to develop expert system: knowledge acquisition, knowledge representation, inference engine development, and database management. The knowledge comes from books, human experts, or historical data and is represented in a production rule format.

Two basic knowledge sources are required in the creation of a knowledge base. The first is the "deep knowledge", which is knowledge of the problem area which is easily obtained. Deep knowledge can be obtained from books, manuals, etc.. The second basic knowledge source is called "shallow knowledge". This is the knowledge, in the form of heuristics, which is used by human experts. For this research, deep knowledge was acquired from several books [DBK88, Cha90, WL91]. These books have been used as references for operation knowledge sources. The machining knowledge of two human experts is the source for the shallow knowledge base for operation knowledge in IIPPS. The first expert who has worked as product design engineer for more than twenty years at Cooper Oil Tool Company. The other expert used to work as a manufacturing engineering for more than ten years for Baker Oil Tool Company. They provided valuable knowledge for machining operations.

In the knowledge representation, IF-THEN rules are used. The IF-THEN rule is the most widely used method of knowledge representation in expert systems. The IF-THEN rules provide a straightforward method of representing the heuristic rules which exist in process planning application. A typical IF-THEN rule used in IIPPS is shown below :

RULE NAME : FINISH MACHINING EXTERNAL CYLINDER

IF TOLERANCE ≥ 0.001 inches
AND TOLERANCE ≤ 0.003 inches
AND SURFACE FINISH ≥ 63 microinches
AND SURFACE FINISH ≤ 250 microinches
AND RAW MATERIAL SHAPE = "ROUND"
AND FEATURE = "CYLINDER"
THEN MACHINE = "NC-LATHE" OR = "MANUAL LATHE"
AND FIXTURE = "THREE JAW CHUCK"
AND OPERATION = "FINISH"

The conditions which are combined with "AND" in the IF parts are called "premise clauses". In this rule, the premise information is derived from drawing. If these four premise clauses are all true, the conclusion in the THEN part can be taken. The machine used for this operation depends on the factory capacity. The machine is not determined until factory capacity is specified. All production rules concerning the final conclusions in the knowledge base were expressed with certainty factors. All the premise clauses were expressed with deterministic facts. The facts are stored in the object-oriented representation. The production rule provides the method to inference the other facts.

As for inference engine development, first of all, the feature mapping knowledge base is developed and inference information such as machine, fixture, and operation roughness. Secondly, the process information can be obtained after the process mapping knowledge base is applied to specific feature and machine information. Third, tool and machining parameter selections can be determined by production rules after the raw material and process are known. Basically, this inference engine is forward chaining. From initial available information (facts) like feature, tolerance, surface finish, etc., final

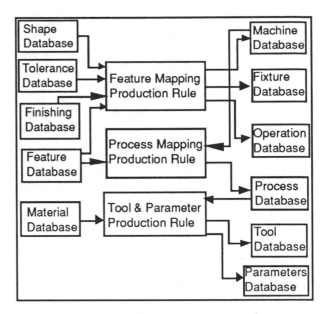

FIGURE 3 The inference engine

information like process, machine used, tool used, machining parameter, etc., is deduced from production rules (FIGURE 3).

As for database management, several databases are created such as material, feature, rule, machine, fixture, tool, and machining parameter. They are created well before they are used. When production rules are applied, the final information is obtained. The database is checked for the information (machine, tool, etc.). If the information is not available, users should be notified. Database maintenance is provided through an editor with ADD, MODIFY and DELETE functions. The editor will provide easy access to the database for users.

5. A SEQUENCING MODULE

This section discusses two material shapes, prismatic and round. The machine, fixture, and tool setups are also considered. Generally speaking, minimizing the setup time is the most economical method to sequence the operation. Constraints are used to determine which machine, fixture, and tool setup should be machined first. If there is no constraint between features in the same tool setup, the shortest path method is used. Otherwise, the constraints can sequence the order in the same tool setup. The minimal machine, fixture and tool setup and the shortest path methods are applied in the sequencing module. Sequence is developed based on operation, geometry, tool selection, tolerance, fixture selection, and machining constraints. Finally the process sequence is generated. In addition, the total machining time can be computed based on the machining parameters, features and available tool information. In this module, the inputs are the operations, machine used, fixture used, tool used, and operation relationships (constraints) generated from the expert system for the machining operation module, feature relationship (constraint) generated from the feature extraction module and other constraints, such as machining constraint, fixture constraint, etc., from manual data entry. The output of this module is the operation sequencing for this part drawing. With detailed operation information such as machine used, tool used, fixture used, machining parameters, and machining time, the output becomes a process plan. This procedure of the sequencing module is as follows:

(1) Group the same machine setup operations
(2) Group the same fixture setup operations.
(3) Group the same tool setup operations.
(4) Determine the natural order based on datum, fixture, tolerance, geometry, operation, tool, and machining constraints.
(5) Apply the shortest path method to the operations in which order cannot be determined by the constraints mentioned above.

6. CASE STUDY

To illustrate the methodology described in the previous sections of this paper, one part is analyzed using this system as one test case for demonstration purposes. This case study is based on FIGURE 4.

The assumed drawing has been done and saved in IGES format. After users input raw material, feature tolerance information, and constraint information, the process planning is generated. The detail process is as follows:

(1) There are two features which are classified. f1, pocket, includes F1, F2, F3, F4, and F5. f2 includes F6, F7, F8, F9, F10, F11, F12, F13, and F14.
(2) The machining direction for f1 is upside down and the number is equal to one. The number of machining

direction for f2 is zero which means more than one machining direction is required to generate this feature. After grouping the machining direction, two machining directions are needed to machine f1 and f2. f2 is a special shape. It is decomposed into one face group, F6, F9, F12, F13, and F14, in one machining direction, which is the same as f1, and the other face group, F6, F7, F8, F10, F11, in the other machining direction. The face group, F6, F9, F12, F13, and F14, can be recognized as one protrusion feature. Based on the feature operation dependence, f1 is the depression feature which should be machined after f2. The face group, F6, F7, F8, F10, and F11, can be decomposed into two keyways. There is no geometrical relationship between these two keyways.
(3) After manufacturing process mapping, f1 should be finish end-milled. The protrusion in f2 should be finish end-milled. In addition, two keyways in f2 should be rough end-milled.
(4) The end mill operation for this protrusion in f2 is grouped with the end mill operation for f1.
(5) Since f2 should be machined before f1, the end-mill for the protrusion should be machined before the end-mill for f1. Then flip the workpiece and end-mill these two keyways in two operations. These two operations can be grouped together.

7. SYSTEM VERIFICATION

The knowledge base has been verified by two human experts. The verification is also based on some principles such as consistency, completeness, etc.

The system is tested for conformance to the human factor design guidelines and the users' need. Usability is considered from a designer's perspective: the designer must ultimately address any design issues that may affect usability. The function's independent variables represent user perception and performance [Mit91].

A simple linear usability function, $U(S_1, ..., S_m, O_1, ..., O_n)$, normalized to the range [0,1] such that $U:[0,1]^{m+n} \to [0,1]$, is considered, i.e.,

$$U = \sum_{j=1}^{m} a_j S_j + \sum_{k=1}^{n} b_k O_k,$$

such that for $j = 1,...,m$ and $k = 1,...,n$,

(1) $0 \leq S_j, O_k \leq 1,$

(2) $0 \leq a_j, b_k \leq 1,$

(3) $\sum_{j=1}^{m} a_j + \sum_{k=1}^{n} b_k = 1.$

The usability score is U and S_j and O_k are function variables measuring opinions and performance. Coefficients are assigned on the basis of pairwise comparisons of function variables, where comparison are

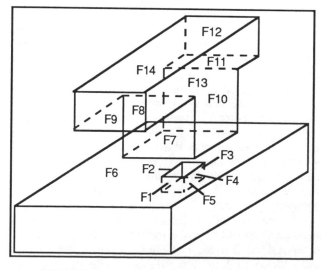

FIGURE 4 An example for case study

made by interface design experts,

$$(4)\quad a_j = \frac{\alpha_j}{\sum_{j=1}^{3}\alpha_j + \sum_{k=1}^{3}\beta_k},$$

$$(5)\quad b_k = \frac{\beta_k}{\sum_{j=1}^{3}\alpha_j + \sum_{k=1}^{3}\beta_k}.$$

In this research, eight students in the Industrial Engineering Department were selected to answer the survey. There are six functions selected: user confidence (S_1), perception of user friendliness (S_2), user perception of system efficiency (S_3), output precision (O_1), output specificity feasibility (O_2), input interactivity (O_3). The coefficients are assigned on the basis of the pairwise comparisons of function variables. The results of pairwise comparison reflect the relative importance of each usability variable, and the results are used as coefficients. The final normalized usability function can be written as $U = 0.23S_1 + 0.14S_2 + 0.09S_3 + 0.27O_1 + 0.18O_2 + 0.09O_3$

Based on the survey answers, the usability score can be computed using the above formula. U = {0.88, 0.8, 0.9, 0.92, 0.77, 0.85, 0.94, 0.91}. The average score for the usability of the system is 0.87.

According to the method mentioned in Mitta's paper, an expert system is useable if the usability score is larger than 0.5, which is the scale midpoint. Since the average U of IIPPS exceeds 0.5, IIPPS is considered as a useable system.

8. SYSTEM VALIDATION

IIPPS is required to perform quick and appropriate process planning just as human experts do. The accuracy of the process plan is the most important factor to evaluate the system. 30 cases are chosen from papers, companies, and books. The human evaluator is one expert who has worked in industry for more than ten years. He is given the same data as is input into IIPPS and he manually generates process plans. Another expert, who has been in design and manufacturing for more than twenty years, makes comparison between the results of IIPPS and the first human expert.

The statistical hypothesis method is used to compare the correctness between human expert and IIPPS. The one-tailed Z test has been selected for comparison since the sample size, 30, is large enough and the hypothesis concerns that the average error number of IIPPS is lower than that of human expert. The error for each evaluator can be feature recognition error, machining operation error, or sequencing error.

If the calculated z value is greater than -2.33, the null hypothesis can not be rejected. In other words, the performance of both is concluded to be the same. If the calculated z is less than -2.33, the null hypothesis must be rejected and the alternative hypothesis is true. In other words, the mean error number of IIPPS is less than that of human expert. The result of the statistical test is calculated as -5.85. Since -5.85 is less than -2.33, the null hypothesis must be rejected and the alternate hypothesis is true. That means the performance of IIPPS is better than that of human expert who has ten years of experience.

9. CONCLUSIONS

The objective of this research is to propose and develop an integrated intelligent process planning system (IIPPS) which prevents human error to the fullest extent and proposes more realistic constraints to direct an operation sequence. In addition, this system proposes a modification concept to strengthen the expert system structure. To achieve these goals, several modules which were mentioned in previous chapters are developed and integrated into IIPPS. The evaluation results, in terms of correctness, indicate that IIPPS performs better than an expert who has ten years of experience.

One possible extension of the work presented in this paper is to develop a system that can examine the machinability of a given part. In other words, the fact that a part can be designed does not mean that it can be manufactured.

REFERENCES

[AZ89]   L. Alting and H. Zhang. Computer-Aided Process Planning: the State-of-the-Art Survey. *International Journal of Production Research*,, 27: 553-585, 1989.

[Cha90]   T. C. Chang. *Expert Process Planning for Manufacturing*,. Addison-Wesley Publishing Company, New York, 1990.

[Chi88]   H. Chiyokura. *Solid Modeling with Designbase*. Addison-Wesley Publishing Company, New York, 1988.

[CWW91]   T. C. Chang, R. A, Wysk and H. P. Wang. *Computer-Aided Manufacturing*, Prentice-Hall, New Jersey, 1991.

[DBK88]   E. P. DeGarmo, J. T. Black, and R. A. Kohser. *Material and Processes in Manufacturing*. Macmillan Publishing Company, New York, 1988.

[HL88]   I. Ham and S. C. -Y. Lu. Computer-Aided Process Planning: The Present and the Future. *Annals of the CIRP*, 37: 591-600, 1988.

[KN93]   T. S. Kang and B. O. Nnaji. Feature Representation and Classification for Automatic Process Planning Systems. *Journal of Manufacturing Systems*, 12: 133-144, 1993.

[Kyp83]   Kyprianou, L. K., Shape Classification in Computer Aided Design, Ph.D. Dissertation, University of Cambridge, UK, 1983.

[Liu94]   S. C. Liu. *An Integrated Intelligent Process Planning System Taking 3D Data Files as Input*. Ph.D. Dissertation, University of Houston, Tx, 1994.

[Mit91]   D. Mitta. A Methodology for Quantifying Expert System Usability. *Human Factors*, 33: 233-245, 1991.

[Sha90]   J. J. Shah. Assessment of Feature Technology. *Computer Aided Design*, 21: 331-343, 1990.

[WL91]   H. P. Wang and J. K. Li. *Computer-Aided Process Planning*. Elsevier Science Publishing Company, New York, 1991.

ENCAPSULATION OF ACTIONS AND PLANS IN CONDITIONAL PLANNING SYSTEMS

Marco Baioletti, Stefano Marcugini, Alfredo Milani
Dipartimento di Matematica, Università di Perugia
Via Vanvitelli, 1 06123 Perugia Italy
E-mail :{marco,gino,milani}@gauss.dipmat.unipg.it
tel:+39(75)585.5049 fax:+39(75)585.5024

Abstract

We introduce a notion of macroaction for a given subplan. A macroaction is an action representing the given plan and it can be considered as an abstraction, neglecting the internal structure of the original plan. The description of the macroaction is sufficiently powerful to represent all the information needed for synchronization and clobbering detection of the given subplan. It is possible to prove that when substituting in a correct plan a macroaction with its real composition, the overall plan remains correct. It is pointed out that it is possible to build large libraries of plans by means of techniques of plan compilation and manage them with fast-selection and plan re-use methods.

A comparison of different kind of macroactions is done in the frameworks of linear, non-linear and conditional planning. It is shown that a conditional macroaction is required even in the case of nonlinear unconditional plans. A conditional planning model based on conditional formula calculus is used. Conditional formulae represent uncertain situations and denote sets of mutually exclusive states.

The method of compilation is defined through a weakest precondition semantics for conditional plans.

1. INTRODUCTION

One of the important problems faced in planning research is the abstraction. Since the early Sacerdoti's works [Sacerdoti], this feature has been mainly used to reduce computational complexity in plan generation because it allows the planner to work in smaller search spaces.

Other existing techniques of plan synthesis based on the use of "precanned" plans, as in [Aylett], solve planning problems by using libraries of previously generated plans, but they have to cope with the problems of fast plan selection and of interactions management between precanned plans in the overall plan under construction.

The problem of detecting and solving interactions between actions arises in hierarchical approaches since abstractions, (as in hierarchical planning of [Sacerdoti] and in the proposals of Tenenberg [Tenenberg]) are obtained by simplifying the description of actions and plans or classifying them into taxonomies.

In those cases plan refinement phases are needed for each level of abstraction, but unfortunately these phases could require heavy reordering of actions and declobbering which can diminish the computational advantages of the approach.

It would be useful to have a method for representing concisely the knowledge embedded in a plan. This representation should have the same power of the abstraction techniques, while keeping a sufficiently detailed description in order to detect and solve conflicts or to check if it can be used to achieve some goals. Some previous approaches in this direction are in [Terragnolo] in a parallel planning framework and in [Sani], a recursive planning approach; where plans are entities which can be used in the plan description itself.

Macroactions

M is a macroaction for a given plan P if the action description M is equivalent to P in that:
1) M is executable in a situation S if and only if P is executable in S
2) the execution of M in any situation S produces the same effects as the execution of P in S

The computation of a macroaction M for a plan P is called **compilation** of P in M. The macroaction M of P is also denoted by M[P]

The inverse of plan compilation is the macroaction **substitution**. A macroaction M[P] can be substituted into a global plan Q by the detailed plan P which originated the macroaction M. The substitution makes each action in the plan P to inherit, in the global plan Q, all the precedence relations which involved macro action M in Q.

There exists an apparent relationship between the expressiveness of plan representation and the expressive power required to the macroactions which represent it.

We shall examine the problem of macroaction compilation with respect to two coordinates: increasing complexity of plan structure (i.e. linearity vs. nonlinearity and conditionals) and increasing complexity of action descriptions (i.e. STRIPS-like operators vs. actions with conditional preconditions and/or effects).

Linear Plans

A Strips-like macroaction M suffices to compile a linear plan P.

A macroaction M for a given linear plan P composed only by STRIPS-like operators A1,...,An can be easily generated by the following constructive steps:
1) the preconditions of M Pre(M[A1;A2;...;An]) are computed by the recursive equations:

$$Pre(M[A])=Pre(A)$$
$$Pre(M[A1;A2;...;An])=Pre(A1) \cup$$
$$(Pre(M[A2;...;An]) - Eff(A1))$$

2) calculate the effects Eff(M[A1;A2;...;An]) by simulating the execution of the linear sequence A1;...;An in the initial situation Pre(M[P]) of step 1.
See example figure 1.

Step 1 calculates the *minimal initial state* of plan P as the minimal preconditions which must be verified in order to guarantee the execution of all elementary actions in P. It consist of all preconditions of all actions in P except those conditions which are *internally* generated by other actions in P. The effects calculated in step 2 are then the minimal ones.

Non Linear Plans

Unfortunately as we increase the complexity of plan structure to allow simple nonlinearity between actions, it is impossible to compile a nonlinear plan (even if composed only by Strips-like operators) into a Strips-like macroaction.

The reason is that a nonlinear plan can encapsulate a set of possibly different final states and a set of possibly different initial states, then the single state structure of Strips-like operators (single input situation and single output situation) is clearly insufficient to represent it.

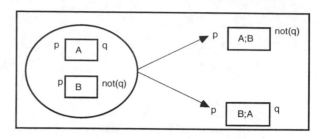

FIGURE 2.a

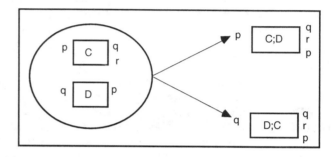

FIGURE 2.b

The example of fig 2a shows as the possible completions of the partial order precedence relation can lead to two different final states. Fig.2b shows the counterpart case for different initial states.

Conditional Plans

The problem of compiling a nonlinear plan into a macroaction requires that macroaction descriptions represent multiple input and multiple output states (i.e. conditional action, for example in [PS92] [Mil94][BMM95a]).

The most general case, which subsumes the preceeding cases, consists of a plan with a conditional structure and elementary conditional actions.

A result of this work is that: a conditional plan P can be compiled into a macroaction M using a conditional action description for M.

In fig.3a and 3b are shown the conditional macroaction solutions to problems of fig.2a and 2b.

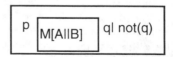

FIGURE 3.a Macroaction for the nonlinear plan in figure 2.a

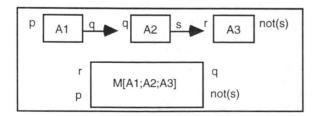

FIGURE 1 Macroaction for a linear plan

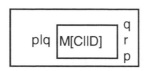

FIGURE 3.b Macroaction for the nonlinear plan in figure 2.b

The conditional planning model appears to be powerful enough to subsume classical linear and nonlinear model. Moreover, with respect to the problem of macroaction compilation, a conditional model is required even if we limite it to nonlinear plans with simple Strips-like operators.

In the following, a conditional model is briefly introduced and a method to compile conditional plans into macroactions is described. The macroaction compilation is proved to be correct with respect to macroaction/plan substitution.

2. THE CONDITIONAL PLANNING MODEL

We briefly recall some definitions and operators which will be used in the following; for a detailed description see [BMM95b]. Our description is somehow similar to SAS formalism introduced by [Bac92] (even if the original model is only for unconditional action) but it is actually an extension of the description given in [BMM95a] to allow the use of multivalued variables.

States

A state is conjunctive proposition of the form $V_1 = v_1 \wedge \ldots \wedge V_n = v_n$, in which the V_i's are state variables and the v_i's are their values, that is each of them is a member of the corresponding domain. A new value U is added to each domain, which represents the lack of knowledge about the real value of that variable.

A state can be viewed as an assignment that gives each variable a value of its extended domain. Given an assignment p and a variable x we denote by p(x) the value of x in the state represented by p.

For instance, in the block world, the state of knowledge "I know that a is over b, b is on the table, but I do not know where c is" can be formulated by the assignment p which is defined as

p(below_a)=b, p(below_b)=table, p(below_c)=U.

Two states are compatible if they do not give two different non U values to the same variable.

We define also an operator for joining two compatible states p and q in a natural way as the aggregation of the two states and a subsumption relation $\sqsubseteq$ between states.

Any description of partial knowledge can be viewed as a set of different mutually exclusive states.

Conditional formulae

Conditional formulae denote a *set of states* or *alternative situations*. Conditional formulae are built according to the following BNF syntax:

Formula ::= φ | Ω | Variable=Value | Formula & Formula | Formula "|" Formula

The special symbols φ and Ω denote respectively the empty set of states and the total ignorance state.

The *alternative formula F|G*, to be read "F alternate G", represents all the states of knowledge belonging either to F or to G (or to both). Its behaviour is similar to disjunction but it allows to represent explicitly tautological situations as p|¬p.

The *conjunctive formula F&G*, to be read "F and G", represents all the possible states of knowledge composed by any state of F joined to any state of G, when these are not incompatible with each other. When all the states in [F] are incompatible with any state in [G] we set by definition that F&G=φ. It reduces to usual conjunction when F and G do not contain alternatives.

Note that & and | are similar to operator "and" and "alt" as defined in [Mil94].

Each formula F denotes a *set of states* [F] according to the obvious definitions given in [BMM95b].

The main properties about conditional formulae are:
1. Completeness: for any set of assignments S there exists a formula F such that [F]=S
2. Normal conditional form: for any formula F there exists an equivalent formula $F_1|F_2|\ldots|F_K$ where the F_i's are conjunctive clauses.

Operator AFTER

The operator $\triangleright$, to be read "after", applied on situation F and G represents the situation obtained when F is updated with G:

$[F \triangleright G] := \{p \triangleright q : p \in [F] \text{ and } q \in [G]\}$.

The operator $\triangleright$ applied to two states p and q is defined as:

$$p(x) \triangleright q(x) = q(x) \text{ if } q(x) \text{ is not U,}$$
$$p(x) \text{ if } q(x) = U$$

for each possible variable.

Necessary consequences

The notion of "Necessary consequence" extends to conditional formulae the concept of formula necessarily true in a situation, analogously as in [Cha87].

We will say that P entails Q, to be read as "Q is necessarily true in P" (or is a necessary consequence of P) if for each state $p \in [P]$ there exists an assignment $q \in [Q]$ such that $q \sqsubseteq p$ (which means that q is not more defined than p, see [BMM95b]).

Minimality

The notion of minimality of a formula is used to eliminate redundancy from formulae.

The minimalization min(P) of a given formula P is a formula denoting the greatest minimal subset of assignments of [P]:

$$[min(P)]=\{p \in [P]: \forall q \in [P]\ q \sqsubseteq p \Rightarrow p=q\}$$

For instance p|p&q|q is not minimal, while p|q is minimal.

The conditional planning model

We give a conditional planning model based on the notions of set of assignments and conditional formulae. We shall denote the set of all the formulae on all the existing atoms by ***Formulae***.

Definition

An action instance I is a couple (P_I, E_I) where P_I and E_I are conditional formulae which denote respectively the preconditions and the effects of the action.

Definition

An action instance $I=(P_I, E_I)$ is executable in a situation F if F *entails* P_I. The situation after the execution of I is $F \triangleright E_I$.

P_I and E_I being formulae, this definition allows us to use conditional preconditions and effects and execution in an uncertain framework which may be not totally defined. If P_I and E_I are not conditional the definition reduces to a classical Strips-like model.

Conditional plans

Conditional nonlinear plans are built composing actions by *precedence operator* ";", which denotes an ordered sequence, operator "||", which means that *any order* is allowed, and the *conditional construct* IF-THEN-ELSE. The syntax of plans is:

 Plan ::= Instance_Action | Plan;Plan | (Plan) | Plan || Plan | IF Condition THEN Plan ELSE Plan
Condition ::= Variable = Value | Variable = Value AND Condition

The semantics of plan execution is described by an execution function $X:***Formulae*** \to [***Plans*** \to ***Formulae***]$ and by an executability predicate $R:***Formulae*** \to [***Plans*** \to \{0,1\}]$. $X(P)_S$ is the formula denoting the situation after the execution of the plan P in the situation S. $R(P)_S$ is true if and only if the plan P is executable in the situation S.

By convention we state that $X(P)_S=\phi$ if $S=\phi$ or $R(P)_S$ is false.

Definition

If P is a plan formed by a single instance $I=(P_I, E_I)$ of elementary action then $R(P)_S \equiv S$ *entails* P_I and

$X(P)_S= S \triangleright E_I$ if S *entails* P_I
ϕ otherwise.

Definition

If P is the ordered sequence $P_1;P_2$ we define $R(P)_S \equiv R(P_1)_S \wedge R(P_2)_{X(P_1)_S}$ and $X(P)_S=X(P_2)_{X(P_1)_S}$.

Definition

If P is the anyorder sequence $P_1 || P_2$ we define $R(P)_S \equiv R(P_1;P_2)_S \vee R(P_2;P_1)_S$ and $X(P)_S=X(P_1;P_2)_S | X(P_2;P_1)_S$.

Definition

If P is the conditional IF c THEN P_1 ELSE P_2 we define $R(P)_S=(c \wedge R(P_1)_S) \vee (\neg c \wedge R(P_2)_S)$ and $X(P)_S=X(P_1)_{S\& c} | X(P_2)_{S\& \neg c}$.

Example

Let us suppose that we have the following planning problem: a robot has to process a piece. If the quality of the partially finished product is good then the robot completes the piece, otherwise the robot sends back the piece to repair. The planner can use the actions described in fig. 4.

According to the previously introduced conditional planning model, a plan to solve the given problem can be developed as shown in fig. 5.

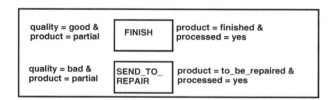

FIGURE 4 Description of actions

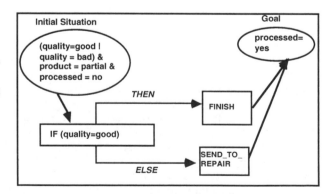

FIGURE 5 The generated plan

3. Weakest preconditions axiomatic semantics

We will define for the actions in a plan a notion equivalent to the weakest preconditions for the instructions in a programming language.

$W(P)_K$ denotes the least situation in which the plan P is executable and after its execution the formula K holds:

Definition

$W(P)_K$ is the minimal formula such that for each formula S, we have S *entails* $W(P)_K$ if and only if $R(P)_S$ and $X(P)_S$ *entails* K.

There are cases in which no such formula can be found: for instance when K requires an atom x to be true, but the effects of P deny it. In such cases we shall use the convention that $W(P)_K = \phi$.

Proposition

For each instance $I = (P_I, E_I)$ of an elementary action
$W(I)_K = min(P_I \& W(I')_K)$
where I' is the instance of an action whose preconditions are empty ($P_{I'} = \Omega$) and effects are the same as I.

Hence we can always replace an action by one (such as I') with empty preconditions.

$W(I')_K$ can be computed by $K \triangleleft E_I$ as described in [BMM95a]. Note that $W(I')_K$ is empty (i.e. Ω) if E_I *entails* K.

Proposition

Weakest preconditions of a sequence
$W(P_1; P_2)_K = W(P_1)W(P_2)_K$.

Proposition

Weakest preconditions of an unordered sequence
$W(P_1 \| P_2)_K = W(P_1; P_2)_K \& W(P_2; P_1)_K$.

Proposition

Weakest preconditions of a conditional
$W(IF\ p\ THEN\ P_1\ ELSE\ P_2)_K = (p\&W(P_1)_K)|$
$(\neg p\&\ W(P_2)_K)$.

4. MACROACTIONS

Macroactions are built by the planner in order to represent a plan or a subplan by a single conditional action. New macroactions augment the set of actions available to the system and can be used in order to rapidly solve already solved problems or subproblems. The conditional action description of the macroaction has preconditions and effects equivalent to preconditions and effects of the original plan.

Definition

A macroaction M[P] derived from plan P can be obtained by defining
$PRE(M[P]) = W(P)_\Omega$ and
$EFF(M[P]) = X(P)_{PRE(M[P])}$

Using the weakest precondition model and the execution function X it is possible to give a method to compile plans into macroactions.

Example

Suppose we want to compile the plan generated in the previous example (see fig. 5). According to the definition (4.6) the first phase consists in computing the preconditions of the candidate macroaction *process_piece* by computing the weakest preconditions of the plan with respect to the unconstrained situation Ω. By WP rules (4.3) and (4.5) it follows that PRE(*process_piece*) are *(quality = good | quality = bad) & product = partial*, see fig.3 and 4.

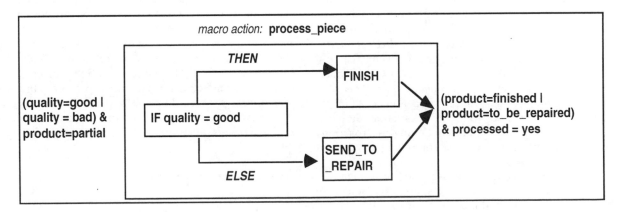

FIGURE 6 The macro action with its internal structure

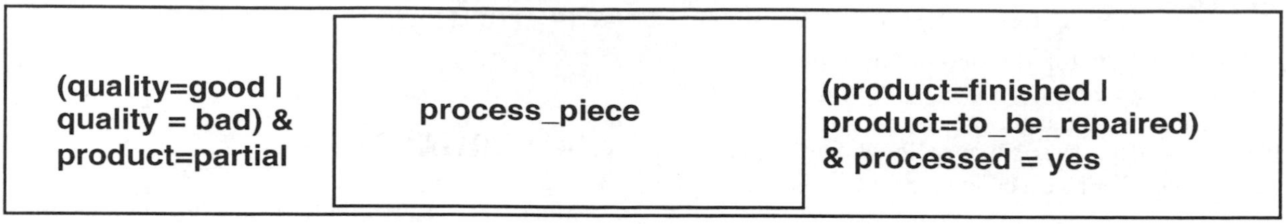

| (quality=good \| quality = bad) & product=partial | process_piece | (product=finished \| product=to_be_repaired) & processed = yes |

FIGURE 7 The final macro action

The second phase consists in computing EFF(*process_piece*), which are easily obtained by calculating the effects of the plan executed in an initial situation equal to PRE(*process_piece*) as shown in fig. 4 and 5.

By applying definitions (3.3), (3.5) and (3.6) of the execution function X, EFF has value *(product = finished | product = to_be_repaired) & processed = yes*. Note that the final macroaction, as shown in fig. 5, contains conditional preconditions and effects, the new action *process_piece* can be used without any regards to the internal structure of its original plan.

5. Conclusions

The advantages of using macroactions in planning are apparent: libraries of solved plans can be compiled into macro actions, which are managed by the same mechanisms used for elementary actions.

For example, if the planner has to solve a new problem, say an initial situation *product = partial & quality = good* and a goal situation *processed = yes*, the generated macro action *process_piece* can be directly used to solve the problem, instead of developing a new ad hoc plan. The action *process_piece* is executable since *product = partial & quality = good entails product = partial & (quality = good | quality = bad)*. The decision between the two solutions is a matter of planning strategy, i.e. fast action selection and plan re-use vs. plan generation.

The compilation of plans into macroactions is a kind of plan abstraction. As other forms of abstraction, they can be compiled at any level: plan containing macroactions and/or elementary actions can be compiled in their turn generating second level macroactions.

In classical approaches to the abstraction in planning [Sac74] [Ten88] actions or plans are abstracted by omitting details or by building taxonomies. Building an abstract plan is certainly easier, but this implies that the detailed plan could be not correct since conflicts can arise due to the omitted information. Some interactions can be hidden at a higher level and only when the planner gets to

a lower level it can detect them and can refine the plan, but that is done at high computational cost.

In the macroaction approach the abstraction is obtained by hiding the internal structure of the plan, while retaining in the macro action the information sufficient to detect conflicts and to use it to solve goals. A similar approach has been proposed in a non conditional framework in [MMT92] to detect conflicts in parallel planning. The given conditional planning semantics guarantees that if a plan built with macroaction achieves a goal G then the detailed plan, obtained by substituting each macro action with the detailed elementary actions, is still correct. Therefore no phases of plan refinement are required.

As noted in introduction, we point out that the conditional model (with conditional formula and alternate operator) is needed to compile plans into actions even if the plan to be compiled does not use conditional actions but it is nonlinear (suppose for example that the plan to be compiled uses only deterministic Tweak-like actions, with no alternatives in preconditions and effects). The reason is that when a nonlinear plan P is compiled into a macro-action M[P], PRE(M[P) and EFF(M[P]) must contain all the information needed to compute the preconditions of execution of P and the effects of execution of P. The internal structure of P is not known at this level but P can contain, for instance, two unordered actions A and B. The order of execution of A and B will be known only at execution time, therefore the effect of A||B can be different in different executions. A conditional formula is then necessarily required to express the effects of nonlinearly ordered actions at planning time, as effects of a macroaction which contains them. A benchmark example of this situation has been given in fig.2.a.

Acknowledgements

This work has been partially supported by Progetto Speciale "Pianificazione Automatica" under contract n. 93.006.27.CT07 of Italian National Research Council (C.N.R.) and by 40% project "Algoritmi, Modelli di Calcolo e Strutture Informative" of M.U.R.S.T.

References

[Ant92] G. Antognoni, "Pianificazione di Azioni con Effetti Alternativi: un Modello di Rappresentazione", Tesi di Laurea, Dipartimento di Matematica, Università di Perugia, Perugia, Italy (1992)

[AFB91] R. Aylett, A. Fish, S. Bartrum, "HELP: A Hierarchical Execution-Led Planner for Robotic Domains", in Proceeding of 1st EWSP, St. Augustin, Germany, (1991)

[Bac92] C. Backstrom, Equivalence and tractability results for SAS planning, Intern. conf. on Principles of Knowledge Represent. and Reasoning KR-92, 126-137,Cambridge, USA (1992)

[BH90] G.Brewka, J.Hertzberg: How To Do Things with Worlds: "On Formalizing Actions and Plans", Tasso-report n.11, GMD, (1990)

[BMM95a] M. Baioletti, S. Marcugini , A. Milani"A Weakest Preconditions Semantics for Conditional Planning", in Proceedings of 3rd Congress of AI*IA, Lecture Notes on Artificial Intelligence 992, 291-302, Springer Verlag (1995).

[BMM95b] M. Baioletti, A. Milani, S. Marcugini "Macroactions and axiomatic semantics for conditional planning", Tech. Report, Dip. di Matem., Univ. di Perugia

[Byl91] T.Bylander, "Complexity Results for Planning", Proc. of IJCAI- 91, 274 (1991)

[Cha87] D.Chapman "Planning for conjunctive goal", Artific. Intell. n.32 (1987)

[Geo86] M.P. Georgeff, "The Representation of Events in Multi-Agents Domains", in Proceedings of AAAI-86, (1986)

[Gin88] M.L.Ginsberg, D.E.Smith, "Reasoning About Action I: A Possible Worlds Approach" Art.Int., n. 35 , 165 (1988)

[Lif87] V.Liftschitz "On the Semantics of STRIPS", Proc. 1986 Workshop Reasoning About Actions and Plans, Timberline, OR, Morgan Kaufmann, 1, (1987)

[Mil94] A. Milani "A Representation for Multiple Situations in Conditional Planning", in Current Trends in AI Planning, EWSP'93 - 2nd European WorkShop on Planning, IOS Press (1994)

[MMT92] S. Marcugini, A. Milani, M. Terragnolo "Plans as Planning Objects", Artficial Intelligence V, Methodology, Systems, Applications, North Holland Elsevier, (1992)

[Mor87] L.Morgensten, "Knowledge Preconditions for Actions and Plans" in Proceeding of IJCAI-87 (1987)

[PS92] M.A.Peot, D.E.Smith, "Conditional Nonlinear Planning", Proc.of the 1st Int. Conf. on A.I.Planning Systems, AIPS92, J.Hendler Ed., Morgan Kaufmann, 189 (1992)

[Sac74] E.D.Sacerdoti, "Planning in a Hierarchy of Abstraction Spaces" Artificial Intelligence 5, 1974

[SS91] R.G. Sani, S. Steel "Recursive Plans", in Proceedings of the 1st European Workshop on Planning EWSP 1991, Sankt Augustin, Germany, Lecture Notes in Artificial Intelligence 522, Springer Verlag, 53 (1991)

[Ten88] J.D. Tenenberg, "Abstraction in Planning", Tech. Rep. 250, Univ. of Rochester, NY, (May 1988)

[War76] D.H.D. Warren, "Generating Conditional Plans and Programs" Proc. of AISB-76 Summer Conference, Edinburgh, 277 (1976)

[War90] D.H.D. Warren, "Warplan: A System for Generating Plans" in Readings in Planning; J. Allen, J.Hendler, A. Tate ed., Morgan Kaufmann (1990)

[Wil83] D.E. Wilkins, "Representation in a Domain-Independent Planner", SRI Intern., Tech. Note N. 266, (May 1983)

[Win88] M.Winslett, "Reasoning about Action Using a Possible Models Approach", Proc. AAAI-88, 89 (1988)

ON-LINE OPTIMIZATION TECHNIQUES FOR
DYNAMIC SCHEDULING OF REAL-TIME TASKS

Yacine Atif & Babak Hamidzadeh[†]
Department of Computer Science
Hong Kong University of Science & Technology
Clear Water Bay, Kowloon, Hong Kong.
Email: hamidzad@cs.ust.hk

ABSTRACT

In this paper, we explore the application of a new class of
on-line optimization techniques, referred to as Self-
Adjusting Real-Time Search (SARTS), to dynamic
scheduling of sporadic real-time tasks on a uniprocessor
architecture. The task model selected is that of non-
preemptable tasks with arbitrary start times and deadlines.
The selected search algorithms address a fundamental
trade-off between the cost of the scheduling process and
the quality of the delivered solution.

Keywords: Real-time search, dynamic scheduling, real-
time tasks, on-line optimization.

Introduction

A scheduling process can be formulated as a search for
a feasible sequence in a space of all partial and/or complete
sequences [ZR87]. The prohibitive optimization cost of
such a search process has limited its applicability to on-line
problem solving. On-line problem solvers are expected to
find the best solution and execute that solution within the
allotted time. These problem solvers are also expected to
improve upon their solutions if additional time is allocated,
while remaining attentive to the changes in the
environment. Two alternatives have been suggested to
cope with these requirements. Anytime algorithms search
for the best complete solution under a time constraint
[DB88] before making the first commitment. An
alternative formulation of the dynamic real-time problem
solving would be to interleave the best partial solutions
with their execution. The latter approach has been adopted
by real-time search algorithms such as RTA* [Kor90] and
DTA* [RW91]. RTA* combines planning and execution
and operates with repetitive plan-execution cycles until a
solution is reached. DTA* is an improved version of
RTA*. Both algorithms involve a fixed depth limit as a

parameter that controls the search. They do not address
explicitly the problem of constraints on response time.

Self-Adjusting Real-Time Search (SARTS)
[SH93][HS95] can be considered as a type of real-time
search algorithm [Kor90] that directly controls the time
allocated for scheduling in order to find solutions and
execute them within a given deadline. SARTS has the
capability to predict deadline violation enabling the
monitoring system to issue alternative actions before the
deadline is reached.

The real-time systems literature has shown that, for
many task models, simple algorithms like Earliest-
Deadline-First perform poorly and fail to provide any type
of guarantee on whether or not the tasks will meet their
deadlines [SSN+95]. Most of the existing scheduling
techniques consider the scheduling costs as negligible
overhead [JSM91] and/or suffice by proposing some ad-
hoc heuristics for reducing scheduling costs without
considering their direct effects on overall quality of the
resulting schedules [ZR87][SZ92][ZRS87a][ZRS87b].

This paper adapts a real-time search strategy belonging
to the SARTS class [SH93] to develop algorithms that
dynamically schedule real-time tasks. The resulting
algorithms make a major contribution to the real-time
scheduling research by incrementally building feasible
schedules while consuming no more than the available time
such that the deadline violations due to scheduling
overhead are minimized. The paper is among the first
attempts to explore the application of an on-line
optimization technique such as SARTS to solving a hard
problem dynamically.

The remaining sections of this paper are organized as
follows. Section 2 provides the problem statement. Section
3 provides the procedures of the proposed algorithm.
Section 4 shows the experimental results and discuss their
behaviour. Finally, section 5 concludes the paper with a
summary of our work and suggests some future extensions.

[†] The order of names is purely alphabetical and has no other significance.

Problem Statement

In this paper, we address the problem of scheduling a set T of n sporadic, non-preemptable, real-time tasks with earliest start times and deadlines on a uniprocessor architecture. Each task T_i in T is characterized by a processing time P_i, arrival time A_i, earliest start time or ready time S_i and a deadline D_i. Coping with tasks' readiness and deadlines simultaneously without preemption on a uniprocessor system has been shown to be *NP-hard*. The scheduling problem is further exacerbated by the unpredictable nature of the sporadic tasks. We selected this problem to evaluate our algorithms' on-line optimization capabilities in domains where simple conventional algorithms (e.g. Earliest-Deadline-First) fail to provide the necessary deadline compliance guarantees [SSN+95].

Our objectives can be formulated as follows:

1) Minimize (to zero) the number of scheduled-and-missed tasks: that is the set of tasks which were predicted to be executed by their deadline but they missed it due to the scheduling overhead.

2) Minimize the scheduling cost.

3) Maximize the deadline compliance or hit ratio. That is the the ratio of tasks which were predicted to be executed by their deadline and they, indeed, met their deadline.

Real-time task scheduling can be regarded as the problem of searching for a permutation of a set of given tasks T such that once executed, they are guaranteed to meet their associated deadlines. Schedules all of whose tasks meet their deadlines when they are executed are referred to as feasible schedules. An example of the search space of partial schedules corresponding to a task set T is shown in figure 1 for a set of four tasks $T = \{T_1, T_2, T_3, T_4\}$. The nodes $\{v_i\} \in V$ in the search space $G(V,E)$ of this problem, represent partial schedules of T and the edges $(v_i, v_j) \in E$ represent transformation functions that extend the partial schedule at one end of the edge by a task T_i. Thus feasible (complete) schedules, if they exist, will be at the leaf nodes of such a search space.

Time Controlled Dynamic Scheduling

Time Controlled Dynamic Scheduling (TCDS) can be regarded as an on-line optimization technique like SARTS which uses a novel parameter tuning and prediction technique to determine the time of a search period. The allocated time to a scheduling period in this algorithm is self-adjusted based on what is known on-line about the nature of the problem and the problem instance. The algorithm continually self-adjusts the search-time based on parameters such as slack[1] and arrival rate or a combination thereof. The larger the slack, the greater is the time allocated to the search process. Also, the lower the arrival rate, the greater is the time allocated to the scheduling process.

An iteration of TCDS within a scheduling period is similar to an iteration of the branch-and-bound algorithm. During one iteration of a scheduling period j, a partial feasible schedule of the set of arrived tasks is developed as follows. The most promising node is removed from a list of candidate nodes (i.e. the open list) in the search space. The immediate children of the removed node are then generated (expanding a node). The decision regarding the feasibility of the partial schedule represented by each child node is evaluated (see following sub-sections) and the feasible child nodes are added to the open list. If a heuristic exists based on which to prioritize the nodes on the open list, the search algorithm will use it to sort the new list with the most promising node in front of the list. A partial schedule implies the possibility of scheduling only a fraction of the arrived tasks. The remaining unscheduled tasks are either rejected (if their deadlines are predicted to be missed) or are postponed to be scheduled in the next scheduling period (if they are not ready yet). Each scheduling period is terminated when there are no more partial or complete feasible schedules to examine or when allocated time to that particular period runs out.

The allocated time of scheduling period j is controlled by different criteria in different versions of the algorithm. One such criterion is:

$$T_s(j) \leq Min\left[Slack(T_l) \,|\, T_l \in Batch(j)\right] \quad \textbf{(SC1)}$$

where $Batch(j)$ denotes the set of tasks considered in scheduling period j. $Slack(T_l)$ is defined as the maximum time by which the start of T_l's execution can be delayed from the beginning of scheduling period j or S_l, without missing its deadline D_l. This criterion is aimed at putting an upper bound on the amount of time allocated to scheduling period j such that none of the deadlines of tasks in the current batch are violated due to scheduling cost. Another

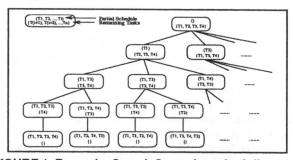

FIGURE 1. Example: Search Space for scheduling

$$T = \{T_1, T_2, T_3, T_4\}$$

---

1. The slack of a task is defined as the maximum time by which the execution of a task can be delayed without violating the task's deadline.

stopping criterion is:

$$T_s(j) \leq \frac{k}{\lambda} \quad \textbf{(SC2)}$$

This criterion is aimed at stopping a scheduling period early in bursty arrivals, in order to account for incoming tasks soon after their arrivals. In the expression of SC2, λ denotes the task arrival rate and k is a coefficient that can be set to control the average number of task arrivals during a scheduling period. Under low arrival rates, SC2 will allow longer scheduling periods to optimize the tasks in the current batch and to allow a reasonable number of tasks to arrive to form the next batch. A criterion using a combination of the previous criteria (i.e. **SC1** and **SC2**) was also designed for another version of TCDS:

$$T_s(j) \leq Min \,(SC1 \,, SC2) \quad \textbf{(SC3)}$$

At the end of the scheduling period, the partial feasible schedule found during the period is delivered to the ready queue for execution. If the period is terminated, due to failure to extend further a feasible partial schedule (say F) to include more tasks, F is delivered for execution as the outcome of scheduling period j.

TCDS employs feasibility tests which are capable of showing whether or not adding the next task to the current partial schedule will lead to a feasible schedule. The feasibility tests also facilitate pruning of the search space, in order to cut down the search complexity and to guide the search process towards plausible (i.e. feasible) paths. The feasibility test takes into account the scheduling time of a period, as well as the current time, deadline, earliest start time and processing time of tasks. Accounting for the scheduling time in the feasibility test ensures that no task will miss its deadline due to the use of resources for scheduling. The test is designed to make sure that a task is feasible at the end of the scheduling period, as well as at the time at which the feasibility test is performed.

IF ($t_c + Rem\_T_s(j) + Start_l \leq S_l$)
THEN F' is **infeasible**
ELSE IF ($t_c + Rem\_T_s(j) + End_l \leq D_l$)
THEN F' is **feasible**
ELSE F' is **infeasible**

FIGURE 2. Feasibility test of TCDS

The decision for adding a task T_l to the current feasible partial schedule F to obtain partial schedule F' in scheduling period j is performed through the feasibility test shown in figure 2. In the test, t_c denotes the current time (or the time at which T_l was scheduled). $Rem\_T_s(j)$ denotes the remaining time of scheduling period j, obtained by deducting the scheduling-time spent sofar from the

allocated scheduling-time to period j. $Start_l$ denotes the scheduled start time of task T_l relative to other tasks in F'. That is the time at which T_l starts its execution providing that the tasks preceeding T_l in the partial schedule F' were executed. End_l denotes the scheduled finish time of task T_l (i.e. $End_l = Start_l + P_l$).

Next, we provide a correctness proof for TCDS. It shows that our objective in minimizing to zero the number of scheduled and missed tasks has been met.

Theorem: The real-time tasks scheduled by TCDS are guaranteed to meet their deadlines, once executed.
Proof: The proof is done by contradiction. Let us assume that a task $T_l \in Batch(j)$, is scheduled during the jth period but once delivered for execution, it misses its deadline. This assumption leads to the following condition: $Delivery(j) + End_l > D_l$ (1). $Delivery(j)$ is the time at which the schedule of period j is delivered for execution. On the other hand, TCDS's bound on scheduling-time allocated to each period ensures that: $Delivery(j) \leq t_c + Rem\_T_s(j)$ (2), where $Rem\_T_s(j)$ is the remaining scheduling-time, once task T_l has been scheduled at time t_c. Thus, combining (1) and (2) leads to: $t_c + Rem\_T_s(j) + End_l > D_l$ (3). The feasibility test (figure 3) performed at time t_c ensures that: $t_c + Rem\_T_s(j) + End_l \leq D_l$, contradicting inequality (3). Therefore, our assumption regarding deadline violation of T_l is false, which concludes the proof of the correctness theorem $\square$.

Note that, according to the specified feasibility test, we mark tasks whose earliest start times are later than their scheduled start time as infeasible. The scheduling of these tasks is postponed until later. The alternative to these decision is to schedule these tasks by introducing delays into the schedule to execute them later when their earliest start times are honored. Such a version of the algorithm is presented in the following section.

Scheduling During the Idle Times

Another version of TCDS, namely TCDS-I, considers the alternative of scheduling with inserted idle-times. TCDS-I consists of relaxing the feasibility test in figure 2 from testing the tasks' readiness (figure 3). The tasks which will succeed the deadline test, will be scheduled to start their execution at their ready time.

IF ($t_c + Rem\_T_s(j) + End_l \leq D_l$)
THEN F' is **feasible**
ELSE F' is **infeasible**

FIGURE 3. Feasibility test of TCDS-I

Thus, the start-time $SWIT_l$ (Start With Idle-Time) of task T_l is: $SWIT_l = Max(S_l, Start_l)$. The advantage of this scheduling strategy is in decreasing the amount of backtracking since few dead-end paths are expected to be encountered using this test. The drawback however, is in introducing delays in the schedule which potentially reduce the CPU utilization. To increase utilization, TCDS-I performs scheduling during the idle periods. The allocated time of scheduling period j is controlled in TCDS-I by the following criterion:

$$T_s(j) \leq [\text{IF } idle\text{-}time \leq L \text{ THEN SC3 ELSE } idle\text{-}time] \text{ (SC4)}$$

Thus the stopping criterion adopted in TCDS (**SC3**) is also applied to TCDS-I. But TCDS-I combines with it another criterion based on the idle-time available in the schedule. Such criterion allows the scheduler to deliver a solution once a task scheduled in previous periods becomes ready for execution. If the algorithm encounters a big enough gap (i.e L units of time or longer) in the current schedule, it switches back to scheduling. In the absence of idle-time gaps in the schedule, TCDS and TCDS-I perform similarly. Finally, we note that the TCDS-I strategy is among the first attempts to take advantage of CPU idle-times in a schedule to perform scheduling.

Experimental Evaluation

In this section, we evaluate the performance of the TCDS algorithms and discuss the results of comparing these algorithms with a set of limited backtracking search algorithms similar to those reported in [ZR87]. These algorithms were selected for comparison with the TCDS algorithms, because they too are based on a search process. The limited backtracking algorithms use the following approach. Assuming no backtracking, the algorithm follows a single path in the solution space attempting to reach a leaf node that schedules all tasks feasibly. If such a leaf node is reached, the algorithm announces success and delivers the tasks for execution. If, on the other hand, the single examined path is deemed infeasible, the algorithm discards all the tasks in the current batch and attempts to schedule tasks in a new batch. With one level of backtracking, the algorithm will have another chance to explore other schedules and so on. These algorithms ignore the effect of scheduling costs on schedule quality and on delivery of feasible schedules as the level of backtracking in these algorithms is a fixed parameter. Some of the tasks scheduled by the limited backtracking algorithms may have missed their deadline by the time they are submitted for execution.

In the experiments, a Poisson job arrival process was used to simulate a natural sequence of sporadic task arrivals. The time window t, within which arrivals are observed, was set to 2000 time units. The arrival rate λ

ranged from 0.1 to 0.5. We modelled the availability of the execution context of a task T_i by that task's earliest start time S_i. S_i's are assigned a value, selected with uniform probability, from the interval (A_i, S_{max}), where $S_{max} = A_i + M^{K_s}$. M^{K_s} is a parameter used to tune the degree to which the S_i's and A_i's are set apart. We chose 3 for M and 7 for K_s as the values of these parameters for our experiments.

The processing times P_i of tasks T_i are uniformly distributed within the interval between 1 and 50 time units. Deadlines D_i are uniformly distributed in the interval (End_i, D_{max}) where End_i is the finish time of T_i assuming it starts at S_i (i.e. $End_i = S_i + P_i$), and D_{max} is a maximum value a deadline can have and is calculated as $D_{max} = End_i + SF^{K_d}$. SF^{K_d} is a parameter that controls the degree of laxity in task deadlines. K_d is fixed to 5 in our experiments whereas the Slack Factor SF[1] ranges from 1 to 10. Larger SF values represent larger slack, whereas small SF values represent tight deadlines. Another one of our algorithm parameters is the constant coefficient k of the term k/λ in the stopping criterion of TCDS (inequality SC2). This parameter implies the expected task batch size for each scheduling period of TCDS and was set to 5 for the experiments. The parameter L in the feasibility test of TCDS-I (figure 3) was set to Zero. This signifies that in our experiments, we use any idle-time gap, for scheduling. In practice, a large enough value for L can be chosen to account for context-switch times and other overheads associated with switching between task execution and scheduling. Finally, the degree of backtracking in the limited-backtracking algorithms constitutes another parameter of our experiments. The value of this parameter was fixed to 2. This level of backtracking was shown to perform well among the other levels for a wide range of parameter values selected in our experiments.

The metric of performance in our experiments was chosen to be deadline compliance. Deadline compliance or hit ratio measures the percentage of tasks which have completed their execution by their deadline.

The entire performance evaluation consists of two major experiments. The first experiment compares the TCDS and the limited-backtracking algorithm. The second experiment investigates the performance of TCDS-I versus TCDS. The remainder of this section discusses the results of these experiments.

Comparison of Algorithm Performance

In this section, we present the results of our experiments in which we compared the performance of the

---

1. In the figures, the terms "laxity" and SF are used interchangeably.

TCDS algorithm with the combination criterion (SC3) and the limited-backtracking algorithm with backtracking level 2 (referred to as BT-2). Figures 4 and 5 show the performance of the two algorithms in terms of the percentage of the task deadlines that were met. As is shown in the figures, TCDS outperforms BT-2 under all parameter configurations.

Figure 4 shows the results as the arrival rate varies for an average laxity value of 6. As is evident from this figure, the gap between TCDS and BT-2 widens by as much as 20% as the arrival rate increases.

Figure 5 shows the results as the degree of laxity varies for arrival rate values 0.1. This figure shows again, that TCDS outperforms BT-2 by wide margins, particularly as the degree of laxity increases. This is due to the fact that as the degree of laxity increases, the scheduling time of BT-2 may increase dramatically due to the fact that now 2 levels of backtracking may mean very long scheduling times.

Performance of TCDS-I Algorithm

In this experiment, we compared the performance of TCDS-I algorithm versus the performance of the original algorithm TCDS. TCDS-I is a modified version of TCDS where the stopping criterion is SC4 and the feasibility test is the one shown in figure 3. By this experiment, we wanted to study the gain of performance in relaxing the feasibility test from checking the tasks' readiness, while allowing the scheduling process to take place during the inserted idle-times. Figures 6 and 7 show the results of this experiment. Figure 6 shows the results of the comparison when the degree of laxity is fixed to 6. In this figure, the balance between tasks' laxities and idle-time gaps inserted in the schedules is favourable to TCDS-I policy.

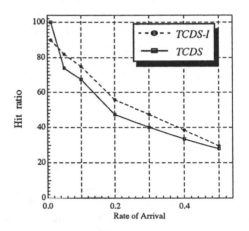

FIGURE 6. Deadline compliance (SF = 6)

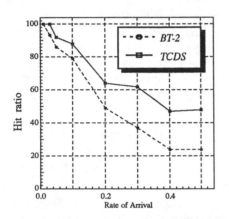

FIGURE 4. Deadline compliance (SF=6)

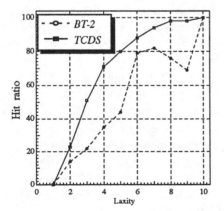

FIGURE 5. Deadline compliance (λ = 0.1)

Figure 7 shows the results of the comparison when the arrival-rate is fixed to 0.1. An interesting crossing at laxity value 5 shows the balance between the idle-times inserted in the partial schedules by TCDS-I and the tasks' laxities. Again for smaller degrees of laxity we can see that TCDS dominates performance due to TCDS-I's small idle-time gaps. We notice, however, that TCDS-I outperforms TCDS for higher degrees of laxity, as the looseness in the deadlines create longer scheduling gaps for TCDS-I.

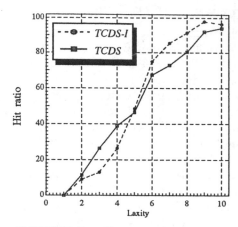

FIGURE 7. Deadline compliance (λ =0.1)

From this experiment, we identified situations where TCDS-I improves the performance of the original algorithm TCDS (see figures 6 and 7). Such situations correspond to less constrained environments (high laxity and low arrival rate).

Conclusion

In this paper, we have proposed a set of dynamic real-time scheduling algorithms called Time-Controlled Dynamic Scheduling (TCDS) based on the SARTS real-time search algorithms. TCDS was designed to explicitly address a fundamental contradiction in dynamic scheduling, namely the trade-off between the time allocated to scheduling and the quality of the resulting schedules. From the results of our experiments we conclude that effective stopping criteria can be designed to adapt the duration of scheduling periods automatically, in order to obtain high deadline compliance. The experiments show that the deadline compliance of TCDS is high compared to other existing algorithms under a wide range of parameter values for a number of parameters. The results also show that introducing idle-time intervals in the schedule can result in improving deadline compliance, if scheduling is performed within the idle-times.

As part of our future research, we plan to investigate the effect of different search heuristics on the performance of our algorithms. We also plan to extend our work to more complicated architectures.

References

[ZR87]. W. Zhao and K. Ramamritham, "Simple ad Integrated Heuristic Algorithms for Scheduling Tasks with Time and Resource Constraints", Journal of Systems and Software, 1987.

[DB88]. T.L. Dean and M. Boddy, "An Analysis of Time-Dependent Planning", In Proceedings of the Seventh National Conference on Artificial Intelligence, pp. 49-54, 1988.

[Kor90]. R.E. Korf, "Real-Time Heuristic Search", Artificial Intelligence vol.42, pp.189-211, March 1990

[RW91]. S. Russel and E. Wefald, "Do the Right Thing", MIT Press, Cambridge, MA, 1991.

[SH93]. S. Shekhar and B. Hamidzadeh, "Self-Adjusting Real-Time Search: A Summary of Results", Proceeding of the IEEE conf. on Tools for Artificial Intelligence, 1993.

[HS95]. B. Hamidzadeh and S. Shekhar, "Deadline Compliance, Predictability and On-Line Optimization In Real-Time Problem Solving", Proceeding of the International Joint Conference on Artificial Intelligence, vol.1, pp. 220-226, 1995.

[SSN+95]. J.A. Stankovic, M. Spuri, M.D. Natale and G.C. Buttazzo, "Implications of Classical Scheduling Results for Real-Time Systems", IEEE Computer, pp. 16-25, June 1995.

[JSM91]. K. Jeffay, D.F. Stanat and C.U. Martel, "On Non-Preemptive Scheduling of Periodic and Sporadic Tasks", Proc. of the 12th IEEE Real-Time Systems Symposium, pp.129-38, 1991.

[SZ92]. K. Schwan and H. Zhou, "Dynamic Scheduling of Hard Real-Time Tasks and Real-Time Threads", IEEE Transactions on Software Engineering, August 1992.

[ZRS87a]. W. Zhao, K. Ramamritham, and J.A. Stankovic, "Preemptive Scheduling Under Time and Resource Constraints", IEEE Transactions on Computers, August 1987.

[ZRS87b]. W. Zhao, K. Ramamritham and J.A. Stankovic, "Scheduling Tasks with Resource Requirements in Hard Real-Time Systems", IEEE Transactions on Software Engineering, vol. SE-12, no. 5, 1987.

Practical Applications

APPLICATION-SPECIFIC CONFIGURATION OF TELECOMMUNICATION SYSTEMS

Andreas Böhm and **Stefan Uellner**
Deutsche Telekom AG, Technology Centre
Am Kavalleriesand 3, D-64295 Darmstadt, Germany
Email: (boehm, uellner)@fz.telekom.de

ABSTRACT

Telecommunication systems are evolving rapidly. The need for new configuration systems in the telecommunication area matches well with advances in the AI domain, especially in the configuration area.
This paper outlines the requirements a new knowledge-based configuration system will have to fulfil to configure complete and ambitious telecommunication systems. It describes the configuration process, embedded in the procurement process, the kind of knowledge needed and the interaction between the configuration system, the operator and the potential user (customer). Furthermore, we discuss our experimental configuration system KIKon.

1. INTRODUCTION

During the last few decades most telecommunication services have been based on plain old telephone systems. Neither hardware nor software technology advanced rapidly in the field of telecommunication. Special features were provided by introducing new user facilities in private branch exchanges or terminal equipment. Different services (e.g. fax) were related to their corresponding terminals. Therefore, telecommunication simply consisted of putting together different services and their corresponding terminal equipment.

New services like mobile communication or even telefax indicate changes. The combination of telephony, data and mobile communication opens up new user groups. New applications in the telecommunication field come into sight. This development causes a demand for individually configured telecommunication scenarios built up from existing services, hardware and software. This process of creating a specific solution for the individual customer has to be supported by software tools which help analyse customer needs and configure the system.

In this paper we will first briefly discuss existing systems used to configure telecommunication systems (chap.2). Secondly the paper describes the requirements to be met by a knowledge-based configurator, which automatically generates a telecommunication solution according to customer needs (chap.3,4). Finally we outline the experimental configuration system KIKon (chap.5).

2. SUPPORT SYSTEMS FOR THE CONFIGURATION OF TELECOMMUNICATION SYSTEMS

In 1988 Deutsche Telekom introduced ExTel. This is a support tool for the analysis and configuration of simple telecommunication systems used by private persons and small companies. ExTel was implemented in PROLOG using SINIX and to be used via a graphical user interface. In 1990 every "Telefonladen" (point of sale for private customers) of Deutsche Telekom was equipped with this tool. With ExTel the user requirements were evaluated in a dialogue between the customer and an operator (e.g. number of telephones, type,...). The operator chose components deemed to match and fulfil the requirements of the customer. Afterwards ExTel checked whether the chosen components fitted together and, if not, gave hints for changing the configuration so that a usable telecommunication solution was built. Some often used solutions were integrated into ExTel. This system was not accepted very well by the operating staff, since the system did not work fast enough and the operator already had the solution in mind, while having to feed the system with all the data. After redesigning ExTel, a new system called VIBS was generated running on PCs under MS Windows, having no graphical user interface but pick lists to select terminals and services easily and fast. ExTel and VIBS are rule-based systems with all the problems of knowledge maintenance of such systems. The scenarios which can be configured with these systems are very limited in size and the type of available components.

Another configuration system used by Deutsche Telekom is IKU-CAS, which is employed for the configuration of private branch exchanges. IKU-CAS only checks whether the configuration chosen by the operator is a usable solution or not.

3. THE DOMAIN KNOWLEDGE

The domain knowledge, which has to be taken into account for the configuration task, depends on the intended capabilities of the system. The following knowledge areas are identified [EOUB 94]: Terminal equipment like telephones or fax machines, PBXs, services like ISDN or a pager service, communication software and hardware and related computer platforms, networks, and interfaces. The knowledge of these fields is of different reliability and it is recommended to develop effective strategies in order to update the knowledge and keep it consistent. The intended configuration solutions embrace scenarios ranging from

- a private customer who only wants the telephone service,
- a free-lancer who likes to have a telephone, an answering machine and a fax and likes to do telebanking from his PC,
- a lawyer office with different work sites and all the typical telecommunication infrastructures perhaps integrated in a computer network to allow online services and fax management, to
- a software company wishing to enable teleworking (e-mail, file transfer, application sharing) for its employees at home and to connect the salespersons to its computer network.

4. FROM USER REQUIREMENTS TO A TELE-COMMUNICATION SOLUTION

The different phases

There is an increasing demand for support for the configuration of telecommunication solutions. In the following we will analyse the complete process of advice in order to specify the requirements to be met by a new configuration system. We identified four phases from retrieving user requirements to presenting a final solution:

- Introduction. During the introduction phase the first contact to the customer is established. For this phase we only expect a minor contribution of the configuration system.
- Specification. In the second phase the identification and specification of the requirements is carried out. We assume the contact to have been initiated by the customer. Therefore, the system is able to start with some customer requirements. These requirements have to be checked for completeness and consistency. An interactive user interface helps obtain the missing information from the customer.
- Solution. With this information the system should be able to develop a configuration and present it to the advisor or the customer. Furthermore, the system should also present alternatives with an increased functionality or more comfortable features and alternatives that only in part fulfil the requirements in order not to exceed a given cost limit. The alternatives

should be presented to the customer in a way that he is able to compare these alternatives with respect to technical and economical aspects. In case of incomplete requirements or unsatisfactory results, there may be multiple iterations through phases two and three.
- Completion. If an acceptable solution is reached, phase four leads to an offer. In case of a positive decision, the system may prepare for ordering the components and for applying for the necessary services and optional value-added services including the appropriate service characteristics.

The requirements specification phase

Phases two and three represent the central process of the configuration system. Therefore, we examine these phases in more detail. Phase two, the specification of the configuration task, may be classified as follows:

- New configuration - There is no existing infrastructure which has to be taken into account. Consequently, there are no restrictions for the configuration.
- Substitution - There are no changes in the requirements, but possibly the new configuration provides a more comfortable solution or lower running costs.
- Modification - There are changes in the requirements and there exists an infrastructure to be taken into account. This infrastructure can be used in the new configuration but it can also be rejected.

In all three cases, a set of requirements has to be analysed being the input to the configuration. Therefore, it must be expressed in an appropriate formalism. In general, we expect the following difficulties:

- Requirements are expressed in a colloquial form.
- Requirements are incomplete.
- Requirements cannot be fulfilled.
- Requirements have to take into account existing or wanted infrastructures.

These points may give rise to two main questions:

1. What kind of formalism is suitable to represent the user requirements?
2. By which means can the non-formal requirements be transformed into such a representation?

One possible representation may be the description of the requirements (or communication functions) as an aggregation of several basic functions. These basic functions describe elementary communication relations (e.g. transmit, receive, remote processing). Their related contents are audio, video, data etc.. For further characterisation of communication functions, more attributes and relations are needed (e.g. quality, bandwidth, security, synchronism, description of relations to applications).

The difficulty of developing a formalism for the representation arises from the question whether the used language is powerful enough to describe all requirements. Moreover, the services to be configured would have to be described in the same language as the requirements. The advantage of such a formalism would be that, for describing the requirements, one needs not to know anything about the existing services. Furthermore, such a general formalism is capable of creating and expressing new value-added services very fast.

An alternative approach is the description of the requirements by a set of services, value-added services and the desired user interface. In addition, the requirements can be expressed by means of terminal equipment that supports the required services and satisfies the requirements of the user interface. The terminal equipment could be divided into dedicated types like telephones and fax terminals and universal platforms like computers. The universal platforms are modular and contain software components, which define the user interface. The interface to other system components has to be described very carefully since system components may transfer some communication functions to other system components (e.g. send a fax through an application). These possibilities have to be described as additional characteristics of this specific component.

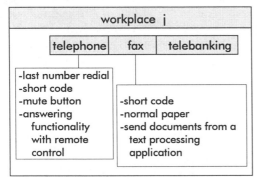

FIGURE 1 Requirements for one workplace

Existing or wanted infrastructure could easily be taken into account by means of this pragmatic approach. Furthermore, the advantage is the immediate access from the requirements to the configuration but it also has the disadvantage that detailed knowledge about services and components is needed during the specification phase.

Specifying the requirements as described above, an example workplace would look as depicted in figure 1. The requirements are expressed for different sites (workplaces) and only common server tasks are defined globally.

To summarise, the following information seems to be sufficient to express the local requirements:
- set of services (with characteristics) out of a pick list,
- requirements to be met by the user interface expressed in terms of the characteristics of the terminal equipment,
- existing and wanted infrastructure.

This second approach of representing the requirements also provides a solution to the problem of non-formal requirements. The possibilities of input are restricted to a selection of services and components held in the knowledge base including the set of attribute values. Consequently, the explicit appearance of non-formal requirements is avoided.

Unsatisfiable and incomplete requirements

Having transformed the imprecise colloquial requirements into formalised requirements, the question arises if these formalised requirements are complete and if they can be fulfilled by the configuration system. This will be decided during the process of configuration. According to the two cases (incomplete / unsatisfiable) the following strategies can be used to cope with the problem.

incomplete requirements:
- the system generates questions which are to be answered by the customer
- the system falls back on defaults for not completely specified details
- the system uses existing solutions from similar scenarios, which are stored in a case library

unsatisfiable requirements:
- the configuration process is cancelled
- the requirements are slightly modified and a new configuration process is initiated
- the system uses existing solutions from similar scenarios, which are stored in a case library

The configuration phase

This phase includes the execution of the configuration, the validation against the requirements and the presentation of the solution to the customer. The configuration system which has to be designed will have to fulfil the following requirements:
- configuration of a telecommunication scenario
- use of existing case libraries
- revision-oriented configuration
- optimisation with respect to costs, performance, reliability, security ...
- proposal of alternatives including more functionality and comfort
- deletion of singular requirements which cannot be fulfilled
- explanation of decisions taken by the configuration system
- provision of information on services and terminal equipment to the operator or customer

Figure 2 summarises the configuration system/process starting with the requirements of the customer, followed by an analysis yielding the formalised requirements which serve as the input to the configurator itself. The formalised requirements are presented graphically in order to enable an easy manipulation.

The knowledge base of the configurator consists of two categories of knowledge, a static and a dynamic one. Knowledge about the strategies is essential for the process of configuration. In combination with the knowledge about the customer and his business, the system chooses the appropriate methods, arranges the order of the steps of the configuration, and decides how to proceed in case of incomplete requirements. The static knowledge base also contains the domain model, which describes how to configure telecommunication scenarios, and knowledge about real components, services, and constraints with the possibility of up-dating from an external product data base.

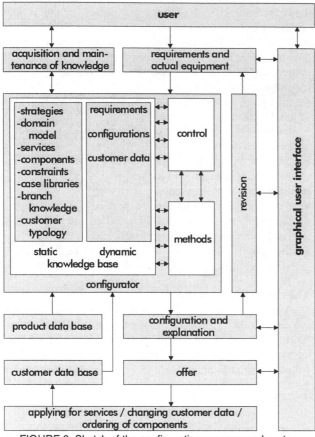

FIGURE 2 Sketch of the configuration process and system

The dynamic knowledge base includes the worked-on configuration(s), contains the (modified) requirements and holds some customer data coming from an external customer data base and / or from some information gained during the phase of the requirement analysis

The configurator builds up a configuration, presents it in a graphical way, and provides the user with explanations. In the case of an unsatisfactory result, the user can revise the configuration with modified requirements. Finally, the system presents an offer to the customer and may prepare to order the components, apply for the services, and modify the customer data in the customer data base.

5. THE EXPERIMENTAL SYSTEM KIKON

The KIKon project [EOUB94] started in 1994 to firstly evaluate the need for a knowledge-based configuration system within Deutsche Telekom for the customer-specific configuration of telecommunication services. This included the collection of information and the analysis of the structure of the related domain knowledge. Secondly it was necessary to find an appropriate

configuration method, which is linked to the choice of the knowledge representation. Then an experimental system had to be built in order to demonstrate the ability of the chosen configuration method within the regarded domain. Furthermore, we also had to focus on the user interface providing an easy input of the requirements, a clear presentation of the configuration results, and a convenient possibility of revising these results.

Domain model

The configuration took place in the telecommunication domain as defined in chapter 3. A taxonomic hierarchical order can be built up in terms of the above mentioned fields such as terminal equipment, PBXs, telecommunication services, communication hardware and software, computer platforms, and networks.

It is not the aim of this paper to discuss the entire domain model but this section will give a short overview of the used service concept. The main difference from other configuration domains lies in the telecommunication services. The general service concept defined for the ISDN [ITU88] distinguishes between bearer services provided by the telecommunication network and teleservices enabled through bearer services and specific terminal equipment (figure 3a).

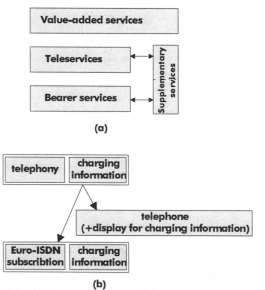

FIGURE 3 a) Scheme of telecommunication services b) example of partial configuration for a teleservice "telephony" provided in conjunction with a supplementary service "charging information" which requires a telephone enabled to display the charging information and an Euro-ISDN subscription which includes the bearer service for speech transmission.

These services are accompanied by what is termed supplementary services which can only be used in conjunction with other services (e.g. "calling line identification presentation"). The services "only" transmit the information. Additionally, value-added services, which require these transmission services, perform a further processing of the information (e.g. telebanking is a value-added service offered by most banks). This concept of different services is represented in our domain model.

At first sight, simple telecommunication scenarios are always similar in structure as for example: a telephone and a fax machine connected via a PBX to a terminal adapter. The functions "making a phone call" and "remote copying" are realised here by terminal equipment, a central "distributor" and the analog telephone service provided by the environment through a socket. But for more complex scenarios most of the sets of functionalities can be realised in quite different ways. This often implies a different use of the components with different connections or dependencies on other components. These connections and dependencies could be modelled differently. One possibility is to express them in a compositional and specialisation hierarchy outlining the possible configuration solutions, the other possibility is to follow a port concept, i.e. to introduce entities called resources for connecting components. The type of modelling to be used depends on the chosen configuration method and vice versa.

Configuration methods

No universal configuration method exists for the solution of general configuration tasks. The well known and often cited XCON system [McDermott82] is a rule-based system which was used by DEC to configure VAX-11/780 computers. But purely rule-based systems often carry problems with the knowledge base maintenance. A set of different configuration methods was developed during the last few years in order to tackle these problems [Tank93].

Skeleton-oriented methods as used in the configuration tools PLAKON [CGS91] and IDAX [Paulokat95] are useful in strongly structured domains, i.e. they are appropriate for the configuration of systems which have always the same basic structure. The domain model is represented by two different hierarchies: a composition (has-parts) and specialisation (is-a) hierarchy. The has-parts relation describes the functionality between the components. The possible configuration solutions are defined through these hierarchies. During the top down configuration process it is necessary to decide how to expand the configuration graph. This means either to specialise a component following the is-a hierarchy or split it up into its sub components, following the has-parts hierarchy, respectively.

The resource-oriented approach as realised in MOKON [SW91] or COSMOS [Heinrich91] is useful in domains structured more weakly. It is based on two different entities: components and resources. The relations between the components are expressed in resources supplied or consumed by the components. Configuration means to look for and add components which supply resources balancing the consumption of other components. The requirements are reflected by a distinct choice of components with a set of demanded resources. A configuration satisfies the requirements if no unbalanced consumed resource of any component remains. This approach

enables a more flexible way of configuring a solution because it is not necessary to define all possible structures in advance. But due to the lack of a predefined structure, it is not obvious which component of the whole scenario fulfils the resource demand of another component. On the other hand a concept of local balances leads to a kind of structure.

The skeleton- as well as the resource-oriented configuration methods are applied to our telecommunication domain. First results are briefly given in this chapter.

User interface

It is well known that the quality of the user interface essentially contributes to the acceptance of a new system. General requirements such as goal-oriented dialogues and presentations, interactive and user-driven order of the operations are necessary to achieve this acceptance. The user interface of KIKon realises two major tasks: (i) to specify the requirements and (ii) to present the configuration solution(s).

(i) There are different possibilities of specifying the requirements. One way is to choose a set of predefined functionalities. Another way is to select a set of components which fulfil the desired functionalities. The selection of components can take place directly through a named component as well as through the choice of a component type and a few determined attribute values. The higher the number of determined attribute values, the smaller the number of remaining components, and finally one component is selected. It is also possible to select several services in order to stimulate the direction in which the solution will be realised. The definition of different work sites with specified requirements structures the scenario. During the configuration process the configuration kernel chooses the components to satisfy the requirements. Some of the requirements are satisfied on the spot others through central "servers". A server could be a computer in a computer network or a PBX providing a central switching service for telephony, for example.

(ii) It is also the task of the user interface to present the configuration results or some alternative configurations to the user. The user chooses one of the alternatives and is able to modify the results, e.g. to change some components or add new requirements. This revision-oriented iterative procedure is quite an interactive way to generate a satisfying configuration.

Figure 4 roughly shows the described relations between user, user interface, configuration kernel and configuration result.

First results

Up to now (end of 1995) such an experimental system has been built up in co-operation with GMD [GMD] and we learned a lot about the modelling of this particular knowledge domain, the fitting configuration methods, and suitable user interfaces. As described above, the domain knowledge was expressed in taxonomic hierarchies. The definite representation depends on the configuration method. The configuration trials include a skeleton- and a resource-oriented configuration method.

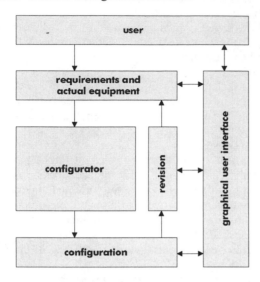

FIGURE 4 Architecture of the experimental system KIKon

The configuration kernel PLAKON [CGS91] was used to test a skeleton-oriented method. The knowledge was encoded in the PLAKON internal knowledge representation. We saw some problems of expressing the domain knowledge in such a representation and realising some strategies. Especially if there is a component with a multiple functionality which could replace two other components or there are components which can be used in conjuncture with different types of other components. PLAKON prefers a top down configuration but allows also a bottom up process. But the requirements which should be handled in KIKon may be quite different. It is conceivable that there is a mixture of functionalities, services and named components a solution for which it is not easy to find.

The resource-oriented method was tried by using COSMOS [Heinrich91] as a configuration kernel. The knowledge was encoded in the COSMOS internal knowledge representation. Due to the additional concept of resources, it is easier to model the domain knowledge and handle the above-mentioned problems of components with multiple functionalities and processing of mixed requirements. On the other hand there is a lack of predefined structures of possible configurations. These structures are modelled in a more implicit way through the supplied or consumed resources. Therefore, the modelling has to be done very carefully to avoid unwished configurations. The first results we obtained with this configuration method are promising for configuring more complex scenarios as well as handling more components.

6. CONCLUSION

We have outlined the primary difficulties experienced in designing a configuration system being able to configure telecommunication equipment and services with respect to customer/user specific demands and preferences. The different phases of the configuration process (introduction, specification, solution, completion) were described and examined. Two different methods for the description of formalised user requirements were discussed, and a proposal for a configuration system was made.

We introduced the experimental KIKon configuration system, which is used to study different configuration methods and strategies as well as the representation of the domain knowledge. First results for small to medium sized telecommunication systems are in favour of a resource-oriented approach.

We intended to enhance the configuration system using case bases and revision-oriented interactive methods in order to scale up the promising behaviour for the configuration of more complex scenarios.

ABBREVIATIONS AND ACRONYMS

AI   Artificial Intelligence
ExTel, IKU-CAS, VIBS configuration systems of Deutsche Telekom
ISDN   integrated services digital network
ITU   International Telecommunications Union
KIKon   Kundenindividuelle Konfiguration von Telekommunikations-Diensten (customer specific configuration of telecommunication services)
PBX   private branch exchange
SINIX   UNIX derivative of Siemens Nixdorf

REFERENCES

[McDermott82] J. McDermott, *AI* **19**, 39-88 (1982)

[Tank93]   W. Tank, *KI* **93**, 7, 1993

[CGS91]   R. Cunis, A. Günter, H. Strecker (Hrsg.) "Das PLAKON-Buch", *Springer, Informatik Fachberichte* 1991

[EOUB 94] Eusterbrock, Orth, Uellner, Böhm "Kundenindividuelle Konfiguration von Telekommunikations-Diensten", internal study, Deutsche Telekom research centre, 1994

[GMD]   Gesellschaft für Mathematik und Datenverarbeitung mbh, Birlinghoven, Darmstadt

[Heinrich91] M. Heinrich, in "Beiträge zum 5. Workshop 'Planen und Konfigurieren' ", Universität Hamburg 1991

[ITU88]   ITU-T Recommendation I.200, 1988

[Paulokat95] J. Paulokat, in Proceedings of "Expertensysteme 95" Kaiserslautern 1995,

[SW91]   B. Stein, J.Weiner, in "Beiträge zum 5. Workshop 'Planen und Konfigurieren' ", Universität Hamburg 1991

CHARACTER DESIGN BASED ON CONCEPT SPACE FORMATION

Takenao Ohkawa, Kaname Kakihara and Norihisa Komoda

Department of Information Systems Engineering
Faculty of Engineering, Osaka University
2-1, Yamadaoka, Suita, Osaka 565, Japan
Tel: +81-6-879-7826, Fax: +81-6-879-7827
E-mail: {ohkawa,kakihara,komoda}@ise.eng.osaka-u.ac.jp

ABSTRACT

The design method based on concept space formation is an approach to designing products from the viewpoint of individual impressions to the products. In this method, a concept space is formed for an individual by evaluating the images of several existing cases of design products on the individual in advance. By modifying an existing case located nearest to the specified image point on the concept space, a new product can be created.

This paper describes an application of the method to the image character design. The characters are represented using the cone connection models. To compare cases, the parts that compose a case are identified using the model pattern, which defines the typical meanings of the parts. A new character is created by applying the operation sequence consisting of several operations derived through comparison of the cases. We confirmed that a new character suited for individual impressions is successfully designed.

1 INTRODUCTION

The trend of products design is that the customers requirements are highly diversifying, and impressions or images of the products on an individual customer, as well as functions of them, are being considered as important design factors. The computer support of the products design that gives consideration to the customers' impressions is also demanded, and several methods that can deal with emotional factors of the customers have been developed[Nagam88]. In these methods, however, the relation between impressions or images of the products on the customers and physical or structural features of them is clarified averagely by enormous statistical data, which are collected by questionnaires to the customers. Therefore, they can only deal with the generalized impressions rather than the individual impressions.

The design method based on concept space formation[Nagat94], which we have developed, is an approach to designing products from the viewpoint of individual impressions to the products. In this method, a concept space is formed for an individual by evaluating the images of several existing cases of design products on the individual in advance. By modifying an existing case located nearest to the specified image on the concept space, a new product can be created. We have clarified some characteristics of the method through the application to design of simple pictures that are constructed with parts placed on the 5 × 5 points grid[Nagat95].

In this paper, we describe an application of the method to a more real target, namely the image character design. The character means a person or an animal in fiction or animations, etc., who is created by deforming a real person or animal. The characters are represented using the cone connection models. To evaluate the similarity between cases, the parts that compose a case are identified using the model pattern, which defines the typical meanings of the parts. A new character is created by applying the operation sequence consisting of several operations derived through comparison of the cases.

2 DESIGN METHOD BASED ON CONCEPT SPACE FORMATION

In our design method, the existing cases that have similar impressions to the specified image point on the concept space are retrieved and are modified in order to create a new product. An outline of the method is shown in Figure 1 and summarized as follows.

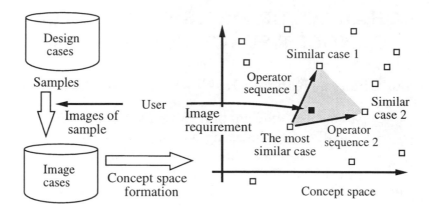

Figure 1: Design method based on concept space formation.

1. A user inputs his images of existing sample cases with some image attributes, and a concept space of the user is formed based on the input images of the sample cases using the principal component analysis.

2. The user inputs a demanded image of product to be designed with the same attributes that is used in the first step, and a point of the specified image on the concept space is calculated.

3. Three cases located around the point of the user's image are retrieved as similar cases, and the operator sequences, which are defined as the manipulations for transforming the most similar case to the other similar cases, are derived.

4. The most similar case is modified by applying the composition of the two operator sequences derived in the previous step. As a result, a new product suited for the user's specified image point is created.

For further details of the procedure of the method, see the reference[Nagat94, Nagat95].

3   REPRESENTATION OF CHARACTER

In the design method by concept space formation, the comparison and the transformation between cases plays a key role. Therefore, the character representing model can be compared or transformed in terms of an external form of the character, such as a skeleton, a physique, a body form, a posture, etc. The requirements for the model is summarized as follows.

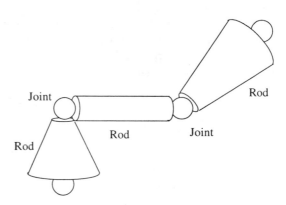

Figure 2: Cone connection model.

- It can represent parts composing the character as not only a set of pictures but a set of objects with meanings.

- It can represent characters which are compared each other and whose difference is calculated relatively.

- It can represent characters which are transformed each other.

We introduce the cone connection model as the character representing model. The cone connection model consists of cone-shaped rods and joints between rods as shown in Figure 2. The parameters which define rods and joints are shown in Table 1.

In order to compare between two cases composed with several parts, which imply sets of rods and joints with meanings, such as, the upper arm, the hand, etc., the model patterns of the character is prepared. The

Table 1: Data structure of rods and joints.

Type	Parameter	Definition
joint	axes	absolute direction of the joint
	position	absolute coordinate of the joint
	scale	absolute scale of the joint
rod	direction	relative direction of the rod
	scale	relative scale of the rod
	length	length of the rod
	top	radius of the top circle of the rod
	bottom	radius of the bottom circle of the rod
	offset	offset to the center of the circle
	joint	identifier of rod that joints the rod

Table 2: Classification of parts.

Group	Parts
Body group	(chest, belly, waist)
Head group	(head, neck)
Backbone group	(neck, chest, belly, waist, tail)
Arm group	(right upper arm, right lower arm, right hand), (left upper arm, left lower arm, left hand)
Leg group	(right upper leg, right lower leg, right foot), (left upper leg, left lower leg, left foot)
Tail group	(tail)

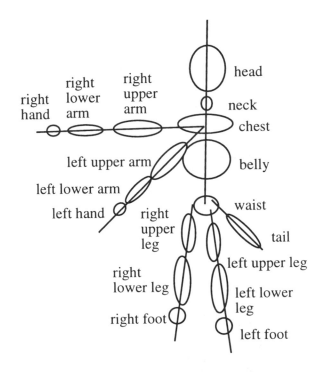

Figure 3: Model pattern of a character.

model pattern is illustrated in Figure 3. Comparing two cases, each part of one case is corresponded with a part of other case according to the model pattern. If a part is not specified with the model pattern, the part is regarded as a part that can be transformed discretely.

4 COMPARISON AND TRANSFORMATION OF CHARACTERS

4.1 Continuous transformation

If both of two parts that compose two cases severally are comparable under the model pattern, the transformational operation can be calculated easily as the difference of parameters of them. The calculation of the operation is performed for each of six groups, namely, the body group, the head group, the backbone group, the arm group, the leg group and the tail group, which are comprised of some parts according to Table2. Table 3 shows the comparison items between cases. For example, the value of the direction of the joint that connects each of four types of parts to the body is compared in two cases, and an operation is derived using the difference of the value. Figure 4 illustrates an example of the operation derivation by the continuous transformation.

4.2 Discrete transformation

Parts that are specific to a case may not appear in other cases. In this situation, the transformation of the part must be regarded as the discrete one, that is, whether it exists, or not. If the numbers of the same parts of two case are different, the discrete transfor-

Table 3: Comparison points between cases.

Case	Parts	Compared data
Connection to body	head, arms, legs, tail	direction
Direction	head, backbone, arms, legs, tail	direction
Skeleton	head, backbone, arms, legs, tail	length
Body form	head, backbone, arms, legs, tail	top,bottom,offset

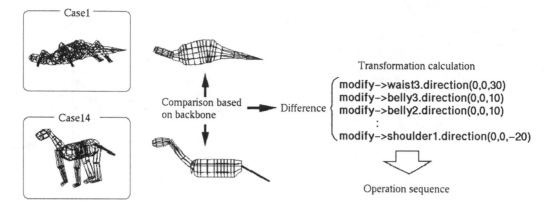

Figure 4: Comparison of parts and the derived operation.

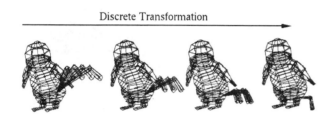

Figure 5: Discrete transformation.

mation should be also considered. The discrete transformation is performed according to the following two manners.

(1) If a part of the case is not registered on the model pattern, there is no way to compare it with the other case. In this case, the possible operation is addition of the part or deletion of the part. In case that there are more than one parts that are not registered, the addition or the deletion is considered one by one according to the distance on the concept space.

(2) If a part of the case is registered on the model pattern but the same parts exist in the case, the only one part is compared with a part of the other case and an operation sequence is derived using with continuous transformation. Both the derived operation sequence and the discrete operation (addition or deletion) are applied at the same time to the other parts.

Figure 5 illustrates an example of the discrete transformation.

5   CHARACTER DESIGN SYSTEM

We have developed the character design system on the Sun Workstation. The overview of the operation flow of the system is shown in Figure 6. The system can display the following three windows.

1. Data window(Figure 7):
   The user can input his impressions of the sampled cases by using sliders on the window.

2. Map window(Figure 8):
   The concept space of the user is formed and displayed with 2D map. When the user inputs the requirement using sliders, the specified image point appears on the map.

3. Transform window(Figure 9):
   The result of design and the similar three cases are displayed. You can also input your requirement on this window.

6   EXPERIMENTS

We made experiments for evaluating the character design based on concept formation. The procedure of the experiments is summarized as follows.

1. Present ten samples of the characters to four test subjects, and get the subjects to input their images of them using sliders. Construct the concept spaces of the subjects.

2. Extract three points on the formed concept spaces. Considering these points as the points of demanded images, create new characters.

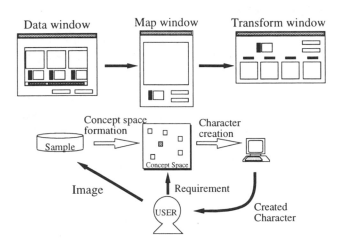

Figure 6: Overview of character design system.

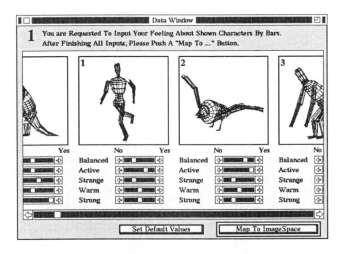

Figure 7: Data window.

3. Show the created characters to the subjects, and have them to answer the images of the characters using the attributes.

4. Locate the answered images on the concept spaces, and calculate the discrepancy between the points that is used to create new characters and the points of answered images.

The calculated discrepancy is regarded as an evaluation value. If the discrepancy is relatively close, we can judge that the method create the characters suited for images of the subjects. We use the distances between any two points of the sample characters on the concept spaces as the reference to evaluate the discrepancy.

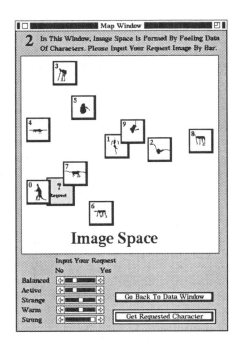

Figure 8: Map window.

Figure 10 illustrates a concept space of a subject. Figure 11 shows the result of experiments. The lines indicates the distribution density of the distance between any two points of the characters on the concept space, and the arrows imply the discrepancies between the random points and the user's points. Most of the discrepancies are smaller than the distance at maximum density. In addition, the discrepancies of one third of the created characters are closer than the minimum distance between cases.

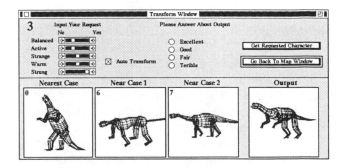

Figure 9: Transform window.

Figure 10: A concept space of a test subject.

REFERENCES

[Nagam88] M.Nagamachi, et al.: "Image Technology Based on Knowledge Engineering and its Application to Design Consultation," *Ergonomics International*, 88, pp.72-74 (1988).

[Nagat94] T.Nagata, T.Ohkawa and N.Komoda: Case-Based Evolutional Design by Primitive Operators, *Proc. 1994 IEEE Int. Conf. on Systems, Man and Cybernetics*, pp. 2492–2497 (1994).

[Nagat95] T.Nagata, T.Ohkawa and N.Komoda: Case-Based Transformational Design Method Based on Personal Specification, *Proc. IEEE Int. Symposium on Industrial Electronics*, pp. 924–929 (1995).

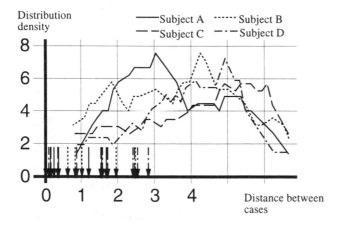

Figure 11: Evaluation based on distance between cases.

7   CONCLUSION

In this paper, we describe the application of the design method based on the concept space formation to the character design. The characters are represented using the cone connection models. To compare cases, the parts that compose a case are identified using the model pattern, which defines the meanings of the parts. A new character is created by applying the operation sequence synthesized with several operations which are derived through the comparison between the cases. We confirmed that a new character suited for a personal image is successfully designed by the system installed the method.

Customizing A* Heuristics for Network Routing

M. Hitz   and   T. A. Mueck

Abteilung Data Engineering
Universität Wien
Rathausstr. 19 / 4, 1010 Vienna, Austria
email: {hitz, mueck}@ifs.univie.ac.at

ABSTRACT

For a given network topology approach built upon the Cayley graph model, A* based routing heuristics are customized to achive fast and fault tolerant routing schemes using network degradation information and path caching.

In particular, the characteristic properties of a given network topology are used to optimize traversal heuristics representing the core parts of the routing algorithm. The configuration phase during network setup is controlled by simulated annealing. At runtime of the routing procedure, i.e., during network operation, the resulting heuristics are part of an A*-path finding algorithm which is executed on demand, e.g., if a standard network path is no longer operational or congested.

1. INTRODUCTION

The construction of network interconnection topologies is a main topic on the parallel and distributed processing research agenda ([1], [2], [3], [9], [11], [15]). With the commercial advent of massively parallel supercomputers, the interest in such proposals has further increased. A particular reseach direction in this area deals with a family of network topologies based on so-called *Cayley graphs* ([1], [6] , [7]). From a graph theoretical point of view, such topologies outperform standard multi-hop topologies (e.g., 2-dimensional mesh, binary hypercube, see [12]) which are current best practice mainly due to attractive routing algorithms compensating for their structural disadvantages.

Our approach is characterized by a generation and evaluation framework based on both, graph theoretical properties and efficient routing algorithms. In a first evaluation step, we select topologies with respect to expected shortest path length. This can be done either application independent, or, if possible, even considering an application dependent estimated communication pattern. This part of the approach which is not within the scope of this paper is described in [6], [7]. In a second evaluation step, we focus on routing heuristics to choose one of the candidate topologies with respect to routing considerations (preliminary results on these issues can be found in condensed form in [8]). The *combined* overall process to find an optimal configuration for a given number of nodes p and a given number of links per node k is outlined in Figure 1.

This paper deals with Step 3 of the overall schema. Technically speaking, we extract the characteristic properties (called *generators*) from particular candidate topologies called *best Forward-Backward* Cayley graphs. Subsequently, in an additional preprocessing step these properties are used to optimize routing heuristics for each

of the topologies. The objective function is the number of node expansions during A*-based path finding at runtime. The optimization procedure is based on simulated annealing.

The paper is organized as follows. In Section 2, we describe Cayley graphs in general and forward-backward Cayley graphs in particular, their properties and their relationship to permutation groups. Section 3 deals with adaptive fault tolerant routing in Forward-Backward-Cayley graphs in general. Heuristic A* path-finding at runtime is discussed in Section 4. Sections 5 and 6 elaborate on appropriate A* heuristics and their customization, respectively. The last part of the paper contains conclusions and references.

2. FORWARD - BACKWARD CAYLEY GRAPHS

This section contains a short outline of a group theoretical approach initially proposed in [1] and subsequently used in [6], [7].

In the following, $S_n \equiv (\Gamma, \bullet)$ with $\Gamma = \{ \pi_1, \pi_2, .. \pi_{n!} \}$ denotes the permutation group of order $n!$. As usual, a specific permutation $\pi \in \Gamma$ is depicted by a string of n digits, the identical permutation $123 .. n$ by I. For example, S_3 corresponds to the group ({ 123, 132, 213, 231, 312, 321 }, $\bullet$), $321 \bullet 231$ is an abbreviation for the composition of the two bijective functions π_1, π_2 to be resolved from right to left yielding 213 as the result.

An arbitrary subset $G = \{ g_1, g_2, .. g_k \}$ of $\Gamma \setminus I$ is called *generator set*, $g_i \in G$ is called *generator*. Informally, a generator set is used to produce a group by iterative application of the group operation $\bullet$ to a set of group members until the result is closed under $\bullet$. Prior to an exact definition of so called *algebraic generations* as used in the sequel, some notational conventions are needed.

Definition 1: Generating function, algebraic generation

Let $\wp(X)$ denote the powerset of X. In the sequel, the *generating function* for a generator set G, denoted by gen_G, is defined as $gen_G: \wp(\Gamma) \to \wp(\Gamma)$ with

$$gen_G = \begin{cases} \{I\} & \text{if } S = \varnothing \\ \{s \bullet g \mid (s \in S \land g \in G)\} & \text{otherwise} \end{cases}$$

Let F_G denote the least fixed-point of gen_G, in other words, F_G is given as the subset of Γ for which $(gen_G (F_G) = F_G \land \neg \exists\ F' \subset F_G: gen_G (F') = F')$ holds. The stepwise computation of F_G starting with $gen_G (\varnothing)$ is called *algebraic generation based on G* ∎

F_G always exists since S_n is a group and therefore $\bullet$ is closed in Γ. Representing the stepwise computation of the least fixed-point for a specific gen_G as a directed graph yields the so called *Cayley graph* (or *group graph*) framework. Informally, the nodes of such a Cayley graph correspond to the permutations yielded by the algebraic generation, whereas each edge corresponds to a particular generation step resulting from the application of a particular generator to a particular permutation (i.e., node).

Definition 2: Cayley graph

A *Cayley graph* corresponds to the interpretation of an algebraic generation yielding a particular F_G as a directed graph (V, E) such that
$V = F_G$ and $e(v_i, v_j) \in E \Leftrightarrow (\exists\ g \in G:\ v_i \bullet g = v_j)$ ∎

A number of well-known interconnection topologies like n-cubes, n-cube-connected cycles and n-star graphs represent special cases in the context of the general Cayley graph framework. A detailed account of the relationship between these special cases and the general Cayley graph approach can be found in [1].

In our application context a physical data transfer step from processing element v_i to processing element v_j is modeled by a node-to-node transition in a Cayley graph, say $v_i \bullet g = v_j$, with g denoting the physical transmission link to be used (e.g., a transmission from node 321 to node 213 is done via physical link 231).

Stating as a basic hardware/firmware assumption that standard communication links are bidirectional (see also Subsection 2.2), we have to focus on those Cayley graphs in which any two adjacent nodes are connected by both a forward and a corresponding backward edge, or, loosely speaking, on Cayley graphs resembling undirected graphs. In the sequel, a Cayley graph with this property is called *Forward-Backward* Cayley graph, or just *FBC* graph. For example, a generator set G_1 = {1243, 1324, 2134} produces a Cayley graph which is also an *FBC* graph, whereas G_2 = {321, 312} results in a Cayley graph which is not an *FBC* graph.

Due to node symmetry, any path, i.e., generator sequence $g_1 \bullet .. \bullet g_x$, between a pair of nodes v_i and v_j is algebraically equivalent to a path between $v_i^{-1} \bullet v_i$ and I. The immediate practical consequence of this theoretical observation is that arbitrary routing requests $v_i \rightarrow v_j$ can be remapped and, in turn, simplified by an algebraic transformation. In particular, an optimal route for $v_i \rightarrow v_j$ corresponds to a minimum length generator action sequence $v_i \bullet g_1 \bullet .. \bullet g_x = v_j$. Now, multiplying both sides of the equation from the left by the inverse of the target node v_j yields $v_j^{-1} \bullet v_i \bullet g_1 \bullet .. \bullet g_x = v_j^{-1} \bullet v_j = I$ which allows to map any routing request $v_i \rightarrow v_j$ to an equivalent request $v_j^{-1} \bullet v_i \rightarrow I$.

This kind of algebraic transformation is used in the context of our systematic evaluation of different routing parameters (see Section) and provides a significant reduction of the evaluation overhead.

3. FAULT TOLERANT DYNAMIC ROUTING

We focus on the processing load for routing in a given *FBC* graph, thus trying to design a routing algorithm which not only finds the shortest path to the destination, but also does so with a minimum amount of work.

In a first step, we present an overall description of a dynamic fault tolerant routing scheme. It is based on congestion information obtained at runtime. This kind of information is used by an A* pathfinding procedure (see next section for details) to be executed on demand at runtime.

Without assuming any stringent hard- or firmware prerequisites, we propose a routing procedure *requestRoute(message, start, goal)* which initiates routing of the given message from *start* to the target node *goal* in the graph defined by the global generator set G. Another global data structure, *taboo*, represents a set of node/link pairs containing all node/link combinations currently out of order which should thus be disregarded by the pathfinding procedure. *requestRoute* calls this procedure (called *SP* in the sequel, discussed in Section 4) which computes the shortest path to *goal* and, as a by-product, returns the subset *effTaboo* ⊆ *taboo* of those node/link pairs which proved to be relevant to the search process.

Both results of *SP* are stored in a fixed sized LRU path cache which is consulted upon the next routing request. The paths stored in this cache are invalidated whenever the topology undergoes a temporary change (see Figure 2).

When a node receives a message concerning the failure of a specific link (addressed as a node/link pair in our context), all paths passing through that link are deleted from the path cache. Conversely, when a formerly faulty link is reported to have recovered, those paths that were computed taking into account (and thus avoiding) this link are deleted from the cache, because they are considered suboptimal. In either case, the invalidated path is being recomputed when a corresponding routing request is issued the next time. The resulting routing scheme is summarized in Figure 2.

4. HEURISTIC PATH FINDING AT RUNTIME

Using an A-algorithm [13] (see Figure 3) to implement the path-finding function appearing in Figure 2, namely $SP((goal^{-1} \bullet start), path, effTaboo)$, it is possible to

---

1. generate a set C of candidate topologies considering particular p and k values;
2. compute a subset bC of C containing topologies with appealing graph theoretical characteristics, e.g., minimal average communication path length;
3. find the best topology in bC with respect to routing considerations as follows:
   for each c ∈ bC
       optimize path-finding heuristics using simulated annealing;
       select best topology w.r.t a minimum number of nodes visited during A* pathfinding

Figure 1: Overall generation and evaluation framework

introduce structural knowledge by means of appropriate *heuristics* which actually control (at least to a certain degree, see discussion below) the search process.

Upon visiting a node u during the search, all yet unvisited neighboring nodes are recorded in a priority queue called "open list" (this step will be called *expansion of u* in the sequel). The next node v to be examined is then extracted from this queue; it is always a node with *minimum total length d(v)* of a shortest path *start* $\rightarrow v \rightarrow$ *goal* from the starting node to the goal via node v . At this point, Dijkstra's standard single source shortest path algorithm (denoted DSS in the sequel) always selects a node v with minimum distance to *start* only. This distance $d(v)$ is obviously composed of $l(start, v)$ and $l(v, goal)$ both of which are unknown and must therefore be estimated at the time of decision. If we denote the corresponding estimating functions by $g(v)$ and $h(v)$, respectively, then d has the form $d(v) = g(v) + w \cdot h(v)$. While $g(v)$ is simply the length of the shortest path from *start* to v the search procedure has found so far, the heuristic $h(v)$ estimating the remaining path length $l(v, goal)$ to the goal represents the essence of the A-algorithm. w tunes the influence of the heuristic on the estimated total distance $d(v)$.

To be able to explore the effects of different heuristics and different weights, the A-algorithm given in Figure 3 has two additional input parameters (h and w) representing the search heuristic and its influence on the A-Algorithm, respectively. On the other hand, as *goal=I* holds in the context of our routing problem (see Subsection 2.2.3), no explicit parameter is necessary to select the target node. The aim is to find a reasonable heuristic h and an appropriate value for w for a given *FBC* graph which reduces CPU time with respect to DSS.

5. CONFIGURABLE ROUTING HEURISTICS FOR *FBC* GRAPHS

The basic idea upon which the construction of heuristics h for *FBC* graphs is founded is as follows: located at a certain permutation (i.e., node of an *FBC* graph) and trying to reach identity I, we inspect all adjacent permutations, some of which - by looking at their displacements - seem to be closer to I than others. The "close" ones are to be assigned a low value for h, while the "distant" ones should yield higher h values.

In an attempt to design easily computable heuristics estimating the remaining distance of a given permutation π to the identity I, we first experimented with two distance functions h_1 and h_2. h_1 counts the number of displacements, e.g. $h_1(\bar{3}21)=2$. It only roughly estimates the number of steps to reach I; in particular, it overestimates the distance if the generator set G contains an appropriate "powerful" permutation for the case at hand. For example, if $321 \in G$, then $l(321, I) = |321 \bullet 321|$ $= 1 < h_1(321)$. The same is true for h_2 which computes the sum of distances of each element to its position in I. Interestingly, in many cases h_2 turns out to be more effective than h_1. Exact definitions of these two distance functions are given as:

$$h_1(\pi) = \sum_{j=1}^{n}(1 - \delta_{\pi(j),j}) \qquad \text{(\# of displacements)} \\ (\delta .. \text{ Kronecker's delta})$$

$$h_2(\pi) = \sum_{j=1}^{n}|\pi(j) - j| \qquad \text{(sum of displacements)}$$

As an example, let us consider $p=120$ processors with $k=3$ links. Starting from the set of *best FBC graphs* for $p=120$

```
FUNCTION requestRoute (message, start, goal)
// Input:          Message to be transmitted, starting node start, target node goal
// Global data:    Generator set G, taboo set taboo, a cache of most recently used paths PC
// Local data:     Subset effTaboo of taboo containing node/link pairs that might have been
//                 part of an optimal route if not contained in taboo.
//                 List path representing the path to goal.
path ← queryPathCache(PC, start, goal);
if (path = ∅) {
     SP((goal⁻¹ • start), path, effTaboo);             // SP routes to I, c.f. Section 2.2.3.
     updatePathCache(start, goal, path, effTaboo); }
execRoute(message, path);                               // hard- and firmware dependent


FUNCTION processLinkStateMess (node, link, type)
// Input:          a node / link (i.e., generator) - pair and its new type ∈ { UP, DOWN }
// Global data:    path cache PC (see above)
if (type = DOWN)                                        // failure detected
     // delete all paths from PC passing through given link via given node:
     deletePathVia(PC, node, link);
else                                                    // type = UP
     // delete all paths from PC having the node/link pair in their effective taboo list:
     deletePathWithTabooContaining(PC, node, link);
```

Figure 2: Overall routing schema

```
FUNCTION SP (start, h, w, path, effTaboo) // finding shortest paths with A-Algorithms
// Input: Starting node start, heuristic h, weight w
// Output: path containing the shortest path from start to I,
//    effTaboo, the effective subset of the taboo list

// Global data: Generator set G for an FBC graph, set of defect node/link pairs taboo

// Mappings l:V→REAL (distance start → v), d:V→REAL (distance start → v → I),
//    and Predecessor:V→V (resulting path)

// Priority queue Open containing nodes already known but not yet expanded
// Set Closed of all nodes already expanded
// EffTaboo contains node/link pairs that are part of an optimal route if not contained in taboo

l(start) ← 0;                          // Initialization:
d(start) ← l(start) + w * h(start);
Predecessor(start) ← NONEXISTING;
Open ← insert(start, ∅, d(start));
Closed ← ∅;
EffTaboo ← ∅;

loop {
v ← first(Open);                       // Fetch node v with minimum d
if (v = I)                             // Target found
break;                                 // Exit loop;
Open ← tail(Open);                     // Remove v from Open
Closed ← Closed ∪ {v};
for each g ∈ G {
if (<v, g> ∈ Taboo)                    // Do not use g starting from v
    EffTaboo ← EffTaboo ∪ {<v,g>}
else {                                 // Expand v:
    s ← v • g;                         // Apply generator g to v
    l' ← l(v) + 1;                     // Compute new distances:
    d' ← l' + w * h(s);
    if (s ∉ Open) {
        if (s ∉ Closed) {// v is yet unknown
            l(s) ← l';
            d(s) ← d';
            Predecessor(s) ← v;
            Open ← insert(s, Open, d(s));
        } else {                       // v has already been expanded
            if (d' < d(s)) // Shortcut found, must re-expand v
                Open ← insert(s, Open, d(s));
        }
    }
    if (l' < l(s))                     // General update of distance mappings:
        l(s) ← l';
    if (d' < d(s)) {                   // Shortcut found
        d(s) ← d';
        Predecessor(s) ← v;            // Update current path
} } } }
path ← makePath(Predecessor);          // Transform mapping Predecessor into a list of routing steps
```

Figure 3 A-Algorithm

and $k=3$ links, i.e., from a set of graphs with identical graph theoretical properties (e.g., average shortest path length), we choose two arbitrary generator sets in order to test the heuristics h_1 and h_2.

The corresponding set of best FBC graphs for $n=5$ and $k=3$ contains 120 graphs with 120 nodes and average path length 4.748. Routing from each of the 119 nodes to I (we exclude $I \rightarrow I$), DSS expands 7259 nodes in absolute terms.

As we are dealing with best FBC graphs, this result holds regardless of the specific graph. The behavior of the A-algorithm when guided by the two heuristics defined above is shown in Table 1 (we identify graphs by their generator sets).

We define $N_C(w)$ as the number of nodes expanded by the above algorithm while searching all shortest paths in a given FBC graph C with a given weight w, compared to the

Graph	Heur.	w	N_C
13245 15432 21543	h_1	0.37	64.29%
	h_2	0.22	54.98%
12354 35142 43215	h_1	0.37	62.85%
	h_2	0.22	61.56%
13254 14325 21435	h_1	0.37	63.15%
	h_2	0.37	28.67%

Table 1: Routing characteristics using h_1 and h_2

respective number of nodes visited during a search with w=0. In other words, $N_C(w)$ is the ratio between the number of nodes expanded during heuristic search and the number of nodes expanded during standard DSS execution. The first two blocks in Table 1 show the performance of heuristics h_1 and h_2 applied to the graphs obtained by generator sets {13245, 15432, 21543} and {12354, 35142, 43215}, respectively, with w set to the largest weight yielding no deviation from the optimal path length. It turns out that h_2 is slightly superior to h_1 for these graphs, and that the second graph is clearly inferior to the first one with respect to routing.

The last block contains the *best* results achieved during an investigation of *all* 120 best *FBC* graphs with n=5 and k=3. The graph yielding minimum N_C (28.67%, i.e., only 2081 nodes are expanded) is generated by {13254, 14325, 21435}. It is interesting to see that, although the average shortest path length is the same for all best *FBC* graphs, their routing behavior may differ considerably. Different graphs C taken from the same family usually yield totally different values for N_C and w_{opt} because of the different properties of their generators[1].

Despite the considerable savings achieved so far, in order to further improve the heuristics found, we may observe that some displacements are more easily resolved than others and construct h_3 by assigning respective weights w_j to each position j:

$$h_3(\pi) = \sum_{j=1}^{n} w_j |\pi(j) - j|$$

(weighted sum of displacements)

Thus, h_3 is an attempt to configure h_2 for a given *FBC* graph. The optimal assignment of weights yields a multidimensional optimization problem, as the vector $\vec{w} = (w_1, w_2, ..., w_n)$ replaces the single independent variable w introduced before. The optimization process (the details of which are deferred to the next subsection) turns out to be a very time consuming process for n≥5. However, results show that it is indeed worth while employing such a "learning" heuristic h_3:

Graph	$\vec{w}$	N_C
13245 15432 21543	(0.47, 0.20, 0.23, 0.55, 0.22)	33.13%
12354 35142 43215	(0.21, 0.17, 0.22, 0.18, 0.21)	54.61%
13254 14325 21435	(0.15, 0.41, 0.41, 0.41, 0.41)	27.50%

Table 2: Routing characteristics using h_3

Table 1 shows the performance of heuristic h_3 applied to

---

1. Even more surprising, also graphs belonging to the same (structural) isomorphism class show dissimilar behavior with respect to routing.

the same graphs as before. The two lines in each block contain the result of the customized heuristic h_3, again under the constraint that the resulting routes have minimal length. We observe that h_3 outperforms the other two heuristics in all cases although the improvement of the best case is only minor (27.5% compared to 28.67%). These figures also justify the assumption that constraining E_C to zero does not make much difference.

The last family of heuristics we have investigated so far again represents a generalization of a former one: While in h_3 any displacement at position j was accounted for equally, in heuristic h_4, each possible displaced value $\pi(j)$ at position j is treated separately:

$$h_4(\pi) = \sum_{j=1}^{n} w_{\pi(j),j} (1 - \delta_{\pi(j),j})$$

$$= \sum_{j=1}^{n} w_{\pi(j),j} \quad \text{with} \quad w_{i,j} = 0$$

To customize this heuristics for a given *FBC* graph, the learning procedure has to assign $n \cdot (n-1)$ values to the weight matrix $w_{i,j}$ which models the specific effort to exchange j with $\pi(j)$. Before turning to the problem of determining an optimal assignment for $w_{i,j}$ in the next subsection, let us briefly present the results achieved with h_4.

For the graph {13254, 14325, 21435}, i.e., the one which yielded the best results together with h_3, the number of expanded nodes could be reduced to 1816 employing heuristic h_4 (=25.02% of DSS).

Elaborating on larger graphs, we choose {123465, 132546, 215436, 543216} with 720 nodes. DSS expands 259559 nodes when determining all 719 shortest paths to I, while the A-algorithm under heuristic h_4 only checked 18.8% or 49042 nodes. However, as we are dealing with a 30-dimensional optimization problem to find all $n \cdot (n-1)=30$ parameters for h_4 necessary to process a graph with n=6, we are by no means sure that we do operate with an optimal set of parameters. Indeed, we had to change our optimization method when we started out with h_4, as we will explain in the following subsection.

6. COMPUTING GRAPH-SPECIFIC CONFIGURATIONS

As explained in the previous section, when given a specific family of heuristics and a certain *FBC* graph, we are faced with the problem to select the optimal configuration for the particular combination of FBC graph and heuristics. w (in the case of h_1 and h_2) in an optimal way, the parameter vector $\vec{w}$ (in the case of h_3) or the parameter matrix $w_{i,j}$ (for h_4) for a certain graph. We started off with Powell's method ([4], [14]), which yielded reasonable results but soon suspected that we might get caught at local minima as the number of parameters was rising. To get an idea of the structure of our optimization problem, we fixed 10 out of 12 parameters in a problem with n=4, k=3, and h_4, varying only $w_{1,4}$ and $w_{2,3}$. In Figure 4, N_C is represented as a two-dimensional function $N_C(w_{1,4}, w_{2,3})$. As depicted in Figure 4, we find at least three valleys in which Powell is likely to get stuck. In this situation, switching to simulated annealing as an optimization procedure [5], [10] seems reasonable.

The main idea is to allow for some uphill movements from

time to time in order to give the system a chance to get away from a local minimum. The decision to go into the "wrong" direction is guided by a random oracle, whose probability to suggest such a move is directly connected with a monotonically decreasing parameter T (called "temperature") via the so-called Boltzmann probability distribution describing the distribution of energy states E in a system in thermal equilibrium at temperature T, that is $P(E) \sim exp(-E/(k \cdot T))$. As T goes down, oszillations of the configuration are "freezing", thus eventually yielding a final, hopefully optimal configuration.

In the case at hand, simulated annealing proves to work quite well. The results yielded were in all cases better than the ones found by Powell's method. Table 1 summarizes some examples to illustrate this fact:

Graph	N_C (Powell)	N_C (S. A.)
1243 1324 3412	38.13%	31.77%
13245 15432 21543	26.20%	25.02%
123465 132546 215436 543216	26.31%	18.80%

Table 3: Routing characteristics obtained by different methods using h_4

7. CONCLUSIONS

We focus on network routing algorithms which are able to deal with fault conditions in an efficient manner at runtime (see Section 3). Structurally promising Cayley graph topologies (see Section 2) are evaluated with respect to routing considerations. In this context, shortest paths are computed at runtime using a modified A*-algorithm based on a set of parameterized path-finding heuristics (see Section 4). For a given graph, these heuristics can be pre-configured by simulated annealing (see Sections 5 and 6).

The following results are shown:

- the generators of specific Cayley network topologies can be used to train routing heuristics yielding attractive A* behaviour;
- the proposed A*-algorithm is able to cope with network degradation in a reasonable way by maintaining a taboo list of network links believed to be out of operation;
- additionally, the routing algorithm is able to use a path

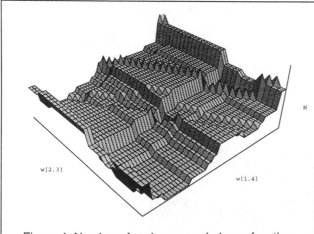

Figure 4: Number of nodes expanded as a function of two parameters

caching scheme to avoid extensive pathfinding overhead; as an optimization strategy for the above mentioned preconfiguration step (i.e., the training phase for the heuristics), simulated annealing is shown to be superior to conventional gradient methods like Powell's algorithm.

REFERENCES

[1] S.B. Akers, B. Krishnamurthy: A Group Theoretic Model for Symmetric Interconnection Networks; Proc. of the 1986 Int. Conf. on Parallel Processing; IEEE CS Press, 1986.

[2] J.-C. Bermond, C. Delorme, J.-J. Quisquater: Strategies for Interconnection Networks: Some Methods From Graph Theory; Journal of Parallel and Distributed Computing vol. 3/#4; Academic Press, 1986.

[3] L.N. Bhuyan: Interconnection Networks for Parallel and Distributed Processing; IEEE Computer Vol. 20/#6, IEEE CS Press, 1987.

[4] R.P. Brent: Algorithms for Minimization without Derivatives; Prentice-Hall, 1973.

[5] L. Davis: Genetic Algorithms and Simulated Annealing; Pitman, 1987.

[6] W. Gutjahr, M. Hitz, T.A. Mueck: Heuristics for the Quadratic Assignment Problem in Cayley Graphs, submitted for publication, 1995.

[7] M. Hitz, T.A. Mueck: Domain Dependent Evaluation of Cayley Graph Topologies; Proc. of the 6th ISCA Int. Conf. on Parallel and Distributed Computer Systems; Louisville 1993, pp. 228-235; ISCA; 1993.

[8] M. Hitz, T.A. Mueck: Routing Heuristics for Cayley Graph Topologies; Proc. of the 10th IEEE Conf. on Artificial Intelligence for Applications (CAIA-94); San Antonio 1994, pp. 474-477; IEEE Computer Society Press, 1994.

[9] M. Jerrum, S. Skyum: Families of fixed degree graphs for processor interconnection; IEEE Transactions on Computers Vol. C-33; IEEE CS Press, 1984.

[10] P.J.M. van Laarhoven, E.H.L. Aarts: Simulated Annealing: Theory and Applications; Mathematics and Its Applications, D. Reidel Publ. Comp., 1987.

[11] Q.M. Malluhi, M.A. Bayoumi: The Hierarchical Hypercube: A New Interconnection Topology for Massively Parallel Systems; IEEE Transactions on Parallel and Distributed Systems Vol. 5/#1; IEEE CS Press, 1994.

[12] L.M. Ni, P.K. McKinley: A Survey of Wormhole Routing Techniques in Direct Networks; IEEE Computer Vol. 26/#2; IEEE CS Press, 1993.

[13] N.J. Nilsson: Principles of Artificial Intelligence; Springer, 1982.

[14] W.H. Press, B.P. Flannery, S.A. Teukolsky, W.T. Vetterling: Numerical Recipes in C - The Art of Scientific Computing; Cambridge University Press, 1988.

[15] D.J. Pritchard, D.A. Nicole: Cube Connected Moebius Ladders: An Inherently Deadlock-Free Fixed Degree Network; IEEE Transactions on Parallel and Distributed Systems Vol. 4/#1; IEEE CS Press, 1993.

FEATURE SELECTION IN AUTOMATIC TRANSMISSION SHIFT QUALITY CLASSIFICATION

Robert Williams[1] and Yi Lu[2]

[1]GM Powertrain Group
General Motors Corporation

[2]Department of Electrical and Computer Engineering
The University of Michigan-Dearborn
Dearborn, MI 48128-1491
USA
yilu@umich.edu

ABSTRACT

We describe an automatic system for classifying the vehicle transimission shift quality. The system is developed by using pattern recognition methodolgy to classify three main types of shift quality; Firm, Smooth, and Run-Through based on torque trace features. The system uses a feedforward neural network for classification. The feature selection procedure consists of two major steps, engineering knowledge elicitation and encoding, and feature refinement based on statistical knowledge. The algorithm has applications in both transmission shift quality control systems and establishing shift quality test metrics.

1. INTRODUCTION

This paper describes our effort in applying pattern classification methods to vehicle transmission shift quality control. Transmission shift can be classified into three classes, firm, smooth and run-through. Different vehicle models may require different shift quality. For example, for a premier luxury vehicle it is important to produce an exceptionally smooth powertrain. However a firm shift, which has an increased magnitude of jerk, is acceptable by a driver who is in a sports/performance vehicle. It is important to develop an automatic algorithm that detects and classifies transmission shift quality in automotive manufacturing.

In general a pattern recognition problem consists of two major problems: feature selection and representation, and pattern classification. Figure 1 illustrates our transmission shift quality system in which a feedforward neural network

classifier is used at the classification stage. Since neural network is a well-known classification technique [AdH95], the emphasis of this paper is to describe our feature extraction procedure.

Feature selection is problem-dependent and is considered the most central to the final result of a recognition system. A good feature set should show little variance within objects in the same class, and show significant difference among the objects in the different classes. Feature selection is a process of a incoporating human knowledge about a particular problem into a computer system. In this project, the feature selection procedure consists of two major steps. The first step is to elicit domain specific knowledge from automotive engineers and encode the knowledge into a feature set. The second step refines the feature set using a statistic method.

The transmission shift quality classification system is developed upon a set of torque trace features and has applications in two major areas, transmission shift quality control systems and establishing shift quality test metrics. In shift quality control systems, the classification algorithm can be used for adaptive shift pressure control. The transmission shift quality classification algorithm can also be used to establish shift quality test metrics. During powertrain development, we need to evaluate shift quality in a standardized fashion[Sch94]. Conventionally, the quality of a shift is determined from the perception of a development engineer which is responsible for the transmission calibrations for a specific powertrain/chassis style combination. This manual method is subjective, and the result can vary from engineer to engineer due to experience and expectation. Therefore it is important to develop an automatic shift quality classification based on a set of sensitive features.

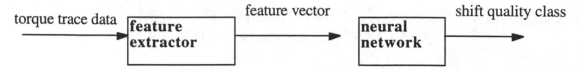

FIGURE 1. Block diagram of an automatic transmission shift quality classification system.

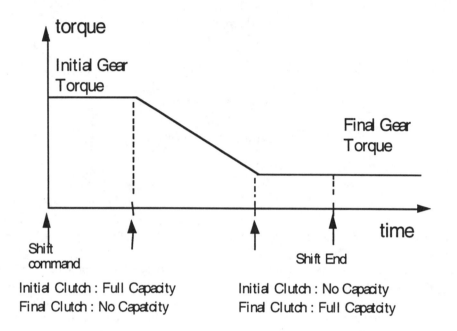

FIGURE 2. Ideal Automatic Transmission Torque Trace for an Upshift.

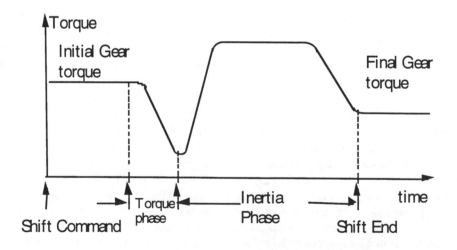

FIGURE 3. Transmission Output Torque Trace for an Upshift in Simplified Inertia Terms.

2. ENGINEERING KNOWLEDGE

In this project we attempt to classify three types of transmission shift, a firm shift, smooth shift and run-through shift. A firm shift has an increased magnitude of jerk. A smooth shift gives a smooth feeling of acceleration. A run-through shift is an indication that not enough pressure was present when applying the acceleration. The transmission shift is affected by a number of factors including the amount of idle and high speed noise and vibration, acceleration continuity, vehicle speed and shift feel. In fact, the output torque of the transmission is directly proportional to the delivered wheel torque, and hence the **transmission output torque is directly proportional to the vehicle fore/aft acceleration**. An ideal transmission output torque trace for an upshift is characterized by the pattern shown in the Figure 2. If we add in simplified powertrain and vehicle inertia parameters, the resulting upshift torque trace is shown in Figure 3. The simplified inertia model does not take into account the torque overshoot and ringing that will actually occur in a real vehicle. The transmission output torque trace for an upshift can be characterized by two distinct phases, the torque phase and the inertia phase.

If the effects of inertia torque transients are added to the torque trace, an actual torque trace for a fairly smooth shift looks more like the trace in Figure 4. Note the overshoot and ringing through out the inertia phase. Figure 5 represents an example of a shift that is objectionably firm. A firm shift is an indication that **too much pressure was present** during the apply of the on-going clutch. Notice the quicker ratio change when compared to the smooth shift trace and the fairly large torque spike that occurs in the inertia phase near the middle of the ratio change. A torque spike which occurs at this point, in conjunction with a quick ratio change, causes firm and abrupt shift feel. Figure 6 shows an example of a shift that **is classified as a run-through** which is classified as **objectionable**. With this torque trace, the ratio change is slow and excessively drawn out. The curve has a low flat area of the torque trace during the torque phase to inertia phase transition. Therefore, a run-through shift can be characterized by a noticeable torque sag followed by a sudden torque spike causing a sizable bump in fore/aft acceleration. By examining a number of shift torque traces and the different grades of shift quality; firm, smooth, and run-through, a predictable pattern was found and an initial set of features was developed.

3. FEATURE EXTRACTION

The pattern feature set was obtained from the shift torque trace in monitoring two types of upshift, 1-2 upshift and 2-3 upshift. The shift torque trace was analysed through the filtered ratio and axle output torque traces. The ratio and axle torque signals were filtered using an equally weighted, three place moving average filter before the data was fed into the feature extractor. The first point of each of the ratio and axle torque

signals is taken at the point that the shift is first commanded.

The first step in the feature extraction is to locate the **pertinent transition points** on the **ratio and axle torque traces.** These points are shown in the below Figures 7 and 8. The seven points illustrated in these two figures are further described in Table 1

Table 1. Description of Feature Points.

point	Description
1	Starting Ratio Change
2	Ending Ratio Change
3	Start of Torque Phase
4	End of Inertia Phase
5	End of Torque Phase
6	Inertia Phase Torque Spike Peak
7	Inertia Phase Torque Sag Valley

Points 1 and 2 are obtained from finding the 10% and 90% ratio change thresholds. Point 3 is found by monitoring the change in torque. In order to be more tolerant to noise, a two point time duration is used in calculating the torque jerk, although torque jerk is calculated for every point of output torque. This two point time duration gives a local average delta torque and will produce more reliable results. Point 3, which is considered the start of torque phase, is found by searching from the first point to the right, and the first point in which the torque jerk value is less than a calibrated threshold (negative value) is captured. Point 4 is determined similar to Point 3, except that the torque jerk values are searched from the last data point and moving to the left. The point in which torque jerk exceeds a calibrated threshold (positive value), is considered the end of the inertia phase. Point 5 are then found by searching for the data point which has the minimum torque value, starting from Point 3 and continuing until the torque jerk exceeds a positive calibrated threshold. The point with the minimum value is considered the end of the torque phase. Point 6 is found similar to Point 5. The torque value is measured for each point between Point 5 and Point 4. The point with the maximum value of torque is flagged as the inertia phase torque spike peak. Finally, Point 7 is found by searching from Point 6 to Point 4 and finding the data point which has the minimum value Based on these seven critical points 13 torque trace features can be obtained(Table 2). The torque$_{initial}$ value used above is calculated by averaging the torque values of each point from the initial data point to Point 3..

Upshift Transmission Output Torque

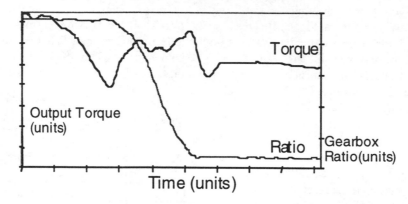

FIGURE 4. Output Torque Trace for a Smooth 1-2 Upshift.

Upshift Transmission Output Torque

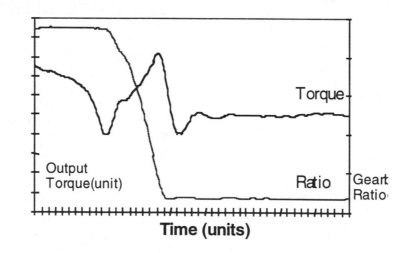

FIGURE 5. Output Torque Trace for a Firm 1-2 Upshift.

Upshift Transmission Output Torque

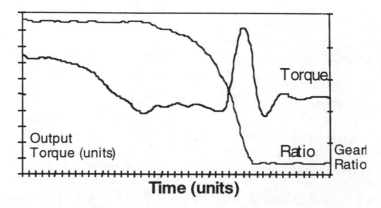

FIGURE 6. Output Torque Trace for a Run-Through 1-2 Upshift.

Ratio Trace Feature Points

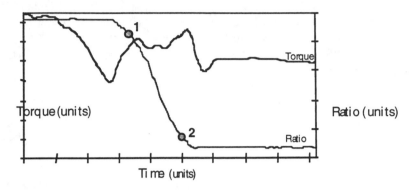

FIGURE 7. Ratio Trace Feature Points.

Torque Trace Feature Points

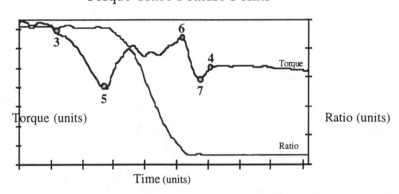

FIGURE 8. Torque Trace Feature Points.

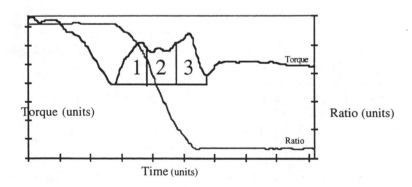

FIGURE 9. Division of Inertia Phase into Thirds.

The torque$_{final}$ value used above is the average of the torque value from the last data point to Point 4. The Area of the Total Inertia Phase is calculated by: $\sum$ (torque$_i$ - torque$_{P5}$), where i goes from Point 5 to Point 7.

The Area of each 1/3 of the inertia phase is done by dividing the Total Inertia Phase Area in three equal divisions (with respect to the time axis), with the 1st 1/3 being closest to Point 5 and the 3rd 1/3 being closest to Point 7(see Figure 9). These features were chosen so that they **would be independent with respect to the magnitude of ratio change** during the shift and to the magnitude of the initial torque levels, which will vary with the magnitude of vehicle acceleration previous to the shift (i.e. light or heavy throttle shifts).

4. FEATURE REFINEMENT

At this step we attempt to evaluate the usefulness of the set of features described in the last section. First we develop the measure of separation between classes. In order to measure separation for this project, the scatter matrix is computed for the feature vectors. The scatter matrix between different shift quality classes is computed by the following equation:

$$Scatter_{xy}[i] = Scatter_{bxy}[i] / Scatter_{wxy}[i]$$

where for feature i

$Scatter_{xy}[i]$: scatter value between class x and y

$Scatter_{bxy}[i]$: between class scatter for classes x and y

$Scatter_{wxy}[i]$ is the within class scatter for classes x and y for feature i. and

$$Scatter_{bxy}[i] = (mean_x[i] - mean_y[i])^2$$

$$Scatter_{wxy}[i] = \frac{\sum_{j=1}^{n}(value_{jx}[i] - mean_x[i])^2}{n} + \frac{\sum_{k=1}^{m}(value_{ky}[i] - mean_y[i])^2}{m}$$

where for feature, $value_{jx}[i]$ is the value of point j of class x, $mean_x[i]$ is the mean of class x, n is the number of data points in class x, $value_{ky}[i]$ is the value of point k of class y , $mean_y[i]$ is the mean of class y, and m is the number of data points in class y. After running a set of upshift data files that contained over 20,000 samples through the feature extractor program described in the last section, we obtained the scatter values shown in Table 3.

Table 3. Summary of Scatter Values.

Shift Type	Shift Qlty	Feat [0]	Feat [1]	Feat [2]	Feat [3]	Feat [4]	Feat [5]	Feat [6]	Feat [7]	Feat [8]	Feat [9]	Feat [10]	Feat [11]	Feat [12]
1-2 Up	1, 2	4.462	4.673	0.070	0.285	0.644	0.184	0.090	0.005	0.675	0.002	1.466	0.037	0.112
1-2 Up	1, 3	7.773	7.841	0.174	0.007	1.046	0.077	0.084	0.026	0.083	0.001	0.359	1.271	0.215
1-2 Up	2, 3	1.353	1.082	0.061	0.213	0.062	0.046	0.296	0.045	0.587	0.005	0.121	1.731	0.909
2-3 Up	1, 2	4.660	6.471	6.792	1.281	0.010	0.596	5.465	0.042	5.726	0.001	0.098	0.356	1.509
2-3 Up	1, 3	3.290	2.578	0.129	0.004	90.45	0.000	0.063	0.037	0.004	0.482	1.341	5.308	5.238
2-3 Up	2, 3	1.760	0.030	0.287	0.085	3.413	0.596	0.391	0.001	1.489	0.489	0.428	2.136	2.722

5. CONCLUSION AND ACKNOWLEDGMENT

We have described an automatic transmission shift quality classification algorithm. The algorithm consists of two stages, feature selection and classification. The feature selection has two processes, elicitng and encoding engineering knowledge and feature refinement. The classification algorithm is developed for luxury vehicles. The methodology described in this paper can be used to develop shift quality classification algorithms for other vehicle models such as performance, economy, and utility vehicles. When examining the results of the scatter values, a high value signifies a large amount of separation between the classes in question for a particular feature. A low value of scatter means that there is a significant amount of overlap of classes, which will make it hard to classify a point as a member of the correct class. A value of 0.5 indicates a significant amount of overlap, and values below that start to make that feature useless in determining the which class a data point belongs too (of the two classes in question).

This project is supported by the CEEP grant from the University of Michigan-Dearborn.

6. REFERENCES

[AdH95] Hojjat Adeli, Shih-Lin Hung, Machin Learning - - Neural Networks, Genetic Algorithms, and Fuzzy Systems, John Wiley & Sons, Inc., 1995.

[Sch94] Leo F. Schwab, "Development of a Shift Quality Metric for an Automatic Transmission," SAE Technical Paper Series, 941009, 1994.

KOA:General Affairs Expert System with Easy Customization

Shigeo KANEDA[*1], Katsuyuki NAKANO[*2], Daizi NANBA,[*3]
Hisazumi TSUCHIDA[*3], Megumi ISHII[*1] and Fumio HATTORI[*3]

[*1]NTT Communication Science Laboratories
[*2]NTT Software Corporation
[*3]NTT Communication & Information Systems Laboratories
[*1]1−2356 Take, Yokosuka−shi, 238−03 Japan   [*1] Email: kaneda@nttkb.ntt.jp

ABSTRACT

In Japan, the General Affairs Section of most large companies has the responsibility of processing applications related to health insurance, housing, various types of allowances, mutual aid, and social welfare benefits. The work involved in this process creates some burdens for two main groups of people: company employees and the General Affairs personnel. Against this background, NTT developed the Knowledge-base Office Automation (KOA) expert system to assist in the preparation of application forms. With this system, the employee (user) chooses an "event" such as "have married" or "child born" and responds to questions from the system. On the basis of these simple actions, the system can prepare forms with all necessary information entered. The KOA expert system has been introduced into NTT offices to assist the 200,000 employees of NTT.

1. INTRODUCTION

In Japan, the processing of applications related to health insurance, housing, various types of allowances, mutual aid, and social welfare benefits in large companies is handle by a Generals Affairs Section ("Soumuka" in Japanese). In the application process, the employee must prepare a number of related forms and submit them to the General Affairs personnel ; some documents seem to be as complicated as a tax form. Filling in these forms is not easy for employees[1] because they have difficulty in determining "What form should I use for this matter?" and "What should I enter for each item on the form?". Errors and omissions frequently occur as a result, and the General Affairs personnel must take a long time to check each form.

To solve the above problems, we developed a Knowledge-base Office Automation (KOA) expert sys-tem for General Affairs work. KOA is a kind of diagnostic expert system. KOA allows users to revise data previously entered, and they can do so whenever they notice such errors. This frequent revision of data by users is a major difference between KOA and common diagnostic expert systems, which do not easily support such modifications.

To allow users to change existing data, additional software coding is needed with conventional procedural or table-driven-schema programming techniques. Furthermore, when rules or regulations related to the General Affairs Section change, this code becomes an object requiring maintenance, which increases the maintenance cost. Please notice that regulations or application forms are very often changed in this domain. Also, it must be emphasized that compared with the main decision-making task of the expert system, the routine that enables the user to return to a prior screen and revise data is undesirable appendage.

In diagnostic expert systems, methods for constructing a user-friendly human-machine interface and techniques for generating questions have long been topics of discussion[GS84, LOWGREN89, KIDD85, KM85]. As far as the authors know, however, there has been no discussion on the code required to support data revision by users. This paper will focus on eliminating this type of code.

In response to the issues described above, the authors constructed the original general-purpose human-machine-interface tool called "Interfacer[KIHT]", and by combining it with object-oriented techniques, developed the KOA system. The most notable feature of Interfacer is that absolutely no additional routine is needed to handle user revisions. Also, the addition and modification of form input items can be performed by simply changing the form layout[2] and revising a few objects (frames) with an expert-system building tool; it is not necessary to modify the graphic user interface.

---

[1]For example, if a young man gets married, at least seven application forms should be selected. Moreover, for a young woman's case, two more forms are required because her name changes.

---

[2]This form layout change is easily realized by a Japanese word processor.

Employee Side	· Omissions in the application arise because "what can be applied for at what time" is not understood.
Omissions in application due to ignorance of system/preparation of forms too complicated	*Example: Age 61, Income ￥1,500,000 (survivor's pension), Mother; what would be the type of dependency? (See Note)* · The way to fill out the form is unclear · The same item must be filled out on several forms.
General Affairs Side	· Much work spent checking and returning incomplete forms
Omissions in application due to independent processing on the General Affairs side	· Because a number of specialized fields exist in General Affairs with separate personnel, a processor receiving a submitted form may not realize that processing by a related field is also necessary, resulting in application omissions and failure to update the database correctly · Because the various fields are highly specialized, a generalist covering all fields is not available for handling rare cases

Note: In Japan, public survivor's pensions are not regarded as income by the tax system, and this case would be a deduction as a dependent relative. However, in terms of dependency allowance for health insurance and wages, the pension would be regarded as a source of income although not income for tax purposes.

TABLE 1: Problems of General Affairs Section.

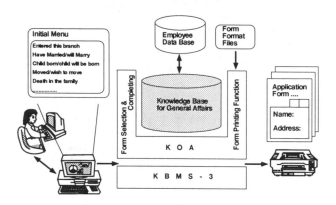

FIGURE 1: System-overview of KOA System.

Section 2 discusses the necessity of KOA. Section 3 describes the problems encountered in system development. Section 4 introduces the internal structure of the system and techniques for solving the above problems, and Section 5 compares KOA with related research. Finally, Section 6 presents the results of a system evaluation, and Section 7 makes some conclusions.

2. PURPOSE OF KOA

2.1 Problem areas in General Affairs work

In this paper, the expression "General Affairs work" refers to the preparation of various types of applications based on the individual life situation of employees making up a company (e.g., applications for dependents' allowance, health insurance, commuting allowance, leave of absence). In the conventional method, the employee first selects the relevant forms and then fills out the selected forms. As selection of the proper forms and filling out the forms require spe-

cialized knowledge, these actions require the assistance of the General Affairs personnel[3].

The selection and filling out forms have always been handled by humans. As a result, omissions, errors and other problems abound. As shown in Table 1, the problems are due to both the employee and the General Affairs personel. On the employee's side, various errors are made due to ignorance of current rules and regulations. On the General Affairs side, any decrease in the number of personnel results in errors of the application form processing task. Considering the present trends, it is obvious that the problems shown in the table will only become worse. Developing a computerized work flow that can conveniently process applications without errors or omissions is greatly needed.

2.2 System overview

Knowledge-base Office Automation (KOA) is an expert system developed to solve the problems described in the previous section by incorporating within a computer, the knowledge possessed by General Affairs experts. The basic configuration of the system is shown in Fig. 1. A company employee, knowing only that an "event" has occurred in his or her life, sits in front of a personal computer. The system displays questions for the employee to answer, and based on the replies, automatically prints out relevant forms with all necessary items filled in. In this process, the employee (user) has no need for specialized domain knowledge, and the forms normally used within the company do not need to be modified for use by KOA.

Table 2 lists "events" currently known by the system. Forms are not fixed for each event but depend on the responses made. For example, in the case of the "marriage" event, notification to the company (application for congratulatory monetary gift on getting married) and application for leave of absence are essential items. However, change of name notification and application for dependent allowance depends on the case in question. The proper forms are decided based on data input by the user after he or she selects an event and follows system instructions.

3. TECHNICAL PROBLEMS

Knowledge acquisition in an expert system is extremely time consuming. In KOA as well, rules and the questioning sequence are determined after a number of interviews with the General Affairs personnel. This technique, in which a Knowledge Engineer (KE) interviews experts, is widely used to build expert systems and will not be mentioned in this paper.

A problem of more concern in KOA development

---

[3]This process can be compared to the work involved in filling out tax forms in the United States.

· Entered this branch
· Have married/will marry
· Child born/child will be born
· Moved/wish to move
· Death in the family
· Wish to add/remove a dependent (health insurance)
· Wish to add/remove a dependent (salary)
· Need extra health-insurance ID
· Lost health-insurance ID
· Need/do-not-need home-owner's allowance
· Need/do-not-need home-rental/apartment allowance
· Setup/change bank account for transferring travel expenses
· Setup/change bank account for transferring wages
· Setup/change bank account for health insurance compensation
· Wish-to-enter/left company housing
· Wish to apply for a parking pass
· Wish to request a leave of absence
· Wish to take a union-related leave
· Need an entrance card
· Will leave this branch
· Name has changed
· Lost company ID
· Need company badge
· Lost company badge

TABLE 2: Initial "Event" Menu.

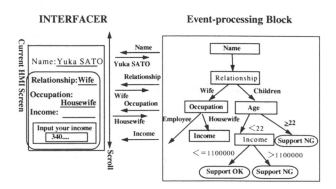

FIGURE 2: Change of Sequence of Query.

is the design of the human machine interface (HMI). Although the target application forms in KOA are limited to the major ones in use at NTT, there are still more than 40 of them and more than 1000 data items. The system must thus be capable of receiving replies to a large number of questions, and as a result, at least the same number of man-hours as used for knowledge acquisition must be allocated to building the HMI.

In the field of diagnostic expert system building, there has been much controversy as to the methods of building an HMI. Most of the discussion, however, has centered around the method of expressing knowledge and of generating questions. In contrast, the following problems are particular to KOA.

Problem 1: Change in question sequence

In KOA, questions and the sequence in which they are displayed on the screen when carrying on a dialog with the user change dynamically and cannot be determined in the design phase. Figure 2 shows a simplified version of part of the process for determining dependents. At first, set questions like name and income are displayed in sequence and the user is prompted to respond. However, questions after the third one in the figure, "relationship," depend on the response given to that question. If instead of "wife" the user replies "child," the sequence of questions displayed on the screen changes in the manner of "name→relationship→age→...." as shown in the figure.

Problem 2: Revision of data at any location

One other problem in KOA is the frequent revision of data by users. Referring to Fig. 2, assume that the

user while entering information in "income", notices that his or her response to "relationship" is wrong. It is desirable for the user to be able to interrupt the current input and correct the erroneous information immediately. If correction is not immediately possible, the user is bound to feel that the system is inconvenient and not user friendly.

The KOA system was originally constructed using a conventional table-driven scheme, but this resulted in an excessive increase in coding to enable such revisions to be made. Thus, a technique was needed that could enable users to revise any data item on the screen and that placed no burden on the programmer.

4. SYSTEM CONFIGURATION

4.1 Basic configuration

Figure 3 shows a block diagram of the KOA system. KOA was developed using KBMS (Knowledge-Base Management System)[KBMS92], NTT's expert system building tool[4], and the C language. KOA consist of the UI Block(User Interface block) and Event-processing Block. A major part of the UI Block is "Interfacer" which will be mentioned later. Another major part of the Event-processing Block are the three layered objects that represent the model of problem solving in the General Affairs Section. The problems described in the previous section were solved by these major parts of KOA. The following sub-sections describe the Interfacer and Object-oriented techniques in the Event-processing block (the other blocks are not described for brevity).

4.2 Interfacer

The two requirements for constructing an HMI in KOA were satisfied by the general-purpose "Interfacer[KIHT]." With Interfacer, the Event-

---

[4]KBMS is a registered trademark of NTT within Japan.

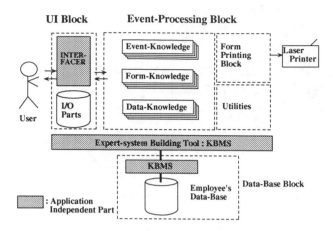

FIGURE 3: Block Diagram of KOA System.

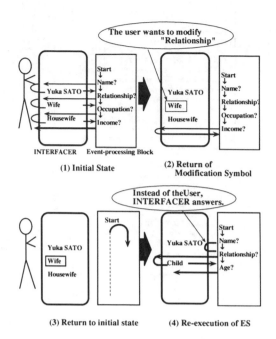

FIGURE 4: Interfacer Response.

processing Block is separated from the HMI. The main feature of Interfacer is that when the user corrects a previously entered datum, KOA internally re-executes all data entry from the top of the sequence in question. The major functions of Interfacer are as follows.

Automatic generation of interactive screens

Based on requests from the Event-processing Block for the display of questions and the capture of answers, Interfacer places questions one by one in a scrolling screen from the top in an interactive manner. The order of displaying questions on the screen, the display positions, and other parameters are recorded by Interfacer and are of no concern to the Event-processing Block. Interfacer also displays messages to prompt the user for input and menus/dialog boxes to handle data input.

Menus and dialog boxes used in acquiring data from the user are defined beforehand as objects (frames) at the time of system development. Similarly, menu titles, selection items, values returned when selecting certain items, data types of input data, inspection functions, default values, etc. are defined as slot values.

No need for data revision code

The most prominent feature of Interfacer is that it re-executes the Event-processing Block using original and revised values. An example of this process is shown in Fig. 4.

Re-execution can occur when the Interfacer returns a special value to the Event-processing Block. In the figure, when the user decides to make a revision after the Event-processing Block issues the request "income," a special symbol with the meaning of "revise" is returned as the return value for the request "relationship" (Fig. 5(2)). On detecting this symbol, the Event-processing Block returns to the top of

this sequence (Fig. 5(3)). Although this requires the Event-processing Block to backtrack with respect to stored data, all stored data in this sequence are simply made to revert to their original values and coding is accordingly simple. The object (frame) state breakpoint-restart function of an expert system building tool[KEA87b, KEA87a, KBMS92] can be used to perform this backtracking.

Backtracking of the Event-processing Block restores stored data to its most recent state. As a consequence, if the Event-processing Block is then executed as such, the same questions as before will be presented to the user. To avoid this problem, the Interfacer saves the data obtained from the user together with the names of the corresponding questions. Then, when the Event-processing Block proceeds to ask the same questions, Interfacer itself returns the saved values without querying the user again.

At the same time, even if the Event-processing Block is being re-executed, Interfacer shows the present screen to the user as it originally was for the sequence of input and display items that were not changed. For those input and display items that were do changed, the new items are shown to the user and input instructions given (Fig. 5(4))[5].

---

[5] After the changing point, each stored item value is displayed as default value on the CRT screen. Thus, the user can easily confirm the stored value.

4.3 Object-oriented techniques in Event-processing block

Knowledge related to the General Affairs Section consists of three levels: event knowledge, form knowledge, and data knowledge, and each item of knowledge consists of multiple objects (event objects, form objects, and data objects). This configuration represents the model of the work performed by the General Affairs Section as follows.

- Event Knowledge :
  This knowledge selects application forms depending relevant to the selected "event". Usually, several questions are required to analyze the user's situation.

- Form Knowledge :
  This knowledge corresponds to each application form and represents what items should be filled-in according to the input data.

- Data Knowledge :
  This knowledge contains data items for the employee and his or her family.

These three types of knowledge consist of multiple event objects, form objects, and data objects, respectively. When an employee selects an event, the event object corresponding to the event becomes activated (as shown in Fig. 5). It then acquires data from data objects and determines the form objects that have to be activated. Each form object, which corresponds to an actual application form, then determines which data items must be filled in, and obtains the necessary data from data objects. Data objects ask Interfacer for the values not yet known. The data object then returns the item value to form objects / event objects. When all of the slots in the form object are filled, the corresponding form is complete.

The query strategy of General Affairs personal and the regulations of General Affairs are represented as slot functions in event objects and form objects. Each "method" is activated when the slot value is queried by another object. Of course, a slot function is used to ask for the other slot values to make decisions. These mechanisms match those in the conventional object-oriented approach, so details are omitted in this paper.

5. RELATED RESERCH

Through KOA, we have uncovered a new application that we can call "General Affairs." The system itself is a type of diagnostic expert system, and rule-expression and other methods that we have adopted exist elsewhere. For this reason, such methods have not been explained in this paper.

Conventional diagnostic expert systems have rule expression and question generation as their main

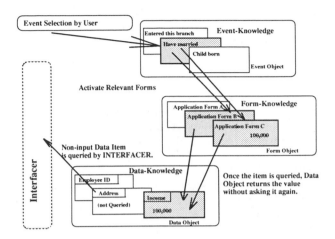

FIGURE 5: Object Layer in Event-processing Block.

objectives[GS84, LOWGREN89, KIDD85, KM85]. The goal of this system, on the other hand, is handling data revision by users. Specifically, Interfacer enables data to be revised by users without any additional code. To achieve this type of data revision, Interfacer backtracks internal system variables using a mechanism that can be found in SmallTalk and various expert system building tools / object-oriented database systems. These conventional backtrack functions, however, return only a slot value at a certain time in the past, while Interfacer shows the past screen as it originally appeared to the user. In addition, questions which come up again due to re- execution are automatically answered by Interfacer.

This ability of Interfacer to automatically reply to questions generated by re-execution is similar to breakpoint restart in conventional DBMSs. Various types of editors also have the ability of restoring data from a transaction file. In these types of breakpoint restarts, only past file states reappear. Interfacer has introduced this idea into the HMI. Unlike a DBMS, Interfacer will reprocess as long as the question sequence is unchanged using the past data, but will halt reprocessing at the point at which the response was changed.

6. EVALUATION

6.1 System descriptive power

Use of both the Interfacer and object-oriented techniques has increased the descriptive power of the KOA system dramatically. Figure 6 compares amount of program description when using conventional procedural description with that when using Interfacer. It should also be mentioned that the KOA program was written using the C language and standard functions provided by KBMS. As shown in the figure, the program amount decreased by more than half, and the beneficial effect of Interfacer combined with object-oriented techniques can easily be seen.

6.2   Decrease of Work

The KOA system was initially installed at the NTT Yokosuka Research and Development Center, which has about 2,000 employees, and the effects of its introduction were measured. This was performed by having employees answer a questionnaire that asked about the events used, the time spent using the system, etc. It was found that the most commonly used events were "entered/leaving branch," "child born," "add/remove a dependent," and "have married," which made up more than half of all events used.

Tabulating the results of the questionnaires also showed that the work performed by the General Affairs Section in handling 2,000 employees could be decreased by one man-year. There was also the opinion that less mistakes were made in entering information through the new system. On the other hand, because users were unfamiliar with the system, there were many time when General Affairs staff had to explain how enter information. A similar problem can be found with the need to use Japanese on a computer; although it is necessary to first input kana (syllabic characters) through an alphabet-based keyboard and then have the kana converted to kanji (Chinese characters), if this "kana-kanji-conversion" method is different from the procedures used in daily life, many users will not understand it at all. From this point on we plan to investigate a system configuration that offers improved ease of use. One approach is to link the system with existing in-house databases.

7. CONCLUSION

With the goal of speeding up the processing of General Affairs forms, we have developed an expert system called Knowledge-based Office Automation (KOA) and introduced it into our General Affairs Section. The two most important considerations in constructing this type of system were making it easy to modifying forms and easy to handle data revision by users.

In light of the above, we decided to express experts' knowledge with rules and developed the original interface tool "Interfacer". These approaches eliminated the need to design screens during system development or to design code specifically for revising data. '

The KOA system was initially in operation at some NTT branch offices. Finally, as a part of NTT's upgraded wage-calculation system, the KOA system has been introduced to all NTT administrative offices (about 4,300 terminals & more than 200,000 users), enabling all entered data to be processed "on-line" with links to in-house databases. We expect this will signif-

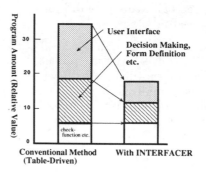

FIGURE 6: Program Amount of KOA System.

icantly raise the level of efficiency in General Affairs work.

References

[KIHK] S. Kaneda, M. Ishii, F. Hattori, and T. Kawaoka : "Interfacer : A User Interface Tool for Interactive Expert-Systems" Journal of Decision Support Systems, (to appear).

[GS84] B. Guchanan and E. Shortliffe : Rule-Based Expert Systems: The MYCIN Experiments of the Stanford Heuristic Programming Project (Addison-Wesley Publ. Comp., 1984)

[KEA87a] J. Kempf et al.,: "Experience with CommonLoops", Proceedings of OOPSLA'87 (1987).

[KEA87b] J. Kempf et al.,: "Teaching Object-oriented Programming with the KEE System", Proceedings of OOPSLA'87 (1987).

[KBMS92] KBMS Reference Manual (in Japanese)(NTT Software Corporation, Yokohama, JAPAN, 1992).

[LOWGREN89] J. Lowgren : "An Architecture for Expert System User Interface Design and Management", Second Annual Symposium on User Interface Software and Technology, Proceedings of the ACM SIGGRAPH Symposium on User Interface Software and Technology, (1989).

[KIDD85] A. Kidd,: The consultative role of an expert system. In P.Johnson and S. Cook, editors, People and Computers: Designing the Interface (Cambridge University Press, Cambridge, 1985).

[KM85] A. Kidd and M. Coope : "Man-machine interface issues in the construction and use of an expert system", Int. Journal of Man-machine Studies, Vol.22, (1985).

USING CONSTRAINT TECHNOLOGY FOR PREDICTIVE CONTROL OF URBAN TRAFFIC BASED ON QUALITATIVE AND TEMPORAL REASONING

Toledo F. [1], Moreno S.[2], Bonet E.[2], Martín G.[2]

[1]Departamento de Informática. Universitat Jaume I, Penyeta Roja, Castellón, Spain 12071
Email: toledo@inf.uji.es
[2]Instituto de Robótica, Universidad de Valencia, Hugo de Moncada 4, Valencia, Spain 46010
Email: lisitt@vm.ci.uv.es

ABSTRACT

In this paper we show the use of constraint technology for building a Urban Traffic (UT) control system able to perform predictive control in real time. This control system is based on a model of UT evolution involving temporal reasoning (an implementation of an extension of the Event Calculus, which is a temporal reasoning formalism we use to represent the concept of problematic situation in UT) and qualitative reasoning (to represent the behavior of a UT control system). Constraint technology has shown particularly efficient in the implementation of the temporal reasoning formalism, and on the other hand it has followed a qualitative simulator (which generates a future state starting from a given initial state) to be transformed in a traffic plan generator (which given an initial state and the representation of the knowledge about what are problematical situations, generates plans aiming to avoid future problematical situations if possible).

The work presented here has been developed inside the ESPRIT CHIC project and implementation has been done with ECL^{1}PS^e.

INTRODUCTION

Urban Traffic Control (UTC) Systems are generally composed of two layers [Am+91], which can also be found in many other control systems: a first one where the decision-taking process is made by an "algorithmic" system, and a second one where the decision-taking process is made by a human expert (using his knowledge about the systems dynamics and control). In UTC, human control acquires a great relevance, due to the fact that algorithmic systems present an acceptable performance within a determinate range of situations (the domain in which they can optimize a given function), but outside this range (congestive situations and so on) they can even cause a worse situation [Rob87] moving problems to another and even more problematic zone. In these critical conditions, the engineers are capable of taking more coherent decisions, because of their global knowledge of the city, than the algorithmic systems will do (because these systems generally focus on a good flow administration, but do not handle global information, thus allowing the problem to be translated to other areas).

One of the most critical aspects of the expert control which is almost impossible to perform in an algorithmic system is the predictive feat, that is, to be able to take corrective measures in advance, preventing traffic collapses. Particularly, trying to avoid that a primary congestion (increase of a queue by an excess of the demand versus the capacity of a street) can turn into a secondary one (blockade due to a growing queue, of a crossing located afterwards, with the consequent extension of the blockade along the network).

These facts make the actuation and supervision of human engineers to be continuos, and this fact, together with other factors [BSI91], makes the UTC systems based on two layers unable to reach an adequate control. This conclusion has made several authors to consider the possibility of putting Knowledge Based Systems (KBS) inside the architecture of the UTC systems, with the objective of introducing with them the knowledge the expert handles in his decision-taking process. Nevertheless, the first generation KBS built starting from shallow knowledge, are not enough to develop many of the funcionalities a UTC system must have. In order to explain such deficiencies, we must add in the UTC application the great complexity of the system and the lack of information which a first generation KBS suffers (information of sensors and causal rules fired by this information). For these reasons, it would be desirable in UTC to include the reasoning structures corresponding to what we call the theories about the application domain [For88]: knowledge about the internal structure, behavior and/or composition of the study objects (which has been name as deep knowledge). This knowledge is structured in two levels: a first level with knowledge explaining the behavior of the system we are reasoning on, and a second level with knowledge about the finding-solution process of a given problem (in our case, what could be called control knowledge).

In previous works [To+90], [MTM94] we have developed a model, which is applied to the prediction of the UTC system dynamics. This model constitutes the kernel of the first knowledge level above mentioned, and it has been developed as a Qualitative Model (using Qualitative reasoning techniques) in the ESPRIT project EQUATOR. In this paper we reformulate the Qualitative model using Constraint Technology, and we develop one module for detecting UTC problems using temporal reasoning implemented by means of Constraints Technology. The set of Qualitative model and temporal reasoning module is named Qualitative Temporal Model (QTM), and its formulation in terms of constraints permits the second knowledge level above mentioned to be built.

AN OVERVIEW OF THE QUALITATIVE MODEL

The basic idea of the Qualitative Model is to represent the spatio-temporal evolution of traffic density in a given street, defining a quantity discrete space (or qualitative values) in which the density takes values. The border line separating regions in which the density takes constant values (from a qualitative point of view), is assumed linear. Figure 1 gives a clearer idea of the representation capacity of the traffic evolution using the density as a bi-dimensional parameter (space versus time) and using the quantity space. The vertex of each region are named events (in the figure are denoted by e1,e2,e3,...), and they represent changes in the dynamic of the system (i.e. appearance or disappearance of a qualitative region). In this way, the vertex or events can represent phenomenon such as appearance or disappearance of a traffic queue, saturation of a street by a traffic queue, a change in the color of a traffic light, etc.).

With the consideration of events as primary entities in the representation of the system evolution, we use entities with cognitive meaning, and we achieve a great level of compacity in the temporal database which state represents the system evolution (we only need the initial state of the street and the events generated). For these

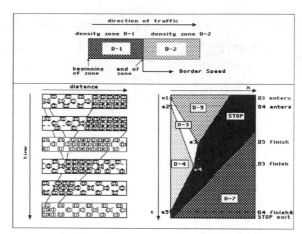

FIGURE 1: Spatio-temporal evolution of traffic density

reasons, we can obtain both high execution speed and easy high level interpretation of the temporal database content.

With this theoretical base, we built a qualitative simulator for UT able to execute in real-time. The system behavior is modeled by means of a set of objects (street, intersections, traffic lights, entries to the city, exits from the city, etc.) each one of them have several parameters (one-dimensional or bi-dimensional parameters, like density) which possible evolution is defined by constraints between them. Recently several authors adopted these approach for practical systems [Wil92].

USING CONSTRAINT TECHNOLOGY FOR TRAFFIC PLAN GENERATION

The events representing the evolution of the UTC system behavior can be obtained reformulating the qualitative model using Constraint Technology. In order to do this we have defined several types of constraints: "instantaneous" constraints (associated to intersections representing the flow conservation principle), "transition" constraints (representing the border line between two adjacent qualitative regions), and constraints for modeling the interaction between objects (traffic lights, intersections, entries and outputs of the city, links between two intersections).

It must be pointed out that, once the traffic state is known, the red time of a traffic light determines the length reached by the queue. This red time is what is named "split", and typically 8 different values (S_1, ..., S_8) for each intersection are handled. The set formed by the splits associated to all the intersections is known as a traffic plan, and its election is the control action a UTC

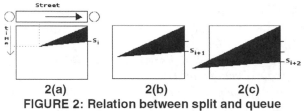

FIGURE 2: Relation between split and queue generation.

system must take. In figure 2 it can be seen the generation process of a traffic queue between two intersections, for three different split values (S_i, S_{i+1}, S_{i+2}). In figure 2(c) the queue is greater than the street length, so it can generates a problematic situation. Therefore, the corresponding traffic light must have a S_{i+1} split or smaller.

This idea can be represented using Constraint Technology, by associating a domain variable to each intersection, formulating the concept of problematic situations as constraints, and using these constraints together with the instantaneous and transition ones, in order to eliminate values of the domain variables. No eliminated values forms one (or several) traffic plans able to avoid (or minimize) future problematic situations (at a short term). For more detailed description, we need a formalism for representing the concept of problematic situation.

REPRESENTATION AND REASONING ON PROBLEMATIC SITUATIONS IN UTC

The reasoning about traffic problems can be performed by means of a spatio-temporal analysis of the database generated by the qualitative model. We use for this task the General Representation Formalism (GRF) [Equ91] developed in the EQUATOR project. It is based on the Event Calculus of Kowalski [KS86], a formalism for reasoning about time using the Horn Clauses subset of the first order logic augmented with negation as failure. The GRF provides several extensions with respect to the Event Calculus (i.e. temporal granularity and continuous change). Due to the detected insufficiencies of Prolog, the GRF implementation has consisted on developing a proof procedure in the Constraint Logic Programming (CLP) paradigm [Ba+94].

GRF concepts

In the GRF, temporal domain knowledge is expressed by event occurrences generally referred to different temporal domains according to the intrinsic time granularity associated to events. The occurrence of an event may be explicitly specified as an uncertain interval. Events are linked to properties by intentionally

defined domain relations, represented by means of the predicates *initiates(Event,Property)* and *terminates(Event,Property)* which describe the effects of events on properties. Instances of events and properties are obtained by attaching one time point/interval to event and property types. Coherently with the persistence properties in GRF, the predicate mholds_for(Property,[Start,End]) finds disjoint time intervals over which a property P, initiated and terminated by two events, maximally holds. State queries about stable properties are allowed by the *holds_at(Property,Time)* predicate. The GRF allows the specification of both ordering and metric constraints over events and properties. The predicate *earlier_than (Event1,Event2)* sets up the corresponding constraints over the involved events. It is also possible to introduce metric constraints by means of the predicate *distance(E1,E2,Distance)* where *Distance* represents the distance between the occurrence times of the events E1 and E2. It introduces also constraints over properties, such as *contradicts(P1,P2)* (properties P1 and P2 can not happen at the same time, i.e. their validity intervals must be disjoint) and *before(P1,P2)* (property P1 holds before of the property P2).

The CLP Implementation

Given a set of events initiating and terminating properties we want to derive in which time intervals a property holds. The key to obtain this is to know when two intervals are equal, i.e. given two events e, e', when *after(e,p)=before(e',p)*. The axiom used by Kowalski is: any two periods associated with the same relationship are either identical or disjoint. It can be expressed as follows:

$$after(e,p) = before(e',p) \; if \; holds(after(e,p))$$
$$and \;\; holds(before(e',p))$$
$$and \;\; e < e'$$
$$and \;\; not(after(e,p)< \; < \; before(e',p))$$

Where *after(e,p)< before(e',p)* means that the intervals *after(e,p)* and *before(e',p)* are disjoint.

The semantics of the GRF temporal features have been expressed by a set of Horn clauses. Therefore the above GRF axiom has been expressed by the following clauses, which directly gives the maximal time intervals over which property holds:

mholds_for(property,[start,end]):-
  happens(e), time(e,start), initiates(e,property),
  happens(e'),time(e',end),terminates(e',property),
  not broken_during(p,[start,end]).
broken_during(property,[start,end]):-
  happens(e), time(e,t), start < t,
  end > t, terminates(e,property).

From a semantic point of view, the GRF combines two conceptually distinct parts: the temporal part, constituted by time points/intervals and their relations, and the logic part, concerning relations over events and properties. Both parts are linked through the assignment of events/properties to time points/intervals.

Our implementation of the GRF axiomatic semantic expresses the logic part of the GRF with the SLDNF procedure provided by Prolog, and the temporal part with a Constraint Solver, in order to avoid a set of problems (floundering, time asymmetry, incompleteness of temporal information, deficiencies with the treatment of arithmetic, redundant computations,...) presented by implementations based on SLDNF, and to obtain a more efficient inference mechanism. The disjointness of time intervals is represented through a constraint which we will call the maximality constraint, and which forces to each two time intervals of a property to be equal, before or after. The constraint solver forces the above constraint to its equal option, if it can do so.

The Constraint Solver we have used is ECL^iPS^e. We represent time points as domain variables and time intervals as two time points start and end which satisfy the interval constraint start < end, in order to handle uncertainty. The Solver reduces the domains by constrain propagation using the Partial Looking Ahead Inference Rule.

<u>Use of temporal reasoning in Urban Traffic Control</u>

We adopt the following classification of problematic situations, given at different abstraction levels: elemental problematic situations, temporal problematic situations, and spatio-temporal problematic situations.

The first level of abstraction is constituted by the elemental problematic situations. The unique situation of this type we will take into account in the UTC system is when a queue reaches an intersection and some links connected to it have their traffic light in the green stage. This situation is called locked_intersection. We introduce a set of properties which associate the locked_intersection situations to the events generated by the qualitative model:

initiates(reaches_queue(Inter,Link),
    blocked_inter(Inter,Link)).
terminates(disappears_queue(Inter,Link),
    blocked_inter(Inter,Link)).
initiates(change_to_green(Link),
    green_stage(Link)).
terminates(change_to_red(Link),
    green_stage(Link).
initiates(new_cycle(Inter), cycle(Inter)).
terminates(new_cycle(Inter),cycle(Inter)).

Where Inter is the intersection identifier and Link is the link identifier. Over the last properties, the constraint contradicts(red_stage(Link), green_stage(Link)) must be satisfied. The property associated to the locked_inter situation is defined by the following axiom, where the

LinkP is any link entrance of the intersection (Inter) perpendicular to Link:

mholds_for(locked_inter(Inter),[Start,End]):-
    mholds_for(blocked_inter(Inter,Link),[Start1,End1]),
    perpendicular(Inter,Link,LinkP),
    mholds_for(green_stage(LinkP),[Start2,End2]),
    inter([Start1,End1],[Start2,End2],[Start,End]).
contradicts(locked_inter(Inter),unlocked_inter(Inter)).

The second level of abstraction consists on recognizing temporal problematic situations, which are the persistency over time of elemental situations (i.e. congestion in a given intersection). The following rules express the form of capturing these properties:

mholds_for(null_congestion(Inter),[Start,End]) :-
    mholds_for(cycle(Inter),[Start,End]),
    mholds_for(unlocked_inter(Inter),[Start,End]).
mholds_for(full_congestion(Inter),[Start,End]) :-
    mholds_for(cycle(Inter),[Start,End]),
    mholds_for(locked_inter(Inter),[Start,End]).
mholds_for(medium_congestion(Inter),[Start,End]):-
    mholds(cycle(Inter),[Start,End]),
    holds_at(locked_inter(Inter),Time1),
    Time1 > Start, Time1 < End,
    holds_at(unlocked_inter(Inter), Time2),
    Time2 > Start, Time2 < End.

Spatio-temporal problematic situations are defined over a set of objects and their temporal problematic situations, and they correspond to the persistency over space of temporal problematic situations. The dead-lock is the most relevant problem of this type (it consists of a circuit of congested streets blocking each other). The rules that express this property are the following ones:

mholds_for(dead_lock(Links),[Start,End]):-
    circuit(Links),
    mholds_for(all_blocked(Links),[Start,End]).
mholds_for(all_blocked([Link\Links]), [Start, End]):-
    intersected([Start, End],[Time1,Time2]),
    mholds_for(blocked_inter(Inter,Link),
        [Time1,Time2]),
    mholds_for(all_blocked(Links),[Start,End]).

Where Links represent all the links that form the dead_lock, circuit/1 obtains the links that form a circuit in the urban traffic network and intersected/2 is a constraint that states that the first interval must be contained in the second interval (we are looking for the situation in which all the links are blocked at the same time).

We have compared the execution of the temporal module in the GRF with an execution of this module in a Prolog implementation of a GRF version without constraints. This execution ran over real data taken from traffic of the city of Valencia. The results, shown in table 1, were obtained on a Sparc-Station10. The results show that the obtention of the first answer is faster with the use of the PROLOG GRF than the ECL^{i}PS^e GRF. This is due to the fact that the ECL^{i}PS^e GRF makes a prune over the search tree, minimizing the failures in the property intervals computations, so the time wasted in stating the constraints is not compensated by the time spent by the backtracking computation. Nevertheless, the obtention of all the answers is much faster in the ECL^{i}PS^e GRF implementation, due to the fast query operations once the constraints have been set.

TRAFFIC PLAN GENERATION WITH A QUALITATIVE TEMPORAL MODEL BASED ON CLP

The set of qualitative model and temporal reasoning module above mentioned, both of them based on CLP, are named Qualitative Temporal Model (QTM) for UTC. The QTM has been implemented using in the framework of our participation in the CHIC project, and we built a demonstrator of UTC based on an area of the city of Valencia with 19 main intersections. In figure 3 a view of this demonstrator is given.

The Traffic Plan Generator we have developed is basically an optimization method which tries to find in real time the minimum of a function. The input domain of the function is the set of all the possible traffic plans, and the output domain is a numerical value named the congestion level of the traffic network, which is a measure of the problematic situations of the traffic network in a given interval of time. The function relates each traffic plan with the congestion level the application of the plan at a given time T will produce, during the interval (T,T+ΔT], (in our case ΔT is five minutes).

The function is the QTM above described. The aim of the qualitative model is to predict the qualitative evolution of the traffic network during the interval (T,T+ΔT], starting from the traffic plan chosen at time T and the state of the traffic network at time T. The aim of the temporal reasoning module is to deduce the presence of problematic situations in the temporal traffic network during the interval (T,T+ΔT]. The database congestion level of the network during the study interval is then calculated from the problematic situations detected. Summarizing, the function calculates the

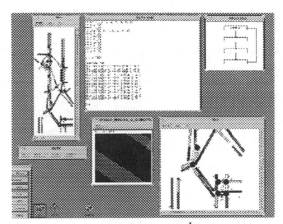

FIGURE 3: A view of the ECL^{i}PS^e demonstrator

congestion level in the next future corresponding to a representing the qualitative evolution of the given traffic plan, and the aim of the Traffic Plan Generator is to find the minimum of the function in real time or at least an acceptable minimum. The minimization method chosen is branch and bound, which is an acceptable method if function can be bounded. Nevertheless, the function is composed of two very complex software modules, and it is not possible to bound it because it is practically a black box, too complex to be bounded.

This is the point where we can obtain a great benefit of Constraint Technology. The Qualitative model has been reformulated in terms of constraints among variables representing the traffic network, and the temporal reasoning module is based on constraints too, so the whole function can be seen as a huge set of domain variables and the constraints among them. This approach allows to handle the function not as a black box, but allowing a treatment of what's inside. The strong point of this approach is that constraint solving techniques can now handle successfully the whole function in a single branch and bound method, bounding automatically the congestion level of high sets of traffic plans through constraint propagation among variable domains, allowing to prune in advance high subsets of traffic plans (when the surviving values in domain variables show that these plans cannot give a better solution than the better one which has been founded through the search) and to find an acceptable plan very fast.

Table 1: Execution time of the temporal module for Quintus and ECL^{i}PS^e

Type of Problematic situation	QUINTUS		ECL^{i}PS^e	
	1st answer	All answers	1st answer	All answers
ELEMENTAL	5.02 sec	366 sec	6.90 sec	7.42 sec
TEMPORAL	18.97 sec	1290 sec	13.75 sec	15.32 sec
SPATIO-TEMPORAL	19.22 sec	1360 sec	14.58 sec	22.42 sec

CONCLUSIONS AND DISCUSSION

The final conclusions reached after the trials to-date are:

The formulation of the problem has gained much from the declarative nature of ECL^{1}PS^e and the use of constraints. Several modules of the original application can now be merged into a single optimization problem with the branch and bound techniques in which the branching process is governed by temporal constraints as determined by the natural temporal branching points of the qualitative model and state of the split domains. This has led to a clear and concise way of representing the proposed control problem.

The branching process has gained in efficiency compared to standard Prolog, due to the efficient propagation of the temporal constraints which prune the split domains of each sub-problem and the search space.

The complexity of the plan generation problem has made it necessary to quantize the possible cost alternatives at sub-problem level too coarsely to apply sensible constraint propagation to reduce the upper cost bound to below the best solution so far. There are two reasons for this: first, the nature of QTM makes it impossible to assign true costs to each sub-problem by using linear approximations, other than for very short intervals, and second, even with this assumption, the linear functions involve all the splits in the sub-problem, resulting in no efficient domain propagation. The solution finally chosen in order to benefit from ECL^{1}PS^e was to build linear functions involving few splits each. However these results in a non-ideal quantization and in the possibility of missing the optimum if it is located on an initially bad branch.

The use of a constraint-based approach covered the inefficiencies arisen from Prolog-like implementation and it allowed the GRF to be used for UTC, maintaining the high level of declarativity. This approach integrates the declarativity and non-determinism of logic programming with the efficiency of the numeric resolution techniques.

The experience on UTC highlighted that ECL^{1}PS^e is particularly suitable for those applications involving both quantitative and qualitative temporal constraints. In fact its mechanisms allow an effective representation and an efficient implementation of temporal reasoning processes.

ACKNOWLEDGMENTS

We wish to thank Alexander Herold from ECRC (München, Germany) and Owen Evans from ICL (London) for their helpful comments during our work in CHIC. The research of this paper has been partially supported by a grant of Fundació Caixa-Castelló.

REFERENCES

[Am+91] Ambrosino B., Bielli M., Boero M., Mastretta M.: "Expert system approach to road traffic control", in Concise encyclopedia of Traffic & transportation systems, M. Papageorgiou, Pergamon Press, pp. 124-130, 1991.

[Ba+94] Barber F., Dondossola G., Berlanga R., Toledo F., Martín G.: "A CLP framework for time-related reasoning in industrial applications", Proc. 2nd. Conf. Intelligent Systems Engineering, IEE-BCS.

[BSI91] Bell M. C., Scename G., Ibbetson L.J.: "CLAIRE: An expert system for congestion management", in Advanced Telematics in Road Transport, Proc. of the DRIVE conference, Elsevier Pub., pp. 596-614, Brussels, 1991.

[Equ91] Equator Team: "Formal specification of the GRF and CRL", Deliverable D123-1, ESPRIT project 2049.

[For88] Forbus K.D.: "Qualitative physics: past, present and future", in Exploring Artificial Intelligence, Morgan Kauffan,1988.

[KS86] Kowalski R., Sergot M.: "A logic-based calculus of events", New generation computers, 4, 1986.

[MTM94] G. Martín, F. Toledo y S. Moreno: "Qualitative Simulation of Traffic Flows For Urban Traffic Control", in Artificial Intelligence Applied to Traffic Engineering, Bielli and Ambrosino (eds), VSP, pp. 201-231, 1994.

[Rob87] Robertson G.D.: "Handling congestion in SCOOT", Traffic Engineering Control, pp. 228-232,1987.

[To+90] Toledo F., Moreno S., Rosich F., Martín G., "Qualitative simulation in urban traffic control: implemetation of temporal features", in Decision Support systems & qualitative reasoning, M. G. Singh and L. Travé-Massuyes eds, Elsevier Science Pub, pp. 395-400, 1990.

[Wil92] Wild B. (1992). SAPPORO: Towards an intelligent integrated Traffic Management system. in Proc. Int. Congress of Artificial Intelligence applied to Transportation. Buenaventura, California, 19-38.

Robotics

FUZZY BEHAVIOR ORGANIZATION AND FUSION FOR MOBILE ROBOT REACTIVE NAVIGATION

Jiancheng Qiu and Michael Walters
Faculty of Design and Technology
University of Luton, Luton LU1 3JU, UK
Email: jqiu@alpha2.luton.ac.uk

ABSTRACT

This paper presents the progress on the development of a navigation system for a construction mobile robot working in indoor environemnts. Fuzzy logic control is used to compose reactive and task-oriented behaviors. The orgazination of reactive behaviors provides efficient support for task-oriented behaviors. Several reactive and task-oriented behaviors have been developed. A task-oriented dynamic activation and inhibition scheme has been devised to fuse behaviors. The effectiveness of the above strategies has been demonstrated in simulation.

1. INTRODUCTION

Our research is to develop a navigation system for a construction mobile robot MARCO(Mobile Autonomous Robot for COnstruction) working in indoor environments. An experimental MARCO testbed is being built. A laser range scanner, a sonar radar and odometry encoders will be the main sensing modality. The testbed will be used to investigate the application of goal-directed reactive control, fuzzy logic and other AI technologies to develop a mobile robot navigation system.

Traditionally, a mobile robot control system is organized based on a top-bottom hierarchical functional decomposition of tasks into sensing, world modelling, planning and execution modules. The drawbacks of the approach are the slow system response to environment changes and incompetence to cope with errors in the system operations. Unlike the traditional method, a behavior-based approach, first advocated by [Brooks86], is to decompose a mobile robot control system into several special purpose task-achieving behaviors[Brooks86] rather than information processing modules. The advantages of this approach are the high real time performance and the tolerance to environment changes and errors. However, this approach is unable to achieve long term goals because high level deliberation is abandoned.

To overcome the limitations of traditional and behavior-based approaches, a number of researchers have seeked to integrate the two approaches [Arkin90] [Firby89] [Gat92] [Saffiotti93] and create goal-directed reactive control systems. In this approach, deliberative plans are no longer applied as strict constraints. Instead, they act as guidance and coordination while a robot is still allowed to maintain reactivity to their environment changes. We adopt the same method and currently focus on the development of the basic mechanism of such a system. This paper describes the progress on the behavior organization and fusion for the reactive control mechanism.

We have chosen fuzzy logic control to compose behaviors. The reasons of this choice is that it allows to accommodate approximate, imperfect and noisy information during mobile robot operations and produces smooth control output.

2. FUZZY LOGIC CONTROL

Fuzzy logic control(FLC) is the application of fuzzy set theory to solve real control problems. Fuzzy set theory was first introduced by [Zadeh65]. In the past several years, fuzzy logic control has been explored for mobile robot navigation by [Saffiotti93] [Goodridge94] [Song92] and other researchers. [Sugeno89] has successfully used fuzzy logic control to guide a car into a garage. Flakey robot developed in SRI using fuzzy logic control successfully completed assigned tasks and won second prize at the first AAAI robot contest[Saffiotti93].

The basic fuzzy set operations defined by [Zadeh65] are: intersection, $A \wedge B = \min(\mu_A(x), \mu_B(y))$; union, $A \vee B = \max(\mu_A(x), \mu_B(y))$ and complement $\sim A = 1 - \mu_A(x)$.

The commonly used fuzzy reasoning method is the minimum operation rule. In this method, a fuzzy rule leads to a control action $\mu_{C'}(z) = w \wedge \mu_C(z)$, where w is weight calculated from antecedent of a fuzzy control rule with the above fuzzy set operations. The membership function μ_C of consequence C is truncated at this weight.

Finally, defuzzification is needed to produce a non-fuzzy control value which best represents the possibility distribution of a fuzzy control action. The widely used strategy is Centroid Calculation. The formula is

$$C = \frac{\sum_{i=0}^{m} \mu_c(z_i) * z_i}{\sum_{i=0}^{m} \mu_c(z_i)} \qquad (1)$$

where m is the number of the quantization level of output , μ_c is the membership function of output fuzzy set.

In order to reduce computation expense, we consider a simplified Centroid Calculation formula. This formula is based on fuzzy singleton representation of output. This representation allows us to use a special form of fuzzy set with only one pair having a value and full degree of truth and zero for the rest pairs. It is illustrated in Figure 1.

The advantage of the representation is obvious. It can greatly reduce computation time to produce the output

$$C = \frac{\sum_{i=0}^{m} w_i * z_i}{\sum_{i=0}^{m} w_i} \qquad (2)$$

fuzzy set, especially when the quantization level of output is big. Furthermore, defuzzification is simplified with formula (1) transformed to formula(2) because $\mu_c(z_i)$ is 1.0 at singleton points and 0 otherwise, where m becomes the number of rules, w_i is the weight of the antecedent of the ith rule and z_i is singleton value for the ith rule output.

3. FUZZY BEHAVIOR ORGANIZATION

To complete a navigation task in a behavior-related method, a mobile robot must have reactive behaviors to provide the ground for the success of task-oriented behaviors. This ground can be obtained through well organizing reactive behaviors.

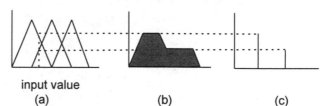

input value
(a)   (b)   (c)

Figure 1: (a) Membership function of input, (b) output fuzzy space in standard representation, and (c) output fuzzy singleton values.

3.1 REACTIVE BEHAVIORS

We have developed three reactive behaviors so far in simulation. The three behaviors are wandering, obstacle avoidance and wall following. They are organized based on the sphere of influence of environments to a robot. Wandering behavior usually is active when there is a large free space around the robot. Therefore, a big free space is its influence area. Obstacle avoidance behavior is active when there is small objects, ill-shaped objects and some moving objects like human nearby. Its influence area is around these objects. Wall following behavior becomes active when the robot is close to larger and relatively shaped objects, such as walls and barriers. To treat wall following as a reactive behavior is different from other methods[Saffiotti93] [Goodridge94] in which walls are taken as specific goals to follow. The capability of this behavior is also different from those introduced in the literature[Saffiotti93][Goodridge94][Gat91]. It can allow the robot to follow not only an inside wall and turn at an inside corner, but also turn at an outside corner and continue to follow an adjacent wall without subgoal positions being planned beforehand. This is the strong support for some task-oriented behaviors, such as goal reaching behavior. Wall following behavior is activated by walls and barriers and keeps active until it guides a robot to an easy position for task execution. Here, we present the implementation of these behaviors.

3.1.1 WANDERING BEHAVIOR

This behavior moves the robot around in an open space and introduce some noise in the movement. The introduction of the noise helps the robot to escape from the immobility possibly caused by some contradictive control rules among or between the behaviors. To do so, we associate some related behaviors with motivational state. Failures in these behaviors can cause wandering behavior be active and the noise helps the robot to move again. We will discuss this later. The following are the three rules to implement the behavior:

if speed is LOW then speed_increase
if speed is HIGH then speed_decrease
if angle_persistence is BIG then angle_change.

3.1.2 OBSTACLE AVOIDANCE

This behavior picks up 40 sample data from a sonar radar. The 40 points cover the front, right and left sides of the robot. The minimum distances to the obstacles in the three directions are calculated and fed into fuzzy rules for evaluation. Escaping speed and heading are produced through the rule synthesis. Currently we implement this behavior with four rules. The first two rules are :

if obs_right is CLOSE and
  obs_left is not CLOSE then left_heading
if obs_left is CLOSE and
  obs_right is not CLOSE then right_heading.

3.1.3 WALL FOLLOWING BEHAVIOR

A well-tuned wall following behaviour can follow a boundary continuously and turn accordingly. This can help the robot to escape from local minima and to dampen a barrier between the robot and a goal position, therefore make it easy to finish tasks. It can also reduce the time in calculating subgoals for guidance and give the system more time to concentrate on high level tasks. In our implementation, a high level planner and an explicit wall model are not needed for the behavior to act. Instead, we use a little local sense about the current shape of a wall it follows. The local sense is provided by an angle between the robot and a wall calculated using angle-histogram [Hinkel89]. With Acuity 3000 laser scanner, it is possible to have a quite accurate estimate of the angle. The angle histogram has been originally developed for world modelling and localization. We use it differently, not calculating over a full scan but a small local section. We call it local angle histogram. The following is the algorithm:

local angle histogram calculation algorithm:
    get_laser_data;
    find_min_distance_point;
    while(not discontinuity)
        extend_to_certain_scope;
        calculate_angle_histogram_over_scope.

The two biggest histograms and related angles are obtained from the calculation. They represent the walls with the maximum reflected points. The shape of a local wall

section can also be inferred from the data, such as a straight wall, an inside corner and an outside corner. An angle is selected according to the data and the angle changing history. This calculation is particularly useful in identifying an outside corner at which an angle is needed for the effective turning. By the local angle histogram calculation, the behavior can have a little more sense about its vicinity without high level reasoning and still maintain its reactivity. Small objects, ill-shaped objects and some moving objects like human are filtered out in the calculation because they cannot accumulate enough consistent reflected points. They are dealt with by obstacle avoidance behavior.

We use seven rules to implement this behavior. The distances to the robot sides are calculated similar to obstacle avoidance behavior and then input to fuzzy control rules together with the angle data. The output of the behavior is heading variation and speed change in order to keep the robot aligned with a wall and turn at corners. The first two rules

if (obs_right is BIG and obs_right is CLOSE) or
  (obs_right is MEDIUM and
    angle is NEGATIVE BIG) then right_heading
if obs_right is SMALL or
  (obs_right is MEDIUM and
    angle is POSITIVE BIG) then left_heading

control the robot to follow a wall at the right and produce a right or left turning recommendation when the robot drifts to the left or right of the path.

3.2 TASK-ORIENTED BEHAVIORS

At present, we have developed goal reaching behavior for the experiment purpose. This is the most common task-oriented behavior as most of mobile robot applications require a robot to move to different positions in order to finish a final task. In analogue to potential field methods, our fuzzy control rules produce a speed change vector given a goal position. The magnitude of the vector is the speed change to move to the goal and the heading is the robot direction change toward the goal position. In the experiments, a goal position is a point in Cartesian coordinates. The robot calculates the position change according to the encoder data and the new position data is fed to goal reaching behavior to create a new speed change vector toward the goal until it reaches the goal. We have developed six rules for the behavior. The first two rules are:

if angle is POSITIVE MEDIUM and
  goal_dist is not SMALL and
  obs_left is not SMALL then left_heading

if angle is NEGATIVE MEDIUM and
    goal_dist is not SMALL and
    obs_right is not SMALL then right_heading.

4. FUZZY BEHAVIOR FUSION

In behavior-related architectures, multiple behaviors usually need to be fused to create one set of output for actuators. Several schemes have been used for this, such as hierarchical switching[Brooks86], weighted averaging [Arkin90], context dependent blending[Saffiotti93] and fuzzy multiplexing[Goodridge94]. Context dependent blending developed by [Saffiotti93] has been effectively used to combine fuzzy reactive behaviors and goal-directed behaviors. In their application, a reactive behavior mainly refers to obstacle avoidance. In order to finish a task, a set of subtasks are exhaustibly planned by an off-line planner. A goal-directed behavior is activated for a subtask sequentially and combined with obstacle avoidance. The weight for goal-directed behavior is subdued with the obstacle avoidance weight and the outputs are defuzzified with the centroid calculation. Although the scheme works efficiently, only a few behaviors can be active at the same time. The success of this scheme depends on a detailed task planning. For example, a set of walls of a concave-shaped barrier must be provided as different subtasks for wall following behavior to execute in order to reach the other side of the barrier. Otherwise, the robot may be trapped in the concave area moving endlessly.

Based on the organization of our fuzzy behaviors, we have developed a different fusion scheme. We call it task-oriented dynamic arbitration. The idea is inspired by [Maes90] in artificial life research. In her method, the selection of a behavior is modelled in a bottom-up way using an activation/inhibition dynamics among different behaviours. This dynamics is both situational and motivational. Our method is similar in terms of activation and inhibition but is very different in definition and design of the dynamics mechanism. Fuzzy rule predicates are exploited to calculate the activation level of a behavior instead of assigned values. The activation level of the behavior changes more naturally and smoothly as the environment changes. Environment conditions are also utilized to alter the activational level of the behaviors in favour of the task-finishing behavior. Through the flexible inhibition/promotion link, more behaviors can be active simultaneously.

4.1 FUZZY BEHAVIOR ACTIVATION

A fuzzy behavior activation level which determines whether or not the behavior can be selected consists of two parts: situational activation and motivational activation/inhibition. Situational activation level depends on its current environment conditions. Motivational activation/inhibition is a way that a behavior subdues or promote other behaviours.

4.1.1 SITUATIONAL ACTIVATION

Each behavior is associated with a situational activation level. It is continuously changed as the robot environment changes and is calculated in a similar way to calculate fuzzy rule antecedent weight upon sensed environment data. It changes smoothly and has 1.0 as full strength. In calculating its situational activation level, goal reaching behavior has more sense about its environment by sensing the approximate obstacle position between the robot and a goal. If it sees the goal position, it has the full activation level. If there is an obstacle between, it has a situational activation level calculated similar to a potential field method. This treatment provides an effective transition of the control from goal reaching behavior to wall following or obstacle avoidance behavior when the robot approaches an object between a goal and the robot.

4.1.2 MOTIVATIONAL ACTIVATION/INHIBITION

We have defined three types of motivation for fuzzy behaviours. They are awareness of a task, safety and frustration of execution. Awareness of a task is mainly defined for task-oriented behaviors. Safety is defined for

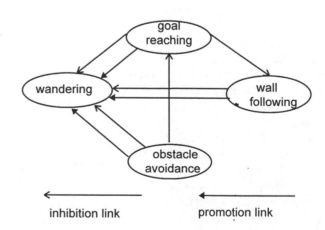

Figure 2: Promotion/inhibition links between behaviors

obstacle avoidance behavior and frustration of execution is for behaviors other than wandering behavior. A fuzzy behavior can subdue other behaviors by decreasing their activation level and promote them by increasing their activation level and decrease its own activation level. The behaviors are connected with promotion/inhibition links depending on their relationships. The current links between our fuzzy behaviors are shown in Figure 2. Goal reaching behavior has an inhibition link to wall following behavior to favour task execution. Obstacle avoidance behavior can subdue goal reaching behavior through their inhibition link when there is the danger of collision. Wandering behavior can subdued and also promoted by all the other behaviors. Appropriate levels of inhibition and activation are selected to maintain the independence of the behaviors and allow the effective transitions of the control between the behaviors toward a task execution. The frustration level of a behavior is calculated for undesirable robot motions under its control. The situations currently defined are: motionless and very slow.

4.1.3 CALCULATION OF BEHAVIOR ACTIVATION LEVEL

We define the following values for the calculation of a behavior activation level:

a_level — activation level;
a_level$_s$ — situational activation level;
f_level — frustration level;
i_value$_i$ — inward inhibition value;
i_value$_o$ — outward inhibition value;
p_value$_o$ — outward promotion value
p_value$_i$ — inward promotion value;
i_factor — inhibition factor;
p_factor — promotion factor.

Behavioral activation level is calculated for the following situations:
   a. outward promotion link;
      a_level = a_level - p_value$_o$
   b. inward inhibition link;
      a_level = a_level - MAX(i_value$_i$ s)
   c. inward promotion link;
      a_level = a_level + MAX(p_value$_i$ s).

Take goal reaching behaviour as an example, the calculation has the following steps:
   (1) inhibition value for wall following behavior
      i_value$_o$ = a_level$_s$ * i_factor;
   (2) promotion value for wandering behavior
      p_value$_o$ = f_level * p_factor;
   (3) inward inhibition link from obstacle avoidance behavior
      a_level = a_level$_s$ - i_value$_i$;

(4) outward promotion link to wandering behavior
   a_level = a_level - p_value$_o$.

The final activation level is truncated at full strength 1.0.

4.2 DYNAMIC BEHAVIOR ARBITRATION

We simply use a winner-take-all principle to select a behavior with the highest activation level. If there are more than one behaviors with the same highest activation level, we randomly select one. The behavior output is then multiplied by its activation level as the final control value.

5. SIMULATIONS AND EXPERIMENTS

A primary navigation system has been developed with the fuzzy behaviors and the behavior fusion scheme described earlier. The examples of the simulation experiments are discussed here.

Figure 3 illustrates the robot to finish a navigation task in a simulation. The robot is given a goal position at the up right corner of a simulated environment. It moves over two big barriers, avoids an small object and reaches the goal point without being told the walls. From its start position, the robot moves toward the first barrier controlled by goal reaching behavior as there is a big free space and goal reaching behavior has enough activation level to subdue wall following behavior. Then it is gradually weak and is overtaken by wall following behavior. Wall following behavior guides the robot to align with the straight wall of the first barrier, turn at the inside corner and the outside corners until the robot reaches the up side of the first barrier where it sees another big free space. The robot is again controlled by goal reaching behaviour until it approaches a small object and goal reaching behaviour is subdued by obstacle avoidance. After the robot escapes from the object, the control is back to goal reaching behaviour but soon handed over to wall following behaviour which leads the robot to the other side of the second barrier. From then, the robot sees the goal point and is quickly guided by goal reaching behavior to the position.

Figure 4 shows another run of our simulated mobile robot from the low side of a concave shaped barrier to the upside. First, goal reaching behaviour guides the robot toward the goal direction. Then wall following behaviour controls the movement from the inside wall to the outside wall until there is no obstacle before the goal position. And goal reaching behaviour completes the final control of the movement.

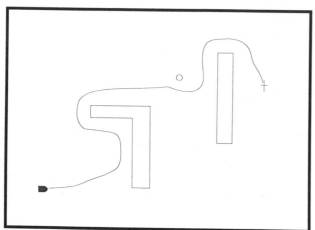

Figure 3: Simulated run to cross two big barriers and avoid a small object and reach a goal.

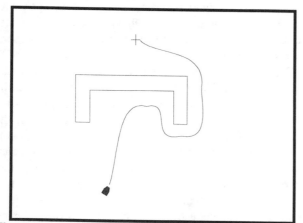

Figure 4: Simulated run to cross a concave-shaped barrier and reach a goal without high level reasoning.

6. CONCLUSIONS

Well-organised reactive behaviours give the efficient support for task-oriented behaviours. Fuzzy logic control provides a convenient engineering method to develop the behaviors for mobile robot reactive navigation. A task-oriented dynamic activation/inhibition mechanism provides an effective control transitions among the behaviors for task finishing. We have implemented some fuzzy behaviors and a fusion scheme and their effectiveness has been demonstrated through simulations. Future research will be in three areas: first, integration of planning into reactive control, second, using evolutionary algorithms to learn effective fuzzy control rules and third, incorporating more reasoning, such as case-based reasoning into the behavior fusion mechanism.

REFERENCES:

[Arkin90] Arkin, Ronald C., "Integrating Behavioral, Perceptual and World Knowledge in Reactive Navigation", *Robotics and Autonomous System*, 6, pp.105-122, 1990.

[Brooks86] Brooks, R.A., "A robust layered control system for a mobile robot", *IEEE J. of Robotics and Automation*, RA-2, pp.14-23, April, 1986.

[Firby89] Firby, R., "Adaptive execution in complex dynamic worlds", Technical Report YALEU/CSD/ RR-#672, Yale University Department of Computer Science, 1989.

[Gat91] Gat, E., "Robust low-computation sensor-driven control for task-directed navigation", *Proceedings of the 1991 IEEE Conference on Robotics and Automation.*

[Gat93] Gat, E., "On the role of Stored Internal State in the Control of Autonomous Mobile Robots", *AI Magazine*, pp.64-73, spring 1993.

[Goodridge94] Goodridge, Steven G.; Luo, Ren C., "Fuzzy Behavior Fusion for Reactive Control of an Autonomous Mobile Robot: MARGE", *Procs. of the 1994 IEEE International Conference on Robotics and Automation*, San Diego, CA, May, 1994.

[Hinkel89] Hinkel, R.; Knieriemen, T.,"Environment perception with a laser radar in a fast moving robot", *Robot Control 1988 (SYROCO88) selected papers from the 2nd IFAC Symposium*, Pergamon, Oxford, UK, 1989.

[Maes90] Maes, P., "A bottom-up mechanism for behavior selection in an artificial creature", *From Animals to Animats*, J.-A. Meyer and S.W. Wilson, Eds.(MIT Press, Cambridge, MA, 1990), pp. 238-246.

[Saffiotti93] Saffiotti, A., "Some notes on the integration of planning and reactivity in autonomous mobile robots", *Procs. of AAAI Spring Symposium on Foundations of Automatic Planning*, Stanford, CA, 1993.

[Song92] Song, K.Y.; Tai, J.C., "Fuzzy navigation of a mobile robot", *Procs. of the 1992 IEEE/RSJ International Conference on Intelligent Robots and Systems*, Raleigh, North Carolina, July, 1992.

[Sugeno89] Sugeno, M.; et.al, "Fuzzy algorithmic control of a model car by oral instructions", *Fuzzy sets and Sys tems*, 31, pp.207-219, 1989.

[Zadeh65] Zadeh, L.A., "Fuzzy Sets", *Information and Control*, 8, pp.338-353, 1965.

LEARNING AND CLASSIFICATION OF CONTACT STATES IN ROBOTIC ASSEMBLY TASKS

Enrique Cervera, Angel P. del Pobil
Department of Computer Science,
Jaume-I University, Penyeta Roja Campus,
Castellón, Spain, E-12071.
{ecervera, pobil}@inf.uji.es

ABSTRACT

The application of connectionist learning techniques, namely unsupervised neural networks, to contact classification is investigated. This approach demonstrates the feasibility and appropriateness of using force sensing to solve this problem. Empirical results are provided for the chamferless two-dimensional peg-in-hole insertion model with friction. The advantages of learning approaches over geometric model-based techniques are discussed. Our neural network approach is simple but robust against unpredictable changes of task parameters, and it exhibits a gracefully degrading behavior and on-line adaptation to new task conditions.

1. INTRODUCTION

In this paper we investigate the classification of contact states in an assembly task by using connectionist techniques, namely a class of unsupervised neural networks called Self-Organizing Maps, which has been applied successfully to other fine motion tasks with uncertainty [CPM+95a].

The two-dimensional peg-in-hole task is used as a canonical assembly problem. In a previous work [CPM+95b] we showed how the Self-Organizing Map was able to identify the relevant contact states in a simple model of this task. In order to make our model more realistic, friction has been added. We also consider the influence of several parameters like the hole clearance and the friction coefficient. We will show how our network is able to generalize, i.e., identify contact states under changes of those parameters. Experiments will show a gracefully degrading behavior of the network as the conditions progressively worsen. As a final result, we demonstrate how the network is able to evolve, in order to adapt itself to changing conditions. This adaptation allows the network to increase dramatically its performance under the new task parameters.

In the rest of the paper, first related work on contact identification is presented, our assembly model is discussed and some basic concepts about unsupervised neural networks and the self-organizing map are introduced. After solving the problem of contact state classification, important properties of the system, like generalization, robustness and adaptation are presented. These properties are extremely important for the application of the system to real-world robot assembly tasks. Together with the simplicity of the system —as it only needs force sensor information— they are major advantages over classical model-based geometric approaches. Finally, a new method for the calculation of the self-organizing map output is presented.

2. RELATED WORK

Recognition of contact states is an important issue in robot fine motion planning, and much research has been recently carried out from different points of view. Hirai & Iwata [HI92] develop a state classifier based on geometric models of the workpieces, which does not include friction. Desai & Volz, [DV89] and Xiao [Xia93] use more accurate geometric models including different sources of uncertainty. They found out that it is not always possible to disambiguate all contact formations due to sensing and geometric uncertainties and approximation of the contact models, and they introduced the concept of *active verification*, i.e. making small test movements to disambiguate contact situations. This technique is also used by Spreng, [Spr93], but it may change the task state in an undesired way.

Due to the inability to solve the problem with geometric model-based techniques, a growing interest about learning approaches in this field is being developed. Asada, [Asa93] uses a neural network to learn a nonlinear compliant mapping for the peg-in-hole insertion task. He argues that the internal layer of the network learns the different contact states of the task. His task model only consists of three contact states. Nuttin et al., [NRS+95] compare different learning approaches to contact estimation in assembly tasks, including two types of neural networks; results are given for the block-in-a-corner task, and both force and position information are needed.

3. THE ASSEMBLY MODEL

Our contact identification method is intended for general fine motion tasks. We assume that positional error is small. Nevertheless, a tiny misalignment of the pieces can greatly complicate or thwart a successful insertion. Based on this premise, we argue that the final phase of the insertion process must rely mainly on force sensing. Our model is intended to identify the contact state with only this type of information, which will be given by a wrist force sensor, a widely used type of contact sensor in real robots. This sensor can measure the three spatial components of the force and torque vectors acting on the robot wrist. For the sake of simplicity, we study the two-dimensional problem, where the sensor output consists only of two force components and one torque value. A schematic view of a gripper grasping a peg and our simplified model is depicted in figure 1.

A quasi-static model is used, i.e. inertia forces are neglected. In order to identify the contact states, only the direction of forces is relevant. We choose an arbitrary modulus, namely the unit vector. The Coulomb friction model is used, and the static and dynamic friction coefficients are considered to be equal.

4. UNSUPERVISED NEURAL NETWORKS: THE SELF-ORGANIZING MAP

The Self-Organizing Map is a type of unsupervised neural networks, introduced by Kohonen. Its theory and applications are thoroughly explained in [Koh95]. There is an input layer which is fully connected to all the units of the network, and each unit has an output. Unlike other multilayer neural networks, the neighboring neurons of the self-organizing map co-operate during training, providing a more powerful system than a usual single-layer neural network.

Unsupervised networks, unlike supervised ones, do not need output information during training. In order to make a supervised network learn to classify contact states, it should be trained with a set of input-output samples, the input being the force signals and the output the desired contact state for those signals. However, an unsupervised network is trained with a set of input data alone.

The main advantage of unsupervised networks is flexibility. We are not constraining the network to learn some a-priori states. The network self-organizes itself discovering regularities or correlations in the input data. Later on, when the training is over, we test the network response on those a-priori states. Hopefully, identification of the states will be possible if the network's response is different for each state. If this does not occur, some states are ambiguous, i.e., they cannot be identified with only the information provided in the input signals. This is an important result which supervised networks are unable to show, and allows the designer to rearrange the input information to overcome those ambiguities. Another interesting advantage is that the network can discover new states that were not considered a-priori. If this occurs, there will be some network response that we cannot associate to any of the known states. Studying the input values which caused that response will allow to identify the unforeseen situation.

5. LEARNING TO IDENTIFY THE CONTACT STATES

The six possible contact states in the peg-in-hole task are depicted in figure 2. Reaction forces are also shown. The appropriate friction forces are chosen from the friction cones with a uniform random probability. The peg angle is kept positive or zero. In [CPM+95b] we showed that negative angles cause ambiguities among states that cannot be solved with only force measurements, and that they are not necessary to successfully perform the insertion task.

Since we are interested in the influence of the task parameters, and not in those of the network, we will keep the network dimensions constant (a lattice of 15x10 hexagonally connected units) as well as the training

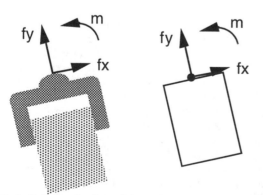

FIG. 1. Schematic view of the gripper and task model.

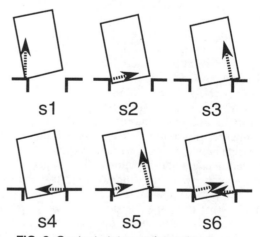

FIG. 2. Contact states and reaction forces.

parameters (learning rate, neighborhood, etc.). We investigate the performance of the network for several combinations of clearance and friction (μ) parameters.

Our experiments consist of three phases. Each phase involves an independent set of samples which are randomly generated. The same number of samples for each contact state is chosen. Any parameter subject to uncertainty is considered to have a uniform probability density function. In the same way, any random choice is equally probable. Each sample consists of the two force components, normalized to unit modulus, and the appropriate torque value.

The three phases are the following:

1) Training. The weights are randomly initialized. The neural network is trained with a set of 1200 random input samples. This process is split into two iterations. The first one (ordering phase) is 3000 training steps long, and the initial parameters are `learning_rate=0.02` and `radius=12`. The second iteration (tuning phase) has 60000 training steps, and initially `learning_rate=0.001` and `radius=5`. The learning rate parameter decreases linearly to zero during training. The radius parameter decreases linearly to one during training. A detailed explanation of the training process is given by [Koh95]

2) Calibration. A set of 600 samples is used. The network response is analyzed and state labels are associated with the network units. A unit will be associated to a contact state if that unit's response is greater with input data of that state than with data of any other state. This can be easily calculated by counting how many times a neuron is selected as the closest to the input samples of the different states. The state that has more 'hits' is selected for that unit's label. A second label (of the state with the second number of hits) will also be used during visualization, in order to highlight the overlapping among states in the map.

3) Testing. The set consists of 600 samples. The performance of the network is tested with an independent set of data. For each sample, the most responsive unit is selected, i.e. the one whose weights are closer to the input signals. The contact state is simply given by that unit's label. An uncertain response occurs when that unit is unlabeled. In order to solve this problem, we will introduce another method for calculating the network's output.

Results for a preliminary experiment are shown in Table 1. Two states, s2 and s5, are perfectly identified. Other two, s1 and s4, are almost perfectly identified. S1 is correctly identified in the 94% of the cases, it is erroneously identified as s3 in the 5% and it is unclassified in the remaining 1%. Meanwhile, s4 is properly classified in the 97% of the cases, but it is unknown in the remaining 3%. The other two states, s3 and s6, are more ambiguous, and the proper classification percentages are smaller. The average network

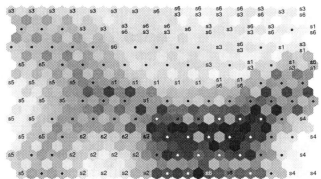

FIG. 3. U-matrix representation of the neural network. Task parameters: μ = 0.2, Clearance = 1%.

performance is very good, a 88% success, and we must take into account that only force information has been used.

This neural network can easily be visualized by using the so called *u-matrix* visualization [Ult93]. This display method has also been described in [KMJ92], and consists in visualizing the distances between reference vectors of neighboring map units using gray levels. The farther the distance, the darker the representation. In this way, one can identify clusters, or groups of neurons with similar response, which should be desirable to belong to the same contact state.

The first map is represented in figure 3, which consists of a big white region on the top with units labeled with states s1, s3 and s6, and three smaller light regions isolated by darker zones, i.e. long distances. This regions are labeled with s2, s4 and s5. This representation reflects the state ambiguities, which are also presented in the table. Some units are labeled twice to show this problem, that occurs with states s1, s3 and s6. This means that those units not only are selected for the first state, but sometimes they are also selected for another state. Unlabeled neurons are displayed as a dot.

Our previous work [CPM+95b] demonstrated that the network is unable to improve unless more information is given to the network by adding new input signals. These could be positional or other kind of useful information. However, tactile information is sufficient to get a good identification of the contact state, according to the classification rate which is obtained in the experiments.

Table 1. Classification percentages for μ = 0.2, Clearance: 1%

States	Right	Wrong	Unknown
s1	94	5	1
s2	100	–	–
s3	59	34	7
s4	97	–	3
s5	100	–	–
s6	78	21	1
Total %	88	10	2

6. ROBUSTNESS AGAINST NEW TASK CONDITIONS

We have argued that an accurate task model is useless if the task parameters change unpredictably. We want to test our neural network against these changing conditions.

The same network that was trained in the previous experiment is now tested with input data which has been generated with other parameters, namely a different clearance or friction.

First, the clearance is decreased to 2 $^0/_{00}$ and 5 $^0/_{00}$. Without further training of the network, its performance is maintained or even improves a little bit. This could be due to the fact that it is easier to identify contacts when the clearance is small. The classification rates are 89% right, 8.5% wrong, and 2.5% unknown, for a clearance of 5 $^0/_{00}$, and 89.3% right, 7.3% wrong, and 3.3% unknown for a clearance of 2 $^0/_{00}$.

Secondly, the clearance is increased up to 2%. The network's performance is only slightly worse: 84.7% right, 11% wrong, and 4.3% unknown.

This demonstrates the robustness of the neural network against clearance changes.

Next we consider some variations in the friction coefficient. If this coefficient is halved (μ=0.1), the network's performance improves up to 90.1% right, 8.1% wrong, and 1.7% unknown. The reason is that friction cones are narrower and they are more unlikely to overlap and to cause ambiguities. However, if the coefficient is increased two and four times, performance progressively worsens down: 74.8% right, 11.5 wrong, and 13.7 unknown (μ=0.4); 54% right, 12.3% wrong, and 33.7% unknown (see Table 2).

It is interesting to note that the network's response gracefully degrades as the friction coefficient increases. The *unknown* value demonstrates how well the network behaves in the sense that states are not randomly misclassified. Instead, the network sends a *warning* to a higher control level, which means that there is something wrong the network cannot cope with and further training or redesigning is required.

Finally, the network also generalizes to a combination of changes of clearance and friction coefficient (μ=0.4, Clerance=2%). Classification rate is slightly lower, 71.7% right, 13.3% wrong, and 15% unknown.

Table 2. Classification percentages for $\mu = 0.8$, Clearance: 1%

States	Right	Wrong	Unknown
s1	57	19	24
s2	74	–	26
s3	24	37	39
s4	37	–	63
s5	70	2	28
s6	62	16	22
Total %	54	12.3	33.7

Table 3. Classification percentages for a new SOM trained with $\mu = 0.8$, Cl=1%

States	Right	Wrong	Unknown
s1	93	7	–
s2	95	2	3
s3	53	41	6
s4	96	–	4
s5	82	15	3
s6	46	54	–
Total %	77.5	19.8	2.7

7. ADAPTATION TO PERMANENT CHANGES

It has been shown how the network gracefully degrades under task parameter changes. However, if the performance falls under a certain level, the output will be absolutely useless, or even harmful if the contact states are confused. Our neural network tends to be conservative, and outputs an *unknown* state in this case, but results show how misclassification of wrong states grows up too.

Considering the difficult problem with a greater friction coefficient, we train a new neural network to demonstrate that the task is still solvable. Results are shown in Table 3, and they are a great improvement over the old network, because training, calibrating and testing have been carried out with data sets generated with the large friction coefficient. Nevertheless, performance is a bit lower than the first problem because the task is more difficult. Now, friction cones are wider and are more likely to overlap and to cause ambiguities between contact states. The successful classification rate is 77.5%, more than 23% better than the network in Table 2.

Frequently, some tasks do not allow to stop the system and train a new neural network in order to adapt to the new conditions. A desirable system should be capable of adapt on-line, while the process is running, to these changes. We demonstrate the ability of our neural network to adapt in this way. The first network is re-trained, i.e. its weights are not initialized randomly but are kept as the initial weights and another training iteration is performed. This process could be done on-line if the training data are collected from the running process. After 20000 training steps, the network's performance is 77% right, 19.8% wrong, and 3.2% unknown. After 60000 training steps, the performance is 78.5% right, 19.2% wrong, and 2.3% unknown. It is even better than the network trained from scratch with new data.

This is a very important result that demonstrates the ability of the network to evolve under new task conditions. It is also an example of learning a difficult task from an initial easier task. The network was first trained with data generated with low friction coefficient, which makes contact classification easier. Afterwards, the network is re-trained with data from the harder task while keeping the previous knowledge of the similar task. Moreover, this transition is smooth, providing robustness and trustfulness to the system to run on-line in demanding processes like assembly tasks.

8. A NETWORK ENHANCEMENT: COLLECTIVE OUTPUT CALCULATION

In the previous experiments, the output of the network was given by the unit whose weights were closest to the input signals. That unit's label was the contact state. This method has two minor drawbacks: the first one is the subset of unlabeled units, i.e. those units that never were selected during the calibration phase. Thus, they have no label at all, and the network's output will be unknown, if one of them is selected. This could be overcome if the search is restricted to the labeled units, but that implies wasting resources, since those units were also trained during the training phase. The second problem is that the network does not give any measure of confidence, i.e. some kind of probability of being right. Our new method will overcome these two problems.

In our method, during the calibrating phase a vector of hits is collected for each unit. This *hit vector* has as many integer components as different contact states. Each component will store the number of times that neuron was selected for that given state during the calibration process. Obviously the maximum component will correspond to the state label that should be assigned by the original procedure.

Now, the network's output is calculated with the hit vectors of all the units in the map. The output vector is a weighted sum of these vectors. The coefficients are functions of the distance from the input signal to each of the unit weights, and a negative exponential function is arbitrarily chosen.

$$\text{Output} = \sum_i \left(\text{hit}_i \cdot e^{-\beta \cdot d_i} \right)$$

where β is an arbitrary coefficient and d_i is the distance from a neuron's weights to the input vector. The greater β, the smaller the influence of the further units. The maximum component of this output vector is selected, and the network's output is the contact state associated to that component. In Table 4 classification rates using this method are shown, for $\mu = 0.8$. The *unknown* column does not appear anymore.

Table 4. Classification percentages for collective output: $\beta = 12$, $\mu = 0.8$

States	Right	Wrong
s1	93	7
s2	99	1
s3	63	37
s4	100	–
s5	88	12
s6	41	59
Total %	80.7	19.3

With this method, a state is always selected. The successful classification rate is improved to 80.7%. For $\mu=0.2$, the successful classification rates are 87.2% and 88.2%, which are very close to the original 88% rate.

Despite this little improvement in the classification rate, the major advantage is the confidence measurement. The magnitude of the vector components can be used for this purpose. If one state is clearly distinguished, its component will be much bigger than the others. However, if there is a confusion between several states, the magnitude of their components will be similar, reflecting this ambiguity. Finally, if no state is confident enough, every component of the output vector will be small. For example, the samples in Table 5 were generated for state s6 with $\mu=0.8$, and the output vectors were calculated with $\beta=12$.

The outputs to the two first samples are confident. The value of the 6th component is far greater than the others, thus the sample will be classified as state s6. In the third example, however, several components share the same order of magnitude. With the original algorithm, the signal would be classified as state s1, the biggest component. However, this classification is not confident due to the similar order of magnitude of the third and sixth components; there is an ambiguity among these states. In the fourth example, the sixth component is at least four orders of magnitude greater than the others, but its absolute magnitude is small. The most likely state is s6, but the network is not very confident about that.

The absolute values of the components depend on the parameter β and the network metrics, which in turn depend on the input signal space metrics. A proper tuning of this parameter is required to maximize the mean classification rate. Further studies are required in order to do this process in an automatic way.

9. CONCLUSIONS

A learning approach to contact classification based on unsupervised neural networks and force sensing has been presented. Despite its simplicity, classification performance in the peg-in-hole insertion task with friction is excellent. Besides that, the network is robust against changes in task parameters, and behaves in a gracefully degrading way, providing a highly confident and reliable output for demanding robot assembly tasks. The neural network is also capable of adapting to these new conditions by means of an on-line retraining process. The performance of the network increases up to the level of those which are specifically trained. In this way, the network is able to adapt to the new conditions of a harder task, by using the knowledge that it previously learned in a simpler problem.

Future work includes the extension of this approach to cope with more complex tasks, e.g. in three dimensions In these tasks, even with a simple geometry, there is a great amount of different contact states and it is more difficult

Table 5. Input and Output vectors for the collective output example ($\mu=0.8$, $\beta=12$).

Input Vector	Output Vector
{-0.09, -0.99, -0.50}	{5.3e-10, 2.3e-7, 2.7e-9, 4.4e-3, 1.3e-11, **15.1**}
{ 0.16, -0.98, -0.19}	{ 2.1e-8, 1.4e-4, 3.1e-9, 5.5e-5, 9.8e-10, **22.7**}
{ 0.31, 0.95, -0.62}	{ **15.25**, 0.01, **12.76**, 0.03, 0.61, **10.8**}
{-0.60, -0.79, 0.47}	{1.8e-11, 8.9e-8, 4.5e-11, 1.4e-7, 7.1e-13, **6.9e-3**}

for the neural network to represent all of them. An implementation of the described approach on a real robot is on the way. A long-term direction is the association of velocity commands instead of state labels to the units of the network. Then the output of the network would provide a motion direction for the task to be performed successfully. New powerful versions of the self-organized map will allow to learn this relationship.

ACKNOWLEDGEMENTS

This paper describes research done in the Robotic Intelligence Laboratory. Support for this work is provided in part by the CICYT under project TAP95-0710, by the Generalitat Valenciana under project GV-2214/94, by Fundacio Caixa-Castello under P1A94-22, and by a scholarship of the FPI Program of the Spanish Department of Education and Science. The neural netowrk simulations have been done with the software package SOMPAK, which was developed at the Helsinki University of Technology.

REFERENCES

[Asa93]   H. Asada, "Representation and Learning of Nonlinear Compliance Using Neural Nets", *IEEE Trans. Rob. Aut.*, vol. 9, no. 6, pp. 863-867, 1993.

[CPM⁺95a] E. Cervera, A. P. del Pobil, E. Marta, M. A. Serna, "Dealing with Uncertainty in Fine Motion: a Neural Approach", in Industrial & Engineering Applications of Artificial Intelligence and Expert Systems, edited by G.F. Forsyth and M. Ali, Gordon and Breach Publishers, Amsterdam, pp. 119-126. ISBN: 2-88449-198-8.

[CPM⁺95b] E. Cervera, A. P. del Pobil, E. Marta, M. A. Serna, "A Sensor-Based Approach for Motion in Contact in Task Planning", *Proc. IEEE/RSJ Int. Conf. on Intelligent Robots and Systems*, IROS'95, vol. 2, pp. 468-473, Pittsburgh, USA, 1995.

[DV89]   R. S. Desai and R. A. Volz, "Identification and verification of termination conditions in fine-motion in presence of sensor errors and geometric uncertainties", *Proceedings of the 1989 IEEE Int. Conf. Rob. Aut.*, Arizona, USA, pp. 800-807.

[HI92]   S. Hirai and K. Iwata. Recognition of Contact State Based on Geometric Model. In *Proceedings of the 1992 IEEE Int. Conf. on Robotics and Automation*, pp. 1507-1512, Nice, France, May 1992.

[Koh95]   T. Kohonen, *Self-Organizing Maps*, Springer Series in Information Sciences, no. 30, Springer, 1995.

[KMJ92]   M. A. Kraaijveld, J. Mao, A. K. Jain, "A non-linear projection method based on Kohonen's topology preserving maps", *Proc. 11th Int. Conf. Pattern Rec. (11ICPR)*, pp. 41-45, Los Alamitos, CA, IEEE Comp. Soc. Press, 1992.

[NRS⁺95]   M. Nuttin, J. Rosell, R. Suárez, H. Van Brussel, L. Basañez, J. Hao, "Learning Approaches to Contact Estimation in Assembly Tasks with Robots", *Third European Workshop on Learning Robots*, Heraklion, Greece, April 1995.

[Spr93]   M. Spreng, "A Probabilistic Method to Analyze Ambiguos Contact Situations", *Proc. 1993 IEEE Int. Conf. Rob. Aut.*, Atlanta, USA, Vol. 3 pp. 543-548.

[Ult93]   A. Ultsch, "Self-Organized feature maps for monitoring and knowledge acquisition of a chemical process", *Proc. Int. Conf. Artif. Neural Nets. (ICANN93)*, pp. 864-867, Springer-Verlag, 1993.

[Xia93]   J. Xiao. Automatic determination of topological contacts in the presence of sensing uncertainties. In *Proc. IEEE Int. Conf. on Robotics and Automation*, pp. 65-70, 1993.

A NEURAL APPROACH FOR NAVIGATION INSPIRED BY THE HUMAN VISUAL SYSTEM

J. Fernández de Cañete and I. García-Moral

Dpt. de Ingeniería de Sistemas y Automática. E.T.S.I. Industriales de Málaga
Plaza El Ejido s/n . 29013 Malaga, SPAIN.
E-mail: canete@ctima.uma.es

ABSTRACT

The application of artificial neural networks to pattern recognition allows path planning in order for a mobile robot to navigate in an environment with obstacles. A backpropagation neural net providing quasi-optimal trajectory for navigating in several environments is presented. Feature extraction to enable learning is achieved using a model of the human visual data acquisition system. Several experiments with different obstacle configurations have been carried out showing how the neural net is capable of generalization to unknown maps as well as over the training map.

1. INTRODUCTION

The aim of developing of controllers for mobile robot guidance is the search for efficient trajectories to navigate from an initial point to a target, whitout colliding with the obstacles placed in the environment.

The path planning task has been approached by several researchers using different techniques, such as graph searching [Loz83], potential fields [Kha86,Tic90] and goal-oriented algorithms [SL92], among others. However, most of the methods described tackle the navigation task in previously known static environments. In practice the environment is dynamic, therefore some limitations arise with these techniques when they are applied, since no previous information concerning the environment is available.

In most situations, the computing time to evaluate the optimal trajectory is excessively high, but in many cases only quasi-optimal trajectories are required, with less computational effort though less accuracy. Several techniques have emerged which are applied to the quasi-optimal path planning task based on artificial neural nets [KV92,PL90].

One possible solution for tackling variable navigation environments could be the use of a neural network working as a controller, to avoid designing intelligent algorithms for path planning [MP92,HF93].

In this paper, a navigation simulator for path planning and obstacle generation is described. A backpropagation neural network working as a controller extracts information about the environment by emulating the human visual data acquisition system, and generating quasi-optimal trajectories for navigation with obstacle avoidance.

2. PROBLEM CHARACTERIZATION

The goal pursued by this application is the design of an artificial neural net for guiding a mobile robot starting from visual information extracted from its own environment, from an initial point to a final one on a two-dimensional navigation map with polygonal obstacles which are in certain cases previously unknown (fig. 1).

Two types of zones of polygonal shaped obstacles and free spaces define the navigation map available for the mobile robot to move in. An environment simulator has been designed to generate variable maps and modify previously defined ones. Moreover, it allows the tracing of quasi-optimal trajectories needed to obtain the training patterns during the learning process, enabling the neural network to decide the appropriate direction in each position (fig. 2).

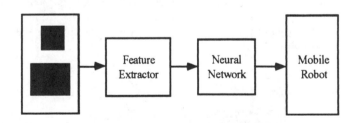

FIGURE 1. Structure of the neural guidance system

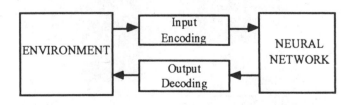

FIGURE 2. Interaction between simulator and neural net.

3. FEATURE EXTRACTION

The feature extraction procedure starts from the knowledge of the map containing initial and target points, as well as obstacle coordinates. Relevant information is extracted enabling the neural network to decide which is the best direction to take. This process is based on the behaviour of the human visual system.

For obstacle recognition and avoidance tasks, the human brain processes incoming visual information and produces a response in the motor organs.

The human visual system has a focusing mechanism based on regulating convexity of the lens, by paying attention to images over a fixed distance, and decreasing perception over the rest. It would be suitable to split the mobile robot's visual field in concentric zones to differentiate between distant and nearby information extracted from the environment.

On the other hand, the geometrical layout of retinal photoreceptors allows discrimination between an image and its surroundings whitin a visual field of 150° width. Therefore, it would be necessary to split the mobile robot's visual field in contiguous sectors covering 360° width.

With the aim of translating this feature extraction process from the human to the mobile robot's visual field, and given the discrete nature of digital computers, the environment is divided into z concentric zones (square shaped for simplicity) centered on the mobile robot's current position, and s circular sectors, corresponding to different focusing planes and directions, respectively (fig. 3).

Once the map is split into subareas or information units, features are extracted for training the neural net, particularly obstacle density per subarea, target direction (i.e. the sector it belongs to), and a dispersion of obstacles index $F(s,z)$ with higher values for the subareas having the best conditions for the mobile robot to choose from.

The disperdion index is a function depending on the following characteristics:

• Obstacle density $D(s,z)$
• Subarea proximity to the mobile robot's current position, weighing zones closer to the mobile robot with a maximum value with a zone modifier $MZ(z)$.
• Target direction, weighing in decreasing order sectors containing the target with a maximum value with a sector modifier $MD(s)$.

therefore, the dispersion index is defined as

$$F(s,z) = f(D(s,z), g(MD (s), MZ(z)))$$

The function f has not arbitrarily chosen, instead it must satisfy certain properties in order to make the discrimination process more efficient.

Considering fixed s and z, and evaluating the f dependence on $d=D(s,z)$, that is $F=f(d)$, is clear that different dispersion values must produce different input values to the net, so $f(d)$ should be monotonic and bounded in [0,1]. A good solution for $f(d)$ would be

$$f(d) = \mu \cdot d$$

being μ the slope of a straight line, that is constant for fixed s and z.. This value just matches the expression of $g(MD(s), MZ(z))$ so each value of μ is related to a pair (s,z). Nevertheless, g defines a set of segments with different slopes included in an area A (fig. 4), so for the most advantageous subarea, the slope μ should be greater than the diagonal to obtain a high value of the net input (scaled in [0,M]). However, this choice produces a set of dispersion values with the same net input values, therefore the best choice of g is the slope of the diagonal for the most adavantageous subarea, while decreasing slopes are assigned for other subareas.

The function g has been heuristically obtained, being the best solution defined by the product of the slopes

$$g(MD(s), MZ(z)) = MD(s) \cdot MZ(z)$$

$MD(s)$ and $MZ(z)$ being functions empirically calculated to achieve maximum efficiency during trajectory planning (fig. 5).

The function f therefore is given by

$$F(s,z) = (1-D(s,z)) \cdot MD(s) \cdot MZ(z)$$

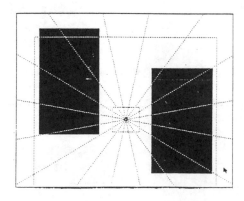

FIGURE 3. Division of the visual field.

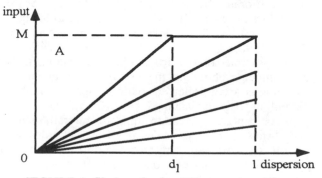

FIGURE 4. Choice of μ for differents situations

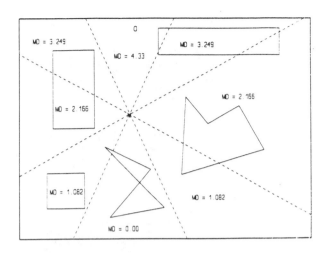

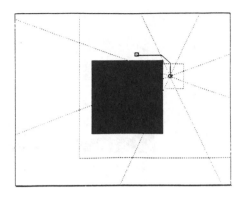

FIGURE 6. Training the net to go round a corner.

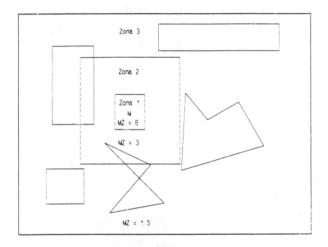

FIGURE 5. Values of MD(s) and MZ(z)

!! Entrenamiento codo2 del laberinto rect POS 351 182 277 130

E			NE			N			NO			O			SO			S			SE			DIR
6	3	0	13	6	0	19	10	0	12	12	0	0	0	0	6	3	0	6	3	0	0	0	0	N
6	3	0	13	6	0	19	10	0	12	13	0	0	0	0	6	2	0	6	3	0	0	0	0	N
6	3	0	13	6	0	19	10	0	14	13	0	0	0	0	6	2	0	6	3	0	0	0	0	N
6	3	0	13	6	0	19	10	0	19	13	0	0	0	0	6	2	0	6	3	0	0	0	0	N
0	0	0	6	3	0	13	6	0	17	10	0	0	0	0	9	3	0	13	5	0	6	3	0	N
0	0	0	6	3	0	13	6	0	18	10	0	0	0	0	0	0	0	13	5	0	6	3	0	NO
0	0	0	6	3	0	13	6	0	19	10	0	0	0	0	0	0	0	12	5	0	6	3	0	NO
0	0	0	6	3	0	13	6	0	19	10	0	9	9	0	0	0	0	10	4	0	6	3	0	NO
0	0	0	6	3	0	13	6	0	19	10	0	16	9	0	0	0	0	6	4	0	6	3	0	NO
0	0	0	6	3	0	13	6	0	19	10	0	22	10	0	0	0	0	0	0	0	6	3	0	NO
! 0	0	0	6	3	0	13	6	0	19	10	0	22	10	0	0	0	0	0	0	0	6	3	0	O
0	0	0	6	3	0	13	6	0	19	10	0	22	10	0	0	0	0	0	0	0	4	3	0	O

FIGURE 7. Patterns obtained during training.

outputs elicited for a 8 sectors by 3 zones map, obtained while the net is trained to go round a corner (fig.6 & 7).

Training was carried out guiding the mobile robot through user-generated quasi-optimal trajectories for a given navigation environment, minimizing the number of changes in direction. Training patterns are elicited for each mobile robot´s current position.

This method can lead to contradictory situations where the same input pattern has different output patterns associated with it. These incoherencies have negative effects on the training quality, so it is necessary to eliminate these uneven patterns to assure coherence.

During neural net execution indecisive situations were encountered owing to the presence of concavities and symmetries, which led to the inefficient tracing of trajectories (fig. 8).

The guidance process tries to avoid zones with greater obstacle density and prevent an obstacle collision, therefore a consistently trained neural net would go round the edges of the obstacles and reach the target without effort.

In situations where the net cannot trace an efficient trajectory a sub-target is computed (defined as a point pertaining to the quasi-optimal trajectory) which is easier to obtain than the initial target (fig. 9).

4. NEURAL CONTROLLER

The guidance process starts by marking the original and target position on the selected navigation environment. In this way, the neural net is presented with the patterns associated with successive positions reached by the mobile robot. The net input is the dispersion index for each subarea, and the most satisfactory direction towards the target is the net output.

A learning method based on the generalized delta rule [RHW96] has been optimized by using both an input patterns discretization process and ternary encoding prior to training. The discretization process obeys to the division of the environment in S sectors by Z zones, and ternary encoding enhances the discrimination capability of the net. Output patterns have been encoded in binary and assigned through a double decreasing distribution centered on the maximum output value, that is, the chosen direction. Below it is shown net input entries and

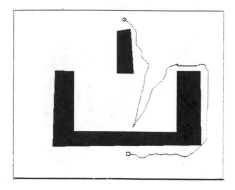

FIGURE 8. Example of indecisive situations

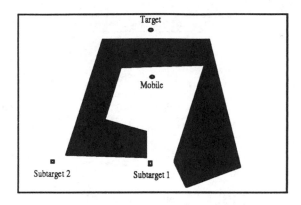

FIGURE 9. Sub-target computation.

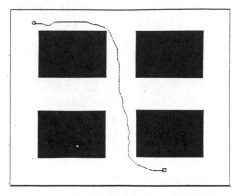

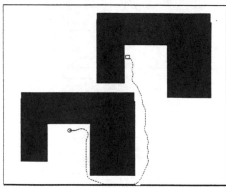

FIGURE 10. Net execution with different maps.

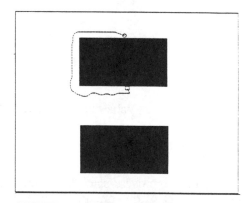

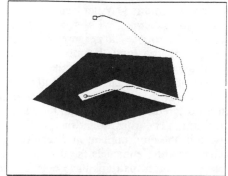

FIGURE 11. Training with sub-targets.

5. RESULTS

The navigation map is considered to be divided into 16 sectors and 3 concentric zones, this choice being a compromise between net size and trajectory tracing efficiency.

The backpropagation net is structured on three levels:
- Input level, 144 neurons (16 sectors x 3 zones x 3 digits).
- Intermediate level, 100 neurons (empirical value).
- Output level, 16 neurons (16 directions).

The net was trained over 20 trajectories in two different navigation environments, with 200 training patterns and 35 epochs. Training trajectories were traced for environments with rectangular shaped obstacles

The net´s ability to trace trajectories not only over the training map but also over unknown maps is shown in fig. 10.

The computation of sub-targets was necessary for navigation maps with concavities, as well as for the removal of limit cycles , as is demonstrated in fig. 11.

The neural net showed a 90% efficiency average in searching for quasi-optimal trajectories.

CONCLUSIONS

A neural network-based solution for navigation in an environment with obstacles has been presented. The net has shown great capability of generalization capability to previously unknown environments.

The best results have been obtained for environments having a large number of obstacles. In the presence of concavities or complex maps the net is provided with computational aids, such as sub-target computations, which avoid situations the net is incapable of tackling giving a satisfactory solution. Nevertheless, extending information about perimeter, size or relative position of the obstacles, would prevent their use.

Future applications will be devoted to tackling navigation environments having mobile obstacles without noticeably modifying the original solution presented here, as well as incorporate the mobile robot dyamics to the navigation process.

REFERENCES

[HF93]   A. Ho and G. Fox. Motion Planning for a Mobile Robot on Binary and Varied Terrains. *IEEE International Workshop on Intelligent Robotics and Systems*, 1993.

[KV92]   A. Kassim and B. Vijaya. A Neural Network Architecture for Path Planning. *Proceedings of the IEEE International Conference on Neural Networks*, 1992.

[Kha86]   O. Khatib. Real-Time Obstacle Avoidance for Manipulators and Mobile Robots. *International Journal on Robotics Research*, 1986.

[Loz83]   T. Lozano-Pérez. Spatial Planning. A Configuration Space Approach ". *IEEE Trans. on Computers*, Vol. 2, 1983.

[MP92]   H. Meng and P. Picton. Obstacle Avoidance Using a Neural Network Controller and Visual Feedback. *Symposium on Intelligent Components and Instruments for Control Applications*, 1992.

[PL90]   J. Park and S. Lee. Neural Computation for Collision-Free Path Planning. *Proceedings of the IEEE International Conference on Neural Networks*, 1990.

[RHW86]   D. Rumelhart, G. Hinton and R. Williams. Learning Internal Representations by Error Propagation. *PDP Group, MIT Press*, 1986.

[SL92]   B. Steer and M. Larcombe. A Goal Seeking and Obstacle Avoidance Algorithm for Autonomous Mobile Robots. *Symposium on Intelligent Components and Instruments for Control Applications*, 1992.

[Tic90]   R. Ticlore. Local Obstacle Avoidance for Mobile Robots Based on the Method of Artificial Potential. *Proceedings of the IEEE International Conference on Robotics and Automation*, 1990.

ULTRASONIC PERCEPTION: TRI-AURAL SENSOR ARRAY FOR ROBOTS USING A COMPETITION NEURAL NETWORK APPROACH

J. Chen, H. Peremans and J. M. Van Campenhout

Department of Electronics and Information Systems
State University of Gent
Sint.-Pietersnieuwstraat 41, B-9000 Gent, Belgium
e-mail: jie.chen@elis.rug.ac.be, peremans@elis.rug.ac.be, jvc@elis.rug.ac.be

ABSTRACT

For truly autonomous behavior of the mobile robot, it is essential to make a mobile robot understand its environment. For this reason a certain measurement device must be installed on the mobile robot. In our system we make use of a tri-aural ultrasonic sensor array to observe the robot's environment. The tri-aural sensor array is composed of one transceiver and two additional receivers. The arrival times of reflected echoes received by the sensor array provide information about the robot environment. From these observations, the robot has to determine the number of objects in the sensor field and their positions. We have developed a competition neural network to process this information. When we send arrival time information to the competition neural network, the number of the objects and their positions in the sensor field can be generated by this network. In this paper, we describe the competition neural network and explain how it solves the correspondence problem. To illustrate how the competition neural network fares in realistic circumstances, we have done a number of simulations.

1 INTRODUCTION

The operation of an intelligent mobile robot requires a device for observing the environment, the purpose of such a device is similar to the human eye. In this paper we propose a new robot vision system which consists of an ultrasonic sensor array combined with a competition neural network. The tri-aural sensor array is used to inspect the robot environment. The competition neural network is used to process the arrival time data extracted from the acoustical signals. From these arrival time information, the competition network must accurately determine the number of objects and their positions (distance and bearing) in the sensor field.

It is well known that the neural network can be trained to perform a great variety of complex tasks. In our system, we could make use of a multi-layered feed-forward neural network to learn the correspondence between input arrival times and output object positions [CPC93]. However, in realistic situations, the reflected echoes are not always detected by the sensor array or sometimes the sensors can measure additional echoes which might be caused by multiple reflections. Consequently, not all straightforward interpretation of the echoes present correspond with the true situation. We need a mechanism to compare the different interpretations and keep the best ones only. Such a mechanism is lacking from a feedforward neural network. A competition neural network however is different, it has lateral interconnections among the output neurons. These lateral interconnections provide a mechanism to force consistency on the output neurons. This is an important motivation for us to study the competition neural network.

In the next section we give a short overview of the tri-aural sensor array and the correspondence problem. In section 3, we then describe the competition neural network. Section 4 contains a performance analysis of this new approach. Finally, we conclude the paper with a discussion of the results and some suggestions for future improvements.

2 ULTRASONIC SENSOR ARRAY

In order to help the reader to understand our robot perception system, we first briefly describe the tri-aural sensor array system and then explain the correspondence problem.

2.1 Tri-aural Sensor Array

In our system, in order to simplify matters, we hypothesize the robot environment to be a specular 2D environment. This environment consists of three primitive types of reflectors only: edges, planes and corners as first proposed by Kuc [KS87]. The object position in the sensor field is described by using polar coordinates (range: r, bearing: θ).

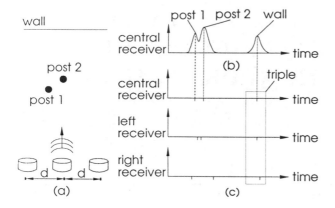

Figure 1: Tri-aural sensor array and its measurement.

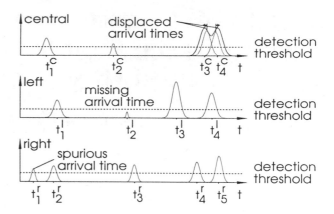

Figure 2: Difficult problems: spurious echoes, missing echoes, displaced echoes.

The tri-aural sensor array [PAC93], as shown in Fig. 1(a), consists of three sensors lined up and spaced $d = 15$ cm apart. The central sensor is used both as a transmitter and a receiver, the two peripheral ones are used only as receivers. The measurement cycle of the tri-aural sensor array is made up of two steps, as shown in Fig. 1: first, the central sensor emits a pulse; next, at each of the three receivers, the precise arrival times of the echoes are extracted from the raw sensor signals [KAC92]. If we denote these arrival times by t_c, t_l and t_r for the central, left and right receiver respectively and the speed of sound by v_s, the measurement vector is given by

$$\mathbf{m} = (m_c, m_l, m_r)^t = v_s(t_c, t_l, t_r)^t.$$

If more than one reflecting object is present in the sensor field, the single emitted pulse results in a number of echoes at each receiver, as shown in Fig. 1(b). These lists of arrival times are to be combined to find those that pertain to the same reflector, resulting in a list of triples. A triple consists of the three arrival times, one from every receiver, of the echoes returned by a single object in the field of view, Fig. 1(c).

2.2   Correspondence Problem

In order to detect the object positions by the tri-aural sensor array, it is necessary that the arrival times from the three receivers are united into triples that pertain to the same object. The problem of how to group the arrival times into triples is very similar to the correspondence problem as described in vision theory.

As we all know, in realistic situations all kinds of complications may occur, such as influence of noise and multiple specular reflections between the objects. Upon reception of the reflected echoes by the tri-aural sensor array, the acoustical signals are processed by a matched filter [KAC92], a possible output from this

filter is shown in Fig. 2. Echoes from objects that are too close together can not be separately detected, and will be perceived as one echo, displaced with respect to the true position. Also, some echoes may be too weak (echo amplitude is less than detection threshold, see Fig. 2) to be found, and other, spurious echoes caused by multiple specular reflections from other objects may be present. These deviations from the ideal situation result in missing echoes, spurious echoes and displaced echoes as we call them. In our robotic sensor system, the most important difficulties faced by any algorithm that tries to solve the correspondence problem are summarized in Fig. 2. Below we briefly describe these problems:

- Spurious echoes might occur when the signal processing software detects false echoes in the outputs of the receivers. This occurs mostly when multiple echoes overlap or when multiple reflections occur.

- Missing echoes would occur when the echo amplitude is too weak to be detected (amplitude value is less than the threshold).

- Displaced echoes may occur when the two echoes overlap considerably, their arrival times are often displaced by an amount which is much larger than the expected measurement noise.

3   NEURAL APPROACH

The question we want to answer in this section is how to determine the number of objects as well as their positions in the sensor field from the reflected echo information.

3.1   Architecture

The architecture of the competition neural network is an one-layer neural network, it is shown in Fig. 3(b).

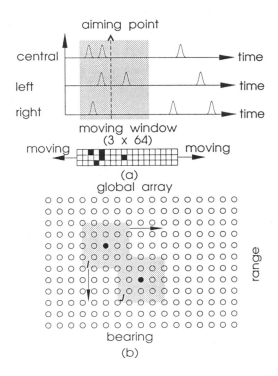

Figure 3: The neural network input array and global competition array.

Note that the moving window at the top serves only to transform the echo arrival time information into the input of the competition neural network (see Fig. 3(a)), it performs no competition and hence would not be considered to be part of the competition neural network array. There is no distinction between input cells and output cells in the global competition array as shown in Fig. 3(b). All these cells make up a single matrix array, each cell in the array represents the exact position of an object in the sensor field. The length of the network array represents the range of the sensor field, the width of the network array represents the angular extent of the sensor field, as shown in Fig. 3(b). The bearing resolution is 0.5° and the range resolution is 0.085 cm. The moving window consists of a binary matrix of (3 × 64) cells as shown in Fig. 3(a). The moving window is used to scan the echo information along the arrival time axis. The resolution of the moving window is 0.085 cm and it is equal to the scanning step of the moving window. We also define an aiming point placed on the 26-th column of the moving window, which is used to indicate the exact position of the moving window. After the moving window scans a certain distance along the arrival time axis, the arrival time information is transformed into the competition neural network.

3.2 Initialization

The objects in a scene give rise to two classes of triples in the arrival time domain: (1) complete triples, consisting of three arrival times, one from each receiver; (2) defective triples or couples, consisting of two arrival times only. In addition we assume that the arrival times are corrupted by measurement noise, the noise model derived in [PAC93] is Gaussian distributed noise. If we suppose the observation vector to be the arrival time vector, we can then represent the hypothesized object positions in the sensor field by the following likelihood functions.

We assume that the triple of arrival times predicted by the object parameter vector $\overline{p}$ is denoted by $\overline{h}(\overline{p}) = [\hat{t}_c, \hat{t}_l, \hat{t}_r]^t$, the measured triple is represented by $\overline{t} = [t_c, t_l, t_r]^t$. The parameter vector of an object consists of the position in polar coordinates and the object type: edge or plane [PAC93]. In this paper we state the results assuming edges only. The results can be easily extended to planes as well. The difference between predicted triple $\overline{h}(\overline{p})$ and measured triple $\overline{t}$ is due to Gaussian noise $\sim \mathcal{N}(0, \sum_{clr})$. The likelihood function for a complete triple, given the object position $\overline{p}$, is then given by

$$p_3(\overline{t}|\overline{p}) = \frac{|\sum_{clr}|^{-1/2}}{(2\pi)^{3/2}} \exp(-\frac{1}{2}(\overline{t}-\overline{h}(\overline{p}))^t \sum_{clr}^{-1}(\overline{t}-\overline{h}(\overline{p}))),$$

where covariance matrix $\sum_{clr}$ is derived in [PAC93]. If the reflection from the object is not detected at the right receiver, the predicted defective triple is given by $\overline{h}(\overline{p}) = [\hat{t}_c, \hat{t}_l]^t$, the measured defective triple is given by $\overline{t} = [t_c, t_l]^t$. The likelihood function for the defective triple is then

$$p_2(\overline{t}|\overline{p}) = \frac{|\sum_{cl}|^{-1/2}}{2\pi} \exp(-\frac{1}{2}(\overline{t}-\overline{h}(\overline{p}))^t \sum_{cl}^{-1}(\overline{t}-\overline{h}(\overline{p}))),$$

where covariance matrix $\sum_{cl}$ is also given in [PAC93].

In order to simplify matters we assume that the arrival times from different receivers are independent, therefore the above two likelihood functions can be simplified. The likelihood function for a complete triple is given by

$$p_3(\overline{t}|\overline{p}) = p_1(t_c|\overline{p})p_1(t_l|\overline{p})p_1(t_r|\overline{p}), \quad (1)$$

the likelihood function for defective triple is given by

$$p_2(\overline{t}|\overline{p}) = p_1(t_c|\overline{p})p_1(t_l|\overline{p}), \quad (2)$$

where $p_1(t|\overline{p}) = \frac{1}{2\pi\sigma} \exp(-\frac{(t - \hat{t})^2}{2\sigma^2})$, the σ is a standard deviation.

We now move the window over the arrival time axis. These neurons corresponding to the range indicated by

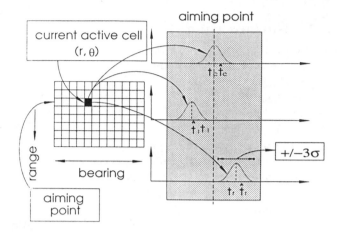

Figure 4: The current active neuron.

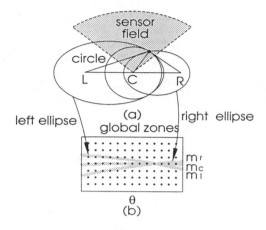

Figure 5: Competition zones: one circle and two ellipse.

the position of the aiming point are initialized with (1) or (2) depending on the situation. The predicted arrival times corresponding with each neuron are compared with the actually measured ones, as shown in Fig. 3. If three arrival times are found within a $\pm 3\sigma$ zone around the predicted arrival times, the equation (1) is used; otherwise, the neural network uses the equation (2) to calculate the likelihood value with which that neuron is initialized.

3.3 Competition Algorithm

The competition neural network is a recurrent feedback network, each cell in the network array is an input neuron receiving activation from its neighbours and it is also an output neuron sending its activation value to its neighbours. During competition among the neurons, the activation of every neuron is always recalculated and used to modify the others neurons in the competition area. This processing is repeated again and again, the successive iterations producing smaller and smaller output changes until the outputs become constant [Fau94]. The entire competition algorithm is divided into two steps: (1) local competition, each neuron competes with its neighbours within the local area; (2) global competition, every peak competes with its neighbouring peaks.

3.3.1 Local Competition

The local competition region is a rectangular array neighbourhood as shown in Fig. 3(b). Each neuron in this rectangular grid (5 × 5 matrix) is fully interconnected to its nearest neighbours without self-connection. Every local competition sub-network in our system is a small Mexican-hat network [Koh89].

The competition principle employed among the neurons is called *Winner Take All* [Fau94]. In our system,

there is no training algorithm for the neural network, the weights between the neurons are fixed. If we denote the link weight between cell j and cell i by w_{ij}, assuming that the output of a cell is equal to its activation value, then the activation value received by cell j is given

$$y_j(t+1) = y_j(t) + \sum_{i \neq j} Sign[y_j(t) - y_i(t)]w_{ij}y_i(t),$$

where $i \in$ local region. With this formula the activations for each cell in the local competition region is recurrently updated ($t = 1, 2, \cdots, n$).

3.3.2 Global Competition

The shape of the global competition area can be defined as follows. We determine the position of the reflector by finding the measured distances m_c, m_l, m_r. The object position is on the intersection of three curves: one circle and two ellipses respectively, as shown in Fig. 5(a). The measurement m_c stands for the distance from the sender to the reflector and back to the central receiver, the locus of such points is of course a circle. The measurements m_l and m_r stand for the distance from the sender to the reflector and back to one of the peripheral receivers, the loci of such points are ellipses (refer to [PAC93]).

For the global competition, we have to add the restriction that every arrival time is contained in exactly one triple. Allowing one explanation for each measured arrival time then reduces to allowing only one peak in each of the three global competition areas as shown in Fig. 5(b). The competition mechanism used by the global competition algorithm is still *Winner Take All*, the way to determine the winner is as follows. After local competition every peak has its own competition

Table 1: Environment: two closely spaced objects.

	range (cm)	bearing (deg)
object 1	150.0	-2.0
object 2	151.0	-9.0

Table 2: The results from the competition network.

situation	σ_r (cm)	σ_θ (deg)	nHit (%)	nFalse
ideal	0.089	0.92	100	0
missing	0.089	0.94	89	2
spurious	0.092	0.94	92	34

zones as mentioned above (see Fig. 5(b)), then we can check the number of neighbouring peaks for each peak. The principle for determining who is the winner peak is that the peak with the fewest neighbouring peaks will be chosen as the winner. During the competition between the peaks, the extent of the competition zones corresponding to each peak is increased. The peak that is closest to the winner will be swallowed first. If we denote the winner peak by j and its neighbouring peaks by i, assuming that the output of a cell is equal to its activation value, then the activation value received by the winner peak j is given

$$y_j(t+1) = y_j(t) + \sum_{i \neq j} \frac{\alpha}{d_i^2} y_i(t),$$

where $i \in$ dynamic region, α is a proportional constant and d_i is the distance between cell j and cell i.

4 SIMULATION RESULTS

Up to now, we have described the architecture and the competition mechanism. In order to see how well the competition network approach is able to solve the correspondence problem, especially in the case of missing and spurious echoes, we use simulation data to evaluate this approach. The simulated robot environment consists of two objects that stand very close together in the sensor field. The true positions are displayed in Table 1.

4.1 Performance Measures

Before we analyse these statistical results, we explain how we evaluate the performance of the competition neural network. In view of the fact that the competition neural network is used for object localization, we make use of localization results to evaluate the neural network. Below we briefly define mathematical expressions for the four performance measures.

Standard deviations of the position estimates. We assume the true object position is represented by (r^*, θ^*), the estimated object position generated by the competition neural network is represented by (r, θ). We can calculate the Mahalanobis distance between these two points to decide if the estimated position falls within the 3σ region (Gaussian distribution) around

the true position. The formulas of calculating standard deviations are given by

$$\sigma_r = \sqrt{\frac{\sum_{i=1}^{n}(r_i - r^*)^2}{n-1}}, \ \sigma_\theta = \sqrt{\frac{\sum_{i=1}^{n}(\theta_i - \theta^*)^2}{n-1}},$$

where n is the number of measurements.

Hit rate and false alarm rate. As we mentioned above, if the estimated position falls within the 3σ region around the true position, the object is assumed to be detected. An object can be seen at most once by a single measurement, i.e., $hit_{obj,i} \in \{0, 1\}$. If the estimated position falls outside all the 3σ regions, we assume that the competition neural network generates a false object. The hit rate for each object and the average number of false alarms are given by

$$hit_{obj}(\%) = \frac{1}{n} \sum_{i=1}^{n} hit_{obj,i} \times 100\%,$$

$$nFalse = \frac{1}{n} \sum_{i=1}^{n} false_i,$$

where n is the number of measurements and $false_i$ is increased for each object not corresponding with a real object, i.e., $false_i$ can take on any positive value.

4.2 Statistical Analysis

Below we describe the different situations that were studied. For each situation we have added realistic measurement noise (Gaussian noise $\sim \mathcal{N}(0, 0.1)$) to the theoretically calculated results. The results from 100 simulation runs for each situation are summarised in Table 2.

Two closely spaced posts (ideal case), no interference. In this situation we assume that the echoes do not interfere with one another. From the third column in the Table 2 we conclude that both objects were seen every time by the competition neural network. The first two columns in the Table 2 give the standard deviations of the position estimate (r, θ). The last column indicates that no ghost objects were created by the competition neural network. From these statistics we conclude that as long as the echoes do not interfere, the competition neural network performs very well. Hence, having cluster of very closely spaced, non-interfering, arrival times does not seem to pose a problem to this approach as such.

Two objects with missing arrival times, no interference. To simulate the occurrence of missing arrival times we have taken the previous data set and eliminated randomly one of the closely spaced arrival times. Furthermore, the remaining arrival times have a fixed probability of 5% of being eliminated too. From the third column in Table 2 we see that the probability that the two objects are still seen by the competition neural network is 89% of the time. The hit rate for this situation has slightly decreased. This is to be expected because sometimes too many arrival times get eliminated and thus the neural network no longer has sufficient information to recognise the presence of an object. From the last column we conclude that the average number of ghost objects created each measurement has risen to 0.02. If arrival times are eliminated it becomes more difficult for the neural network to figure out which arrival times belong together and thus it starts making mistakes.

Two objects with spurious echoes added, no interference. We now add spurious arrival times to the data set of the first experiment, every two consecutive arrival times that are close together have a probability of 50% of generating an additional spurious arrival time. Notice that the uncertainty on the position estimate is still similar to the one in experiments described above. From the third column in Table 2, we see that the hit rate is close to the ideal situation. However, from the last column, we notice that an average 0.34 ghost objects were created each measurement. This rise in the number of false objects is caused by the neural network trying to explain every arrival time. The neural network has no notion of spurious echoes, it is designed to propose an object to explain each arrival time.

5 CONCLUSIONS

From the analysis in the previous section, we can derive a number of conclusions. First of all and perhaps most importantly, the performance of the competition neural network approach described in this paper is very good when compared with other approaches [Per94], especially when missing echoes and spurious echoes occur. The position accuracy is very good when confronted with scenes consisting of multiple objects as long as the echoes do not interfere with each other.

From Table 2 we realise that the competition neural network method generates too many ghost objects

when the data include spurious arrival times. To solve this problem, we are currently investigating the following remedies: (1) One possibility would be that the radiation patterns of the transducers could be used to evaluate the likelihood of proposed triples and couples by the competition neural network; (2) Since the information currently extracted from a single measurement by the tri-aural sensor array is not sufficient, we have to utilize data from different measurements, collected at different viewpoints.

Acknowledgements

This work was supported by the Interuniversity Attraction Pole No.50 initiated by the Belgian State, Prime Minister's office, Science Policy Programme.

References

[CPC93] J. Chen, H. Peremans, and J. Van Campenhout. Application of an artificial neural network in an object localization system using ultrasonic sensors. In *Proc. of Robotics and Manufacturing*, pages 20–23, Oxford, England, September 1993.

[Fau94] Laurene Fausett. *Fundamentals of neural networks*. Prentice Hall International, Inc., 1994.

[KAC92] Y. Kawahara K. Audenaert, H. Peremans and J. Van Campenhout. Readings in computer vision: issues, problems, principles, and paradigms. In *IEEE Int. Conf. on Robotics and Automation*, pages 1733–1738, Nice, France, 1992.

[Koh89] T. Kohonen. *Self-organization and associative memory*. Springer-Verlag, Berlin, 1989.

[KS87] R. Kuc and M. Siegel. Physically based simulation model for acoustic sensor robot navigation. *IEEE Transactions on Pattern Recognition and Machine Intelligence*, pages 766–778, 1987.

[PAC93] H. Peremans, K. Audenaert, and J. Van Campenhout. A high-resolution sensor based on tri-aural perception. *IEEE Transactions on Robotics and Automation*, pages 36–48, 1993.

[Per94] H. Peremans. A maximum likelihood algorithm for solving the correspondence problem in tri-aural perception. In *IEEE Int. Conf. on Multisensor Fusion and Integration for Intelligent Systems*, pages 485–492, Las Vegas, USA, 1994.

Vision

A Genetic Algorithm For Image Segmentation

A. Calle   W.D. Potter   S.M. Bhandarkar
Artificial Intelligence Center, University of Georgia
Athens, Georgia 30602–7415, USA
potter@pollux.cs.uga.edu

ABSTRACT

In this paper we present a genetic algorithm for image segmentation. Instead of adopting the chromosome-like encoding of the classical genetic algorithm we propose a more "natural" representation that adheres to the basic building blocks philosophy underlying the genetic algorithm. As a result, we devise novel crossover and mutation operators. We also devise a hill-climbing procedure to speed up the convergence of the genetic algorithm. The basic philosophy underlying our algorithm is to first identify the building blocks, and then to choose a representation that allows us to manipulate the building blocks by means of the genetic operators. The results are quite illustrative of this philosophy.

1 INTRODUCTION

The purpose of this paper is to introduce a new technique for the image segmentation problem. It is based on an evolutionary technique known as the genetic algorithm. Genetic algorithms have been thoroughly researched and used, often with very good results, in a variety of application domains over the past decade. The genetic algorithm proposed in this paper, for the purpose of image segmentation, introduces several novel ideas, and is in some respects, a departure from some of the ideas and concepts underlying the traditional genetic algorithm.

1.1 Image Segmentation

Image segmentation is the process by which an image is partitioned into disjoint regions such that each region is homogeneous with respect to some predefined homogeneity criterion. Image segmentation is a very important problem in computer vision. A segmented image can be looked upon as the highest domain-independent abstraction or description of the raw (i.e. pixel-based) input image. A segmented image is the typical input to high-level vision processes which then analyze and interpret the image contents using domain-specific knowledge. It is critical that the image be properly segmented because the performance of the high-level vision processes is greatly affected by the quality of the segmented image.

There are several ways in which to tackle the problem of image segmentation [Hara85]. Most image segmentation techniques described in the literature can be categorized into three classes: (i) segmentation techniques based on feature vector clustering, (ii) segmentation techniques based on edge detection and, (iii) segmentation techniques based on region extraction [Fu81]. In this paper, we adhere to the image segmentation approach based on feature vector clustering, where the goal is to group the image pixels into clusters so as to optimize an objective function. In this paper, the objective function depends entirely on the pixel gray level although it can be extended to incorporate any multidimensional set of features derived from the pixel gray level values. Due to the sheer volume of pixel-level data, it is not possible to tackle the clustering problem without using a sensible search technique that is capable of exploring the search space of probable clusters and, at the same time, capable of guiding the algorithm towards the optimal solution.

Deterministic techniques such as thresholding rely on a procedure to label the set of image pixels. They are generally based on a feature histogram that is used to identify the feature range values that are used to partition the image. The resulting search for the feature range values can be done in a reasonable amount of time but this technique is unable to explore alternative solutions. Thus, the segmentation algorithm could be easily biased towards a sub-optimal clustering in the absence of exploration. On the other hand, an exhaustive enumeration process that explores all possible clusterings of the pixel data is computationally prohibitive because of the large volume of the pixel data involved.

Our approach uses a genetic algorithm-based technique that initially performs an exploration of the search space and then increasingly focuses on the most promising clusterings. Consequently our approach is much less likely to be trapped in a sub-optimal clustering. Furthermore, it is not unduly time consuming as the exploration phase utilizes useful information such as the spatial location of pixels and the similarity of pixel gray values. Our technique adheres to the core ideas of natural selection rather than adhering to the typical approach that has been used in the implementation of most genetic algorithms. Other applications of the genetic algorithm in image processing have been reported in [Bhan94] and [Fitz84].

1.2   Genetic algorithms

Genetic algorithms were first introduced by Holland [Holl75]. Genetic algorithms have their roots in the Darwinist concept of natural evolution. In a given environment any point in time, nature tends to favor the presence of the fittest members of a certain species and the features that characterize these members. This is the well-known process of natural selection, also termed as *survival of the fittest*. The fitter an individual, the greater his chances of living longer and reproducing. Consequently, the features that characterize fitter individuals (and in fact make them fitter), tend to prevail and even increase among the members of subsequent generations. At the same time, random changes in the form of mutations also take place which, in case they are positive, also survive into subsequent generations.

The key idea behind this evolutionary approach to optimization lies in the manipulation of the *environmental information* about how to survive (i.e. how to be fitter) in the context of a given environment. This information is encoded within the representation of each individual in the population. To manipulate this information we rely on the selection mechanism and on the genetic operators, *crossover* and *mutation*. With crossover, we provide individuals with a mechanism to "talk" among themselves and thereby exchange information. The selection mechanism encourages the crossover of the fittest individuals and thereby the exploitation of their characteristic features. The mutation mechanism introduces diversity in the population and enables exploration of alternative solutions.

The classical implementation of these ideas are to be found in Holland's genetic algorithm. The kernel of the algorithm is the chromosome-like representation of the solutions and the genetic operators: crossover and mutation. Individuals (solutions) are represented in the form of binary strings, which serve as chromosomes. One possible implementation of the crossover mechanism takes a pair of individuals (i.e. parents) and produces two offspring by selecting a crossover point and

interchanging the substrings of the parents. The probability of crossover is proportional to the fitness values of the parents which is evaluated using a *fitness* function. Crossover does not always take place i.e there exists a probability (1 - *pcross*), that the mating pair avoids crossover in which case both parents are simply copied into the next generation. The mutation mechanism flips the value of a randomly chosen bit in the binary string. The probability of a bit being mutated is indicated by the mutation parameter, *pmut*.

In the classical genetic algorithm the environmental information is encoded within the small substrings of a solution whose fitness (i.e. the average fitness of the different solutions containing this substring) is very high. These substrings are called the *building blocks* or *schemas* and they are the key to the fitness of the individuals and to the performance of the genetic algorithm as a whole. The building block hypothesis states that the proper juxtaposition of these substrings will tend to yield near-optimal solutions. The purpose of the genetic algorithm is to promote the presence of these building blocks through successive generations [Gold90].

2   GA APPROACH

When the building-block hypothesis fails, we are faced with what is termed in the the genetic-algorithm literature as a *deceptive* problem. There are several problems that are deceptive and therefore hard to solve using a genetic algorithm. One of the limitations of the classical genetic algorithm philosophy is that it is constrained to interpret, preserve and propagate useful environmental information in the form of substrings or schemas with high average fitness. This is due mainly to the fact that we are tied to a binary string-like chromosome representation.

Rather than tailoring the algorithm to the problem, the classical genetic algorithm approach tends to tailor the problem to the algorithm. In our genetic algorithm-based approach to image segmentation, the representation of the solutions needs to follow as naturally as possible from the problem definition. In other words, the building blocks, i.e. the solution structures that are to be preserved and propagated over successive generations, need to embody the valuable information contained in a solution that makes it better than the alternative solutions.

What are the building blocks in the image segmentation problem? Obviously, a solution is better than another one if it contains better clusters. So, it is natural to consider the set of clusters within a given solution as the environmental information contained in this solution. Consequently, a solution is represented by a "string" of clusters. By means of the genetic operators i.e. crossover and mutation, we need to promote the presence of good environmental information, that is, good clusters over successive generations of the genetic algorithm. Also note that the building block hypothesis is naturally satisfied i.e. the juxtaposition of good clusters will yield good solutions.

The genetic operators must carry on a search over the space of building-blocks in order to identify the most promising clusters. In the image segmentation problem it is clear that compact regions in the image whose pixels are homogeneous (based on some predefined homogeneity criterion) would constitute some of these building blocks. It is therefore natural to introduce a procedure prior to the genetic algorithm, so that these compact homogeneous regions can be identified. When employing genetic algorithms we need to be very careful about the incorporation of heuristics as they could bias the search towards a local optimal solution. It is therefore important not to be very greedy or opportunistic during the initial phases of exploration where the building blocks of the solution are being identified else the genetic algorithm would be subsequently limited to a very small portion of the search space. Once the building blocks are identified, the initial population can be generated.

3 GA IMPLEMENTATION

As pointed out earlier, prior to the genetic algorithm itself, we need to carry out an exploration the image to identify the building blocks. We have to bear in mind, however, that the kernel of the overall application must be the genetic algorithm. The genetic operators are ultimately responsible for providing a good trade-off between exploration and exploitation of the search space of possible ways to segment the image.

3.1 The test image

To illustrate the performance of the genetic algorithm we have chosen a small image containing a small portion of a computer monitor screen against a fairly uniform background (Figure ??).

3.2 The atomic clusters

The first step in the genetic algorithm-based segmentation is the identification of very compact homogeneous regions in the image which we term as *atomic clusters*. The atomic clusters serve as building blocks and help in the initialization of the first generation.

The atomic clusters are required not contain an edge nor be close to any edges in the image. The atomic clusters constitute the very inner portion of the objects that we wish to segment in the image. Any object must contain at least one of these atomic clusters, otherwise it would be considered so noisy that it should not be worthy of any further attention. In addition, the definition of the atomic clusters would also help the genetic algorithm to focus on on the type of objects that the user wants to extract, since the user would be required to determine the degree of compactness required by these atomic clusters.

To identify the atomic clusters we make use of the pixel gray value histogram. The peaks of the histogram and their neighborhood, correspond to the gray level range values of pixels that are likely to be contained within these atomic clusters. These pixels, that are termed as *atomic pixels* are used as starting points for the first generation of the atomic clusters. Before employing a region growing procedure, we test whether the atomic pixels would result in an atomic cluster. Possible ways to test the compactness of the neighborhood around an atomic pixel could be based on the gray level variance of the pixels in the neighborhood, the derivative of the gray level values of the pixels in the neighborhood (which must be close to zero in "smooth" areas), the number of pixels in the neighborhood with a "similar" gray value (i.e. a sort of nearest-neighbor test), etc.

Once an atomic pixel has passed the initial test, we apply a region-growing technique that ensures the compactness feature of an atomic cluster. In our case, we have focused on preventing the region-growing algorithm from straying into "rough" areas i.e. areas in the image with lot of gray level detail. The region-growing procedure is as follows: We treat each of the atomic pixels as the first member of an atomic cluster and incorporate the adjacent pixels whenever they satisfy the region-growing test. This test is positive if the variance in the local neighborhood of the pixel (where the local neighborhood is window of size 3×3 centered at the atomic pixel, for instance) is lower than some predefined threshold. The process is repeated by enlarging the search window around the atomic pixel. The process is halted when we reach a window in which it is not possible to label a new pixel as belonging to the atomic cluster. The region-growing procedure outlined above ensures the connectivity of the atomic clusters.

Obviously, the larger the value of the threshold, the larger the resulting atomic cluster will be and the less compact the final region will be. In order to set up a proper exploration of the possible clustering of the image it is better to start with very compact clusters and let the random labeling procedure (described in the following subsection) come up with a diversity of initial images. Figure **??** shows the atomic clusters generated for the test image for two different threshold values.

3.3   Initial population

After the initialization of the atomic clusters, we proceed to label the rest of the pixels. To do this, we randomly go over the boundaries of the atomic clusters, incorporating the unlabeled pixels that fall within a neighborhood window centered at the boundary pixel. The size of this window is determined by the user. The larger the size, the less "homogeneous" the expansion and also less the time needed to generate an image. Figure **??** shows some members of the initial population.

3.4   Selection mechanism

The selection mechanism determines the mates that will undergo crossover to generate the next population. First, we establish a fitness function to compare the solutions. It is clear that this fitness function must encourage the compactness of the clusters that constitute a solution. The fitness function must also encourage a proper definition of the cluster boundaries. We consider well defined cluster boundaries to be important because they correspond to edge pixels that constitute object boundaries.

The fitness of a solution is defined to be the sum of the individual fitness values of its constituent clusters. The fitness of a cluster C, is given by:

$$f(C) = W_1 * (\sigma_{max} - \sigma_{tot}(C)) + W_2 * \sigma_{edge}(C).$$

where $\sigma_{tot}(C)$ is the variance of the pixel gray level values in C, $\sigma_{edge}(C)$ is the average variance of pixel gray level values computed in a local neighborhood along the boundary of C, σ_{max} is a constant and W_1 and W_2 are weights used to balance the two terms in the fitness function. The examination of the fitness function shows that the more compact the cluster (i.e. the lower the cluster variance), the higher the fitness function. At the same time, the higher the variance along the cluster boundary pixels, the better the cluster (i.e. better the contrast between the pixels within the cluster and those outside the cluster).

In order to create a list of potential mates, we have adopted a rank-based selection method. As the fitness function is an "artificial" way to evaluate the solutions, we reject the typical roulette wheel method of genetic algorithms. The rationale for this choice is that a solution whose fitness is twice that of another does not necessarily deserve to generate twice as many offspring.

The rank-based selection method assigns a certain percentage of the total number of offspring to a certain percentage of the population. In our experiments, we let the top 20% (in terms of fitness values) of the solutions contribute towards the creation of 40% of the individuals in the following generation. The remaining 60% of the individuals in the following generation come from the next 60% of the solutions in the present generation. The bottom 20% of the solutions in the present generation do not contribute any offspring for the following generation.

3.5   Crossover mechanism

The crossover mechanism is the kernel operator of any genetic algorithm. As previously mentioned, the crossover operation must provide a means of interchange of the possible building blocks within the mating solutions. Given a pair of solutions to be mated, one of them is selected as the dominant parent. The dominant provides half of its clusters, which are randomly chosen, to one of its offspring. We then scan the set of atomic pixels that have not yet been labeled in the offspring. To these atomic pixels, we assign the corresponding cluster from the second parent which the atomic pixel belongs to. In the case of overlap, the cluster from the dominant parent takes precedence. To account for pixels that remain unlabeled, a random expansion of the clusters is carried out.

3.6   Mutation mechanism

The interchange of information between clusters via the crossover operation may not be enough to reach the global optimum. It is necessary to maintain a certain degree of diversity in the population by means of a mutation mechanism. While crossover represents the exploitation of the best information at hand, mutation can be regarded as an exploration of alternatives in the solution space. As is the case with nature, mutations must not be very disruptive. Mutated solutions must "resemble" the original ones as the mutations aim to introduce incremental improvements in the current population.

Mutation is performed over any pixel on the cluster boundary with probability $pmut$, which is set by the user. Since we are dealing with an cluster boundary pixel, some of its adjacent pixels will belong to a different cluster. The mutation operation at a cluster boundary pixel essentially incorporates of all of the pixels in a local neighborhood of the cluster boundary pixel into the cluster. The local neighborhood is defined as the set of pixels that fall within a window centered on the cluster boundary pixel. The size of the window is defined by the user. Our mutation procedure can be regarded as a random expansion of the set of clusters.

3.7 Hill-climbing mechanism

Genetic algorithms are very good at detecting the most promising regions within the space of solutions. The final solution will be in the vicinity of the optimum solution with a high probability. Nevertheless, the genetic algorithm may lack the ability to incorporate the precise information needed to tune the best solution to the global optimum. This motivates the inclusion of a deterministic procedure i.e. the hill-climbing mechanism which consists of a local optimization procedure that attempts to improve the quality of the solution.

Our hill-climbing mechanism works over the set of cluster boundary pixels of a particular solution. Its goal is to slightly reshape the inter-cluster boundaries so that more accurate edges are obtained. To accomplish this task, we scan the set of cluster boundary pixels. For any cluster boundary pixel, we compute the variance of the gray level values of the pixels that fall within a local neighborhood of the cluster boundary pixel (the window size is decided by the user). If moving the cluster boundary pixel within its neighborhood results in a higher variance value, then the cluster boundary is so adjusted. In other words, we "move" the cluster boundary towards areas that are more likely to be the "right" inter-cluster boundaries [Gref87] [Liep86].

4 RESULTS

Some experimental runs of the genetic algorithm-based segmentation algorithm were performed in order to test it and also illustrate the role of the different genetic operators involved in the algorithm. These runs where all performed for a population of 10 solutions with a probability of crossover, $pcross = 0.6$. The window sizes were 7×7 for the region-growing procedure, 3×3 for mutation operator and 3×3 for the hill climbing procedure. The weights in the fitness function W_1 and W_2, were both chosen equal to 1.

4.1 The role of mutation

Whereas crossover tends to reduce the variance in the population, mutation introduces diversity by attempting to find "alternative information", i.e. mutation explores new ways of clustering the input pixel data. Nevertheless, an "excess" of diversity could prevent the crossover operation from boosting the presence of the fittest clusters in successive generations, thus delaying the convergence of the algorithm. A high mutation rate results in very frequent random changes, thereby affecting the convergence of the genetic algorithm.

4.2 The role of hill climbing

To evaluate the impact of hill climbing we have run the algorithm with and without it. The results are very much in accordance with the characteristic features of a conventional genetic algorithm. Figure ??a shows the best image obtained after 90 generations, without making use of a hill-climbing technique. The solution is close to a globally optimum segmentation of the image, but the algorithm is unable to tune the solution to the global optimum. The situation is different when we employ hill climbing. Since this heuristic is quite powerful, it speeds up the convergence of the algorithm and the final result is a very good segmentation of the original image.

Genetic Algorithms are good at identifying the most promising portions of the search space but they sometimes lack the ability to reach the globally optimal solution. However, we have to bear in mind that whenever we include some heuristic such as hill-climbing we, in effect, bias the algorithm. The exploration of alternative solutions is restricted to a smaller portion of the search space which entails a risk of not being able to reach the globally optimal solution. As a rule of thumb, a deterministic procedure should be included only if it has been proven to be useful when used by itself.

5 CONCLUSIONS

In this work, we have presented some novel ideas about evolutionary algorithms. Instead of adopting the chromosome-like encoding of the classical genetic algorithm we propose a more "natural" representation that adheres to the basic building blocks philosophy underlying the genetic algorithm. The point is to first identify the building blocks, i.e. the useful environmental information, and then to choose a representation that allows us to "play" with the building blocks by means of the genetic operators. The results are quite illustrative of this philosophy.

We would like to stress that this is an open algorithm. By selecting specific tests and window sizes in some of the procedures described in this paper we can tailor the genetic algorithm to the specific features and requirements of an application domain underlying the set of images such as brain CAT scans, satellite images, etc.

In our future work, we intend to improve the performance of the genetic algorithm. First, we intend to devise a hill-climbing procedure to merge clusters that the user, typically, would not want separated. The procedure could be performed at the end of an iteration (generation cycle) and it would test, for any cluster, the possibility of merging it with one of its adjacent clusters. The test would be based on the uniformity of the two regions (gray variance, almost zero derivative of pixel gray level values, statistical test of gray level uniformity, absence of a real edge between them, etc.). In our experimental runs the atomic pixels where able to map onto the single objects that we wanted to extract as the final clusters. But this fact cannot always be ensured. Second, we have to come up with a more general approach in order to tackle different types of images. That means, either a procedure to identify the proper tests or more general tests.

6 Bibliography

[Bhan94] S.M. Bhandarkar, Y. Zhang and W. D. Potter. "An edge detection technique using genetic algorithm-based optimization". *Pattern Recogn.* 27(9), 1994, 1159–1180.

[Fitz84] J.M. Fitzpatrick, J.J. Grefenstette and D. Van Gucht. "Image registration by genetic search". *Proc. of IEEE Southeastern Conf.* 1984, 460–464.

[Fu81] K.S. Fu and J.K. Mui. "A survey on image segmentation". *Pattern Recogn.* 13, 1981, 3–16.

[Gold90] D. Goldberg. *Genetic Algorithms in Search, Optimization and Machine Learning.* Reading, MA: Addison-Wesley, 1990.

[Gref87] J. Grefenstette. "Incorporating problem specific knowledge into genetic algorithms". in *Genetic Algorithms and Simulated Annealing.* (L. Davis, Ed.) London: Pittman, 1987.

[Hara85] R.M. Haralick and L.G. Shapiro. "Image segmentation techniques". *Comp. Vision Graphics and Image Proc.* 29, 1985, 100–132.

[Holl75] John Holland. *Adaptation in Natural and Artificial Systems.* Ann Arbor, MI: University of Michigan Press, 1975.

[Liep86] G. Liepins. "Greedy genetics". *Genetic Algorithms and their Applications, Proc. Second Int. Conf. Genetic Algorithms.* 1986, 90–99.

Figure 1: The test image

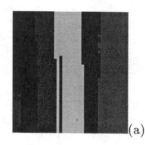 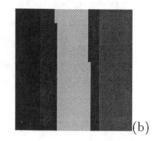

Figure 2: Atomic clusters generated from three atomic points by setting different variance thresholds: (a) 100 (b) 400

Figure 3: Some members of the initial population (window size for random expansion is 5 × 5).

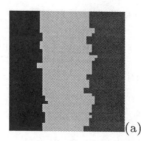 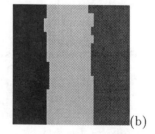

Figure 4: Best solution obtained after different runs: (a) No hill-climbing is used; 90 generations (b) Hill-climbing is used; 10 generations

A HIERARCHY OF DETAIL FOR REPRESENTING NON-CONVEX CURVED OBJECTS

Begoña Martínez and **Angel P. del Pobil**
Computer Science Dept., Jaume-I University
Campus Penyeta Roja, E-12071 Castellón, Spain
{bmartine, pobil}@inf.uji.es

ABSTRACT

Spatial representations are a bottleneck for real-world applications involving increasingly more complex geometries, in such a way that more efficient ways of representing objects are called for. Hierarchies of detail based on spheres seem a powerful approach to overcome this problem. A representation for non-convex objects with curved surfaces is presented, which is independent of the number of features used for the polyhedral model of the object. It easily lends itself to parallel implementations and can also be used in connection with computer vision approaches that use generalized cylinders.

1. MOTIVATION

Spatial representation is a fundamental problem in different domains within Artificial Intelligence [Dav90], [dPS95]. Some areas in which it is used are computer vision, intelligent robotics and spatial reasoning among others. In general, an adequate spatial representation would be necessary in those intelligent systems that deal in one way or another with the real world by perceiving or acting.

Artificial Intelligence has been studied for more than four decades and many important contributions have been made. However, some researches feel that it has made few inroads into real industrial applications: current AI models perform poorly when applied to real-world problems more often than it would be desired, they are too slow, complex, specialised or simplistic. A fact that partly accounts for this situation is that current approaches tend to use simplistic geometric models for the involved physical objects. For example, in the case of robot motion planning the examples reported in the literature [BL90] involve such simple geometries that many computation costs can be ignored. When these approaches are applied to real-world problems with complex geometric models, dealing with geometric details becomes a bottleneck: as reported by Chang [Cha95], relatively simple applications took 23 hours, and even in some cases almost 100 hours. More efficient ways of representing objects are called for, if AI techniques are to be used in realistic situations.

Several different models for 3D objects have been used in Computer Science and Robotics: Constructive Solid Geometry, Boundary Representation, spatial occupancy enumeration, cell decomposition, swept volumes, octrees or even prism-trees. In the domain of Intelligent Robotics, where previous work has focused on convex objects which have been used extensively in robotics, much research has been directed at finding efficient algorithms ([Can88] uses a quaternion representation, [Cam90] uses S-Bounds and a four-dimensional extrusion, [FH93] computes a convex approximation that guarantees to encompass the real swept volume). Faverjon and Tournassoud [FT88] have used an octree representation; octrees, however, present important drawbacks when dealing with motion [Hay86], as the involved transformations (rotations and translations) are computationally expensive, since the complete representation tree has to be computed anew for each placement of the robot. The hierarchical nature of CSG has made possible some interesting results: Cameron [Cam89] uses it for intersection detection and Faverjon [Fav89] for collision avoidance.

Most approaches are based on the properties of convex polyhedra. For non-convex objects however, few efficient algorithms exist. Non-convex objects can be divided into a set of convex objects and the solution for these ones is computed, but after decomposing a non-convex object into a set of convex ones, many new polygons are introduced, resulting in an increment in computation times for real applications. In addition, most of current approaches work with the features (vertices, edges or faces) of a polyhedron. Their complexity obviously depends on the number of these features. If objects defined by curves and curved surfaces are modeled as polyhedra in a realistic way, the number of features involved will make the computation burden too big.

Generally, a trade-off between accuracy and simplicity is necessary, since a too detailed model may often be too complex, while a too simple model may be too inaccurate. A good representation in AI must be one that is simple enough but, at the same time, one that contains all the necessary information to deal with the problem at hand and can be refined as required. There are two possibilities to simplify the representation of an object:

decomposing the object as a combination of simpler parts, on the one hand, or computing an approximation of the shape of the object, on the other.

The notion of hierarchy of detail was used by [dPSL92] to overcome this problem. By using this concept, the decomposition of the object and the approximation of its shape are obtained simultaneously and at different levels of detail which are used only when needed.

Spherical representations have proved to be very useful in different domains (see [dPS95] for a thorough review). Just to mention a few recent examples: Ranjan and Fournier [RF94] use a representation of volumetric data based on the union of spheres for visualization; Hebert et al. [HID95] introduce a spherical representation for recognizing curved objects in the domain of computer 3D vision. In both cases, only one description is used to represent the shape of an object. An algorithm by Quinlan [Quin94] builds a hierarchical bounding representation with spheres; it is used for distance computation between non-convex polyhedra which are described as a set of convex components.

Many researchers in different areas do not define a spherical representation, but they just *assume* that such a representation exists, in this way the problem at hand becomes much simpler. Such is the case of an algorithm for intelligent robot motion called *collision prediction* by Hayward et al. [HAF+95] who assume that 3D moving objects are modeled as collections of spheres. A few dozens of similar examples are reported by [dPS95].

In this paper, we present an approach to physical object representation for AI applications. The objects can be non-convex and they are not decomposed into convex ones. The objects can also be limited by curves and curved surfaces, the complexity of the approach is independent of the number of vertices, edges and faces in the underlying polyhedral model, so that curves and curved surfaces can be represented with as many details as desired without impairing the performance of the algorithm.

The main drawback of previous related approaches was that spherical approximations either tend to be coarse, or a unique representation is available with an excessive number of spheres to represent outer shape. Our model combines the simplicity of dealing solely with spheres with the power of a hierarchy of detail that permits us to quantify and control the accuracy of the representation. For this purpose several heuristic techniques were implemented. The process to select the number and locations of the spheres of both the exterior and interior representations is based on a rule-based system, and some results from the mathematical theory of packing and covering are necessary.

Presently, objects are limited to those that can be represented as generalized cylinders, or that can be decomposed in a set of generalized cylinders. This model is very powerful and well-known in computer vision and artificial intelligence, and has been used to represent many kinds of every-day objects ([Bro81], [RN90] [AB76], [CN95]). The current implementation of the algorithm is limited to straight generalized cylinders with constant cross-section, although this is not a fundamental limitation of the method. Our approach is based on the notion of hierarchy of detail and uses the sphere as representation primitive. It offers many additional advantages: for instance, it easily lends itself to parallel implementations and can also be used in connection with computer vision approaches that use generalized cylinders.

2. MAIN FEATURES OF THE SPHERICAL REPRESENTATION

The spatial representation that supports our method is based on an extension of the representation developed by del Pobil and Serna, [dPS94a], [dPSL92]. The original model used a double hierarchy of exterior bounding spheres and interior enclosed spheres, but was limited to prisms having convex polygonal cross-sections. It was used in a robot motion planning algorithm described in [dPS94b].

The present extension aims at objects that can be represented as generalized cylinders, that is, by sweeping a cross-section along an axis in such a way that: (1) the cross-section can be any arbitrary planar shape, (2) the axis may either be a straight line or a curve and it may be contained on a plane or in general 3D space, (3) the size and shape of the cross-section can be constant or it can vary along the axis, according to a certain cross-section function and (4) the angle between the section and the axis may or may not be straight.

In the exterior representation two algorithms are used to cover the cross-section: one to completely cover it with spheres and another to cover its boundary with disks. The quality of a representation is measured by using the difference between the volume or surface of the representation and that of the actual object. As the problem of finding an optimal solution is NP-complete [dPS94a], a heuristic approach is called for. Once a solution is obtained for a given number n of disks for the boundary, sweeping the cross-section will give rise to a set of circular generalized cylinders obtained by sweeping the disks. These cylinders are then covered by using spheres to give a complete covering of the object with spheres. A rule-based expert *spherizer* is used to select the best action to obtain a refined representation from a given one, the rules implement several heuristics that try to adapt the spheres to the shape of the object by minimizing the global error [dPS95].

Starting from a unique outer sphere the sequence of coverings that correspond to each representation must be generated. A general representation for an object is built using five different coverings which are themselves 3D approximations: one for the side surfaces obtained by sweeping, two for the swept cross-section at the initial and final positions, and two more —in certain cases— called tips around the ends of the objects. This fact yields five different regions that can be selected for modification; in addition, each one can be changed in several possible ways. Consequently, the expert spherizer is in charge of the decision process involved in the definition of the sequence of representations. To summarize, we have the following coverings:

(1) The covering for the side region is expressed as the system of translates (see [Rog64] for a definition of translate in the mathematical theory of packing and covering):

$$\{K_s + \mathbf{a}^s_j;\ j=1, 2, ..., n_s\},$$

K_s is the set of spheres: $K_s = \{S^s_j,\ i=1,..., m_s\}$.

Vector $\mathbf{a}^s_j$ is defined as $\mathbf{a}^s_j = (2j-1)\mathbf{a}_s + h_{ti}\mathbf{u}$, being $\mathbf{a}_s = h_s / 2n_s\ \mathbf{u}$, where $\mathbf{u}$ is a unit vector with the direction of segment H (generator of the object for sweeping), and h_{ti} and h_s are the heights of the tip and side boundary.

Moreover, the set of circles $\{C^s_i,\ i=1,..., m_s\}$, is the covering for the boundary of the base polygon that has served to build up the 3D covering.

(2) Similarly, we have the equivalent definitions for the coverings for the regions called tips:

$$\{K_{ti} + \mathbf{a}^{ti}_j;\ j=1, 2, ..., n_{ti}\},\ K_{ti} = \{S^{ti}_i,\ i=1,..., m_{ti}\}.$$

Vector $\mathbf{a}^{ti}_j$ is different for both tips; at the initial position it is:

$$\mathbf{a}^{ti}_j = (2j-1)\mathbf{a}_{ti},\ \text{being } \mathbf{a}_{ti} = h_{ti}/2n_{ti}\ \mathbf{u},$$

and for the final position it is:

$$\mathbf{a}^{ti}_j = (2j-1)\mathbf{a}_{ti} + (h_{ti} + h_s)\mathbf{u}.$$

And the set of circles is $\{C^{ti}_i,\ i=1,..., m_{ti}\}$.

(3) Finally, for both cross-sections the covering system is $\{K_{to} + \mathbf{a}_{to}\}$,

since in this case $n_{to} = 1$.

$$K_{to} = \{S^{to}_i,\ i=1,..., m_{to}\}$$

Vector $\mathbf{a}_{to}$ is $\mathbf{a}_{to}=0$ for the lower base, and for the upper it is:

$$\mathbf{a}_{to} = (2h_e + h_l)\mathbf{u}.$$

The set of circles is $\{C^{to}_i,\ i=1,..., m_{to}\}$.

The basic premise, upon which the selection of the next action relies, is that a balanced and quasi-optimal improvement to a representation can be obtained by detecting the zone with the worst local quality and subsequently refining its representation. It has to be noticed that a certain action can improve the local representation for a zone, while, at the same time, it makes the quality of another zone worse. Several heuristics have been employed to deal with this kind of situation. For this problem we have chosen to equip the Expert Spherizer with the structure of a *rule-based system*.

The expert spherizer works by following the fundamental criterion of detecting the representation area whose local quality is the worst, in order to later refine its representation by means of a certain action. Of course, it must be taken into account that the applied action may have side effects that must be corrected. The quality coefficients that will serve to make the main decisions are: δ_s and ε_s for the side covering; δ_{ti} and ε_{ti} for that of the tips; δ_{to}, δ_{fs}, δ_{ft} for the cross-sections; and finally, δ_{tot} which characterizes the complete exterior representation.

The error set is composed of those points Q that belong to a certain sphere of the covering but are not contained inside any real object. That is,

$Q \in E$ if and only if $\exists\ S_j \in S\ |\ Q \in S_j$ and $\forall\ i\ (i=1, 2,...)\ Q \notin O_i$. Where: $O = \{O_i,\ i=1,..., m\}$ is the set of all objects and $S = \{S_j,\ j=1,..., n\}$ is the set of all spheres in the representation.

The δ quality coefficient for a covering may be described as the ratio of the volume of its error set to the surface of the covered boundary of the object: $\delta = \text{Vol}\ (E_i)\ /\ \text{Surf}\ (BO_i)$, where E_i is the error set corresponding to object BO_i, and BO_i is its outer boundary. This definition corresponds to the average error distance when we approximate the exterior boundary of the true object by the boundary of the representation. Local coverings have their corresponding quality coefficients.

Next, we will comment on the main heuristic rules that the expert spherizer uses. From the relationships between the previous quality coefficients it will decide which is the best action to be applied.

Rule (0). If the specific top coverings $\{K_{to}+\mathbf{a}_{to}\}$ have not been defined, and it is necessary that they exist to guarantee a correct representation, then the first covering is defined $\{K_{to}+\mathbf{a}_{to}\}$ with $m_{to}=1$. We must point out two cases:

Rule (0.1). If the covering $\{K_{ti}+\mathbf{a}^{ti}_j\}$ has not yet been defined, verify if $\{K_s+\mathbf{a}^s_j\}$ guarantees that the top and bottom are covered; if this is not so then define $\{K_{to}+\mathbf{a}_{to}\}$.

Rule (0.2). If the covering $\{K_{ti}+\mathbf{a}^{ti}_j\}$ has already been defined, verify if it guarantees that the top and bottom are covered; if this is not so then define $\{K_{to}+\mathbf{a}_{to}\}$.

Rule (1). When the covering $\{K_{ti}+\mathbf{a}^{ti}_j\}$ exists and it contributes to the top and bottom, it may be necessary to correct it in certain cases. We must distinguish between two cases:

Rule (1.1). If $\{K_{to}+\mathbf{a}_{to}\}$ has been defined, verify if $\{K_{ti}+\mathbf{a}^{ti}_j\}$ surpasses it, that is, $\delta_{ft}>\delta_{to}$. This rule tries to avoid anomalous situations in which the tip covering, whose mission is to cover part of the side boundary, introduces an error by exceeding the top and bottom coverings. Then, the set K_{ti}, must be modified, so in this way, upon decreasing the radius of the spheres of the generating set, its contribution to the error set on the top and bottom will be less.

Rule (1.2). If $\{K_{to}+\mathbf{a}_{to}\}$ does not exist, in this case $\{K_{ti}+\mathbf{a}^{ti}_j\}$ is in charge of covering the top and bottom. If its error set on the top and bottom generates an average error distance $\delta_{ft}>\delta_s+\varepsilon_s$, that is, worse than the one due to the side covering, then it will be necessary to correct K_{ti} by introducing more spheres.

Rule (2). This rule considers the effect produced when the side covering $\{K_s+\mathbf{a}^s_j\}$ in some way affects the top and bottom. Logically, this case will come up only if the specific tip covering has not been defined; otherwise $\{K_s+\mathbf{a}^s_j\}$ could never surpass the top and bottom ($\delta_{fs}=0$). We should consider different cases:

Rule (2.1.1). If $\{K_{to}+\mathbf{a}_{to}\}$ has been defined, verify if

$\{K_s+\mathbf{a}^s{}_j\}$ surpasses it, that is, $\delta_{fs}>\delta_{to}$. This rule tries to avoid anomalous situations in which the side covering introduces an error as it exceeds the top and bottom coverings. A first possibility to solve this situation is to introduce an initial $\{K_{ti}+\mathbf{a}^{ti}{}_j\}$ with $m_{ti}=n_{ti}=1$. This solution is not always convenient. Whether it is convenient or not to define these coverings depends on a certain heuristic coefficient γ.

Rule (2.1.2). If in the previous situation it is not possible to define the tip covering, the only possibility left to improve δ_{fs} is by modifying K_s so that the spheres that make it up will have smaller radii.

Rule (2.2.1). If $\{K_{to}+\mathbf{a}_{to}\}$ has not been defined, then the top and bottom must be covered by $\{K_s+\mathbf{a}^s{}_j\}$. If the error set on them is such that $\delta_{fs}>\delta_s+\varepsilon_s$, that is, with a quality coefficient worse than that of the local covering of the side surfaces by $\{K_s+\mathbf{a}^s{}_j\}$ itself, then it will be necessary to define the tip covering if it is possible.

Rule (2.2.2). If in the last case it is not convenient to define an initial $\{K_{ti}+\mathbf{a}^{ti}{}_j\}$, it will be necessary to correct K_s by introducing more spheres.

Rule (3). After having considered in the previous rules the side effects of $\{K_s+\mathbf{a}^s{}_j\}$ and $\{K_{ti}+\mathbf{a}^{ti}{}_j\}$ on the top and bottom, now we are going to search out which is the worst partial covering among those corresponding to the sides, tips, and top (and bottom), so as to try to correct it. In this rule we look at the less frequent case in which the tip covering is the worst, that is, $\delta_{ti}+\varepsilon_e>\delta_s+\varepsilon_s$ and furthermore $\delta_{ti}+\varepsilon_{ti}>\delta_{to}$. We have two possible actions to improve the two coefficients —δ_{ti} and ε_{ti}— which contribute to the quality of $\{K_{ti}+\mathbf{a}^{ti}{}_j\}$. We will give precedence to the first of these actions over the second by introducing a coefficient $\kappa<1$ which will multiply ε_{ti}. The reason is that it does not make sense to try to adjust set K_{ti} until the cylinders are well approximated by spheres. Furthermore, the first of these operations always gives good results, while the second is more uncertain. The concrete value used in the implementation was $\kappa=0.5$.

Rule (3.1). If $\delta_{ti}>\kappa\varepsilon_{ti}$, we modify the sequence $\{\mathbf{a}^{ti}{}_j\}$ in order to improve δ_{ti} so that when n_{ti} increases by 1 the cylinder approximations improve.

Rule (3.2). If, on the other hand, $\delta_{ti}\leq\kappa\varepsilon_{ti}$ then, to improve ε_{ti}, we modify the set K_{ti}, in such a way that the error set of the cylinders in relation to the tips of the object decreases.

Rule (4). Let us now consider the case in which the covering $\{K_s+\mathbf{a}^s{}_j\}$ is the worst, that is, $\delta_s+\varepsilon_s\geq\delta_{ti}+\varepsilon_{ti}$ and furthermore $\delta_s+\varepsilon_s\geq\delta_{to}$. As in rule (3), we have two possible actions to improve this covering, either improve δ_s, or ε_s. Likewise, we will favor δ_s over ε_s, and we will use the κ factor defined in the exact manner as in the previous case. The reason is obvious, since the coverings for the tips and sides are conceptually the same although they are applied to different situations.

Rule (4.1). If $\delta_s>\kappa\varepsilon_s$, to improve δ_s we modify the sequence $\{\mathbf{a}^s{}_j\}$ so that when n_s increases by 1, the approximation of the corresponding cylinders improves.

Rule (4.2). If, on the other hand, $\delta_s\leq\kappa\varepsilon_s$ then, to improve ε_s, we modify the set K_s, in such a way that the error set of the cylinders in relation to the tips of the object decreases.

Rule (5). Finally, the last case to be dealt with will be that in which the covering $\{K_{to}+\mathbf{a}_{to}\}$ is the worst, that is, $\delta_{to}>\delta_s+\varepsilon_s$. In this situation we only have one possible action to improve the top and bottom coverings, namely, increasing m_{to} by modifying the set K_{to} (since the sequence $\{\mathbf{a}_{to}\}$ is fixed).

A similar approach —but with the help of Voronoi diagrams— is used to fill the object with inner spheres, these will be used in the applications as a lower bound of the actual object volume.

3. RESULTS FOR STRAIGHT GENERALIZED CYLINDERS

In the present extension, straight generalized cylinders with a constant cross-section can be represented. The cross-section can be non-convex and can be delimited by curves. Non-convex objects are not decomposed into convex ones, but are dealt with as a whole entity. This is a fundamental property of this approach: when the boundary of a concavity in such a *generalized polygon* is curved, then it is said not to be *locally non-convex* (see Fig. 1), because there exists an infinite number of points p on its boundary such that the intersection of the generalized polygon with a sufficiently small convex neighborhood of p is non-convex. If a generalized polygon is not locally non-convex, then it cannot be decomposed into a finite number of convex parts. As a consequence, the traditional approach to deal with a non-convex object by partitioning it into convex parts cannot be applied.

Two important features of the representation are that first, it does not rely on the properties of convex objects. Second, the number of spheres in the representation is

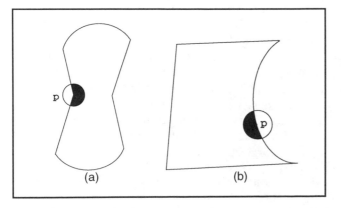

FIGURE 1 (a) A locally non-convex polygon. (b) This generalized polygon is not locally non-convex

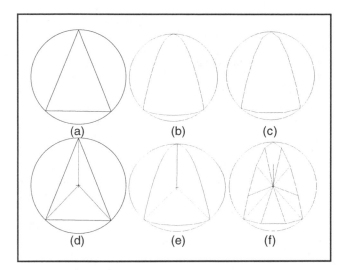

FIGURE 2   Efficient edge heuristics

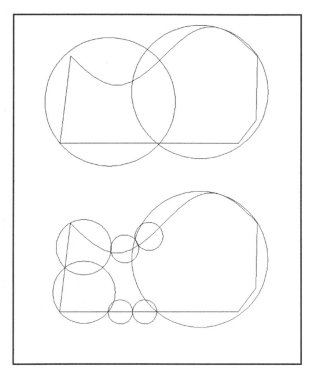

FIGURE 3   Representation of a generalized polygon in 2D

independent of the number of vertices, edges and faces in the underlying polyhedral model, so that curves and curved surfaces can be represented with as many details as desired without impairing the performance of the algorithm. This is so due to the use of the so-called *efficient edge heuristics*, by which two objects having the same global shape will give rise to the same spherical representation independently of the number of vertices and edges that are used to model them. In Fig. 2, polygons (a), (b) and (c) have similar shapes, but they differ in the number of vertices of their polygonal model: obviously, (a) has three sides but (b) is made of curved sides and (c) is polygonal representation of (b). The efficient edge heuristics groups points by means of curvature or the angle between two consecutive segments, so that it internally represents a set of vertices defining a side or lateral zone as a unique efficient edge. Thus, polygons (b) and (c) have three efficient edges (figure (e)) like the triangle (d), instead of the 11 real edges of (f).

The curves are then modeled as bounding polygonal lines and the algorithm can be applied giving the correct results. Similarly, the extension for non-convex polygons with curved boundaries is obtained by modifications of the data structures and the algorithms at geometric level. A detailed description of these computational geometry methods can be found in [MP95a] and [MP95b]. These include algorithms to compute the error measure in terms of surface differences when non-convex regions are involved and to keep track of the different concave portions of the boundary as a representation evolves. Figure 3 shows typical results in two dimensions; in this case polylines with 40 vertices were used for the curve.

3.1. Representation of Locally Non-Convex Polygons

Figure 4 (a) shows a locally non-convex polygon with three non-convex angles. If we number the vertices of the polygon (b), only vertices 3, 7 and 13 are non-convex. Note that we could easily divide it into three convex polygons (b). The traditional solution would be to partition the polygon and to spherize each one of the subpolygons separately. By using the new extended spherizer, a remarkable result is obtained: Figures 4 (c) and (d) show how the spherizer, in the third representation, divides the original polygon into these three parts without using any algorithm for the partition of the polygon into convex regions.

We will analyze this example in depth in order to better understand the actions that are carried out with locally non-convex polygons. We shall start with the first representation corresponding to Figure 4 (a). In this initial representation, we have, as was previously stated, three non-convex angles, which will be defined in the following way: α (vertices 2, 3 and 4), β (vertices 6, 7 and 8), and γ (vertices 12, 13 and 1). Initially, the efficient edges of the unique list are made up of the following vertices:
- efficient edge A: vertices 1 and 2;
- efficient edge B: vertices 2, 3 and 4;
- efficient edge C: vertices 4 and 5;
- efficient edge D: vertices 5, 6, 7 and 8;
- efficient edge E: vertices 8, 9, 10 and 11;
- efficient edge F: vertices 11 and 12;
- efficient edge G: vertices 12, 13 and 1.

We can verify that the edges that make up a non-convex angle belong to the same efficient edge. Furthermore, we may have the case of other convex edges belonging to the same efficient edge, as occurs in efficient edge D.

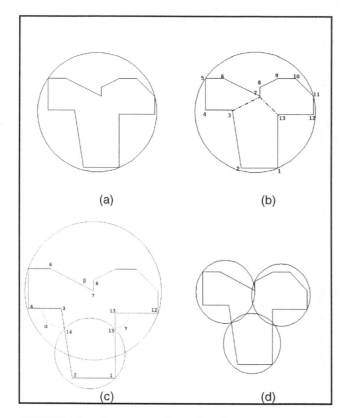

FIGURE 4   Representation of a locally non-convex polygon

After computing the error areas that are associated to each efficient edge, the system decides to cut the list at efficient edges B and G; more concretely, it decides to cut at the edges whose vertices are 2-3 and 13-1, respectively (Figure 4 (c)). In this way we obtain two sublists delimited by the new vertices, which we will denoted as vertex 14 and vertex 15. The following vertices will make up the first sublist: 14, 3, 4, 5, 6, 7, 8, 9, 10, 11, 12, 13 and 15; vertices 15, 1, 2 and 14 will make up the second sublist.

The cut edges belong to non-convex angles. This implies that the description of these angles vary in terms of data structures in the following way: vertices 14-3-4 will make up angle α and vertex 2 will no longer belong to this angle. In fact, vertex 2 belongs to a different list altogether. The same thing occurs with angle γ, now formed by vertices 12-13-15 and, as before, vertex 1 no longer makes up part of a non-convex angle. This is the information which will be used for the correct functioning of the algorithms.

To obtain the third representation (Figure 4 (d)) note that the efficient edge with greatest error belongs to the first list. For this reason, edge 6-7 will be cut obtaining a new vertex. Again, this vertex, will delimit two new sublists and furthermore, vertex 6 —which defined a non-convex angle— is replaced by the new vertex in the definition of the non-convex angle.

In the three cuts carried out, a non-convex vertex has ceased to define a non-convex angle while a newly created vertex takes its place. We may also have the case where the vertex that has been cut coincides with the central vertex of the non-convex angle. In this case, this angle *disappears* since the vertices that made it up now belong to different lists. In any case, these are the two possible situations that we find when working with isolated non-convex angles. We will see that in the general case there are more possible results.

To end this example up, we will insist on the fact that the three lists of representation number 3 tend to divide the polygon into three convex subpolygons, as if an algorithm for this purpose had been used.

The data structures and algorithms implemented for this extension can be reviewed in [dPS95] and [MP95b].

3.2. Polygons with Curved Concavities

Algorithms get more complicated in the general case, when consecutive non-convex angles are found or the polygon is not locally non-convex. Let us remember that in this kind of polygon we have an infinite number of non-convex points, which is equivalent to speaking of non-convex curves.

Let us assume that we are working with a concavity of n vertices, $v_1, v_2, ..., v_n$, enumerated clockwise. We will use v_k to denote the cutting point or new vertex. Now, we will state how the general case is treated when the cut to obtain the next representation takes place on the edge of vertices v_i, v_{i+1}. The concavity is divided into two parts by the new vertex. Each one of these two parts will belong to one of the two new sublists obtained by the cut:

Part 1.- i+1 vertices ($v_1, v_2, ..., v_i, v_k$).

Part 2.- n-i+1 vertices ($v_k, v_{i+1}, v_{i+2}, ..., v_n$).

The cutting point v_k will belong to both new concavities.

Briefly, the main actions to be carried out in this case are:

a) Compute the number of vertices of the first new concavity.

b) Store the vertices of the first new concavity in a vector.

c) Take into account that v_k will belong to both new concavities, therefore, we must link them.

d) Compute the number of vertices of the second new concavity.

e) Store the vertices of these angles.

f) Update the links for the rest of the vertices.

Figure 5 shows an example of different steps in the process of the boundary covering for a polygon that is not locally non-convex. The polygon is made up of a combination of curves and segments. Furthermore, we have a non-convex curve as well as a convex one. Initially the figure has four efficient edges, three of them convex and one non-convex. In (a), the result of the second spheration step is shown. This result was obtained after having cut the initial polygon at both of the curved efficient edges. In this step, the non-convex concavity has

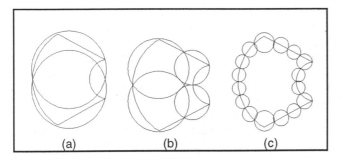

FIGURE 5 Representation of the boundary of a non-convex generalized polygon. This example combines curved boundaries with concavities

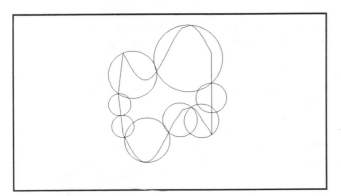

FIGURE 6 Representation of the side of a generalized polygon, non-convex and with curved sides

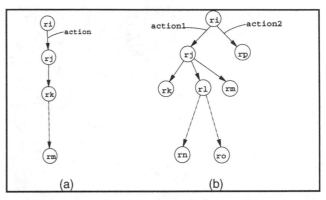

FIGURE 7 (a) The linear structure resulting from the global operation mode to obtain the representations. (b) Tree generated with the local operation mode, starting in the same representation, r_j, three different representations can be obtained

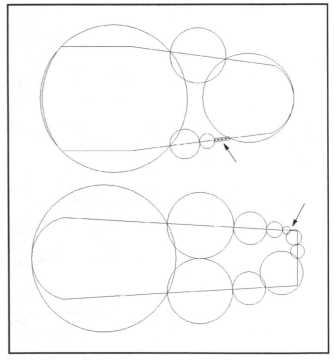

FIGURE 8 Two examples of local improvement in the zones nearest to an hipotetical obstacle

been divided into two new portions. Finally, (c) shows the result after going through more steps.

Figure 6 represents the planar outer representation of another generalized polygon for the covering of its boundary.

4. THE HIERARCHY OF DETAIL

An additional extension to the original spherical representation is intended for more efficient applications. Two operation modes for the spherizer are proposed: the global mode and the local mode. For a given representation, the expert spherizer is used to choose the best action —among several possible ones— to obtain a refined representation from a given one. In the global operation mode, this selection was made automatically in order to obtain a globally improved shape representation, so that the error regions were distributed homogeneously around the surface of the object. Now, in the local mode, the spherizer will modify only those spheres that the particular application requires, the resulting representation will be only locally improved, precisely over the surface portions that are of interest in a certain moment for the application.

The underlying data structure is a tree having as root node a unique sphere that encloses the whole object (Fig. 7). At each node there is a representation. The deeper the

level is, the more refined a representation will be. A branch from a node corresponds to an action of the spherizer. Different branches and their associated subtrees will correspond to representations that are locally refined at different regions. The selection of one action/branch will depend of the spheres that the particular application passes on to the spherizer to be refined. Since the number of actions will in general be high at lower levels, as the number of spheres increases, a trade-off between efficiency and space suggests pre-computing some first levels of the tree and generating the rest levels on-line only as needed.

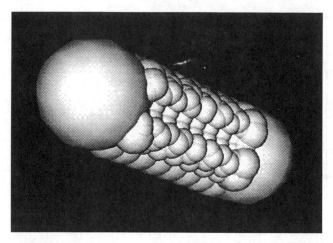

FIGURE 9   Spheration of an object in the local mode

Figure 8 shows an example for the cross-sections of two limbs of a Puma robot, the arrows show the position of a hypothetical obstacle. Figure 9 corresponds to an object defined as a generalized cylinder by sweeping the generalized polygon shown in Fig. 5. In this case, its whole central part has been refined. It can be seen how the concavity is approximated by sets of spheres with the shape of a croissant.

The question of evaluating the efficiency of the present algorithm is not a trivial one. This approach tries to apply to real-life situations, where the solids are of a complex geometry, so we are forced to use approximations and heuristics due to the inherent complexity of the problem. Worst-case behaviour is obviously of no interest, since the aim of all the employed heuristics is just to avoid dealing with this kind of situation. Instead, we will informally discuss the expected average-case performance on the basis of our experiments.

As we have already stated, there are some quality coefficients for a given representation. However, for the local mode, the total error of an approximation makes no sense since the resulting representation will be only locally improved. In this way, going back to the first example of Figure 8, the minimum distance we are able to discern is 0.54 units, which corresponds to the radius of one of the smallest circles. If we define a relative error as the ratio of the smallest distance that can be discerned with respect to a characteristic size for the object (for example, the distance between the farthest points of the cross-section, 71.72 units in this case), then we obtain a relative error of 0.76% for this figure, and using only 9 circles for the representation.

The first levels of the tree can be computed once and stored for later use. As the spheration process may be included as part of the complete design or image processing stages, the computer time to automatically generate all representations of a set of objects will always be negligible when compared with the time and effort spent on the rest of the process.

Figures 10 and 11 show some examples of our latest experiments in the global mode. Figure 10 diplays different stages in the spheration process of a particular case of a non-convex cross-section with a hole. The hole is treated as a limit case of non-convexity. Figure (a) is the first representation with only 1 circle. The approximation has notably been improved in Figure (b), specially the concave zone, using 7 spheres. Finally, Figure (c) shows a representation refined in a limit case, since the approach tends to an non-zero error approximation.

Figure 11 shows the results of the spheration of a generalized polygon with a concavity, which is itself non-convex.

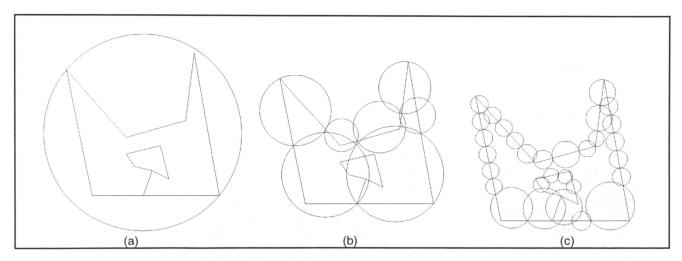

(a)   (b)   (c)

FIGURE 10   Different steps in the representation of a generalized polygon with a hole

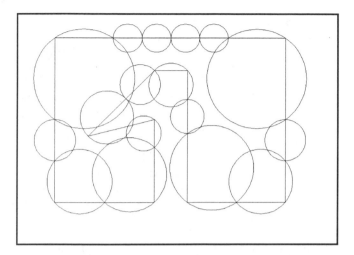

FIGURE 11 Representation of a non-convex generalized polygon whose concavity is itself non-convex

5. CONCLUSIONS

Spatial representations are a bottleneck for real-world applications involving increasingly more complex geometries, in such a way that more efficient ways of representing objects are called for. Hierarchies of detail based on spheres seem a powerful approach to overcome this problem. A representation for non-convex objects with curved surfaces is presented, which is independent of the number of features used for the polyhedral model of the object. It easily lends itself to parallel implementations and can also be used in connection with computer vision approaches that use generalized cylinders.

ACKNOWLEDGEMENTS

This paper describes research done in the Robotic Intelligence Laboratory. Support for this work is provided in part by the CICYT under project TAP95-0710, by the Generalitat Valenciana under project GV-2214/94, by Fundacio Caixa-Castello under P1A94-22 and by a scholarship of the FPU Program of the Spanish Department of Education and Science

REFERENCES

[AB76]   Agin, G.J., Binford, T.O., "Computer Description of Curved Objects", *IEEE Trans. on Computers* 1976, Vol. C-25, No. 4, pp. 439-449.

[BL90]   Barraquand, J., Latombe, J. C., 1990, "A Monte-Carlo Algorithm for path planning with many degrees of freedom", *Proc. IEEE Conference on Robotics and Automation*, pp. 1712-1717.

[Bro81]   Brooks, R.A. 1981, "Symbolic Reasoning Among 3-D Models and 2-D Images", *Artificial Intelligence*, 17, 285-348.

[Cam89]   Cameron, S., 1989, "Efficient intersection tests for objects defined constructively", International Journal of Robotics Research, Vol 8, pp. 3-25.

[Cam90]   Cameron, S., 1990, "Collision detection by four-dimensional intersection testing", IEEE Transactions on Robotics and Automation, 6(3).

[Can88]   Canny, J., 1988, "The complexity of robot motion planning", Cambridge MA, MIT Press.

[Cha95]   Chang, H., 1995, "Motion Planning in Virtual prototyping: Practical considerations", *Proc. IEEE International Symposium on Assembly and Task Planning*, pp. 427-428.

[CN95]   Chung, J.M., Nagata T., 1995, "Reasoning Simplified Volumetric Shapes for Robotic Grasping", *Proc. of the IEEE/RSJ Intl. Conference on Intelligent Robots and Systems IROS'95*, Vol. 2, pp. 348-353, Pittsburgh, Pennsylvania.

[Dav90]   Davis, E., 1990, *Representations of Commonsense Knowledge*, Morgan Kaufmann Publishers, San Mateo, CA.

[dPS92]   del Pobil, A.P., Serna, M.A., 1992, "Solving the Find-Path Problem by a Simple Object Model", *Proc. 10th European Conference on Artificial Intelligence ECAI-92*, Vienna, Austria, pp. 656-660.

[dPS94a] del Pobil, A.P., Serna, M.A., 1994a, "A New Object Representation for Robotics and Artificial Intelligence Applications", *International Journal of Robotics & Automation,* vol. 9, no. 1, pp. 11-21.

[dPS94b] del Pobil, A.P., Serna, M.A., 1994b, "A Simple Algorithm for Intelligent Manipulator Collision-Free Motion", *Journal of Applied Intelligence*, vol. 4, pp. 83-102.

[dPS95] del Pobil, A.P., Serna, M.A., 1995, *Spatial Representation and Motion Planning*, Springer-Verlag, Berlin.

[dPSL92] del Pobil, A.P., Serna, M.A., Llovet, J., 1992, "A New Representation for Collision Avoidance and Detection", *Proc. IEEE International Conference on Robotics and Automation*, Nice, France, pp. 246-251.

[Fav89] Faverjon, B., 1989, "Hierarchical Object Models for Efficient Anti-Collision Algorithms", *Proc. IEEE Intl. Conf. on Robotics and Automation*, pp. 333-340.

[FT88] Faverjon, B., Tournassoud, P., 1988, "A Practical Approach to Motion-Planning for Manipulators with Many Degrees of Freedom", INRIA, Rapport de Recherche No. 951.

[FH93] Foisy, A., Hayward, V., 1993, "A safe swept volume method for collision detection", Proc. 6th International symposium on Robotic Research, pp.61-68.

[Hay86] Hayward, V., 1986, "Fast Collision Detection Scheme by Recursive Decomposition of A Manipulator Workspace", Proc. IEEE Intl. Conf. on Robotics and Automation, pp. 1044-1049.

[HAF+95] Hayward, V., Aubry, S., Foisy, A., Ghallab, Y., 1995, "Efficient collision prediciton among many moving objects", *International Journal of Robotics & Automation*, 14(2), pp. 129-143.

[HID95] Hebert, M., Ikeuchi, K., Delingette, H., 1995, "A Spherical Representation for Recognition of Free-Form Surfaces", *IEEE Trans. on Pattern Analysis and Machine Intelligence*, Vol. 7, No. 7, pp. 681-689.

[MdP95a] Martínez, B., del Pobil, A.P., 1995a, "Covering Generalized Polygons with Disks: an Application to Collision Detection", *Proc. of the 6th National Symposium on Computational Geometry*, pp. 247-253, Barcelona, Spain.

[MdP95b] Martínez, B., del Pobil, A.P., 1995b, "Covering Non-Convex Polygons with Disks", *Proc. of the 6th National Symposium on Computational Geometry*, pp. 238-246, Barcelona, Spain.

[Qui94] Quinlan, S., 1994, "Efficient distance computation between non convex objects", *Proc. IEEE International Conference on Robotics and Automation,* pp. 3324-3329.

[RF94] Ranjan, V., Fournier, A., 1994, "Volume Models for Volumetric Data", *IEEE Computer,* Vol. 27, No. 7, pp. 28-36.

[RN90] Rao, K., Nevaitia, R., 1990, "Computing Volume Descriptions from Sparse 3-D Data", *in Advances in Spatial Reasoning*, edited by S. Chen, Ablex, Norwood, New Jersey.

[Rog64] Rogers, C.A., 1964, *Packing and Covering*, Cambridge University Press, Cambridge, England.

A QUALITATIVE TRAFFIC SENSOR BASED ON THREE-DIMENSIONAL QUALITATIVE MODELING OF VISUAL TEXTURES OF TRAFFIC BEHAVIOR

E. Bonet[1], S. Moreno[1], F. Toledo[2], G. Martín[1]

[1]Instituto de Robótica, Universidad de Valencia, Hugo de Moncada 4, Valencia, Spain 46010
Email: lisitt@vm.ci.uv.es
[2]Departamento de Informatica. Universitat Jaume I, Penyeta Roja, Castellón, Spain 12071
Email: toledo@inf.uji.es

ABSTRACT

This paper introduces a camera-based traffic sensor which uses qualitative image processing to provide a qualitative description of the state of traffic. The video input is processed by means of a temporal Gabor transform, and qualitative traffic textures are recognized in the temporal Gabor space. The resulting qualitative traffic textures are matched with a qualitative model of urban traffic behavior in video sequences, in order to obtain the qualitative description of traffic behavior which better explains the observed textures. The matching process is performed focusing only in discriminant regions of the video space, which are obtained from the envision graph of the qualitative model, resulting in a great gain in performance. The proposed method obtains an exceptionally good response to low or null movement congestive situations, which most state-of-the-art camera traffic sensors based on movement analysis fails to perform.

INTRODUCTION

During the last ten years, the traffic control rooms of many big and medium-sized cities have been equipped with TV-based traffic monitoring systems, thus allowing traffic controllers to take benefit of visual monitoring instead of traffic detector based one. Nevertheless, automatic traffic control systems cannot handle this visual information, making necessary to have human traffic experts continuously monitoring the TV cameras, so that they can take corrective actions on the traffic control system when they see conflictive situations which the control system does not detect. In addition, the continuous supervision of human experts is not able to reach an adequate level of performance, due to several

reasons [BSI91]. As a consequence, it is necessary to develop Urban Traffic Image Recognition Techniques (UTIR) able to take benefit of these visual information for its use in traffic control systems.

Current approaches to UTIR are mainly based on numerical techniques which focus on individual vehicles. Each vehicle has to be individually recognized, and desired traffic parameters must be calculated from data of each vehicle. The most commonly used techniques classical UTIR approaches are based on movement segmentation and object segmentation, both based on the subtraction of the current image and a previously obtained reference image [Iñi85], [Ku+90] and [Ma+94]. Nevertheless, the usefulness of these techniques is generally limited by the fact that congestive or queue states are characterized by low or null car movement, so they are often confused with an empty street state.

The technique presented in this paper is based in the obtention of qualitative textures of vehicles instead of individual vehicle processing, which not only reduces computational effort by reducing the level of detail but also overcomes the problem of confusing low movement states because of the higher differences among their textures. In addition, the interpretation of qualitative textures by the qualitative model of traffic allows to focus the image processing effort only on small regions of the qualitative video textures, thus obtaining an even greater performance.

ARCHITECTURE

The overall architecture of the qualitative image traffic sensor can be seen in figure 1. It is based on the following three modules, which are connected sequentially:

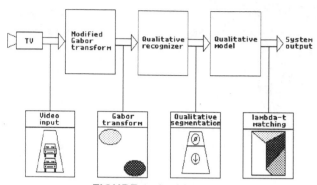

FIGURE 1: Architecture

- Video Transform Module.
- Qualitative Recognizer Module.
- Qualitative Modeling Module.

The video transform (VT) module gets the video signal from the TV camera and performs a video transform whose coefficients are arranged as a 3D sequence of texture vectors.

The Qualitative Recognizer (QR) module performs a pattern recognition process aimed at recognizing qualitative states of traffic from the VT output. The final result is a 3D sequence of qualitative traffic states, each one corresponding to the quantification of a single texture vector.

The Qualitative Modeling (QM) module tries to model the qualitative behavior of traffic which better explains the QR output, using a three-dimensional qualitative model of image traffic behavior. It generates a complete representation of the street in the form of a qualitative and temporal database which can be queried directly by an expert system, and which allows an easy implementation of advanced features, such as problematic queue detection (to be able to discriminate between queues which can cause problems and those which can not) and so on.

VIDEO TRANSFORM

The transform used is based on the knowledge obtained at the beginning of the 90's of the visual preprocessing that the brain performs at the primary visual cortex. Although the complete processes are still unclear, a lot of information on this processing has been obtained and several theoretical models have been developed trying to explain the data. The key elements are: representation of receptive fields for simple cells as 2D Gabor functions, discretization of these Gabor functions in direction each 15°, phase shift of 90°, near-octave discretization of the spatial frequency,

overlapping of about 5:1 among the receptive fields, combinations of non-temporal 2D Gabor functions along time to detect movement in complex cells [Law89], and so on. These facts have served as basis to develop several theoretical models as [PZ88] and [OP94]. Several transforms have been developed in the literature using this knowledge, but most of them have been developed for still image analysis [PZ88], [NW93], [Shu94], and cannot be applied for our application; others can be used for video analysis [Gr+89], but they are too complex to be included in our system, which needs a real-time response. So we have developed a real time video transform for our application focusing on speed, while retaining the key elements above listed.

The VT we have developed uses 3D Gabor functions as basic functions (Gabor functions defined on both spatial directions and also along time, which can measure both static texture features and dynamic ones), half symmetric and the other half anti symmetric (because of the 90° phase shift), with angle discretization along space of 15°, an overlapping of 5:1 (which causes the texture representation to be greater than video space, but which allows a better classification of textures with a winner-take-all procedure), and a octave discretization of the frequency in three levels (both spatial and temporal, it allows to perform multirresolution texture analysis at three different levels, thus allowing to make a Qualitative Recognizer able to recognize the same textures scaled from 1 to $2^3 = 8$ times). The following equation transforms the I video sequence in the M transformed sequence through a discrete convolution with the G base of 3D Gabor tensors. Each Gabor tensor extends Nx, Ny and Nt in the spatial and temporal axis, and Sx, Sy and St are related with the Ni's and the overlapping of the Gabor tensors on I. M can be interpreted as a texture vector associated to each point in the original video sequence, in a similar way to [PZ88], but using video instead of still images. The VT calculates three different M sequences on the same I, by using the same G base scaled by octaves, to perform multirresolution texture analysis.

$$M_{xyt}^{\alpha} = \sum_{i=1}^{N_x} \sum_{j=1}^{N_y} \sum_{k=1}^{N_t} G_{ijk}^{\alpha} I_{(S_x x+i)(S_y y+j)(S_t t+k)}$$

QUALITATIVE RECOGNIZER

The aim of the QR is to obtain a 3D sequence of qualitative traffic states (I'), which associates to each (x,y,t) point in the I video sequence the qualitative traffic state from the quantic domain {Empty, Free, Congestive,

Queue, Non-relevant} which better matches the three M texture vectors calculated on the same (x,y,t) point at the three different spatio-temporal resolutions. It can be represented by the equation:

$$I'_{xyt} = QR\left(M^{\alpha}_{(S_x x)(S_y y)(S_z z)}, M'^{\alpha}_{(S'_x x)(S'_y y)(S'_z z)}, M''^{\alpha}_{(S''_x x)(S''_y y)(S''_z z)} \right)$$

where M, M' and M" represent the three M sequences, and the Si's represent scale factors for the x, y and t coordinates, to adequate the different spatio-temporal resolutions of the M sequences to the lower one of I'.

The obtention of the qualitative traffic state for each (x,y,t) point has been performed considering only six elements, which correspond to the coefficients of two different 3D Gabor tensors seen in M, M' and M" for the given point. One of these two tensors is symmetric on time, and we will name it the static component, and the other one is anti symmetric on time, and we will name it the dynamic component. The six chosen elements can be interpreted as a three-stage multirresolution analysis focused on the static and the dynamic component, which makes the recognition of qualitative traffic states scale-invariant possible.

The static component has been chosen as the 3D Gabor tensor which is the best approximation to a car and its surrounding space when the car is part of a queue at a given car size. The corresponding coefficient in the texture vector will therefore be maximum on a street when the voxel (volume element) of the texture vector is centered on a car. The averaging of the static component at voxels whose size is greater than a car, behaves therefore as a car density function at a given car size. The addition of the three averages based on M, M' and M" allows to obtain a Car Density function which eliminates the dependence of the car size. We have proven in practice the close relation of our function with the real car density, quite independently from its size.

A similar treatment is performed on the dynamic component, and the function built by adding the averages on the three dynamic components is related to the Car Flow, quite independently from the car size.

The vector formed by these two functions takes value in a flow-density plane, as shown in figure 2. A street mask is used on I' to set all (x,y,t) points located outside the street to the Non-relevant qualitative state, and the points located on the street can only take value on the {Empty, Free, Congestive, Queue} regions at figure 2, because flow and density are related in traffic. The qualitative state of street points are assigned as the name of the region where the (Density, Flow) vector is located, which is recognized through the use of two threshold values, Ud and Us.

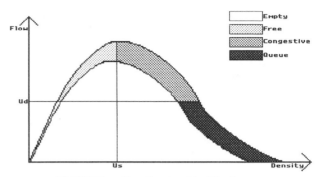

FIGURE 2: Qualitative Traffic States

QUALITATIVE MODELING

The QM module is aimed at analyzing the I' sequence by using a three-dimensional qualitative and temporal model of traffic texture behavior in the 3D video space. It tries to find out the temporal sequence of the traffic model which better matches the I' sequence, with the objective of understanding in traffic terms what it is happening in the video input. The obtained temporal sequence describes completely the traffic behavior deduced for the street in traffic terms. The model used is composed of a qualitative representation method for 3D qualitative functions, a knowledge representation method based on restrictions (which is used to represent traffic knowledge), an envision procedure similar to classical qualitative reasoning formalisms, and a monitoring process which is the working cycle for the QM.

Qualitative Representation

The representation of qualitative discrete functions of type I'(x,y,t), where x and y are spatial parameters, t is time, and I' ranges on the quantic space {Empty, Free, Congestive, Queue, Non-relevant} is performed in the following way: The I'(x,y,t) divides the video 3-D space in a set of disjoint and conexe 3-D regions, being each one formed by a set of points having the same value of I'(x,y,t). The surfaces which separates the regions are limited by lines, and the lines are limited by points. The set of regions, surfaces, lines and points can be used to represent qualitatively the behavior of the I'(x,y,t). Nevertheless, the surfaces are not necessarily planar and the lines are not necessarily straight, so the representation should include extra information about shapes which can make the representation prohibitive. Fortunately, we have seen that the surfaces in the I' sequence obtained in the QR are almost planar, and the lines are almost straight, due to traffic knowledge [MTM94], so we introduce the assumption that, in our model, lines will always be straight and surfaces will always be planar. The final representation of the I'(x,y,t) sequence is composed of a set of polyhedra (having each one a qualitative value), a set of planar regions, a set of straight lines, and a set of points, which we name events (having each one a (x,y,t) coordinate set associated), and the relationships among them (i.e. line2 is limited by event3, planar_surface3 is limited by line2, and so on).

The elements of the representation are finally grouped along the temporal axis in a sequence of snapshots and video sequences, where snapshots are defined for each time point when an event happens, and video sequences are defined as video intervals between two consecutive snapshots.

Figure 3 shows the qualitative representation of a video input taken from a street according to the explained procedure. The video space is divided in five polyhedra: two corresponding to the video volume occupied by the two non-street areas at the left and the right of the street, and the other three corresponding to evolution of traffic states along time in the street area.

The qualitative representation of the video input is organized as a temporal sequence of three elements: snapshot_1 (S1), video_sequence_1 (VS1), and snapshot_2 (S2), which can be seen in figure 3. S1 is the initial state, so it must be defined completely (the following snapshots are defined only from changes on the former ones). S1 is defined by 5 planar regions, 16 straight lines, and 12 events, from which only four events are important: e1a, e1b, e2a and e2b, because they start the only 4 moving lines along time, according to traffic knowledge. These moving lines will finally converge generating e3a and e3b at S2 (see figure 3). VS1 represents the ending process of the polyhedra at "free" state whose vertex are e1a, e1b, e2a, e2b, e3a and e3b. S2 represents a qualitative change of behavior: the

ending of the free region, and it is represented only by two events (e3a and e3b), and a set of relationships indicating which lines they terminate, which lines they generate, and so on. Both VS1 and S2 can be deducted from S1 and traffic knowledge.

Restrictions

The knowledge on traffic is expressed as a set of constraints indicating the vector of direction of the lines started by events, and knowledge about line convergence. For example, the line started by e1a must be contained in a plane perpendicular to S1 which contains the line e1a-e2a, and it must move always in the sense of cars (from the upper side to the bottom side) because of the two traffic polyhedra it separates: empty and free. And the line started by e2a must move on the same plane but in the opposite direction to the cars, as it separates a "Free" state from a "Queue" one. Clearly both lines must converge in a point which defines e3a. Indicating the disappearing of the "Free" state at the left side of the street. The rest of knowledge is similar.

Envisionment Graph

The envisionment graph, in qualitative reasoning, represents a graph containing all possible qualitative states in which the system can be as nodes, and all possible transitions among these states as arcs of the graph. In our model, the states of the system can be snapshots or video sequences, and the arcs indicates which states can follow to each state. A general restriction is that a snapshot can only be followed by a video sequence, and a video sequence can only be followed by a snapshot. This is similar to the Point-Interval-Point-... sequence which can be found in qualitative reasoning paradigms like [Kui86], [DB84], and so on.

Figure 4 shows the structure of the envision graph. The

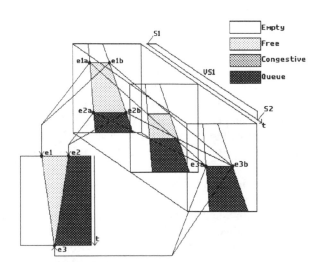

FIGURE 3: Qualitative representation of the stabilization of a traffic queue in a street

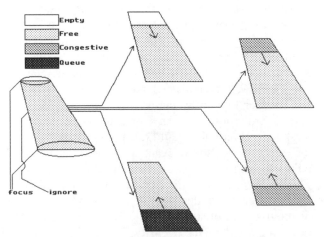

FIGURE 4: Transitions among qualitative states in the envision graph

video sequence at the left represents a view of a street in an homogeneous free state. It will necessarily end in one of the four snapshots shown at the right, due to the application of traffic knowledge as restrictions on event generation (In fact, combinations among them are also possible, but we do not consider them for the sake of simplicity). The two upper ones represent the ending of the video sequence caused by the entrance at the top side of the street of an empty state or a congestive state, and the two lower ones represent the ending of the video sequence caused by the entrance at the end side of the street of a queue or congestive state propagating backwards.

Monitoring process

The working cycle of the Qualitative Module consists on going through the envision graph, selecting the proper arc in each transition by using the qualitative video of traffic textures provided by the QR. The obtained path across the graph will represent the qualitative traffic behavior of traffic in the street which better matches the qualitative video, given in terms of a sequence of snapshots and video sequences which can be easily used to extract meaningful events (i.e. a sequence containing the first transition of the four ones in figure 4 would be interpreted as an empty state region entering at the beginning of a street in a homogeneous free state).

Two advantages of this process are filtering and focusing: the monitoring process filters the noisy input from the QR, as its output is always composed of a noise-free, well-formed traffic behavior sequence. In addition, it allows to direct the matching process by focusing only on discriminant regions of the qualitative

video input, instead of processing the whole of it. For example, if the current state of the QM corresponds to the homogeneous free state at the left of figure 4, the envision graph ensures that changes can only arrive from the beginning or ending of the street, so it will only be necessary to focus on these areas (which we call discriminant regions) on the I'(x,y,t) sequence to determine which one of the four transitions is the right one.

RESULTS

We have implemented the proposed architecture on a HP-735. The input of the system is provided by digitizing VCR recordings of real TV-cameras, which we have recorded in the Traffic Control Centre of the city of Valencia. The digitized video sequence is rated at 15 frames per second, and the proposed architecture is currently able to process 4 frames per second, due to the processing effort required by the VT module. We hope to reach real time processing speed by doing the VT on a proper image processing board.

Figure 5 shows the windows that monitorize our system. The left window represents the digitized video sequence (I) from the VCR, which shows a light flow of vehicles. The middle window shows in real time the qualitative analysis of the Video Input (the I' sequence), and it has correctly represented the street state by a sequence of free-empty-free states. The right window represents the qualitative behavior of traffic obtained by the QM, represented as a space-temporal

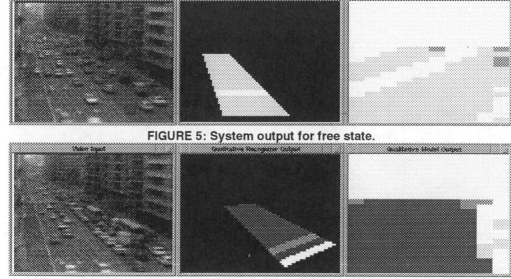

FIGURE 5: System output for free state.

FIGURE 6: System output for queue state.

map where the temporal axis is vertical, and the horizontal axis represents space along the street. The bottom line represents the present state, which is the free-empty-free sequence along the spatial axis already seen in I'.

Figure 6 represents the generation of a queue in the same street, but located in the opposite direction of movement. An example of noise filtering can be seen here: the QR output in the middle window should show only two states, empty-queue, indicating a stabilized queue, but recognition errors have originated a new congestive state in the middle of the empty one, so the QR sequence is empty-queue-congestive-queue. The error is corrected by the QM, which matches this input as the empty-queue sequence.

CONCLUSIONS

The main advantages obtained with the proposed architecture have been:

The use of video traffic textures to solve the problem of discrimination among traffic states with low or null car movement, as queue and empty states.

The use of a qualitative three-dimensional traffic model to filter erroneous traffic recognitions with the aid of traffic knowledge.

The use of a qualitative three-dimensional traffic model to focus the processing effort in discriminant regions of the video space.

The description of the Qualitative behavior of traffic given by the QM allows to model the traffic behavior using a very reduced set of meaningful entities, providing the visual input of a Second Generation Knowledge-based Traffic Control System developed during the ESPRIT EQUATOR P2409 project [Equ94].

ACKNOWLEDGMENTS

We thank the strong support received from D. Victoriano Sanchez, Head of the Traffic Control Room of the city of Valencia. This work has been performed thanks to the economical support of CICYT TIC95-0704.

REFERENCES

[BSI91]   Bell M.C., Scenama G., Ibbetson L.J.: "CLAIRE: An expert system for congestion management", Advanced Telematics in Road Transport, DRIVE conference proc., Elsevier Pub., pp. 596-614, Brussels, 1991.

[DB84]   De Kleer J., Brown J.S.: "A Qualitative Physics based on Confluences", Artificial Intelligence 24, pp. 7-83, 1984.

[Equ94]   EQUATOR Team: "EQUATOR Final Report", Esprit Project N°2409 of the EEC. February 1994.

[Gr+89]   Grossberg S., et al.: "A neural network architecture for preattentive vision", IEEE Trans. in Biomedical Engineering, ISSN 0018-9294, vol 36, N°1, pp. 65-84, 1989.

[Iñi85]   Iñigo R.M.: "Traffic Monitoring and Control using Machine Vision: a Survey", IEEE Transactions on Industrial Electronics, August 1985.

[Ku+90]   Kudo Y. et al.: "Traffic Flow Measurement System using Image Processing", Syst. and Comput., Japan. Vol. 17, 1990.

[Kui86]   Kuipers B.: "Qualitative Simulation", Art. Intelligence 29, pp. 289-338, 1986.

[Law89]   Lawton Teri B.: "Outputs of Paired Gabor Filters Summed Across the Background Frame of Reference Predict the Direction of Movement", IEEE Trans. on Biomedical Engineering, vol. 36, N° 1, January 1989.

[MTM94]   Martín G., Toledo F., Moreno S.: "Qualitative simulation of traffic flows for urban traffic control", in Artificial Intelligence Applied to Traffic Engineering, Bielli and Ambrosino (eds), VSP, pp. 201-231, 1994.

[Ma+94]   Martinez J.J., et al.: "Evaluation of a Computer Vision based Automatic Incident Detection System in an urban context". First World Congress on ATT&IVHS, Paris, December 1994.

[NW93]   Niemann Heinrich, Wu Jian-Kang: "Neural Network Adaptive Image Coding", IEEE Transactions on Neural Networks, vol 4, N° 4, July 1993.

[OP94]   Oram Mike W., Perret David I.: "Modeling Visual Recognition From Neurobiological Constraints", Neural Networks, vol 7, Nos 6-7, pp. 845-972, 1994.

[PZ88]   Porat Moshe, Zeevi Yehoshua Y.: "The Generalized Gabor Scheme of Image Representation in Biological and Machine Vision", IEEE Transactions on Pattern Analysis and Machine Intelligence, vol 10, N° 4, July 1988.

[Shu94]   Shustorovich Alexander: "A Subspace Projection Approach to Feature Extraction: The Two-Dimensional Gabor Transform for Character Recognition", Neural Networks, vol 7, N° 8 pp. 1295-1301, 1994.

Towards an Automatic Determination of Grasping Points Through a Machine Vision Approach

P.J. Sanz, J.M. Iñesta and A.P. del Pobil
Departamento de Informática, Universitat Jaume I, Campus de Penyeta
Roja, 12071-Castellón, Spain.
E-mail: sanzp,inesta,pobil@inf.uji.es

ABSTRACT

A system is presented which coordinates a parallel-jaw gripper and a vision system with the aim of choosing grasping points taking into account stability conditions.

The choice of grasping points is done by a new heuristic approach that tries to manage all kind of planar parts with or without internal centre of gravity, possibly containing holes. It needs only two moment-based features, centroid and the direction of the main axis, easily obtained from a Freeman chain-code representation of the boundary. The results are exclusively visual parameters, expressed in visual coordinates (pixels).

These algorithms have been implemented in a 4DOF robot arm with one camera fixed over the hand. Visual feedback was used in the control system to improve precision up to the limits of the available visual system.

1. INTRODUCTION

The system can be divided into two different, but closely connected parts: the visual algorithms and the visuo-motor control. The main objectives of the first part are to detect each object isolating it from the background by an appropriate choice of a global threshold, and then to extract the contour to determine the features that will be used to obtain suitable grasping points for a given hand configuration. Objects are supposed planar and homogeneous, and they are not known in advance: no trial of shape identification is done. The only previous knowledge is the geometry of the fingers.

Planar grasping can be described considering only the transmission of forces from a set of contact points to the object [MLS94]. In a first approximation to the problem, we assume that the contact between the fingers and the object will take place around contact points with friction. In order to preserve stability, the grasping point candidates to should have to basically fulfill two conditions [KFE94]: The grasping line, defined as the line segment that links the contact points, must be close to the centroid, and must be approximately normal to the contour at both points. These conditions are in correspondence with the approach followed by other authors [Mon91]. The way to connect these conditions with visual parameters assuring the possibility of the automatic process in a simple and robust manner will be described in this paper. The visual parameters that we show to be sufficient for the task are two moment-based features: centroid and direction of the main axis. The developed algorithm shows great efficiency with a more general set of planar parts than other research works in this area.

As we point out above, our algorithm returns visual coordinates. The main objective of the visuo-motor system is then to position the arm at any given point as to be seen with such coordinates. In our case, the camera is attached to the arm, that we need to position the arm so as to have the relevant visual feature at a given image point, by appropriately moving the arm with the camera. Our approach constructs a mapping between input (visual) coordinates and output (motor) ones, which can be prestored or learnt on line through the knowledge of the visuo-motor transformation and its derivatives at a small set of points. It is inspired by the observation of the Vestibulo-Ocular reflex, since a scientific objective is to use a biological model of an existing reflex (behaviour) as an inspiration for a real robotic system.

2. RELATED WORK

The objective of this section is to give a little description of a few interesting recent works closer to ours. Some relevant work in the automatic planar grasping determination based on vision can be divided into the following approaches:

• Systems that attempt to recognize the objects from some kind of representation before grasping them, and using prestored data about the object to choose grasping points [Sta88].

• Systems that do not attempt recognition. They may act either on a complete representation of the boundary [KFE94] or work without it [Jar88]: "*It is not necessary for the robot to recognize an object in order to grasp it; it need only apprehend that object*" [Sta91].

The latter approach is more suitable for unstructured environments, where parts may be faulty or incomplete and sometimes overlapping. In these cases, the sub approach of working with a complete boundary representation would be unfeasible, but it is very interesting in normal conditions, because the representation of the part would allow its recognition to be used in further reasoning, and the grasp generation task becomes only one stage in the most general assembly problem. This paper does not try to recognize shape, but it works with a complete boundary representation able to be used for further recognition.

A key issue for the visual positioning of a robot arm relies on the ability of the programmer to find a mapping between visual and motor coordinates, or to provide the system with the capability of constructing it by itself. The approach adopted to face this problem classifies systems for positioning a robot arm at some place seen by TV cameras into two main categories: those based on the knowledge of an approximate analytical form of the transformation from some cartesian reference frame to the camera frame, and those based on acquiring knowledge about the camera-robot direct transformation (pixels-joints). Our approach, fully explained in [Dom91], fits well into this latter group, but in it, knowledge is not acquired or deduced, but introduced through the measurement of the transformation value and its Jacobian at a set of selected points.

3. STRATEGY FOR GRIP DETERMINATION

The basic steps to assess the grasp will be:
1.- Image Segmentation.
2.- Boundary Extraction.
3.- Moment-Based Features: only computation of the moments up to order two is needed, and from them, the centroid and orientation of the main axis.
4.- Choose the Candidates to Grasping Points.
5.- The Supervisor Mechanism: after the two grasping point candidates are found, we need to evaluate them according to certain conditions of stability.
6.- Safety Conditions.

Details on steps 4 and 5 can be found elsewhere [SDP* 95]. Steps 1, 2 and 3 are related with de image processing part of this strategy and it is explained better in the following section (4). Other steps are explained in [SDP* 95], so here only a short review is presented. After step 3 is completed, the contour of the object is represented in a dynamic array, saving the pixels belonging to the 3-sampling of the boundary (see below). Step 4 guarantees the existence of two unique grasping point candidates: $(P1,P2) = P$. Step 5 evaluates these points, according to three thresholds related with: the curvature at $P_i = \alpha$, the angle between $P1P2$ (the grasping line) and the normal to P_i (^NP_i) = μ and the distance between $P1P2$ and the centroid = λ. We can summarize this strategy in the following expression:

$$Stable\ (\vec{P}) = \prod_{i=1}^{3} a_i(\vec{P})$$

where $a_i(P) \in [0,1]$ is the computation of α_i (when $i = 1$), μ_i (when $i = 2$) and λ_i (when $i = 3$). So,

$$Stable\ (\vec{P}) = \left\{ \begin{array}{ll} 0 & \text{iff } ((a_1(\vec{P}) \vee a_2(\vec{P}) \vee a_3(\vec{P})) = 0 \\ 1 & \text{iff } a_1(\vec{P}) = a_2(\vec{P}) = a_3(\vec{P}) = 1 \end{array} \right\}$$

The algorithm would be as follows:
Compute_α_i
If $a_1(P) = 0$ **Then** $Stable(P) \leftarrow 0$
Else
    Compute_μ_i
    If $a_2(P) = 0$ **Then** $Stable(P) \leftarrow 0$
    Else
        Compute_λ_i
        If $a_2(P) = 0$ **Then** $Stable(P) \leftarrow 0$
        Else $Stable(P) \leftarrow 1$
    EndElse
EndElse

Finally, step 6, adapted from [WVW85] and [KFE94], takes into account two conditions:
1) The distance between the grasping points does not exceed the aperture of the gripper.
2) The fingers of the jaw gripper do not intersect with the object in its trajectory of grasping.

4. THE IMAGE PROCESSING STAGE

Once the image has been captured and the thresholding has been done, the boundary is found by an algorithm adapted from [GW87], and it is represented as a sequence of n integer-coordinate points, $\mathbf{C} = \{p_i = (x_i, y_i), i = 1,...,n \}$ where p_{i+1} is a neighbor of p_i (module n).

The Freeman chain-code of $\mathbf{C}$ consists of the n vectors $c_i = p_{i-1}p_i$ each one represented by an integer $f = \{0, ..., 7\}$ [Fre61]. Using this representation, the approach followed for moment computation is based on the extended delta rule [SIB* 94]. This is an extension of the work of Zakaria [ZVZ*87], that takes into account non convex objects, and parts with holes.

The segmentation process is based in an adaptative thresholding that assumes that the images are bimodal. We will suppose that illumination conditions are not strictly controlled, so objects are distinguishable from the background by their reflective properties. If this holds, two modes appear in the histogram, and from their characterization, a look up table function can be defined that transforms the original image into a new one with the highest possible contrast that can be already used to easily extract the object contours using an optimal threshold. Details can be found elsewhere [Iñe94].

After the segmentation procedure, it is often important to find a suitable representation of the object contours in order to gain efficiency in the descriptor computation, but keeping the object morphology for performing geometric reasoning. This is specially true in our case, since we need high speed computation for visual feedback, and we are trying to select grasping points, so if the shape is damaged by a data reduction procedure, the algorithms can lead to wrong conclusions. So, we need a fast and reliable method to perform data reduction in the contours. This topic is usually known as curve segmentation, and a number of algorithms have been proposed to carry out this task, but all of them are time consuming, since they try to make an *intelligent* segmentation based on the study of geometric properties at each point of the contour, and then apply a selection procedure to choose a suitable subset of points with a high informative content. But our visual feedback system can not afford such an approach, so a less *intelligent* but effective enough method, should be applied for data reduction. One way to achieve this is called *m*-sampling, and was described by Proffit and Rosen [PR79] for estimating the length of digital curves. We are going to justify this methodology and the choice of a suitable value for *m*. The technique of the *m*-sampling is based on the selection of one point out of every m points of the curve. Thus, if $m = 1$ the effect on the curve is leave it unchanged and no data reduction is achieved. On the other hand, a high value for *m* implies an important rate of data reduction, but the morphology may suffer a serious damage. We are going to try to assess the effects of *m*-sampling on a simple shape descriptor such as the perimeter for different values of *m*. We will also study the reliability of the taken measurements. The length of the sampled curve is evaluated as

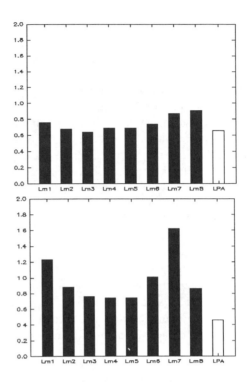

Figure 1. Percentages found for the dispersions with respect to the mean values for the perimeter estimations with different values for *m* and a polygonal approximation. The smallest dispersions are found for the central values of *m* and for L_{PA}. These results have been reproduced for other contours tested.

$$L_m(m) = \sum_{j=0}^{N/m} d(m)_j$$

that is dependent on the value chosen for *m*, being $d(m)_j$ the euclidean distance between consecutive *m*-sampled points. We will also measure the length of a polygonal approximation computed by a dominant point selection procedure [TC89], to keep it as reference value, and for comparison purposes. This value has validity only if the criterion for point selection is related with the keeping of the morphological content of the shape, and the method used in that paper really does it. Its length will be the summation of the lengths of the straight segments that build the polygonal approximation:

$$L_{PA} = \sum_{j=0}^{n} l_j$$

where n is the number of segments in the resulting polygon and l_j is the euclidean distance between the extremes of the jth segment.

Two points of view are considered for assessing the quality of the *m*-sampling for data reduction: 1) reliability of the measurements, or which is the error interval expected for a perimeter value extracted through *m*-sampling. We will use a set of 16 digitizations of a number of objects trying to analyze dispersion values. 2) proximity of the obtained value to the "true" measured value (see figure 1). Other experiments show that with a value for *m* equal to 3 or 4, the estimation of perimeters yields values close to the real ones, hand measured on real parts. Between these values for *m*, 3 represents a better choice for preserving original shape if sharp angles are presented. After all these discussions we have two good estimators for the perimeter length of digital curves. The best of all seems to be L_{PA}, since it combines good reliability and precision, as expected taking into account the *quality* of the points selected for constructing the polygonal approximation. But, as we have stated above, when the computation time is a constraint, an alternative less intelligent data reduction procedure is needed, and, in this regard, L_{m3} seems to give good performances, since its results are very acceptable and its computation is straightforward.

5. SYSTEM DESCRIPTION

The experimental set up is composed of a semieducational SCARA robot arm with low repeatability, (± 0.5mm.) and a CCD microcamera attached to the gripper. The robot has 6 DOF but only four will be used since they are sufficient for planar part grasping: *zed* height, shoulder, elbow and yaw angles. Pitch will be kept at -90°. These will be sent to the robot as counts for the encoders. The height over the floor, variable with the *zed* coordinate, will only take three possible values which will be explained below. The hardware to capture and process images consists of a CCD microcamera (3×3×5 cm) with resolution 256×256 pixels. The video signal is digitalised and stored by a Matrox MVP-AT board. 64 gray levels have been used. All the programs are executed by a 486DX computer at 33 MHz with 8Mb RAM. A sketch of the whole system is shown in figure 2.

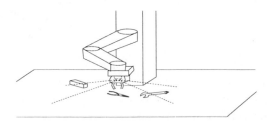

Figure 2. A sketch of the work area

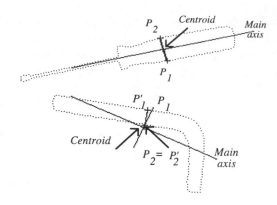

Figure 3. (Top) grasping points in the screwdriver. (Bottom) grasping points candidates (P1,P2) and corrected ones (P'1,P'2), in the allen key.

Details followed to position the arm ere described in [SDP* 96]. The approaching phase was divided into two parts: gross positioning, that works on the image taken by the camera with the arm at its maximum height ("high position"); and fine positioning, that operates with the image taken at a height appropriate to see the part occupying most of it ("middle position"). The grasping is executed with the fingers almost touching the floor ("low position"). The system works in two phases: a previous phase of measurement taking (off-line), storing a list with pairs that connect visual coordinates of the centre seen from high position and (*sho, elb, yaw*) the motor coordinates of the arm when it is just over the object; and a second phase: the on-line process.

The on-line process is as follows:

1) Go to the high position and take a global image of the work area.

2) Process that image to find blobs for the objects and obtain an approximate visual center for each of them.

3) Choose an object to grasp (out of the scope of this paper).

4) Look up for the closest visual point of those stored in first step of measurement, the high visual point we called (*vxh,vyh*). Recover its associated middle visual point (*vxm, vym*), motor point (*sho, elb, yaw*) and its derivative matrix.

5) Send the arm to (*sho,elb,yaw*) and take image again. Use the visual algorithms explained before to get visual centre and orientation.

6) While the centroid of the object does not match the point (*vxm, vym*), calculate the difference between both of them and multiply it by the derivative matrix. Add the resulting joint (motor) vector to the present motor position and go there.

7) Iterate this procedure (visual feedback) until the centroid of the object coincides with (*vxm, vym*). This

will have led the gripper to the appropriate point to grasp the object, since (vxm, vym) is the point in which the target object was seen in the measurement phase just after releasing it.

(8) Finally, rotate the yaw the same angle that the orientation line makes with the x axis.

Finally, we must point out that the visual module behaves as an autonomous part that takes an image and returns visual coordinates when invoked by the motor control module. In the same way, visuo motor module takes visual coordinates and then moves the robot in a quite independent way. Even these cannot be really called behavioural modules, they are a first attempt to build a more flexible architecture for a restricted class of assembly systems.

6. EXPERIMENTAL RESULTS

In order to visualize the aforementioned conditions to determine the grasping points, we will show some convenient results:

Case 1: The method finds a stable two-fingered grasp (see figure 3).

1.1. Trivial example: screwdriver. Here the Supervisor Mechanism (step 5) evaluates satisfactorily the grasp delivered from step 4.

1.2. Non-trivial example: allen key. Here the Supervisor Mechanism (step 5) evaluates and corrects the grasp delivered from step 4. Finally, it finds that all conditions are true and it concludes that the grasp is stable.

Case 2: The method cannot find a stable two-fingered grasp (see figure 4).

Example: Scissors. In this case the algorithm recovers the grasping points candidates from step 4, and evaluates, in first place, the curvature conditions making use of the k-cosine method [RJ73]. Figure 5 shows why these points (the effect over $P2$ is similar to $P1$) fails the necessary smooth conditions required to get a satisfactory preshaping.

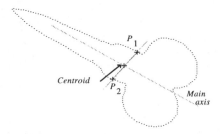

Figure 4. Scissors. The contour and the two moment-based features are found. After 3-sampling the algorithm finds the grasping point candidates.

Initial state		Final state	
Coords.	Curvat.	Coords.	Curvat.
P1 (171,115)	133.66º	P1 (166,111)	177.78º
P1-k (160,107)	166.59º	P1-k (154,104)	154.51º
P1+k (182,113)	130.56º	P1+k (177,118)	93.51º

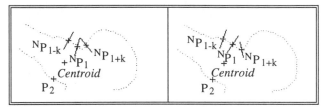

Figure 5. Scissors. Evolution of the curvature analysis from the initial state (left) to the final state (right). The table (above), shows it in degrees. The pictures with the normals in the points around P1, are showed below.

Another important question is related to the impact of the 3-sampling on the efficiency of the visual algorithms. In figure 6, we show a sketch of this. The response of the 3-sampling is very good due to the linear dependence of the algorithms with the number of points.

The positioning algorithm was evaluated by making the arm go to each point of a grid of 64 (4×16) points, chosen at the centre of the original grid where measurements were taken, so these are the points where maximum error is expected. The average error without visual feedback was 1.6 mm. There were no significant differences in the behaviour of the error depending on the part of the workspace. This average error was reduced to 1.2 mm with one visual feedback cycle, 0.9 with two cycles, and 0.7 with three. More cycles did not give substantial improvement; this is due to the limited precision of the visual system, since the size of a pixel at our distances is about 0.6 mm. The total time for the execution on our computer is between 0.3 and 0.8 seconds for the visual part, depending on the length of the boundary, and 0.1 sec. for the positioning algorithm. Since up to three visual feedback cycles can be needed, the total time would be 3.6 seconds. This is not so due to the slow arm we use, which consumes most of the time obeying the positioning commands.

7. DISCUSSION AND CONCLUSIONS

Among the contributions of this paper we can mention an efficient way of representing the contours of the parts

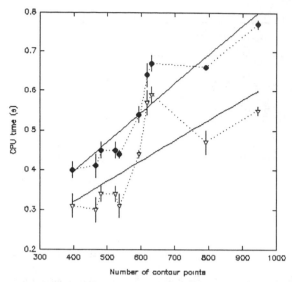

Figure 6 Relation between the number of points in the contour and the CPU time, in seconds for 1-sampling and 3-sampling The regression line shows which are the tendencies in each case when the values are incremented.

via 3-sampling, improving the efficiency of the algorithms that make an extensive use of geometric reasoning from them.

Other important contribution goes in the direction of guaranteeing the uniqueness of the grasp found, avoiding possible conflicts.

Moreover, the system lets an open way to computing other grasps, relaxing the conditions imposed by the initial thresholds.

The global system has shown to work with a quite general kind of planar parts, and positioning algorithms have shown to be precise enough for the task.

ACKNOWLEDGMENTS: This work has been funded by a CICYT (TAP 95-0710-C-01), by the Generalitat Valenciana (GV-2214/94), and Fundació Caixa-Castelló (P1A94-22) project grants.

REFERENCES

[Dom91] Domingo J. "*Stereo part Mating*". Master's thesis, Department of Artificial Intelligence, Univ. of Edinburgh, Sept. 1991.

[Fre61] Freeman H. "*On the encoding of arbitrary geometric configurations*". IRE Trans. Electronic Computers, vol. EC-10(2): 260-268. 1961.

[GW87] Gonzalez RC, Wintz P. "*Digital Image Processing*". 2nd Ed., Addison-Wesley, Reading, Massachussets. 1987.

[Iñe94] Iñesta JM. "*A suitable framework to automatically achieve morphometric measurements from vertebral pieces based on image analysis*".

Tech.Rep. DI 05-11/94, Dep. de Informática, Univ. Jaume I, Spain. 1994.

[Jar88] Jarvis RA "*Automatic Grip Site Detection for Robotics Manipulators*". Australian Computer Science Comm., Vol 10, No 1, pp. 346-356. 1988.

[KFE94] Kamon I, Flash T, Edelman S. "*Learning to Grasp Using Visual Information*". Int. Report. Dept. of Applied Mathematics and Computer Science. The Weizmann Institute of Science. Israel. 1994.

[Mon91] Montana DJ. "*The Condition for Contact Grasp Stability*". Proceedings of the IEEE Int. Conf. on Robotics and Automation, pp. 412-417. Sacramento, California. April 1991.

[MLS94] Murray RM, Li Z y Sastry SS. "*A Mathematical Introduction to Robotic Manipulation*". CRC Press.1994.

[PR79] Proffit D, Rosen D. "*Metrication errors and coding efficiency of chain-encoding schemes for the representation of lines and edges*". Comp. Graph. Image Processing, 10: 318-32. 1979.

[RJ73] Rosenfeld A. and Johnston E. "*Digital Picture Processing*". Academic Press, Inc.1973.

[SIB* 94] Sanz PJ, Iñesta JM, Buendia M and Sarti MA. "*A Fast and Precise Way for Computation of Moments for Morphometry in Medical Images*". In Proc. of the V Int. Symposium on Biomedical Engineering, pp. 99-100. Santiago de Compostela, Spain, Sept. 1994.

[SDP* 96] Sanz PJ, Domingo J, del Pobil AP, and Pelechano J. "*An Integrated Approach to Position a Robot Arm in a System for Planar Part Grasping*". Advanced Manufacturing Forum, (in press).

[Sta88] Stansfield SA "*Reasoning about grasping*". In Proc. of the AAAI-88 Conference, pp. 768-773. Morgan Kaufmann Publishers. 1988.

[Sta91] Stansfield SA "*Robotic Grasping of Unknown Objects: A Knowledge-based Approach*". The International Journal of Robotics Research, Vol 10, No 4, pp. 314-326. August 1991.

[TC89] Teh C-H, Chin RT. "*On the Detection of Dominant Points on Digital Curves*". IEEE Transactions on Pattern Analysis and Machine Intelligence. Vol II. No.8. August, 1989.

[WVW85] Wolter JD, Volz RA, and Woo AC. "*Automatic generation of gripping positions*". IEEE Transactions on Systems, Man, and Cybernetics, SMC-15. 1985.

[ZVZ*87] Zakaria MF, Vroomen LI, Zsombor-Murray PJA, van Kessel JMHM. "*Fast Algorithm for the Computation of Moment Invariants*". Pattern Recog., 20(6): 639-643. 1987.

Abstracts for Poster Session

A CASE-BASED REASONING APPROACH IN INJECTION MOLDING PROCESS DESIGN

K. Shelesh-Nezhad, E. Siores

School of Mechanical and Manufacturing Engineering, Queensland University of Technology,
2 George Street, GPO Box 2434, Brisbane Q 4001, Australia.
Email: k.sheleshnezhad@student.qut.edu.au

ABSTRACT

An AI System for obtaining the magnitude of process parameters in plastic injection molding operation has been implemented. The system is user interactive and can be used at shop floor. This system applies two techniques, Rule-Based and Case-Based Reasoning. The system reduces optimization time and human expert dependency. Keywords: Case-Based Reasoning (CBR); Rule-Based System (RBS); Injection Molding

1. CBR IN INJECTION MOLDING DESIGN

An expert in injection molding often goes through his previous works to find a molding design similar to the current molding and uses successfully tested molding parameters with intuitive adjustments and modifications as a start for a new molding application. This approach saves a substantial amount of time and cost in experimental based corrective actions to reach the optimum molding conditions. A CBR system performs the same task by retrieving the most similar case to the new case from the case library and uses the modification rules to adapt a solution to the new case. Therefore, a CBR system can simulate human expert strategy in injection molding process design.

2. CBR SYSTEM IMPLEMENTATION

The first stage of implementation was to determine an adequate number of basic successful cases called frame cases for the system library development. Each frame case includes the molding features (i.e. type of material, flow pattern, flow length and thickness) and corresponding optimum operating parameters (i.e. melt and mold temperature, required pressure and injection time). A simulation FEM analysis software called Moldflow [Aus88] was used to determine these preliminary cases through the iterative trial and error actions.

The second stage of implementation was ranking the frame cases to relate the features of the new problem to the existing solutions. To search for the existing solutions, IF - THEN rules were applied to find out the similarities between new problem features and existing solutions. The condition part of similarity rules included the material and flow type as well as the flow length and thickness ranges. The action part of similarity rule was a set of required operating parameters for the existing similar cases.

The third stage of implementation entailed the use of existing solutions (corresponding frame cases) to derive the new solution by taking into account the differences between the newly defined problem and the frame cases.

After determining the required molding parameters for the cavity utilizing CBR system, the melt flow analysis was carried out by the Hybrid Expert System to find the pressure drop, temperature difference and shear rate in the mold feed system and subsequently, the system determined the required processing parameters (barrel temperature, injection speed, switch over time and machine hydraulic pressure). Following, these parameters were set on the machine and tried out. If there was any molding defects, the RBS conducted inquiries and suggestions procedures to deal with the variations presented.

REFERENCE

[Aus88]   C. Austin, " Moldflow Analysis and Simulation Program ", Australia, 1988.

SMARTUSA: A CASE-BASED REASONING SYSTEM FOR CUSTOMER SERVICES

Pradeep Raman, Kai H. Chang*, W. Homer Carlisle, and James H. Cross
Department of Computer Science and Engineering
Auburn University
Auburn, AL 36849-5347   USA
* Email: kchang@eng.auburn.edu

ABSTRACT

Case-Based Reasoning (CBR) is the process of solving a given problem based on the knowledge gained from solving precedents. This paper describes the implementation results of a particular helpdesk system, namely SmartUSA, developed for the Union Camp Corporation. This system solves a customer's problem by filtering the problem description through an alias table to generate a brief description and then matching the brief description with the cases in the database. It has proved to be an effective and user-friendly system that has successfully handled different descriptions of the same problem and allowed for the casebase to be built in free-format (plain) text. This system has significantly reduced the workload and the response time in the customer services department of the Union Camp Corporation.

1. SMARTUSA

In theory, CBR can be considered as a five step problem solving process [AL94]: *Representation, Retrieval, Adaptation, Validation, and Update*. **SmartUSA** [RA95] implements these five steps. It provides its user a direct access and assists the user in locating a solution for his computer operation related problems. It was developed for the Union Camp Corporation.

SmartUSA stores the problem description of a precedent in a free-format text and transforms the description into a brief "key-words-filled" description. The key to a high rate of case retrieval success is the presence of *keywords*, which uniquely identify the precedent [HA86]. It generates the brief description by stripping out all non-key words such as "the", "when", and "how", and abbreviating words in the original problem description.

SmartUSA improves the retrieval efficiency by using the indexing words in the brief description to match cases. The candidate cases are ranked according to the number of words matched and then filtered through some expert provided rules that take into account factors such as problem category, user's job title, and problem expiration date to further modify the rankings of the matched cases. The case that retains the highest ranking after this analysis is termed the *closest-matching precedent* and its associated solutions are provided to the user for validation.

SmartUSA was developed in Microsoft FoxPro, a PC-based database product, and runs in a client-server environment under the MS-Windows and Novell Netware operating environments. Its primary benefit is to eliminate or drastically reduce the number of calls placed to the helpdesk personnel. Union Camp Corporation is providing this self-assisting tool to the end users in an attempt to funnel helpdesk resources into other much needed areas. The system has been used by the helpdesk personnel for some time and has amply demonstrated its capability for increasing productivity. While not comprehensive, in the sense that not all queries may be resolved successfully by the system, SmartUSA is certainly a practical, user-friendly, and intelligent solution to handling large volume of repetitive cases. The technique of word-aliasing serves a dual purpose in the sense that it allows for the system to process two different user descriptions in the same manner and also recognizes the precedent of a given problem as all non-key words are stripped out.

2. REFERENCES

[AL94] B.P. Allen, "Case-Based Reasoning: Business Applications," Communications of the ACM, Vol. 37, No. 3, March 1994, pp. 40-42.

[HA86] C. Hammond, "Case-Based Planning: An Integrated Theory of Planning, Learning, and Memory," Ph.D. Thesis, Yale University, 1986.

[RA95] P. Raman, "Case-Based Reasoning: An Implementation Methodology," Technical Report 95-04, Department of Computer Science and Engineering, Auburn University, 1995.

MODELING AND IMPLEMENTING ASPECTS OF HOLISTIC JUDGMENT IN INDUSTRIAL DECISION MAKING

Philip A Collier[1] and Stewart A Leech[2]

[1]Department of Computer Science, [2]Department of Accounting and Finance

University of Tasmania, GPO Box 252C, Hobart, 7001; e-mail: pac@cs.utas.edu.au

KEY WORDS

Descriptive model, judgment, weighted-additive model.

ABSTRACT

We outline a method for acquiring a model of industrial decision making when aspects of the process are poorly understood. We argue that in this context it is often necessary to model holistic judgments, and for these it is appropriate to use a weighted additive model with equal weights. This model is simple and has advantages over more complex forms of uncertain reasoning and overcomes the problem of unknown values in simple production rules.

1 INTRODUCTION

Our aim is to create a validated *descriptive model* of decision-making in insolvency (DMI). This requires various judgments to be made across a spectrum of industrial settings. Our descriptive model will assist parties concerned with the insolvency to better understand the ensuing process. This includes the directors of the company, trading creditors who are often innocent by-standers, and government agencies.

In industry it is frequently necessary to form judgments about various matters. For example, when assessing the risk of taking on or continuing work on a contract. One of our chief concerns with modeling DMI, is assessing the background, attitudes and abilities of several classes of human agents, including suppliers, customers, bankers and the directors. We also need to reason about the business and the relevant industry sector. This can be as diverse as a building contractor, a mine or a magazine publisher.

Frequently this type of judgment is intangible, for two significant reasons:

- the factors which underlie the judgment are unclear, and are often not recorded reliably; and
- how these factors combine together is usually unclear.

Many decisions in an insolvency depend upon how the above two points are resolved. According to experts, such judgments are often holistic in nature. We argue that weighted-attribute methods are suitable for modelling these judgments, and that the weights in these models can be chosen more or less arbitrarily. The judgments can then be combined in relatively simple *process models*, represented as production rules or decision trees, to produce a final outcome.

In the context of the very sparse literature on insolvency decisions and few if any case descriptions, there is little hope of attempting to find a model that improves on existing practice. Instead we focus on modelling human DMI. This model has the potential to improve future decisions by being objective and consistent.

To model human DMI, we need to know the processes that occurs in the minds of human experts. In the methodology section of this paper, we outline some of the psychology of human decision making and methods for discovering and recording the processes in the human mind. There are some practical problems with applying some of the more powerful methods in this domain.

2 SELECTED BIBLIOGRAPHY

Bailey, AD, K. Hackenbrack, P. De and J. Dillard 1987, Artificial Intelligence, Cognitive Science and Computational Modeling in Auditing Research: A Research Approach, *Journal of Information Systems*, **1**, 2, 20–40.

Carroll, J. S. and E. J. Johnson 1990, *Decision Research—A Field Guide*, Sage Publications, Newbury Park, CA.

Dawes, R. M. and B. Corrigan 1974, Linear Methods in Decision Making, *Psychological Bulletin*, 81, 95–106.

Dungan, C. W. 1983, A Model of Auditing Judgment in the Form of an Expert System, PhD dissertation, University of Illinois at Urbana-Champaign, University Microfilms International, Ann Arbor.

Leech, S. A. and Sexton, T-L 1994, Liquidate or Trade Out? An Expert System to Explain the Decision Process, *Australian Accounting Review* Vol 4 No 1, May, 28–36.

Nisbett R. E. and T. DeC. Wilson 1977, Telling more than we can know: Verbal Reports on Mental Process, *Psychological Review*, **84**, 3, 231–259.

Computer Intrusion Detection and Incomplete Information

Mansour Esmaili   **Reihaneh Safavi-Naini**   **Josef Pieprzyk**
Center for Computer Security Research
Computer Science Department
University of Wollongong
Wollongong, NSW 2522
Australia.
{mansour,rei,josef}@cs.uow.edu.au

Abstract

Intrusion Detection Systems (IDS) have previously been built by hand. These systems have difficulty in successfully classifying intruders, and require a significant amount of computational overhead making it difficult to create robust real-time IDS systems. Artificial Intelligence techniques can reduce the human effort required to build these systems and can improve their performance. AI has recently been used in Intrusion Detection (ID) for anomaly detection, data reduction and induction, or discovery, of rules explaining audit data [Fra94].

The use of Expert System technology allows certain intrusion cenarios to be specified much more easily and naturally than in the case using other technologies. However, expert system technology alone provides no support for developing models of intrusive behavior and encourages the development of ad hoc rules.

1 Intrusion Detection Example

Consider the following scenario:

```
ln <file> -<anystring>    %Creating a link to
-<anystring>              %<file> where file
                          %is a user's setuid
                          %script with #!/bin/sh
                          %or #!/bin/csh in
                          %the first line.
```

The network in Figure 1 can represent the above scenario. Suppose the current audit record contains a record showing that user has created a file. It is a definite piece of evidence, which corresponds to `Create` node in the network in Figure 1. Receiving this evidence, will change the posterior probabilities of `Type(Link)` and `Type(Symb_Link)` nodes.

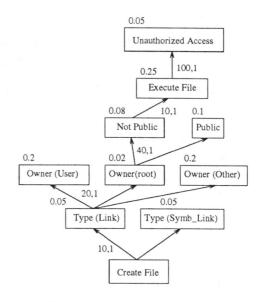

Figure 1: Network representing intrusion scenario

The prior Probability of `Type(Link)` is 0.05. Converting it to Odds we have

$$Odds(Type) = \frac{0.05}{1 - 0.05} = 0.0526$$

Propagating up the network posterior probability of `Unauthorized Access` would be 0.55. It means an increase of 0.5 in the probability of an intrusion to the system.

References

[Fra94] J. Frank. Artificial intelligence and intrusion detection: Current and future directions. In *Proceedings of 17th National Computer Security Conference*, volume 1, pages 22–33, Baltimore, Meryland, 11 - 14 oct 1994.

TURBOLID: TIME USE IN A RULE BASED ON LINE INDUSTRIAL DIAGNOSER

Alonso González, C.; Pulido Junquera, B.; Acosta Lazo, G. G.

Department of Computer Science, Faculty of Science, University of Valladolid,Spain

Prado de la Magdalena s/n, 47011 Valladolid, Spain

Phone + 34 83 423162; fax + 34 83 423161; e-mail: carlos@infor.uva.es

KEY WORDS: Knowledge Based System , Diagnosis, Process Supervision, Continuos Processes.

ABSTRACT

A Knowledge Based System was developed performing automatic diagnosis in a continuous process and in real time. The main contributions of this work were the presence of a diagnosis protocol and the ability to follow plant evolution during the diagnosis process. The system was developed using G2 and nowadays it comprises up to 2500 rules.

TIME MANAGEMENT APPROACH

This applications arises as a consequence of previous systems: AEROLID [ALON94] and TEKNOLID [ACOS94]. The first one analyses current plus historical data with a set of rules encoding expert diagnoser's knowledge. The second one models in a precise way temporal relationships between causes and effects, using an "at any time" diagnosis algorithm. Although TEKNOLID is able to detect faults with rather complicated dynamics, users preferred AEROLID due to its easier maintenance. Hence TURBOLID tries to keep some TEKNOLID capabilities preserving AEROLID simplicity.

Symptoms were split into two sets: fast and slow symptoms. Fast symptoms included past and current effects. They allowed to obtain all the diagnoses covered by AEROLID. Slow symptoms might appear after the diagnosis had started and they covered some of the diagnoses obtained by TEKNOLID. Specifically those faults with slow symptoms overlapping in time. This was a strong assumption that seemed to work well for this application.

DIAGNOSIS PROTOCOL

These family of industrial diagnosers tried to find physical faults (diagnosis) and also operation mode errors. As these systems assumed nominal working conditions, operation mode errors just checked that working hypotheses held, before using expert knowledge to seek a diagnosis.

Diagnosis comprised four stages: *initial*, looking for operation mode errors; *fast,* examining fast symptoms; *slow*, waiting to observe slow symptoms; and *discard*, where weaker information was supplied to the control room operator. Diagnosis started when a monitored variable [ALON94] changed its state to critical; immediately diagnosis stage became *initial.*

The system assumed single fault hypothesis. Hence when any stage found a fault source, diagnosis finished; otherwise it proceeded to the next stage.

CONCLUSIONS

First of all, TURBOLID models operators reasoning strategy by mean of diagnosis protocol.

Second one, it includes temporal information in a simple and effective way.

TURBOLID is undergoing field test, but has already been accepted as the final diagnoser system.

REFERENCES

[ACOS94] Acosta, G. C., Alonso González, C. J., Acebes, L. F., Sanchez, A., de Prada, C. Knowledge Based Diagnosis: Dealing with Fault Modes and Temporal Constrains, Proc. of IECON´94, Vol 2, pp. 1419-1424, Bologna, Italy. September 1994.

[ALON94] Alonso González, C.J., Acosta Lazo G.G., de Prada Moraga, C., Mira Mira J. A knowledge base approach to Fault Detection and Diagnosis in Industrial Processes: a case study, Proc. of the ISIE '94, Santiago, Chile, pp. 397-402. May 94.

An Approach to a Multi-agent Based Scheduling System Using a Coalition Formation

Takayuki Ito and Toramatsu Shintani
Department of Intelligence and Computer Science
Nagoya Institute of Technology, Gokiso, Showa-ku, Nagoya, 466, Japan
Email: itota@ics.nitech.ac.jp

ABSTRACT

In this paper, we propose methods for multi-agent scheduling based on a coalition game and two mechanisms called "nemawashi" and "settoku." The "nemawashi" and "settoku" mechanisms facilitate reaching a consensus among agents. We show the effect of these mechanisms in a practical scheduling system.

1. SCHEDULING SYSTEM

In this paper, we propose a multi-agent scheduling system based on the coalition game[Zl93] in game theory and the mechanisms of "nemawashi" (maneuvering behind the scenes) and "settoku" (persuasion). In the coalition game, an agent calculates his utility by means of a characteristic function, and participates in a coalition that increases his utility. We define the characteristic function $v(S)$ as being based on event size $|S|$, private events U_e, human relations $\mathcal{H}$, and agents relations $\mathcal{A}$ as follows.

$$v(S) = |S| - U_e + \mathcal{H} + \mathcal{A}$$

A scheduling system is used to make a schedule which is agreed on by most members in an organization. In this system, an agent is assigned to an user, and the agent knows his/her user's private schedule. Instead of a human user, agents negotiate by playing the coalition games. In practical applications, it is difficult to reach a consensus if only conventional negotiation protocols are used. To solve this problem, we use "nemawashi" and "settoku" mechanisms to facilitate reaching a consensus among agents.

2. NEGOTIATION PROCESS

In a negotiation process, the agents' preferences are collected by a circulation board protocol[Ki89] used to collect preferences of agents by sending a notice in circulation. In order to reach a consensus effectively in the process, we divide a negotiation process to a fundamental negotiation and a soft "settoku" mechanism. These are done by turns until agents reach a consensus or make no change in their preferences. The fundamental negotiation is done by free agents (who have no private events scheduled on candidate days). The soft "settoku" mechanism is done by busy agents (who do have a private events scheduled on candidate days).

In the fundamental negotiation, the free agents declare preferences based on the characteristic function $v(S)$ (in this case, agents consider human relations $\mathcal{H}$and agent relations $\mathcal{A}$ only). Next, based on these preferences, the busy agents declare their preferences based on the characteristic function $v(S)$ in the process of the soft "settoku" mechanism.

In the process of the "nemawashi" mechanism, one agent collects information of the other agents' preferences for candidate days. According to the information, the agent can narrows down candidate days effectively. When the agents' preferences do not tend to reach a consensus, a forced "settoku" mechanism forces the agents in the minority to unselect their most preferable candidate days. These agents must then reselect an alternative day from the preferable candidate days, and can receive rewards that make the agents more influential in the next negotiation process. The system improves the relations $\mathcal{A}$ for the agents received rewards. And, the agents get higher priority to access information of the notice in the circulation board protocol.

3. CONCLUSIONS

In this paper, we propose a method for negotiation among agents based on a coalition game, as well as "settoku" and "nemawashi" mechanisms. By using this method and these mechanisms, we promote the reaching of a consensus among agents.

REFERENCES

[Zl93] G. Zlotkin and J.S. Rosenschein: A Domain Theory for Task Oriented Negotiation. *IJCAI-93*, pp. 416–422, 1993.

[Ki89] Y. Kitamura, T. Okumoto: Optimal Task Allocation by Circulation Board Protocol. *9th WorkShop on DAI*, pp. 163–177, 1989.

INTEGRATING AGENT AND OBJECT-ORIENTED PROGRAMMING PARADIGMS FOR MULTI-AGENT SYSTEMS DEVELOPMENT

Agostino Poggi, Giovanni Adorni and Paola Turci
Dipartimento di Ingegneria dell'Informazione
University of Parma, Viale delle Scienze, 43100 Parma, Italy
Email: {poggi,bambi,turci}@CE.UniPR.IT

ABSTRACT

This paper presents an environment for the development of multi-agent systems that integrates agent and object-oriented programming paradigms. In fact, this environment, called HOMAGE, offers two different programming levels: object and agent. The object level allows the development of agent models and systems on the basis of two traditional object-oriented programming languages (C++ and Common Lisp). The agent level allows to specialize the agent models defined at the object level and to develop real multi-agent systems through an agent oriented language, called ALL.

HOMAGE

HOMAGE (Heterogeneous Object Multi AGent Environment) [Pog96a] is an environment that allows the development of multi-agent systems composed of agents distributed on internet and communicating with each other through different protocols. HOMAGE allows the user to program at two different levels of abstraction: at a "high level" through an agent-oriented language, ALL (Agent Level Language), and at a "low level" through two well-known object-oriented languages (C++ and Common Lisp).

ALL [Pog96b] is an evolution of MAPL++ and offers a dedicated set of operators for agent-oriented programming. In particular, ALL offers an agent model based on an engine, a set of variables, and a set of rules. The behavior of an agent is driven by the engine that serves input messages through rules. A rule has the following form:

rule_name msg_pattern precondition service
[enable *enable_list*] [disable *disable_list*]

where: *msg_pattern* is a list that has the same form of a message, but contains the special symbols "?" and "+"; the symbol "?" matches any element belonging to an ALL datatype and the symbol "+" matches any sequence of zero or more elements belonging to an ALL datatype; *precondition* is a condition on agent state; *service* is a piece of code allowing the agent both to modify its state and to interact with the other agents; "[" and "]" denote optionality of the enclosed form; *enable_list* and *disable_list* are lists containing the names of two sets of rules.

When an agent receives a message, the engine checks whether the message matches the *msg_pattern* of one of its rules; if the matching fails for all the rules, then it rejects the message, else it checks the *precondition* of the "matched" rule; if *precondition* is false and there is not another rule whose *msg_pattern* matches the message, then the message is put into the queue again, else the *service* code of the "matched" rule is executed. After the execution of *service*, if the enable code is present, then the rules contained in *enable_list* are enabled, that is, the next messages can activate them; if the disable code is present, then the rules contained in *disable_list* are disabled, that is, the next messages cannot activate them until a rule enables them again.

The object level allows the definition of new agent models and systems on the basis of C++ and Common Lisp languages. In particular, these two languages have been used to implement the kernel of HOMAGE: i) the set of libraries implementing ALL operators and supporting object distribution and communication, ii) two parsers translating ALL code, and iii) a set of predefined agents.

The user can develop multi-agent systems using and specializing the agent model presented above. Usually, a new agent model can be obtained by specializing this agent model at the agent level by introducing and modifying some rules and some variables; however, sometimes it is not sufficient because, for example, the user will define a more complex agent model. In these case, the user can specialize the agent model at the object level by modifying the engine, by introducing some methods and by introducing some other passive or active objects that cooperate with the engine to define the behavior of this new agent model.

The work has been partially supported by MURST 40% through the grant "Sistemi Intelligenti".

REFERENCES

[Pog96a] A. Poggi. HOMAGE a heterogeneous object based environment to develop multi-agent systems. In *Proc. HICSS '96*, Maui, HI, 1996.

[Pog96b] A. Poggi. A multi language environment to develop multi-agent applications. Technical Report DII-CE-TR001-96, DII - Università di Parma, Parma, 1996.

NECESSARY KNOWLEDGE SHARING
FOR COOPERATIVE LEARNING AMONG AGENT

Akira Namatame and Yoshiaki Tsukamoto
Dept. of Computer Science, National Defense Academy,
Yokosuka, 239, JAPAN
E-mail : nama@cc.nda.ac.jp

We consider the problems of knowledge exchange, transfer, and sharing in cooperative learning among agents. Each learning agent is specialized to interact with the different learning environment and encapsulates a specific set of knowledge independently obtained from his own experiences. Each individual learning agent learns locally as a precursor to cooperation with other learning agents. We consider two types of cooperation, the cooperative process model and cooperative learning. In the cooperative model, learning agents put forward their decisions based on their own learnt knowledge, and the negotiation among agents is made in order to reach a group or global decision at the cooperative stage. In cooperative learning, learning agents put forward their learnt knowledge. They exchange and share learnt knowledges. They modify their own reflecting other agents' learnt knowledge until they reach some consistent group decision. We show cooperative learning has many attractive features that not found in the cooperative model

Many real world situations are currently being modeled as a set of cooperating intelligent agents, and which can provide effective methodologies for supporting cooperation among databases and knowledge bases. Many agent models proposed so far provide a set of pre-determined representation and inference capability, and that may not be amended. In this paper, we explore the concept of learning agent. For the individual agents, each agent, endowed with his own learning capability, is specialized to interact with a specific problem domain, and it learns to improve his problem-solving skills based on his observation and experiences. At the cooperative stage, learning agents put forward their learnt knowledge. We are especially concerned with so called emergent properties induced from the interactions among agents in which agents actively learn from both their own experiences and the experiences of other agents. We consider connectionist learning for specializing the learning capability in a multi-agent environment. In multi-agent setting, two types of learning may occur: each agent can learn as an autonomous entity based on its own experiences, while at the same time, agents can learn as a group by sharing their learnt knowledge and by adjusting

their views and actions. As a group, learning takes effect in the form of group decision. The improved coordination can be also achieved by knowledge sharing or more efficient knowledge transfer among learning agents. Learning as a group can be also improved by learning the specialization of agents. Cooperation is supported by instantiating various cooperation strategies. We consider two types of cooperation strategies, the cooperative process model and cooperative learning. In the cooperative process model, learning agents put forward their decisions based on their own learnt knowledge, and the negotiation among agents proceeds until they reach a group decision. In the cooperative learning model, on the other hand, learning agents put forward their learnt knowledge, and they exchange and share those knowledge, and each learning agent modifies his own learnt knowledge reflecting other agents' learnt knowledge until they reach some coincide group decision as shown in Figure 1.

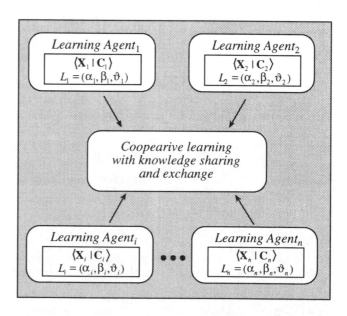

Figure 1 The comparison of cooperative model and cooperative learning agent.

782

OPTIMAL SELECTION OF CUTTING LENGTH OF BARS BY GENETIC ALGORITHMS

Toshihiko Ono and Gen Watanabe
Dept. of Communication and Computer Eng.,
Fukuoka Institute of Technology
3-30-1 Wajiro-higashi, Higashi-ku,Fukuoka 811-02, Japan
e-mail: ono@cs.fit.ac.jp

ABSTRACT

The optimal selection method of cutting lengths of bars by genetic algorithms is explained. When the raw bars of various length are cut into finished products of predetermined various length, the combination of cutting lengths in a raw bar is determined by genetic algorithms to achieve the minimum length of scrap produced, at the same time, keeping the amount of each size of products in accordance with the order of customers at the end of production.

OUTLINE OF SYSTEMS

The purpose of the system is to determine the combination of cutting length in a raw bar, hence the chromosome of each individual which corresponds to a raw bar consists of genes expressing the number of cuts to each finished bar length.

To make the length of scrap minimum and keep balance among the number of finished bars, a fitness function which is one of the most important factors is defined as follows:

$$F(k) = a(k)F_1(k) + b(k)F_2(k)$$

The function $F_1(k)$ is to make the scrap length minimum and reduced as follows:

$$F_1(k) = f_1(\Delta L(k))$$

$$\Delta L(k) = L(k) - \sum_{i=1}^{N} L_i n_i(k)$$

where, $L(k)$ is the length of each raw bar, N is the number of different kind of finished bar lengths, L_i and $n_i(k)$ are the length and the number of cuts for i-th finished bar length. Hence $\Delta L(k)$ stands for the length of each scrap and should be non-negative. In order to make the scrap length minimum, we determined the

function $f_1(x)$ as a non-negative decreasing function whose value becomes maximum at $x = 0$ and is always non-negative in the operating range.

The function $F_2(k)$ is to keep balance among the number of finished bars and is determined as follows:

$$F_2(k) = \sum_{i=1}^{N} f_2(\Delta r_i(k))$$

$$\Delta r_i(k) = \frac{r_i}{\sum_{j=1}^{N} r_j} - \frac{p_i(k) + n_i(k)}{\sum_{j=1}^{N}(p_j(k) + n_j(k))}$$

where, r_i stands for the target of production ratios among the finished bars and p_i is the summarized number of the finished bars produced until this time for each size. Hence $\Delta r_i(k)$ is the deviation of production ratios from the object values. The function $f_2(*)$ is similar to $f_2(*)$.

CONCLUSIONS

We can get the genetic algorithms which can solve two optimal requirements of minimum scrap length, while keeping balance among the amounts of finished bars and we have confirmed their performance by simulation studies.

REFERENCES

[1] T.Ono & G.Watanabe : Application of Genetic Algorithms to Optimizing Problems(in Japanese), Language and Information Processing(Fukuoka Institute of Technology), Vol.6, pp.89/96, 1995

[2] T.Ono & G.Watanabe : Application of Genetic Algorithms to Optimal Selection of Cutting Length of Bars, Proceedings of The Fifth FIT-AJOU University Joint Seminar, pp.25/31, 1995

A DESCRIPTIVE MODEL OF STUDENT PROGRAM GENERATION BASED ON A PROTOCOL ANALYSIS

Tsuruko Egi and Kazuoki Osada

Department of Computer Science, Ube College,

Ube, Yamaguchi, Japan, 755.

Email:egi@cs.ube-c.ac.jp

Faculty of Engineering in Kyushu, Kinki University,

Iizuka, Fukuoka, Japan, 820.

Email:osada@fuk.kindai.ac.jp

ABSTRACT

The proposed model represents student programmers' behaviors using the *subject goals* and the *tactical goals* selected during program generation. The above goals are based in a cognitive analysis of the protocol data.

1. DATA

Four types of data were collected from the six subjects.

(1) *Online protocol* consisted of all versions of C programs generated by subjects. (2) *Reports from the subjects* were written by each subject subjectively. (3) *Retrospective protocol* consisted of accurate records of interviews with each subject. (4) *Investigation reports* were the records of the program generation processes written by an observer based on the online protocol and the retrospective protocol.

The investigation reports of program generation are helpful in understanding and in analyzing the many kinds of protocol data.

2. GOALS AND PLANS

There were two kinds of goals using by students. Subject goals are (1) pure goals specified by the assignment; and (2) goals used to extend some functions. Tactical goals are; (3) goals used to confirm the execution; and (4) goals used to investigate the causes of bugs. Goals (1) and (2) are goals derived from the assignment. Goals (3) and (4) mainly occur as subgoals of Goals (1) and (2), but are not used to fulfill a requirement of the assignment directly. A tactical goal is used to help smooth generation of programs or to take opportunities to get out of an impasse.

The programming knowledge includes those goals and plans. Plans are collect plans, buggy plans, coded plans and abstract plans. There may be a number of plans for each goal, and also subgoals for each plan. These subgoals in turn may require a number of plans. The model is represented by a tree structre with hierarchies of goals and plans [John 86] [Spo 92].

3. MODEL OF PROGRAM GENERATION

The model has five phases: preparation, goal schedul-

```
program  generation
(PREPARATION  phase;
 repeat until goals exhausted from the goal agenda
  (Generation  phase
   ( GOAL-SCHEDULE    phase;
     PLAN-GENERATION  phase;
     if the plan  is a tactical  plan
       (Generation  phase;)
     else
       (EXECUTION   phase;
        EVALUATION   phase;) ) ) )
```

FIGURE 1: The basic model.

ing, plan generation, execution and evaluation (Figure 1). If a tactical plan is generated in the plan generation phase, then the four phases, with the exception of the preparation phase are carried out recursively. Student programmers executed their program either on a computer, in their minds or in other's minds, and then evaluated the generated plan. The results trigger off goal scheduling. The cycle of four phases is the basic composition of the processes of program generation.

According to our model using two kinds of goals, it becomes possible to simulate flexibly novice programmers.

4. DISCUSSION

Most student programmers acquire new programming knowledge and form some concepts of programming during their program generations. To establish a student model for programming ITS, the processes of concept formation and program generation are needed. We think that an evaluation of tactical goals used in the processes of program generation influences to start up the processes of concept formation.

REFERENCES

[John 86]   W. L. Johnson. Intention-Based Diagnosis of Novice Programming Error, *Morgan Kaufmann Publishers*, Los Altos (1986).

[Spo 92]   J. C. Spohrer. MARCEL: Simulating the Novice Programmer, *Ablex Publishing Corporation*, New Jersey (1992).

AN INTELLIGENT TUTORING SYSTEM FOR THE TEACHING OF THE INDUSTRY EQUIPMENT AND ITS STUDENT MODELLING

MSc. Lidia L. Elías Hardy and Jose M. Yáñez Prieto
Instituto Superior de Ciencias y Tecnologia Nucleares,
Ave Salvador Allende esq Luaces, Quinta de los Molinos,
Plaza,Apdo 6163, C. Habana, Cuba
Email: isctn@ceniai.cu

ABSTRACT

This paper shows a model of Tutor-Help, an Intelligent Tutoring System oriented to train on functioning and operation of industrial equipment. Written in C++ v 3.1, the system uses the OOP facilities, mainly the inheritance and polymorphism.

The construction of the new industry and its subsequent operation with quality and efficiency, performing the established norms, requires an appropriate qualification of personnel who will work in it.

Nowadays, for training and certification of operation personnel different computing systems are in use. These systems are very expensive and complex. The main disadvantage of them is that they are unique.

During the last few years simulators and Computer Based Systems without adaptability for the user have been developed [DEF92]. Furthermore, the Artificial Intelligence techniques have been introduced to make these systems more powerful [DEF92,MDL92].

Starting from the idea that each ITS can be divided in two parts: a general and an specific part, we built our model.
The fundamental characteristics of our model are:

- It solves the tasks with expertise.

- It selects the more adequate pedagogical strategies taking into consideration the learner's knowledge.

- Graphical environments allowing the communication of learner with tutoring system by means of several learner's actions on equipment graphic.

- It contains the student modelling.

The general part of the system consists of the: Controller, Tasks, Expert System, Data Structure, Utilities

In this ITS the knowledges are represented using production rules. The rules are used by Expert System in the solution of the tasks presented to the student and allows it the direct student-used knowledge relation.

The Student Modelling consists of two levels: provisional level and permanent level.

In provisional level, the system has not any information about student preceeding work.

In permanent level, the teaching process is strongly entailed to individual characteristics of each student.

The Student Modelling in provisional level works independent at the Expert System, but keeps up the strongly relation with it. At Expert System it obtains the possible series of rules that carry to correct answers using the same rules in the detection of student misconceptions.

The used deviation model is the knowing as Overlay Model. In this model the student knowledge is represented as a subgroup of dominated knowledges by Expert. Hence, we can find four states of knowledges: not used, unknown, forgotten and known.

This model allows to make a valuation of the student depending on dominated knowledges by himself inside the expert system knowledges.

To facilitate the work with knowledges we designed a data structure: KNOWLEDGES GRAPH. This structure gives the knowledges-subknowledges relations.

REFERENCES

[DEF92] Díaz A., Elorriaga J., Fernández I., Gutiérrez J., Vadillo J. **Diagnóstico y estrategias de recuperación en sistemas ICAI aplicados a entornos industriales**. Memorias del 3er Congreso Iberoamericano de Inteligencia Artificial, La Habana, Febrero, 1992.

[MDL92] Mercier V., Delmas D., Lonca P., Moreau J. **SEPIA: An intelligent training system for French Nuclear Power Plant operators**. NATO Advanced Research Workshop "The use of computer models for explication, analysis and experiental learning". Bonas (France), October , 1992.

[PL89] Prieto, M. Manuel, Loret de Mola, Gustavo **Consideraciones sobre la construcción de Entrenadores basados en Sistemas Expertos**. Universidad de la Habana, 1989.

[AS] Alessi M., Stanley R. **Computer-Based Instruction (Methods and Development)**. Prentice Hall.

SKILL TRAINING AT A DISTANCE

Shuichi Fukuda, Yoshifusa Matsuura and Premruedee Wongchuphan
Tokyo Metropolitan Institute of Technology
6-6, Asahigaoka, Hino, Tokyo, 191, JAPAN
Email: fukuda@mgbfu.tmit.ac.jp

ABSTRACT

The conventional distance learning system are more oriented toward conveying messages and not too much attention is paid to the transfer of knowledge in terms of behaviors. And using these system, the images of behaviors can be transmitted only one way from a teacher to a student, and a teacher can not correct the behavior of a student who is following his or her teacher's example.

We developed a preliminary interactive system for distance learning the behaviors. In this system, the teacher watches the motion of this student and can correct it by means of editing.

1. MOTIVATION OF THE RESEARCH

As the world is quickly globalizing and the infrastructure for the computer network is being developed rapidly, distance learning is getting wide attention. But it seems that most of the present distance learning systems are more oriented toward conveying messages and not too much attention is paid to the transfer of knowledge in terms of behaviors. What we are trying to do in this research is to develop a methodology to cope with the situation where a teacher and a student may be located far apart and they can educate and learn in a very interactive manner using a computer network. In the conventional distance learning system such as the one using a TV , a video, etc. , the images of behaviors can be transmitted only one way from a teacher to a student, and a teacher can not correct the behavior of a student who is following his or her teacher's example. If a teacher could show his or her student how he or she should behave when he or she is following the example, the student would know more exactly what he or she is supposed to do and would learn the skill more effectively and much faster.

Therefore, this kind of skill education with emphasis on behavioral learning will be very important if we consider the fact that manufacturing is very quickly globalizing. It is expected that the behavior-based education will fill the gap to a large extent. Because the messages alone will not be sufficient to fully master the skill.

It should also be mentioned that this kind of system can be utilized for a teacher to show his or her student where he or she should focus his or her attention to.

2. OUTLINE OF THE SYSTEM

We developed a preliminary interactive system for distance learning the behaviors. In this system, a student watches the video of the model behavior of a teacher and follows the model motion. The computer takes in these motion images using a camera and it sends them to another computer. The teacher watches the motion of this student and corrects it by means of editing.

Mac Quadra950 was used for the system development. The outline of the system is that a student receive the model motion video of his or her teacher and the copied motion images are taken in using QuickTime in first step. Second step, the teacher receive the copied motion images from his or her student, and cut out the inadequate motion image frames using Premier and the corrected behaviors are generated based upon these original images using Morph. Last step, the student receive the corrected motion images again and paste it back into their original positions, and is able to watch the teacher's exemplary behavior by a video at the same time. The corrected images can also be turned into a video and shown at the same time if another window will be opened. These application softwares are linked together using HyperCard.

K-TREE: AN EFFICIENT STRUCTURE FOR VERIFICATION AND INFERENCE IN RULE-BASED SYSTEMS

T. Rajkumar and H. Mohanty

Artificial Intelligence Laboratory, University of Hyderabad, Hyderabad, India, 500 046.
Email: hmcs@uohyd.ernet.in

ABSTRACT

A K-Tree based approach for verifying completeness and consistency in rule-based systems is discussed. A structure called Parameter Dependency Network (PDN) is used to perform the global level verification. The K-Tree approach has the potential to carry out verification, in an efficient manner than alternative approaches, and it can be generally implemented in any rule-based expert system shell. A K-Tree Based Inference Algorithm (KTBIA) for rule-based systems is also presented. KTBIA makes use of the K-Tree and PDN structures generated during the verification process.

1. VERIFICATION AND INFERENCE IN RULE-BASED SYSTEMS USING K-TREES

Most of the existing approaches for rule base verification are not feasible for verifying large rule bases. A K-Tree based verification approach [SM94] overcomes the major defects of existing approaches in verifying a rule base at the local level (rules that have the same consequent parameter). However, the main drawback of this approach is that it cannot identify errors at the global level (errors in chained inference).

We propose algorithms for verifying a rule base at the global level using K-Trees. A structure called Parameter Dependency Network (PDN) is used to perform the global level verification. The K-Tree based global level verification is discussed in [Raj96]. The main computational advantage of the global level verification method proposed by us, is that it identifies global errors without traversing all the inference chains in a rule base. The K-Tree approach is a generic shell independent rule verification methodology, that can be implemented in any rule-based expert system shell.

We argue that the K-Tree and PDN structures generated during the verification process, should be made use of for further activities in a rule-based system, since these are compiled knowledge. We propose a K-Tree Based Inference Algorithm (KTBIA) for rule-based systems, which makes use of K-Tree and PDN structures generated during the verification process. KTBIA is similar to LFA [Wu93], but it is better than LFA since it requires fewer pattern matches with working memory during inferencing [Raj96].

2. CONCLUSIONS

We have shown the efficiency of K-Tree for verification and inferencing in rule-based systems. We are currently working towards developing a parallel inference algorithm for rule-based systems, using K-Tree and PDN structures.

REFERENCES

[Raj96] T. Rajkumar. K-Tree: An Efficient Structure for Verification and Inference in Rule-Based Systems. M.Tech Thesis, Department. of Computer and Information Sciences, University of Hyderabad, Hyderabad, India, January 1996.

[SM94] Y. H. Suh and T. J. Murray. A Tree-Based Approach for Verifying Completeness and Consistency in Rule-Based Systems. Expert Systems with Applications, volume 7, number 2, pages 199-220, 1994.

[Wu93] X. Wu. LFA: A Linear Forward Chaining Algorithm for AI Production Systems. Expert Systems, volume 10, number 4, November 1993.

SOFTWARE FAULT REDUCTION METHODOLOGY
FOR RELIABLE KNOWLEDGE BASES

Yasushi Shinohara
Central Research Institute of Electric Power Industry (CRIEPI)
2-11-1 Iwado-Kita, Komae-Shi, Tokyo 201, Japan
E-mail: sinohara@denken.or.jp

ABSTRACT

We propose a systematic methodology named "software fault reduction methodology" to design software development/V&V plan for a target knowledgebase. It focuses on the types of faults in the outputs (documents, programs etc.) of each development stage and selects proper fault-detection methods or prevention methods by their detective / preventive power and coverage.

1. INTRODUCTION

There have been proposed many V&V methods for knowledge-based systems, but few to organize them in a systematic way for a target knowledge-based system. So, we developed "software fault reduction methodology". It organizes not only fault-detective V&V methods but also some preventive methods based on the analysis of the relationship among representation of products (documents or programs), types of possible faults of products and V&V methods and others.

2. OVERVIEW OF THE METHODOLOGY

The basic procedure of the proposed methodology is:
(1) To design a skeleton of development process, to select one of basic process models and modify it in response to demanded level of software reliability for a target system.
(2) For the products created at each stage, to determine their representation (to prevent faults) and fault-detection V&V methods by selection metrics, in order to achieve the products' target level of reliability. Also, to select development support tools to create products with the representation and to detect faults by the V&V methods.

3. SUPPORT ENVIRONMENTS

To implement the proposed procedure, we developed 3 databases: representation database(RDB), V&V method database (MDB), and a support tool database (TDB) and a set of worksheets in Excel. RDB has a list of representation (texts, flow charts, rules, etc.) and their expressive power in several aspects. MDB has a list of V&V methods and its detection power to types of faults that can occur in products (contradictions in descriptions, incomplete data, etc.). TDB has information of development / V&V support tools.

5. CONCLUSION

A trial application to a sample knowledge-base system demonstrated the proposed methodology can design a V&V plan with balanced combination of analytic methods and dynamic testing methods. However, the methodology heavily depends on the databases, especially, on the analysis of method's power. We need to increase the number of methods and objectiveness of the analyses.

REFERENCES

[YS95] Yasushi Shinohra and Lance Miller, et.al., "Software Fault Reduction Using CASE TOOLS", CRIEPI-REPORT, March 1995.

The project was sponsored by CRIEPI and EPRI(Electirc Power Research Institute).

MACRO SCALE OBJECT COMPLEXITY MEASUREMENT AND ITS RELATION TO A SEMANTIC MODEL

John W. Gudenas, Department of Computer Science, Aurora University,
Aurora, IL, US, 60506, Email: jgudenas@admin.aurora.edu
and
C. Robert Carlson, Department of Computer Science, Illinois Institute of Technology,
Chicago, IL, US, 60601

ABSTRACT

A metric of macro scale object complexity defined by an adaptive minimal object is presented. A mapping of these minimal objects is demonstrated to a semantic model based upon a case grammar structure.

MACRO SCALE OBJECT COMPLEXITY

The granularity of object complexity is user defined by the necessity to satisfy requirements, not the ultimate or natural complexity of an object. Using an enterprise to bound the object domain, an appropriate unit of complexity measurement must include the behavior as well as static descriptive data about the enterprise.

Using concepts of Kolmogorov complexity from an observer's view[ZURE90], real system complexity is the result of (remaining ignorance) + (algorithmic entropy).

Let $O_i = [E_i(a_1,a_2...a_n) R_i E_j(a_1,a_2...a_n)]$ be a minimal object where E represents an entity with attributes that have a crisp relational strength and R_i is the association between E_i and E_j that has a varying strength indicated by a range of 0 to 1.

Define UE be a universal enterprise that accepts the set O consisting of all O_i needed to describe UE and R $=\{u/R_i \mid \forall R_i \exists uR_i[x]$ that defines the certainty of $R_i\}$. Entity complexity is $|O| = |E| + |a|$. I is the remaining ignorance defined by the Kosko fuzzy entropy of R, where $I = M(R \cap R^c)/ M(R \cup R^c)$. Therefore the macro system complexity of UE is given by $|O| + (I \times |O|)$.

Complexity of an enterprise can be derived by decomposing an entity relation diagram into minimal objects and applying the macro scale complexity metric.

Linguistic meaning is added to the minimal object by associations of words in UE produced by a case grammar structure[FILL68]. Utilizing the description of roles as case labels and their relation to predicates a simple case structure of O_i is developed. In First Level Canonical Form (FLCF) the predicate must produce unambiguous association and usually but not necessarily indicates possession. E exhibits case roles of Agent, Instrument, Experiencer, or Source. The attributes exhibit case roles of Object, Result, or Goal. The FLCF structure is:
<ENTITY> <PREDICATE> <ATTRIBUTES> or reverse.

The Second Level Conical Form(SLCF) links E_i and E_j with a predicate that may inherit ambiguity from the roles of E. The ambiguity of the predicate is expressed by a hedge or a value in a fuzzy set. E may exhibit roles of Agent, Instrumenter, Experiencer, Object, Result, Source or Goal. The Structure is given by:
<ENTITYi > <QUALIFIED PREDICATE> <ENTITYj >.

Consider this example: Faculty have courses and classrooms assigned to them. Classrooms always have overhead projectors, boards, and room numbers and faculty usually find their assigned classrooms.

FLCF is   <faculty> <assigned> <classrooms>
<classrooms> <have> <room#,board,projector>

SLCF is   <faculty> <find/0.8> <classroom>

A direct mapping to minimal objects O is evident.

CONCLUSIONS

A macro scale object complexity metric can be defined from an adaptive minimal object. A complete set of minimal objects can define an enterprise and can map to a semantic model that exhibits canonical form.

REFERENCES

[FILL68] Fillmore C.J., "Lexical Entries for Verbs", *Studies in Descriptive Linguistics*, ed. R. Driven and G. Radden , Julius Groos, Verlag, Heidleberg , pp. 35-45

[ZURE90] Zurek, W.H., *Complexity, Entropy and the Physics of Information*, Addison Wesley, Redwood City, CA, 1990, pp. 73-88

B-PROLOG: A HIGH PERFORMANCE PROLOG COMPILER

Neng-Fa Zhou, Keiichi Katamine
Isao Nagasawa, Masanobu Umeda, and Toyohiko Hirota

Faculty of Computer Science and System Engineering
Kyushu Institute of Technology,
Iizuka, Fukuoka, Japan, 820.
E-mail: zhou@mse.kyutech.ac.jp

ABSTRACT

In this poster presentation, we will demonstrate a system called B-Prolog and present its features, implementation and programming techniques, and applications.

1. INTRODUCTION

B-Prolog is a fast, portable, and robust Prolog implementation. It consists of an emulator of the ATOAM (yet Another matching Tree Oriented Abstract Machine) [Zho95] written in C, a compiler written in canonical-form Prolog, and a library of built-in predicates. B-Prolog is available by anonymous ftp from

ftp.kyutech.ac.jp(131.206.1.101)

in the directory

pub/Language/prolog.

We have confirmed that the system works on SPARC runing SunOS and Solaris, HP runing UX, and 486 runing Linux and MSDOS (Windows).

B-Prolog is significantly faster than WAM-emulator-based systems. The ATOAM retains many good features of the WAM, but differs from it in (1) predicate arguments are passed directly in stack frames; (2) only one frame is used for each predicate; and (3) predicates are translated into trees, called matching trees, if possible, and clauses are indexed on all input arguments. B-Prolog (2.0) is 40% faster than emulated SICStus Prolog (3.0), 50% faster than Bin-Prolog (4.0), more than twice as fast as XSB (1.40) and SWI-Prolog (2.1).

Besides Edinburgh-style clauses, B-Prolog also accepts canonical-form programs. In canonical-form Prolog, each predicate consists of a sequence of *matching clauses* in which input and output unifications are separated, and determinism is denoted explicitly. The compiler is able to translate predicates in this form into matching trees and index them using all input arguments. The compiler can compile programs quickly because it is mostly written in canonical-form Prolog.

B-Prolog is a complete implementation. Besides the compiler, It also has an interpreter written mostly in canonical-form Prolog. It provides an interactive interface through which the user can consult, list, compile, load, debug and run programs. In addition, it provides an interface through which C functions can be called from Prolog. With this interface, the user can integrate other software such as X Windows, sockets, and DBMSs, etc. into the system.

B-Prolog is compatible to ISO standard for Prolog, but also provides several new facilities. It provides a special data structure called state tables that can be used to represent graphs, simple and complex domains in constraint satisfaction problems, and situations in various combinatorial search problems. It also includes a finite-domain constraint solver with which constraint satisfaction problems can be specified declaratively but solved efficiently.

Some functionalities including big integers, delay, module system, and garbage collection are not supported now and are to be implemented in the future.

REFERENCES

[Zho95]   N.F. Zhou: Parameter Passing and Control Stack Management in Prolog Implementation Revisited, available through anonymous ftp from ftp.kyutech.ac.jp/pub/Language/prolog, an earler version appears in Proc. International Conference on Logic Programming, MIT Press, 159-173, 1994.

A CASCADE OF NEURAL NETWORKS FOR COMPLEX CLASSIFICATION

David Philpot and Tim Hendtlass
Centre for Intelligent Systems,
School of Biophysical Sciences and Electrical Engineering,
Swinburne University of Technology.
P.O. Box 218 Hawthorn 3122. Australia.
Email: dnp@bsee.swin.edu.au, tim@bsee.swin.edu.au.

ABSTRACT

Cascades of neural networks (CONN) is a technique that uses several different neural networks to find a relationship in a set of training data. It can be thought of as a divide and conquer technique. Each neural network learns a particular region in the training space, and all the learned regions of all the networks fit together to cover the entire training space in a similar way to pieces of a jigsaw puzzle.

1. INTRODUCTION

CONN is a technique that allows a cascade of several separate networks to work together to form a single relationship. The separate networks can be of any type, and need not be all the same type. There is nothing new in the actual training of the individual networks, this paper shows how they can all work together in a cascade to solve a common problem. The main advantages of the CONN technique is stability, and its ability to learn a relationship in the training data that often fits the training data exactly, providing the training data has a many to one relationship between the inputs and outputs.

This technique removes a lot of the guess work which is often involved when finding the values of various parameters in the neural networks, such as structure, number of layers, learning coefficients and so on. Often these parameters can be varied and a solution will still be found. However, better parameters will produce a solution faster and often with fewer networks.

The main disadvantage of the CONN technique is that the more complex the relationship between the inputs and outputs in the training data, the greater the number of networks required. This takes longer to train

2. A BASIC IMPLEMENTATION

The training process involves training the first network on all of the training data. Any points that haven't been learned adequately by the first network, or which lie on the boundary between a region that has been learned and one that has not been learned, are passed to a second network to learn. Relearning the boundary points improves generalisation in regions which contain no training points. The third network is only trained on the points that the second network passes to it, and so on. This process is continued until all of the training data has been learned.

Finding the outputs of testing data involves putting the test data inputs to the first network that was trained. If this produces an output in its unlearned region, the input is passed to the second network, and so on until it falls into a network's learned region. When this happens, the output from that network is the final output for the test point.

3. RESULTS

A relationship that fits the training data perfectly has been found for every training set tried. The most complex training set attempted has been Frey and Slate's Letter Recognition Problem [FS91]. The 16000 point training set was learned with 100% accuracy by a cascade of 241 backpropagation networks. The networks had between one and nineteen hidden nodes with an average of four.

The 4000 point test set was classified with an accuracy of 88.55%, compared with Frey and Slate's best result of 82.7% using Holland Style Adaptive Classifiers with exemplar-based induction.

REFERENCE

[FS91]   Frey, P.W. & Slate, D.J., Letter recognition using holland-style adaptive classifiers, *Machine Learning*, 6, 161-182

PAC-LEARNING OF WEIGHTS IN MULTIOBJECTIVE FUNCTION BY PAIRWISE COMPARISON

(EXTENDED ABSTRACT)

Ken Satoh

Division of Electronics and Information Engineering, Hokkaido University

North 13, West 8 Sapporo 060, Japan

Email:ksatoh@db.huee.hokudai.ac.jp

In engineering domain, It is frequent that there are many objectives required to be optimal. However, it is rare that an optimal solution is obtained in which every objective takes an optimal value; we often encounter situations where some of the objectives conflict each other.

One approach for this problem is to learn multiple objective functions. It is very useful especially when we would like to transfer expert's preference into an expert system doing the above optimization problem in place of the expert. This paper gives a theoretical analysis based on the PAC-learning framework[Valiant84] for a learning method of multiple objective functions by pairwise comparisons of solutions[Srinivasan73].

Let $A \in R^n$ be a solution and $u_i(A)(1 \leq i \leq t)$ be objective functions which can be calculated in the time proportional to n. And preference function $F(A, W)$ is assumed to be of the form:

$$F(A, W) = \sum_{i=1}^{m} W_i * F_i(u_1(A), ..., u_t(A)),$$

where W is a weight vector $(W_1, ..., W_m)$ and F_i is a polynomially evaluatable function of $u_1(A), ..., u_t(A)$ and m is a polynomial of n. We also assume that there exists a true weight vector $W^* = (W_1^*, ..., W_m^*)$.

Then, the learning problem is to find a hypothetical weight vector W which approximates W^* as possible.

Let $\mathbf{P}$ be any probability distribution over n-dimensional Euclidean space, R^n. Then, a set of different pairs between W and W^* is defined as follows:

$$diff(W, W^*) \overset{def}{=} \{\langle A, B \rangle \in R^n \times R^n |$$
$$(F(A, W) \geq F(B, W) \wedge F(A, W^*) < F(B, W^*)) \vee$$
$$(F(A, W) < F(B, W) \wedge F(A, W^*) \geq F(B, W^*))\}$$

The above set consists of solution pairs (A, B) such that (1) a solution A is actually preferable to the other solution B, but from the hypothesis weight, B is preferable to A or, (2) vice versa.

W is said to be an ϵ-approximation of W^* w.r.t. different pairs for $\mathbf{P}^2$, if the probability of $\mathbf{P}^2(diff(W, W^*))$ is at most ϵ. We call ϵ an error rate.

The following theorem shows that this framework is polynomially PAC-learnable.

Theorem 1. *There exists a learning algorithm which satisfies the following conditions for any probability*

distribution over R^n, $\mathbf{P}$, and an arbitrary constants ϵ and δ in the range $(0, 1)$:

1. *The teacher selects a true weight vector W^* from $[0, \infty)^m$.*

2. *The teacher gives the definition of a preference function $F(A, W)$ with W unknown and gives N pairs according to $\mathbf{P}^2$ with the results of pairwise comparison defined by W^* to the algorithm.*

3. *The algorithm outputs a hypothetical weight vector W and the following hold.*

• *The probability that W is not an ϵ-approximation of W^* w.r.t. different pairs for $\mathbf{P}^2$ is less than δ. We call δ a confidence.*

• *The size of required pairs N for learning is bounded by a polynomial in n, ϵ^{-1} and δ^{-1}, and its running time is bounded by a polynomial in the size of required pairs.*

The following shows a learning algorithm of weights by binary comparison used in the proof of the theorem.

Learn(ϵ, δ, m)

ϵ: accuracy, δ: confidence,

m: the number of weights in the preference function

begin

  Receive the definition of the preference function $F(A, W)$ with W unknown

  and $max(\dfrac{4}{\epsilon} log_2 \dfrac{2}{\delta}, \dfrac{8m}{\epsilon} log_2 \dfrac{13}{\epsilon})$ pairs of solutions and the results of comparison from the teacher.

  for every pair (A, B)

    if $A \leq B$ **then**

      add $F(A, W) \leq F(B, W)$ to the constraint set:

    if $B < A$ **then**

      add $F(B, W) + 1 \leq F(A, W)$ to the constraint set:

  Get consistent values for the above constraint set by linear programming and output W.

end

RERERENCES

[Srinivasan73] Srinivasan, V. and Shocker, A., Linear Programming Techniques for Multidimensional Analysis of Preferences, *Psychometrika*, Vol.38, No.3, pp. 337-369 (1973).

[Valiant84] Valiant, L. G., A Theory of the Learnable, *CACM*, **27**, pp. 1134-1142 (1984).

A MACHINE TRANSLATION APPROACH FROM JAPANESE TO TAMIL

Sivasundaram Suharnan and Hiroshi Karasawa
Department of Electrical Engineering and Computer Science,
Faculty of Engineering, Yamanashi University,
Yamanashi, JAPAN
E-mail(suharnan, karasawa)@opal.esi.yamanashi.ac.jp

ABSTRACT

This paper presents a rule-based approach to Japanese-Tamil Machine Translation(MT) system. Our approach proves the possibility of the MT between Japanese and Tamil, and we give special attention and describe the Tamil synthesis in this paper.

1. INTRODUCTION

From the viewpoint of the linguistic typology, Tamil and Japanese languages are both in such an agglutinated, flexible word order where European languages are not. Both Tamil and Japanese are of Subject-Object-Verb (SOV) order languages which shares many grammatical characteristics such as a same word order, postposition location, rich inflections for verbs, and some other features. However in Tamil every addition of an suffix undergoes a complex phonological rule application[1]. These Tamil suffix creation rule depends not only on the suffix of the word but also the postposition.

2. SYSTEM OVER VIEW

Our system consists of three phases ie. (1) Building the Modified Japanese Functional Structure(MJFS) from Japanese sentence. (2) Creating Modified Tamil Functional Structure(MTFS) from MJFS. (3) Tamil synthesis. First through the Morphological analysis, to Syntax analysis, to Semantic analysis, Japanese sentences are changed into MJFS. Then this MJFS is transformed into MTFS. In this part Japanese morphemes and postposition will be changed into Tamil with the help of a preliminary Tamil-Japanese dictionary and the Tamil-Japanese Postpositions table. Finally MTFS will be changed to Tamil sentence, by considering the syntactic and morphological generation rules[2].

3. THE SYSTEM PROCESSING AND PRELIMINARY RESULT

Our system can translate complex sentence, compound sentence as well as mixed sentences. This example is one of the many experimental results of our system.

Japanese : かれは昨日私に自分が持っ
ていた新しいシャツをくれた。

Tamil : அவன் நேற்று எனக்கு தான்வைத்
திருந்த புதிய சட்டையை தந்தான்.
(Yesterday he gave me his new shirt.)

4. CONCLUSION

Because of the thick characteristics relationship between Tamil and Japanese. The machine translation is possible by appling transfer method in the level of Lexical Functional Structure. We proved by providing experiments from the translation.

REFERENCE

[1] Renganathan Vasu. A Lexical Phonology Approach to Processing Tamil words by Computer. University of Michigan, 1994.

[2] Raman, S and Alwar, N. An AI- Based Approach to Machine Translation in Indian Languages Comm. ACM, vol. 3, pp. 521-527(1990)

ANNEALED HOPFIELD NETWORK APPROACH FOR BOUNDARY-BASED OBJECT RECOGNITION

Jung H. Kim[*], Eui H. Park[**] and Celestine A. Ntuen[**]
*Dept. of Electrical Enigineering, North Carolina A&T State University, Greensboro, NC 27411
**Dept. of Industrial Engineering, North Carolina A&T State University, Greensboro, NC 27411

ABSTRACT

The annealed Hopfield network which is derived from the mean field annealing(MFA)[BS90] has been developed to find global solutions of a non-linear system. In our early work, we developed the hybrid Hopfield network(HHN) on the purpose of fast and reliable matching. In this paper, we present the annealed Hopfield network(AHN) for occluded object matching problems. In AHN, the mean field theory is applied to the hybrid Hopfield network in order to improve computational complexity of the annealed Hopfield network and provide reliable matching under heavily occluded conditions. AHN provides near global solutions without initial restrictions and provides less false matching than HHN. In conclusion, a new algorithm based upon a Neural Network approach was developed to demonstrate the feasibility for the object recognition with the robustness in occlusion environments.

SUMMARY AND CONCLUSION

A set of 12 models is obtained and used in the matching procedure to compare the performance of the HHN with that of AHN in finding objects in occluded images. The models consist of a set of tools and a couple of guns. The number of boundary points for the models ranges from 225 to 514. For experiments, fifty images which are combinations of models are created. The number of boundary points in occluded images ranges from 421 to 816. The number of nodes ranges from 6 to 26. The boundary segmentation algorithm is very reliable in the sense that it is not noise dependable and thus it keeps detecting the same corner points from the object in different scenes.

The average matching scores of HHN and AHN are .80 and .86, respectively. It means that AHN goes nearer to global optimal solution than HHN does. We know from the average matching scores that the performance of AHN is superior to that of HHN. The results show the desired matchings are successfully obtained[KPN95].

We also experiment on the critical temperature to save run time. Annealing through the higher temperature wastes time since it has no effect on the energy function. Instead, most of the optimization occurs near the critical temperature. There is a steepest drop in the energy function around the critical temperature. Annealing through the low temperature does not improve the solution but serves only to saturate the neurons at 1 or 0. The experiment result was 0.45 of T_c while the T_c was 0.7. This discrepancy may result from the small number of neurons of the example since we assume that the number of neurons are very large for mean field approximation. Therefore, we have the T_c be the half of the estimated T_c to get a valid solution. Nevertheless, the estimation of T_c gives us an information of annealing temperature range in the process. This significantly reduces the operation time.

To get the best performance the temperature needs to be decreased by 5%. By 5% decrement of temperature, AHN has pretty good solutions. First, AHN gives a reliable matching of the corresponding segments between two objects. The method eliminates possibility for a part of an object to be matched to similar segments in a different object. Second, there may be a critical solution.

REFERENCES

[BS90] Griff L. Bilbro and Wesley E. Snyder," Applying Mean Field Annealing to Image Noise Removal," *Journal of Neural Network Computing*, Fall,1990

[KPN95] Jung H. Kim,Eui H. Park and C. Ntuen, *"Robust Algorithms for Threat Object Recognition in the Presence of Occlusion and Scale Variations"*, FAA Tech Center, Atlantic City, Under Grant #: 93-G-012, Final Report, August 1995.

* This research has been supported by McDonnell Douglas under Grant No. Z50038.

BETTER NEURONS FOR BIOLOGICAL SIMULATIONS

Howard Copland and Tim Hendtlass
Centre for Intelligent Systems,
School of Biophysical Sciences and Electrical Engineering,
Swinburne University of Technology,
P.O. Box 218 Hawthorn 3122. Australia.
E-mail hxc@brain.physics.swin.oz.au and tim@bsee.swin.edu.au

ABSTRACT

New types of artificial neurons incorporating more aspects of their biological counterparts are proposed that will better model biological neural systems.

1. PROPERTIES OF ARTIFICIAL NEURONS

For more accurate modeling of neural systems, such as spinal reflex systems, artificial neurons incorporating more of the biophysical properties of real neurons are required [Nic94]. Models of single neurons are often not readily implemented in networks to any great degree of biological plausbility, and it is at the network level that information is processed. For networks, simpler models of neurons will suffice and are cheaper computationally. When simulating a network, the activation pattern of the neurons is the important aspect of their operation. To model reflex processing requires artificial interneurons (AIN) and artificial motor neurons (AMNs). In specialized forms, these largely form the spinal cord regions of interest.

Conventional neurons generally sum the products of the input signals with their weights to generate an internal activation, which is non-linearly transformed. In addition to this, AINs and AMNs possess activation and supramaximal thresholds, which relate to input signal strength, an exponentially decaying short term memory which emulates charge decay, refractory periods, and an adaptive transfer function. The single-neuron learning rule applied changes the weights, the transfer function and the thresholds. Both AINs and AMNs trained using this heuristic demonstrated good convergence and stability, with correlations better than .998 in under 200 training cycles. Peak error was between 2% and 5%. Random data presentation is not possible. A strictly sequential data set must be used.

2. ARTIFICIAL NEURAL NETWORKS

Spinal networks are irregular in architecture, though the spinal cord is highly organized [DH91]. The networks contain extensive feedback loops and inhibitory meaures. Significant temporal delays between neurons result from the architecture. Information flow through the network is not uniform and across a large temporal space. A modified Directed Random Search (DRS) [SW81] with a metabolic costing has been used to determine neural parameters within the network. The lowest energy solution is considered best. Over 200 point data sets, the network achieved an output-ideal correlation not less than .96 in under 30,000 DRS training cycles.

REFERENCES

[Nic94]   Nicholls, David G., *Proteins, Transmitters and Synapses*, Blackwell Scientific Publications, Oxford, 1994, pp66-100.

[DH91]   Davidoff, R.A. & Hackman, J.C., Aspects of Spinal Cord Structure and Reflex Function, *Neurologic Clinics*, Vol. 9, No. 3, August 1991, pp533 - 548.

[SW81]   Solis, F. J. & Wets, J.-B., Minimization by Random Search Techniques, *Mathematics of Operations Research*, vol. 6, no. 1, 1981, pp19 - 30.

CANONICAL FORM OF RECURRENT NEURAL NETWORK ARCHITECTURE

N. Selvanathan,
Mashkuri Hj. Yaacob
Faculty of Computer Science & Information Technology
University of Malaya
59100 Kuala Lumpur
Email: selva@fsktm.um.edu.my

ABSTRACT

A recurrent neural network (RCNN) allows feedback from the output of a neuron to its own input, input to neurons of the same layer or neurons of other layers. Feedforward network allows output of neurons in a particular layer to be fed to inputs in the forward layers, hence the excitation is propagated forward. Recursive network allows feedforward and feedback and its recursive nature allows sustained activity, without any external influence. RCNN are able to store information and are suitable for time series prediction. In addition, it is applied for speech processing especially for sequence to label or vice versa application. The mathematical analysis of RCNN is complex in virtue of its feedforward and feedback nature and various researchers have made a attempts to model its dynamics [NC83]. This paper is based on the Nerrand et al model [A94] and is the most general type of recurrent neural network that has been proposed. An attempt is made to obtain the canonical representation of the Nerrand et al architecture and its dynamics written in the state space formulation.

A generalised recurrent neural network (RCNN) is defined and the dynamic equations are written in a canonical form. The canonical form of the RCNN is ideal for simulation using the Matlab software. Stability of recurrent neural network is important when data is used to train the network. The canonical form of RCNN together with the extensive theory of stability of control systems will be helpful in obtaining a generalised stability criteria for RCNN.

REFERENCES

[NC83] Nerrand et al, Cohen MA, Grossberg S.G., 1983, Absolute stability of global pattern formation and parallel memory storage by competitive neural networks. IEEE Transactions on Systems, Man and Cybernetics,13,pp 815-826.

[A94] Ah Chung Tsoi, 1994, Notes on Recurrent Neural Networks, University of Queensland, Queensland, Australia.

EQUALISATION OF DIGITAL COMMUNICATION CHANNEL USING HARTLEY-NEURAL TECHNIQUE

Jitendriya K. Satapathy*, Ganapati Panda ** and Laxmi N. Bhuyan***

* Department of Electrical Engineering

** Department of Applied Electronic & Instrumentation Engineering

Regional Engineering College, Rourkela, India 769 008

*** Department of Computer Science, Texas A&M University, Texas 77843-3112, USA

ABSTRACT

A highly efficient adaptive neural equaliser for a digital communication channel has been developed. The proposed transformed based equaliser has outperformed the conventional multilayer feedforward neural network based equaliser both in terms of rate of convergence and steady state noise floor level.

1. PROPOSED STRUCTURE

Adaptive equalisation is primarily intended at effectively compensating for distortion introduced into the transmitted data sequence by the communication channel due to intersymbol interference alongwith the additive noise. But they exhibit poor performance when the channel has a deep spectral null in its bandwidth. The present work is emphasised at developing an equaliser structure employing ANN & DHT which overcomes the problems of slower convergence rate due to squeezing of the EVR and has resulted in significant enhancement in the overall performance.

2. PROBLEM SIMULATION AND RESULTS

The proposed Hartley-Neural equaliser shown in Fig.1 is simulated having a channel impulse response

$$H(z) = 0.4 + z^{-1} + 0.4 z^{-2}$$

The simulation study as detailed in [SGC90] is followed.

3. CONCLUSION

The proposed hybrid type Hartley-Neural equaliser has shown excellent performance over its conventional counterpart in all departments (Fig.2). The proposed scheme has come out as a clear winner in terms of fast learning characteristics and minimum steady state error.

REFERENCE

[SGC90] S. Siu, G.J. Gibson and C.F.N. Cowan. Decision Feedback Equalisation Using Neural Network Structures and Performance Comparison with Standard Architecture. In *Proceeding IEE, Pt. I*, 137 : 221-225, 1990)

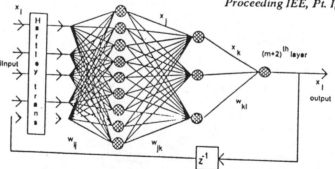

Fig. 1 Proposed hybrid Hartley-Neural equaliser

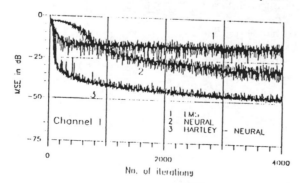

A NONLINEAR OPTIMIZATION NEURAL NETWORK FOR OPTIMUM MULTIUSER DETECTORS

Guiqing He   Puying Tang
Lab504, University of Electronic Science and Technology of China,
Chengdu, Sichuan 610054, P. R. China

ABSTRACT

A nonlinear optimization neural network for the implementation of optimum multiuser detector (MUD) is proposed. By theoretical analyses and computer simulations, it is shown that the proposed neural network detector has the dynamic performance of real time application and the ability of anti-interference of multiple access, and has the less error bit rate than the traditional methods.

1. INTRODUCTION

Code-division multiple-access (CDMA) can make the effective communication among the great number of users and obtains the widespread application. In CDMA communiation systems, optimum MUD is key problem.

The main idea of optimum MUD is find the supposed vector $b^* = [b_1^* \ b_2^* \cdots b_k^*]^T$ which makes the likelihood function maximum on the basis of the received signal's waveform $r(t)$ for the viewing interval $[0, T_b]$. [Ver86]

$$\hat{b}^* = arg_{b \in \{-1, +1\}^k}^{min} \int_0^{T_b} (y(t) - \sum_{j=1}^k b_j c_j(t))^2 dt$$
$$= arg_{b \in \{-1, +1\}^k}^{max} - \frac{1}{2} b^T Hb + Y^T b \qquad (1)$$

where Y is the sample vector of output $y(t)$ at time $t = T_b$, which is the output of the matching filter passed by the received signal $r(t)$. The sample vector is called the viewing vector, and can be written as.

$$Y = Hb + z = RWb + z = F^T FWb + F^T n \quad (2)$$

where $H = RW$ is called the equivalent transfer matrix in CDMA system. W is called the energy matrix, and is the diagonal matrix.

According to (1), the computing complexity increases exponentially with K (the number of users), which is an NP-Complete problem. Such computing complexity is unacceptable in many applications. A neural network opproach to the optimum MUD will be disscused in the next section.

2. NEURAL NETWORK MUD

The optimum multiuser detection problem of (1) can be changed as the following optimization problem.

$$Min : G(b) = \frac{1}{2} b^T Hb - Y^T b + \gamma \sum_{i=1}^K (b_i^2 - 1)^2$$
$$= \frac{1}{2} \sum_{i=1}^K \sum_{j=1}^K h_{ij} b_i b_j - \sum_{i=1}^K Y_i b + \gamma \sum_{i=1}^K (b_i^2 - 1)^2 \qquad (3)$$

where γ is the penalty factor for b_i should take $+1$ or -1. From (3), the neural network for optimum MUD can be derived as:

$$\left. \begin{array}{l} RC \dfrac{dv_1}{dt} = -\dfrac{dG(v)}{dv_1} = -\sum_{j=1}^k h_{1j} v_j + y_1 - 4\gamma(v_1^2 - 1)v_1 \\[2mm] RC \dfrac{dv_2}{dt} = -\dfrac{dG(v)}{dv_2} = -\sum_{j=1}^k h_{2j} v_j + y_2 - 4\gamma(v_2^2 - 1)v_2 \\[2mm] RC \dfrac{dv_k}{dt} = -\dfrac{dG(v)}{dv_k} = -\sum_{j=1}^k h_{kj} v_j + y_k - 4\gamma(v_k^2 - 1)v_k \end{array} \right\} (4)$$

It is shown theoretically that the above neural network is asymptotically stable.

3. SIMULATION RESULTS ANALYSIS

Computer simulation shows that, the neural network MUD is whole asymptotically stable; Its convergence time is less than 0.4τ (the time constant $\tau = RC$ could be made sufficently small to meet the need of real-time application) and its bit error-rate (BER) performance obviously take advantages over the conventional detectors such as single user detectors and decorrelating detectors.

REFERENCES

[Ver89]S. Verdu, "Optimum multiuser asymptotic efficiency" *IEEE Trans. Commun.* Vol. COM-34, No. 9, Sept. 1986, 89

PARALLEL COMPUTER IMPLEMENTATION FOR FEATURE EXTRACTION VIA ANSWER-IN-WEIGHTS NEURAL NETWORK

Iren Valova and Yukio Kosugi

Tokyo Institute of Technology, Department of Precision Machinery Systems,
4259 Nagatsuta, Midori-ku, Yokohama 226, Japan.

E-mail: iren@pms.titech.ac.jp

ABSTRACT

An inverse neural network - Answer-in-Weights [VKK95] - is applied to parallel neurocomputer (MY NEUPOWER) for the purposes of image decomposition and compression, the latter being based on omitting image components which are not significant for the image presentation. Results from workstation and parallel neurocomputer simulations are given. The advantages and shortcoming of the method are presented. Possible further improvements are discussed.

DISCUSSION

A set of Hadamard orthogonal functions is used for image expansion [VKK95]. The images are two-dimensional 128x128 pixel grey scale pictures. The correlation between neighboring pixels should be considered therefore two-variable functions are needed. The image is represented by the weight coefficients for each Hadamard function. The compression strategy is based on removing the weights with small absolute values from the final image calculation. The original images are partitioned into subimages due to quantity of processing neurons (512) in MY NEUPOWER [SSA93], where the subimages are fed one after another and each is processed in parallel. Original, followed by compressed images are shown in Fig.1. The percentages should be regarded as percentage of omitted weights from the final image calculation.

CONCLUSIONS

The advantages of the proposed neural network application include: simple architecture, fast convergence, high compression rates while maintaining good image quality, easy adjustment to a particular task due to orthogonal functions based image expansion. The main advantage over the wavelet compression methods is that the edges of separately processed image parts are not blurred, therefore no information is lost upon final image restoration, which is conducted by combining the processed parts. The processed images can be used as source for neural network implementing image segmentation or understanding algorithms.

REFERENCES

[SSA93] Y. Sato, K. Shibata, M. Asai, M. Ohki, M. Sugie, T. Sakaguchi, M. Hashimoto and Y. Kuwabara.Development of a High Performance, General Purpose Neuro-Computer Composed of 512 Digital Neurons. *Proc. of 1993 International Joint Conference on Neural Networks - Japan, pp 1967-1970, 1993.*

[VKK95] I. Valova, K. Kameyama and Y. Kosugi. Image Decomposition by Answer-in-Weights Neural Network. *IEICE Trans.Inf&Syst., No9:1221-1224, 1995.*

FIGURE 1   Original followed by 60% and 90% compressed by MY NEUPOWER images

SERIAL AND PARALLEL NEURAL NETWORKS FOR IMAGE COMPRESSION

Ryuji Hamabe and Ho Chun Kuo

Department of Communication and Computer Engineering,
Fukuoka Institute of Technology, Higashi-ku, Fukuoka-shi, 811-02 Japan
E-mail:hamabe@fit.ac.jp

ABSTRACT

This paper presents the experimental results of the compression performance concerning the serial and parallel NNs intended for the stationary images. The accuracy of reproduction image quality and the generalization property affected by the some constituent elements of NNs are clarified.

1. NN FOR IMAGE COMPRESSION

The main object of image compression is to store or transmit a compressed data which keeps the reasonable image quality. In a 3-layer NN, the input-to-hidden layer performs the compression and the hidden-to-output layer performs the reproduction. The compression is to induce input units N to hidden units n, and the reproduction is to expand the compressed hidden n to output N. The back-propagation method and sigmoid function are used for the learning of NNs. The SSE(Sum Square Error) between original $T_k(k = 1 \sim N)$ and reproduction image O_k indicates with $SSE = (T_K - O_k)^2$.

2. SIMULATION RESULTS

The original images (256×256, $256 levels$) are four SIDBA's standard images(Couple, Girl, Moon Surface, Aerial). These images are divided into 1024 blocks, and each block size is $N = 8 \times 8$ pixels. To evaluate the quality of the reproduction images, the following SNR(Signal to Noise Ratio) is used, in here E_v is the average. $SNR = 10 log_{10} Ev(T_k)^2/E_v(T_k - O_k)^2$
Performances of serial/parallel NNs ; We examined the reproduction quality about every input image afected with the unit number n of hidden layer, iteration number L, and weight modification gain coefficient α. One of the results of SNR(Girl) relative to the L is presented in Fig. 1. In serial NN, the better score

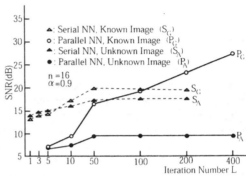

Fig. 2 SNR for known/unknown input images

son is considered that the characteristic of the last few learning blocks give stronger effect to the learning result because of an only NN for an image. And then, it may be seen that the serial NN performs better than parallel NN in the range of smaller L, but the high score may be obviously obtained in parallel NN for large L.

Generalization property ; The known image(Girl) and unknown image(Aerial) are inputted to the both NNs, which are designed by using the three learning set(Girl, Moon, Couple). In Fig. 2. the examples against the both inputs(S_G, S_A), the serial NN is better than parallel one in the smaller L, but the image quality is not revised so much(lower than 20 dB) for large L. As there is an only NN for three images, the characteristics will be learned in the direction of average according to increasing of L. In parallel NN, the known input(P_G) can be seen the good improvement in large L. Although the oposite result is obtained against the unknown input(P_A), it will be that the just three blocks are used for learning a NN.

3. CONCLUSIONS

The better performances of serial NN in lower L may be usefull to solve the problem of data storing because of high compression rate. Hereafter, we are going to study about the adaptive applying of NNs which may be obtained more high quality.

REFERENCES

[SKM88] N. Sonehara, M. Kawato, and S. Miyake : "Neuro-CODEC:Image Data Compression", IEICE, IE88-62, pp.57-64,1988.

[CPS91] S. Carrato, A.Premori, and G.I.Sicuranza: "Linear and Nonlinear Neural Networks for Image Compression", Proc. of Int. Conf. on Digital Signal Processing, pp.526-531, September 1991.

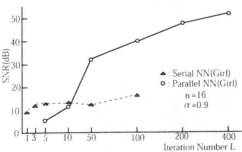

Fig. 1 SNR based on serial/parallel NNs

of SNR did not obtain for increasing of L. The rea-

EXPLORE THE OPERATIONAL PROBLEMS OF FMSs THROUGH ANALYTICAL HIERARCHY PROCESS AND SIMULATION APPROACH

Felix T. S. Chan[+] and Ralph W. L. Ip[*]

[+] School of Manufacturing and Mechanical Engineering, University of South Australia, Ingle Farm, PO Box 1, SA5098, Email: metsc@levels.unisa.edu.au

[*]Department of Manufacturing Engineering, City University of Hong Kong.

ABSTRACT

An attempt is made to look at the operational problems of FMS through simulation. The purpose of the present simulation is to analyse various combinations of scheduling rules in the FMS system, i.e. given a set of scheduling rules applied at the workstations, the effect of different scheduling at the input loading buffer to the system performance are studied. Various measures, such as throughput, utilisation, lateness, etc., are examined. An Analytical Hierarchy Process(AHP) multi-attribute analysis is performed by using a decision support software package, AUTOMAN. This package evaluates and combines the qualitative and quantitative factors for different scheduling rules.

Looking at the operational problems of FMS, such as scheduling and loading strategies, simulation methodology seems to be useful to address these issues. Many authors[G+92] have used various criteria for the generation of optimum schedules, such as number of tardy jobs, number of completed jobs in process inventory and machine utilisation. Montazeri and Wassenhove[MW90]stress the need for simulation prior to actually setting up the FMS. They use a user-oriented discrete event simulator to mimic the operation of a real life FMS. Stecke and Solberg[S+81]have carried out a detailed simulation of a real life system. They have tested various alternatives and evolved loading and control methods which significantly improved the systems production rate.

For the given configuration and data, the FMS was simulated with thirty different combined scheduling rules. From the results, it is clear that the combination of rules plays an important role while evaluating the overall system performance. Thus it must be emphasised that, in an FMS environment, a single rule aimed at a single objective is not desirable. Depending on the scheduler's objective (i.e. different people may have different perspectives in assigning the weight on each evaluation criterion), performance of various combined rules may vary on attributes such as lateness, utilisation, throughput, etc.

In the AHP modelling, the weight assigned on each evaluation criterion is very subjective. Here we had only considered one set of rules. It needs further investigation upon the significance of various weights on the performance of scheduling rules. A fuzzy logic approach applied in assigning various weights is under investigation.

The work reported here is only a pilot study carried out on the FMS configured as a part of pedagogical material being developed to understand the strength of simulation and AHP as a design tool for FMS. The motivation for this approach stemmed from the fact that very few studies have been reported in this direction. Using the simulator together with the developed AHP models, a feasible combined scheduling rule can be found to give an optimum overall performance measure. The simulation model incorporates certain assumptions regarding transportation times, breakdown of machines or AGV's, and due dates may need close scrutiny. However, for the present simulation and AHP exercises, these assumptions are justified on the grounds that they do not seriously affect the relative effectiveness of the scheduling rules.

REFERENCES

[G+92]   Y. P. Gupta and S. K. Goyal. Flexibility tradeoffs in a random flexible manufacturing system, a simulation study, *International Journal of Production Research*, Vol. 30, No. 3, pp. 525-57, 1992.

[MW90]   M. Montazeri and L. N. Wassenhove. Analysis of scheduling rules for an FMS, *International Journal of Production Research*, Vol. 28, No. 4, pp. 785-802, 1990.

[S+81]   K. E. Stecke and J. J. Solberg. Loading and control policies for a flexible manufacturing system, *International Journal of Production Research*, Vol. 19, No. 5, pp. 650-55, 1981.

SPECTACLE DESIGN AND ADVICE COMPUTER GRAPHICS SYSTEM USING ARTIFICIAL INTELLIGENCE

Ryuto Fujie, Hiroyuki Fujie, Kunie Takeuchi *, Oskar Bartenstein **, Kosaku Shirota ***

*Paris-MIKI. Inc *http://www.paris-miki.co.jp* **IF Computer *http://www.biz.isar.de/ifcomputer/* ***NEC

ABSTRACT

The user requirement towards eyeglasses is first the medical function of eyesight correction and second fashion aspects, ie. lens shape and color. By nature, every single person has unique eye problems and a unique face. Thus, only individual design can fully satisfy these requirements.

In this paper we describe an integrated design, sales support, and CAD/CAM system that proposes and produces custom made spectacles based on customer requirements on-site in the store.

The system employs computer vision to measure face features, artificial intelligence for the concurrent engineering task to solve the design problem satisfying medical, fashion, and manufacturability requirements, and high quality computer graphics to visualize the proposed eyeglasses by rendering them on a digital image of the customer's face.

The system covers a sense of beauty reflecting personal preferences and a regular level of functional design. The system is in large scale field operation in optical stores, integral part of the daily business.

AI BASED SPECTACLE DESIGN

A typical user session evolves along the steps:
Input of Personal Data → Digital Photograph → Extraction of Cardinal Points → Edit Cardinal Points → Automatic Make-Up → Color Measurements → Customer Hearing → Lens Shape Design and Editing → Lens Color Suggestion → Composition with Metal Parts → Presentation → Manufacturing

The words from customer hearing describing the final image to be projected by the new eyeglasses influence both color and shape of the designed lens. For example, if you specify *"modern businesslike"* the AI system will suggest both a *modern businesslike* color and a *modern businesslike* shape.

To write the knowledge base we interviewed highly experienced sales people and expert designers, did extensive preliminary studies, literature research, and lots of cross checks.

To propose lens shape we use a special purpose method that covers natural face features and fashion preferences. Figure 1 shows an example of the result of shape design.

To propose lens colors we also developed a special purpose method. which covers personal natural colors, age, sex, personal preferences and climate.

The knowledge base obviously holds universally applicable notions like concepts of beauty and the theory of color harmony. It is implemented in Prolog.

USER INTERACTION

The system supports design and editing of eyeglasses.
It is used in optical stores by the sales staff together with the customer. Therefore the system is designed for easy interactive operation and has many features to edit proposed designs and entertain the audience.

We tried to stick to an artistically appealing display throughout all phases of use. The user interface was planned with careful consideration of the sales process and the customer's attention during the session. Major presentation features are: asymmetrical design; symmetrical face; memory function; automatic make-up; background scene composition; special make-up effects; lens color variation; lens thickness simulation; morphing and warping.

CONCLUSION

The system is implemented in C and IF/Prolog. The user interface is realized in OSF/Motif and X. The system is now widely deployed and integral part of the daily business at optical stores. At the time of this writing about fifty-thousand customers experienced the use of the system as outlined in this paper. Under consideration for future extensions are: - to increase 2D rendering speed; - integration with external PCs, point-of-sales systems, and networks; - extension to 3D graphics; - feedback of experience to the knowledge base.

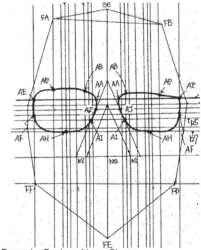

Figure 1 Example: Design of Lens Shape

REFERENCES

[AKA] Goro Akagi; The Eyeglasses; Medical Aoi Publishing Co; (in Japanese)

[JIS] Japan Industrial Standard; JIS Color Handbook; (in Japanese)

[CHI] Hideaki Chijiiwa; The Color Planning Handbook; Visual Design Research Institute; (in Japanese)

Study of a Consultation System for Railway Safety Countermeasure

Hisaji FUKUDA
Fundamental Research Division, Railway Technical Research Institute
2-8-38 Hikari-cho, Kokubunji-shi, Tokyo 185, Japan
Email: Fukuda@rtri.or.jp

ABSTRACT

An advanced safety management system in non-rail areas embodied from a scientific approach; a new concept of a consultation system for railway; and its prototype applying an expert system based on PC technology combining frame theory and production rule are studied. The system models a doctor's process of inquring a patient for diagnosis, and includes functions such as i) process analysis, ii) detailed factor analysis, iii) proposal of countermeasure and automatic reporting.

BASIC CONCEPT OF CONSULTATION SYSTEM FOR RAILWAY SAFETY COUNTERMEASURE

The consultation system for railway safety we indicate is an expert system supporting the proposal of a countermeasure, applying an expert's knowledge and judgement to accident information or situation, by interactively consulting the personal computer in time of accident occurrence or planning a countermeasure. This system is aimed at serving for prevention of occurrence or reccurrence of an accident through standardization of input/output information about accident or safety, and speedy, elaborate pursuit of cause and proposing a countermeasure. The consultation system for safety countermeasure, modeling medical diagnostics, has the following functions and features:

(1) diagnostics about outline of accident situation;
When, Where, Who, What (Who met What, When and Where)
(2) analysis of accident occurrence process: confirm the facts in time series
  (phases I∼IV:HOW) and principal causes(WHY)
(3) Detailed factors analysis: inference of fundamental cause/factors

Detailed factors leading to an accident can be traced as follows: that is to say, explore not only the immediate cause of accident occurrence, but also infer fundamental cause or factors such as human error, weather and physical background to the accident. Here, we can derive knowledge about the background from DBS (database system) without questioning any person concerned.
(4) Proposing countermeasures and writing an accident report automatically

We can propose a countermeasure corresponding to the above (2)and (3), making the most of KB(knowledge base) about countermeasures, and we can automatically draft an accident report as follows:
TYPE0: accident statistics(date, time and place of occurrence)
TYPE1: accident outline(description of accident within 50

words, mainly about the cause, the same as accident reporting now in practice)
TYPE2: accident outline(description of accident within 150 words, the same as newspaper article by 5W1H)
TYPE3: accident outline(TYPE1+proposal of countermeasure)
TYPE4: accident outline {TYPE3+image data(photo of the spot, reference drawing,etc.)}

DEVELOPMENT OF EXPERT SYSTEM FOR SAFETY COUNTERMEASURE AT RAILWAY CROSSING

We developed a prototype " consultation system for prevention of railway crossing accidents " as a concrete application of an expert system based on personal computer, which combines frame theory and production rule. It is available for retrieving features or image data relating to an accident from the crossing database, based on an information of accident occurrence, analyzing an accident in detail, and automatically drafting an accident report. Moreover, it can conduct an interactive dialogue, make a detailed analysis, and eventually output a proposal for countermeasure.

Acquisition of expert knowledge is at present one of the difficult problems in construction of an expert system. Here we retrieve articles about crossing accident and countermeasure from recent professional journals and papers, arrange them according to the rule "if∼then∼", and add an expert's judgement, extract about 150 pieces of knowledge, and store them into knowledge base for a computer.

Information from professional journals is likely to contain duplication or partially contradiction. In that case, this system adopts an information strategy which gives priority to a newer information.

CONCLUSIONS

The system, providing standardization of accident and safety information, accommodation to the changing situations and speedy, elaborate information and proposal of countermeasures, is available for prevention of railway crossing accidents. It is theoretically applicable for general operational accidents other than railway crossing accidents.

REFERENCE

[Fuk+94] Fukuda, H et al.: Development of human error database in the railway, RTRI report, Vol.8, No.7, 1994 (in JPN).

A Step Toward Human - Robot Cooperative System

Masaru Ishii   and   Hironori Saita

Depart. of Commun. and Computer Eng., Fukuoka Institute of Technology

Wajiro-Higashi 3-30-1, Higashi-Ku, Fukuoka, 811-02 Japan

Email: ishii@fit.ac.jp

ABSTRACT

In near future, we will have to live with many machines such as home robots, hospital robots and so on. User friendly communication between man and those machines is very important for cooperative system utilizing their features and abilities. As an example of human - robot cooperative system, we propose an intuitive approach to robot teaching method with multi-media tools.

A ROBOT TEACHING WITH MULTIMEDIA TOOLS

As a first step for constructing human - robot cooperative system, robot teaching system with multi-media tools is described in this paper. Clicking buttons on the display by a mouse device mean controlling a robot system. After we make many primitives of robot's operations using this method, we can create any necessary tasks easily by a macro operation that combined those primitives. Also, we can select a wanted task from robot macro operation's database. One of our aims is to make such a simple command for executing complicated robot tasks. Actually, a complicated task of a robot system consists of several primitive robot operations. So we can replace a robot complex work with a sequence of these primitive operations.

In robot macro card in *Fig.2*, we can give a name to a macro operation. Left lower part of this card shows macro operation's names such as **put-into-video-deck**, **put-back-video-tape** for example. Operation **put-into-video-deck** means that an operator teaches 3D positions and orientation of a robot system for putting a tape into video-deck by our teaching method. As soon as we push

a **put-into-video-deck** button, robot system plays back the same operation quickly. As this system has a hierarchy structure, we can make a higher macro operation such as handling a **Video Tape** operation by selecting and collecting optimal operations from lower macro ones. This operation has manipulating sequences as follows: grasping a video tape, inserting it and so on. So we can make a simple command for such a complicated task. In this way, a robot operation's sequence is created as one card stack by this teaching method. A registered task is performed automatically by clicking a button of above operation on the main menu card.

As an experiment example, we make macro operations about handling a video tape from teaching cards mentioned above. An experimental task is to grasp a video tape, insert and push it into a video deck. Next, these systems pull a video tape from a deck and put it back to the shelf. Tthis system can insert a video tape into a deck successfully.

This paper describes an intuitive robot teaching method with multimedia tools for user friendly interfaces.

The goal of our system is making human - robot cooperative system for overcoming common operational tasks with mutual conversation and collaboration. Sharing and learning ability of robot system by operator's advice are expected to solve common problems in our system.

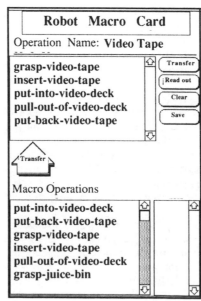

Fig.2. *A robot macro operation card*

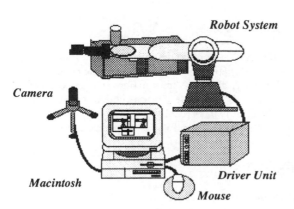

Fig.1. *A robot teaching system*

INDEX OF AUTHORS

Abhary, K.	491	Devedzic, V.	267
Adorni, G.	781	Ding, R.	141
Aguilar, G.D.	283	Durand, N.	585
Akashi, T.	317	del Pobil, A.P.	725, 751, 767
Akutsu, T.	455		
Ali, F.E.A.F	235	Egi, T.	784
Ali, M.	455	Egresits, Cs.	469
Alliot, J-M.	585	Esmaili, M.	778
Almuallim, H.	431		
Alpigini, J.J.	195	Filipic, B.	461
Anh, D.T.	395	Fimbell, E.	553
Atif, Y.	673	Fouet, J-M.	327
		Fox, D.	511
Babaguchi, N.	127	Frasson, C.	261
Baioletti, M.	665	Frasson, M.C.	261
Bartenstein, O.	802	Fujie, H.	802
Bhandarkar, S.M.	745	Fujie, R.	802
Bhuyan, L.N.	797	Fukuda, H.	803
Bohm, A.	681	Fukuda, S.	786
Bonet, E.	711, 761	Furukawa, K.	389
Braithwaite, N.	593		
Burgard, W.	511	Garcia-Moral, I.	731
		Gero, J.S.	369
Cain, G	241	Ghwanmeh, S.H.	213
Calle, A.	745	Gonzalez, C.A.	779
Campenhout, J.M.V.	737	Gudenas, J.W.	789
Canete, J.F.d.	731		
Carlisle, W.H.	776	Hamabe, R.	800
Carlson, C.R.	789	Hamamoto, Y.	247, 559
Casey, M.G.	161	Hamidzadeh, B.	673
Cavalieri, S.	479	Han, W.	141
Cervera, E.	725	Hardy, L.L.	785
Chan, F.T.S.	223, 491, 801	Hase, T.	559
Chang, K.H.	503, 776	Hashimoto, M.	71
Chen, J.	737	Hattori, F.	705
Chen, M-C.	333	He, G.	798
Chen, Y-W.	235	He, L.	175
Chuang, T-N.	201	Heidelbach, M.	511
Collier, P.A.	161, 777	Hendtlass, T.	253, 537, 791, 795
Copland, H.	795	Higuchi, T.	77
Cremers, A.B.	511	Hirahara, M.	65
Cross, J.H.	776	Hirota, T.	71, 790
		Hitz, M.	693
Damski, J.C.	369	Ho, T.	413
Davidsson, P.	403	Honda, M.	149

Hong, T-P.	201	Lee, Y.H.	339
Hopgood, A.A.	593	Leech, S.A.	777
Hori, M.	647	Lim, E.N.	95
Horinouchi, N.	317	Lim, S.	547
Hu, L.	189	Lim, S.S.	95
Huarng, K.	333	Liu, H.	419
		Liu, S-C.	659
Inesta, J.M.	767	Lu, Y.	167, 521, 699
Ip, R.W.L.	223, 491, 801		
Ishii, Ma.	804	Maddouri, M.	295
Ishii, Me.	431, 705	Maeda, A.	275
Ishitobi, M.	653	Maeda, H.	89, 117
Ito, K.	135	Marcugini, S.	665
Itoh, H.	57, 175	Martin, G.	711, 761
Itoh, T.	780	Martinez, B.	751
Iwazume, M.	305	Mase, H.	627
		Masuda, G.	449
Jaoua, A.	295	Matsuo, T.	639
Jerinic, L.	267	Matsuura, Y.	786
Jinno, K.	135	McDermott, D.	25
Jones, K.O.	213	McElyea, R.	503
Junquera B.P.	779	Mckinney, M.	521
		Mikami, I.	89
Kadar, B.	469	Miki, T.	43
Kakihara, K.	687	Milani, A.	665
Kamakura, C.	57	Mitani, Y.	559
Kanchanasut, K.	395	Mohanty, H.	787
Kaneda, S.	431, 705	Monostori, L.	469
Kaneyama, C.	247	Moreno, S.	711, 761
Kappel, A.M.	511	Morgenstern, B.	605
Karasawa, H.	793	Morita, C.	425
Katamine, K.	71, 790	Mueck, T.A.	693
Kato, S.	57		
Kawamura, A.	135	Nagamatsu, M.	563
Kazerooni, A.	491	Nagasawa, I.	71, 77, 790
Kennedy, K.	521	Naka, Y.	207
Kim, J.H.	794	Nakajima, N.	3
Kim, S.C.	547	Nakanishi, K.	317
Kinukawa, H.	627	Nakano, K.	705
Kitahasi, T.	127	Nakao, Z.	235
Kobayashi, A.	89	Nakatani, Y.	103
Komoda, N.	687	Namatame, A.	575, 782
Kosugi, Y.	799	Namioka, Y.	363
Koujitani, K.	117	Nanba, D.	705
Koyama, T.	283	Natter, M.	611
Kuo, H.C.	800	Ngoi, K.A.	95
		Nguyen, T.	413
L-Kappel, S.	511	Nikovski, D.	531
Lazo, G.G.A.	779	Nishida, T.	117, 181, 305, 639
Lee, B.H.	95	Nkambou, R.	261
Lee, G.	633	Nouira, R.	327
Lee, J-H.	633	Ntuen, C.A.	794
Lee, T.	633		

Obeid, N.	357	Siores, E.	775
Ohara, K.	127	Staudt, M.	345
Ohkawa, T.	687	Stewart, B.	439
Ohsuga, S.	9	Straten, G.V.	569
Oka, N.	65	Stumptner, M.	155
Okuma, A.	377	Suharnan, S.	793
Okuno, H.G.	47		
Ono, T.	783	Takaai, M.	181
Osada, K.	784	Takasu, A.	455
		Takata, O.	317
Panda, G.	797	Takeda, H.	181, 305
Park, E.H.	794	Takeuchi, K.	802
Peremans, H.	737	Tanaka, Hide.	47
Perpar, M.	461	Tanaka, Hiro.	207
Phillips, H.J.	593	Tanaka, S.	89
Philpot, D.	791	Tanaka, Ta.	35
Picton, P.D.	593	Tanaka, To.	363
Pieprzyk, J.	778	Tang, P.	798
Podlena, J.R.	253	Taudes, A.	611
Poggi, A.	781	Tegoshi, Y.	71
Potter, W.D.	745	Tien, B.T.	569
Prieto, J.M.Y.	785	Toledo, F.	711, 761
		Tomita, S.	247, 559
Qin, A.	621	Trcka, M.	611
Qiu, J.	719	Tsuchida, H.	705
		Tsuji, H.	627
Rajkumar, T.	787	Tsukamoto, Y.	575, 782
Raman, P.	776	Tsukimoto, H.	425
Rogers, M.	503	Tsukiyama, M.	103
Russell, D.W.	195	Tsuneta, Y.	247
		Turci, P.	781
Safavi-Naini, R.	778		
Saita, H.	804	Uellner, S.	681
Sakai, H.	377	Umeda, M.	77, 790
Sakamoto, N.	449	Ushijima, K.	449
Salami, M.	241		
Sanz, P.J.	767	Valova, I.	799
Satapathy, J.K.	797		
Satoh, K.	792	Wake, T.	103
Satoh, Y.	275	Walters, M.	719
Schimpe, H.	345	Wang, J.	289
Seki, H.	57, 175	Wang, S-L.	111
Selvanathan, N.	796	Watanabe, G.	783
Setiono, R.	419	Watanabe, T.	317
Shelesh-Nezhad, K.	775	Wen, C-Y.	543
Shimazu, K.	389	Wilk, E.	605
Shimodaira, H.	599	Wilk, J.	605
Shimokuni, O.	47	Williams, D.	213
Shinohara, Y.	788	Williams, R.	699
Shintani, T.	780	Wong, W.S.	621
Shiqi, Z.	487	Wongchuphan, P.	786
Shirota, K.	802	Wotawa, F.	155
Simono, M.	43		

Yaacob, M.Hj.	796	Yoshida, T.	647
Yamada, K.	149	Yu, S.	487
Yamakawa, T.	43		
Yamato, H.	283	Zargham, M.	189, 531
Yanaru, T.	563	Zhigang, Y.	487
Yano, H.	317	Zhou, N-F.	383, 790
Yoo, S.I.	339	Zongxue, X.	135
Yoshida, K.	65, 563	Zun, I.	461

GREYSCALE

BIN TRAVELER FORM

Cut By _Yadira_ Qty _15_ Date _4/19/24_

Scanned By _____ Qty_____ Date_____

Scanned Batch IDs

_____ _____ _____

Notes / Exception

_____

_____

_____

_____

_____